Matsl Coating

ELECTROPLATING ENGINEERING
Handbook

ELECTROPLATING ENGINEERING
Handbook

Fourth Edition

EDITED BY
Lawrence J. Durney

VAN NOSTRAND REINHOLD
New York

Copyright © 1984 by Van Nostrand Reinhold

Library of Congress Catalog Card Number 84-3548
ISBN 0-442-22002-2

All rights reserved. No part of this work covered by the copyright hereon may be reproduced or used in any form or by any means—graphic, electronic, or mechanical, including photocopying, recording, taping, or information storage and retrieval systems—without written permission of the publisher.

Printed in the United States of America

Van Nostrand Reinhold
115 Fifth Avenue
New York, New York 10003

Van Nostrand Reinhold International Company Limited
11 New Fetter Lane
London EC4P 4EE, England

Van Nostrand Reinhold
480 La Trobe Street
Melbourne, Victoria 3000, Australia

Nelson Canada
1120 Birchmount Road
Scarborough, Ontario, M1K 5G4, Canada

16 15 14 13 12 11 10 9 8 7 6 5 4 3

Library of Congress Cataloging in Publication Data

Main entry under title:

Electroplating engineering handbook.

    Includes index.
    1. Electroplating. I. Durney, Lawrence J.
TS670.E4614 1984   671.7′32   84-3548
ISBN 0-442-22002-2

CONTRIBUTORS

David M. Anderson*
Bethlehem Steel Corporation

Reginald K. Asher Sr.
Principal Staff Engineer
Motorola Inc.

Leslie C. Borchert*
Houdaille Industries

Allen D. Brandt*
Bethlehem Steel Corporation

George E. F. Brewer
Consultant

Peter Cambria
Chief Engineer
Rapid Electric Company Inc.

Edward Clegg
Army Materials and Mechanics Research Ctr.

William Cochran
Consultant

Richard W. Crain
Vice President
Industrial Filter and Pump Mfg. Co.

Isidore Cross
Isidore Cross Associates

John V. Davis
The Udylite Corporation

H. K. DeLong*
Dow Chemical Company

George A. DiBari
Senior Program Manager
Inco Research and Development Labs. Inc.

Alfred Douty*
Amchem Products Inc.

Lawrence J. Durney
Vice President for Technology
Frederick Gumm Chemical Company Inc.

Robert Duva
Consultant
Catholyte Inc.

Anthony O. Facciolo
Alexandria Metal Finishers, Inc.

Charles L. Faust*
Batelle Memorial Institute

Simon Gary
President
Scientific Control Laboratories

Kenneth J. Gatchel
Lea Michigan Inc.

A. Kenneth Graham*
Deceased

Juan B. Hajdu
Enthone Inc.

J. Bernard Hignett
Vice President
The Harper Company

Jack Hyner
Whyco Chromium Corporation

William Innes
Vice President and Technical Director
MacDermid Inc.

William H. Jackson*
The Udylite Corporation

Samuel S. Johnson*
Consultant

William Jones
Research Director
The Heatbath Corp.

CONTRIBUTORS

HENRY L. KELLNER
The Lea Manufacturing Co.

JAMES K. LONG
Consultant

IAN MCDONALD
Frederick Gumm Chemical Company Inc.

MARTVIG J. MOLL*
Former Hanson Van Winkle Munning Company

JOHN MORICO
Clearwater Analytical Labs.

LELAND M. MORSE*
Chrysler Corporation

FIELDING OGBURN
Corrosion and Electrodeposition Group National Bureau of Standards

ROBERT R. PIERCE*
Pennsalt Chemical Corporation

H. L. PINKERTON*
Westinghouse Electric Corporation

CLARENCE H. ROY
Rainbow Research Inc.

WILLIAM SAFRANEK
Technical Editor Plating and Surface Finishing

RICHARD SALTONSTALL*
The Udylite Corporation

KEITH SHEPPARD
Assistant Professor of Materials and Metallurgical Engineering Stevens Institute of Technology

DONALD L. SNYDER
Metal Finishing Dept., The Harshaw Chemical Co.

ELLSWORTH A. STOCKBOWER*
Amchem Products Inc.

KEN SURPRENNANT
Dow Chemical USA

KENNETH TATOR*
Deceased

HERB TILTON
Tilton Rack Basket Mfg. Co.

JAMES D. THOMAS
General Motors Research Laboratories

JOHN A. THORNTON
Department of Metallurgy and Mining Engineering, and The Coordinated Science Laboratory, University of Illinois

DEAN WARD
President Haward Corp.

ROLF WEIL
Professor of Metallurgy Stevens Institute of Technology

BRYAN E. WINTER
Alexandria Metal Finishers, Inc.

*Indicates contributors retained from third edition.

PREFACE

As an instructor in various finishing courses, I have frequently made the statement over the years that "In the field of metal finishing there is very little black and white, just a great deal of grey. It is the purpose of the instructor to familiarize the student with the beacons that will guide him through this fog."

To a very considerable extent, a handbook such as this serves a similar purpose. It is also subject to similar limitations. Providing all the required information would result in a multi-volume encyclopedia rather than a usable handbook. In the pages that follow, you will therefore find frequent references to other sources where more detailed explanations or information can be found. The present goal is proper guidance and the provision of the most frequently required facts, not everything that is available.

In the 13 years since the last edition, changes in the finishing industry have been profound but in one sense have resulted in simplifying matters rather than complicating them. Because technology has advanced to a level of complexity rendering "home brew" impractical in many cases, dependence on proprietary compounds has become common. Therefore, detailed solution compositions are often no longer significant or even practical. It is thus more important to provide instruction about the factors that affect the choice of the most suitable type of proprietary material.

In a few instances, technology has rendered some procedures or equipment obsolete, or nearly so. Such, for example, is the case with generators; these have almost universally been replaced by rectifiers. The chapter on generators has therefore been dropped, and the chapter on bus-bars and connections, which formerly emphasized generator connections, has been altered to reflect the change to rectifiers (and the almost complete elimination of rheostats as well).

The discussion of design for electroplating has been expanded to include the importance of proper material selection; that is, plating needs to be considered as an actual contributor to the engineering qualities of the final product rather than as a purely cosmetic treatment to improve customer appeal. As a result, the discussion has necessarily changed a great deal both in presentation and emphasis.

Because of their interdependence, the chapter on water requirements and the chapter on waste disposal have been combined into a single chapter that emphasizes the interaction of the two subjects. Since the relevant technology has advanced so far and so rapidly, this chapter provides guiding instructions on how to go about gathering the required information, whom to consult, and what methods to consider, not just a cookbook list of ready-made answers. Such answers, incidentally, no longer exist. Only by completely analyzing the entire layout, work load, and solution use can a proper answer be obtained.

P.R. plating has also essentially disappeared, since the emerging technology of pulse plating has replaced it. The related chapter reflects this change.

The technology of electropolishing has changed in a relatively minor way, but its applications have become more widespread. The chapter on this subject has therefore been rewritten by the owner and operator of a successful electropolishing job shop to emphasize the practical aspects of the technique.

Since hi-speed plating is another process that is apparently growing, a chapter on this subject has been added.

Analytical techniques have shifted considerably. Instrumentation has become more prevalent, and the need for analytical results in the ppm range for waste control have forced other changes. The chapter on solution analysis has therefore been completely altered.

The chapter on troubleshooting has been completely rewritten to emphasize an organized procedure based on a logical process. This discussion, however, has been kept general in nature. Detailed troubleshooting information has been added to the chapter on plating bath composition so that it will be readily available. The chapter on cleaning has been altered to include information that should be of help when problems arise. In short, wherever possible, troubleshooting information has been located as closely as possible to the item affected or subject involved.

All told, 44 chapters (and/or subchapters) have been either completely rewritten or reviewed and revised.

A great deal of the material contained in previous editions has been retained. The authors who labored so long and so industriously to build this body of information leave us forever in their debt. We have therefore adopted a policy of differentiating between those chapters that are unaltered, those that have been reviewed and revised, and those that have been mostly or entirely rewritten. For unaltered chapters, the original authors are listed. For chapters reviewed and revised, the original authors are listed along with the reviewer. For major revisions or complete rewrites, only the new author is listed.

The editor owes an equally massive debt to the authors who took time out of their very busy schedules to prepare the new or revised chapters. Without their expertise and cooperation, the task of preparing this new edition would have been insurmountable. My sincerest thanks are extended to them all.

Thanks, too, to Fred Gumm, who permitted the use of his office photocopier for the multitude of photocopies required for this project.

Special thanks go to my family, and, in particular, to my wife Marie. The time needed to produce this new edition has of necessity been taken from the time normally spent with them or in carrying out the usual household chores—raking the lawn, painting the house, and so forth. To my youngest sons Martin and Thomas, my thanks for filling in for me in this area while I pounded my typewriter or the keyboard of the computer used in the later stages of progress. And to my wife, my deepest thanks for tolerance, understanding, and support during the two years that it took to put this book together.

Lawrence J. Durney
No. Caldwell, N.J.

CONTENTS

Contributors / v
Preface / vii

PART I. GENERAL PROCESSING DATA

1. Tables of Data, *Edwin R. Bowerman, Jr.* / 3
2. Design for Plating, *J. Hyner* / 50
3. Metal Surface Preparation and Cleaning / 58
   A. Basis Metal Surface, *Keith Sheppard* / 58
   B. Polishing, Brushing and Buffing, *Henry L. Kellner* and *Kenneth J. Gatchel* / 61
   C. Mass Finishing Methods, *J. Bernard Hignett* / 78
   D. Electropolishing, *Dean E. Ward* / 100
   E. Solvent Cleaning, *Kenneth S. Surprenant* / 121
   F. Alkaline Cleaning, *Lawrence J. Durney* / 147
   G. Oxide Removal, *Cloyd A. Snavely* and *Charles L. Faust* / 159
4. Typical Processing and Operating Sequences / 174
   A. Metals, *Donald L. Synder* and *James K. Long* / 174
   B. Plastics, *J. B. Hajdu* and *Gerald Krulik* / 202
5. Wastewater Control and Treatment, *Clarence H. Roy* / 206
6. Plating Bath Compositions and Operating Conditions, *Lawrence J. Durney* / 226
7. Plating Standards and Specifications, *Edward T. Clegg* and *Fielding Ogburn* / 263
8. Troubleshooting, *Lawrence J. Durney* / 285
9. Analysis of Plating Solutions, *John Morico* / 289
10. Testing Electrodeposited Coatings / 309
    A. Thickness Tests, *Isidore Cross* / 309
    B. Corrosion Tests, *James Douglas Thomas* / 328
    C. Inspection, *Leland M. Morse* / 332
11. Industrial Hygiene and Safety, *Bryan E. Winter* and *Anthony O. Facciolo, Jr.* / 341
12. Relevant Materials Science for Electroplaters, *Rolf Weil* / 352
13. Surface Protection and Finishing Treatments / 366
    A. Phosphate Coating Processes, *Alfred Douty* and *Ellsworth A. Stockbower*, revised by *William C. Jones* / 366
    B. Chromate Conversion Coatings, *William P. Innes* / 390
    C. Sulfuric and Chromic Acid Anodizing of Aluminum, *W. C. Cochran* / 396
    D. Anodizing and Surface Conversion Treatments for Magnesium, *H. K. DeLong* / 410

14. Nonelectrolytic Metal Coating Processes / 420
    A. Non-Catalytic Chemical Methods, *H. L. Pinkerton* / 420
    B. Vapor-Phase Methods, *John A. Thornton* / 426
    C. Catalytic Methods, *Ian McDonald* / 438
15. Current and Metal Distribution, *H. L. Pinkerton*, revised by
    Lawrence J. Durney / 461
16. Electroforming, *George A. DiBari* / 474

PART II. ENGINEERING FUNDAMENTALS AND PRACTICE

17. Plant Location and Layout, *Simon P. Gary* / 493
18. Floors—Plans and Construction, *Robert R. Pierce* / 503
19. Tanks—Design, Construction Installation and Maintenance, *John V. Davis*
    and *A. Kenneth Graham*, revised by *Lawrence J. Durney* / 516
20. Linings, *Kenneth Tator* / 525
21. Heating and Cooling Equipment, *H. L. Pinkerton* and
    A. Kenneth Graham / 537
22. Design, Construction, Insulation and Maintenance of Plating Racks,
    Herbert Tilton / 559
23. Manually Operated Installations, *Lawrence J. Durney* / 567
24. Barrels, *William H. Jackson* and *A. Kenneth Graham*, revised by
    Reginald K. Asher, Sr. / 573
25. Semi- and Fully Automatic Plating Machines, *Martvig J. Moll*, revised by
    Lawrence J. Durney / 606
26. Continuous Plating Equipment for Steel Mill Products, *A. Kenneth Graham*
    and *Samuel S. Johnston* / 617
27. Anode and Cathode Rod and Bus Systems, *A. Kenneth Graham* and
    H. L. Pinkerton, revised by *Lawrence J. Durney* / 631
28. Exhaust Systems, *Allen D. Brandt* and *David M. Anderson* / 638
29. Electrode Materials and Design, *H. L. Pinkerton*, revised by
    Lawrence J. Durney / 656
30. Rectifiers, *Peter Cambria* / 669
31. Pulse Plating, *Robert Duva* / 684
32. Rinsing, *H. L. Pinkerton* and *A. Kenneth Graham* / 691
33. Drying Practices and Equipment, *H. L. Pinkerton* / 710
34. Filtration, *Richard W. Crain* / 716
35. Auxiliary Equipment for the Plating Room, *H. L. Pinkerton*, revised by
    Lawrence J. Durney / 728
36. General Maintenance, *Richard B. Saltonstall* and *Leslie C. Borchert* / 739
37. Electrochemical Machining, *Charles L. Faust* / 744
38. Electrodeposition of Organic Coatings, *George E. F. Brewer* / 755
39. High Speed Plating / 767
    A. Applications, *W. H. Safranek* / 767
    B. Principles, *Robert Duva* / 772

Index / 775

1 / GENERAL PROCESSING DATA

1. TABLES OF DATA

EDWIN R. BOWERMAN, JR.

Manager, Power Sources, Energy Conversion Department, General Telephone and Electronics Laboratories, Bayside, N. Y.

TABLE 1. ELECTROCHEMICAL EQUIVALENTS AND RELATED DATA*
(All figures in this table are based on 100% current efficiency)

Element	Atomic weight	Valence	mg/coulomb	g/amp hr	oz/amp hr	lb/1,000 amp hr	sp gr	oz/sq ft for 0.001 in.	amp hr to deposit 0.001 in./sq ft	Symbol of element
Aluminum	26.97	3	0.0932	0.3354	0.0118	0.7394	2.7	0.225	19.05	Al
Antimony	121.76	5	0.2523	0.9085	0.032	2.0028	6.68	0.557	17.4	Sb
		3	0.4206	1.5141	0.0535	3.3380	6.68	0.557	10.4	
Arsenic	74.91	5	0.1525	0.5589	0.0197	1.2322	5.73	0.475	24.1	As
		3	0.2587	0.9315	0.0328	2.0536	5.73	0.475	14.5	
Bismuth	209.0	5	0.4332	1.5594	0.055	3.4378	9.8	0.816	14.8	Bi
		3	0.7219	2.5990	0.0915	5.7297	9.8	0.816	8.93	
Cadmium	112.4	2	0.5824	2.097	0.074	4.6226	8.64	0.72	9.73	Cd
Chromium	52.01	6	0.0898	0.323	0.0114	0.7129	7.1	0.591	51.8	Cr
		3	0.1796	0.646	0.0228	1.4258	7.1	0.591	25.9	
Cobalt	58.94	2	0.3054	1.099	0.0388	2.4236	8.9	0.74	19.0	Co
Copper	63.57	2	0.3294	1.186	0.0418	2.6142	8.92	0.74	17.7	Cu
		1	0.6588	2.372	0.0837	5.2283	8.92	0.74	8.84	
Gallium	69.72	3	0.2408	0.8670	0.0306	1.9114	5.9	0.491	16.0	Ga
Germanium	72.60	4	0.1881	0.6771	0.02388	1.4928	5.35	0.445	18.6	Ge
		2	0.3762	1.3542	0.04776	2.9855	5.35	0.445	9.31	
Gold	197.2	3	0.6812	2.4522	0.0865	5.406	19.3	1.61	18.6	Au
		2	1.0218	3.6783	0.1297	8.1094	19.3	1.61	12.4	
		1	2.0435	7.3567	0.2595	16.2187	19.3	1.61	6.2	
Indium	114.76	3	0.3964	1.4271	0.05033	3.1461	7.31	0.608	12.1	In
Iridium	193.1	4	0.5003	1.8001	0.06349	3.9704	22.42	1.869	29.4	Ir
		3	0.6670	2.4012	0.08469	5.2938	22.42	1.869	22.1	
Iron	55.84	2	0.2893	1.042	0.0368	2.2963	7.9	0.66	17.9	Fe
Lead	207.2	2	1.074	3.865	0.1362	8.5210	11.3	0.94	6.91	Pb
Manganese	54.93	2	0.2846	1.0246	0.0362	2.2588	7.2	0.598	16.5	Mn
Mercury	200.61	2	1.0394	3.7420	0.1320	8.2496	13.55	1.129	8.55	Hg
		1	2.0788	7.4839	0.2640	16.4992	13.55	1.129	4.27	
Nickel	58.69	2	0.3041	1.095	0.0386	2.4135	8.9	0.742	19.0	Ni

TABLE 1 (continued)

Element	Atomic weight	Valence	mg/coulomb	g/amp hr	oz/amp hr	lb/1,000 amp hr	sp gr	oz/sq ft‡ for 0.001 in.	amp hr to deposit 0.001 in./sq ft□	Symbol of element
Palladium	106.7	4	0.2764	0.9951	0.035	2.1939	12.0	0.998	28.6	Pd
		3	0.3686	1.3268	0.0467	2.9252	12.0	0.998	21.4	
		2	0.5528	1.9903	0.0703	4.3878	12.0	0.998	14.2	
Platinum	195.23	4	0.5058	1.8208	0.0645	4.0142	21.4	1.78	27.6	Pt
		2	1.0115	3.6416	0.1284	8.0283	21.4	1.78	13.85	
Polonium	210	4	0.5440	1.958	0.0691	4.318	—	—	—	Po
Rhodium	102.9	4	0.2666	0.9598	0.0338	2.1160	12.5	1.04	30.8	Rh
		3	0.3555	1.2797	0.0451	2.8213	12.5	1.04	23.1	
		2	0.5332	1.9196	0.0677	4.2319	12.5	1.04	15.37	
Rhenium	186.31	7	0.2758	0.9929	0.0350	2.1890	20.53	1.710	48.8	Re
Selenium	78.9	4	0.2046	0.7364	0.0260	1.6235	4.81	0.400	15.4	Se
Silver	107.88	1	1.1179	4.0245	0.142	8.8726	10.5	0.875	6.16	Ag
Tellurium	127.61	4	0.3306	1.1901	0.0420	2.6238	6.25	0.520	12.4	Te
		2	0.6612	2.3803	0.0840	5.2476	6.25	0.520	6.19	
Thallium	204.39	1	2.1180	7.6249	0.2583	16.8000	11.85	0.986	3.82	Tl
Tin	118.7	4	0.3075	1.1070	0.039	2.4406	7.3	0.61	15.63	Sn
		2	0.6150	2.2141	0.078	4.8812	7.3	0.61	7.82	
Zinc	65.38	2	0.3387	1.2195	0.043	2.6886	7.1	0.59	13.7	Zn

* Portions of this table are taken from "Modern Electroplating," 1st Edition, Appendix Table VIII, by permission of The Electrochemical Society, publisher.
‡ One side only.
□ Equals Sp. Gr. × 0.08323.

TABLE 2. COATING WEIGHTS FOR ONE-MIL THICK DEPOSITS

Metal	oz/sq ft	mg/sq in.	mg/sq cm	Metal	oz/sq ft	mg/sq in.	mg/sq cm
Aluminum	0.225	44.7	6.9	Lead	0.94	185	28.7
Antimony	0.557	111	17.0	Manganese	0.598	118	18.2
Arsenic	0.475	94.3	14.5	Nickel	0.742	146	22.6
Bismuth	0.816	162	24.9	Palladium	0.998	196	30.4
Cadmium	0.72	143	22.0	Platinum	1.78	350	54.3
Chromium	0.591	117	18.0	Rhodium	1.04	205	31.8
Cobalt	0.74	147	22.6	Rhenium	1.71	336	52.2
Copper	0.74	147	22.6	Selenium	0.400	78.7	12.2
Gallium	0.491	97.5	15.0	Silver	0.875	172	26.7
Germanium	0.445	88.3	13.6	Tellurium	0.520	102	15.9
Gold	1.61	320	49.2	Thallium	0.986	194	30.1
Indium	0.608	121	18.6	Tin	0.61	120	18.6
Iridium	1.869	368	57.0	Zinc	0.59	116	18.0
Iron	0.66	130	20.1				

TABLE 3. CHEMICAL COMPOUNDS

Name	Formula	Formula weight	Metal content % (Theoretical)
Acid			
Acetic	CH_3COOH	60.05	
Arsenious	As_2O_3	197.82	75.8
Boric	H_3BO_3	61.84	
Chloroauric	$HAuCl_4 \cdot 4H_2O$	412.10	47.9
Chloroplatinic	$H_2PtCl_6 \cdot 6H_2O$	518.08	37.7
Chromic	CrO_3	100.01	52.0
Citric	$H_3C_6H_5O_7 \cdot H_2O$	210.14	
Formic	$HCOOH$	46.03	
Fluoboric	HBF_4	87.83	
Hydrobromic	HBr	80.92	
Hydrochloric	HCl	36.46	
Hydrofluoric	HF	20.01	
Molybdic	MoO_3	143.95	66.6
Nitric	HNO_3	63.02	
Oxalic	$H_2C_2O_4 \cdot 2H_2O$	126.07	
Perchloric	$HClO_4$	100.46	
Perrhenic	$HReO_4$	251.32	74.4
Phosphoric	H_3PO_4	98.00	
Pyrophosphoric	$H_2P_2O_7$	177.99	
Silicic	$SiO_2 \cdot XH_2O$		
Sulfamic	$HSO_3 \cdot NH_2$	97.09	
Sulfuric	H_2SO_4	98.08	
Tartaric	$(CHOHCOOH)_2$	150.09	
Tungstic	WO_3	231.92	79.3
Aluminum	Al	26.97	
Chloride, anhyd.	$AlCl_3$	133.34	20.2
Chloride	$AlCl_3 \cdot 6H_2O$	241.43	11.1
Oxide	Al_2O_3	101.94	52.9
Potassium sulfate	$AlK(SO_4)_2$	258.19	10.4
	$AlK(SO_4)_2 \cdot 12H_2O$	474.38	5.7
	$AlK(SO_4)_2 \cdot 24H_2O$	690.57	3.9
Sulfate, anhyd.	$Al_2(SO_4)_3$	342.12	15.8
Sulfate	$Al_2(SO_4)_3 \cdot 18H_2O$	666.41	8.1
Ammonium			
Acetate	$NH_4C_2H_3O_2$	77.08	
Aluminum sulfate	$(NH_4)_2Al_2(SO_4)_4 \cdot 24H_2O$	906.64	5.9
Bifluoride	NH_4HF_2	57.05	
Carbonate	$(NH_4)_2CO_3 \cdot H_2O$	114.09	
Chloride	NH_4Cl	53.50	
Fluoborate	NH_4BF_4	104.86	
Hydroxide	NH_4OH	35.05	
Molybdate	NH_4MoO_4	196.08	48.9
Nitrate	NH_4NO_3	80.05	
Persulfate	$(NH_4)_2S_2O_8$	228.20	
Sulfamate	$NH_4(SO_3 \cdot NH_2)$	114.13	
Sulfate	$(NH_4)_2SO_4$	132.14	
Sulfide	$(NH_4)_2S$	68.14	
Thiocyanate	NH_4CNS	76.12	
Thiosulfate	$(NH_4)_2S_2O_3$	148.20	
Antimony	Sb	121.76	
Potassium antimonyltartrate	$2K(SbO)C_4H_4O_6 \cdot H_2O$	677.80	35.9
Trichloride	$SbCl_3$	228.13	53.4
Trifluoride	SbF_3	178.76	68.1
Trioxide	Sb_2O_3	291.52	83.5
Sulfide, tri	Sb_2S_3	339.70	71.7
Sulfide, penta	Sb_2S_5	403.82	60.3

TABLE 3 (*continued*)

Name	Formula	Formula weight	Metal content % (Theoretical)
Arsenic	As	74.91	
Trichloride	$AsCl_3$	181.28	41.3
Trioxide	As_2O_3	197.82	75.7
Barium	Ba	137.36	
Carbonate	$BaCO_3$	197.37	69.6
Chloride	$BaCl_2 \cdot 2H_2O$	244.31	56.2
Cyanide	$Ba(CN)_2 \cdot 2H_2O$	225.42	60.9
Hydroxide	$Ba(OH)_2 \cdot 8H_2O$	315.50	43.7
Nitrate	$Ba(NO_3)_2$	261.38	52.6
Sulfate	$BaSO_4$	233.42	58.9
Sulfide	BaS	169.42	81.2
Bismuth	Bi	209.00	
Trichloride	$BiCl_3$	315.37	66.3
Trioxide	Bi_2O_3	466.00	89.7
Cadmium	Cd	112.41	
Cyanide	$Cd(CN)_2$	164.43	68.4
Fluoborate*	$Cd(BF_4)_2$	296.05	37.9
Nitrate	$Cd(NO_3)_2 \cdot 4H_2O$	308.49	36.4
Oxide	CdO	128.41	87.5
Sulfate	$CdSO_4$	208.47	54.0
Calcium	Ca	40.08	
Carbonate	$CaCO_3$	100.09	
Chloride, anhyd.	$CaCl_2$	110.99	
Chloride	$CaCl_2 \cdot 2H_2O$	147.03	
Cyanide	$Ca(CN)_2$	92.12	
Fluoride	CaF_2	78.08	
Hydroxide	$Ca(OH)_2$	74.10	
Nitrate	$Ca(NO_3)_2 \cdot 4H_2O$	236.16	
Oxide	CaO	56.08	
Sulfate	$2CaSO_4 \cdot H_2O$	290.30	
	$CaSO_4 \cdot 2H_2O$	172.17	
Chromium	Cr	52.01	
Potassium sulfate	$Cr_2(SO_4)_3 \cdot K_2SO_4 \cdot 24H_2O$	998.84	
Trioxide	CrO_3	100.01	52.0
Cobalt	Co	58.94	
Ammonium sulfate	$Co(NH_4)_2(SO_4)_2 \cdot 6H_2O$	395.24	14.9
Carbonate	$CoCO_3$	118.95	49.5
Chloride	$CoCl_2 \cdot 6H_2O$	237.95	24.8
Fluoborate*	$Co(BF_4)_2$	232.58	25.3
Sulfate	$CoSO_4 \cdot 7H_2O$	281.11	21.0
Copper	Cu	63.57	
Cupric			
Acetate	$Cu(CH_3COO)_2 \cdot H_2O$	199.67	31.8
Carbonate, basic	$CuCO_3 \cdot Cu(OH)_2$	221.17	67.4
Chloride	$CuCl_2$	134.48	47.3
Cuprosulfite	$CuSO_3 \cdot Cu_2SO_3 \cdot 2H_2O$	386.86	49.3
Fluoborate*	$Cu(BF_4)_2$	237.21	26.8
Nitrate	$Cu(NO_3)_2 \cdot 3H_2O$	241.63	26.3
Oxide	CuO	79.57	79.9
Pyrophosphate	$Cu_2P_2O_7 \cdot 3H_2O$	355.15	35.8
Sulfate	$CuSO_4 \cdot 5H_2O$	249.71	25.4
Cuprous			
Chloride	CuCl	99.03	64.2
Cupricsulfite	$Cu_2SO_3 \cdot CuSO_3 \cdot 2H_2O$	386.86	49.3
Cyanide	CuCN	89.59	70.9

* Note: Compounds available only in water solutions.

TABLE 3 (continued)

Name	Formula	Formula weight	Metal content % (Theoretical)
Gallium	Ga	69.72	
Hydroxide	Ga(OH)$_3$	120.75	57.7
Gold	Au	197.2	
Auric			
Acid chloride	HAuCl$_4$·4H$_2$O	412.10	47.9
Chloride	AuCl$_3$	303.57	65.0
Chloride, dihydrate	AuCl$_3$·2H$_2$O	321.59	61.3
Aurous			
Cyanide	AuCN	232.66	84.8
Potassium cyanide, anhyd.	KAu(CN)$_2$	288.33	68.3
Potassium cyanide	KAu(CN)$_2$·2H$_2$O	324.36	60.7
Sodium cyanide	NaAu(CN)$_2$	272.21	72.5
Hydrogen	H$_2$	2.016	
Peroxide	H$_2$O$_2$	34.02	
Indium	In	114.76	
Chloride	InCl$_3$	221.13	51.9
Fluoborate*	In(BF$_4$)$_3$	375.22	30.6
Hydroxide	In(OH)$_3$	165.78	69.5
Sulfamate	In(SO$_3$·NH$_2$)$_3$	403.03	28.5
Sulfate	In$_2$(SO$_4$)$_3$	517.70	66.5
Iridium	Ir	193.10	
Tetrachloride	IrCl$_4$	334.93	57.6
Trichloride	IrCl$_3$	299.47	64.5
Iron	Fe	55.85	
Ferric			
Chloride, anhyd.	FeCl$_3$	162.22	34.4
Chloride	FeCl$_3$·6H$_2$O	270.32	20.7
Fluoride	FeF$_3$	112.85	49.6
Fluoborate*	Fe(BF$_4$)$_3$	316.31	17.6
Nitrate	Fe(NO$_3$)$_3$·9H$_2$O	404.02	13.8
Oxide	Fe$_2$O$_3$	157.70	69.9
Potassium ferricyanide	K$_3$Fe(CN)$_6$	329.25	16.9
Ferrous			
Ammonium sulfate	FeSO$_4$(NH$_4$)$_2$SO$_4$·6H$_2$O	392.16	14.2
Chloride, dihydrate	FeCl$_2$·2H$_2$O	144.78	38.6
Chloride, tetrahydrate	FeCl$_2$·4H$_2$O	198.83	28.1
Fluoborate	Fe(BF$_4$)$_2$	230	24.2
Potassium ferrocyanide	K$_4$Fe(CN)$_6$	422.39	13.4
Sulfate	FeSO$_4$·7H$_2$O	278.03	20.1
Lead	Pb	207.21	
Acetate	Pb(CH$_3$COO)$_2$·3H$_2$O	379.35	54.6
Basic carbonate	2PbCO$_3$·Pb(OH)$_2$	775.67	80.1
Dioxide	PbO$_2$	239.21	86.6
Fluoborate*	Pb(BF$_4$)$_2$	380.85	54.3
Fluosilicate*	PbSiF$_6$	463.27	44.7
Monoxide	PbO	223.31	92.8
Nitrate	Pb(NO$_3$)$_2$	331.24	62.6
Sulfamate	Pb(SO$_3$NH$_2$)$_2$	399.39	51.9
Magnesium	Mg	24.32	
Carbonate	MgCO$_3$	84.33	28.8
Chloride	MgCl$_2$	95.23	25.5
Fluoride	MgF$_2$	62.32	39.0
Hydroxide	Mg(OH)$_2$	58.34	41.7
Sulfate	MgSO$_4$·7H$_2$O	246.50	9.8
Manganese	Mn	54.93	
Potassium permanganate	KMnO$_4$	158.03	34.7
Manganic			
Dioxide	MnO$_2$	86.93	63.2

* Note: Compounds available only in water solutions.

TABLE 3 (continued)

Name	Formula	Formula weight	Metal content % (Theoretical)
Manganous			
Chloride	$MnCl_2 \cdot 4H_2O$	197.91	27.7
Sulfate	$MnSO_4 \cdot H_2O$	169.01	32.5
Mercury	Hg	200.61	
Mercuric			
Chloride	$HgCl_2$	271.52	74.0
Cyanide	$Hg(CN)_2$	252.65	79.4
Nitrate	$Hg(NO_3)_2 \cdot H_2O$	342.65	58.6
Oxide	HgO	216.61	92.7
Sulfate	$HgSO_4$	296.67	67.7
Mercurous			
Chloride	$HgCl$	236.07	84.8
Nitrate	$HgNO_3 \cdot H_2O$	280.63	71.5
Sulfate	Hg_2SO_4	497.28	80.5
Nickel	Ni	58.69	
Acetate	$Ni(CH_3COO)_2 \cdot 4H_2O$	248.84	23.6
Ammonium sulfate	$NiSO_4(NH_4)_2SO_4 \cdot 6H_2O$	394.99	14.9
Carbonate, basic	$2NiCO_3 \cdot 3Ni(OH)_2 \cdot 4H_2O$	587.58	50.0
Chloride	$NiCl_2 \cdot 6H_2O$	237.70	24.7
Cyanide	$Ni(CN)_2 \cdot 4H_2O$	182.79	32.1
Fluoborate*	$Ni(BF_4)_2$	232.33	25.2
Formate	$Ni(HCOO)_2 \cdot 2H_2O$	184.76	31.8
Hydroxide	$Ni(OH)_2$	92.71	63.3
Oxide	NiO	74.69	78.6
Sulfamate	$Ni(SO_3 \cdot NH_2)_2$	250.85	23.4
Sulfate			
Hexahydrate	$NiSO_4 \cdot 6H_2O$	262.85	22.3
Heptahydrate	$NiSO_4 \cdot 7H_2O$	280.87	20.9
Palladium	Pd	106.7	
Chloride	$PdCl_2 \cdot 2H_2O$	213.65	49.9
Chloride, sodium	$PdCl_2 \cdot 2NaCl \cdot 3H_2O$	348.57	30.6
Cyanide	$Pd(CN)_2$	158.72	67.1
Nitrite, potassium	$K_2Pd(NO_2)_4$	368.92	28.9
Platinum	Pt	195.23	
Chloride	$PtCl_4 \cdot 2HCl \cdot 6H_2O$	518.08	37.7
Diamminonitrite	$Pt(NH_3)_2(NO_2)_2$	321.31	60.8
Potassium	K	39.10	
Acetate	$K(CH_3COO)$	98.14	
Acid tartrate	$KHC_4H_4O_6$	188.18	
Bifluoride	KHF_2	78.10	
Bisulfate	$KHSO_4$	136.16	
Carbonate, anhyd.	K_2CO_3	138.20	
Carbonate	$K_2CO_3 \cdot 1\frac{1}{2}H_2O$	165.23	
Chlorate	$KClO_4$	122.56	
Chloride	KCl	74.55	
Chromate	K_2CrO_4	194.20	
Copper cyanide	$K_2Cu(CN)_3$	219.78	29.0
Cyanide	KCN	65.11	
Dichromate	$K_2Cr_2O_7$	294.21	
Ferricyanide	$K_3Fe(CN)_6$	329.25	
Ferrocyanide	$K_4Fe(CN)_6$	422.39	
Fluoborate	$K(BF_4)$	125.92	
Fluoride, anhyd.	KF	58.10	
Fluoride	$KF \cdot 2H_2O$	94.13	
Gold cyanide, anhyd.	$KAu(CN)_2$	288.33	68.8

* Note: Compounds available only in water solutions.

TABLE 3 (continued)

Name	Formula	Formula weight	Metal content % (Theoretical)
Gold cyanide	KAu(CN)$_2 \cdot$2H$_2$O	324.36	60.7
Hydroxide	KOH	56.10	
Hypophosphite	KH$_2$PO$_2$	104.09	
Nickel cyanide	K$_2$Ni(CN)$_4 \cdot$H$_2$O	258.97	22.7
Nitrate	KNO$_3$	101.10	
Permanganate	KMnO$_4$	158.03	
Perrhenate	KReO$_4$	289.41	64.4
Phosphate			
Monobasic	KH$_2$PO$_4$	136.09	
Dibasic	K$_2$HPO$_4 \cdot$3H$_2$O	228.23	
Tribasic	K$_3$PO$_4$	212.27	
Pyrophosphate	K$_4$P$_2$O$_7 \cdot$10H$_2$O	510.59	
Silver cyanide	KAg(CN)$_2$	199.01	54.0
Stannate	K$_2$SnO$_3 \cdot$3H$_2$O	298.94	39.7
Thiocyanate	KCNS	97.17	
Mercury cyanide	KHg(CN)$_3$	317.73	63.2
Rhenium	Re	186.31	
Perrhenic acid	HReO$_4$	251.32	74.4
Potassium, perrhenate	KReO$_4$	289.41	64.4
Sodium perrhenate	NaReO$_4$	273.31	68.7
Rhodium	Rh	102.91	
Hydroxide	Rh(OH)$_3$	153.94	66.8
Phosphate	RhPO$_4$	197.89	52.0
Sulfate	Rh$_2$(SO$_4$)$_3$	494.01	20.8
Selenium	Se	78.96	
Dioxide	SeO$_2$	110.96	71.2
Silver	Ag	107.88	
Chloride	AgCl	143.34	75.3
Cyanide	AgCN	133.90	80.5
Nitrate	AgNO$_3$	169.89	63.5
Potassium cyanide	KAg(CN)$_2$	199.01	54.2
Sodium cyanide	NaAg(CN)$_2$	182.91	58.9
Sodium	Na	23.00	
Acetate	Na(CH$_3$COO)$\cdot$3H$_2$O	136.09	
Bicarbonate	NaHCO$_3$	84.02	
Bifluoride	NaHF$_2$	62.00	
Bisulfate	NaHSO$_4$	120.07	
Bisulfite	NaHSO$_3$	104.06	
Carbonate, anhyd.	Na$_2$CO$_3$	106.00	
Carbonate	Na$_2$CO$_3 \cdot$H$_2$O	124.02	
Chromate	Na$_2$CrO$_4 \cdot$4H$_2$O	234.07	
Chloride	NaCl	58.45	
Chlorite	NaClO$_2$	90.46	
Cyanide	NaCN	49.02	
Dichromate	Na$_2$Cr$_2$O$_7 \cdot$2H$_2$O	298.05	
Fluoborate	NaBF$_4$	109.82	
Fluoride	NaF	42.00	
Formate	Na(HCOO)	68.02	
Hydrosulfite	Na$_2$S$_2$O$_4$	174.11	
Hydroxide	NaOH	40.00	
Metaborate	NaBO$_2 \cdot$4H$_2$O	137.88	
Metaphosphate	NaPO$_3$	102.02	
Nitrate	NaNO$_3$	85.00	
Nitrite	NaNO$_2$	69.00	

TABLE 3 (*continued*)

Name	Formula	Formula weight	Metal content % (Theoretical)
Phosphate			
Monobasic	$NaH_2PO_4 \cdot H_2O$	138.01	
Dibasic	$Na_2HPO_4 \cdot 7H_2O$	268.09	
Tribasic	$Na_3PO_4 \cdot 12H_2O$	380.16	
Potassium tartrate	$NaK(C_4H_4O_6) \cdot 4H_2O$	282.23	
Pyrophosphate, anhyd. (tetrasodium)	$Na_4P_2O_7$	266.02	
Pyrophosphate (tetrasodium)	$Na_4P_2O_7 \cdot 10H_2O$	446.11	
Sesquicarbonate	$Na_2CO_3 \cdot NaHCO_3 \cdot 2H_2O$	226.04	
Sesquisilicate	$Na_6Si_2O_7 \cdot 11H_2O$	504.28	
Silicate			
Meta	Na_4SiO_3	122.05	
Ortho	Na_4SiO_4	184.05	
Stannate	$Na_2Sn(OH)_6$	266.74	44.5
Sulfamate	$NaSO_3 \cdot NH_2$	119.09	
Sulfate	$Na_2SO_4 \cdot 10H_2O$	322.21	
Sulfide	$Na_2S \cdot 9H_2O$	240.20	
Tetraborate	$Na_2B_4O_7 \cdot 10H_2O$	381.43	
Tetraphosphate	$Na_6P_4O_{13}$	470.09	
Thiocyanate	$NaCNS$	81.08	
Thiosulfate	$Na_2S_2O_3 \cdot 5H_2O$	248.21	
Tripolyphosphate	$Na_5P_3O_{10}$	367.93	
Tungstate	Na_2WO_4	293.91	62.6
Zincate	Na_2ZnO_2	143.38	45.5
Tin	Sn	118.70	
Stannic			
Potassium stannate	$K_2Sn(OH)_6$	298.94	39.7
Sodium stannate	$Na_2Sn(OH)_6$	266.74	44.5
Stannous			
Chloride	$SnCl_2 \cdot 2H_2O$	225.65	52.7
Fluoborate*	$Sn(BF_4)_2$	292.34	40.6
Sulfate	$SnSO_4$	214.76	55.3
Oxide	SnO	134.70	88.1
Chloride, anhyd.	$SnCl_2$	189.61	62.6
Zinc	Zn	65.38	
Ammonium chloride	$ZnCl_2 \cdot 2NH_4Cl$	243.29	26.9
Chloride	$ZnCl_2$	136.29	47.9
Cyanide	$Zn(CN)_2$	117.39	55.7
Fluoborate*	$Zn(BF_4)_2$	239.02	27.4
Oxide	ZnO	81.38	80.4
Sulfate	$ZnSO_4 \cdot 7H_2O$	287.56	22.8
Tungsten	W	183.92	
Sodium tungstate	Na_2WO_4	293.91	62.6
Trioxide	WO_3	231.92	79.3

Note: Compounds available only in water solutions.

TABLE 4. COMMONLY USED NAMES FOR CHEMICALS

Commonly Used Names	Chemical Name or Compound	Commonly Used Names	Chemical Name or Compound
Alum	Potassium aluminum sulfate, $K_2Al_2(SO_4)_4 \cdot 24H_2O$ (varies)	Gypsum	Calcium sulfate, $CaSO_4 \cdot 2H_2O$
"Aloxite" Alumina "Alundum"	Aluminum oxide, Al_2O_3	Hypo	Sodium thiosulfate, $Na_2S_2O_3 \cdot 5H_2O$
		Lime	Calcium oxide, CaO
Aqua fortis	Nitric acid	Litharge	Lead monoxide, PbO
Aqua regia	Nitric acid − 1 part + hydrochloric acid − 3 parts	Liver of sulfur	Potassium polysulfide, K_2S_x
		Marble	Calcium carbonate, $CaCO_3$
Baking soda	Sodium bicarbonate, $NaHCO_3$	Metso	Sodium metasilicate, $Na_2SiO_3 \cdot 5H_2O$
Bleaching powder	Calcium hypochlorite, $CaOCl_2$	Milk of lime	Calcium hydroxide, $Ca(OH)_2$
Blue dip	Solution of mercury salt	Muriatic acid	Hydrochloric acid, HCl
Blue stone Blue vitriol	Copper sulfate, $CuSO_4 \cdot 5H_2O$	Oil of vitriol	Acid sulfuric, H_2SO_4
		Pearl ash	Potassium carbonate, K_2CO_3
Boracic acid	Boric acid		
Borax	Sodium tetraborate, $Na_2B_4O_7 \cdot 10H_2O$	Plaster of Paris	Calcium sulfate, $2CaSO_4 \cdot H_2O$
Butter of antimony	Antimony trichloride, $SbCl_3$	Quick lime	Calcium oxide, CaO
		Quicksilver	Mercury
Calcite	Calcium carbonate, $CaCO_3$	Rochelle salt	Sodium potassium tartrate, $NaK(C_4H_4O_6) \cdot 4H_2O$
Calomel	Mercurous chloride, Hg_2Cl_2		
		Rouge	Iron oxide, Fe_2O_3
"Carborundum"	Silicon carbide, SiC	Sal ammoniac	Ammonium chloride, NH_4Cl
Caustic potash	Potassium hydroxide, KOH	Single nickel salt	Nickel sulfate, $NiSO_4 \cdot 6H_2O$ or $NiSO_4 \cdot 7H_2O$
Caustic soda	Sodium hydroxide, NaOH		
Chalk	Calcium carbonate, $CaCO_3$	Slaked lime	Calcium hydroxide, $Ca(OH)_2$
Chevereul salt	Cupro cupricsulfite, $Cu_2SO_3 \cdot CuSO_4 \cdot 2H_2O$	Soda ash	Sodium carbonate, Na_2CO_3
Chloride of lime	Calcium hypochlorite, $CaOCl_2$	Sour water	Dilute sulfuric acid
		Sugar of lead	Lead acetate, $Pb(CH_3COO)_2$
Common salt	Sodium chloride, NaCl		
Copperas	Ferrous sulfate, $FeSO_4 \cdot 7H_2O$	TSP	Trisodium phosphate, Na_3PO_4
Cream of tartar	Potassium acid tartrate, $KH(C_4H_4O_6)$	Washing soda	Sodium carbonate, $Na_2CO_3 \cdot 10H_2O$
Double nickel salt	Nickel ammonium sulfate, $NiSO_4(NH_4)_2SO_4 \cdot 6H_2O$	Water glass	Sodium silicate solution
		Wood's metal	Low melting temperature alloy, e.g., 4 pts Bi, 2 pts Pb, 1 pt Sn, 1 pt Cd
Emery	Aluminum oxide, Al_2O_3		
Epsom salt	Magnesium sulfate, $MgSO_4 \cdot 7H_2O$		
Flowers of sulfur	Sulfur, S	Vienna lime	Calcium carbonate, $CaCO_3$
Fluorspar	Calcium fluoride, CaF_2		
Glauber's salt	Sodium sulfate, $Na_2SO_4 \cdot 10H_2O$		

TABLE 5. SOME PHYSICAL PROPERTIES OF THE ELEMENTS†

The following table has been compiled from many sources.

For the less common metals and for the non-metallic elements, most of the data are from National Bureau of Standards Circular C447, 1943, page 459.

Atomic weights are from the 1947 Report of the International Commission on Atomic Weights. Nearly all the melting points are those most recently quoted by the National Bureau of Standards. Most of the boiling points were selected by K. K. Kelley (Bureau of Mines Bulletin 383, 1935) from smoothed curves of vapor pressure versus temperature.

All entries in this table are for elements of high purity. Impurities may have a great effect on certain properties, for example, on the melting point and electrical resistivity. Most of the data do not apply to commercial metals, which contain greater amounts of impurities than the elements considered here. For metals, most of the properties refer to annealed specimens. Cold working will change certain properties—for example, density and electrical resistivity. Grain size may also affect some properties.

It will be noted that the number of significant figures contained in the various values ranges from one to six. Such differences in the uncertainties of the quoted values may reflect the greater or lesser care that has been taken in measuring the properties. Certain properties are much more difficult to measure than others; for example, the moduli of elasticity of soft and ductile metals and the boiling points of many metals are difficult to determine precisely. It is not uncommon for different investigators to report different values for a given property, depending on the method of measuring; examples are thermal conductivity and latent heat of fusion. Because of these factors, some of the values are probably less accurate than would be inferred from counting the number of digits contained in the values listed.

In using this table, the reader should be guided by the above considerations concerning the probable accuracy of the numbers.

Element	Symbol	Atomic number	Atomic weight (1947)	Density at 20°C (68°F), g/cu cm	Density at 68°F (20°C), lb/cu in.	Atomic volume, cu cm/g-atom	Melting point, °C	Melting point, °F	Boiling point, °C	Boiling point, °F	Specific heat at 20°C, cal/g/°C	Heat of fusion, cal/g	Heat of fusion, Btu/lb
Aluminum	Al	13	26.97	2.699	0.09751	9.993	660.2 ± 0.1	1220.4 ± 0.2	2060	3740	0.215	94.6	170
Antimony	Sb	51	121.76	6.62	0.239	18.4	630.5 ± 0.1	1166.9 ± 0.2	1440	2620	0.049	38.3	68.9
Arsenic	As	33	74.91	5.73	0.207	13.1	814ᵈ	1497ᵈ	610ⁱ	1130ⁱ	0.082	—	—
Barium	Ba	56	137.36										
Boron	B	5	10.82										
Bromine	Br	35	79.916										
Cadmium	Cd	48	112.41	8.65	0.313	13.0	320.9 ± 0.1	609.6 ± 0.2	765	1409	0.055	13.2	23.8
Calcium	Ca	20	40.08										
Carbon (graphite)	C	6	12.010										
Cerium	Ce	58	140.13										
Chlorine	Cl	17	35.457										
Chromium	Cr	24	52.01	7.19	0.260	7.23	1890 ± 10	3430 ± 20	2500	4500	0.11	75.6	136
Cobalt	Co	27	58.94	8.9	0.32	6.6	1495 ± 1	2723 ± 2	2900	5250	0.099	58.4	105
Copper	Cu	29	63.54	8.96	0.324	7.09	1083.0 ± 0.1	1981.4 ± 0.2	2600	4700	0.092	50.6	91.1
Fluorine	F	9	19.00										
Germanium	Ge	32	72.60	5.36	0.194	13.5	958 ± 10	1760 ± 20	—	—	0.073	—	—
Gold	Au	79	197.2	19.32	0.698	10.2	1063.0 ± 0.0	1945.4 ± 0.0	2970	5380	0.031	16.1	29.0
Hydrogen	H	1	1.0080	0.08375 × 10⁻³	0.3026 × 10⁻⁴	—	−259.4 ± 0.1	−434.6 ± 0.2	−252.7	−422.9	3.45	15.0	27.0

Element	Symbol	Z	At. Wt.										
Indium	In	49	114.76	7.31		0.264		156.4 ± 0.1	313.5 ± 0.2		0.057		—	—
Iodine	I	53	126.92										—	—
Iridium	Ir	77	193.1	22.5		0.813		2454 ± 3	4449 ± 5	5300	0.031	65	117
Iron	Fe	26	55.85	7.87		0.284		1539 ± 3	2802 ± 5	2740	0.11	6.3	11.3
Lead	Pb	82	207.21	11.34		0.4097		327.4 ± 0.1	621.3 ± 0.2	1740	0.031		
Lithium	Li	3	6.940										
Magnesium	Mg	12	24.32	1.74		0.0628		650 ± 2	1202 ± 4	1110	0.25	89	160
Manganese	Mn	25	54.93	7.43		0.268		1245 ± 10	2273 ± 20	2150	0.115	64	115
Mercury	Hg	80	200.61	13.55		0.4896		−38.87 ± 0.02	−37.97 ± 0.04	357	0.033	2.7	4.9
Molybdenum	Mo	42	95.95	10.2		0.369		2625 ± 50	4760 ± 90	4800	0.061	70	126
Nickel	Ni	28	58.69	8.90		0.322		1455 ± 1	2651 ± 2	2730	0.105	74	133
Nitrogen	N	7	14.008										
Oxygen	O	8	16.0000	1.3318 × 10⁻³	0.048118 × 10⁻³			−218.8 ± 0.1	−361.8 ± 0.2	−183.0	−297.4	0.218	3.3	5.9
Palladium	Pd	46	106.7	12.0		0.434	8.89	1554 ± 1	2829 ± 2	4000	7200	0.058ᵐ	34.2	61.6
Phosphorus (yellow)	P	15	30.98										
Platinum	Pt	78	195.23	21.45		0.7750	9.102	1773.5 ± 1	3224.3 ± 1.8	4410	7970	0.032	27	49
Potassium	K	19	39.096	20		0.72	9.3	3170 ± 60	5740 ± 110			0.033		
Rhenium	Re	75	186.31											
Rhodium	Rh	45	102.91	12.44		0.4495	8.273	1966 ± 3	3571 ± 5	4500	8100	0.059		
Ruthenium	Ru	44	101.7	12.2		0.441	8.33	2500 ± 100	4500 ± 180	4900	8850	0.057ᵐ		
Selenium	Se	34	78.96	4.81		0.174	16.4	220 ± 5	428 ± 9	680	1260	0.084	6.6	11.9
Silicon	Si	14	28.06											
Silver	Ag	47	107.880	10.49		0.3790	10.28	960.5 ± 0.0	1760.9 ± 0.0	2210	4010	0.056ᵐ	25	45
Sodium	Na	11	22.997											
Sulfur (yellow)	S	16	32.066											
Tantalum	Ta	73	180.88	16.6		0.600	10.9	2996 ± 50	5425 ± 90			0.036ᵐ		
Tellurium	Te	52	127.61	6.24		0.225	20.5	450 ± 10	840 ± 20	1390	2530	0.047	7.3	13.1
Thallium	Tl	81	204.39	11.85		0.4281	17.24	300 ± 3	572 ± 5	1460	2660	0.031	7.2	13.0
Tin	Sn	50	118.70	7.298		0.2637	16.26	231.9 ± 0.1	449.4 ± 0.2	2270	4120	0.054	14.5	26.1
Titanium	Ti	22	47.90	4.54		0.164	10.6	1820 ± 100	3300 ± 180		10,700	0.126		
Tungsten	W	74	183.92	19.3		0.697	9.53	3410 ± 20	6170 ± 35	5930		0.032	44	79
Uranium	U	92	238.07	18.7		0.676	12.7	1130	2065			0.028		
Vanadium	V	23	50.95	6.0		0.217	8.5	1735 ± 50	3150 ± 90	3400	6150	0.120		
Zinc	Zn	30	65.38	7.133ʳ		0.258ʳ	9.17	419.46	787.03	906	1663	0.0915	24.09	43.36
Zirconium	Zr	40	91.22	6.5		0.23	14	1750 ± 700	3200 ± 1300			0.066		

TABLE 5 (*continued*)

Element	Coefficient of Linear Thermal Expansion near 20°C, micro-in./°C	Coefficient of Linear Thermal Expansion near 68°F, micro-in./°F	Thermal Conductivity near 20°C, cal/sq cm/cm per °C/sec	Electrical Resistivity, microhm-cm	Modulus of Elasticity in Tension, 10⁶ psi	Crystal Structure
Aluminum	23.9*	13.3*	0.53	2.655 (20 C)	10	Face-centered cubic
Antimony	8.5 to 10.8ᵇ,ᶜ	4.7 to 6.0ᵇ,ᶜ	0.045	39.0 (0 C)	11.3	Rhombohedral
Arsenic	4.7	2.6	—	35 (0 C)	—	Rhombohedral*
Cadmium	29.8	16.6	0.22	6.83 (0 C)	8	Close-packed hexagonal
Chromium	6.2	3.4	0.16	13 (28 C)	36	Body-centered cubic*
Cobalt	12.3	6.8	0.165	6.24 (20 C)	30	Close-packed hexagonal*
Copper	16.5	9.2	0.94	1.673 (20 C)	16	Face-centered cubic
Germanium	—	—	—	89,000 (0 C)	—	Diamond cubic
Gold	14.2	7.9	0.71	2.19 (0 C)	12	Face-centered cubic
Hydrogen	—	—	4.06×10^{-4}	—	—	Hexagonal
Indium	33	18	0.057	8.37 (0 C)	—	Face-centered tetragonal
Iridium	6.8	3.8	0.14	5.3 (20 C)	75	Face-centered cubic
Iron	11.7	6.5	0.18	9.71 (20 C)	28.5	Body-centered cubic*
Lead	29.3*	16.3*	0.083	20.65 (20 C)	2.6	Face-centered cubic
Magnesium	26ᵃ	14ᵃ	0.38	4.46 (20 C)	6.5	Close-packed hexagonal
Manganese	22	12	—	185 (20 C)	23	Cubic* (complex)
Mercury	—	—	0.0201	94.1 (0 C)	—	Rhombohedral
Molybdenum	4.9*	2.7*	0.35	5.17 (0 C)	50	Body-centered cubic
Nickel	13.3°	7.4°	0.22	6.84 (20 C)	30	Face-centered cubic
Nitrogen	—	—	0.000060	—	—	Hexagonal*
Oxygen	—	—	0.000059	—	—	Cubic*
Palladium	11.8	6.6	0.17	10.8 (20 C)	17	Face-centered cubic
Platinum	8.9	4.9	0.17	9.83 (0 C)	21	Face-centered cubic
Rhenium	—	—	—	—	—	Close-packed hexagonal
Rhodium	8.3*	4.6*	0.21	4.5 (20 C)	42	Face-centered cubic*
Ruthenium	9.1	5.1	—	7.6 (0 C)	60	Close-packed hexagonal*
Selenium	37	21	—	—	—	Hexagonal*

Silver	19.7[o]	10.9[o]	1.0[m]	11	Face-centered cubic
Tantalum	6.5	3.6	0.13	12.4 (18 C)	27	Body-centered cubic
Tellurium	16.8	9.3	0.014	2×10^5 (19.6 C)	6	Hexagonal
Thallium	28	16	0.093	18 (0 C)	—	Close-packed hexagonal*
Tin	23	13	0.16	11.5 (20 C)	6	Body-centered tetragonal*
Titanium	8.5	4.7	—	80 (0 C)	16.8	Close-packed hexagonal*
Tungsten	4.3	2.4	0.48	5.5 (20 C)	50	Body-centered cubic*
Uranium	—	—	0.064	60 (18 C)	—	Orthorhombic*
Vanadium	7.8	4.3	—	26 (20 C)	—	Body-centered cubic
Zinc	39.7[b]	22.1	0.27	5.916*(20 C)	t	Close-packed hexagonal
Zirconium	5	3	—	41.0 (0 C)	11	Close-packed hexagonal*

NOTES. * Ordinary form; other modifications known or probable.

[a] Computed [b] For polycrystalline zinc; in single crystals, varies from 61.5 (parallel to hexagonal axis) to 15 (perpendicular to hexagonal axis) [c] From 20 to 60°C (68 to 140°F) [d] At 36 atmospheres pressure [e] From 20 to 100°C (68 to 212°F) [f] At 30 atmospheres pressure [g] At minus 62°C (minus 80°F) [h] From 20 to 50°C (68 to 122°F) [i] Sublimes [j] From 20 to 200°C (68 to 392°F) [k] From 20 to 750°C (68 to 1382°F) [l] At 0°C (32°F) [m] At 40°C (104°F) [o] From 0 to 100°C (32 to 212°F) [p] At 100°C (212°F) [q] Monoclinic sulfur [r] At 25°C (77°F) [s] For polycrystalline zinc; in single crystals, varies from 6.16 (parallel to hexagonal axis) to 5.89 (perpendicular to hexagonal axis) [t] Pure zinc has no clearly defined modulus of elasticity [u] The solid helium listed is stable only under pressure.

† Reprinted from the "Metals Handbook," 1948 Edition, by permission of the publisher, The American Society for Metals, Cleveland 3, Ohio.

TABLE 6. DENSITIES OF METALS AND ALLOYS*

(Values listed here are for as-cast, as-wrought or as-annealed material)

Metal and Composition	Density g/cu cm	lb/cu in.	Metal and Composition	Density g/cu cm	lb/cu in.
Aluminum and Aluminum Alloys			**Aluminum and Aluminum Alloys** (continued)		
Wrought			A108—5.5 Si, 4.5 Cu	2.79	0.101
Aluminum (99.996 Al)	2.699	0.09751	C113—7 Cu, 3.5 Si	2.91	0.105
1100 (2S)—99.0+ Al	2.71	0.0979	A132—12 Si, 2.5 Ni, 1.2 Mg, 0.8 Cu	2.68	0.0968
3003 (3S)—1.2 Mn	2.73	0.0986	Red X-13—1.5 Cu, 0.7 Mg, 0.7 Mn, 12 Si	2.7	0.098
3004 (4S)—1.25 Mn, 1.0 Mg	2.72	0.0983	B195—4.5 Cu, 2.5 Si	2.78	0.100
2011 (11S)—5.5 Cu, 0.5 Pb, 0.5 Bi	2.82	0.102	A214—3.8 Mg, 1.8 Zn	2.65	0.0957
2014 (14S)—4.4 Cu, 0.8 Si, 0.8 Mn, 0.4 Mg	2.80	0.101	750—6.5 Sn, 1 Cu, 1 Ni	2.89	0.104
2017 (17S)—4 Cu, 0.5 Mg, 0.5 Mn	2.79	0.101	13—12 Si	2.66	0.0961
2018 (18S)—4 Cu, 2 Ni, 0.5 Mg	2.80	0.101	85—5 Si, 4 Cu	2.78	0.100
2024 (24S)—4.5 Cu, 1.5 Mg, 0.6 Mn	2.77	0.100	218—8 Mg	2.53	0.0914
2025 (25S)—4.5 Cu, 0.8 Mn, 0.8 Si	2.79	0.101	360—9.5 Si, 0.5 Mg	2.68	0.0968
4032 (32S)—12.5 Si, 1.0 Mg, 0.9 Cu, 0.9 Ni	2.69	0.0972	380—8.5 Si, 3.5 Cu	2.76	0.0997
6151 (A51S)—1.0 Si, 0.6 Mg, 0.25 Cr	2.69	0.0972			
5052 (52S)—2.5 Mg, 0.25 Cr	2.68	0.0968	**Cobalt and Cobalt Alloys**		
5053 (53S)—1.3 Mg, 0.7 Si, 0.25 Cr	2.69	0.0972	Pure Cobalt	8.9	0.32
5056 (56S)—5.2 Mg, 0.1 Mn, 0.1 Cr	2.64	0.0954	*Cast*		
6061 (61S)—1.0 Mg, 0.6 Si, 0.25 Cu, 0.25 Cr	2.70	0.0975	"61" Alloy—23 Cr, 5 W, 2 Ni	8.54	0.309
7075 (75S)—5.5 Zn, 2.5 Mg, 1.5 Cu	2.80	0.101	"Vitallium"—27 Cr, 6 Mo, 2 Ni	8.30	0.300
2014 (R301)—4.5 Cu, 0.8 Si, 0.5 Mn, 0.4 Mg, clad	2.81	0.101	"X-40" Alloy—23 Cr, 10 Ni, 7 W	8.61	0.311
Cast			"422-19" Alloy—23 Cr, 16 Ni, 6 Mo	8.31	0.300
43—5 Si	2.69	0.0972	"S-816" Alloy—20 Ni, 19 Cr, 4 W, 4 Mo, 4 Cb, 3 Fe	8.59	0.310
108—4 Cu, 3 Si	2.79	0.101	"6059" Alloy—32 Ni, 23 Cr, 6 Mo	8.21	0.297
113—7 Cu, 2 Si, 1.7 Zn	2.91	0.105	**Copper and Copper Alloys**		
122—10 Cu, 0.2 Mg	2.95	0.107	Pure copper	8.96	0.324
142—4 Cu, 2 Ni, 1.5 Mg	2.81	0.101	Electrolytic tough pitch copper (99.92 Cu-0.04 O)	8.92[a]	0.322[a]
195—4.5 Cu	2.81	0.101	Deoxidized or phosphorized copper (99.94 Cu-0.02 P)	8.94	0.323
214—3.8 Mg	2.65	0.0957	*Wrought*		
220—10 Mg	2.58	0.0932	Primer gilding metal—97 Cu, 3 Zn	8.89	0.321
355—5 Si, 1.3 Cu, 0.5 Mg	2.70	0.0975	Gilding metal, 95%—95 Cu, 5 Zn	8.86	0.320
A355—5 Si, 1.4 Cu, 0.5 Mg, 0.7 Mn, 0.7 Ni	2.74	0.0990	Commercial bronze, 90%—90 Cu, 10 Zn	8.80	0.318
Allcast—5 Si, 3 Cu	2.76	0.0997	Red brass, 85%—85 Cu, 15 Zn	8.75	0.316
356—7 Si, 0.3 Mg	2.68	0.0968	Low brass, 80%—80 Cu, 20 Zn	8.67	0.313
Red X-8—8 Si, 0.3 Mg, 0.3 Mn, 1.5 Cu	2.73	0.0986			
40E—5.5 Zn, 0.6 Mg, 0.5 Cr, 0.2 Ti	2.81	0.101			

Copper and Copper Alloys (continued)

Alloy	Density	—
Brazing brass, 75%—75 Cu, 25 Zn	8.59	0.310
Cartridge Brass, 70%—70 Cu, 30 Zn	8.53	0.308
Yellow brass, 65%—65 Cu, 35 Zn	8.47	0.306
Muntz metal, 60%—60 Cu, 40 Zn	8.39	0.303
Leaded commercial bronze—89 Cu, 1.75 Pb, 9.25 Zn	8.83	0.319
Leaded red brass—78.5 Cu, 1.5 Pb, 20 Zn	8.70	0.314
Low-leaded brass—67 Cu, 0.5 Pb, 32.5 Zn	8.50	0.307
Low-leaded brass—64.5 Cu, 0.5 Pb, 35 Zn	8.47	0.306
Medium-leaded brass—64.5 Cu, 1.0 Pb, 34.5 Zn	8.47	0.306
High-leaded brass—62.5 Cu, 1.75 Pb, 35.75 Zn	8.47	0.306
Extra-high-leaded brass—62.5 Cu, 2.5 Pb, 35 Zn	8.50	0.307
Free-cutting brass—61.5 Cu, 3 Pb, 35.5 Zn	8.50	0.307
Leaded Muntz metal—60 Cu, 0.5 Pb, 39.5 Zn	8.41	0.304
Free-cutting Muntz metal—60.5 Cu, 1.1 Pb, 38.4 Zn	8.41	0.304
Forging brass—60 Cu, 2 Pb, 38 Zn	8.44	0.305
Architectural bronze—57 Cu, 3 Pb, 40 Zn	8.47	0.306
Admiralty metal—71 Cu, 1 Sn, 28 Zn	8.63	0.308
Naval brass—60 Cu, 0.75 Sn, 39.25 Zn	8.41	0.304
Leaded naval brass—60 Cu, 0.75 Sn, 1.75 Pb, 37.5 Zn	8.44	0.305
Manganese bronze—58.5 Cu, 1 Sn, 1.4 Fe, 39 Zn, 0.1 Mn	8.53	0.308
Aluminum brass—76 Cu, 22 Zn, 2 Al	8.33	0.301
Phosphor bronze, 1.25% E—98.75 Cu, 1.25 Sn	8.89	0.321
Phosphor bronze, 5% A—95 Cu, 5 Sn	8.86	0.320
Phosphor bronze, 8% C—92 Cu, 8 Sn	8.80	0.318
Phosphor bronze, 10% D—90 Cu, 10 Sn	8.78	0.317
Cupro-nickel, 30%—70 Cu, 30 Ni	8.94	0.323
Nickel silver, 18% A—65 Cu, 17 Zn, 18 Ni	8.73	0.315
Nickel silver, 15%—64 Cu, 21 Zn, 15 Ni	8.69	0.314
Nickel silver, 10%—65 Cu, 25 Zn, 10 Ni	8.68	0.313
Nickel silver, 18% B—55 Cu, 27 Zn, 18 Ni	8.70	0.314
Nickel silver, 15%—57 Cu, 28 Zn, 15 Ni	8.63	0.312
Nickel silver, 25%—55 Cu, 20 Zn, 25 Ni	8.72	0.315
Nickel silver, 30%—47 Cu, 23 Zn, 30 Ni	8.74	0.316
Silicon bronze, Type A—97 Cu, 3 Si	8.53	0.308
Silicon bronze, Type B—98.5 Cu, 1.5 Si	8.75	0.316

Copper and Copper Alloys (continued)

Alloy	Density	—
Aluminum bronze, 5%—95 Cu, 5 Al	8.17	0.295
Cupro-nickel, 15%—85 Cu, 15 Ni	8.94	0.323
Cupro-nickel, 20%—80 Cu, 20 Ni	8.94	0.323
Constantan—55 Cu, 45 Ni	8.9	0.32
Cunife—60 Cu, 20 Fe, 20 Ni	8.61	0.311
Beryllium copper—98 Cu, 2 Be	8.23	0.297

Cast

Alloy	Density	—
Leaded tin bronze—88 Cu, 6 Sn, 1.5 Pb, 4.5 Zn	8.7	0.315
Leaded tin bearing bronze—87 Cu, 10 Sn, 1 Pb, 2 Zn	8.80	0.318
High-leaded tin bronze—85 Cu, 5 Sn, 9 Pb, 1 Zn	8.87	0.320
High-leaded tin bronze—83 Cu, 7 Sn, 7 Pb, 3 Zn	8.93	0.323
High-leaded tin bronze—80 Cu, 10 Sn, 10 Pb	8.80	0.318
High-leaded tin bronze—78 Cu, 7 Sn, 15 Pb	9.25	0.334
High-leaded tin bronze—70 Cu, 5 Sn, 25 Pb	9.30	0.336
Ounce or composition metal—85 Cu, 5 Sn, 5 Pb, 5 Zn	8.80	0.318
Leaded red brass—83 Cu, 4 Sn, 6 Pb, 7 Zn	8.6	0.311
Leaded semi-red brass—81 Cu, 3 Sn, 9 Pb, 7 Zn	8.70	0.314
Leaded semi-red brass—76 Cu, 3 Sn, 6 Pb, 15 Zn	8.6	0.311
Leaded yellow brass—71 Cu, 1 Sn, 3 Pb, 25 Zn	8.50	0.307
Leaded yellow brass—66 Cu, 1 Sn, 3 Pb, 30 Zn	8.4	0.303
Leaded yellow brass—60 Cu, 1 Sn, 1 Pb, 38 Zn	8.40	0.303
High-strength yellow brass—62 Cu, 26 Zn, 3 Fe, 5.5 Al, 3.5 Mn	7.9	0.285
High-strength yellow brass—58 Cu, 39.25 Zn, 1.25 Fe, 1.25 Al, 0.25 Mn	8.2	0.296
Leaded manganese bronze—59 Cu, 0.75 Sn, 0.75 Pb, 37 Zn, 1.25 Fe, 0.75 Al, 0.5 Mn	8.2	0.296
Nickel silver—66 Cu, 5 Sn, 1.5 Pb, 2 Zn, 25 Ni	8.85[b]	0.320[b]
Nickel silver, dairy bronze—64 Cu, 4 Sn, 4 Pb, 8 Zn, 20 Ni	8.85	0.320
Nickel silver, Benedict metal—57 Cu, 2 Sn, 9 Pb, 20 Zn, 12 Ni	8.95	0.323
Leaded nickel silver—60 Cu, 3 Sn, 5 Pb, 16 Zn, 16 Ni	8.95	0.323
Aluminum bronze—89 Cu, 1 Fe, 10 Al	7.6	0.274
Aluminum bronze—87.5 Cu, 3.5 Fe, 9 Al	7.4	0.267
Aluminum bronze—86 Cu, 4 Fe, 10 Al	7.5	0.271

TABLE 6 (continued)

Metal and Composition	g/cu cm	Density lb/cu in.	Metal and Composition	g/cu cm	Density lb/cu in.
Iron and Iron Alloys			ASTM-10 bearing metal—83 Pb, 15 Sb, 2 Sn	10.07	0.3638
Electrolytic iron	7.874	0.2845	Lead-base babbitt—80 Pb, 15 Sb, 5 Sn	10.04	0.3627
Ingot iron	7.866	0.2842	Lead-base babbitt—75 Pb, 15 Sb, 10 Sn	9.73	0.351
Wrought iron	7.7	0.28	ASTM-6 bearing metal—63.5 Pb, 15 Sb, 20 Sn, 1.5 Cu	9.33	0.337
Gray cast iron	7.15c	0.258c	Tin-lead solder—95 Pb, 5 Sn	11.0	0.397
Malleable iron	7.27d	0.262d	Tin-lead solder—80 Pb, 20 Sn	10.20	0.3685
0.06 Carbon steel	7.871	0.2844	50-50 half and half—50 Pb, 50 Sn	8.89	0.321
0.23 Carbon steel	7.859	0.2839	Magnesium and Magnesium Alloys		
0.435 Carbon steel	7.844	0.2834	Magnesium (99.80)	1.74	0.0628
1.22 Carbon steel	7.830	0.2829	A10—10 Al, 0.1 Mn (w, sc, and pmc)	1.81	0.0654
18 Cr, 8 Ni stainless steel	7.93	0.286	AZ91—9 Al, 0.7 Zn, 0.2 Mn (die cast)	1.81	0.0654
25 Cr, 20 Ni stainless steel	7.98	0.288	AZ92—9 Al, 2 Zn, 0.1 Mn (sc and pmc)	1.82	0.0658
12.5 Cr stainless steel	7.75	0.280	A8—8 Al, 0.2 Mn (sand cast)	1.80	0.0650
25 Cr stainless steel	7.60	0.273	AZ61X—6Al, 1 Zn, 0.2 Mn (wrought)	1.80	0.0650
18 W, 4 Cr, 1 V high speed steel	8.67	0.313	AM244—4 Al, 0.2 Mn (sand cast)	1.76	0.0636
20 W, 4 Cr, 2 V, 12 Co high speed steel	8.89	0.321	AM11—1.25 Al, 1 Mn (die cast)	1.70	0.0616
6 W, 5 Mo, 2 V high speed steel	8.16	0.265	AZ80X—8.5 Al, 0.5 Zn, 0.15 Mn (wrought)	1.80	0.0650
8 Mo, 1.5 W, 1 V high speed steel	7.88	0.285	AZ63—6 Al, 3 Zn, 0.2 Mn (sand cast)	1.84	0.0665
Invar 36 Ni	8.00	0.289	AZ51X—5 Al, 1 Zn, 0.25 Mn (wrought)	1.79	0.0647
"Hipernik 50" Ni	8.25	0.298	AZ31X—3 Al, 1 Zn, 0.3 Mn (wrought)	1.78	0.0643
4 Si	7.6	0.27	M1—1.5 Mn (wrought)	1.76	0.0636
10.27 Si	6.97	0.252	TA54—5 Sn, 3 Al, 0.5 Mn (wrought)	1.84	0.0665
Lead and Lead Alloys			Mg-Al Alloys		
Pure lead (99.73+ Pb)	11.34	0.4097	98 Mg, 2 Al	1.75	0.0632
Chemical lead (99.90+ Pb)	11.34	0.4097	96 Mg, 4 Al	1.77	0.0639
Cable sheath alloy—99.8 Pb, 0.028 Ca	11.34	0.4097	94 Mg, 6 Al	1.78	0.0643
1% antimonial lead—99 Pb, 1 Sb	11.27	0.4071	92 Mg, 8 Al	1.80	0.0650
Hard lead—96 Pb, 4 Sb	11.04	0.3989	90 Mg, 10 Al	1.81	0.0654
Hard lead—94 Pb, 6 Sb	10.88	0.3931	88 Mg, 12 Al	1.82	0.0658
8% antimonial lead—92 Pb, 8 Sb	10.74	0.3880	Nickel and Nickel Alloys		
Grid metal—91 Pb, 9 Sb	10.66	0.3848	Nickel (99.95 Ni + Co)	8.902	0.3216
ASTM-12 bearing metal—90 Pb, 10 Sb	10.67	0.3855	"A" Nickel (99.4 Ni + Co)	8.885	0.3210
ASTM-11 bearing metal—85 Pb, 15 Sb	10.28	0.3714	Cast nickel—97 Ni, 1.5 Si, 0.5 Mn, 0.5 C	8.34	0.301
Lead-base babbitt—85 Pb, 10 Sb, 5 Sn	10.24	0.3700	"D" Nickel—95.2 Ni, 4.5 Mn	8.78	0.317
"G" Lead-base babbitt—83 Pb, 12.75 Sb, 3 As, 0.75 Sn	10.1	0.365	"Duranickel"—94 Ni, 4.5 Al	8.75	0.316
"S" Lead-base babbitt—83 Pb, 15 Sb, 1 Sn, 1 As	10.1	0.365	Monel—67 Ni, 30 Cu, 1.4 Fe, 1.0 Mn, 0.15 C	8.84	0.319

Material	Density	Value
Nickel and Nickel Alloys (*continued*)		
Cast monel—63 Ni, 32 Cu, 1.6 Si, 0.2 C	8.63	0.312
"K" monel—66 Ni, 29 Cu, 3 Al	8.47	0.306
"S" monel—63 Ni, 30 Cu, 4 Si, 2 Fe	8.36	0.302
"Hastelloy A"—60 Ni, 20 Mo, 20 Fe	8.80	0.318
"Hastelloy B"—65 Ni, 30 Mo, 5 Fe	9.24	0.334
"Hastelloy C"—58 Ni, 17 Mo, 15 Cr, 5 W, 5 Fe	8.94	0.323
"Hastelloy D"—85 Ni, 8 to 11 Si, 3 Cu	7.8	0.28
"Illium G"—58 Ni, 22 Cr, 6 Cu, 6 Mo, 6 Fe, 0.2 C	8.58	0.310
"Inconel"—80 Ni, 14 Cr, 6 Fe	8.51	0.307
Cast "Inconel"—77.5 Ni, 13.5 Cr, 6 Fe, 2 Si	8.3	0.30
"Chromel A"—80 Ni, 20 Cr	8.4	0.30
"Nichrome"—60 Ni, 16 Cr, 24 Fe	8.25	0.298
"Chromax"—35 Ni, 15 Cr, 50 Fe	7.95	0.287
Constantan		
45 Ni, 55 Cu (wrought)	8.9	0.32
45 Ni, 55 Cu (cast)	8.6	0.31
Ni-Fe alloys		
90 Ni, 10 Fe	8.8	0.32
80 Ni, 20 Fe	8.6	0.31
70 Ni, 30 Fe	8.5	0.31
60 Ni, 40 Fe	8.35	0.302
Permalloy—78 Ni, 22 Fe	8.6	0.31
Mumetal—76 Ni, 16 Fe, 6 Cu, 2 Cr	8.6	0.31
Tin and Tin Alloys		
Pure tin	7.298	0.2637
Soft solder—70 Sn, 30 Pb	8.32	0.301
Soft solder, eutectic solder—63 Sn, 37 Pb	8.42	0.304
Tin babbitt, ASTM-1—91 Sn, 4.5 Sb, 4.5 Cu	7.34	0.265
Tin babbitt, ASTM-2—89 Sn, 7.5 Sb, 3.5 Cu	7.39	0.267
Tin babbitt, ASTM-3—83 Sn, 8.3 Sb, 8.3 Cu	7.46	0.269
Tin babbitt, ASTM-4—75 Sn, 12 Sb, 10 Pb, 3 Cu	7.75	0.280
Tin babbitt, ASTM-5—65 Sn, 15 Sb, 18 Pb, 2 Cu	7.52	0.272
Zinc and Zinc Alloys		
Pure zinc	7.133	0.2577
"Zamak 2," ASTM XXI (dc) 92 Zn, 4 Al, 3 Cu, 0.03 Mg	6.7	0.24
"Zamak 3," ASTM XXIII—95 Zn, 4 Al, 0.04 Mg	6.6	0 24
"Zamak 5," ASTM XXV—94 Zn, 4 Al, 1 Cu, 0.04 Mg	6.7	0.24
Zinc and Zinc Alloys (*continued*)		
SAE 63, T-11 (cast)—86 Zn, 10 Cu, 4 Al	6.9	0.25
Commercial rolled zinc—99 Zn, 0.08 Pb	7.14	0.258
Commercial rolled zinc—99 Zn, 0.06 Pb, 0.06 Cd	7.14	0.258
Commercial rolled zinc—99 Zn, 0.3 Pb, 0.3 Cd	7.14	0.258
"Zilloy 40" (rolled)—98 Zn, 1 Cu, 0.08 Pb	7.18	0.259
"Zilloy 15" (rolled)—98 Zn, 1 Cu, 0.1 Pb, 0.01 Mg	7.18	0.259
Sintered Carbides		
97 WC, 3 Co	15.25	0.5510
95.5 WC, 4.5 Co	15.05	0.5438
94 WC, 6 Co	14.85	0.5365
91 WC, 9 Co	14.60	0.5275
87 WC, 13 Co	14.15	0.5112
80 WC, 20 Co	13.55	0.4896
Predominantly WC with TaC and 13 Co	13.90	0.5022
Predominantly WC with TaC and 6 Co	14.70	0.5311
Predominantly WC with TiC and 6 Co	11.20	0.4047
Predominantly WC, less TiC and 8 Co	12.80	0.4625
Predominantly WC, TaC, TiC, 8 Co	11.7	0.423
Predominantly WC, TaC, TiC, 11 Co	11.6	0.419
Predominantly WC, TaC, TiC, 15 Co	11.4	0.412
Precious Metals		
Gold	19.32	0 6984
69 Au, 25 Ag, 6 Pt	16.11	0.5820
10-Carat		
41.5 Au, 11.7 Ag, 46.8 Cu	11.2	0.405
41.6 Au, 6.8 Ag, 44.3 Cu, 7.3 Ni	11.55	0.4173
41.6 Au, 4.6 Ag, 43.4 Cu, 5.0 Ni, 5.4 Zn	11.60	0.4190
12-Carat		
50.0 Au, 4.0 Ag, 37.1 Cu, 4.3 Ni, 4.6 Zn	12.17	0.4397
50.0 Au, 42.0 Cu, 6.0 Ni, 2.0 Zn	12.19	0.4404
14-Carat		
58.3 Au, 3.3 Ag, 31.0 Cu, 3.5 Ni, 3.9 Zn	12.37	0.4469
58.3 Au, 4.9 Ag, 31.6 Cu, 5.2 Ni	13.08	0.4726
Iridium	22.5	0.813
Osmium	22.5	0.813
Palladium	12.02	0.4343
Pd, 4 to 5 Ru	12.0	0.434
Pd, 40 Ag	11.3	0.408
Pd, 40 Cu	10.6	0.383

TABLE 6 (*continued*)

Metal and Composition	g/cu cm	Density lb/cu in.	Metal and Composition	g/cu cm	Density lb/cu in.
Platinum	21.45	0.7750			
Pt-Ru Alloys			Ag-Mo Alloys		
Pt, 5 Ru	20.67	0.7468	Ag, 20 Mo	10.42	0.3765
Pt, 10 Ru	19.94	0.7204	Ag, 40 Mo	10.38	0.3750
Pt-Ir Alloys			Ag, 60 Mo	10.35	0.3739
Pt, 5 Ir	21.49	0.7764	Ag, 80 Mo	10.27	0.3710
Pt, 10 Ir	21.53	0.7779	Ag-W Alloys		
Pt, 25 Ir	21.66	0.7826	Ag, 10 W	10.95	0.3956
Pt-Rh Alloys			Ag, 20 W	11.52	0.4162
Pt, 3.5 Rh	20.9	0.755	Ag, 30 W	12.2	0.441
Pt, 18.74 Rh	18.74	0.6771	Ag, 40 W	12.87	0.4650
Pt-Ni Alloys			Ag, 50 W	13.6	0.491
Pt, 1 Ni	21.1	0.762	Ag, 70 W	15.4	0.556
Pt, 2 Ni	20.8	0.751	Ag, 90 W	17.8	0.643
Pt, 5 Ni	20.0	0.723	Silmanal—86.75 Ag, 8.8 Mn, 4.45 Al	8.996	0.3250
96 Pt, 4 W	21.3	0.770	Miscellaneous Pure Metals		
84 Pt, 10 Pd, 6 Ru	18.1	0.654	Antimony	6.62	0.239
Rhodium	12.44	0.4495	Beryllium	1.816	0.0658
Ruthenium	12.2	0.441	Bismuth	9.8	0.35
Silver	10.49	0.3790	Cadmium	8.65	0.313
Ag-Cd Alloys			Calcium	1.55	0.056
Ag, 2.5 Cd	10.43	0.3768	Chromium	7.19	0.260
Ag, 5.0 Cd	10.387	0.3753	Cobalt	8.9	0.32
Ag, 10 Cd	10.29	0.3718	Columbium	8.57	0.3096
Ag, 15 Cd	10.18	0.3678	Gallium	5.91	0.214
Ag-CdO Materials			Lithium	0.53	0.019
Ag, 2.5 CdO	10.38	0.3750	Manganese	7.43	0.268
Ag, 5.0 CdO	10.24	0.3700	Molybdenum	10.2	0.369
Ag, 10 CdO	9.97	0.3602	Rhenium	20	0.72
Ag, 15 CdO	9.75	0.352	Selenium	4.81	0.174
Ag-graphite			Silicon	2.33	0.0842
Ag, 1 graphite	9.9	0.36	Strontium	2.6	0.094
Ag, 2 graphite	9.65	0.349	Tantalum	16.6	0.600
Ag, 4 graphite	9.0	0.325	Tellurium	6.24	0.225
Ag, 7 graphite	8.0	0.29	Thallium	11.85	0.428
			Thorium	11.5	0.415

Miscellaneous Pure Metals (continued)

Material	Density	Value
Titanium	4.54	0.164
Tungsten	19.3	0.697
Uranium	18.7	0.676
Vanadium	6.0	0.217
Zirconium	6.5	0.234

Iron

Material	Density	Value
Electrolytic iron	7.874	0.2845
Ingot iron	7.866	0.2842
Carbonyl iron	7.860	0.2840
Wrought iron	7.7	0.28

Cast Iron

Material	Density	Value
Gray cast iron	6.95 to 7.35	0.251 to 0.265
White cast iron	7.7	0.28
Malleable iron (ASTM grade 35018)	7.20 to 7.45	0.260 to 0.269
Malleable iron (ASTM grade 32510)	7.25 to 7.34	0.262 to 0.265

Carbon Steel

Composition	Density	Value
0.06 C, 0.01 Si, 0.38 Mn (ann 1700°F)	7.871	0.2844
0.23 C, 0.11 Si, 0.635 Mn (ann 1700°F)	7.859	0.2839
0.435 C, 0.20 Si, 0.69 Mn (ann 1580°F)	7.844	0.2834
1.22 C, 0.16 Si, 0.35 Mn (ann 1470°F)	7.830	0.2829

Low-Chromium Steel

Composition	Density	Value
0.31 C, 1.00 Cr, 0.74 Mn (o–q 1650°F, tem 1350°F)	7.84	0.283
0.315 C, 1.09 Cr, 0.69 Mn (ann 1580°F)	7.84	0.283
0.35 C, 1.56 Cr, 0.24 Mn (ann)	7.83	0.283
1.73 C, 1.65 Cr, 0.30 Mn (ann)	7.80	0.282
0.80 C, 1.67 Cr, 0.28 Mn (ann)	7.82	0.282
0.62 C, 1.67 Cr, 0.22 Mn (ann)	7.82	0.282
0.98 C, 1.68 Cr, 0.28 Mn (ann)	7.81	0.282
0.20 C, 1.85 Cr, 0.14 Mn (o–q 1650°F, tem 1380°F)	7.84	0.283
0.22 C, 2.80 Cr, 0.10 Mn (o–q 1650°F, tem 1380°F)	7.82	0.282
0.21 C, 3.88 Cr, 0.19 Mn (o–q 1650°F, tem 1380°F)	7.81	0.282
0.30 C, 5.54 Cr, 0.08 Mn (o–q 1650°F, tem 1380°F)	7.79	0.281
0.35 C, 0.88 Cr, 0.20 Mo, 0.59 Mn (ann 1580°F, tem 1185°F)	7.845	0.2834

Low-Alloy Nickel-Chromium Steel

Composition	Density	Value
0.33 C, 3.38 Ni, 0.80 Cr, 0.53 Mn (ann 1580°F, tem 1185°F)	7.85	0.283
0.325 C, 3.41 Ni, 0.71 Cr, 0.55 Mn (ann 1580°F, tem 1185°F)	7.85	0.283

Low-Alloy Nickel-Chromium Steel (continued)

Composition	Density	Value
1.28 C, 3.46 Ni, 1.80 Cr, 0.24 Mn (b–q 2190°F)	7.92	0.286
(ann 1435°F)	7.82	0.282
1.50 C, 3.46 Ni, 1.80 Cr, 0.25 Mn (b–q 2190°F in iced brine)	7.91	0.286
(ann 1435°F)	7.82	0.282
0.325 C, 3.47 Ni, 0.17 Cr, 0.55 Mn (ann 1580°F)	7.855	0.2838
0.51 C, 3.52 Ni, 1.72 Cr, 0.22 Mn (b–q 2190°F)	7.79	0.281
(b–q 2190°F, tem 375°F)	7.80	0.282
(b–q 2190°F, tem 690°F)	7.82	0.282
(b–q 2190°F, tem 1110°F)	7.835	0.2831
(ann 1435°F)	7.835	0.2831
0.34 C, 3.53 Ni, 0.78 Cr, 0.55 Mn, 0.39 Mo (ann 1580°F, tem 1185°F)	7.86	0.284

Wrought Stainless and Heat-Resisting Steel

Composition	Density	Value
302—0.10 C, 18 Cr, 9 Ni	7.93	0.286
303—18 Cr, 9 Ni, 0.5 Mo, Zr	7.93	0.286
309—23 Cr, 13 Ni	7.98	0.288
310—25 Cr, 20.5 Ni	7.98	0.288
316—17 Cr, 12 Ni, 2.25 Mo	7.98	0.288
321—18 Cr, 10.5 Ni, Ti	8.02	0.290
410—12.5 Cr	7.75	0.280
416—13 Cr, 0.5 Mo, Zr	7.73	0.279
420—13 Cr	7.70	0.278
430—16 Cr	7.70	0.278
440—17 Cr, 0.6 Mo	7.68	0.277
446—25 Cr	7.60	0.273
0.07 C, 17.88 Cr, 8.26 Mn	7.77	0.281
0.07 C, 17.55 Cr, 10.48 Mn	7.76	0.280
0.09 C, 18.40 Cr, 5.33 Mn, 4.07 Ni, 0.78 Cu	7.91	0.286
0.06 C, 18.50 Cr, 6.79 Mn, 4.06 Ni	7.90	0.285
0.06 C, 18.04 Cr, 7.90 Mn, 2.06 Ni	7.78	0.281
0.07 C, 17.70 Cr, 9.40 Mn, 0.68 Cu	7.77	0.281

Corrosion-Resistant Castings

Composition	Density	Value
CA—12 Cr	7.612	0.2750
CB—20 Cr	7.529	0.2720
CC—28 Cr	7.529	0.2720
CE—29 Cr, 9 Ni	7.667	0.2770
CF—19 Cr, 9 Ni	7.750	0.2800

TABLE 6 (*continued*)

Metal and Compositions	g/cu cm	Density lb/cu in.
Corrosion-Resistant Castings (*continued*)		
CG—21 Cr, 11 Ni	7.778	0.2808
CH—24 Cr, 13 Ni	7.723	0.2790
CK—25 Cr, 20 Ni	7.750	0.2800
Tool Steel		
1.0 C, 0.71 Cr (ann)	7.84	0.283
(w-q 1470°F)	7.758	0.2803
(w-q 1470°F, tem 350°F)	7.765	0.2805
(w-q 1470°F, tem 600°F)	7.786	0.2813
(w-q 1470°F, tem 1000°F)	7.808	0.2821
18 W, 4 Cr, 1 V (ann)	8.67	0.313
18 W, 4 Cr, 2 V (ann)	8.67	0.313
8.24 Mo, 1.64 W, 3.68 Cr, 1.00 V, 0.80C (q 2200°F)	7.925	0.2863
(q 2200°F, tem 1150°F)	7.87	0.284
4.11 Mo, 5.20 W, 4.60 Cr, 4.00 V, 1.32C (hardened)	7.93	0.286
7.75 Mo, 4.39 Cr, 4.10 V, 1.20C (hardened)	7.76	0.280
20 W, 4 Cr, 2 V, 12 Co (ann)	8.89	0.321
18 W, 4 Cr, 1 V, 5 Co (ann)	8.68	0.314
6 W, 5 Mo, 2 V (ann)	8.16	0.295
8 Mo, 1.5 W, 1 V (ann)	7.88	0.285

Metal and Composition	g/cu cm	Density lb/cu in.
Permanent Magnet Alloys		
Alnico 1—20 Ni, 12 Al, 5 Co	6.892	0.2490
Alnico 2—17 Ni, 10 Al, 12.5 Co, 6 Cu (cast)	7.086	0.2560
Alnico 3—25 Ni, 12 Al	6.892	0.2490
Alnico 4—28 Ni, 12 Al, 5 Co	7.003	0.2530
Alnico 5—24 Co, 14 Ni, 8 Al, 3 Cu	7.307	0.2640
Alnico 12—35 Co, 18 Ni, 6 Al, 8 Ti	7.197	0.2600
Miscellaneous Ferrous Alloys		
28.37 Ni, 0.28 C (q 1740°F)	8.16	0.295
Invar—36 Ni	8.00	0.289
Radiometal—45 Ni	8.3	0.30
Hipernik—50 Ni	8.25	0.298
Austenitic manganese steel—1.2C, 13 Mn (1920°F, a–c)	7.87	0.284

The following abbreviations have been used to indicate the condition of the metal: ann, annealed; q, quenched; b–q, quenched in brine; w–q, quenched in water; o–q, quenched in oil; a–c, air cooled; f–c, furnace cooled; tem, tempered. All values of density measured at or near room temperature.

* Reprinted from the "Metals Handbook," 1948 Edition, by permission of the publisher; The American Society For Metals, Cleveland 3, Ohio.

[a] 8.89 to 8.94 g/cu cm and 0.322 to 0.323 lb/cu in.
[b] 8.80 to 8.90 g/cu cm and 0.318 to 0.322 lb/cu in.
[c] 6.95 to 7.35 g/cu cm and 0.251 to 0.265 lb/cu in.
[d] 7.20 to 7.34 g/cu cm and 0.260 to 0.265 lb/cu in.

TABLE 7. STEEL COMPOSITIONS*
Composition Limits of Standard Steels
Society of Automotive Engineers and American Iron and Steel Institute

Carbon Steels (Revised April 16, 1947)

SAE Number	C	Mn	P Max	S Max	AISI Number	SAE Number	C	Mn	P Max	S Max	AISI Number
—	0.06 max	0.35 max	0.040	0.050	C1005	—	0.70–0.80	0.40–0.70	0.040	0.050	C1075
1006	0.08 max	0.25–0.40	0.040	0.050	C1006	1078	0.72–0.85	0.30–0.60	0.040	0.050	C1078
1008	0.10 max	0.25–0.50	0.040	0.050	C1008	1080	0.75–0.88	0.60–0.90	0.040	0.050	C1080
1010	0.08–0.13	0.30–0.60	0.040	0.050	C1010	—	0.80–0.93	0.60–0.90	0.040	0.050	C1084
—	0.10–0.15	0.30–0.60	0.040	0.050	C1012	1085	0.80–0.93	0.70–1.00	0.040	0.050	C1085
						1090	0.85–1.00	0.60–0.90	0.040	0.050	C1090
—	0.11–0.16	0.50–0.80	0.040	0.050	C1013	1095	0.90–1.05	0.30–0.50	0.040	0.050	C1095
1015	0.13–0.18	0.30–0.60	0.040	0.050	C1015						
1016	0.13–0.18	0.60–0.90	0.040	0.050	C1016	—	0.08 max	0.45 max	0.07–0.12	0.060	B1006
1017	0.15–0.20	0.30–0.60	0.040	0.050	C1017	—	0.13 max	0.30–0.60	0.07–0.12	0.060	B1010
1018	0.15–0.20	0.60–0.90	0.040	0.050	C1018	—	0.43–0.50	0.50–0.80	0.050	0.050	D1049
						—	0.50–0.60	0.50–0.80	0.050	0.050	D1054
1019	0.15–0.20	0.70–1.00	0.040	0.050	C1019	—	0.55–0.65	0.50–0.80	0.050	0.050	D1059
1020	0.18–0.23	0.30–0.60	0.040	0.050	C1020	—	0.60–0.70	0.50–0.80	0.050	0.050	D1064
—	0.18–0.23	0.60–0.90	0.040	0.050	C1021	—	0.65–0.75	0.40–0.70	0.050	0.050	D1069
1022	0.18–0.23	0.70–1.00	0.040	0.050	C1022	—	0.70–0.80	0.40–0.70	0.050	0.050	D1075
—	0.20–0.25	0.30–0.60	0.040	0.050	C1023						
1024	0.19–0.25	1.35–1.65	0.040	0.050	C1024						
1025	0.22–0.28	0.30–0.60	0.040	0.050	C1025						
—	0.22–0.28	0.60–0.90	0.040	0.050	C1026	Resulfurized Carbon Steels (Free-Cutting Steels)					
1027	0.22–0.29	1.20–1.50	0.040	0.050	C1027	(Revised April 16, 1947)					
—	0.25–0.31	0.60–0.90	0.040	0.050	C1029						
1030	0.28–0.34	0.60–0.90	0.040	0.050	C1030						
1033	0.30–0.36	0.70–1.00	0.040	0.050	C1033						
1034	0.32–0.38	0.50–0.80	0.040	0.050	C1034						
1035	0.32–0.38	0.60–0.90	0.040	0.050	C1035	SAE Number	C	Mn	P Max	S Max	AISI Number
1036	0.30–0.37	1.20–1.50	0.040	0.050	C1036	—	0.08 max	0.30–0.60	0.045	0.08–0.13	C1106
						—	0.08–0.13	0.50–0.80	0.045	0.07–0.12	C1108
1038	0.35–0.42	0.60–0.90	0.040	0.050	C1038	1109	0.08–0.13	0.60–0.90	0.045	0.08–0.13	C1109
—	0.37–0.44	0.70–1.00	0.040	0.050	C1039	—	0.08–0.13	0.30–0.60	0.045	0.08–0.13	C1110
1040	0.37–0.44	0.60–0.90	0.040	0.050	C1040	—	0.08–0.13	0.60–0.90	0.045	0.16–0.23	C1111
1041	0.36–0.44	1.35–1.65	0.040	0.050	C1041						
1042	0.40–0.47	0.60–0.90	0.040	0.050	C1042	—	0.10–0.16	1.00–1.30	0.045	0.24–0.33	C1113
1043	0.40–0.47	0.70–1.00	0.040	0.050	C1043	1114	0.10–0.16	1.00–1.30	0.045	0.08–0.13	C1114
1045	0.43–0.50	0.60–0.90	0.040	0.050	C1045	1115	0.13–0.18	0.60–0.90	0.045	0.08–0.13	C1115
1046	0.43–0.50	0.70–1.00	0.040	0.050	C1046	1116	0.14–0.20	1.10–1.40	0.045	0.16–0.23	C1116
1050	0.48–0.55	0.60–0.90	0.040	0.050	C1050	1117	0.14–0.20	1.00–1.30	0.045	0.08–0.13	C1117
—	0.45–0.56	0.85–1.15	0.040	0.050	C1051	1118	0.14–0.20	1.30–1.60	0.045	0.08–0.13	C1118
1052	0.47–0.55	1.20–1.50	0.040	0.050	C1052	1119	0.14–0.20	1.00–1.30	0.045	0.24–0.33	C1119
—	0.50–0.60	0.50–0.80	0.040	0.050	C1054	—	0.18–0.23	0.70–1.00	0.045	0.08–0.13	C1120
1055	0.50–0.60	0.60–0.90	0.040	0.050	C1055	—	0.22–0.28	0.60–0.90	0.045	0.08–0.13	C1125
—	0.50–0.61	0.85–1.15	0.040	0.050	C1057	1126	0.23–0.29	0.70–1.00	0.045	0.08–0.13	C1126
—	0.55–0.65	0.50–0.80	0.040	0.050	C1059						
						1137	0.32–0.39	1.35–1.65	0.045	0.08–0.13	C1137
1060	0.55–0.65	0.60–0.90	0.040	0.050	C1060	1138	0.34–0.40	0.70–1.00	0.045	0.08–0.13	C1138
—	0.54–0.65	0.75–1.05	0.040	0.050	C1061	1140	0.37–0.44	0.70–1.00	0.045	0.08–0.13	C1140
1062	0.54–0.65	0.85–1.15	0.040	0.050	C1062	1141	0.37–0.45	1.35–1.65	0.045	0.08–0.13	C1141
1064	0.60–0.70	0.50–0.80	0.040	0.050	C1064	1144	0.40–0.48	1.35–1.65	0.045	0.24–0.33	C1144
1065	0.60–0.70	0.60–0.90	0.040	0.050	C1065	1145	0.42–0.49	0.70–1.00	0.045	0.04–0.07	C1145
1066	0.60–0.71	0.85–1.15	0.040	0.050	C1066	1146	0.42–0.49	0.70–1.00	0.045	0.08–0.13	C1146
—	0.65–0.75	0.40–0.70	0.040	0.050	C1069	1151	0.48–0.55	0.70–1.00	0.045	0.08–0.13	C1151
1070	0.65–0.75	0.60–0.90	0.040	0.050	C1070	1111	0.13 max	0.60–0.90	0.07–0.12	0.08–0.15	B1111
—	0.65–0.76	0.75–1.05	0.040	0.050	C1071	1112	0.13 max	0.70–1.00	0.07–0.12	0.16–0.23	B1112
1074	0.70–0.80	0.50–0.80	0.040	0.050	C1074	1113	0.13 max	0.70–1.00	0.07–0.12	0.24–0.33	B1113

Alloy Steels (Revised August 10, 1947)

AISI Number	C	Mn	P Max	S Max	Si	Ni	Cr	Other	SAE Number
1320	0.18–0.23	1.60–1.90	0.040	0.040	0.20–0.35	—	—	—	1320
1321	0.17–0.22	1.80–2.10	0.050	0.050	0.20–0.35	—	—	—	—
1330	0.28–0.33	1.60–1.90	0.040	0.040	0.20–0.35	—	—	—	1330
1335	0.33–0.38	1.60–1.90	0.040	0.040	0.20–0.35	—	—	—	1335
1340	0.38–0.43	1.60–1.90	0.040	0.040	0.20–0.35	—	—	—	1340
2317	0.15–0.20	0.40–0.60	0.040	0.040	0.20–0.35	3.25–3.75	—	—	2317
2330	0.28–0.33	0.60–0.80	0.040	0.040	0.20–0.35	3.25–3.75	—	—	2330
2335	0.33–0.38	0.60–0.80	0.040	0.040	0.20–0.35	3.25–3.75	—	—	—
2340	0.38–0.43	0.70–0.90	0.040	0.040	0.20–0.35	3.25–3.75	—	—	2340
2345	0.43–0.48	0.70–0.90	0.040	0.040	0.20–0.35	3.25–3.75	—	—	2345
E 2512	0.09–0.14	0.45–0.60	0.025	0.025	0.20–0.35	4.75–5.25	—	—	2512
2515	0.12–0.17	0.40–0.60	0.040	0.040	0.20–0.35	4.75–5.25	—	—	2515
E 2517	0.15–0.20	0.45–0.60	0.025	0.025	0.20–0.35	4.75–5.25	—	—	2517
3115	0.13–0.18	0.40–0.60	0.040	0.040	0.20–0.35	1.10–1.40	0.55–0.75	—	3115
3120	0.17–0.22	0.60–0.80	0.040	0.040	0.20–0.35	1.10–1.40	0.55–0.75	—	3120
3130	0.28–0.33	0.60–0.80	0.040	0.040	0.20–0.35	1.10–1.40	0.55–0.75	—	3130
3135	0.33–0.38	0.60–0.80	0.040	0.040	0.20–0.35	1.10–1.40	0.55–0.75	—	3135
3140	0.38–0.43	0.70–0.90	0.040	0.040	0.20–0.35	1.10–1.40	0.55–0.75	—	3140
3141	0.38–0.43	0.70–0.90	0.040	0.040	0.20–0.35	1.10–1.40	0.70–0.90	—	3141
3145	0.43–0.48	0.70–0.90	0.040	0.040	0.20–0.35	1.10–1.40	0.70–0.90	—	3145
3150	0.48–0.53	0.70–0.90	0.040	0.040	0.20–0.35	1.10–1.40	0.70–0.90	—	3150
E 3310	0.08–0.13	0.45–0.60	0.025	0.025	0.20–0.35	3.25–3.75	1.40–1.75	—	3310
E 3316	0.14–0.19	0.45–0.60	0.025	0.025	0.20–0.35	3.25–3.75	1.40–1.75	—	3316
								Mo	
4017	0.15–0.20	0.70–0.90	0.040	0.040	0.20–0.35	—	—	0.20–0.30	4017
4023	0.20–0.25	0.70–0.90	0.040	0.040	0.20–0.35	—	—	0.20–0.30	4023
4024	0.20–0.25	0.70–0.90	0.040	0.035–0.050	0.20–0.35	—	—	0.20–0.30	4024
4027	0.25–0.30	0.70–0.90	0.040	0.040	0.20–0.35	—	—	0.20–0.30	4027
4028	0.25–0.30	0.70–0.90	0.040	0.035–0.050	0.20–0.35	—	—	0.20–0.30	4028
4032	0.30–0.35	0.70–0.90	0.040	0.040	0.20–0.35	—	—	0.20–0.30	4032
4037	0.35–0.40	0.70–0.90	0.040	0.040	0.20–0.35	—	—	0.20–0.30	4037
4042	0.40–0.45	0.70–0.90	0.040	0.040	0.20–0.35	—	—	0.20–0.30	4042
4047	0.45–0.50	0.70–0.90	0.040	0.040	0.20–0.35	—	—	0.20–0.30	4047
4053	0.50–0.56	0.75–1.00	0.040	0.040	0.20–0.35	—	—	0.20–0.30	4053
4063	0.60–0.67	0.75–1.00	0.040	0.040	0.20–0.35	—	—	0.20–0.30	4063
4068	0.63–0.70	0.75–1.00	0.040	0.040	0.20–0.35	—	—	0.20–0.30	4068
—	0.17–0.22	0.70–0.90	0.040	0.040	0.20–0.35	—	0.40–0.60	0.20–0.30	4119
—	0.23–0.28	0.70–0.90	0.040	0.040	0.20–0.35	—	0.40–0.60	0.20–0.30	4125
4130	0.28–0.33	0.40–0.60	0.040	0.040	0.20–0.35	—	0.80–1.10	0.15–0.25	4130
E 4132	0.30–0.35	0.40–0.60	0.025	0.025	0.20–0.35	—	0.80–1.10	0.18–0.25	—
E 4135	0.33–0.38	0.70–0.90	0.025	0.025	0.20–0.35	—	0.80–1.10	0.18–0.25	—
4137	0.35–0.40	0.70–0.90	0.040	0.040	0.20–0.35	—	0.80–1.10	0.15–0.25	4137
E 4137	0.35–0.40	0.70–0.90	0.025	0.025	0.20–0.35	—	0.80–1.10	0.18–0.25	—
4140	0.38–0.43	0.75–1.00	0.040	0.040	0.20–0.35	—	0.80–1.10	0.15–0.25	4140
4142	0.40–0.45	0.75–1.00	0.040	0.040	0.20–0.35	—	0.80–1.10	0.15–0.25	—
4145	0.43–0.48	0.75–1.00	0.040	0.040	0.20–0.35	—	0.80–1.10	0.15–0.25	4145
4147	0.45–0.50	0.75–1.00	0.040	0.040	0.20–0.35	—	0.80–1.10	0.15–0.25	—
4150	0.48–0.53	0.75–1.00	0.040	0.040	0.20–0.35	—	0.80–1.10	0.15–0.25	4150
4317	0.15–0.20	0.45–0.65	0.040	0.040	0.20–0.35	1.65–2.00	0.40–0.60	0.20–0.30	4317
4320	0.17–0.22	0.45–0.65	0.040	0.040	0.20–0.35	1.65–2.00	0.40–0.60	0.20–0.30	4320
4337	0.35–0.40	0.60–0.80	0.040	0.040	0.20–0.35	1.65–2.00	0.70–0.90	0.20–0.30	—
4340	0.38–0.43	0.60–0.80	0.040	0.040	0.20–0.35	1.65–2.00	0.70–0.90	0.20–0.30	4340
4608	0.06–0.11	0.25–0.45	0.040	0.040	0.25 Max	1.40–1.75	—	0.15–0.25	4608
4615	0.13–0.18	0.45–0.65	0.040	0.040	0.20–0.35	1.65–2.00	—	0.20–0.30	4615
—	0.15–0.20	0.45–0.65	0.040	0.040	0.20–0.35	1.65–2.00	—	0.20–0.30	4617
E 4617	0.15–0.20	0.45–0.65	0.025	0.025	0.20–0.35	1.65–2.00	—	0.20–0.27	—
4620	0.17–0.22	0.45–0.65	0.040	0.040	0.20–0.35	1.65–2.00	—	0.20–0.30	4620
X 4620	0.18–0.23	0.50–0.70	0.040	0.040	0.20–0.35	1.65–2.00	—	0.20–0.30	X 4620
E 4620	0.17–0.22	0.45–0.65	0.025	0.025	0.20–0.35	1.65–2.00	—	0.20–0.27	—
4621	0.18–0.23	0.70–0.90	0.040	0.040	0.20–0.35	1.65–2.00	—	0.20–0.30	4621
4640	0.38–0.43	0.60–0.80	0.040	0.040	0.20–0.35	1.65–2.00	—	0.20–0.30	4640
E 4640	0.38–0.43	0.60–0.80	0.025	0.025	0.20–0.35	1.65–2.00	—	0.20–0.27	—
4812	0.10–0.15	0.40–0.60	0.040	0.040	0.20–0.35	3.25–3.75	—	0.20–0.30	4812
4815	0.13–0.18	0.40–0.60	0.040	0.040	0.20–0.35	3.25–3.75	—	0.20–0.30	4815

AISI Number	C	Mn	P Max	S Max	Si	Ni	Cr	Other	SAE Number
4817	0.15–0.20	0.40–0.60	0.040	0.040	0.20–0.35	3.25–3.75	—	0.20–0.30	4817
4820	0.18–0.23	0.50–0.70	0.040	0.040	0.20–0.35	3.25–3.75	—	0.20–0.30	4820
5045	0.43–0.48	0.70–0.90	0.040	0.040	0.20–0.35	—	0.55–0.75	—	5045
5046	0.43–0.50	0.75–1.00	0.040	0.040	0.20–0.35	—	0.20–0.35	—	5046
—	0.13–0.18	0.70–0.90	0.040	0.040	0.20–0.35	—	0.70–0.90	—	5115
5120	0.17–0.22	0.70–0.90	0.040	0.040	0.20–0.35	—	0.70–0.90	—	5120
5130	0.28–0.33	0.70–0.90	0.040	0.040	0.20–0.35	—	0.80–1.10	—	5130
5132	0.30–0.35	0.60–0.80	0.040	0.040	0.20–0.35	—	0.80–1.05	—	5132
5135	0.33–0.38	0.60–0.80	0.040	0.040	0.20–0.35	—	0.80–1.05	—	5135
5140	0.38–0.43	0.70–0.90	0.040	0.040	0.20–0.35	—	0.70–0.90	—	5140
5145	0.43–0.48	0.70–0.90	0.040	0.040	0.20–0.35	—	0.70–0.90	—	5145
5147	0.45–0.52	0.75–1.00	0.040	0.040	0.20–0.35	—	0.90–1.20	—	5147
5150	0.48–0.53	0.70–0.90	0.040	0.040	0.20–0.35	—	0.70–0.90	—	5150
5152	0.48–0.55	0.70–0.90	0.040	0.040	0.20–0.35	—	0.90–1.20	—	5152
E 50100	0.95–1.10	0.25–0.45	0.025	0.025	0.20–0.35	—	0.40–0.60	—	50100
E 51100	0.95–1.10	0.25–0.45	0.025	0.025	0.20–0.35	—	0.90–1.15	—	51100
E 52100	0.95–1.10	0.25–0.45	0.025	0.025	0.20–0.35	—	1.30–1.60	—	52100
								V	
6120	0.17–0.22	0.70–0.90	0.040	0.040	0.20–0.35	—	0.70–0.90	0.10 Min	—
6145	0.43–0.48	0.70–0.90	0.040	0.040	0.20–0.35	—	0.80–1.10	0.15 Min	—
6150	0.48–0.53	0.70–0.90	0.040	0.040	0.20–0.35	—	0.80–1.10	0.15 Min	6150
6152	0.48–0.55	0.70–0.90	0.040	0.040	0.20–0.35	—	0.80–1.10	0.10 Min	—
								Mo	
8615	0.13–0.18	0.70–0.90	0.040	0.040	0.20–0.35	0.40–0.70	0.50–0.60	0.15–0.25	8615
8617	0.15–0.20	0.70–0.90	0.040	0.040	0.20–0.35	0.40–0.70	0.40–0.60	0.15–0.25	8617
8620	0.18–0.23	0.70–0.90	0.040	0.040	0.20–0.35	0.40–0.70	0.40–0.60	0.15–0.25	8620
8622	0.20–0.25	0.70–0.90	0.040	0.040	0.20–0.35	0.40–0.70	0.40–0.60	0.15–0.25	8622
8625	0.23–0.28	0.70–0.90	0.040	0.040	0.20–0.35	0.40–0.70	0.40–0.60	0.15–0.25	8625
8627	0.25–0.30	0.70–0.90	0.040	0.040	0.20–0.35	0.40–0.70	0.40–0.60	0.15–0.25	8627
8630	0.28–0.33	0.70–0.90	0.040	0.040	0.20–0.35	0.40–0.70	0.40–0.60	0.15–0.25	8630
8632	0.30–0.35	0.70–0.90	0.040	0.040	0.20–0.35	0.40–0.70	0.40–0.60	0.15–0.25	8632
8635	0.33–0.38	0.75–1.00	0.040	0.040	0.20–0.35	0.40–0.70	0.40–0.60	0.15–0.25	8635
8637	0.35–0.40	0.75–1.00	0.040	0.040	0.20–0.35	0.40–0.70	0.40–0.60	0.15–0.25	8637
8640	0.38–0.43	0.75–1.00	0.040	0.040	0.20–0.35	0.40–0.70	0.40–0.60	0.15–0.25	8640
8641	0.38–0.43	0.75–1.00	0.040	0.040–0.060	0.20–0.35	0.40–0.70	0.40–0.60	0.15–0.25	8641
8642	0.40–0.45	0.75–1.00	0.040	0.040	0.20–0.35	0.40–0.70	0.40–0.60	0.15–0.25	8642
8645	0.43–0.48	0.75–1.00	0.040	0.040	0.20–0.35	0.40–0.70	0.40–0.60	0.15–0.25	8645
8647	0.45–0.50	0.75–1.00	0.040	0.040	0.20–0.35	0.40–0.70	0.40–0.60	0.15–0.25	8647
8650	0.48–0.53	0.75–1.00	0.040	0.040	0.20–0.35	0.40–0.70	0.40–0.60	0.15–0.25	8650
8653	0.50–0.56	0.75–1.00	0.040	0.040	0.20–0.35	0.40–0.70	0.50–0.60	0.15–0.25	8653
8655	0.50–0.60	0.75–1.00	0.040	0.040	0.20–0.35	0.40–0.70	0.40–0.60	0.15–0.25	8655
8660	0.50–0.65	0.75–1.00	0.040	0.040	0.20–0.35	0.40–0.70	0.40–0.60	0.15–0.25	8660
8720	0.18–0.23	0.70–0.90	0.040	0.040	0.20–0.35	0.40–0.70	0.40–0.60	0.20–0.30	8720
8735	0.33–0.38	0.75–1.00	0.040	0.040	0.20–0.35	0.40–0.70	0.40–0.60	0.20–0.30	8735
8740	0.38–0.43	0.75–1.00	0.040	0.040	0.20–0.35	0.40–0.70	0.40–0.60	0.20–0.30	8740
8742	0.40–0.45	0.75–1.00	0.040	0.040	0.20–0.35	0.40–0.70	0.40–0.60	0.20–0.30	—
8745	0.43–0.48	0.75–1.00	0.040	0.040	0.20–0.35	0.40–0.70	0.40–0.60	0.20–0.30	8745
8747	0.45–0.50	0.75–1.00	0.040	0.040	0.20–0.35	0.40–0.70	0.40–0.60	0.20–0.30	—
8750	0.48–0.53	0.75–1.00	0.040	0.040	0.20–0.35	0.40–0.70	0.40–0.60	0.20–0.30	8750
—	0.50–0.60	0.50–0.60	0.040	0.040	1.20–1.60	—	0.50–0.80	—	9254
9255	0.50–0.60	0.70–0.95	0.040	0.040	1.80–2.20	—	—	—	9255
9260	0.55–0.65	0.70–1.00	0.040	0.040	1.80–2.20	—	—	—	9260
9261	0.55–0.65	0.75–1.00	0.040	0.040	1.80–2.20	—	0.10–0.25	—	9261
9262	0.55–0.65	0.75–1.00	0.040	0.040	1.80–2.20	—	0.25–0.40	—	9262
E 9310	0.08–0.13	0.45–0.65	0.025	0.025	0.20–0.35	3.00–3.50	1.00–1.40	0.08–0.15	9310
E 9315	0.13–0.18	0.45–0.65	0.025	0.025	0.20–0.35	3.00–3.50	1.00–1.40	0.08–0.15	9315
E 9317	0.15–0.20	0.45–0.65	0.025	0.025	0.20–0.35	3.00–3.50	1.00–1.40	0.08–0.15	9317
9437	0.35–0.40	0.90–1.20	0.040	0.040	0.20–0.35	0.30–0.60	0.30–0.50	0.08–0.15	9437
9440	0.38–0.43	0.90–1.20	0.040	0.040	0.20–0.35	0.30–0.60	0.30–0.50	0.08–0.15	9440
9442	0.40–0.45	1.00–1.30	0.040	0.040	0.20–0.35	0.30–0.60	0.30–0.50	0.08–0.15	9442
9445	0.43–0.48	1.00–1.30	0.040	0.040	0.20–0.35	0.30–0.60	0.30–0.50	0.08–0.15	9445
9747	0.45–0.50	0.50–0.80	0.040	0.040	0.20–0.35	0.40–0.70	0.10–0.25	0.15–0.25	9747
9763	0.60–0.67	0.50–0.80	0.040	0.040	0.20–0.35	0.40–0.70	0.10–0.25	0.15–0.25	9763
9840	0.38–0.43	0.70–0.90	0.040	0.040	0.20–0.35	0.85–1.15	0.70–0.90	0.20–0.30	9840
9845	0.43–0.48	0.70–0.90	0.040	0.040	0.20–0.35	0.85–1.15	0.70–0.90	0.20–0.30	9845
9850	0.48–0.53	0.70–0.90	0.040	0.040	0.20–0.35	0.85–1.15	0.70–0.90	0.20–0.30	9850

Wrought Stainless Steels

Composition Limits of Austenitic Stainless Steels

AISI Type No.	Carbon	Composition, percentage Chromium	Nickel	Other Elements	General Properties and Uses
301	0.08 to 0.20	16.0 to 18.0	6.00 to 8.00	Mn 2.00 max	A general-utility stainless steel, easily worked; for trim, household utensils, structural purposes
302	0.08 to 0.20	17.0 to 19.0	8.00 to 10.00	Mn 2.00 max	A readily fabricated stainless steel for decorative or corrosion-resistant use
302B	0.08 to 0.20	17.0 to 19.0	8.00 to 10.00	Si 2.00 to 3.00 Mn 2.00 max	Silicon added to increase resistance to scaling at high temperatures
303	0.15 max	17.0 to 19.0	8.00 to 10.00	P, S, Se 0.07 min Zr, Mo 0.60 max	A free-machining grade 18-8 stainless steel
304	0.08 max	18.0 to 20.0	8.00 to 11.00	Mn 2.00 max	A low-carbon 18-8 steel, weldable with less danger of intercrystalline corrosion
308	0.08 max	19.0 to 21.0	10.00 to 12.00	Mn 2.00 max	For use when corrosion resistance greater than that of 18-8 is needed
309	0.20 max	22.0 to 24.0	12.00 to 15.00	Mn 2.00 max	For use at elevated temperature, combining high scaling resistance and good strength
309S	0.08 max	22.0 to 24.0	12.00 to 15.00	Mn 2.00 max	Low carbon permits welded fabrication with a minimum of carbide precipitation
310	0.25 max	24.0 to 26.0	19.0 to 22.0	Mn 2.00 max	Similar to 25-12 stainless, with higher nickel content for greater stability at welding temperatures
316	0.10 max	16.0 to 18.0	10.0 to 14.0	Mo 2.00 to 3.00	Superior resistance to chemical corrosion
317	0.10 max	18.0 to 20.0	11.0 to 14.0	Mo 3.00 to 4.00	Higher alloy content than 316 for increased corrosion resistance
321	0.08 max	17.0 to 19.0	8.0 to 11.0	Ti min 5 times C	An 18-8 type, stabilized against intercrystalline corrosion at elevated temperatures
347	0.08 max	17.0 to 19.0	9.00 to 12.00	Cb 10 times C	A stabilized 18-8 steel for service at elevated temperatures

Composition Limits of Martensitic Stainless Steels

AISI Type No.	Carbon	Chromium	Nickel	Other Elements	General Properties and Uses
403	0.15 max	11.5 to 13.0	—	—	Widely used for forged turbine blades
410	0.15 max	11.5 to 13.5	—	—	A low-priced general-purpose heat treatable stainless steel
414	0.15 max	11.5 to 13.5	1.25 to 2.50	—	For springs, knife blades, tempered rules
416	0.15 max	12.0 to 14.0	—	P, S, Se 0.07 min Zr, Mo 0.60 max	A free-machining grade
420	more than 0.15	12.0 to 14.0	—	—	For cutlery, surgical instruments, valves, ball bearings, magnets
420F	more than 0.15	12.0 to 14.0	—	P, S, Se 0.7 min Zr, Mo 0.60 max	A free-machining variation of 420
431	0.20 max	15.0 to 17.0	1.25 to 2.50	—	For high mechanical properties
440A	0.60 to 0.75	16.0 to 18.0	—	Mo 0.75 max	For instruments, cutlery, valves; high hardness obtainable in high-carbon grades
440B	0.75 to 0.95	16.0 to 18.0	—	Mo 0.75 max	
440C	0.95 to 1.20	16.0 to 18.0	—	Mo 0.75 max	
501	more than 0.10	4.00 to 6.00	—	—	Corrosion-resistant rather than stainless; good strength at elevated temperatures
502	0.10 max	4.00 to 6.00	—	—	Corrosion-resistant rather than stainless; good strength at elevated temperatures

Composition Limits of Ferritic Stainless Steels

AISI Type No.	Carbon	Chromium	Nickel	Other Elements	General Properties and Uses
405	0.08 max	11.5 to 13.5	—	Al 0.10 to 0.30	Nonhardening when air cooled from high temperatures
406	0.15 max	12.0 to 14.0	—	Al 3.50 to 4.50	For electrical resistances; aluminum also reduces air hardening
430	0.12 max	14.0 to 18.0	—	—	An easily formed stainless alloy, much used for automobile trim and chemical equipment
430F	0.12 max	14.0 to 18.0	—	P, S, Se 0.07 min Zr, Mo 0.60 max	A free-machining variety of 430
442	0.20 max	18.0 to 23.0	—	—	For high-temperature service when ease of fabrication is not required
443	0.20 max	18.0 to 23.0	—	Cu 0.90 to 1.25	An easily worked steel of high corrosion resistance, for chemical equipment, and use at elevated temperatures
446	0.35 max	23.0 to 27.0	—	—	High resistance to corrosion, and scaling resistance up to 2150° F

Wrought Heat-Resisting Alloys

Composition Limits of Wrought Heat-Resisting Steels[a, b]

Type No. (AISI)	Carbon	Manganese (max)	Silicon (max)	Phosphorus (max)	Sulfur (max)	Chromium	Nickel	Other Elements
				Composition, percentage				
			Wrought Chromium Steels					
501	0.10 min	1.00	1.00	0.040	0.030	4.00 to 6.00	—	—
7 Cr	0.15 max	0.60	0.50 to 1.00	0.030	0.030	6.00 to 8.00	—	Mo 0.45 to 0.65
9 Cr	0.15 max	0.60	1.00	0.030	0.030	8.00 to 10.00	—	Mo 0.90 to 1.10
403	0.15 max	1.00	0.50	0.040	0.030	11.50 to 13.00	—	—
410	0.15 max	1.00	1.00	0.040	0.030	11.50 to 13.50	—	—
430	0.12 max	1.00	1.00	0.040	0.030	14.00 to 18.00	—	—
442	0.25 max	1.00	1.00	0.040	0.030	18.00 to 23.00	—	—
446	0.35 max	1.50	1.00	0.040	0.030	23.00 to 27.00	—	N 0.25 max
			Wrought Chromium-Nickel Steels					
302	0.08 to 0.20	2.00	1.00	0.040	0.030	17.00 to 19.00	8.00 to 10.00	—
302B	0.08 to 0.20	2.00	2.00 to 3.00	0.040	0.030	17.00 to 19.00	8.00 to 10.00	—
304	0.08 max	2.00	1.00	0.040	0.030	18.00 to 20.00	8.00 to 11.00	—
309	0.20 max	2.00	1.00	0.040	0.030	22.00 to 24.00	12.00 to 15.00	—
310[c]	0.25 max	2.00	1.50	0.040	0.030	24.00 to 26.00	19.00 to 22.00	—
316	0.10 max	2.00	1.00	0.040	0.030	16.00 to 18.00	10.00 to 14.00	Mo 2.00 to 3.00
321	0.08 max	2.00	1.00	0.040	0.030	17.00 to 19.00	8.00 to 11.00	Ti 5 × C min
347	0.08 max	2.00	1.00	0.040	0.030	17.00 to 19.00	9.00 to 12.00	Cb 10 × C min
330[d]	0.25 max	2.00	1.00	0.040	0.030	14.00 to 16.00	33.00 to 36.00	—

[a] AISI Steel Products Manual, Section 24, Part 10.

[b] Not all the compositions in this table conform strictly with those established for seamless tubing. Composition ranges for tubing are listed in AISI Steel Products Manual, Section 18, and in the following ASTM specifications: A 158, A 213, A 268, A 269 and A 271.

[c] A modification of this type, known as 310B, contains 2.0 to 2.5% Si and is used to a considerable extent for caburizing boxes and furnace parts.

[d] This material is not listed as a standard AISI grade, but it has been used to some extent in wrought form.

Chemical Compositions of Wrought Heat-Resisting Alloys

Alloy	C	Mn	Si	Cr	Ni	Co	Mo	W	Cb	Ti	N	Fe	Other
19-9 W-Mo	0.12 max	0.50	0.60	19.00	8.50	—	0.35	1.25	0.50	0.35	—	Rem	—
19-9 DL	0.25	0.50	0.60	19.00	9.00	—	1.25	1.20	0.30	0.20	—	Rem	—
234-A-5	0.38	4.17	0.30	18.52	4.55	—	1.35	1.34	0.57	—	—	Rem	—
Timken 16-25-6	0.10	1.35	0.70	16.72	25.23	—	6.25	—	—	—	0.147	Rem	—
Low-carbon N-155	0.15	1.50	0.50	21.3	20.0	20.0	3.0	2.0	1.0	—	0.15	Rem	—
N-155	0.36	1.54	0.59	21.08	20.8	20.54	3.0	2.18	0.98	—	0.11	Rem	—
S-495	0.40–0.50	1.0 max	1.0 max	13.0–15.0	19.0–21.0	—	3.50–4.50	3.50–4.50	3.50–4.50	—	—	Rem	—
S-497	0.40–0.50	—	—	—	—	19.0–21.0	—	—	—	—	—	Rem	—
S-590	0.40–0.50	2.0 max	1.0 max	18.50–20.5	19.0–21.0	19.0–21.0	3.50–4.00	3.50–4.00	3.50–4.00	—	—	Rem	—
S-816	0.35–0.45	1.0 max	1.0 max	18.50–20.50	19.0–21.0	42.0–44.0	3.50–4.00	3.50–4.00	3.50–4.00	—	—	Rem	—
Inconel "X"	0.04	0.50	0.40	15.0	73.0	—	—	—	1.0	2.5	—	7.0	Al 0.7
K-42-B (Type 5)	0.06	0.70	0.34	18.0	42.0	22.0	—	—	2.56	—	—	13.0	Al 0.59
Refractaloy 26	0.03	0.70	0.65	17.9	37.0	20.0	3.03	—	2.99	—	—	Rem	Al 0.25
Hastelloy B	0.05	0.59	0.19	—	65.10	—	28.63	—	—	—	—	4.71	—
Vitallium	0.22	0.66	0.53	27.42	2.84	62.20	5.53	—	—	—	—	0.70	—
25 Cr-20 Ni-1.5 Si (Type 310)	0.25	2.0 max	1.5 max	24.0–26.0	19.0–22.0	—	—	—	—	—	—	Rem	—

[*] Reprinted from the "Metals Handbook," 1948 Edition, by permission of the publisher, The American Society For Metals, Cleveland 3, Ohio.

TABLE 8. COMPOSITION OF NONFERROUS ALLOYS*

Name	Composition	Liquidus, °F	Solidus, °F
Aluminum and Aluminum Alloys			
1100 (2S) Alloy	99.0+ Al	1215	1190
3003 (3S) Alloy	Al-1.2Mn	1210	1190
2011 (11S) Alloy	Al-5.5Cu-0.5Pb-0.5Bi	1190	995
2014 (14S) Alloy	Al-4.4Cu-0.8Si-0.8Mn-0.4Mg	1180	950
2017 (17S) Alloy, "Dural"	Al-4Cu-0.5Mg-0.5Mn	1185	955
2018 (18S) Alloy	Al-4Cu-2Ni-0.5Mg	1180	945
2024 (24S) Alloy	Al-4.5Cu-1.5Mg-0.6Mn	1180	935
2025 (25S) Alloy	Al-4.5Cu-0.8Mn-0.8Si	1185	970
4032 (32S) Alloy	Al-12.5Si-1.0Mg-0.9Cu-0.9Ni	1060	990
6151 (A51S) Alloy	Al-1.0Si-0.6Mg-0.25Cr	1200	1026
5052 (52S) Alloy	Al-2.5Mg-0.25Cr	1200	1100
6053 (53S) Alloy	Al-1.3Mg-0.7Si-0.25Cr	1205	1075
5056 (56S) Alloy	Al-5.2Mg-0.1Mn-0.1Cr	1180	1055
6061 (61S) Alloy	Al-1.0Mg-0.6Si-0.25Cu-0.25Cr	1205	1080
7075 (75S) Alloy	Al-5.5Zn-2.5Mg-1.5Cu-0.3Cr-0.2Mn	1180	890
13 Alloy	Al-12Si	1085	1065
43 Alloy	Al-5Si	1165	1070
85 Alloy	Al-5Si-4Cu	1135	970
108 Alloy	Al-4Cu-3Si	1170	970
Allcast	Al-5Si-3Cu	—	970
A108 Alloy	Al-5.5Si-4.5Cu	1140	960
113 Alloy	Al-7Cu-2Si-1.7Zn	1165	975
C113 Alloy	Al-7Cu-3.5Si	1165	975
122 Alloy	Al-10Cu-0.2Mg	1160	1005
A132 Alloy	Al-12Si-2.5Ni-1.2Mg-0.8Cu	1095	1000
Red X-13	Al-12Si-1.5Cu-0.7Mn-0.7Mg	—	—
142 Alloy	Al-4Cu-2Ni-1.5Mg	1165	995
195 Alloy	Al-4.5Cu	1195	1020
B195 Alloy	Al-4.5Cu-2.5Si	1160	980
214 Alloy	Al-3.8Mg	1185	1075
218 Alloy	Al-8Mg	1150	1005
220 Alloy	Al-10Mg	1150	840
319 Alloy	Al-6Si-3.5Cu	1120	950
355 Alloy	Al-5Si-1.3Cu-0.5Mg	1160	1075
356 Alloy	Al-7Si-0.3Mg	1130	1075
Red X-8	Al-8Si-1.5Cu-0.3Mg-0.3Mn	—	—
360 Alloy	Al-9.5Si-0.5Mg	1095	1050
380 Alloy, 4-9 Alloy	Al-8.5Si-3.5Cu	1090	970
750 Alloy	Al-6.5Sn-1Cu-1Ni	1200	450
40E Alloy	Al-5.5Zn-0.6Mg-0.5Cr-0.2Ti	1140	1060
Copper and Copper Alloys			
OFHC Copper	99.9+ Cu	1981	—
Electrolytic tough pitch copper	99.92Cu-0.04O$_2$	1981	1949
Deoxidized copper	99.94Cu-0.02P	1981	—
Gilding metal	Cu-5Zn	1950	1920
Commercial bronze	Cu-10Zn	1910	1870
Red brass	Cu-15Zn	1880	1810
Low brass	Cu-20Zn	1830	1770
Cartridge brass	Cu-30Zn	1750	1680
Yellow brass	Cu-35Zn	1710	1660
Muntz metal	Cu-40Zn	1660	1650

TABLE 8 (*continued*)

Name	Composition	Liquidus, °F	Solidus, °F
Leaded commercial bronze	Cu-9.25Zn-1.75Pb	1900	1850
Low-leaded brass	Cu-35Zn-0.5Pb	1760	1650
Low-leaded brass, tube	Cu-32.5Zn-0.5Pb	1720	1660
Medium-leaded brass	Cu-34.5Zn-1.0Pb	1700	1630
High-leaded brass, clock or engraver's brass	Cu-35.75Zn-1.75Pb	1670	1630
Extra high-leaded brass, block or rule brass	Cu-35Zn-2.5Pb	1660	1630
Free-cutting brass (yellow), free-turning brass	Cu-35.5Zn-3Pb	1650	1630
Leaded muntz metal	Cu-39.5Zn-0.5Pb	1650	1630
Free-cutting muntz metal, half-leaded muntz	Cu-38.4Zn-1.1Pb	1650	1630
Forging brass	Cu-38Zn-2Pb	1640	1620
Architectural bronze	Cu-40Zn-3Pb	1630	1610
Admiralty metal	Cu-28Zn-1Sn	1720	1650
Naval brass (Grade A)	Cu-39.25Zn-0.75Sn	1650	1630
Leaded naval brass (Grade C)	Cu-37.5Zn-1.75Pb-0.75Sn	1650	1630
Manganese bronze (Type A)	Cu-39Zn-1.4Fe-1Sn-0.1Mn	1630	1590
Aluminum brass	Cu-22Zn-2Al	1780	1710
Phosphor bronze, 5% Grade A	Cu-5Sn-trace P	1920	1750
Phosphor bronze, 8% Grade C	Cu-8Sn-trace P	1880	1620
Phosphor bronze, 10% Grade D	Cu-10Sn-trace P	1830	1550
Phosphor bronze, 1.25% Grade E	Cu-1.25Sn-trace P	1970	1900
Cupro-nickel, 30%	Cu-30Ni	2260	2140
Nickel silver, 18%-Alloy A	Cu-18Ni-17Zn	2030	1960
Nickel silver, 18%-Alloy B	Cu-27Zn-18Ni	1930	—
(High) silicon bronze-Type A, copper-silicon alloy	Cu-3Si	1880	1780
Silicon bronze-Type B	Cu-1.5Si	1940	1890
5% Aluminum bronze	Cu-5Al	1940	1920
8% Aluminum bronze	Cu-8Al	1900	1890
10% Aluminum bronze	Cu-10Al	1905	1890
Aluminum bronze	Cu-10Al-5Ni-2.5Fe	1930	1895
Beryllium copper	Cu-2Be-0.25Co (or 0.35Ni)	1750	1587
Leaded tin bronze, Navy M, steam or valve bronze	Cu-6Sn-1.5Pb-4.5Zn	—	1800
Leaded tin bearing bronze	Cu-8Sn-1Pb-4Zn	1830	1570
High-leaded tin bronze	Cu-5Sn-9Pb-1Zn	—	1750
High-leaded tin bronze	Cu-7Sn-7Pb-3Zn	—	1750
High-leaded tin bronze	Cu-10Sn-10Pb	—	—
High-leaded tin bronze, anti-acid bronze	Cu-7Sn-15Pb	—	1700
High-leaded tin bronze, semi-plastic bronze	Cu-5Sn-25Pb	—	1650
85-5-5-5, Ounce metal, composition metal	Cu-5Sn-5Pb-5Zn	—	1810
Leaded red brass, hydraulic bronze	Cu-4Sn-6Pb-7Zn	—	1800
Leaded semi-red brass valve metal	Cu-3Sn-7Pb-9Zn	—	1750
Leaded semi-red brass, plumbing goods brass	Cu-3Sn-6Pb-15Zn	—	1725
Leaded yellow brass, high-copper yellow brass	Cu-1Sn-3Pb-25Zn	—	1700

TABLE 8 (*continued*)

Name	Composition	Liquidus, °F	Solidus, °F
Leaded yellow brass, No. 1 yellow brass	Cu-1Sn-3Pb-30Zn	—	1700
Leaded yellow brass, die-cast brass	Cu-1Sn-1Pb-38Zn	—	1675
High-strength yellow brass, high-strength manganese bronze	Cu-26Zn-3Fe-5.5Al-3.5Mn	1650	—
High-strength yellow brass, manganese bronze	Cu-39.25Zn-1.25Fe-1.25Al-0.25Mn	1660	—
Leaded manganese bronze, stem manganese bronze	Cu-0.75Sn-0.75Pb-37Zn-1.25Fe-0.75Al-0.5Mn	1675	—
Nickel silver, leaded nickel bronze	Cu-5Sn-1.5Pb-2Zn-25Ni	—	—
Nickel silver, dairy bronze	Cu-4Sn-4Pb-8Zn-20Ni	—	—
Nickel silver, Benedict metal	Cu-2Sn-9Pb-20Zn-12Ni	—	—
Leaded nickel brass, nickel silver	Cu-3Sn-5Pb-16Zn-16Ni	—	—
(89-1-10) Aluminum bronze	Cu-1Fe-10Al	—	—
Aluminum bronze	Cu-3.5Fe-9Al	—	—
(86-4-10) Aluminum bronze	Cu-4Fe-10Al	—	—
(79-5-11-5) Aluminum bronze	Cu-5Fe-11Al-5Ni	—	—
Gold and Gold Alloys			
Gold	99.9+ Au	1945	—
Green gold	Au-25Ag	1970	1900
Yellow gold	Au-12Ag-13Cu	—	1623
15-Carat white gold	Au-16Cu-17Ni-8.65Zn	—	—
18-Carat white gold	Au-3.5Cu-16.5Ni-5.0Zn	—	—
14-Carat pink gold	Au-4.9Ag-31.6Cu-5.2Ni	—	1710
12-Carat pink gold	Au-4Ag-37Cu-4.3Ni-4.6Zn	—	1720
10-Carat pink gold	Au-4.6Ag-43.4Cu-5Ni-5.4Zn	—	1715
Telephone relay contact metal	Au-25Ag-6Pt	2030	1886
Lead and Lead Alloys			
Chemical lead	99.9+ Pb	618	—
Cable lead	Pb-0.028Ca	621.5	621.3
1% Antimonial lead	Pb-1Sb	608	595
Hard lead	Pb-4Sb	570	486
Hard lead	Pb-6Sb	545	486
8% Antimonial lead	Pb-8Sb	520	486
Grid metal	Pb-9Sb	509	486
Silver-lead solder	Pb-1.5Ag-1Sn	589	—
5-95 Soft solder	Pb-5Sn	594	518
20-80 Soft solder	Pb-20Sn	531	361
40-60 Soft solder	Pb-40Sn	460	361
50-50 Soft solder	Pb-50Sn	421	361
Lead-base Babbitt	Pb-10Sb-5Sn	493	464
Lead-base Babbitt	Pb-15Sb-5Sn	522	464
Lead-base Babbitt	Pb-15Sb-10Sn	514	464
Arsenical lead-base Babbitt, "S" Babbitt	Pb-15Sb-1Sn-1As	667	595
Arsenical lead-base Babbitt, "G" Babbitt	Pb-12.75Sb-3As-0.75Sn	595	477
Electrotype metal	Pb-2.5Sb-2.5Sn	578	475
Stereotype metal	Pb-13Sb-6.5Sn	485	462
Linotype metal	Pb-11Sb-3Sn	477	462
Monotype metal	Pb-15Sb-7Sn	503	462
Foundry-type metal	Pb-20Sb-10Sn	553	462
Hard foundry-type metal	Pb-20Sb-20Sn-1.5Cu	—	—

TABLE 8 (*continued*)

Name	Composition	Liquidus, °F	Solidus, °F
Magnesium and Magnesium Alloys			
Pure magnesium	99.8+ Mg	1202	—
"Mazlo AM244"	Mg-4Al-0.2Mn	1165	1045
"Dowmetal A," "Mazlo AM241"	Mg-8Al-0.2Mn	1125	918
"Dowmetal G," "Mazlo AM240," "Mazlo AMC59S"	Mg-10Al-0.1Mn	1101	867
"Dowmetal H," "Mazlo AM265"	Mg-6Al-3Zn-0.2Mn	1130	850
"Dowmetal C," "Mazlo AM260"	Mg-9Al-2Zn-0.1Mn	1100	830
"Dowmetal R," "Mazlo AM263"	Mg-9Al-0.7Zn-0.2Mn	1105	875
Eclipsaloy 130	Mg-1.25Al-1Mn	1190	1165
"Dowmetal M," "Mazlo AM403," "Mazlo AM3S"	Mg-1.5Mn	1200	1198
"Dowmetal FS-1," "Mazlo C52S"	Mg-3Al-1Zn-0.3Mn	1170	1050
"Dowmetal JS-1"	Mg-5Al-1Zn-0.25Mn	1150	990
"Dowmetal J-1," "Mazlo AMC57S"	Mg-6Al-1Zn-0.2Mn	1140	950
"Dowmetal 0-1," "Mazlo AMC58S"	Mg-8.5Al-0.5Zn-0.15Mn	1115	900
"Dowmetal D," "Mazlo AM65S"	Mg-5Sn-3Al-0.5Mn	—	—
Nickel and Nickel Alloys			
Electrolytic nickel	99.95 (Ni + Co)	2651	—
"A" nickel	99.4 (Ni + Co)	2635	2615
"D" nickel	Ni-4.5Mn	—	—
"Duranickel"	Ni-4.5Al	2635	2615
Cast nickel	Ni-1.5Si	—	—
Monel	Ni-30Cu-1.4Fe-1Mn	Melts 2370 to 2460	
Cast monel	Ni-32Cu-1.6Si-1.5Fe	Melts 2400 to 2450	
"K" monel	Ni-29Cu-2.75Al-0.9Fe-0.75Mn	2460	2400
"S" monel	Ni-30Cu-4Si-2Fe	Melts 2300 to 2350	
"Hastelloy A"	Ni-20Mo-20Fe	2425	2370
"Hastelloy B"	Ni-30Mo-5Fe	2460	2410
"Hastelloy C"	Ni-17Mo-15Cr-5W-5Fe	2380	2320
"Hastelloy D"	Ni-10Si-3Cu	2050	2030
"Inconel"	Ni-14Cr-6Fe	2600	2540
Cast "Inconel"	Ni-13.5Cr-6Fe-2Si	Melts 2500 to 2550	
"Illium G"	Ni-22Cr-6Cu-6Mo-6Fe	—	2370
ASTM B82-46, "Nichrome V"	Ni-20Cr	Melts 2550	
ASTM B83-46	Ni-24Fe-16Cr	Melts 2750 to 2800	
	Ni-50Fe-15Cr	Melts 2775 to 2825	
Constantan	Ni-55Cu	2355	2230
"Inconel X"	Ni-15Cr-7Fe-2.5Ti-1Cb		
ACI type HW alloy	Ni-28Fe-12Cr		
ACI type HX alloy	Ni-17Cr-16Fe		
K-42B (Type 5)	Ni-22Co-18Cr-13Fe-2.5Ti		
78.5 Permalloy	Ni-21.5Fe		
56 Permalloy	Ni-55Fe		
3.8-78.5 Cr-Permalloy	Ni-17.7Fe-3.8Cr		

TABLE 8 (*continued*)

Name	Composition	Liquidus, °F	Solidus, F
3.8-78.5Mo-Permalloy	Ni-17.7Fe-3.8Mo		
5-79Mo-Supermalloy	Ni-16Fe-5Mo		
45-25 Perminvar	Ni-30Fe-25Co		
7-45-25Mo-Perminvar	Ni-23Fe-25Co-7Mo		
70-7.5 Perminvar	Ni-22.5Fe-7.5Co		
"Conpernik," "Hipernik"	Ni-50Fe		
"1040" alloy	Ni-14Cu-11Fe-3Mo		
Mumetal	Ni-21Fe-5Cu		
Silver and Silver Alloys			
999 Fine silver	99.9+ Ag	1761	—
Sterling silver	Ag-7.5Cu	1652	1430
Coin silver	Ag-10Cu	1616	1435
BT alloy (eutectic)	Ag-28Cu	1435	—
"Easy-Flo"	Ag-15.5Cu-16.5Zn-18Cd	1175	1160
"Sil-Fos"	Ag-80Cu-5P	1300	1185
RT alloy	Ag-25Cu-15Zn	1325	1260
RT-SN alloy	Ag-30Cu-10Sn	1445	1385
Hard solder	Ag-20Cu-5Zn	1450	1365
Dental amalgam	Ag-52Hg-12.5Sn-2Cu-0.5Zn	Hg sweats at 167	
Tin and Tin Alloys			
Grade A Straits tin	99.8+ Sn	449.4	—
Hard tin	Sn-0.4Cu	446	441
Antimonial tin solder	Sn-5Sb	464	452
Tin-silver solder	Sn-5Ag	473	430
70-30 soft solder	Sn-30Pb	378	361
Eutectic soft solder	Sn-37Pb	361	—
60-40 soft solder	Sn-40Pb	370	361
Tin Babbitt No. 1	Sn-4.5Sb-4.5Cu	700	433
Tin Babbitt No. 2	Sn-7.5Sb-3.5Cu	669	466
Tin Babbitt No. 3	Sn-8.3Sb-8.3Cu	792	464
Tin Babbitt No. 5	Sn-18Pb-15Sb-2Cu	565	358
Tin foil	Sn-8Zn	395	390
White metal	Sn-8Sb	500	475
Pewter, Britannia metal	Sn-7Sb-2Cu	563	471
Zinc and Zinc Alloys			
Pure zinc	99.9+ Zn	787	—
"Zamak-3"	Zn-4Al-0.04Mg	730	717
"Zamak-5"	Zn-4Al-1Cu-0.04Mg	727	717
"Zamak-2"	Zn-4Al-3Cu-0.03Mg	734	715
Zinc-base slush-casting alloy	Zn-4.75Al-0.25Cu	734	716
Zinc-base slush-casting alloy	Zn-5.5Al	743	716
Deep-drawing zinc	Zn-0.08Pb	786	—
Commercial rolled zinc, battery sheet	Zn-0.06Pb-0.06Cd	786	—
Commercial rolled zinc, photo-engraver's sheet	Zn-0.3Pb-0.3Cd	786	—
"Zilloy-40"	Zn-1Cu	792	786
"Zilloy-15"	Zn-1Cu-0.01Mg	792	786

* Reprinted from the "Metals Handbook," 1948 Edition, by permission of the publisher, The American Society for Metals, Cleveland, Ohio.

TABLE 9. UNITS AND FACTORS

Acre foot	= 43,560 cubic feet
	= 325,851 gallons
Barrel, U.S., liquid, (bbl.)	= 31.5 gallons
Base box (Double base-box = double quantities)	= 112 sheets— 14 inches × 20 inches
	= 435.5 sq ft surface area (both sides)
Board foot	= 144 cubic inches
Boiler (hp)	= 33,479 Btu/hr
British thermal unit (Btu)	= 0.293 watt-hour
Carat (c)	= 200 milligrams
Centimeter (cm)	= 0.3937 inch
	= 0.03281 foot
	= 393.7 mils
Circular mil (cir mil)	= 1 mil dia. circular area (mil = 0.001 in.)
Cubic centimeter (cc)	= 0.0610 cubic inch
	= 0.99997 milliliter
	= 0.0338 U. S. fluid oz
Cubic foot per second	= 0.6463 million gallons per day (mgd)
	= 448.8 gallons per minute (gpm)
Cubic inch (cu in.)	= 16.3872 cubic centimeters
	= 16.3868 milliliters
	= 0.5541 U.S. fluid oz
Decimeter (dm)	= 3.9370 inches
	= 0.3281 feet
	= 10 centimeters
Dram, avoirdupois (dr av)	= 0.0625 avoirdupois ounce
	= 1.7718 grams
Dram, fluid (dr fl)	= 0.125 fluid ounce
	= 3.6966 milliliters
Dram, troy (dr t)	= 0.1371 avoirdupois oz
	= 3.8879 grams
Foot (ft)	= 30.4801 centimeters
Gallon, British (gal)	= 1.2009 U.S. gallons
Gallon, U.S. (gal)	= 231.00 cubic inches
	= 3.7853 liters
	= 3785.4 cubic centimeters
Grain	= 0.0648 gram
	= 0.002286 avoirdupois ounce
Gram (g)	= 0.002205 avoirdupois pound
	= 0.03527 avoirdupois ounce
Grams per cubic centimeter	= 62.43 avoirdupois pounds per cubic foot
	= 8.3452 avoirdupois pounds per U.S. gallon
Horse power (hp)	= 745.7 watts
Inch (in.)	= 2.5400 centimeter
	= 0.0833 foot
Karat (k)	= unit of gold purity, a 1/24 part
Kilogram (kg)	= 2.2046 avoirdupois pounds
Kilowatt-hour (kwhr)	= 1000 watt-hours
	= 3413 Btu
Liter (l)	= 0.03532 cubic foot
	= 1.05671 U.S. liquid quarts
	= 33.8147 U.S. fluid oz
	= 1000.027 cubic centimeter
Meter (m)	= 3.2808 U.S. feet
	= 100 centimeters
Micron (μ)	= 3.937 × 10^{-5} inch
	= 0.001 millimeter
Mil	= 0.001 inch
	= 0.002540 centimeter
	= 25.4 microns
Milligram (mg)	= 0.001 gram
Milliliter (ml)	= 1 cubic centimeter (approx)
Ounce, avoirdupois (oz av)	= 28.3495 grams
Ounce, fluid, U.S. (fl oz)	= 0.007813 U.S. gallon
	= 29.5729 milliliters
Ounce, troy (oz tr)	= 1.09714 avoirdupois ounce
	= 31.1035 grams
Pennyweight, troy (dwt)	= 0.003429 avoirdupois pound
	= 1.5552 grams
Pint, U.S. liquid (pt)	= 473.167 milliliters
Pint, British liquid	= 568.25 milliliters
Pound, avoirdupois (av lb)	= 453.5924 grams
Square centimeter (sq cm)	= 0.1550 square inch
	= 0.001076 square foot
	= 0.0100 sq decimeter
Square decimeter (sq dm)	= 15.500 square inch
	= 0.1076 square foot
Square foot (sq ft)	= 929.0341 square centimeters
	= 9.2903 square decimeters
	= 144 square inches
square inch (sq in.)	= 6.4516 square centimeters
	= 0.006944 square foot
	= 0.06452 square decimeter
Square mil	= 1.2732 circular mil
Ton, long	= 2240 avoirdupois pounds
	= 1016.0470 kilograms
Ton, metric	= 2204.6 avoirdupois pounds
	= 1000 kilograms
Ton, short	= 2000 avoirdupois pounds
	= 907.1850 kilograms
Ton, refrigeration	= 12,000 Btu/hr

Table 10. Conversion Factors

Concentration:

Multiply	by	to obtain
Ounces (avoirdupois) per gallon (U. S.)	7.489	grams per liter
Ounces (avoirdupois) per gallon (British)	6.236	grams per liter
Ounces (troy) per gallon (U. S.)	8.217	grams per liter
Ounces (troy) per gallon (British)	6.842	grams per liter
Grams per liter	0.1335	ounces (avoirdupois) per gallon (U. S.)
Grams per liter	0.1217	ounces (troy) per gallon (U. S.)
Pennyweight (troy) per gallon (U. S.)	0.4108	grams per liter
Pennyweight (troy) per gallon (British)	0.3421	grams per liter
Grams per liter	2.434	pennyweight per gallon (U. S.)
Fluid ounces per gallon	7.8125	milliliters per liter
Pints per gallon	125	milliliters per liter
Parts per million	1.336×10^{-4}	ounces (avoirdupois) per gallon (U.S.)
Parts per million	8.345	pounds per million gallons (U.S.)

Current Density:

Multiply	by	to obtain
Amperes per square decimeter	0.06452	amperes per square inch
Amperes per square decimeter	9.29	amperes per square foot
Amperes per square foot	0.006944	amperes per square inch
Amperes per square foot	0.1076	amperes per square decimeter
Amperes per square inch	144	amperes per square foot
Amperes per square inch	15.50	amperes per square decimeter

Approximate Conversion Factors for Brass, Bronze, Cadmium, Copper, and Zinc Cyanide Baths

Multiply	by	to obtain
Cadmium, wt	1.14	weight of CdO
	1.46	weight of $Cd(CN)_2$
	0.875	weight of NaCN equivalent to (CN) in $Cd(CN)_2$
	1.75	weight of NaCN equivalent to (CN) in $Na_2Cd(CN)_4$
CdO, wt	0.875	weight of Cd
	1.62	weight of $Cd(CN)_2$ equivalent
	1.53	weight of NaCN to convert to $Na_2Cd(CN)_4$
	0.625	weight of NaOH formed by adding NaCN to form $Na_2Cd(CN)_4$
$Cd(CN)_2$, wt	0.683	weight of Cd
	0.693	weight of CdO equivalent
	0.596	weight of NaCN equivalent to (CN) in $Cd(CN)_2$
	0.479	weight of NaCN to convert to $Na_2Cd(CN)_4$
	0.940	weight of NaCN equivalent to (CN) in $Na_2Cd(CN)_4$
Copper, wt	1.43	weight of CuCN equivalent
	0.77	weight of NaCN equivalent to (CN) in CuCN
	2.31	weight of total NaCN equivalent to (CN) in $Na_2Cu(CN)_3$
	1.02	weight of KCN equivalent to (CN) in CuCN
	3.07	weight of total KCN equivalent to (CN) in $K_2Cu(CN)_3$
CuCN, wt	0.70	weight of Cu
	0.546	weight of NaCN equivalent to (CN) in CuCN
	0.725	weight of KCN equivalent to (CN) in CuCN
	1.09	weight of NaCN to convert to $Na_2Cu(CN)_3$
	1.45	weight of KCN to convert to $K_2Cu(CN)_3$
Zinc, wt	1.82	weight of $Zn(CN)_2$ equivalent
	1.24	weight of ZnO equivalent
	1.50	weight of NaCN equivalent to (CN) in $Zn(CN)_2$
	3.0	weight of total NaCN equivalent to (CN) in $Na_2Zn(CN)_4$
$Zn(CN)_2$, wt	0.55	weight of Zn
	0.83	weight of NaCN to convert to $Na_2Zn(CN)_4$
	0.687	weight of ZnO equivalent

TABLE 10 (*continued*)

Multiply	by	to obtain
ZnO, wt	0.80	weight of Zn
	1.44	weight of Zn(CN)$_2$ equivalent
	2.41	weight of NaCN to convert to Na$_2$Zn(CN)$_4$
	0.985	weight of NaOH formed by adding NaCN to form Na$_2$Zn(CN)$_4$

NH$_3$ or NH$_4$OH *volume relationships*

$$1 \text{ ml/l } (28\% \text{ NH}_4\text{OH}) = \begin{array}{l} 3.78 \text{ ml/gal} \\ 0.25 \text{ g/l NH}_3 \\ 0.0147 \text{ Normal} \end{array}$$

$$1 \text{ ml/gal } (28\% \text{ NH}_4\text{OH}) = \begin{array}{l} 0.264 \text{ ml/l} \\ 0.066 \text{ g/l NH}_3 \\ 0.0039 \text{ Normal} \end{array}$$

$$0.1 \text{ Normal NH}_3 \text{ or NH}_4\text{OH} = \begin{array}{l} 6.8 \text{ ml/l } (28\% \text{ NH}_4\text{OH}) \\ 25.7 \text{ ml/gal } (28\% \text{ NH}_4\text{OH}) \end{array}$$

TABLE 11. TEMPERATURE OF SATURATED STEAM

This table gives the gage pressure of saturated steam in pounds per square inch corresponding to the stated temperature. Zero gage pressure is equal to 14.70 pounds per square inch absolute.

°F	psi	°F	psi	°F	psi
212	0.00	242	11.19	272	28.56
214	0.59	244	12.13	274	29.99
216	1.31	246	13.10	276	31.45
218	1.84	248	14.10	278	32.96
220	2.49	250	15.13	280	34.51
222	3.17	252	16.19	282	36.09
224	3.86	254	17.28	284	37.72
226	4.58	256	18.40	286	39.39
228	5.32	258	19.55	288	41.10
230	6.08	260	20.73	290	42.86
232	6.87	262	21.95	292	44.66
234	7.68	264	23.20	294	46.51
236	8.52	266	24.49	296	48.39
238	9.38	268	25.81	298	50.33
240	10.27	270	27.16	300	52.32

TABLE 12. CAPACITY OF ROUND TANKS
Basis: Inside Measure, 1 cubic foot = 7.481 gallons
Table Gives Gallons Per Foot of Height

Diameter	gal/ft*	Diameter	gal/ft*	Diameter	gal/ft*
1'-0"	5.88	3'-0"	52.88	9'	475.92
1'-3"	9.18	3'-6"	71.98	10'	587.56
1'-6"	13.22	4'-0"	94.01	11'	710.94
1'-9"	17.99	4'-6"	118.98	12'	846.08
2'-0"	23.50	5'	146.89	15'	1322.00
2'-3"	29.75	6'	211.52	20'	2350.23
2'-6"	36.72	7'	287.90	25'	3672.23
2'-9"	44.43	8'	376.04	30'	5288.01

* gal/ft = 5.8756 (diameter in feet)^2

TABLE 13. CAPACITY OF RECTANGULAR TANKS
Basis: Inside Measure, 1 cubic foot = 7.481 gallons (U. S.)
Table Gives Gallons Per Foot of Depth

Length	1'-0"	1'-6"	2'-0"	2'-6"	3'-0"	4'-0"	5'-0"	6'-0"	7'-0"	8'-0"	9'-0"	10'-0"
1'-0"	7.48	11.22	14.96	18.70	22.44	29.92	37.41	44.89	52.37	59.85	67.33	74.81
1'-3"	9.35	14.03	18.70	23.38	28.05	37.41	46.76	56.11	65.46	74.81	84.16	93.51
1'-6"	11.22	16.83	22.44	28.05	33.66	44.89	56.11	67.33	78.55	89.77	100.99	112.22
1'-9"	13.09	19.64	26.18	32.73	39.28	52.37	65.46	78.55	91.64	104.73	117.83	130.92
2'-0"	14.96	22.44	29.92	37.41	44.89	59.85	74.81	89.77	104.73	119.70	134.66	149.62
2'-3"	16.83	25.25	33.66	42.08	50.50	67.33	84.16	100.99	117.83	134.66	151.49	168.32
2'-6"	18.70	28.05	37.41	46.76	56.11	74.81	93.51	112.22	130.92	149.62	168.32	187.03
2'-9"	20.57	30.86	41.15	51.43	61.72	82.29	102.86	123.44	144.01	164.58	185.15	205.73
3'-0"	22.44	33.66	44.89	56.11	67.33	89.77	112.22	134.66	157.10	179.54	201.99	224.43
3'-6"	26.18	39.28	52.37	65.46	78.55	104.73	130.92	157.10	183.28	209.47	235.65	261.84
4'-0"	29.92	44.89	59.85	74.81	89.77	119.70	149.62	179.54	209.47	239.39	269.32	299.24
4'-6"	33.66	50.50	67.33	84.16	100.99	134.66	168.32	201.99	235.65	269.32	302.98	336.65
5'-0"	37.41	56.11	74.81	93.51	112.22	149.62	187.03	224.43	261.84	299.24	336.65	374.05
6'-0"	44.89	67.33	89.77	112.22	134.66	179.54	224.43	269.32	314.20	359.09	403.97	448.86

TABLE 14. SIZE OF HANGING WIRES

Where wire is used only once for hanging work in a plating tank, the cost of wire should be minimized by selection of the smallest wire diameter that will support the weight of the work and that will adequately carry the current. A recommended procedure of size selection is to select first the smallest wire gage that will carry the current without burning the worker and then to increase the gage if necessary to obtain the required physical strength. The following table gives bare copper wire sizes which have been found adequate for use in air.

Required Current, amp	Copper Diameter, mil	Nearest B & S Gage	Required Current, amp	Copper Diameter, mil	Nearest B & S Gage
5	21.6	23	30	71.1	13
10	34.2	19	35	78.7	12
15	44.2	17	40	86.1	11
20	54.2	15	45	93.5	10
25	62.8	14	50	100	10

Formula:
  Copper diameter, mils = 7.35 (amperes)$^{2/3}$
  Good electrical contact to work rod is assumed.

TABLE 15. SIZE OF BUS BARS

The size of bus bars for a required current capacity depends upon the total length of run and the permissible voltage drop in the bus bar. For copper bus bars the following formulas apply:

$$A = 11.5 \, LI/E$$
$$a = 9.03 \times 10^{-6} \, LI/E$$

where
A = cross section area in circular mils
a = cross section area in square inches
I = current in amperes
L = total length of bus in feet
E = permissible voltage drop in volts

The *area* of 1100(2S) aluminum bus bar for equivalent service is larger by a factor of 1.8 due to the lower conductivity of aluminum.

TABLE 16. SPECIFIC GRAVITY AND DEGREES BAUMÉ OF AQUEOUS HYDROCHLORIC ACID SOLUTIONS*

Sp Gr $\frac{20°}{4°}$ C	°Baumé	% HCl	HCl Content	
			g/l	oz/gal
1.0032	0.5	1	10.03	1.34
1.0082	1.2	2	20.16	2.68
1.0181	2.6	4	40.72	5.43
1.0279	3.9	6	61.67	8.23
1.0376	5.3	8	83.01	11.08
1.0474	6.6	10	104.7	13.95
1.0574	7.9	12	126.9	16.9
1.0675	9.2	14	149.5	19.9
1.0776	10.4	16	172.4	23.0
1.0878	11.7	18	195.8	26.1
1.0980	12.9	20	219.6	29.25
1.1083	14.2	22	243.8	32.5
1.1187	15.4	24	268.5	35.8
1.1290	16.6	26	293.5	39.1
1.1392	17.7	28	319.0	42.5
1.1493	18.8	30	344.8	46.0
1.1593	19.9	32	371.0	49.5
1.1691	21.0	34	397.5	53.0
1.1789	22.0	36	424.4	56.6
1.1885	23.0	38	451.6	60.2
1.1980	24.0	40	479.2	64.0

* Reprinted from "Modern Electroplating" by permission of the publisher The Electrochemical Society, New York City.

TABLE 17.* SPECIFIC GRAVITY AND DEGREES BAUMÉ OF AQUEOUS SULFURIC ACID SOLUTIONS

Sp Gr $\frac{20°}{4°}$ C	°Baumé	H_2SO_4 Content			Sp Gr $\frac{20°}{4°}$ C	°Baumé	H_2SO_4 Content		
		% H_2SO_4	g/l	oz/gal			% H_2SO_4	g/l	oz/gal
1.005	0.7	1	10.05	1.4	1.445	44.7	55	794.9	106
1.018	2.6	3	30.55	4.07	1.466	46.1	57	835.7	111.5
1.032	4.5	5	51.59	6.88	1.487	47.5	59	877.6	117
1.045	6.3	7	73.17	9.76	1.509	48.9	61	920.6	122.7
1.059	8.1	9	95.32	12.7	1.531	50.3	63	964.5	128.7
1.073	9.9	11	118.0	15.73	1.553	51.7	65	1,010	134.8
1.087	11.7	13	141.4	18.85	1.576	53.0	67	1,056	141.0
1.102	13.4	15	165.3	22.05	1.599	54.3	69	1,103	147
1.117	15.2	17	189.9	25.3	1.622	55.6	71	1,152	153.8
1.132	16.9	19	215.0	28.7	1.646	56.9	73	1,201	160
1.147	18.6	21	240.9	32.1	1.669	58.1	75	1,252	167
1.163	20.3	23	267.4	35.6	1.693	59.3	77	1,303	174
1.178	21.9	25	294.6	39.3	1.716	60.5	79	1,355	181
1.194	23.6	27	322.4	43.0	1.738	61.6	81	1,408	188
1.210	25.2	29	351.0	46.8	1.759	62.6	83	1,460	195
1.227	26.8	31	380.3	51.0	1.779	63.5	85	1,512	201.6
1.243	28.4	33	410.3	54.7	1.795	64.2	87	1,562	208.4
1.260	29.9	35	441.0	58.8	1.809	64.8	89	1,610	214.6
1.277	31.4	37	472.5	62.9	1.819	65.3	91	1,656	220.6
1.294	33.0	39	504.7	67.3	1.828	65.7	93	1,700	226.5
1.312	34.5	41	537.8	71.6	1.834	65.9	95	1,742	232.5
1.329	35.9	43	571.6	76.1	1.836	66.0	97	1,781	237.5
1.348	37.4	45	606.4	80.8	1.834	65.9	99	1,816	242
1.366	38.9	47	642.2	85.5	1.830	65.8	100	1,831	244.5
1.385	40.3	49	678.8	90.5					
1.405	41.8	51	716.5	95.5					
1.425	43.2	53	755.1	100.8					

* Reprinted from "Modern Electroplating" by permission of the publisher, The Electrochemical Society, New York City.

TABLE 18. SPECIFIC GRAVITY AND DEGREES BAUMÉ OF CHROMIC ACID SOLUTIONS AS A FUNCTION OF THE CrO_3 CONTENT*

Sp Gr $\frac{15°}{4°}$ C	Degrees Baumé	CrO_3 Content			Sp Gr $\frac{15°}{4°}$ C	Degrees Baumé	CrO_3 Content		
		Molarity	g/l	oz/gal			Molarity	g/l	oz/gal
1.01	1.44	0.15	15	2.0	1.18	22.1	2.57	257	34.4
1.02	2.84	0.29	29	3.9	1.19	23.2	2.72	272	36.4
1.03	4.22	0.43	43	5.8	1.20	24.2	2.88	288	38.6
1.04	5.58	0.57	57	7.6	1.21	25.2	3.01	301	40.3
1.05	6.90	0.71	71	9.5	1.22	26.2	3.16	316	42.3
1.06	8.21	0.85	85	11.4	1.23	27.1	3.30	330	44.2
1.07	9.5	1.00	100	13.4	1.24	28.1	3.45	345	46.2
1.08	10.7	1.14	114	15.3	1.25	29.0	3.60	360	48.2
1.09	12.0	1.29	129	17.3	1.26	29.9	3.75	375	50.2
1.10	13.0	1.43	143	19.1	1.27	30.8	3.90	390	52.2
1.11	14.4	1.57	157	21.0	1.28	31.7	4.06	406	54.5
1.12	15.5	1.71	171	22.9	1.29	32.6	4.22	422	56.5
1.13	16.7	1.85	185	24.8	1.30	33.5	4.38	438	58.7
1.14	17.8	2.00	200	26.8	1.31	34.3	4.53	453	60.7
1.15	18.9	2.15	215	28.8	1.32	35.2	4.68	468	62.7
1.16	20.0	2.29	229	30.6	1.33	36.0	4.84	484	64.8
1.17	21.1	2.43	243	32.6	1.34	36.8	5.00	500	67.0

* From data compiled by J. A. Beattie for "International Critical Tables," McGraw-Hill Book Company, Inc., New York.

By permission from "Principles of Electroplating and Electroforming" by Blum and Hogaboom. Copyright, 1924, 1930, 1949, by the McGraw-Hill Book Co., Inc.

TABLE 19. TOTAL CONCENTRATIONS OF COPPER SULFATE PLUS SULFURIC ACID IN SOLUTIONS OF GIVEN SPECIFIC GRAVITY*

Sp Gr $\frac{25°}{4°}$ C	Copper Sulfate plus Sulfuric Acid		Sp Gr $\frac{25°}{4°}$ C	Copper Sulfate plus Sulfuric Acid	
	g/l	oz/gal		g/l	oz/gal
1.01	20	2.7	1.13	217	29.1
1.02	36	4.8	1.14	234	31.3
1.03	52	7.0	1.15	251	33.6
1.04	68	9.1	1.16	268	35.9
1.05	84	11.3	1.17	286	38.3
1.06	100	13.4	1.18	303	40.6
1.07	117	15.7	1.19	321	43.0
1.08	133	17.8	1.20	339	45.4
1.09	150	20.0	1.21	357	47.8
1.10	166	22.3	1.22	375	50.2
1.11	183	24.5	1.23	393	52.6
1.12	200	26.8			

* By permission from "Principles of Electroplating and Electroforming" by Blum and Hogaboom. Copyright, 1924, 1930, 1949 by the McGraw-Hill Book Co., Inc.

Reprinted from "Modern Electroplating" by permission of the publisher, The Electrochemical Society, New York City.

TABLE 20. SURFACE DATA FOR FLAT METAL PRODUCTS
U.S. Standard Gage (a Weight Gage)

Used for: iron and steel sheet, galvanized iron, tinned plate, terne plate, black iron, hot and cold rolled sheet steel, monel and nickel sheet.

Gage	oz/sq ft	Approximate thickness (in.)	Area of 100 linear in. of edge expressed in sq ft	Surface in sq ft/100 lb of steel without edges *
7/0	320	0.4902	0.340	10.0
6/0	300	0.4596	0.319	10.7
5/0	280	0.4289	0.298	11.4
4/0	260	0.3983	0.276	12.3
3/0	240	0.3676	0.255	13.3
2/0	220	0.3370	0.234	14.5
0	200	0.3064	0.213	16.0
1	180	0.2757	0.191	17.8
2	170	0.2604	0.181	18.8
3	160	0.2451	0.170	20.0
4	150	0.2298	0.166	21.3
5	140	0.2145	0.144	22.9
6	130	0.1991	0.138	24.6
7	120	0.1838	0.128	26.7
8	110	0.1685	0.117	29.3
9	100	0.1532	0.106	32.0
10	90	0.1379	0.096	35.6
11	80	0.1225	0.085	40.0
12	70	0.1072	0.074	45.7
13	60	0.0919	0.063	53.3
14	50	0.0766	0.053	64.0
15	45	0.0689	0.048	71.2
16	40	0.0613	0.043	80.0
17	36	0.0551	0.038	88.9
18	32	0.0490	0.034	100
19	28	0.0429	0.030	114
20	24	0.0368	0.026	133
21	22	0.0337	0.023	146
22	20	0.0306	0.021	160
23	18	0.0276	0.019	178
24	16	0.0245	0.017	200
25	14	0.0214	0.015	229
26	12	0.0184	0.013	267
27	11	0.0169	0.012	291
28	10	0.0153	0.011	320
29	9	0.0138	0.010	356
30	8	0.0123	0.009	400

* Total surface, both sides.

TABLE 20 (*continued*)

Brown & Sharpe (B. & S.) or American (A. W. G.)

Used for: aluminum sheet; brass and phosphor bronze sheet, German silver (nickel brass, nickel silver)

Gage	Thickness (in.)	Area of 100 lineal in. of edge in sq ft	Surface in sq ft/100 lbs of metal (without edges)*			
			Aluminum (0.100 lb/cu in.)	Yellow brass (0.306 lb/cu in.)	Phosphor bronze (0.320/lb/cu in.)	German silver (0.313 lb/cu in.)
6/0	0.5800	0.403	23.9	7.81	7.46	7.63
5/0	0.5165	0.358	26.9	8.77	8.38	8.57
4/0	0.4600	0.320	30.2	9.85	9.32	9.63
3/0	0.4096	0.284	33.9	11.1	10.6	10.8
2/0	0.3648	0.254	38.0	12.4	11.9	12.1
0	0.3249	0.226	42.7	13.9	13.3	13.6
1	0.2893	0.201	48.0	15.7	15.0	15.3
2	0.2576	0.179	53.9	17.6	16.8	17.2
3	0.2294	0.159	60.5	19.8	18.9	19.3
4	0.2043	0.142	68.0	22.2	21.2	21.7
5	0.1819	0.126	76.4	24.9	23.8	24.4
6	0.1620	0.113	85.6	28.0	26.7	27.4
7	0.1443	0.100	96.2	31.4	30.0	30.7
8	0.1285	0.089	108	35.3	33.7	34.5
9	0.1144	0.079	121	39.6	37.8	38.7
10	0.1019	0.071	136	44.5	42.5	43.5
11	0.0907	0.063	153	50.0	47.7	48.8
12	0.0808	0.056	172	56.1	53.6	54.8
13	0.0720	0.050	193	63.0	60.2	61.5
14	0.0641	0.044	228	70.6	67.5	69.1
15	0.0571	0.040	243	79.4	75.9	77.6
16	0.0508	0.035	273	89.1	85.2	87.2
17	0.0453	0.032	306	100	95.5	97.8
18	0.0403	0.028	344	112	107	110
19	0.0359	0.025	387	126	121	123
20	0.0320	0.022	434	142	135	139
21	0.0285	0.020	487	159	152	156
22	0.0253	0.018	548	179	171	175
23	0.0226	0.016	614	201	192	196
24	0.0201	0.014	690	225	216	221
25	0.0179	0.012	775	253	242	248
26	0.0159	0.011	873	285	272	279
27	0.0142	0.010	977	319	305	312
28	0.0126	0.009	1103	359	344	352
29	0.0113	0.008	1230	402	383	392
30	0.0100	0.007	1388	453	433	443

* Total surface, both sides.

TABLE 20 (*continued*)

Copper is Rolled to Weight and the Gage is Ounces or Pounds per Square Foot

oz/sq ft	lb/sq ft	Approximate thickness (in.)	Area of 100 lineal in. of edge expressed in sq ft	Surface in sq ft/100 lb without edges*
1	1/16	0.0014	0.0010	3200
2	1/8	0.0027	0.0019	1600
3	3/16	0.0041	0.0028	1067
4	1/4	0.0054	0.0038	800
5	5/16	0.0068	0.0047	640
6	3/8	0.0081	0.0056	533
8	1/2	0.0108	0.0075	400
10	5/8	0.0135	0.0094	320
12	3/4	0.0162	0.0112	267
14	7/8	0.0189	0.0131	229
16	1	0.0216	0.0150	200
18	1 1/8	0.0243	0.0169	178
20	1 1/4	0.0270	0.0187	160
24	1 1/2	0.0323	0.0224	133
28	1 3/4	0.0377	0.0262	114
32	2	0.0431	0.0299	100
36	2 1/4	0.0485	0.0337	88.8
40	2 1/2	0.0539	0.0374	80.0
44	2 3/4	0.0593	0.0412	72.7
48	3	0.0647	0.0449	66.7
56	3 1/2	0.0755	0.0514	57.1
64	4	0.0863	0.0599	50.0
72	4 1/2	0.0970	0.0674	44.4
80	5	0.1078	0.0748	40.0
	5 1/2	0.1186	0.0823	36.3
	6	0.1294	0.0898	33.3
	6 1/2	0.1402	0.0973	30.8
	7	0.1510	0.105	28.6
	7 1/2	0.1617	0.112	26.7
	8	0.1725	0.120	25.0
	8 1/2	0.1833	0.127	23.5
	9	0.1941	0.135	22.2
	9 1/2	0.2049	0.142	21.1
	10	0.2157	0.150	20.0
	11	0.2372	0.165	18.2
	12	0.2588	0.180	16.7
	13	0.2804	0.195	15.4
	14	0.3019	0.210	14.3
	15	0.3235	0.224	13.3
	16	0.3451	0.240	12.5

* Total surface, both sides.

TABLE 21. SURFACE DATA FOR ROUND METAL PRODUCTS

Fraction, inches	Decimal, inches	Area of cross section (sq in.)	Surface area per 100 in. length (sq in.)
1	1.0000	0.7854	314.16
15/16	0.9375	0.6903	294.53
7/8	0.8750	0.6013	274.89
13/16	0.8125	0.5185	255.26
3/4	0.7500	0.4418	235.62
11/16	0.6875	0.3712	215.99
5/8	0.6250	0.3068	196.35
9/16	0.5625	0.2485	176.72
1/2	0.5000	0.1964	157.08
7/16	0.4375	0.1503	137.45
3/8	0.3750	0.1105	117.81
5/16	0.3125	0.0767	98.18
1/4	0.2500	0.0490	78.54
3/16	0.1875	0.0276	58.91
1/8	0.1250	0.0123	39.27
1/16	0.0625	0.0030	19.64

Steel Wire Gage, Washburn & Moen, American Steel & Wire Co., Roebling, National Wire Gage Used for Iron and Steel Wire

Gage	Decimal, inches	Area of cross section (sq in.)	Surface area per 100 in. length (sq in.)
7/0	0.4900	0.1886	153.9
6/0	0.4615	0.1673	145.0
5/0	0.4305	0.1456	135.2
4/0	0.3938	0.1218	123.6
3/0	0.3625	0.1032	113.9
2/0	0.3310	0.0860	104.0
0	0.3065	0.0738	96.3
1	0.2830	0.0629	88.9
2	0.2625	0.0540	82.4
3	0.2437	0.0467	76.5
4	0.2253	0.0398	70.8
5	0.2070	0.0337	65.0
6	0.1920	0.0290	60.3
7	0.1770	0.0246	55.6
8	0.1620	0.0206	50.9
9	0.1483	0.0173	46.6
10	0.1350	0.0143	42.4
11	0.1205	0.0114	37.8
12	0.1055	0.0087	33.1
13	0.0915	0.0066	28.8
14	0.0800	0.0050	25.1
15	0.0720	0.0041	22.6
16	0.0625	0.0031	19.64
17	0.0540	0.0023	16.97
18	0.0475	0.0018	14.92
19	0.0410	0.0013	12.88
20	0.0348	0.0010	10.93

TABLE 21 (*continued*)

Gage	Decimal, inches	Area of cross section, (sq in.)	Surface area per 100 in. length (sq in.)
21	0.0318	0.0008	9.99
22	0.0286	0.0006	8.99
23	0.0258	0.0005	8.11
24	0.0230	0.0004	7.23
25	0.0204	0.0003	6.41
26	0.0181	0.0003	5.69
27	0.0173	0.0003	5.44
28	0.0162	0.0002	5.09
29	0.0150	0.0002	4.71
30	0.0140	0.0002	4.40

Brown & Sharpe (B. & S.) or American (A. W. G.) Used for Nonferrous Wires

Gage	Decimal, inches	Area of cross section (sq in.)	Surface area per 100 in. length (sq in.)
6/0	0.5800	0.2642	182.0
5/0	0.5165	0.2095	162.2
4/0	0.4600	0.1662	144.2
3/0	0.4096	0.1317	128.6
2/0	0.3648	0.1045	104.6
0	0.3249	0.0829	102.0
1	0.2893	0.0657	90.8
2	0.2576	0.0520	80.9
3	0.2294	0.0413	72.0
4	0.2043	0.0328	64.1
5	0.1819	0.0260	57.1
6	0.1620	0.0206	50.8
7	0.1443	0.0164	45.4
8	0.1285	0.0130	40.4
9	0.1144	0.0103	35.9
10	0.1019	0.0082	32.0
11	0.0907	0.0065	28.5
12	0.0808	0.0051	25.3
13	0.0720	0.0041	22.6
14	0.0641	0.0034	20.14
15	0.0571	0.0024	17.94
16	0.0508	0.0020	15.96
17	0.0453	0.0016	14.23
18	0.0403	0.0013	12.66
19	0.0359	0.0010	11.28
20	0.0320	0.0008	10.05
21	0.0285	0.0006	8.95
22	0.0253	0.0005	7.95
23	0.0226	0.0004	7.10
24	0.0201	0.0003	6.31
25	0.0179	0.0003	5.62
26	0.0159	0.0002	5.00
27	0.0142	0.0002	4.46
28	0.0126	0.0001	3.96
29	0.0113	0.0001	3.55
30	0.0100	0.0001	3.14

TABLE 21 (*continued*)

Birmingham or Stubs Iron Wire Gage Used Occasionally for Iron and Copper Wire

Gage	Decimal, inches	Area of cross section (sq in.)	Surface area per 100 in. length (sq in.)
5/0	0.500	0.1964	157.1
4/0	0.454	0.1619	142.6
3/0	0.425	0.1419	133.5
2/0	0.380	0.1134	119.4
0	0.340	0.0908	106.8
1	0.300	0.0707	94.2
2	0.284	0.0633	89.2
3	0.259	0.0527	81.4
4	0.238	0.0445	74.8
5	0.220	0.0380	69.1
6	0.203	0.0324	63.8
7	0.180	0.0254	56.5
8	0.165	0.0214	51.8
9	0.148	0.0172	46.5
10	0.134	0.0141	42.1
11	0.120	0.0113	37.7
12	0.109	0.0093	34.2
13	0.095	0.0071	29.8
14	0.083	0.0054	26.1
15	0.072	0.0041	22.6
16	0.065	0.0033	20.4
17	0.058	0.0026	18.22
18	0.049	0.0019	15.39
19	0.042	0.0014	13.20
20	0.035	0.0010	11.00
21	0.032	0.0008	10.05
22	0.028	0.0006	8.80
23	0.025	0.0005	7.85
24	0.022	0.0004	6.91
25	0.020	0.0003	6.28
26	0.018	0.0003	5.65
27	0.016	0.0002	5.03
28	0.014	0.0002	4.40
29	0.013	0.0001	4.08
30	0.012	0.0001	3.77

TABLE 22. STANDARD SCREW THREADS

	National Coarse (U. S. Standard)			National Fine (SAE)	
Nominal size	Tap drill	Clearance drill	Nominal size	Tap drill	Clearance drill
1–64	53	47	0–80	$3/64''$	51
2–56	50	42	1–72	53	47
3–48	47	36	2–64	50	42
4–40	43	31	3–56	45	36
5–40	38	29	4–48	42	31
6–32	36	25	5–44	37	29
8–32	29	16	6–40	33	25
10–24	25	$13/64''$	8–36	29	16
12–24	16	$7/32''$	10–32	21	$13/64''$
$1/4''$–20	7	$17/64''$	12–28	14	$7/32''$
$5/16''$–18	F	$21/64''$	$1/4''$–28	3	$17/64''$
$3/8''$–16	$5/16''$	$25/64''$	$5/16''$–24	I	$21/64''$
$7/16''$–14	U	$29/64''$	$3/8''$–24	Q	$25/64''$
$1/2''$–13	$27/64''$	$33/64''$	$7/16''$–20	$25/64''$	$29/64''$
$9/16''$–12	$31/64''$	$37/64''$	$1/2''$–20	$29/64''$	$33/64''$
$5/8''$–11	$17/32''$	$41/64''$	$9/16''$–18	$33/64''$	$37/64''$
$3/4''$–10	$21/32''$	$49/64''$	$5/8''$–18	$37/64''$	$41/64''$
$7/8''$–9	$49/64''$	$57/64''$	$3/4''$–16	$11/16''$	$49/64''$
$1''$–8	$7/8''$	$1\text{-}1/64''$	$7/8''$–14	$13/16''$	$57/64''$
			$1''$–14	$15/16''$	$1\text{-}1/64''$

TABLE 23. DIMENSIONS AND AREAS OF STEEL PIPE

				Standard-Weight		Extra-Heavy Weight		Double Extra-Heavy Weight	
Nominal size (in.)	Threads per in.	External diameter (in.)	External area per lineal foot (sq ft)	Thickness (in.)	Internal area/linear ft (sq ft)	Thickness (in.)	Internal area/linear ft (sq ft)	Thickness (in.)	Internal area/linear ft (sq ft)
$1/8$	27	0.405	0.106	0.068	0.070	0.095	0.056	—	—
$1/4$	18	0.540	0.141	0.088	0.095	0.119	0.079	—	—
$3/8$	18	0.675	0.177	0.091	0.129	0.126	0.111	—	—
$1/2$	14	0.840	0.220	0.109	0.163	0.147	0.143	0.294	0.066
$3/4$	14	1.050	0.275	0.113	0.216	0.154	0.194	0.308	0.114
1	$11\frac{1}{2}$	1.315	0.344	0.133	0.274	0.179	0.251	0.358	0.157
$1\frac{1}{4}$	$11\frac{1}{2}$	1.660	0.435	0.140	0.361	0.191	0.334	0.382	0.235
$1\frac{1}{2}$	$11\frac{1}{2}$	1.900	0.498	0.145	0.421	0.200	0.393	0.400	0.288
2	$11\frac{1}{2}$	2.375	0.622	0.154	0.541	0.218	0.507	0.436	0.394
$2\frac{1}{2}$	8	2.875	0.752	0.203	0.646	0.276	0.607	0.552	0.464
3	8	3.500	0.917	0.216	0.803	0.300	0.759	0.600	0.602
$3\frac{1}{2}$	8	4.000	1.048	0.226	0.929	0.318	0.886	—	—
4	8	4.500	1.179	0.237	1.055	0.337	1.002	0.674	0.826
5	8	5.563	1.458	0.258	1.321	0.375	1.259	0.750	1.064
6	8	6.625	1.736	0.280	1.590	0.432	1.508	0.864	1.282
8	8	8.625	2.254	0.322	2.092	0.500	2.000	0.875	1.802

TABLE 24. TWIST DRILL DIAMETERS

Drill no.	Diameter (in.)	Drill no.	Diameter (in.)	Drill letter	Diameter (in.)
1	0.2280	41	0.0960	A	0.2340
2	0.2210	42	0.0935	B	0.2380
3	0.2130	43	0.0890	C	0.2420
4	0.2090	44	0.0860	D	0.2460
5	0.2055	45	0.0820	E	0.2500
6	0.2040	46	0.0810	F	0.2570
7	0.2010	47	0.0785	G	0.2610
8	0.1990	48	0.0760	H	0.2660
9	0.1960	49	0.0730	I	0.2720
10	0.1935	50	0.0700	J	0.2770
11	0.1910	51	0.0670	K	0.2810
12	0.1890	52	0.0635	L	0.2900
13	0.1850	53	0.0595	M	0.2950
14	0.1820	54	0.0550	N	0.3020
15	0.1800	55	0.0520	O	0.3160
16	0.1770	56	0.0465	P	0.3230
17	0.1730	57	0.0430	Q	0.3320
18	0.1695	58	0.0420	R	0.3390
19	0.1660	59	0.0410	S	0.3480
20	0.1610	60	0.0400	T	0.3580
21	0.1590	61	0.0390	U	0.3680
22	0.1570	62	0.0380	V	0.3770
23	0.1540	63	0.0370	W	0.3860
24	0.1520	64	0.0360	X	0.3970
25	0.1495	65	0.0350	Y	0.4040
26	0.1470	66	0.0330	Z	0.4130
27	0.1440	67	0.0320		
28	0.1405	68	0.0310		
29	0.1360	69	0.0292		
30	0.1285	70	0.0280		
31	0.1200	71	0.0260		
32	0.1160	72	0.0250		
33	0.1130	73	0.0240		
34	0.1110	74	0.0225		
35	0.1100	75	0.0210		
36	0.1065	76	0.0200		
37	0.1040	77	0.0180		
38	0.1015	78	0.0160		
39	0.0995	79	0.0145		
40	0.0980	80	0.0135		

TABLE 25. Weight and Volume Percent Relationship of Commercial Acids and Ammonia*

	Acetic Acid		Phosphoric Acid				Sulfuric Acid	
Designation	Glacial		75%	80%	85%		66° Bé	
Sp Gr	1.056		1.583	1.639	1.695		1.835	
g/L	1056		1188	1311	1440		1762	
Wt %	Sp Gr	Vol %	Sp Gr	Vol %	Vol %	Vol %	Sp Gr	Vol %
10	1.015	9.6	1.055	8.9	8.1	7.3	1.069	6.3
20	1.029	19.5	1.116	18.8	17.0	15.5	1.143	13.4
30	1.042	29.6	1.184	29.9	27.1	24.6	1.222	21.5
40	1.053	39.9	1.257	42.3	38.4	34.9	1.307	30.6
50	1.062	50.3	1.339	56.3	51.0	46.3	1.399	40.8
60	1.069	60.7	1.430	72.2	65.3	59.5	1.502	52.7
70	1.074	71.2	1.531	90.2	82.1	74.3	1.615	66.2
80	1.072	81.5	1.639		100.0	91.0	1.733	81.2
90	1.072	91.3					1.821	95.8
100	1.056	100.0						

	Aqua Ammonia		Hydrochloric Acid				Nitric Acid			
Designation	26°Bé		18° Bé	20° Bé	22°Bé		36° Bé	38° Bé	40° Bé	42° Bé
Sp Gr	0.897		1.142	1.160	1.179		1.330	1.355	1.381	1.408
g/L	264		319	360	413		696	766	848	945
Wt %	Sp Gr	Vol %	Sp Gr	Vol %	Vol %	Vol %	Sp Gr	Vol %	Vol %	Vol %	Vol %
10	0.060	36.4	1.050	32.9	28.7	25.3	1.056	15.2	13.7	12.5	11.1
20	0.926	70.3	1.100	69.1	60.3	53.8	1.120	32.2	29.1	26.4	23.6
30			1.153		93.7	83.3	1.185	51.1	46.2	41.9	37.5
40							1.252	71.9	65.2	59.1	52.8
50							1.317	94.6	85.8	77.7	69.5
60										97.3	86.9

TABLE 26. Flow in Pipes

For aid in estimating pipe sizes and flows for non-critical applications, the following approximate values are given.

Nominal Pipe Size Schedule 40	Water Capacity* US gal/min at 1 ft/sec	Air Capacity† cfm at 1 atm. 70 F	15 psig	Saturated Steam lb/hr at 5000 ft/min 30 psig	50 psig	100 psig
⅛	0.2	1.4	5	8	11	20
¼	0.3	3.2	10	15	23	40
⅜	0.6	7	20	30	45	75
½	0.9	13	35	50	75	120
¾	1.6	28	65	95	140	240
1	2.7	54	110	160	230	400
1¼	4.5	110	190	300	400	700
1½	6.3	170	260	400	550	950
2	10	330	450	650	900	1600
2½	14	525	640	900	1350	2300
3	22	950	1000	1450	2100	3600
4	38	2000	1750	2500	3700	6300
5	60	3500	2750	4000	6000	10000
6	85	6000	4000	5800	8500	14000

* The water velocities shown are often exceeded in short runs of pipe, especially in diameters below 3 in.

† High pressure air at 80 psig conveyed with 1 psig drop/100 ft run.

This table is adapted freely from data supplied by courtesy of H. L. Pinkerton.

TABLE 27. THE ELECTROMOTIVE SERIES FOR METALS AND ALLOYS*

(ANODIC END)

Lithium (-2.959 v)
Rubidium
Potassium (-2.92 v)
Strontium
Barium
Calcium (-27 v)
Sodium (-2.71 v)
Magnesium and its alloys (-1.55 v)
Aluminum (-1.33 v)
Beryllium
Uranium
Manganese
Tellurium
Zinc (-0.76 v)
Chromium (-0.56 v)
Sulfur
Gallium
Iron (Fe^{++}) (-0.44 v)

A (Note 1)
- Galvanized steel ([1])
- Galvanized wrought iron
- Al Zn-Mg alloys (eg: 7075) ([2]) ([3])
- Al-High Mg. alloys (eg: Al-220)
- Al-Low Mg. alloys (eg: 3004)
- Aluminum plus very low percentage of alloying constituents (eg: 6053)
- Alclads

Cadmium (-0.40 v)
Al-Si-Mg alloys (e.g.: Al-356)
Al-Cu alloys (with or without small additions of Mg) (e.g.: 2024)
Al-Cu alloys (with or without small additions of Zn) (e.g.: Al-113)
Al-Cu-Si alloys (e.g.: Al-108)
Mild steel
Copper steel
S.A.E. 4140
S.A.E. 3140
Wrought iron
Cast iron
4-6% chromium steel, Type 501 or 502 ([4]) (active)
12-14% chromium steel, Types 403, 410, 416 (active)
16-18% chromium steel, Type 440 (active)
23-30% chromium steel, Type 446 (active)
Indium ([5])
Thallium
Cobalt (-0.28 v)
Ni-Resist cast iron
50-50 lead tin solder
17% Cr-7% Ni steel, Type 301 (active)
18% Cr-8% Ni steel, Types 302, 303, 304, 321, 347 (active)
23% Cr-14% Ni steel, Type 309 (active)
25% Cr-20% Ni steel, Type 310 (active)
18% Cr-12% Ni-3% Mo steel, Type 316 (active)
Hastelloy "C" (59% Ni, 17% Mo, 5% Fe, 14.6 Cr, 5% W, 0.1% C)
Lead (-0.12 v)
Tin
Iron (Fe^{+++})

Hydrogen (0.00 v)
Antimony
Bismuth
Arsenic

B (Note 6)
- Muntz metal (60% Cu, 40% Zn([6]))
- Manganese bronze (66.5% Cu, 19% Zn, 6% Al, 4% Mn)
- Naval brass (Add. of ¾% Sn to Muntz metal)
- Nickel (active)
- 60% Ni-15% Cr (active)
- Inconel (78% Ni, 13.5% Cr, 6% Fe) (active)
- 80% Ni-20% Cr (active)
- Hastelloy "A" (60% Ni, 20% Mo, 20% Fe, 0.1% C)
- Hastelloy "B" (65% Ni, 30% Mo, 5% Fe, 0.1% C)
- Yellow Brass (58-70% Cu, 0.50-1.5% Sn, 0.75-3.5% Pb, balance Zn)
- Admiralty brass (71% Cu, 28% Zn, 1% Sn)
- Aluminum bronze (Add. of 2¼% Al to 75% Cu-25% Zn alloy)
- Red brass (85% Cu, 15% Zn)

Copper (-ic) ($+0.522$ v)
Oxygen
Polonium ($+0.34$ v)
Copper (-ous) ($+0.52$ v)
Iodine
Tellurium (Te^{++++})

C (Note 7)
- Silicon bronze (1.0 to 3.0% Si) ([7])
- Nickel-silver
- Ambrac (5.0% Zn, 20% Ni, balance Cu)
- 70% Cu-30% Ni
- Comp. G-bronze (88% Cu, 2% Zn, 10% Sn)
- Comp. M-bronze (88% Cu, 3% Zn, 6.5% Sn, 1.5% Pb)
- Silver solder
- Nickel (passive)
- 60% Ni-15% Cr -25% Fe (passive)
- Inconel (passive)
- 80% Ni-20% Cr (passive)
- Titanium
- Monel (70% Ni, 30% Cu)
- 12-14% Cr steel, Types 403, 410, 416 (passive)
- 16-18% Cr steel, Type 440 (passive)
- 17% Cr-7% Ni steel, Type 301 (Passive)
- 18% Cr-8% Ni steel, Types 302, 303, 304, 321, 347 (passive)
- 23% Cr-14% Ni steel, Type 309 (passive)
- 23-30% chromium steel, Type 446 (passive)
- 25% Cr-20% Ni steel, Type 310 (passive)
- 18% Cr-12% Ni-3% Mo steel, Type 316 (passive)

Mercury ($+0.85$ v)
Silver ($+0.80$)
Lead (Pb^{++++})
Palladium
Bromine
Chlorine ($+1.36$ v)
Graphite ([8])
Gold (-ic) ($+1.3$ v)
Gold (-ous)
Platinum
Fluorine

(CATHODIC END)

[1] Members of Group (A) are listed, relative to each other, but the position of the group with respect to the group between chromium and ferrous iron is uncertain. The group, however, is listed correctly between zinc and cadmium.

[2] Aluminum types are the standard designations of the Aluminum Association for Wrought Alloys, other types are Alcoa designations.

[3] The heat treatment of aluminum and its alloys has an effect upon their galvanic potentials. They fall in the aluminum group, however, regardless of the heat treatment.

[4] Stainless Steel types are standard American Iron and Steel Institute designations.

[5] Indium, thallium, and cobalt are listed relative to each other, but their position relative to the stainless steels is uncertain. They are properly placed between aluminum and lead.

[6] The members of Group (B) are listed relative to each other, but the position of the group relative to ferric iron to arsenic is uncertain. The group is, however, properly placed between tin and copper.

[7] The members of Group (C) are listed relative to each other, but the position of this group with respect to iodine, tellurium, and mercury is uncertain. However, the group is properly placed between copper and silver.

[8] Graphite lies between lead and gold but its position relative to palladium, bromine, and chlorine is uncertain.

* Reprinted from *Metal Finishing* by permission of the publisher, Metal Finishing, One University Plaza, Hackensack, N.J. 07601.

In the Electromotive Series, metals are arranged in order of their standard electrode potentials. The metals at the anodic end tend to form ions while ions of the metals at the cathodic end are easily reduced to the metallic state. The series is useful in a general way in predicting the tendency of one metal to displace another or to protect another against corrosion. The arrangement of the series is reasonable if the solutions contain simple ions at moderate concentrations. If complex ions are formed, corrections for actual activity must be made with consideration for the particular solution.

2. DESIGN FOR PLATING

J. HYNER

An article should not be designed for plating. It should be designed to do a job. If the part is successfully designed, provision will have been made for corrosion resistance, hardness, lubricity, decorative appeal or whatever required property the finishing process is to supply. Therefore, design for plating cannot be thought of as a separate and specific operation, step or process. The designer of the part must know what the part should do, how strong it is to be, how hard, how heavy, how bright, how precise and, also, what qualities finishing should add. The designer of the part should know the properties of the various plated coatings, how they function, what to expect of them and where they are best employed, or he should work with someone who can provide this information.

Fifty years ago the designer did not have a wide choice in the coatings he could specify. If his part required corrosion resistance, it was usually manufactured from brass with a nickel topcoat or steel plated with zinc or cadmium to provide sacrificial protection. As years passed and the manufacturer became more dependent on imported copper, the increased cost and uncertainty of supply predicated that less brass was specified as basis metal, and steel, at much lower cost, substituted. Chromium was just beginning to be used as a hard surfacing tool, and gold and silver were plated for decorative purposes only. Since World War II, plating technology has exploded and become a valuable tool used to enhance and improve the properties of the basis metal, giving the entire article many superior properties through coating with a minute top layer.

A few of the ways of improving the properties of the basis metal through plating are listed below:

1. *Protection from Corrosion*
   Copper, nickel and chromium on steel and zinc die castings
   Zinc or cadmium on steel
2. *Appearance*
   Copper, nickel and chromium on steel
   Nickel and gold on brass
   Silver on brass
3. *Superior Hardness and Better Wear Resistance*
   Chromium on steel
   Electroless nickel on steel
   (This hardness improvement is gained without sacrifice of ductility; the plating allows a hard surface while maintaining a softer, ductile core.)
4. *Lower Contact Resistance and Increased Reliability for Electrical Contacts*
   Gold on brass or copper
5. *Improved Solderability and/or Weldability*
   Tin on brass
   Electroless nickel on steel
6. *Better Base for Other Finishes*
   Nickel under gold or chromium
   (The nickel inhibits the migration of the gold into the brass basis.)
7. *Improved Lubricity Under Pressure*
   Silver on bronze
8. *To Strengthen the Base and Render it More Temperature Resistant*
   Copper, nickel and chromium on plastics
9. *To Act as a Stop-off in Heat Treating*
   Copper on steel for carburizing
   Bronze on steel for nitriding

These and numerous other properties can be incorporated onto steel, brass and other substrates. Combinations of different plates will yield properties different and often superior to heavier thicknesses of a single plate. For example, both sacrificial and barrier protection can be given to the same plated part. Modern instrumentation, such as the atomic absorption spectrophotometer, has enabled us to control alloy deposition in a manner not dreamed of 25 years ago, allowing us to plate many new and useful alloys. These new systems are available for incorporation into an improved product. It is therefore obvious that the designer cannot consider plating as a separate treatment to be decided upon *after* the other properties of the part are determined and the part manufactured. When the chemical and mechanical properties are defined, the need or desirability of a plated finish should also be established. Cost/benefit analyses should be carried out, and the proper finish selected. Once this has been done, bearing in mind the characteristics of the finishing system to be used, the geometry of the part to be plated should be considered. Sharp edges, ridges, recesses, blind holes, etc. introduce added difficulty and, therefore, increased cost into the finishing operation. They should therefore be minimized or eliminated. Good design for plating requires knowledge of plating specifications, the characteristics of the plated coatings, some knowledge of basis materials, heat treating, corrosion testing and all of the other bits of knowledge which, when put together, produce the satisfactory product.

The designer need not be experienced in any or all of the above skills. He should, however, recognize the need to consult with knowledgeable people, both in and out of his company, prior to finalizing a designed part. The design stage is the time to avoid and eliminate problems and minimize costs. It is the time to go over previous problems and to come up with improvements. Sometimes the performance of the part needs improvement. Have corrosion or excessive wear been problems in the past? The design stage is the time to rectify past mistakes. It is surprising how often serious design blunders involving plating occur in factories that have captive plating operations run by competent engineers. All too often the skill and knowledge of the plater is not utilized. A half century ago plating technology may have been very archaic and simple. This does not hold true today. Plating is an essential tool to almost all industries from the manufacture of pins to the construction of the space shuttle. We now are capable of plating numerous alloys; we deposit composite plates with diamonds, fluorocarbons and other particulates occluded to improve wear resistance and lubricity. We now plate copper circuits onto plastic sheets, and we spot plate gold on a continuously moving strip. No one can expect the designer to be expert in everything. However, it is expected that a good designer will discuss his problem with a consultant, be he an in-plant plating engineer, a good outside supplier or a professional finishing consultant. The consultant can serve an important function even if only to verify what is already known.

Much of the literature covering design for plating seems to deal primarily with the shape of the part. Does the part adapt itself to satisfactory rack or barrel plating? These are important considerations, and shape, size and form characteristics must be reviewed.

The following general principles will serve as a partial guide for this review:

1. Sharp edges and right angles should be avoided. Every sharp, protruding edge will draw extra current and build-up with extra plate. Conversely, the part will receive very little plate in an acute angle. All sharp edges and angles should be rounded to the greatest degree design allows.
2. Holes should be either countersunk or counterbored, because build-up on the sharp edges may exceed tolerance allowance.
3. Deep recesses should be avoided. The recess will receive a lesser thickness of plate than the adjoining area, and either require heavier average thickness on the overall part to meet the minimum specification or receive too little plate if the average area is used to compute thickness.
4. If the article is a threaded fastener or threaded screw machine part, special care should be taken to account for build-up of at least four times the plating thickness on the pitch diameter.
5. Formed tubular articles will often trap and carry over solution if drainage holes are not provided in the design.
6. Larger parts to be rack plated must be provided with some way to rack the part either in hole, lug or rim. Since the contact point will be poorly plated, the electrical contact should be in a non-significant area.
7. Blind holes, rolled edges, seams and other crevices will trap solution unless special plating techniques are used.

8. Bolted assemblies should be avoided because of possible unplated areas subject to subsequent corrosion.
9. Dissimilar metals are difficult to plate, because cleaning methods vary for different basis materials.

(The next six illustrations and the text accompanying them are excerpts from the previous edition of *Electroplating Engineering Handbook*, Chapter 2, "Design for Plating," by Bigge and Graham.)

In the design of a part to be plated, it is well to utilize all the advantages that may be incorporated to permit as uniform a distribution of plating thickness as possible and still retain the basic design desired. Each sharp corner (recessed or protruding) should be provided with as large a radius as possible. Figure 1 illustrates the distribution of nickel plating thickness on a formed part. Note the lack of adequate radii at the ends of the central recessed section. The ratio of 9.0 illustrates the effect of these sharp corners.

The design shown in Fig. 2 is a modification of that shown in Fig. 1. The elimination of the sharp corners at the ends of the central recessed section has improved the distribution of plating thickness, indicated by the more favorable ratio of 5.5.

If greater latitude of design is possible, it is well to avoid deep recesses and substitute convex surfaces for those which are concave. Figure 3 illustrates distribution of plating thickness on a part having a recessed surface. The ratio of 6.5 on this part may be compared to the ratio of 2.0 obtained on a part whose cross-section is shown in Fig. 4.

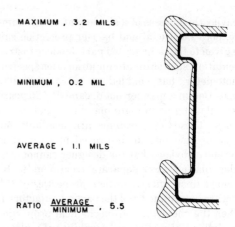

FIG. 2. Distribution of (nickel) plate; thickness is exaggerated.

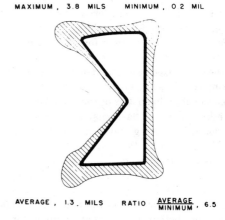

FIG. 3. Distribution of (nickel) plate; thickness is exaggerated.

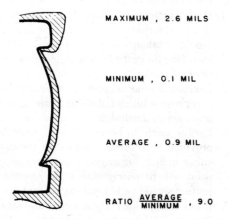

FIG. 1. Distribution of (nickel) plate; thickness is exaggerated.

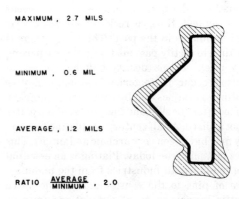

FIG. 4. Distribution of (nickel) plate; thickness is exaggerated.

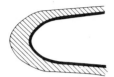

MAXIMUM, 2.7 MILS   MINIMUM, 0.7 MIL

AVERAGE, 1.4 MILS   RATIO AVERAGE/MINIMUM, 2.0

FIG. 5. Distribution of (nickel) plate; thickness is exaggerated.

Figure 5 also illustrates the advantage of a convex surface for plating, but the relation of this convex surface to the shape of the part being processed has a pronounced effect on the distribution of plating thickness. The section shown in Fig. 5 is favorable from the plating standpoint but if this shaped section is part of a larger surface to be plated, then the distribution of plating thickness can be greatly changed, as shown in Fig. 6. Compare the ratio of 2.0 in Fig. 5 with the ratio of 8.0 in Fig. 6.

If reasonably uniform distribution of plating thickness is desired, every effort should be made to avoid recesses, to fillet all sharp corners, and to use convex instead of concave surfaces wherever possible. This will improve distribution of plating thickness and, by so doing, will provide a finish having better corrosion resistance at a reduced cost of plating.

The current distribution principles outlined above also apply to barrel plating, usually

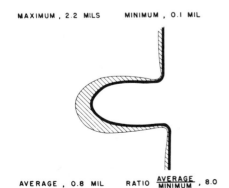

MAXIMUM, 2.2 MILS   MINIMUM, 0.1 MIL

AVERAGE, 0.8 MIL   RATIO AVERAGE/MINIMUM, 8.0

FIG. 6. Distribution of (nickel) plate; thickness is exaggerated.

chosen because of potential cost savings made possible by bulk handling. The smaller the part barrel plated instead of rack plated, the greater the savings. Figure 7 shows some ideal shapes for barrel plating, size varying from ¼ in. in diameter to approximately 1½ in. long and not too heavy. Figure 8 illustrates stampings and castings of 2½ to 4 in. long. These can be plated very successfully in the larger production barrels but at questionable or small savings over rack plating. In designing for barrel plating, care should be taken to keep parts from interlocking. Figure 9 is a curtain pin that normally plates in the closed position. If a few inadvertently open, a ball will form, with rather poor

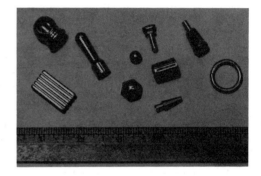

FIG. 7. Ideal parts.

FIG. 8. Marginal parts.

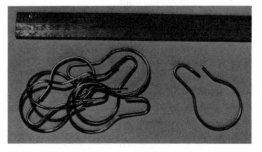

FIG. 9. Interlocked.

plating and bent parts resulting. Male and female threads should never be plated together—they will find each other and mate during plating. Truncated, cone-shaped parts will also get together and form long, poorly plated "sticks." Delicate threads are very often damaged during the longtime rotation of barrel plating.

However, it is not enough to just consider the geometry; there are other factors that must be taken into account. The thought processes used in the course of the design and manufacture of several different parts can be used to illustrate proper technique.

If the part being designed requires good corrosion resistance, the designer has the choice of a corrosion-resisting alloy steel, a non-ferrous material or steel plated to provide the needed corrosion resistance. The third choice usually represents the lowest cost. There are numerous published corrosion specifications available to all interested designers. These are published by the American Society of Testing Materials, the military, the aircraft industry, the automotive industry, the electronic industry and numerous trade organizations. Plating standards and specifications are discussed more fully by Ogburn in Chapter 7.

Table 1, reproduced from MFSA's *Quality Metal Finishing Guide*, is a typical guide.

The specification that most closely meets the designer's service requirements should be chosen. Good plating will then assure satisfactory life and utility.

Unfortunately, overspecification is all too common in the plating industry. One common engi-

TABLE 1. SUGGESTED STANDARDS FOR QUALITY ZINC COATINGS ON IRON AND STEEL PRODUCTS

Degree of Exposure	Min. Thick (micrometre) (1)	Chromate Finish	Salt Spray Hrs. to White Corr.	Typical Applications
MILD-Exposure to indoor atmospheres with rare condensation & subjected to minimum wear or abrasion	5	none clear iridescent yellow olive drab	— 12-24 24-72 72-100	screws, nuts and bolts, buttons, wire goods, fasteners
MODERATE-Exposure mostly to dry indoor atmospheres but subjected to occasional condensation, wear or abrasion	8	none clear iridescent yellow olive drab	— 12-24 24-72 72-100	tools, zipper pulls, shelves, machine parts
SEVERE-Exposure to condensation perspiration, infrequent wetting by rain, cleaners	13	none clear iridescent yellow olive drab	— 12-24 24-72 72-100	tubular furniture, insect screens, window fittings, builders' hardware, military hardware, washing machine parts, bicycle parts
VERY SEVERE (2) Exposure to bold atmospheric conditions, and subject to frequent exposure to moisture, cleaners and saline solutions plus likely damage by denting, scratching or abrasive wear	25	none		plumbing fixtures, pole line hardware

25.4 Micrometres = 1.0 Mil.

(1) Thickness of the coating after chromate treatment.
(2) Although there exist some applications for heavy electrodeposited coatings for very severe service they are most usually satisfied by hot dipped or sprayed coatings.

neering design failing is to assume that if 12.5 μm (0.0005 in.) of plate is good, 25 μm (0.001 in.) should be better. This is not necessarily a valid assumption. The extra plate thickness could be superfluous for the service conditions and may cause tolerance problems. Overspecification is the lazy man's way out of an unsure situation. It is safer and more cost-effective to spend the added time in research and consultation so the correct specification for the service requirements is determined.

To illustrate the kind of thinking that will help establish the proper plating and finishing specification, let us go through the process of designing a very common article—the body of a two-slice, pop-up toaster. The design calls for a bright, highly polished chromium part.

The toaster is normally used in a mild, dry, indoor atmosphere. Since corrosion is not a great problem, carbon steel should be chosen for the basis metal. ASTM B456 and MFSA *Quality Metal Finishing Guide*, Volume IV, No. 1, both indicate that 8-μm (0.0004-in.) bright nickel and 0.25-μm (0.00001-in.) chromium should be satisfactory. The designer now has the choice of either rack nickel and chromium plating the individual part, stamped out of cold rolled steel, or of forming the toaster body from strip steel, polished and preplated to the chosen specification. The sheared edges exposed after fabrication from preplated steel would be bare and a site for corrosion. However, in the mild, dry application of the toaster body, the corrosion of folded under edges would not be too severe or detrimental. Of more concern would be marks left by the forming dies and the ductility of the nickel on the strip. Will it be able to take the bend of the forming without cracking? High-quality, preplated strip can be purchased with a protective, removable coating that will protect the chromium surface during forming and that is ductile enough to be formed without cracking.

A completely coated part would theoretically be better than one with exposed edges. There would also be no concern about cracked plating caused by forming. Cost is a serious and most important consideration and must be weighed when designing and fabricating. The designer should strive for best value, which is achieved by choosing the least expensive specification that will do the work. In practice both methods of plating toaster bodies have been used. In the most expensive models that require a mirror-bright, highly buffed surface, the toaster bodies are formed, polished and sometimes buffed to achieve the most perfect, flawless surface. Less expensive models have been produced from preplated stock with very satisfactory results.

An interesting example of how the part designer and plater have worked together to solve a very serious cost problem is illustrated by the story of the evolution of the gold plated electronic contact. Gold is often specified as the final coating on electronic contacts because of its conductivity, solderability and resistance to attack by the elements, yielding unmatched reliability.

When gold was selling at $36.00 per ounce, contacts were made on a screw machine or multiple stamping machine and then barrel gold plated all over—often with an undercoating of nickel to prevent migration of the gold. When the cost of gold began to esculate, the method of manufacture was changed. Instead of forming completely and then plating, the contact was partly formed on the strip, and the strip was partially plated to cover just the contact tips. Eventually this technique evolved to the point where gold is spot plated, through a mask, onto a small but significant area of the contact. The evolution of the spot gold plated contact is a classic example of what can be accomplished when skilled designers, engineers and platers combine their talents to achieve a common goal. Literally, millions of dollars will be saved without any sacrifice in reliability.

There are also times when designers may think they have done everything according to best practice but miss an important point and, thereby, still get into trouble. The part shown (approximately full size) in Fig. 10 is a hub for a four-wheel drive mechanism used by most of the

FIG. 10

major American automotive companies. The hub is held onto the wheel by five machine screws, which were subject to great stress. Since the fasteners would pass through a polished chromium-plated hub, zinc or cadmium, which could meet the corrosion requirements, were ruled out. Before choosing the plating finish, the basis metal of the fastener had to be picked. Because of the great strength required on a relatively small fastener, either a case hardened carbon steel alloy or a hardened 400 series stainless steel could be used. A 300 series stainless steel could not be considered, because it was not strong enough for the application. All of the automobile companies have very complete plating specifications. A GM exterior specification, GM6112M (20), was chosen to be applied to a hardened alloy steel screw coated with a sequence of five layers of plate composed of 5 μm (0.0002 in.) of a micro-throwing nickel to seal the laps in the cold-headed part, 7.5 μm (0.0003 in.) cadmium, 5 μm (0.0002 in.) copper, 7.5 μm (0.0003 in.) nickel and 0.25 μm (0.00001 in.) chromium. The plating sequence was required to pass a 22-hour CASS test and a 96-hour of 5% neutral salt spray in electrolytic test.

The other basis material possible was an inert atmosphere, hardened 410 stainless steel screw plated with 8 μm (0.0003 in.) nickel plus 0.25 μm (0.00001 in.) chromium to pass 96 hours 5% neutral salt spray. The hardened 410 stainless steel without plating would not pass the salt spray test.

Since both the different fasteners with their respective plating systems passed the automotive strength and corrosion requirements, the fastener that would do the required job at the lowest cost was chosen—the five-layer plated steel fastener at 30% lower cost than the nickel and chromium plated 410 stainless steel.

Because the fastener had to resist vibration and was to be removed several times in adjustment, the thread required a patch which was fabricated after plating by applying a nylon-based compound and then induction heating the fastener to cure the patch.

The decision making process seemed to be good and made sense. All factors apparently were taken into account: corrosion resistance, color and cost.

Unfortunately, after over one million parts were ordered from three sources, a serious problem arose. Everyone had overlooked the consequences of applying the nylon patch, which was sprayed onto the part and then cured by induction heating. The temperature of the heating was over 600°F and melted the cadmium in the second layer, causing balls and blisters on the threads, making torquing pressure too high—sometimes even preventing proper mating.

The specification had to be changed to 410 stainless steel basis metal followed by 8 μm (0.0003 in.) nickel plus 0.25 μm (0.00001 in.) chromium. This gave the corrosion resistance, appearance, ability to take the patching temperature without blistering and, unfortunately, costing 30% more.

All too often the electroplating specification is chosen after the article is designed and blueprinted. An engineer then consults a book of specifications and chooses one that will give the necessary corrosion resistance. Hidden in the blueprint is a hole drilled to tight dimensions to take a mating component. The hole requires a 5-μm (0.0002-in.) diameter tolerance. The specification calls for 8-μm (0.0003-in.) zinc plate.

This simple example illustrates a typical problem. Easy to rectify in the design state—impossible after manufacture. The solution to the problem is usually to plate with 2.5-μm (0.0001-in.) zinc, turning out an inferior product that will fail in the field. Poor plating will inevitably be blamed! Those of us who have plated for many years have been faced with countless variations of the above example.

Design engineers have taken on the qualities of Renaissance men—they must know something about everything.

References

1. Safranek, W. H. and A. A. Underwood. *The Influence of Design on Electroplating of Zinc Die Castings.* New York: American Zinc Institute, 1966.
2. Gonas, A. J. and J. H. Lindsay. The Influence of Plating Requirements on Design of Decorative Zinc Die Castings. Society of Die Casting Engineers, Technical Paper No. C-42, 1964.
3. Barton, H. K. Plating Requirements and the Design of Zinc Alloy Die Castings. *Machinery* (December 25, 1963).
4. Safranek, W. H. and A. A. Underwood. The Influence of Design on Electroplating Zinc Die Castings. *Precision Metal Molding*, Industrial Publishing Company, pp. 75–80 (September 1962).
5. Layton, D. N. *Design for Electroplating.* London: The International Nickel Company (Mond) Limited, 1962.
6. *Design for Metal Finishing.* London: Institute of Metal Finishing, 1960.
7. Christie, J. J. and D. J. Thomas. Prediction of

Nickel Plate Distribution by Plating Simulation. *Plating* **52**:855–859 (1965).
8. Safranek, W. H. The Relationship of Shape to Effectiveness of Plating. *ASME Design Engineering Conference and Show*, New York, May 15–18, 1967.
9. *MFSA Quality Metal Finishing Guide.*
10. ASTM Specification B456-71.
11. ASTM Specification B450.
12. Recommendations for the Design of Metal Articles that are to be Coated, BS4779 (1969).
13. Silman, H., G. Isserlis and A. F. Averill. Protective and Decorative Coatings for Metals. *Finishing Publications* (1978).
14. Goldberg, S. and F. Ogburn. Plating Standards and Specifications. *Electroplating Engineering*, Third Edition. New York: Van Nostrand Reinhold, 1971.

3. METAL SURFACE PREPARATION AND CLEANING

A. Basis Metal Surface

Keith Sheppard

Assistant Professor of Materials and Metallurgical Engineering, Stevens Institute of Technology, Hoboken, N.J.

Most electrodeposition takes place on metallic substrates. It is important that the electroplater have an understanding of the influence that may be exerted by the basis metal surface on the integrity, appearance and properties of electrodeposits. Many of the basis metal effects can be influenced by the electroplater's actions, particularly during surface preparation and cleaning. Many of the metallurgical concepts described in this section are discussed in more detail in Chapter 12.

NATURE OF THE SURFACE

Atomically the surface differs from the bulk of metal in that surface atoms have fewer nearest neighbors and can therefore be considered as being more active (energetic) than bulk atoms. The need for near neighbors can to some extent be satisfied by either chemical reaction—e.g., to form oxides—or by adsorption of foreign substances, such as gas molecules or greases and soils, which generally are organics. These surface films may be held tenaciously, often making their removal difficult.

Metal surfaces under normal circumstances are not atomically smooth. Crystal defects such as dislocations, twins and grain boundaries, emerging at the surface can give rise to steps and ledges that can be many atoms high. Surface treatments prior to electroplating can lead to further enhancement of this surface roughness. Atoms in ledges and steps are even more energetic than those in a smooth surface and are sites of strong adsorption. Different crystal phases within the surface of alloys as well as impurities and nonmetallic inclusions such as entrapped slag, create additional surface inhomogeneities.

DEPOSIT GROWTH MODE

Consider a basis metal surface free of oxides or adsorbed films and one that is not mechanically damaged from the surface preparation procedure. The first atoms to deposit on this surface would prefer to take up positions that continue the substrate crystal structure. It is easier for this to occur if the two metals comprising the deposit and substrate have the same crystal structure and similar interatomic spacings—e.g., gold on silver, nickel on copper. Growth is then said to be epitaxial and the deposit reproduces the crystallographic orientation of the substrate. Epitaxy can also occur for a deposit/substrate combination of different crystal structures, but one in which the atomic arrangement in a certain crystallographic direction in the deposit structure matches one in the substrate. Unless the interatomic spacings are identical in the substrate and deposit, the initial atomic layers of an epitaxial deposit will be in a state of stress, having been stretched or compressed from the interatomic spacing they would otherwise assume. Dislocations can form at the interface to relieve these misfit stresses and may subsequently influence the mechanical properties of the deposit and its corrosion resistance.

Few practical electroplating systems exhibit epitaxial growth to great thickness. Those that do, such a acid-copper on copper, do so only in the absence of addition agents and impurities. Both low temperatures and high current densities limit the ability of depositing atoms to diffuse to epitaxial growth sites on the surface, such conditions therefore favoring a non-epitaxial growth mode. There are many systems where the initial deposit layers are epitaxial, but subsequent growth is influenced by the presence of growth inhibiting species which are adsorbed at the surface during plating, blocking epitaxial growth sites. These species may result from deliberate additives, (e.g., leveling agents), impurities or chemicals formed by the electrodeposition process itself (e.g., nickel hydroxide in Watts nickel). The adsorption of inhibiting species leads to growth twin formation and the ensuant development of non-epitaxial growth, often having a fiber texture which is a function of bath composition and plating conditions. Plating from Watts nickel onto copper is an example, as is pure acid gold onto copper.

In the presence of large quantities of growth inhibitors it becomes impossible for even the initially depositing atoms to follow the basis metal crystal structure. Under these conditions, epitaxial growth does not occur, growth being determined solely by plating conditions and bath composition, often resulting in a fine-grained, randomly oriented deposit structure.

If the basis metal surface has been subjected to abrasion during a preparation procedure such as grinding or mechanical polishing, epitaxial growth is generally not observed in plating systems that otherwise produce such growth. This is due to severe distortion of the crystal structure in the basis metal surface layer. Such a distorted layer can be removed by a chemical or electrochemical polishing procedure prior to plating. Similarly, epitaxial growth is not seen if the crystal structure of the deposit and an undeformed basis metal have a large mismatch in interatomic spacings (generally greater than 12%).

ADHESION

Adhesion is one of the most important considerations in electroplating, as without adequate adhesion the coating will generally not perform the task for which it was intended, such as impart corrosion protection or wear resistance to a component.

One might expect that the best conditions to promote adhesion are those that favor epitaxial deposition, particularly where the deposit and basis metal have the same crystal structure and small atomic misfit. At one time epitaxy was thought to be a prerequisite for good adhesion. In reality, strong adhesion can also be obtained with a non-epitaxial deposit. One of the major factors influencing adhesion is the presence of suface contamination due to oxides or soils. Adhesion is thus very much determined by the surface preparation techniques used by the elecroplater. It is desirable to remove all surface films during the preplating treatment to obtain maximum adhesion. There are, however, cases—e.g., Watts nickel—on a copper basis, where good adhesion can still be achieved even though a thin (less than 1000 Å) oxide film remains. It is believed that the plating bath dissolves this thin residual copper oxide layer on immersion. Thicker oxides are not fully dissolved and lead to reduced adhesion.

Basis metals such as nickel that develop a passive oxide layer under open circuit in the plating bath must be introduced directly after cleaning with a cathodic potential already applied to prevent passivation.

Adhesion can be affected by mechanical damage to the basis metal. Severe plastic deformation due to grinding or polishing can produce a relatively brittle surface layer which may crack and break away under the influence of high internal stress in the subsequently applied electrodeposit or an externally applied load to the plated component. Embrittlement of the surface by abrasion may also result from a change in structure due to overheating when grinding or polishing has been too aggressive. It is possible, particularly at elevated temperature, for alloying to occur at the interface between the deposit and basis metal. If this alloy phase is brittle, cracking at the interface may occur under stress. Examples are copper plated on zinc and tin deposit on copper or steel.

Certain basis metals may contain non-conducting or poorly conducting phases. Where such phases are exposed on the surface, plating will not occur, although coverage may be produced by bridging of the deposit from neighboring areas that are conducting. The same effect is produced by abrasive particles such as silicon carbide that become embedded in the surface during a polishing or grinding operation and are not subsequently removed by electropolishing or etching. If the basis metal contains a phase that has a low hydrogen overvoltage—e.g., graphite in cast-iron—such areas evolve hydrogen rather than deposit metal. Again, bridging over may produce a continuous deposit, but one containing areas with no adhesion.

Another effect is seen when a soft second phase exists—e.g., lead in leaded brass. Abrasion can produce a poorly adherent smeared layer of the soft phase across the basis metal surface. This smeared layer must be removed by chemical or electrochemical means prior to plating if good adhesion is desired.

Adhesion of electrodeposits to nonmetallic substrates almost always requires some mechanical keying of the deposit to a rough or porous surface. Pores may be created for example in plastics by conditioning in chromic acid. The relative importance of mechanical keying compared to chemical bonding of electrodeposits to plastic surfaces remains unresolved.

POROSITY

Many of the factors which cause poor adhesion also produce porosity in the electrodeposit. Areas of the basis surface that have soils remaining or contain non-conducting phases such as slag particles are not plated and pores are formed as the deposit bridges over them. Likewise, low hydrogen overvoltage phases produce hydrogen bubbles which may be occluded by the growing deposit or create channels through the deposit as the bubbles evolve from the surface. Aggressive mechanical treatment of the surface may produce fine surface cracks that are not plated, again leading to porosity. Some of these causes of porosity may be removed by treatments prior to plating such as adequate degreasing, chemical or electropolishing. However, there is little that can be done about non-conducting and low hydrogen overvoltage phases except to avoid them where possible.

BRIGHTNESS

The basis metal effect on brightness is essentially that due to surface topography. This topography is very much dependent on surface treatments. Surface roughness greater than the wavelength of visible light—i.e., about 0.15 micrometers (6 microinches) causes diffuse scattering of light and a dull appearance. If such a surface is plated with a deposit which is thin, has good microthrowing power and follows the basis metal surface without producing significant leveling, the deposit will also appear dull. At greater thickness, such a deposit will tend to exhibit increased brightness due to geometric leveling. In many plating baths, leveling additives are used which promote preferential deposition in valleys rather than on the hills of the growing surface. Consequently, the surface roughness due to the basis surface is reduced. Often the growth structure (i.e., grain size and crystal morphology) of the deposit itself is the major determinant of brightness at the microscopic level. Deposit structure is often controlled by the plating bath composition and plating conditions rather than the basis metal surface.

STRESS

It was earlier mentioned that stresses could result from the atomic mismatch between an epitaxial deposit and the basis metal surface. If the interatomic spacing of the deposit is smaller than that of the basis metal, the crystal structure of the initially depositing atoms will be stretched and consequently be in a state of tensile stress. Compressive stress arises when the deposit has a larger spacing than the basis metal. These stresses can be significant in thin deposits, often leading to the formation of dislocations which influence the mechanical properties and corrosion resistance of the deposit.

Stresses already present in the basis metal surface due, for example, to grinding, may become superimposed upon those that develop in the deposit during the plating process to an extent that cracking or peeling of the deposit occurs. Stresses can also develop between deposit and basis metal due to a difference in thermal expansion coefficient upon cooling after plating at elevated temperature.

Stresses may be produced in the deposit and/or basis metal by gases, particularly hydrogen. In some plating processes—e.g., nickel—hydrogen is co-deposited. Diffusion of this hydrogen into the basis metal can leave the deposit in a state of tensile stress and the basis metal in compression. The ease with which hydrogen can enter the basis metal depends on its crystal structure and surface condition. Diffusion through an etched surface is many times faster than through a polished one. Any oxide film on the basis metal will further reduce hydrogen diffusion. The extent of dislocations and grain boundaries in the basis metal has an influence, these being considered preferred paths for diffusion.

Apart from co-deposition during plating, atomic hydrogen may be introduced into the basis metal during preplating treatment processes such as acid pickling or cathodic cleaning. This hydrogen can collect as molecules in voids and produce considerable internal pressures, leading to brittle cracking of the basis metal under relatively low stresses. Heat treatment is the only way to remove the entrapped hydrogen, but this cure may itself cause problems if it reduces desired properties such as hardness.

References

1. Baker, E. M., *J. Soc. Automotive Engrs.*, **15**, 461 (1924).
2. Phillips, W. M., *Proc. Am. Electroplaters' Soc.*, **24**, 249 (1936).
3. Pinner, W. L., *Proc. Am. Electroplaters' Soc.*, **28**, 136 (1940); *ibid.* **40**, 83 (1953).
4. Steer, A. T., Proc. 3rd Int. Electrodeposition Conf. (London), 165 (1947).
5. Westerman, A. E. R., and Mornheim, F. A., *Proc. Am. Electroplaters' Soc.*, **40**, 138 (1953).
6. Jones, M. H., and Zajdowski, J., *Proc. Am. Electroplaters' Soc.*, **45**, 45 (1958).
7. Jones, M. H., Chih-Yeu Lu, Mohrnheim, A. F., and Zajdowski, J., *Proc. Am. Electroplaters' Soc.*, **46**, 113 (1959).

B. Polishing, Brushing and Buffing

Henry L. Kellner

Vice-Chairman, The Lea Manufacturing Company, Waterbury, Conn.

AND

Kenneth J. Gatchel

Technical Director, Lea-Michigan, Inc., Grand Rapids, Mich.

POLISHING

In the metal finishing industry, the abrading operation which follows grinding and precedes buffing is called polishing. The purpose of polishing is to remove a considerable amount of metal and to effect a preliminary smoothing of the surface preparatory to more refined finishing procedures. As there is considerable resiliency in the media used for polishing, and much more in those used for buffing, neither of these operations can be considered precision processes, as are grinding and lapping. Buffing, which usually follows polishing, smooths the metal surface to improve its appearance but removes very little metal in comparison to the preceding operations.

Polishing Wheels

Polishing wheels are generally made of muslin, canvas, felt or leather. Variations in construction make available wheels of varying flexibility that best suit the shape and surface condition of the individual object to be finished. The polishing wheels in widest use are made of woven cotton fabrics. The hardest wheel of this type is made of individual discs of canvas cemented together and the softest is composed of discs of muslin sewn together. Between these extremes the most popular wheels are formed from sewed sections of muslin discs glued or cemented together by adhesives. For reasons of economy these sewed sections are often made of balanced pieces of muslin rather than full discs of cloth. Cotton fabric wheels as a class are the most commonly used medium for general all-around polishing because of their versatility and their relatively moderate cost.

Pressed felt wheels, available in densities from rock hard to extra soft, are indicated where the face of the wheel must be kept true and be absolutely uniform in density over its entire surface. The face of a felt wheel can be easily shaped to fit irregular contoured articles.

Because of the higher initial cost the use of felt wheels is generally restricted to the finer abrasive grit sizes.

Leather-faced wooden wheels are popular for use on flat surfaces where a minimum of flexibility is desired. Solid leather wheels of Walrus or Bull Neck leather are tough but resilient, with a springy open grain, hence are widely used for the fine polishing required in the cutlery and gun field. Wheels made of sheepskin discs are used where the need for greater flexibility and less density is indicated. Varying the construction of the sheepskin wheel will determine the degree of hardness. When a harder wheel is necessary the discs are cemented together. A softer wheel is produced by hand stitching the discs together.

Nearly all the materials described which are assembled in disc form for polishing wheels can be used in a different manner in the production of the so-called "compress" wheel. In this type small pieces of leather or woven fabric are attached endwise as segments to a rigid center section of steel or plastic so that the cloth edges are perpendicular to the side of the wheel. Accordingly, there are no seams following the direction of rotation; hence more precise polishing can be done than with any other type of built-up wheel. Various degrees of flexibility are available for each type of material used, and with the lower-density woven fabric compress wheel in particular, a degree of fine polishing can be obtained that cannot be duplicated with wheels of different construction.

Although no hard-and-fast rules can be made for the selection of polishing wheels for any particular job, in general the more rigid wheels are more frequently used for the coarser polishing operations requiring rapid metal removal or where the surface is relatively regular and this regularity of surface is to be maintained. The more flexible wheels are used on irregular surfaces and for fine polishing where rapid removal of metal is less important.

Adhesives

The adhesives commonly used for fastening the abrasive grains to the surface of the polishing wheel are hide glue and silicate-base cements. Glue, when used, should be high-grade material selected for melting point, jelly strength, viscosity, and flexibility. It should be as free as possible from bacteria which will spoil the glue and decrease its strength.

Covered, sterilized containers should be used to soak a one-day supply of flake or ground glue. Melted glue should never be made up for more than one day's production because bacteria will be picked up from the air. Ground glue is usually preferred to flake glue, since one hour of soaking is generally sufficient. Soaked glue should be melted in a water-jacketed glue pot (of copper or aluminum preferably) at 140° to 150°F. Overheating or prolonged heating even at low temperatures quickly reduces glue strength. It is advisable to have the glue pot thermostatically controlled and to determine accurately the amount of glue necessary for a day's production in order to minimize the time that the glue is kept in the melted condition. At the end of the day the glue pot and brush should be cleaned and sterilized so that bacteria do not remain to contaminate the next day's batch.

The size of the abrasive to be used and the strength of the dry glue will determine the proportion of glue to water. When the type of hide glue has been selected, the exact proportion of glue and water should be decided upon for the size of abrasive to be used and this proportion should be rigidly maintained. The glue and water are measured by weight rather than by volume or guess-work. Although the exact proportion of glue to water will depend upon the type of glue selected, Table 1 will give approximate proportions which will serve as a guide in determining the final figures.

Before setting up the wheel, it and the abrasive should be heated to approximately 120°F to prevent chilling when applying the melted glue. The first or sizing coat should be applied to the clean surface of the polishing wheel by means of a bristle brush and allowed to dry before putting on the heading coat that is to hold the abrasive.

TABLE 1. PROPORTION OF DRY GLUE TO WATER FOR VARIOUS GRIT SIZES

Grit size of abrasive	Dry glue (wt. %)	Water (wt. %)
30	50	50
36	45	55
46	40	60
60	35	65
80	33	67
100	30	70
150	25	75
220	20	80

New wheels should be sized on both the face and sides. After the coating for the first head is brushed on, the wheel should be rolled in the proper abrasive.

When wheels are headed by hand the equipment consists of an abrasive trough to hold the grain and a round stick or piece of pipe to slip into the arbor hole of the wheel. After the wheel is coated with the adhesive the operator puts the round stick or pipe into the arbor hole and rolls the wheel in the abrasive trough. Moderate pressure is sufficient, since the weight of the wheel itself will press the abrasive into the adhesive and provide a uniform coating. In larger plants, a wheel heading machine (Fig. 1) is generally used instead of the hand method. Such a machine consists of a moter-driven pan containing the abrasive, a scraper to keep the grain level, and a live spindle which carries the wheel. This spindle is raised or lowered by means of a lever bringing the wheel in contact with the abrasive grain when wanted.

After the wheel is dried in air for 1 to 2 hr, a second coating is applied in the same manner, if required. Separate brushes and separate glue pots should be used for each grit size to avoid contamination of wheels with abrasive grains of other sizes. After the final coating, the wheels should be dried in a well-ventilated room at about 80°F with a relative humidity of about 50%. The drying time should be 24 hr for each abrasive head applied. After drying, the surface is broken up to provide resilience and free cutting by hitting it with a round bar at about

Fig. 1. Wheel header. (*Courtesy of Divine Brothers Co.*)

Fig. 2. Floor-type Balancing tool. (*Courtesy of Divine Brothers Co.*)

45° to the axis of the arbor hole from each side of the wheel so that an "X" pattern is finally produced on the face of the wheel.

The wheel must now be balanced (see Fig. 2) before it will be ready for use. If balancing tubes have been installed in the wheel by the manufacturer, balancing is easily done by the insertion of lead wire in these tubes. An alternate method is to attach plates to the side of the wheel. However, unless a polishing wheel has a metal bushing to support the arbor hole, it will be extremely difficult to balance the wheel so that it will run true on the spindle of the polishing lathe.

Before recoating a used polishing wheel the old head is removed, preferably by means of an abrasive stick. In many plants this is done while the polishing wheel is on the lathe on which it is used. In larger plants a modern wheel dressing machine (Fig. 3) designed for this purpose is used instead. Such a machine has a motor-driven arbor to hold the wheel, with infeed and cross-feed slides for proper movement of the abrasive tool to ensure that the face of the wheel is trued up properly for reheading.

Rotating the wheel on a pair of wet rolls will remove the glue and abrasive, but this method

Fig. 3. Wheel dressing machine. (*Courtesy of Divine Brothers Co.*)

should be restricted to those wheels that are not harmed by water soaking. If the polishing wheel is lubricated, the grease must be removed before reheading; it is sometimes necessary to use an organic solvent, particularly if the fabric of the polishing wheel has been exposed and has become grease-impregnated during use.

Silicate-base polishing wheel cements offer an attractive alternative to glue. One big advantage of these cold cements is that they can be applied at ordinary room temperatures as received from the manufacturer, without the need for special equipment and the careful precautions and elaborate preparation required for the proper and efficient use of hide glue. Polishing wheels set up with cement can stand higher temperatures, both in drying and in use. They can be dried more rapidly and also stand a higher frictional heat caused by higher speeds and greater pressures. Cements are particularly favored for fast, tough, and coarse polishing, although they have been used successfully for some fine polishing operations. However, for fine work, where flexibility of head is desired, most operators still prefer hide glue. Most silicate-base cements used in polishing are proprietary mixtures and it is therefore best to follow the manufacturer's recommendations in each case rather than to attempt here to give general directions to cover all of them.

Abrasives

Both natural abrasives such as emery and corundum and artificial abrasives such as aluminum oxide and silicon carbide are widely used for polishing metals. The artificial abrasives are harder and more uniform than the natural, and are therefore of greater importance in polishing. The most widely used abrasive in polishing metals is an artificial aluminum oxide grain possessing high capillarity and with an etched surface so that it will be properly bonded by the adhesive. This material is hard, sharp, tough and fast-cutting; it is recommended for tough metals such as carbon steels, alloy steels, high-speed steels, anealed malleable iron, wrought iron, and certain bronzes. Artificial silicon carbide is harder than aluminum oxide, fractures more easily when dull, and presents new sharp cutting edges; for this reason it is selected for cutting materials which are of low tensile strength, such as brass, aluminum and copper, and for hard brittle substances, such as hard alloys, chilled iron, gray iron and cemented carbide tool stock. Silicon carbide grain, however, is more difficult to bond to the wheel by adhesives and its scope is therefore limited.

Of the natural abrasives, Turkish emery was once the standard polishing grain. Emery is a natural compound of aluminum and iron oxide containing between 57 and 75% aluminum oxide. Because of the iron oxide, the edges of Turkish emery wear smooth rapidly which decreases the degree of cutting but at the same time makes this abrasive highly desirable for certain fine polishing operations, particularly in the high grade cutlery field.

The Abrasive Grain Association standard grain sizes for the artificial abrasives are as follows:

Aluminum Oxide Abrasive

Screened Sizes: 4, 6, 8, 10, 12, 14, 16, 20, 24, 30, 36, 46, 54, 60, 70, 80, 90, 100, 120, 150, 180, 220, 240.

Unclassified Flours: F, 2F, 3F, 4F, XF.
Classified Flours: 280, 320, 400, 500, 600.

Silicon Carbide Abrasive

Screened Sizes: 8, 10, 12, 14, 16, 20, 24, 30, 36, 46, 60, 70, 80, 90, 100, 120, 150, 180, 220, 240.

Unclassified Flours: F, 2F, 3F, 4F, XF.
Classified Flours: 280, 320, 400, 500, 600.

Although it is seldom that two polishers will agree on the polishing method for any given article, Tables 2 and 3, compiled by the Abrasive Grain Association, give approximate procedures which will serve as a guide in determining the best sequence of operations for a particular job.

Lubrication

Lubrication of the cutting face of a polishing wheel with oil or grease is desirable in many cases to prevent gouging or tearing when a fine polished surface is required. It is also used to

TABLE 2. TABLE OF GRAIN SELECTION
(Naxos emery and Turkish emery)

Parts	Polishing Operation				
	First	Second	Third	Fourth	Fifth
Axes	46–60	70–90	120	150–180*	—
Aluminum, sand-cast (Inside-bottom)	36–46	—	—	—	—
Aluminum, sand-cast (Outside)	60–80	120–180	Buff	—	—
Aluminum, die-cast	150*	Buff	—	—	—
Aluminum, sheet	120*	180*	Buff	—	—
Auto bumpers	60–90	120	150–180*	220*	—
Auto headlights	180–220*	Buff	—	—	—
Band saw steel	60–80	120–150	—	—	—
Brass, sand-cast	60–80*	150–180*	—	—	—
Brass, sheet	180–220*	Buff	—	—	—
Electric irons	80	120*	150*	180–240*	—
Glass beveling	70–90	120–150	220	Pumice	Rouge
Granite polishing	60–90	120–150	F	3F	Buff tin oxide
Gray iron, pickled	80	120–150	—	—	—
Gray iron, not pickled	70	120–150	—	—	—
Hammer heads	46–60	100–120*	—	—	—
Knives, table and steel blades	80–90	120–150*	220F	Buff (special machines)	—
Knives, table backs	46–60	—	—	—	—
Knives, machete, edges	46–60	—	—	—	—
face	80	120*	—	—	—
Lenses	—	—	—	—	—
Lenses, prescription	60–80	180–220	{ Optical flour	Rouge	—
Lenses, telescope	60–80	180–220	{ Optical flour	{ Optical flour	Rouge
Locomotive side rods	36	60–70	120	—	—
Monel metal, deep-drawn	120	150	180*	Buff	—
Monel metal, cast	80	120	150	150*	Buff
Monel metal, full-finish sheet	180	180*	220*	Buff	—
Plows	24–36	80	120–150	180–220*	—
Plow shares	36–46	—	—	—	—
Plow disks	30–46	70–90	—	—	—
Shears, tinsmith	46	60	120–150	180	—
Shovels, blades	30–46	120	—	—	—
Shovels, straps	36–70	—	120 (belt)	—	—
Stainless steel:					
Mirror finish	60–80	100–120*	150*	220–3F*	Buff
Commercial finish	80	100*	120*	150*	—
Wrenches	30–60	80–90	120*	—	—

*Denotes grease or oil wheel. (*Source*: Heywood, *Abrasive Grains and their Uses*.)

TABLE 3. TABLE OF GRAIN SELECTION
(Aluminum oxide unless otherwise specified)

Parts	Polishing Operation					
	First	Second	Third	Fourth	Fifth	Sixth
Axes	46–60	70–90	120	150–180*	—	—
Aluminum, sand-cast (Inside-bottom)	36–46	—	—	—	—	—
Aluminum, sand-cast (Outside)	60–80	120–180	Buff	—	—	—
Aluminum, die-cast	150*	Buff	—	—	—	—
Aluminum, sheet	120*	180*	Buff	—	—	—
Auto bumpers	60–90	120	150–180*	—	220*	—
Auto headlights	180–220*	Buff	—	—	—	—
Automobile fenders and sheet stock for enameling	90	120–150	—	—	—	—
Automotive hardware	36–54	90	120	220*	—	—
Band saw steel	60–80	120–150	—	—	—	—
Brass, sand-cast	60–80*	150–180*	220–3F*	—	—	—
Brass, sheet	180–220*	Buff	—	—	—	—
Cutlery	80	120	Pumice	Buff with compound		
Electric irons	60–80	120*	150*	180–240*	—	—
Forks, hay	60–70	100–120	—	—	—	—
Forks, spade	24	—	—	—	—	—
Glass beveling	70–90 SiC	120–150	220	Pumice	Rouge	—
Granite polishing	60–80 SiC	120–150 SiC	F SiC	3F SiC	Buff tin oxide	—
Gray iron, pickled	80	120–150	—	—	—	—
Gray iron, not pickled	70	120–150	—	—	—	—
Hammer heads	46–60	100–120*	—	—	—	—
Hoes, first quality	36–46	70	100–120	—	—	—
Hoes, second quality	36–46	—	—	—	—	—
Knives, table and steel blades	80–90	120–150*	220–F	Buff	—	—
			(special machines)			
Knives, table backs	46–60	—	—	—	—	—
Knives, machete, edges face	46–60 80	120*	—	—	—	—
Lenses	60–80	180–220	Optical flour	Rouge	—	—
Lenses, prescription	60–80 SiC	180–220	Optical flour	Rouge	—	—
Lenses, telescope	60–80	180–220	320	Optical flour	Optical flour	Rouge
Locomotive side rods	36	60–70	120	—	—	—
Monel metal, deep-drawn	120	150	180*	Buff	—	—
Monel metal, cast	80	120	150	150*	Buff	—
Monel metal, full-finish sheet	180	180*	220*	Buff	—	—
Plows	24–36	80	120–150	180–220*	—	—
Plow shares	36–46	—	—	—	—	—
Plow disks	30–46	70–90	—	—	—	—
Shears, tinsmith	46	60	120–150	180	—	—
Shovels, blades	36–46	120	—	—	—	—
Shovels, straps	36–60	120 (belt)				
Stainless steel:						
Mirror finish	60–80	100–120*	150*	220–3F*	Buff	—
Commercial finish	80	100*	120*	150*	—	—
Tools, small hand†	(See note.)					
Wrenches	30–46	80	120*	—	—	—

*Denotes grease or oil wheel. †Small tool polishing operations are too numerous to enumerate in detail. Apply to any abrasive grain manufacturer for recommendations. (*Source:* Heywood, *Abrasive Grains and their Uses.*)

minimize frictional heat when polishing some of the softer metals, particularly aluminum. Although applying a tallow-grease mixture in bar form to the rotating surface is the popular method of lubricating polishing wheels on hand lathes, liquid lubricants sprayed onto the polishing wheels are used almost exclusively on automatic equipment.

Speeds

Speeds for the efficient operation of polishing wheels generally fall between the limits of 5000 to 8000 surface ft/min when glue is the adhesive used (see Table 4). At higher speeds glue tends to break down because of overheating, although polishing wheels set up with cement can safely operate up to 9000 surface ft/min. The lowest speeds are used for the soft metals and the highest speeds are used for the hard steels. With too low a speed the abrasing operation is slowed down and, in addition, there is a tendency for the abrasive to be ripped out of the wheel. On certain metals susceptible to undesirable physical changes because of overheating the limiting factor is heat tolerance of the metal rather than the ability of the adhesive to withstand heat.

Abrasive Belts

Endless belts coated with adhesives and abrasive grain by the user or the belt manufacturer are used for numerous polishing operations instead of conventional polishing wheels. Abrasive belts of cloth or paper precoated by the manufacturer are available in a full range of grit sizes and ready for immediate use in polishing metals without the necessity for the maintenance of a set-up room. Cloth-backed belts are favored for polishing on all types of metals, whether manual or automatic, while paper-backed belts are used mostly for sanding wood. Provided the backing is not torn, these belts can be reheaded several times by the user, or by companies offering this service. Factory-coated belts are offered with the same fast-cutting abrasives found popular for conventional set-up wheels. These belts have the abrasive bonded to the cloth surface with glue, resin or resin over glue and then are recoated with the bonding material—thus the name *coated abrasives*. The glue has the most resiliency, while the waterproof resin has the least. The unslit belt is drawn over a sharp edge or run between a large diameter rubber drum and a small diameter steel bar to decrease the stiffness. If this flexing is perpendicular to its length, it is called single flexed. If this flexing is at 45° to each side, then it is double flexed. If the single and double flexing are combined, it is triple flexed. When these flexings and bonding materials are combined with the choice of the softer jean cloth (J designation) backing or the harder drill cloth (X designation) backing, the selection becomes quite varied, and something should be found to suit most needs. They have the advantage of providing an even, controlled coating which will consistently give uniform, polished surfaces.

Metal surfaces are polished on abrasive belts running over two or more wheels or rolls, proper tension being maintained to prevent slippage. One of the wheels or rolls is the driving roll and the others run free. One of the wheels or rolls is the cushioned contact wheel against which the metal surface being polished is held.

The field open to abrasive belt polishing has been greatly increased by the development of contact wheels with a wide variation in resiliency. Compressed leather, compressed canvas, felt, bias-type buffs, rubber-covered fiber, rubber-covered steel and solid urethane are being used successfully as contact wheels for belt polishing. The contact wheel should provide the resiliency necessary to permit one to polish contoured surfaces without sacrificing production speed.

Abrasive belts operate to best advantage on relatively flat surfaces when the part is held against the periphery of the contact wheel and on irregular convex surfaces when the slack of the belt is used. With concave and irregularly con-

Table 4. Surface Speed of Polishing Wheel for Various Diameters
(ft/min to nearest 10 ft)

Speed of Arbor (RPM)	Diameter of Wheel								
	2"	4"	6"	8"	10"	12"	14"	16"	18"
800	420	840	1260	1680	2100	2510	2930	3350	3770
1000	520	1050	1570	2100	2620	3140	3670	4190	4710
1200	630	1260	1880	2510	3140	3770	4400	5030	5650
1400	730	1470	2200	2930	3670	4400	5130	5860	6600
1600	840	1680	2510	3350	4190	5030	5860	6700	7540
1800	940	1890	2830	3770	4710	5650	6600	7540	8480
2000	1050	2100	3140	4190	5240	6280	7330	8380	9420
2200	1150	2300	3450	4600	5760	6910	8060	9220	10370
2400	1260	2510	3770	5030	6280	7540	8800	10060	11310
2600	1360	2720	4080	5450	6800	8170	9530	10890	12250
2800	1470	2930	4400	5860	7330	8790	10260	11730	13190
3000	1570	3140	4710	6280	7850	9430	11000	12570	14140
3200	1680	3350	5020	6700	8380	10050	11730	13410	15080
3400	1780	3560	5340	7120	8900	10680	12430	14250	16020
3600	1880	3770	5650	7540	9430	11310	13200	15080	16960

toured surfaces requiring the softest possible cushion to the polishing medium, conventional polishing wheels work to better advantage. Abrasive belts offer an enormous abrasive-coated area at all times, with cooler operation and less danger of burning the work, especially when the pressure is increased in an effort to speed up production. Also, because of the greater abrasive surface on a belt, the individual grains do not lose their sharp edges as rapidly. With polishing wheels the abrasive-coated area is smaller but much thicker and more economical on jobs requiring removal of considerable metal. When extremely cool polishing is essential, wet belt polishing is employed using abrasive belts bonded with a waterproof synthetic adhesive, applied under controlled factory conditions and run over a contact wheel unaffected by water such as one that is rubber-covered.

Paste Wheels and Flexible Polishing

In the polishing coatings described above, whether the abrasive is coated as in the instance of the belt or uncoated as in the case of the set-up wheel, the abrasive particle protrudes from the bonding surface. This is highly desirable in many instances, particularly when fast cutting is required. However, in a number of fine-polishing operations, it is more desirable to have a closed surface, which is obtained by having intimate mixtures of the abrasive grain and the adhesive. Paste wheels, in which the melted glue is thoroughly mixed with abrasive and the mixture then is troweled onto the face of the polishing wheel in multiple coatings, have long been used in the cutlery industry, particularly for "double header" polishing on steel knife blades. Silicate cements have also been used with abrasives for similar purposes.

A special technique has been developed so that wheels of this type can be built up by frictional transfer of greaseless compound applied to the polishing wheel as it revolves on the spindle of the polishing lathe. These flexible polishing wheels are headed up by the following procedure: A glue-base sizing material in bar form is brought to the revolving wheel, pressure is applied and the power is shut off. During deceleration the heavy sizing coat is melted and flowed onto the surface of the wheel. The lathe is then turned on and off to start drying at less than full speed and is finally run at full speed for about 2 min or until the sizing coat is no longer sticky to the touch. A bar of a glue-base greaseless compound is then applied in the same manner after the power is shut off. During deceleration the heavy coating of greaseless compound is melted and transferred to the wheel with sufficient frictional heat so that the layer of greaseless compound is firmly bonded to the sizing coat. This coating is dried in the same manner as mentioned for the sizing coat and for approximately the same time. A second coating of greaseless compound can then be applied without additional sizing and heads can be built up to $3^{1}/_{6}$ in. thickness.

In the formation of flexible polishing wheels of this type it is essential that the stop-and-start technique be followed closely, because if the lathe were allowed to run at full speed and the compound were applied at the necessary pressure, centrifugal force would throw off much of the material. During deceleration, however, the greaseless compound is flowed evenly onto the face of the wheel. With this procedure, true polishing wheels 80 grit and finer can be produced ready for use in less than 10 min without the necessity of ever removing the wheel from the spindle.

When a cloth polishing wheel of proper resiliency is selected, flexible polishing wheels can be made up with greaseless compound that have many advantages over broken-down conventional polishing wheels which have long been used for fine-polishing operations. Such a wheel can be promptly produced with the proper surface condition, and consequently the wheel inventory can be reduced significantly. The technique in producing a flexible polishing wheel with greaseless compound is such that much softer cloth wheels can be employed than can be used by the conventional polishing wheel set-up procedure; hence such wheels utilize the resiliency of the cloth to cushion the cutting action of the abrasive grain. With the conventional polishing wheel, on the other hand, this effect can only be obtained by the use of excessive lubrication. Two complete cleaning cycles, each with alkali cleaning, rinses, and acid dips, are frequently used with greased wheels. One complete cleaning cycle is safely being eliminated in many plants when flexible polishing with greaseless compound is substituted for the final grease conventional polishing wheel procedure. Flexible polishing wheels operate most efficiently at 5000 to 6000 surface ft/min.

BRUSHING

In metal finishing there are a number of operations accomplished by brushing which cannot be placed correctly under either polishing or buffing. While abrasive compounds of one type or another are often applied to a rotating brush, such abrasives do not adhere to the surface of the brush as well as they do to a polishing or buffing wheel. Brushing compounds, when used, are back-transferred to the work and subsequently moved along when bristles or wires come in contact with them, and a surface finish of a somewhat different type is produced. Fine finishing usually requires the use of nonmetallic brushes, while coarser finishing may be done with wire brushes.

Brushes of steel wire approximately 0.01 in. thick are used without compositions for removing rust, scale, paint and other incrustations from metallic surfaces. Speeds vary from 4200 to 6500 surface ft/min, depending on the brush construction and the finish required. Similar wire wheels, as well as Tampico brushes, have been used successfully with a special grease-base emery paste for removing burrs and rounding edges when the contour of the part is such that maximum flexibility is required. Brushing with emery paste on Tampico brushes was once the standard procedure between the final polishing operation and subsequent cut-down buffing. In many cases this method has been replaced by flexible polishing with greaseless compounds because of the increased cleanliness possible.

For final scratch brush finishing for an extremely dull effect, fine-wire brushes of brass or nickel-silver wire are still employed, even though an additional cleaning operation is required. This application is particularly adaptable to parts having a thin electroplate, since it minimizes the danger of cutting through on sharp edges. For such work, the wire brush is generally run wet and for dull finishes pumice and water are used. In such cases speeds are very low, averaging around 1500 surface ft/min. In the jewelry industry bristle brushes are used with buffing compositions to produce effects similar to those produced by true buffing. They have the advantages here of more readily getting into crevices and delicate designs and, in addition, minimize the danger of cutting through on exposed edges.

String wheels are sometimes referred to as string brushes. They are made of soft cotton yarn, fastened to a hub, and used with greaseless compounds to produce wiped finishes varying from dull satin to bright butler.

Cord wheels or brushes are similar to string wheels in construction and use. Special cords, natural and synthetic, are fastened to hubs, forming a more dense brush than string and are generally used to produce coarser finishes. Brushing or buffing compounds must be used with nonmetallic brushes to achieve the desired finish.

Except for those cases requiring the special properties of the brushes, the final finishing is generally accomplished by buffing methods, to reduce the cleaning problem, which is always a factor after brush finishing.

BUFFING

Buffing can be divided into four classifications: (1) satin finishing for producing satin, brushed or butler finishes, (2) cut-down buffing for producing a preliminary smoothness, (3) cut and color buffing for producing smoothness and some luster, and (4) color buffing for the production of a high-gloss or mirror finish.

Buffing Wheels

The wheels most extensively used in buffing are made of muslin. In order of decreasing density or rigidity buff construction may be sewed, pocketed or folded, full-disc or packed, the latter being made up of alternate discs of large and small diameters. A softer wheel is the string brush which has no cross threads common to sheeting. Although there is some overlapping in the case of sewed buffs, in general, buffing wheels are much more flexible than those used in polishing.

The grades of muslin sheeting commonly used for buffing wheels are 48 x 48, 64 x 68, 80 x 92 and 86 x 82. In general, lower count sheeting is preferred for coloring operations and the higher count sheeting is used for heavy cut-down buffing and wherever a hard surface wheel is required. A cloth with medium count such as 80 x 92 is a popular all-around buffing material. Although buffs made with other than cotton or sisal have always failed, a muslin type fabric made with a blend of polyester and cotton has proven to be effective in a bias-type buff for cut-down on non-ferrous metals.

For fast cut-down buffing sewed buffs are

commonly used. Full-disc sections of cloth are sewed spirally, concentrically, radially, square stitched or in a crescent pattern from the center to the edge. In general, the closer the rows of sewing the denser is the buff. Spiral sewing is the most common type and is used almost exclusively on pieced buffs and on most of the sewed full-disc buffs, although other types of sewing show distinct advantages in many cases.

Pocketed or folded buffs are widely used where fast cutting is wanted with a greater degree of flexibility. These buffs are made by sewing folded pieces of cloth together with the pockets open at the periphery, or by taking long strips of cloth, which are shirred in on one end and sewed to a hard center causing folds to form radially toward the periphery. Quite often these strips of cloth are cut on the bias which minimizes unraveling during use. In some cases a perforated metal center is used, so designed that in operation air will enter near the arbor and be thrown by centrifugal force through the folds of the buff, thus giving additional ventilation to the surface. The pockets on buffs of this type and those with radial or crescent sewing tend to collect and hold buffing compositions increasing the efficiency of tough cutdown buffing operations.

Finger buffs also are used where a great deal of flexing is necessary. These buffs are made from folded strips of cloth which are sewed along their length. Segments are cut, folded in half and secured in a steel clinch ring.

The bias buffs are probably the work-horse in the buffing field. These are made with cloth that has been cut on a 45° bias, wrapped around a split drum and squeezed into a steel clinch ring with an iris type device. This creates puckered buffs where the thread ends are in a weaving pattern, which helps to cool the buff and hold composition. The density of these buffs depends upon the diameter of the split drum and the final diameter of the buff.

If greater flexibility is needed, the biased cloth is shirred and sewed, and then clinched into a steel ring the same as the bias type buffs. This results in flexible buffs with the no streaking properties of the bias type buffs, referred to as open-face buffs.

Loose buffs are employed where greater flexibility and less cutting are required, such as in final coloring operations. When greater flexibility is needed in special cases, such as in the production of fine satin finishes on sheet metal parts, discs of sheeting are packed or separated by smaller discs of cloth or cardboard, the ratio of large discs to small discs will determine the degree of flexibility. The string brush is also used for this type of work.

In addition to muslin, buffs made of canton and domet flannel, felt, wool cloth or sheepskin discs are used for the final coloring of precious metals. For fast cut-down buffing, particularly on ferrous metals of regular shape, buffing sections made of muslin and sisal, as well as all sisal, have high cutting properties when used with coarse, greasy compositions. The scratches left by these sisal buffs are usually taken out with the same compositions on all cotton muslin buffs.

Flap polishing and buffing wheels, where hinged segments of polishing belts, sisal, cloth or combinations of these are mounted around a hub with the ends of these segments perpendicular to the side of the wheel, combine flexibility and aggressiveness. Even the polishing belt types can be used in conjunction with buffing compositions to increase wheel life and change the cutting effect.

A non-woven sponge like fabric, incorporating aluminum oxide, silicon carbide or garnet abrasives in it, is used to make bias, full disc, flap and solid type buffs. The effect is similar to greaseless compositions, although it does more scouring than surface removal. These buffs are used for deburring and for brushed finishes.

Compositions in Bar Form

Satin finishing operations are performed by means of greaseless compounds combining fast-cutting abrasives with a glue-base binder. Numerous grades are available, employing abrasives 80 and finer, for varying degrees of dullness of finish on all base metals and electrodeposits. In the most popular all-around grades, artificial aluminum oxide and silicon carbide abrasives are used in grit sizes from 180 to 220. Silicon carbide grades are widely used for finishing aluminum, cast brass, stainless steel and titanium, and the aluminum oxide grades are favored for brass and other nonferrous metals, as well as for carbon steel prior to plating. For the best effects on brass and electrodeposits, finer sizes of emery and hard silica are employed. For butler finishes on silver plate and sterling, fine buffing powders of unfused aluminum oxide and soft silica are used. Bright butler finishes on silver that approach the luster produced by grease-base coloring bars can be obtained with extremely fine greaseless compositions made with a specially lubricated binder.

Greaseless compounds are used at speeds of between 4000 and 6000 surface ft/min. Higher speeds waste the composition without a proportionate increase in production rate. Greaseless compositions do not penetrate into the buff, as do grease-base compositions, but adhere to the surface, which favors buff life. The buff should be allowed to run for 20 to 30 sec after a greaseless compound has been applied before the work is brought to a wheel, so that there will be no back-transfer of the compound. When the greaseless compound wheel is used correctly the work will leave the wheel clean, dry, and in proper condition for inspection and packing. Certain metals, such as aluminum, are susceptible to dragging by dry abrading, but a light top dressing of a low free-grease content buffing bar will sufficiently lubricate the surface and produce an even finish. In this case care should be used so that the lubricant does not penetrate the layer of greaseless compound, saturate the cloth, and prevent the proper adherence of subsequent additions of greaseless compound. The necessary lubrication can also be obtained in certain cases by the use of a bar of greaseless composition employing a special binder with sufficient lubricating properties, so that top dressing becomes unnecessary.

A great variety of fats and waxes of animal, vegetable, and mineral origin are used in binders for grease-base buffing compositions. The more commonly used include stearic acid, hydrogenated fatty acids, tallow, hydrogenated glycerides, and petrolatum. Under the conditions of buffing, the fatty acids appear to have a chemical effect in the formation of metallic stearates which is beneficial, in addition to their physical properties.

Grease-base buffing bars for cut-down buffing are composed of fast-cutting buffing powders in a binder formulated to produce a great amount of "drag" between the buffing wheel and the work. The proportion of powder to bind will vary with the oil sorption of the buffing powder, the type of work, and the particular conditions under which the work must be done. Final decision can be made only after an examination of the job in question.

The buffing powders to be used in cut-down buffing vary with the metal being buffed. Once-ground rose Tripoli is the most popular buffing powder for compositions for cutting down non-ferrous metals. It is relatively inexpensive and performs very efficiently. Fused and unfused aluminum oxide powders are most widely used in compositions for cut-down buffing of carbon and stainless steels and to a lesser degree are indicated for some of the aluminum alloys, particularly for cast, extruded, and rolled shapes.

Cut-and-color buffing is a step which is often passed over, but in other cases is used alone in place of the faster cut-down and the higher coloring operations. In general, a binder is selected of the same general type as above but with less drag, since fast cut is not the prime requisite. Abrasive powders are selected which will give some brilliance with moderate cut, sacrificing both cutting properties and coloring properties to produce a general-purpose composition. For the nonferrous metals cut-and-color compositions employ white silica powders or a blend of these powders with Tripoli. Similarly, for the ferrous metals, coarse unfused aluminum oxide powders are used for a combination of these with some fused aluminum oxide powder. Crocus bars made of a coarse grade red iron oxide powder are used in some cases for cut-and-color buffing on certain nonferrous articles, as well as on some steel cutlery.

Compositions for producing a color, luster, or mirror finish on metals are composed of the finest abrasives selected with special attention to their cutting properties, so that a minimum of scratches will appear in the final finish. As fast cutting is not required in this operation, binders are formulated mainly to hold the selected abrasive powder to the wheel without building a buffing head so that removed particles are not embedded in the buff surface to cause unwanted scratches. The result enables the wheel to color or burnish out the scratch marks left by previous buffing. Powdered lime is used in compositions for coloring nickel plate and to a somewhat lesser degree for the coloring of brass. Fine unfused aluminum oxide powders are used in coloring nickel, chromium plate, stainless steel and, on occasion, aluminum and brass. Soft white silica powders of finer sizes are generally used in compositions for coloring brass and aluminum. Compositions containing fine chromium oxide powder are used for coloring chromium and stainless steel. Rouge, compounded of the finest red iron oxide powder; is widely used in coloring bars for silver and gold because it has the proper burnishing qualities to bring the soft, precious metals to the required high luster.

TABLE 5. APPROXIMATE SURFACE SPEEDS FOR
HAND-BUFFING VARIOUS MATERIALS

Material	Surface speeds (ft/min) Cutting down	Color buffing
Carbon and stainless steel	8000–9000	7000–9000
Brass	6000–9000	6000–9000
Nickel	6000–9000	6000–8000
Aluminum	6000–9000	6000–7000
Zinc and other soft metals	5000–8000	6000–7000
Chromium		7000–8000
Plastics	3000–5500	3000–5500

Although the speeds for buffing with grease-base bars will vary greatly from job to job and operator to operator, the data in Table 5 on surface ft/min will serve as a guide for hand buffing operations. Automatic buffing speeds may be somewhat higher, as the contact pressure of the work to the wheel can be more definitely fixed without depending upon the physical ability of the hand buffer to maintain the proper position and pressure.

The speeds in the Table 5 for various metals are those suggested for manual buffing and those automatic buffing systems where the wheel follows the shape of the surface of the parts—sometimes called contour buffing. Much lower speeds are needed for automatic equipment designed for "mush buffing." In this case, the buffing operation runs much slower to cut down centrifugal force so that the surface of the cloth can be pushed into the irregularities in the surface of the parts being buffed. Contact of the face of the buff with all sections of the surface of the metal part is of prime importance even at the expense of efficiency of the buffing operation. With the diminished frictional heat, variations in the composition formula can recover some of the lost efficiency caused by the subnormal speed. To repeat, speeds are set to insure contact of buff to all of the surface to be buffed with little or no regard for the most efficient speed for the metal involved.

Spray Buffing Compositions

Liquid-spray compositions use the same type of abrasive powders used in the bars, but the big difference lies in the binder. Instead of using fats, greases, waxes, and such combinations that set up solid at room temperatures, similar materials are used in oil solutions or water emulsions so that the binder and the resulting spray compositions are fluid at ordinary temperatures. These mixtures are fed either by gravity or pressure to a spray gun similar to that used in organic spray finishing, and then sprayed by means of compressed air onto the revolving buffing wheel. Automatic spray guns are used in the majority of installations on rotary indexing, continuous, straight line, and other automatic buffing machines. Where buffing machines require hand application of buffing compounds, spray buffing application may be performed with a manual spray gun.

With automatic low-pressure spray gun installations, if narrow buffing wheels not exceeding 9 in. wide are used, one automatic spray gun will give sufficient coverage of the buff with proper fluid and fan spray control. Wider-faced buffs will require the use of multiple gun installations or use of a spray gun moving mechanism.

Spray guns are usually mounted in convenient positions near the buffing wheels so that they will not interfere with the operator or the work being buffed, and to permit efficient operation of the guns.

An important factor to consider in proper positioning and operation of the spray guns is the effect of the revolving "blanket" of air generated by the rotating buff and its effect on efficient coating of the buff with the liquid buffing compounds. The counteracting force of air varies with different types of buffs and with their speed. This must be considered in determining the proper angle of spray, fluid and atomizing pressures, to penetrate this "blanket" of revolving air with the liquid buffing composition.

The low-pressure liquid buffing compositions range from 20,000 to 50,000 centipoises and tend to seek their own level. Liquid compositions thicker than these are used in airless, high-pressure systems. The pump for these thick compositions sucks the composition from the top of the drum. This makes it necessary for the plate containing the pump and the hose gaskets around the plate to fit tight inside the drum. A small pressure on the plate ensures that the pump sucks up the composition instead of air. The composition is pushed around the system at pressures ranging from 600 to 3000 psi.

The airless eliptical spray gun orifice fans the composition out so that it can cover quite uniformly a buffing wheel 24 to 30 in. wide. The composition is propelled at the revolving buffing wheel at high speeds, enabling it to go through the revolving air barrier and penetrate the top inch of buff surface with little or no loss. This means more of the composition is put to work with less buff wear and more reproducible buffing results. Only the softer abrasives, such as Tripoli, can be used in this system since the harsh abrasives quickly reduce the eliptical shape of the spray gun orifice and the width of the fan pattern. Operators should not clean or adjust guns when there is a possibility that the spray gun can discharge composition which can penetrate the skin and cause complications.

Compressed air, controlled by a pressure regulator, is supplied to actuate the spray guns. Automatic machines employ cam-operated or electrically operated timer air valves. In some cases knee or foot valves control the supply of air. The airless system, because it requires small increments of time, necessitates fast, electronic timers.

The principal advantages of the liquid spray buffing method are as follows:

1. Optimum quantity of composition is always maintained on the buff, the composition being supplied regularly rather than intermittently. For practical purposes, automatic bar compound application or manual bar application invariably results in an excess of compound being applied to the buffing wheel. Where very wide buffs are used, bar compound application becomes cumbersome and uneconomical in some cases. In bar compound application insufficient amounts of compound are present for the last piece buffed before another application of the bar. Using the spray buffing method the desired amount of composition is administered for each piece buffed.
2. With a deficiency of composition and the natural abrasion of the bar against the wheel the cloth is worn excessively. Spray buffing compositions, eliminating this deficiency of bar compound application, eliminate unnecessary buff wear. Buffs have been made to last four or five times longer.
3. Solid buffing dirt is packed into the crevices of the work when an excess of buffing composition is present. The serious cleaning problem presented by this dirt is well known. The use of spray buffing method helps cleaning in two ways. First, there need be no excess of composition since spraying the compounds allows better control of application, i.e., normally one applies less compound per application but applications are more frequent. Secondly, by their very nature liquid compounds are easier to clean. This is particularly true of water emulsions. In many cases costly cleaning equipment and compounds have been eliminated.
4. Production is increased due to less down time than would be required for changing of used or broken bars.
5. There are savings in compound consumption. All the spray buffing composition brought to the lathe can be used; there are no unusable nubbins left over.
6. In the conventional buffing procedures using high pressures of the work to the wheel, a deficiency of composition on the buff has often resulted in such frictional heat that the muslin buff catches fire. The spray method eliminates this hazard by keeping the buff properly coated at all times. However, a spray buffing composition must be selected which does not constitute a fire hazard, as is the case when it is composed of volatile, combustible fluids.

Recognizing the numerous advantages of this method, the finishing industry has accepted the spray finishing method as a standard procedure for high production buffing.

DEBURRING

Deburring, unlike some of the other terms listed herein, does not refer to an individual procedure, but rather to any finishing method which can remove burrs and break sharp edges. Hand filing, polishing, flexible polishing, satin finishing, brushing, and tumbling, properly used, become deburring operations.

For decorative products, such smoothing of edges is usually a consequence of the polishing and buffing operations necessary to achieve an attractive, salable finish. When functional parts are deburred, this is usually the final mechanical finish.

Although burrs can be removed by hand methods, such as filing, the labor cost is usually so high that mechanical means are preferred in most cases. Where only restricted areas can be touched, set-up polishing wheels and muslin buffs coated with a greaseless compound are the ideal mediums. These methods are described in some detail above under the headings Polishing Wheels and Flexible Polishing and later under Satin Finishing in the section on Compositions in Bar Form. The method for any one job will be determined by the regularity of the edge and the amount of metal that must be removed. Where edges are straight and the burrs heavy, rigid set-up wheels are indicated. Where the contours are irregular and the burrs not excessive, sewed, loose, or packed muslin buffs coated with a fast-cutting grade of greaseless compound work more efficiently. Where maximum flexibility is required, a string brush with greaseless compound or a Tampico brush with emery paste is used.

BUFFING AND POLISHING EQUIPMENT

Basic equipment for buffing and polishing must include a spindle to which the buffing or polishing wheel can be fastened securely in such a position that it can make contact with the surface of the metal being finished. A source of power must be available which will rotate the wheel at the proper speed for the operation in question in spite of the friction between the wheel and the work.

Manually operated buffing lathes generally consist of a double-ended spindle running between suitable bearings and belt-driven by means of an electric motor usually mounted in the base of the pedestal holding the spindle (see Fig. 4). The advantage of the belt-driven lathe is that by the use of pulleys of the proper size any arbor speed can be selected, either higher or lower than the speed of the driving motor. Variable-speed drives can also be used between the motor and the spindle which allow the arbor speeds to be changed quicky when shifting from one job to another. There are also manual buffing lathes with a direct-driven spindle, but these are restricted to the single speed of the motor.

For abrasive belt operations, a backstand with an idler pulley is used with the manual buffing lathe on which is mounted the contact wheel which drives the belt. Backstand equipment is designed so that the operator can readily regulate belt tension and control the tracking of the belt.

Semiautomatic attachments are available for use with manual buffing lathes (see Fig. 5). Such devices provide fixtures or work-holders which carry the parts to the buffing or polishing wheel

Fig. 4. Manual buffing lathe-variable speed. (*Courtesy of Hammond Machinery, Inc.*)

FIG. 5. Semiautomatic attachment—four spindle. (*Courtesy of Acme Manufacturing Co.*)

in a uniform manner eliminating the natural variation inherent in the manual holding of the work. Such semiautomatic devices carry the parts through only one step of the finishing operations and not progressively from one wheel to another.

Automatic buffing or polishing machines (see Fig. 6, 7 and 8) consist of chucks or work-holders fastened to a rotary table or a straight-line conveyor which carries the parts to each of a series of polishing or buffing wheels arranged in the proper order and mounted in the proper position so that all required surfaces of the article to be finished are abraded. With round objects the spindles rotate as they pass under or in front of the wheels; in some cases they progress continuously and in others of the indexing type, they halt for a predetermined period in contact with each wheel. With other objects of more angular design the work-holders can be made to give a partial turn as they are brought to each new station so that a different surface can be abraded at each step on the automatic machine. Such mechanisms can be extremely complex and, of course, are designed especially for the particular article to be finished. Another approach to finishing these difficult to buff parts is to mush buff them (see Fig. 9) with long buffs at slow speeds.

Other objects lend themselves to finishing on a table passing back and forth under wide-faced buffing or polishing rolls (see Fig. 10). With

FIG. 6. Continuous rotary automatic polishing and buffing machine. (*Courtesy of Hammond Machinery Inc.*)

Fig. 7. Straight-line automatic polishing and buffing machine. (*Courtesy of Divine Brothers Co.*)

certain parts, such as knife blades, flatware and metal sheets, the work, properly fastened at one edge, can be passed between two rolls so that both sides can be finished at one time. In common with the semiautomatic attachments previously mentioned, only one finishing operation can be performed at a time.

Regular round stock such as bar and tubing can be finished automatically on machines employing the centerless grinding principle (see Fig. 11). At each station from the coarse to the fine, rods or tubes are passed between a driving roll and a cutting wheel, the angles of which are so set that the work progresses from one station to the next propelled by the force of the revolving wheel. Set-up polishing wheels or abrasive belts running over contact wheels are used for the coarse work and buffing wheels with considerable rigidity are found at the buffing or coloring end of the series of operations.

The decision to install automatic polishing and buffing equipment depends upon a number of

Fig. 8. Indexing type rotary automatic polishing and buffing machine. (*Courtesy of Acme Manufacturing Co.*)

Fig. 9. Mush buffing machine (*Courtesy of The Harper Co.*)

factors, the most important of which is whether the shape of the article is such that it can be handled on an automatic machine. If the part is round, hexagonal or square, it can generally be handled satisfactorily. An automatic machine can produce uniform surface finishes consistently and such uniformity is difficult, if not impossible, to maintain manually. If this uniformity of finish is of sufficient importance the automatic machine may well be installed for this reason alone. The obvious factor to be considered next is the volume of work to be polished or

Fig. 10. Reciprocating table machine. (*Courtesy of Olean Finishing Machine Co., Inc.*)

Fig. 11. Centerless polishing machine. (*Courtesy of Production Machine Co.*)

buffed. If it is great enough, even though the required quality can be produced manually, it then becomes economically sound to make the capital investment necessary for automatic machines. When the uniformity of finish is paramount the quality to be produced is secondary.

References

1. Heywood, Johnson. *Abrasive Grains and their Uses.* Cleveland: Abrasive Grain Association, 1943.
2. Coated Abrasives Manufacturers' Institute. *Coated Abrasives, Modern Tool of Industry* New York: McGraw-Hill, 1958.
3. Illustrated Lecture Number 43, "Polishing and Buffing." American Electroplater's Society, Winter Park, Fla. 1978.
4. Burkart, W. and K. Schmotz. *Grinding and Polishing, Theory and Practice.* Redhill, Surrey, England: Portcullis Press, 1981.
5. *Quality Metal Finishing Guide, Mechanical Finishing.* Birmingham, Mich.: Metal Finishing Suppliers' Association, 1983.
6. *Metal Finishing Guidebook and Directory.* Hackensack, N.J.: Metal and Plastics Publications, Annual Editions.

C. Mass Finishing Methods

J. Bernard Hignett

*Vice President, The Harper Co.,
East Hartford, Conn.*

The processes available for deburring and surface conditioning are:

1. Manual
2. Mass finishing
3. Buff, brush and polish
4. Thermal energy
5. Abrasive flow
6. Abrasive blast and non-abrasive blasting
7. Chemical
8. Electrochemical.

The mass finishing processes comprise a group of deburring, edge and surface finishing techniques. When considering the best means of improving the mechanical edge and surface condition of any product, the mass finishing processes should normally be the first to be considered, being intrinsically low cost, versatile and mechanized.

By definition, the mass finishing processes are those where components to be finished are placed into a container with some form of medium, and motion of the container is provided to cause the medium to rub against components (and perhaps for parts to rub against one another), hence producing the edge and surface refinement. Most generally parts are loaded loosely into the container, but fixturing or individual compartmenting of parts can be used. The mass finishing processes include:

1. Tumbling barrels
2. Vibratory finishing machines
   A. Tub vibrators
   B. Bowl vibrators
3. Spindle finishing machines (not always considered mass finishing equipment)
4. Centrifugal barrel machines
5. Centrifugal disc machines
6. Reciprocal finishing machines.

The mass finishing processes are used for both decorative and functional finishing applications and have the following capabilities:

1. Deburring
2. Edge and corner radiusing
3. Surface improvement
4. Cleaning
5. Derusting and descaling
6. Surface texturing.

The basic advantages of the mass finishing processes are:

1. Low cost
2. Absolute consistency from part to part
3. Basic consistency from batch to batch
4. The capability of handling all metals and many nonmetals
5. The capability of handling most sizes and shapes of parts
6. Broad process capability.

The limitations of mass finishing processes are that:

1. During normal processing, action will be effective on all surfaces, edges and corners of parts exposed to media. It is not normally possible to give preferential treatment to one area compared with another.
2. Action will be greater on corners than on similarly exposed edges, and will be more on edges than similarly exposed surfaces.
3. Action in holes and recesses is significantly less than on exposed areas.
4. Some of the mass finishing processes are very slow, resulting in substantial work-in-progress and inventory.
5. Process techniques have to be finalized on a trial and error basis. This limitation, of course, applies to other mechanical finishing methods.

This article outlines the advantages and limitations of the principal mass finishing processes available today, discusses the media and compounds used in those processes and offers some guidelines for selection and use of the equipment and materials.

TUMBLING BARRELS

The original mass finishing process was barrel tumbling. Barrel tumbling, which is one of the earliest mechanized manufacturing techniques, was used many centuries ago to achieve smooth finishes on jewelry and weapons.

The barrel tumbling process comprises a container, the barrel, which is loaded with components to be deburred and finished together with media and most generally water and some form of compound. The container is rotated to cause parts and media to rub against one another as they tumble down the slope so formed by the action.

Barrel tumbling equipment (Fig. 1) is low cost, quite versatile and easy to use, and it still has many applications within industry. Because the more modern mass finishing processes are faster and frequently better, tumbling should no longer be the first mass finishing process to consider.

There are two types of conventional barrels, the horizontal and the oblique. In the oblique barrel, parts can be inspected during the process; water and compound additions can be made while the machine is in operation. For reloading, the barrel can be tilted to allow the work to flow out from the open end. The horizontal barrel is faster and more versatile; there is less possibility

1 / GENERAL PROCESSING DATA

Fig. 1. Typical tumbling barrel machine.

of parts and media migrating away from one another during the process.

Process Method

The typical operation is to load barrels approximately 60% full with a mixture of parts and media (30 to 40% full for oblique barrels). Higher loading slows down the action; lower loading wastes space and energy. For most applications, water is added to cover the load. Increasing water level provides gentler action, but slows down the process. Reduced water level increases action, but reduces cleanliness and consistency of result. The compound is normally added as a means of increasing the abrading or the polishing action while keeping the load clean, and inhibiting corrosion.

The finishing action within the tumbling barrel results from parts and media sliding down the slope formed by barrel rotation (hence rubbing against each other). Useful work is being done only during the time the components are sliding down the surface of that slope; hence the action is very slow. Speed of rotation of the barrel depends upon quality of surface and edge finish required; fifty surface feet per minute is considered very gentle, suitable for delicate parts, while 250 surface feet per minute is about maximum. Anything above this results in heavy water falling and impingement.

There are a number of modifications to the basic tumbling barrels for special applications, and to somewhat automate the process. Fitting a scroll into one end of a horizontal tumbling barrel enables it to be automatically reloaded. The drum rotates in one direction for the processing cycle, and then reverses for parts and media to be fed out through the scroll onto a screener. With the oblique barrel, abrasive compound may be rinsed out and finishing compound added during the process cycle. Perforated horizontal barrels may be used to achieve this same capability by immersing the barrels into tanks of suitable mixtures of abrasives and cleaning compounds.

Barrel tumbling equipment is low in initial investment and low in maintenance cost. Results of this fairly versatile process are consistent and operation is simple. However, barrel tumbling is time consuming and space consuming, requiring high levels of inventory and work-in-progress. More modern mass finishing processes offer faster and better means of achieving the results desired.

VIBRATORY FINISHING EQUIPMENT

Vibratory deburring and surface finishing is the standard mechanical finishing technique used by the metalworking industry. The first vibratory finishing machines were introduced in 1957, and there are more than 13,000 machines in operation in the U.S. today.

Two basic benefits of vibratory finishing are simplicity and low cost. The process is readily automated, and is suited for handling virtually any size component. Once optimum process parameters are established, quality and cost control are easily maintained.

Vibratory finishing machines are almost invariably open-topped so that the parts are readily accessible for easy checking and inspection. Addition of compound can be made during the process.

Vibratory Machines

Vibratory finishing machines consist of a container mounted on springs which is vibrated. The entire load of parts and media within the container is in motion, resulting in much faster action than that achieved in tumbling barrels. There are two basic types of vibratory equipment, the tub type and the bowl type.

Tub Type Vibrators

A tub type machine has a rectangular tub with either a "U" shaped or "inverted keyhole" shape cross-section. The vibratory motion is created either by a vibratory motor attached to the tub bottom with its shaft running along the tub, or

METAL SURFACE PREPARATION AND CLEANING

FIG. 2. Typical tub type vibratory machine.

FIG. 4. Bowl or "donut" vibratory machine.

by one or two shafts mounted eccentrically and driven by a motor (Fig. 2). Tub type equipment can handle parts of over 100-ft length, and up to 5-ft cross-section. Long tub vibrators (Fig. 3) are readily automated by loading at one end and unloading at the other. In general, tub type vibrators are considered the more versatile design with faster process capability.

Round Bowl Machines

Round style or "donut" vibrators (Fig. 4) have toroidal shaped containers positioned horizontally with the vibratory motor or vibratory shaft positioned vertically at the center of the bowl. Parts and media move around the container as they are vibrated against one another. There are a number of configurations of such machines where the base of the toroid might be stepped and it is possible to insert a dam into the bowl enabling the parts and media to be lifted out across a screen. With such an arrangement, the bowl type vibrator is lower cost and simpler for applications where such separation is effective. Bowl vibrators cannot handle parts as large as those that can be processed in tub machines, and they are somewhat slower in action. However, for applications where they are suited, they are of lower cost, more convenient and require less space. For these reasons, the bowl vibrator should be the first style of machine to be considered for any application. If frequent changes of media are required, it is important to maintain close size classification of media; if the integral separation unit is not fully effective, then tub type machines may be required.

Vibratory Operation

Process conditions for vibratory equipment are similar to those for conventional tumbling barrels. Media are selected to be sufficiently aggressive to remove burrs and radius corners and edges while leaving a sufficiently smooth surface finish. Ultimately, selection of the type of media, their size and shape are on a trial and error basis. Water level can have a radical effect: too much water dampens the action; too little will result in a dirty finish, or perhaps glazing of the media, and hence give inconsistent results. With vibratory equipment, it is desirable to have continuous flow-through of water and compound to maintain complete consistency. Compounds are used in the same way as for tumbling barrels; to enhance the abrasive action, improve surface color, inhibit corrosion and, of course, maintain cleanliness of the load during processing. Other basic process variables may be ampli-

FIG. 3. Long tub with material handling equipment.

tude of vibration and frequency of vibration, the shape and size of container and, of course, parts and media ratio.

Vibratory equipment offers a convenient and faster means of deburring and surface finishing components than does tumbling barrels. Larger components can be effectively handled and more action can be achieved in recesses. The tub type vibratory machine generally has more vigorous action, but on some styles there is a tendency for parts and media to migrate away from one another. This increases the likelihood of part on part impingement. With the donut style machine, this separation of parts from media does not happen.

The basic action of vibratory machines is a fast rubbing and tapping action of short amplitude. This precludes the ability of such equipment to handle very fragile parts or parts requiring very fine surface finishes. As with all mass finishing equipment, action is greater on edges and protrusions than on basic surfaces and recesses. Vibratory equipment is now the most common and general purpose mass finishing technique in use within industry today. When considering mechanized mechanical finishing requirements, vibratory machines should be considered first.

Vibratory Process Methods

As with conventional tumbling barrels, the standard means of deburring and finishing components is to load them into the vibratory container together with media, compound and water. There are some occasions when it is possible to finish components without the use of media. There are some components where it is necessary to place the parts in individual containers within the machine, or to fixture them.

For normal processing, the ratio of parts to media for typical applications can be along the following basis.

Ratio by Volume	Typical Application
0:1	No media. Part on part action. Used for beating off burrs. Only suitable for small regular shaped parts.
1:1	Equal volumes of media and parts. Crude operation for forgings and sand castings, applications where no precision needed and rough surfaces are acceptable.
2:1	Still severe impingement as a result of part on part action, but able to achieve adequate results on fairly hard metals.
3:1	Minimum ratio for any reasonable application on nonferrous parts, but significant part on part impingement. Fairly good surface finishes can be achieved at this ratio on hard parts.
4:1	Average ratio for deburring nonferrous parts, good for ferrous.
5:1	Good results on nonferrous mid-sized components with little part on part impingement action.
6:1	Good preplate finishes can be obtained on zinc based die castings.
8:1	High quality preplate finishes.
10:1 or more	Larger, irregularly shaped parts where fine finishes are needed.

Some Typical Applications for Various Types of Vibratory Machines

1. Hand tool forgings (Fig. 5). The finishing requirement for wrenches, sockets, plier halves, etc. is to deflash and radius edges while improving surface finish; usually prior to plating. A two-step process cycle is

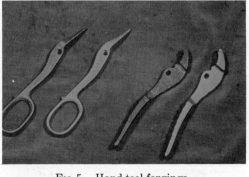

Fig. 5. Hand tool forgings.

normal, typically about four hours for a grinding operation followed by about one hour to refine surfaces. For the preliminary operation, parts to media ratio of 1:3 is acceptable, for the finishing operation about 1:6. If the variety of parts during a given day is small and no frequent change of media is needed, then two bowl type vibrators might be considered ideal, perhaps coupled with a bowl cob dryer.

2. Small plumbers brassware components (Fig. 6), where a wide variety of components is to be finished during a single shift. Tub vibratory equipment is more convenient as quick media changes can be made. These parts may require single or double operations, typically one to two hours to smooth and 15 to 20 minutes for fine surface finishing. Parts to media ratio for the first operation 1:4, for the second operation 1:8. The tub type machine shown in Fig. 2, coupled with a parts/media separator is well suited for comparatively small batches, say 1 to 2 ft^3 of parts per load.

3. Golf clubs (Fig. 7), both forged and cast, are finished in vibratory equipment. Bowl type and tub type vibrators are used depending on individual companies requirements. Parts are deflashed before vibratory finishing. Early operations are carried out with parts run loose, ratios may be as high as 1:3. There may be some intermediate polishing or buffing stages, but the final, very high quality smoothing for plating can be achieved in vibratory equipment, generally by fixturing parts within the container.

4. Die castings (Fig. 8), both zinc and aluminum, of all sizes are deflashed and deburred in vibratory equipment. Large die castings such as tractor engine components are deburred with 15-min process cycles in automated continuous flow tub type vibratory machinery.

5. Gears (Fig. 9) and sprockets of most materials and types are suited to deburring in vibratory equipment. For batch production of a limited variety of gears, bowl vibrators may be best. For large quantities of a multiplicity of different types of gears, automated tub equipment with ability to select from a number of different types of media offers considerable versatility.

The Future for Vibratory Equipment

Vibratory deburring and surface finishing will remain the standard method of deburring fairly regular shaped parts, and improving surfaces where extremely high standards are not required. As production becomes more sophisticated and labor costs increase, demands for automation will increase. There are a number of CNC vibratory machines now in operation as well as many continuous feed units. To meet requirements for improved quality of parts, improved media selection and reclassification will become more readily available. Precise control of compounds will be more generally used.

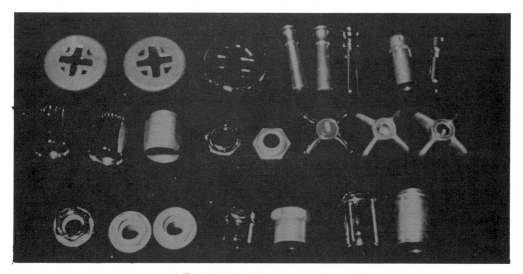

FIG. 6. Plumbing components.

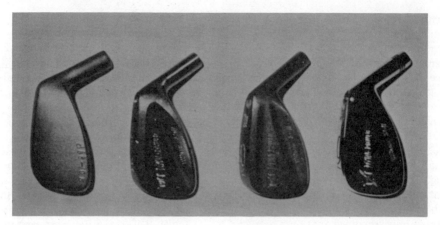

Fig. 7. Golf club heads.

While vibratory machines will remain the standard process, the share of the total mass finishing market will decrease somewhat as the high speed and high precision mass finishing processes take over. As consumer demand becomes more sophisticated, the average product becomes more complex. This may result in mass finishing having a somewhat smaller proportion of the total deburring and finishing market, ultimately, vibratory finishing will become a less important part of the technology.

CENTRIFUGAL BARREL FINISHING

Like all other forms of mass finishing, centrifugal barrels normally use abrasive media, compound and water to deburr and surface finish components. The differences between the centrifugal process and those considered so far is that CBF is a very high speed, high precision technique achieving more controllable and better finishes and having the capability to impart very high compressive stresses to the surface of components.

Principle of Operation (Fig. 10)

The centrifugal barrel machine comprises a number of drums mounted on the periphery of a turret. The turret rotates at a high speed in one direction while the drums rotate at a slower speed in the direction opposite to that of the turret. Drums are loaded in a manner similar to that for normal tumbling equipment; that is with the parts, media, water and some form of compound. Rotation of the turret generates high centrifugal force which compacts the load of parts and media into a tight mass, the load having effective weight of as much as 50 times normal gravitational weight. Rotation of the drums in opposite direction to turret rotation causes an activity of the abrasive media sliding against the parts removing burrs and refining the surfaces.

The abrading action (Fig. 11), under high centrifugal force, results in very short process cycles. A product that is lightly rubbed by an abrasive is not much effected, but the same abrasive when rubbed against components with pressure increased by 50 times will have far greater effect. Further, the increase of pressure permits finer abrasives to be used to get far more

Fig. 8. Typical die casting.

Fig. 9. Gears.

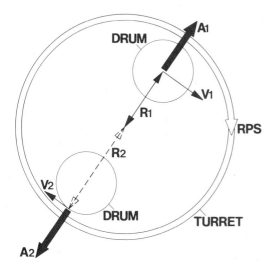

Fig. 10. Principle of centrifugal barrel process.

banging against one another. In tumbling barrels, the action is, of course, one of tumbling; in vibratory equipment, there is a short stroke rub combined with a short stroke impingement action. Because centrifugal barrels have a completely smooth sliding action, very high tolerances can be maintained on components and even the most fragile parts can be processed achieving very fine surface finishes.

Centrifugal Barrel Equipment

Modern centrifugal barrel equipment (Fig. 12), together with its material handling system is customized to meet the individual company's requirements, although there are many small standard model units. CBF machines may have turret and drums rotating in a horizontal plane or in a vertical plane. Only on rare occasions does the different configuration have any effect on process results. Rotation in a horizontal plane is used where the work containers are removable from the machine for reloading. Machines with turrets and drums rotating in a vertical plane need less floor space and are normally better suited for handling of large parts, mid-sized equipment of this type is more versatile.

uniform action over all corners, edges and surfaces as well as achieving finer finishes.

Substantially reduced process cycles are achieved in centrifugal barrels, typically between five and ten minutes compared with one to four hours in vibratory machines. This results in easier justification of capital equipment, reduced floor space in the finishing department, reduced inventories, and improved work flow. The main value of centrifugal barrel finishing extends beyond just accelerating the process cycles. The effect of rotating drums in the opposite direction to turret rotation is to create a completely smooth sliding action of media against components with no possibility of parts

Fully automated centrifugal barrel equipment (Fig. 13) is available with capacities of up to 3000 lb weight of components per load. Between four and six loads per hour from this type of equipment is quite normal. Systems are

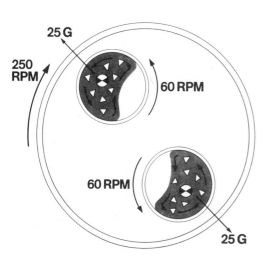

Fig. 11. Action within a centrifugal barrel machine.

Fig. 12. Typical centrifugal barrel machine.

FIG. 13. Automated centrifugal barrel machine.

available to handle a multiplicity of parts. Equipment which will automatically switch from substantial stock removal to fine finishing is normal. Loading, unloading, separation of parts from media, media reclassification and storage and the subsequent drying and passivating of parts can all be incorporated into a pushbutton or numerically controlled system.

Most centrifugal barrel equipment is built to generate forces of up to 50 times gravitational weight for very high-speed processing and total smoothness of action. For the less sophisticated deburring and surface conditioning tasks, such as edge and surface finishing of stampings, castings and die castings; optimum results can be achieved with process cycles of 10 to 15 min with centrifugal force of only about 5G. For this reason, there is a range of centrifugal barrel machines of large capacity arranged to run at comparatively slow speeds. Choice between such equipment and vibratory machines depends on size of component, batch quantity and throughput.

Special purpose CBF machines are built to meet individual deburring and finishing applications, unusual production control needs, and component size and shape. Machines have been built to finish parts weighing up to 750 lb each and as long as 4 ft. For very large parts, CBF equipment is practical only when very complex shapes or very high standards of finish and tolerance are needed; otherwise vibratory machines are more economical.

Economic considerations frequently dictate the choice between CBF and other types of mass finishing. Of course, if CBF is the only means to achieve satisfactory results because of the precision of the part, its delicacy or the finish needed; then the choice is easy. If process cycles in other mass finishing machines are lengthy or a variety of parts must be handled in a single machine, then these are applications where CBF will be more efficient and more economical.

Capabilities of CBF

The main functions are similar to those for other mass finishing processes:

Deburring
Edge radiusing
Corner radiusing and smoothing of edges
Smoothing of surfaces
Brightening of surfaces and edges
Stock removal
Cleaning
Changing stress in the surface of parts
Rust inhibiting, passivating or activating of surfaces.

CBF is a versatile process. Where a variety of components must be processed in a fairly short time and where conventional mass finishing techniques do not meet the requirement for precision control or surface finish quality, CBF may well prove uniquely suited. A unique asset of CBF is its capability to generate high compressive stresses uniformly in the surface of parts which results in improved performance of many highly stressed components. Metal fatigue is the most common cause of fracture in metal parts. Numerous repeated applications of low stress cause metal fatigue failure. This stress is much lower than that needed for fracture in a single application. The higher the stress, however, the fewer applications are needed to cause failure. It follows that fatigue failure is most common in components that are highly stressed and subjected to repeated applications of stress while functioning. A fatigue crack usually starts because the tensile stress component at the surface of the material is too high. It is thus beneficial to impart compressive stress of components to oppose any tensile stress to which the component may be subjected in service. The CBF process imparts the high compressive stress uniformly in the surface of the components while improving edge and surface condition. CBF is then able to improve performance of parts such as springs, spring clips, flapper valves, aircraft blades, bearing and instrument components to a degree that cannot be reached by other means.

Some Typical Applications for CBF

1. Precision Bearing Races (Fig. 14). Centrifugal barrel equipment is capable of generating surfaces of better than two microinches AA on bearing raceways maintaining tolerances, smoothing edges and imparting multi-directional scratch pattern for longer life and smoother running. (Typical process cycles 5 to 15 min, 3 to 6 loads/hr, which in the machine shown here is 6 to 12 batches each of $\frac{1}{4}$ ft^3 of parts, per hour.)

2. Bearing Balls and Rollers. Centrifugal barrel machines are used for both heavy grinding and superfinishing. On bearing balls, prior to hardening, stock removal of 0.001 in. within 7 to 15 min is normal, and 50% of this after hardening. On cylindrical rollers, edge radii of 0.020 in. can be achieved in 20 min, 0.030 in. in 50 to 60 min, depending on size. Needle rollers are processed in this equipment to generate the special knob-shape on the end and to grind excess metal. Superfinishing operations achieve a surface of better than four microinches AA while at the same time, generating very high compressive stress on the surface of parts and multidirectional scratch pattern for optimum performance. Equipment shown here handles 3000 lb of parts per load for the grinding operations and usually in excess of 1000 lb/load for the finishing.

3. CBF machinery processes precision machined castings, complex shaped castings of iron and steel; aluminum and magnesium. Shown here (Fig. 15) is an aircraft fuel pump body, four parts being processed at a time process cycle of 15 min. CBF achieves better finish, more uniform edge and corner radii, better results in recesses than in other mass finishing equipment. This type of CBF machine is well-suited for short production runs.

4. Jewelry (Fig. 16). CBF replaces many manual polishing and buffing operations of fine jewelry and offers very fast processing for costume jewelry. Typical process cycles run 5 to 20 min. Capable of maintaining intricate patterns and detail, smoother finishes and more versatile than conventional mass finishing; equipment shown handles six batches of parts per load.

5. Instrument Parts (Fig. 17). CBF is an ideal means of finishing small precision machined components such as those in watches and instruments, precision deburring and fine finishing. Imparting high compressive stresses results in better components than can be obtained when finishing in conventional barrels and vibrators. Typical process cycles run 5 to 15 min, 4 to 10 loads/hr with two batches of work being processed simultaneously in equipment such as that shown.

6. Other important applications for finishing by CBF include aircraft turbine (Fig. 18) and compressor blades; surgical and medical components including instruments, implants, ear molds; dental and orthodontal parts (Fig. 19); ceramic capacitors; watch components; precision stampings, castings and forgings; springs; spring clips; reed and flapper valves and carbide tooling.

The Future of CBF

As industry increases requirements for closer tolerance, finer finishes and more highly stressed components; the capabilities of CBF become increasingly important. High-energy mass finishing will have an increasing share of the equipment market. Developments of machines include smaller units, very large units, and more automated equipment including CNC. While CBF machinery is available with advanced automation including precise control and handling of media and compounds, development of continuous flow machines is unlikely in the forseeable future. Fully automatic CBF equipment is expensive in comparison to vibratory machines, and automated fast batch equipment is generally more expensive than centrifugal disc.

CENTRIFUGAL DISC MASS FINISHING PROCESS

The centrifugal disc (CD) process is a high-energy mass finishing method. It is the newest significant type of mass finishing process.

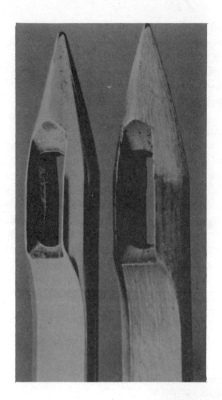

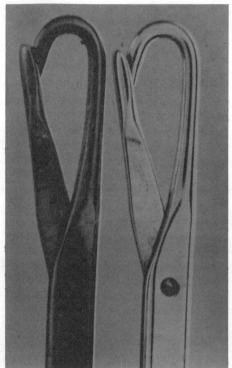

Fig. 14. Precision bearing race, before and after CBF.

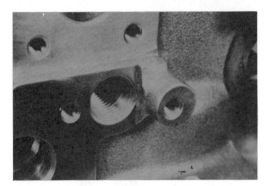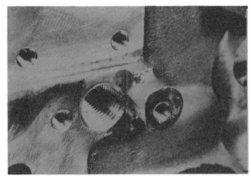

Fig. 15. Before and after precision castings.

Equipment comprises a bowl shaped container, similar to a bowl type vibratory unit but without the center column. The base of the machine is separate from the sides and forms a disc which is driven to rotate at a high speed while the side walls of the machine are stationary (Fig. 20). The side walls of CD machines are lined with urethane; the disc may be urethane or very hard wearing steel.

The gap between the disc and the sides of the machine is kept small and water (with compound) is pumped through the gap and into the machine itself; this gap must be clean and lubricated; contents of the machine must not enter into this gap.

Parts and media are loaded into the machine in similar fashion to the method of loading vibratory equipment. Rotation of the disc at high speed forces the load out against the walls of the container. The wall acts as a brake to create a sliding action between different layers of the load while high centrifugal force is exerted. Thus compared to barrel and vibrators, CD machines achieve fast processing and smoother action for most applications.

Process cycles are short, about one-tenth those in vibratory machines (generally double those of CBF). This results in the equipment being highly cost-effective and versatile.

Centrifugal Disc Equipment

Centrifugal disc machines are available with capacities of up to 20 ft^3, this being total media and parts capacity of the equipment. A major advantage of CD processing is that the bowl may be open topped so that inspection may be made while the machine is running. Of greater importance, simple, economical and highly effective automation is available.

For reloading, a door in the wall of the container may be opened, manually or automatically. With the disc rotating at a fairly slow speed; the total load is fed smoothly and quickly out of the machine where parts may be separated and media returned to storage or back into the equipment for the next load. The machine shown has the total load raised to a separating unit placed above the bowl so that media drops straight back into the work container while parts are fed out for drying or further processing.

Fig. 16. Samples of before and after CBF of jewelry.

Applications for Centrifugal Disc Equipment

CD equipment can be used for a wide variety of processes, but primarily for the following:

1. Descaling
2. Deflashing
3. Deburring
4. Radiusing
5. Metal removal
6. Surface improvement.

1. Descaling. Heat treatment and forging scale can be removed from parts in process cycles of 3 to 5 min duration. With the machine set to give an aggressive cut, all component parts are scale-free in this short process time. A big advantage is that it is not necessary to use an acid based compound to achieve this result.

2. Deflashing. The machine can easily remove flash (and sprue) from zinc based and aluminum die cast parts. A typical application is for die cast toy automobiles. The body of the automobile had the sprue and some of the heavier flash removed by hand, then run in a vibrator to remove the remainder of the flash. This operation involved the use of

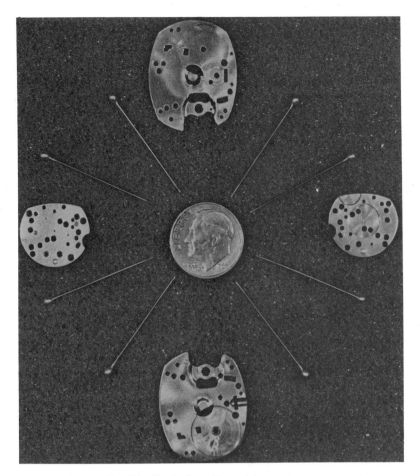

Fig. 17. Typical small precision parts.

eight operators and a 10 ft³ capacity bowl vibratory machine. Today, one operator and a 6 ft³ centrifugal disc machine will do the job with production capacity to spare.

3. Deburring. The CD machine will remove tough burrs in faster time cycles than vibratory finishing machines, and in some cases, will remove burrs that cannot be removed by those machines. For example, deburring of the milled out area between the prongs on universal joints—a 45-min process; a further 15 min produces a preplate finish.

4. Radiusing. A measurable, controlled radius can be easily produced from CD machines. On hardened steel parts, it is possible to produce a 0.007 in. radius in a 15-min cycle time. On aircraft brake parts, a radius of 0.030 in. is achieved in a process time cycle of 25 min. Burner ring fingers used in domestic cooking stoves are CD processed for 45 min to achieve uniform 0.040 in. radius.

5. Metal Removal. The removal of metal from the surface of a part is not a common use of CD machines; however, it can be achieved. The tool industry frequently requires removal of metal from oversized, hardened drills; 0.005 in. can be removed from the diameter of the drills in a 2-hr cycle.

6. Surface Improvement. Surface improvement includes both preplate finishing and microinch improvement. Zinc based die cast bathroom fittings are made ready for plating with a 15-min process cycle. A turbine blade has surface improved from 45 microinch AA to 10 to 12 microinch AA with a 40-min cycle.

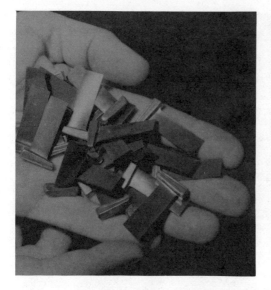

FIG. 19. Dental/orthodontal parts.

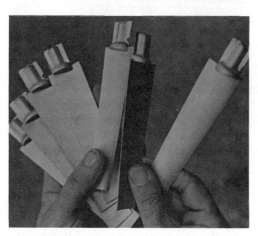

FIG. 18. Aircraft turbine blades.

Parts To Media Ratios

Ratio by Volume
- 1:5 Standard ratio for good results without any impingement.
- 1:3 For heavy burr removal and radiusing of small nonferrous parts.
- 1:1 For heavy burr removal and radiusing of small ferrous parts.

The Future

The centrifugal disc process is new to the U.S. metalworking industry. It is versatile and can prove highly economical compared with both centrifugal barrel and vibratory machinery. In Europe and Japan, it has found acceptance for a broad range of applications and, thus, deserves greater attention than it is currently receiving.

SPIN FINISH OR SPINDLE FINISHING PROCESS

Spin finishing is frequently not included as one of the mass finishing processes as components are always fixtured to deburr and finish. However, it is proper to consider the spindle method of deburring with vibratory, CBF and CD techniques, for they all rely on parts to be immersed in mass media.

Spin finishing is a high-speed pressure type process characterized by the workpiece being fixtured and held within a high-speed flow of abrasive media. Work cycles are short and the possibility of part-on-part impingement damage is nil.

Equipment comprises a bowl containing the abrasive media, and one or more heads with spindles at the end of them. Components are mounted on fixtures at the ends of the spindles.

The most standard type of machine has the bowl rotating at high speed to compact the media mass around its periphery. Fast processing results. Other equipment is available with stationary bowls, some with vibratory ones. The head may be stationary, rotating or oscillating, as may the spindle. The spin finishing machine

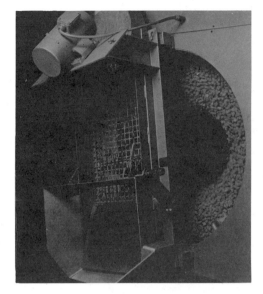

FIG. 20. Action within a centrifugal disc, high-energy machine.

may have one or more heads, and each head may have one or more spindles to hold work pieces.

Spindle Finishing Equipment (Fig. 22)

Standard equipment is built with bowl diameters ranging up to 6 ft. Heads and spindles are built to handle work loads up to 100 lb.

Bowls should rotate to create as much as 1800 surface ft/min for heavy deburring operations, although speeds as low as 200 surface ft/min are used for fine edge break and finish applications. For large work pieces, vibratory motion of the bowl may help avoid excess resistance to work piece movement and strong directional action.

Manipulation of the work piece within the mass of media depends upon shape and size of parts being finished and, of course, the finish requirement. Action with more than one component on a spindle or more than one spindle on a head effects action, so some degree of trial and error testing is required. Equipment with variable speed rotation and capability to oscillate and present work at different angles within the bowl are all beneficial.

FIG. 21. A centrifugal disc machine and control panel.

FIG. 22. Typical spindle finishing equipment.

Process cycles ranges from as little as 15 sec to several minutes.

Applications

Spindle finishing equipment is only suited to processing parts that can be readily fixtured to expose all significant surfaces and edges to the finishing media. Parts generally have to be regularly shaped. Production quantities must be sufficient to justify the cost of fixturing.

Ideal applications are for gears and sprockets. Equipment is capable of handling high-precision parts and will improve the surface condition of gear teeth at the same time as removing burrs. Carbide tool tips can have edges radiused in a spindle machine; aircraft turbine blades can have edges radiused and surfaces finished.

The Future

Spindle finishing machines in several sizes and types have been available for many years, and applications have been carefully studied and established. Compared with other mass finishing processes, the basic disadvantage is that parts must be fixtured. Further, fixtures are subject to wear by the abrasive media.

It seems reasonable to expect then that spindle finishing equipment will maintain its share of the mass finishing business, but is unlikely to become one of the standard and versatile deburring and finishing methods.

OTHER MASS FINISHING PROCESSES

Processes not yet discussed include reciprocal finishing, chemically accelerated vibratory and CBF processing, combination action equipment such as vibratory tumbling barrels, and electrochemically accelerated mass finishing. These methods are highly specialized and unlikely to be of any general significance in the short term.

MASS FINISHING MEDIA AND COMPOUNDS

The choice of consumable materials used in any mass finishing process is critical for optimum results. The choice of media ranges from natural stones, manufactured abrasives which may be bonded into either plastic or ceramic matrices, and randomly shaped manufactured abrasive products, to metallic shapes and agricultural products. Unfortunately, the choice of media must ultimately be determined on a trial and error basis. Some basic guidelines should be followed before starting trials, but no exact criteria exist to select best type, size or shape of media. To complicate the situation, few standards exist for any type of medium, so that selection of similar types from different manufacturers requires extensive testing, followed by checks on the quality control of the media during its manufacture. Manufacturing variations of those media account for vast differences in performance.

While a simple reference chart is not feasible

to guide the user of mass finishing equipment to the best materials for his purpose, an outline of the different available media with a description of their properties can offer some direction. Fortunately, during the past three years, not only have significant improvements in performance and consistency of manufactured media been developed; but efforts to establish universal standards have been made. Many quantitative evaluations of media have been carried out and results have been published. Several major suppliers offer more precise and comprehensive data to assist in choice of the ideal media. A brief description of the more important types of media is as follows.

Natural Abrasives (Fig. 23)

Most natural stones have been used for tumbling media, but are of significantly reduced importance as mass finishing techniques become increasingly sophisticated. Natural media are less consistent than their manufactured counterparts and, or course, occur only in random shapes. Natural stones, such as limestone and granite, are usually softer and, therefore, faster wearing than the manufactured materials. Many are liable to fracture and create expensive lodging problems. As quarrying costs increase and use of manufactured materials increase, much less economic incentive exists to find the suitable natural stone.

Nevertheless, natural materials still have valuable applications. Corundum, a natural aluminum oxide, is available at prices comparable to standard manufactured fused aluminum oxide nuggets and offers an alternative where slightly improved surface finish are desired and less abrasion is acceptable. As it wears down, novaculite maintains a sharp and flaky particle shape which can be valuable for removing burrs from fine holes and grooves. Limestone and granite may still have use in conventional tumbling barrels. Natural abrasive grain still has economic value for use in the manufacture of bonded preforms.

Agricultural Products

Products such as ground corncobs, ground walnut shells and sawdust are used in heated tumbling barrels for drying purposes. The wiping action can apply final maximum luster to some parts. When mixed with fine abrasive, such as rouge and alumina, these materials are capable of smoothing surfaces and polishing intricately detailed components with little corner radiusing. Extremely high luster on parts such as jewelry can be achieved.

Manufactured Abrasive Media

Fused aluminum oxide is the most standard random-shaped tumbling medium used today by industry for mass finishing. Most of these media have substantially greater abrasion properties than any natural stone, while having good wear characteristics. Consistent quality, uniformity and size can be obtained. This enables consistent quality finishing more economically and, generally, faster than can be achieved with natural stones. Some harder forms of these media are available and capable of producing smooth finishes and high color.

Fused silicon carbide abrasive media in small nugget size are also available, but are too fragile to be used in sizes above about $\frac{1}{4}$ in. diameter. When components have to be brazed or welded, aluminum oxide cannot be used because it may leave residual particles in the surface of the components which can have a detrimental effect on the final component. Silicon carbide media are mandatory for such applications.

FIG. 23. Natural abrasive media.

Fig. 24. Preform media.

Preform Media (Fig. 24)

The synthetic preformed shaped media have wider cutting range than natural media and fused aluminum oxide. The bonds may be ceramic or plastic with various types of abrasive added and may be formed into a wide range of shapes and sizes for fast or slow cutting. The versatility of synthetic media together with the development of equipment has resulted in mass finishing processes becoming precision tools for metal removal.

Ceramic Bonded Media

Ceramic bonded media is manufactured by vitreous or fusion bonding. Porcelain or other vitreous material is used by itself or as a matrix into which the abrasive grains are bonded. The abrasives are quartz, aluminum oxide or silicon carbide, abrasive content in the media is up to 50% and the size of abrasive grains is between 600 and 60 mesh.

Resin Bonded or Plastic Bonded

Resin and plastic bonded media contains up to 70% by weight of abrasive bonded with either polyester or urea formaldehyde resin. Plastic media are formed by casting—sometimes by extrusion—and are available in similar shapes to those of the ceramic media.

The major difference between ceramic and plastic bonded media is that the plastics are softer and have lower bulk density. Because of the softness, plastic media give greater uniformity of metal removal on corners compared with edges and surfaces. The low density results in a good cushioning effect, resilience and a milder abrasive action. Plastic media are used for deburring soft alloys, such as zinc based die castings and aluminum, and for precision finishing of intricate and complicated delicate parts. Plastic media have bulk density between 50 and 65 lb/ft^3 compared with ceramic bonded media at 85 to 95 lb/ft^3. Ceramic media, being harder, will have greater abrasion and are more economical because the ratio of metal removal to rate of wear of media is better.

Shape and Size Considerations

The first consideration for proper media selection is to choose material which reaches into all areas of components to be finished without wedging into the work pieces. The second rule is that the larger the media, the faster the abrasive action, but the greater variation between metal removal at edges, corners and surfaces. The smaller the media, the greater uniformity of metal removal over all surfaces and edges.

When selecting a specified sized medium; the user should remember that different manufacturers have different nomenclature, not only for the type and shape of the media they produce, but also size. One-half inch triangles from one manufacturer may be significantly different in size from another's one-half inch triangles.

The shape of preformed media is of great importance. Shapes available include triangles, cones, cylinders, diamonds, arrow heads, stars, spheres, some with edges cut square and some with edges cut at angles to the faces. Cylinders, cones and spheres roll over surfaces and, therefore, should have more effect on edges and corners than would triangles. These shapes are chosen to deburr parts with a variety of holes. Triangular shapes give greater action on the surfaces of parts relative to edges and corners and will deburr slots better than other shapes. An important variable which is shape dependent is the effective pack density. Media are purchased by the pound, and denser media are liable to be expensive to use. Sometimes when media are of different density from the parts, parts will migrate together, causing part on part impingement.

Triangles are the most standard preformed media because they keep their shape during the useful life and reach into corners well. The angle cut triangle produces deeper penetration into remote areas and allows the media to retain a sharp edge throughout their life.

Metallic Media

Hardened steel balls and steel shapes are widely used for burnishing operations helping achieve maximum luster for primarily decorative finishing purposes. Steel media are of considerable value having uniform size and shape without fracturing in use and, therefore, without lodging in recesses. Tacks and nails have a unique ability to pick out burrs from small holes and recesses when used as media with a loose abrasive compound. Zinc shapes are used for burnishing applications as an alternative to steel.

Considerations When Choosing Media

1. Capability to remove burrs, achieve required radius, achieve required surface finish and produce desired action on edges, corners and surfaces relative to one another.
2. Optimum parts to media ratio. This is likely to be a function of media density and particle size as well as type.
3. Short process cycle.
4. Size, type and shape that will not jam in holes and recesses.
5. Non-fracturing.
6. Minimum wearing rate for lowest cost and minimum amount of reclassification to avoid lodging.
7. Ready availability of consistent quality.
8. Capability of handling the range of products to be used within a given machine.
9. Correct density to avoid parts migration, minimum part on part impingement and maximum economy. Capability, with appropriate compound changes, of combining an abrasive operation followed by a surface refinement operation.
10. Requiring minimum break-in requirements, that is, achieving desired degree of abrasion and surface refinement when new and when somewhat worn.
11. Shape and size to reach into all areas needing to be processed.
12. Available from a supplier who is able to advise on capabilities of the material to avoid minimum trial-and-error testing.
13. Economical price per pound.

Testing of Media

When testing media to establish the best type for a specific purpose, the following variables should be understood and perhaps measured:

1. Initial condition of media. Some media are likely to have very sharp edges and a number of fragments. Tested media should always have a preliminary run and then be screened to establish the amount of waste and amount of inconvenience this initial trash can be.
2. Quality control of media. Excessive porosity or bubbles, cracks or sharp edges detract from the quality of media. Plastic media may be undercured resulting in poor cut, or overcured resulting in irregular results. Variations of ceramic media vitrification creates variable results of those media. For these reasons, testing samples should be taken from a number of batches. It is worth checking specific gravity of each batch and making visual inspection for cracks, pores or other defects.
3. Equipment in which media are to be used is important. Media that perform well in high-energy equipment might have no action in a bowl vibrator, while that which is excellent in a bowl vibrator might fracture and break down in some high energy machines.
4. Rate of cut of media.
5. Rate of wear of media.

The combination of the above is the measure of media efficiency. When comparing media efficiency, compare the same shape, size and stage of wear. A number of excellent articles describing how media basic efficiency measurements should be made to obtain proper comparisons are available. These basic guidelines might need modification for specific applications.

6. Speed of deburr and edge radius. Generally, the rate of cut and the speed of deburring are similar, although it depends upon type of burr. Sone parts may be deburred by knocking off the burr rather than grinding off.
7. Capability to improve surface finish. Each type of media will have some specific surface finish improvement capability, depending upon the type of equipment, speed of equipment, size of media and compound used. When deciding the surface finish improvement needed, it is normally not sufficient to consider only in terms of micro inches, RMS or AA; usually consideration of the particular application will be rele-

vant. Brightness, smoothing of defect and the type of scratch all have to be considered.

Mass Finishing Compounds

For mass finishing operations where parts are deburred and finished with media or by part-on-part action, the process is carried out in an aqueous solution, and a suitable compound with the water allows for better control of the process and can enhance the action. Compounds may be powder or liquid. They may be abrasive or non-abrasive. The most important function of the compound is to keep both parts and abrasive media clean, to prevent the media from becoming loaded with contaminant or glazing, and to insure consistency of action throughout the whole process as well as uniformity of result from one load to the next.

Compounds are needed to maintain cleanliness for deburring and surface finishing. With the development of improved equipment and these compounds, use of mass finishing has become attractive for many cleaning operations. A good cleaning compound coupled with extremely efficient scrubbing action, no requirement to heat solution and parts makes most mass finishing more effective than other forms of scrubbing and cleaning.

Other purposes for compounds are:

1. *Conditioning of water.* Compounds should contain water softening ingredients. Most commercially available compounds are suited to any standard water supply in the U.S. For hard water, it may be worth considering modifying standard compounds to increase the level of water softening agents. If this is not done, overdose the solution with compound to get sufficient water softening.
2. *Cushioning parts.* Some compounds create high foamimg action which forms a cushion. High foaming compounds must not be used in vibratory equipment, but in conventional tumbling barrels and occasionally in centrifugal barrel equipment. The cushioning effect improves fine finishing capabilities.
3. *Corrosion inhibition.* Some compounds offer temporary rust inhibition after processing, sufficient for temporary storage or movement to a subsequent operation. More important, compounds must protect parts from corrosion during the process cycle.
4. *Improving luster or color.* This is largely a function of cleanliness, ability to remove oil from parts. Some compounds have chemical coloring and brightening action.
5. *Increasing abrasion.* Compounds may have added free abrasive (powder compound only). Some non-abrasive compounds can enhance the cut of media, action being primarily one of cleaning to keep the abrasive bonded in the media open.
6. *Descaling.* Both strong acid and strong alkaline compounds can be used, the alkaline for ferrous metals only. Acid compounds should not be used in the closed containers, because a chemical action could lead to an explosion. Modern equipment relies on mechanical action to descale rather than chemical.
7. *Burnishing.* Compounds for burnishing are mildly alkaline mixtures with good lubricating action. Different compounds may be used for different metals.
8. *Defoaming in compounds.* These are desirable for processing in vibratory equipment where heavy foaming is both messy and reduces or stops the action.

Choice of compounds should be made considering OSHA and EPA requirements. Changing of compound from an abrasive cutting type to fine finishing can permit two operations to combine without need for unloading parts nor changing media.

Some Operating Considerations

1. Equipment. The finishing department in every metalworking plant will be used as the general rectification shop. Therefore, versatility of equipment should be considered as for most other fabrication equipment. Most mass finishing machines are inherently versatile.
2. Material Handling. Equipment should separate parts from media. The most economical means of doing this is with screening units. Media will need reclassifying as it wears down, and the screening equipment used for media reclassification may also be suitable for parts separation.
3. Washing and Drying. After unloading from the mass finishing equipment, it may, on occasion, be necessary to rinse parts, then dip in a rust preventative or oil. Parts may be dried in centrifugal dryers, hot air blowers, corncobs or sawdust; but it

is unlikely that air drying will be acceptable.
4. Media and Compound Storage. Clean storage facilities must be available for the consumable materials. Accidental mixup should be impossible.
5. Process Instructions. In most organizations, the operator no longer decides proper process technique nor judges quality. Proper process instructions governing all variables can and should be issued. This is still not normal practice, even in some high-technology corporations. Having established the type of equipment to be used, the media and the compounds and the other operating variables are not difficult to list. These are as follows:
   a. Speed. Most modern equipment offers variable speed, amplitude, frequency, stroke or centrifugal force. Equipment manufacturers list reasons for speed change and, for most equipment, precise determination of this variable is essential for optimum process efficiency.
   b. Water level. This is important in mass finishing equipment because too little water causes dirty work, glazing of media and incorrect action within the machine. Too much water slows down the action. In general terms, the more water, the more gentle the action.
   c. Load in the Machine. In conventional tumbling barrels and centrifugal equipment, the level of the total load in the drums effects the process time and smoothness of action. For consistent results, it is necessary to specify the proportion of parts to media. For maximum economy and efficiency, the more parts the better. Too many parts cause damage to the components. Other than cost considerations, rarely is there any disadvantage in having too few components in a load. It is preferable to mix parts and media before loading into some equipment, or to load abrasive and components in layers. Initial mixing by running the equipment slowly may be accomplished.
6. Control of Incoming Quality. Few manufacturers control the size of burrs produced by their cutting machinery. Therefore, inspection of parts before loading into mass finishing equipment is recommended. Control of the amount of foreign matter, such as oil and swarf, is also important.
7. Cycle Time. This is self-explanatory. For some operations, adding water or compound during the process cycle is desirable and this too should be specified.
8. Compound Quantity. For many mass finishing machines, continuous flow of compound through the load is practical. If there is a flow system, maintaining a careful check of the compound quantity in the reservoir is important. For batch addition of compound, quantity must be specified to avoid variation in quality of components processed and to control costs. In continuous flow, the flow rate must be specified.
9. Masking and Fixturing. It is sometimes advisable to plug holes in components with rubber or plastic to avoid media being jammed into those holes during processing. Edges and corners that must be left sharp may be masked. Some components are best processed on fixtures in mass finishing equipment and, of course, fixturing is an essential part of spindle finishing equipment. Design of fixture and position of the component within the machine are variables which must be considered.
10. Cleanliness. Most processes tolerate some oil, scale and dirt on components; mass finishing is used to clean parts. Some operations, particularly fine finishing applications demand utmost cleanliness and, on occasion, even the hardness of local water creates problems in achieving the brightest surfaces. It is necessary to know what type of dirt settles on components coming into the finishing department as proper compounds and process methods can be selected. The finishing department must be notified of any changes in prior operations (for example, changes of heat treatment method) which effect the amount and type of soil that must be handled.
11. Maintenance of Records. Because standards and specifications are not clear, many mass finishing operations frequently have poorly maintained standard operating procedures compared to those for metal-forming operations. Process sheets are, of course, important. Because of the variety of variables, it is not wise to leave this task to the equipment operator or his immediate supervision.

FUTURE TRENDS AND REQUIREMENTS

Mechanical edge and surface conditioning technology is still somewhat neglected. The cost of deburring and surface finishing most metal components exceeds 5%, yet the average component manufacturer still invests only 1% of total capital into improved deburring and finishing methods. Thus, greater opportunities for improvement in the field of mechanical finishing exist than in any other phase of the production cycle.

Of the deburring and surface conditioning processes, mass finishing is the most used. It is also where developments are fastest. Not only is equipment for efficient and high quality finishing more available, automated systems are being designed and built for highly sophisticated purposes, and the consumable materials are becoming ever more consistent, categorized and efficient.

Education of the manufacturing engineer in the field of mechanical finishing, particularly in the field of mass finishing, is still somewhat neglected. Improved education will enable the metalworking industry to take substantial strides forward. This will create greater incentive for research and development in the industry. Mass finishing processes will remain a vital part of the industry for several decades.

Reference

1. Weiss, H. J., *Electrodepositors' Tech. Soc.*, **24**, 171–180 (1949).

General References

Mitchell, R. W., "Metals Handbook," 301–305 (1948), American Society for Metals, Cleveland, Ohio.

Bregman, A., "Economic Possibilities in Tumble Finishing," American Society of Tool Engineers preprint, presentation at Nineteenth Annual Meeting March 16, 1951, New York, N. Y.

Stone, Wayne, *Iron Age*, **172**, 117–121 (Sept. 24, 1953).

Chase, H., *Metal Finishing*, **46**, 56–60 (Apr. 1948).

Wingate, J. E., *Metal Finishing*, **46**, 67–74 (Oct. 1948).

Wingate, J. E., *Metal Finishing*, **47**, 55–61 (May 1949).

Charleson, J. E., *Metal Finishing*, **47**, 51–52 (Nov. 1949).

Enyedy, R., "Handbook of Barrel Finishing," Reinhold Publishing Corporation, New York, 1955.

Letner, H. R., "Special Report on Abrasive Tumbling," Mellon Institute of Industrial Research, Pittsburgh, Dec., 1956.

D. Electropolishing

Dean E. Ward, President

*Haward Corp.
No. Arlington, N.J.*

Electropolishing, an electrochemical process, is the reverse of electroplating. Therefore, metal is removed rather than deposited. The article to be electropolished is made the anode in an electrolyte which, when low voltage is applied, forms a polarized film over the entire surface. This film is thickest over the microdepressions and thinnest over the microprojections. Where the polarized film is the thinnest, electrical resistance is the least and therefore the rate of metallic dissolution is the greatest. Electropolishing selectively removes the microscopic high points much faster than the rate of attack on the "valleys" or microdepressions. Stock is removed as a metallic salt. Stock removal is controllable and can be held to 0.0001 to 0.0025 in. However, stock removal can be much higher if desired, and under stringent conditions can be held to a lesser amount. It is believed that the polarized film is responsible, under proper operating conditions, for brightening and smoothing the metal surface. It must be pointed out, however, that brightness and smoothness do not always go together. For instance, a welded or rough ground piece may be bright but not smooth. Conversely, a lapped surface may be smooth but not particularly bright.

When a metal surface is polished, the light rays are reflected in parallel lines, thus the surface acts like a mirror. On a rough or unpolished surface, the light rays are reflected in a random pattern; therefore, no image can be seen (Fig. 1).

To obtain a smooth, highly reflective surface on various metals, several factors come into play. The degree of successful electropolishing is determined, to a large extent, on the surface conditions of the base metal. Poor conditions for electropolishing include nonmetallic inclusions, overpickling, heat scale, large grain size, directional roll marks, insufficient cold reduction or excessive cold working. Several of these conditions may be inherent in the metal as it comes from the mill. During electropolishing, metal is removed, revealing these flaws. When metal is removed, smoothing can take place. Other factors contributing to a smooth, highly reflective surface are change in bath chemistry and precipitation of metal salts. When bath chemistry is out of balance, poor polishing or no polishing can result or an extended time may be necessary to obtain a satisfactory finish. When salts build up in the bath, higher voltages are required to maintain the desired current density. This can result in burned contact points.

The Electropolished Surface

Comparison With Mechanically Finished Surface. The absence of scratches, strains, metal debris, and embedded abrasive characterizes the electropolished surface. It has the "true" crystal structure of the metal undistorted by the cold working that accompanies mechanical finishing methods. The comparison is shown by Figs. 2 through 14.

The unaided eye does not see the difference shown by Fig. 2 and 3. (Both have a microinch finish to 0.1 and 0.5 rms.) It sees the two surfaces in terms of light reflection as shown in Table 1. Therein, a surface like that of Fig. 2 is shown to give the expected sharp clear image reflection and directionality; whereas the surface like that of Fig. 3 gives greater total reflectivity, has no directionality, and gives almost as good image definition.

The surfaces of Figs. 2 and 3 differ in other respects. Finishing by abrasives or other cutting or burnishing action, regardless of how small the amount of work, always distorts the surface. Figure 4 shows a carefully ground surface with a microinch finish of 3 to 5 rms. The important fact is the unavoidable presence of those rough

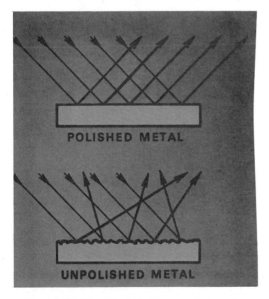

FIG. 1. Light reflected from metal surfaces. (*Courtesy of American Electroplaters' Society, from AES Illustrated Lecture Series on electropolishing.*)

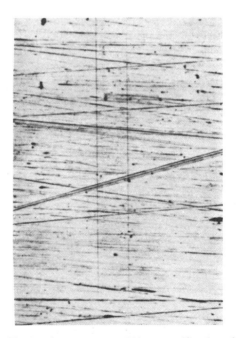

FIG. 2. Appearance at 100× magnification of a good mirror finish by abrasive-finishing methods. (Faust, C. L., *Proc. Am. Electroplater's Soc.*, 137-150, 1950).

FIG. 3. Appearance at 100× magnification of a good mirror finish by electropolishing. (Faust, C. L., *Proc. Am. Electroplater's Soc.*, 137–150, 1950.)

FIG. 5. The same surface as Fig. 3 after electropolishing. Also shown at 10,000× magnification. Typical faintly wavy surface, showing tiny crystalline protuberances.

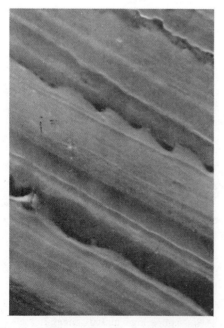

FIG. 4. Surface-ground steel at 10,000× magnification. Note loosely attached fragment in left foreground and rough ridges.

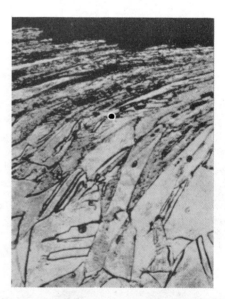

FIG. 6. Shear and tearing produced by mechanical action on a metal surface. (Steer, A. T., Electrodepositors' Tech. Soc., Birmingham Symposium, October 4, 1949.) 200×
(a)—depth of distortion

FIG. 7. The surface of Fig. 4 with insufficient electropolishing to remove all surface damaged metal. Shown at 10,000× magnification.

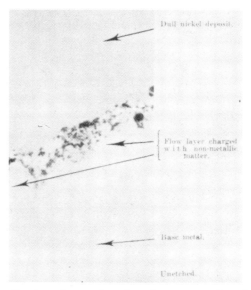

FIG. 9. Abnormal surface layer contaminated with nonmetallic matter retained during polishing and cleaning. (Steer, A. T., *loc. cit.*) Shown at 3,000× magnification.

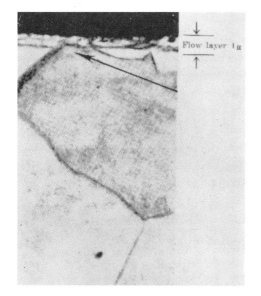

FIG. 8. Flow layer from polishing still intact after 300 hrs annealing at 500C. The flowed layer was originally striated by severe burnishing action and has recrystallized but is not absorbed. (Steer, A. T., *loc. sit.*) Etched by immersion technique. Shown at 3,000× magnification.

FIG. 10. A ground surface of steel after some burnishing. Shown at 10,000× magnification. Evidence of metallic debris in the grooves.

FIG. 11. Surface view of 8-min. nickel plate on the metallographically polished surface of stainless steel. (Pray, H. A., and Faust, C. L., *Iron Age*, April 11, 1940.) Shown at 100× magnification.

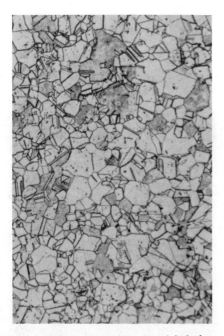

FIG. 12. Surface view of 8-min. nickel plate on the electropolished surface as shown in Fig. 3. (Pray, H. A., and Faust, C. L., *loc. cit.*) Shown at 1000× magnification.

FIG. 13. Nickel plate showing fine, abnormal structure because of the disturbed surface on mechanically (metallographic) polished stainless steel. (Faust, C. L., *J. Electrochem. Soc.*, 95, 62c, March, 1949.) Shown at 500× magnification.

ridges and loose clods of metal. Those defects are absent after electropolishing, as seen in Fig. 5.

The differences shown by Figs. 2 versus Figs. 3 and 4 versus Fig. 5 are much more than topographical. The cold-work damage penetrates into the metal and the abrasive is embedded in the surface. The shearing and tearing is shown by Fig. 6. It reveals in cross-section what are probably the rough ridges seen in Fig. 4. The depth of damage is also revealed in Fig. 7, which shows the surface of Fig. 4 without enough electropolishing to remove all the scratches and rough-ridge metal.

Much is known about heat treating metals to relieve them of cold-work effects. Figure 7 shows

FIG. 14. Coarser structured nickel plate on electropolished stainless steel. The nickel continues some of the stainless steel crystals and reveals total absence of mechanically disturbed surface. (Faust, C. L., *loc. cit.*) Shown at 500× magnification.

TABLE 1. LIGHT REFLECTION FROM POLISHED STAINLESS STEEL (TYPE 302)

Finish for Smoothness	Reflection			
	Diffuse		Specular	
	Transverse	Longitudinal	Transverse	Longitudinal
No. 8 Mill finish (Fig. 1)	0.164	0.068	103	104
No. 7 Mill finish	0.241	0.020	117	118
No. 2 Mill finish after electropolishing[b]	0.072	0.042	120	121
No. 2 Mill finish after electropolishing[b] (Fig. 2)	0.043	0.023	124	123

[a] Averages of five readings.
[b] Two different electropolishing methods.

TABLE 2. SMOOTHNESS VERSUS POLISHING METHOD AND KIND OF STEEL

Kind of Steel	Polishing Method		Profilometer Reading[b] (RMS)	
	Mechanical, abrasive belts[a]	Electro (time, min)	As mechanical polished	As electropolished
High-carbon SAE 1080	150[c]	3	16.5	21.0
	150[c]	5	16.5	13.0
	150[c]	10	19.0	5.7
	150[c]	10	13.5	4.0
	Sandblast	20	140.0	16.5
Low-carbon SAE 1020	180[d]	3	23.5	19.0
	180[d]	4	27.0	21.5
	180[d]	10	29.5	12.5
	180[c]	3	14.5	12.0

[a] Number shown is screen size of grit on the belt.
[b] Same locations as accurately as possible before and after mechanical and electropolishing. Average of several readings. All RMS units, where stated in this paper, are microinches.
[c] Hand polishing.
[d] Automatic-machine polishing.

that the superficial damage done to a metal surface by mechanical finishing does not necessarily disappear as a result of heat treatments known to restore gross properties. Nor will heat treatment (or any other method) remove embedded and smeared-over abrasive particles, which are revealed in Fig. 13.

Burnishing by lapping, buffing, or coloring decreases the microinch roughness and improves the image-defining quality of a surface, but never completely removes the debris and damaged metal. Figure 10 shows a burnished surface with debris still in the remaining, but shallower, scratches. The minimum of cold-work and mechanical damage is done in metallographic polishing, yet the effects are there, as shown by Figs. 11, 12, 13 and 14. The fine abnormal structure of nickel plate reveals the presence of the cold-worked surface. The nickel plate in Figs. 12 and 14 clearly reveals the true, undisturbed metal in the electropolished surface.

The mechanical strength of the surface metal is lowered by cold work accompanying simple cutting operations. Machining of steel having 100,000 psi tensile strength can leave a surface skin of worked metal having only 35,000 psi tensile strength[9]. This might be expected from an examination of Figs. 4 and 6.

Advantages and Limitations. The preceding discussion shows some of the *positive technical advantages of electropolishing*. It shows why electropolishing produces a surface with good properties for: receiving electroplates having better smoothness, better appearance, and, because of fewer voids, better corrosion protection; resisting corrosion when there is no plate or other coating; more uniform anodizing, phosphating, black oxidizing, and other conversion coatings; greater reflectivity of light and heat; better emissivity in electronic tubes; and wear against other metal surfaces without loose metal fragments to cause fretting.

The surface damage and cold-work effects are overlooked if only smoothness is specified as a criterion of quality in a metal finish. As the preceding figures show, smoothness is not an independent variable in surface definition. It is only a part of an important subject that can be called "surface metallurgy." Smoothness specification, according to gages, can be met by electropolishing (with or without plating as an adjunct) just as well as by abrasive polishing. This was shown to some degree by Table 1. It is further shown by the data in Tables 2, 3, and 4. Electropolishing, according to these data, is not influenced by the hardness or plasticity of the metals, whereas abrasive methods are.

The principal limitations of electropolishing are: the process cannot smear over and cover up defects such as seams and nonmetallic inclusions in metals; multiphase alloys in which one phase is relatively resistant to anodic dissolution are

TABLE 3. SMOOTHNESS OF STEEL (SAE 1010) SURFACE BY MECHANICAL POLISHING, ELECTROPOLISHING, AND ELECTROPLATING

Polishing Method				Profilometer reading (RMS)			
Mechanical abrasive set	Nonmechanical (mils metal)			As belt polished	After electropolish	After nickel plate	After electrobuff
	Electropolish	Nickel plate	Electrobuff[a]				
120	0.3	2.0[b]	0.5	48	37	33	27
120	0.6	2.0	0.5	52	34	26	21
120	0.9	1.8	0.3	48	25	24	22
150	0.3	2.0	0.5	38	27	22	20
150	0.6	2.0	0.5	43	28	25	21
150	0.9	1.8	0.3	47	19	17	16
180	0.3	2.0	0.5	33	23	19	15
180	0.6	2.0	0.5	37	25	20	16
180	0.9	1.8	0.3	36	20	18	17
240	None	1.8	0.3	20	None	19	17
120	0.6	2.0[c]	0.5	45	28	28	22
150	0.6	2.0[c]	0.5	37	23	27	22
180	0.6	2.0[c]	0.5	26	17	19	16
240	None	2.0[c]	0.5	19	None	25	20

[a] Electrobuff is a term used to designate the electropolishing of an electroplate as distinct from the electroplishing of the basis metal. Both electrobuffing and electropolishing remove metal. The nickel-plating operation adds metal.

[b] The group of plates was bright nickel; a mirror-like smooth electrodeposit.

[c] The group of plates was dull nickel, appearing smooth, but lacking luster.

usually not amenable to electropolishing; heavy orange peel, mold-surface texture, and rough scratches are not removed by a practical amount of electropolishing and require "cutting down" first, just as needed before buffing and coloring. This last situation can be reversed, and electropolishing is used as a "roughing operation" before color buffing on wheels.

Seams and nonmetallic inclusions impose a limitation. In showing up such conditions, electropolishing is a good inspection tool. Although unpassable by established inspection habits, an electropolished and plated surface can be superior in performance. Figure 15 shows an example. Chromium plate has filled a hole left by electropolishing out a nonmetallic inclusion. The depression in the chromium plate at the site of the filled-in hole would be termed a "pit" and the part would be rejected. Yet, the inclusion would

TABLE 4. SMOOTHING OF BRASS AND ALUMINUM

Metal	Profilometer reading (RMS), after			
	Mechanical polish, 180 grit belt	Electropolishing (mils removed)		
		0.2	0.35	0.5
Brass	31	26	24	21
Brass	27	20	17	16
Aluminum	48	37	37	35
Aluminum	90[a]	80	75	65

[a] The belt "loaded" during polishing of previous sample.

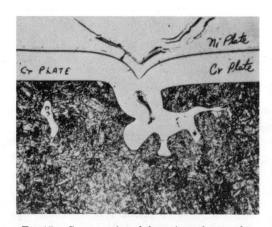

FIG. 15. Cross-section of chromium plate on electropolished steel, at 150× magnification. The chromium has filled the hole left by the nonmetallic inclusion removed by electropolishing. (Faust, C. L., *Proc. Am. Electroplater's Soc.*, 37, 137–150, 1940.)

remain in an abrasively polished surface and be "bridged over" by the chromium plate. A point of weakness would exist, since the chromium would not adhere to the nonmetallic. The plated product would be passed by inspection.

Clearly, an informed approach to the use of electropolishing can give improved products, and processing costs can be reduced in many applications. By electropolishing, surface appearance and quality can be as easily reproduced as can an electroplate. Electropolishing baths are not so critical as plating baths to control and are less subject to contamination.

Types of Metal Electropolished

Most metals can be electropolished successfully but best results are obtained with those having fine grain boundaries and free of nonmetallic inclusions and seams. Also, those comprising a high content of silicon, lead or sulfur are usually troublesome.

Stainless steels are the most frequently electropolished alloys and all can be electropolished. Castings will polish to a bright finish but not to the same brightness or smoothness as wrought alloys.

Besides stainless steel, other commercially electropolished metals include high and low carbon steel, copper, brass, beryllium copper, phosphor bronze and many grades of aluminum. Other metals which can be electropolished are columbium, gold, silver, tantalum, titanium, tungsten and vanadium.

Table 5 lists various metals and solutions which are suitable for electropolishing. There are numerous suppliers of proprietary solutions on the market today. Some are complete solutions while others are concentrates which are mixed with acids purchased locally. Before purchasing a solution, each should be evaluated to obtain the best operating conditions for a particular application. Type of solution, operating temperature, ventilation requirements and polution problems are some of the things to be considered.

Table 6 shows electropolishing results and properties of automotive trim parts made from regular cold-rolled stainless steel strip, obtained without prior knowledge of whether the stainless steel would electropolish.

Metallographic. Electropolishing is widely used for metallographic examination. Discussion of techniques and results are not within the scope of this chapter. They are described adequately in the technical literature. Table 7 briefly lists methods for many metals. Essential details are discussed in the following references selected from the technical literature: metallographic specimens, references 12, 17 and 18; electropolishing steel specimens, reference 20; copper, brass, bronze, aluminum, tin, zinc, nickel, cobalt, reference 21; and electron microscope samples, reference 5.

Applications

Stampings, spinnings, weldments, castings, drawings, forgings and wire goods are all suitable candidates for electropolishing. Typical items fabricated from stainless steel and electropolished include hospital, medical and surgical equipment; dairy, food and beverage processing and handling equipment; bone and joint implants, vacuum equipment; paper mill equipment; automotive and truck parts; electronic and communication parts; tubing, pipe, valves and fittings; fasteners and all types, sizes and shapes of wire goods. These parts can range in size from a small nut to tanks with thousands of gallons capacity.

Steel items are electropolished for microinch improvement, burr removal and to obtain a better surface for electroplating. For instance, the I.D. of gun barrels are electropolished prior to hard chrome plating. Electropolishing is used on brass costume jewelry items, electronic parts and as a finish prior to electroplating.

Electropolishing, by removing a small amount of metal from the surfaces, also removes any contamination on or just under the surface. This enables a much stronger subsequent weld or braze to be achieved, many times with less heat or brazing material being used.

Many times electropolishing has been used as an inspection tool. For example, during the fabrication of heavy walled stainless steel elbows, tees, crosses, etc., which are used in nuclear applications, it has been found that electropolishing was a superior method of crack detection than was the standard dye check. Cracks can develop in the surface which are not detectable by visual inspection and are even suspect for xyglo inspection. Electropolishing, by removing the top surface, reveals material flaws such as cold shots, cracks and inclusions. This is not only valuable for nuclear applications but also for high-pressure systems. Electropol-

TABLE 5. COMMERCIAL ELECTROPOLISHING METHODS

Metal	Type of Bath	U. S. Patent No.	Assignee
Aluminum	Sodium phosphate-carbonate	2,096,309	Aluminum Company of America
	Fluoboric acid	2,108,603	Aluminum Company of America
	Phosphoric-sulfuric-chromic acid	2,550,544	Battelle Development Corporation
	Phosphoric acid-carbitol	—	Aluminum Company of America
Brass	Phosphoric-chromic acid	2,407,543	Battelle Development Corporation
Copper	Phosphoric-chromic acid	2,347,039	Battelle Development Corporation
	Phosphoric acid-modified	2,366,714	Battelle Development Corporation
	Phosphoric acid	—	—
Monel	Phosphoric-sulfuric-hydrochloric acid	2,440,715	Battelle Development Corporation
Nickel	Sulfuric acid	2,145,518	ElectroKemiska Actiebolaget
	Phosphoric-sulfuric-chromic acid	2,334,699	Battelle Development Corporation
	Phosphoric-sulfuric acid	2,429,676	Battelle Development Corporation
	Phosphoric-sulfuric-hydrochloric acid	2,440,715	Battelle Development Corporation
Nickel-Silver	Phosphoric-chromic acid	2,407,543	Battelle Development Corporation
Silver	Potassium cyanide	2,416,294	Arthur D. Little, Inc.
	Potassium cyanide	2,610,143	Oneida, Ltd.
Stainless steel 300 series	Phosphoric acid	1,658,222	Western Electric Company, Inc.
	Phosphoric-sulfuric acid	2,334,698	Battelle Development Corporation
	Phosphoric-sulfuric-chromic acid	2,334,699	Battelle Development Corporation
	Sulfuric-citric acid	2,335,354	Armco
	Phosphoric-chromic acid	2,366,712	Battelle Development Corporation
	Phosphoric acid-butanol	2,424,674	Armco
	Sulfuric-glycolic acid	2,607,722	Armco
	Sulfuric-phosphoric + aniline	2,594,124	Electropol, Ltd.
	Sulfuric-phosphoric + morpholine	2,820,750	Electropol, Ltd.
Stainless steel 400 series	Phosphoric-sulfuric acid	2,334,698	Battelle Development Corporation
	Phosphoric-sulfuric-chronic acid	2,334,699	Battelle Development Corporation
	Sulfuric-citric acid	2,335,354	Armco
	Phosphoric-chromic acid	2,366,712	Battelle Development Corporation
	Phosphoric butanol	2,424,674	Armco
	Sulfuric-glycolic acid	2,607,722	Armco
Steels	Phosphoric-sulfuric acid	2,334,698	Battelle Development Corporation
	Phosphoric-sulfuric-chromic acid	2,338,321	Battelle Development Corporation
	Phosphoric-chromic acid	2,347,040	Battelle Development Corporation
	Phosphoric-sulfuric acid-sodium salts	2,493,579	Standard Steel Spring Company
Zinc	Caustic soda + additive	Pending	Diamond Alkali Company

TABLE 6. RESULTS OF ELECTROPOLISHING AND METALLOGRAPHIC EXAMINATION OF STAINLESS STEEL AUTOMOTIVE TRIM PARTS*

Specimen No.	Article Shape	Alloy	Electropolish Result							Structural Features (at 500× Mag.)				Strain Lines
			Luster[1]	Orange Peel	Pitting	Roll Marks	ASTM Grain Size	Slag	Magnetism	Inclusions Number	Size	Distribution	Type	
41A	Deep	301	F-G	None	None	Slight	7	None	Very slight	Moderate	Small	Singly and in broken stringers	Oxide; sulfide	No
41B	Deep	301	F	Moderate	Severe	None	6	None	Very faint	Many	Small	Singly and in broken stringers	Oxide; sulfide	No
41C	Deep	301	G	Slight	None	Moderate	7	None	Slight	Moderate	Moderate	Singly	Oxide	No
41E	Deep	301	G	Slight	None	None	7	Moderate	Slight	Moderate	Small	Singly	Oxide	No
41G	Deep	301	Ex.	None	None	None	9	None	Slight	Moderate	Small	Singly	Oxide	No
41J	Flat	430	P[2]	—	None	—	9	—	Very magnetic	Moderate	Small	Singly	Oxide	No
41K	Flat	430	G	—	Moderate	—	9	—	Very magnetic	Many	Moderate	Singly	Oxide	No
41M	Strip	301	P	—	Moderate	Severe	6	Moderate	None	Moderate	Small	Singly and in small stringers	Oxide; sulfide	No
40A	Deep	301	—[3]	Moderate	—	Slight	6	None	Slight	Moderate	Small	Singly	Oxide	Yes
40B	Deep	301	G	Moderate	None	Slight	7	Moderate	Very slight	Moderate	Small	Singly	Oxide	No
40I	Deep	301	F-G	Moderate	None	Moderate	6	None	Very faint	Many	Small	Singly and in broken stringers	Oxide; sulfide	No
40J	Deep	301	G	Slight	None	Slight	7	None	Slight	Many	Small	Singly	Oxide; sulfide	Yes
40K	Deep	301	G	Slight	Moderate	Moderate	7	None	Very slight	Moderate	Small	Singly and long broken stringers	Oxide; sulfide	Yes
40L	Deep	301	F	Moderate	Severe	Slight	6	None	Very faint	Moderate	Small	Singly and long broken stringers	Oxide	No
40M	Deep	301	F-G	None	Slight	Moderate	8	None	Very slight	Moderate	Small	Singly and long broken stringers	Oxide	No
40N	Deep	301	F	Moderate	None	Moderate	6	None	None	Few	Small	Singly	Oxide	No

* Faust, C. L., and Graves, E. E., *Proc. Am. Electroplaters' Soc.*, **35**, 223–239 (1948).

TABLE 7. ELECTROLYTIC POLISHING OF METALLOGRAPHIC SAMPLES
(Compiled (1942) by G. E. Pellissier, Jr., Harold Markus, and Robert F. Mehl)

Metal	Solution	C.D.§	Voltage	Temp. (°C)	Time (Min)	Remarks	Ref.
All carbon steels, martensitic, pearlitic, and sorbitic; Armco and white cast iron; 3% silicon steel	Acetic anhydride, 765 cc Perchloric acid*, 185 cc sp gr 1.61 (65%) Distilled water 50 cc 0.5% Al	4 to 6	50†	<30	4 to 5	Prepare solution 24 hr before using. Use moderate agitation. Al increases viscosity which permits more vigorous agitation and current density of 3. Can use at current density of 10 for austenitic steels. Prepare samples to 000 paper. Fe or Al cathode.	1
Steels (except low carbon)	Perchloric acid*, 54 cc of 70% concentration Water, 146 cc Alcohol with 3% ether, 800 cc	2 to 6	70†	Room	½	Preliminary smoothing on 60-grit alundum wheel; stirring paddle set close to specimen operates at all times.	2
Austenitic steels‡	Acetic anhydride, Perchloric acid (65%) 2 parts to 1 part*	6	50†	<30	4 to 5	Same as items at top of column.	1
Iron and silicon-iron	Orthophosphoric acid sp gr 1.316	0.6	0.75 to 2.0			Iron cathode.	3
Tin	Perchloric acid* (sp gr 1.61), 194 cc Acetic anhydride, 806 cc	9 to 15	25 to 40†	15 to 22	8 to 10	Stir solution if length of electrolysis is over 10 min. Polish to 000 paper. Tin cathode. Electrodes 2 cm apart.	4
Copper‡	Orthophosphoric acid sp gr 1.3 to 1.4	0.65 to 0.75	2	Room	±5	Polish to 0000. Copper cathode. Electrodes 2.2 cm apart.	5
Copper‡	Pyrophosphoric acid 530 g	8 to 10	1.6 to 2.0	15 to 22	10 to 15	Polish to 00000 paper. Copper cathode.	6
Cobalt	Orthophosphoric acid sp gr 1.35		1.2			Rough metallographic polish. Cobalt cathode.	7
Aluminum‡	Perchloric acid* (sp gr 1.48) Acetic anhydride, 2 parts to 7 parts	3.0 to 5.0	50 to 100†	<50	15	Allow 4 to 5 g/l to enter solution. Polish to 000 paper. Aluminum cathode.	8
Zinc‡	Potassium hydroxide 25% solution	16	6	Room	15	0000 paper. Solution agitated by air or nitrogen. Copper cathode. Electrodes 2.5 to 15 mm apart.	9

METAL SURFACE PREPARATION AND CLEANING 111

Table 7. *Continued*

Metal	Solution	C.D.§	Voltage	Temp. (°C)	Time (Min)	Remarks	Ref.
Lead	Acetic acid, 650 to 750 cc Perchloric acid*, 350 to 250 cc	1 to 2			3 to 5	0000 paper, horizontal anode. Use current density of 20 to 25 for 1 to 2 min to remove flowed layer. Copper cathode.	6
Pb-Sn alloy	Same as above	2					6
Tin + 3% Sb	Same as for tin, above	9 to 15	25 to 40†	15 to 22	8 to 10	Same as for tin, above.	4
Copper + 3.2% Co	Orthophosphoric acid sp gr 1.35	0.07	2		5 to 10	Polish to 000 paper. Copper cathode. Electrodes horizontal and ½ in. apart.	
Copper + 2.4% Fe							10
Brass, 70-30‡ (1 constituent)	Orthophosphoric acid 430 g/l	13 to 15	1 to 2	Room	10 to 15	File, swirl in 40% HNO₃ 45 sec. Large size copper anode horizontal; electrodes ½ to 2 in. apart.	11
66.7-33.3 Brass‡	Orthophosphoric acid 990 g/l	2.5 to 3					6
Two-constituent 60-40 brass‡	Pyrophosphoric acid 530 g/l	9 to 11	1.9				6
Aluminum bronze; Leaded bronze (85 Cu, 10 Sn, 3 Zn, 2 Pb)	Orthophosphoric acid 990 g/l	1 to 2					6
Phosphor bronze, silicon bronze, monel, nichrome, nickel, and‡ metals	Methyl alcohol (abs.), Nitric acid (conc.). 2 parts to 1 part		40 to 50†	20 to 30	Seconds	Cathode of stainless steel cloth in bottom of dish. Distance between electrodes ½ to 1 in.	12

* Explosive mixture! §Current density in amperes per square decimeter. †External applied voltage.
‡See last item in table.

1. Jacquet, P., and Rocquet, P., *Compt. rend.*, **208,** 1012 (1939).
2. Parcel, R. W., *Metal Progress*, p. 209 (August 1942).
3. Elmore, W. C., *J. Applied Phys.* (Oct. 1939).
4. Jacquet, P., International Tin Research and Development Council, Bull. 90.
5. Lowery, Wilkerson and Smare, *Phil. Mag.*, **22,** 769, (1936).
6. Jacquet, P., *Bull. soc. chim. France*, **3,** 705 (1936).
7. Elmore, W. C., *Phys. Rev.*, **53,** 757 (1938).
8. Jacquet, P., *Comp. rend.*, **205,** 1232 (1937); *The Metallurgist*, Supplement to *The Engineer*, p. 116 (April 1938).
9. Vernon and Stroud, *Nature*, **142,** 477, 1161 (1938).
10. Gordon and Cohen, American Society for Metals Preprint No. 39, 1939; ASM Symposium on "Age Hardening of Metals," p. 161.
11. *Metal Progress*, pp. 756, 771 (Dec. 1939.)
12. Communication from Hugh E. Brown.

ishing is currently being used to decontaminate equipment used in nuclear power plants.

Reactor vessels, pipes, fittings, valves, heat exchangers, tanks, etc. for the food chemical and pharmaceutical and polymer industries are currently being electropolished to provide a smooth, clean surface which provides greater anti-stick qualities coupled with easier cleanability. In many cases an electropolished stainless steel fabrication provided anti-stick qualities equal to glass lining. Other applications for electropolishing include decorative purposes and removing heat tint from spot welds, heli-arc welds and other types of welding. This is of great value for wire fabrications which have been spot welded.

SOLUTION OPERATION

Over 90% of electropolishing is performed on stainless steel. All electropolishing is done in the same manner, therefore, our comments will be based primarily on stainless steel.

Various solutions are used for electropolishing, both acid and alkaline. The preponderance of solutions used commercially are acid and are based on one of the grades of phosphoric acid and one or more additional acids. Basically there are two types of electropolishing solutions—infinite and finite. When a part is being electropolished, a small amount of metal is removed. This metal in combination with components of the bath form a metallic salt which drops to the bottom of the tank and forms a sludge. This bath has an infinite life. A finite solution is one in which metallic salts remain dissolved in the solution and do not settle to the bottom of the tank. To obtain optimum results these metallic salts must be held within limits. When the upper limit is reached, the solution must be decanted and replaced with new solution to keep the metallic salt concentration within satisfactory operating limits. This can be expensive because of the Resource Conservation and Recovery Act regulations for disposing of this used acid.

Infinite life solution requires that periodically the solution be removed from the tank and stored. The sludge in the bottom of the tank is then removed and disposed of according to RCRA regulations. The solution is then pumped back into the tank and fresh solution added to proper operating level.

In electropolishing, the parts are made the anode in the appropriate solution which will dissolve the oxides of the metal (Fig. 16). As these metal ions are removed from the metal surface more metal atoms are exposed, thus polishing occurs. As in plating, a DC rectifier is used for the power source. A typical set-up is shown in Fig. 17. As work is being electropolished oxygen is liberated at the anode and hydrogen is liberated at the cathode. Thus no hydrogen can be imparted to the part. There can be times when an imbalance of solution occurs and the oxygen liberated at the anode and present in air can cause an explosion of the hydrogen. Such explosions will be loud but not dangerous. This condition is preceded by a larger than normal foam blanket on the solution

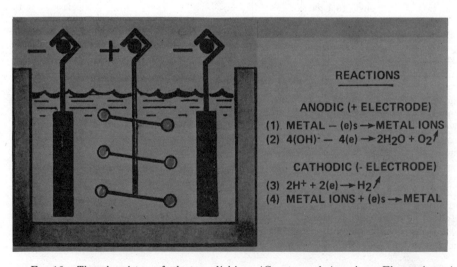

FIG. 16. The chemistry of electropolishing. (*Courtesy of American Electroplaters' Society, from AES Illustrated Lecture Series on electropolishing.*)

METAL SURFACE PREPARATION AND CLEANING 113

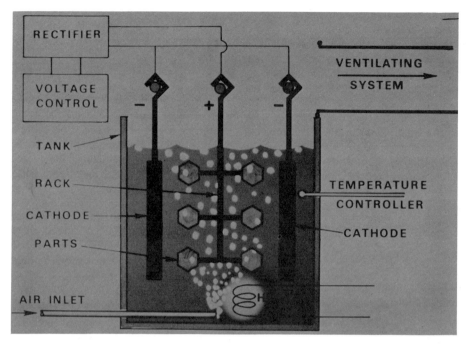

Fig. 17. Electropolishing equipment. (*Courtesy of American Electroplaters' Society, from AES Illustrated Lecture Series on electropolishing.*)

surface and is ignited by a spark from a work holder.

Electropolishing is usually done in concentrated acid solutions. Therefore, protective clothing, gloves and glasses should be worn to prevent injury from splashes. When parts are hung in the solution and current is applied, polishing will occur only when the current density is in the proper range (see Fig. 18). As the voltage is increased the current will increase along the straight line on the left and etching will take place. When polarization occurs, the

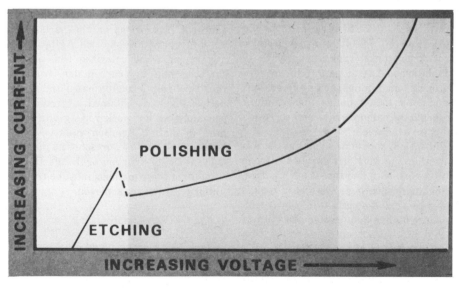

Fig. 18. Current-voltage curve. (*Courtesy of American Electroplaters' Society, from AES Illustrated Lecture Series on electropolishing.*)

current will drop slightly and the voltage will increase as shown by the dotted line.

The curved line on the center portion of the graph represents the range in which electropolishing takes place. Voltage is the variable that is adjusted so the proper current flows through the bath to enable the desired polish to be obtained. Current, time, electrode efficiency and bath temperature are the factors which determine the amount of metal removed from the part. As voltage and current are increased further, large volumes of gas are liberated which can cause gas streaks from holes and edges on the piece. This high current density, right side of graph, is used for deburring and, at times, for microinch improvement.

ELECTROPOLISHING EQUIPMENT

Since electropolishing is a reverse procedure of plating, similar types of equipment may be used. Tanks may be constructed of steel which are lined with rubber, Koroseal, acid brick, plastics or lead. Type of lining to be used is dictated by the kind of solution used.

Stainless steel tanks are not generally used because stray currents can cause pitting of the tank and shorting out of the tank due to poor insulation and can cause a hole and perhaps loss of the bath. The same holds true for a lead lined tank. If lead or stainless is used it is advisable to insert wire reinforced glass or plastic panels along the sides and bottom of the tank to protect them from these stray currents.

Electropolishing baths are very heavy, weighing about 12 to 15 lb/gal. Steel tanks should be fabricated from 3/16 in. plate using heavier gauges for larger tanks. Tank linings of Koroseal or rubber should be at least 3/32 in. on sides and 3/16 in. on bottom. Linings should extend over, around and under the top flange. The outside of the tank should be coated with the best corrosion resistant paint available.

Tank dimensions should be such that there is room for either one work bar and two cathode bars or two work bars and three cathode bars; and cooling and/or heating coils. Size of tank is determined by size of objects to be electropolished and production requirements. Depth of tank should be 6 to 12 in. deeper than that required for parts to leave space for collecting sludge, should an infinite type solution be used. This will enable the sludge to accumulate in the bottom of the tank without interfering with production.

Buss bars are usually copper. Because of higher current densities for electropolishing than plating, they generally have a heavier cross-section. Work bars and cathode bars should be round to ensure the best electrical contact.

Cathodes can be either lead, copper, stainless steel (302 or better) or carbon. Strips of these materials are preferred over sheets and should be 2 to 4 in. wide and 1/8 to 1/4 in. thick. For internal work, sheets or wire cloth conforming to the piece are satisfactory. The composition of the electrolyte is a determining factor of the cathode material. Sides of a metal tank should not be used as the cathode. Cathodes can become encrusted with metallic salts and must be kept clean by periodic cleaning and scrubbing in water. Cathodes which are not clean can cause an increase in polishing time for a satisfactory finish. Conductivity is seriously decreased; voltage may need to be increased or no polishing will take place.

Whenever possible, cathode to part ratio should be about 2:1. Obviously this ratio cannot be maintained when doing internal work. Cathodes placed too close to one area during internal work, can cause extreme high current densities and excessive material removal at that location.

Agitation

During electropolishing oxygen is liberated at the anode. When parts are being electropolished at high current densities or long time cycles, this liberated oxygen can cause gas streaks at holes, edges or protrusions on the parts. Agitation tends to relieve this condition. Agitation continually replaces the polarized film at the part surface. Thus, high current density burns from rack tips tend to be eliminated and generally a better finish is obtained. Agitation can be accomplished by work rod agitation, such as used in plating; solution agitation by air or mechanical stirring, or solution pumping. To prevent the destruction of the electropolishing film, all of these methods must be controlled so uniform results are achieved.

Racks

Rack tips may be made of copper, brass or phosphor bronze. However, rack tips should be compatible with the electrolyte being used. If racks are used to electropolish steel or copper alloy parts which are then to be nickle or chrome plated, the same racks may be used. Titanium tips may also be used and although they have a higher initial cost they may be justified for production runs because they will not be attacked by the electrolyte. Racks should be plastisol coated and only that portion of the tips touching the part to be electropolished should be exposed.

Rack tips as well as splines shoud have sufficient cross-section to carry the required amperage. Table 8 gives the conductivity values of various materials used for racks. It must be remembered that these values are measured in air. When contacts are in solution, a cooling effect takes place, thus increasing conductivity. Overheating and burning of contacts are less likely to occur. Tips must make positive contact with the part being electropolished. If the contact is not firm, the parts will not receive the proper current density. This can cause burning at the contact point. Due to the density of the solution and gas action, which exerts a lifting effect, "J" hooks cannot be used for electropolishing unless the parts have enough weight to make a firm contact.

Heating/Cooling

Because of the high electrical resistance of most electropolishing solutions a heating effect generally occurs depending on the operating temperature of the tank. Solutions may need to be heated before starting the operation but may need cooling during the day. Heating is accomplished through coils of carbon, tantalum, titanium or type 316 stainless steel plate-coils.

Low watt density electric immersion heaters can also be used as long as the material of the heater is compatible with the solution. Automatic temperature controls are desirable. There are times when, with the proper controls, a heat exchanger can be used for both heating and cooling. As a tank is being run and sludge develops, it also will build up on the surface of heating and cooling equipment. These items will need to be cleaned to maintain efficiency more often than the periodic cleaning of the tank. If using a heat exchanger the tubes and hoses will also have to be kept clean.

Stainless steel rinse tanks after electropolishing should be used whenever possible. As a second choice, coated tanks can be used. Plain steel tanks will have a very limited life. After the first rinse tank an acidulated rinse tank may be beneficial to remove any cloud or stain which might develop on the finished part.

Power Supply

Power requirements are determined by the surface area to be electropolished at one time. As in plating, it is calculated in amps/in.^{2}, amps/ft^2 or amps/decimeter^2. Voltage requirement for most popular metals is between 5 and 12 volts, although there have been cases where voltages have been under 5 volts and up to 18 volts for a specific application.

This low-voltage direct current is generally

TABLE 8. METAL CONDUCTIVITY CHART IN AMPERES

Size	Copper	Aluminum	Brass	Steel	Phosphor Bronze	Stainless Steel	Titanium
1 × 1	1000	600	250	120	180	23	31
½ × 1	500	300	125	60	90	12	16
⅜ × 1	375	225	94	45	68	9	12
¼ × 1	250	150	63	30	45	6	8
¼ × ¾	187	112	47	22	33	4.5	6
½ × ½	250	150	63	30	45	6	8
¼ × ½	125	75	31	15	22	3	4
⅜ × ⅜	140	84	35	17	27	4	5
½⌀	200	120	50	24	36	5	6
¼⌀	50	30	13	6	9	1	1.5
³⁄₁₆⌀	28	16	7	3.5	5	.6	.8
⁵⁄₃₂⌀	20	12	5	2.5	3.7	.5	.6
⅛⌀	12	7	3	1.5	2.2	.3	.4
³⁄₃₂⌀	7	4	1.7	.8	1.2	.2	.2
¹⁄₁₆⌀	3	1.8	.7	.4	.5	.1	.1

supplied by rectifiers. Voltage control is almost always required. Most electropolishing is done at current densities from 50 to 500 amps/ft^2, depending on material being electropolished and electrolyte used.

Ventilation

Many solutions require ventilation for various reasons such as objectionable odor and misting from the tank. It has been noted that the higher the operating temperature, the more desirable ventilation becomes. However, there are solutions that do not give off an objectionable odor and do not throw up a mist. When ventilation equipment is required or recommended it can be a fume hood, or slot type draft boxes along one or more sides of the tank depending on tank size. Materials of construction can be plastic, stainless steel or a plastic coated steel. Exhaust capacity would be between 200 and 300 ft^3/min/ft^2 of bath surface.

PROCEDURES FOR ELECTROPOLISHING

Electropolishing operating sequence is similar to that of plating. The work is racked and processed through the following sequence:

Rack parts
Clean
Rinse
Electropolish
Rinse
Acid dip
Rinse
Hot rinse
Dry.

If the parts are to be plated, the acid dip is an acid activator dip, then rinse and plate.

As seen earlier, there are many variables to obtain good electropolishing results. One which the operator can control is good cleaning methods. These methods include soak cleaning, electrocleaning, vapor degreasing, solvent cleaning or a combination of these. If oil films are not removed, in all probability they will come off in the electropolishing tank. However, this oil floats to the surface and forms a scum on the solution surface which will have to be skimmed off. If it is not skimmed off, this scum may be redeposited on the finished parts, which will give a poor appearance and will be difficult to remove in subsequent rinses.

Heat tint, such as that produced from spot welding, will not be removed by cleaning but will generally be removed in the electropolishing cycle. However, if the oil on the metal surface was not removed before spot welding, the heat generated from the spot weld on the adjacent area causes the oil to carbonize, making it extremely difficult to remove by normal cleaning procedures.

Drawing compounds, lead and heat treat scale must be completely removed. If not, etching, pitting or a non-uniform polish will result. Heat treat scale and welding scale can be removed by pickling, salt bath descaling or various abrasive methods. Overpickling can cause etching of the part, which will have a deleterious effect on the final finish. Trying to remove scale by electropolishing will cause the scale to be removed—if at all—in an uneven manner and the part will have an etched or mottled appearance.

Thorough rinsing after cleaning is mandatory so that no alkaline materials, wetting agents or noncompatible materials will be dragged into the electropolishing solution.

Time of electropolishing varies with the solution temperature, material being polished, distance of part from the cathode, voltage and amperage used and degree of polishing desired. Times are usually in the range of 1 to 20 min.

Due to today's emphasis on pollution control, it is advisable to use double or triple counter flow rinses. This will enable thorough rinsing to be obtained with up to 85% less water being used. These tanks should be designed to accommodate the largest part to be processed. Tanks which are too large will waste water. This is very important if rinse waters must be treated prior to discharging to the sewer. The less water treated, the less cost for treatment. Water treatment can be a significant cost factor.

Parts are placed in the electropolishing bath on the anode rod between and parallel to the two rows of cathodes. The parts are arranged so that the greatest area is exposed parallel to the cathodes. The distance from the cathode to anode bar can be from a very few inches to perhaps 2 ft. The closer the parts are to the cathode, the higher the current density and metal will be removed rapidly on those edges facing the cathode. On irregularly shaped pieces, a greater distance from anode to cathode may give a more uniform finish. Higher voltage will be required to overcome resistance if there is too great a distance between anode and cathode.

Care must be taken so gas is not trapped in deep recesses or inside corners. Racks are designed so that gas is either trapped in the least objectionable area, or internal conforming cathodes may be used to overcome this phenomenon.

Deburring

During the electropolishing process, edges become high current density areas; therefore, metal is removed at a higher rate from sharp edges and protrusions than other areas (Fig. 19). Thus electrolytic deburring occurs and sharp edges can be broken. This does not mean that all burrs can be removed. Two things that determine the size of the burr that can be removed are the size of the burr and the tolerance of the piece. Obviously, if the burr is 0.005 and tolerances are held to 0.0005, it is unlikely that the burr can be removed and still hold tolerance. Many times burrs located on the I.D. of a part can be removed by use of strategically located internal cathodes. Sometimes it is necessary to mechanically remove a large chip, leaving a small burr which can then be removed by electrodeburring.

Electrodeburring requires 500 to 2000 amperes/ft^2 (ASF) rather than 50 to 500 ASF for electropolishing.

Figure 20 shows perforated stainless steel aircraft fuel filters, beryllium-copper contact strips, berrylium-copper connector and a stainless steel mouse feeder (the latter used in laboratories for germ-free mice), all of which were electropolished for burr removal.

Passivation

Because a small amount of metal will be removed during electropolishing, any imbedded contamination either on the surface or just under the surface will be removed. Electropolishing passivates stainless steel to a greater extent than does any other passivation treatment. Storage life of carbon steel parts, from the standpoint of corrosion, is aided by electropolishing. For example, electropolished steel parts have been stored at 60 to 70% relative humidity for over six months without visible rust.

Brass electropolished in phosphoric-chromic acid solution tarnishes only slightly, if at all, whereas buffed brass turns dark in the same atmospheric environment. Although electropolishing does not make brass or steel tarnishproof or rustproof, the improvement in resistance to corrosion and oxidation can be a practical advantage during manufacturing.

Electromachining

At times parts have been made to improper tolerances or tolerances need to be changed slightly. Electropolishing becomes a tool to change tolerances by closely controlling the removal of a small amount of metal. Figure 21 shows nuts which needed a small amount of material removed so that the threads would fit the mating part. The turbine blades were machined oversize. By controlling time and temperature, up to 0.008 material was removed on the spline, holding tolerances to ±0.0005. There were three diameters on the I.D. of the connector and all were 0.001 undersize. Electropolishing, with an internal cathode, was used to enlarge the bore. An air gauge was used to check each part before and after electropolishing so that metal removal would be accurate to ±0.0001.

Surface Roughness

Surface roughness is commonly classified as RMS, root mean square, which denotes smoothness of ground or machined surfaces. This measurement has no real relationship as to how easily an electropolished part can be cleaned after use or on its anti-stick properties.

Surface roughness measurement is usually

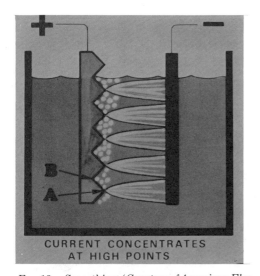

FIG. 19. Smoothing. (*Courtesy of American Electroplaters' Society, from AES Illustrated Lecture Series on electropolishing.*)

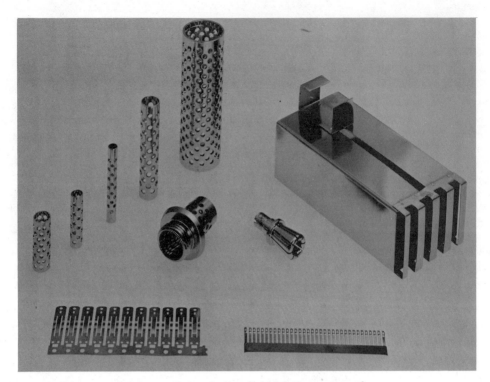

FIG. 20. Parts electropolished for burr removal.

made with a profilometer which cannot measure peak to valley distance accurately. Such peaks may be reduced by electropolishing from substantial points to insignificant mounds without changing peak to peak distance at the same ratio. However, a microscopic examination will show up to a 90% reduction in surface area. Such surfaces will now exhibit superior smoothness and anti-stick properties. On some work the profilometer reading will show a 50% improvement. Here again the cleanability and anti-stick properties will be enhanced many times.

FIG. 21. Electromachining.

To obtain the lowest microinch finish, the best surface to begin with is a 2B finish. Figure 22 shows two panels of stainless steel. The one on the left is a magnified view of a mechanically polished surface. The other shows an electropolished surface. Both appear polished, but when magnified, the mechanically polished surface shows a series of fine scratches while none are apparent on the electropolished surface.

Figure 23 shows two curves measured by a surface roughness meter that records, on moving graph paper, an electronically magnified view of surface irregularities. By this graph it is readily seen that a mechanically polished surface is considerably rougher than a mechanically finished surface which has been electropolished.

Friction

Electropolishing reduces the coefficient of friction of metals because the small asperities on the surface have been either removed or rounded. The coefficient of friction of an electropolished surface is about one-fourth that of a mechanically polished surface.

Hydrogen Embrittlement

As was discussed earlier, during the electropolishing cycle oxygen is liberated at the part being electropolished. Hydrogen is discharged at the cathode; hence, electropolishing cannot cause hydrogen embrittlement.

Castings

Castings are generally not good candidates for electropolishing because of the various alloying materials used in the castings. An exception is stainless steel. These castings will brighten but will not become as smooth or obtain a mirror finish in comparison to strip stock. Electropolishing will remove contamination from the surface and provide a good passivated surface. During electropolishing, the surface of the casting will be removed and may expose subsurface porosity.

An investment casting is a better candidate for electropolishing than a sand casting, because of its inherently smoother surface. However, sand castings will electropolish to a clean, bright surface.

Pollution Control

Due to EPA and RCRA regulations it is important that thorough rinsing be accomplished with as little water as possible to provide a clean, neutral part, free of staining and cloudy appearance. A good cycle after electropolishing would include a drain-off tank. When enough solution is saved it can be put back in the electropolishing tank. Next tank can be a still rinse with air agitation. When this rinse tank

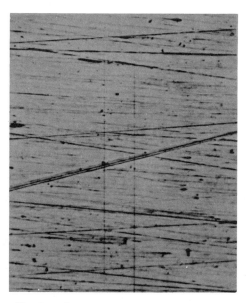 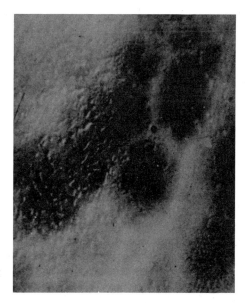

FIG. 22. Stainless steel panels. (*Courtesy of American Electroplaters' Society, from AES Illustrated Lecture Series on electropolishing.*)

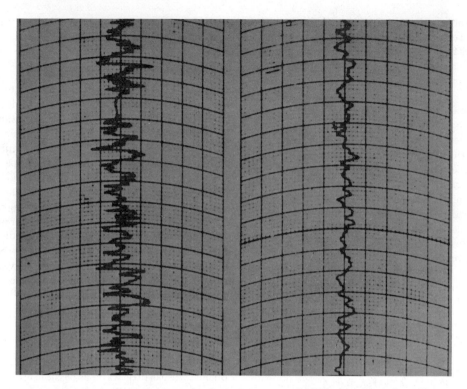

Fig. 23. Curves measured by surface roughness meter. (*Courtesy of American Electroplaters' Society, from AES Illustrated Lecture Series on electropolishing.*)

reaches a certain level of concentration it can be pumped into another tank, heated to drive off water and then put back in the electropolishing tank. The next rinse can be either spray or counterflow rinsing using either two or three tanks, depending on part configuration and as little water as practical to obtain a thoroughly clean part. Should there be a problem with cloudy work, an acidified or alkaline tank followed by a clean rinse, preferably counterflow, and then a hot water rinse to facilitate drying. It cannot be stressed enough to use as little water as possible. If city water is being used it has to be purchased. After the water is used it goes to a sewer where there is usually a sewer charge by the gallon. Between the inflow and the outflow there will probably be pollution control equipment, and the more water used the more chemicals are needed to neutralize and treat the effluent to acceptable standards.

Inasmuch as electropolishing removes metal, it is necessary to recapture these heavy metals, chrome, nickel, copper, etc., to a level acceptable to the EPA. To date, the most economical system is neutralization to a ph where these metals form a hydroxide and precipitate. A flocculating agent is then added, which speeds up the precipitation. This water is then pumped into a clarifier, where settling takes place and a wet sludge is formed. This sludge is pumped into a centrifuge or filter press, where the sludge emerges at about a 20 to 25% solids level. This sludge is now neutral but it must be hauled away at substantial cost to a properly approved disposal area.

This method is one way to control pollution discharge, particularly for the infinite solutions. Solutions which are of the finite variety require the tank to be decanted and replenished with new solution. The discarded solution is not only highly acid but it also has heavy metals in suspension. This solution must also be disposed of in an approved manner. Pollution control is a major consideration in any present or future electropolishing installation.

Reference

1. Faust, Charles L. "Electropolishing." *Metals Handbook*, Vol. 2, 8th Ed. Taylor Lyman, Ed. American Society for Metals, 1964, p. 487.

E. Solvent Cleaning

Kenneth S. Surprenant

*Research Leader, Dow Chemical Co. U.S.A.
Midland, MI*

Cleaning is the removal of undesirable material from a base material and is normally limited to the surface. The unwanted material may be rust or oxide films, metal fines, shop dirt such as dust or paper, rust preventatives, buffing or polishing compounds, wax, oil, fingerprints, grease, asphalt and even water in some cases. Earlier sections of this chapter have described buffing, polishing, barrel finishing and electropolishing. These processes are valuable in removing rust or oxide films or metal surface imperfections. Salts and other water-soluble soils, soaps and some buffing and polish suspending agents can be cleaned from parts by alkali cleaning as described in Section F. The presence of organic compounds such as grease or oil can foul these cleaning processes and add to their cost, maintenance and waste. These organic compounds are easily solubilized by solvent and removed from the work parts.

In some cases, solvent cleaning before other surface preparations can extend the life of those cleaning operations; reduce their costs. In other cases, solvent cleaning provides work parts in a condition ready for the next operation such as assembly, painting, inspection, further machining or packaging. Prior to plating, solvent cleaning is usually followed by an alkaline wash or other similar process which provides an hydrophilic surface. Further, solvent cleaning can be employed to remove water from electroplated parts, as is common in the jewelry industry.

SOLVENT CLEANING

Solvent cleaning can be accomplished in room temperature baths or by employing vapor degreasing techniques. Room temperature solvent cleaning is commonly referred to as cold cleaning. Vapor degreasing is the process of solvent cleaning parts by condensing solvent vapors of a non-flammable solvent on a work part(s).

COLD CLEANING

Without a doubt, the first solvent cleaning was done in a small pail or bucket. It is still done today and it is the simplest kind of cold cleaning. These simple operations are mobile as opposed to fixed or stationary cleaning operations. Such operations require a minimum of equipment capital, energy and space. Generations of cold cleaning have yielded much more complex cousins as well.

Cold cleaning is the usual choice for infrequent maintenance cleaning of locomotives, manufacturing machines, electric motors, watches and thousands of other items large and small. Many of these are spray cleaning or wipe cleaning, while dip or immersion cleaning is used for others.

The prime properties of a solvent for simple cold cleaning are good solvency, minimum flammability, low toxicity and a moderately fast evaporation rate. Carbon tetrachloride was regarded as a nearly ideal cold cleaning solvent before its highly toxic properties were recognized. Cold cleaning solvent properties are summarized in Table 1. A relatively rapid evaporation rate is preferred for dip or wipe cleaning operations, whereas much slower evaporation is more desirable for spray or flush cleaning. In the latter operations, the solvent must reach the object to be cleaned and must have an opportunity to drain off the work part to remove the soil. Too rapid evaporation in the latter operation redeposits the soil on the work part. Further, the risk of fire is increased by spraying. The choice of a spray cleaning solvent should lean toward solvents which are non-flammable* or low in flammability. Where distillation is practiced, as an effective means to reduce solvent costs, non-

---

*The accepted definition of the word flammable is given in 40 CFR 60030, December 31, 1975, and is determined by means of TAG OPEN CUP (ASTM D56-70), or SETAFLASH (ASTM D3278-73).

TABLE 1. COMMON METAL CLEANING SOLVENTS

Type of Solvent Solvent	Solvency for Metal Working Soils	Toxicity (ppm)	Flash Point	Evaporation Rate**	Water Solubility (% wt)	Boiling Point (Range)	Pounds per Gal
Alcohols							
Ethanol (95%)	poor	1000*	60°F	24.7	∞	165–176°F	6.76
Isopropanol	poor	400*	55°F	19	∞	179–181°F	6.55
Methanol	poor	200*	58°F	45	∞	147–149°F	6.60
Aliphatic hydrocarbons							
Heptane	good	500*	<20°F	26	<0.1	201–207°F	5.79
Kerosene	good	500	149°F	0.63	<0.1	354–525°F	6.74
Stoddard	good	200	105°F	2.2	<0.1	313–380°F	6.38
Mineral spirits 66	good	200	107°F	1.5	<0.1	318–382°F	6.40
Aromatic hydrocarbons							
Benzene***	good	10*	10°F	132	<0.1	176–177°F	7.36
SC 150	good	200*	151°F	0.48	<0.1	370–410°F	7.42
Toluene	good	200*	45°F	17	<0.1	230–232°F	7.26
Turpentine	good	100*	91°F	2.9	<0.1	314–327°F	7.17
Xylene	good	100*	81°F	4.7	<0.1	281–284°F	7.23
Chlorinated solvents							
Carbon tetrachloride***	excellent	10*	none	111	<0.1	170–172°F	13.22
Methylene chloride	excellent	500*	none	363	0.2	104–105.5°F	10.98
Perchloroethylene	excellent	100*	none	16	<0.1	250–254°F	13.47
1,1,1-trichlohroethane	excellent	350	none	103	<0.1	165–194°F	10.97
Trichloroethylene	excellent	100*	none	62.4	<0.1	188–190°F	12.14
Fluorinated solvents							
Trichlorotrifluoro-ethane (FC-113)	good	1000*	none	439	<0.1	118°F	13.16
Ketones							
Acetone	good	1000*	<0°/F	122	∞	132–134°F	6.59
Methyl ethyl ketone	good	200*	28°F	45	27	174–176°F	6.71

*Federal Register, June 27, 1974, Vol. 39, No. 125.
**Evaporation rate determined by weight loss (in mg/m^2/min) of 50 ml in a 125-ml beaker on an analytical balance (Dow Chemical Co. method).
***Not recommended or sold for metal cleaning (formerly standard in industry).

flammable and relatively low boiling solvents are preferred. Solvents with good electrical insulating properties and low residues on evaporation are preferred for electrical and electronic cleaning.

Nearly all cold cleaning operations involve some exposure to people. Therefore, high priority should be given to selecting a solvent with minimal toxic properties. For this reason, *carbon tetrachloride and benzene should not be used for cleaning solvents.*

SOLVENT BLENDS

Compromises in solvent properties can be obtained by blending two or more of the appropriate solvents. For example, the cost of a cold cleaning solvent can be reduced by a solvent blend which contains quantities of a low-cost solvent. Such blends are all often referred to as safety blends, which is misleading to many users. Without evidence to the contrary, solvent blends must be considered to have the toxic properties of the solvents in their composition. Estimates of their toxicity are made based on their relative volatility (vapor pressure) and percent composition. However, these estimates assume that the toxicity of one solvent in the composition does not aggravate the toxic reaction of other portions of the blend.

Solvent blends can serve useful purposes by offering special solvent properties, decreasing the cost per gallon or reducing the flammability of the final composition. For example, alcohols can be added to chlorinated solvents to increase their cleaning effectiveness for water-soluble solder fluxes or fingerprints. Solvent blends based on methylene chloride are popular as paint, varnish and lacquer stripping formulas. In instances where the solvent cannot be recaptured and distilled for reuse, a solvent blend containing 25% methylene chloride by volume, 5% perchloroethylene and 70% stoddard solvent or mineral spirits has been used to reduce cleaning costs.

The design of cold cleaning solvent blends should be done under the guidance of solvent experts to avoid problems in performance, toxicity, stability and flammability. Although methylene chloride is nonflammable, small quantities (less than 25% by volume) added to flammable solvents will often increase the flammability by decreasing the flash or fire point of the composition. Blends formulated to reduce costs nearly always result in compositions where the less volatile and cheaper solvent is flammable. When these formulas are evaporated, sometimes by as little as 10% by volume, they may become as flammable as the low-cost solvent.

The chlorinated solvents contain stabilizers or inhibitors to protect them from degradation from sunlight, oxidation or reactive metals such as aluminum. Mixing other solvents or even two chlorinated solvents together can result in diluting the stabilizers to the extent that they will not protect the new solvent composition.

Crude appliances such as pails are often used in cold cleaning; however, they can result in unnecessary fire and vapor exposure risks. Examples of cold cleaning equipment for flammable solvents include safety transfer cans with flame arrestors (Fig. 1), dip or soak tanks with fusible link cover supports (Fig. 2), plunger type safety cans with drain screens (Fig. 3) and covered sinks with a solvent inventory tank below (Fig. 4). These devices can be used by the nonflammable solvents as well as flammable solvents and are effective in reducing evaporative losses. Mild steel and galvanized or zinc metallized steel are common materials of construction. Cast iron pump housings with stainless or brass impellers and stainless shafts are suitable for most solvents. Gasketing and sealing compounds should be chosen according to the solvent used. Aluminum parts are commonly cleaned by solvents in both maintenance and manufacturing operations. However, aluminum is a highly reactive metal, particularly when the aluminum oxide film is removed and can react with a number of solvents, including the alcohols and the chlorinated solvents. For this reason, aluminum is not recommended as a material of construction for pumps, meters, spray nozzles, valves, storage tanks, mixing blades (impellers) or piping for solvent cleaning operations.

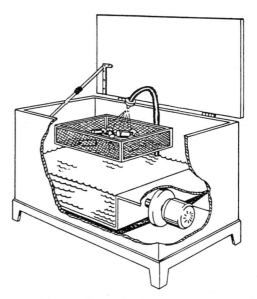

FIG. 2. Spray cleaning equipment.

More sophisticated cold cleaning operations are commonly found in the electronics industry. In the manufacture of printed circuit boards, the raw circuit board consists of a fiberglass epoxy sheet with a copper film laminate. A photosensitive plastic film is bonded to the copper. This

FIG. 1.

FIG. 3.

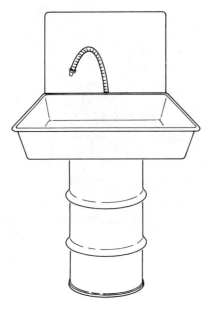

Fig. 4.

photo resist film is exposed to ultraviolet light through a photographic tool in the pattern of the desired circuitry.

The ultraviolet light exposed portions of the film further polymerizes to form a solvent-insoluble plastic. The circuit board is developed by dissolving the unexposed portion of the film with 1,1,1-trichloroethane, exposing the copper circuit. Note that additives or inhibitors added to 1,1,1-trichloroethane can have a significant effect on the quality of the developing operation. Some inhibitors have been known to attack the polymerized film during development, causing quality defect. The exposed copper circuit is plated or protected with solder. Then, the remaining polymerized film is stripped, using methylene chloride or methylene chloride base formulations. The newly exposed copper surface is removed by etching with ferric chloride, for example. Both the solvent developing operation and the stripping operation are in effect sophisticated cold cleaning techniques. Figure 5 illustrates typical equipment used for both developing and stripping.

More recently a water base system for printed circuit board manufacture has been introduced but the solvent base system offers finer line quality, requires less energy and equipment capital and avoids concerns for water pollution. The solvent system, to date, requires less process control and is faster also. Solvent developing and stripping equipment are enclosed and well ventilated, posing minimal exposure to vapor inhalation.

Obviously, the water base system avoids potential solvent vapor exposure entirely. Both systems are competitive economically.

DIPHASE COLD CLEANING

Frequently, solvent cleaning is chosen to avoid exposing the processed work to water. In some instances, such as before plating, aqueous cleaning after solvent cleaning is desirable. In such situations, diphase cleaning can offer distinct advantages. In this cleaning operation, a water layer which may contain surfactants is placed on top of one of the chlorinated solvents or fluororcarbon 113. The chlorinated and fluorocarbon solvents are heavier than water and remain below the water surface. Generally, the petroleum solvents are not used in diphase cleaning because they are lighter than water and float. Water soluble solvents cannot be used in diphase cleaning; fortunately, the chlorinated and fluorocarbon solvents are essentially water-insoluble. Not only are the parts cleaned in both solvent and water in this process, but the evaporation rate of the solvent is suppressed by the water blanket above it, increasing the solvent economy in such operations. Corrosion of the tank at the water solvent interface and in the area contacted by the water phase can be held to a minimum by maintaining a pH of approximately 10 with buffer salts (e.g., Na_2HPO_4) and by employing 304 or 316L stainless steel. Strong alkalies such as sodium hydroxide or potash should not be used in the water phase of a diphase cleaning operation. These highly alkaline materials can react to degrade the chlorinated solvents. Diphase cleaning of aluminum is not recommended for 1,1,1-trichloroethane because the aluminum inhibitor system in the solvent may be extracted into the water phase, leaving the solvent inadequatly protected. Diphase cleaning is competitive with normal cold solvent cleaning operations and both are frequently economically attractive cleaning processes when compared with alkaline washing operations. Diphase cleaning obviously suffers the handicap of leaving the cleaned parts wet with water.

VAPOR DEGREASING

Vapor degreasing is the process of cleaning articles, usually metal parts, by condensing pure

METAL SURFACE PREPARATION AND CLEANING 125

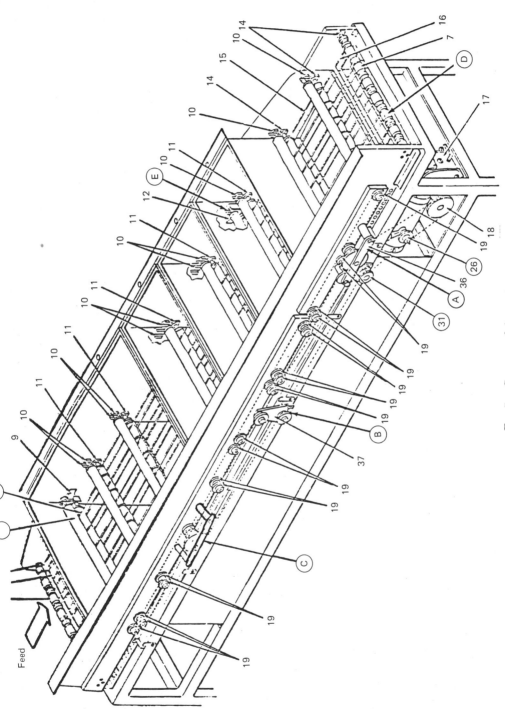

FIG. 5. Conveyor drive system.

solvent vapors on them. In this process, a nonflammable solvent is brought to a boil within a vapor degreaser (Fig. 6). The solvent vapors are three or more times heavier than air; so they displace the air as they are generated and create a pure solvent vapor zone. The solvent vapors are prevented from escaping by condenser coils located below the upper rim of the degreaser. Parts being cleaned are lowered into this vapor zone and condense solvent vapor until they reach the approximate boiling temperature of the solvent. The soils removed from the work parts boil at much higher temperatures than the solvent which results in essentially pure solvent vapors being formed, even though the boiling solvent may be quite contaminated with soil from previous work parts. In this regard vapor degreasing is an improvement over cold solvent cleaning because the parts are always washed with pure solvent. By contrast, in cold cleaning the solvent bath becomes more and more contaminated as repeated workloads are processed and redeposition of soil increases. In vapor degreasing, the parts are heated to the boiling temperature of the degreasing solvent and they dry instantly as they are withdrawn from the vapor zone. Cold cleaned parts dry more slowly.

The demands made on a vapor degreasing solvent are much more severe than those imposed on cold cleaning solvents. Thus, relatively few solvents qualify for use in vapor degreasing. The primary vapor degreasing solvents and their properties are summarized in Table 2. A discussion of the critical vapor degreasing solvent properties follows.

Flammability

The flammability of the solvent is determined by standard flash point test methods such as Tag Open Cup. All flammable solvents have a flash point below their boiling point. At temperatures above this flash point, the vapors from the given flammable solvent can be readily ignited and will burn at least for a brief interval. A fire point is similar except that the vapors continue to burn. Vapor degreasing solvents do *not* have a flash or fire point by standard tests. The fire hazard associated with using a flammable solvent in vapor degreasing would be wholly unacceptable because this process demands the use of the solvent at its boiling point. This would result in highly flammable air and solvent vapor mixtures at the air-solvent vapor interface.

Although all of the nonflammable solvents present essentially no fire hazard in vapor degreasing, some of these solvents do have limited flammable properties. These flammable properties can be measured by determining the ignitable or flammable limits in air. All flammable solvents have ignitable limits in air as well as flash and fire points. Perchloroethylene and fluorcarbon 113 have no flammable limits in air. Although nonflammable methylene chloride, trichloroethylene and 1,1,1-trichloroethane also have flammable limits. The flammable limits of the latter three chlorinated solvents are at much higher vapor concentrations than those typical for truly flammable solvents. Also, the minimum ignition energy for these solvents are 1000 to 3000 times the minimum ignition energy required for flammable solvents (Table 3). These differences account for the minimum fire risk experienced with methylene chloride, trichloroethylene and 1,1,1-trichloroethane. Fires are virtually nonexistent with all of the nonflammable halogenated solvents.

Toxicity

The Occupational Safety and Health Administration publishes the legal worker exposure health standards. However, a more convenient and up-to-date reference is published annually by the American Conference of Governmental Industrial Hygienists (ACGIH). Values from this reference have been at least as conservative as those required by OSHA and represent recommended maximum time-weighted average (TWA) concentration exposures acceptable for a normal 8-hr work day (40-hr work week) to which nearly all workers may be repeatedly exposed without adverse affect. Higher exposures are permissible for short periods if the average is not exceeded. Generally, higher ac-

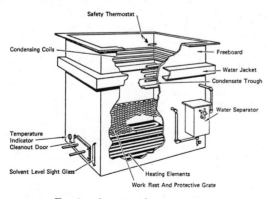

FIG. 6. Open top degreaser.

TABLE 2. VAPOR DEGREASING SOLVENTS

Parameter	Trichloroethylene	Trichloroethane	Perchloroethylene	Trichlorotrifluoroethane Fluorocarbon 113	Methylene Chloride
Flash point	None	None	None	None	None
Toxicity—TLV*	50 ppm	350 ppm	50 ppm	1000 ppm	100 ppm
Solvency	Strong	Moderate	Moderate	Mild	Strong
Photochemical reactivity	Yes	No	No	No	No
Vapor Density (Air = 1.0)	4.5	4.6	5.7	6.5	2.9
Volume of condensate (gals)	1.00	0.86	1.57	0.54	0.19
Stabilization	Yes	Yes	Yes	No	Yes
Boiling point	189 F	165 F	250 F	118 F	104 F
Molecular weight	131	133	166	187	85

*American Conference of Governmental Industrial Hygienists (1981)

1 / GENERAL PROCESSING DATA

TABLE 3. FLAMMABILITY OF SOLVENTS*

Solvent	Flash Point Open Cup, °C	Flash Point Closed Cup, °C	Flammable Limits in Air (20°C), Vol%		Minimum Ignition Energy (Millijoules)	
			Upper	Lower	25°C	50°C
Methylene chloride	NF	NF	22.0	14.8	3,200	
1,1,1-trichloroethane	NF	NF	10.5	8.0	**	600
Trichloroethylene	NF	NF	10.5	8.0	**	280
Perchloroethylene	NF	NF	NONE		NONE	NONE
Acetone	−9	−17	13	2.15	1.15	
Cyclohexane		2.5	8.35	1.35	0.22	
Ethyl acetate	7	−5	11	2.25	1.42	
Ethyl benzene	54		6.7	1.0		
n-hexane	−28	−22	6.9	1.25	0.24	
Methanol	15.6	6	36.5	6.72	0.14	
Methyl ethyl ketone	−5.6	−7	11.5	1.81	0.53	
Toluene		4	7.0	1.3		

*Values derived from:
1. Archer, W. L. and V. L. Stevens, "Comparison of chlorinated, Aliphatic, Aromatic and Oxygenated Hydrocarbons as Solvents." *Industrial Eng. Chem. Prod. Res. and Dev.* **16**, No. 4 (1977).
2. Explosion Venting. National Fire Protection Association, 1978.
3. *Electrostatic Hazards: Their Evaluation and Control.* Verlag Chemic Weinheim, 1977.
4. Unpublished research by R. Tait, and The Dow Chemical Company.
** No ignition up to 20,000 millijoules.

ceptable exposure values indicate less toxic materials.

Solvency

Kauri butanol values and other systems intended to quantify solvent power tend to give misleading information for cleaning operations. The subjective descriptions of solvency offered in Table 2 represent experience in industrial cleaning applications.

Vapor Density

More dense solvent vapors are less easily disturbed by room air turbulence than lighter vapors. The density of each solvent's vapors can be estimated by dividing the molecular weight of the solvent by 29 (the average molecular weight of air).

Solvent Condensate Volume

The volume of solvent which condenses on the work parts controls how much washing action occurs from simple solvent vapor condensation. The solvent condensate volume is expressed in gallons of solvent which form on cleaning 100 lb of steel or 1/100 kg. It is calculated by the following formula:

$$\frac{100 \text{ lb} \times 0.11 \text{ BTU/F}° \times (\text{boiling point-room temp.})}{\text{Solvent latent heat of vaporization}} = \text{gal}/100 \text{ lb.}$$

Thus, a low heat of vaporization results in a greater quantity of solvent condensing, as does a higher boiling point.

Boiling Point

A low boiling vapor degreasing solvent permits easy handling of cleaned parts due to their lower temperature and avoids possible damage to heat sensitive parts; e.g., thermometers. However, these solvents may require refrigeration for effective vapor condensation and control. Conversely, a high boiling solvent like perchloroethylene provides the maximum solvent condensate rinsing on parts which have a high surface to weight ratio such as radiators but it requires high pressure steam (50 to 60 psig) for heating.

Stability

The stresses which must be withstood by vapor degreasing solvents include heat, water, metal fines, air oxidation, light and the accumulation of soluble soils. All of the vapor degreasing solvents except fluorocarbon 113 require chemical additives to maintain their stability in this application. However, each of the stabilized degreasing solvents will tolerate all but the most abusive treatment with appropriate stabilization.

In addition to the primary vapor degreasing solvents, some co-solvent blends with the basic solvents are available to increase or enhance the solubility characteristics of the base solvent. To be effective, these co-solvent systems must boil together with one of the primary degreasing solvents so that they can be distilled and will have essentially the same composition in the vapor zone as in the liquid. These special mixtures which boil together are called azeotropes. Again, to be useful and safe in vapor degreasing operations, these azeotropic blends must meet all of the requirements of a vapor degreasing solvent. The preparation of azeotropic vapor degreasing solvent blends should not be attempted without expert advice and laboratory testing. Azeotropic mixtures may have unacceptable fire or toxic properties.

Vapor Degreasing Equipment

As in the case of cold cleaning, vapor degreasing equipment varies in complexity. All of the vapor degreaser designs provide for an inventory of solvent, a heating system to boil the solvent and a condenser system to prevent loss of solvent vapors and to control the upper level of the vapor zone within the equipment (Fig. 6).

Heating of the degreaser is usually accomplished by steam. However, electrical resistance heaters, gas combustion tubes and hot water can be used. Gas combustion heaters with open flames located below the vapor degreaser are not recommended and are prohibited by OSHA regulations. Specialized degreasers are designed to use a heat pump principle for both heating and vapor condensation. In this instance, the compressed gases from the heat pump are used for heating the vapor degreasing solvent and the expanded refrigeration gases are used for vapor condensation. Such a degreaser offers mobility that permits movement without connecting to water, steam or gas for operation.

Normal vapor control is achieved with plant water circulation through the condensing coils. Refrigeration-cooled water or direct expansion of the refrigeration gases in the condenser coils are effective means of vapor control where a sufficient cool water supply is not available; a low-boiling vapor degreasing solvent is chosen (such as methylene chloride or fluorocarbon 113) or plant water is excessively warm. A less common vapor control employing a vapor level control thermostat(s) can be used with the higher boiling solvents. In this vapor control method, the heat is intermittently turned on and off by one or more thermal sensors which detect the heat of the solvent vapors as they rise to a specific level.

For safety, economy and, in some cases, to comply with regulations, degreasers are usually equipped with a number of auxiliary devices. These devices include:

1. A Water Separator. This is a chamber designed to separate and remove water contamination from the degreaser (Fig. 7). Solvent and water condensate collected by the condenser coils is carried by the condensate collection trough and exterior plumbing to the water separator. The water separator is designed to contain five to six minutes of solvent and water condensate. This provides for nonturbulent flow and flotation of the insoluble water. This water is discharged from the equipment while the solvent condensate is returned to the degreasing equipment.
2. Safety Vapor Thermostat. This device, located just above the condensing coils, detects the solvent vapors if they rise above

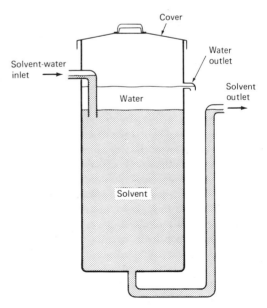

Fig. 7. Water separator.

the designed level in the equipment. This could occur with inadequately cool condensing water or condenser water flow interruption. When solvent vapors are detected, the heat input to the degreaser is turned off automatically. Manual resetting is preferred and commonly used. This demands attention and alerts the operator to a malfunction.

3. Boiling Sump Thermostat. In the cleaning operation, high-boiling oils and greases are removed and collect in the boiling chamber. These contaminants elevate the boiling temperature of the solvent and could cause solvent decomposition if left to accumulate without control. This device is located in the boiling chamber solvent and like the safety vapor thermostat, turns off the heat to the degreaser if it senses temperatures higher than those appropriate for the solvent being used.
4. Condenser Water Thermostat and/or Flow Control Switch. Again, these devices turn off the degreaser heat if the water flow through the condenser system is inadequate or the water temperature is insufficiently cool to control the solvent vapors in the degreaser.
5. Solvent Spray Thermostat. This temperature sensing device is located just below the normal vapor air interface in the degreaser and is designed to prevent manual or automatic spraying if the vapor zone is not at or above the thermostat level. This device has been required by some regulations, although no available devices are known to be effective. Spraying above the vapor zone can exaggerate solvent losses by causing air and solvent vapor mixing.

Modifications in this basic vapor degreaser are designed to permit various cleaning cycles including spraying of the work parts or immersing them in boiling or cool solvent (Fig. 8). Further vapor degreaser designs are available to provide various conveyor and transport means through the cleaning cycles. Common conveyor systems include the monorail vapor degreaser, the crossrod vapor degreaser, the vibratory conveyorized degreaser and the elevator degreaser. Open top degreasers constitute over 80% of the vapor degreasers used in industry. Their size ranges from benchtop models with perhaps 2 ft² (0.2 m²) of open top area to tanks over 100 ft long (30 m). The most common sizes range between 4 and 8 ft (1 and 3 m) long and 2 and 4 ft (0.5 and 1.5 m) wide. The most frequently used cleaning cycle is vapor-solvent spray-vapor. Among the conveyorized vapor degreasers, the monorail is the most prevalent. Generally, open top degreasers are much lower in cost, permit great flexibility in cleaning different workloads, occupy much less floor space and are adaptable to both maintenance and production cleaning. Due to their relative low cost and minimum space requirements, they are preferred for intermittent operations and for decentralized cleaning where transport of parts to be cleaned to a centralized location would add substantially to the cleaning cost.

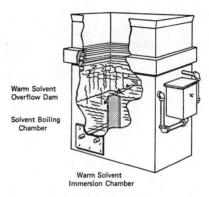

FIG. 8. Two compartment degreaser.

Monorail conveyor degreasers (Fig. 9) lend themselves particularly well to plants which use a monorail conveyor to transport parts from one operation to another. In such cases, the plant monorail system carries the parts down into the vapor zone for initial cleaning. The parts are moved through a stationary spray zone in the vapor zone and held in the vapor zone until they are heated to the vapor temperature. The monorail then carries them out of the equipment to the next operation. Often, no manual handling of the

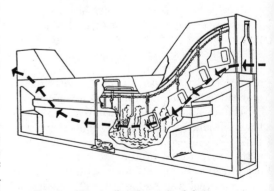

FIG. 9. Monorail conveyorized degreaser.

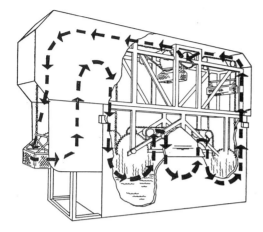

FIG. 10. Cross-rod conveyorized degreaser.

parts is needed. This system is often used prior to dip or spray painting.

A crossrod degreaser (Fig. 10) depends on two chains which carry work parts suspended from a rod between the chains through the process. Unlike the monorail conveyor which can only transport parts at a descending angle of about 45°, the crossrod degreaser can lower the parts vertically. This enables the crossrod degreaser to lower parts into sequential chambers of solvent for immersion cleaning. The parts carrier can be a simple pendant basket or can be a wire mesh cylinder. Where a wire mesh cylinder is used, the parts container is often rotated by a gear and pinion mechanism within the liquid baths to improve removal of metal fines and chips from recesses in the parts. The conveyor system can be designed to load on one end of the equipment and unload at the opposite end or the load and unload station can be located at the same end. When the parts workload is sufficiently high, an automatic loading and unloading conveyor can be used to save the labor of loading and unloading the parts. Since each basketload of parts is maintained discretely separate from the other loads, varieties of small parts can be cleaned in this equipment without intermixing one with another. This equipment is particularly useful in cleaning a variety of small parts with cavities or holes and removing metal fines and chips from the parts.

The ferris wheel (Fig. 11) degreaser is similar in its parts handling features to a crossrod. However, the parts are carried through the cleaning cycle on a rotating wheel which again may cause the parts to be tumbled in the solvent immersion stage. This equipment is smaller and

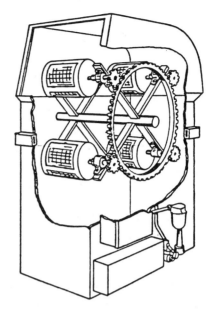

FIG. 11. Ferris wheel degreaser.

less costly than the crossrod but is able to process the work parts through only one immersion cycle. Parts loading and unloading is usually done manually in a manner similar to that of a crossrod degreaser.

Although a vibratory degreaser (Fig. 12) can handle a large variety of small parts, it is particularly well suited to cleaning parts coming from one or more high-volume manufacturing machines making the same part. Since this conveying means is essentially continuous, it does not keep different small parts separate from one another. Parts are fed into a chute which directs them to a circular pan flooded with solvent. The parts are moved by a vibratory motion through the pan and up a spiral where they are exposed first to solvent vapor condensation and then dried in air before being dis-

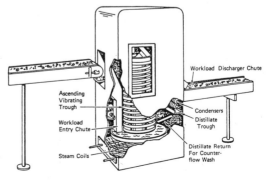

FIG. 12. Vibra degreaser.

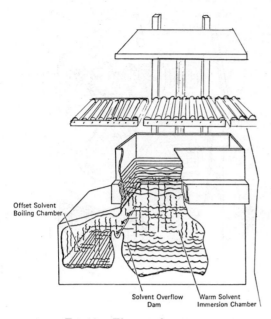

Fig. 13. Elevator degreaser.

charged through the exit chute. Of the conveyorized degreasers, this equipment occupies a minimum of floor space. However, the vibratory action is quite noisy and often requires noise insulation. Further, this vibratory action causes considerable wear and requires somewhat more maintenance. This conveyor system is especially well suited to the cleaning of rapidly manufactured cold headed parts such as nails or the cleaning of high alloy metal chips and turnings prior to recycling.

The elevator degreaser (Fig. 13) is very much like an open top degreaser. The parts can be transferred onto the elevator by belt conveyor or roller conveyor. The elevator lowers the parts into the vapor zone and can be indexed to hold them there for a specific interval. They may be further lowered into a solvent liquid chamber or sprayed or both before being held in the vapor zone again prior to removal. This kind of degreaser can be designed so that the cover is maintained in place on the equipment for all operations except loading and unloading. As such, it would meet the definition of "an enclosed degreaser design" as described in the section on regulations. Loading and unloading can be manually controlled or automated and the elevator can be designed to maintain the cover on the equipment when no parts are available for cleaning. The cost of an elevator degreaser will normally be considerably less than other conveyorized vapor degreasing systems. As in the case of other conveyorized degreasers, an elevator degreaser can reduce the labor costs of cleaning to a minimum. Although conveyorized degreasers are usually larger than open top degreasers, their solvent consumption per ton of parts cleaned is usually much less than in open top equipment.

MATERIALS OF CONSTRUCTION

Most vapor degreasing equipment is constructed of 316L or 304 stainless steel. Phenolic resin coating such as Heresite® over mild steel can be employed but is subject to cracking or chipping and subsequent corrosion. Galvanized and zinc metallized equipment is no longer recommended. Stainless steel or copper tubing are used for condenser coils. Centrifugal pumps are used for most solvent transfer and spraying operations. Cast iron construction with a stainless steel shaft and stainless or brass impeller and mechanical seals are recommended for this use. Aluminum pumps and/or impellers should not be used.

Gasketing materials must be solvent resistant. Impregnated cellulose fiber base materials or compressed asbestos impregnated with graphite, or Viton® or Teflon® polytetrafluoroethylene are suitable gasketing. Rubber gasketing will swell and lose strength in contact with the chlorinated solvents. Glycerin or ethylene glycol can be used as gasket lubricants. Graphite, molybdate and lithium stearate can be used as general solvent resistant lubrication on vapor degreasing machinery when needed. Stainless steel pipe and fittings are recommended for solvent plumbing. Standard black iron piping and storage tank construction is adequate for dry solvent storage and transfer. Teflon® tape is preferred over standard pipe dope for plumbing joints.

DESIGN CONSIDERATION

Heat Supply

Steam is a preferred heating means for vapor degreasing and distillation equipment because control of the steam pressure limits the temperature to which the solvent will be exposed. Allowing the solvent level to drop below a portion or all of the steam heating coil will usually not cause solvent decomposition because of the limited temperature exposure. By contrast, exposing electrical resistance heaters or gas fired immer-

®Registered Trademark E. I. duPont deNemours & Co.

sion tubes will often exceed the solvent thermal degradation temperature and result in the formation of hydrochloric acid in quantities which may be corrosive to the equipment and parts cleaned. Steam may not be available at some sites and does not lend itself to mobility of the vapor degreasing operation. Electrical heating is also common and contributes to permitting the degreasing operation to be moved from one place to another. When combined with refrigeration condensation of the solvent vapors, a small degreaser may become completely mobile. Electrical resistance heaters should have a watt density of 20 watts/in.^{2} or less to avoid excessively hot surface temperatures.

The final common heating means is by gas fired immersion combustion tubes. The exhaust of the combustion gases must be constructed of acid-proof insulating materials such as asbestos cement. Without insulation, this exhaust duct will be heated by the combustion gases and could cause solvent vapor decomposition on its surface. The required heat input per hour can be estimated by multiplying the maximum weight load to be cleaned in three minutes, times the specific heat of the metal involved, times the temperature rise from ambient to the boiling point of the solvent chosen, times 20. An estimate of the radiation heat losses from the degreaser plus the quantity of heat required to supply fresh distillate solvent for spraying or diluting clean solvent rinse chambers must be added. This value converted to kilowatt hours can be used directly for electrical heat or steam calculations. Approximately twice this energy input is required for gas fired combustion tubes due to the heat loss with the combustion gases.

Condensing System

The condenser coils and freeboard jacket must contain the solvent vapors in the degreaser. Usually, the condensing system will be designed to maintain the upper level of the vapor zone at or below the mid-point of the condenser coils. Therefore, the condensing coils should be calculated to handle approximately twice the heat input needed for a steam or electrically heated degreaser or roughly the same heat input for a gas fired degreaser. Control of the condenser water effluent temperature is achieved by locating a thermostat in the discharge pipe. The regulating valve and the manual control valve for the condenser coils should be located on the inlet side of the equipment to prevent line pressure build-up within the condenser system when the equipment is shut down.

Size

Open top degreasers should be designed to have an open top area 50% larger than the largest basket or single part to be processed through the equipment. If immersion cleaning is intended, the immersion tank must permit parts to be submerged within it.

Conveyor or Hoist Speed

The vertical component of travel for an overhead hoist or conveyor system should not exceed 11 ft/min. Thus, a monorail conveyor which descends at an angle of 45° could have a conveyor line speed of approximately 15½ ft (4.7 m) per min.

Freeboard

The freeboard on a vapor degreaser is the distance from the top of the vapor zone to the lip of an open top degreaser or to the lowest point of a part's entrance or exit on conveyorized equipment. This portion of the equipment is designed to prevent unnecessary disturbance of the vapor zone by air turbulence in the workroom. Current open top equipment is manufactured with a freeboard height equal to 75% of the width of the equipment or greater. Exceptions are made for extraordinarily wide equipment and the freeboard height may be increased if room air disturbance is expected to be greater than normal or if the equipment is long and narrow.

Ventilation

Normal room ventilation is frequently adequate to avoid excessive solvent vapor exposure to workers using or nearby the small degreasing operations. Where ventilation is needed to protect against vapor exposures or comply with regulations, the solvent loss rate from the equipment can be expected to be increased by as much as 100% of the normal solvent loss rate without ventilation. Open top degreasers are customarily vented by means of a lip exhaust on three sides of a degreaser. The design standard is 50 cfm/ft^2 of open top area. Degreasers equipped with a lip exhaust ventilation system should have covers which close the degreaser below the lip exhaust during shut down intervals. Ventilation for conveyorized degreasers usually provide for the removal of 50 to 100 cfm/ft^2 (15 to 30 m^3/min/m^2) of open area. When a degreaser is located in a pit, ventilation of the pit should be provided which will permit a minimum of two air changes per minute. Pit ventilation should be located at or near the bottom of the pit.

SELECTING VAPOR DEGREASING EQUIPMENT

Some of the key considerations involved in choosing vapor degreasing equipment are as follows:

1. The size and quantity of parts to be cleaned.
2. The variety and need for segregation of parts cleaned from one another.
3. The degree of cleanliness required.
4. The space available.
5. The advantages of cost and efficiency in centralized cleaning versus the handiness of multiple cleaning operations located near their need.
6. The type of soil to be removed.
7. The utilities available.
8. Capital cost considerations.

As discussed earlier, open top degreasers are less costly to purchase. This leads to their use where multiple cleaning stations are desirable or where the frequency of cleaning is less continuous. They also occupy a minimum of floor space.

Conversely, conveyorized equipment will often consume only half as much solvent per unit of parts production. This may amount to as little as $\frac{1}{2}$ to 1 gal/ton of free draining work parts cleaned. Open top degreasers tend to be labor-intensive, whereas conveyorized vapor degreasing equipment frequently requires little or no operator attention. The choice in vapor degreasing equipment should be made carefully. Advice can be obtained from solvent and/or equipment manufacturers.

LOCATION OF VAPOR DEGREASER

Space Needed

Whenever possible, a degreaser should be located in a space that allows a walkway on all sides of the equipment. Particular attention should be paid to providing the space necessary to open and remove clean-out doors or ports. Clean-out doors frequently house the heating coils. Space must be available to totally withdraw the heating coils from the vapor degreaser. Overhead room must be available on open top equipment to permit the workloads to be raised over the lip of the vapor degreaser. Where inadequate headroom is available, the degreaser may require installation in a pit. The top of an open top degreaser should extend at least 42 in. above floor level or be provided with a railing at this height to avoid a risk of personnel falling into the equipment. As noted earlier, pits should be equipped with ventilation to exhaust the air volume within the pit at least twice per minute. Drainage and a sump pump should be provided in the pit to remove any solvent or water spillage. If possible, this pump should transfer the solvent or water to a separate vessel where the water and solvent can be permitted to separate so that quantities of solvent will not be pumped to a sewer in case of a spill. The pit should be provided with adequate space to permit clean-out and maintenance and should be covered with a protective grating or other removable flooring.

Room Size

Experience has shown that degreasers located in a room having at least 2000 ft^3 of air/ft^2 (610 m^3/m^2) of open top degreaser area will usually provide sufficient ventilation to protect workers from excessive vapor exposure. Actual measurement of solvent vapor concentrations in the work area can determine the need for additional mechanical ventilation.

Gas Heated Degreasers

The combustion products from gas heated degreasers must be ventilated from the work area. Where a negative pressure exists in the operating room, positive exhaust of the combustion products must be provided by an exhaust fan.

Heat, Open Flame or Welding

Excessively hot surfaces such as ovens or open flames will decompose vapor degreasing solvents. Further, both the heat and the ultraviolet light emitted from welding operations can also cause solvent vapor decomposition. Solvent decomposition will form hydrochloric acid fumes and other corrosive and potentially toxic products. Usually, the corrosive products pose a greater risk of corrosion to exposed metal surfaces rather than a threat to workers. Consequently, vapor degreasers should not be located where the solvent vapors are likely to be exposed to these conditions. Examples of operations which should not be located near vapor degreasing operations are heat treating ovens, flame annealing, gas fired or radiant space heaters and infrared ovens, and welding.

Drafts

Normal room air circulation is favorable toward maintaining low solvent vapor concentrations in the work area. However, excessive room drafts caused by air intakes, fans, doors or windows, etc. can dramatically increase the solvent loss rate from the equipment whether conveyorized or open top. Location of the equipment in non-drafty areas is preferable. The use of baffles or partial enclosures can be used when the location would otherwise cause unnecessary vapor disturbance.

STARTUP PROCEDURE

1. A check should be made that clean-out ports are in place and sealed and all drain valves closed.
2. The ventilation should be turned on if available.
3. The condenser water should be turned on.
4. Electric power to the safety devices should be turned on if separate from the ventilation power.
5. Solvent should be added to all chambers within the equipment to their operating levels. A solvent transfer pump is recommended to avoid splashing.
6. Check and adjust all thermostat settings.
7. Turn on heat.
8. Observe vapor generation until assured that the condenser coils will maintain control of the vapor zone at or below the midpoint of the coils.
9. Start conveyor and spray pump if continuous spraying is employed. If spraying is manually controlled, check operation of spray pump.
10. Ensure solvent condensate flow through the water separator and return to the proper vapor degreaser compartment.
11. Check safety vapor thermostat and water flow control detector for proper operation.
12. Adjust condensor water discharge temperature to approximately 100° to 110°F (a temperature of approximately 80°F is preferred for fluorocarbon 113 and methylene chloride).
13. Work processing can begin. The workload should not cause excessive collapse of the solvent vapor zone.

SHUTDOWN PROCEDURE

Shutdown of a vapor degreasing operation follows essentially the reverse order of steps provided in the startup procedure. Care should be maintained that the condenser system is not turned off until the equipment has cooled sufficiently to collapse the vapor zone. The degreaser ventilation can be turned off when the degreaser cover(s) is positioned in place.

SOLVENT CONSERVATION

Economies in solvent consumption can be achieved by employing improved operating techniques, by equipment design and by the use of supplementary equipment. Good operating techniques are discussed as part of solvent regulations. Repetition of these techniques will be avoided in this section, with one exception. The well-disciplined use of covers on open top equipment is recognized as an outstanding means for economically controlling solvent losses. By placing a cover on the degreaser during non-use periods, solvent consumption has been reduced by 24 to 50% in actual using operations.

Equipment Design

Prior to the mid-1970's, the industrial standard for freeboard height on vapor degreasers was approximately 50% of the width of the degreaser. Since that time the freeboard design has been increased to 75% of the width as the result of experimentation showing that between 27 and 46% of the solvent loss could be reduced by this technique. Increasing the freeboard from 75% of the width to 100% of the width has been shown to reduce the solvent losses experienced with a 75% freeboard by as much as 39% where air turbulence in the room was present.

Covers for open top degreasers can be modified so that the equipment can be covered even while the cleaning operation is taking place. With this modification an open top degreaser can simulate the advantages of an elevator conveyor degreaser. This design modification is particularly useful when cleaning massive workloads requiring a prolonged cleaning cycle. Cover designs which slide horizontally across the open surface disturb the vapor zone less than covers that are hinged on one or both sides of the equipment. Covers can also be automated so that they will close without operator control after a workload has been cleaned and removed from the equipment.

Although most degreaser operators are diligent, they are subject to inattention, haste and distractions which can and do result in greater solvent consumption. Automated parts handling through open top equipment has recently been introduced. Like the automated cover operation, these mechanisms can remove operator error and substantially reduce the manual costs associated with open top vapor degreasing. These automated systems can be programmed to process parts through spray, vapor and solvent immersion cycles where the dwell time in each phase of the cleaning is controlled. Further, the parts baskets can be designed so that they can be rotated within the liquid or vapor or freeboard zones to prevent excess solvent dragout on the workparts.

Supplementary Conservation Equipment

Solvent stills (Fig. 14) are usually attached directly to large vapor degreasing operations such as the conveyorized degreasers. When connected directly to degreasing operations, the solvent stills can be operated by a float control or a level detector which will automatically transfer dirty solvent from the degreaser boiling chamber to the still. As in the degreaser, the solvent is boiled, producing pure solvent vapors. The solvent vapors are condensed on condensing coils, collected in a trough, routed through a water separator and returned to the degreasing operation or to clean solvent storage. Degreasing operations equipped with a still are constantly replenished with clean solvent and require far less frequent cleaning to remove accumulated soils. Alternatively, a centrally located still can be used to recover dirty solvent from several small degreasing operations. Distillation will recover the solvents or azeotropic solvent systems unless the solvent has been exposed to abusive conditions in the basic cleaning operation. The recovered solvent can usually be used without further treatment in the original cleaning operation. The value of the solvent recovered and reused will often recover the capital investment in distillation equipment in a short period of time. In addition, the labor costs associated with cleaning the vapor degreasing operation are reduced to a minimum and the degreasing operation can be conducted on a more continuous basis without production interruptions.

Refrigerated freeboard chillers (Fig. 15) have been used to reduce solvent losses. A second set of coils located above the condenser coils are installed in the vapor degreaser. Expansion of refrigerant gases within these coils or the circulation of water glycol refrigerated solutions

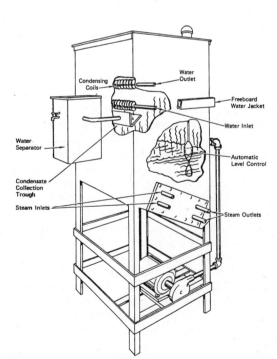

Fig. 14. Still.

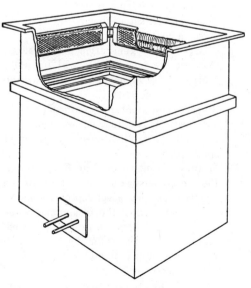

Fig. 15. Refrigerated freeboard chiller.

maintain these coils at a low temperature which chills the air above the vapor zone. This chilled air creates a secondary barrier to the loss of solvent vapors. Some atmospheric moisture or frost can be expected to build up on the refrigerated freeboard chiller coils and these coils must be equipped with a suitable refrigeration system. Where the coils are operated at a temperature below the freezing point of water, frost accumulates. A defrost cycle is incorporated on a time-scheduled basis to remove the frost. A separate collection trough is the preferred method of removing the water collected on these coils. By this means, a minimum of water contamination occurs within the equipment. Water contamination can result in equipment corrosion and water spotting of the cleaned parts. Although some solvent accumulates on the refrigerated coils, the primary means of reducing solvent vapor losses is through increasing their confinement within the equipment. In some instances, the introduction of water makes the application of this conservation device impractical for azeotropic co-solvents that are water soluble, such as the alcohols. Reductions in solvent consumption up to 60% have been measured in industrial operations using this equipment. A more conservative 40% reduction in solvent is recommended for evaluating a potential application of this device.

Carbon adsorption (Fig. 16a) is the only true solvent recovery process widely used. Vapors as dilute as 200 ppm in air can be collected by this process. A vapor degreasing operation using this recovery process must be equipped with a ventilation system to collect the solvent vapors in air. In the adsorption phase, the solvent vapor-air mixture is filtered and blown down through a bed of carbon granules or pellets (Fig. 16b). The activated carbon captures the solvent vapor from the air. The relatively pure air is exhausted through the bottom of the carbon bed and usually outside the plant. The carbon bed is said to be saturated when it begins to lose its effectiveness in collecting the solvent vapor.

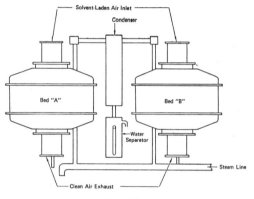

Fig. 16a.

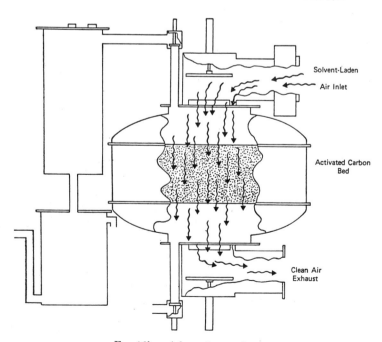

Fig. 16b. Adsorption cycle.

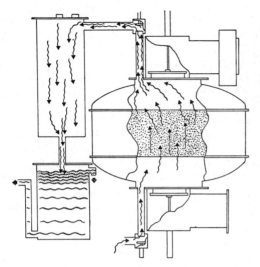

Fig. 16c. Desorption cycle.

The adsorption phase ends when bed saturation occurs or on a time cycled basis established by experience. Dampers close off the air inlet and outlet, and the desorption cycle begins (Fig. 16c). Steam is injected through the bottom of the bed, causing the solvent to vaporize from the bed. The uncondensed steam and solvent vapors are condensed to liquid solvent and water in a condenser. The solvent and water mixture is separated in a water separator and the solvent is returned to the operating process or to solvent storage. Most industrial carbon adsorption systems are time sequence automated dual bed systems so that one carbon bed can be on the adsorption phase of the cycle at all times.

Field studies of user carbon adsorption systems indicate that approximately 7 gal of steam condensate are collected per gal of solvent recovered in the desorption cycle. This results in a very large exposure of the solvent to water. This water exposure causes any water-soluble additives to be substantially removed from the solvent. Thus, the recovered solvent is very often stripped of its stabilizer system. In spite of this, this recovery system has been applied to perchloroethylene and trichloroethylene degreasing operations in reasonable numbers.

Solvent usage can be reduced by 40 to 50% and even higher by this system. Laboratory and field studies have shown carbon bed systems to be effective in capturing in excess of 95% of the vapors. However, ordinary ventilation systems are not nearly as effective in delivering the solvent-air mixture to the carbon bed system and increase solvent losses by disturbing the vapor zone. In addition to steam, water and electrical power, carbon adsorption systems require compressed air to operate the various valves and dampers. Experience has shown that these systems require moderately high maintenance to maintain good operating efficiency. The recovery of 1,1,1-trichloroethane by carbon adsorption is not generally recommended due to severe stabilizer removal. Likewise, methylene chloride is not well suited to carbon adsorption recovery due to its 2% solubility in water. Again, azeotropic blends containing water-soluble solvents cannot be recovered intact. The presence of large amounts of water creates serious corrosion potential. Phenolic resin coatings such as Heresite® over mild steel have been adequate for perchloroethylene and trichloroethylene. Hasteloy construction and titanium clad equipment has been offered for the exceptional applications recovering 1,1,1-trichloroethane or methylene chloride.

Of the methods to conserve solvent, carbon adsorption has the highest capital and utility costs, demands more maintenance for efficient operation, and severely depletes the solvent stabilizer systems. Large degreasing systems operating on a multi-shift basis provide the most economic application of this solvent recovery system, particularly where other solvent loss controls are impractical. The outstanding advantage of carbon adsorption is its ability to recover dilute solvent vapors from air.

CHOOSING A VAPOR DEGREASING SOLVENT

All of the vapor degreasing solvents or commercial azeotropic solvent blends possess the required characteristics of low flammability, low toxicity, good solvency, a relatively high vapor density, a low heat of vaporization (which provides a relatively high volume of solvent condensate), an intermediate boiling point and chemical stability. However, they are not all equally suited to all vapor degreasing operations and some possess special advantages to meet particular cleaning needs. Although trichloroethylene was the dominant vapor degreasing solvent prior to about 1970, 1,1,1-trichloroethane and trichloroethylene now share this position. Perchloroethylene is the third most popular vapor degreasing solvent followed by methylene chloride and fluorocarbon 113.

Some of the special considerations which influence the choice of vapor degreasing solvent are described below.

Regulations

More information on regulations of solvent cleaning operations is provided later. The most important change in the use pattern of vapor degreasing solvents was caused by air pollution control regulations. Trichloroethylene and perchloroethylene* have been identified and regulated as contributors to the formation of "smog" in the ambient atmosphere whereas 1,1,1-trichloroethane, methylene chloride and fluorocarbon 113 have been excluded from these regulations as non-smog forming agents. These regulations, introduced in the late 1960s and early 1970s, have resulted in the gradual replacement of trichloroethylene by 1,1,1-trichloroethane primarily and the other solvents to a lesser extent. Another regulatory influence has been Occupational Safety and Health Regulations to control the worker exposure to defined vapor concentrations. This has resulted in an increased use of vapor degreasing solvents with higher acceptable time weighted average exposure health standards, specifically 1,1,1-trichloroethane and fluorocarbon 113.

Available Steam Pressure

Many manufacturing facilities have low pressure steam (approximately 15 psig), suitable for trichloroethylene and 1,1,1-trichloroethane. Fewer plants have high-pressure steam in the range of 50 to 100 psig. Since steam is the preferred means of heating vapor degreasing equipment, perchloroethylene frequently cannot be used due to its requirement of 50 to 60 psig steam. The control of steam heat for low-boiling solvents (methylene chloride and fluorocarbon 113) is somewhat more sensitive and discourages their application to some extent. However, steam regulation and equipment designs can overcome the sensitivity of steam heating these solvents. All of the solvents can be heated with electrical resistance heaters or gas immersion combustion tubes.

*A recently published research study by the U.S.E.P.A. has shown that perchloroethylene does not contribute to "smog" or ozone and the EPA has proposed excluding it from ozone regulations.

Available Water Cooling

When plant cooling water is not available year around at temperatures of 50°F or lower, refrigerated condenser systems are required for methylene chloride and fluorocarbon 113. These refrigeration systems add to the overall cost of the vapor degreasing equipment but this cost may be offset at least partially by the improved portability of these vapor degreasers. Normal plant cooling water is usually adequate for controlling the vapors of 1,1,1-trichloroethane, trichloroethylene and perchloroethylene.

Special Solvency

Fluorocarbon 113 is recognized as being less aggressive on paints or identification markings for parts or assemblies. 1,1,1-Trichloroethane can often be used in operations where damage to markings for identification must be avoided. Conversely, methylene chloride and, to a lesser extent, trichloroethylene, can be effective in stripping lacquer or some uncured resin coatings prior to refinishing. The particular solvency and boiling temperature of trichloroethylene is used in electric motor rebuilding operations. Brief exposure to the trichloroethylene vapors performs simple cleaning while prolonged exposure to the solvent vapors softens the insulating varnish on the wire windings and permits easy removal for rewinding.

Water Removal

Each of the solvents used in vapor degreasing will displace water from parts mechanically and will azeotrope with the water so that it may be removed from the vapor degreaser. Perchloroethylene is the only solvent having a boiling point higher than that of water. This high-boiling solvent has been used in plating and jewelry operations as a rapid means of drying small parts. Special surfactant additives to fluorocarbon 113 have also been employed effectively to remove water from parts.

High Melting Soils

Heavy deposits of asphalt or creosote as encountered in maintenance cleaning of road equipment, the removal of cosmoline rust preventative and the removal of high melting waxes are improved by the high boiling point of perchloroethylene. Both the higher temperature and the prolonged bathing action provided by the increased volume of solvent condensate on the parts improve its cleaning efficiency. In some instances, these characteristics also yield better cleaning results on parts with a high surface to weight ratio such as radiators.

Heat Sensitive Parts

Where high-temperature exposure may cause parts damage, methylene chloride or fluorocarbon 113 are preferred vapor degreasing solvents. Some electronic components, thermometers and temperature switches fall in this category. Each of the vapor degreasing solvents can be used for calibration of thermal devices, depending on the temperature desired for calibration.

Price/Cost

The chlorinated solvents tend to be roughly price competitive on a volume basis. Fluorocarbon 113 usually costs three to four times as much as the chlorinated solvents. However, many other aspects of the cleaning operation can affect the overall cost of vapor degreasing with one or another of the solvents. If the wrong solvent choice has been made, damage to work parts or reprocessing due to poor cleaning can rapidly shift an apparent savings in purchase price of solvent. Generally, low-boiling solvents evaporate much more rapidly than the higher boiling solvents. But surprisingly, the solvent loss rate from an operating vapor degreaser tends to be lower for the lower boiling solvents. Differences in equipment costs, the costs of heating and cooling, the ease of parts handling after degreasing and the need for ventilation or emission control equipment are important considerations in determining the overall costs of vapor degreasing.

Heat Pump Degreasers

Degreasers of this design concept have been developed to operate for the lower boiling solvents including 1,1,1-trichloroethane. Heat pump application to trichloroethylene and perchloroethylene vapor degreasing is not available.

ULTRASONIC AGITATION

Ultrasonic agitation of the degreasing solvents can assist in the removal of insoluble soils, speed the removal of soluble matter and diminish the time necessary for cleaning both in cold cleaning operations and vapor degreasing operations. Ultrasonic agitation is achieved by transmitting high-frequency sound waves through the solvent. This high-frequency energy causes small voids or cavities to be formed in the fluid which collapse with great energy on the surface of the parts to be cleaned as well as in the general fluid. The formation of and collapse of these bubbles provides a strong scrubbing action which will remove insoluble matter such as dust, metal powder, metal fines, etc. Normal 60-cycle electrical current is transformed to current at 20 to 40 kilohertz (20,000 to 40,000 cycles/sec). This energy activates a transducer which may be magnetostrictive or piezoelectric crystals such as barium titanate or lead zirconate titanate. The transducer energized by this high-frequency electrical energy vibrates to transmit the ultrasonic energy through the cleaning fluid. Ultrasonic cleaning can be used in heated cold cleaning tanks or in clean solvent chambers of vapor degreasers. To operate efficiently, the solvent must be degassed. That is, the air dissolved in the solvent must be removed from it. Degassing occurs when the solvent is heated and/or boiled.

Each solvent has a temperature range in which ultrasonic cleaning or cavitation provides maximum effectiveness. Generally, this maximum effectiveness occurs in a range of 10° to 25° F below the solvent boiling temperature. If sonic energy is transmitting into a boiling solvent, the cavities formed do not collapse and the energy results in increased boiling but the aggressive cleaning action is reduced. Due to these conditions, the application of ultrasonic agitation is well-suited to vapor degreasing operations to enhance cleaning. Ultrasonics can be used in heated cold cleaning baths, however, the solvent loss rate from such systems can be expected to be substantial. Although this supplementary cleaning action is not inexpensive, it often permits the removal of insoluble soils which can cause product rejects or hand cleaning which would ultimately be more expensive.

METAL CLEANING COSTS

As noted in the introduction, the soils to be removed, type and size of parts, frequency of expected use, and the operation following cleaning all enter into the choice of cold cleaning, vapor degreasing or other cleaning processes, particularly alkaline washing. The wrong choice will result in poor cleaning, parts failure or high rejection rates which will overcome any anticipated costs savings projected. Cold cleaning is usually chosen for low volumes or intermittent uses or small wide spread multiple stations. Vapor degreasing fills the need for more regular cleaning, higher cleaning quality and more centralized cleaning, especially for larger conveyorize degreasers. Versus alkaline washing, vapor degreasing is preferred for: 1) removing solvent soluble soil; e.g., cutting lubricants; 2) cleaning of high precision small parts; 3) removing metal chips and fines; 4) concentrating waste for disposal rather than diluting in waste water; 5) high cleaning quality; 6) lower utility requirements; and, 7) often lower total operating costs.

The last two reasons may surprise even experienced metal cleaning people. Water would appear to be cheaper but it requires 10 or more times as much energy to dry it from metal parts than the solvents. To properly forecast cleaning costs all of the costs must be included. For example, chemical costs, energy, capital, maintenance, water treatment, waste disposal, building space, insurance and labor must be considered.

One cost comparison was reported by M. E. Baker and G. H. Hetrick of E. I. duPont de Nemours & Company in the third edition of this handbook. Although the unit costs have escalated for both vapor degreasing and alkaline washing, this study is presented in Table 4 as originally developed and is still a valid comparison.

Another, more recent, comparison of costs for cleaning was the subject of an article, "Vapor Degreasing or Alkaline Cleaning?" by K. S. Surprenant, Dow Chemical. The labor costs were the same for both operations in this evaluation. A summary of the comparison is shown in Table 5.

Additional cost studies are available from the following literature references:

1. "Metal Cleaning Costs," *Metal Progress* (August 15, 1955).
2. "How to Get the Most from Solvent Vapor Degreasing," *Metal Progress* (May 1960), p. 96.
3. Appendix, S. "The Control of Organic Solvent Emissions to the Atmosphere," *Report to Aerospace Industries Association of America* (March 1968), p. S-20.

TABLE 4. COMPARISON OF TOTAL CLEANING COSTS FOR VAPOR DEGREASING AND ALKALI WASHING*

	Vapor Degreaser	Washer-Dryer
Investment		
Cleaning equipment	$20,400	$26,000
Installation	4,000	5,000
Floor space @ $15/sq ft	8,400	11,500
General facilities @ 5% of above totaled	1,600	2,100
Total†	$34,400	$44,600
Hourly Operating Costs		
Solvent @ $1.37/gal	$4.16	$—
Chemicals @ $0.07/lb	—	0.53
Utilities		
Gas @ $0.625/1,000 cu ft	—	1.64
Steam @ $0.78/1,000 lb	1.56	4.22
Water @ $0.10/1,000 gal	0.24	0.01
Electricity @ $0.0125/KWH	0.03	0.63
Operating labor @ $6.00/hr including overhead	0.75	0.75
Maintenance (including equipment cleanouts)	0.59	2.13
Depreciation, taxes & insurance (based on installed equipment costs)	1.38	1.45
Total	$8.71	$11.36
Investment Carrying Cost		
@ 30% of total investment†/year	$5.16	$6.69
Total Hourly Cost	$13.87	$18.05
Total Annual Cost (2,000 hr/year)	$27,740	$36,100

* Basis: For cleaning 48,000 lb/hr of heavy sheet steel automotive parts prior to dip painting. Line speed 14 ft/min; operation 2,000 hr/yr. Equipment—Stainless clad monorail vapor-spray vapor degreaser with auxiliary still. Work space—700 sq ft. Monorail, steam-heated washer, rinse tank and gas-fired dryer. Work space—960 sq ft.
† The investment in gas and water distribution facilities, steam boilers ($6–15/lb/hr of steam), power distribution ($100–200/KW) and working capital (usually estimated at 100 calendar days' operating costs) are normally included as part of the total investment in any plant process Because of the variance in accounting procedures in the many metal fabricating plants, these items have been deleted from this comparison.

TABLE 5. OPERATING COST COMPARISON:
ALKALINE WASHING VERSUS VAPOR DEGREASING
(Cost per ton)

	Degreaser	Alkaline Washer
Equipment	$0.78	$0.83
Building	0.03	0.09
Insurance	0.12	0.14
Maintenance	0.93	0.12
Chemicals	0.79	1.21
Steam	0.49	1.03
Electric power	0.09	0.24
Water	0.10	Neg.
Waste water	Neg.	0.10
Make-up air heat		0.03
Totals	$3.33	$3.79

VAPOR DEGREASING PROCESS QUALITY CONTROL

Most vapor degreasing solvents require stabilizers to provide assured cleaning quality and protect against solvent decomposition resulting in corrosive acid formation (hydrochloric acid). These stabilizers boil with the solvents and are present in quantities sufficient for normal use. When solvent conservation extends the solvent use period, soils or parts are changed, distillation recovery or carbon adsorption recovery is started, or when a different degreasing solvent is used, a solvent analysis program is particularly advisable. Routine operations often can benefit by less frequent analysis.

The most sophisticated and thorough analysis is accomplished by gas chromatography (GC). This method can measure all of the stabilizers. GC analysis requires expensive equipment usually not available at user locations. However, this analysis is often supplied by local consulting laboratories and may be offered by the solvent producer.

Vapor degreasing stabilizer systems comprise multiple chemical additives designed for each particular solvent. In each case, one of those additives has the ability to neutralize small amounts of acid. This type of additive is referred to as an "acid acceptor." If the solvent is abused or a deficiency occurs in the other stabilizers, small amounts of acid will form and be neutralized by the acid acceptor. Thus, analysis of these acid acceptor chemicals generally indicate the condition of the entire stabilizer system. The acid acceptance value, expressed as equivalent sodium hydroxide, is determined by adding a known amount of acid to a solvent sample and titrating the residual acid after reaction with the stabilizer. This procedure is described in detail in ASTM. D 2942.

Acid acceptor test kits are available from some solvent suppliers also. These methods apply equally to cold cleaning operations, although cold cleaning places less stress on the solvents.

SOLVENT REGULATIONS

This section is not intended to be a comprehensive guide to compliance with solvent regulations. On occasion, federal, state and local regulations will apply. These regulations may require licensing, reporting solvent use and volume, equipment design, operating procedures, plumbing and electrical codes, etc. Although seemingly complex, compliance with these requirements becomes relatively simple when users or potential users avail themselves of the information resources. Regulatory information can be obtained for a specific location from equipment manufacturers, solvent suppliers, governmental agency people, local building contractors and tradesmen and corporate or local attorneys. The information presented here is intended to acquaint the reader with an overview of regulations as they apply to solvent cleaning.

Occupational Safety and Health Administration (OSHA)

OSHA has the primary responsibility for protecting worker health. Numerous general regulations exist which apply to open tanks or heated equipment. Examples of these include: 1) providing a cover, 2) guardrails for platforms or walkways, 3) an open-top edge or guardrail 42 in. high, 4) enclosed combustion heaters with corrosion resistant exhaust ducts. Where flammable solvents are employed, special requirements such as explosion-proof equipment and fusible link cover supports are required. Solvent spraying in general must be conducted in an enclosure to prevent spray discharge into the working area. Spraying in a vapor degreaser should be done only below the solvent vapor zone to prevent forcing air into the vapor zone. Welding and chlorinated solvent cleaning operations must be located separately so that the solvent vapors will not be drawn into welding areas. Exposure of the chlorinated solvent vapors to the ultraviolet light radiated by welding can cause solvent decomposition to corrosive and toxic products.

Aside from the risk of fire, the primary health hazard associated with solvent cleaning is the inhalation of excessive vapor concentrations. Acceptable Time-Weighted Average (TWA) vapor exposure standards have been adopted by OSHA for 8 hr/day. OSHA requires that worker exposures be maintained at or below these concentration limits. Mechanical ventilation may be required to control exposures below these concentrations. The measurement of actual exposures to vapor concentrations can be accomplished by industrial hygiene surveys using activated carbon collection tubes and calibrated air pumps, continuous reading vapor detectors and detector tubes. Additional information can be found in the May 29, 1971 *Federal Register*, p. 10,466, and June 27, 1974, p. 23,540.

Environmental Protection Agency (EPA) Air Regulations

The EPA air regulations were developed to limit the formation of ozone or "smog" in the ambient atmosphere. Nearly all hydrocarbon solvents react with the oxides of nitrogen in the presence of sunlight to form ozone. Three particular solvents have been found to cause essentially no ozone formation. They are methylene chloride, 1,1,1-trichloroethane and fluorocarbon 113. For this reason, these three solvents, along with methane and ethane, have been excluded from air regulations in most states. The EPA requires a state regulation to control solvent emissions to the atmosphere if one or more localities within a state exceeds the national ambient air quality standard for ozone (0.12 ppm). Most industrialized states have at least one area that exceeds this standard. These regulations vary from state to state. However, some understanding of these regulations in general can be obtained by reviewing the regulation guidelines to the states provided by the U.S. EPA. These guidelines are presented below.

The Resource Conservation and Recovery Act (RCRA)

The Resource Conservation and Recovery Act, also known as the Solid Waste Disposal Act, has as its objectives to promote the protection of health and the environment and to conserve valuable material and energy resources. Foremost in meeting these objectives was the regulating of treatment, storage, transportation and disposal of hazardous wastes which have adverse effects on health and the environment. Virtually all chemical wastes have the potential to be defined as hazardous since the EPA defined solid waste as any solid, liquid, semisolid or contained gaseous material resulting from industrial, commercial, mining or agricultural operations or from community activities. There are exceptions and a good background document is the May 1980 *Federal Register*, Vol. 45, No. 98, "Identification and Listing of Hazardous Waste."

These regulations may be administered federally or by the individual states. Wastes are categorized hazardous by individual chemical listing, residues from specified processes (e.g., electroplating) or based on a property of the waste. The characteristics of a hazardous waste are ignitability, corrosivity, reactivity or EP (Extraction Procedure) toxicity.

The hazardous waste regulations, in general, require the use of a manifest system to record the handling beginning with the source (generators), and tracing through storers, transporters, treaters and/or disposers. A generator is required to obtain an identification number to start the process. Each stage of the waste handling must be approved by the regulating agency.

Most electroplating wastes, including solvent residues, require disposal according to these regulations. Quantity exemptions (such as less than 1000 kg/mo) exist for some wastes, providing relief from paper work; however, proper waste disposal is still required.

Solvent distillation can reduce the quantity of waste to a minimum, particularly with the nonflammable vapor degreasing solvents. Under some circumstances, still bottoms (residues) can be used as a fuel in industrial boilers. Nonhazardous waste such as paper should be segregated from hazardous wastes to minimize disposal costs. Incineration is the best known ultimate disposal method for wastes from solvent cleaning operations. However, wastes containing quantities of solvent may be salable to local reclaimers.

The Clean Water Act

The Federal Water Pollution Act, known as the Clean Water Act, has the objective to restore and maintain the chemical, physical and biological integrity of the nations's surface waters. The

TABLE 6. CONTROL SYSTEMS FOR COLD CLEANING

Control System A
Control Equipment:
1. Cover
2. Facility for draining cleaned parts
3. Permanent, conspicuous label, summarizing the operating requirements.

Operating Requirements:
1. Do not dispose of waste solvent or transfer it to another party, such that greater than 20% of the waste (by weight) can evaporate into the atmosphere.* Store waste solvent only in covered containers.
2. Close degreaser cover whenever not handling parts in the cleaner.
3. Drain cleaned parts for at least 15 sec or until dripping ceases.

Control System B
Control Equipment:
1. Cover: Same as in System A, except if (a) solvent volatility is greater than 2 kPa (15 mm Hg or 0.3 psi) measured at 38°C (100°F),** (b) solvent is agitated, or (c) solvent is heated, then the cover must be designed so that it can be easily operated with one hand. (Covers for larger degreasers may require mechanical assistance, by spring loading, counterweighting or powered systems.)
2. Drainage facility: Same as in System A, except that if solvent volatility is greater than about 4.3 kPa (32 mm Hg or 0.6 psi) measured at 38°C (100°F), then the drainage facility must be internal, so that parts are enclosed under the cover while draining. The drainage facility may be external for applications where an internal type cannot fit into the cleaning system.
3. Label: Same as in System A.
4. If used, the solvent spray must be a solid, fluid stream (not a fine, atomized or shower type spray) and at a pressure which does not cause excessive splashing.
5. Major control device for highly volatile solvents: If the solvent volatility is > 4.3 kPa (33 mm Hg or 0.6 psi) measured at 38°C (100°F), or if solvent is heated above 50°C (120°F), then one of the following control devices must be used:
    a. Freeboard that gives a freeboard ratio*** ≥ 0.7
    b. Water cover (solvent must be insoluble in and heavier than water)
    c. Other systems of equivalent control, such as a refrigerated chiller or carbon adsorption.

Operating Requirements:
Same as in System A

*Water and solid waste regulations must also be complied with.
**Generally solvents consisting primarily of mineral spirits (Stoddard) have volatilities < 2 kPa.
***Freeboard ratio is defined as the freeboard height divided by the width of the degreaser.

---

primary impact of this legislation will be on aqueous processes like alkaline washing. Solvent operations are rarely designed in such a way that they discharge water containing solvent. The greatest risk of water contamination from solvents arises from improper disposal practices (controlled by RCRA) and spills. Even when contamination occurs, the halogenated solvents are rapidly volatilized from water if aerated or turbulent in flow.

Hazard Evaluation and Prevention

Total hazard evaluation is seldom practiced. It demands the assessment of all risks of an operation such as fire, exposure to toxic chemicals or electricity or radiation, falling, falling objects, being struck by moving objects, noise, etc. These risks are present in every industrial activity and they are often much greater in non-solvent using operations. For example, a comprehensive statistical study* (1957–1959) covering the U.S. industrial safety experience with trichloroethylene as a vapor degreasing solvent for the 1948–1957 period was conducted.

This study, which reviewed injury statistics, types and causes, preventive measures and plant conditions, produced the following facts:

1. Relative to other industrial processes, vapor degreasing with trichloroethylene exhibited a low-injury-frequency rate. This rate in the range of 0.3 injury per million man hours indicates the use of trichloroethylene in a vapor degreaser imposes a negligible additional hazard upon the operation

*Hargarten, J. J., G. H. Hetrick, and A. J. Fleming. *13th National Congress on Occupational Health,* 1960.

TABLE 7. COMPLETE CONTROL SYSTEMS FOR OPEN TOP VAPOR DEGREASERS

Control System A
Control Equipment:
  1. Cover that can be opened and closed easily without disturbing the vapor zone.
Operating Requirements:
  1. Keep cover closed at all times except when processing work loads through the degreaser.
  2. Minimize solvent carry-out by the following measures:
     a. Rack parts to allow full drainage.
     b. Move parts in and out of the degreaser at less than 3.3 m/sec (11 ft/min).
     c. Degrease the work load in the vapor zone at least 30 sec or until condensation ceases.
     d. Tip out any pools of solvent on the cleaned parts before removal.
     e. Allow parts to dry within the degreaser for at least 15 sec or until visually dry.
  3. Do not degrease porous or absorbent materials, such as cloth, leather, wood or rope.
  4. Work loads should not occupy more than half of the degreaser's open top area.
  5. The vapor level should not drop more than 10 cm (4 in) when the work load enters the vapor zone.
  6. Never spray above the vapor level.
  7. Repair solvent leaks immediately, or shut down the degreaser.
  8. Do not dispose of waste solvent or transfer it to another party such that greater than 20% of the waste (by weight) will evaporate into the atmosphere. Store waste solvent only in closed containers.
  9. Exhaust ventilation should not exceed 20 m^3/min/m^2 (65 cfm/ft^2) of degreaser open area, unless necessary to meet OSHA requirements. Ventilation fans should not be used near the degreaser opening.
  10. Water should not be visually detectable in solvent exiting the water separator.

Control System B
Control Equipment:
  1. Cover (same as in system A).
  2. Safety switches.
     a. Condenser flow switch and thermostat (shuts off sump heat if condenser coolant is either not circulating or too warm.)
     b. Spray safety switch (shuts off spray pump if the vapor level drops excessively, about 10 cm (4 in).)
  3. Major Control Device.
     Either: a. Freeboard ratio greater than or equal to 0.75, and if the degreaser opening is > 1 m^2 (10 ft^2), the cover must be powered,
             b. Refrigerated chiller,
             c. Enclosed design (cover or door opens only when the dry part is actually entering or exiting the degreaser.),
             d. Carbon adsorption system, with ventilation ≥ 15 m^3/min/m^2 (50 cfm/ft^2) of air/vapor area (when cover is open), and exhausting < 25 ppm solvent averaged over one complete adsorption cycle, or
             e. Control system, demonstrated to have control efficiency, equivalent to or better than any of the above.
  4. Permanent, conspicuous label, summarizing operating procedures #1 to #6.
Operating Requirements:
  Same as in System A

and maintenance of equivalent mechanical equipment.
2. Those injuries which occurred were largely associated with specific episodes of poor safety practices rather than from daily exposure. The most serious hazards were from entry into an operating degreaser and acute exposure while working alone during a degreaser cleanout.
3. Based on a large amount of air test data, solvent concentrations in the working area of the average degreaser were found to be well below recommended limits.

The most accessible reference to evaluate hazards associated with solvents is the Material Safety Data Sheets supplied by solvent producers. These data sheets discuss flammability, toxicity, first aid, fire protection recommendations, etc. Some other sources include:

1. *Handbook of Organic Industrial Solvents.*

TABLE 8. CONTROL SYSTEMS FOR CONVEYORIZED DEGREASERS*

Control System A
Control Equipment: None
Operating Requirements:
1. Exhaust ventilation should not exceed 20 m^3/min/m^2 (65 cfm/ft^2) of degreaser opening, unless necessary to meet OSHA requirements. Workplace fans should not be used near the degreaser opening.
2. Minimize carry-out emissions by:
   a. Racking parts for best drainage.
   b. Maintaining vertical conveyor speed at <3.3 m/min (11 ft/min).
3. Do not dispose of waste solvent or transfer it to another party such that greater than 20% of the waste (by weight) can evaporate into the atmosphere. Store waste solvent only in covered containers.
4. Repair solvent leaks immediately, or shut down the degreaser.
5. Water should not be visibly detectable in the solvent exiting the water separator.

Control System B
Control Equipment:
1. Major control devices; the degreaser must be controlled by either:
   a. Refrigerated chiller,
   b. Carbon adsorption system, with ventilation ≥ 15 m^2/min/m^2 (50 cfm/ft^2) of air/vapor area (when down-time covers are open), and exhausting < 25 ppm of solvent by volume averaged over a complete adsorption cycle, or
   c. System demonstrated to have control efficiency equivalent to or better than either of the above.
2. Either a drying tunnel, or another means such as rotating (tumbling) basket, sufficient to prevent cleaned parts from carrying out solvent liquid or vapor.
3. Safety switches.
   a. Condenser flow switch and thermostat (shuts off sump heat if coolant is either not circulating or too warm.)
   b. Spray safety switch (shuts off spray pump or conveyor if the vapor level drops excessively, e.g. >10 cm (4 in.).)
   c. Vapor level control thermostat (shuts off sump heat when vapor level rises too high.)
4. Minimized openings: Entrances and exits should silhouette work loads so that the average clearance (between parts and the edge of the degreaser opening) is either < 10 cm (4 in.) or < 10% of the width of the opening.
5. Down-time covers: Covers should be provided for closing off the entrance and exit during shutdown hours.

Operating Requirements:
1 to 5. Same as for System A.
6. Down-time cover must be placed over entrances and exits of conveyorized degreasers immediately after the conveyor and exhaust are shut down and removed just before they are started up.

*Conveyorized degreasers include both room temperature and vapor degreasing conveyorized equipment.

Alliance of American Insurers, 20 N. Wacker Drive, Chicago, IL 60606.
2. *Fire Hazard Properties of Flammable Liquids, Gases, Volatile Solids.* National Fire Protection Association, 470 Atlantic Avenue, Boston, MA 02210.
3. *Manual of Hazardous Chemical Reactions.* National Fire Protection Association.
4. *Threshold Limit Values for Chemical Substances and Physical Agents in the Workroom Environment.* American Conference of Governmental Industrial Hygienists, 65 Glenway Avenue, Bldg. D-5, Cincinnati, OH 45211.

Each year about 12,000 deaths, 123,000 injuries and hundreds of thousands of dollars in industrial property damage* result from fire. In evaluating solvent risk, the potential for fire must be the first order of business. Solvents with lower flash and fire points pose greater risk than higher flash point or non-flashing solvents. Where flammable solvents must be used, the fire risk can be reduced by ventilation of their vapors to levels below 25% of their lower flammable limit, the use of explosion-proof equipment and switches and the installation of automatic fire

*Statistical Abstract of the United States, 1980.

detection and extinguishing equipment. Although less risk exists with higher flash point solvents, spraying or cleaning with shop rags can significantly increase their ease of ignition at room temperature. All cleaning and transfer equipment should be grounded to avoid static sparks. Even in the use of the practically nonflammable chlorinated solvents or fluorocarbon 113, welding or torch cutting of equipment or drums which may contain vapors of these solvents is extremely dangerous. These same operations where flammable solvent vapors may be present can only be described as suicidal.

Contrary to popular belief, it is not necessary to choose between a fire hazard with flammable solvents and a toxic hazard with the common vapor degreasing solvents. All solvent vapors possess some toxic properties. Many of the petroleum or flammable solvents are equally or more toxic than the chlorinated solvents or fluorocarbon 113. With proper design of the cleaning operation and its ventilation, exposure to vapor concentrations is readily controlled to safe levels for both cold cleaning and vapor degreasing. Highly toxic solvents such as carbon tetrachloride or benzene are not recommended and should not be used!

Aside from fire, entering a confined or poorly vented area such as a pit or tank with high solvent vapors present poses the gravest risk from solvents. All volatile solvents (flammable and nonflammable) present this potential problem. Such exposures account for most non-fire related solvent deaths. To avoid this threat the following procedures are recommended.

1. Obtain written permission from supervision.
2. Prepare the tank or confined area by:
   a. Locking off electrical power.
   b. Shutting off any heat supply to the equipment.
   c. Draining all solvent.
   d. Venting to remove all solvent vapors and liquid.
   e. Open all clean out ports, removing sludge from outside the tank.
   f. Posting "Worker in Tank" signs in approaches.
3. Disconnect and block off all fill and drain pipes.
4. Continue ventilation throughout tank entry.
5. Check air in tank or area for:
   a. A minimum of 19.5% oxygen.
   b. Absence of flammable vapors.
   c. Vapor concentrations below the acceptable weighted average standard.
6. Wear a harness and lifeline attached to a manual or powered hoist, protective gloves and clothing, and carry an approved canister respirator (for escape only).
7. Station an observer with an auxiliary air supply by entrance continuously and maintain voice communication between observer and worker.
8. Be certain other workers are available within calling distance.

Solvent cleaning operations are cost-effective and experience excellent safety records when properly designed, operated and supplied with a reasonable solvent choice.

F. Alkaline Cleaning

Lawrence J. Durney

Although convenience of discussion requires that the subject of part preparation be separated into distinct areas, it is a mistake to think that these areas are not interrelated and interdependent. Rather than to think in terms of any single step, the plater is encouraged to think in terms of a preplating, or part preparation cycle, every step of which is affected by what went before, and in turn affects that which comes after.

The object of this preplating cycle is to remove those surface films which can be characterized as soils, and replace them with films which will be compatible with the solutions being used to apply the final finish. When the sequence is properly selected and operated, the parts will enter the final processing solution with a surface in an activated or receptive state for the finish to be applied. To accomplish this preparation, four basic steps are required:

1. Gross cleaning—the removal of heavy soil.
2. Fine cleaning—the removal of residues from gross cleaning, along with fine particulate matter.
3. Oxide removal—the removal of the thin layer of oxide which covers every metallic surface.
4. pH adjustment—to bring the residual surface film close to the same pH as the processing solution.

These basic steps constitute the objectives of each stage of the preplating cycle. The actual processing sequence may be considerably more elaborate. Any stage may require more than a single processing solution; some soils may require that a previous stage be repeated; the rinsing steps must be considered as part of each stage, not merely incidental; if a multi-component plate is to be applied, intermediate activating or preparatory steps may be required. The complete process, therefore, can become extensive.

In addition to preplating cycles, there also may be less demanding cycles for other manufacturing operations; cleaning prior to applying rust preventives; cleaning prior to conversion coating; and specialized cleaning operations.

Factors which affect all of these processes include:

1. The nature of the soil
2. The base metal
3. The finish to be applied.

SOILS

The definition of a soil may be compared to the definition of a weed. A weed is a plant that is out of place. A rose bush in a wheat field is a weed. A wheat stalk in a rose garden is a weed. Similarly, a soil is matter out of place. A rust-proofing oil on a part in storage is not a soil. Only when the part moves to the finishing room does it become a soil. The same is true for cutting oils used in machining; drawing or stamping lubricants; buffing compounds, etc. The air is filled with particulate matter, oil sprays and various fumes, all of which can settle out on parts in storage and which are lumped together as shop dirt. A part cannot be made without contaminating it to some degree with some sort of soil.

Soils not only vary in their basic nature, but the same soil may present varied cleaning problems, depending on the method of application and its history. Some soils are particularly susceptible to these effects.

Buffing Compounds

Buffing compounds are mixtures of lubricating materials (usually fatty acids), abrasives (complex silicates, carbides or metal oxides) and materials to control the melting point (often high-melt parafinnic compounds or waxes). Since the buffing process is a friction related process, very high temperatures may be generated at the point of contact, and all the ingredients can react with each other and the metal surface. These temperatures can vary widely with buffing conditions; the reactions can vary as well. It is not unusual, therefore, to find that the same parts buffed with the same buffing compound by two different buffers clean somewhat differently. Since the buffing residues often react with the base metal to form metal-organic compounds which then bind the residues to the surface, storage or transfer time between buffing and cleaning operations will also affect the cleaning process. In general, the shorter the delay between operations, the easier will be the cleaning process. In extreme cases of delay, it is possible for the buffing residues to react so extensively with the surface, that when they are removed, an etch pattern will remain.

Rust-proofing Compounds

Rust-proofing compounds can roughly be placed in three categories:

1. Inorganic, water-soluble compounds for protection between operations, or short-term protected storage. These normally do not present any cleaning problems.
2. Emulsifiable organic mixtures cut back with water to form the required emulsion. When the emulsion "breaks" due to a change of temperature or the evaporation of water, the organic portion is left on the surface as a protective film. The formulation usually contains one or more

volatile constituents which evaporate with the water during drying so the protective film is no longer emulsifiable. Protection is adequate for long-term protected storage, or interplant transfer. Cleaning problems are similar to the next category.

3. Solvent cutback organic mixtures provide a wide degree of protection, depending on composition and degree of cutback. Protection may be adequate to permit outdoor storage for reasonably extended periods. They may be formulated with water-displacing characteristics so parts to be protected may be immersed wet. The organic protective materials generally contain an oil base, a highly protective material such as a fatty acid, a metallic soap, or a polar material with an affinity for the substrate. If they are not fully dry to the touch they become magnets for shop dirt. Dryness or lack of tack is usually imparted by incorporating a wax, a drying oil or a film forming resin. Since these materials are designed to protect by preventing the penetration of moisture to the metal surface, they are often difficult to clean in aqueous systems. Solvent or vapor degreasing before aqueous cleaning is often helpful. A solvent dip to penetrate the film and reduce its viscosity also helps. If waxes are used for dryness, the temperature of the cleaning solution must be higher than the melting point of the wax.

Age of the film can be an important factor. Some of the polar materials may react with the metal surface. Unsaturated compounds may polymerize to form varnish-like materials. Evaporation of the solvent used for cutback will alter the viscosity. Coiled stock is particularly susceptible to these effects. Depending on the tightness of the coiling, these variations may occur at different rates in various areas of the coil. Hence, differences in cleaning requirements from point to point on the coil are not unusual.

Machining and Forming Oils

Increasingly often, these oils are being fortified with additives providing extreme pressure lubrication. Since these adhere strongly to the substrate, aggressive, high-alkalinity cleaners may be required. Machining and forming conditions by generating locally high temperatures can affect the cleaning process. Additionally, poorly designed or poorly maintained tooling can introduce surface conditions that complicate the cleaning process. Double-cleaning cycles may be required to compensate for these defects. Post-plating defects, including roughness, plate porosity, and spotting out can result from these causes.

Certain chlorinated or sulfonated oils may be gelled by high alkalinity cleaners, and low alkalinity materials may be more effective. Once gelled, they can be very difficult to remove. Where the presence of these materials is suspected, a sequence of low alkalinity followed by high alkalinity is the safest procedure.

Smuts

A smut is defined as finely divided particulate matter strongly adherent to the metal surface. It may be conductive or non-conductive. The non-conductive smuts consist of inorganic residues including carbon from acid treatment of high-carbon steels, or from heat treating operations such as oil quenching or controlled atmosphere heat treatment; pigments from the use of pigmented drawing compounds; insoluble constituents of an alloy brought to the surface by previous chemical treatment; e.g., silicon in aluminum alloys, beryllium in beryllium copper, etc.; abrasive compounds from buffing or mass finishing operations; mold residues from casting operations; and certain types of shop dirt.

This type of smut usually responds well to reverse current treatment in electrocleaners or alkaline descalers.

The conductive smuts usually consist of metallic fines or finely divided metallic oxides from a previous operation such as polishing, mechanical finishing, machining or forming. This type of smut does not usually respond well to electrolytic treatment, since the gas is generated at the surface of the smut rather than at the metal surface. Much of the lifting action of the gas is therefore lost. Relatively strong or specialized acid treatment is often the only effective procedure. It is not uncommon for oil films to be trapped under these smuts so their removal results in the reappearance of a "water break." Double cleaning cycles may be required.

Both types of smut can occasionally be held to the surface by either electrostatic or magnetic attraction. Ultrasonic or spray cleaning may be required to overcome these forces.

Base Metal Effects

The nature of the base metal has a critical bearing on the type of cleaning system selected. Materials must be selected to provide the required cleaning action without undue or selective attack on the base metal. Since metals vary greatly in reactivity, allowable limits of pH, temperature and concentration and the type and concentration of inhibiting agents are dictated by the base metal. Cleaners for aluminum or zinc will generally be quite different from those for brass or steel.

Finish Effects

Some finishes are applied from solutions which either have cleaning and deoxidizing action by virtue of their composition, or which are very tolerant of marginal part preparation. Cyanide zinc plating solutions fall into this category, and it has been common practice to use very condensed, and often marginal preparatory cycles ahead of these solutions. The wisdom of the approach is highly questionable, but acceptable, if not high-quality, work can be produced in this way. Nickel plating, on the other hand, is highly susceptible to improper part preparation, so more extensive and effective cycles must be used. Numerous other instances can be found to illustrate this point.

Tests for Cleaning

The literature on various tests for the effectiveness of cleaning procedures is extensive.[1-5] Unfortunately, with few exceptions, these remain research tools rather than production control methods. Among the methods suggested are analysis and control of numerous characteristics of the cleaning solution; the waterbreak test,[40,56,57,63] spray pattern test,[74a,b,c,d] atomizer test,[52] residual soil measurements,[19,34,56] residue pattern evaluation,[56] fluorescent dye evaluation,[61] radioactive tracer tests[41,42,83,85] and the copper sulfate test.[45,56,57,76]

In production, the cleaning process is usually controlled by a combination of solution analysis and the waterbreak test.

Solution Analysis

The increasing complexity of cleaning blends, and the widespread use of proprietaries, has reduced control by solution analysis to a simple test for solution concentration or total alkalinity, occasionally supplemented by an analysis for total surfactant concentration. These may be carried out by laboratory procedures, but most often are monitored with simple test kits provided by the suppliers. Dilute solutions, such as used in spray washing equipment, may be monitored by concentration measurements based on simple conductivity meters, or hand-held refractometers.

More extensive breakdowns involving measurements of pH, complete titration curves, emulsification characteristics, surface tension, colloidal suspension properties, etc. are generally reserved for use by the suppliers laboratory in trouble shooting, cleaner selection or the development of new materials.

The Waterbreak Test

This is based on the ability of a properly cleaned metal surface to retain an unbroken film of water. The test is subject to possible misinterpretation due to retained alkali from inadequate rinsing of cleaner residues, or the presence of hydrophilic smuts with oil trapped under the smut. These difficulties can be avoided by using a suitable acid treatment before making the observation. In production, parts are inspected at various stages of the preplating cycle for any evidence of "waterbreak" or failure to retain a continuous film of water.

The other tests listed for determining the cleanliness of a surface are used as research tools for evaluating cleaning mechanisms and developing new materials. Occasionally one or more of these techniques may be used as a control procedure in very demanding applications; e.g., sophisticated electronic manufacture, or space applications.

CLEANER OPERATION

Emulsifiable Solvents

These are mixtures of suitable solvents and sufficient concentrations of surfactants to cause the solvent to emulsify when added to water. They may be used full strength followed by a water rinse; or as a prepared emulsion, generally at a concentration of 5 to 10% by volume in water. A residual film is always left on the metal surface. They therefore are generally used as precleaners. Cleaning is by dissolving action of the solvent on oily soils present on the surface, although the surfactants used to form the emulsion provide additional cleaning action. The prepared emulsions are frequently used in spray units. Increasingly stringent restrictions on the levels of hexane solubles in effluents are having a negative influence on the use of these materials since they contribute heavily to this type of effluent contamination.

Buffing Compound Removers

These are essentially highly specialized forms of soak cleaners, designed for the effective removal of buffing compound residues. They fall into three basic categories:

1. Neutral detergent—usually liquids; mixtures of surfactants; pH close to neutral with buffering provided by the surfactants used. Concentrations in the range of 1 to 10% by volume
2. Enhanced detergent—similar to neutral detergent but fortified with organic alkalies which can react with the fatty acid in the buffing compound to form organic soaps. Concentrations 2 to 10% by volume.
3. Modified soak cleaners—similar to soak cleaners (q.v.) but modified to be especially effective on buffing compounds. Concentrations 45 to 120 g/l (6 to 16 oz/gal).

Types 1 and 2 often show poor performance on oily soils other than buffing compounds. Temperature of operation should be above the melting point of the buffing compound, 60° to 80°C (140° to 180°F). Use of ultrasound, or vigorous agitation will often permit operation at lower temperatures.

Alkaline Cleaners

Alkaline cleaners are blends of various inorganic alkaline salts with deflocculants, inhibitors and surfactants as required to provide the various cleaning mechanisms and functions discussed below.

Saponification. The chemical action by which a fatty acid, a fatty oil or other reactible soil is converted to a water-soluble compound such as a soap. Elevated temperature, concentration and pH promote the speed and completion of the reaction. The main advantage is that cleaning will proceed in the absence of surfactants, and that the reaction products may function as additional cleaning agents to improve the performance of the cleaner. Disadvantages include the fact that at least initially only reactible soils will be affected; the reaction products may build up to levels that cause rinsing and drying-on problems; incomplete rinsing may result in redeposition of the soils in a subsequent acid treatment; the solubilized soils unless separated will contribute heavily to hexane solubles in the effluent, and such separation is not always easy to attain.

Emulsification. The chemical process by which surfactants penetrate oily soils and break them down into globules sufficiently small to allow dispersion and suspension in the solution. Advantages include the fact that the reaction is often independent of pH; temperatures and concentrations required can be somewhat lower than with saponification; all types of oily soils will be removed; and rinsing will generally be somewhat better than for saponified soils. Disadvantages are similar to those for saponification except as noted, and with the added possibility that the surfactant concentration may be depleted at a rate different from the alkali depletion. The cleaner may therefore drift out of balance and fail to perform even when concentrations appear to be within limits.

Deflocculation. The process by which special chemical compounds surround particles of solid soil, removing them from the surface and dispersing them in solution. The process is generally improved by mechanical action and/or the development of gas by electrolysis. Elevated temperatures may also be helpful. Different deflocculants may be specific to certain solids, so complex soils may require mixtures of several agents for effective action.

Displacement. The process by which surfactants lift oily soils from the surface of the parts to be cleaned. A film of surfactant and solution is left on the part surface. The oily soil floats to the surface of the cleaning bath. Advantages include longer solution life and the possibility of operating at lower concentrations and temperatures. The main disadvantage is the need to continually skim the solution surface to remove the displaced oil. Failure to keep the solution surface properly cleared may result in the redeposition of the oily soil as the parts are removed from the solution. When properly operated, hexane solubles in the effluent are reduced, since the oily soil is constantly separated from the cleaning solution.

Spray Cleaners

Cleaning solutions which are sprayed on the parts, sometimes under considerable pressure. Any of the mechanisms previously discussed, including emulsified solvents may be used. Careful attention must be given to choosing materials with low-foaming characteristics. The combination of chemical action and the mechanical action of the spray produces effective cleaning. Spray patterns must be designed to provide complete coverage of the parts, and the units given periodic maintenance to insure that nozzles are not plugged. Except for the foaming requirement, alkaline spray cleaners are similar to soak cleaners. Concentrations and temperatures, however, are generally much lower, in the range of 15 to 30 g/l (2 to 4 oz/gal) and 35° to 60°C (100° to 140°F). The newer low-temperature spray cleaners often operate at 4 to 15 g/l (½ to 2 oz/gal) and 20° to 30°C (70° to 90°F). Liquid forms of the materials are sometimes available and operate at ½ to 2% by volume.

Ultrasonic Cleaning

The use in the cleaning process of ultrasonic energy provided by an ultrasonic generator which produces the necessary signal, and transducers which convert the signal to mechanical energy within the solution. The most commonly used frequency is probably 20 KHz, although higher frequencies (40 to 100 KHz) are sometimes used. The ultrasonic energy alternately compresses and expands the solution and produces several concurrent effects.

Cavitation. The alternate pressure effects can literally tear the solution apart to produce "cavitation bubbles." When these "bubbles" are collapsed on the compression portion of the cycle, high-pressure mechanical effects are created, blasting solid soil away from the surface. Pressures as high as 180,000 psi may be generated. In extreme cases, the substrate being cleaned may be etched or otherwise damaged. Cavitation occurrence will be increased, and cavitation pressures reduced by increasing temperature. It disappears completely at the boiling point.

Electrical Effects. Very high voltages may be developed across opposite faces of the cavitation bubble. These can neutralize electrostatic charges holding particles to the substrate being cleaned, or even produce oxidizing effects by generating ozone from oxygen dissolved in the solution.

Transmission Effects. Since any relatively rigid material will transmit and reradiate ultrasonic energy with only about 5% loss at each interface, ultrasonic cleaning is highly effective for blind holes, and internal threads or bores.

Soft, resilient materials such as rubber and plastic are energy absorbers and therefore do not respond well to ultrasonic cleaning.

Cleaning Materials. Any of the standard cleaning materials may be used with proper adjustment for the ultrasonic effect. Since much of the ultrasonic energy is converted to heat, solvents with low flash points or low boiling points may require cooling to avoid the possibility of fire or excessive evaporation. Alkaline cleaners will generally require better inhibition for sensitive metals, since the ultrasonic effects will increase the chemical action of the cleaner on the substrate. Cleaners can be specially formulated for ultrasonic use and/or to control cavitation pressures.

Equipment Considerations. The energy levels are generally calculated in terms of watt density/in.^{2} of surface to be cleaned. Common values are 5 to 10 W/in.^{2} since at higher levels a layer of heavy cavitation may form immediately adjacent to the transducer and prevent proper transfer of the energy into the solution. Only one side of the rack or part is used to calculate this "cleaning window." Transducer area must be adequate to provide the necessary energy without exceeding the 10 W/in.^{2} limit mentioned above. Provision should be made to remove solid materials from the active area by filtration, or by settling, to avoid attenuation of the ultrasonic energy, or marking of the substrate by ultrasonically agitated abrasive particles.

Soak Cleaners

The work-horses of the industry, they remove the major portion of the heavy oily soils, and often some of the solid soils. They are generally used at 60 to 120 g/l (8 to 16 oz/gal) with an average level of 75 to 90 g/l (10 to 12 oz/gal). Temperatures range from 50° to 95°C (120° to 200°F); more commonly 60° to 70°C (140° to 160°F). The newer, low-concentration, low-temperature materials will operate at 15 to 30 g/l (2 to 4 oz/gal) with a maximum of 45 g/l (6 oz/gal) at temperatures of 20° to 40°C (70° to 100°F) in the displacement mode and 30 to 60 g/l (4 to 8 oz/gal) with a maximum of 75 g/l (10 oz/gal) at temperatures of 25° to 40°C (80° to 110°F) in the emulsification mode. Soak cleaners consist of blended alkalies to establish the desired pH range and reserve alkalinity, surfactants for detergency, and often deflocculants for the removal of solid soil. Inhibitors for specific base metals may be included. The pH range is usually established by the reactivity of the metal to be cleaned; strongly alkaline materials often being used for steel, magnesium and copper alloys, and mild materials for zinc, aluminum, brass and other sensitive alloys. Generally only relatively mild materials can be considered for all-purpose use.

Soak Cleaning Equipment. Plain steel equipment is usually satisfactory. Tanks should be equipped with a bottom drain, a dam type overflow for skimming action, a grease trap and a circulating pump if displacement type cleaners are used, and a heating coil (see Fig. 1).

Electrocleaners

Alkaline blends for use with current. Work can be either cathodic or anodic, although general practice now emphasizes anodic use. Cathodic electrocleaning has the advantage that twice as much gas is developed to provide scrubbing action to remove solid soils. Since the gas developed is hydrogen, sensitive metals may be subject to hydrogen embrittlement. Additionally, permissible levels of chromate contamination are considerably lower with cathodic cleaning, and metallic contaminants may be plated out onto the work. Good practice, therefore, dictates that when cathodic cleaning is used, special precautions are taken. These include use of sufficient anodic cleaning to remove plated on contaminants; prevention of hexavalent chromium contamination; and special care in regard to hydrogen embrittlement possibilities.

Electrocleaners are commonly used at concentrations of 60 to 120 g/l (8 to 16 oz/gal) with an average of 75 to 90 g/l (10 to 12 oz/gal); temperatures from 50 to 95 C (120 to 200 F), more often 60 to 70 C (140 to 160 F), and current densities of 5 to 15 Å/dm^2 (50 to 150 ASF). The voltage required to develop the desired current density is dependent on the type of electrocleaner used, the tank configuration and the temperature. The normal voltage range is 6 to 12 volts. The newer, low-concentration, low-temperature materials will operate at 45 to 75 g/l (6 to 10 oz/gal) with a maximum of 90 g/l (12 oz/gal); temperatures of 45 to 75 C (80 to 110 F); current densities of 3 to 8 Å/dm^2 (30 to 80 ASF). For similar equipment configurations, required voltages will be somewhat higher than for heated materials.

Electrocleaners consist of blended alkalies to establish the desired pH range and reserve alkalinity (generally at least some free caustic or other highly conductive salt is present to promote conductivity); inhibitors to prevent attack of the base metal (frequently silicates of one type or another); deflocculants and/or complexers for solid soil and smut removal. Since precleaning has usually been carried out, surfactants are frequently limited to the amount needed to provide a mist suppressing foam blanket. Compromise materials with sufficient surfactant to act as both soak and electrocleaner are available. Foam levels must be watched carefully. The hydrogen and oxygen resulting from electrolysis will be trapped in the foam, and if foam levels

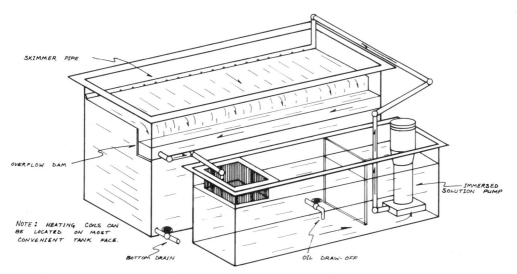

Fig. 1. Diagram of method of constructing a soak cleaning tank showing an automatic skimmer and an outboard separator tank for removing grease and oil.

are excessive, any sparks from poor contacts, vibration of racks, etc., can produce ear-shattering, although seldom dangerous, explosions.

Electrocleaners may be contained in plain steel equipment under normal conditions. In addition to the equipment listed for soak cleaning, the tank should be equipped with bus-bars, ventilation and a power source (see Fig. 2).

Alkaline Descalers

These are highly alkaline, heavily chelated or complexed materials, sometimes fortified with surfactants, used in a variety of different ways for the removal of rust, scale and smut.

For rust removal these descalers are used by simple immersion at concentrations of 90 to 360 g/l (12 to 48 oz/gal); temperatures from 60 to 90 C (140 to 190 F) for the removal of simple corrosion. When fortified with surfactants, they will also be effective in removing rust preventives, marking paints, etc.; e.g., pipe from outdoor or partially protected storage. Simple immersion is not generally effective on scales.

For smut removal they are used with reverse current (generally) for the removal of difficult smuts. Conditions of use and restrictions are similar to electrocleaners (*q.v.*). Since alkaline descalers do not contain inhibitors, excess current densities or dwell times may occasionally result in mild etching of high-current density areas of susceptible steels. Not generally recommended for non-ferrous metals. Not usually used cathodically since dissolved metals will plate out at high-current densities.

For scale removal they are used at 90 to 240 g/l (8 to 32 oz/gal); temperatures 50 to 80 C (120 to 180 F), current densities 5 to 15 Å/dm^2 (50 to 150 ASF). Sodium cyanide is often added at concentrations of 15 to 120 g/l (2 to 16 oz/gal) frequently in the ratio of two parts descaler to one part cyanide. Periodic reversal of current on a 7-sec direct, 7-sec reverse, or 10-10, cycle is effective on most scales. When cyanide is used, operating temperature should be limited to 60 C (140 F) to avoid destruction of cyanide and build-up of carbonates.

When periodic reversal is used, parts should only be removed from the tank on the reverse cycle to insure that plated out metals are removed from the high-current density areas. Cycles are sometimes set at 7–10 or 10–15, as additional protection against plating out. When used with periodic reversal, hydrogen embrittlement of susceptible steels can still occur, but severity appears to be less pronounced than with acidic scale removal (heavy pickling). Cyanide-free materials are available but their application and success appear to be limited to specific conditions.

Alkaline descalers generally consist of sodium hydroxide with high concentrations of complexers and/or soft chelating agents. Hard chelating agents are sometimes used at lower percentages. Rinses are usually tolerated by waste disposal systems, but heavy dumps are often disrup-

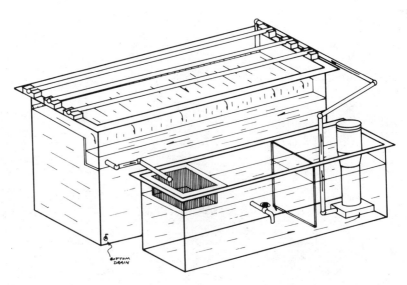

Fig. 2. Diagram of method of constructing an electrocleaner tank showing placement of anode and cathode bars.

tive. Solutions fortunately have excellent life, so dumps are relatively infrequent. When a solution is discarded, it should be transferred to a storage tank and bled into the waste disposal system slowly to avoid disruption.

Equipment required depends on use, similar to soak or electrocleaning q.v.

Aluminum Cleaners

Aluminum has high reactivity, an oxide film of variable thickness is always present, and many of the finishing processes for aluminum are especially sensitive to surface residues. The cleaning formulations are, therefore, restricted and because of this are grouped in a separate category. Cleaning materials developed for aluminum often will work quite well on other substrates; the reverse is not always true.

Aluminum cleaners are subdivided into non-etch and etch types. The non-etch types are generally used for processing buffed parts, or parts made from prefinished or bright rolled stock. The etch cleaners are used to produce matte finishes, and in particular are widely used to produce finishes which mask surface imperfections such as die marks, rolling imperfections, orange peel, score marks and other defects produced during, or as a result of, processing procedures.

The non-etch cleaners fall into two general categories: inhibited and non-inhibited.

The inhibited cleaners use the principle that alkaline silicates will react with aluminum to form insoluble aluminum silicate. Silicates, therefore, are incorporated in the cleaner as inhibitors against etching. The protective film has a finite time of formation, so some gassing may be evident on initial immersion. This rapidly diminishes and stops, however, and the resulting etch is microscopic and essentially invisible. The slight gassing that does occur helps to dislodge solid particles and speed cleaning.

Inhibited alkaline cleaners of this type can be formulated over a wide range of composition but can be split into two subclasses based on pH: those with a pH of 9.5 to 11.0, and those with a pH over 11.0. The higher-pH materials will develop heavier inhibiting films with corresponding problems of removal later. They do offer the advantage, however, that they are more effective on soils that can be removed by saponification. The lower-pH materials produce lighter inhibiting films more easily removed and are particularly effective on sulfurized oils. The time required for film formation is generally considerably shortened with low-pH materials and the microscopic etch during the induction period is correspondingly reduced.

Since the inhibiting action used in this type of cleaner is dependent on the formation of aluminum silicate, it is generally necessary to remove this film in a deoxidizing solution containing a controlled percentage of fluoride ion. Typical would be:

42 Be' nitric acid	50 to 75% by volume
Fluoride salt	15 to 120 g/l (2 to 16 oz/gal)

Powdered mixtures are also available either for use alone, or with an addition of nitric acid. Nitrate-free mixtures are also available.

Chemical polishing and certain conversion coatings are particularly susceptible to the presence of these silicate films. For these operations, consideration must be given to the use of the so-called non-silicated, non-etch cleaners.

These compounds are not in the strictest sense non-etching. They are actually carefully buffered mild etchants operated under controlled conditions to hold the etch to a microscopic level. These solutions are usually operated in the range of 23 to 60 g/l (3 to 8 oz/gal) 60 to 82 C (140 to 180 F). At the lower concentrations and temperatures, the etch will be essentially unnoticeable. At the more severe conditions, a definite "frosting" will usually occur, although the etch will not be so severe that it cannot be eliminated by chemical polishing or a bright plating operation. Careful attention to operational conditions is a must with these solutions.

While these cleaners, properly operated will leave the surface generally free of any film, it is common practice to follow their use with a simple nitric acid deoxidizing step—25 to 75% by volume of 42 Be' nitric acid—to insure the removal of any trace of alloying elements from the surface.

Etching cleaners for aluminum are based on the rapid reaction of aluminum with alkalies according to the reaction:

$$2Al + 2OH^- + 4H_2O \rightarrow 2H_2AlO_3^- + 3H_2.$$

Unless these solutions are modified by the addition of weak complexing and deflocculating agents, such as sodium gluconate, a hard scale of hydrated aluminum oxide quickly forms on tank walls, coils, etc. It may also form on the aluminum surface and cause non-uniform etch patterns. The presence of other alkalies such as carbonate or phosphate can also affect the appearance of the etch. Where a characteristic etch pattern must be maintained with unusual uniformity frequent analysis and close control of free alkalinity, carbonate concentration, complexer concentration and concentration of dissolved aluminum may be necessary.

Etching solutions of this type are relatively poor cleaners. The amount of gas evolved limits the amount and type of surfactant which can be employed without developing excessive foaming. Variable etch patterns delineating the soil distribution can occur easily. It is common practice, therefore, to preclean critical parts in a non-etch cleaner prior to etching. All of these etching solutions have relatively high rates of attack and leave the surface covered with a "smut" consisting of undissolved alloying elements. The degree of smut formation naturally varies with the alloy, and the degree of etch, from barely discernible, to heavy black, but can be removed in one of the acid deoxidizers mentioned above.

Cleaner Maintenance

Cleaner compositions have become sufficiently complicated that they are most often supplied as proprietary materials. In most instances, the formulations have been adjusted to compensate for losses and changes during use so that the only control needed is concentration and operating conditions (temperature, voltage, etc.). Concentration control is by titration, or in the case of certain spray cleaner or buffing compound removers by conductivity meters or simple refractometers to measure the refractive index. Simple test kits for operator use are often available from the supplier. Cleaner life will be dependent on the amount of work processed, the soil load and good housekeeping practice. Housekeeping should include:

1. Periodic settling and removal of solid soils and metallic particles, including, in particular, parts which have fallen to the bottom of the tank.
2. Periodic oil removal: for displacement types by frequent or continuous skimming; for emulsifying types by periodic cooling (to allow emulsified oil to discharge to the surface) followed by skimming.
3. Proper control of concentration with careful attention to proper techniques to insure complete solution of additions.
4. For electrocleaners, periodic cleaning of busbars, tank electrodes, exhaust ducts, etc.
5. Proper control of cleaner contaminants; e.g., reduction of hexavalent chrome in all cleaners, control of dissolved copper in electrocleaners, etc.

Cleaner Life

Cleaner life is dependent upon a number of factors, some of which are not controllable. These include 1) the formulation of the cleaner; 2) the nature and the amount of soil introduced; 3) the amount of dragout; and 4) the maintenance procedures used.

Formulation. Cleaner formulas can be varied to provide different degrees of cleaning capability and tolerance for soils. Increasing either capability generally increases cleaner cost more or less in direct proportion to the increased capability. As a rule of thumb, more expensive materials either do a better cleaning job, or last longer, or both. Purchasing the best material available, however, is not necessarily the best approach. Makeup and dragout cost may be unnecessarily high. (See cleaner cost calculation, which follows.) The ideal compromise is the least expensive material which will do a satisfactory job with a slight margin of safety.

Soil. Heavily soiled materials contaminate a cleaner more quickly than lightly soiled parts. A minor ingredient in a soil may selectively remove cleaner ingredients and reduce life.

Dragout. High dragout conditions may result in a complete replacement of the cleaner solution in a relatively short time span. In this case, a steady state may be established and a cleaner will have almost an indefinite life. Low dragout ratios conversely may result in rapid loading of the cleaner with a correspondingly shorter life.

Maintenance. The procedures listed above will often considerably lengthen the life of a cleaner. Similarly, complete lack of maintenance may shorten life.

Because of all these factors, prediction of cleaner life in a production situation is virtually impossible. A production test is the only reliable method of establishing the proper operating period.

Once the practical life of a cleaner has been established, periodic rechecks are essential. Any change in production load, soil load, soil nature or end requirements can change the life expectancy.

Cleaner Cost

The delivered cost of the cleaning compound and the total consumption do not by themselves give a true value for the cost of cleaning. Other factors are involved as well. The true cost includes:

Operating cost (S) includes the cost of chemicals, both makeup and replenishment, plus the energy costs for operating the process (heat and, in the case of electrocleaners, power). From this should be subtracted any reclaim value; e.g., the value of oil skimmed from the cleaner and added to the fuel oil used in boilers, etc.

Maintenance cost (M) includes labor and/or downtime required to dump the solutions, clean the tanks and build new solutions. It should also include the labor required to analyze the solutions, make additions, skim oil, etc.

Reprocessing costs (R) include the cost of reprocessing any rejects attributable to this part of the cycle. Certain highly critical applications may require zero reject assurance. In these cases, all costs become secondary to the accomplishment of the objective, and are therefore ignored. Examples could include space applications, some electronic applications, and parts for critical nuclear reactors. Most operations, however, can tolerate a low but finite percentage of rejects to reduce costs. The cost of reprocessing these rejects, however, must be factored into the overall cleaning cost. These reprocessing costs should include not only the cost of stripping, both labor and materials, but also an allowance for the production lost while reprocessing the rejected parts.

Waste disposal costs (W) include the cost of neutralizing and/or removing heavy metals from the cleaning solution before or while dumping.

If the production (P) is expressed in some meaningful units, generally thousands of square feet processed, and the life of the solution (D) in number of days, or shifts of production, the actual cleaning cost (C) is given by

$$C = S/PD + M/PD + R/P + W/PD.$$

Use of this formula will permit the evaluation of the savings from reduced energy costs by using low temperature materials, and other factors that will be neglected if only material costs are considered. It gives a true picture of the cleaning cost.

References

An excellent source for a comprehensive review of the literature on metal cleaning in all its phases may be obtained from Jay C. Harris, "Metal Cleaning Bibliographical Abstracts, 1842–1951" Special Technical Publication No. 90-B, American Society for Testing Materials, Philadelphia 3, Pa. (1953).

1. A.S.T.M. Standard Method Chemical Analysis Industrial Metal Cleaning Composition, D800–45.
2. A.S.T.M. Committee D-12, "A.S.T.M. Standards on Soaps and Other Detergents," American Society for Testing Materials, Philadelphia (Nov. 1947).
3. A.S.T.M. Standard Method of Total Immersion Corrosion Test of Water-Soluble Aluminum Cleaners, D930–49.
4. A.S.T.M. Standards on Soaps and Other Detergents, March 1953, Appendix I.
5. A.S.T.M. Standards on Soaps and Other Detergents, March 1953, Appendix II.
6. A.S.T.M. Standards on Soaps and Other Detergents, March 1953, Appendix III.
7. A.S.T.M. Standards on Soaps and Other Detergents, March 1953, Appendix IV.
8. Baker, E. M., and Schneidewind, R., *Trans. Electrochem. Soc.*, **45**, 327–352 (1924).
9. Blum, W., and Hogaboom, G. B., "Principles of Electroplating and Electroforming," pp. 200–211, New York, McGraw-Hill Book Co., 1949.
10. Burg, A., *Emaille Technische Monats-Blaetter*, **7**, No. 5, 51–54 (1931); *J. Inst. Met.*, **50**, 192 (1932).
11. Cowles Detergent Co., Cleveland, Ohio, "Cowles Metal Cleaning Tips" (June, Sept., Nov., 1940).
12. Crowther, H. G., *Metal Ind. (London)*, **56**, 23–25 (1940).
13. Dinley, C. F., *Products Finishing*, **6**, No. 10, 24–26, 28 (1942).
14. Dobbs, E. J., *J. Electrodepositors' Tech. Soc.*, **7**, 161–162 (1932).
15. Federal Specifications, Bur. of Ships 51 C-20 (INT).
16. Federal Specifications, Fed. Spec. P-D-236.
17. Federal Specifications, Navy Aer. Spec. RM-70 (1938).
18. Federal Specifications, Air Corps Spec. No. 20015 B (1940).
19. Federal Specifications, Navy Aer. Spec. C-109a (1942).
20. Federal Specifications. Navy Aer. Spec. C-113 (1941).

21. Federal Specifications, Navy Aer. Spec. C-114, Am. 1 (1942).
22. Federal Specifications, Army-Navy Aer. Spec. AN-O-T-631a, Am. 1 (1942).
23. Federal Specifications, Navy Aer. Spec. M-363a (1941).
24. Federal Specifications, Fed. Spec. P-C-576 (1942).
25. Federal Specifications, Ord. Dept. Tent. Spec. TM-38-305 (1943).
26. Federal Specifications, TAC ES-No. 382b (1943).
27. Federal Specifications, TAC ES-No. 452b (1943).
28. Federal Specifications, TAC ES-No. 542b (1943).
29. Federal Specifications, TAC ES-No. 645a (1943).
30. Federal Specifications, TAC ES-No. 398b (1943).
31. Ferguson, A. L., et al., *Monthly Rev. Am. Electroplaters' Soc.*, **32**, 894, 1006, 1116, 1237 (1945); **33**, 45, 166, 279, 620, 1285 (1946); *Plating*, **35**, 724–728 (1948).
32. Gavin, F. J., *Metal Cleaning Finishing*, **1**, 585, 586 (1929).
33. General Motors Approved Standards, Section A: Materials and Processes Cleaners Specs. O-11 to O-80. Cleaner Testing Procedures W-151 to W-153.
33a. Graham, A. K. Trans. Inst. Met. Finishing, Advance Copy 17 (1954).
34. Greulich, E., *Korrosion u. Metallschutz*, **14**, 340–345 (1938).
35. Griggs, F. E. P., *Can. Chem. Met.*, **20**, 258–260, 316–318 (1936).
36. Groggins, P. H., and Scholl, W., *Ind. Eng. Chem.*, **19**, 1029–1030 (1927).
37. Gwathmey, A. T., Leidheiser, H., Jr., and Smith, G. P., Nat. Advisory Comm. Aeronaut. Tech. Notes 1460–1461 (1948).
38. Halls, E. E., *Metal Treatment*, **6**, 131–133 (1940).
39. Harris, J. C., *Proc. Am. Electroplaters' Soc.*, **28**, 59–63 (1940).
40. Harris, J. C., Special Technical Publication No. 90-B, A.S.T.M., 1916 Race St., Phila. 3, Pa. (1953).
41. Harris, J. C., *A.S.T.M. Bull.*, **158**, 49–52 (1949).
42. Hensley, J. W., Skinner, H. A., and Suter, H. R., A.S.T.M., Spec. Tech. Pub. No. 115, 18–32 (1951).
43. Hermann, C. C., and Mitchell, R. W., *Iron Age*, **144**, 48–50 (July 6, 1939).
44. Hess, W. F., Wyant, R. A., and Averbach, B. L., *Welding J. (N.Y.)*, **23**, 402s–413s (1944).
45. Hogaboom, G. B., *Proc. Am. Electroplaters' Soc.*, **35**, 215–221 (1948).
46. Jesson, W. F., *Metallurgia*, **25**, 177–178 (1942).
47. Juraschek, F., *Iron Age*, **143**, 21–28, 48 (1939).
48. Kaye, A. L., *Metal Cleaning Finishing*, **3**, 9–12, 40, 71–72, 93–94, 179–182, 311–312 (1936).
49. Kushner, J. B., *Plating*, **36**, 798–801, 915–18 (1949).
50. Kushner, J. B., *Plating*, **38**, 933–5 (1951).
51. Liddiard, P. D., *Chemistry & Industry*, **60**, 480–482, 684–688, 713–716 (1941).
52. Linford, H. B., and Saubestre, E. B., *Plating*, **37**, 1265–1269 (1950); **38**, 60–65, 254 (1951); **38**, 367–375 (1951); **38**, 713–717 (1951); **38**, 847–855 (1951); **38**, 1157–1161 (1951); **38**, 1263–1266 (1951); **39**, 55–63 (1952); **40**, 379–386 (1953); **40**, 489–496 (1953); **40**, 633–634, 639–645 (1953); **40**, 1269–1271 (1953).
53. Lyons, E. H., Jr., *Trans. Electrochem. Soc.*, **80**, 367–386 (1941).
54. Macnaughtan, D. J., *Trans. Faraday Soc.*, **26**, 465–481 (1930).
55. Mabb, P., *Metallurgia*, **23**, 81–83 (1941).
56. Mankowich, A., *Metal Finishing*, **45**, No. 12, 77–78, 88 (1947).
57. Mantell, C. L., *Metal Cleaning Finishing*, **4**, 25–26, 33–34 (1932).
58. Merrill, R. C., Jr., *Ind. Eng. Chem., Anal. Ed.*, **15**, 743–746 (1943).
59. Mitchell, R. W., *Metal Cleaning Finishing*, **2**, 13, 111, 207, 299, 389, 485, 585, 673, 759, 839, 935, 1025 (1930).
60. Mitchell, R. W., *Proc. Am. Electroplaters' Soc.*, **26**, 238–241 (1938).
61. Morgan, O. M., and Lankler, J. G., *Ind. Eng. Chem., Anal. Ed.*, **14**, 725–726 (1942).
62. Myers, E. W., *Rev. Current Lit. Paint, Colour, Varnish and Allied Inds.*, **14**, 166 (1941).
63. Nielsen, C., *Steel*, **113**, No. 14, 106–108, 132–136 (1943).
64. Pinner, W. L., *Plating*, **40**, 1115–1118, 1123–1125 (1953).
65. Quadland, H. P., *Metal Finishing*, **41**, 463–465 (1943).
66. Rogner, H., *Korrosion u. Metallschutz*, **19**, No. 4, 113–116 (1943).
67. Sanders, R., *S.A.E. Journal*, **51**, 23–30 (1943).
68. Sanigar, E. B., *Monthly Rev. Am. Electroplaters' Soc.*, **20**, No. 3, 26–29 (1933).
69. Savage, F. K., *Monthly Rev. Am. Electroplaters' Soc.*, **25**, 445–450 (1938).
70. Scott, G. W., Jr., and Charles, E. B., *Welding J. (N.Y.)*, **23**, 1s–7s (1944).
71. Seligman, R., and Williams, P., *J. Inst. Metals*, **28**, 297–298 (1922).
72. Simmie, W. S., *Welding J. (N.Y.)*, **11**, 462–463 (1943)
73. Sizelove, R., *Monthly Rev. Am. Electroplaters' Soc.*, **30**, 54–56 (1943).
74. Spring, S., Forman, H. I., and Peale, L. F., *Ind. Eng. Chem., Anal. Ed.*, **18**, 201–204 (1946).
74a. Spring, S., and Peale, L. F., *Ind. Eng. Chem.*, **38**, 1063–66 (1946).

74b. Spring, S., and Peale, L. F., *Metal Prog.*, **51**, 102–6 (1947).
74c. Spring, S., and Peale, L. F., *Ind. Eng. Chem.*, **40**, 2099–2102 (1948).
74d. Spring, S., *Metal Finishing*, **48**, No. 3, 67–72, 74 (1950).
75. Strow, H., *Metals & Alloys*, (now Materials & Methods), **18**, 503–505 (1943).
76. Tartakovskaya, and Ivanova, M. B., *C.A.*, **34**, 3222 (1940).
77. Walker, W. H., et al., *J. Am. Chem. Soc.*, **29**, 1256 (1907).
79. A.S.T.M. "Designation B322—58T Tentative Recommended Practices for Cleaning Metals Prior to Electroplating," American Society for Testing Materials, Philadelphia (1958).
80. Engel, E., *Products Finishing*, **20**, No. 4, 38–50 (1956).
81. Frey, S. S., *Ultrasonic News*, **2**, No. 5, 8–10, (1958).
82. Harris, J. C., "Supplement to the Metal Cleaning Bibliographical Abstracts," Special Tech. Publication No. 90-D, American Society for Testing Materials, Philadelphia (1957).
83. Harris, J. C., Kamp, R. E., and Yanko, W. H., *J. Electrochem. Soc.*, **97**, 430–432 (1950).
84. Harris, J. C., Stericker, W., and Spring, S., A.S.T.M. Bull. No. 204, 31–34 (1955).
85. Hensley, J. W., and Ring, R. D., *Plating*, **42**, 1137–1143 (1955).
86. Hightower, F. W., *Plating*, **43**, 359–362 (1956).
87. Meyer, W. R., *Trans. Inst. Metal Finishing, Fourth International Conf.*, London, England, **31**, 290–292 (1954).
88. Meyer, W. R., Process for Cleaning and Plating Ferrous Metals, U.S. Patent 2,915,444 (Dec. 1, 1959).
89. Myers, B. S., *Metal Progr.*, **76**, No. 5, 108–116; No. 6, 103–113, 172 (1959).
90. Osipow, L., Segura, G., Jr., Snell, C., and Snell, F. D., *Ind. Eng. Chem.*, **45**, 2779–2782 (1953).
91. Rubinstein, M., "Metal Finishing Guidebook" (1961) 107–109, Metals and Plastics Publications, Inc., Westwood, N. J.
92. Scheifele, B. F. H., *Deut. Farben-Z.*, **8**, 8–18 (1954).
93. Spring, S., "Metal Finishing Guidebook" (1961) 232, Metals and Plastics Publications, Inc., Westwood, N. J.
94. Tesser, K., *Processes Metalware Industrie and Galvanotechnik*, **47**, No. 12, 552; Abstract—*Metal Finishing*, **56**, 75 (1958).
95. Ultrasonic Manufacturers Association, Engineering Standards Committee, *Ultrasonic News*, **3**, No. 1, 5–7 (1959).

G. Oxide Removal

Cloyd A. Snavely

Manager, Development Department, Battelle Development Corporation, Columbus, Ohio

and

Charles L. Faust

Formerly Chief, Electrochemical Engineering Division, Battelle Memorial Institute, Columbus Laboratories, Columbus, Ohio

Revised by Lawrence J. Durney

Oxide-removal operations represent the greatest single use of chemicals in metal working. During 1952, such operations consumed 5% of all sulfuric acid produced, 25% of the hydrochloric acid and most of the hydrofluoric acid.[3] In addition, large quantities of nitric acid, phosphoric acid, and ferric sulfate are used. Oxide removal in the metal-finishing trade is far overshadowed by the large-scale pickling operations necessary during metal production. In general, heavy oxide scales are produced and must be removed during metal-production operations.

Oxide removal prior to plating is usually necessary because heat treating, welding, or other similar operations have oxidized surfaces which are to be plated. Occasionally, work is allowed to rust and corrode between fabrication steps, necessitating oxide-removal treatments. Reclaiming old machinery and fabricated metal parts requires special oxide-removal operations. Bright dipping and removal of superficial oxide films are an important part of metal-finishing procedures.

Character of Oxides and Scales on Metals

The character of oxides and scales on metals varies widely depending on how they are produced. An oxide produced on iron by simple rusting under atmospheric conditions is somewhat different from a scale produced during heat treating. Most oxide layers vary in composition from the outside to the inside, the outside being more fully oxidized. For instance, heat-treated scales on steel consist largely of Fe_2O_3 on the outside, Fe_3O_4 as an intermediate phase, and approximately FeO next to the metal.[3] The same general condition is sometimes noted for copper alloys with CuO on the outside and more or less Cu_2O on the inside.

Oxide layers are sometimes cracked and permeable because of volume changes which take place during their formation or on cooling from elevated temperatures. Some of the most difficult metals to descale are those forming tightly adhering, dense layers of acid-resistant oxide. Some of the AISI 400 series steels (stainless alloys) are of this type.

Figure 1 shows a tight oxide scale on hot-rolled steel. The surface shown is a taper section. The region of scale shown is about 0.001 in. wide.

Figure 2 is a taper section of a similarly scaled surface after pickling.

Reactions Between Oxides and Acids

Most pickling operations simply involve solution of oxide scale in acid. For steel, the following reactions apply:

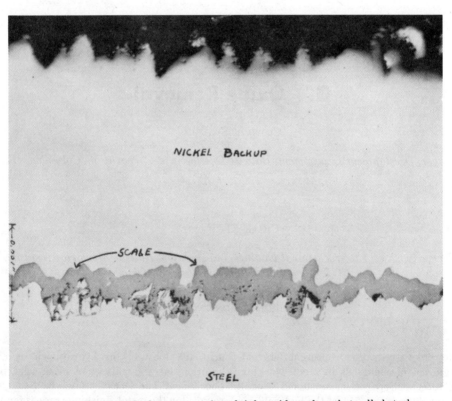

FIG. 1. Photograph of a taper section of tight oxide scale on hot-rolled steel.

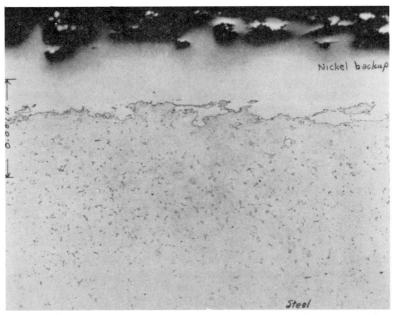

FIG. 2. Photograph of a taper section of a hot-rolled steel surface after pickling.

$$Fe_2O_3 + 2H_2SO_4 + H_2 \rightarrow 2FeSO_4 + 3H_2O$$
$$Fe_3O_4 + 3H_2SO_4 + H_2 \rightarrow 3FeSO_4 + 4H_2O$$
$$FeO + H_2SO_4 \rightarrow FeSO_4 + H_2O.$$

Hydrogen may be supplied by the reaction of acid with steel:

$$Fe + H_2SO_4 \rightarrow FeSO_4 + H_2.$$

The same general types of reactions can be written for hydrochloric, phosphoric, and other acids, as follows:

$$MO_2 + 2HA \rightarrow MA_2 + H_2O$$

(M represents the metal).

These reactions are usually speeded up by an increase in acid concentration, temperature, or agitation. They decrease in rate as the solution becomes more concentrated in the metal being dissolved.

Oxide Removal from Low and Medium Alloy Steels

Oxide removal, pickling, or descaling operations are practiced to a far greater extent on low and medium alloy steels than on all other classes of metallic materials combined. This is true for total tonnages as well as for the number of descaling installations. Most steel in fabricated articles has undergone at least one descaling operation and in many cases three or four at various stages of manufacture. Because of this commercial interest, there is more detailed technical information available on oxide removal from steels than for most other metals.

Pickling with Sulfuric Acid. *Pickling Process.* Sulfuric acid is the most common pickling agent. This is largely because it is the lowest cost acid available for the job. The operation is simple. The steel is immersed in tanks of acid in such a fashion that the acid can reach all surfaces. The acid first attacks the scale and exposes the clean metal surface. Attack on the metal then begins. When the scale is uniform and all surfaces are equally exposed to the acid, the work can be removed as soon as it is scale free, thus avoiding any serious attack on the metal. When the descaling action is spotty because of nonuniform scale or other factors, overpickling often occurs in localized areas. This produces an extremely rough surface and spoils the work for many applications.

An inhibitor is usually added to the pickling bath to reduce the rate of attack on the metal. Inhibitors have little effect on the rate of descaling. Their effectiveness in protecting the metal varies. If used in large amounts, some will almost completely stop the attack on low carbon steel but are less effective for high carbon steel.

Specially built racks are used for sheets or open coils. Much steel is pickled in continuous strips or wires reeled through long tanks or series of tanks. The most awesome equipment arrangements of this sort are used in electrogalvanizing or electrotinning lines.[38] The operation simply involves immersion of the steel in sulfuric acid solution. Only the equipment is complex.

Small parts are usually placed in baskets of acid-resistant alloy for pickling. Occasional shaking or turning may be necessary to obtain even pickling on all surfaces.

Effect of Acid Concentration and Temperature. As the temperature or acid concentration of the pickling bath is increased, the pickling time is reduced as illustrated in Fig. 3 for a hot-rolled plain-carbon steel. Benefits from increasing the acid concentration above 15% (by weight) are not significant at temperatures in the range shown. However, the higher boiling point of more concentrated acid allows its use at even higher temperatures than 100C (212F), with an increase in pickling rate. A few continuous strip-pickling installations are operated under these extreme conditions to attain desired strip speeds in the length of tanks available.

The acid concentration and temperature to be selected are related to economic factors. Where tank space is ample, and the steel need not be rushed through to the next operation, a 5% (by weight) acid solution operated at 65C (150F) is often considered most economical. The rate of attack is slow enough to minimize danger of overpickling. The acid consumed during operation may be replenished from time to time. The ferrous sulfate produced in the pickling operation gradually reduces the pickling rate until the bath must be discarded.

Discarding the 5% acid bath represents less loss of unused acid than is the case for more concentrated baths. Sometimes more concentrated baths are used and are allowed to deteriorate to low acid contents just before they are discarded. By good operation, less than 20% of the acid is wasted.

In general, continuous strip-pickling lines are operated at 10% acid concentration or above. The extra investment in longer tanks to use lower acid concentration is not justified. These installations must operate with the pickling rate reasonably constant. Therefore, the acid level cannot be allowed to go down appreciably before discarding the baths. In such installations, over half of the acid is usually wasted in the spent pickle liquor.

One method for eliminating spent pickle-liquor waste involves concentrating the acid and using the low solubility of ferrous sulfate at high temperatures to precipitate the dissolved iron. Such practices are usually uneconomical, but are becoming necessary to comply with stream pollution regulations.

Effect of Dissolved Iron. Figure 4 shows the effect of dissolved iron on the pickling rate. In practice, this effect is most serious where high-speed pickling is essential, and the operation cannot be slowed down to match the declining activity of the acid.

Effect of Scale Breaking. Most high-speed continuous-strip pickling lines have a scale-

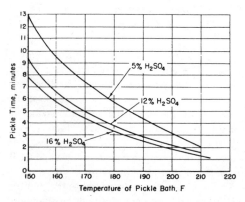

Fig. 3. Time-temperature relationship for pickling hot-rolled plain-carbon steel in 5, 12, and 16 per cent sulfuric acid solutions (b weight). (*Courtesy of National Electric Products Corp.*)

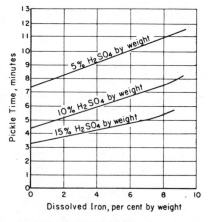

Fig. 4. Effect of dissolved iron on the time of pickling hot-rolled boxing strap at 140 F.

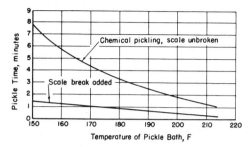

FIG. 5. The effect of scale breaking on pickling time for a hot-rolled plain-carbon steel. (16 per cent H_2SO_4, by weight.) (*Courtesy of National Electric Products Corp.*)

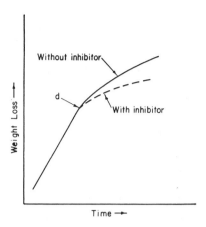

FIG. 6. Effect of a pickling inhibitor in reducing the attack of acid on steel. "d" indicates point where scale removal is complete.

breaking operation immediately before pickling.[26, 38] Scale breaking may be done either by bending the strip around relatively small diameter rolls or by a light pass through a conventional rolling mill[13]. In some cases, a major portion of the scale can be removed by the scale breaker. The scale remaining on the work is cracked and fragmented so that the pickling rate is greatly increased. Figure 5 illustrates this effect. The advantages of scale breaking are less at the higher pickling temperatures.

Scale breaking has several disadvantages. Both scale breaking methods work-harden the steel. The flexing method has a greater hardening effect than the simple-role reduction method. The change in physical properties may be great enough to make the steel unsuitable for subsequent forming operations. In addition, the scale particles may be "rolled into" the steel surface, thus roughening it. The importance of these factors depends on the condition of the steel and scale, and the intended use of steel.

The Effect of Inhibitors. Inhibitors are added to pickling baths to reduce the amount of attack on the steel in areas where the scale is removed. This has been a fertile field for commercialization and a considerable number of different materials are marketed under various trade names. These materials, liquid or soluble solid, are added to the pickling bath in amounts up to approximately 10 per cent. The efffect of a standard commercial product is illustrated in Fig. 6. The inhibitor normally has no appreciable effect on the rate of scale removal.

The exact mechanism by which inhibitors function may not be conclusively established. It is agreed, however, that they form adsorbed films on the clean metal surface.[9,41] If these adsorbed films remain during later operations such as plating, they may affect the operation of the bath or the adherence of the plates. Therefore, inhibitors should be tested for compatibility with operations subsequent to pickling.

Electrolytic Pickling. Electrolytic action can be used to increase the speed of acid pickling. The electric current may be alternate or direct, with the work anodic or cathodic. As shown in Fig. 7, cathodic pickling is the most rapid method. However, the advantage decreases greatly with a rise in bath temperature and acid concentration. For most installations, acid and heat are less costly than direct current, thus electrolytic pickling is not economical. Cathodic pickling is sometimes considered objectionable because it charges the steel with hydrogen at a faster rate than chemical pickling.

The Bullard-Dunn[15, 20] process, a cathodic electropickling process in which lead or tin deposits are used to protect exposed descaled areas, is no longer used.

It has been replaced by proprietary powdered acid salts, utilizing an immersion deposit of tin to provide inhibition. In many cases it is unnecessary to remove the deposit before further processing. Where removal is desirable, however, it is readily accomplished by reverse current treatment in high-alkalinity electrocleaners, or by short immersion in an alkaline tin stripping solution.

Pickling with Hydrochloric or Phosphoric Acids. *Hydrochloric Acid.* Hydrochloric acid is a good pickling agent. It is used less than sulfuric acid because it is more costly and the fumes are a greater problem. However, many small establishments prefer hydrochloric acid

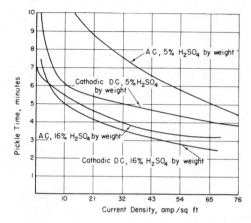

FIG. 7. Effect of current density on pickling time for anodic, cathodic, and alternating current electrolytic pickling. (*Courtesy of National Electric Products Corp.*)

because it gives a good pickling rate at atmospheric temperatures. Therefore, a crock or tank of hydrochloric acid can be used intermittently as needed. Because of its quick action on cold work, hydrochloric acid also is favored for light acid dips immediately preceding plating. Another advantage is the high solubility of the iron chloride salts.

The pickling rate for hydrochloric acid mixtures can be increased by raising the temperature. However, loss of HCl as vapor and corrosion of equipment present serious problems at temperatures above about 130°F.

Phosphoric Acid. Phosphoric acid is a good pickling agent, intermediate in cost between sulfuric and hydrochloric acids. It leaves the pickled steel in a somewhat passivated state having formed iron phosphates at the descaled surface. This is helpful in some cases where rust-resistance is required but unsatisfactory if the steel is to be electroplated after pickling. The pickling rate is slower than with hydrochloric- or sulfuric-acid solutions. Heating is necessary for satisfactory pickling rates.

The advent of the newer types of ion exchange resins makes it technically feasible to pickle with phosphoric acid, and then regenerate the acid by passing the spent liquor through an ion exchange bed[29]. The ion exchange bed would be reactivated with sulfuric acid. Such procedures are not yet in common use.

Gas-Phase Pickling. Gas-phase pickling involves conversion of the iron oxides to ferrous chloride and water and driving off these products as vapors[39]. The work to be pickled is heated in a special furnace. Dry HCl diluted with combustion gases is fed to the furnace. The $FeCl_2$ is collected in a condenser.

This method has been tried in at least one commercial installation but has not been sufficiently successful to warrant continued use. One of the most logical applications appears to be in hot-dip galvanizing where continuous strip could be fed from the gas-pickling furnace directly to the galvanizing pot.

Other Descaling Methods

Alkaline Electrolytic Descaling. Alkaline electrolytic descaling methods have found wide acceptance. (See *Alkaline Descalers*, Section 3F). The major drawback of these methods is the need for the use of sodium cyanide for certain types of scale removal.

Abrasive Descaling Methods. Abrasive descaling is usually somewhat more costly than chemical descaling but may be used for special purposes. In some cases, the abraded finish is desirable[4, 13, 24, 32]. Blasting with sand, steel grit or shot, or with wet abrasive may be employed for special effects. Wet abrasive blasting, in particular, is increasing in use as a preparation for plating. A slurry of fine abrasive in water is pumped or aspirated to a blast nozzle and discharged against the work. A fine mat finish is produced. The smaller sizes of steel grit, used dry, also produce surfaces suitable for many types of electroplating.

Oxide Removal from Stainless Steels and Corrosion-Resistant Alloys

Oxide removal from stainless steels and corrosion-resistant alloys is greatly complicated by the differences in chemical behavior of this class of materials. In general, sulfuric-acid pickling alone is unsatisfactory, because of the black smut left on the work and/or length of pickling time required. A 10% nitric acid (sp gr 1.42)—2% hydrofluoric acid (sp gr 1.24) (by vol.) pickle (at 150 F) following a 10% (by vol.) sulfuric acid (sp gr 1.84) pickle (at 180 to 200 F) removes the smut and provides a fairly white surface. Scale breaking before pickling is often necessary for satisfactory results.

The AISI "400" series steels, known as the "straight chromium" steels, are seldom successfully handled by acid treatments alone. Fused-salt descaling baths are used to perform the descaling or to alter the scale so that it is amenable to acid treatment.

New sulfuric acid baths for pickling stainless steels should be "activated" by dissolving some iron in them before use; otherwise, they may give very low pickling rates until a certain amount of iron is dissolved. Some operators save a small amount of the discarded bath to activate the new one.

The Sodium Hydride Bath. The sodium hydride bath[*1] consists of molten caustic (750 F) containing from 0.75 to 2.5% of sodium hydride. The sodium hydride is formed in a special compartment of the caustic pot by reaction of molten sodium and cracked ammonia. The work to be descaled is immersed in the hydride bath for periods up to 15 min and then water-quenched. The hydride bath reduces most of the metal oxides in the scale, and the steam formed during quenching blasts off most of the powdery residue. A final dip in a sulfuric acid and/or a nitric-hydrofluoric acid solution may be required to produce a satisfactory surface.

The Oxidizing Salt Baths. Oxidizing salt baths are used in conjunction with acid baths. Their function is to alter the scale to a form more readily attacked in the sulfuric and nitric acid baths. The major constituent is usually caustic with various other salts added to improve the effect on the scale. These baths are held in steel tanks, usually gas-fired. They are operated in the range of 900 to 1000 F. The time of immersion may be up to 15 min. The work is then quench rinsed and successively pickled in baths containing 10% sulfuric (sp gr 1.84) (by vol.) and 10% nitric (sp gr 1.42)—2% hydrofluoric (sp gr 1.24) (by vol.) acids. A mechanical brushing along with a steam blast is often necessary to remove any last traces of smut. Box-annealed type 430 stainless steel is one of the most difficult to descale. It is often necessary to send work through the descaling procedure two or three times to achieve satisfactory descaling. The bath compositions in most common use are proprietary.*

Electrolytic Pickling. Electrolytic pickling in fused-salt baths and in nitric acid baths[6] has been practiced commercially. Neither method has become widely accepted, however.

---

*A proprietary process. E. I. du Pont de Nemours & Company.

*Hooker Electrochemical Company, Diamond Alkali Company, and Kolene Company.

Much more commonly used is cathodic electropickling in dilute hydrochloric (15 to 30% by vol.) or sulfuric (10 to 25% by vol.) acids. Parts are made cathodic at 1 to 2 volts. Lead or carbon anodes are used. Baths are normally operated at room temperature.

Powdered acids (proprietary) at 60 to 120 g/l (8 to 16 oz/gal) are often substituted for the dilute mineral acids. Operating temperatures up to 50 C (120 F) are sometimes used. Voltage is 1 to 2 volts. Where these mixtures contain fluoride salts or other additives, lead anodes may not be acceptable and carbon anodes must be used. The procedure is particularly effective for the removal of light spot welding scales or for activating nickel prior to replating.

Abrasive Descaling. Abrasive descaling has become competitive with chemical methods for descaling corrosion-resistant alloys.[32] The type of equipment which propels steel shot abrasive by means of a rotating wheel is favored for this application. The surface finish obtained, however, is often less suitable for subsequent polishing operations than a chemically pickled surface.

Descaling Special Alloys. In recent years, a new group of alloys for high temperature or other special applications has been under development. Some of these, such as the more common alloys used in turbine buckets, can be descaled by one or several of the methods already described, i.e., the oxidizing salt-bath treatment, or the sodium hydride method followed by acid dips. Titanium alloys also appear to be amenable to these treatments.[5, 19] An acid dip in cold 10% nitric acid (sp gr 1.42) plus 0.25% hydrofluoric acid (50%) for about 15 sec, followed by light scrubbing, is recommended as a followup to salt-bath treatments. It is important not to overheat the oxidizing baths (maximum temperature is 950 F) when descaling titanium because rapid oxidation of the titanium may occur. Zirconium may be pickled in a bath containing 10% nitric acid (sp gr 1.42) and 40% hydrofluoric acid (sp gr 1.24) (by vol.).[27] For some applications of special alloys, electropolishing is the most satisfactory descaling method. The procedures are discussed elsewhere.

Oxide Removal from Copper Alloys

Pickling. The copper-rich alloys are usually resistant to attack by sulfuric acid while copper oxide is readily attacked. For this reason, sulfuric acid is most often used as the pickling agent for heavy scales. A solution containing 5 to 10% H_2SO_4 by volume often gives a good pickling rate at atmospheric temperatures. If the scale is heavy, the bath may be heated to about 175 F as necessary to obtain the desired pickling rate. Pickling time may range from 1 to 15 min. Dichromate may be added to increase the rate of pickling by aiding in solution of cuprous oxide and some of the base metal.

Bright Dipping

Bright-dipping solutions are used for ferrous and nonferrous alloys and usually involve mixtures of two or more of the acids, sulfuric, phosphoric, chromic, nitric, and hydrochloric. The acid ratios and concentration vary widely. Dipping times range from 5 sec to 5 min or more. Some bath compositions and dipping conditions give simply a bright, clean appearance; others give a smoothing action with a brilliant appearance. Thus, the processes range in action from bright dipping to chemical polishing. Whenever nitric acid is used, special precautions are necessary. The nitrogen oxide fumes are poisonous and must be removed by proper ventilation.

Bright dipping of many nonferrous metals and alloys has been standard commercial practice for many years. Solutions for ferrous alloys are more recent. Because of the great benefit to anodizing (and subsequent dyeing, if used), bright dipping of aluminum is most extensive with many millions of square feet of aluminum surface treated annually.

The concentration and composition of bright dips must be adapted to the particular work being processed. Fast-acting bright dips can be used for individual parts of simple shape which can be easily removed from the bath and quickly rinsed. For bulk lots and complicated shapes, slower-acting and free-rinsing dips must be used. Because of these factors and the wide variety of metal compositions to be treated, many bright-dip formulas are in use. Aside from other factors, the results in bright dipping are largely dependent on the operator's skill in manipulating the work. No bright-dipping procedure is satisfactory if unskillfully used. Good results can be obtained with a wide variety of procedures.

The general types of baths are shown in the following sections. Most of those producing smooth as well as brilliant surfaces are proprietary.

Bright Dipping and Tarnish Removal for Copper and Copper Alloys. An oxidizing agent must be added to sulfuric acid if a bright dip or removal of light tarnish films is desired because these operations require some attack on the copper alloy. Chromic acid, dichromates, ferric sulfate, and nitric acid are the most commonly used oxidizing agents. The rate of attack on the metal is controlled by the relative amounts of sulfuric acid and oxidizing agent present in the bath. The rate of dissolving also varies with temperature. Table 1 shows the rates for various sulfuric-nitric acid solutions at 160 F. These solutions brighten, but do not appreciably smoothen the surface. If immersion is too long, etching occurs.

The action is rapid. It may be slowed down by adding small amounts of "inhibiting materials" such as 10 g/l of sodium nitrite plus 5 g/l of sodium chloride to baths 50/50 or greater in nitric acid, or by adding chromic acid to those of lower ratio. Additions of up to 50% phosphoric acid (by vol.) improves the smoothing action and diminishes the tendency to tarnish.

TABLE 1. APPROPRIATE RATES OF DISSOLVING COPPER AND COPPER ALLOYS IN SULFURIC-NITRIC ACID SOLUTIONS

Bath Compositions*			Dissolution, inch per 30 sec						
			Brass[30]		18 Per Cent Nickel Silver[30]		Phosphor Bronze[30]		Copper 160F
Sulfuric	Nitric	Water	Cold	160F	Cold	160F	Cold	160F	
0	100	0	0.002	—	0.001	0.012	0.003	0.007	0.007
0	25	75	0.002	—	0.002	0.009	0.001	0.004	<0.0001
17	17	66	—	—	<0.001	—	<0.0001	—	0.0007
25	75	0	0.002	0.002	<0.001	<0.001	<0.001	<0.001	—
50	38	12	<0.001	<0.001	—	—	—	—	—
75	25	0	<0.0001	<0.001	<0.001	<0.001	<0.001	<0.001	nil

* Per cent by vol. of 1.84 sp gr sulfuric and 1.42 sp gr nitric acids.

Concentrated baths are most satisfactory for individually racked work. For bulk dipping, the baths are usually diluted with water to slow down the action and minimize staining. Cold solutions with cooling coils to prevent any temperature rise are preferred by many.

For greater smoothing action so that a measureable amount of polishing takes place, the dissolution process must be changed. Phosphoric-nitric-acetic acid solutions provide brillance and smoothing.[31] Preferred compositions range from:

10 to 80% by wt. phosphoric acid (1.70 sp gr)
10 to 50% by wt. nitric acid (1.42 sp gr)
10 to 80% by wt. glacial acetic acid (1.05 sp gr).

An example of a typical bath is:

    20% nitric acid
    25% acetic acid
    55% phosphoric acid
    0.5% hydrochloric acid.

The bath is operated at 190 F, and the dipping time ranges up to 5 min. Another example is:

    40% nitric acid
    30% phosphoric acid
    30% acetic acid
    1.0% sodium chloride.

This bath is operated at 150 F, and the dipping time ranges up to 4 min.

The hydrochloric acid additions are particularly suitable for treating alloys containing nickel, monel and nickel silver.

The metal removal rate ranges from 0.0002 to 0.0007 in./min. Dipping usually is long enough to remove 0.0005 to 0.0002 in. of metal. Thereafter, no further increase in brilliance occurs. Longer dipping may give some additional smoothing and thus improve surfaces which were initially very rough. An important fact is that etching does not occur regardless of the immersion time.

Dull copper plate (such as from acid sulfate bath) and dull brass plates cannot be made bright by a practical amount of chemical polishing. Semibright and off-luster bright plates can be made uniformly bright by short-time dips.

Table 2 illustrates the amount of smoothing obtained by chemical polishing.[14]

A typical bright-dipping sequence for racked work is as follows[18]:

1. Dip in a scaling bath containing three parts concentrated nitric acid and one part concentrated sulfuric acid (by vol.).
2. Cold water rinse.
3. Dip in a bath containing one part concentrated nitric acid and three parts concentrated sulfuric acid (by vol.).
4. Double cold water rinse or spray rinse.
5. Rinse in cold 5 per cent sodium cyanide solution.
6. Final cold and hot water rinses.

The sodium cyanide dip helps eliminate stains resulting from the acid dips. The scaling bath may be made from a spent bright-dip bath (No. 3) by addition of sulfuric acid.

For bulk work, the action of the bath should be slowed down by the addition of sodium nitrite or sodium chloride, as mentioned previously, and by dilution with about 50 per cent (by vol.) of water. Changing the sulfuric acid/nitric acid ratio from 3/1 to 6/1 is helpful in addition to dilution with water.

A fumeless bath is available for use where ventilation is not adequate for the copiously fuming baths. It contains[18]:

3 parts sulfuric acid
1 part nitric acid
$\frac{1}{20}$ part hydrochloric acid
6 parts water
$\frac{1}{2}$ lb chromic acid per gallon of solution.

TABLE 2. SURFACE FINISH ON NICKEL, BRASS, AND ALUMINUM BY CHEMICAL POLISHING AFTER DRY-BELT POLISHING

			Profilometer Measure—μ in, (RMS)				
			After Chemical Polishing for				
Meta	Dry-Belt Grit	As Belt Polished	1 min	1.5 min	2 min	3 min	6 min
Dull-nickel plate*	120	25	24	—	21	—	—
	150	27	23	—	18	—	—
	180	17	14	—	14	—	—
Aluminum (2S 1/2H)	180	40	28	—	23	22	—
	240	80	60	—	50	50	—
Brass (70–30)	180	30	—	30	—	24	16
	180	30†	—	27	—	20	16
	240	26	—	21	—	17	15
	240	26†	—	21	—	18	16

* Dull-nickel plate was over steel, belt polished as shown in second column. Third column shows nickel-plate smoothness. No mechanical polishing on the nickel plate.

† Higher temperature than other brass specimens for chemical polishing.

The bath is used cold and followed by cold and hot water rinses. It is not recommended for work to be plated, as a passive film is formed which is difficult to remove uniformly. Electroplates over such films are likely to be nonadherent.

The rate of solution of zinc in a bright dip may be increased by increasing the amount of hydrochloric acid.[16] Solution of copper may be increased by increasing the amount of nitric acid. This provides a general method for adjusting bright-dip compositions for optimum results with brass. If the dipped brass is dull and pale yellow, there is an excess of zinc, and hydrochloric acid should be added to the bath. Too much hydrochloric acid causes the brass to stain and take on a brown color indicating that too much zinc has been removed and that copper remains on the surface. Low nitric acid concentrations yield similar results. An increase in temperature increases the rate of attack on the copper but has little effect on the rate of attack on the zinc. A dip that is too hot (115 F) gives brass a white, smoky color because of rapid solution of copper leaving an excess of zinc.

The following formulas are suitable for bulk dipping brass.[16]

Chemical	Sp Gr	Scaling Dip (cc)	Bright Dip (cc)
H_2SO_4	1.84	380	435
HNO_3	1.38	72	72
HCl	1.17	4	2
H_2O.	—	444	491

Two recent developments have greatly reduced the dependence of bright dipping and chemical polishing solutions on nitric and/or chromic acid and its salts. A highly stabilized hydrogen peroxide is now available and can be successfully used as the oxidizer in some of these solutions.

Considerable progress has also been made in the use of iron salts as oxidizing agents in bright dips. Both developments are usually supplied as proprietary processes, and in many cases are used in combination to provide not only brightening, but also smoothing or polishing action.

Dips for Aluminum and Aluminum Alloys. Aluminum is very active in solutions with nonoxidizing conditions and inert in oxidizing solutions. Acid mixtures represent both of these conditions. The most widely used are mixtures of phosphoric and nitric with or without acetic acid[31] or with sulfuric acid.[36]

Compositions range from:

0 to 95% phosphoric acid (1.70 sp gr) (by vol.)
5 to 85% nitric acid (1.42 sp gr) (by vol.)
0 to 95% glacial acetic acid (1.05 sp gr) (by vol.).

Typical bath compositions are in the range[31]:

50 to 85% phosphoric acid
5 to 15% nitric acid
0 to 20% acetic acid
140 to 170 F and up to 5 min dipping time.

Phosphoric-nitric acid and phosphoric-nitric-acetic acid chemical polishing solutions have been discussed in some detail.[33] Thickness removal versus time charts is shown for the following solutions:

	Percent by Volume	
	1	2
Phosphoric acid (1.70 sp gr)	89	76
Nitric acid (1.42 sp gr)	3.5	3
Acetic acid (glacial)	—	15
Water	7.5	5

Brightening with slight, if any, polishing is reported for dipping in 89 phosphoric acid—11 water (% by vol.) or in a solution containing 80 phosphoric acid—14.5 glacial acetic—5.5 water. Other compositions range from:

30 to 92% phosphoric acid
5 to 60% sulfuric acid
3 to 10% nitric acid.

The metal-removal rate in aluminum-dipping solutions ranges from 0.00004 to 0.00002 in./min. The speed varies considerably with the type of solution and temperature. Commercial treatments generally range from 2 to 10 min and remove a total of about 0.0002 to 0.002 in. in producing the optimum finish.

Hydrogen peroxide is used in aluminum bright dips. One such dip consists of[22]:

40% phosphoric acid (1.70 sp gr) by vol.
50% nitric acid (1.25 sp gr) by vol.
10% hydrogen peroxide (100 vol.).

The stability of such a solution is open to question and a violent reaction would be hazardous.

Another dip consists of[7]:

75% phosphoric acid (1.70 sp gr) by wt.
3.5% hydrogen peroxide (30%) by wt.
21.5% water by wt.

This bath is recommended for low silicon aluminum alloys. It is used at 195 F. The dipping time is from 15 sec to 4 min. The water content should be controlled within several per cent, as it is critical.

Very extensive use is made of acid deoxidizing or desmutting baths to remove the residues present on aluminum after alkaline etching. Nitric acid (25 to 75% by vol 42 Be′) is common. Hydrofluoric acid, or a fluoride salt, may be added if the alloy contains silicon. Sulfuric acid may be included for alloys containing magnesium. A so-called "universal" acid dip for use on all alloys consists of

Nitric acid 42 Be′	50% by vol.
Sulfuric acid 98%	25
Water	25
Fluoride salt	60 g/l (8 oz/gal)

There are also available a wide range of proprietary powdered and/or liquid deoxidizers. They are available both with and without fluorides and utilize three different oxidizer systems.

Chromate based deoxidizers provide very low-surface resistance plus a residual film that retards oxide regrowth. Excellent for preparing aluminum for spot welding, and ahead of chromate conversion coating.

Peroxygen acid salt based deoxidizers present minimum waste disposal problems, but no residual protection. Some versions may decompose if contaminated with copper or silver. These materials are considered excellent ahead of chemical polishing, anodizing, or zincate preplating treatments.

Iron salt based deoxidizers increase the sludge content of waste disposal systems, but may beneficially remove phosphates and acts as coagulants for some metal hydroxides that normally do not settle well. Suitable for all uses, they do require more attention to rinsing. No residual film exists to prevent oxide regrowth.

Bright Dips for Ferrous Metals. Solutions that provide brightness on *steel* contain hydrogen peroxide. The luster obtainable will depend on the carbon content and alloying constituents. One solution giving brightness and smoothness is composed of oxalic acid and hydrogen peroxide.[25] It contains:

2.5% oxalic acid (by wt.)
1.3% hydrogen peroxide (30%) (by vol.)
0.007-0.010% sulfuric acid (1.84 sp gr)(by vol.).

The operating temperature is 65 F. Metal dissolution rates are about 0.0004 in./hr on mild steel. This can be increased to 0.0008 by agitation and to about 0.001 by raising the temperature to 105 F. The bath removes grinding scratches in about 30 min. It reduces the rms microinch finish from 39 to 21 and 19 to 12 in 90 min.

To smoothen and to prevent pitting corrosion *stainless steel* is treated in a solution consisting of[39]:

29.5% HCl (sp gr 1.16) (by vol.)
40.0% H_2SO_4 (sp gr 1.84) (by vol.)
0.5% HNO_3 (sp gr 1.42) (by vol.)
5.5% $TiCl_4$
24.5% H_2O.

The solution is operated at about 165 F for dipping times of 2 to 5 min. Metal removal is about 0.0004 in.

Another solution[2] for austenitic types of stainless steel consists of:

2.5% nitric acid (sp gr 1.42) (by vol.)
7.5% hydrochloric acid (sp gr 1.16) (by vol.)
1.0% Rodine 45.*

To the bath is added 10 lb of Toner No. 1* per 100 gal of bath. Dipping time is about 3 min at 205 to 210 F.

Dips for Nickel and Monel. Phosphoric-nitric-acetic solutions for copper alloys[30] containing small additions of sulfuric acid and hydrochloric acid, brighten and smoothen nickel and monel.

Sulfuric-phosphoric-nitric acid solutions brighten nickel.[41] They contain:

45 to 60% phosphoric acid (by wt.)
8 to 15% nitric acid (sp gr 1.42) (by wt.)
15 to 25% sulfuric acid (sp gr 1.84) (by wt.)
10 to 20% water (by wt.).

A typical example is:

60% phosphoric acid (sp gr 1.70) by vol.
20% sulfuric acid (sp gr 1.84) by vol.
20% nitric acid (sp gr 1.42) by vol.

*Marketed by Amchem Products, Inc.

Dipping times are 1 to 3 min. At about 180 F, the metal-removal rate is about 0.0001 to 0.0003 in./min.

Dull nickel plate (such as Watts) cannot be made bright by chemical polishing under practical conditions. Some semibright and off-luster bright nickel plates can be made uniformly bright.

Dips for Cadmium and Zinc. Dilute nitric acid, chromate, and hydrogen peroxide solutions are often used for these metals. The nitric acid solutions are 0.3 to 1% by volume of 1.42 specific gravity acid. They are used at about 70 F.

Chromate solutions are more widely used. Protective films form on the zinc and cadmium so a nonmetallic color may result. Chromic acid-sulfuric acid solutions are reported for micro-smoothing. [10,11,12] Typical solutions are [27]:

For zinc plate—
300 g/l CrO_2 (flake)
5 g/l $H SO_4$ (sp gr 1.84)
For racked cadmium plate—
100 g/l CrO_3 (flake)
1.67 g/l H_2SO_4 (sp gr 1.84)
For barrel-dipping cadmium plate—
100 g/l CrO_2 (flake)
1 g/l H_2SO_4 (sp gr 1.84)

These dips are used cold. Ordinarily colored films do not form. When they occur, they may be removed by dipping the work in weak acids (pH > 2.5) or in caustic solutions (> 4 normal).

Hydrogen peroxide dip solutions for cadmium contain [21]:

3 to 4.5% H_2O_2 (30%) (by wt.)
0.3 to 0.5% H_2SO_4 (sp gr 1.84) (by wt.)
balance water.

The removal rate for cadmium is about 0.00005 in./min. For *zinc*, the composition is changed slightly to about 4% hydrogen peroxide and 0.25% sulfuric acid. The zinc and cadmium are usually electrodeposited and, if from an alkaline bath, the bright dip is preceded by a dip in a weak acid (about 1% sulfuric) and rinsed.

Progress in chromate coatings (see Chap. 13B) has been extensive, and proprietary compounds are now available with a wide range of appearance and corrosion resistance, including clear, clear-blue, iridescent, olive drab, black, and clear with pastel dyes.

Dip for Silver. An aqueous sodium cyanide-hydrogen peroxide bath is described for silver [28]:

1.5% sodium cyanide (by wt.)
8.5% hydrogen peroxide (30%) (by vol.).

This bath should be mixed at 80 F and used below 90 F. When gassing commences, the silver is removed and dipped in a solution of 37.5 g/l sodium cyanide. The procedure is repeated until the desired brightness is obtained.

Dip for Lead. An aqueous solution [27] of

3.5% hydrogen peroxide (30%) (by vol.)
3.5% glacial acetic acid (sp gr 1.05) (by vol.)

causes periodic brightening and darkening of lead. This may serve as the basis for a bright dip with a somewhat lower acid content.

Dip for Beryllium. Beryllium is chemically smoothed and brightened in [35]:

5% sulfuric acid (sp gr 1.84) (by vol.)
75% phosphoric acid (sp gr 1.70) (by vol.)
7% (by wt.) chromic acid (CrO_2 flake)
balance water.

At about 120 F, beryllium dissolves 0.00005 in./min. The surface is semipassive and can be activated by dipping 0.5 to 1 min in 6% (by wt.) sulfuric acid.

Dips for Zirconium and Titanium. A brightened and somewhat smoothed surface is obtained by dipping zirconium in a solution containing [33]:

100 g/l ammonium bifluoride ($NH_4F \cdot HF$)
400 ml/l nitric acid (sp gr 1.42)
200 ml/l fluosilicic acid (30–31%)
balance water.

At 70 to 80 F, this dip removes about 0.0006 in. of metal per minute. At 90 to 95 F, the rate is 0.0018 in./min.

The surface of titanium can be similarly benefited by this solution. Some alloys are only smoothened and others are polished and brightened. In some cases the results are improved if the nitric acid concentration is increased to 500 ml/l.[34]

Dip for Uranium. Uranium can be bright pickled in a solution containing:

TABLE 3. MATERIALS OF PICKLING EQUIPMENT

Pickling Agent	Tanks	Heating	Pumps and Valves	Baskets and Racks
H_2SO_4 (for steel)	Wood Stoneware Lead Organic coatings including rubber Carbon Brick (over membrane)	Lead "Karbate" } heat exchangers Tantalum "Duriron"-steam injectors	"Worthite" "Duriron" "Karbate" Rubber	Rubber-coated steel Monel "Inconel" Acid-resisting bronze
H_2SO_4 (for copper alloys) or H_2SO_4 + $Na_2Cr_2O_7$	Same as above plus Types 347, 321, or E.L.C. stainless for welded tanks	Lead Type 316 stainless } heat exchangers "Karbate" "Duriron-steam injectors	Same above plus Types 302, 304, or 316 stainless	Same as above plus Types 302, 304 or 316 stainless
HCl	Wood Stoneware Organic coatings including rubber Carbon or graphite	"Karbate" Tantalum } heat exchangers "Durichlor" "Chlorimet 3" } steam injectors	"Karbate" Rubber "Durichlor" "Chlorimet 3"	Rubber-coated steel Monel Nickel "Inconel" "Hastelloy B"
H_3PO_4	Wood Stoneware Lead Carbon or graphite Type 304 stainless steel Organic coatings including rubber Monel	Lead "Karbate" Type 316 stainless } heat exchangers "Duriron" Types 304 and 316 stainless } steam injectors Monel	"Karbate" Rubber "Duriron" Types 304 and 316 stainless Monel	Types 304 and 316 stainless Monel
Nitric acid	Stoneware Carbon Butyl rubber Types 304, 321 or 347 stainless	"Karbate" Tantalum "Duriron" Types 304, or 316 stainless	"Karbate" "Duriron" "Worthite" Types 304 or 316 stainless	Rubber-(butyl) coated steel Types 304 or 316 stainless
Nitric-hydrofluoric acid	Carbon and graphite Rubber (butyl) Neoprene	"Karbate"—heat exchangers "Durimet 20"—steam injectors	"Durimet 20" "Worthite" "Karbate" Rubber (butyl)	Rubber-(butyl) coated steel Neoprene "Inconel"
Sodium hydride or oxidizing fused salts	Steel	Steel	Cast iron or steel	Steel
CrO_3 + H_2SO_4 (Zn, Cd, bright dip)	Stoneware Lead Types 304, 321, or 347 stainless			Stainless steel
H_2SO_4 + HNO_3* + HCl or $Na_2Cr_2O_7$ (Cu alloy bright dip)	Stoneware Glass Types 316 or 317 stainless	"Karbate" Type 316 stainless	Type 316 stainless	Type 316 stainless

* Note: H_2SO_4-HNO_3-(HCl) bright dips attack some stainless alloys when fresh, but may be safely used with these materials when inhibited with copper.

HNO$_3$ (sp gr 1.42) 500 ml/l
balance water.

The solution temperature should be 120 F or higher.

Dip for Thorium. Thorium can be bright pickled in a solution containing:

HNO$_3$ (sp gr 1.42) 500 ml/l
H$_2$SiF$_6$ (30%) 20 ml/l
balance water.

The solution temperature should be 120 F or higher.

Bright-Dipping and Chemical-Polishing Costs. (Costs for bright dipping or chemical polishing vary widely, depending on the solution, the kind of metal, the temperature, and the original finish. Practical operations involve costs ranging from 0.1 to 10 cents. In general, the lower figures apply for bright dipping and the higher ones for chemical polishing.

All chemical solutions "wear out" because metal dissolves in them and slows them up, and because constituents are decomposed or neutralized by reaction with the metals. An operator will need to establish empirically the quantity of work that can be processed while the bath builds up to its finite limit of dissolved metal.

The life of nitric acid-containing baths can be extended by replacing nitric acid.

Equipment for Pickling and Bright Dipping

Regardless of the size and nature of the pickling or bright-dipping operation, the corrosiveness of the pickling solutions requires special tank materials, special heating arrangements, and adequate ventilation.

Table 3 shows some of the materials which can be used satisfactorily. The table is incomplete, especially as regards some of the newer organic materials which have not been fully proven in service.

Where a tank surface may suffer mechanical damage, as in continuous pickling where the strip may rub on the tank bottom, a brick lining backed with an organic membrane is favored. The brick also serves to insulate the organic from high bath temperatures.

Heating may be accomplished by conventional heat exchanger coils of lead tubing, "Karbate," or tantalum carrying steam. Another popular method involves the discharge of live steam through nozzles submerged in the bath. Still another innovation is the use of submerged combustion burners which discharge hot gases directly into the pickling bath. These are normally gas fired, but oil has also been used.

Submerged combustion tubes are often used for heating the fused-salt descaling baths. These are gas-fired and the exhaust gases are drawn off where the tubes emerge from the tank. Small fused-salt baths may be heated with gas or oil burners directly under the tanks, by electric immersion heaters, or by electrical resistance heating.

The fused-salt baths may be pumped with ordinary cast iron centrifugal pumps, provided the conventional bronze bearings are replaced with steel or cast iron bushings. The bearing clearances should be increased to 0.010 to 0.020 in. The fused salt serves as a lubricant.

Small-scale bright dipping is conveniently handled in large crocks set in steel tanks. The space around the crocks is filled with water, which serves to cool the bright dip and is also used as a rinse.

Wood is one of the most readily available materials which is resistant to acids. Its use in tanks, floor grilles, and structures may be most practical if it is inspected regularly and replaced periodically. A life of at least two years can be expected of a properly constructed wood pickling tank. Acid-resistant brick and tile set in acid-resistant cement can provide a long-wearing floor area around pickling tanks where acid dripping is a problem.

References

1. Alexander, H. L., *Iron Steel Engr.*, **24**, 44–51 (1947).
2. American Chemical Paint Co., "Efficient Pickling with Rodine," Bul. No. 13, Ambler, Pa., 1952.
3. Anon, *Chem. Eng.*, **61**, No. 1, 379 (January, 1954).
4. Anon., *Met. Ind. (London)*, **73**, 471–472 (Dec. 10, 1948).
5. Anon., Modern Metal Finishing, **11** (2), 4, publ. by Electrochemicals Div., Du Pont Co. (Nov., 1953).
6. Anon., *Modern Metal Finishing*, **10** (2), 2–3, publ. by Electrochemicals Div., Du Pont Co. (June, 1952).
7. Cochran, W. C., "Chemical Brightening of Aluminum," U. S. Patent No. 2,613,141 (Oct. 7, 1952).

8. Davis, M. H., Simnad, M. T., and Birchenall, C. E., *J. Metals*, **3**, No. 10, 889-896 (Oct., 1951).
9. Douty, Alfred, *Plating*, **37**, 269-271 (Mar., 1950).
10. Dubpernell, G., and Soderberg, K. G., "Method of Brightening Metals Electronegative to Iron," U.S. Patent 2,021,592 (Nov. 19, 1935).
11. Dubpernell, G., and Soderberg, K. G., "Method of Brightening Metals Electronegative to Iron," U.S. Patent No. 2,186,579 (Jan. 9, 1940).
12. Dubpernell, G., and Soderberg, K. G., "Method of Brightening Metals Electronegative to Iron," U.S. Patent No. 2,194,498 (March 26, 1940).
13. Farrell, J. F., *Metal Finishing*, **47**, No. 2, 69-75 (February, 1949).
14. Faust, C. L., *37th Annual Proc. Am. Electroplaters' Soc.*, 137-150 (1950).
15. Fink, C. G., and Wilber, T H., *Trans. Electrochem. Soc.*, **66**, 381-392 (1934).
16. Graham, A. K., *Trans. Electrochem. Soc.*, **52**, 289-300 (1927).
17. Greenberger, J. I., *Iron and Steel Engr.*, **28**, No. 6, 95 (June, 1951).
18. Halls, E. E., *Metal Finishing*, **51**, (5), 68-73 (May, 1953).
19. "Handbook of Titanium Metal," publ. by Titanium Metals Corporation of America, Third Printing (March, 1951).
20. Hanley, H. R., *Metal Progr.*, **41**, 179-183 February, 1942).
21. Hull, R. O., "Bright Dip," U. S. Patent No. 2,154,451 (Apr. 18, 1939).
22. Klein, B., British Patent No. 679,078.
23. Meyer, W. R., and Brown, S. H., *37th Ann. Proc. Am. Electroplaters' Soc.*, 163-188 (1949).
24. Marks, B. H., *Materials & Methods*, **29**, No. 4, 64-67 (April, 1949).
25. Marshall, W. A., *J. Electrodepositors Tech. Soc.*, **50**, 78-83, 85 (1953).
26. Martin, E. D., Yearbook of the American Iron and Steel Institute, 602-666 (1948).
27. Miller, G. L., *Iron Age*, **167**, No. 20, 83-87 (May 17, 1951).
28. Parker, E. A., *Trans. Electrochem. Soc.*, **88**, 297-305 (1945).
29. Paulson, C. F., and Gilwood, M. E., *Iron Age*, **166**, 97-99 (Dec. 14, 1950).
30. Pinner, R., *Electroplating*, **6** (10), 360-367 (1953).
31. Pray, H. A. H., Igelsrud, I., and Simard, G. L., "Chemical Polishing of Metal Surfaces," U.S. Patent No. 2,446,060 (July 27, 1948).
32. Rose, K., *Materials & Methods*, **28**, No. 5, 72-75 (November, 1948).
33. Schickner, W. C., Beach, J. G., and Faust, C. L., *Trans. Electrochem. Soc.*, **100**, 289-291 (June, 1953).
34. Beach, J. G., "Polishing Metals" U. S. Patent No. 2,711,364 (June 11, 1955).
35. Schickner, W. C., Beach, J. G., and Faust, C. L., *Trans. Electrochem. Soc.*, **100**, 276-279 (June, 1953).
36. "Soc. Trefileries et Laminoirs du Hovre," French Patent No. 984,863.
37. Soderberg, K. G., *Trans. Electrochem. Soc.*, **88**, 297-305 (1945).
38. Stuck, W. H., and Abrams, J. H., *Iron and Steel Engr.*, **28**, No. 11, 101-108 (February, 1951).
39. Turin, J. J., *Iron Age*, **153**, 64-70 (April 20, 1944).
40. Uhlig, H. H., "Corrosion Resistant Ferrous Alloys," U. S. Patent No. 2,172,421 (Sept. 12, 1939).
41. Warner, J. C., *Metal Cleaning Finishing*, **10**, No. 2, 104 (Feb., 1938).
42. Zelley, W. G., "Brightening Nickel," U.S. Patent No. 2,638,410 (May 12, 1953).

4. TYPICAL PROCESSING AND OPERATING SEQUENCES

A. Metals

Donald L. Snyder

Research Associate, Metal Finishing Department, The Harshaw Chemical Company, Cleveland, Ohio.

and

James K. Long

Consultant

In plating, as in many other manufacturing processes, "the chain is no stronger than the weakest link." While the plating may usually be considered the most complicated stage in an electroplating sequence, it may frequently be surpassed in importance by the "simple" step of rinsing.

Although many processing cycles are presented in this chapter, seldom will a given cycle meet the requirements of a particular case. Therefore, it is hoped that from the cycles given and the accompanying discussion, the reader will gain a sufficient understanding of the requirements in order to be able to plan the electroplating sequence which will be most suitable for each particular situation. It is of utmost importance for the reader to refer to other chapters to understand fully the various steps in a sequence. In addition, the cited references* will be of great aid.

The most elementary cycle prior to the electroplating step is a simple sequence of cleaning and rinsing. In the majority of cases, one or more additional steps, such as acid dipping, striking, activating, conditioning, etc., are required. Even in the above elementary sequence the cleaning may involve several steps.

In order to understand and be able to project an electroplating sequence, the following fundamental factors and questions should be considered:

1. The work should be free of foreign matter such as oil, grease, dirt, and oxide before attempting to electroplate. Long transfer times allowing contamination and oxidation should be avoided.
2. Rinsing should be adequate. In many cases, it may be necessary to use distilled, deionized or A.S.T.M. type reagent grade water.†

---

*The references for this chapter are separated according to subject headings and subheadings.

†"Standard Specification For Reagent Water," A.S.T.M. Designation D 1193, *1981 Book of A.S.T.M. Standards*, Part 9, p. 546.

3. Acid dips, cleaning solutions, and water should be of a composition so as not to leave films which cannot be rinsed off. This means that the composition of the acid dip and cleaning solution will usually depend on the composition of the basis metal.
4. Does the basis metal require a special conditioning treatment prior to plating? Examples are the activation of stainless steel, immersion or anodic treatments for aluminum, and silver striking before silver electroplating.

In the solution formulas given in this chapter, the concentrations are expressed in metric units followed by English units in parenthesis, unless otherwise stated.

Liquids: Percent by volumes, %, as for example, 5% is equivalent to 5 liters in a total of 100 liters; grams per liter, g/L; (avoirdupois ounces per gallon, oz/gal; or fluid ounces per gallon, fl oz/gal).

Solids: Grams per liter, g/L; (avoirdupois ounces per gallon, oz/gal).

Unless otherwise stated, the acids used in the solution formulas are common technical or commercial grade concentrated acids that have the following approximated compositions:

Sulfuric Acid: Commercial 66 Be' (93% by weight H_2SO_4). Sometimes expressed as g/L (oz/gal) of the 66 Be' acid.

Hydrochloric Acid (Muriatic Acid): Commercial 20 Be' (31% by weight HCl).

Nitric Acid: Commercial 42 Be' (67% by weight HNO_3).

Phosphoric Acid, Orthophosphoric Acid: (75% by weight H_3PO_4).

Fluoboric Acid: (42% by weight HBF_4).

Acetic Acid—As Glacial: (99.5% by weight CH_3COOH).

Hydrofluoric Acid: (48% by weight HF).

Many of the chemicals referred to in this chapter are of a potentially hazardous nature. Before attempting to use them, the safe and proper methods for storing, handling, and disposing them should be thoroughly investigated.

PRETREATMENTS

Pretreatment is the preparation of the article for the actual electroplating step. For clarity in discussion, pretreatment is divided into two stages—preliminary and final. The preliminary treatment removes heavy surface soils, such as grease, buffing compound, drawing compound, scale, heavy rust, and burnt oil. Although it can be a part of the plating-room cycle, it frequently precedes it.

The final treatment removes only the last traces of oil and grease, and conditions the surface for electroplating. Acid dips in this final stage should not be expected to remove scale or heavy rust. They are only to neutralize the last traces of alkaline cleaner remaining after rinsing, and to activate the surface for electroplating.

In any preparatory cycle where the parts have both oil and oxide contamination, it is better practice to remove the oily material before attempting to remove rust or scale unless the oxide scale is to be removed by mechanical means. This will facilitate uniform removal of the latter.

Preliminary Treatment

This involves one or both of two basic steps: (1) Removal of heavy amounts of oil, grease, buffing compound, drawing compound, etc., and (2) removal of scale, heavy rust, burnt-in oil, etc.

These steps are only followed when required, as determined by the kind and degree of contamination. The methods for accomplishing step (1) may in most cases be used for all basis metals. The methods for accomplishing step (2) vary depending on the metal and type of article.

Removal of Grease, etc. (See Chapter 3-F and Reference 1.) This may be accomplished by one or more of the following methods:

1. *Spray cleaning.* Alkali or emulsion-type cleaners are used with a nozzle pressure of 0.2 to 0.4 MPa (30 to 60 psi.) The temperature, alkalinity, and time should be adjusted to suit the material being cleaned.

2. *Solvent degreasing.* Trichloroethylene or perchloroethylene is ordinarily used. The parts may be washed by agitation in a heated or cold liquid, or by vapor degreasing by suspending the cooled article in the vapor. Since the latter often leaves solid matter on the article, a multiphase operation is frequently used. This involves washing in a heated liquid, followed by dipping in a

cool condensed liquid, and finally vapor degreasing. Proper health precautions should be taken, and because of fire hazard, the use of mineral spirits is not recommended. Solvent degreasing is usually followed by very thorough washing or preferably soak cleaning.

3. *Soak cleaning.* This is done by immersion, preferably with agitation, in alkali or emulsion cleaners, and the same precautions should be taken as with spray cleaning.

4. *Electrolytic alkali cleaning.* This may be cathodic or anodic depending on the metal being cleaned and other conditions. Generally, the same conditions for a given metal are used as are given in the final cleaning cycle.

5. *Cleaning of parts in bulk.* Parts in bulk may be cleaned by:

 a. Spray cleaning or solvent degreasing, holding the parts in suitable baskets or trays.
 b. Electrolytic cleaning in a high-conductivity cleaner using an insulated barrel cyliner will effect more complete removal of grease, etc., but may not be a necessary step in the preliminary cleaning cycle.
 c. The parts may be tumble-cleaned without current in a suitable cleaner.

6. *Ultransonic cleaning.* Used in conjunction with cleaning solutions to produce or increase agitation which in turn facilitates the cleaning process.

Thorough rinsing should follow aqueous cleaning. Where there is a time lapse or storage period after cleaning, an alkaline cleaner film should remain on the part.

Removal of Scale, etc. Scale may be removed by the following methods:

1. *Mechanical treatment.* This consists of polishing, tumbling, and sand, grit, or vapor blasting. Frequently this precedes or makes unnecessary any cleaning in the preliminary treatment cycle. If not complete, the mechanical treatment at least reduces the alkaline treatment or acid pickling to a minimum. Where the scale is both heavy and tenacious, as with hot-formed, high-carbon steel parts, the pratice is first to remove most of the scale by pickling. The polishing then removes the remainder of the scale and smooths the surface, which has been pitted and roughened by the pickling operation. On softer metals such as lead alloys, zinc and aluminum castings, and brasses, the thin surface oxide can be removed by fine mesh abrasive polishing or buffing, followed by cleaning the surface by passing it over a clean, dry buffing wheel.

2. *Alkaline treatment.* Alkaline descaling may be conducted on ferrous metal parts. A typical bath, which may be used as a soak or with current contains:

Sodium hydroxide	180 g/L (24 oz/gal)
Sodium cyanide	120 g/L (16 oz/gal)
Chelating agents	80 g/L (10.5 oz/gal)
Temperature	40 C (104 F)

This method minimizes the effect of acid attack on the ferrous metal and will not produce smut, both of which could be a problem with acid pickling.

For removal of heavy scale, periodic reverse current (NOTE 1) may be utilized.

NOTE 1—During periodic reverse (PR), the part is made alternately cathodic and anodic at intervals of a few seconds using DC current. Barrels or racks may be used.

3. Pickling (see NOTE 2 and Chapter 3-G):

(1)[2] Hydrochloric acid	20-85%
Temperature	Room
(2)[2] Sulfuric acid	5-15%
Inhibitor	Consult supplier
Temperature	50-80 C (120-150 F)
(3)[2] Sulfuric acid	4-6%
Temperature	50-65 C (120-150 F)
Parts as anode at	3.2-6.5 A/dm^2 (30-60 ASF)
(4)[2] Hydrochloric acid	20-85%
Nitric acid	1-5%
Temperature	Room
(5)[3] Sulfuric acid	22.5 g/L (3 oz/gal)
Potassium nitrate	22.5 g/L (3 oz/gal)
Temperature	70 C (160 F)
(6)[3] Sulfuric acid	45 g/L (6 oz/gal)
Ferrisul[4]	37.5 g/L (5 oz/gal)

NOTE 2—Proprietary acid salts, many which contain fluorides, may be used along with or in place of the acid pickles.

a. *Low-carbon steel.* Solution (1) is used where the rust or scale is not heavy or tenacious. Solutions (2) and (3) are used for heavy and tenacious scale. Solutions (4), (5), and (6) may be used for brown, glazed or burnt-in oil surfaces.

b. *High-carbon, case-hardened, and low-alloy steels.* High-carbon steel is usually hot-formed, producing heavy scale; low-alloy steels are cold-formed, producing little scale. These steels are susceptible to hydrogen embrittlement, and any pickling, except anodic, should be eliminated where possible or held to a minimum, especially for spring-temper and case-hardened parts. Mechanical methods frequently eliminate the need for acid pickling.

Solutions (1), (2), and (3) for pickling low-carbon steel are used for removing scale from high-carbon steel. Oxidizing acids, as in solutions (4) and (5), are sometimes used for case-hardened or spring-temper parts.

c. *Cast irons.* As with high-carbon steel, acid pickling of cast irons should be kept to a minimum since cast irons are susceptible to a low-hydrogen overvoltage surface condition that causes low electroplating efficiency, and acid treatment will frequently intensify this effect. Formation of smudge can also result from overpickling; therefore, scale, rust, and sand should be removed wherever possible by mechanical means. If it is necessary to acid pickle, two solutions have been recommended[3]:

(1) Sulfuric acid   125 ml (16 fl oz)
    Hydrofluoric
    acid   125 ml (16 fl oz)
    Water   1 L (1 gal)
    Temperature   Room or elevated

(2) Sulfuric acid   90 g/L (12 oz/gal)
    Nitric acid   37.5 g/L (5 oz/gal)
    Zinc   7.5 g/L (1 oz/gal)

d. *Stainless steels.* As with some other types of steel, mechanical methods such as blasting, shot peening, tumbling, and wheel abrading will prove economical prior to pickling. Iron-containing abrasive should not be used. Frequently, pickling of stainless steels consists of two steps: scale softening and final scale removal.

Scale Softening[5]

(1) Sulfuric acid   8–11%
    Temperature   65–70 C (150–160 F)
    Time   10–45 min

(2) Hydrochloric
    acid   10–15%
    Temperature   50–60 C (120–140 F)
    Time   30–90 min

Inhibitors should be used and the parts rinsed thoroughly before going into the scale removal solution.

Scale Removal[5]

(1) Nitric acid   6–10%
    Hydrofluoric
    acid   1.5%
    Temperature   Room

(2) Nitric acid   9–10%
    Hydrofluoric
    acid   1.5%
    Temperature   60–70 C (140–160 F)

Solution (2) is used only for the austenitic stainless steels (except for type 303) to shorten the time required. The high-carbon grades (types 420, 440A, 440B, and 440C) should be mechanically descaled, if possible. Pickling in the fully hardened condition should be avoided to prevent pickling cracks. Other solutions for scale removal have been used[3]:

(3) Nitric acid   10%
    Hydrofluoric
    acid   2.4%
    Hydrochloric
    acid   1.2%
    Temperature   55–60 C (130–140 F)

(4) Hydrochloric
    acid   25%
    Nitric acid   5%
    Temperature   50–70 C (120–160 F)

e. *Copper and copper-base alloys.*[6] In addition to copper, the following solutions may be used for brasses, bronzes, and nickel-silver. If buffed, descaling is seldom necessary.

(1) Sulfuric acid   10–40%
    Temperature   Room to 80 C (176 F)

(2) Sulfuric acid   60–70%
    Temperature   50 C (120 F)
(3) Sulfuric acid   0–30%
    Nitric acid   10–60%
    Temperature   Room
(4) Sulfuric acid   60–70%
    Nitric acid   20–35%
    Water   5–10%
    Hydrochloric
      acid   0.1%
    Temperature   Room
(5) Nitric acid   25%
    Phosphoric acid   75%
    Temperature   Room

Solutions (2) and (5) and the so-called fire-off dip (3) can be used for removing heavy oxide, solution (3) sometimes being used after (1). For beryllium-copper, solution (2) is used. The bright dip solution (4) may be part of the preliminary or final pretreatment cycle, and it should be followed by very thorough rinsing. Solution (5) may be used for lighter oxides, but the parts should be clean and dry when immersed.

Final Treatment

It is assumed that the work going into the final cycle will be reasonably free of heavy contamination. As previously discussed, however, there is no clear demarcation between the preliminary and final treatments. Therefore, some of the cycles which follow will essentially be duplications of each other, the differences only being to allow for different degrees of cleaning necessary.

(I) Low-Carbon Steel[7]

Cycle I-1

(1) Anodic clean
(2) Rinse
(3) Acid dip (NOTE 3)
(4) Rinse
(5) Plate

Cycle I-2

(1) Preliminary treatment
(2) Rinse
(3) Anodic clean
(4) Rinse
(5) Acid dip (NOTE 3)
(6) Rinse
(7) Plate

Cycle I-3

(1) Anodic clean
(2) Rinse
(3) Acid dip (NOTE 3)
(4) Rinse
(5) Anodic clean
(6) Rinse
(7) Acid dip (NOTE 3)
(8) Rinse
(9) Plate

Cycle I-5

(1) Anodic clean
(2) Rinse
(3) Acid dip (NOTE 3)
(4) Rinse
(5) Cyanide copper strike
(6) Rinse
(7) Acid dip— 5% sulfuric acid
(8) Rinse
(9) Plate

Cycle I-4

(1) Preliminary pretreatment
(2) Rinse, tumble in cleaner, no current
(3) Warm rinse
(4) Cold rinse
(5) Acid dip (NOTE 3)
(6) Rinse
(7) Rinse
(8) Plate

NOTE 3—The acid dip may depend on the plating solution but is usually 4 to 10% sulfuric acid or 5 to 25% hydrochloric acid. Aklaline derusters are sometimes used in place of the acid dips. They are usually operated at 40 C (104 F) and 2 to 5 A/dm² (20 to 50 ASF), with and without periodic reverse current (see NOTE 2). They would be followed by a rinse, acid dip, and rinse.

For nickel electroplating. Cycles I-1 through I-5 may be used.

For copper (see also Cycles II-8, II-9), zinc, cadmium, and tin electroplating. The above cycles may be used, but very frequently a room temperature, 30 to 75 g/L (4 to 10 oz/gal) sodium cyanide dip or alkali dip for alkaline tin is used just prior to plating in the cyanide or alkaline solution.

For silver electroplating. Proceed after the last rinse in the above cycles as follows:

Cycle I-6

(1) First silver strike

Cycle I-7

(1) Copper strike
(2) Rinse

(2) Second silver Strike (NOTE 4)
(3) Silver

(3) Silver strike (NOTE 4)
(4) Silver plate

Cycle I-8

(1) Nickel strike or plate
(2) Rinse
(3) Silver strike
(4) Silver plate

NOTE 4—A rinse may be desirable after the silver strike. Its use will prevent a carry-over of impurities into the silver plating solution but will result in some loss of silver and cyanide.

For electroplating directly over steel, two silver strikes are necessary (see Chapter 7), as in Cycle I-6. Cycle I-8 is usually preferred over I-7.

For gold electroplating. Proceed after the last rinse in Cycles I-1 through I-4 as follows:

Cycle I-9[8]

(1) Bright nickel strike
(2) Rinse
(3) Rinse
(4) Active 37.5 g/L (5 oz/gal) KCN, cathodic at 4.0 A/dm^2 (40 ASF)
(5) Rinse in distilled or pure water
(6) Gold plate

Cycle I-10

(1) Copper, brass, silver, or nickel strike
(2) Warm rinse
(3) Rinse (distilled or pure water)
(4) Gold plate (frequently a potassium cyanide dip and another rinse are used prior to gold plating in the above cycle)

For lead electroplating. Proceed after the last rinse in Cycles I-1 through I-4 as follows:

Cycle I-11

(1) Lead plate (NOTE 5)

Cycle I-12

(1) Copper strike
(2) Rinse
(3) Acid dip (see NOTE 6): 2–10% Fluoboric acid, 2–10% acetic acid, or 2–10% hydrochloric acid
(4) Rinse
(5) Lead plate

Cycle I-13

(1) Nickel flash
(2) Rinse
(3) Rinse
(4) Lead plate

NOTE 5—Lead directly over steel usually lacks good covering power, and unless a heavy deposit is to be used, Cycles I-12 and I-13 are preferred.

NOTE 6—The acid dip may be omitted, sometimes advantageously, if extra rinsing is used.

For chromium electroplating. Chromium directly on steel is usually for functional purposes and is known as hard chromium plate.[9]

Cycle I-14

Proceed after the last rinse in Cycle I-1, I-2, or I-3 as follows:
(1) Anodic in chromic acid solution[10]
(2) Chromium plate

Cycle I-15[10]

(1) Stress relieve at 175 C (350 F) if necessary
(2) Anodic clean
(3) Rinse
(4) Anodic in chromic acid solution
(5) Chromium plate
(6) Rinse
(7) Bake at 175 C (350 F) to remove occluded hydrogen

(II) High-Carbon and Low-Alloy Steels[11]

As stated previously, these steels are susceptible to hydrogen embrittlement. Therefore, the acid treatments are either very mild, of short duration, used with anodic current, or contain oxidizing agents.

Cycle II-1

(1) Anodic clean, 90 C, 5 A/dm^2 (195 F, 50 ASF), less than 2 min.
(2) Rinse, 50 C (120 F) (NOTE 7)
(3) Rinse, room temp.
(4) Acid dip, hydrochloric acid, 1–10%

Cycle II-2

(1) Preliminary pretreatment
(2) Rinse
(3) Anodic clean
(4) Rinse
(5) Anodic etch (as in II-1)
(6) Rinse
(7) Rinse
(8) Plate

(5) Rinse
(6) Smut removal (NOTE 8)
  (a) Dip in NaCN, 22 g/L (3 oz/gal) or
  (b) Anodic in NaCN, 22-45 g/L (3-6 oz/gal) at 1.5 to 2 A/dm^2 (15-20 ASF) for ½-1 min
(7) Rinse
(8) Anodic etch (NOTE 9, sulfuric acid, 250-1000 g/L (33-133 oz/gal), 30 C (86 F), 10-40 A/dm^2 (100-400 ASF)
(9) Rinse
(10) Plate

NOTE 7—This includes the use of immersion rinses followed always by spray rinses at the recommended temperature.

NOTE 8—Alkaline derusters with or without periodic reverse current (see NOTE 1) or chelated alkaline cleaners can be used in place of the NaCN treatments.

NOTE 9—A high acid content, low temperature, and high current density will minimize smut formation. Carry-over of water into the anodic etch solution should be held to a minimum. Frequently this is accomplished by the use of a hot rinse. Long transfer times after the anodic etch should be avoided.

Cycle II-3

(1) Preliminary pretreatment
(2) Rinse
(3) Anodic clean
(4) Rinse
(5) Nitric acid dip, 5%
(6) Rinse
(7) Anodic in NaCN (as in Cycle II-1-6b)
(8) Rinse
(9) Acid dip, 1% hydrochloric acid (optional depending on plating solution)
(10) Rinse
(11) Plate

Cycle II-4

(1) Preliminary pretreatment
(2) Rinse
(3) Anodic clean
(4) Rinse
(5) Electropolish or anodic etch
(6) Rinse
(7) Rinse
(8) Anodic clean
(9) Rinse
(10) Acid dip, 20% hydrochloric

Cycle II-5[12]

(1) Preliminary pretreatment
(2) Rinse
(3) Anodic clean
(4) Rinse—warm
(5) Spray rinse
(6) Alkaline deruster with PR current (See NOTE 1)
(7) Rinse—cold
(8) Acid dip—10% H$_2$SO$_4$
(9) Rinse—cold
(10) Spray rinse
(11) Plate

Cycle II-7 (For Spring Temper Parts)

(1) Anodic clean
(2) Rinse
(3) Acid dip: sulfuric acid, 0-20%, plus nitric acid, 2-10%
(4) Rinse
(5) Rinse
(6) Plate

Cycle II-6 (For Spring Temper Parts)

(1) Anodic clean
(2) Rinse
(3) Anodic in NaCN (as in Cycle II-1-6b—optional)
(4) Rinse
(5) Rinse
(6) Plate

Cycle II-8

(1) Preliminary pretreatment
(2) Rinse
(3) Anodic clean
(4) Rinse—hot
(5) Anodic etch
(6) Rinse
(7) Rinse (optional)
(8) Cyanide copper strike
(9) Rinse
(10) Acid dip, 5% sulfuric (occasionally omitted)
(11) Rinse
(12) Plate

While every consideration should be given to prevention of hydrogen embrittlement, trial will determine to what extent this factor is important. Sometimes, a seemingly harsh treatment will give satisfactory results, an example being the use of a full strength hydrochloric acid dip for certain case-hardened parts.

For nickel electroplating. Cycles II-1 through II-8.

For copper electroplating. Proceed after the last rinse in Cycles II-1 through II-7 as follows:

Cycle II-9

(1) Cyanide copper plate (if bright copper, a cyanide copper flash usually precedes the bright copper)

Cycle II-10

(1) Cyanide copper strike
(2) Rinse
(3) Acid dip
(4) Rinse
(5) Acid copper plate

Cycle II-11

(1) Nickel strike
(2) Rinse
(3) Acid copper plate

For zinc, cadmium, and tin electroplating. Although Cycles II-1 through II-7 may be used, frequently a cyanide dip or alkali dip is employed just prior to plating in the cyanide or stannate solution. Some high-carbon steels are difficult to plate in the cyanide zinc solution (see Section IV on plating cast irons).

For silver electroplating. Proceed after the last rinse in Cycles II-1 through II-7 with Cycle I-6, I-7, or I-8.

For gold electroplating. Proceed after the last rinse in Cycles II-1 through II-7 with Cycle I-9 or I-10.

For lead electroplating. Proceed after the last rinse in Cycles II-1 through II-7 with Cycle I-11, I-12, or I-13.

For chromium electroplating. Proceed as in Cycle I-15.

(III) Stainless Steels[13]

Cycle III-1

(1) Preliminary treatment
(2) Rinse
(3) Electroclean (anodic cleaning is usually preferred, and when brightness is important the severity of cleaning should be kept to a minimum). A short electropolishing treatment may be used in place of cleaning, providing the preliminary cleaning is adequate.
(4) Rinse
(5) Activate. This step is most important except in the case of chromium electroplating. The following activating treatments have been used, and the best treatments will depend on the type of steel and the particular circumstances:

(a) Immersion Treatments
  (1) Immerse in 20-50% sulfuric acid at 63-80 C (150-180 F) for at least 1 min after gassing commences. It may be necessary to start the gassing by touching the stainless steel with a piece of carbon steel.
  (2)[14] Immerse for 26 sec at room temperature in:

Hydrochloric acid	0.1%
Sulfuric acid	1.0%

(b) Cathodic treatments
  (1)

Sulfuric acid	5-50%
Temperature	Room
Current density	0.54 A/dm² (5 ASF)
Time	1-5 min

  (2)[14]

Hydrochloric acid	5-50%
Temperature	Room
Current density	2.15 A/dm² (20 ASF)
Time	1-5 min

  (3) Immerse in 10-30% hydrochloric acid at room temperature for 30-60 sec, treat cathodically in:

Sulfuric acid	5-50%
Temperature	Room
Current density	0.54-2.7 A/dm², (5-25 ASF)

  (4) Acid salts containing fluorides: 120 g/L (16 oz/gal)

Temperature	45 C (110 F)
Current density	0.54-1.6 A/dm² (5-15 ASF)
Time	1-5 min

(c) Simultaneous activation-plating treatments
  (1)[15]

Nickel chloride	240 g/L (32 oz/gal)
Hydrochloric acid	85 ml/L (11 fl oz/gal)
Temperature	Room
Electrodes	Nickel

First, treat anodically at 2.2 A/dm² (20 ASF) for 2 min, followed by cathodic treatment at 2.2 A/dm² (20 ASF) for 6 min.

  (2)[16]

Nickel chloride	240 g/L (32 oz/gal)

Hydrochloric acid	126 ml/L (16 fl oz/gal)
Temperature	Room
Anodes	Nickel

First, treat cathodically at 5.4–21.5 A/dm² (50–200 ASF) for 2–4 min, then reduce to 1.6–5.4 A/dm² (15–50 ASF) for 15–30 min.

(3)
Nickel chloride	30–300 g/L (4–40 oz/gal)
Hydrochloric acid	15–160 ml/L (2–20 fl oz/gal)
Temperature	Room
Anodes	Nickel
Current density—cathodic	0.55–10.75 A/dm² (5–100 ASF)
Time	½–5 min

(4) Hydrochloric acid, concentrated
Copper sulfate	0.4 g/L (0.05 oz/gal)
Temperature	Room
Anodes	Nickel
Current density—cathodic	4.5–6.6 A/dm² (40–60 ASF)
Time	1–5 min

In the above activation-plating treatments using nickel anodes, the nickel content of the solution gradually increases due to the low cathode efficiency. This can be compensated for by removing a portion of the solution and replenishing the hydrochloric acid. Nickel anode material containing greater than 0.01% sulfur should not be used in a nickel strike bath operated at a pH of 0.5 or lower. Carbon anodes have been used but the resultant chlorine fumes create a ventilation problem.

(5)[17]
Nickel sulfate	240 g/L (32 oz/gal)
Sulfuric acid	50 g/L (6.7 oz/gal)
Temperature	30–40 C (86–104 F)
Anodes	Lead
Current density—cathodic	16.2 A/dm² (150 ASF)
Time	5–10 min

(6) See also Refs. 18–25.

After activation, the work is rinsed and transferred to the plating solution as quickly as possible. Where possible, the rinse water should be slightly acid (pH 2.5–3.5). Frequently carry-over of acid from the activation treatments will suffice. Depending on the type of stainless steel and activating treatment, adherent deposits of the common metals can usually be deposited directly on the activated surface. If possible, the current should be on as the parts are immersed into the plating solution. Except in the case of chromium plating, the use of a simultaneous activation-plating treatment usually gives more consistent good adhesion. In some cases, however, the thin intermediate layer of nickel reduces the corrosion resistance of the finished article.

Gold electroplating on stainless steel. This may be accomplished by the above activation-plating treatments, but special treatments have been proposed and used successfully to deposit gold directly on stainless steel.[14,26-29]

Chromium electroplating on stainless steel. The immersion-activating treatment Cycle III-1, 5a(2), followed by rinsing, has been used with success for chromium plating on stainless steel automobile parts. The above activating treatments can frequently be eliminated and the following cycles used:

Cycle III-2[30]

(1) Cathodic clean
(2) Rinse
(3) Treat as anode at 10.8 A/dm² (100 ASF) in chromium plating solution
(4) Chromium plate

Cycle III-3[25]

(1) Soak clean
(2) Rinse
(3) Anodic clean at 5.4 A/dm² (5 ASF)
(4) Rinse
(5) Acid dip, 45 sec: sulfuric acid, 1%, hydrochloric acid, 0.1%
(6) Rinse
(7) Rinse
(8) Chromium plate

(IV) Cast Irons[31,32]

Cycle IV-1

(1) Cathodic clean
(2) Rinse—hot 60 C (140 F)
(3) Rinse—cold
(4) Acid dip, hydrochloric acid, 20%, or 5–10% sulfuric acid, 3–10 sec
(5) Rinse
(6) Plate

Cycle IV-2[33-35]

(1) Soak clean or degrease
(2) Cathodic clean or cathodic followed by a short anodic cleaning
(3) Rinse—hot
(4) Rinse—cold
(5) Treat anodically in sulfuric acid

TYPICAL PROCESSING AND OPERATING SEQUENCES 183

25–30%, 10.8–21.6 A/dm² (100–200 ASF) (black film should be absent at finish of anodic treatment)
(6) Rinse
(7) Plate

Cycle IV-3[32]

(1) Soak clean or degrease
(2) Anodic clean
(3) Rinse—hot
(4) Rinse—cold
(5) Treat anodically in sulfuric acid, 25–35%, 10.8–21.6 A/dm² (100–200 ASF)
(6) Rinse—cold
(7) Anodic clean
(8) Acid dip 5–10% sulfuric acid
(9) Rinse—cold
(10) Plate

For nickel electroplating. As in Cycle IV-1, IV-2, or IV-3. For a bright nickel finish the casting is sometimes given a preliminary 4-min strike in a dull Watts bath.

For copper electroplating. Proceed as in Cycle II-9, II-10, or II-11 after the last rinse in Cycles IV-1, IV-2, and IV-3.

For cadmium and tin electroplating. As in Cycle IV-1, IV-2, or IV-3, although it is advisable to use a preliminary cyanide or alkali dip prior to alkaline electroplating.

For zinc electroplating. Frequently, zinc electroplating of cast irons in a cyanide zinc solution is avoided because of the difficulty in obtaining good coverage by the zinc, but it has been done successfully using Cycle IV-3. Castings can be plated in the acid baths using Cycle IV-1, IV-2, or IV-3 providing the throwing power is adequate. Intermediate deposits may be used as follows:

Cycle IV-4

(1) Cyanide copper strike, cadmium strike or alkaline tin strike

Cycle IV-5

(1) Acid zinc strike
(2) Rinse
(3) Cyanide zinc plate

Methods[31,36] have been proposed for zinc electroplating directly on castings from a cyanide bath, but their application has been limited. Cycle IV-3 has been most successful.

For silver electroplating. After the last rinse in Cycle IV-1, IV-2, or IV-3, proceed as in Cycles I-6 through I-8.

For chromium electroplating.

Cycle IV-6[37]

(1) Anodic clean
(2) Rinse
(3) Anodic etch in chromic acid, 3–5 sec
(4) Flash in plating bath at 155 A/dm² (1440 ASF) for about 30 sec
(5) Drop current to normal

Cycle IV-7[37]

(1) Buff well
(2) Wash with solvent
(3) Dry
(4) Rub down the surface with slaked lime or alumina
(5) Flash and plate as in Cycle IV-6

Cycle IV-8[38]

(1) Abrasive polish, light pressure, with 600 mesh emery paper or belt
(2) Anodic clean, 40 min at 5.4 A/dm² (50 ASF) in 12% sodium hydroxide at 43–46 C (110–115 F)
(3) Cold rinse
(4) Dip in 5% hydrochloric acid, 5 sec
(5) Cold rinse
(6) Reverse current at 10.8 A/dm² (100 ASF) in chromium plating solution, then plate starting at no less than 81 A/dm² (750 ASF) for 5 min, then lower to normal current density (do not use steps 3 and 4 until ready to plate)

(V) Copper and Copper-Base Alloys[39]

Cycle V-1

(1) Preliminary pretreatment (NOTE 10)
(2) Rinse
(3) Electroclean (NOTE 11)
(4) Rinse
(5) Acid dip (NOTE 12)

Cycle V-2

(1) Preliminary pretreatment
(2) Rinse
(3) Electroclean (NOTE 11)
(4) Rinse
(5) Sodium cyanide solution 15–45 g/L (2–6 oz/gal)

(6) Rinse
(7) Plate in acid solution

(6) Rinse, or copper strike and rinse
(7) Plate in alkaline solution

Cycle V-3 (For Soldered Parts or Leaded Brass)

(1) Preliminary pretreatment
(2) Rinse
(3) Cathodic clean
(4) Rinse
(5) Acid dip, 10–20% fluoboric acid
(6) Rinse
(7) Cyanide solution dip
(8) Rinse, optional
(9) Cyanide copper strike
(10) Rinse
(11) Acid dip (see NOTE 12)
(12) Rinse
(13) Plate

Cycle V-4

(1) Preliminary pretreatment
(2) Rinse
(3) Electroclean
(4) Rinse
(5) Bright dip[39]
(6) Rinse
(7) Rinse
(8) Plate

NOTE 10—See Chapter 3-G for cleaning and pickling.

NOTE 11—In many cases, proprietary cleaners are used in accordance with the supplier's recommendations. Anodic cleaning of copper and alloys has in recent years become preferred. Occasionally, cathodic followed by anodic cleaning in a separate tank is used. If the parts are tarnished and cathodic cleaning is to be used, it is advisable to remove the tarnish by an acid or cyanide dip prior to cathodic cleaning.

NOTE 12—The acid dip for copper or brass may be 5 to 10% by volume sulfuric acid, 10 to 20% by volume hydrochloric acid or 10 to 20% by volume fluoboric acid, the latter usually used when a fluoborate plating solution follows. Because of the interaction of the various steps in plating on leaded brass, it is difficult to prescribe a cycle which will always give adherent electrodeposits. Cathodic cleaning is usually preferred, but very mild anodic cleaning is sometimes used. In many cases, fluoboric acid functions no better than sulfuric acid.

Occasionally it has been found necessary to add carbonate to a new copper cyanide strike solution.

For nickel electroplating. Use Cycle V-1, V-3, or V-4.

For copper electroplating. Use Cycle V-1, V-2, V-3, or V-4.

For cadmium, zinc, and tin electroplating. Use Cycle V-1, V-2, V-3, or V-4.

For silver electroplating. After the last rinse in Cycle V-1, V-2, V-3, or V-4, proceed as in Cycles I-6 through I-8, except for duplication of the copper strike.

For gold electroplating. After the last rinse in Cycle V-1, V-2, V-3, or V-4, proceed as in Cycles I-9 and I-10.

For lead electroplating. Use Cycle V-1, V-2, V-3, or V-4.

For chromium electroplating (directly on copper or alloy). Use Cycle V-1. For hard chromium the cycle is as follows:

Cycle I-5[40]

(1) Cathodic or soak clean
(2) Rinse
(3) Acid dip, hydrochloric acid, 50%
(4) Rinse
(5) Plate

(VI) Zinc-Base Die Castings[41,42]

Cycle VI-1

(1) Preliminary pretreatment
   (a) Solvent degreasing
   (b) Emulsion cleaning
   (c) Power spray wash
(2) Rinse (NOTE 13)
(3) Electroclean (NOTE 14)
(4) Rinse
(5) Acid dip, sulfuric acid, 0.25–1%, or hydrochloric acid, 0.25–1%, or hydrofluoric acid, 0.25–1%, or acid salts with fluorides, 0.5 g/L (2 oz/gal)
(6) Rinse
(7) Copper strike, optional (NOTE 15)
(8) Copper plate

NOTE 13—A cycle including a warm water rinse, a cold water rinse, followed by a spray rinse is recommended.

NOTE 14—The choice of cathodic or anodic cleaning depends on the cleaner composition.

The temperature, current density and time should be adjusted to minimize attack on the casting.

NOTE 15—Customary prior to high concentration copper cyanide baths.

Virtually all zinc die castings are first copper plated, as in Cycle VI-1. Usually other metals such as nickel, chromium, silver, gold, and brass are deposited over the copper plate. In the case of bright nickel over bright copper it is sometimes necessary or advisable to follow the copper plating with special activating or cleaning treatments prior to going into the bright nickel solution. In this case, the suppliers should be consulted. The minimum thickness of copper is about 5 µm (0.2 mil) and it is necessary to use a copper strike prior to most proprietary copper solutions. For simple shapes not requiring high throwing power, nickel may be deposited directly on zinc die castings with good results from solutions formulated for that purpose.[43,44] An improvement in appearance and corrosion protection may be obtained on porous castings by the use of a leveling zinc deposit prior to the copper.[45]

(VII) Aluminum and Aluminum Alloys[46]

In general, aluminum and aluminum alloys do not respond to the surface conditioning treatments employed for the preparation of other metals for electroplating. The alloys which can be plated with the use of various cycles have been published.[46,47,48]

Modifications of the cycles can be made to fit the circumstances if the composition of the alloy is considered, particularly regarding silicon, copper, and magnesium. There are four major steps in the finishing of aluminum and aluminum alloys:

A. Preliminary cleaning
B. Conditioning
C. Immersion-strike treatment
D. Final electrodeposits

Depending on the alloy, various cleaning and conditioning treatments should provide a surface with uniform activity for the deposition of the initial immersion-strike layer.

A. *Preliminary cleaning.* The surface must be free from grease, oil, or buffing compounds. This can be accomplished by vapor degreasing, solvent washing, or solvent emulsion cleaning. A 60-80 C (140-175 F) mild etching type cleaner for 1-3 min may also be used.

Sodium carbonate, anhydrous	25 g/L (3.3 oz/gal)
Trisodium phosphate, anhydrous	25 g/L (3.3 oz/gal)

B. *Conditioning.* Following the preliminary cleaning, the surface should be conditioned. Several methods have been successful for various alloys.

1. For most alloys:

Sodium hydroxide	50 g/L (6.7 oz/gal)
Time	30 sec to 1 min
Temperature	50 C (125 F)

followed by a water rinse and immersion for about 45 seconds in a desmutting solution of:

Nitric acid	75%
Ammonium bifluoride	120 g/L (16 oz/gal)
Temperature	20-25 C (65-75 F)

2. Heat treated (T-Temper) alloys:

Sulfuric acid	10%
Chromic acid	35 g/L (4.7 oz/gal)
Temperature	70-80 C (160-175 F)
Time	2-5 min

3. Wrought alloys of the 1100 and 3003 type: treat with the carbonate-phosphate solution in A above. Follow by a dip in a 50% nitric acid solution.

4. All wrought and cast aluminum-magnesium alloys:

Sulfuric acid	15%
Temperature	80 C (175 F)
Time	2-5 min

5. High-silicon casting alloys:

Nitric acid	75%
Hydrofluoric acid	25%
Temperature	Room
Time	3-5 sec

C. *Immersion—strike treatment.* Following the preliminary cleaning and conditioning steps, it is necessary to further treat the surface to obtain adequate adhesion of the final electrodeposited metals.

1. *The zinc immersion process*[46,49,50]
   Depending upon the alloy involved, differ-

Alkaline Zincate Solutions

	1	2	3	4
Zinc oxide	100 (13)	100 (13)	5 (0.67)	20 (2.7)
Sodium hydroxide	525 (70)	525 (70)	50 (6.7)	120 (16)
Ferric chloride		1 (0.13)	2 (0.27)	2 (0.27)
Rochelle salts		10 (1.3)	50 (6.7)	50 (6.7)
Sodium nitrate			1 (0.13)	1 (0.13)
Temperature, C	15–27	15–27	20–25	20–25
(F)	(60–80)	(60–80)	(70–75)	(70–75)
Time, sec	30–60	30–60	30 or less	30 or less

Acid Zincate Solutions

	5	6
Zinc sulfate·7 H_2O	720 (96)	
Zinc fluoborate		75 (10)
Hydrofluoric acid	3.5%	
pH (with fluoboric acid)		3.0
Temperature C	25	20–27
(F)	(77)	(70–80)
Time, sec	30–60	30

Formulations 1 through 4 are alkaline zincate solutions. Formulations 5 and 6 are acid zinc solutions. Formulations 2, 3, and 4 give more uniform and satisfactory results on wrought and cast alloys. In addition to iron being added to zincate solutions, nickel and copper have also been used with repeatedly good results.[51,52,53]

Also available are proprietary zincates which are modified with copper, copper/nickel, and copper/nickel/iron. These deposit alloy zinc films instead of pure zinc, and permit somewhat greater latitude in subsequent plating conditions and applications. They also claim advantages in reduced porosity, and improved corrosion resistance. The pretreatment procedure may be affected as well. The suppliers should be consulted for details on the processes and procedures.

Prior to depositing any subsequent metallic layers it is usually advisable to apply a copper strike over the zinc immersion layer. The parts are transferred to a low-pH, low-temperature tartrate-type cyanide copper solution. An initial high-current density strike of about 2.6 A/dm² (24 ASF) is applied for the first 2 min followed by a reduced current density of 1.3 A/dm² (12 ASF) for 3 to 5 min.

ent acid dips and zinc immersion processes are used in the following basic cycle:

Cycle VII-1[46,48]

(1) Preliminary cleaning
(2) Rinse
(3) Soak clean
(4) Rinse
(5) Acid dip(s) (see below, *a*)
(6) Rinse
(7) Zinc immersion (see below, *b*)
(8) Double rinse
(9) Plate

a. *Acid dips*

The following acid dips have been used in Cycle VII-1, step 5 above. The choice depends upon the aluminum surface.

(1) Single acid: 50% nitric acid, room temperature
(2) Double acid: 15% sulfuric acid, 2 min at 82 C (180 F), rinse, 50% nitric acid
(3) Double Zincate: zinc-immersion (see below: b. Zinc-Immersion), rinse, 50% nitric acid
(4) Mixed acid: 75% nitric acid with 25% hydrofluoric acid

b. *Zinc immersion*

Depending upon the nature of the aluminum surface, the following basic formulations in g/L (oz/gal) can be used in the zinc immersion step in Cycle VII-1 step 7 and in the double zincate process given above (*a*).

2. *Zinc immersion—neutral nickel strike*[54]

After the proper precleaning and conditioning, the parts are given a double zinc immersion treatment similar to Cycle VII-1 using acid dip, *a*-3. After the final rinse the surface is given a nickel strike in a solution with the following compositions:

Nickel sulfate	142 g/L (19 oz/gal)
Ammonium sulfate	34 g/L (4.5 oz/gal)
Nickel chloride	30 g/L (4.0 oz/gal)
Sodium citrate	142 g/L (19 oz/gal)
Sodium gluconate	30 g/L (4.0 oz/gal)
Temperature	57–66 C (135–150 F)

pH at 140°F   6.8–7.2
Agitation   Mechanical

After an initial strike at 9.5–13 A/dm² (95–130 ASF) for 30–45 sec, reduce the cathode current density to 4–5.5 A/dm² (40–55 ASF) and electroplate for an additional 3–5 min.

3. *Zinc immersion—nickel glycolate strike*[55,56]

After the proper precleaning and conditioning, as in Cycle VII-1, the parts are given either a single or a double zinc immersion treatment. This is followed by a rinse and a nickel strike in a solution of the following composition:

Nickel acetate	65 g/L (8.7 oz/gal)
Boric acid	45 g/L (6.0 oz/gal)
70% glycolic acid	60 ml/L (7.7 fl oz/gal)
Saccharin	1.5 g/L (0.2 oz/gal)
Sodium acetate	50 g/L (6.7 oz/gal)
Temperature	Room
pH	5.5–6
Current density	2.7 A/dm² (25 ASF)
Time	2–5 min
Anodes	Nickel or inert
Agitation	Mechanical

4. *Tin immersion—bronze strike*[57]

The parts should be precleaned as described in Section A and then conditioned as in Section B-1. After cleaning and conditioning the parts are subjected to a tin activation treatment. This can be accomplished either by simple immersion, or electrolytically, with cathodic current in a proprietary,[57] aqueous stannate bath for 30 sec at 26–30 C (75–85 F). Following the stannate treatment, without rinsing and with a minimum time delay, the parts are transferred into a proprietary, aqueous bronze bath,[57] where they are given a strike of 3–4 min at 26–30 C (75–85 F), with a cathode current density of 3.2–5.4 A/dm², (32–54 ASF).

5. *Anodic process (anodizing)*[58-62]

The anodic process forms an oxide film on the surface of the aluminum over which other metals may be electrodeposited. Two typical cycles are as follows:

Cycle VII-2[63-64]

(1) Preliminary cleaning
(2) Rinse
(3) Soak clean and etch
(4) Rinse
(5) Nitric acid dip
(6) Double rinse
(7) Phosphoric acid anodize:
    phosphoric acid (20–60%),
    temperature 27–35 C (80–95 F),
    voltage 5–30 volts,
    time 5–15 minutes
(8) Rinse
(9) Plate

Cycle VII-3

(1) Soak clean
(2) Rinse
(3) Anodize
(4) Rinse
(5) Nickel plate

D. *Final electrodeposits.* Following the immersion or strike deposits listed in Section C, other metals can be electroplated on the conditioned aluminum parts using conventional electroplating solutions. A copper strike, as stated in Section C-1, is usually preferred following either the zinc immersion strike or the phosphorus anodized strike described in Section C-5. Two precautions should be taken when plating over the zinc immersion or anodized surfaces:

(1) The current should be on as the work is immersed in the plating solution.
(2) The plating solution should not be too acid or too alkaline.

For copper electroplating. After the copper strike, the work can be transferred to any conventional copper plating bath.

For nickel electroplating. After the copper strike, the work is rinsed and transferred to a conventional nickel electroplating bath. When nickel electroplating directly on the zinc immersion coating, the pH of the nickel solution should be 4.0 to 5.0; whereas, for electroplating over the anodized surface the pH may be 3.5 to 4.5. Nickel can be electroplated directly on the nickel strike and bronze immersion strikes.

For zinc and tin electroplating. These metals can be applied over any of the immersion strike metals or copper strike films.

For cadmium electroplating. Cadmium can be applied over copper, nickel, or bronze strike deposits or directly on the zinc immersion film. If applied directly on the zinc film, a cadmium strike prior to cadmium electroplating of the following compositions should be used:

Cadmium oxide	7.5 g/L (1 oz/gal)
Sodium cyanide	60 g/L (8 oz/gal)
Temperature	Room
Current density	2.7 A/dm^2 (25 ASF)
Time	1 min

For silver electroplating. Silver may be deposited over the copper, nickel, or bronze strikes as in Cycle I-7 or I-8, or on the anodized surface as in Cycle I-6.

For gold electroplating. Gold may be applied over a copper strike by use of Cycle I-9 or I-10.

For lead electroplating. Lead may be applied over a heavy copper, nickel, or bronze strike after rinsing.

For chromium electroplating:

Cycle VII-4

(1) Soak clean
(2) Rinse
(3) Etch in: sodium hydroxide, 45 g/L (6 oz/gal); temperature, 40.5 C (150 F); time, 1-2 min
(4) Rinse
(5) Nitric acid dip
(6) Rinse
(7) Chromium plate in standard bath

Cycle VII-5

(1) Cycle VII-1
(2) Chromium plate at 15.5-21 C (60-70 F)
(3) Transfer to higher-temperature bath and hold until part reaches bath temperature
(4) Gradually increase current density from 16-33 A/dm^2 (150-300 ASF)

Cycle VII-6

(1) Cycle VII-1
(2) Chromium plate at 18-21 C (65-70 F) in standard or Bornehauser solution
(3) Buff

Cycle VII-7

(1) Cycle VII-1
(2) Copper strike, 3-5 min
(3) Rinse
(4) Hard chromium plate

(VIII) Magnesium and Its Alloys[65]

Magnesium requires a zinc immersion deposit prior to plating with other metals, which is somewhat different from that used on aluminum.

Solution for Zinc Immersion

Zinc sulfate (ZnSO$_4$·H$_2$O)	30 g/L (4 oz/gal)
Tetrasodium pyrophosphate	120 g/L (16 oz/gal)
Sodium fluoride	5 g/L (0.67 oz/gal)
or lithium fluoride	3 g/L (0.40 oz/gal)
Sodium carbonate	5 g/L (0.67 oz/gal)
pH (measured colorimetrically)	10.2-10.4
Temperature	79-85 C (175-185 F)
Time	
Aluminum-containing alloys	5-7 min
Aluminum-free alloys	3-5 min
Unalloyed magnesium	3-5 min
Agitation	Mild

Cycle VIII[66-70]

(1) Preclean and pickle if necessary (NOTE 16)
(2) Rinse
(3) Cathodic clean 7.5-13 A/dm^2 (75-130 ASF) 85 C (185 F)
(4) Rinse
(5) Pickle if necessary
(6) Rinse
(7) Activate by immersion in: phosphoric acid, 20%, and sodium, potassium, or ammonium bifluoride, 105 g/L (14 oz/gal); temperature, 16-38 C (60-100 F); time 0.5-2 min
(8) Rinse
(9) Zinc immersion treatment (NOTE 17)
(10) Rinse
(11) Copper strike

NOTE 16—See Chapter 13 E.

NOTE 17—For some alloys, a double zinc immersion treatment is required after step 10. Repeat steps 7 through 10.

Copper strike. The following two baths have been found suitable for electroplating over zinc immersion coating.

Bath A[70]

Copper cyanide	41 g/L (5.5 oz/gal)
Potassium cyanide	68 g/L (9.0 oz/gal)
Potassium fluoride	30 g/L (4.0 oz/gal)
Free cyanide	7.5 g/L (1.0 oz/gal)

Bath B

Copper cyanide	41 g/L (5.5 oz/gal)
Sodium cyanide	52.5 g/L (7 oz/gal)
Rochelle salt	45 g/L (6 oz/gal)
Free cyanide	4 g/L (0.5 oz/gal)

Both solutions are operated at 54 to 60 C (130 to 140 F) and a pH of 9.6 to 10.4 (colorimetric). Cathode rod agitation is used. In solution, make electrical contact quickly at 5 to 10 A/dm² (45 to 90 ASF) then lower current to 1 to 2.5 A/dm² (9 to 23 ASF). Use periodic reverse current (see NOTE 1) after 2.5 µm (0.1 mil) if deposits thicker than 25 µm (1 mil) are needed.

Racks. Magnesium alloy racks may be used but are not advantageous after the first plating cycle. Although the use of conventional racks is operational, rack coatings should be used on the entire rack except for small contact areas. When used for other than copper, brass, zinc, or cadmium electroplating, the rack should be given a copper strike before recycling.

For copper electroplating. Heavier copper deposits may be applied by additional time in the copper strike solution, transferring to another alkaline bath, or electroplating over nickel.

For nickel electroplating. Nickel may be applied after copper electroplating to about 7.6 µm (0.3 mil) minimum thickness. Preferably, the bright nickel solution should have a pH of 4.0 or above. For recessed parts where the copper may be thin or for electroplating directly over the zinc, the following solution, which does not seriously attack magnesium, can be used:[70,71]

Hydrofluoric acid (70%)	60%
Citric acid	30 g/L (4 oz/gal)
Basic nickel carbonate (2NiCO₃·3Ni(OH)₂·4H₂O)	120 g/L (16 oz/gal)
Sodium lauryl sulfate	0.9 g/L (0.125 oz/gal)

The chemicals should be added in the above order to approximately one-tenth of the final volume of water; the nickel carbonate should be added slowly to prevent foaming. The pH should be held between 1.0 and 3.0, the temperature at 50 to 60 C (120 to 140 F), and the current density at 3.0 to 10 A/dm² (30 to 100 ASF). Cathode rod agitation is used.

A thin layer of electroless nickel[72] applied directly to the magnesium surface has been used as a base for further electroplating.

For zinc and brass electroplating. These may be applied from cyanide solutions directly on the zinc immersion coating, but preferably after copper or nickel electroplating.

For cadmium electroplating. This is applied over the copper coating of Cycle VIII.

For silver electroplating. After copper electroplating the usual nickel strike, silver strike, silver plate cycle is used as in Cycle I-8.

For chromium electroplating. Chromium may be deposited directly over copper or nickel deposits.

(IX) Nickel Silver

Nickel silver is a copper-base alloy containing 17 to 32% zinc and 10 to 30% nickel.

Cycle IX-1

(1) Cathodic clean
(2) Rinse
(3) Hydrochloric acid dip, 30%
(4) Rinse
(5) Plate

Cycle IX-2

(1) Cathodic clean
(2) Anodic clean
(3) Rinse
(4) Sodium cyanide dip, 45–60 g/L (6–8 oz/gal)
(5) Rinse
(6) Sulfuric acid dip, 5%
(7) Rinse
(8) Cyanide copper strike
(9) Rinse
(10) Sulfuric acid dip, 2%
(11) Rinse
(12) Nickel strike, gray or bright nickel up to 0.25 µm (0.01 mil)

Cycle IX-3[73] *(Hollow Ware)*

(1) Soak clean
(2) Rinse
(3) Cyanide dip where tarnished
(4) Rinse
(5) Cathodic clean

(6) Rinse
(7) Brush, scour, or sponge
(8) Rinse
(9) Cathodic clean
(10) Rinse
(11) Hydrochloric acid dip, 4–30%
(12) Rinse
(13) Nickel strike

Cycle IX-4[73] (Flatware)

(1) Preclean
(2) Hot rinse
(3) Swirl in hot soap solution
(4) Soak clean
(5) Cathodic clean
(6) Rinse
(7) Cyanide dip
(8) Rinse
(9) Hydrochloric acid dip
(10) Rinse
(11) Nickel strike

For silver electroplating. Most nickel-silver is silver electroplated, and the usual nickel strike, silver strike, and silver plate are used as in Cycle I-8. If the nickel strike is omitted, a cycle similar to I-6 or I-7 is used.

For gold and rhodium electroplating. A nickel strike is preferred. Proceed as in Cycle I-9 to I-10 after the last rinse in Cycles IX-1 through IX-4.

For brass, nickel-chromium, tin electroplating. The usual copper or nickel undercoat is first applied.

(X) Lead and Lead Alloys[74]

Electroplating on lead or lead alloys falls in two main categories: electroplating on soldered parts and electroplating on lead alloys such as in jewelry and hollow ware. Two complications arise: (1) usually the lead or lead alloy is only part of the surface being electroplated as in soldered brass parts or brass hollow ware with white metal spouts; (2) certain fluxes and stains from hot flux can be troublesome from the cleaning standpoint. The cycles which follow are to be used only as a guide in determining the best cycle for a given circumstance. In many cases, hand scrubbing and soak cleaning have been found more reliable than electrocleaning. One of the most important steps in plating hollow ware made up of soldered parts is to remove both the inside and outside stain caused by the hot flux. Buffing will remove the outside stain, but the outside surface will frequently become stained or etched by the time the inside stain is removed by cleaning. Therefore, immediately after soldering, the inside should be thoroughly scratch brushed.

Cycle X-1 (As for Jewelry)

(1) Degrease
(2) Cathodic clean-low voltage, 70 C (160 F), 30 sec
(3) Rinse
(4) Sodium cyanide dip, 15–22 g/L (2–3 oz/gal)
(5) Rinse
(6) Fluoboric acid dip, 5–25% or 8% glacial acetic acid and 4.5% hydrogen peroxide
(7) Water rinse
(8) Plate

Cycle X-2 (As for Percolators)

(1) Preclean
(2) Rinse
(3) Cathodic clean
(4) Rinse
(5) Soak clean, swabbing may be used
(6) Rinse
(7) Anodic in 10% fluoboric acid at 1.6 A/dm^2 (15 ASF) for 15–20 sec
(8) Rinse
(9) Plate

Cycle X-3[75] (for Antimonial Lead Alloys)

(1) Cathodic clean in trisodium phosphate, 30–45 g/L (4–6 oz/gal)
(2) Anodic clean in soda ash solution, 30 g/L (4 oz/gal), 15 sec
(3) Rinse
(4) Dip in caustic soda solution, 15 g/L (2 oz/gal), 65 C (150 F)
(5) Rinse
(6) Acid dip, hydrochloric acid, 20%
(7) Rinse
(8) Dip in 3% sodium cyanide solution
(9) Copper flash

Cycle X-4 (For Lead and Lead Alloys)

(1) Soak clean
(2) Scour with pumice
(3) Rinse
(4) Cathodic in mild cleaner
(5) Rinse

(6) Dip in 20% fluoboric acid
(7) Rinse
(8) Plate

For nickel electroplating. The plating step in Cycle X-1, X-2, or X-4 may be used for nickel plating. In cycle X-3, a copper flash may be used, followed by a rinse, mild acid dip, rinse, and nickel.

For zinc, cadmium, tin, brass, silver, gold, and rhodium electroplating. These may be deposited over a nickel or copper flash (preferably nickel) as in other previously described cycles.

(XI) Powder Metal Compacts

Customarily, articles are made by compacting a metal powder in a die at slightly elevated temperature to produce a weak, "green" molding. It is then sintered at a temperature below the melting point in an inert or reducing atmosphere.

There are no general cycles that can be given which apply to all powder compacts and for all deposits. Good rinsing may be the most important step due to the porosity of the surface. This cannot be overemphasized in order to prevent spotting out. The degree of spotting out varies also with the thickness of the deposits, which may be grouped as follows: (1) thin deposits can be rinsed more easily and give little trouble, (2) medium weight deposits, 25 to 50 μm (1 to 2 mil), because of the partial bridging of the voids, give considerable trouble, (3) heavy deposits, 150 to 300 μm (5 to 12 mil), where the bridging is completed give very little trouble. The typical cycles given here have been used with success on certain powder metal parts. Other parts will require modification to fit the paticular circumstance. Application of the principles involved in these cycles, and the realization of the high degree of porosity of such parts, will usually lead to a satisfactory cycle.

Cycle XI-1[76] (Nickel on Iron Powder Compacts)

(1) Unsintered green molding
(2) Pickle in warm 10% hydrochloric acid
(3) Rinse thoroughly in warm water
(4) Nickel plate to 25 μm (1 mil) in a solution containing:
Nickel sulfate
($NiSO_4 \cdot H_2O$)   120 g/L (16 oz/gal)
Ammonium
chloride   15 g/L (2 oz/gal)
Boric acid   15 g/L (2 oz/gal)
pH   5.2–5.8
Current
density   1.9 A/dm^2 (18 ASF)
(5) Rinse thoroughly
(6) Sinter in hydrogen at 1090 C (2000 F) for 1 hr taking 30 min each for heating and cooling.

Cycle XI-2[77] (Copper, Nickel, Chromium on Brass Compacts)

(1) Coarse, followed by fine greaseless buffing
(2) Emulsion or soak cleaning
(3) Anodic clean in brass cleaner
(4) Hot rinse
(5) Cold rinse
(6) Repeat steps 4 and 5
(7) Dip in sodium cyanide solution, 7.5 g/L (1 oz/gal)
(8) Repeat steps 4, 5, and 6.
(9) Dip in sulfuric acid, 3–5%
(10) Repeat steps 4, 5, and 6
(11) Copper electroplate to 25 μm (1 mil) in acid copper solution
(12) Repeat steps 4, 5, 6, 7, and 8
(13) Hot rinse and dry
(14) Copper buff (Use greaseless compound prior to coloring)
(15) Clean as in steps 2 through 10.
(16) Nickel electroplate
(17) Repeat steps 4 and 5
(18) Chromium electroplate
(19) Hot rinse and dry

The porosity and high surface area of powder metal parts may be reduced by rolling or other type of cold working, or by impregnation with copper, lead, tin, waxes, or resins. The excess impregnating material is removed by cleaning, polishing, or tumbling.[78-81].

Cycle XI[77] (Nickel-Chromium on Brass Compacts)

(1) Proceed as in steps 1 through 10 in Cycle XI-2.
(2) Proceed as in steps 16 through 19 in Cycle XI-2.

Cycle XI-4[77] (Copper or Brass on Brass Compacts)

(1) Coarse, followed by fine greaseless buffing
(2) Color with dry lime (if bright final finish is desired)
(3) Emulsion or soak clean
(4) Anodic clean
(5) Hot then cold rinse
(6) Repeat step 5
(7) Dip in sulfuric acid, 3–5%
(8) Repeat steps 5 and 6
(9) Dip in sodium cyanide solution, 7.5 g/L (1 oz/gal)
(10) Repeat steps 5 and 6
(11) Copper or brass plate, cyanide solution
(12) Repeat steps 5 and 6
(13) Dip in citric acid solution, 1% by wt
(14) Treat cathodically at high voltage for 1 min in chromic acid solution, 1.1–2.3 g/L (0.15–0.3 oz/gal) (optional)
(15) Repeat steps 5 and 6
(16) Agitate in dewatering liquid
(17) Vapor degrease
(18) Buff with fine greaseless compound, then with dry lime
(19) Degrease
(20) Lacquer

Cycle XI-5[82] (Cadmium Plating)

(1) If the parts are filled with oil, degrease (If not oiled, they may be degreased or electrocleaned)
(2) Rinse thoroughly
(3) Dip in 20% sulfuric acid solution for 15 to 30 sec
(4) Rinse thoroughly; for barrel plating soak in warm water for 15 min
(5) Cadmium electroplate in cyanide solution at as high a current density as possible without burning
(6) Rinse in cold water
(7) Rinse in hot water
(8) Dry in hot air blast

(XIII) Less Common Metals

Electroplating on the following metals is done to a much less extent than the proceeding ones discussed in this chapter. They are generally more difficult to plate and therefore require very specific preplate treatments which in themselves demand greater control and expertise. The electroplating of these metals is more often done for specific engineering and industrial purposes rather than for purely decorative reasons.

A. Beryllium

Cycle 1[83]

(1) Clean
(2) Anodic pickle in solution containing 10% phosphoric acid and 2% hydrochloric acid at 20–27 C (70–80 F) using 5.2–16 A/dm^2 (50–150 ASF) for 2 min
(3) No rinse
(4) Chemical pickle in concentrated nitric acid (70%) at 20–27 C (70–80 F) for 2 min
(5) Rinse
(6) Plate in baths suitable for electrodeposition on active metals such as zinc, aluminum, and magnesium
(7) Rinse and dry
(8) Outgas: e.g., outgassing of iron-plated beryllium is necessary before subsequent nickel plating to avoid blistering when the composite is to be heated. Heating for about 1 hr at 500 C (932 F) in an inert gas atmosphere is sufficient

Cycle 2[83] (Zinc Immersion Method)

(1) Clean
(2) Immerse in solution containing $Na_4P_2O_7$, 120 g/L (16 oz/gal); $ZnSO_4 \cdot 7H_2O$, 40 g/L (5.3 oz/gal); NaF 7.5 g/L (1 oz/gal); K_2CO_3, 5.2 g/L (0.7 oz/gal); pH 7.5–8 (adjusted with sulfuric or phosphoric acid) at 82 C (180 F) for about 5 min
(3) Rinse
(4) Plate with copper, iron, zinc, or nickel in solutions suitable for electroplating on zinc
(5) Rinse and dry
(5a) Or rinse and outgas; e.g., copper-plated beryllium is outgassed by soaking for 30 min in boiling water prior to subsequent electroplating so as to avoid blistering
(6) Dip in 10% sulfuric acid
(7) Rinse

B. Chromium[84]

Cycle 1 (Chromium on Chromium)

(1) Preclean
(2) Abrasive cleaning (if necessary)
(3) Alkaline cleaning (always cathodic)
(4) Activating
 Etch in 225 g/L (30 oz/gal) CrO_3
 Temperature room
 Time 5–6.0 sec at 6 volts

(5) Electroplate in chromium solution containing 250-400 g/L (33-54 oz/gal) CrO_3 and a sulfate ratio of 100 to 1. Make part cathodic up to 3.0 volts, then slowly increase the current so that electroplating and gassing start after 30-60 sec and full current is reached in 5 min

C. Germanium[85]

Cycle 1 (for rectifying contacts)

(1) Clean
(2) Chemical polish in solution containing:

Nitric acid	5 parts by vol
Hydrofluoric acid	3 parts by vol
Acetic acid	3 parts by vol
Liquid bromine	10 drops per 50 cc of the above mixture

(3) Thorough water rinse
(4) Place under special nozzle which at its tip contacts germanium with plating bath flowing continuously
(5) Plate generally in cyanide type bath. Exception: inidium sulfate; chromic acid-sulfate; bismuth perchlorate; tellurium fluoride-sulfate
(6) Rinse
(7) Acid etch in suitable acid to remove the feather edge around the deposit to ensure good rectifying contact
(8) Rinse

D. Molybdenum

Cycle 1[86] (Chromium plus Nickel)

(1) Clean—cathodic preferred
(2) Anodic etch in solution containing equal volumes of concentrated sulfuric acid and phosphoric acid at 21-32 C (70-90 F) using 1.9-9.7 A/dm² (18-90 ASF) for 2-3 min
(3) Rinse
(4) Plate 25 μm (1 mil) of chromium in a low-contraction chromium bath containing 247 g/L (33 oz/gal) of CrO_3 and 2.47 g/L (0.33 oz/gal) sulfuric acid at 85 C (185 F) using 81-108 A/dm² (750-1000 ASF)
(5) Rinse
(6) Etch in 50% hydrochloric acid
(7) Nickel strike, all chloride baths using 19.5-58 A/dm² (180-540 ASF) for 1-3 min
(8) Nickel plate—watts or all-chloride type nickel

Cycle 2[87] (Alloy Coating of 20 Cr-80 Ni)

(1) Carefully grind to remove any surface imperfections
(2) Heat 2 hr in hydrogen atmosphere at 980 C (1800 F) to outgas
(3) Degrease in alkaline solution
(4) Rinse
(5) Pickle in 50% nitric acid for 1 min at 21-27 C (70-80 F)
(6) Rinse
(7) Anodic etch in solution containing 99 g/L (13.2 oz/gal) CrO_3 plus 1.3 g/L (0.173 oz/gal) $Na_2Cr_2O_7$ at 21-27 C (70-80 F) using 38 A/dm² (350 ASF) for 2 min
(8) Plate with 37.5 μm (1.5 mil) of chromium in a solution of 247 g/L (33 oz/gal) CrO_3 and 2.47 g/L (0.33 oz/gal) H_2SO_4 at 85 C (185 F) using 81 A/dm² (750 ASF)
(9) Rinse
(10) Dip in concentrated hydrochloric acid
(11) Rinse
(12) Dip in sulfuric acid solution of pH 1.0 t 82 C (180 F) for 1 min
(13) Rinse
(14) Nickel strike electroplate in solution containing 450 g/L (60 oz/gal) $NiSO_4 \cdot 6H_2O$ and 50 g/L (6.7 oz/gal) H_2SO_4 at 38 C (100 F) using 10.8 A/dm² (100 ASF) for 1 min
(15) Rinse
(16) Nickel electroplate 19 μm (0.75 mil) in a sulfamate bath
(17) Rinse
(18) Repeat step 12
(19) Rinse
(20) Plate with 6.35 μm (0.25 mil) of chromium in a low-contraction bath such as that given in step 8
(21) Repeat steps 9 through 20 for as many alternate layers of Cr-Ni-Cr as desired
(22) Rinse
(23) Dry
(24) Heat-treat 4 hr at 980 C (1800 F) in hydrogen. This forms a Ni-Cr alloy coating by diffusion

Cycle 3[88] (Chromium plus Nickel)

(1) Degrease in 10% NaOH solution with 60-cycle alternating current at 25 volts for 20 sec
(2) Rinse
(3) Anodic treat in 70% sulfuric acid at 21-32 C (70-90 F) using 10.8-32.3 A/dm² (100-300 ASF), approximately 10 volts, for 30 sec

(4) Rinse in water which removes most of the blue oxide film formed in step 3
(5) Dip in alkaline solution such as a proprietary cleaner or trisodium phosphate-sodium hydroxide solution
(6) Rinse
(7) Dip in 10% sulfuric acid
(8) Rinse
(9) Chromium strike electroplate at a temperature not lower than 49 C (120 F) using 16.2 A/dm^2 (150 ASF) for 60 sec
(10) Rinse
(11) Nickel strike electroplate in a Wood's strike using 5.4 A/dm^2 (50 ASF) (NOTE 18)
(12) Rinse and apply any plate that can be plated over nickel

NOTE 18—This is the name generally given to a solution containing nickel chloride and a high concentration of hydrochloric acid. See References 15, 16 and 18 under Stainless Steels.

Cycle 4[89] (Al-Ni-Cr Alloy)

(1) Same as Cycle 1 through nickel electroplate step 8
(2) Rinse and dry
(3) Aluminum plate in molten cryolite bath at 1010 C (1850 F) using 9–19 A/dm^2 (90–180 ASF) to deposit 63.5 μm (2.5 mil) aluminum; e.g., 2.5 hr at 19 A/dm^2 (180 ASF). The aluminum alloys with the nickel by diffusion during deposition.

E. "Inconel" X and "Hastelloy" C

Cycle 1 (Chromium Plate[90])

(1) Degrease
(2) Cold rinse
(3) Dip in 20% hydrochloric acid at 21–27 C (70–80 F) for 1 min
(4) Cold rinse
(5) Anodic etch in Wood's nickel strike (see NOTE 18) at 43 C (110 F) using 5.4 A/dm^2 (50 ASF) for 20 sec
(6) Plate in Wood's bath at 43 C (110 F) using 5.4 A/dm^2 (50 ASF) for 2 min
(7) Cold rinse
(8) Dip in chromium plating bath for 30 sec
(9) Chromium plate in bath containing 247 g/L (33 oz/gal) CrO_3 and 2.5 g/L (0.33 oz/gal) H_2SO_4 at 54.5 C (130 F) using 31 A/dm^2 (288 ASF)

F. "Hastelloy" B, "Stellite" 21, and 19-9 DL Stainless Steel

Cycle 1 (Chromium Plate[90])

(1) Degrease
(2) Rinse
(3) Dip in 20% hydrochloric acid at 21–32 C (70–80 F) for 1 min
(4) Cold rinse
(5) Dip in 67% nitric acid at 71 C (160 F) for ½ min
(6) Cold rinse
(7) Dip in chromium plating bath for ½ min
(8) Chromium plate in same bath as in preceding cycle

G. Nimonic Alloys, "Incoloy T," "Inconel," and 18-8 Stainless Steels "Uniloys" 1 and 2

Cycle 1 (Nickel Plate[91])

(1) Degrease
(2) Rinse
(3) Anodic etch in Wood's-nickel strike bath at 21–27 C (70–80 F) using 2.3 A/dm^2 (30 ASF) for 2 min (see NOTE 18)
(4) Cathodic strike in Wood's bath at 21–27 C (70–80 F) using 3.2 A/dm^2 (30 ASF) for 6 min
(5) Without rinsing, transfer to nickel electroplating bath and plate

H. "Invar" (Selenium-Free) and 40 to 50% Ni-Fe Alloys and "Monels"

Cycle 1 (Nickel Plate[91])

(1) Degrease
(2) Rinse
(3) Anodic treat in 34% sulfuric acid at 21–27 C (70–80 F) using 21.6 A/dm^2 (200 ASF) for 3–5 min after onset of passivity
(4) Rinse
(5) Nickel plate

I. Incoloys W, 901, and X[91]

Cycle 1 (Nickel Plate)

(1) Degrease
(2) Rinse
(3) Dip in 25% hydrofluoric acid at 21–27 C (70–80 F) for 10 sec
(4) Rinse
(5) Nickel plate

J. "Duranickel", "Ni-Span-C", "Ni-Resist" II, III, IV, V

Cycle 1 (Nickel Plate)

(1) Degrease
(2) Rinse
(3) Anodic treat in 34% sulfuric acid at 21-27 C (70-80 F) using 21.6 A/dm² (200 ASF) for 2-10 min
(4) Cathodic treat in same bath as step (3) using 21.6 A/dm² (200 ASF) for 2-3 sec
(5) Rinse
(6) Nickel plate

K. Niobium

Cycle 1[92] (Nickel or Iron Plated)

(1) Soak in alkaline cleaning solution at 90 C (195 F) for 2 min (any strong ferrous cleaner)
(2) Rinse
(3) Anodic etch in 10% sodium hydroxide solution at 21-27 C (70-80 F) using 5 volts to produce a yellow straw color (colorless—too thin; blue to purple—too thick)
(4) Rinse
(5) Plate with nickel in a Wood's-nickel strike bath at 29.5 C (85 F) using 3.2 A/dm² (30 ASF) for a sufficient time to cover with nickel (see NOTE 18)
(6) Rinse
(7) Nickel plate in Watts-type bath of pH 4 at 65.5 C (150 F) using 5.4 A/dm² (50 ASF)
(7a) Alternative—iron plate in solution containing 300 g/L (40 oz/gal) of $FeSO_4 \cdot 7H_2O$, plus 39.7 g/L (5.3 oz/gal) of $FeCl_2 \cdot 4H_2O$ and 15 g/L (2 oz/gal) of $(NH_4)_2SO_4$ at a pH of 4 at 60 C (140 F) using 1.0-1.2 A/dm² (10-20 ASF)
(8) Rinse
(9) Dry

Cycle 2[93] (Iron Plate)

(1) Degrease
(2) Activate in hydrofluoric acid (48%) at 21-27 C (70-80 F) using alternating current at current density of about 21.6 A/dm² (200 ASF) for 1-3 min
(3) Remove smut in solution containing 1% hydrofluoric acid and 50% nitric acid at 21-27 C (70-80 F) by immersion for ¼-½ min
(4) Rinse

(5) Iron plate at a pH of 4 at 60 C (140 F) and 3.2 A/dm² (30 ASF) in:
   300 g/L (40 oz/gal), $FeSO_4 \cdot 7H_2O$
   42 g/L (5.6 oz/gal), $FeCl_2 \cdot 4H_2O$
   15 g/L (2 oz/gal) $(NH_4)_2SO_4$
   15 g/L (2 oz/gal), sodium formate, $NaHCO_2$
   30 g/L (4 oz/gal), boric acid, H_3BO_3
   0.98 g/L (0.13 oz/gal) sodium lauryl sulfate (wetting agent)
(6) Bake for 4 hr or longer at 205 C (400 F) to prevent blistering of plate during subsequent heating
(7) Heat for 1 hr at 705 C (1300 F) to improve as-plated adhesion

L. Silicon

Cycle 1[94]

(1) Degrease
(2) Rinse
(3) Chemical etch in solution containing 44% nitric acid, 18% hydrofluoric acid, balance H_2O
(4) Rinse
(5) Chemical etch in concentrated hydrofluoric acid for 10 sec
(6) Rinse
(7) Nickel coat in alkaline electroless nickel bath[95]

Cycle 2[85] (Nickel Plate)

(1) Clean
(2) Soak 1 min in a nickel plating solution containing 7.8% hydrofluoric acid, 30 g/L (4 oz/gal) citric acid, 130 g/L (17.4 oz/gal) basic nickel carbonate, and 20 drops per liter (75 drops/gal) sodium lauryl sulfate
(3) Apply current to electrodeposit nickel
(4) Rinse
(5) Plate additional nickel or other metal

M. Titanium and Titanium Alloys

Due to the many alloys of titanium it is impossible to list one cycle to accommodate all alloys. The ones listed below are for the specific alloys stated. It is possible that all will work on pure titanium and all its alloys; however, the suitability of any cycle for a given alloy should be experimentally determined first.

Cycle I⁹⁶ (Pure Titanium and 6AL-4V, 4AL-4Mn, 3AL-5Cr)

(1) Preliminary cleaning
(2) Rinse
(3) Pickle in solution of 25% hydrofluoric acid and 75% nitric acid until red fumes are evolved
(4) Rinse
(5) Etch at 82-100 C (180-212 F) in one of the following aqueous solutions for 20 min

	Standard	3AL-5Cr
$Na_2Cr_2O_7 \cdot 2H_2O$	250 g/L (33 oz/gal)	390 g/L (52 oz/gal)
HF	4.8%	2.5%

(6) Rinse
(7) Electroplate with chromium, with copper from an acid bath or with nickel from either a Watts or sulfamate bath (electroless nickel can also be deposited; if metals other than the above are desired, use a nickel strike first)

Cycle II⁹⁶ (4AL-4Mn)

(1) Preliminary cleaning
(2) Rinse
(3) Pickle (see Cycle I-3)
(4) Rinse
(5) Etch at 54-60 C (130-140 F) for 15-30 min and 5.8 A/dm² (54 ASF) in a solution containing 20% hydrofluoric acid and 80% ethylene glycol
(6) Rinse
(7) Copper strike under the following conditions:
Copper cyanide, 23 g/L (3.1 oz/gal)
Total sodium cyanide, 33.7 g/L (4.5 oz/gal)
Sodium carbonate (Na_2CO_3), 15 g/L (2.0 oz/gal)
Room temperature; no agitation; 5.4 A/dm² (50 ASF) for a few seconds, then 1.5 A/dm² (15 ASF) for 3-5 min
(8) Rinse
(9) Copper electroplate in fluoborate solution

Cycle III⁹⁶ (Pure Titanium, 3AL-5Cr, 5AL-2Cr-2Mo, 7AL-5Cr, 2.5AL-16V, 4AL-4Mn, 2Fe-Cr-2Mo, 28Cr-1.5Fe, 6AL-4V, 3AL-13V-11Cr)

(1) Preliminary cleaning
(2) Rinse
(3) Blast all surfaces with a water-abrasive slurry. The grit may be as coarse as 100 mesh or as fine as 1250 mesh.
(4) Rinse

(5) Plate with either electrolytic or electroless nickel
(6) Hot-treat in an inert gas atmosphere for 1-4 hr at 540-800 C (1000-1472 F)

N. Thorium[97]

Cycle 1

(1) Descale mechanically to remove oxide
(2) Cathodic alkaline clean at 82 C (180 F) using 2.7 A/dm² (25 ASF) for 2 min
(3) Cold rinse
(4) Anodic pickle in 10% hydrochloric acid at 21-32 C (70-90 F) using 5.4 A/dm² (50 ASF) for 5 min without rinsing
(5) Chemical pickle in 10% sulfuric acid at 21-32 C (70-90 F) for 5 min
(6) Cold water rinse
(7) Electroplate; immerse with current on (most plating solutions are satisfactory except low-pH, chloride-containing solutions, and high-temperature chromium solutions)

O. Tungsten[98,99]

Cycle I

(1) Preliminary cleaning
(2) Anodic treatment in either 5-25% sodium hydroxide solution at 71 C (160 F) and 16-25 A/dm² (160-250 ASF) *or* etch in 10% hydrochloric acid solution at 24 C (85 F) and 5.4 A/dm² (54 ASF)
(3) Activate in 5-50% hydrofluoric acid solution using alternating current (60 cycle) at 1-5 volts for 1-2 min
(4) Rinse
(5) Electroplate in acid electrolyte

Cycle II

(1) Preliminary cleaning
(2) Anodic etch in a 30% potassium hydroxide solution at 30 A/dm² (300 ASF) for 2-5 min and 48-60 C (120-140 F)
(3) Rinse
(4) Acid dip in 6% sulfuric acid at room temperature for 1 min
(5) Rinse
(6) Chromium strike in bath containing 247 g/L (33 oz/gal) chromic acid plus 2.5 g/L (0.33 oz/gal) sulfuric acid at 15-25 A/dm² (150-250 ASF) for 1-3 min at 60-72 C (140-160 F)
(7) Rinse

(8) Acid activate in 20% hydrochloric acid at room temperature for 2-5 sec
(9) Rinse
(10) Electroplate with desired metal

P. Uranium

Cycle I[100] (Nickel or Iron)

(1) Descale mechanically
(2) Clean by immersion in alkaline cleaner at 82 C (180 F) for 2 min
(3) Cold rinse
(4) Pickle in 50% nitric acid at 21-32 C (70-90 F) for 5-10 min
(5) Warm rinse at 32-38 C (90-100 F)
(6) Anodic pickle in 50% phosphoric acid and 2% hydrochloric acid at 38C (100 F) using 2.7-8.0 A/dm^2 (25-75 ASF) for 2-10 min
(7) Cold rinse
(8) Pickle as in step 4 for 3-5 min (until anodic film is removed)
(9) Cold rinse
(10) Immerse in plating bath and apply current within ½ min

Nickel

NiSO$_4$·7H$_2$O	150 g/L (20 oz/gal)
NiCl$_2$·6H$_2$O	15 g/L (2 oz/gal)
MgSO$_4$·7H$_2$O	75 g/L (10 oz/gal)
NH$_4$Cl	15 g/L (2 oz/gal)
H$_3$BO$_3$	15 g/L (2 oz/gal)
Sodium lauryl sulfate	
pH	5.5
Temperature	38 C (100 F)
Current density	1.6 A/dm^2 (15 ASF)

Iron

Same as for niobium	Cycle 2, step 5

(11) Rinse and dry
(12) Heat-treat after 12.7-25.4 μm (0.5-1.0 mil) of nickel or iron plate in vacuum (<10 microns) at 650 C (1200 F) for 1 hr to alloy partially the deposit with the uranium and improve the corrosion resistance of the coating

Q. Zirconium

Cycle 1[101] (Nickel or Iron Plate)

(1) Descale
(2) Clean by immersion or treat cathodic in hot alkaline cleaner
(3) Rinse
(4) Chemical etch by immersion in a solution containing 35 g/L (4.7 oz/gal) NH$_4$F and 30 g/L (4 oz/gal) HF (as 100%) in deionized or distilled water at 38 C (100 F) (mole ratio NH$_4$F/HF = 1.2 to 4.1) for 1-3 min
(5) Rinse
(6) Nickel or iron plate, 50 μm (2 mil) or less. *Nickel plate* in Watts-type bath at pH 2.5, 60 C (140 F), 4.3 A/dm^2 (40 ASF), and using 0.03% hydrogen peroxide (30%) antipit (no organics of any kind). *Iron plate:* same as for niobium, Cycle 2, step 5
(7) Rinse and dry
(8) Bake 2-4 hr at 204 C (400 F)
(9) Vacuum heat-treat ½ hr at 705 C (1300 F) for nickel and 815 C (1500 F) for iron plate

Cycle 2[102] (Platinum Plate)

(1) Clean by soaking in alkaline cleaner, 82-93 C (180-200 F), 3-5 min
(2) Rinse
(3) Chemical polish at 27 C (80 F) in solution containing 100 g/L (13.2 oz/gal) NH$_4$F·HF; 79 g/L (10.5 oz/gal) H$_2$SiF$_6$ (as 100%); 405 g/L (54 oz/gal) HNO$_3$ (as 100%); 3-4.5 g/L (0.4-0.6 oz/gal) Zr to remove worked metal surface layer (approximately 50 μm (2 mil) [NOTE: Action is too violent in fresh bath without the dissolved zirconium]
(4) Rinse immediately
(5) Activate, nickel plate, and heat-treat as detailed in steps 4 through 9 of Cycle 1
(6) Clean and pickle in 50% hydrochloric acid at 27 C (80 F) for 1 min
(7) Rinse
(8) Platinum plate; diamminonitrate bath containing 4-9 g/L (0.53-1.2 oz/gal) of platinum maintained slightly ammoniacal by additions of ammonium hydroxide and operated at 93 C (200 F). Periodic reverse cycle of 5 sec cathodic plus 2 sec anodic using a current density of 5.4-6.5 A/dm^2 (50-60 ASF) produces sound platinum coatings at a plating rate of 5-6.5 μm/hr (0.2-0.3 mil/hr)
(9) Rinse and dry

INTERMEDIATE ELECTRODEPOSITED COATING AS BASIS METAL SURFACE

For the most part, the procedures given previously will be applicable when the basis metal is a previously deposited metal; i.e., Cycle V-1 would apply to that case where a copper deposit has been buffed and is to be plated with another metal such as nickel. An understanding of the reasons for each step in the cycles will enable one to establish a satisfactory cycle for most of the common metals and deposits. However, certain intermediate deposits are so frequently used as basis surfaces for other deposits that these will be covered separately.

I. Copper Deposits (Cyanide or Acid)

A. Buffed

Buffed copper deposits may be cleaned and plated with other metals as in Cycles V-1 and V-2.

B. Unbuffed

Cycle 1 (For Acid Copper Going into Acid Plating Bath)

(1) Rinse thoroughly
(2) Plate or strike as in Cycle V-1 and V-2

Cycle 2 (For Acid Copper Going into Cyanide Plating Bath)

(1) Rinse
(2) Cyanide or alkali dip
(3) Rinse
(4) Plate or strike as in Cycles V-1 and V-2

Cycle 3 (For Cyanide Copper Going into Acid Plating Solution)

(1) Rinse
(2) Acid dip
(3) Rinse
(4) Plate or strike as in Cycles V-1 and V-2

Cycle 4 (For Cyanide Copper Going into Cyanide Plating Solution)

(1) Rinse thoroughly
(2) Plate or strike as in Cycles V-1 and V-2

In cycles such as 2 and 3 above, it is frequently possible, and in some cases desirable, to omit the cyanide, alkaline, or acid dip and to rinse thoroughly before going into the next plating solution. However, for some proprietary bright copper deposits, it is necessary to treat the surface more thoroughly, e.g., by a cleaning step or cathodic cyanide treatment to remove surface films, especially if bright nickel is to follow. The supplier's recommendations should be followed.

II. Nickel Deposits

Bright and Unbuffed. Other than plating with chromium, the cycles for plating of nickel on fresh nickel surfaces have been sufficiently covered. In plating bright nickel with chromium, it is only necessary to rinse after nickel plating prior to chromium plating; however, very seldom is it this simple. In some cases, better coverage by the chromium is obtained if the nickel is "activated." Where the work is nickel and chromium plated on the same rack, it is necessary to free the rack of all chromium solution before it is returned to the cycle. The above cases are illustrated in the following cycles:

Cycle 1

(1) Bright nickel plate
(2) Rinse and dry
(3) Rerack
(4) Clean: soak or cathodic in mild cleaner, or "activate" cathodically in 30–45 g/L (4–6 oz/gal) of sodium cyanide
(5) Rinse
(6) Acid dip
(7) Rinse
(8) Chromium plate

Cycle 2 (Nickel and Chromium Plating on Same Racks)

(1) Bright nickel plate
(2) Rinse
(3) Chromium plate
(4) Water rinse
(5) Neutralize in alkaline solution
(6) Rinse
(7) Hot rinse, dry, and unrack
(8) If necessary, the racks are stripped of chromium by anodic treatment in alkaline solution

Cycle 3

(1) Bright nickel plate
(2) Rinse
(3) Electropolish
(4) Rinse
(5) Dip in 5–10% sulfuric acid
(6) Rinse
(7) Chromium plate

A. Buffed Nickel

Cycle 4

(1) Soak or spray clean
(2) Rinse
(3) Clean cathodically or activate cathodically in 30-45 g/L (4-6 oz/gal) of sodium cyanide
(4) Rinse
(5) Acid dip (NOTE 19)
(6) Rinse
(7) Chromium plate

Steps 1 and 2 may be omitted if the parts are free of visible buffing compound.

NOTE 19—The usual acid dip used is a 5% sulfuric acid solution. Due to surface film, passivity, etc., frequent trouble has been encountered in obtaining a desirable chromium deposit or maximum plating range. Methods have been proposed for testing and controlling the chromium and acid solutions to obtain optimum results.[103,104]

Cycle 5

(1) Cathodic clean
(2) Rinse
(3) Dip in solution of 5-10% sulfuric acid and 0.75-1.5 g/L (0.01-0.2 oz/gal) of potassium iodide
(4) Rinse
(5) Chromium plate

Cycle 6[105]

Buffing of dull nickel may be eliminated by use of the following cycle:
(1) Nickel plate, dull
(2) Rinse
(3) Electropolish
(4) Rinse
(5) Dip in 20% hydrochloric acid
(6) Rinse
(7) Bright nickel strike
(8) Rinse
(9) Chromium plate

B. Nickel on Aged or Passivated Nickel[106]

Nickel surfaces, which have been aged or treated anodically, as in a chromium stripping solution, become passivated and are not easily nickel plated. An anodic etch in sulfuric acid followed by rinsing will usually suffice to activate the surface prior to nickel plating. In some cases, it may be necessary to use a low-pH and high-chloride nickel strike.

References

(The references for this chapter are numbered consecutively and divided under main and secondary subject headings.)

Preliminary Treatment

1. "Standard Practice for Cleaning Metals Prior to Electroplating," A.S.T.M. Designation B322, *1981 Book of A.S.T.M. Standards*, Part 9, p. 135.
2. "Standard Practice for Preparation of Low-Carbon Steel for Electroplating," A.S.T.M. Designation B183, *1981 Book of A.S.T.M. Standards*, Part 9, p. 28.
3. Hall, N. "Pickling and Acid Dipping," *Metal Finishing Guidebook & Directory*. New Jersey: Metals and Plastics Publications, pp. 173-182, 1981.
4. Dockray, T. "Resurfacing of Metals," U.S. Patent 2,295,204 (September 8, 1942).
5. "Standard Practice for Preparation of and Electroplating on Stainless Steel," A.S.T.M. Designation B254, *1981 Book of A.S.T.M. Standards*, Part 9, p. 80.
6. "Standard Practice for Preparation of Copper and Copper-Base Alloys for Electroplating," A.S.T.M. Designation B281, *1981 Book of A.S.T.M. Standards*, Part 9, p. 94.

Final Treatment

I. Low-Carbon Steel

7. See Ref. 2.
8. Shapiro, M. *Metal Finishing* 48, 7:46-49 (1950).
9. "Standard Practice for Chromium Electroplating on Steel for Engineering Use," A.S.T.M. Designation B177, *1981 Book of A.S.T.M. Standards*, Part 9, p. 23.
10. Logozzo, A. "Preparation of Metals for Hard Chrome Plating," *Metal Finishing Guidebook & Directory*. New Jersey: Metals and Plastics Publications, pp. 215-220, 1981.

II. High-Carbon Steel and Low-Alloy Steel

11. "Standard Practice for Preparation of High-Carbon Steel for Electroplating," A.S.T.M. Designation B242, *1981 Book of A.S.T.M. Standards*, Part 9, p. 52.
12. *Canning Handbook on Electroplating*, 22nd Edition, 1978, p. 345.

III. Stainless Steels

13. See Ref. 5.
14. Underwood, J. E. (to C. Howard Hunt Pen Co.). "Art of Gold Plating," U.S. Patent 2,133,996 (October 25, 1938).
15. Wesley, A. (to the International Nickel Co.). "Process for Electrodepositing an Adherent Coating of Copper on Chromium-Containing Alloys of Iron and/or Nickel," U.S. Patent 2,285,548 (June 9, 1942).
16. Tucker, W. M. (to Eastman Kodak Co.). "Activating and Electroplating Stainless Steel," U.S. Patent 2,437,409 (March 9, 1948).
17. Gardam, G. E. *J. Electrodepositors' Tech. Soc.* **13**, *13:*1–8 (1937).
18. Wood, D. *Metal Ind. (N.Y.)* **36**:330–331 (1938).
19. Rogers, R. R. "Process of Electroplating Copper," U.S. Patent 2,133,255 (October 11, 1938).
20. McGill, E. F. "Electroplating," British Patent 575,309 (February 12, 1946).
21. Batcheller, C. "Copper Plating Stainless Steel," Canadian Patent 448,565 (May 18, 1948).
22. Weiner, R. *Electroplating and Metal Finishing* **3**:455–456 (1950).
23. Batcheller, C. "Method of Electroplating Stainless Steels and Iron," U.S. Patent 2,528,717 (November 7, 1950).
24. Haas, J. *Metal Finishing* **49**, *6*:50–54 (1951).
25. Head, H. *Plating* **37**:260–264 (1950).
26. Lukens, H. S. "Pen," U.S. Patent 2,039,326 (May 5, 1936).
27. Lukens, H. S. "Proc. for Gold Plating Chromium Alloy Steels," U.S. Patent 2,133,995 (October 25, 1938).
28. Kushner, J. "Method of Gold Plating Steel and Ferrous Alloys," U.S. Patent 2,227,464 (January 7, 1941).
29. Kushner, J. *Products Finishing* **6**, *6*:34–39 (1942).
30. See Ref. 10.

IV. Cast Irons

31. Burgess, C. C. *Metallic and Non-Metallic Coatings for Gray Iron* Cleveland, Ohio: Gray Iron Founders' Soc., 1950.
32. "Standard Recommended Practice for Preparation of Iron Castings for Electroplating," A.S.T.M. Designation B320, *1981 Book of A.S.T.M. Standards*, Part 9, p. 130.
33. Orbaugh, N. H. *Metal Finishing* **47**, *11*:53–55, 59 (1949).
34. Wernick, S. and F. Willetts. *Metal Finishing* **46**, *11*:76–81 (1948).
35. Wesley, W. and W. Prine. *Practical Nickel Plating*, New York: The International Nickel Co.
36. Diggin, M. *Metal Finishing* **41**:277–281 (1943).
37. See Ref. 10.
38. Communication from C. L. Faust, Battelle Memorial Institute, Columbus, Ohio.

V. Copper and Copper-Base Alloys

39. See Ref. 6.
40. See Ref. 10.

VI. Zinc-Base Die Castings

41. "Standard Recommended Practice for Preparation of Zinc-Alloy Die Castings for Electroplating," A.S.T.M. Designation B252, *1981 Book of A. S. T. M. Standards*, Part 9, p. 65.
42. Burns, R., M. Caldwell, R. Wagner, E. Anderson, and C. Reinhard, (a Symposium). *Plating* **35**, *2*:131–152 (1948).
43. Tuttle, R. "Nickel Plating Solutions and Processes," U.S. Patent 2,069,566 (February 2, 1937).
44. Pinner, W., G. Soderberg, and E. Baker, *Trans. Electrochem. Soc.* **80**:550 (1941).
45. Safranek, W. and C. Faust. *Plating* **45**, *10*: 1027–1037 (1958).

VII. Aluminum and Its Alloys

46. "Standard Practice for Preparation of and Electroplating on Aluminum Alloys by the Zincate Process," A.S.T.M. Designation B253, *1981 Book of A.S.T.M. Standards*, Part 9, p. 70.
47. "Electroplating of Aluminum Alloys," Aluminum Company of America, *Bull. 7* (1952).
48. Raymond, W. A. "Preparation of Various Base Metals for Plating," *Metal Finishing Guidebook—Directory.* New York: Finishing Publications, pp. 220–222, 1951.
49. Hewitson, E. U.S. Patent 1,627,900 (1927).
50. Keller, F. and W. Zelley. *J. Electrochem. Soc.* **87**:143–151 (1950).
51. Wyszynski, A. E. *Trans. IMF* **58**:34. (1980).
52. Lashmore, D. S. *Plating and Surface Finishing*, **65**, *4*:44 (1978).
53. Heiman, S. *J. Electrochem. Soc.* **95**, *5*:205–225 (1949).
54. U.S. Patent 3,417,005, assigned to General Motors Corp.
55. Leo Missel, IBM Corp. *Plating and Surface Finishing* **64**, *7*:32–35 (1977).
56. Missel, L. and G. Kishi. *Metal Finishing* **79**, *8*:37–42. (August 1981).
57. "Proprietary Process," Available from M & T Chemicals, Rahway, NJ 07065.
58. Travers, W. "Protection of Metals," U.S. Patent, 1,971,761 (August 28, 1934).
59. Travers, W. J. *Trans. Electrochem. Soc.* **75**: 201–208 (1939).
60. Travers, W. J. *Proc. Am. Electroplaters' Soc.*, pp. 119–120 (1942).
61. Yates, R. F. *Proc. Am. Electroplaters' Soc.*, pp. 118–119 (1943).

62. Bengston, H. *Trans. Electrochem. Soc.* **88**: 307-324 (1945).
63. *Electroplating on Aluminum and Its Alloys*, New Kensington, Pa.: Aluminum Co. of America, 1946.
64. O'Keefe, J. "Surface Preparation of Aluminum for Plating," *Metal Finishing Guidebook—Directory.* New York: Finishing Publications, pp. 222-227, 1951.

VIII. *Magnesium and Its Alloys*

65. "Standard Practice for Preparation of Magnesium and Magnesium Alloys for Electroplating," A.S.T.M. Designation B480," *1981 Book of A.S.T.M. Standards*, Part 9, p. 274.
66. DeLong, H. K. *Metal Finishing* **46**, *7*:46-49 (1948).
67. DeLong, H. K. "Method of Producing a Metallic Coating on Magnesium and Its Alloys," U.S. Patent 2,526,544 (October 17, 1950).
68. Diggin, M. "Electroplating on Magnesium Alloys," *Metal Finishing Guidebook—Directory.* New York: Finishing Publications, pp. 218-224, 1953.
69. DeLong, H. K. *Metal Ind.* **82**, *17*:327-329 (1953).
70. *Magnesium Finishing*. Midland, Mich.: The Dow Chemical Co., 1958.
71. DeLong, H. K. "Method of Producing an Electroplate of Nickel on Magnesium and Magnesium Base Alloys," U.S. Patent 2,728,720 (December 27, 1955).
72. "Chemical Plating of Nickel on Magnesium," The Dow Chemical Co., British Patent 830,597 (March 16, 1960), U.S. Patents 2,983,634, 3,121,644, 3,152,009, and 3,211,578.

IX. *Nickel Silver*

73. Gray, D. *Electroplating and Metal Finishing* **3**, *16*:605 (1950).

X. *Lead and Lead Alloys*

74. "Standard Practice for Preparation of Lead and Lead Alloys for Electroplating," A.S.T.M. Designation B319, *1981 Book of A.S.T.M. Standards*, Part 9, p. 126.
75. Meyer, W. *Metal Cleaning Finishing* **8**, *11*:751-756 (1936).,

XI. *Powder Metal Compacts*

76. *Electroplating and Metal Finishing* **5**:201 (1952).
77. Graham, A. K. *et al. Plating* **36**:702-709 (1949).
78. Hausner, H. *Metal Finishing* **48**, *3*:73-74 (1950).
79. Bonanno, J. *Proc. Metal Powder Assoc.* **5**:74-93 (1949).
80. Schwarzkopf, P. *Powder Metallurgy, Its Physics and Production.* New York: The Macmillan Co., 1947.
81. Anderson, E. *Materials and Methods (Materials in Design Engineering)* **30**, *2*:55-57 (1949).
82. "Cadmium Plating-Unbuffed-Oilite," Chrysler Corp., Engineering Div., Process Standard #930.

XII. *Less Common Metals*

83. Beach, J. G. and C. L. Faust. *J. Electrochem Soc.* **100**:276 (1953).
84. "Standard Recommended Practice for Preparation of Chromium for Electroplating with Chromium," A.S.T.M. Designation B 630, *1981 Book of A.S.T.M. Standards*, Part 9, p. 473.
85. Turner, D. R. *J. Electrochem. Soc.* **106**:786 (1959).
86. Couch, D. E., H. Shapiro, J. K. Taylor, and A. Brener. *J. Electrochem. Soc.* **105**:450 (1958).
87. Safranek, William H. and G. R. Schaer. *Proc. Am. Electroplaters' Soc.* **43**:105 (1956).
88. Korbelak, A. *Proc. Am. Electroplaters' Soc.* **40**:90 (1953).
89. Couch, D. T. H. Shapiro, and A. Brenner. *J. Electrochem. Soc.* **105**:485 (1958).
90. Levy, C. *Proc. Am. Electroplaters' Soc.* **43**:219 (1956).
91. Sellers W. W. and C. B. Sanborn. *Proc. Am. Electroplaters' Soc.* **44**:36 (1957).
92. Saubestre, E. B. *J. Electrochem. Soc.* **106**:305 (1959).
93. Beach, John G. and Charles L. Faust. "Electroplated Coatings on Niobium," May 24, 1955, Report No. BMI-1004, *Metallurgy and Ceramics*, TID-4500, 9th Edition, Contract No. W-7505-eng-92, Office of Technical Services, U.S. Dept. of Commerce, Washington, D.C.
94. Sullivan, M. V. and J. H. Eigler. *J. Electrochem. Soc.* **104**:226 (1957).
95. Brenner, A. and Grace Ridell. "Deposition of Nickel and Cobalt by Chemical Reduction," *Proc. Am. Electroplaters' Soc.* **34**:156 (1947); U.S. Patents 2,532,283; 2,532,284 (December 5, 1950).
96. "Standard Practice for Preparation of Titanium and Titanium Alloys for Electroplating," A.S.T.M. Designation B481, *1981 Book of A.S.T.M. Standards*, Part 9, p. 277.
97. Beach, J. G. and G. R. Schaer. *J. Electrochem. Soc.* **106**:392 (1959).
98. Cannizaro, J. J. "Process of Electroplating on Tungsten," U.S. Patent 2,443,651 (June 1948).
99. "Standard Practice for Preparation of Tungsten and Tungsten Alloys for Electroplating," A.S.T.M. Designation B482, *1981 Book of A.S.T.M. Standards*, Part 9, p. 280.

100. Beach, J. G. and C. L. Faust. *J. Electrochem. Soc.* **106**:654 (1959).
101. Schickner, W. C., J. G. Beach, and C. L. Faust. *J. Electrochem. Soc.* **100**:289, (1953).
102. Tripler, A. B., J. G. Beach, and C. L. Faust. Report No. BMI 1097 "Electrocladding of Zirconium with Platinum," June 21, 1956. *Metallurgy and Ceramics*, TID-4500 11th Edition, Contract No. W-7405-eng.-92, Office Technical Services, U.S. Dept. of Commerce, Washington, D.C.

Intermediate Coating as Basis Metal Surface

103. Tucker, W. and R. Flint *Trans. Electrochem. Soc.* **88**:335–355 (1945).
104. Winters, J. and R. Hull. *Proc. Am. Electroplaters' Soc.*, pp. 93–101 (1949).
105. Pray, H. A. "Method of Metal Electroplating," U.S. Patent 2,336,568 (December 14, 1943).
106. "Standard Practice for Preparation of Nickel for Electroplating with Nickel," A.S.T.M. Designation B343, *1981 Book of A.S.T.M. Standards*, Part 9, p. 158.

B. Plastics

J. B. Hajdu

Vice President, Technology, Enthone, Inc., New Haven, Conn.

AND

Gerald Krulik

Supervisor, Technical Services, Enthone, Inc. Chicago, Ill.

The term *plated plastics* generally refers to plastic parts finished with bright chrome electroplate. Plated plastic parts are used in a variety of automotive, appliance, and hardware applications. This section will review the general procedures for electroplating plastics.

SELECTION AND MOLDING OF THE PLASTIC

At present, about 85% of plated plastic parts are injection molded of acrylonitrile-butadiene-styrene (ABS) terpolymer. Special plating grade ABS molding compounds are generally preferred for better-quality plating. Also, for successful plating, basic design criteria should be observed avoiding blind holes, large flat surfaces, and sharp corners. During the molding special parameters recommended by the resin manufacturer should be observed. Finally, silicone mold releases should be avoided in all cases.

PLATING

The typical plating cycle for ABS plastics is described on Fig. 1. Other plastics, based on polyphenylene oxide, nylon, polysulfone, fluoropolymers, and alloys of ABS and polycarbonate have been commercially plated. Special plating cycles are available for these materials, either from the manufacturer of the resin or from the suppliers of plating chemicals. These cycles differ mainly by requiring special cleaning, solvent treatment, or etching procedures.

The steps indicated in Fig. 1 can be carried out as one single plating line or divided into two lines, the first comprising all steps needed to render the plastic conductive and the second containing the electroplating operations. In either approach, automatic plating machines, in general programmed hoist units, are used for better economy and consistent quality. Adequate water rinses should be included after each step.

The main characteristics of the steps of Fig. 1 are as follows:

Cleaning. Cleaning of the plastics may require a separate treatment, particularly when molding and plating are being done at different locations. Extra care should be taken in the handling, packaging, and transportation of

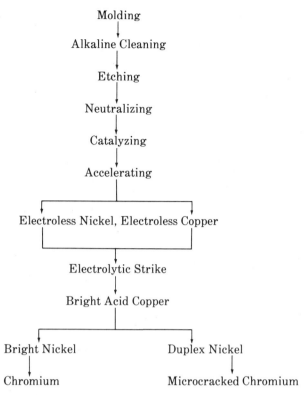

FIG. 1. Typical electroplating process for ABS plastics. Rinses have been omitted.

molded parts to eliminate such dirts as airborne dust, oils, and fingerprints. Some of these materials could absorb into the plastic and affect subsequent processing. The purpose of cleaning is to remove materials that might interfere with uniform chemical attack in the next steps. Alkaline detergent solutions are commonly used.

Etching. The etching solutions are specifically designed to render the plastic surface hydrophilic, and to produce a micro-etch of the surface by selectively attacking components in the plastic. It is this selective micro-etch that supplies the required bond between the plastic and the conductive first coating. The etchants usually are strongly oxidizing solutions, often containing high concentrations of chromic acid and sulfuric acid. These etching solutions are used for ABS at 60 to 70 C (140 to 160 F), for 5 to 10 minutes.

Neutralizing. Neutralizers normally are used following the chromium containing etchants so as to eliminate as completely as possible the carryover of hexavalent chromium compounds which are detrimental to operation of all other treatments in the cycle. The elimination of chromium salts from the plastic surface also improves the absorption of the catalyst or activator in the next step.

Catalyzing. In this step a catalyst is adsorbed on the plastic to initiate the electroless deposition process. In general, catalysts or activators are strongly acidic mixtures of tin and palladium salts designed to be adsorbed onto the plastic surface in limited and controllable amounts. The tin compound is strongly attracted to the organic surface and bonds the palladium to produce a catalytic surface. These catalysts are generally supplied as proprietary concentrates which are suitably diluted (1 to 10% by volume) at 15 to 50 C (60 to 120 F) for immersion times of 2 to 10 minutes.

Accelerating. An accelerator or post-activator is used in the next step to remove excess tin compounds to expose the palladium. Dilute acid or alkaline solutions are used at 20 to 50 C (70 to 120 F) for 1 to 2 minutes immersion times.

Electroless Plating. Electroless nickel or electroless copper deposits without electric current on the catalyzed surface to produce the conductive coating required for electroplating. Most present operations utilize electroless copper because of specific outdoor performance requirements. Electroless nickel baths are more stable and are acceptable for many less severe applications.

In general, proprietary electroless solutions are used. These baths will deposit sufficient metal at room temperature in 5 to 10 minutes to render the parts conductive. Present proprietary baths are highly stabilized. Bath lives of several months for electroless copper and over a year for nickel can be expected. Since the usual thickness of these electroless deposits will only be 0.12 to 0.75 μm (0.005 to 0.030 mil), it is very important that the parts now be handled with extreme care.

Electroplating. The previous steps can be carried out having the parts racked or in bulk using barrels or baskets. Most of the electroplating is done on racks, and only small parts plated in specially designed barrels, since plastic parts float in the plating bath. For the same reason, the racks should make strong positive contact with the parts to avoid floating and give adequate electric conductivity.

Although the conductivity of the electroless deposit may be sufficient to carry the required electroplating current on small and medium sized parts, it often is insufficient for large parts such as automotive grilles when electrical rack contacts may be quite far apart. In general, it is advisable to use a nickel or copper strike deposit from a high-efficiency solution at low current density in order to avoid burn-off of the thin electroless layer. Thickness of the strike deposit usually is less than 2.5 μm (0.1 mil).

The next electrodeposit is bright, leveling ductile copper obtained from acid copper sulfate baths. This layer of copper serves both to improve surface appearance prior to other metal deposits and to act as a stress-absorbing layer both for the stresses in the following nickel deposits and the stresses which may be set up because of the differences in the thermal expansion of plastic and the plated metals. Thickness of this deposit will vary with the size and design of the plastic part and with its intended use. For example, 15 μm (0.6 mil) usually will be sufficient for decorative knobs and frames on radios and television sets and decorative parts of small appliances. A minimum 20 μm (0.8 mil) may be necessary on larger appliance parts subject to some temperature variations and on most interior automotive parts. Exterior automotive parts, which usually are fairly large and which must withstand extreme temperature changes, require thicker deposits with the average usually being 20 to 25 μm (0.8 to 1 mil). In all cases, greater thicknesses may be applied in order to improve final surface finish.

Over the bright copper, any of the other plated finishes may be applied. However, the most common finish is bright nickel and chromium, usually to meet a specific specification. Most appliance and interior automotive specifications will call for only bright nickel and chromium while exterior automotive specifications require a deposit of a sulfur-free, semi-bright nickel, a

TABLE 1. ELECTRODEPOSIT THICKNESSES FOR PLATING OF PLASTICS.

Type of Serivce	Minimum Thickness									
	Copper		Semi-bright Nickel		Bright Nickel		Special* Nickel		Chromium	
	μm	mils	μm	mils	μm	mils	μm	mils	μm	mils
Mild indoor exposure; warm dry, minimum wear	15	0.6	0	0	7.5	0.3	0	0	0.125	0.005
Indoor exposure with moderate temperature and moisture changes, medium wear	15	0.6	0	0	15	0.6	0	0	0.25	0.01
Indoor or outdoor exposure, periodic wetting, possible major temperature changes; some abrasive wear	15	0.6	15	0.6	5	0.2	2.5	0.1	0.25	0.01
Very severe outdoor exposure such as exterior automotive	15	0.6	19	0.75	12.5	0.5	2.5	0.1	0.25	0.01

*Special deposit of nonmetallic particles in nickel matrix. Used to produce microporous chromium structure. Special microcracked chrome may be an acceptable substitute for some applications.

full-bright nickel, and a special microcracked or microporous chromium. Table 1 shows the minimum plate thickness normally considered advisable for various service conditions. Major manufacturers will specify thicknesses and specific performance tests. These may include thermocycling, adhesion tests, and corrosion tests.

References

These references are of a general nature for the reader interested in studying the subject matter in more detail.

1. American Society of Electroplated Plastics. Standards and Guidelines for Electroplated Plastics. Washington, D.C.
2. Goldie, W. *Metallic Coating of Plastics*. Hatch End, England: Electrochemical Publications, 1969.
3. Kirk-Othmer. *Encyclopedia of Chemical Technology*, 3rd Edition, Vol. 8. New York: Wiley, 1979.
4. Krulik, G. *J. Chem. Education* 55:361–365 (1978).
5. Wedel, R. G. *International Metals Review No. 217*:97–118 (1977).

5. WASTEWATER CONTROL AND TREATMENT

Clarence H. Roy, Ph.D.

Rainbow Research, Inc., Port St. Lucie, Fla.

There have been substantial changes in federal, state, and local regulations since the previous edition of this book, while few dramatic advances have been made in the chemistry of industrial wastewater treatment. New laws and regulations have forced suppliers and electroplaters to improve technology selection and application.

The American Electroplater's Society and other concerned professional organizations have made serious efforts over the past ten years to inform electroplaters of the many aspects of wastewater treatment. Numerous publications, seminars and technical sessions have provided the opportunity to acquire essential information. The U.S. EPA has published several booklets which also contribute to the electroplater's knowledge of the field.

Despite these educational opportunities, most shops, both captive and job, have not mastered various aspects of waste management problems. Management desire is an element vital to the success of wastewater technology at the applied level. Educational services are but tools of the motivated and concerned professional.

WATER SUPPLY

An effective water pollution control program is initiated by examining all factors, beginning with the water supply itself. Tap water is often unsuitable for use in plating baths and has become a major source of impurities in closed loop water recovery systems. A chemical analysis of the incoming water supply is a necessary investment. Iron, calcium, magnesium, manganese, and chloride are impurities responsible for many plating problems; use of contaminated water shortens bath life, reduces bath efficiency, precludes rinse water recovery, causes rejects and contributes to the waste treatment load. Water composition will vary with the seasons. Routine analysis monitors those fluctuations and enables platers to assess and compensate for contamination levels.

Discussion of the effects of impurity build-up in plating baths discourages many from applying recovery technology to ideal situations. Even a relatively good water supply is subject to evaporation. Daily additions of makeup water can add substantial quantities of impurities to the bath.

Demineralized water was primarily used in the past by the more sophisticated plating shops. Equipment was costly and troublesome to maintain and required frequent attention. Today's picture has improved in both equipment and technology. Suppliers have formulated complete "package systems," thereby eliminating confusion in the selection of components.

Most electroplaters do not need the extremely high-purity water used in semiconductor manufacture. Water demineralization by ion exchange or reverse osmosis is adequate without further polishing. Acidic and alkaline regeneration waters from ion exchange columns, including approximately twenty bed-volumes of rinse water, will generally require treatment by the plant pollution system. If the system has an averaging or equalizing tank capable of holding these flows, in addition to the normal rinse water flows, the acidic and alkaline regeneration waters will nearly offset each other. The acidity will usually dominate and a relatively small amount of excess caustic will be required to restore the drained discharge pH.

In the case of reverse osmosis plants, the permeate to concentrate (or reject) ratio most likely will be around one to one. Therefore, for each gallon of purified water produced, a gallon of reject water goes down the sewer. This reject water may also require treatment in some localities.

The prospect of adding chemicals and/or a hydraulic burden to the existing or planned wastewater system can be a source of dismay. A careful analysis demonstrates that the benefits derived from the use of purified water frequently outweigh the disadvantages.

Factors to be considered are:

1. Money saved by producing fewer rejects— including lower stripper consumption and waste disposal costs.
2. Reduction in scrap parts from repeated stripping and plating.
3. Reduced labor in reject salvage.
4. Extended plating bath life.
5. Potential for reclaim of rinses.

The project—to select demineralization equipment—should not be initiated until a comprehensive water conservation program has been completed. While this word of caution may be obvious, the priorities are frequently reversed and the plant purchases excess capacity in the pure water systems. Thousands of words have been written about water conservation, yet most plating shops have done little or nothing in this area. Time, effort, and money spent on an intensive water conservation program are the best investments a plating shop can make in terms of the cost/benefit relationship.

WATER AND CHEMICAL CONSERVATION

A water conservation program begins with careful determination of where the water goes. Water bills cannot always be reconciled with actual sources and flows. Leaks can be discovered and excess flows can be found everywhere. A tank-by-tank inventory of the entire plating and finishing area is the starting point. Determining flows for each rinse can be done using a bucket and a stopwatch or by measuring the rate at which an empty rinse tank fills. An assessment of the possibilities for adding rinse tanks in series with existing tanks can then be made.

Counterflow rinsing is the most desirable method of conserving water. The water required for effective rinsing will be reduced by a factor of ten for each added counterflow rinse tank. The factor is actually nearer to 20 to 1, though skeptics subscribe to a 10 to 1 ratio (see also Chapter 32).

Platers are advised to find large, old rinse tanks that can be replaced by a two-compartment counterflow tank. The rinsing of small parts, barrels, or racks does not necessitate large rinse tanks. Sizing rinse tanks to suit the work will improve tank turnover time and efficiency. Counterflow water should ideally enter the top of the tank and exit at the bottom or the reverse. If external piping is used, the pipe diameter should be as large as practicable to ensure an unobstructed flow of water. Larger pipe sizes are particularly important in longer hookups, as well as in bypassing process tanks such as pickle rinses flowing to cleaner rinse tanks or plating rinses flowing to preplate tanks. Since the effectiveness of counterflow rinsing depends on complete mixing, agitation of the rinse tank with air, or by mechanical means is important.

Dragout or still rinse tanks belong in virtually every electroplating line. Not only do they provide water conservation but they are the fulcrum of chemical recovery projects as well. In some applications, these tanks may be termed "reclaim" tanks and their contents returned to the plating bath without concentration. A nickel bath operating at 140 F will evaporate 0.1 gal water/hr/ft^2 of surface area. With air agitation, this rate is considerably higher.

A large surface area is not necessary to allow return of substantial amounts of dragout solution. Dragout recovery is most satisfactorily carried out by frequent or hourly return of dragout solution to the plating tank(s) accompanied by an equivalent addition of purified water to the reclaim tank. Two counterflow reclaim tanks may be used to capture over 80% of the metal dragout. In the case of nickel, the figure is close to the total of all metal added to the bath as salts. Some anode metal is also lost as dragout.

For optimum efficiency, the entire dragout recovery process can be automated by level controllers in the plating and reclaim tanks. A low level in the plating tank calls in a transfer pump from the reclaim tank while a low level in the reclaim tank calls for fresh water. This type of packaged unit is available commercially.

The employment of drip trays between process

and rinse tanks is advisable. Their utilization prevents potential pollutants from reaching the floor. A marked chemical savings can be achieved with a slight tilt toward the process tank. The thesis that dragout is the sole method of maintaining bath cleanliness cannot be substantiated. Meticulous housekeeping manifests stronger results. Retrieving dropped parts frequently makes a significant difference. Maintaining racks and barrels will contribute as well.

In cyanide plating on steel, when pickle tanks are consistently blue, post-plating rinsing is inadequate and/or the plating line is badly designed. If the work must pass back across the pickle tank for post-plate rinsing, the pickle life is shortened and the waste treatment picture needlessly complicated. Untreated cyanide-bearing pickles will show some cyanide in the effluent even in the best wastewater systems.

Work patterns and flows should be made as direct as possible in the early stages of the project, even taking some major tank relocations under consideration. Although in the short term there are problems with production interruption and expense, the long-term values are marked. Several additional adjustments may insure greater results:

1. Replace worn-out lines.
2. Automate hand lines.
3. Avoid return type machines wherever possible. Although loading and unloading at the same line end is attractive from a labor position, the continual bath cross-contamination by dribbles from racks and barrels could overbalance the convenience factor.
4. Install a carousel type designed operation with the advantage of a single load/unload area.
5. Specify the number of counter-flow rinse tanks desired as well as any other water and chemical saving features deemed suitable.

If these features are not specified, the low bidder will provide single rinse tanks and only the bare essentials. Tank and barrel design call for in-depth examination. Improper rack configuration and part orientation increase dragout and associated waste. Platers are advised to become acquainted with the drainage characteristics of their parts. They must read literature reports as well as conduct experiments with the parts to determine in which of various orientations a part drains best. The results may lead to unexpected savings in water, metals, chemicals, and waste treatment costs, both capital and operating.

There are a variety of water and chemical devices and practices to be considered for new or existing installations. Perhaps the most effective and least practiced is spray rinsing. Spray rinse tanks can be designed to rinse almost any part, including cup configurations. Sprays can succeed where multiple immersions are only marginal. Ingenuity and persistence are prerequisites in obtaining proper spray distribution. A novel spray rinse employs a shallow sump in the bottom of the rinse tank created by a 6- to 8-in.-high standpipe overflow. By connecting a recirculating pump between the sump and spray headers, a substantial amount of water can be saved while using the high rinse rate of the pump. By connecting the first standpipe to the bottom of a second spray tank in counterflow series, having a 1-in. shorter overflow standpipe, enormous water savings can be realized.

Simple water conservation devices such as flow restrictors and aerators rate highly in cost saving. These inexpensive devices increase agitation in rinse tanks while reducing flow. The net effect is improved rinsing with less water in virtually every application.

Timer activated solenoid valves are recommended for rinse tanks with sporadic use patterns. A fixed duration timer holds a solenoid valve open for a preset duration to restore the rinse quality of the tank before it shuts off. A water activated timer, similar to those used to water lawns, can also serve the purpose. Conductivity controls can be installed in these tanks but are a more expensive method to accomplish the same task. These instruments' essential function is to monitor rinse water quality by measuring the ionic content (conductivity) of the water until it reaches a predetermined purity set point. At set point, the control closes a solenoid valve in the fresh water line.

Timer and conductivity devices rarely find application in rinsing situations involving frequent tank use. High production hand lines or automatic machines produce so sustained a level of contamination that rinse water flow, particularly if flows are restricted, can only cease when the operation is idle. These savings seldom justify the expense. A single master valve can suffice for all rinse tanks in the line.

CHEMICAL AND WATER RECOVERY

Once all possible water and chemical conservation measures have been implemented, a comprehensive review of all sources and flows will serve as basis for the design of a wastewater treatment system. Before embarking upon such a project, recovery technology demands serious evaluation. Justification for investment in recovery is based upon three factors.

First, the potential for metal value or plating solution recovery is the primary consideration in the application of recovery technology. Various metal and/or salt values lost can be readily checked by a review of purchasing and consumption records for a representative period of activity. Exceptionally slow or active times should not be heavily weighed. A dollar loss value can be obtained using these figures and the ratio of metal applied to the parts versus the amount lost. For example, all nickel salts and approximately 10% of anodes used are lost to rinsing; therefore, virtually the entire value of the nickel salt lost can be recovered. Similarly, 10 to 20% of chromium purchased is plated out on the work; the balance is lost to rinsing and the air scrubber. It is reasonable to expect recovering nearly all losses with a system capable of handling both the rinse water and the scrub water.

If loss factors are unknown, they can be estimated by determining the amount of plating bath dragged out by a single rack or barrel. Rinse the rack or barrel thoroughly in a clean, still rinse tank after electroplating. Analysis of rinse water for metal content will provide the basis of actual losses experienced by multiplying the units produced per day, week, or year. If plating a variety of parts, a few pieces representative of shape, configuration, and surface area should be subjected to the aforementioned procedure. A profile of losses can be developed with knowledge of production figures for each. When tabulated, the findings indicate the recovery potential of the process. A valid recovery technology can recover 90% of the value lost.

Second, concerning recovery of water itself, since numerous recovery systems reclaim water, a credit for water savings may be taken in those cases. The derived credit may initially be small but could increase in significance with water rate increases.

Third, when assessing the merits of metal recovery, the cost of treating rinses involved as well as the removal of sludge generated must be taken into account.

The savings and benefits obtainable by implementing a conservation and/or recovery program can be estimated by referring to Table 1. This table indicates the value (make-up cost) of various plating bath losses on the basis of 1 gal dragout/hr for an 8-hr shift. With multiplication or division, the figures can be corrected to match actual dragout rates and costs.

TABLE 1. TREATMENT COSTS FOR 8 hr AT 1 G.P.H.D.O. (NaOH Used for Neutralization)

Plating Bath	Make-Up Cost	Treat Cost	Disposal Cost at 30% Solids	Total Cost
Cu	$14.08	$1.12	$1.87	$17.07
CuCN	11.92	7.87	0.67	20.46
Ni	29.44	1.84	2.74	34.02
Cr	18.80	3.61	4.27	26.68
Zn	16.16	1.20	1.73	19.09
ZnCN	8.16	7.31	0.53	16.00
Cd	8.88	12.86	0.60	22.34

These costs were calculated in 1982, using handbook formulations and generic chemicals. Use of proprietary formulations and inflation of the dollar will increase these costs. The value of bath losses, the cost of treating the effluent for these bath losses, and the cost of disposing of the sludge created by the treatment are never-ending. They must be perceived as dollars lost from the bottom line.

On the other side of the ledger are capital costs which, if profits permit the tax deduction, can be depreciated over five years. Operating and maintenance costs cannot be ignored. These costs vary considerably with the technology selected but seldom outweigh the recovery benefits. The small plater will have the greatest difficulty in justifying an investment in recovery systems. The recovery technology prospects are further complicated by the fact that application seldom eliminates the necessity for conventional wastewater treatment. While the double cost may be initially discouraging, the anticipated benefits will often justify the expenditure.

EVAPORATIVE RECOVERY

Recovery technologies may be classified according to their operating principles. The oldest, evaporative recovery, has been in use since the discovery of distilled beverages. When applied to electroplating rinses, both distilled water and concentrated residue (plating solution) are valuable. The basic process involves two heat exchangers: one to heat the dilute rinses and the other to condense the water vapor. If water recovery values are small, there is little reason to have the second heat exchanger.

Numerous suppliers of packaged evaporative recovery systems have historically favored the double heat exchanger concept; however, the number of atmospheric evaporators in the market is growing. Atmospheric evaporators have the added advantage of operating without the complications and expense of a vacuum system. Thermally stable solutions do not require vacuum systems. The application of a vacuum lowers boiling temperatures and aids in preventing the decomposition of reclaimed plating solutions such as zinc and cadmium cyanide. Evaporative recovery is often viewed as energy-intensive and vacuum capability increases the energy demand by an additional 10 to 20%. A more elaborate cooling system will also be required to condense water vapors at temperatures below the atmospheric boiling point.

Evaporative recovery is best suited to extremely low flow applications where corrosive solutions are involved. While vacuum evaporative recovery equipment is more costly on a dollar/gal recovered per hour basis than other recovery methods, it has the advantage of offering a variety of corrosion-resistant materials. Commercially packaged systems are constructed of glass, titanium, fiberglass, CPVC, and stainless steel. In the case of chromium rinse recovery, evaporation is one of the most viable options.

Insuring the proper installation of evaporative equipment is vital. Adequate cooling is essential; sufficient pump or aspirator capacity for proper performance of vacuum systems is a necessity. Without attention to the engineering of these details, efficient operation is impossible.

Application of atmospheric systems can circumvent the complexities or expense of vacuum systems. Atmospheric systems employ towers filled with irregularly shaped plastic or ceramic packing. The heated solution is sprayed at the top of the tower and air (or hot air) is introduced at the bottom. As the solution cascades over the surfaces of the packing, water evaporates and concentrated solution returns to the heated sump tank. When the sump concentration reaches plating strength, the solution is transferred to the process tank. Because of the possibility of contamination being concentrated as well as the desired chemical constituents, recovery systems of this type require special attention to contamination control. Rinsing prior to entering the plating tank should receive special attention to hold drag-in of contaminants to a minimum. It may still be desirable, or even necessary, to include a purification stage in the recovery system. The composition of the solution, the nature of the contaminant, and the flow rate of the dilute solution for recovery will determine whether this purification step should be placed ahead of the recovery unit to process the dilute solution; or after recovery before the concentrated solution is returned to the plating tank.

REVERSE OSMOSIS

Reverse osmosis (RO) ranks next to evaporative recovery in terms of recovery rate (gal/hr) versus investment dollars. It will process approximately 180 gal/hr for almost the same investment as a 25 gal/hr vacuum evaporative unit. RO functions in the opposite manner from osmosis. Osmosis is a natural phenomenon involving the inclination of water to pass through a semipermeable membrane from the weak solution side to the strong solution side. The pressure generated by this activity can be measured; stronger solutions will produce greater pressure. If pump pressures are applied to overcome natural osmotic pressure and water is forced through a membrane from the concentrated side to the pure (permeate) side, reverse osmosis occurs.

Reverse osmosis functions most efficiently on very dilute solutions. The greater the ionic strength of a solution, the higher the pump pressure and horsepower required to produce reasonable permeate flows. RO is therefore well-suited to the concentration of dilute nickel plating rinse waters. For this reason, nickel rinse recovery using cellulose acetate membranes has been a frequently successful application of reverse osmosis in the metal finishing industry. Even this advantageous application is not without difficulties. It is impractical for extracting water from still rinses and treated plant effluents where total desolved solids often exceed 3000 ppm.

A major drawback is the plugging of fine tubular-type membranes with an insoluble yellow powder, thought by some to be nickel boride. The spiral wound sheet configuration has not manifested this problem, possibly due to the spacer between the winding facilitating cleaning of the membrane surface. Some membranes are now guaranteed for two years, but even these can be damaged by heat. High-pressure pumps can generate sufficient heat levels in closed loop systems to cause membrane compression. A heat exchanger (chiller) and a thermostat can prevent such problems.

Cellulose acetate membranes are not durable in chromic acid and cyanide applications. Strong oxidizing agents and high and low pH conditions cause membrane deterioration. Newer membrane materials, including the polyamides and the polysulfones, extend the range of pH and the chemical resistance in RO applications. The growth of RO technology should depend on the development of new membrane materials.

ELECTRODIALYSIS

Electrodialysis, or ED, is taking its place in the ranks of proven recovery technology. Early laboratory applications were hampered by a lack of practical membrane materials. With the advent of suitable membranes in the marketplace, ED is a valuable commercial option. Electrodialysis functions according to Faraday's Law by using an ion exchange type membrane to permit ion migration from the dilute side of the membrane to the concentrated side.

Flat, square cells are stacked on top of each other creating a module with considerable membrane area in a compact configuration. The membrane spacing is about 6mm (¼ in.), in order to create a short migration path through the solution, as well as improve removal efficiency. Rinse waters can be pumped through the cell stack at relatively high flow rates, placing it next in the ranking of gal/min/investment dollar.

Solution conductance is proportional to concentration. In contrast to RO, ED is better suited to more concentrated solutions. Electrical power requirements and resistance (heating) increase as concentration decreases, leading to inefficiency and possible overheating of the stack. It would, therefore, not be commercially practical to attempt to extract metal ions from dilute rinse water. ED units are sized in terms of pounds of metals to be recovered per hour. The dragout rates and the concentrations determined earlier in the project are valuable in assessing the merits of ED. Concentrated solution generated by electrodialysis may be suitable for return directly to the plating bath; the dilute stream is usually returned to the dragout tank or rinse tanks.

ION EXCHANGE

Ion exchange (IX) systems generally demonstrate the highest hydraulic capacity per dollar invested. Ion exchange units employ cylindrical columns of short or tall design, filled with polymeric beads with a chemical affinity for either cations (metallic) or anions. As the rinse is passed through the column packing (resin bed), the desired ions are extracted from the flow together with other ions of similar charge which may also be present in the stream.

The flow must eventually be interrupted and the resin bed regenerated. Duty and regeneration cycles are shortened considerably in the short or reciprocating bed.* The relatively small bed resin volume allows quicker regeneration and shorter duty cycles. An advantage of the short bed is its decreased down-times for regeneration; it demands smaller sumps and/or holding tanks for containment of flow during these periods. An additional short bed asset is that when automated, the duty cycle can be shortened to prevent "breakthrough." Most resins allow the escape of ions they are designed to retain at a time well before they are loaded to 100% capacity. Ion leakage in recovery situations is costly and should be minimized. The more classical, tall column systems can also be operated on a shortened cycle to prevent leakage or breakthrough of recoverable ions. Down-time is still greater than for "short bed" columns and thus requires correspondingly larger holding tanks for regeneration during shut-down periods.

When IX is applied to metal recovery, the metal values obtained from the regeneration process may not be suitable for return directly to the plating bath. Concentration and pH adjustment are frequently required before reuse. Much of the regeneration rinse waters are too dilute or otherwise unsuitable for reclaim; they will often require treatment in the wastewater system prior to discharge. The most recent chelating or ion specific resins are contributing to expansion of IX applications in metal finishing and printed circuit manufacture.

ION MIGRATION

Traditional technologies can attribute improved recovery performance to the integration of innovative materials. Thick polymeric barriers, such as porous polypropylene, have made ionic migration a practical methodology to concentrate chromic acid rinse waters. Strong chromic acid solution can be produced in the anode compartment by using an electrolytic cell or cells with these porous membranes. Trivalent chromium can also be oxidized to the hexavalent state under controlled conditions.

ELECTROWINNING

Electrowinning, an established technology, finds its roots in the early 1900's use of carbon granules and charred wood shavings as substrates for the electrowinning of gold. Modern processes employ mats or cloth composed of

---

*Patented.

carbon fibers to enhance efficiency. Nearly any electroplatable metal can be extracted from rinse water by this method under proper conditions. On an applied level, metal is electrodeposited on carbon fiber mats which serve as cathodes until the resistance falls (or amperage rises) to a preset value. The process is then interrupted while the metal is removed from the carbon by acid, cyanide, or reverse current. The entire cycle can be automated in a closed loop manner. Reclamation of the concentrated metal extract may require some imagination, but, undeniably, the values are not lost to the sewer or do not appear in the waste treatment sludge. Cyanide in the rinse is destroyed by oxidation at the anode.

Simple, familiar technology can also be a valuable source for the electroplater in his/her quest to recover metal and reduce pollution. One time-honored application calls for the employment of closely spaced alternating rows of anodes and cathodes in ordinary plating tanks. Numerous patents and papers describe the treatment of spent cyanide plating baths using electrolysis in this way. As with any electroplating application, metal concentration, temperature, agitation, and electroplate concentration are important factors. A large successful segment of the electroplating industry is subscribing to electrowinning as metal recovery process and/or waste treatment procedure. A variety of electrowinning devices are commercially available for those preferring the pre-engineered approach. The more readily deposited metals such as copper, silver, and gold do not require complicated equipment to function satisfactorily. The metal is often of such purity that it can be used as anode material in the plating bath rather than sold for scrap or sent to a refiner.

Upon completion of the recovery technology survey pertaining to prevailing conditions, the plater must then determine which process or processes best meet his or her requirements. The final selection(s) may depend upon cash flow and budgetary considerations, both calling for a clear, comprehensive review of the complete wastewater program. Investing in recovery equipment before acquiring a keen perception of its actual impact on wastewater treatment costs is counterproductive. Recovery will often justify itself on the basis of chemical and water values saved. Review possible credits, such as reduced sludge disposal costs. The cost saving levels can be substantial, depending upon the magnitude and the nature of the complete wastewater treatment requirements.

It would be appropriate to review what has been discussed to this point before proceeding with wastewater treatment:

1. A comprehensive survey of all waste sources.
2. A completed intensive conservation program, including changes or modifications of existing equipment, processes and practices.
3. Investigation of available recovery technology.
4. Preliminary judgments made concerning specific applications.

WASTEWATER TREATMENT—SEGREGATION AND COLLECTION

The majority of industrial wastewater treatment systems in the metal finishing industry currently utilize technology predicated upon removal of metal ions from the waste stream by formation and precipitation of relatively insoluble metal hydroxides. A smaller number use sulfide precipitation. The only essential chemical requirements of hydroxide precipitation are that the metal ions in the waste stream are free (uncomplexed) and that the pH is high enough to insure optimum hydroxide precipitation. The requirement that metal ions be free creates additional problems for those shops doing cyanide-bath electroplating, chromium plating, chromating, chromic acid bright dripping, electroless plating, and printed circuit manufacture. It is vital at the outset to insure that these rinses are segregated from all other acidic and alkaline (A/A) rinses because they require special treatment. The hexavalent chromium rinses must be fastidiously segregated from the cyanide rinses. In addition to the piping of rinse tanks, attention must be given to floor areas where spills, splashes, and washdown waters can be mingled. Mixed waters often pass through treatment and produce effluent violations. After segregation of chromium and cyanide rinses at the source, they must flow to segregated holding tanks, sumps, pits, or whatever collecting basins are envisioned. The collection basin is normally at or below floor level to allow gravity flow of rinses from their point of origin.

Collection basins should generally be as large as practicable with due regard for space and budget limitations. The collection basin for the acidic and alkaline rinses should be sized to receive these waters as well as the combined flows from the cyanide and chromium treatments. In other words, the A/A basin should be sized as if it were receiving the total wastewater flow directly. If space is not available inside the plant, outside locations for in-ground designs could be considered. Properly designed in-ground concrete basins will not normally freeze during winter months in most states. Good sense and experience are needed for designs found in northernmost locations. Plant shut-down periods, such as weekends and holidays must be taken into account when determining freezing potential.

When sizing collection basins, their role as holding tanks during shut-down periods must be examined. If the plant intends to stock spare parts, the plant's location with respect to spare part suppliers, proximity to service centers for pumps, instrumentation, etc., may all influence holding capacity requirements.

A nominal sized basin should have at least a 4-hr retention capacity. Larger operations require a 24-hr capacity to avoid interrupting manufacturing when the waste treatment system is inoperative or not meeting discharge standards.

The role of collecting tanks or basins can be greatly enhanced by the addition of a mixer. Mixing of the contents averages the composition of the flows as well as equalizing the surges with lower flow conditions. Hence, these tanks and basins are more properly termed averaging tanks. Averaging tanks are well-suited to to serve as sumps for the transfer pumps that feed the treatment process. Level controls can be installed in the tanks to automate the transfer.

Twin alternating transfer pumps are preferred for reliability. The pumps employed should be sized to match the specific design flow rating of the waste treatment process with spare capacity to compensate for pump wear. Centrifugal pumps with either open or closed impellers are frequently selected for this application. Pumps should be securely mounted on the floor next to the tank or on a bridge platform over the tank. Vertical submerged suction pumps are preferred by some, but the cardinal rule is never mount pumps, horizontal or vertical, where the motor can be submerged during a power failure. This rule eliminates the cost and time lost in replacing ruined motors. The use of a foot valve is recommended even when the pump is self-priming. The exception is when the averaging tanks will be exposed to freezing weather. Wastewater should not be retained in the pump's suction line.

A low level shut-off point should be selected for the transfer pumps to preserve some averaging capacity and to keep the mixer blades submerged at all times. Out-of-water mixer operation may cause wobble or whipping that could damage the drive, the shaft, or both, depending upon the mixer design, shaft speed, and blade type. In large tanks, low-speed turbine blade mixers are preferred; smaller tanks call for faster three-blade propeller types. "Props" are often used, with the bottom prop having a stabilizing ring to dampen shaft whipping.

CYANIDE OXIDATION

With the completion of the wastewater collections and transfer system design, treatment of cyanide and chromium rinse waters are the next order of business. All treatment processes used in the United States involve the oxidation of cyanide to cyanate or carbonate (CO_2) and nitrogen. Chlorine or sodium hypochlorite are commonly employed. Table 2 presents the chemical equations for these reactions. Chlorine gas is preferred for larger systems because of operating cost, while hypochlorite can be justified in smaller systems. Comparative costs are shown in Table 3.

The initial cost of a vacuum feed system for chlorine gas quickly pays for itself where flows and/of cyanide levels are high. Smaller systems with less than 20 gal/min flows are better suited to simple chemical feed pumps supplying a 15% solution of sodium hypochlorite. Sodium hypochlorite decomposes in storage, and hot weather accelerates the process. It may be difficult, in

TABLE 2. CYANIDE OXIDATION EQUATIONS

1. $NaCN + 2NaOH + Cl_2 \rightarrow$
   $NaCNO + NaCl + H_2O$
2. $2NaCNO + 4NaOH + 4Cl_2 \rightarrow$
   $2CO_2 \uparrow + N_2 \uparrow + 8NaCl + 3H_2O$
3. $2NaCN + 10NaOH + 5Cl_2 \rightarrow$
   $2NaHCO_3 + N_2 \uparrow + 10NaCl + 4H_2O$
4. $2NaCN + 5NaOCl + 2NaOH \rightarrow$
   $2Na_2CO_3 + N_2 \uparrow + 5NaCl + H_2O$

TABLE 3. TREATMENT COSTS. 1-gal CYANIDE PLATING BATH

Plating Bath	1st Stage Using Cl_2	1st Stage Using NaOCl	2nd Stage Using Cl_2	2nd Stage Using NaOCl
CuCN	$0.25	$0.36	$0.63	$0.94
ZnCN	0.21	0.31	0.54	0.80
Cd	0.41	0.60	1.05	1.55

warm climates, to justify the use of hypochlorite even for small flows.

Chemical equations are but the starting point for determining actual chlorine demand. Some cyanide complexes, such as copper and zinc, are oxidized with relative ease; others require substantial (100%) excesses of oxidant. Cyanide complexes of iron, cobalt, and nickel are extremely stable and may never oxidize completely. Yet these complexes decompose, at least partially in the total cyanide analytical procedure producing detectable levels of cyanide. It is important, therefore, to avoid combinations of cyanide with these metals, especially in those areas where the total cyanide discharge limits are particularly low (0.1 ppm or less). A blue color in a steel pickle tank indicates that iron and cyanide are complexed and gives indication of potential trouble in meeting low total cyanide limits.

The initial reaction or contact time required is actually rather short in the design of cyanide treatment process equipment. With good mixing, the first stage reaction can be completed with 10- to 15-min residence time on a flow-through basis. Due to the intermediate formation, of cyanogen chloride, a toxic gas, safe practice dictates that a 20- to 30-min detention tank follow this first stage. Some states require as much as 1-hr detention. When metal contents are high, a mixer may be needed in this tank to prevent sludge build-up. Although theoretically, mixing could defeat the purpose of this tank, on an applied level, high sludge levels will shorten detention times.

Overflow from the detention tank is piped to the acid-alkali collection/averaging area if an installation is but a single stage. If a two-stage oxidation is required by state or federal regulation, the same two-tank format of the first stage is repeated using the same retention times.

The instrumentation required to control the alkaline chlorination (first stage) is identical for chlorine and sodium hypochlorite use. A reliable pH controller must be used to maintain a pH of 11.5 in the reaction tank. Some designers omit pH control if hypochlorite is used. When pH control is lost, a health hazard occurs due to the evolution of cyanogen chloride gas. The pH is often maintained by relatively small chemical feed pumps activated by a signal from the pH controller. Using a drum wand, a 50% solution of sodium hydroxide ("liquid caustic") can be drawn directly from the drum, without dilution, and injected into the reaction tank. Larger systems may employ bulk storage of liquid caustic with the same chemical feed pump concept. One system supplier uses bulk caustic storage in conjunction with an air pressurized feed system which relies upon solenoid valves to convey caustic to all demand points in the treatment plant. This method of chemical feeding avoids the maintenance problems associated with feed pumps, but applied without engineering expertise it could be dangerous. Liquid caustic will almost instantly destroy the cornea of the eye. Wearing face and eye protection when working with or around this chemical is mandatory. Other protective clothing such as gloves, aprons, and boots is recommended here as well as elsewhere in the waste treatment area.

The addition of chlorine gas or sodium hypochlorite to the reaction tank is controlled by an oxidation-reduction (ORP) potential controller. As the name implies, these instruments measure the actual voltage generated by contact of the effluent with two electrodes in the probe assembly. Acting much like a battery, a millivolt signal is generated between the reference cell filled with a saturated solution of potassium chloride and a platinum or gold measuring electrode. In the first stage, an oxidizing potential of about 350 millivolts (Au-KCl) should suffice. This setting may change to suit the actual character of the effluent with fine tuning. A high organics loading may increase chlorine demand. The chlorine addition may be effected through the use of chemical feed pumps working from 55-gal drums of 15% sodium hypochlorite solution. Alternatively, the signal from the controller can be used to activate a normally closed solenoid valve in a chlorine line. The chlorine gas is pulled into the system by means of a vacuum eductor located in a recirculating loop with the treatment tank. The recirculating pump in this loop must be capable of producing a pressure of at least 40 psi to assure adequate vacuum from the eductor.

In a two-stage chlorination system, the instrumentation required is identical to the first stage. The ORP meter scale should read to 1000 millivolts to allow for an operating range of 500 to 800 millivolts. The pH controller should be set up to add sulfuric acid from a carboy by means of a chemical feed pump. The pH must be maintained between 8.0 to 8.5 to assure complete oxidation of the intermediate cyanate to nitrogen and carbon dioxide (carbonate). The pH must not be allowed to fall below 8.0 because of the potential release of cyanogen chloride gas, chlorine gas, or even hydrogen cyanide gas in malfunctioning systems. It is a wise option to provide for the automatic addition of sodium hydroxide to eliminate low-pH excursions. Audible alarms and lights should also be used to signal the operator when either pH or ORP measurements are not correct. Alarms for mixer malfunction are desirable in any waste treatment process. Inadequate mixing allows improper treatment and substandard effluent quality. A plater must plan ventilation to cover contingencies of the process. Exhaust fans or hoods near or over the installation are the best means of preventing a hazardous accumulation of various toxic gases associated with the treatment process.

HEXAVALENT CHROMIUM REDUCTION

The anionic hexavalent chromium (chromate, dichromate, etc.) must be chemically reduced to the trivalent cationic state so that it can be precipitated. This reduction in valence also reduces the associated toxicity by a factor of 1000.

Table 4 gives the chemical consumption and costs for the common methods of reduction. Ferrous sulfate is seldom used because of its inefficiency, expense, and high sludge generation rate. The most economical reagent for chromium reduction is sulfur dioxide gas. At a pH of 2 or lower, the reduction is almost instantaneous. A reaction tank having 10 to 15 min retention time is therefore quite adequate. However, at pH values above 2, the reaction rate slows markedly; at pH 4, it is too slow to be effective. This is unfortunate, since at pH 4 there is much less SO_2 released to the atmosphere than at pH 2. Proper ventilation is essential with this process.

Instrumentation for the sulfur dioxide (SO_2) reduction of chromium is similar to that described for the chlorination of cyanide. A reliable pH controller is needed to maintain the optimum pH by addition of sulfuric acid upon demand. Because SO_2 in water is naturally acidic (sulfurous acid), the acid demand is quite low. An ORP controller is employed to maintain a reducing potential with about 100 millivolts being a good starting point. pH and potential should be adjusted to obtain the optimum conditions for the waste stream involved. Rinse waters from chrome plating will be more demanding than those from the chromating of zinc or aluminum.

Reduction of chromium with sodium bisulfite or sodium metabisulfite is widely used in place of SO_2 gas, particularly in smaller systems. While more expensive by a factor of almost 4, it avoids the handling of gas cylinders. Bisulfite solution is applied from a solution make-up tank by means of a chemical feed pump, activated by the ORP controller. Acid demands will be higher with bisulfite than with SO_2 because of the slight alkalinity of the material. As with SO_2, the optimum pH for reducing hexavalent chromium is in the vicinity of 2.

Sodium hydrosulfite may be used for chromium reduction in special circumstances. It offers the advantage of efficiently reducing chromium in the pH range of 8.0 to 9.0. Treating small batches and spills is, perhaps, its best application. In addition to its higher cost, hydrosulfite has the disadvantage of being pyrophoric and can self-ignite when damp. Care in storage and handling is required with all reducing agents such as sulfite, bisulfite, metabisulfite and hydrosulfite. They should be stored in a dry area well away from oxidants such as dichromate and chromic acid. Serious explosions and fires have resulted from mixing such incompatible materials.

Iron or steel scrap, in the form of punchings, is now being widely used for the reduction of hexavalent chromium. Early efforts to apply this straightforward chemistry proved impractical because of high sludge product and the extremely large quantities of steel needed to treat even very low flows.

In 1970, a patented process using small multiple reactors and careful automatic pH control resulted in an economical and efficient way to treat chromium bearing rinses. The process uses set points between pH 1.5 and 2.5 to obtain complete reduction. Because most rinse waters are acidic to begin with, the addition of sulfuric acid is generally less than 1 gal/hr for a modular unit capable of treating 25 gal/min of chrome

TABLE 4. TREATMENT COSTS, 1-gal CHROME PLATING BATH

Treatment Chemical	Amount (lb)	Cost	Total Cost
Ferrous Sulfate + Sulfuric Acid	6.76 + 2.38	$1.49 + $0.11	$1.60
Iron + Sulfuric Acid	0.45 + 2.38	0.05 + 0.11	0.16
Sodium Bisulphite + Sulfuric Acid	1.26 + 0.60	0.55 + 0.03	0.58
Sulphur Dioxide	0.78	0.17	0.17

rinses. This process ranks a close second to SO_2 gas in operating cost while avoiding the health hazards of the gas.

The electrolytic generation of ferrous sulfate is available in commercial form but is not widely used in the electroplating industry because its best application is to very low hexavalent chromium levels such as might be found in cooling tower waters.

Hydrazine solutions reduce hexavalent chromium but their use has declined with growing knowledge of the health hazards they present. Not only are hydrazine inhalation and ingestion hazardous, but the chemical's absorption through the skin is dangerous as well. Hydrazine's effects are cumulative; the liver and the kidneys are particularly vulnerable to exposure. If hydrazine is employed, avoid storage or use in basements or other poorly ventilated enclosures. The old but unhealthy practice of workers removing chrome stains from their hands with the chemical should be forbidden as well as any other direct contact with hydrazine solutions.

PRETREATMENT

Following oxidation of cyanide and reduction of hexavalent chromium and a combination of these waters with other acidic and alkaline flows, pH adjustment and metal ion removal still remain to be accomplished. In practice, the processes involved are referred to as neutralization and precipitation or sedimentation. Neutralization is not an accurate description because the actual pH is almost never 7. For the sake of conformity, the word *neutralization* will be used here to describe the pH adjustment process.

For those situations where discharge limits of most metals are in the range of 3 ppm or less, it may be necessary to introduce a pretreatment stage before neutralization. Some suppliers of complete systems have adopted the pretreatment stage as routine practice. In the pretreatment process, aluminum sulfate solution (27%) and/or calcium chloride solution (40%) are added to the flow in a separate mixing tank having a residence capacity of about 20 min or less. The pH may or may not be controlled at this stage, depending upon the nature of the effluent. If heavy loadings of emulsified oils and greases or wide pH fluctuations are anticipated, automatic pH control of this stage would be worthwhile. By using both calcium and aluminum at this stage at a pH of 5 or 6, the emulsified oils can be readily removed while dampening the pH swings. In addition to removing oils by absorption, the introduction of calcium and aluminum ions antagonizes surfactants and complexing agents that might hinder subsequent settling; insolubilizes phosphates, fluorides and silicates; and aids in the flocculation of suspended solids.

NEUTRALIZATION

Following pretreatment, the effluent will require pH adjustment to a pH of between 8.5 to 9.5. A pH over 9 will usually be required to obtain optimum insolubilities of zinc, cadmium, nickel, lead, and possibly copper. The actual operating pH selected will depend upon the mix of metals present. Graphs presented in the literature are often misleading in showing the theoretical insolubility curve for each individual metal. In actual fact, there is a strong tendency for metal ions to co-precipitate as hydroxides or a mixed hydroxide-carbonate species. When calcium and aluminum ions are present, this synergism is further enhanced and residual metal levels in the effluent will often fall below the theoretical solubility limits.

The reagent chemicals generally used in the neutralization process are 96% sulfuric acid for high-pH conditions and 50% sodium hydroxide solution ("liquid caustic") for pH elevation. Lime slurry has also been used as a source of alkalinity in the neutralization process. While it is less

TABLE 5. THEORETICAL CONSUMPTION (lb) OF ALKALIES FOR THE NEUTRALIZATION OF 100 lb OF ACID

Alkali	Acid to be Neutralized				
	H_2SO_4 (Conc)	HCL 35%	HNO_1	HF	H_3BO_3
$Ca(OH)_2$	76	36	59	185	180
(Cost)	($2.28)	($1.08)	($1.77)	($5.55)	($5.40)
NaOH	82	39	64	200	194
(Cost)	($12.30)	($5.85)	($9.60)	($30.00)	($29.10)

TABLE 6. THEORETICAL CONSUMPTION (lb) OF ALKALIES FOR THE PRECIPITATION OF HEAVY METAL PER SHIFT

Alkali	Heavy Metal to be Precipitated				
	Cu	Ni	Cr	Zn	Cd
Ca(OH)$_2$	7.60	11.91	25.22	6.96	3.16
(Cost)	($0.23)	($0.36)	($0.76)	($0.21)	($0.09)
NaOH	4.56	7.04	14.97	4.08	1.84
(Cost)	($0.68)	($1.06)	($2.25)	($0.61)	($0.28)

Metal Hydroxide Formation
(Neutralization)

$$M^{+2}L^{-2} + Ca(OH)_2 \rightarrow M(OH)_2 + CaL$$
$$M^{+2}L^{-2} + NaOH \rightarrow M(OH)_2 + Na_2L$$

expensive than liquid caustic, it tends to be messy, clogs pumps and lines, and significantly increases sludge accumulation rates.

Table 5 compares the quantities and costs of lime and "liquid caustic" for acid neutralization. Table 6 presents the comparisons for metal precipitation. In both cases, lime appears to be the reagent of choice; when sludge solids contents and sludge disposal costs are considered, as in Table 7, "liquid caustic" becomes most attractive. Sodium carbonate solutions may be used on occasions where the solubility of a particular hydroxide is not low enough to meet discharge standards. The mixed hydroxide-carbonates (also known as basic carbonates) are often less soluble.

Of primary importance in the neutralization stage is the careful control of pH. A difference of 0.5 pH units is equal to a tenfold difference in the theoretical solubility of a divalent metal hydroxide. There are several good pH controllers commercially available, but they all require the application of sound engineering practices to obtain the desired result—pH stability.

Tank design is the starting point. Ten to fifteen minutes of retention time should be allowed for the flow. Three or four vertical baffles are necessary in round tanks to obtain turbulent mixing. Adequate mixing is essential for homogeneous blending of the treatment reagents with the effluent flow. If a mixer is viewed as a pump without a housing, then pumping or turnover rates are easily understood. It is important that the pH-sensing probe in this tank is "seeing" the actual pH condition prevailing. If the contents are not homogeneous, or nearly so, then the pH controller cannot be expected to perform its function. Flow-through pH sensors also have the disadvantage of not "seeing" the true condition prevailing in the treatment tank at any given moment. This sensing delay leads to instability that is particularly exaggerated by fluctuations in the incoming pH. Locate the injection point for the reagents so that they can be blended quickly with the tank contents as they are driven down by the mixer (usually about one-half the radius). Position the pH probe near the tank edge where it will register the return or up-flow from the mixer.

Most mixer manufacturers will assist in selecting the proper mixer for a given application, but a good rule-of-thumb requires one horsepower for each 750 gal of tank volume, if direct drive "prop"-type mixers are employed. Gear drive

TABLE 7. SLUDGE VOLUME AND DISPOSAL COST PER SHIFT AT 30% SOLIDS (wt.)

Type of Plating	NaOH TMT (lb)	Disposal At 0.10/lb	Ca(OH)$_2$ TMT (lb)	Disposal At 0.10/lb
Acid Cu	18.7	$1.87	30.7	$3.07
Cyanide Cu	6.7	0.67	9.3	0.93
Nickel	27.4	2.74	35.4	3.54
Chromium	42.7	4.27	61.4	6.14
Acid Zn	17.3	1.73	24.7	2.47
Cyanide Zn	5.3	0.53	8.4	0.84
Cadmium	6.0	0.60	8.7	0.87

mixers of less power can be used to turn larger blades at slower speeds; the turnover rate must be preserved.

The pH control philosophy employed in neutralization, being more demanding than those described previously for cyanide and chromium treatment, requires special attention. Simple on-off control of a single caustic or acid pump will develop a saw-tooth graph of the tank discharge pH. With a large flocculation and settling tank, these swings may smooth out to where discharge quality will be acceptable in some locations for sewer (POTW) discharge. Adjustment of acid and caustic pumping rates to better match demand will further help to dampen the tendency of these pumps to fight each other and reduce the magnitude of the swings.

Maximum pH control requires that responses to change in pH follow quickly and precisely. Proportional control has the potential for achieving this objective. Electronic interfacing of the pH controller output with valve positioning motors can regulate the flow of sulfuric acid or liquid caustic to compensate for fluctuations in pH. There is some delay in valve actuation that slows response speed in some cases.

Another proportional feed format involves converting the pH controller output to a variable frequency signal that controls the stroke of a chemical feed pump. The response is almost instantaneous and provides good stability, particularly in small feed applications.

An effective alternative to proportional control is known mathematically as multilevel suboptimal control. This control has been in use longer than the mathematical theory supporting it. On an applied level, control by this method is simple. The pH controller must have two independent set points for controlling two independent chemical feed pumps or solenoid valves. The first or lead set point is the actual pH control point, while the second or back-up point is slightly lower, as in the case of caustic addition. It is slightly higher for acid addition. The pumping rate for the lead pump is set to closely match the demand rate. The back-up pump is set at a higher rate.

If the pH falls below set point by the margin between the two set points, the second pump quickly corrects the situation. By carefully adjusting the pumping rates to match the routine characteristics of the flow, a nearly straight line trace will be generated. This control format will rapidly correct large-pH excursions caused by spills. If solenoid valves are used instead of pumps, throttling valves in the feed line permit the same control format. When both acid and caustic are employed, a four set point controller will be required.

FLOCCULATION

Flocculation is an essential process in every industrial wastewater treatment system. Following the establishment of the optimum precipitation pH, the waters should flow (preferably by gravity) to a separate tank or compartment specifically designed to achieve effective flocculation of suspended metal hydroxides and other particles in the flow. The flocculation process involves coating the small particles with polyelectrolyte, followed by agglomeration of the small particles into larger groupings. These larger groupings will have greatly improved settling characteristics, thus insuring proper settling and a clear discharge.

In the flocculation process, an organic polyelectrolyte is added to the waste stream at a rate of from 1 to 20 ppm. The feed rate required will depend, in part, upon the particle population of the flow. Higher particle densities and dispersing agents in the flow will call for the higher dose rates.

The flocculation tank should provide for between 3 and 10 min of residue time. It should be fitted with a mixer having a speed of from 60 to 400 rpm. The mixing times and speeds are selected to obtain the maximum particle collision probability without degrading the particles. The agglomerated particles are extremely fragile and will break up under shear or turbulent conditions. Once broken up, these "flocs" will not reassociate without added polymer. For this reason, the transition from flocculation to settling should be made, if at all possible, by gravity flow. If design elevations do not allow for such flow, then a pumping stage should precede flocculation.

An alternative flocculation format that may be used to avoid elevating large tanks, involves providing about 30 sec of rapid mixing of polyelectrolyte with the flow, followed by approximately 3 min of very slow mixing in a separate compartment or tank.

While cationic polyelectrolytes are often used in sewerage treatment, the anionic species are far superior for the treatment of metal finishing waste streams. The more highly charged materials usually work best. A series of parallel beaker or jar tests, with various polyelectrolytes and application rates will quickly reveal the most advantageous selection. The plater should obtain a gallon or more of the plant's effluent and treat it through the neutralization stage. Using this stock solution and one or more beakers with slow mechanical or magnetic mixing, the plater should then apply the polyelectrolyte(s) to be evaluated.

He/she should then, after a minute or so, halt the mixing action in the beakers and time the rate of settling. Note the sludge volume after an hour or two to ascertain the sludge accumulation rate. The observed sludge volumes will be higher in a beaker than in reality, but they will reveal potential problems before the plant is built. The polyelectrolyte that produces the best clarity in the upper layer in the shortest settling time should be selected.

Most anionic polyelectrolytes are sold by the manufacturer as granular solids. There are few 100% active liquid materials on the market. Water solutions of polyelectrolytes tend to degrade and to lose their activity; it is recommended when using granular products to make up working solutions on a daily or not more than weekly basis.

Make-up of the working solution can be difficult without a disperser to prevent clumping and to wet each granule individually. A disperser is a jet mixer actuated by a garden hose water supply. Most suppliers can provide these devices for under $100 or furnish names of manufacturers. More sophisticated apparatus selling in the $5000 to $10,000 range is not necessary.

A "make-up" of 1 lb of polyelectrolyte in 50 gal of water should be suitable for a 50 to 100 gal/min flow when applied at the rate of 5 gal/hr or less. Adjustable stroke piston type chemical feed pumps are well-suited for this application. The working solution should be allowed to age 12 to 24 hr to provide time for complete hydration of the polyelectrolyte. This aging allows most efficient use of the polyelectrolyte and avoids pump problems caused by partially hydrated granules (fish eyes). The most efficient method of solution preparation is to provide two make-up tanks and to alternate from one to the other on an every-other-day basis.

In cases where particularly low-discharge limitations prevail, supplementary filtration may be required. In a well-designed and operated system, the discharge from settling will contain under 20 ppm suspended solids and 1 ppm or less of each of the commonly electroplated metals. This performance may not be acceptable in some localities. It will then be necessary to take such measures as needed to meet the discharge standards. Supplementary sand filtrations will normally reduce the suspended solids content to the 3 to 5 ppm range and lower the metal content to approximately 0.1 ppm for each metal. Zinc, cadmium, and lead may prove troublesome and laboratory treatability studies are recommended to confirm that the metals will be in a suspended form that responds to filtration. (Other methods of addressing the problem of dissolved metal will be described later.)

Commercially available sand filters are of two general designs. The first, and perhaps the best known, is the mixed media type where sand, as well as various other media, is employed in a packed column configuration. Flow may be either up or down the column, but the densities of the media must be such that the waters flow through the coarser materials first and the finer material last. In this mode of operation, the filter cycle is extended. When particulates emerge from the filter or the pressure drop through the bed reaches a predetermined limit, the filtration must be halted and the bed backwashed to remove accumulated solids. The backwash water must be discharged to the A/A tank for reprocessing through the treatment system. This type of filtration has the disadvantage of demanding system shutdown during backwashing or two filters for alternating duty. A holding tank could be used to store unprocessed water, but if the receiving sumps are large enough to accept the flow, system shutdown may be a better choice.

The other type of sand filter currently enjoying some popularity is a patented continuous upflow filter featuring a continuous sand washer. The dirty sand from the bottom of the filter is lifted by air to a washing chamber in the top. Wash water bearing dirt from the sand overflows the washer section and is returned to the A/A sump for reprocessing while the cleaned sand falls back on the top of the sand column. The filtrate overflows at the top of the filter, having passed through the cleanest sand last.

SPECIAL TREATMENT METHODS

In some instances, the wastewater treatment processes described thus far may not remove metals to the desired levels. This problem occurs when large amounts of metal complexing agents are employed in the manufacturing processes. The effluents from the manufacture of printed wiring (circuit) boards (PWBs) are representative of this situation. Special treatment methods must be employed to extract the metal from the complex. Until recent times, PWB effluents were treated either with calcium chloride, hydrogen peroxide and sulfuric acid in a patented process that broke the complex and reduced copper to cuprous form prior to conventional treatment or by the addition of ferrous sulfate prior to conventional treatment. The ferrous sulfate and the hydrogen peroxide (at pH 9) apparently reduced copper to cuprous while occupying the complex with either ferrous or ferric ions. Newer systems employ insoluble starch xanthate (patented), ferrous sulfide (patented) and an axial flow horizontal reactor packed with steel scrap (patent pending).

While ferrous sulfate treatment of complexed effluents is widely used, its main disadvantage, apart from the cost of the relatively high dose rate needed, is an extremely high sludge accumulation rate. The calcium chloride and lime at high-pH processes are also in use, but they also suffer from the high sludge problem. The sludge problem carries over to ferrous sulfide applications since it involves the substitution of copper (or other metal) ions for ferrous ions in the sulfide molecule. The ferrous ions must be subsequently precipitated as hydroxide along with the metal sulfides.

Sodium polysulfide and sodium hydrosulfide have been used in similar applications, but, while avoiding the excess sludge problem, they increase the potential for generating and releasing the very toxic hydrogen sulfide gas. The odor of rotten eggs is most deceptive and should not be taken lightly. Deaths occur each year from hydrogen sulfide poisoning. Careful pH control and good ventilation are essential to any sulfide application.

Insoluble starch xanthate (ISX) functions essentially as a throw-away ion exchange resin, which releases magnesium ions and takes up heavy metals. The sulfide-based treatments, including ISX, can reduce the heavy metal content of effluents to the very low levels consistent with the solubilities of metal sulfides. Invariably the sulfide solubilities are lower than the hydroxides. Sulfide-based treatment processes have also been applied to effluents other than those from PWB manufacture. However, for reasons mentioned previously, they are best reserved for smaller flows or low-metal-content effluents. Polishing applications where discharge limits are particularly low have been reported to produce good results with ISX treatment followed by filtration.

The axial flow horizontal reactor functions to remove copper from complexed effluents by cementing the copper as copper metal while releasing ferrous ions. The ferrous ions serve to antagonize or form complexes with the free complexing agents. While ferrous ion activity in this technology resembles the ferrous sulfate treatment described previously, sludge generation is only 10 to 20% of that from ferrous sulfate use. When the horizontal reactor is followed by calcium chloride, aluminum sulfate pretreatment pH adjustment, flocculation and settling, the residual copper in the effluent will be less than 1 ppm.

New chelating ion exchange resins belong in the category of special treatments because they do not function as conventional resins. The regeneration of these resins is accomplished in the conventional manner by sodium hydroxide to yield the sodium form or hydrochloric acid to produce the hydrogen form. These resins, however, have a selective affinity for heavy metals. When a wastewater flow is passed through the resin bed, the hydrogen ions are displaced by sodium ions, the sodium ions by calcium ions, calcium ions by copper ions, etc. This progressive displacement permits chelating resins to function in applications not favored by conventional resins. Final effluent polishing, for example, is now possible without most of the interference usually caused by sodium and calcium ions. One particular resin has a preference for copper over calcium of 2000 to 1. Recovery of copper from PWB waters by a chelating resin has been patented and described in a recent paper. While few full-scale installations using chelating resins in metal finishing wastewater treatment have been described in the literature, prospects are good that more successful applications will be reported.

SOLIDS MANAGEMENT

In general, solids management can be divided into categories, beginning with suspended solids removal, followed by settled solids handling, possibly thickening, and finally dewatering and disposal. Following flocculation of the particulates in the wastewater stream, they can be removed by a variety of mechanisms. While filtration may appear to be an ideal approach, it is often difficult to achieve. Most successful applications employ diatomaceous earth coated filters. The downtimes required to clean and precoat such filters necessitate considerable holding capacity for the interrupted flow or alternating filters. Consequently, the most frequently employed technology uses gravity settling in either clarifier tanks or with tube/plate settlers.

A number of clarifier designs are commercially available. The circular clarifier is perhaps the oldest design, having its origins in sewerage treatment and the mining industry. Many circular clarifiers have been used for suspended solids removal from metal finishing effluents. The results have been mixed largely because of the application of engineering criteria from sewerage treatment situations. When more conservative design specifications are used to compensate for the lighter hydroxide particles, circular clarifiers can produce acceptable effluent.

The circular clarifier features a center-well injection of the effluent. If the effluent must be pumped to the clarifier, the flocculation process should be conducted in a tank above the center well. The flocculated waters then flow down the well and emerge near the bottom. The upward and outward flow slows the particulate velocity and permits settling. The clarified waters overflow a circumferential vee-notched wier. It is essential that the wier be perfectly horizontal to assure symmetrical distribution of flow. Improper distribution of flow will cause short circuiting and unsatisfactory solids removal.

Circular clarifiers are often fitted with rotating scrapers that move the sludge slowly down the slope of the bottom and toward the center of the tank for removal. Scrapers may be fitted with a rake resembling a short picket fence which aids in thickening the sludge. When the settled sludge is disturbed, the interstitial water rises and the particles move close together to produce thicker sludge.

A more modern clarifier design employs a rectangular tank configuration with an overflow wier at one end and an inlet baffle or wier at the other. Studies show the design to be quite efficient with only about 150-min residence time required to produce less than 20 ppm of suspended solids in the effluent. One supplier of this type of clarifier has the added feature of a self-propelled thickener and collector mechanism that operates above the tank. The bridge of the device carries a sludge pump that draws the sludge from the bottom through a slotted pipe extending the width of the tank. The sludge is discharged into a holding tank that runs the length of the settling tank. It is not uncommon to obtain sludges with 3 to 5% solids with this design.

Tube or plate settling type clarifiers are predicated upon providing an inclined plane upon which suspended particles can settle. An incline of about 60° will cause the settled particles to slide downward against the flow and fall into a collection chamber below the tubes or plates. When tubes or plates are stacked to form an inclined bundle, the particles have only a few inches to fall before reaching a tube or plate floor. Square or rectangular tubes provide a better floor surface for the particles to settle upon than round tubes. Parallel plates can present even more effective settling in a given area without the sidewall drag effect which sometimes causes sludge to adhere and clog the tubes.

Settling devices of the plate or tube types require less floor space for any given flow than would a comparable circular clarifier or rectangular clarifier. However, tube and plate settlers will require more head room because of the configuration of the collection chamber beneath the tube or plate array. If the settling unit can be installed in a pit in the floor, it can receive a gravity overflow from the flocculation stage; if not, then pumping must precede flocculation at the feed elevation of the settler.

Tube or plate settlers must be operated in such a manner that the accumulated sludge level is kept well away from the bottom of the tubes or the plates. If sludge builds up too much, it can be caught up in the flow and appear in the discharge. Either a slow, steady sludge withdrawal or a frequent batch withdrawal will insure proper performance. The sludge withdrawn will frequently be too low in solids (0.5% or less) for efficient dewatering. A thickening device or secondary settling tank may be needed. To avoid this complaint, at least one supplier offers a thickening device in the base of the settler.

SLUDGE DEWATERING

Enactment of the Resource Conservation and Recovery Act (RCRA) at the federal level and passage of many regulations at the state and local levels have led to extremely serious problems for the metal finishing industry. Sludges from wastewater treatment systems have traditionally been buried in landfills and dumps. Now many of these residues have been categorically classified as hazardous because of their origin or fall into that category by virtue of failure to pass extraction tests (federal or state). The stigma of this classification, together with the regulations concerning the transportation, storage, and disposal of metal finishing wastewater treatment sludges have made it increasingly difficult to obtain legitimate arrangements. Some states have no legal hazardous waste landfill sites. Out-of-state transportation and disposal can be very expensive. The situation could become increasingly difficult as more and more landfills are closed and new ones become harder to find.

Recent studies of metal finishing sludges show that many sludges can pass the U.S. EPA extraction procedure (EP) toxicity test. These studies indicate that "good" sludges come from "good" systems: i.e., systems that are well-designed, well-operated, and well-maintained. Such results are consistent with the fact that there would be virtually no hexavalent chromium, no free cyanides, and minimal metal ions in the interstitial water of such sludges. It also shows that the sludges with the least residual water content usually give better EP test results. These findings, together with the high cost and difficulty of sludge disposal, cause sludge dewatering to become a vital aspect of wastewater treatment.

Commercially available sludge dewatering devices fall into three essential categories, the first being of the centrifugal type. When sludges are spun at high speed, suspended particles are driven by gravitational force to the outer perimeter of the bowl. Most metal hydroxides are highly hydrated and have specific gravities only slightly greater than water itself. Hence, the high G-forces are necessary to separate the suspended particles from the water. The most efficient centrifuges generate G-forces in excess of 2100 and the best of these exceed 3000 G^s.

Even at high speeds, centrifugal dewatering is never complete. The centrate or dilute aqueous phase always retains suspended particles that must be considered in the design of flow-through centrifugal dewatering facilities. The centrate may be returned to the sludge holding tank for reprocessing or discharge to acid-alkali averaging tank for recycling through the treatment system. There is an element of awkwardness to both options. In no event, however, can the centrate from continuous centrifugal dewatering be discharged with the treated effluent. Batch centrifuges can produce a clear centrate but their size and cost are usually prohibitive for most metal finishing applications. The maintenance of high-speed centrifuges is another serious consideration, regardless of whether they are of the continuous or batch type. If lime, sand, or tumble finishing residues are in the sludge, wear of exposed surface will be accelerated. Even tungsten carbide clad sludge scoops will erode quickly under abrasive conditions. Support bearings for the bowl and shaft assembly are also critical and must be shrouded from exposure to abrasives.

Spare parts should be stocked for all critical components of any centrifugal dewatering operation. Even the best wastewater system can be virtually brought to a halt by a breakdown of the dewatering facility. When settling tanks and clarifiers begin to back up with sludge, discharge quality will quickly deteriorate. Sludge continues to accumulate until there is no place to store it. (While these remarks are made with respect to centrifugal dewatering, they pertain to the critical nature of any and all dewatering methods.)

Perhaps the most vital consideration of centrifugal dewatering technology is the residual water content of the processed sludge. Very often sludges with 10% solids and 90% water will be produced, and 20% solids must be considered as outstanding performance. When the centrifuge operates exclusively on a heavy metal sludge such as lead, the percentage of solids could rise to 30, reflecting the higher atomic weight of the metal.

The second method of sludge dewatering employed in the metal finishing industry is vacuum filtration. This technology usually employs a rotating vacuum drum upon which a filter media is applied. A vacuum is applied through a swivel assembly on the drum axel and is distributed to the perforated drum face through a series of radial tubes. The media most often used is either a string or cloth endless belt. The string belts have not been very successful in dewatering the finely divided, gelatinous metal finishing sludges. Nylon or polypropylene cloth belts are used with better results. The best results are achieved when the belt is coated with diatomaceous earth (DE) or paper. DE is applied as a thick precoat to the cloth covered drum while rotating in a vat configured to provide about 40% drum submergence. As the drum rotates, the portion (about 60%) exposed to the atmosphere provides drying time for the adherent coating. When the drum has been suitably coated to a thickness of 2 to 4 in. or more, depending upon filter size and design, the sludge suspension is introduced to the vat. The sludge particles coat the DE precoat (heel) and are dewatered during the atmospheric exposure period. The drum submergence can usually be adjusted to obtain optimum dewatering. Dewatered sludge is scraped or cut from the drum with an adjustable knife mechanism called a doctor blade. Some diatomaceous earth is cut from the drum with each revolution. The blade must be set to cut deep enough to remove sludge that has penetrated the DE coating. If significant sludge remains in the pores of the coating, dewatering will be poor. The solids content of sludges dewatered by filters of this type will normally run from 20 to 35% solids. The solids content will also include the DE used. Likewise, the volume of sludge generated will reflect the DE content.

The paper precoat vacuum filter works much like the DE precoat filter except that rolls of newsprint paper are employed. The sludge is collected upon the paper as it rotates with the drum. The sludge coated paper continuously separates from the drum and discharges into a hopper or bin for disposal. The solids content will be about the same as that produced with a DE filter and will reflect the paper content. The normal operating cost for the paper filter will be less than for diatomaceous earth. Both types of filters have substantial energy demands when compared with other dewatering methods.

Other types of filters belonging in the vacuum category are those employing cloth covered discs or plates. The filter cloths are generally precoated with diatomaceous earth to improve particle retention. The hollow discs or plates are arranged in series connecting to a vacuum manifold. The array of discs or plates may be submerged in an open or closed vessel, depending upon the manufacturer and design preferences. Filters of this type are essentially batch filters and must be shut down for cleaning and recoating.

While these filters may be regarded as sludge dewatering devices, their best application is most likely in effluent polishing. In locations where stringent effluent limits are established, these filters can be used to capture finely divided particles that escape other suspended solids removal methods. A body feed of additional DE will be needed to extend filter cycles to a reasonable duration (one shift, one day, etc.). The DE and sludge residue may be captured in a suitable bin after dewatering or directed as a slurry to the A/A sump for return through the treatment system. The choice depends upon filter design, solids content, and quantities involved. The added capital and operating expense of this filter calls for careful examination of the benefits and professional design experience with the specific application.

A third category of solids dewatering employs pressure. Pressure filters have been improved considerably in the past decade and are now the most widely used dewatering devices in the metal finishing industry.

Newer materials of construction and improved designs account for much of the pressure filter's success. Early filter designs consisted of a plate and frame configuration with cloth filters on two opposing plates separated by a spacer or frame. A series of plates and frames were held together by a high-pressure ram while the sludge slurry was fed under pressure through a center core into the cavities created by the frames. Grooves in the plates allowed the filtrate to escape through the cloth to discharge ports. When the filter was filled with sludge, the ram pressure would be released and the sludge removed from each frame individually.

While the concept of the pressure filter has not changed, the advent of rigid polypropylene recessed plates and woven polypropylene cloths has certainly simplified construction and operation of these filters. The recessed plate design eliminates the frame component by having a recessed cavity on both sides of the plate. When a number of these plates are compressed together, they form a series of cavities into which the sludge slurry is forced as in the plate and frame instance. The sludge does not adhere to the newer filter cloths, thus providing quick release of the sludge cake when the ram pressure is released and the plates shifted apart. The sludge cakes simply fall, with little or no prodding, into a bin under the filter or onto a conveyor belt or screw conveyor leading to a storage bin. The addition of an automatic plate shifting mechanism that pulls the plate stack apart one plate at a time permits automation of virtually the entire dewatering process. An operator may be needed to make the transition from the filter cycle to the unload cycle and to prod a few cakes which may be reluctant to drop freely.

In practice, there are essentially two operating pressures to choose from in the recessed plate pressure filter design. There are the low-pressure designs that employ 80 to 100 psi sludge pumps to fill the filter cavities; there are high-pressure types that use 150 to 225 psi for the same purpose.

The low-pressure filters employ an air actuated diaphragm type pressure pump that serves to transport the sludge slurry from a sludge holding tank to the filter, as well as supply the pressure for sludge dewatering. The simplicity of this design is ideal for small to medium sized sludge generators. The solids content of the

TABLE 8. SLUDGE VOLUME AND DISPOSAL COST PER SHIFT AT 20% SOLIDS (wt.)

Type of Plating	NaOH TMT (lb)	Disposal At 0.10/lb	Ca(OH)$_2$ TMT (lb)	Disposal At 0.10/lb
Acid Cu	27.4	$2.74	44.7	$4.47
Cyanide Cu	10.4	1.04	14.6	1.46
Nickel	41.4	4.14	52.7	5.27
Chromium	64.1	6.41	92.1	9.21
Acid Zn	25.4	2.54	36.7	3.67
Cyanide Zn	8.8	0.88	12.7	1.27
Cadmium	8.7	0.87	12.7	1.27

sludge produced will usually be in the 25 to 35% range when the filter is operated properly.

Low-pressure filters are generally used with plates in the 1- to 2-ft^2 range. Because plates in this range are relatively light in weight, a single operator can shift the plates with little exertion. Automated plate shifting seldom justifies the additional expense, unless the filter has a large number of plates and quick dump cycles are desired. The small manual, low-pressure filters have few moving parts, are easy to maintain, and are competitive in price with alternative dewatering methods as well as producing markedly drier sludge. Only high-pressure filtration will produce drier sludge cakes. A few manufacturers have recognized this fact and supply a low-pressure press that can be converted to high pressure if the need arises. The need might arise in those locations that legislated landfill ordinances banning sludge of less than x% solids. In a case where the limit was 40% solids, numerous plants in the area were caught with centrifuges, vacuum filters, and low-pressure filter presses that could not produce an acceptable sludge on a regular basis. A filter capable of upgrading to meet new demands would have been ideal in this situation.

Apart from the sludge pumping and feeding arrangements, there is not a great difference between low-and high-pressure filters. The side rails, end plate, and ram assembly are of more durable construction in high-pressure applications, but the plates are often identical. The major difference in high-pressure filters is the pressure pump itself. Progressive cavity and multiple piston pumps are commonly used in conjunction with a pressure accumulator tank. Pressure regulating switches on the accumulator tank control the operating pressure and the cycle of the pressure pump. By introducing a timer to monitor the duration of "off" time, the filtration cycle can be automated. Thus, for example, if it has been determined that when the feed pump remains off for 1 min the sludge has 35% solids, this time can be used to indicate that the filter cycle is finished. If a drier sludge is desired, then a longer "off" time or a higher pressure may be needed. Sludges with 40 to 50% solids are not uncommon.

While high-pressure recessed plate filters are more expensive than the lower pressure models, it is important to consider sludge hauling costs when evaluating them. It may be that the cost of transporting and disposing of the added water in the low-pressure sludge cake may justify application of higher pressures. Comparison of Table 8 with Table 9 will illustrate this point.

MAINTENANCE

In closing this chapter, the single most important factor in wastewater treatment—maintenance—must be addressed. Despite the professional integrity of the design, regardless of the materials of construction, and the quality of the

TABLE 9. SLUDGE VOLUME AND DISPOSAL COST PER SHIFT AT 40% SOLIDS (wt.)

Type of Plating	NaOH TMT (lb)	Disposal At 0.10/lb	Ca(OH)$_2$ TMT (lb)	Disposal At 0.10/lb
Acid Cu	14.0	$1.40	22.7	$2.27
Cyanide Cu	5.5	0.55	7.3	0.73
Nickel	20.7	2.07	26.7	2.67
Chromium	32.0	3.20	46.0	4.60
Acid Zn	12.7	1.27	18.7	1.87
Cyanide Zn	4.0	0.40	6.0	0.60
Cadmium	4.7	0.47	6.7	0.67

components, no wastewater treatment system can function indefinitely without a proper maintenance program. The first step toward such a program is the preparation of a comprehensive operating manual. Every technical operating parameter and detail should be fully described with explicit instructions for troubleshooting malfunctions.

Each pump, mixer, motor, gear drive, etc. must be scheduled for lubrication with the correct oil and grease specified for each component. Instrumentation must be calibrated routinely and probe assemblies cleaned regularly. Although not every contingency can be anticipated, effort must be made to cover as much as possible. A wastewater system, after all, is simply an organized assortment of mechanical, electromechanical, and electronic devices.

Management must assume the responsibility for encouraging operator concern and interest in the system. Neglect will degrade a system rapidly while conscientious operation and maintenance can sustain specification performance year after year. Without thoughtless dumps and spills, there should be no "bad days." "Bad days" is a term excusing a multitude of sins. Invariably, the sins relate to indifference, carelessness, neglect, and ignorance. When these conditions are corrected, any wastewater system will function as it was designed to perform.

Bibliography

The published literature on waste disposal and waste treatment is so extensive, and proliferating so rapidly, that preparation of a meaningful bibliography is almost impossible. It would not come close to being current, and a critical review of the literature could be a major undertaking.

The following sources are suggested for those seeking additional information:

The American Electroplaters Society, 1201 Louisiana Ave., Winter Park, Fla. 32789. 1. Reports on AES Research Projects; 2. Proceedings of AES/EPA Seminars.

Suppliers: Suppliers of specific types of equipment often have selected bibliographies available on their equipment.

Monthly journals: Consulting current copies, or the most recent annual index for the following journals is suggested:

Plating and Surface Finishing, American Electroplaters Society

Products Finishing, Gardner Publications, Inc., 6600 Clough Pike, Cincinnati, Ohio 45244

Metal Finishing, Metals and Plastics Publications, One University Plaza, Hackensack, N.J. 07601

Industrial Finishing, Hitchcock Publishing Co., Hitchcock Bldg., Wheaton, Ill. 60187.

References

1. Kushner, J. B., *Metal Finishing*, **47**, No. 12, 52–58 (1949).
2. Nordell, E., "Water Treatment," New York, Reinhold Publishing Corp., 1951; 2nd ed. 1961.
3. Shaw, G. V., and Loomis, A. W., "Cameron Hydraulic Data," Ingersoll-Rand, 1951.
4. Smith, J. B., *Metal Finishing*, **51**, 2, 69–72, 74 (1953), **51**, 3, 65–69 (1953).
5. "Betz Handbook of Industrial Water Conditioning," 5th Ed., Betz Laboratories, Inc., 1957.
6. Milne, D., and Foulke, D. G., "Water—A Global Problem for Electroplaters," *Plating*, **44**, No. 8, 859–863 (1957).
7. Engel, E., "Notes on the Use of Water for Cleaning and Electroplating," *Products Finishing*, **20**, No. 1, 38–50 (1956).
8. Doria, A., "Water and Waste," *Plating*, **43**, No. 8, 975 (1956).
9. Fair, G. M., and Geyer, J. C., "Water Supply and Waste Water Disposal," New York, John Wiley & Sons, Inc., 1954.
10. Doria, A., "Water and Waste," *Plating*, **42**, No. 8, 1005 (1955).
11. Kovatis, P. P., "Water! Water!! Water!!!," *Plating*, **42**, No. 5, 463 (1955).
12. Helbig, W. A., "Water Purification With Activated Carbons," *Plating*, **42**, No. 8, 1044–1045 (1955).
13. Korbelak, A., "From Waste to Profits—In Water," *Plating*, **40**, No. 12, 1361 (1953).
14. Bueltman, C., and Mindler, A. B., "Rinse Water Re-Use by Ion Exchange," 42nd Annual Technical Proceedings, Electroplaters' Soc., 83–89 (1955).
15. Spencer, L. F., "Water—Its Use in the Plating Shop," *Metal Finishing*, **56**, No. 7, 44–51 (1958).
16. Keating, R. J., "Ion Exchange—A Practical Tool in the Plating Room," *Metal Finishing*, **52**, No. 12, 52–55 (1954).
17. Smith, J. B., 'Water Impurities and Electroplating," *Metal Finishing*, **51**, No. 2, 69–74; **51**, No. 3, 65–69 (1953).
18. Engel, E., "Notes on the Use of Water for Cleaning and Electroplating," *Products Finishing*, **20**, No. 4, 38–50 (1956).

6. PLATING BATH COMPOSITIONS AND OPERATING CONDITIONS

Lawrence J. Durney

*Vice President for Technology, Frederick Gumm Chemical Co., Inc.
Lyndhurst, NJ*

DISCUSSION OF TABLES

The purpose of this chapter is to present in brief table form information on plating baths, for the most part commonly used in production,* essential for intelligent handling of the engineering problems that may arise in connection with their installation and operation. It is not intended to cover all the metals that can be electrodeposited, nor, for the metals covered, to discuss any of the fundamental theory. Such information is very adequately covered in other texts.[1,2]

Names of proprietary processes and information on their patent status are omitted from the tables, but included in the references. Similarly, readers are referred to the suppliers for information on addition agents and other details of proprietary processes.

The data given in the tables are confined to the chemical composition of typical plating baths, and to their engineering requirements, such as operating temperature, current density, agitation, anodes, ratio of anode-to-cathode area, voltage, filtration and materials of construction. In a few instances purification is mentioned, but detailed information on this subject should be obtained from the references. The uses and applications of the baths are also noted.

Following each set of formulas is a brief outline of the major problems that can be encountered with each bath—a troubleshooting chart to serve as a guide when problems develop. These charts have been developed from published information on both proprietary and non-proprietary baths. They are not intended to be definitive and/or all-inclusive. In the case of proprietary solutions, they may be materially altered and/or changed by the nature and quantity of addition agents. They are offered as helps and guides for preparing more specific and all-inclusive charts for the particular baths in use. As suggested in the chapter on troubleshooting it is strongly recommended that specific, more extensive charts be prepared when any new proprietary bath is placed in operation. The required information should be available from the supplier of the additives for the bath.

Bath ingredients are given by name rather than by chemical formulas. The latter are only given where clarification is considered necessary. For example, copper sulfate, $CuSO_4 \cdot 5H_2O$, is the commercial salt corresponding to the weight or concentration given.

Concentrations are given in both metric and English equivalents. English equivalents are avoirdupois oz/gal except in the case of the noble metals. Noble metal concentration English equivalents are troy oz/gal.

The three variables of temperature, current density, and agitation are interrelated in their

---

*The use of platinum and paladium plating has increased in recent years. However, since these baths are operated with insoluble anodes and are only used in small volume, there is little that need be known about their use from the engineering point of view. They have, therefore, not been included in the tables, and information with respect to bath compositions and plating conditions can be obtained from references 85–89 and 90–96.

effect on a plating operation. It is customary to control all three closely by standardization and instrumentation. Increasing temperature and agitation usually enables one to obtain higher current efficiencies and to use higher current densities. It is good practice to operate at the highest values of temperature and current density consistent with the limitation imposed by the quality of deposit required, or by the equipment and materials of construction. One thus obtains the maximum production from the available facilities.

The operating temperatures in the tables represent the best average, if a single value is given, or the usual temperature range, if two values are given. The operating temperature is an important consideration in the selection of suitable linings and protective coatings. It also determines what provision must be made for heating or cooling.

In continuous operation a bath will attain an equilibrium temperature; this is determined by such factors as resistivity, the anode-to-cathode spacing, the amperes per gallon employed, and the heat losses. With these factors fixed, facilities must be provided for (a) cooling if the bath is to be operated below the equilibrium temperature or (b) heating if the bath is to be operated above the equilibrium value. Since the equilibrium temperature varies with the conditions employed, it can only be stated in general terms that a bath operated above 32 C (90 F) usually requires some provision for rapidly heating to the operating temperature, and both heating and cooling may be necessary to control the temperature. Below 32 C (90 F) cooling only may be required, especially when using high current densities with agitation, as commonly practiced with bright acid copper plating processes.

Single values of current density in the tables are good average values for the conditions of temperature and agitation shown. A practical operating range of current density is indicated by a high and low value. It is well to allow for somewhat higher values when estimating current requirements. However, when using these current density values to calculate the plating time to give a deposit of required thickness, or to determine the size of a plating tank to give a specified number of processed parts per hour, other factors must always be considered. In Chapter 22 the effect of racking and current distribution on the metal distribution is discussed. The calculation of the size of a plating tank to produce a definite number of parts per hour with a deposit of a specified thickness is discussed in Chapters 23 and 25.

Agitation may be of two kinds—cathode movement and solution movement. The former is obtained by a cathode rod reciprocating mechanism, a rotating plating barrel, the use of semi- or fully automatic plating equipment and sometimes by movement of the part being plated. Solution movement is sometimes obtained by pumping, with continuous filtration or heat exchangers; sometimes by a separate circulating pump, with motor-driven propeller agitators in the bath; and sometimes with low-pressure air. Frequently both cathode and solution movement are employed. There is a degree of agitation which if exceeded produces no further advantages. Excessive agitation may be detrimental, and any agitation should be as uniformly distributed as possible to avoid nonuniformities in the structure, appearance and thickness of the electrodeposited coating. For this reason, propeller agitation is the least satisfactory and air requires both experience and care in its use for successful results. Further information on agitation is given in Chapter 35.

The values of cathode efficiency given in the tables are only approximate, since the efficiency may vary widely with changes in the bath composition and operating conditions. In practice, one has to determine the metal distribution by trial runs and thickness measurements in order to find the ampere minutes per square foot required to meet a plated thickness specification. While doing this, it is essential to control the operating conditions and plating bath composition at specified values.

The various types of anodes to be used with the different plating baths are mentioned in the tables. Soluble and insoluble anodes of various types and compositions are discussed in Chapter 29. Most modern plating processes and specifications require the use of high purity soluble anodes. The ratio of anode-to-cathode area is also important in determining the anode current density. Too high an anode current density may cause objectionable polarization when using soluble anodes with resultant loss in anode efficiency, formation of solid particles, or sludge, rough deposits and adverse changes in the composition of the plating bath.

The use of anode bags to prevent solid particles from getting into the plating bath and causing rough deposits is common practice in the deposition of certain metals. Anode bags are mentioned in the tables. Both anode bags and the use of diaphragms in plating tanks are discussed in Chapters 19 and 29 respectively.

The bath voltages given in the tables are only to indicate what voltage direct current source is required for the baths and operating conditions given. The actual voltage at the plating tank cannot be given since it depends on such factors as the bath composition, area of anodes and cathodes, anode-to-cathode spacing, temperature, degree of agitation and amperes per gallon of solution. The control of the tank voltage is discussed in Chapters 30 and 35.

In barrel plating the concentration of the plating bath is frequently greater than for still or automatic plating. This applies chiefly to the cyanide baths which are most commonly used in barrel plating because of their better throwing power. Higher voltages are usually necessary for such applications, i.e., 9, 12, and sometimes, 15 volts.

Most plating baths require purification and filtration when first prepared. Some processes require continuous filtration and many require periodic purification. This is especially true of the bright plating processes. Filters are therefore essential in any modern plating installation, especially when plating to quality specifications. The subject of filtration is discussed in Chapter 34.

The purification methods generally employed are outlined below:

(1) Filtration through filter aid.

Purpose: To remove solid particles.

Conditions:

(a) Precoat filter with about 2 oz filter aid per ft^2 of filter area. See also Chapter 34.

(b) Average rate for batch filtration recirculating back into the same tank is not critical, but filtration should continue until about ten times the solution volume has passed through the filter. Agitation of the solution during filtration is necessary for best results.

(c) Average rate for continuous filtration should be sufficient to turn over the solution volume once per hour.

(2) Carbon treatment and filtration through filter aid.

Purpose: To remove solid particles and most organic contamination.

Conditions:

(a) Carbon dose for batch treatment is 1 to 4 lb activated carbon per 100 gal of solution, depending on degree of contamination.

(b) Carbon dose for continuous treatment is ½ to 3 lb activated carbon per 100 gal of solution, but is also limited by filter area available, as the carbon should be limited to about 3 oz/ft^2 of filter area.

(c) Precoat filter with 2 oz/ft^2 filter aid, then apply the carbon, together with half its weight of filter aid.

(3) Electrolytic purification.[8]

Purpose: To remove dissolved metallic contamination.

Conditions:

(a) Low current density is used, from 0.05 to 0.8 A/dm^2 (0.5 to 8 asf) depending on the metal to be removed. The cathodes are frequently in the form of corrugated sheet or louvers, either of which if operated at an average current density of about 0.5 A/dm^2 (5 asf) will give actual current densities covering the desired range.

(b) Batch dummying in the plating tank is used to remove relatively large amounts of metallic impurities; this can usually be accomplished by passing from ½ to 5 amp-hr/gal of solution. Dummy cathodes must be removed as soon as the current is shut off to avoid redissolution of the impurities. The solution should be agitated and filtered continuously during purification to remove any loose metallic particles which may fall from the cathodes.

(c) Continuous electrolytic purification is used to maintain a low concentration of metallic impurities. A continuous current of from 0.01 to 0.1 amp/gal of solution is usually adequate for this purpose. At the specified average current density on the dummy cathodes of 5 asf, there will thus be required 0.2 to 2.0 ft^2 of cathode per 100 gal of solution. A separate tank is used for continuous purification (see Chapter 35). The solution is circulated through this tank and passed through a filter before returning to the plating tank to remove any loose metallic particles that may fall off the dummy cathodes.

(d) Agitation is advantageous, not only because it brings all the solution in contact with the cathode, but it also assists in removal of most impurities by lowering their deposition potentials which occurs with agitation.

(e) In a few isolated cases, impurities are best removed by medium to high current density dummying. Such cases will be readily evident from consulting the troubleshooting charts. Where such instances occur, continuous treatment results from the actual plating operation and normally keeps the problem under control. In some cases, however, a periodic supplemental treatment may be required. Where this is required, follow the procedures outlined in (a) and (b) above, but increase the current densities recommended in (a) and (b) by the necessary factor (required average current density from the troubleshooting chart divided by 5).

(4) Chemical purification.

Purpose: To remove dissolved ions of specific impurities.

Conditions: These methods are specific for each bath, but the most commonly used are:

(a) Displacement, e.g., zinc dust is added to a zinc bath to precipitate the nobler metals, which are filtered off.

(b) Sodium sulfide is used to precipitate zinc and lead from cyanide baths.

(c) High pH precipitation of heavy metals from a nickel bath. Since this method is widely used, it will be described in some detail. As the pH of a solution containing heavy metal ions is increased, the hydroxides of these metals start to precipitate, each at some critical pH value. At a fairly high pH, say 5.2 electrometric, heavy metal impurities may exist partly as the soluble salt and partly as flocculated or colloidally dispersed hydroxides. Because heat accelerates flocculation of the colloidal particles, purification is preferably carried out at elevated temperatures. In addition, at this pH iron is more completely precipitated in the ferric condition than the ferrous, thus it is advantageous to add hydrogen peroxide to oxidize the iron. Treatment with activated carbon to remove organic impurities is also desirable, although wetting agents and many organic brightening agents will be removed at the same time. If the concentration of organic impurities or organic addition agents is high, it is sometimes advantageous to carbon treat prior to the addition of hydrogen peroxide since the latter will react with most organic matter. In any case, all traces of hydrogen peroxide must be removed by heating before replacing wetting agents and brighteners, otherwise objectionable oxidation products may be formed.

Under no condition should the purification be carried out in the plating tank itself; a suitably lined purification tank of the same volume must be provided, with means for heating and agitation. Because traces of activated carbon in the plating solution will cause roughness, it is preferable to locate this tank at some distance from the plating tank, or in a separate enclosure. Many deviations from the details of the procedure given below are possible, but the cycle described is preferred for best results:

(1) Heat the solution to about 65 C (150 F).

(s) Add sufficient nickel carbonate to raise the pH to 5.0 to 5.3, measured after stirring for about 30 min. Use of nickel carbonate at the rate of 250 to 500 g/100 liters (2 to 4 lb/100 gal) will usually suffice. Lime is sometimes used to attain higher pH values, but this is not generally recommended because of the introduction of the undesirable calcium ion.

(3) Add 0.1 to 1.0 gal of 30% by weight (100 vol) hydrogen peroxide per 100 gal of solution, depending on degree of contamination. Hold at temperature with agitation for 1 hr or more.

(4) Add 1 to 4 lb activated carbon per 100 gal of solution, and continue to agitate for 1 hr or more. Sometimes the carbon and peroxide are added concurrently to save time, but since hydrogen peroxide is decomposed catalytically by carbon, this is not considered the best practice.

(5) Add $\frac{1}{4}$ to 1 lb filter aid per 100 gal of solution, continue to agitate and transfer the solution to the clean plating tank through a filter precoated with a filter aid. In handling large volumes of solution, where the total load of carbon, filter aid, and precipitates is such that the filter would have to be cleaned several times during the transfer, some labor may be saved if the solution is allowed to settle 6 hr or more after adding the filter aid. In this case, the suction hose intake should initially be placed in the relatively clear upper layer of solution and slowly lowered as filtration progresses.

The plating bath composition and the conditions recommended for its operation are important factors in determining the materials of construction to be used. The choice is further limited by the necessity of considering the effect of impurities which may be dissolved or extracted by the plating bath. As little as 5 to 50 ppm of certain "foreign" metals or organic contamination from some rubber or plastic lining may cause serious plating defects and adversely affect the properties of electrodeposited coatings. Lead, for example, may be used with many plating baths as a corrosion-resistant material, but from a compatibility viewpoint, the small amount of lead that will dissolve in a bath frequently restricts its use in plating equipment. For similar reasons, it is not sufficient to specify just any rubber or plastic for a lining. Only "approved" formulations can safely be used; this term implies that compatibility has been established by a suitable test.

Materials of construction commonly used for electroplating equipment are listed with key numbers in Table 1. The key numbers have been

TABLE 1. MATERIALS OF CONSTRUCTION
(Limited to materials coming in direct contact with the electroplating bath, but excluding anodes, electrical conductors, anode bags and diaphragms)

Key No.	Material	Uses
1	Steel, low carbon	Tanks, filters, pumps, pipe, fittings, heating coils
2	Cast iron	Pumps, filters, valves, fittings
3	Stainless steel[a]	Tanks, pumps, filters
4	High silicon cast iron ("Duriron")	Pumps, pipe, fittings, heat exchanges
5	Lead, usually 6% antimony alloy	Linings, pipe
6	Copper	Heating coils
7	Nickel	Heating coils
8	Carbon ("Karbate")	Heaters and heat exchangers primarily, pumps, air diffusers
9	Glass ("Pyrex" or tempered)	Heat exchangers, tanks, pumps
10	Chemical stoneware	Tanks, tower concentrators and tower packing
11	"Haveg"	Tanks, pipe
12	Hard rubber[b]	Pipe, fittings, pumps
13	Rubber (approved compositions)[b]	Linings, hose
14	Plastics (approved compositions)[b]	Linings, hose, pipe, fittings, barrels, heating coils and exchangers
15	Acid resisting brick	Linings
16	Wood	Tanks

[a] Limited to special alloys.

[b] Compatability tests are necessary in the absence of data on bath contamination and effect on deposits.

used in the other tables to designate materials which are suitable for specific plating baths. If steel or cast iron is satisfactory, more expensive and possibly less available materials have not always been indicated. A certain degree of judgment must always be exercised when deciding what materials of those recommended should be considered for specific items, such as pumps, pipe fittings and linings, even though general information on uses is given in Table 1. Unless experience has indicated the choice of materials, it is best to check any uncertainties which might arise by consulting with suppliers.

Under the heading "Uses" in the tables, the purpose for which a bath may be used is stated, such as for a decorative, protective or rust preventative coating, for coloring or for dull, semi-bright or bright deposits. The usual coating thicknesses for these uses are also given. In some instances, the thickness value corresponds to the upper practical value for satisfactory deposits from a given bath.

Under the heading "Applications," an attempt has been made to indicate the type of basis metal to which the plated coating may be applied, the kind of manufactured item for which the plating bath is commonly used, and how the coating is used, i.e., as an undercoat or otherwise. Engineering applications of plated coatings have also been mentioned. The above distinction between uses and applications is unfortunately not without some inconsistencies.

The references at the end of this chapter are classified according to the metals and types of plating baths. The number assigned each reference is the same as that used in the tables under the heading "References." Both literature and proprietary references are cited, but no attempt has been made to prepare a complete bibliography. Instead, the objective has been to give the most pertinent references, including important engineering applications, from which the reader should be able to obtain additional information.

The necessity for exhausting fumes in connection with the operation of any of the plating baths mentioned in the tables is not indicated. Chromic acid plating baths are the only acid-type plating processes requiring tank ventilation and most large chromium installations are equipped with water spray washers for exhaust gases. Of the alkaline-type plating processes, most all cyanide plating baths require an exhaust system. With the latter baths, the temperature, current density, and efficiencies of the process are contributing factors. A low temperature, high efficiency silver cyanide bath is commonly operated without such a system. Also, the high temperature, high efficiency, high concentration copper cyanide baths may be operated without ventilation, but such facilities are frequently provided to remove the water vapor from the working environment. Ventilation and plating room hazards are discussed elsewhere in this book.

EFFECTS OF HYDROGEN

Embrittlement of metals by hydrogen has been recognized for many years but, as two authors have pointed out[124], some other defects in electrodeposits, such as blistering, cracking, gas pits, peeling and poor adhesion, may also be related to hydrogen in ways not yet identified. Their conclusions, quoted below, have added greatly to our understanding of these matters.

"1. Many electroplating processes favor the absorption of large quantities of hydrogen by metals. Embrittlement of the metals may be due to hydride formation, to strain imposed by occlusion of molecular hydrogen in submicroscopic rifts and to deposition of hydrated ions— phenomena that are already recognized by electroplaters.

"2. Less recognized, perhaps, are the functions of hydrogen occluded within the steel. The extremely low solubility of steel for hydrogen at ordinary temperature, combined with the extraordinary ability of this metal to absorb huge quantities of the gas when presented atomically at the surface during pickling and cathodic electrolysis, causes important effusion of this gas from the steel when the atomic layer is removed by the presence of an applied coating or by cessation of the hydrogen-producing process.

"3. When the effusion occurs beneath a coating of any material, including metals, whose permeability is unsuited to the quantity of the effusion, the coating may be (1) lifted from the steel or ruptured; or both, by the pressure of the accumulating gas, or (2) blistered, depending upon the qualities of the coating.

"4. If the plated ware is heated, the effusion of hydrogen is accelerated and the coating is weakened, so that occurrence of the above defects is favored.

"5. Effusion of the gas from the steel base during electroplating leads to the formation of gas pits in the coating.

"6. Cathodic cleaning in either acid or alkaline solutions provides large quantities of hydrogen for absorption."

According to Faust[125] the minimum amount of hydrogen that may harm the physical properties of steel is dependent on the type, physical condition and past history of the steel. Also, this minimum amount may be called a threshold quantity, which must be exceeded before objectionable effects appear. The preplating cleaning and pickling must preferably be carried out so as to reduce hydrogen absorption. Even then, further hydrogen may be absorbed during plating and, if the threshold value of hydrogen is exceeded, embrittlement of the steel basis metal and defects associated with hydrogen absorptions such as blistering, pitting and poor adhesion may result.

With respect to the removal of hydrogen from metals, Zapffe and Faust[124] point out that the behavior of hydrogen in steel is complex, and gas cannot be successfully removed by any one broadly defined treatment. The future service largely determines the preliminary treatment necessary to prevent the occurrence of hydrogen-caused defects. For an item that is to be subjected to an elevated temperature in service, especial treatment must be applied prior to electroplating to remove the fraction of hydrogen that would appear at the temperature in question, and care must again be taken to prevent subsequent absorption. The solubility curves of hydrogen in nonferrous metal may serve as a guide in heat treating electrodeposits for relieving embrittlement as a result of absorbed hydrogen. The references listed in the above reference will be of value to those interested in this subject.

Hydrogen embrittlement is frequently experienced when plating on high-carbon, high-alloy or stainless steels from copper, zinc or cadmium cyanide baths. Baking the plated parts for 1 to 23 hr at 150 to 190 C (300 to 375 F) is commonly recommended as a means of eliminating embrittlement.

According to Valentine[126], however, this aftertreatment was only successful for lockwashers having a hardness of Rockwell C 47 to 49. For hardnesses of Rockwell C 50 to 56, heating at a subcritical temperature prior to the hardening and tempering heat treatment was beneficial, in addition to low temperature baking after plating. Ritzenthaler also reported difficulties when plating spring-type metals with hardness above Rockwell C 45. These difficulties frequently can be eliminated by baking at 150 to 190 C (300 to 375 F). Baking should be done again after plating.

Other work has indicated that certain high tensile steels may require temperatures as high as 260°C (500 F) with times of up to 8 hr for proper elimination of absorbed hydrogen.

Cadmium and zinc coatings which require chromate coatings, must be baked before application of the chromate, since the high temperatures would dehydrate the chromate and destroy its effectiveness. If parts are handled carefully to avoid contamination during baking, only a light activation is required before the application of the chromate.

These and other examples could be cited to support the argument that not infrequently the plating operation does not provide the hydrogen which causes embrittlement. It is certain that in some cases plating only provides a barrier coating which prevents the removal, naturally or by baking, of hydrogen introduced in some preceding operation. Where the hydrogen concentration of the as-received steel is high, slight additions in cleaning, pickling or plating may bring the total concentration into the critical range for embrittlement. It is apparent, therefore, that an operation which may be safe on some occasions may cause trouble on others because of differences in residual hydrogen concentration in the workpieces. Baking of suspicious material will usually be effective in bringing residual hydrogen to a safe level prior to plating.

Another manifestation of the effect of hydrogen may be the commonly observed fact that a plated steel sheet will frequently break in forming at a much lower degree of deformation than the unplated steel. This applies to copper, nickel, zinc and chromium (and possibly other metals) even when these metals are deposited from acid-plating baths. Plating with chromium has a great tendency to impart brittleness to certain alloy steels and even to nickel electrodeposits because of the large amount of hydrogen associated with its electrodeposition.

Although hydrogen embrittlement may be revealed by the behavior described above, in many instances failure of the plated part does not occur immediately upon bending or loading. Instead, it may be delayed for days or even weeks. Temperature is a factor also, and, in opposition to the usual effect of temperature on brittle-ductile transitions in metals, the higher the temperature the shorter will be the time to failure. The fundamental reasons for this behavior have been summarized by Read[127]. Because of these factors a bend test or any other quick-fracture test often does not reveal serious damage for certain types of load applications, particularly when workpieces are highly stressed. There is no satisfactory fast routine test for damage of this kind; only a time-consuming, expensive stress-rupture test of some sort will serve.

The nuisance, cost, and uncertainties of baking as a means of avoiding hydrogen embrittlement have led to many attempts to devise plating processes which do not cause embrittlement. Cadmium plating processes have been a favorite target, and success has been claimed by several investigators. There is not much on which to base a selection of one or another process either for cadmium or for other metals, but some of those which have been well described in the literature have been summarized so far as claims and results are concerned[128].

STRIPPING AND SALVAGING OF DEFECTIVE PLATED ITEMS

Stripping of deposits is an effective method of determining the average coating thickness by a measurement of the weight difference between the plated and stripped part. However, the improvements in equipment and precision for spot measurements have made the procedure almost obsolete.

A great deal of stripping, however, is carried out either for maintenance purposes, or to salvage defective parts.

Maintenance procedures usually involve stripping of racks and barrel contacts and danglers. With the increased use of stainless steel or titanium for the contacts of plastisol coated racks and for barrel components, the use of accelerated nitric acid strippers for this purpose has become common. These involve the use of 50 to 90% by volume of 42° Be' nitric acid along with a suitable proprietary accelerator which sometimes incorporates fume control. Stripping rates can be accelerated to the point where it is sometimes necessary to provide cooling or to limit the survace area per gallon processed in order to hold the temperature down.

More recently, the increased restrictions on air pollution, and the problems of disposing of the metal saturated nitric acid has led to the increased use of electrolytic strippers, many of which are proprietary.

Where racks and contacts are of copper alloys, the usual strippers for these alloys are used for maintenance stripping.

Salvaging of defective plated parts is generally expensive, and the quality of the replated part frequently is affected by the process of stripping and replating. The decision to strip and replate is therefore affected by whether it is a captive or job plating shop.

To justify stripping and replating, one or more of the following criteria must be met:

1. The cost of stripping and replating is less than the cost of scrapping and replacing the defective parts;

2. The parts are from a limited run and would require extensive setup costs to produce replacements;
3. The parts are needed to satisfy a production requirement which cannot be satisfied by 1 or 2 above.

Obviously, these alternatives are not as easily available to job shops as to captive shops. Job shops therefore are more likely to be involved in stripping for salvage purposes.

Copper-nickel-chrome plated zinc-base die castings are frequently scrapped when defective, and remelted without stripping the deposit. The nickel and chromium are removed in the dross and the small amount of copper has no significant effect on the zinc alloy composition. It is generally conceded that repolishing and buffing a zinc-base die casting after stripping the deposit may cut below the chilled skin surface of the casting, resulting in a more porous and less satisfactory surface for plating.

The development of both electrolytic and chemical strippers, based on electrolytic and chemical polishing technology, has greatly alleviated this condition by reducing the amount of mechanical rework required after stripping. These solutions are usually available as proprietary materials, although a few formulations may be found in the expired patent literature.

When defects are minor, copper and/or nickel chromium plated steel items are frequently replated without stripping the nickel. Similarly, large items, such as automobile bumpers, are repolished as required and replated without removing the remaining nickel. Special attention must be given to the preplating cycle to insure that the unstripped nickel is sufficiently activated to insure good adhesion of the new deposit. A Wood's nickel strike consisting of 30 to 60 g/L (4 to 8 oz/gal) of nickel in 10 to 15% by volume hydrochloric acid is quite effective for this purpose. The strike is operated at room temperature with nickel anodes at 3 to 10 A/dm^2 (30 to 100 ASF) for 15 sec to 1 min. Parts may be transferred directly to the nickel bath without rinsing. If a rinse is used, the time should be kept as short as possible to avoid passivation of the strike layer. Because of the difference in the anode and cathode efficiency, nickel tends to build up in the bath. Periodic withdrawal of solution must be made to maintain the nickel content below 75 g/L (10 oz/gal) and hydrochloric acid added to maintain the free acidity. If the defects are major, the parts may be sold as scrap steel, or stripped in a proprietary nickel stripper, either by immersion or electrolytically.

Barrel plated small parts are usually stripped in a proprietary immersion nickel stripper and reprocessed.

Chromium may be stripped easily in 10% by volume or stronger hydrochloric acid. It may also be stripped by making it the anode in an alkaline cleaning solution, or a solution of sodium hydroxide. Hydrochloric acid generally leaves the nickel sufficiently active so that replating of the chromium poses no problems. Reverse current stripping generally passivates the nickel so that a good activation step is required before rechroming. Suitable methods include the use of a strong hydrochloric acid dip, cathodic treatment in an electrocleaner followed by a strong acid dip, or cathodic treatment in sulfuric acid. There are also very effective proprietary activating compounds available, some of which can be used with current, others by immersion.

Heavy deposits of chromium are usually stripped in a reverse current alkali. Solutions of sodium hydroxide, well-buffered with sodium carbonate or metasilicate may be used. Proprietary high current density electrocleaners or alkaline descalers are also effective. Agitation of parts and/or solution is recommended since depleted solution containing dissolved chromium may drop markedly in pH and if allowed to collect in pockets or blind holes can cause severe etching of the steel substrate.

Immersion coatings of copper are used extensively as an aid in the drawing of steel. This thin layer can be removed by tumbling in specially designed barrel finishing compounds which will also bring up a lustrous finish. For larger parts, special electrocleaners which also strip the copper have been developed, so the stripping operation becomes part of the preplating cycle. Alternatively, one of the methods listed below may be used.

Heavy layers of copper are used as stopoffs in selective carburizing for surface hardness. These layers can be removed by the same methods used for defectively plated parts. Many of the methods previously used for the removal of copper have been outmoded. These include the lime sulfur/sodium cyanide method; stripping in chromic acid solutions; and some of the less effective electrolytic methods. The more common methods involve proprietary compounds, and include both cyanide-containing and cyanide-free immersion strippers, and electrolytic methods. A non-proprietary electrolytic stripper consists of copper cyanide, 5.5 oz/gal; sodium cyanide, 11.0 oz/gal; trisodium phosphate, 8 oz/gal; and sodium hydroxide to give pH of 12.2. An anode current density of 20 asf and a temperature of 180 F are satisfactory. Additions of 6.5 oz/gal of sodium sulfite and cyclohexylamine prolong the life of the bath. Agitation is beneficial, increasing the stripping rate. The baths may be regenerated by the removal of carbonate and the addition of cyanide.

A true bronze consisting of a plated copper-tin alloy is similarly used as a stopoff for nitriding. These coatings can be removed by one of the proprietary strippers listed above or by one of the following methods.

1. Treat anodically in a solution of 12.2 oz/gal sodium cyanide and 2.7 oz/gal sodium hydroxide at room temperature using 6 volts.
2. Immerse in solution of 13.4 oz/gal sodium hydroxide, 2.7 oz/gal sodium cyanide, and 2.7 oz/gal sodium chlorite at a temperature of 80-90 C (175-195 F). Note that sodium chlorite is a hazardous chemical and instructions should be followed implicitly.

Lead and lead alloys have been successfully stripped from steel as follows[133]:

1. Anodically treating at 20 to 40 asf and 82 C (180 F) in a solution consisting of sodium hydroxide 13 oz/gal, sodium metasilicate 10 oz/gal and Rochelle salts 6.7 oz/gal.
2. Anodically treating at 20 to 200 asf and 20-82 C (70-180 F) in a sodium nitrate solution of 67 oz/gal at a pH of 6 to 10.
3. Immersion in a solution of acetic acid 10 to 85% by volume and hydrogen peroxide (100 vol) at 5% by volume.

Proprietary strippers for lead are also available.

Silver-plated household items of pewter, britannia metal, steel or copper alloys are successfully stripped, repaired if necessary, and refinished, but this can only be justified on the basis of sentimental value rather than cost. The manufacturer of silver-plated ware, however, can partly justify the cost of salvaging defective items on the basis of the value of the silver recovered. Conditions for stripping silver deposits are as follows:

1. Reverse current in a sodium cyanide solution (4 to 8 oz/gal) at room temperature and an anode current density of 10 to 20 asf.
2. Same as (1) but at high pH, i.e., with an addition of 4 to 8 oz/gal of sodium hydroxide to the stripping bath.
3. Immersion in an acid mixture consisting of 95 and 5% by volume of concentrated sulfuric and nitric acids, respectively, and operated at about 80 C (175 F). This bath should be covered when not in use to prevent dilution by moisture from the air; parts should be dry when immersed, otherwise the basis metal attack will be excessive.

Procedure 1 is used for lead alloys, such as pewter and britannia metal, and for steel items. These basis metals are attacked and buffing or refinishing is necessary prior to replating. Procedure 2 is preferred for steel since the high pH eliminates the attack on the basis metal and the item can usually be replated without refinishing the surface. Procedure 3 is used for basis metals of copper, brass and nickel brass, but in all cases the surface is etched and requires refinishing before replating.

The cyanide-containing strippers are effective for removing silver from steel, and are also available in inhibited forms for stripping silver from copper alloys and nickel. These are of particular use in the electronics industry since they generally will also strip gold. Gold may be stripped by most of the methods suitable for silver. In addition, stainless steel or titanium racks may be stripped with some of the accelerated nitric acid strippers. The present high cost of gold has made stripping of racks an economic necessity. In fact, it has even become desirable to strip the copper wires so widely used for stringing parts in the jewelry business. The cost of recovering the gold from the stripping solution is far less than the reduction in scrap value of the unstripped copper wires.

Zinc or cadmium plated steel parts can be stripped without injuring the surface finish by using a weak sulfuric or hydrochloric acid dip with an inhibitor. Salvaging and replating of such parts are relatively quick and simple operations.

Zinc may also be stripped readily in a highly alkaline reverse current electrocleaner, particularly if it is "chelated." There are also available combined zinc stripper/cleaners for stripping parts and racks in the preplating line. Unless the cleaning cycle is relatively long, however, the use of these materials is generally limited to thin deposits.

Tin plate can often be stripped in the same type of solution used for lead. For stripping stainless steel or titanium rack tips, or aluminum, 25% nitric acid with a proprietary accelerator may be used. This same solution is recommended for stripping solder plate (lead/tin alloy) from racks, and rates as high as 0.5 mm (0.022 in)/hr are claimed.

INDEX TO TABLES

Plating Bath	Table No.	Plating Bath	Table No.
Cadmium	3, 8	Black Nickel	17
Copper	4, 5, 7	Nickel	18, 19
Copper-Cyanide	10	Rhodium	2
Copper-Zinc	9	Silver	20, 21
Chromium	12	Tin	25, 26
Gold	11	Tin-Copper	22
Indium	13	Tin-Nickel	23
Iron	14	Tin-Zinc	24
Lead	16	Zinc	6, 28, 29
Lead-Tin	15	Zinc-Cyanide	27

TABLE 2. RHODIUM BATHS[a]

Composition	1	2	3[c]
Rhodium metal[b], g/L (dwt/gal)	2 (5)	2 (5)	
Rhodium metal[b], g/L (oz/gal)			10–20 (1.3–2.7)
Sulfuric acid, c.p., %/vol	2.0		2.5
Phosphoric acid, 85%, %/vol		2.0	
Conditions:			
Temperature, C (F)	40–45 (104–113)	40–45 (104–113)	50 (122)
Current density, A/dm² (asf)	1–10 (10–100)	1–10 (10–100)	1–2 (10–20)
Agitation	as required	as required	vigorous
Cathode efficiency, (15 asf), %	80	80	85
Ratio area platinum anode to cathode	1:1	1:1	1:1
Nickel undercoat recommended except for gold or platinum group metals			
Volts:	6	6	6

Suitable materials of construction: 8, 9, 12, 13, 14
Use: For thin bright decorative and tarnish-resistant coatings (0.001–0.006 mil). Heavier coatings for high reflectivity (78% incident light) and for scratch and corrosion resistance (0.06 to over 1.0 mil).
Applications: For jewelry, silverware, reflectors, electrical contact surfaces and electronic applications.
References: 2, 71–80   2, 71–80   78, 82, 83, 84

[a] Patented proprietary additions for crack-free, stress-free deposits are available.
[b] Metal is added as required in form of concentrate.
[c] For heavy deposits.

Table 3. Cadmium-Cyanide Bath

Composition	g/L	oz/gal
Cadmium oxide	30.0	4.0
Sodium cyanide	100.0	13.1
or	or	or
(Cadmium cyanide)	40.0	(5.1)
(Sodium cyanide)	75.0	(10.1)
Sodium hydroxide	20.0	2.5
By analysis:		
Cadmium, Cd	26.5	3.5
Total cyanide, NaCN	100.0	13.1
Wt. ratio = $\dfrac{\text{Total NaCN}}{\text{Cd}}$		3.74
Conditions:		
Temperature, C (F)	20–35	(70–95)

Current density, 0.5–5 A/dm² 5.0–50 asf
Agitation, cathode   Preferred
Cathode efficiency, %   90–95
Ratio anode to cathode area   2:1
Anodes   Cadium and steel

Filtration: As required
Volts: 6
 (Barrel plating)   6–12
Suitable materials of construction: 1, 2, 13, 14.
Use: Primarily for rust preventive coating.
 Bright deposits when using brighteners[a] and bright dips. Thicknesses 0.15–0.5 mil.
Applications: Over all ferrous basis metals as rust preventive coating. On moving parts where white corrosion products of zinc could be objectionable, especially on communications equipment and instruments.
References: 1, 2, 7, 8.

Low cyanide and/or low caustic can cause poor plating rates and a tendency to blister. High cyanide will improve throwing power, but reduce efficiency. Low cyanide frequently causes anodes to polarize. Cadmium metal tends to climb under normal circumstances. Balance the ratio of steel to cadmium anodes to control metal content. Contamination with copper, zinc, tin, or hexavalent chromium will cause problems with brightness and/or efficiency. Chrome can cause failure to plate or loss of adhesion. Copper can affect salt spray resistance and cause problems in chromating before effects are noticeable in the plate. Excess brightener concentration can cause brittleness and sometimes laminated peeling plate when barrel plating heavy parts.

---
[a]For proprietary and patented bright processes, contact suppliers.

Table 4. Acid Copper Sulfate Bath

Composition	g/L	oz/gal
Copper sulfate, $CuSO_4 \cdot 5H_2O$	195–248	26–33
Sulfuric acid	30–75	4–10
By analysis:		
Copper	40–50	5.2–6.6
Sulfuric acid	30–75	4–10
Conditions:		
Temperature, C (F)	21–49	(70–120)
Current density, A/dm² (asf)	2–10	(20–100)
Agitation, cathode and/or air	Preferred	
Cathode efficiency, %	95–100	
Ratio anode to cathode area	1:1	
Anodes	Copper	

Filtration: Continuous preferred, especially for heavy deposits.
Voltage: Less than 6 volts generally. Higher for some applications.
Suitable materials of construction: 5, 8, 13, 14, 15.
Use: Heavy copper deposits to any required thickness. Also as a bright easily buffed deposit using proprietary and patented processes (sometimes for printed circuits).
Applications: To all ferrous base metals over copper strike. Under nickel-chromium deposits for protective coatings. For electroforming-electrotypes, printing rolls for textiles and rotogravure, other applications.
References: 1, 2, 10, 11, 12, 56.

Problems:
Coarse or powdery deposit: Check for low acid content; excess current; low temperature; low chloride content.
Soft deposit: Check for low acid content; high temperature; low current density.
Hard or brittle deposit: Check for excess acid; low temperature; organic contamination; high iron contamination.
Poor throw: Check for low acid; low brightener.
Low conductivity: Check for low temperature; low acid; low copper.
Polarized anodes: Check for low acid; metallic contamination; high chloride.
Bright acid copper solutions require very carefully controlled chloride concentrations, generally in the range of 60 to 100 ppm, although some brightener systems operate to 180 ppm. Hexavalent chromium contamination can be especially troublesome, causing brightness, efficiency and adhesion problems

TABLE 5. Copper-Fluoborate Baths[a]

Composition	Low Copper g/L	Low Copper oz/gal	High Copper oz/gal	High Copper g/L
Copper fluoborate, $Cu(BF_4)_2$	225.0	30	60	450.0
Fluoboric acid, HBF_4		to pH	to pH	
By analysis:				
Copper	60.0	8	16	120.0
Sp gr, °Be' at 80 F		21–22	37.5–39	
Conditions:				
pH, colorimetric		0.8–1.7	<0.6	
Temperature, C(F)	27–77	(80–170)	(80–170)	27–77
Current density, A/dm^2 (asf)	7.5–12.5 (75–125)		12.5–30.0 (125–300)	
Cathode efficiency, %	95–100		95–100	
Ratio anode to cathode area	1:1		1:1	
Anodes	Copper		Copper	
Filtration:	As required		As required	
Volts:	6		6–12	

Agitation, cathode preferred, also air
Suitable materials of construction: 3, 8, 12, 13, 14.
Use: Heavy copper deposits to any required thickness.
Applications: To all base metals over prior copper strike. Coating steel wire and electroforming generally (sometimes for printed circuits).
References: 1, 8, 13, 14, 56.

Fluoborate copper is generally free of problems if the free acidity is maintained, and organic contamination avoided.

[a]Sold as aqueous concentrate (specific gravity of 1.54 at 68 F) weighing 12.8 lb/gal and containing approximately 25 oz/gal of copper and 6 oz/gal of fluoboric acid and boric acid respectively (Ref. 8 and 9).

TABLE 6. Acid Zinc Bath*

Composition	Sulfate oz/gal	g/L
Zinc sulfate, $ZnSO_4 \cdot 7H_2O$	32–48	240–360
Ammonium chloride, NH_4Cl	2–4	15–30
Aluminum sulfate, $Al_2(SO_4)_3 \cdot 18H_2O$	4	30
or	or	
(Sodium acetate, $NaC_2H_3O_2 \cdot 3H_2O$)	(2)	15
Additions, Optional		
Glucose	16	120
or	or	
(Licorice)	(0.13)	1.0
By analysis:		
Zinc	7.3–11	55–83
pH, colorimetric	3.5–4.2	
Conditions:		
Temperature, C (F)	(75–85)	24–30
Current density, A/dm^2 (asf)	(10–30)	1.0–3.0
Volts:		6
Cathode efficiency, %		95–100
Ratio anode to cathode area		1:1

Anodes, prime wester, intermediate and high purity zinc.
Filtration: Preferred for continuous plating of steel-mill products.
Agitation, cathode and solution, optional but recommended, especially with high current density
Suitable materials of construction: 3, 4, 5, 8.
Use: As a rust preventive coating (0.1–2.0 mil).
Applications: To ferrous basis metals as rust preventive coating for wire, strip, sheet conduit, wire cloth, also on cast iron and malleable iron parts as an undercoat on steel prior to painting.
References: 1, 2, 123.

[a]With good agitation higher current densities may be used.
*Two zinc baths have been developed by steel companies for plating strip and wire at high current densities in the range of 100 to 600 asf and temperatures in the range of 100 to 150 F which contain sodium sulfate and magnesium sulfate or ammonium sulfate, according to private communications. (See Ref. 2, page 453)

Table 7. Copper-Pyrophosphate Bath

Composition	g/L	oz/gal
Copper-pyrophosphate, proprietary compound	345.0	46
Potassium hydroxide	18.0	2.4
Ammonia (29%)	1 ml/L	1.2 fl oz/gal
By analysis:		
Copper	22.5	3.0
Pyrophosphate	157.5	21.0
Ammonia	2.0	0.25
Ratio P_2O_7/Cu		7.0
Conditions:		
pH, electrometric		8.2–9.2
Temperature, C (F)	43–60	(110–140)
Current density, A/dm^2 (asf)	1–7.5	(10–75)
Voltage, volts		2–5 at tank
Agitation, cathode and/or air		
Cathode efficiency, %	95–100	
Ratio anode to cathode area		1:1–2:1
Anodes: Copper, electrolytic or rolled electrolytic (not bagged).		

Filtration: Intermittent or continuous as required. Filter aid and carbon recommended.

Suitable materials of construction: 1, 2, 3, 13, 14.

Use: Undercoat for bright nickel and chromium, carburizing and nitriding stopoff, heavy plate for industrial use.

Applications: To any basis metal over copper strike. Usually as undercoat for bright nickel and chromium for bright nickel and chromium for protective coatings. Also, gears, printing, paper and textile rolls (sometimes for printed circuits).

References: 1, 2, 18, 19, 20.

Except for printed circuit work, pyrophosphate solutions are usually operated with additives to provide a bright deposit. Loss of brightness can be due to brightener imbalance, or low ammonia content. Dullness in the low current density areas only may be due to too high an operating temperature; excess ammonia; or low pyrophosphate content. Burning in the high current density areas may be due to inadequate agitation; low temperature; low pyrophosphate content; low copper content; or excessive current. Hexavalent chromium at low levels may cause patchy dullness or even bare spots. At higher levels, it may stop plating completely. Dull, brittle, or stressed deposits may also be due to organic contamination. The pH of the solution will normally fall due to the loss of ammonia by vaporization. It may be adjusted with potassium hydroxide, or ammonium hydroxide as required to maintain the proper ammonia content. For those unusual cases where the pH rises excessively, small amounts of sulfuric acid may be used for adjustment.

Table 8. Cadmium-Fluoborate Bath[a]

Composition	g/L	oz/gal
Cadmium fluoborate, Cd(BF$_4$)$_2$	240.0	32.2
Fluoboric acid, HBF$_4$		to pH
Boric acid	22.5	3.0
Ammonium fluoborate	60.0	8.0
Licorice	1.0	0.134
By analysis:		
Cadmium	95.0	12.6
pH, colorimetric		3.0–3.5
Conditions:		
Temperature, C (F)	20–38	(70–100)
Current density, A/dm^2	3–6	30–60 asf
Agitation, cathode		Preferred
Cathode, efficiency, %		100
Ratio anode to cathode area		2:1
Anodes		Cadmium

Filtration: As required

Volts: 4–6

(Barrel plating) 6–12

Suitable materials of construction: 3, 8, 12, 13, 14.

Use: For rust preventive coating.

Applications: Over all ferrous basis metals. Of particular value where freedom from hydrogen embrittlement is required.

Reference: 8.

Excess free acid can reduce efficiency and lead to hydrogen embrittlement. Contamination with copper, lead or zinc can affect color and/or corrosion resistance of deposit.

[a] Sold as liquid concentrate, specific gravity 1.60, 13.3 lb/gal, cadmium 42 oz/gal and fluoboric acid 6.4 oz/gal, to be diluted with water as required (Ref. 8, 9).

TABLE 9. COPPER-ZINC ALLOY BATHS

Composition	#1 Brass g/L	oz/gal	#2 Brass g/L	oz/gal	#3 Brass g/L	oz/gal	Brass #4 Hi speed g/L	oz/gal	#5 White Brass g/L	oz/gal	#6 Bronze g/L	oz/gal
Copper cyanide	27.0	3.6	52.0	7.0	42.7	5.7	60.0	8.0	15.7-20.2	2.1-2.7	53.5	7.2
Zinc cyanide	9.0	1.2	30.0	4.0	13.8	1.8	3.75	0.5	34.4-39.7	4.6-5.3	3.8	0.5
Sodium cyanide	56.0	7.5	90.0	12.0	67.7	9.0	99.0	13.2	47.7-53.6	6.4-7.2	62.2	8.3
Sodium carbonate	35.5	4.5	30.0	4.0	15.0	2.0	15.0	2.0	37.5	5.0	30.0	4.0
Rochelle salts			45.0	6.0			45.0	6.0	1.5-2.2	0.2-0.3	45.0	6.0
Ammonium hydroxide	As needed		0.5-1.5% By volume		12.5-25.0	1.5-3.0					1.0-5.0	0.1-0.7
Sodium hydroxide			15.0	2.0			60.0	8.0	30.0-37.5	4.0-5.0		
Sodium bicarbonate												
By analysis												
Copper	20.0	2.5	37.5	5.0	30.0	4.0	42.5	5.7	11-14	1.5-2.0	38.0	5.1
Zinc	5.0	0.65	15.0	2.0	7.7	1.0	2.25	0.3	29-33.5	3.9-4.5	2.1	0.28
Cu/Zn ratio	4.0		2.5		4.0		19.0		0.38-0.44		18.0	
Free cyanide	17.5	2.3	7.5	1.0	7.3	1.0	36.0	4.8	4.7-6.4	0.6-0.85	4.5	0.6
Conditions												
pH	10.3-10.7		10.3-10.7		9.8-10.3		Not given		Not given		10.3	
Temperature	28-50 C	95-120 F	43-60 C	110-140 F	35-44 C	95-110 F	60 C	140 F	21-29 C	70-85 F	38-60 C	100-140 F
Current density	0.5A/dm²	5.0asf	0.5-3.5A/dm²	5-35asf	0.5-1.6A/dm²	5-16asf			4.3A/dm²	40asf	0.5-3.2A/dm²	5-30asf
Anodes Cu/Zinc	80-20		70-30		80-20		100		35-65		95-5	

Filtration: Intermittent or continuous as required. For high speed and bright baths, check suppliers' recommendations.
Voltage: Usually less than 6 volts for rack; 9 to 12 volts for barrels.
Suitable materials of construction: 1, 2, 13, 14 for all baths.
Use: For color mainly. Usual thickness 1 to 15 μm (0.05 to 0.3 mil)
Also used for rubber adhesion. Sometimes as a strike on zinc die castings. White brass used instead of copper as a base for nickel in some cases.
Problems:
Lack of adhesion: Check cleaning; free cyanide—excess or lack can both interfere.
Red color: Check for excess temperature level; lack of ammonia; proper zinc content; proper free cyanide; low current density.
Pale color or "zinc red": Check for low temperature; proper copper content; low free cyanide; high current density.
Polarized anodes: Check for low cyanide; low ammonia content; adequate anode area; lead contamination.
Spotting out: Check rinsing; part preparation; solids in solution; proper drying.
For proprietary bright baths, check supplier for effects of additives.

TABLE 10. COPPER-CYANIDE BATHS

Composition	1 Copper		2 Rochelle Salt		3 High Concentration[a] (Sodium)		4 (Potassium)	
	g/L	oz/gal	g/L	oz/gal	g/L	oz/gal	g/L	oz/gal
Copper cyanide	22.5	3.0	26.0	3.5	120.0	16.0	60.0	8.0
Sodium cyanide	34.0	4.5	35.0	4.6	135.0	18.0	94.0	12.5
Sodium carbonate	15.0	2.0	30.0	4.0	15.0	2.0	15.0	2.0
Sodium hydroxide	to pH		to pH			4.0	42.0	5.6
Rochelle salt	—		45.0	6.0	—		—	
By analysis:								
Copper	16.0	2.1	19.0	2.5	10.0	11.3	42.0	5.6
Free cyanide	7.5	1.0	6.0	0.8	4.0	0.5	7.5	1.0
Conditions:								
pH, colorimetric	12–12.6		12–12.6		>13.0		>13.0	
Temperature, C (F)	32–43 (90–110)		54–71 (130–160)		77–82 (170–180)		77–82 (170–180)	
Current density asf	10–15		20–40		30–60		30–60	
Agitation, cathode preferred, also air in No. 2 and No. 4								
Cathode efficiency, %	30		50		100		100	
Ratio anode to cathode area	2:1		2:1		2:1		2:1	
Anodes	Copper and steel		Copper		Copper		Copper	
Filtration: Continuous preferred for all baths								
Volts:	6		6		6		6	
Suitable materials of construction: 1, 2, 3, 13, 14 for all baths.								
Use:	Strike or thin deposit, 0.05–0.1 mils		Strike or thin deposit, 0.05–0.3 mils		Bright or heavy deposit, 0.3–2.0 mils		Bright or heavy deposit, 0.3–2.0 mils	

Applications: Baths No. 1 and No. 2 as a strike over all basis metals. Baths No. 3 and No. 4 require prior copper strike and produce easily buffed or bright copper undercoat for nickel—chromium protective coatings using proprietary addition agents and also "PR" current in No. 4. No. 3 and No. 4 also used to prevent case hardening of selected surfaces of steel basis metal.

References:	1, 2	1, 2, 15	1, 2, 16	1, 2, 16, 17

Problems:
Poor adhesion: Check cleaning; high or low cyanide concentration; contamination with hexavalent chromium; failure to use strike when required by high-efficiency solution.
Solution turbid or blue: Check for low cyanide.
Poor efficiency: Check for high cyanide concentration; hexavalent chromium contamination.
Anodes polarized: Check for low cyanide; low Rochelle salts; lead contamination.
Rough deposits: Check for solids in solution; excessive current; poor brightener balance.
Spotting out: See copper alloy section.
Specific brightener systems may require control of sodium hydroxide, temperature, etc. within different limits than given above. Check supplier for specifics. Brightness can be affected by solution imbalance, brightener imbalance, metallic contamination or improper operating conditions. Consult with brightener supplier for specifics.

[a]Commonly used with addition agents as proprietary and patented processes.

PLATING BATH COMPOSITIONS AND OPERATING CONDITIONS

Composition	Alkaline Cyanide				Neutral Gold				Acid Gold			
	Matte		Bright		Matte		Bright		Matte		Bright	
	g/L	oz/gal	g/L	oz/gal	g/L	oz/gal	g/L	oz/gal	g/L	oz/gal	g/L	oz/gal
Gold metal as cyanide	2–12	0.25–1.5 tr.	4–16	0.5–2.0	4–16	0.5–2.0	2–10	0.25–1.0	2–16	0.25–2.0	4–12	0.5–1.5
Potassium cyanide	15–45	2–6	15–90	2–12	—	—	—	—	—	—	—	—
Potassium carbonate	0–45	0–6	0–30	0–4	—	—	—	—	—	—	—	—
Potassium phosphate	0–45	0–6	0–45	0–6	0–90	0–12	0–90	0–12	0–100	0–13.5	0–100	0–13.5
Potassium hydroxide	10–30	1.3–4	10–30	1.3–4	—	—	—	—	—	—	—	—
Brighteners	0	0	0.1–10	0.01–1.0	0	0	0.1–30	0.01–4.0	0	0	0.1–20	0.01–2.5
Chelates	—	—	—	—	15–90	2–12	50–150	6.7–20	10–200	1.3–27	10–150	1.3–20
Secondary brighteners	—	—	—	—	—	—	—	—	0–10	0–1.3	0.1–10	0.01–1.3
Temperature, C (F)	50–70	(120–160)	15–25	(60–75)	25–70	(75–160)	35–60	(95–140)	40–70	(105–160)	20–50	(70–120)
pH	11–13		9–13		6–8		6–8		3–5		3–5	
Current density, A/dm² (asf)	0.1–0.5	(1–5)	0.3–1.5	(3–15)	0.2–1.0	(2–10)	0.5–2.0	(5–20)	0.1–0.5	(1–5)	0.8–2.0	(8–20)
Agitation	Moderate		Rapid		Moderate		Rapid		Moderate		Rapid	
Anodes	Platinum, stainless steel, gold				Platinum, stainless steel, Pt/Ti				Carbon, platinum, Pt/Ti			

Most gold is now plated from proprietary solutions, except in the case of some decorative work and jewelry. Alkaline golds have good throwing power and will retard the deposition of base metals. In the jewelry industry, they are frequently modified with alloying metals to give the desired color: 14-Karat gold—copper or nickel approximately 25% of total metal; green gold—silver approximately 5% of total metal, with small amounts of arsenic or lead to intensify the green color (ternary alloys of cadmium/copper/gold may also be used); red gold—copper as required for color; white gold—nickel up to two times the gold content. The neutral and acid golds are used to produce alloys with special characteristics for electronic work. The acid golds produce the purest gold deposits when needed.

Suitable materials of construction: 3, 9, 14

Applications: Jewelry, ornamentation of various items, flatware, numerous electronic applications.

Aside from the usual problems of maintaining proper solution composition and operating conditions, problems with gold plating generally fall into the following areas:

Problems with contamination: Gold solutions are particularly sensitive to both metallic and organic contamination. In many cases, practical economical purification methods do not exist, so a contaminated solution must be discarded and reclaimed. Strike solutions are often used, not only to promote adhesion, but to serve as buffers or barriers against the introduction of contaminants into the main tank. Particular attention should be given to anode and cathode bars, etc. to ensure that salts from these sources do not contaminate the solutions. It is common practice to plate the copper bus bars with nickel, or tin-nickel. Alternatively, titanium or stainless steel may be used.

Problems with activation: In many cases, gold is applied over a barrier layer to prevent diffusion of the gold into the substrate, or of the substrate through the gold. Silver in particular will diffuse through pores in the gold layer and seriously raise contact resistance. Nickel is frequently used as the barrier layer, and if improperly activated may result in peeling, blistering, or even exfoliation of the gold layer.

Problems with salt build-up: Since many gold solutions use insoluble anodes and replenish the gold with prepared gold salts, build-up of these salts may require solution treatment of some type. The recommendations of the suppliers of the replenishing salts must be followed carefully.

TABLE 12. Chromium (Baths)[a]

Composition	1		2		3[d]		4	
	g/L	oz/gal	g/L	oz/gal	g/L	oz/gal	g/L	oz/gal
Chromic acid, CrO_3	248	33.0	398	53.0	338	45.0	180–225	24–30
Sulfuric acid, H_2SO_4	2.5	0.33	4.0	0.53	2.2	0.29	0.9–1.1	0.12–0.15
Ratio[b] CrO_3/H_2SO_4	100:1		100:1		155:1		200:1	
Catalyst	—	—	—	—	—	—	0.5–1	0.07–0.13
Conditions[c]								
Temperature, C (F)	43–49	(110–120)	43–49	(110–120)	55	(130)	38–49	(100–120)
Current density, A/dm^2 (asi)	10–22	(0.7–1.5)	10–22	(0.7–1.5)	15–36	(1.0–2.5)	15–36	(1.0–2.5)
Cathode efficiency (approximate)	13–18		13–18		13–18		16–24	
Volts	6–12		6–12		6–12		4–12	

Agitation is not generally used. Filtration is uncommon. Anodes are usually lead or a lead alloy used in the ratio of 1:1 to 2:1. Catalysts are usually fluoride compounds of one type or another. Self regulating baths use a fluoride with limited solubility present in excess so baths must be stirred occasionally to ensure solution of the required catalyst. Silicofluoride catalysts give the highest efficiency. Trivalent chromium is usually required for proper operation of the bath and is usually maintained at about 1% of the chromic acid concentration.

Suitable materials of construction: 1, 5, 9, 10, 14, 15. (Also see Ref. 2.)

Use: Bright decorative deposits 0.01–0.03 mil in thickness. Hard deposits for wear resistance 0.1 mil and heavier.

Applications: Decorative coating over copper and/or nickel protective coatings on all basis metals. Industrial or hard coatings over ferrous basis metals for tools, dies and gauges. Also for surface protection of electrotypes, engraving plates and other items. Porous coating for improved lubrication, wear and corrosion resistance of piston rings and cylinder liners.

References: 1, 2, 22, 23, 24, 28, 56; 25, 26, 27, 28

Problems with chromium plating generally revolve around proper brightness and covering power for decorative baths. These problems can usually be ascribed to one of three classes of causes:

Improper operating conditions: This includes deviations in current density, temperature, bath composition, particularly bath ratio, trivalent chromium levels or contamination with iron, copper or nickel. At the present time there is no generally simple way of removing metallic contamination. Specially constructed ion exchange units using carefully selected resins may be used in some cases, but it is far better to avoid contamination.

Improper current control: This includes inadequate or excessive current densities, excess ripple in the current supply, interruption of the current during plating, and stray currents.

Improper preparation of substrate: This includes inadequate activation of nickel or nickel iron, inadequate cleaning, etc. Nickel plated parts from solutions with excessive brightener levels can cause "rainbow" in the low current areas; stripped parts improperly activated before rechroming will not cover properly, etc.

Because of the high current densities involved, and the generally poor throwing power of the chromium solutions, racking of parts is particularly important. Racks must be of adequate cross-section, and use of auxiliary anodes, or thieves and/or shields may be necessary for complete coverage. Proper maintenance of racks is also important for the same reasons. Additionally, chromium is a serious contaminant for most plating solutions, and improperly maintained racks can carry solution from tank to tank, causing severe problems.

[a]Proprietary baths are also available (see Ref. 2, Chapter 6).
[b]Bath ratios of 80:1 to 120:1 are not uncommon (see Ref. 2 for discussion of F^- and SiF_6^- catalysts).
[c]Temperatures from 70 to over 150 F with variations in cathode current densities from 1.5–8.0 asi and cathode efficiencies between 10 and 20% are possible. For bright deposits the exact values must be carefully controlled and related (see References for detailed information).
[d]Non-proprietary crack-free process.

PLATING BATH COMPOSITIONS AND OPERATING CONDITIONS 243

TABLE 13. INDIUM (BATHS)

Composition	1 Sulfate chloride		2 Chloride		3 Fluoborate	
	g/L	oz/gal	g/L	oz/gal	g/L	oz/gal
Indium sulfate, $In_2(SO_4)_3$	45.0	6				
Sodium sulfate, Na_2SO_4	9.75	1.3				
Indium chloride, $InCl_3$			58.0	7.7		
Potassium cyanide			158.0	21		
Potassium hydroxide			40.0	5.3		
Dextrose, $C_6H_{12}O_6$			30.0	4		
Indium fluoborate,[a] $In(BF_4)_3$					109	14.5
Ammonium fluoborate, NH_4BF_4					50.0	6.7
Fluoboric acid, HBF_4					15.0	2.0
Boric acid					26.0	3.4
By analysis:						
Indium	20.0	2.7	30.0	4.0	33.0	4.4
Total cyanide, KCN			140.0	18.7		
pH, colorimetric		2.0–2.7		>13.0		1.5–20
Conditions:						
Temperature, C (F)	21–60	(70–140)	21	(70)	21	(70)
Current density, A/dm² (asf)	0.5–5.0	(5–50)	1.5–3.0	15–30	0.5–2.5	5–25
Agitation, cathode	Preferred		Preferred		Preferred	
Cathode efficiency,%	50–80		70–95		80–90	
Ratio anode to cathode area	1:1		1:1		1:1	
Anodes	Indium and insoluble (platinum)		Platinum, graphite, steel or stainless steel		Indium and insoluble (platinum or graphite)	
Filtration:	As required		As required		As required	
Volts:	6		6		6	
Suitable materials of construction:	3, 4, 8, 13, 14		1, 2, 3, 13, 14		3, 8, 12, 13, 14	

Use: Primarily used as thin coating (0.1 mil) to alloy by subsequent heat treatment with various bearing metals.

Applications: Applied over electrodeposited bearing metals such as silver, cadmium, lead, tin, etc., and heat treated to produce alloy surface layer and impart superior corrosion resistance and other properties.

References: 1, 34   1, 35, 36   8, 9

In the acid baths, excessive free acidity will reduce plating efficiency. This may be corrected by removing the insoluble anodes and using only indium anodes for a suitable period. Using only insoluble anodes will increase the acidity when needed.

Numerous indium alloy plating baths have been reported in the literature, including alloys with antimony, arsenic, bismuth, cadmium, copper, gallium, lead, tin and zinc. Alloy properties include wear resistance, hardness and high corrosion resistance.

[a] Sold as concentrate specific gravity 1.60, 13.4 lb/gal, indium 32.8 oz/gal, fluoboric acid 4.9 oz/gal and boric acid 5.1 oz/gal (Ref. 8).

TABLE 14. IRON BATHS

Composition	1 Sulfate/Chloride		2 Chloride		3 Fluoborate	
	g/L	oz/gal	g/L	oz/gal	g/L	oz/gal
Ferrous sulfate, $FeSO_4 \cdot 7H_2O$	248.0	33				
Ferrous chloride, $FeCl_2 \cdot 4H_2O$	30.0	4	300.0	40		
Ferrous fluoborate,[a] $Fe(BF_4)_2$					225.0	30
Calcium chloride, $CaCl_2$			338.0	45		
Ammonium chloride	22.5	3				
Sodium chloride					10.0	1.3
Boric acid					22.5	3.0
By analysis:						
Iron	58.0	7.7	84.0	11.2	55.5	7.4
pH, colorimetric		4.5–6.0		1.0–1.5		3.0–3.7
Conditions:						
Temperature, C (F)	38	(100)	90	(195)	55–90	(130–190)
Current density, A/dm^2 (asf)	5.0–10.0	(50–100)	6.0	(60)	2.0–9.0	(20–90)
Agitation[b], cathode and/or solution	Required		Required		Required	
Cathode efficiency, %	95–100		95–100		95–100	
Ratio anode to cathode area	1:1		1:1		1:1	

Anodes: Low carbon steel, bagged, (African blue asbestos best for baths 1 and 2, cotton duck or high temperature "Vinyon" for bath 3).

Filtration: Continuous, with activated carbon recommended.

Volts:	6	6	6
Suitable materials of construction:	4, 8, 9, 13, 14, 15	8, 9, 15, 16	3, 8, 12, 13, 14

Use: For heavy deposits in building up worn parts, electroforming.

Applications: Printing plates, parts for electric irons and molds. Salvage of worn parts. Plating stereotypes as substitute for nickel.

References:	1, 2, 37, 56	1, 2, 37	8, 9, 38, 56

Presence of excess ferric iron (III) causes brittleness, pitting and low efficiency. Excessive temperatures promote oxidation of Fe(II) to Fe(III) and therefore should be avoided except in the chloride bath, where higher temperatures improve the appearance and the ductility of the deposit. Control of Fe(III) may be by reduction with iron filings or steel wool, dummying at low current density or operating at the higher pH range where Fe(III) will precipitate. In other respects, iron baths are similar to nickel baths. Organic contamination causes stress, pitting and brittleness. Activated carbon can be used for removal. Hydrogen peroxide or permanganate must not be used.

[a] Sold as concentrate, sp gr 1.52, 12.7 lb/gal, iron 22.4 oz/gal, fluoboric acid 1.4 oz/gal and boric acid 6.2 oz/gal (Ref. 8).

[b] Cathode movement is primarily to jar gas bubbles loose and prevent pitting.

TABLE 13. LEAD-TIN ALLOY BATHS

Bath make-up	1 g/L	1 oz/gal	2 g/L	2 oz/gal	3 g/L	3 oz/gal	4 g/L	4 oz/gal	5[d] g/L	5[d] oz/gal
Lead fluoborate, $Pb(BF_4)_2$[a]	345.0	46	330.0	44	217.5	29	146.0	19.5	848-765	113-102
Tin fluoborate, $Sn(BF_4)_2$[a]	30.0	4	50.0	6.7	202.5	27	300.0	40	75-188	10-25
Boric acid, H_3BO_3[b]	16.5	2.2	16.5	2.2	16.5	2.2	16.5	2.2	40	5.3
Fluoboric acid 48%, HBF_4[b]	30.0	4	30.0	4	30.0	4	30.0	4	25	3.3
Glue	0.4	0.05	0.4	0.05	0.4	0.05	0.4	0.05		
Hydroquinone[e]									10	1.3
By analysis:										
Lead	95.0	12.6	91.0	12.1	60.0	8.0	40.0	5.4	230-210	31-28
Tin	6.0	0.8	10.0	1.3	40.0	5.4	60.0	8.0	15-37.5	2-5
Boric acid	22.5	3.0	22.5	3.0	22.5	3.0	22.5	3.0	30.0	4
Fluoboric acid	30.0	4.0	30.0	4.0	30.0	4.0	30.0	4.0	45.0	6
Ratio Pb:Sn	16:1		9.3:1		1.5:1		0.68:1		15-5.6:1	
pH, colorimetric	>0.5		>0.5		>0.5		>0.5		>0.5	
Deposits, Sn % (approx.)	6		10		40		60		5-15	
Conditions:										
Temperature, C (F)	21-38	(70-100)	21-38	(70-100)	21-38	(70-100)	21-38	(70-100)	55-70	(130-160)
Current density, A/dm^2 (asf/av)	2.5	(25)	2.5	(25)	2.5	(25)	2.5	(25)	20-35	(200-350)
Agitation, cathode	mild		mild		mild		mild		75-250[f]	
Cathode efficiency, % (approx.)	100		100		100		100		100	
Area, anode to cathode	2:1		2:1		2:1		2:1		1:1	
Anodes: Same alloy as deposit										
Filtration: As required										
Volts:	6		6		6		6		12-18	

Suitable materials of construction: 3, 8, 12, 13, 14.
Use: Over various basis metals for improved solderability.
Applications: Bearings and for solderability. Also see Refs. 48 and 49 for coating properties and performance.
References: 1, 2, 41, 42, 43, 44, 46, 47, 48, 49, 146

Low concentration of glue causes a granular deposit with loss of throwing power and a decrease of tin in the deposit. Change in the metal ratio will affect composition of the deposit. Current density will also affect composition, tin increasing with an increase in current density. The effect of metallic impurities has not been clearly defined.

[a]Lead and tin liquid concentrates (15.3 and 13.5 lb/gal, respectively) sold separately and diluted and mixed as required to give bath composition. (See Refs. 8 and 9.)
[b]Allows for free acid content of the concentrates.
[c]Glue and resorcinol may also be used in various proportions with different bath compositions to obtain the desired deposit composition. (See Refs. 42 and 43.)
[d]For strip plating steel mill products.
[e]See Refs. 46 and 47.
[f]Surface velocity, fpm.

TABLE 16. Lead-Fluoborate Baths[a]

Composition	1 g/L	1 oz/gal	2[b] g/L	2[b] oz/gal
Lead fluoborate,[a] $Pb(BF_4)_2$	202 or 404	27 or 54	470	62.5
Fluoboric acid, HBF_4	20 or 40	2.6 or 5.2	45	6
Boric acid, H_3BO_3	20 or 60	2.6 or 8.0	45	6
Gelatin or glue	0.15 or 0.38	0.02 or 0.05	—	—
Hydroquinone	—	—	10	1.3
By analysis:				
Lead	113 or 218	15 or 29	255	34
pH, colorimetric	<1.5		<1.5	
Conditions:				
Temperature, C (F)	25–45	(75–110)	30–70	(90–160)
Current density, A/dm^2 (asf)	0.5–7	(5–70)	10–60	(100–600)
Cathode efficiency, %		100		100
Ratio anode to cathode area		2:1		1:1
Anodes, lead				c

Filtration: as required
Volts: 6
Suitable materials of construction: 3, 8, 12, 13, 14.
Use: For protective coatings (0.25–1.0 mil) including copper strike on ferrous basis metals.
Applications: See Refs. 39a & c for coating properties and performance.
References: 1, 2, 39, 40, 41, 45, 47, 49.
Lead baths are relatively free of problems, provided the proper free acid is maintained, and the bath is kept free of undesirable organics by carbon treatment. Proper concentration of colloid is essential to avoid powdery, crystalline or treed deposits.

[a]Sold as a liquid concentrate to be diluted with water as required. Sp. gr. 1.85, 15.4 lb/gal, lead 67 oz/gal, boric acid 9.9 oz/gal, fluoboric acid 1.7 oz/gal (Ref. 8 and 9).
[b]For plating strip steel mill products.
[c]High purity, 99.99% pb.

TABLE 17. BLACK NICKEL BATHS

Composition	1 g/L	1 oz/gal	2 g/L	2 oz/gal
Nickel ammonium sulfate, $NiSO_4(NH_4)_2SO_4 \cdot 6H_2O$	60	8	—	—
Nickel sulfate, $NiSO_4 \cdot 6H_2O$	—	—	75	10
Zinc sulfate, $ZnSO_4 \cdot 7H_2O$	7.5	1	30	4
Ammonium sulfate	—	—	37.5	5
Sodium thiocyanate, NaCNS	15	2	15	2
By analysis:				
Nickel	9	1.2	16.5	2.2
pH, colorimetric		5.5–6.0		5.0–6.2
Conditions:				
Temperature,	24	(75)	21	(70)
Current density, A/dm^2 (asf)	0.2	(2)	0.08–0.2	(0.75–2.25)
Agitation, mild cathode or none				
Ratio anode to cathode area		1:1		1:1
Anodes: Carbon and/or nickel				
Filtration: Not ordinarily required				
Volts: (at tank)		1–2		1–2

Suitable materials of construction: 3, 4, 5, 10, 13, 14.

Use: To produce black, nonlight reflecting, gun metal finish. Thin coating. Always protected with organic coating.

Application: To typewriter parts, military hardware and novelties.

References: 1, 64

For additional corrosion resistance, black nickel is usually applied over a nickel layer. This underlayer of nickel should not be bright, but rather dull or lustrous, or the desired non-reflective characteristics in the black nickel will be difficult to maintain. Proper activation of the nickel layer is essential. Current density must be carefully controlled. High currents will produce grey or even pure nickel deposits.

TABLE 18. SPECIAL NICKEL BATH

Composition	1 Sulfamate		2 Sulfamate		3 Ni-P		4 Ni-P	
	g/L	oz/gal	g/L	oz/gal	g/L	oz/gal	g/L	oz/gal
Nickel sulfate, $NiSO_4 \cdot 6H_2O$					330	44	150	20
Nickel chloride, $NiCl_2 \cdot 6H_2O$		[a]	30	4	45	6	45	6
Nickel sulfamate, $Ni(SO_3NH_2)_2$	450	60	300	40				
Boric acid H_3BO_3	30	4	30	4	30	4		
Phosphoric acid, H_3PO_4							50	6.7
Phosphorous acid, H_2PO_3					0.2-2	0.03-0.3		5.3
Stress-reducing agents	[b]		[b]					
By analysis:								
Nickel	105	14	77	10	88	11.5	45	6
pH (Electrometric)	3-5		3.5-4.2		17.-3.0		0.5-1.0	
Non-pitter	0.38	0.05			0.15	0.02		
Conditions:								
Temperature, C (F)	49	(120)	49	(120)	63	(145)	85	(185)
(practical range)	38-60	(100-140)	24-71	(75-160)	60-65	(140-150)	75-95	(167-203)
Current density, A/dm^2 (asf)	5-30	(50-300)	2-14	(20-140)	2-5	(20-50)	2-5	(20-50)
Agitation: cathode and solution. Air in absence of wetting agents								
Cathode efficiency, %	98-100		95-100					
Ratio of anode to cathode area	1:1		1:1		1:1		1:1	
(for high-current density)	3:1		2:1					
Anodes: Nickel, bagged	depolarized		cast carbon		carbon type		carbon type	

PLATING BATH COMPOSITIONS AND OPERATING CONDITIONS 249

Filtration: continuous. Turn over once in 1 to 4 hr				
Volts	6–12	6–12	6–12	6–12
Suitable materials of construction: 3, 4, 8, 12, 13, 14 and 15 for baths 1 to 4.				
Use: Bath 1 and 2 give hard, low stress deposits. Bath 3 and 4 give hard deposits 2–3% and 12–15% phosphorous, respectively.				
Applications: Engineering purposes usually where hardness is important.				
References:	1, 2, 58, 61	1, 2, 59, 60, 61	1, 2, 62	1, 2, 63

Woods nickel strike:
Nickel chloride, $NiCl_2 \cdot 6\,H_2O$ 180 g/L 24 oz/gal
Hydrochloric acid (commercial) 10–12% by volume.
Room temperature to 35C (100F)
Current density 2–10 A/dm^2 (20–100 asf)
Anodes, nickel

This widely used strike is effective on stainless steels, nickel, high-alloy steels and selectively carburized and hardened steels. Time is as required to obtain good coverage, generally 30 sec to 2 min. The anode efficiency greatly exceeds the cathode efficiency so metal content will tend to climb. This should be maintained at 45–60 g/L (6–8 oz/gal) by decanting part of the solution and replacing it with hydrochloric acid. The decanted portion may be used as a replenisher for chloride and nickel, and for pH control in standard plating baths.

Nickel Problems:
No plate: Check electrical system; check for heavy chrome or nitrate contamination.
Blistering or peeling: Check preplating cycle; pH too low or too high; check for metallic contamination, excess brightener, contamination with hexavalent chromium or nitrates.
Plate peeling from plate: Check for current interruption, replate without activation, replating with chromium incompletely stripped, excess iron in solution, excessive stress due to high brightener or metallic contamination.
Highly stressed deposit: Check for contamination with zinc or cadmium, excess iron content, organic contamination, brighteners out of balance, excess brightener, pH out of range, chloride levels too high.
Dark deposits: Especially in low current density area, check for copper contamination; at high levels, zinc and cadmium may produce dark, streaky deposits.
Streaky deposits: See above; check chrome or nitrate contamination; inadequate or wrong wetting agent for the type of agitation used (air or mechanical).
Hazy deposits: Check precleaning cycle for smut removal, immersion deposits in acids, etc; brighteners out of balance; pH out of range, iron levels too high.
Pitting: Check for solids in solution; lack of non pitter; oil or grease drag-in; low temperature; lack of agitation.
Roughness: Check for solids in solution; excess boric acid at low temperature; excessive current.

[a] Addition of 0.44 oz/gal of nickel chloride improves anode efficiency without significantly increasing internal stress of deposits.
[b] Proprietary stress lowering agents are usually recommended.

TABLE 19. NICKEL BATHS

Composition	1 Watts Type		2 High Chloride		3 All Chloride		4 Fluoborate[c]	
	g/L	oz/gal	g/L	oz/gal	g/L	oz/gal	g/L	oz/gal
Nickel sulfate, $NiSO_4 \cdot 6H_2O$	300	40	240	32				
Nickel chloride, $NiCl_2 \cdot 6H_2O$	45	6	90	12	240	32		
Nickel fluoborate, $Ni(BF_4)_2$							220	29.3
Boric acid	30–37.5	4–5	30–37.5	4–5	30	4	30	4
By analysis: Nickel	77.0	10.3	75	10	75	10	55	7.3
pH range (electrometric)								
Low	2.0–2.5		2.0–2.5		0.9–1.1			
Medium	3.5–4.0						3.0–4.5	
High	4.8–5.2							
Non-pitter: I or II								
I. Approved wetting agent[a] to give dynes/cm at 21 C (70 F)	35–45		34–45		35–45		35–45	
II. Hydrogen peroxide[b] to give free oxygen, ppm For engineering applications as required	5–10		5–10		5–10		5–10	
Conditions:								
Temperature, C (F) av.	55	(130)	55	(130)	55	(130)	5	(130)
(practical range)	32–71	(90–160)	38–71	(100–160)	38–63	(100–145)	32–71	(90–160)
Current density, A/dm^2 (asf)	1–6	(10–60)	1–6	(10–60)	5–10	(50–100)	5–10	(50–100)
(possible range)					<50	(500)	<50	(500)
Agitation: Cathode and (mild) solution preferred. Air effective in absence of wetting agents.								
Cathode efficiency, %	95–100		95–100		90–100		95–100	
Ratio anode to cathode area (for high current density)	1:1		1:1		1:1–(3:1)		1:1–(3:1)	
Anodes: Nickel, bagged, cast or rolled, depolarized or carbon type.								
Filtration: Continuous, turn over once every 1 to 4 hr. Filter aid and activated carbon commonly used.								
Volts:	6–12		6–12		6–12		6–12	
Suitable materials of construction:	3, 4, 5, 8, 13, 14		3, 4, 8, 13, 14, 15		8, 9, 12, 13, 14, 15		3, 8, 12, 13, 14	

Use: Baths 1 and 2 (dull nickel, 0.1–2.0 mil and higher. For semi- and full-bright nickel, 0.1–2.0 mil, using proprietary addition agents[d]). Bath 3 (hard dull nickel, any thickness). Bath 4 (dull nickel, any thickness)
Applications: For decorative and protective coating, copper and/or nickel-chromium, on most basis metals use baths 1 or 2. For corrosion resistant coating on steel for chemical equipment use baths 1 or 2. For building up worn parts or electroforming use baths 1 or 4. For a nickel strike, especially over steel, copper and copper alloys prior to silver plating, use baths 1 or 3 (less concentrated frequently)

| References | 1, 2, 50, 55, 56, 57 | 1, 2, 50, 57 | 1, 2, 50, 51 | 8, 52, 53, 54, 56 |

[a] Sodium lauryl alcohol sulfate, one of few acceptable.
[b] Cannot be used in bath containing organic addition agents or wetting agent.
[c] Sold as liquid concentrate to be diluted with water as required. Specific gravity 1.58, 13.2 lb/gal nickel, 23.6 oz/gal, boric acid 6.3 oz/gal, fluoboric acid nil (Refs. 8 and 9).
[d] See suppliers for patented proprietary processes.

PLATING BATH COMPOSITIONS AND OPERATING CONDITIONS 251

TABLE 20. STRIKING BATHS* PRIOR TO SILVER PLATING

Composition	1 g/L	1 oz/gal	2 g/L	2 oz/gal	3 g/L	3 oz/gal	4 g/L	4 oz/gal	5 g/L	5 oz/gal
Silver cyanide	1.7	0.23	7	0.9			1.0	0.13	5.0	0.67
Potassium cyanide	75	10.0	75	10.0			45	5.0	15	2.0
Potassium carbonate							30	4.0	30	4.0
Copper cyanide	15	2.0								
Nickel chloride, $NiCl_2 \cdot 6H_2O$					240	32.0				
Hydrochloric acid, sp gr 1.18						10.0%/vol				
By analysis:										
Silver, troy oz/gal	1.6	0.2	6.0	0.7			0.8	0.1	4.0	0.5
Copper	10.5	1.4								
Nickel					60	8.0				
Free cyanide, KCN	53	7.0	71	9.5			44	5.9	13	1.7
Conditions:										
Temperature, C (F)	24	(75)	24	(75)	24	(75)	24	(75)	24	(75)
Current density, A/dm^2 (asf)	3	(30)	3	(30)	15	(150)	3	(30)	0.4	(4)
Time, min		0.33		<1.0		1–2		0.33		1.0
Anodes	Steel		Steel		Nickel, bagged		Steel		Steel	
Filtration:	As required		As required		Constant and carbon		Constant and carbon		Constant and carbon	
Volts:	<6		<6		6		<6		<6	
Suitable materials of construction: For baths 1, 2, 4, and 5 use 1, 2, 3, 9, 10, 13, 14. For bath 3 use 8, 9, 10, 12, 13, 14, 15.										
Use:	First strike on steel.		Second strike on steel. Silver strike over nickel or other basis metal.		First strike on steel or other basis metal.		Second strike over No. 3 strike.		Third strike over No. 4 strike.	

Applications: Strikes 1, 2 and 3 for regular silver plating applications, such as tableware, hollow ware, novelties, jewelry and decorative coatings. Strikes 3, 4 and 5 for engineering application, such as silver plating of bearings.

| References: | 1, 2, 66 | 1, 2, 66 | 68, 70 | 68, 70 | 68 |

*No. 3 nickel strike is used where necessary prior to plating any metal. See Wood's nickel strike Table 16.

TABLE 21. SILVER BATHS

Composition	1 Low Metal[a]		2 High Metal		3 Cyanide Nitrate	
	g/L	oz/gal	g/L	oz/gal	g/L	oz/gal
Silver cyanide	36	4.8	105	14.0	22–45	3–6
Cyanide potassium (sodium)	60	8.0	113	15.0	33–45	4.4–6
Carbonate, potassium (sodium)	45	6.0	15–115	2–15	22–45	3–6
Potassium hydroxide			30	4.0		
Potassium nitrate					45–120	6–20
Carbon disulfide	0.00075	0.0001	[b]		0.00075	0.0001[b]
By analysis:						
Silver,	28.7	3.5 (Troy)	84	10.2 (TR)	16–37	2–4.5 (Troy)
Free potassium cyanide	43	5.7	62	8.2	19–23	2.5–3.1
Conditions:						
Temperature, C (F)	24–32	(75–90)	43–55	(110–130)	27–52	(809–125)
Current density, A/dm² (asf)	0.5–1.5	(5–15)	6–15	(60–150)	5–15	(5–150)
Agitation, cathode	Preferred		Plus solution		Preferred	
Cathode efficiency, %	100		100		100	
Ratio anode to cathode area	1:1		2:1–4:1		1:1	
Anodes	Silver		Silver		Silver	

Filtration: As required for 1 and 3, continuous through paper for 2.
Volts: <6   <6   <6
Suitable materials of construction: 1, 2, 3, 9, 10, 13, 14.
Use: Baths 1 and 3 for regular bright silver deposits following silver strike, 0.1–1.3 mil. Bath 2 for heavy silver deposit after prior silver strike, up to 60 mil.
Applications: Baths 1 and 3 for decorative and protective silver deposits on novelties, jewelry, lighting fixtures, tableware and hollow ware. Baths 2 and 3 for engineering applications including plated bearings.
References   1, 2, 66, 67   1, 2, 66, 67, 68, 69   1, 2, 65, 66, 70

Problems:
Lack of adhesion: Check preplating cycle; check strikes for proper choice and use; check for passivity of nickel layer if used.
Porous deposits: Check for solids in solution; check current.
Dull and/or thin deposits: Check for low temperature; low current; low metal; polarized anodes.
Polarized anodes: Free cyanide too low; insufficient anode area.
Brown stains or spots: Poor rinsing; spotting out due to porosity.
Patchy brightness: Organic contamination; brightener imbalance.
Yellowish or pink deposits: High copper contamination and/or low silver.
Since many solutions now use proprietary additive to control brightness and/or other properties, the troubleshooting information given above should be checked with the supplier and modified and/or supplemented per his instructions.

[a] Proprietary full bright silver processes are available from suppliers.
[b] Ammonium thiosulfate as brightener added at regular intervals in amount equivalent to 1.5 mg/l/hr. Hydrogen peroxide is also recommended as an anti-pit agent, 0.2 ml/gal/30 min (Ref. 68).

TABLE 22. TIN-COPPER ALLOY BATHS[a]

Composition	1 g/L	1 oz/gal	2 g/L	2 oz/gal
Copper cyanide, CuCN	28.5	3.8	13	1.7
Potassium stannate, $K_2SnO_3 \cdot 3H_2O$[b]	35	4.7	100	13.3*
Potassium cyanide, KCN	62	8.3	25	3.3*
Potassium hydroxide, KOH	10	1.3	15	2.0*
Rochelle salt, $KNaC_4H_4O_6 \cdot 4H_2O$	45	6.0		
By analysis:				
Copper	20.25	2.7	9	1.2
Tin	14.25	1.9	45	6.0
Free potassium cyanide	21	2.8	11	1.4*
Free potassium hydroxide	10	1.3		
Deposit, tin, %		10–12		45
Conditions:				
Temperature, C (F)	65–71	150–160	65	150
Cathode current density A/dm² (asf)	2.5–5.0	(25–50)	2.5	(25)
Agitation	cathode		mild	
Cathode efficiency, %	50–60		60	
Anode to cathode area	1.5:1			
Anode C. D.[c], A/dm² (asf)				
Copper			0.5–1.0	(5–10)
Tin			1.5–2.0	(15–20)
Anode, volts				
Copper			2–3	
Tin			3–5	

Filtration: Continuous or at intervals as required.
Suitable materials of construction: 1, 2, 3, 13, 14.
Use: Bath 1 recommended for low tin-bronze coatings (0.1–0.5 mil). Bath 2 is recommended for tarnish-resistant high-tin white alloy such as speculum. Thickness range, 0.5–1.0 mil.
Applications: Primarily for protective coatings. Low tin alloy suggested use as undercoat for bright nickel-chromium coatings on objects of brass, steel or zinc die castings.
References: 2, 93–104.

* Sodium salts.
[a] Features of these baths are patented. (See references.)
[b] Hydrogen peroxide or sodium peroxide is added as required to oxidize stannous tin to stannate.
[c] Parallel circuits for copper and tin anodes have been recommended for bath 2. Bath 1 may be operated with soluble copper and some small area of insoluble carbon anodes balanced to control the metal concentration, in which case tin may be added as copper stannate. Soluble alloy anodes have also been operated with similar baths with difficulty. Also a pyrophosphate stannate bath containing fluoride can be operated with soluble copper-tin anodes to produce most any composition of copper-tin deposit. (See US Patents Nos. 2,658,032; 2,854,388 and 2,886,500.)

TABLE 23. TIN-NICKEL ALLOY BATHS

Composition	1 g/L	1 oz/gal	2 g/L	2 oz/gal
Stannous chloride, $SnCl_2 \cdot 2H_2O$	50	6.7	50	6.7
Nickel chloride, $NiCl_2 \cdot 6H_2O$	300	40.0	250	33.3
Ammonium fluoride			50 or 33	6.7 or 4.4
Ammonium bifluoride	35	4.7		
Sodium fluoride	28	3.7	0 or 20	0 or 2.7
Hydrochloric acid (36%)				1.0
By analysis:				
Tin	27	3.6	27	3.6
Nickel	75	10.0	63	8.4
Bath ratio, tin-nickel %		73.5:26.5		70:30
Deposit ratio, tin-nickel %		65:35		65:35
Ammonium bifluoride	35	4.7		
Total fluorine	39	5.2		
Conditions:				
pH		2.5		2.5
Temperature, C (F)	65	(150)	65	(150)
Current density, A/dm^2 (asf)	2.5	(25)	2.5	(25)

Agitation, cathode preferred. Air cannot be used.

| Cathode efficiency, % | 100 | | 100 | |

Anodes: Bagged alloy anodes of 72% tin-28% nickel are preferred. Separate anodes of tin and nickel with dual circuits can be used if tin anodes are bagged and not left in the bath with current off.

Filtration: Continuously through active carbon is preferred to control pitting. Wetting agents cannot be used to control pitting.

Purification: Metallic impurities can best be removed by using a nickel powder.

Voltage: 6–12

Suitable materials of construction: 7, 12, 13, 14

Precaution: A suitable exhaust system must be provided to take care of hydrofluoric acid vapor.

Use: Recommended as a substitute for bright nickel-chromium coatings for bright decorative and protective coating applications.

Applications: To all ferrous basis metals, preferably over a copper undercoat or directly to nonferrous basis metals.

References: 2, 105, 106, 107, 108, 109, 110, 111, 112.

TABLE 24. TIN-ZINC ALLOY BATHS

Composition	1		2		3[a]	
	g/L	oz/gal	g/L	oz/gal	g/L	oz/gal
Potassium stannate, $K_2SnO_3 \cdot 3H_2O$	120	16	95	12.7	100	13.3
Zinc cyanide, $Zn(CN)_2$	10.5	1.4	13.5	1.8	27	3.6
Potassium cyanide, KCN	33	4.4	30	4.0	17	2.3
Potassium hydroxide, KOH	7	0.9	10	1.3	60	8.0
By analysis:						
Tin	30	4.0	35	4.7	25	3.3
Zinc	6	0.8	7.5	1.0	15	2.0
Total cyanide (as KCN)	45	6.0	45	6.0	47	6.3
Deposit:						
Tin, %		75–85		75–85		5–25
Zinc, %		15–25		15–25		75–95
Conditions:						
Temperature, C (F)	65	(150)	65	(150)	71	(160)
Current density:						
Cathode, A/dm^2 (asf)	3–5	(30–50)	3–5	(30–50)	0.5–4	(5–40)
Anode, A/dm^2 (asf)	1.5–2.5	(15–25)	1.5–2.5	(15–25)	2–5	(20–48)
Agitation	mild		mild		good	
Ratio anode to cathode area	2:1		2:1		1:1	
Anodes[b]:						
Tin, %	80		80		25	
Zinc, %	20		20		75	
Cathode efficiency, %	80–100		80–100		>85	
Filtration: Continuous preferred.						
Volts:	6		6		6	
Suitable materials of construction: 1, 2, 3, 13, 14						
Use: Baths 1 and 2 for high tin alloy for still and barrel plating, respectively. Bath 3 for low tin alloy applications.						
Applications: Used as a coating for solderability. High tin alloy used for anti-friction property or as a protective coating (0.5–1.0 mil).						
References:	2, 113, 114, 115, 116, 117, 118		2, 113, 114, 115, 116, 117, 118		119, 120, 121	

[a] Patented. Additions for bright-plate gelatin 0.25–0.5 oz/gal and ammonium molybdate or thiourea 1–2 oz/gal. (See Ref. 121.)

[b] Anodes must be polarized at high current density after each interruption of current just prior to plating to ensure tin dissolving as stannate. (See Ref. 120.)

TABLE 25. Acid Tin Baths[a]

Composition	1 Fluoborate[b]		2 Sulfonate		3 Bright Proprietary	
	g/L	oz/gal	g/L	oz/gal	g/L	oz/gal
Tin fluoborate, $Sn(BF_4)_2$[b]	200	27				
Fluoboric acid, HBF_4	53	7				
Boric acid	30	4				
Stannous sulfate			55	7.3	60–120	4.0–16.0
Sulfuric acid			100	13.4	7% by vol.	
Cresol sulfonic acid, $C_7H_6OH \cdot HSO_3$			100	13.4	3% by vol.	special additive
Beta naphthol, $C_{10}H_7OH$	1.0	0.13	1.0	0.13		
Gelatin	6.0	0.8	2.0	0.27		
By analysis:						
Tin	83.0	11	30	4	28–62	3.75–8.3
pH		0.2			chloride	max. 200 ppm
Conditions:						
Temperature, C (F)	24–38	(75–100)	25	(77)	15–30	(60–90)
Current density, A/dm^2 (asf)	2.5–12.5	(25–125)	2.5 > 30	(25–> 300)	1.5–2.0	(15–20)
Agitation, cathode and solution	Optional		Required		Required	
Cathode efficiency, %	100		100		100	
Ratio anode to cathode area (at higher current densities)	2:1		1:1		1:1	
Anodes	Tin		Tin		Tin	
Filtration: As required.						
Volts:	6		6–12		6–12	
Suitable materials of construction:	3, 8, 12, 13, 14		2, 3, 12, 14		8, 9, 13	

Uses: Tin plate (0.01–0.06 mil), thicker coating as required.
Applications: Piston cylinders and rings, electro-tin plate. (Frequently with modification of bath 2. See Ref. 65a.) Wire, soldering, electrotypes (backing shells), protective coating for refrigerator parts and kitchen ware.
References: 1, 8, 56   1, 2, 80, 90

Proprietary acid solutions:
Dull low current density areas: Chloride content too high; lack of brightener; temperature too high.
Dull high current density areas: Brightener levels too low; suspended solids in solution; temperature too high.
Pitting: Not enough agitation; not enough wetter; too much current; solution needs dummying.
Poor coverage and/or throwing power: Metal content too high; acid content too low; inadequate current flow.
Dark streaks: Excess brightener levels.
Rapid tarnishing of deposit: Inadequate rinsing; low brightener levels; with alloys containing zinc—may be due to migration of the zinc into the tin—use copper or nickel barrier layer.
Acid sulfate solution:
Shelf roughness: Suspended solids.
Slow deposition rate: Low temperature, low metal, low acid.
Coarse grained deposit: Addition agents too low. Normally this condition is accompanied by a reduction of the voltage required to produce the normal operating current. This voltage drop will signal the problem before it reaches serious proportions.
Alkaline tin solutions:
Rough, dark, or spongy deposits: Stannous tin in solution—anodes must be kept properly polarized. Use hydrogen peroxide for quick correction.
Poor conductivity, anodes grey: Low caustic, low metal or both.

PLATING BATH COMPOSITIONS AND OPERATING CONDITIONS

TABLE 25. (*Continued*)

Poor conductivity, anodes black: Anode current density too high.

Anodes loose polarization, or will not polarize properly: Check caustic level; if caustic level is right, anode area is too high. High caustic will cause similar problem when anode area is right.

[a] For information on patented "Halogen Tin" process see Refs. 91 and 92. See suppliers for proprietary bright tin process.
[b] Sold as tin concentrate, 1.60 sp gr, 13.3 lb/gal, tin (as fluoborate) 40.7 oz/gal, fluoboric acid 8.6 oz/gal and boric acid 6.4 oz/gal (Ref. 8 and 9).

TABLE 26. STANNATE TIN BATHS

Composition	1 Sodium		2 Potassium		3 Sodium	
	g/L	oz/gal	g/L	oz/gal	g/L	oz/gal
Sodium stannate, $Na_2SnO_3 \cdot 3H_2O$	90	12	80	10.6	140	19
Sodium hydroxide (potassium)	7.5	1	30	4.0	15	2
Sodium acetate	15	2				
Hydrogen peroxide, 100 vol.	As required (0.07)					
By analysis:						
Tin	38	5.0	29	3.8	60	8
Conditions:						
Temperature, C (F)	60	(140–180)	85	(185)	93–99	(200–210)
Current density, A/dm^2 (asf)	1–2.5	(10–25)	4.0	(40)	4.5–6.5	(45–65)
Agitation, cathode and solution:	Optional		Preferred			
Cathode efficiency, %	60–90		80–100		90	
Ratio anode to cathode area	1:1		1:1		1:1	
Anodes[a]	Tin		Tin		Tin	
Filtration: as required						
Volts:	6		6		6	

Suitable materials of construction: 1, 2, 3, 13, 14

Use: Average tin coating thickness 0.03–0.3 mil. Greater thickness required for protective coatings, 1.0–3.0 mil.

Applications: Baths 2 and 3 used for producing electro tin plate. Bath 1 is commonly used for tin plating all basis metals in applications requiring ease of soldering, protective coatings or good throwing power. Also used to prevent nitriding of selected steel surfaces.

References: 1, 2, 85   1, 2, 85, 87, 88

[a] Anodes must be filmed with oxide for successful operation of the baths. (Ref. 2, pp. 338–339).

TABLE 27. ZINC-CYANIDE BATHS

Composition	1[a] g/L	oz/gal	2 g/L	oz/gal	3[b] g/L	oz/gal
Zinc oxide, ZnO	45	5.8	66	8.8	10-22	1.3-3.0
Sodium cyanide, NaCn	100	13.6	120	16.0	2.2-19	0.3-2.5
Sodium hydroxide, NaOH	50	7.0	50	6.7	7.5-112	10-15
or						
Zinc cyanide, Zn(CN)$_2$	40	5.5	97	12.9		
Sodium cyanide, NaCN	50	6.6	40	5.4		
Sodium hydroxide, NaOH	95	12.7	115	15.4		
By analysis:						
Zinc	35	4.7	53	7.1	7.5-15	1-2
Total cyanide	102	13.6	120	16.0	2.2-19	0.3-2.5
Total NaOH	95	12.7	115	15.4	75-112	10-15
Ratio, total NaCN:Zn		2.9		2.3		
Conditions:						
pH, colorimetric		>13.0		>13.0		>13.0
Temperature, C (F)	20-30	(70-85)	40-60	(104-140)	20-35	(70-95)
Current density, A/dm^2 (asf)	1-6	(10-60)	9-17	(90-170)	1-3.5	(10-35)
Agitation, cathode		Yes		Yes		+bath
Cathode efficiency, %		65-85		75-95		65-80
Ratio, anode to cathode area		2:1		2:1		3:1

Anodes: Steel and zinc
Filtration: As required

Volts:		6-15		12-18		6-15

Suitable materials of construction: 1, 2, 3, 13, 14

Use:		Rack or Barrel		Conduit or strip		Rack or Barrel

Applications: For bright finish and anti-rust. Special precautions for plating cast or malleable iron (see Ref. 2).
References: 1, 2, 122

Poor adhesion: Contamination with chromium or nitrates; solution out of balance.
Deposit dark: Metallic contamination (copper, cadmium, lead).
Hazy deposits: Solution out of balance, organic contamination, lack of rinsing, lack of brightener.
Poor throwing power: Cyanide too high, caustic too high, metal too high.
Anodes polarized: Low cyanide, low caustic, insufficient anode area.
Staining in storage: Poor rinsing; failure to use proper post-treatment.

[a] Typical high cyanide bright zinc bath formulation.
[b] Typical range for low-cyanide (so called non-cyanide) baths. See suppliers for proprietary brighteners required for these baths.

TABLE 28. ALKALINE NON-CYANIDE ZINC SOLUTIONS[a]

	Barrel g/L	oz/gal	Rack g/L	oz/gal
Zinc metal	8-15	1-2	6-12	0.8-1.6
Sodium hydroxide	80-150	10-20	60-120	8-16
Temperature, C (F)	18-35		(65-95)	
Current density	0.5-1.2 A/dm^2 (5-12 asf)		1-2.5 A/dm^2 (10-25 asf)	

Problems with these solutions are essentially the same as with the cyanide solutions, but they are more susceptible to all effects. Cleaning, contamination, brightener balance, are all more critical. Equipment is the same as for cyanide solutions.

[a] These solutions are generally proprietary and are sometimes supplied as ready-made concentrates. It is also possible to make them from zinc oxide and caustic soda.

TABLE 29. Acid Chloride Zinc Solutions

Amoniated Solutions	Rack g/L	Rack oz/gal	Barrel g/L	Barrel oz/gal
Zinc chloride	74–95	10–12.5	34–64	4.5–8.5
Ammonium chloride	90–120	12–16	112–225	15–30
Sodium chloride	45–67.5	6–9	—	—
Boric acid or ammonium acetate	19–26	2.5–3.5	19–26	2.5–3.5
operating conditions				
Zinc metal	36–45	4.8–6	16.5–30	2.2–4.0
Chloride ion	126–170	16.8–22.6	93–183	12.4–24
Boric acid or ammonium acetate	19–26	2.5–3.5	19–26	2.5–3.5
pH	4.4–5.6		4.4–5.6	
Temperature, C (F)	15–45 (60–110)		15–45 (60–110)	
Potassium Formulations				
Zinc chloride	75–85	10.0–11.3	50–85	6.7–11.3
Potassium chloride	202–270	27–36	180–270	24–36
Boric acid or	22.5–34	3.0–4.5	22.5–34	3.0–4.5
potassium acetate	7.5–12.0	1.0–1.6	7.5–12.0	1.0–1.6
operating conditions				
Zinc metal	36–40	4.8–5.4	24–40	3.2–5.4
Chloride ion	135–175	18.0–23.0	112–175	15–23
Boric acid or	22.5–34	3.0–4.5	22.5–34	3.0–4.5
Potassium acetate	7.5–12	1.0–1.6	7.5–12	1.0–1.6
pH	4.4–5.6		4.4–5.6	
Temperature, C (F)	15–45 (60–110)		15–45 (60–110)	

Equipment: Linings of PVC or polypropylene. The same for all auxiliary equipment. Continuous filtration is recommended.
Anodes: Zinc—high purity required. Anode current density: 0.5–3.0 A/dm^2 (5–30 asf)
Cathode current density: 0.2–6.0 A/dm^2 (2–60 asf).
Deposits dull: Lack of brightener.
Deposits brittle: Excess brightener or solution out of balance.
Pitting: Solution out of balance, or insufficient wetting agent.
Burning in high current density area: Excess current, or low metal.
Brownish deposit: Chloride too high; temperature too low; brighteners not properly balanced.
Poor throwing power: pH too low; metal content too high.
Deposit darkens when chromated: Metallic contamination; iron content too high.
Shelf roughness: pH too high; suspended matter in solution.
Barrel work shows print of barrel holes: Current density too high; barrel speed too slow; high iron concentration.
Spongy dark deposit: Acid content too low.
Low efficiency: Temperature too low; metal too low; solution out of balance.

Acknowledgment

We wish to thank the vendors of proprietary processes for their cooperation in supplying data.

References

General Plating

1. Blum and Hogaboom, "Principles of Electroplating and Electroforming," Third Ed., New York, 1949.
2. Lowenheim, F. A., "Modern Electroplating," Second Ed., John Wiley & Sons, New York, 1963.
3. Case, B. C., 34th Proc. Am. Electroplaters' Soc., 228-248 (1947).

Brass or Bronze Plating

4. Coats, H. P., Trans. Electrochem. Soc., 80, 445-457 (1941).
5. Graham, A. K., Proc. Am. Electroplaters' Soc., 35, 143-156 (1948).
6. Graham, A. K., Plating, 36, 1120-1126 (1949).

Cadmium Plating

7. Soderberg, G., and Westbrook, L. R., Trans. Electrochem. Soc., 80, 429-444 (1941).
8. Data received from General Chemical Div., Allied Chemical and Dye Corp., New York.
9. Harshaw Chemical Co., Cleveland, Ohio.

Copper Plating, Acid

10. Clifton, F. L., and Phillips, W. M., Proc. Am. Electroplaters' Soc., 30, 92-99 (1942).
11. Phillips, W. M., and Clifton, F. L., U.S. Patent 2, 489, 538 (Nov. 29, 1949).
12. Phillips, W. M., and Clifton, F. L., U.S. Patent 2,563,360 (Aug. 7, 1951).
13. Struyk, C., and Carlson, A. E., Monthly Rev., Am. Electroplaters' Soc., 33, 923-934 (1946).
14. Harshaw Chemical Co., Cleveland, Ohio.

Copper Plating, Alkaline

15. Graham, A. K., and Read, H. J., Trans. Electrochem. Soc., 80, 341-354 (1941).
16. Benner, H. L., and Wernlund, C. J., Trans. Electrochem. Soc., 80, 355-360 (1941).
17. Jernstedt, G. W., Proc Am. Electroplaters' Soc., 36, 63-78 (1949); 37, 151-170 (1950).
18. Coyle, T. C., Proc. Am. Electroplaters' Soc., 29, 113-117 (1941).
19. Stareck, J. E., Monthly Rev., Am. Electroplaters' Soc., 30, 25-27 (1943).
20. Fahy, E. R., and Sternberger, W. V., Metal Progr., 47, 278-279; 48, 1311-1312 (1945).
21. Metal & Thermit Corp., Rahway, N. J.

Chromium Plating

22. Dubpernell, G., Trans. Electrochem. Soc., 80, 589-615 (1941).
23. Stareck, J. E., Passal, F., and Mahlstedt, H., Proc. Am. Electroplaters' Soc., 37, 31-49 (1950).
24. Winters, J. B., and Hull, R. O., Proc. Am. Electroplaters' Soc., 36, 93-101 (1949).
25. Safranek, W. H., Miller, H. R., Hardy, R. W., and Faust, C. L., Plating, 47, 41 (1960).
26. Caldwell, M. R., and Sperry, L. B., Plating, 47, 397 (1960).
27. Safranek, W. H., Miller, H. R., and Faust, C. L., Plating, 47, 513-519 (1960).
28. Morisset, P., Oswald, J. W., Draper, C. R., and Pinner, R., "Chromium Plating," England, Robert Draper, Ltd., 1954.

Gold Plating

29. Weisberg, L., and Graham, A. K., Trans. Electrochem. Soc., 80, 509-520 (1941).
30. Parker, E. A., Plating, 36, 448-451, 516, 744 (1949).
31. Parker, E. A., Plating, 38, 1256-1259 (1951); 39, 43-46, 50 (1952).
32. Parker, E. A., Plating, 45, 631-5 (1958).
33. Parker, E. A., Plating, 46, 621-7 (1959).

Indium Plating

34. Linford, H. B., Trans. Electrochem. Soc., 79, 443-452 (1941).
35. Mohler, J. B., Metal Finishing, 43, 60, 77, (1945).
36. Martz, W. M., U.S. Patent 2,409,983 (Oct. 22, 1946).

Iron Plating

37. Thomas, C. T., Trans. Electrochem. Soc., 80, 499-507 (1941).
38. Hoessler, K, L., and Sloan, R. R., Electrotypers and Stereotypers Mag., 38, 7-16, July, 1952.

Lead and Lead-Tin Plating

39. Blum, W., and Haring, H. E., Trans. Electrochem. Soc., 40, 287-306 (1921).
40. Gray, A. G., and Blum, W., Trans. Electrochem. Soc., 80, 645-657 (1941).
41. DuRose, A. H., Trans. Electrochem. Soc., 89, 417-432 (1946).
42. Harshaw Chemical Co., Cleveland, Ohio.
43. General Chemical Div., Allied Chemical Corp., New York, N. Y.
44. Carlson, A. E., and Kame, J. M., Monthly Rev., Am. Electroplaters' Soc., 33, 255-261 (1946).
45. Graham, A. K., and Pinkerton, H. L., Trans. Inst. Metal Finishing, 51, 249-258 (1964).
46. Graham, A. K., and Pinkerton, H. L., Plating, 52, 309-314 (1965).
47. Graham, A. K., and Pinkerton, H. L., Am. Electroplaters' Soc., 50th Annual Technical Proceedings, 139-146, 1963.
48. Graham, A. K., and Pinkerton, H. L., Plating, 54, 367-377 (1967).

49. International Lead-Zinc Research Organization, Inc., Project LE-36, Final Report No. 28, High Speed Lead Plating, (coating properties and performance), December, 1968.

Nickel Plating

50. Pinner, W. L., Soderberg, G., and Baker, E. M., *Trans. Electrochem. Soc.*, **80**, 539-578 (1941).
51. Wesley, W. A., *Monthly Rev., Am. Electroplaters' Soc.*, **33**, 504-506 (1946).
52. Struyk, C., and Carlson, A. E., *Plating*, **37**, 1242-1246, 1263-1264 (1950).
53. Roehl, E. J., and Wesley, W. A., *Plating*, **37**, 142-146, 171 (1950).
54. Koessler, K. L., and Sloan, R. R., *Proc. Am. Electroplaters' Soc.*, **37**, 85-90 (1950).
55. Wesley, W. A., and Carr, D. S., *Plating*, **38**, 1243-1250, 1259 (1951).
56. Peters, E. L., *Proc. Am. Electroplaters' Soc.*, **38**, 69-84 (1950).
57. Savage, F. K., Graham, A. K., and Strothman, E. P., *Materials & Methods (Materials in Design Engineering)*, **36**, 94-97 (1952).
58. Barrett, R. C., *Plating*, **41**, 1027-1033 (1954).
59. Diggin, M. B., *Trans. Inst. Met. Finishing*, **31**, 243 (1954).
60. Diggin, M. B., *Metal Progr.*, **66** (Oct. 1954).
61. Fanner, D. A., and Hammond, R. A. F., *Trans. Inst. Met. Finishing*, **36**, Part 2, 32-43 (1958/1959).
62. Knap, B. B., and Carr, D. S., U.S. Patent 2,594,933 (Apr. 29, 1952).
63. Brenner, A., Couch, D. E., and Williams, E. K., *Plating*, **37**, 36, 161 (1950).

Black Nickel Plating

64. Moline, W. E., *Proc. Am. Electroplaters' Soc.*, **31**, 21-24 (1943).

Silver Plating

65. Wood, D., *Proc. Am. Electroplaters' Soc.*, **25**, 150-155 (1937).
66. Promisel, N., and Wood, D., *Trans. Electrochem. Soc.*, **80**, 459-487 (1941).
67. Schaefer, R. A., and Mohler, J. B., *Proc. Am. Electroplaters' Soc.*, **31**, 29-33 (1943).
68. Hart, J. S., and Heussner, C. E., *Monthly Rev., Am. Electroplaters' Soc.*, **33**, 142-149 (1946).
69. Mesle, F. C., *Monthly Rev., Am. Electroplaters' Soc.*, **33**, 937-942, 1056-1065 (1946).
70. Schaefer, R. A., *Monthly Rev., Am. Electroplaters' Soc.*, **33**, 1176-1178 (1946).

Rhodium Plating

71. Fink, C. G., and Lambros, G. C., *Trans. Electrochem. Soc.*, **63**, 181-185 (1932).
72. Atkinson, R. H., and Draper, A. R., *J. Electrodepositors' Soc.*, **9**, 77-82 (1933).
73. Schumpelt, K., *Trans. Electrochem. Soc.*, **80**, 489-490 (1941).
74. Brenner, A., and Olson, W. A., *Proc. Am. Electroplaters' Soc.*, **33**, 29-31 (1946).
75. Stuart, A. H., *Electroplating*, **1**, Jan., 88-90, 102 (1948).
76. Schumpelt, K., *Plating*, **37**, 1052 (1950).
77. Reid, F. H. (to The International Nickel Co.), U.S. Patents 2,577,364-5 (Dec. 4, 1951).
78. Wiesner, H. J., *Proc. Am. Electroplaters' Soc.*, **39**, 79-99 (1952).
79. Pinner, R., *Electroplating*, **6**, 276-281 (1953).
80. Laister, E. H., and Benham, R. R., *J. Electrodepositors' Tech. Soc.*, **29**, 1-25 (1953).
81. Parker, E., *Plating*, **42**, 882 (1955).
82. Weisner, H. J., *Plating*, **43**, 347 (1956).
83. Reid, F. H., *J. Electrodepositors' Tech. Soc.*, **33**, 105-140 (1956).
84. Reid, F. H. (to The International Nickel Co.), U.S. Patent 2,866,740 (Dec. 30, 1958).

Tin Plating, Alkaline

85. Oplinger, F. F., and Bauch, F., *Trans. Electrochem. Soc.*, **80**, 617-629 (1941).
86. Sternfels, M. M., and Lowenheim, F. A., *Trans. Electrochem. Soc.*, **82**, 77-100 (1942).
87. Lowenheim, F. A., *Trans. Electrochem. Soc.*, **84**, 195-209 (1943).
88. Lowenheim, F. A., Tin Plating from Potassium Stannate Baths, U.S. Patent 2,424,472 (July 22, 1947).

Tin Plating, Acid

89. Pine, P. R., *Trans. Electrochem. Soc.*, **80**, 631-644 (1941).
90. Anon., *Can. Metals Met. Inds.*, **13**, 8-11 (1950).
91. Schweiker, E. W., (to E. I. duPont de Nemours & Co., Inc.), Electrodeposition of Tin, U.S. Patent 2,407,579 (Sept. 10, 1946).
92. Schweiker, E. W., (assigned to E. I. duPont de Nemours & Co., Inc.), Tin Electrodepositing Composition and Process, U.S. Patent 2,402,185 (June 18, 1946).

Tin-Copper Alloy Plating, Cyanide

93. Angles, R. M., Jones, F. V., Prive, J. W., and Cuthbertson, J. W., *J. Electrodepositors' Tech. Soc.*, **21**, 19-44 (1946).
94. "Working Instructions for Speculum Plating," Tin Research Institute, 1947.
95. Cuthbertson, J. W., *J. Electrodepositors' Tech. Soc.*, **23**, 143-150 (1948).
96. Sawyer, W. H., *J. Electrodepositors' Tech. Soc.*, **23**, 151-162 (1948).
97. Bennett, P. S., *J. Electrodepositors' Tech. Soc.*, **26**, 107-118 (1950).
98. Lowenheim, F. A., U.S. Patent 2,528,601 (Nov. 7, 1950).
99. Bair, S. W., and McNaughton, D. J., U.S. Patent 1,970,548-9 (1950).
100. Rooksby, H. P., *J. Electrodepositors' Tech. Soc.*, **26**, 119-124 (1950).

101. Heymann, E., and Schmerling, G., U.S. Patent 2,722, 508 (Nov. 1, 1955).
102. Lowenheim, F. A., *Metal Finishing*, **53**, 51 (1955).
103. Heymann, E., and Schmerling, G., U.S. Patent 2,793,990 (May 28, 1957).
104. Lee, W. T., *Trans. Inst. Met. Finishing*, **36**, Part 2, 51–57 (1958/59).

Tin-Nickel Alloy Plating

105. Parkinson, N., *J. Electrodepositors' Tech. Soc.*, **27**, 129–151 (1951).
106. Cuthbertson, J. W., and Parkinson, N., *J. Electrochem. Soc.*, **100**, 107–119 (1953).
107. Britton, S. C., and Angles, R. M., *Trans. Inst. Met. Finishing*, **29**, 26–39 (1953).
108. Britton, S. C., and Michael, D. G., *Trans. Inst. Met. Finishing*, **29**, 40–58 (1953).
109. Lowenheim, F. A., Sellers, W. W., and Carlin, F. X., *J. Electrochem. Soc.*, **105**, 338–348 (1958).
110. Davies, A. E., *Trans. Inst. Met. Finishing*, **31**, 401–415 (1954).
111. Lowenheim, F. A., *Trans. Inst. Met. Finishing*, **31**, 386–397 (1954).
112. Ramanathan, V. R., *Trans. Inst. Met. Finishing*, **34**, 1–7 (1957).

Tin-Zinc Alloy Plating

113. Angles, R. M., *J. Electrodepositors' Tech. Soc.*, **21**, 45–64 (1946).
114. Cuthbertson, J. W., and Angles, R. M., *J. Electrochem. Soc.*, **94**, 73 (1948).
115. MacIntosh, R. M., and Lowenheim, F. A., *J. Electrodepositors' Tech. Soc.*, **27**, 115 (1951).
116. "The Plating of Tin-Zinc Alloys," Metal & Thermit Corp. Technical Data Sheet 107 (1952).
117. Forrester, P. G., and Lowinger, V. A. (to Glacier Metal Co.), British Patent 669,344 (Apr. 2, 1952).
118. Lowenheim, F. A. (to Metal & Thermit Corp.), U.S. Patent 2,675,347 (Apr. 13, 1954).
119. Saubestre, E. B., and Arnaut, A. D. (to Sylvania Electric Products, Inc.), U.S. Patent 2,898,274 (Aug. 4, 1959).
120. Arnaut, A. D., and Saubestre, E. B. (to Sylvania Electric Products, Inc.), US.S Patent 2,907,702 (Oct. 6, 1959).
121. Saubestre, E. B. (to Sylvania Electric Products, Inc.), U.S. Pat. 2,900,314 (Aug. 18, 1959).

Zinc Plating, Alkaline

122. Hull, R. O., and Wernlund, C. J., *Trans. Electrochem. Soc.*, **80**, 407–427 (1941).

Zinc Plating, Acid

123. Lyons, E. H., *Trans. Electrochem. Soc.*, **80**, 387–405 (1941). "Hydrogen Embrittlement."
124. Zapffe, C. A., and Faust, C. L., *28th Proc. Am. Electroplaters' Soc.*, 1–25 (1940).
125. The Electrochemical Society, "Modern Electroplating," 1st Ed., 114–115 (1942).
126. Valentine, K. B., Method of Processing Electroplated Spring Steel Parts, U.S. Patent 2,572,219 (Oct. 23, 1951).
127. Read, H. J., *Proc. Am. Electroplaters' Soc.*, **47**, 110 (1960).
128. "Hydrogen Embrittlement in Metal Finishing," edited by H. J. Read, Reinhold Publishing Corporation, N. Y., in preparation.
129. Lawrence, Samuel C., Jr., *Proc. Am. Electroplaters' Soc.*, **47**, 135 (1960).

Stripping Copper

130. Mathers, F. C., Landwerlen, C. E., and Martin, E. L., *Monthly Rev., Am. Electroplaters' Soc.*, **32**, 268–270, 672–679, 719 (1945).
131. Mathers, F. C., and Martin, E. L., *Monthly Rev., Am. Electroplaters' Soc.*, **33**, 958–962 (1946).
132. Mathers, F. C., and Martin, E. L., *Plating*, **35**, 569–570, 575–576 (1948).
133. *Iron Age*, **155**, 56, May 24, 1945.

Palladium Plating

134. Atkinson, R. H., and Raper, A. R., *J. Electrodepositors' Tech. Soc.*, No. 10, 1–15 (1933).
135. Schumpelt, K., *Trans. Electrochem. Soc.*, **80**, 493–4 (1941).
136. Lambros, G. C., *Metal Finishing*, **4**, 73, (1943).
137. Schumpelt, K., *Plating*, **37**, 1055 (1950).
138. Wise, E. M., *J. Electrochem. Soc.*, **97**, No. 3, 57c–64c (1950).

Platinum Plating

139. Schumpelt, K., *J. Electrodepositors' Tech. Soc.*, **13**, No. 24, 1–6 (1937).
140. Atkinson, R. H., *J. Electrodepositors' Tech. Soc.*, **13**, No. 25, 1–14 (1937).
141. Davies, E. C., and Powell, A. R., *J. Electrodepositors' Tech. Soc.*, **13**, No. 26, 1–7 (1937).
142. Kushner, J. B., *Metal Finishing*, **37**, 131, 182 (1939).
143. Schumpelt, K., *Trans. Electrochem. Soc.*, **80**, 489–493 (1941).
144. Young, C. B. F., and Davis, J. F., *Metal Finishing*, **40**, 252 (1942).
145. Atkinson, R. H., *Trans. Inst. Met. Finishing*, **36**, 7–16 (1958/59).
146. Rothchild, B. F., and Saunders, D., *Plating*, 1363–1369 (1969).

7. PLATING STANDARDS AND SPECIFICATIONS

Edward T. Clegg

*Engineering Standardization Division,
Army Materials and Mechanics Research Center*

AND

Fielding Ogburn

Materials Chemistry Division, National Bureau of Standards

The term *plating standards*, as used here, includes a variety of documents, including specifications, test methods, practices, definitions, guidelines, etc. They are developed and issued by many industrial, commercial and government organizations and cover a wide variety of products and applications. To keep the scope of this chapter to manageable proportions, covering only the most important information, only those standards of the broadest significance will be treated. These are plating specifications, test methods and practices of the International Organization for Standardization (ISO), the American Society for Testing and Materials (ASTM), the Aerospace Material Specifications (AMS) of the Society of Automotive Engineers and the Federal and Military Specifications of the U.S. government. Coating specifications for specific items are not included, though lists of related standards are.

SPECIFICATION REQUIREMENTS

For the purposes of this chapter, a plating specification is considered to be a set of requirements that provides a detailed description of a plating process or of a plated coating. Ideally the coating is specified and the supplier is allowed to use any process he wishes. In practice, however, we find a mixture of process requirements and product requirements, but with a strong emphasis on the requirements for the final coating.

Most product specifications for plated coatings include requirements pertaining to workmanship, preparation of surfaces for plating, appearance, adhesion, thickness, resistance to accelerated corrosion and post-plating treatment. They also include methods of testing to determine compliance.

Workmanship is a description of the quality of all the processing results. It is intended to cover those attributes not covered by the other requirements.

Since the condition of the substrate influences both the characteristics of the coating and the mechanical performance of the plated part, it is important that the condition of the substrate be characterized and controlled. Accordingly, specifications usually require that the substrate shall be free from defects that will be detrimental to the appearance and protective value of the coating. Since the damaging effect of hydrogen can occur during processing with solutions and under operating conditions which generate hydrogen, they often require that residual tensile stresses be minimized by a suitable stress-relief heat treatment or that the surface be mechanically treated to introduce residual compressive stresses, such as shot-peering prior to preplate processing.

The requirements for appearance are usually described in terms of bright, dull, matte or satin finishes. There are no practical means of quantitatively defining and measuring these conditions. Appearance requirements also prohibit visible flaws such as pits, nodules, blisters, stress cracks and indications of burning.

Good adhesion of the coating to the substrate is essential. Except for electrodeposited coatings on plastics, there is no suitable method for its quantitative determination; thus, it is defined in terms of qualitative tests.

Thickness of the coating is important since the protective value of the coating is dependent on its thickness. Because it is important and readily defined and measured, virtually every plating specification will state minimum local thicknesses or minimum average thickness. Since a coating is most likely to fail where it is thinnest, most specifications state the minimum thickness. The average thickness may be specified if it is more readily measured than the minimum thickness if it conforms with commercial practices, or if, for some reason, it is more significant than the minimum thickness. The actual thickness specified, however, depends on the basis metal, the coating material and the service to which the article will be subjected.

Thickness requirements of ISO, ASTM, AMS and government plating specifications are summarized in the accompanying tables. The ISO and ASTM specifications use metric units; generally, AMS and government specifications are still using the conventional units. The tables give the thicknesses as given in the specifications with approximate metric or English equivalents in parentheses.

Though the requirements discussed above are important for determining the quality of a coating, factors such as porosity, internal stress and impurities may determine the corrodibility and protective value of coatings. Whether a coating is poor with respect to these conditions can often be determined by submitting the coated part or specimen to an accelerated corrosion test; hence, many specifications require such a test.

Requirements for post-plating treatments may include supplementary chromate or phosphate coatings (see Chapter 13) and may also include chemical and corrosion tests for verification of quality (see Chapter 10). The specifications may also include a hydrogen embrittlement relief treatment for plated steel parts (Rockwell C "40" and higher hardness) since, as discussed in Chapters 6 and 13, hydrogen introduced during the plating process affects the mechanical behavior of plated steel parts.

A coating specification may also include a requirement for use of a sampling plan for acceptance testing. This would set forth a definition of a lot, the number of items to be tested, and the criteria for rejection of a lot.

Usually the specification will classify or designate the coatings according to a system of classes and types based on coating thickness, appearance, supplementary coatings, etc. ASTM and ISO are now using a system comprised of service condition numbers and a classification number comprised of letters and numbers to indicate the substrate material, the coating material and the coating thickness. For example; the ASTM specification for nickel-chromium coatings, B 456, (summarized in Tables 9 through 13) defines four service condition (SC) numbers as follows:

SC 4, very severe. Service conditions that include likely damage from denting, scratching and abrasive wear in addition to exposure to corrosive environments.

SC 3, severe. Exposure that is likely to include occasional or frequent wetting by rain or dew or possibly strong cleaners and saline solutions.

SC 2, moderate. Exposure indoors in places where condensation of moisture may occur.

SC 1, mild. Exposure indoors in normally warm, dry atmospheres with coating subject to minimum wear or abrasion.

For each service condition, the specification calls for one or more alternate coating systems. To designate the coating system, a classification number such as "Cu/Ni30d Cr mp" is used. Cu indicates a copper alloy substrate, Ni30d refers to 30 μm of a double-layer (duplex) nickel coating, and Cr mp designates 0.25 μm (0.01 mil) of micro-porous chromium. Chemical symbols are used to designate the various metals, and lowercase letters the kind of electrodeposit:

b—bright nickel
p—dull nickel requiring polishing to give full brightness
d—double- or triple-layer nickel
r—regular (conventional) chromium; 0.25 μm (0.010 mil) thick

for SC 4, SC 3 and SC 2, and 0.13 μm (0.005 mil) for SC 1

mc—micro-cracked chromium; thickness same as for r.

mp—micro-porous chromium; thickness same as for r.

Each of these is defined in more detail in the specification.

When using this specification, the coating system is specified by the classification number. Alternatively, the service condition number may be specified and then any of the coatings systems given for that service condition number may be used by the plater.

The following list of specifications includes most of the plating and anodizing specifications of ISO, ASTM, AMS and the government. Those ASTM and AMS specifications which have been accepted for use by the Department of Defense are indicated by an asterisk preceding the number. The thickness requirements are summarized in the tables indicated in the list.

Anodized Aluminum (Table 1)
  ASTM *B 580 Anodic Oxide Coatings on Aluminum
  AMS 2468 Hard Coating Treatment of Aluminum Alloys
  AMS *2469 Process and Performance Requirements for Hard Coating Treatment of Aluminum Alloys
  AMS 2470 Anodic Treatment of Aluminum Base Alloys, Chromic Acid Process
  AMS 2471 Anodic Treatment of Aluminum Base Alloys, Sulfuric Acid Process, Undyed Coating
  AMS 2472 Anodic Treatment of Aluminum Base Alloys, Sulfuric Acid Process, Dyed Coating
  MIL-A-8625 Anodic Coatings, for Aluminum and Aluminum Alloys
  MIL-C-60536 Coating, Anodic, Hard, for Aluminum and Aluminum Alloys
  MIL-C-60639 Coating, Anodic, Conventional, for Aluminum and Aluminum Alloys

Anodized Magnesium (Table 2)
  AMS 2476 Electrolytic Treatment for Magnesium Base Alloys, Alkaline Type, Full Coat
  AMS 2478 Anodic Treatment of Magnesium Base Alloys, Acid Type, Full Coat
  AMS 2479 Anodic Treatment of Magnesium Base Alloys, Acid Type, Thin Coat
  MIL-M-45202 Magnesium Alloy, Anodic Treatment of

Anodized Zinc
  MIL-A-81801 Anodic Coatings for Zinc and Zinc Alloys

Anodized Titanium
  AMS 2488 Anodic Treatment of Titanium and Titanium Alloys

Black Coatings
  MIL-C-14538 Chromium Plating, Black (Electrodeposited)
  MIL-P-18317 Plating, Black Nickel (Electrodeposited) on Brass, Bronze or Steel

Cadmium (Table 3)
  ISO 2082 Metallic Coatings—Electroplated Coatings of Cadmium on Iron or Steel
  ASTM *A 165 Electrodeposited Coatings of Cadmium on Steel
  AMS 2400 Cadmium Plating
  AMS 2401 Cadmium Plating, Low Hydrogen Content Deposit
  Fed. QQ-P-416 Plating, Cadmium (Electrodeposited)

Cadmium-Titanium Alloy
  AMS *2419 Cadmium-Titanium Alloy Plating

Chromium (Table 4)
  ASTM B 650 Electrodeposited Engineering Chromium Coatings on Ferrous Substrates
  AMS 2406 Chromium Plating, Hard Deposit
  AMS 2407 Chromium Plating, Porous
  Fed. QQ-C-320 Chromium Plating (Electrodeposited)
  MIL-C-20218 Chromium Plating, Electrodeposited Porous

Copper (Table 5)
  AMS 2418 Copper Plating
  MIL-C-14550 Copper Plating (Electrodeposited)

Gold (Table 6)
  ASTM B 488 Electrodeposited Coatings of Gold for Engineering Uses
  AMS 2422 Gold Plating for Electronic Applications
  AMS *2425 Gold Plating for Thermal Control
  MIL-G-45204 Gold Plating, Electrodeposited

Lead (Table 7)
  ASTM B 200 Electrodeposited Coatings of Lead and Lead-Tin Alloys on Steel and Ferrous Alloys
  AMS 2414 Lead Plating
  MIL-L-13808 Lead Plating (Electrodeposited)

Lead-Indium Alloy
  AMS *2415 Lead and Indium Plating

Lead-Tin Alloy
  ASTM B 200 Electrodeposited Coatings of Lead and Lead-Tin Alloys on Steel and Ferrous Alloys

MIL-L-46064 (MR) Lead-Tin Alloy Coating, Electrodeposited

Nickel (Table 8)
  ISO 1458 Metallic Coatings—Electrodeposited Coatings of Nickel
  ASTM B 689 Electroplated Engineering Nickel Coatings
  Fed. QQ-N-290 Nickel Plating (Electrodeposited)
  AMS 2403 Nickel Plating, General Purpose
  AMS 2423 Nickel Plating, Hard Deposit
  AMS *2424 Nickel Plating, Low-Stressed Deposit
  MIL-P-27418 (USAF) Plating, Soft, Nickel (Electrodeposited, Sulfamate Bath)

Nickel plus Chromium (Tables 9–13)
  ISO 1456 Electroplated Coatings of Nickel Plus Chromium
  ISO 1457 Electroplated Coatings of Copper Plus Nickel Plus Chromium on Iron or Steel
  ASTM B 456 Electrodeposited Coatings of Copper Plus Nickel Plus Chromium and Nickel Plus Chromium

Nickel-Cadmium
  AMS 2416 Nickel-Cadmium Plating, Diffused

Nickel-Zinc Alloy (Table 14)
  AMS *2417 Nickel-Zinc Alloy Plating

Palladium (Table 15)
  ASTM B 679 Electrodeposited Coatings of Palladium for Engineering Use
  MIL-P-45209 Palladium Plating, Electrodeposited

Rhodium (Table 16)
  ASTM B 634 Electrodeposited Coatings of Rhodium for Engineering Use
  MIL-R-46085 Rhodium Plating, Electrodeposited

Silver (Tables 17, 18)
  AMS *2410 Silver Plating, Nickel Strike, High Bake
  AMS 2411 Silver Plating, for High-Temperature Applications
  AMS *2412 Silver Plating, Copper Strike, Low Bake
  Fed. QQ-S-365 Silver Plating (Electrodeposited), General Requirements for

Silver Plus Rhodium (Table 19)
  AMS *2413 Silver and Rhodium Plating

Tin (Table 20)
  ISO 2093 Metallic Coatings—Electroplated Coatings of Tin
  ASTM *B 545 Electrodeposited Coatings of Tin
  AMS 2408 Tin Plating
  MIL-T-10727 Tin Plating: Electrodeposited or Hot Dipped, for Ferrous and Nonferrous Metals

Tin Plate (Table 21)
  ASTM *A 624

Tin-Cadmium Alloy
  MIL-P-23408 Plating, Tin-Cadmium (Electrodeposited)

Tin-Lead Alloy (Table 22)
  ASTM B 579 Electrodeposited Coatings of Tin-Lead Alloy (Solder Plate)
  MIL-P-81728 Plating, Tin-Lead (Electrodeposited)

Tin-Nickel Alloy (Table 23)
  ISO 2179 Electroplated Coatings of Tin-Nickel Alloy
  ASTM B 605 Electrodeposited Coatings of Tin-Nickel Alloy

Zinc (Table 24)
  ISO 2081 Metallic Coatings—Electroplated Coatings of Zinc on Iron or Steel
  ASTM *B 633 Electrodeposited Coatings of Zinc on Iron and Steel
  AMS 2402 Zinc Plating
  Fed. QQ-Z-325 Zinc Coatings (Electrodeposited), Requirements for

TEST METHODS

Although the test methods for compliance with the above requirements may vary in detail from one specification to another, they are fairly well standardized. The designations and titles of ISO and ASTM test methods are given below.

Adhesion Tests

Tests of adhesion often involve bending or grinding the sample and subsequently examining the coating-basis metal interface for evidences of detachment.

ISO 2819. Metallic coatings on metallic substrates—electrodeposited and chemically deposited coatings—review of methods available for testing adhesion.

ASTM B571. Test methods for adhesion of metallic coatings.

Thickness Tests

ASTM B 554. Measurement of thickness of metallic coatings on nonmetallic substrates.

ASTM B 659. Measuring thickness of electrodeposited and related coatings.

TABLE 1. SPECIFICATIONS FOR ANODIZED ALUMINUM

Specification	Classification	Thickness
ASTM *B 580-79	Type A hard coat	50 µm (2.0 mil)
	B architectural I	18 µm (0.7 mil)
	C architectural II	10 µm (0.4 mil)
	D auto. exterior	8 µm (0.3 mil)
	E interior a	5 µm (0.2 mil)
	F interior b	3 µm (0.1 mil)
	G chromic acid	1 µm (0.04 mil)
AMS 2468C	hard coat	2 mil (50 µm)
AMS *2469C	hard coat	2 mil (50 µm)
AMS 2470G	chromic acid	no thickness requirement
AMS 2471C	sulfuric acid, undyed	no thickness requirement
AMS 2472B	sulfuric acid, dyed	no thickness requirement
MIL-A-8625C	chromic acid	0.05–0.3 mil (1.3–8 µm)
	sulfuric acid	0.1–1 mil (3–30 µm)
	hard coat	0.5–4.5 mil (13–114 µm)
MIL-C-60536(MU)		nominal 50 µm (2.0 mil)
MIL-C-60539(MU)	chromic acid	no thickness requirement
	sulfuric acid	no thickness requirement

TABLE 2. SPECIFICATIONS FOR ANODIZED MAGNESIUM

Specification	Classification	Thickness
AMS 2476A	alkaline type, full coat	no thickness requirement
AMS 2478B	acid type, full coat	0.9–1.6 mil (23–41 µm)
AMS 2479A	acid type, thin coat	0.1–0.5 mil (3–13 µm)
Mil-M-45202C	light-HAE	0.1–0.3 mil (3–8 µm)
	light-Dow 17	0.1–0.5 mil (3–13 µm)
	heavy-HAE	1.3–1.7 mil (33–43 µm)
	heavy-Dow 17	0.9–1.6 mil (23–41 µm)

ISO 3882. Metallic and other non-organic coatings—review of methods of measurement of thickness.

ASTM B 137. Measurement of *weight of coating* on anodically coated aluminum.

ISO 2106. Surface treatment of metals—anodisation (anodic oxidation) of aluminum and its alloys—measurement of the mass of the oxide coatings—*gravimetric method.*

ASTM B 244. Measurement of thickness of anodic coatings on aluminum and of other nonconductive coatings on nonmagnetic basis metals with *eddy current* instruments.

ISO 2360. Nonconductive coatings on nonmagnetic basis metals—measurement of coating thickness—*eddy current method.*

ASTM B 487. Measuring metal and oxide coating thicknesses by *microscopical* examination of a cross-section.

ISO 1463. Metal and oxide coatings—measurement of thickness by *microscopical* examination of cross-sections.

ASTM B 499. Measurement of coating thicknesses by the *magnetic method:* nonmagnetic coatings on magnetic basis metals.

ISO 2178. Nonmagnetic metallic and vitreous or porcelain enamel coatings on magnetic basis metals, measurement of coating thickness, *magnetic method.*

ASTM B 504. Measurement of the thickness of metallic coatings by the *coulometric method.*

ISO 2177. Metallic coating—measurement of coating thickness—*coulometric method* by anodic dissolution.

ASTM B 530. Measurement of coating thickness by the *magnetic method:* electrodeposited nickel coatings on magnetic and nonmagnetic substrates.

TABLE 3. SPECIFICATIONS FOR CADMIUM PLATING ON STEEL

Specification	Service Condition Number	Class or Type	Thickness, Minimum	
			μm	(mil)
ISO 2082-83	4	Fe/Cd25	25	(1.0)
	3	Fe/Cd12	12	(0.5)
	2	Fe/Cd8	8	(0.3)
	1	Fe/Cd5	5	(0.2)
ASTM *A 165-80		NS	12	(0.5)
		OS	8	(0.3)
		TS	5	(0.2)
Fed QQ-P-416C		1	(13)	0.50
		2	(8)	0.30
		3	(5)	0.20

			Externally Threaded Parts			
			External Threads		Unthreaded Surfaces	
Class	μm	mil	μm	mil	μm	mil
AMS 2400R	— (8–13)	0.3–0.5	(3–10)	0.1–0.4	(5–13)	0.2–0.5
AMS 2401C	1 (3–8)	0.1–0.3	(3–8)	0.1–0.3	(5–10)	0.2–0.4
Low hydrogen	2 (5–10)	0.2–0.4	(3–10)	0.1–0.4	(5–10)	0.2–0.4
content	3 (8–13)	0.3–0.5	(5–13)	0.2–0.5	(8–13)	0.3–0.5
AMS 2419A	4 (10–15)	0.4–0.6	(8–15)	0.3–0.6	(10–15)	0.4–0.6
0.1–0.2 Ti	5 (13–18)	0.5–0.7	(10–18)	0.4–0.7	(13–18)	0.5–0.7

ISO 2361. Electrodeposited nickel coatings on magnetic and nonmagnetic substrates—measurement of coating thickness—*magnetic method*.

ASTM B 555. Measurement of electrodeposited metallic coating thickness by the *dropping test*.

ASTM B 556. Measurement of thin chromium coatings by the *spot test*.

ASTM B 567. Measurement of coating thickness by the *beta backscatter method*.

ISO 3543. Metallic and nonmetallic coatings—measurement of thickness—*beta backscatter method*.

TABLE 4. SPECIFICATIONS FOR CHROMIUM PLATING

Specification	Class or Type	Minimum Thickness
(See tables for nickel specifications)		
Fed. QQ-C-320B	1. Decorative	0.01 mil (0.3 μm)
	2. Engineering	2 mil (50 μm)
AMS 2406F	Hard Deposit	specified on drawings 2 mil (50 μm) recommended when used for corrosion protection
AMS 2407C	Porous	4–6 mil (100–150 μm) Porosity shall be of pinpoint type and depth shall be 1–1.5 mil (25–38 μm)
MIL-C-20218	Porous	5 mil (130 μm)

TABLE 5. SPECIFICATIONS FOR COPPER PLATING

Specification	Class	Minimum Thickness mil	(μm)
MIL-C-14550A	0	1.0-5.0	(25-127)
	1	1.00	(25)
	2	0.50	(13)
	3	0.20	(5)
	4	0.10	(2.5)
AMS 2418C		0.5-0.7	(12.7-17.8)
	flash	0.1	(3)

ASTM B 568. Measurement of coating thickness by *X-ray spectrometry*.

ISO 3497. Metallic coatings—measurement of coating thickness—*X-ray spectrometric methods*.

ASTM B 588. Measurement of thickness of transparent or opaque coatings by double-beam *interference microscope* technique.

ISO 3868. Metallic and other non-organic coatings—measurement of coating thickness—Fizeau *multiple-beam interferometry* method.

ISO 4518. Metallic coatings—measurement of coating thickness—*profilometric method*.

TABLE 6. SPECIFICATIONS FOR GOLD PLATING FOR ENGINEERING USE

Specification	Class	Minimum Thickness (μm)	(mil)
ASTM B 488-80			
Type 1 99.90% gold, minimum			
Type 2 99.70% gold, minimum			
Type 3 99.00% gold, minimum			
Hardness code A: 90 HK25 max			
Hardness code B: 91-200 HK25			
Hardness code C: 130-250 HK25			
suggested minimum thicknesses of gold:		0.25	(0.01)
		1	(0.04)
		2.5	(0.1)
		5	(0.2)
		10	(0.4)
		20	(0.8)
		40	(1.6)
MIL-G-45204B			
Type I 99.7% gold			
Type II 99.0% gold			
Type III 99.9% gold			
	00	(0.5)	0.02
	0	(0.8)	0.03
	1	(1.3)	0.05
	2	(2.5)	0.10
	3	(5)	0.20
	4	(8)	0.30
	5	(13)	0.50
	6	(38)	1.50

*When undercoat is required on steel and zinc alloys, the minimum total thickness is 1 mil (25 um).

AMS 2422B		(1.3)	0.05
AMS *2425B	Minimum thickness of:		
	copper:	(<2.54)	<0.1
	nickel:	(10.2-22.9)	0.4-0.9
	gold:	(2.03)	0.08

TABLE 7. SPECIFICATIONS FOR LEAD AND LEAD-TIN PLATING ON STEEL

Specification	Type	Class	Copper		Lead	
			μm	(mil)	μm	(mil)
ASTM B 200-76 <15% tin		suggested thicknesses:			38	1.5
					25	1.0
					12	0.5
					6	0.24
ASTM B 200-proposed revision <15% tin		4			40	(1.5)
		3			25	(1.0)
		2			12	(0.5)
		1			6	(0.24)
			mil	(μm)	mil	(μm)
MIL-L-13808B 0% tin	I	1	—	—	1.0	(25)
	I	2	—	—	0.50	(13)
	I	3	—	—	0.25	(6)
	II	1	0.015	(0.4)	1.0	(25)
	II	2	0.015	(0.4)	0.50	(13)
	II	3	0.015	(0.4)	0.25	(6)
AMS 2414B 0% tin		flash			0.5–0.7	(13–18)
					0.1	(2.5)
MIL-L-46064(MR) 6% tin			as specified in contract			

ISO 3892. Conversion coatings on metallic materials—determination of mass per unit area—*gravimetric methods*.

Corrosion Tests

ASTM B 117. Salt spray (fog) testing.

ISO 3768. Metallic coatings—neutral salt spray test (NSS test).

ASTM B 287. Acetic acid-salt spray (fog) testing.

ISO 3769. Metallic coatings—acetic acid-salt spray test (ASS test).

ASTM B 368. Copper-accelerated acetic acid-salt spray (fog) testing (CASS test).

ISO 3770. Metallic coatings—copper-accelerated acetic acid salt spray test (CASS test).

ASTM B 380. Corrosion testing of decorative chromium electroplating by the Corrodkote procedure.

ISO 4541. Metallic and other non-organic coatings—Corrodkote corrosion test (CORR test).

ASTM B 538. FACT (Ford anodized aluminum corrosion test) testing.

ASTM B 627. Electrolytic corrosion testing (EC test).

ISO 4539. Electrodeposited chromium coatings—electrolytic corrosion testing (EC test).

ISO 4538. Metallic coatings—thioacetamide corrosion test (TAA test).

Microhardness Tests

ASTM B 578. Microhardness of electroplated coatings.

ISO 4516. Metallic and related coatings—Vickers and Knoop microhardness tests.

Ductility Tests

ASTM B 489. Practice for bend test for ductility of electrodeposited and autocatalytically deposited metal coatings on metals.

ASTM B 490. Practice for micrometer bend test for ductility of electrodeposits.

Solderability Test

ASTM B 678. Solderability of metallic coated products.

TABLE 8. Specifications for Nickel Plating Without Overcoat of Chromium

Specification	Grade	Classification Number	Basis Metal	Minimum Thickness Copper Undercoat µm	(mil)	Nickel µm	(mil)
ISO 1458-74	high	Fe/Ni30b	steel	8	(0.3)	30	(1.2)
		Zn/Cu Ni25b	zinc alloy	8	(0.3)	25	(1.0)
		Cu/Ni20b	copper or copper alloy			20	(0.8)
	medium	Fe/Ni20b	steel	8	(0.3)	20	(0.8)
		Zn/Cu Ni5b	zinc alloy	8	(0.3)	5	(0.8)
		Cu/Ni10b	copper or copper alloy			10	(0.4)
	low	Fe/Ni10b	steel	8	(0.3)	10	(0.4)
		Zn/Cu Ni8b	zinc alloy	8	(0.3)	8	(0.3)
		Cu/Ni5b	copper or copper alloy			5	(0.2)

p or d nickel may be substituted for b nickel.
Brass may be substituted for copper undercoat.

For additional requirements pertaining to p, d, and b nickel, see Table 13.

ASTM B 689-81 Engineering Coatings

		5				5	(0.2)
		25				25	(1.0)
		50				50	(2.0)
		100				100	(3.9)
		200				200	(7.9)

Fed. QQ-N-290A
Class 1 Corrosion Protective Plating

Class	Basis Metal	Cu (µm)	(mil)	Ni (µm)	(mil)
A	steel, zinc or zinc alloy	(5)	0.2	(41)	1.6
B	steel, zinc or zinc alloy	(5)	0.2	(30)	1.2
	copper and copper alloy	*(8)	0.3	(30)	1.2
C	steel, zinc or zinc alloy	(5)	0.2	(25)	1.0
	copper and copper alloy	*(8)	0.3	(25)	1.0
D	steel, zinc or zinc alloy	(5)	0.2	(20)	0.8
	copper and copper alloy	*(8)	0.3	(20)	0.8
E	steel, zinc or zinc alloy	(5)	0.2	(15)	0.6
	copper and copper alloy	*(8)	0.3	(15)	0.6
F	steel, zinc or zinc alloy	(5)	0.2	(10)	0.4
	copper and copper alloy	*(8)	0.3	(10)	0.4
G	copper and copper alloy	*(8)	0.3	(5)	0.2

*Copper undercoat required only if copper alloy is greater than 40% zinc.

Class 2 Engineering Plating   (76)   3 or as specified— by purchase order

AMS 2403G   as specified on drawings
If nickel flash is specified, nickel thickness shall be approximately 0.1 mil (3um).
AMS 2423A (hard deposit)   as specified on drawings
AMS *2424B (low stress deposit)   as specified on drawings
MIL-P-27418 (USAF) (soft deposit)   2 + 0.3 mil (51 + 8 µm)

TABLE 9. SPECIFICATIONS FOR NICKEL PLUS CHROMIUM PLATING ON STEEL

Grade or Service Condition Number	Classification Number		Minimum Thickness					
			Nickel			Chromium		
			Kind	μm	(mil)	Kind	μm	(mil)
ISO 1456-74								
4	Fe/Ni40b Cr	mc	bright	40	(1.6)	micro-cracked	0.3	(0.01)
	Fe/Ni40b Cr	mp	bright	40	(1.6)	micro-porous	0.3	(0.01)
	Fe/Ni40p Cr	r	dull	40	(1.6)	regular	0.3	(0.01)
	Fe/Ni30p Cr	mc	dull	30	(1.2)	micro-cracked	0.3	(0.01)
	Fe/Ni30p Cr	mp	dull	30	(1.2)	micro-porous	0.3	(0.01)
	Fe/Ni40d Cr	r	layered	40	(1.6)	regular	0.3	(0.01)
	Fe/Ni40d Cr	f	layered	40	(1.6)	crack-free	0.8	(0.03)
	Fe/Ni30d Cr	mc	layered	30	(1.2)	micro-cracked	0.3	(0.01)
	Fe/Ni30d Cr	mp	layered	30	(1.2)	micro-porous	0.3	(0.01)
3	Fe/Ni40b Cr	r	bright	40	(1.6)	regular	0.3	(0.01)
	Fe/Ni30b Cr	mc	bright	30	(1.2)	micro-cracked	0.3	(0.01)
	Fe/Ni30b Cr	mp	bright	30	(1.2)	micro-porous	0.3	(0.01)
	Fe/Ni30p Cr	r	dull	30	(1.2)	regular	0.3	(0.01)
	Fe/Ni25p Cr	mc	dull	25	(1.0)	micro-cracked	0.3	(0.01)
	Fe/Ni25p Cr	mp	dull	25	(1.0)	micro-porous	0.3	(0.01)
	Fe/Ni30d Cr	r	layered	30	(1.2)	regular	0.3	(0.01)
	Fe/Ni30d Cr	f	layered	30	(1.2)	crack-free	0.8	(0.03)
	Fe/Ni25d Cr	mc	layered	25	(1.0)	micro-cracked	0.3	(0.01)
	Fe/Ni25d Cr	mp	layered	25	(1.0)	micro-porous	0.3	(0.01)
2	Fe/Ni20b*Cr	r	bright	20	(0.8)	regular	0.3	(0.01)
1	Fe/Ni10b*Cr	r	bright	10	(0.3)	regular	0.3	(0.01)

All these coatings shall have an undercoat of at least 8 um (0.3 mil) of copper or brass.

*For service conditions numbers 2 and 1, p or d nickel may be substituted for b nickel; f, mc or mp chromium may be substituted for r chromium.

Additional requirements for the various kinds of nickel and chromium are summarized in Table 13.

			ASTM B 456-79					
4	Fe/Ni40d Cr	r	layered	40	(1.6)	regular	0.25	(0.01)
	Fe/Ni30d Cr	mc	layered	30	(1.2)	micro-cracked	0.25	(0.01)
	Fe/Ni30d Cr	mp	layered	30	(1.2)	micro-porous	0.25	(0.01)
3	Fe/Ni30d Cr	r	layered	30	(1.2)	regular	0.25	(0.01)
	Fe/Ni25d Cr	mc	layered	25	(1.0)	micro-cracked	0.25	(0.01)
	Fe/Ni25d Cr	mp	layered	25	(1.0)	micro-porous	0.25	(0.01)
	Fe/Ni40p Cr	r	dull	40	(1.6)	regular	0.25	(0.01)
	Fe/Ni30p Cr	mc	dull	30	(1.2)	micro-cracked	0.25	(0.01)
	Fe/Ni30p Cr	mp	dull	30	(1.2)	micro-porous	0.25	(0.01)
2	Fe/Ni20b*Cr	r	bright	20	(0.8)	regular	0.25	(0.01)
	Fe/Ni15b*Cr	mc	bright	15	(0.6)	micro-cracked	0.25	(0.01)
	Fe/Ni15b*Cr	mp	bright	15	(0.6)	micro-porous	0.25	(0.01)
1	Fe/Ni10b*Cr	r	bright	10	(0.4)	regular	0.13	(0.005)

When a dull or satin finish is required, unbuffed p nickel may be substituted for a b nickel or for the bright layer of d nickel.

*p or d nickel may be substituted for b nickel in service condition numbers 2 and 1, and mc or mp chromium may be substituted for r chromium in service condition number 1.

Additional requirements for the various kinds of nickel and chromium are summarized in Table 13.

TABLE 10. SPECIFICATIONS FOR NICKEL PLUS CHROMIUM PLATING ON COPPER AND COPPER-BASE ALLOYS

Classification Number	Nickel Kind	Nickel μm	Nickel (mil)	Chromium Kind	Chromium μm	Chromium (mil)
ISO 1456-74						
Service condition number 4						
Cu/Ni25b Cr mc	bright	25	(1.0)	micro-cracked	0.3	(0.01)
Cu/Ni25b Cr mp	bright	25	(1.0)	micro-porous	0.3	(0.01)
Cu/Ni30p Cr r	dull	30	(1.2)	regular	0.3	(0.01)
Cu/Ni25p Cr mc	dull	25	(1.0)	micro-cracked	0.3	(0.01)
Cu/Ni25p Cr mp	dull	25	(1.0)	micro-porous	0.3	(0.01)
Cu/Ni30d Cr r	layered	30	(1.2)	regular	0.3	(0.01)
Cu/Ni30d Cr f	layered	30	(1.2)	crack-free	0.8	(0.03)
Cu/Ni25d Cr mc	layered	25	(1.0)	micro-cracked	0.3	(0.01)
Cu/Ni25d Cr mp	layered	25	(1.0)	micro-porous	0.3	(0.01)
Service condition number 3						
Cu/Ni20b*Cr r	bright	20	(0.8)	regular	0.3	(0.01)
Service condition number 2						
Cu/Ni10b*Cr r	bright	10	(0.4)	regular	0.3	(0.01)
Service condition number 1						
Cu/Ni5b*Cr r	bright	5	(0.2)	regular	0.3	(0.01)

*For service condition numbers 3, 2 and 1, p or d nickel may be substituted for b nickel; f, mc or mp chromium may be substituted for r chromium.
Additional requirements for the various kinds of nickel and chromium are summarized in Table 13.

Classification Number	Nickel Kind	Nickel μm	Nickel (mil)	Chromium Kind	Chromium μm	Chromium (mil)
ASTM B 456-79						
Service condition number 4						
Cu/Ni30d Cr r	layered	30	(1.2)	regular	0.25	(0.01)
Cu/Ni25d Cr mc	layered	25	(1.0)	micro-cracked	0.25	(0.01)
Cu/Ni25d Cr mp	layered	25	(1.0)	micro-porous	0.25	(0.01)
Service condition number 3						
Cu/Ni25d Cr r	layered	25	(1.0)	regular	0.25	(0.01)
Cu/Ni20d Cr mc	layered	20	(0.8)	micro-cracked	0.25	(0.01)
Cu/Ni20d Cr mp	layered	20	(0.8)	micro-porous	0.25	(0.01)
Cu/Ni25p Cr r	dull	25	(1.0)	regular	0.25	(0.01)
Cu/Ni20p Cr mc	dull	20	(0.8)	micro-cracked	0.25	(0.01)
Cu/Ni20p Cr mp	dull	20	(0.8)	micro-porous	0.25	(0.01)
Cu/Ni30b Cr r	bright	30	(1.2)	regular	0.25	(0.01)
Cu/Ni25b Cr mc	bright	25	(1.0)	micro-cracked	0.25	(0.01)
Cu/Ni25b Cr mp	bright	25	(1.0)	micro-porous	0.25	(0.01)
Service condition number 2						
Cu/Ni15b Cr r	bright	15	(0.6)*	regular	0.25	(0.01)
Cu/Ni10b Cr mc	bright	10	(0.4)*	micro-cracked	0.25	(0.01)
Cu/Ni10b Cr mp	bright	10	(0.4)*	micro-porous	0.25	(0.01)
Service condition number 1						
Cu/Ni5b Cr r	bright	5	(0.2)*	regular	0.13	(0.005)

When a dull or satin finish is required, unbuffed p nickel may be substituted for b nickel or for the bright layer of d nickel.

*p or d nickel may be substituted for b nickel in service condition numbers 2 and 1 and mc or mp chromium may be substituted for r chromium in service condition number 1.

Additional requirements for the various kinds of nickel and chromium are summarized in Table 13.

TABLE 11. SPECIFICATIONS FOR NICKEL PLUS CHROMIUM PLATING ON ZINC AND ZINC-BASED ALLOYS

Classification Number	Nickel Kind	Nickel μm	Nickel (mil)	Chromium Kind	Chromium μm	Chromium (mil)
ISO 1456-74						
Service condition number 4						
Zn/Cu Ni35b Cr f	bright	35	(1.4)	crack-free	0.8	(0.03)
Zn/Cu Ni35b Cr mc	bright	35	(1.4)	micro-cracked	0.3	(0.01)
Zn/Cu Ni35b Cr mp	bright	35	(1.4)	micro-porous	0.3	(0.01)
Zn/Cu Ni35p Cr r	dull	35	(1.4)	regular	0.3	(0.01)
Zn/Cu Ni35p Cr f	dull	35	(1.4)	crack-free	0.8	(0.03)
Zn/Cu Ni25p Cr mc	dull	25	(1.0)	micro-cracked	0.3	(0.01)
Zn/Cu Ni25p Cr mp	dull	25	(1.0)	micro-porous	0.3	(0.01)
Zn/Cu Ni35d Cr r	layered	35	(1.4)	regular	0.3	(0.01)
Zn/Cu Ni35d Cr f	layered	35	(1.4)	crack-free	0.8	(0.03)
Zn/Cu Ni25d Cr mc	layered	25	(1.0)	micro-cracked	0.3	(0.01)
Zn/Cu Cr mp	layered	25	(1.0)	micro-porous	0.3	(0.01)
Service condition number 3						
Zn/Cu Ni35b Cr r	bright	35	(1.4)	regular	0.3	(0.01)
Zn/Cu Ni25b Cr f	bright	25	(1.0)	crack-free	0.8	(0.03)
Zn/Cu Ni25b Cr mc	bright	25	(1.0)	micro-cracked	0.3	(0.01)
Zn/Cu Ni25b Cr mp	bright	25	(1.0)	micro-porous	0.3	(0.01)
Zn/Cu Ni25p Cr r	dull	25	(1.0)	regular	0.3	(0.01)
Zn/Cu Ni25p Cr f	dull	25	(1.0)	crack-free	0.8	(0.03)
Zn/Cu Ni20p Cr mc	dull	20	(0.8)	micro-cracked	0.3	(0.01)
Zn/Cu Ni20p Cr mp	dull	20	(0.8)	micro-porous	0.3	(0.01)
Zn/Cu Ni25d Cr r	layered	25	(1.0)	regular	0.3	(0.01)
Zn/Cu Ni25d Cr f	layered	25	(1.0)	crack-free	0.8	(0.03)
Zn/Cu Ni20d Cr mc	layered	20	(0.8)	micro-cracked	0.3	(0.01)
Zn/Cu Ni20d Cr mp	layered	20	(0.8)	micro-porous	0.3	(0.01)
Service condition number 2						
Zn/Cu Ni15b Cr r	bright	15	(0.6)*	regular	0.3	(0.01)
Service condition number 1						
Zn/Cu Ni8b Cr r	bright	8	(0.3)*	regular	0.3	(0.01)

For each classification number, a copper or brass undercoat of at least 8 μm (0.3 mil) is required.

*For service condition numbers 2 and 1, p or d nickel may be substituted for b nickel and f, mc or mp chromium may be substituted for r chromium.

Classification Number	Nickel Kind	Nickel μm	Nickel (mil)	Chromium Kind	Chromium μm	Chromium (mil)
ASTM B 456-79						
Service condition number 4						
Zn/Cu5 Ni40d Cr r	layered	40	(1.6)	regular	0.25	(0.01)
Zn/Cu5 Ni30d Cr mc	layered	30	(1.2)	micro-cracked	0.25	(0.01)
Zn/Cu5 Ni30d Cr mp	layered	30	(1.2)	micro-porous	0.25	(0.01)
Service condition number 3						
Zn/Cu5 Ni30d Cr r	layered	30	(1.2)	regular	0.25	(0.01)
Zn/Cu5 Ni25d Cr mc	layered	25	(1.0)	micro-cracked	0.25	(0.01)
Zn/Cu5 Ni25d Cr mp	layered	25	(1.0)	micro-porous	0.25	(0.01)
Zn/Cu5 Ni40p Cr r	dull	40	(1.6)	regular	0.25	(0.01)
Zn/Cu5 Ni30p Cr mc	dull	30	(1.2)	micro-cracked	0.25	(0.01)
Zn/Cu5 Ni30p Cr mp	dull	30	(1.2)	micro-porous	0.25	(0.01)
Service condition number 2						
Zn/Cu5 Ni20b Cr r	bright	20	(0.8)*	regular	0.25	(0.01)
Zn/Cu5 Ni15b Cr mc	bright	15	(0.6)*	micro-cracked	0.25	(0.01)
Zn/Cu5 Ni15b Cr mp	bright	15	(0.6)*	micro-porous	0.25	(0.01)
Service condition number 1						
Zn/Cu5 Ni10b Cr r	bright	10	(0.4)*	regular	0.13	(0.005)

For each classification number, a copperr undercoat of at least 5 μm (0.2 mil) is required.

When a dull or satin finish is required, unbuffed p nickel may be substituted for b nickel or for the bright layer of d nickel.

*p or d nickel may be substituted for b nickel in service condition numbers 2 and 1, and mc or mp chromium may be substituted for r chromium in service condition number 1.

Additional requirements for the various kinds of nickel and chromium are summarized in Table 13.

TABLE 12. SPECIFICATION FOR COPPER PLUS NICKEL PLUS CHROMIUM PLATING ON STEEL

Classification Number	Copper		Nickel			Chromium		
	μm	(mil)	Kind	μm	(mil)	Kind	μm	(mil)

ISO 1457-74

Classification Number	μm	(mil)	Kind	μm	(mil)	Kind	μm	(mil)
Service condition number 4								
Fe/Cu20 Ni25b Cr mc	20	(0.8)	bright	25	(1.0)	micro-cracked	0.3	(0.01)
Fe/Cu20 Ni25b Cr mp	20	(0.8)	bright	25	(1.0)	micro-porous	0.3	(0.01)
Fe/Cu20 Ni30p Cr r	20	(0.8)	dull	30	(1.2)	regular	0.3	(0.01)
Fe/Cu20 Ni25p Cr mc	20	(0.8)	dull	25	(1.0)	micro-cracked	0.3	(0.01)
Fe/Cu20 Ni25p Cr mp	20	(0.8)	dull	25	(1.0)	micro-porous	0.3	(0.01)
Fe/Cu20 Ni30d Cr r	20	(0.8)	layered	30	(1.2)	regular	0.3	(0.01)
Fe/Cu20 Ni30d Cr f	20	(0.8)	layered	30	(1.2)	crack free	0.8	(0.03)
Fe/Cu20 Ni25d Cr mc	20	(0.8)	layered	25	(1.0)	micro-cracked	0.3	(0.01)
Fe/Cu20 Ni25d Cr mp	20	(0.8)	layered	25	(1.0)	micro-porous	0.3	(0.01)
Service condition number 3								
Fe/Cu20 Ni30b Cr r	20	(0.8)	bright	30	(1.2)	regular	0.3	(0.01)
Fe/Cu20 Ni20b Cr mc	20	(0.8)	bright	20	(0.8)	micro-cracked	0.3	(0.01)
Fe/Cu20 Ni20b Cr mp	20	(0.8)	bright	20	(0.8)	micro-porous	0.3	(0.01)
Fe/Cu15 Ni25p Cr r	15	(0.6)	dull	25	(1.0)	regular	0.3	(0.01)
Fe/Cu15 Ni20p Cr mc	15	(0.6)	dull	20	(0.8)	micro-cracked	0.3	(0.01)
Fe/Cu15 Ni20p Cr mp	15	(0.6)	dull	20	(0.8)	micro-porous	0.3	(0.01)
Fe/Cu15 Ni25d Cr r	15	(0.6)	layered	25	(1.0)	regular	0.3	(0.01)
Fe/Cu15 Ni25d Cr f	15	(0.6)	layered	25	(0.1)	crack free	0.8	(0.03)
Fe/Cu15 Ni20d Cr mc	15	(0.6)	layered	20	(0.8)	micro-cracked	0.3	(0.01)
Fe/Cu15 Ni20d Cr mp	15	(0.6)	layered	20	(0.8)	micro-porous	0.3	(0.01)
Service condition number 2								
Fe/Cu20 Ni10b*Cr r	20	(0.8)	bright	10	(0.4)	regular	0.3	(0.01)
Service condition number 1								
cFe/Cu10 Ni5b*Cr r	10	(0.4)	bright	5	(0.2)	regular	0.3	(0.01)

*For service conditions 2 and 1, p or d nickel may be substituted for b nickel and f, mc or mp chromium may be substituted for r chromium.

Additional requirements for the various kinds of nickel and chromium are summarized in Table 13.

ASTM B 456-79

Classification Number	μm	(mil)	Kind	μm	(mil)	Kind	μm	(mil)
Service condition number 4								
Fe/Cu15a Ni25d Cr mc	15	(0.6)	layered	25	(1.0)	micro-cracked	0.25	(0.01)
Fe/Cu15a Ni25d Cr mp	15	(0.6)	layered	25	(1.0)	micro-porous	0.25	(0.01)
Service condition number 3								
Fe/Cu12a Ni20d Cr mc	12	(0.5)	layered	20	(0.8)	micro-cracked	0.25	(0.01)
Fe/Cu12a Ni20d Cr mp	12	(0.5)	layered	20	(0.8)	micro-porous	0.25	(0.01)

When a dull or satin finish is required, unbuffed p nickel may be substituted for the bright layer of d nickel. "a" designates ductile copper deposited from acid-type baths containing additives that promote leveling by the copper deposit that has an elongation not less than 8%.

Peel Test

ASTM B 533. Peel strength of metal-electroplated plastics.

Porosity Test

ASTM B 583. Porosity in gold coatings on metal substrates.

Tests for Anodized Aluminum

ASTM B 136. Measurement of stain resistance of anodic coatings on aluminum.

ASTM B 137. Measurement of weight of coating on anodically coated aluminum.

IS 2106. Surface treatment of metals—anodisation (anodic oxidation) of aluminum and its

TABLE 13. ISO AND ASTM Supplementary Requirements for Decorative Nickel and Chromium Plating

Coating	Designation	Description	Elongation	Sulfur Content
Nickel	b	Full bright		
Nickel	p	semi-bright	>8%	<0.005%
Nickel	d	layered		
		bottom layer:		<0.005%
		thickness >60% of total nickel for double layer, *>75% for double layer on steel, >50% for triple layer nickel.		
		top layer:		>0.04%
		thickness >20% of total nickel		
		intermediate layer:		>top layer
		thickness <10% of total nickel		
Chromium	Cr f	crack-free		
Chromium	Cr mc	micro-cracked	>250 cracks/cm	
Chromium	Cr mp	micro-porous	>10 000 pores/cm	

*For double layer nickel on steel of ASTM B 456

These requirements apply to ISO 1456, ISO 1457, ISO 1458 and ASTM B 456.

TABLE 14. Sification for Nickel-Zinc Alloy Plating AMS *2417C

Composition	Coating	Thickness (μ)	(mil)
16–30% nickel balance zinc	alloy:	(8–18)	0.3–0.7
		(5–10)	0.2–0.4
	nickel strike, minimum:	1.3	0.05

TABLE 15. Specification for Palladium Plating for Engineering Use

Specification	Class	Minimum Thickness μm	(mil)
ASTM B679-80	5.0	5.0	(0.20)
	2.5	2.5	(0.10)
	1.2	1.2	(0.05)
	0.6	0.6	(0.02)
	0.3	0.3	(0.01)
	F	0.025	(0.0010)
MIL-P-45209		(1.3)	0.05 unless otherwise specified

TABLE 16. Specification for Rhodium Plating for Engineering Use

Specification	Class	Minimum Thickness μm	(mil)
ASTM B 634-78	0.2	0.2	(0.008)
	0.5	0.5	(0.02)
	1	1	(0.04)
	2	2	(0.08)
	4	4	(0.16)
	5	6.25	(0.25)
MIL-R-46085A	1	0.05	(0.002)
	2	0.3	(0.01)
	3	0.5	(0.02)
	4	2.5	(0.10)
	5	6.4	(0.25)

alloys—measurement of the mass of the oxide coatings—gravimetric method.

ASTM B 457. Measurement of impedance of anodic coatings on aluminum.

ISO 2931. Anodizing aluminum and its alloys—assessment of quality of sealed anodic oxide coatings by measurement of admittance or impedance.

ASTM B 680. Seal quality of anodic coatings on aluminum by acid resistance.

ISO 2932. Anodizing of aluminum and its alloys—assessment of sealing quality by measurement of the loss of mass after immersion in acid solution.

ISO 3210. Anodizing of aluminum and its alloys—assessment of sealing quality by measurement of the loss of mass after immersion in phosphoric-chromic acid solution.

ISO 2085. Surface treatment of metals—anodisation of aluminum and its alloys—check of continuity of thin coatings—copper sulfate test.

ISO 2376. Anodisation (anodic oxidation) of

TABLE 17. Specifications for Silver Plating

Specification	Basis Metal or Type, Grade, and Class		Nickel or Copper Minimum Thickness		Silver	
			μm	mil	μm	mil
ASTM B 700-81	Type 1	99.9% silver				
	Type 2	99.0% silver				
	Type 3	98.0% silver				
	Grade A	Mat finish				
	Grade B	Bright finish (brighteners used in electroplating bath)				
	Grade C	Bright finish (mechanically or chemically polished)				
	Class N	no tarnish-resistant treatment				
	Class S	tarnish-resistant (chromate) treatment				
					Suggested thicknesses of silver:	
					1	(0.04)
					2.5	(0.10)
					5	(0.2)
					10	(0.4)
					20	(0.8)
					>=40	(1.6)
Fed. QQ-S-365C	Nonferrous				(13)	0.5
	Ferrous		(13)	0.5	(13)	0.5
AMS 2410F					as specified on drawings	
	flash				(3)	0.1
AMS 2411C					as specified on drawings	
	flash				(3)	0.1
AMS 2412					as specified on drawings	
	flash				(3)	0.1

aluminum and its alloys—insulation check by measurement of breakdown potential.

ISO 3211. Anodizing of aluminum and its alloys—assessment of resistance of anodic coatings to cracking by deformation.

ISO 2767. Surface treatment of metals—anodic oxidation of aluminum and its alloys—specular reflectance at 45°—total reflectance—image clarity.

ISO 2135. Surface treatment of metals—colored anodisation of aluminum and its alloys—determination of the light-fastness of colored anodized aluminum.

ISO 2143. Surface treatment of metals—anodisation of aluminum and its alloys—estimation of the loss of absorption power by colorant drop test with prior acid treatment.

ASTM B 681. Measurement of thickness of

TABLE 18. Commercial Grades of Silver-Plated Flatware
Basis Metal: Nickel Silver

Grade	Weight of Silver, troy oz/gross of teaspoons*	Average Thickness Approximate	
		(μm)	mil
Half-plate	1	(4)	0.14
Standard plate	2	(7)	0.28
Double plate	4	(14)	0.56
Triple plate	6	(21)	0.83
Quadruple plate	8	(28)	1.11
Federal specification plate (QQ-T-51)	9	(32)	1.25

For flatware articles other than teaspoons, the weight of coating is in direct proportion to the area and the approximate thickness is the same as given above.

*Area is approximately 9 ft^2.

TABLE 19. Specifications for Miscellaneous Composite Coatings

Specification	Layer	Coating	Thickness (μm)	mil
AMS 2413B	1st	copper or nickel	minimum: (2.5)	0.1
	2nd	silver	minimum: (13)	0.5
	outer	rhodium	minimum: (0.5)	0.02
AMS 2415D	1st	lead	as specified	
	outer	indium	as specified	
AMS 2416E	1st	nickel } diffused	(5–10) to word	0.2–0.4
	outer	cadmium	(3–5) to word	0.1–0.2

TABLE 20. Specifications for Tin Plating

Specification	Service Condition Number	Basis Metal	Classification Number	Thickness, Minimum μm	(mil)	
ISO 2093-73	4	steel	Fe/Sn30	30	(1.2)	
		copper or copper alloy	Cu/Sn30	30	(1.2)	
	3	steel	Fe/Sn20	20	(0.8)	
		copper or copper alloy	Cu/Sn15	15	(0.6)	
	2	steel	Fe/Sn12	12	(0.5)	
		copper or copper alloy	Cu/Sn8	8	(0.3)	
	1*	steel	Fe/Sn4	4	(0.16)	
		copper or copper alloy	Cu/Sn4	4	(0.16)	8 μm (0.3 mil) max.
	1f	steel	Fe/Sn4	4	(0.16)	
		copper or copper alloy	Cu/Sn4	4	(0.16)	

*Undercoat of 2 μm (0.08 mil) of copper, bronze, or nickel on brass basis metal.
f = flow-brightened coating

Specification	Service Condition Number	Basis Metal	Classification Number	Thickness, Minimum μm	(mil)
ASTM B 545-72	4	steel	Fe/Sn30	30	(1.2)
		copper and copper alloy	Cu/Sn30	30	(1.2)
	3	steel	Fe/Sn20	20	(0.8)
		copper and copper alloy	Cu/Sn15	15	(0.6)
	2	steel	Fe/Sn10	10	(0.4)
		copper and copper alloy	Cu/Sn8	8	(0.3)
	1	steel	Fe/Sn5	5	(0.2)
		copper and copper alloy	Cu/Sn5	5	(0.2)
MIL-T-10727B				as specified in drawings	
AMS 2408D				as specified by drawings	
		Flash		0.1 mil (3 μm)	

TABLE 21. SPECIFICATION FOR TIN PLATE

Specification	Designation Number	Nominal Weight Each Surface, lb/base box	Minimum Average Coating Weight, lb/base box
ASTM * A 624	10	0.05/0.05	0.04/0.04
	20	0.10/0.10	0.08/0.08
	25	0.125/0.125	0.11/0.11
	35	0.175/0.175	0.16/0.16
	50	0.25/0.25	0.23/0.23
	75	0.375/0.375	0.35/0.35
	100	0.50/0.50	0.45/0.45
	50/25	0.25/0.125	0.23/0.11
	75/25	0.375/0.125	0.35/0.11
	100/25	0.50/0.125	0.45/0.11
	100/50	0.50/0.25	0.45/0.23
	135/25	0.675/0.125	0.62/0.11

anodic coatings on aluminum and of other transparent coatings on opaque surfaces using the light-section microscope.

ISO 2128. Surface treatment of metals—anodisation (anodic oxidation) of aluminum and its alloys—measurement of thickness of oxide coatings—nondestructive measurement by light section microscope.

Also see thickness tests, above.

Tests for Conversion Coatings

ASTM B 201. Practice for testing chromate coatings on zinc and cadmium surfaces.

ISO 3613. Chromate conversion coatings on zinc and cadmium—test method.

ISO 3982. Conversion coatings on metallic materials—determination of mass per unit area—gravimetric methods.

TABLE 22. SPECIFICATION FOR TIN-LEAD ALLOY PLATING

Service Condition	Substrate	Classification Number	Minimum Thickness	
			μm	(mil)
Specification: ASTM B 579-73				
Alloy composition: tin—50% min, 70% max				
SC4	Steel	Fe/SnPb30	30	(1.2)
	copper*	Cu/SnPb30	30	(1.2)
SC3	steel	Fe/SnPb20	20	(0.8)
	copper*	Cu/SnPb15	15	(0.6)
SC2	steel	Fe/SnPb10	10	(0.4)
	copper*	Cu/SnPb8	8	(0.3)
SC1	steel	Fe/SnPb5	5	(0.2)
	copper*	Cu/SnPb5	5	(0.2)

*Substrate may be copper alloys or nonmetal.
If substrate is brass with more than 15% zinc or is beryllium copper, an undercoat of at least 2.5 μm (0.1 mil) of copper or nickel is required.
Specification: MIL-P-81728A

Application		Thickness	
		mil	(μm)
Other than electronic components		0.3-0.5	(8-13)
Electronic components	minimum average:	0.3	(8)
	minimum:	0.2	(5)
Undercoat of copper or nickel on copper alloys with more than 15% zinc and on beryllium copper:		0.1	(3)

TABLE 23. SPECIFICATIONS FOR TIN-NICKEL ALLOY PLATING

				Minimum Thickness			
	Service			Copper		Sn-Ni	
Specification	Condition Number	Basis Metal	Classification Number	µm	(mil)	µm	(mil)
ISO 2179-72	3	steel	Fe/Cu Sn-Ni25	8	(0.3)	25	(1.0)
		copper and copper alloy	Cu/Sn-Ni25			25	(1.0)
		zinc alloy	Zn/Cu Sn-Ni25	8	(0.3)	25	(1.0)
	2	steel	Fe/Cu Sn-Ni15	8	(0.3)	15	(0.6)
		copper and copper alloy	Cu/Sn-Ni15	—		15	(0.6)
		zinc alloy	Zn/Cu Sn-Ni15	8	(0.3)	15	(0.6)
	1	steel	Fe/Cu Sn-Ni8	8	(0.3)	8	(0.3)
		steel	Fe/Sn-Ni8	—		8	(0.3)
		copper and copper alloy	Cu/Sn-Ni8	—		8	(0.3)
		zinc alloy	Zn/Cu Sn-Ni8	8	(0.3)	8	(0.3)
ASTM B 605-75	4	steel	Fe/Cu Sn-Ni 45	4	(0.16)	45	(1.8)
		copper or copper alloy	Cu/Sn-Ni 45	—		45	(1.8)
		zinc alloy	Zn/Cu Sn-Ni 45	4	(0.16)	45	(1.8)
	3	steel	Fe/Cu Sn-Ni 25	4	(0.16)	25	(1.0)
		copper or copper alloy	Cu/Sn-Ni 25	—		25	(1.0)
		zinc alloy	Zn/Cu Sn-Ni 25	4	(0.16)	25	(1.0)
	2	steel	Fe/Sn-Ni 15	—		15	(0.6)
		copper or copper alloy	Cu/Sn-Ni 15	—		15	(0.6)
		zinc alloy	Zn/Cu Sn-Ni 15	4	(0.16)	15	(0.6)
	1	steel	Fe/Sn-Ni 8	—		8	(0.3)
		copper or copper alloy	Cu/Sn-Ni 4	—		4	(0.16)
		zinc alloy	Zn/Cu Sn-Ni 8	4	(0.16)	8	(0.3)

Sampling Plans

ASTM B 602. Attribute sampling of electrodeposited metallic coatings and related finishes.

ISO 4519. Electrodeposited metallic coatings and related finishes—sampling procedures for inspection by attributes.

MIL-STD-105. Sampling procedures and tables for inspection by attributes.

Tests for Plating Processes

ASTM B 519. Mechanical hydrogen embrittlement testing of plating processes and aircraft maintenance chemicals.

ASTM B 636. Measurement of internal stress of plated metallic coatings with the spiral contractometer.

SPECIFICATIONS FOR RELATED MATERIALS AND PROCESSES

Closely associated with electroplating are shot peening, mechanical plating, electroless plating, metal spraying, vacuum deposition and anodizing. Corresponding specifications and test methods include the following:

Mechanical Plating

ASTM B 635. Cadmium-tin mechanically deposited on iron and steel.

ASTM B 454. Mechanically deposited coatings of zinc on ferrous materials.

MIL-C-81562A. Coating—cadmium, tin-cadmium and zinc (mechanically deposited).

PLATING STANDARDS AND SPECIFICATIONS

TABLE 24. SPECIFICATIONS FOR ZINC PLATING ON STEEL

Specification	Service Condition Number	Classification Number	Thickness, Minimum μm	mil
ISO 2081-83	4	Fe/Zn25	25	(1.0)
	3	Fe/Zn12	12	(0.5)
	2	Fe/Zn8	8	(0.3)
	1	Fe/Zn5	5	(0.2)
ASTM *B 633-78	4	Fe/Zn25	25	(1.0)
	3	Fe/Zn13	13	(0.5)
	2	Fe/Zn8	8	(0.3)
	1	Fe/Zn5	5	(0.2)
Fed. QQ-Z-325C		1	(25)	1.0
		2	(13)	0.50
		3	(5)	0.02

AMS 2402F

Class	μm	mil	Externally Threaded Parts — External Threads μm	mil	Untreaded Surfaces μm	mil
—	8–13	0.5–0.7	3–10	0.1–0.4	5–13	0.2–0.5
1	3–8	0.1–0.3	3–8	0.1–0.3	5–10	0.2–0.4
2	5–10	0.2–0.4	3–10	0.1–0.4	5–10	0.2–0.4
3	8–13	0.3–0.5	5–13	0.2–0.5	8–13	0.3–0.5
4	10–15	0.4–0.6	8–15	0.3–0.6	10–15	0.4–0.6
5	13–18	0.5–0.7	10–18	0.4–0.7	13–18	0.5–0.7

Immersion Deposits

AMS-2409. Plating—tin, immersion.

AMS-2420. Plating—aluminum for solderability, zincate immersion process.

AMS-2421. Plating—magnesium for solderability, zinc immersion process.

Metal Spraying

MIL-M-6874. Metal spraying, process for.

AMS-2435. Flame deposition, tungsten carbide.

AMS-2450. Sprayed metal finish—aluminum.

Shot Peening

AMS-2430. Shot peening.

MIL-S-13165. Shot peening of ferrous metal parts.

MIL-R-81841(AS). Rotary flap peening of metal parts.

MIL-P-81985(AS). Peening of metals.

Vacuum Deposition

MIL-C-23217B. Coating, aluminum, vacuum deposited.

MIL-C-8837. Coating, cadmium, vacuum deposited.

AMS 2426. Plating, cadmium, vacuum deposition.

SPECIFICATIONS FOR PLATING SUPPLIES

As discussed in Chapter 29 for anode compositions and in Chapter 6 for bath composition limits, the compositions of anodes and chemicals used in electroplating are critical. Government

TABLE 25. SPECIFICATIONS FOR ANODES

Fed QQ-A-671 Cadmium

Cd (min) %	Ag + Pb + Sn (max) %	As + Sb + Tl (max) %	Zn + Cu %
99.90	0.045	0.007	0.050

Fed. QQ-A-673 Copper

Type:	I Non-Deoxidized	II Oxygen-free	III Deoxidized, Phosphorus Bearing
Cu % min	99.90	99.95	99.90
O % max	0.045		
Metallic inpurities % max	0.01	0.01	0.01
Fe % max	0.003	0.002	0.003
S % max	0.003	0.003	0.003
Pb % max	0.003	0.002	0.002
Sb % max	0.002	0.002	0.002
Ni % max	0.003	0.002	0.002
P % max		0.001	0.015–0.040

Fed. QQ-A-677 Nickel

Type	Ni + Co % min	Fe % max	Cu % max	S % max	Si %
I Carbon	99.00	0.15	0.08	0.02	0.20–0.35
II Oxide	99.00	0.15	0.10	0.01	0.02 max

specifications exist for anodes and chemicals which indicate purity limits as summarized in Tables 25 and 26.

A related specification is Federal Specification P-C-535, "Cleaning Compound, Platers' Electrocleaning, for Steel." The specified composition is 10 to 35% silicate as SiO_2, 5% phosphate as P_2O_5, 0.4% organic detergent and the remainder caustic soda.

ASTM PRACTICES

To meet specification requirements and provide a high-grade product, good plating procedures must be followed. Factors such as the quality and preparation of the basis metal, the composition and purity of the plating baths, the temperatures and current densities used and the preparatory plating cycle employed are most important in determining the quality of the final product. They all have a direct bearing on the porosity, wear resistance, corrosion resistance, adhesion and appearance.

ASTM has prepared a series of "Practices" to provide methods and procedures found to produce good-quality products. These Practices are:

B 656. Autocatalytic nickel deposition on metals for engineering use.
B 449. Chromate treatments on aluminum.
B 177. Chromium plating on steel for engineering use.
B 322. Cleaning metals prior to electroplating.
B 507. Design of articles to be plated on racks.

Electroforming:
B 503. Use of copper and nickel electroplating solutions for electroforming.
B 450. Engineering design of electroformed articles.
B 431. Processing of mandrels for electroforming.

Preparation of metals for electroplating:
B 253. Aluminum alloys.
B 630. Chromium for electroplating with chromium.
B 281. Copper and copper base alloys.
B 320. Iron castings.
B 319. Lead and lead alloys.
B 480. Magnesium and magnesium alloys.

TABLE 26. SPECIFICATIONS FOR CHEMICALS FOR PLATING BATHS

MIL-C-6151   Cadmium oxide

Cd % min	Ag, Pb, Sn, Cu, Hg % max	As, Sb, Tl % max	Volatiles at 105 C %
87.0	0.05	0.005	0.5

Fed. O-C-828   Cupric sulfate

Copper sulfate as $CuSO_4 \cdot 5H_2O$ min %	max %	Insoluble Matter %	Fe % max
98.0	105.0	0.1	0.2

Fed. O-C-303   Chromic acid

CrO_3 % min	SO_4 % max	Cl % max	Insoluble Matter % max
99.5%	0.20	0.10	0.10

Fed. O-N-335   Nickel salts

Salts:	$NiSO_4 \cdot 6H_2O$ or $NiSO_4 \cdot 6H_2O + NiSO_4 \cdot 7H_2O$	$NiSO_4 \cdot (NH_4)_2SO_4 \cdot 6H_2O$	$NiCl_2 \cdot 6H_2O$
Ni + Co % min	21.4	14.6	24.3
Fe % max	0.02	0.02	0.02
Zn % max	0.025	0.025	0.025
Cu % max	0.005	0.005	0.005
Free acid % max	0.10	0.10	0.10
Insoluble matter % max	0.05	0.05	0.05

B 629. Molybdenum and molybdenum alloys.
B 558. Nickel alloys.
B 343. Nickel for electroplating with nickel.
B 242. High-carbon steel.
B 183. Low-carbon steel.
B 254. Stainless steel.
B 481. Titanium and titanium alloys.
B 482. Tungsten and tungsten alloys.
B 252. Zinc-base die castings.

ISO STANDARDS

The International Organization for Standardization (ISO) issues specifications and test methods for electrodeposited and related coatings that are recognized all over the world. Some countries, such as England, often adopt the ISO Standards where applicable rather than maintain their own standards. Other countries may use the ISO Standards as a basis for writing their own standards. It is intended that these standards be used to facilitate international trade.

The requirements of the ISO Standards for electrodeposited and related coatings are almost the same as those of the ASTM. That is to say, anyone plating to ASTM Standards will usually conform to the corresponding ISO Standards and no hardships will be encountered because of conflicting requirements.

Copies of ISO Standards may be purchased from the American National Standards Institute, 1430 Broadway, New York, NY 10018.

ASTM STANDARDS

The American Society for Testing and Materials (ASTM) has prepared and published plating specifications for a number of coating-basis metal combinations. General in nature, these specifications do not refer to any specific application or product, but they provide for several types or thicknesses for applications in a variety of environments. The accompanying tables summarize the thickness requirements for the existing ASTM plating specifications.

The ASTM also publishes test methods and practices for electroplating.

These ASTM Standards are prepared by cooperating representatives of the plating industry, suppliers, consumers and the government, and have received wide recognition in the United States and foreign countries. They often serve as a basis for the specifications prepared by industrial concerns and other organizations and many have been approved as American National Standards by the American National Standards Institute.

The ASTM electroplating standards appear in Vol. 02.05 (1983) of the *Annual Book of ASTM Standards*. The book or the individual standards may be purchased from ASTM, 1916 Race Street, Philadelphia, PA 19103.

GOVERNMENT SPECIFICATIONS

Government specifications are issued to meet the needs of the procurement activities of the federal government. Private organizations sometimes use the government specifications, but such use is voluntary except when selling to the government. Most government specifications covering plated coatings are written by the Department of Defense (DOD). If a nonmilitary agency regularly uses a specification, it is issued as a federal specification even though it is also prepared and/or used by the DOD. If, on the other hand, it is of interest only to the DOD, the specification will appear as a military specification. These government specifications are designated by numbers of the form QQ-S-365B and MIL-G-45204B. Federal specifications begin with a group of one, two or three capital letters, and military specifications begin with MIL. The middle letter in the specification designation is the first letter of the specification title. The letter at the end of the specification serial number indicates the revision. For example, "B" indicates the second revision.

The number of specifications used by the DOD is tremendous and growing every day. Developing, issuing and maintaining these specifications have become extremely expensive and time consuming for the DOD. As a consequence, the DOD has adopted a policy of using specifications written by nationally recognized technical and professional societies and organizations (ASTM, ANSI, etc.) which promulgate standards. The policy of the DOD is to use any such specification as long as it meets the requirements of the DOD.

Each non-government document adopted for use within the DOD is listed in the *DODISS (Department of Defense Index of Specifications and Standards)*. The *DODISS* identifies each such document by industry group symbol, number, title, date of the DOD-adopted issue and military coordinating activity within the DOD. In this chapter, the identifying number of such a specification is preceded by an asterisk.

The *DODISS*, military specifications and federal specifications prepared by the military may be obtained from the Naval Publications and Forms Center, 5801 Tabor Ave., Philadelphia, PA 19120. Other federal specifications and the *Index of Federal Specification and Standards* may be obtained from the General Services Administration's Regional Office Building, Specification Section, Room 6039, 7th and D Streets SW, Washington, DC 20407.

INDUSTRIAL SPECIFICATIONS

Generally, industrial specifications are based on ASTM and government specifications with additional requirements and modifications peculiar to product lines and processes used.

References

1. Uhlig, H. H., "Corrosion Handbook," p. 829, John Wiley & Sons, Inc., New York, 1948.
2. "Metals Handbook," 8th Ed., Vol. 2, pp. 110, 128, 144, 160, ASM, Metals Park, Ohio, 1964.
3. Uhlig, H. H., "Corrosion Handbook," pp. 420–426, 481, John Wiley & Sons, Inc., New York, 1948.

8. TROUBLESHOOTING

Lawrence J. Durney

*Vice President for Technology, Frederick Gumm Chemical Co., Inc.
Lyndhurst, NJ*

Troubleshooting has historically been treated as direct cause and effect process, without emphasis being placed on the procedural logic involved. Much of the published literature has been written from this point of view. If you encounter a specific defect, correct the listed cause or causes.

Experience has therefore assumed a highly important position in the troubleshooting procedure. Under these circumstances, a difficulty not previously experienced posed special problems, required a search for someone who had experienced it or resulted in an empirical hit or miss approach which frequently either delayed the solution or caused additional troubles.

If, however, we define troubleshooting as the logical process by which a defective procedure is corrected, the troubleshooting process can be organized and outlined so it can be applied to any problem, whether previously experienced or not.

Under this definition, we must consider two different applications: 1) the correction of an established, previously successful process which no longer produces acceptable results; and 2) the analysis and correction of a new process which does not produce acceptable results.

The logic used in both cases is essentially the same, but the starting assumptions, and therefore the investigative procedure, will be somewhat different.

In the first case, since the process had been successful, it can safely be assumed that the difficulty is associated with some change in conditions, either in the parts or in the process. The investigation will therefore emphasize the detection of the nature and degree of this change.

The literature listing cause and effect relationships becomes an important part of the background information needed to carry out the search, but is not a substitute for the logical procedure.

In the second case, since the object is to develop a successful process, the investigation will emphasize the determination of the changes needed in the selected process, or the control parameters needed, to ensure that the process functions properly and reproducibly. Again, the cause and effect literature becomes an important aid to the procedure.

GENERAL TROUBLESHOOTING PROCEDURE

The first and essential step in either procedure is to define the problem. The common habit of referring to the defect as the problem must be discarded. The defect is a symptom, not the problem. And just as medical symptoms point to possible diseases or causes, so do the symptomatic defects encountered in plating point to problems or sources of the defects.

Close and intelligent observation of the defect is necessary in any attempt to define the problem. Peeling or blistered plating is a commonly encountered defect. This can point to several different areas of the cycle, depending on whether the peeling is occurring from the substrate, or from an intermediate plated layer. Failure to determine the exact nature of the peeling encountered can seriously delay the troubleshooting process. Failure to accurately assess the symptom has left the troubleshooter with a very broad area of investigation. Proper analysis will greatly narrow the area that must be investigated, simplifying the procedure, and reducing the time required. For example: If the peeling is from the substrate, attention will be focused on the preplating cycle; if the peeling is from another plated layer, or because of laminated plate, the attention will be centered on the plating tanks, or the steps between the plating tanks.

In most cases, if the problem can be adequately defined, the appropriate remedy or correction is reasonably obvious. This is true, however, only if there is sufficient background information available. It does little good to recognize that the defect results from metallic contamination of the plating solution unless the method or methods of removing the contamination are known. Similarly, it does little good to recognize the existence of a smut, unless methods of smut removal are known.

This points up the need for a constant program of self-education. The more that is known about the nature of a process, its shortcomings, the required areas of control and possible areas of difficulty, the easier it will be to solve a problem when it arises.

Any complex process consisting of a series of interdependent steps includes the possibility of additive or domino effects. An additive effect system is one in which minor deviations in several steps add up to a defect producing problem. A domino effect system is one in which a deviation in one step produces a deviation in a subsequent step, which in turn affects still another step until the end result is a defective system. In both cases, the actual source of the problem may be masked or hidden, or attention may be focused on the wrong area as the problem source.

If it is recognized that in such a complex system, each step is affected by what went before, and in turn affects that which follows, the system will be considered as a whole. If each step is examined from this viewpoint, additive and domino effects will be recognized, and the troubleshooting process expedited. Once again, observation, logic and background knowledge are the essential ingredients.

TROUBLESHOOTING AN ESTABLISHED PROCESS

Experience has shown that in an established and tested process, the development of a defect problem will in the majority of cases be caused by a deviation from standard conditions. As high as 90% of the problems encountered are attributed to this source. Yet these problems should never occur. Proper attention to solution control, equipment maintenance and operating conditions would prevent their happening.

Two things are required to implement a preventive program aimed at avoiding problems of this type: a complete listing of the proper conditions for each tank; and a program of analysis, maintenance and inspection to ensure adherence to the required conditions. Not only solutions, racks and processing equipment, but also all ancillary equipment, such as rectifiers, filters, meters, air blowers, exhaust systems, etc., must be included in the program. Some sort of prefinishing inspection of parts received for processing will ensure that such parts are within the quality limits that can be processed under the established conditions.

In the event that a defect problem does occur, one of the first steps in the procedure is to check each step of the process for adherence to the established conditions. Any deviation from the standard should be corrected. Conditions of particular importance include temperature, concentration, voltage and/or amperage, rinse water flow, immersion times, solution purity and solution age. For particular processes, other or additional conditions may assume importance. Many, if not most, problems will be corrected when proper conditions are reestablished.

For those problems which are not corrected in this way, careful observation and analysis of the defect pattern are important steps in arriving at a solution. Each step of the process should be carefully observed for evidence of any deviation from normal. In order to accomplish this, however, the observer must be completely familiar with the appearance of each step when it is functioning properly. This familiarity can only be attained by repeated careful observation of the process when quality acceptable work is being produced. This observation should not be limited to the process steps. It should include the general area, the ancillary equipment and the condition of the parts being processed.

Observations of deviations noted in this careful inspection are valuable indicators of the cause of the problem, provided their meaning and effect is understood. It does little good to observe that a cleaner is contaminated with hexavalent chromium unless the effect of such contamination and its possible relation to the defect problem is understood. The importance of background information and process understanding cannot be overemphasized in this regard. Suppliers of equipment and process chemicals are valuable sources of this type of information. If at all possible, the information should be obtained before a defect problem develops.

In Chapter 6, as part of the solution composition parameters, basic troubleshooting charts for the solutions are provided. These charts are not intended to be absolute or all-inclusive. They will be altered, modified and/or extended due to the effects of solution additives, the preplating sequences, plating procedures (rack or barrel), dwell times and other factors. They will, however, serve as a model and be a base for a more detailed chart specifically drawn up for the process being utilized. Such a detailed chart should be prepared whenever a new process is installed, or an already installed process is modified or altered. Suppliers of the specific materials being used in the process should be consulted for their input and advice on the chart preparation.

The defective parts must be carefully examined to determine the defect pattern. This consists of three components:

1. The nature of the defect—peeling, blistering, pitting, hazing, failure to obtain proper coverage, etc.
2. The defect location pattern—location of the defect on the part, the rack, etc.
3. The defect occurrence pattern—the percentage of rejects, the time distribution, spatial distribution, etc.

Information on all three aspects of the defect pattern when carefully assembled and analyzed will frequently provide the clue necessary for the solution.

TROUBLESHOOTING THE NEW PROCESS

The technique for troubleshooting a new process differs from an established process only in where the emphasis is placed. With an established process, the emphasis is on the reestablishment of proven successful conditions. With a new process, the emphasis is on determining the conditions needed to produce successful results. Concentration, therefore, is on the defect and the defect pattern. In observing the process, the emphasis is on the condition of the parts rather than on the condition of the solutions; the object of the observational process is to determine the changes needed to consistently produce quality-acceptable parts. In examining parts received for processing, the emphasis is on the expected treatment needed, rather than on conformance to standards known to be acceptable in the process. The difference can be illustrated by the example of an oil-quenched heat treated bolt:

	Established Process	New Process
Oil contamination	Is the amount normal?	What cleaning is needed to remove this soil?
Carbon smut	Is the smut within standards?	Will this smut come off in an electrocleaner?
Scale	Within limits?	What acids will be needed?
Threads	Normal?	Any special conditions to be seen?

In short, in troubleshooting an established process, you look for the change that has occurred and needs to be corrected. With the new process, you look for the changes that have to be made to make the process successful and consistent.

SUMMATION OF THE TECHNIQUE

The basic troubleshooting technique can be broken down into four steps, each of which must be taken in logical progression.

1. Analysis—an evaluation of the data available to determine the probable cause of the problem.
2. Synthesis—the development of an explanation for the problem and the corrective measures required.
3. Testing—small-scale or production line changes to prove the accuracy of the explanation.
4. Implementation—the actual corrective procedure.

If the testing must be carried out on the production line, Steps 3 and 4 may be combined into a single step.

The final step is to make a record of the corrective steps taken and to set up procedures to prevent the recurrence of the same problem.

It should be noted that experienced troubleshooters will often appear to shortcut or eliminate some of the listed steps. What has actually happened is that a previously experienced situation has recurred and been recognized. In recalling the memory of the previous situation, the required steps are also recalled, along with the answer. The steps are taken, but by instantaneous memory, and are not therefore apparent to the observer. Thoroughly experienced troubleshooters have such a storehouse of experience, and the ability to recall it, that they often effectively solve problems by phone without seeing parts or equipment.

ADDITIONAL SOURCES OF INFORMATION

All through the previous discussions, it has been emphasized that while technique and logic are essential to the troubleshooting process, experience and process knowledge are almost equally important. Experience may be firsthand or vicarious. Many of the suggested sources discuss problems in considerable detail. This type of discussion and analysis can constitute vicarious experience, which, remembered and recalled when needed, is an excellent substitute for personal acquaintance with the problem. Information is readily available provided the proper sources are known. Such sources include:

1. Plating courses: The American Electroplaters Society offers several types of courses, ranging from intensive one-week sessions given regionally, to evening courses offered by local A.E.S. branches. In some cases, local branches through local colleges may offer somewhat more extensive evening courses.[1] Correspondence courses are also available, both through the A.E.S.[1] and independently.[2]
2. Many monthly publications have specialized articles containing valuable information. Several have question-and-answer columns which provide answers to specific problems.[3,4,5,6]
3. Numerous technical books and manuals are available.
4. Literature from proprietary suppliers frequently contains very detailed troubleshooting information. Additionally, suppliers have experts available for discussion of process parameters and suggested conditions for special situations.

With sufficient study and application of the proper techniques, a plater can prevent a major portion of the problems that normally develop, and solve most of those that cannot be prevented.

References

1. American Electroplaters Society, 1201 Louisiana Ave., Winter Park, FL 32789.
2. Joseph B. Kushner Electroplating School, 732 Glencoe Street, Sunnyvale, CA 94087.
3. Plating and Surface Finishing—see 1 above.
4. Products Finishing, 600 Main Street, Cincinnati, OH 45202.
5. Metal Finishing, One University Plaza, Hackensack, NJ 07601.
6. Industrial Finishing, Hitchcock Bldg., Wheaton IL 60187.
7. Durney, L. J. *Handbook for Solving Plating Problems*. Gardner Publications, Cincinnati, 1983.
8. Lowenheim, F. A. *Modern Electroplating*, New York: Wiley/Interscience. 1974.

9. ANALYSIS OF PLATING SOLUTIONS

John Morico

Clearwater Analytical Laboratories, Inc.

INDEX

Finishing Solution	Analytical Procedure
(1) Anodizing Solution, Chromic	
a. Chromic Acid	6, 20
b. Aluminum	5
(2) Anodizing Solution, Sulfuric	
a. Sulfuric Acid	28
b. Aluminum	5, 29
(3) Brasses	
a. Copper Metal	1, 12
b. Zinc Metal	1, 13
c. Cyanide, Total and Free	8, 9
d. Rochelle Salt	19
e. Carbonate	11
(4) Bronzes	
a. Copper Metal	1, 12
b. Tin metal	2, 12
c. Cyanide, Total and Free	8, 9
d. Rochelle Salt	19
e. Carbonate	11
(5) Cadmium	
a. Cadmium	1, 13
b. Cyanide, Total	9
c. Sodium Hydroxide	10
d. Carbonate	11
(6) Copper, Acid	
a. Copper Metal	23
b. Sulfuric Acid	24
c. Fluoboric Acid	24
d. Chloride	25

Finishing Solution	Analytical Procedure
(7) Copper, Cyanide	
a. Copper Metal	1, 12
b. Cyanide, Total and Free	8, 9
c. Hydroxide	10
d. Rochelle Salt	19
e. Carbonate	11
(8) Copper Electroless	
a. Copper Metal	1, 12, 23, 32
b. Hydroxide	32
c. Formaldehyde	32
(9) Copper, Pyrophosphate	
a. Copper Metal	1, 12, 23
b. Pyrophosphate	35
(10) Chromium	
a. Chromic Acid	6, 20
b. Chromium, Trivalent	20
c. Sulfate	21
d. Fluoride	22
(11) Gold, Acid	
a. Gold Metal	3, 12
b. pH	*
(12) Gold, Cyanide	
a. Gold Metal	3, 12
b. Cyanide, Total and Free	8, 9
c. Carbonate	11
(13) Indium, Acid	
a. Indium Metal	1, 13
b. Sulfate	21
c. Fluoboric Acid	24

*See *Important Notations.*

Finishing Solution	Analytical Procedure
(14) Indium, Cyanide	
a. Indium Metal	1, 13
b. Cyanide, Total and Free	8, 9
c. Hydroxide	10
(15) Iron, Chloride	
a. Ferrous Iron	27
b. Chloride	15
(16) Iron, Sulfate	
a. Ferrous Iron	27
b. Sulfate	21
(17) Lead, Acid	
a. Lead Metal	7, 34
b. Fluoboric Acid	24
(18) Nickel, Black	
a. Nickel Metal	1, 14
b. Zinc Metal	1, 26
c. Chloride	15
d. Boric Acid	18
e. Thiocyanate	36
(19) Nickel, Chloride	
a. Nickel Metal	1, 14
b. Hydrochloric Acid	17
(20) Nickel, Electroless	
a. Nickel Metal	1, 14
b. Hypophosphite	30
(21) Nickel, Fluoborate	
a. Nickel Metal	1, 14
b. Fluoboric Acid	24
c. Boric Acid	18
(22) Nickel, Sulfamate	
a. Nickel Metal	1, 14
b. Boric Acid	18
(23) Nickel, Watts	
a. Nickel Metal	1, 14
b. Chloride	15
c. Boric Acid	18
(24) Nickel, Iron Alloy	
a. Nickel Metal	1, 14
b. Iron Metal	1, 27
c. Boric Acid	18
d. Chloride	15
(25) Rhodium	
a. Rhodium Metal	3, 31
b. Acid, Total	24
(26) Silver, Cyanide	
a. Silver Metal	4, 12
b. Cyanide, Total and Free	8, 9
c. Carbonate	11

Finishing Solution	Analytical Procedure
(27) Tin, Acid	
a. Stannous Metal	7, 33
b. Fluoboric Acid	24
(28) Tin, Alkaline	
a. Tin Metal	2, 12
b. Hydroxide	10
c. Carbonate	11
(29) Tin/Lead and Tin/Lead Copper	
a. Tin Metal	7, 33
b. Lead Metal	7, 34
c. Copper Metal	7, 32
d. Fluoboric Acid	24
(30) Zinc, Cyanide	
a. Zinc Metal	1, 13
b. Hydroxide	10
c. Cyanide	9
d. Carbonate	11
(31) Zinc, Non-Cyanide	
a. Zinc Metal	1, 26
b. Chloride	15
c. Boric Acid	18

ANALYTICAL PROCEDURES

(1) Cadmium, Copper, Indium, Iron, Nickel and Zinc by Atomic Absorption Micro-method

(2) Tin by Atomic Absorption Micro-Method

(3) Gold and Rhodium by Atomic Absorption Micro-Method

(4) Silver by Atomic Absorption Micro-Method

(5) Aluminum by Atomic Absorption Micro-Method

(6) Chromium by Atomic Absorption Micro-Method

(7) Tin-Lead and Tin-Lead-Copper by Atomic Absorption Micro-Method

(8) "Free" Cyanide

(9) Total Cyanide

(10) Hydroxide, Sodium or Potassium

(11) Carbonate, Sodium or Potassium

(12) Copper, Gold and Tin in Cyanide Solutions including Preliminary Procedures

(13) Cadmium, Indium and Zinc by EDTA Method

(14) Nickel by EDTA

(15) Chloride by Silver Nitrate Titration

(16) Nickel Sulfate Calculation

(17) Hydrochloric Acid in Nickel Solution
(18) Boric Acid
(19) Rochelle Salt
(20) a. Chromic Acid
    b. Chromium, Trivalent
(21) Sulfate
    a. Gravimetric
    b. Centrifugation
(22) Fluoride Using Specific Ion Electrode
(23) Copper in Acid Solution
(24) Acid Content
(25) Chloride in Acid Copper Solution
(26) Zinc in Non-Cyanide Solution
(27) a. Ferrous Iron
    b. Alternate Ferrous Iron
(28) Sulfuric Acid in Anodizing Solution
(29) Aluminum in Anodizing Solution
(30) Hypophosphite in Electroless Nickel Solution
(31) Rhodium
(32) Electroless Copper
    a. Copper Method
    b. Sodium Hydroxide and Formaldehyde
(33) Stannous Tin
(34) Lead
    a. Gravimetric
    b. EDTA
(35) Pyrophosphate in Copper Solution
(36) Thiocyanate in "Black Nickel"

REAGENTS

Primary Standard Solutions

PS-1	0.1 N KCl
PS-2	0.05 M EDTA
PS-3	0.05 M Zn
PS-4	0.1 N KIO_3
PS-5	0.1 N $K_2Cr_2O_7$
PS-6	0.1 N As_2O_3
PS-7	Atomic Absorption Standards
PS-8	20 g/L Pyrophosphate Standard

Standard Solutions

S-1	1.0 N HCl
S-1a	0.1 N HCl
S-2	1.0 N H_2SO_4
S-2a	0.1 N H_2SO_4
S-3	0.1 N NaOH
S-3a	1.0 N NaOH
S-4	0.1 N $AgNO_3$
S-5	0.1 N $Na_2S_2O_3 \cdot 5H_2O$
S-6	0.1 N $Ce(SO_4)_2 \cdot 2(NH_4)_2SO_4 \cdot 2H_2O$
S-7	0.1 N I_2
S-8	0.1 N NaSCN
S-9	100 mg/L CN^{-1}
S-10	0.1 N Ferrous Ammonium Sulfate
S-11	0.005 N $Hg(NO_3)_2$
S-12	100° mg/L Fluoride

Non-Standard Solutions

NS-1	10% $BaCl_2$ or $Ba(NO_3)_2$
NS-2	20% Ammonium Acetate
NS-3	10% Na_2CO_3
NS-4	KI (Solid)
NS-5	NH_4OH/NH_4Cl Buffer
NS-6	10% H_2SO_4 or 20% HCl
NS-7	10% H_2O_2
NS-8	NH_4OH, Concentrate
NS-9	20% triethanolamine
NS-10	Mannitol (Solid)
NS-11	10% HCHO
NS-12	10% NaCN
NS-13	0.25% OsO_4
NS-14	Nitrobenzene
NS-15	TISAB Buffer (Total Ionic Strength Acid Buffer)
NS-16	20% Trichloracetic Acid
NS-17	10% Hydrazine
NS-18	1 M Na_2SO_3
NS-19	4% Guanylurea Sulfate
NS-20	12.5% $ZnSO_4$

Indicators

I-1	1% Starch
I-2	Sodium Chromate
I-3	20% KI
I-4	Methyl Yellow
I-5	Caustic Blue
I-6	Diphenylamine
I-7	Murexide
I-8	Eriochrome Black T
I-9	2% Ferric Ammonium Sulfate
I-10	Ferroin
I-11	Bromcresol Purple
I-12	1% Phenolphthalein
I-13	Methyl Orange—Xylene Cyanol
I-14	Diphenylcarbazone
I-15	PAN

EQUIPMENT NEEDED

125 ml Erlenmeyer flasks
250 ml Erlenmeyer flasks
500 ml Erlenmeyer flasks

100 ml volumetric flasks
250 ml volumetric flasks
500 ml volumetric flasks
1000 ml volumetric flasks
250 ml beakers
500 ml beakers
1000 ml beakers
1500 ml beakers
50 ml burettes
1, 2, 5, 10, 50 and 100 ml volumetric pipettes
1, 5, and 10 ml serological pipettes, graduated 0.1 ml
Stirring rods
Watch Glasses
Filter funnels, glass or polyethylene
Filter funnel, Buchner, 15 cm
Whatman No. 40 (or equivalent) filter paper, 15 cm
Test tubes, 16 mm × 125 mm
Glass and polyethylene bottles, 500 and 1000 ml
Glass and polyethylene dropper bottles, 50 and 100 ml

INSTRUMENTS REQUIRED

Electric magnetic heater-stirrer
Electric magnetic stirrer
Teflon stir bars, 1 in., 1.5 in., 2 in.
pH meter capable of recording 1 mv
Drying oven
Analytical balance sensitive to 0.1 mg
Triple-beam balance sensitive to 0.1 g

OPTIONAL INSTRUMENTATION

0–50 ul adjustable micropipette
50–250 ul adjustable micropipette
250–1000 ul adjustable micropipette
Atomic absorption spectrophotometer
Micro-titrator and accessories

IMPORTANT NOTATIONS

It is good practice to properly label *all* reagents with the normality or molarity, reagent concentration in grams per liter (g/L), date and initials of the person making up the reagents. The labeling of indicator solutions should also include the pH range. Standard solutions require the use of an analytical balance accurate to 0.1 mg. The pH meter is also an essential piece of laboratory equipment. It should be standardized frequently in accordance with manufacturer's instructions.

Generally speaking, alkaline reagents should be stored in polyethylene bottles and acid reagents in either glass or polyethylene. Reagents should *never* be stored in volumetric flasks. Prompt rinsing of volumetric glassware will help to prevent contamination and will maintain good laboratory practice. Atomic absorption working standards (0.1 to 5 ppm) should be made fresh daily from concentrated standards (20 to 100 ppm). Working standards are not stable. Concentrated standards can be kept for six months.

Some reagents should be standardized more frequently than others depending on usage. Examples are: silver nitrate, sodium thiosulfate, hydrochloric acid, sodium hydroxide, sodium cyanide, potassium permanganate, sodium thiocyanate and iodine. Many indicators are not stable and should be prepared in small quantities, i.e., quantities sufficient to last about one month.

For convenience sake, abbreviations will be used to denote primary standards not requiring standardization (PS-1, PS-2, etc.); standard solutions which are usually standardized (S-1, S-2, etc.); non-standard reagents used in various analyses (NS-1, NS-2, etc.); and indicators (I-1, I-2, etc.).

When sampling, always wear protective equipment to prevent injury. It is best to use a long, rigid ¼-in. ID (internal diameter) plastic tube for sampling solutions. It should be long enough to almost reach the bottom of the tank. Slowly immerse the tube into the solution and when immersed far enough, cover the open end of the tube with the thumb. Withdraw the tube from solution, hold it over an open 500-ml plastic beaker which has been clearly marked to identify the solution; release the thumb, which then releases the liquid in the tube. Repeat this procedure in various parts of the tank to obtain a representative sample. Pour into a plastic bottle which has been labeled to identify the solution and mix thoroughly prior to analysis. If necessary, the solution can also be filtered. After analyses are completed, return the balance of solutions to the tanks.

All plating tanks should be numbered to avoid confusion. A diagram of the plating room would be helpful. Each tank should be listed by physical size and volume (gallons and/or liters) so that additions can be calculated more easily. To determine gallons, multiply length by width by depth of solution (in inches) and divide by 231. To determine liters, multiply length by width by depth of solution (in centimeters) and divide by

1000. To change gallons to liters, multiply by 3.785, and to change liters to gallons, multiply by 0.2642. Solution level should be at the same height each time sampling takes place to avoid errors in addition. Air agitated solutions can be sampled directly by filling the plastic bottle rather than using the plastic tube procedure.

Occasionally, acid treatment of cyanide solutions, treatment of cyanide solutions with acid or treatment of a solution by fuming with acids or by charring may be required during the course of analysis. It is strongly recommended that a fume-scrubber hood be utilized for these purposes in order to prevent toxic fumes from entering the atmosphere, and for the safety of all personnel. These wastes are more easily treated by the existing waste treatment system.

Experience has shown that most methods of analysis can be reduced to micro-methods. This can be accomplished by reducing all volumes ten-fold or greater with the use of micropipettes, which are very accurate, easy to calibrate and durable. The advantage of micronization is that the sample size is smaller, the reagents last longer because of diminished volume usage, there is less use of expensive glassware and analysis is usually more rapid. For titrations, a microtitrator such as the Oxford model is necessary to give accurate and reproducible results. More sophisticated models are available but not necessary unless plating solution quality control is carried out often during the course of a day. When applied to atomic absorption, micro-methods may eliminate the use of large volumetric flasks and eliminate the destruction of large volumes of cyanide solution. For example, using a sample size of 0.1 ml instead of 10 ml in analyzing brass solution can eliminate the necessity of using a fume hood when destroying cyanide.

In order to analyze waste effluent, it is necessary to obtain the latest editions of *Standard Methods for the Examination of Water and Waste Water* or comparable ASTM and EPA manuals. Atomic absorption is the preferred instrument for analysis of metals and there are several inexpensive models (or used models) on the market. It is strongly recommended that the investment in an atomic absorption spectrophotometer be made. It will also aid in quality control of trace impurities as well as analyses of binary and tertiary alloy solutions, metal cyanide solutions and alloy composition of electroplated metal.

Preparation and Standardization of Standard Solutions

PS-1   0.01 N KCl (7.46 g/L)—Potassium Chloride

Place 30 g potassium chloride, A.R., into a weighing bottle and dry in the oven at 105 C for 2 hr. Cool in desiccator. Weigh out 7.5 g dried KCl and record the exact weight. Transfer into 1 L volumetric flask and dissolve in 500 ml distilled water. Dilute to the mark, mix well and store in glass or polyethylene bottle which has been properly labeled with normality of KCl, concentration of KCl in g/L, date and initials.

$$N = \frac{\text{wt KCl (g)}}{74.6}$$

PS-2   0.05 M EDTA (18.6 g/L EDTA-2)—Ethylenediamine Tetraacetic Acid, Disodium

Place 10 g NaOH, A.R., into a 1 L volumetric flask. Add 200 ml distilled water and dissolve the NaOH. Weigh 18.6 g disodium ethylenediamine tetraacetic acid, A.R., and record the exact weight. Transfer this weighed amount of EDTA to the flask and stir to dissolve completely. Dilute to the mark with distilled water, mix well and store in a polyethylene container which has been properly labeled with molarity of EDTA, concentration of EDTA in g/L, date and initials. (1 ml = 3.27 mg zinc, 8.18 mg copper, 2.93 mg nickel, 5.62 mg cadmium, 10.36 mg lead)

$$M = \frac{\text{wt EDTA-2 (g)}}{372}$$

PS-3   0.05 M (93.27 g/L Zinc)—Zinc Solution

Add 50 ml concentrated HCl, A.R., and 50 ml distilled water to a 1 L volumetric flask. Weigh 3.27 g zinc metal, A.R., granular, and record the exact weight. Transfer the weighed amount of zinc to the flask and stir to dissolve completely. Gentle warming in hot water will aid this process. (Note: Hydrogen gas will be given off, and precautions against explosion should be taken.) Dilute to the mark with distilled water. Mix well and store in glass or polyethylene bottle which has been properly labeled with normality of zinc, concentration of zinc in g/L, date and initials.

$$N = \frac{wt\ Zn\ (g)}{65.4}$$

PS-4 0.1 N KIO_3 (3.567 g/L)—Potassium Iodate

Place 30 g potassium iodate, A.R., into a weighing bottle and dry in oven at 180 C for 2 hr. Cool in desiccator. Transfer 200 ml distilled water into a 400-ml beaker and add 1 g NaOH and 10 g KI; stir to dissolve. Weigh out 3.5 to 3.6 g KIO_3 and record the exact weight. Transfer the weighed amount of dried KIO_3 to the beaker and stir to dissolve. Transfer to a 1 L volumetric flask and dilute to the mark with distilled water. Store in a glass bottle properly labeled with normality of KIO_3, concentration of KIO_3, date and initials.

$$N = \frac{wt\ KIO_3\ (g)}{35.67}$$

PS-5 0.1 N $K_2Cr_2O_7$ (4.9035 g/L)—Potassium Dichromate

Place 30 g potassium dichromate, A.R., into weighing bottle and dry in oven at 110 C for 2 hr. Cool in desiccator. Weigh out about 4.9 g $K_2Cr_2O_7$ and record the exact weight. Add the weighed amount of $K_2Cr_2O_7$ to the volumetric flask; add 700 ml distilled water, stir to dissolve and dilute to the mark with distilled water. Mix well. Store in a glass bottle and label properly with normality of $K_2Cr_2O_7$, concentration of $K_2Cr_2O_7$ in g/L, date and initials.

$$N = \frac{wt\ K_2Cr_2O_7\ (g)}{49.035}$$

PS-6 0.1 N As_2O_3 (4.9460 g/L)—Arsenic Trioxide

Place 30 g arsenic trioxide, A.R., into weighing bottle and dry in the oven at 110 C for 2 hr. Cool in desiccator. Transfer 50 ml distilled water into a 250-ml beaker. Add 4 g potassium hydroxide and stir to dissolve. Weigh out 4.9 to 5.0 g As_2O_3 and record exact weight. Transfer the weighed amount of As_2O_3 to the warm KOH solution and stir to dissolve. When completely dissolved, carefully add 50% H_2SO_4 dropwise until just neutral to phenolphthalein (Reagent I-12) endpoint (colorless). Cool, transfer to 1 L volumetric flask and dilute to the mark with distilled water. Store in a glass or polyethylene bottle properly labeled with normality of As_2O_3, concentration of As_2O_3 in g/L, date and initials.

$$N = \frac{wt\ As_2O_3\ (g)}{49.460}$$

PS-7 Atomic Absorption Standards

These standards should be made up according to manufacturer's specifications or purchased from a reliable source which would ensure accuracy. They are usually available at 500 mg/L and 1000 mg/L, from which dilutions can be made to give the range of metals being analyzed. A 50 mg/L standard containing several metals can be made up as a "stock working standard" which is much easier to use for dilutions.

Pipet 5.0 ml of each 1000 mg/L standard (or 10 ml of each 500 mg/L standard) into a 100-ml volumetric flask. Add 10 ml concentrated HNO_3, A.R., and dilute to volume with distilled water. Mix and store in a glass bottle properly labeled with concentration, date and initials. Daily "working standards" of 0.2 to 5.0 mg/L can then be made from the "stock working standard." Label "stock working standard" bottle with concentration of each metal in mg/L, date and initials. Make up new "stock working standard" every three months. The following metals should not be mixed with other standards: silver, rhodium, tin, lead, gold, titanium, barium and silicon.

PS-8 51.25 g/L $Na_4P_2O_7 \cdot 10H_2O$ [20.0 g/L $(P_2O_7)^{-4}$ Pyrophosphate]

Place 750 ml distilled water into a 1 L volumetric flask. Weigh 51.25 g sodium pyrophosphate, A. R., and record exact weight. Transfer to the 1 L volumetric flask and stir to dissolve. Dilute to the mark with distilled water and mix well. Store in a polyethylene bottle labeled properly with concentration of pyrophosphate in g/L, date and initials. Calculation: Exact wt, g × 0.39 = g/L $(P_2O_7)^{-4}$, Pyrophosphate

S-1 1.0 N HCl (83.3 ml/L)—Hydrochloric Acid

Place 500 ml distilled water into a 1 L volumetric flask. Add 83.3 ml concentrated HCl, A.R., and mix. Dilute to the mark, mix well, and store in a glass or polyethylene bottle which has been labeled with normality of HCl, concentration of HCl in g/L, date and initials. Pipet 2.0 ml into a 250-ml Erlenmeyer flask, add 50 ml distilled water, 2 to 3 drops phenolphthalein indicator (Reagent I-12) and titrate while swirling with standard 0.1 N NaOH (Reagent S-3) to first pink endpoint.

$$N = \frac{\text{ml NaOH} \times \text{N NaOH}}{\text{sample size (2 ml)}}$$

S-1A 0.1 N HCl—Hydrochloric Acid

Pipet 100.0 ml of standardized 1.0 N Hydrochloric Acid (reagent S-1) into a 1 L volumetric flask. Dilute to the mark with distilled water and mix well. Store in a polyethylene bottle labeled properly with normality of acid, concentration, date and initals. No stardardization is necessary.

S-2 1.0 N H_2SO_4 (27.8 ml/L)—Sulfuric Acid

Place 500 ml distilled water into a 1 L volumetric flask. With mixing, add 27.8 ml concentrated H_2SO_4, A.R. Cool, dilute to the mark, mix well and store in glass or polyethylene bottle which has been properly labeled with normality of H_2SO_4, concentration of H_2SO_4, date and initials. Pipet a 2 ml sample of H_2SO_4 into a 250-ml Erlenmeyer flask, add 50 ml distilled water, 2 to 3 drops phenolphthalein indicator (Reagent I-12) and titrate while swirling with standard 0.1 N NaOH (Reagent S-3) to first pink endpoint.

$$N = \frac{\text{ml NaOH} \times \text{N NaOH}}{\text{sample size (2 ml)}}$$

S-2A 0.1 N H_2SO_4—Sulfuric Acid

Pipet 100.0 ml of standardized 1.0 N sulfuric acid (reagent S-2)—into a 1 L volumetric flask. Dilute to the mark with distilled water and mix well. Store in a polyethylene bottle properly labeled with normality of acid, concentration, date and initials. No standardization is necessary.

S-3 0.1 N NaOH (4.0 g/L)—Sodium Hydroxide

Add 4.0 g NaOH, A.R., to a 1 L volumetric flask and add 50 ml boiled and cooled distilled water, stir to dissolve and dilute to the mark with boiled, cooled distilled water and store in a polyethylene container which has been properly labeled with normality of NaOH, concentration of NaOH in g/L, date and initials.

Dry 50 g potassium acid phthalate, A.R. (KHP), in oven at 110 C for 2 hr. Cool in desiccator. Weigh 0.5 g KHP into a tared 250-ml Erlenmeyer flask. Record exact weight of KHP. Add 50 ml warm distilled water and 2 to 3 drops phenolphthalein indicator (Reagent I-12). Titrate while swirling with NaOH solution to first pink endpoint.

$$N = \frac{\text{wt KHP (g)}}{\text{ml NaOH} \times 0.2042}$$

S-3A 1.0 N NaOH (40 g/L)—Sodium Hydroxide

Add 40 g NaOH, A.R., to a 1 L volumetric flask and add 50 ml boiled and cooled distilled water; stir to dissolve and dilute to the mark with boiled, cooled distilled water. Mix well and store in a polyethylene container which has been properly labeled with normality of NaOH, concentration of NaOH in g/L, date and initials.

Dry 50 g potassium acid phthalate, A.R. (KHP), in oven at 100 C for 2 hr. Cool in desiccator. Weigh 5.0 g KHP into a 250 ml Erlenmeyer flask. Record exact weight of KHP. Add 50 ml warm distilled water, and 2 to 3 drops phenolphthalein (Reagent I-12). Titrate while swirling with NaOH solution to first pink endpoint.

$$N = \frac{\text{wt KHP (g)}}{\text{ml NaOH} \times 0.2042}$$

S-4   0.10 N $AgNO_3$ (17.0 g/L)—Silver Nitrate

Add 700 ml distilled water and 5 ml concentrated HNO_3, A.R., to a 1 L volumetric flask. Add 17 g silver nitrate, A.R., and stir to dissolve completely. Dilute to the mark with distilled water, mix well and store in an amber glass bottle which has been properly labeled with normality of $AgNO_3$, concentration of $AgNO_3$ in g/L, date and initials. (Note: If solution is cloudy, chloride contamination is indicated. Wait several days and then filter through very fine filter. Proceed with standardization.)

Pipet 25 ml 0.1 N KCl (PS-1) into a 250 ml Erlenmeyer flask. Add 25 ml distilled water, 5 ml of 20% ammonium acetate solution (NS-2) and 1 ml sodium chromate indicator (I-2). Titrate while swirling with silver nitrate solution prepared above. Endpoint is reached when silver chloride precipitate formed turns pink in the flask.

$$N = \frac{\text{ml KCl} \times \text{N KCl}}{\text{ml AgNO}_3}$$

S-5   0.1 N $Na_2S_2O_3 \cdot 5H_2O$ (24.818 g/L)—Sodium Thiosulfate

Boil 1300 ml distilled water for 5 min. Cool. Add 24.8 g sodium thiosulfate, A.R., to 700 ml boiled, cooled distilled water in a 1 L volumetric flask. Stir to dissolve and dilute to the mark with boiled, cooled distilled water. Mix well and transfer to a glass or polyethylene bottle properly labeled with normality of $Na_2S_2O_3$, concentration of $Na_2S_2O_3 \cdot 5H_2O$ in g/L, date and initials.

Pipet 25.0 ml potassium iodate standard (PS-4) into a 125 ml Erlenmeyer flask. Add 30 ml distilled water, 1 g KI and 10 ml 25% H_2SO_4. Titrate with swirling motion using standard (S-5) to a light yellow color. Add 2 ml starch reagent (I-1) and continue titration until blue color disappears. Record ml (S-5) thiosulfate standard used.

$$N = \frac{\text{ml KIO}_3 \times \text{N KIO}_3}{\text{ml Na}_2\text{S}_2\text{O}_3}$$

S-6   0.1 N $Ce(SO_4)_2 \cdot 2(NH_4)_2SO_4\ 2H_2O$ (63.256 g/L)—Ceric Sulfate

Weigh out 64 g ceric ammonium sulfate, A.R., $Ce(SO_4)_2 \cdot 2(NH_4)_2SO_4 \cdot 2H_2O$. Transfer 500 ml distilled water to a 1 L volumetric flask and with stirring add 28 ml concentrated H_2SO_4. Dissolve the ceric salt in the flask and dilute to the mark with distilled water. Mix well and store in a glass or polyethylene bottle properly labeled with normality ceric sulfate, concentration of ceric sulfate in g/L, date and initials.

Pipet 25 ml aliquot of arsenious acid (reagent PS-6) into 250 ml Erlenmeyer flask. Add 50 ml distilled water and 20 ml concentrated H_2SO_4. Add 2 drops osmium reagent (NS-13) and 2 drops ferroin (reagent I-10). Titrate while swirling with ceric sulfate from pink to pale blue.

$$N = \frac{\text{ml As}_2\text{O}_3 \times \text{N As}_2\text{O}_3}{\text{ml ceric sulfate}}$$

S-7   0.1 N I_2 (12.69 g/L)—Iodine

Transfer 30 ml distilled water to a 100 ml beaker. Add 40 g KI and dissolve. Weigh out 12.7 g iodine, A.R., crystals and add to the KI solution. Stir until completely dissolved. Transfer to a 1 L volumetric flask and dilute to the mark with distilled water. Store in a dark glass bottle properly labeled with normality of I_2, concentration of I_2 in g/L, date and initials.

Pipet 25.0 ml As_2O_3 (reagent PS-6) to a 250-ml Erlenmeyer flask. Add about 2 g sodium bicarbonate, 50 ml distilled water and 2 to 3 drops 1% starch solution (Reagent I-1). While swirling, titrate with iodine (Reagent S-7) to first blue endpoint that remains for at least 30 to 60 seconds.

$$N = \frac{\text{ml As}_2\text{O}_3 \times \text{N As}_2\text{O}_3}{\text{ml I}_2}$$

S-8   0.1 N NaSCN (8.11 g/L)—Sodium Thiocyanate

Transfer 700 ml distilled water to a 1 L volumetric flask. Weight out 10 to 12 g NaSCN and transfer to the 1 L volumetric flask. Add 1 ml formaldehyde solution (37% formalin). Stir to dissolve and dilute to the mark with distilled water. Mix well and store in a glass or polyethylene bottle properly labeled with normality of NaSCN, concentration of NaSCN in g/L, date and initials.

Pipet 25.0 ml silver nitrate (reagent S-4) to a 250-ml Erlenmeyer flask, add 50 ml distilled water and 1 ml ammonium hydroxide (reagent I-9). Titrate with NaSCN to a pink endpoint.

$$N = \frac{ml\ AgNO_3 \times AgNO_3}{ml\ NaSCN}$$

S-9   1000 mg/L Cyanide (2.5 g/L KCN) Note: POISON

Transfer 700 ml distilled water to a 1 L volumetric flask. Add 5 g Na_2CO_3 and 1 g NaOH. Stir to dissolve. Weigh out about 2.5 g KCN, add to flask and stir to dissolve. Dilute to the mark with distilled water, mix well and transfer to a polyethylene bottle properly labeled with concentration of KCN in mg/L, date and initials and POISON warning. Storing at refrigeration temperature will prolong stability.

Transfer 50.0 ml KCN solution (reagent S-9) to a 250 ml Erlenmeyer flask, add 50 ml distilled water, 10 ml ammonium hydroxide (reagent NS-9) and 2 ml 20% potassium iodide (reagent I-3). Titrate while swirling with silver nitrate solution (reagent S-4) until solution becomes light pearlescent yellow.

Concentration Cyanide, mg/L
$$= \frac{ml\ AgNO_3 \times AgNO_3 \times 52{,}000}{ml\ potassium\ cyanide}$$

S-10   0.1 N $Fe(NH_4)_2(SO_4) \cdot 6H_2O$ (39.2 g/L)—Ferrous Ammonium Sulfate (F.A.S.)

Transfer 700 ml distilled water to a 1 L volumetric flask. Add 50 ml concentrated H_2SO_4 with mixing. Cool. Weigh out 30 to 40 g ferrous ammonium sulfate, A.R. (also known as F.A.S.), and transfer to the 1 L volumetric flask. Stir to dissolve and dilute to the mark with distilled water. Mix well and store in glass or polyethylene bottle properly labeled with normality F.A.S., concentration of F.A.S. in g/L, date and initials. Add a pinch of sodium carbonate to exclude air contact with the solution by forming a CO_2 atmosphere.

Pipet 25.0 ml potassium dichromate (reagent PS-5) into a 250 ml Erlenmeyer flask, add 50 ml distilled water and while mixing, add 10 ml concentrated H_2SO_4. Cool, add 4 drops orthophenanthroline (reagent I-10). Titrate while swirling with ferrous ammonium sulfate (reagent S-10). Color changes will progress from yellow to green through grey and finally sharp red endpoint.

$$N = \frac{ml\ K_2Cr_2O_7 \times N\ K_2Cr_2O_7}{ml\ Fe(NH_4)_2SO_4 \cdot 6H_2O}$$

S-11   0.005 N $Hg(NO_3)_2 \cdot H_2O$—Mercuric Nitrate

Transfer 700 ml distilled water to 1 L volumetric flask; add 0.86 g mercuric nitrate, A.R., 2 ml concentrated HNO_3 and stir to dissolve. Dilute to the mark and store in a polyethylene bottle labeled properly with normality of $Hg(NO_3)_2$, concentration of $Hg(NO_3)_2$ in g/L, date and initials.

Pipet 1.0 ml 0.1N KCl (reagent PS-1) into a 250-ml Erlenmeyer flask, add 100 ml distilled water and 10 ml 20% TCA (reagent NS-16). Titrate while swirling with mercuric nitrate (reagent S-11) to the first purple color change.

$$N = \frac{ml\ KCl \times N\ KCl}{ml\ Hg(NO_3)_2}$$

S-12   1000 mg/L Fluoride

Place 700 ml distilled water into a 1 L volumetric flask. Weigh out 2.2105 g sodium fluoride, record the exact weight and transfer to the 1 Liter volumetric flask. Add 1 ml 1 N NaOH (reagent 3-3a), stir to dissolve and dilute to the mark with distilled water. Store in a polyethylene bottle labeled properly with concentration of fluoride, date and initials.

$$\text{Concentration Fluoride, mg/L} = \text{NaF (g)} \times 452.4$$

Non Standard Reagents

NS-1   10% Barium Chloride, A.R., or 10% Barium Nitrate, A.R.
Add 50 g of either salt to a beaker of 500 ml distilled water; stir to dissolve. Store in a glass or polyethylene bottle properly labeled with concentration, date and initials.

NS-2   20% Ammonium Acetate
Add 100 g ammonium acetate A.R., to a beaker of 500 ml distilled water. Stir to dissolve. Store in a glass or polyethylene bottle properly labeled with concentration, date and initials.

NS-3   10% Sodium Carbonate
Add 50 g sodium carbonate, A.R., to a beaker of 500 ml distilled water. Stir to dissolve. Store in a polyethylene bottle properly labeled with concentration, date and initials.

NS-4   Potassium Iodide Crystals, A.R.

NS-5   10% Ammonium Chloride in 50% Ammonium Hydroxide Buffer
In fume hood, place 500 ml distilled water into a 1500-ml beaker. Add 500 ml concentrated NH_4OH, A.R., stir and add 100 g NH_4Cl, A.R. When dissolved, store in a polyethylene bottle which has been properly labeled with concentration, date and initials.

NS-6   Acids
Always add acids to water with mixing when making up acid dilutions from concentrated acids. NEVER pour water into concentrated acids when making up an acid dilution.

NS-7   10% Hydrogen Peroxide
Place 100 ml distilled water into a 250-ml beaker. Add, with stirring, 50 ml 30% H_2O_2, A.R. Store in a polyethylene bottle, keeping the cap loose in order to prevent swelling of the bottle due to oxygen pressure. Label properly with concentration, date and initials.

NS-8   Ammonium Hydroxide, A.R.

NS-9   20% Triethanolamine
Place 400 ml distilled water into 600-ml beaker. Add 100 ml triethanolamine, A.R., and stir well. Store in a polyethylene bottle and label properly with concentration, date and initials.

NS-10 Mannitol, A.R.
This can also be purchased through a medical laboratory supply house as Difco brand D-Mannitol No. 0170-17, 500-g bottle.

NS-11 10% Formaldehyde
Place 450 ml distilled water into a 600-ml beaker. Add 50 ml 37% HCHO, A.R., and mix well. Store in a glass or polyethylene bottle and label properly with concentration, date and initials.

NS-12 10% Sodium Cyanide (POISON)
Place 1000 ml distilled water in a 1500-ml beaker. Add 2 g NaOH and stir to dissolve. Add 100 g sodium cyanide and stir to dissolve. Store in polyethylene bottle labeled properly with concentration, date and initials. Also place POISON label on bottle.

NS-13 0.25% Osmium Catalyst
Place 100 ml distilled water into a 250-ml beaker and add 0.3 ml concentrated H_2SO_4. Mix. Add 0.25 g osmium tetroxide, A.R., and stir to dissolve. Store in a glass or polyethylene dropper bottle labeled properly with concentration, date and initials.

TABLE 1. DATA ON COMMONLY USED ACIDS

	HCl	HNO_3	H_2SO_4	NH_4OH	$HC_2H_3O_2$
Mol wt	36.46	63.02	98.08	35.04	60.03
Av sp gr of conc. reagent	1.19	1.42	1.84	0.90	1.06
Av wt % present in conc. reagent	36.00	69.50	96.00	58.60	99.50
Active ingredient, (g/ml)	0.426	0.985	1.76	0.527	1.055
Normality of conc. reagent	11.7	15.6	35.9	15.1	17.6
Ml conc. reagent/Liter N solution	85.5	64.0	28.4	66.5	56.9
Ml reagent to be diluted to 500 ml for use as shelf reagent (5N)	214	161	70	166	143

NS-14 Nitrobenzene, A.R.
NS-15 TISAB Buffer
Prepare solution of 5 molar NaOH by dissolving 20 g NaOH in 100 ml distilled water. Cool. Place 500 ml distilled water into a 1500-ml beaker, add 57 ml glacial acetic acid, 58 g sodium chloride and 4 g cyclohexylene dinitrilo tetraacetic acid (CDTA). Stir to dissolve and place in cold water bath. Immerse pH electrodes to record pH, and while stirring slowly, adjust pH to 5.0 to 5.5 with the 5 molar NaOH. Cool and dilute to 1000 ml with distilled water. Store in a polyethylene bottle labeled properly with contents, date and initials.

NS-16 20% Trichloracetic Acid
Transfer 900 ml distilled water to a 1500-ml beaker, add 200 g trichloracetic acid, A.R., stir to dissolve and dilute to 1000 ml. Store in polyethylene bottle labeled properly with concentration, date and initials.

NS-17 10% Hydrazine
Place 480 ml distilled water into a 1000-ml beaker, add 20 ml 54% hydrazine (N_2H_4). Mix and store in a polyethylene bottle labeled properly with concentration, date and initials.

NS-18 1 M Sodium Sulfite
Place 500 ml distilled water into a 1000-ml beaker, add 126 g Na_2SO_3, A.R., and adjust pH to 10.0 with 0.1 N NaOH (reagent S-3). Store in a polyethylene bottle labeled properly with concentration, date and initials. Makeup fresh monthly.

NS-19 4% Guanylurea Sulfate
Place 1000 ml distilled water into a 1500-ml beaker and heat to 60 C. Add 40 g guanylurea sulfate and stir to dissolve. While still hot, add 1 to 2 g activated carbon and stir. Filter through Buchner funnel with vacuum. Store in a polyethylene bottle labeled properly with concentration, date and initials. NOTE: When cold, the reagent will crystallize. Warming the reagent will redissolve the crystals.

NS-20 12.5% Zinc Sulfate
Place 1000 ml distilled water into a 1500-ml beaker. Add 125 g $ZnSO_4$, A.R., and stir to dissolve. Add 5 drops methyl orange-xylene cyanol (reagent I-13) and titrate with 0.1 N HCl or 0.1 N NaOH until color changes to pink endpoint. Store in a polyethylene bottle labeled properly with concentration, date and initials.

Indicators

I-1 1% Starch
Boil 500 ml of distilled water in 1000 ml beaker. Make slurry of 5 g soluble starch A.R. with a small amount of cold water. Slowly add to the boiling water and boil for 3 to 5 minutes. Cool and store in a glass or polyethylene bottle which has been properly labeled with concentration, date and initials. A pinch of mercuric oxide can be added as a preservative to prevent mold growth.

I-2 Sodium Chromate
Place 100 ml distilled water in a 250 ml beaker. Add 2 g Na_2CRO_4, A.R. and stir to dissolve. Store in glass or polyethylene and label properly with concentration, date and initials.

I-3 20% Potassium Iodide
Place 100 ml distilled water in a 250 ml beaker. Add 20 g KI, A.R., and stir to dissolve. Store in a glass or polyethylene bottle and label properly with concentration, date and initials.

I-4 Methyl Yellow—Transition pH 2.9 Red, pH 4.0 Yellow
Place 50 ml isopropyl alcohol in a 100 ml beaker. Add 0.5 g methyl yellow (dimethylaminoazobenzene) and stir to dissolve. Store in a glass dropper bottle and label properly with concentration, date and initials.

I-5 Caustic Blue—Transition pH 11.4 Blue, pH 12.0 Yellow
Place 50 g sodium chloride, A.R., in mortar and grind to fine powder. Add 0.5 g 5,5′ indigo disulfonic acid, disodium salt (indigo carmine) and mix well. Store in a dark glass or polyethylene bottle. Label properly with contents, date and initials.

I-6 1% Diphenylamine
Place 40 ml distilled water in a 250 ml beaker. Add 60 ml acetic acid, A.R., mix well. Weigh 1 g diphenylamine and dissolve in the acetic acid. Store in a glass dropper bottle and label properly with concentration, date and initials.

I-7   Murexide
Place 100 g sodium chloride, A.R., in mortar and grind to a fine powder. Add 0.2 g murexide indicator and mix well. Store in a glass or polyethylene bottle and label properly with contents, date and initials.

I-8   Eriochrome Black T
Place 100 g sodium chloride, A.R., in a mortar and grind to a fine powder. Add 1 g Eriochrome Black T powder and mix well. Store in a glass or polyethylene bottle and label properly with contents, date and initials.

I-9   2% Ferric Ammonium Sulfate
Place 100 ml distilled water into a 250 ml beaker. Add 2 g ferric ammonium sulfate, A.R., and stir to dissolve. Add 2 drops concentrated sulfuric acid with stirring. Store in a glass or polyethylene dropper bottle and label properly with concentration, date and initials.

I-10 0—Phenanthroline—Ferrous Complex (Ferroin)
Place 100 ml distilled water into a 250 ml beaker. Add 7 g ferrous sulfate, A.R., and mix well to dissolve. Add 1.5 g 1,10 phenanthroline and mix well. Store in a glass dropper bottle and label properly with contents, date and initials.

I-11 0.05% Bromcresol Purple Transition pH 5.2 Yellow, pH 6.8 Purple
Place 200 ml distilled water in a 250 ml beaker. Add exactly 2.0 ml 0.1 N NaOH (reagent S-3) and 0.1 g bromcresol purple, A.R. Mix well and store in a glass or polyethylene dropper bottle properly labeled with concentration, date and initials.

I-12 1% Phenolphthalein Transition pH 8.3 Colorless, pH 10.0 Red
Place 100 ml isopropyl alcohol in a 250 ml beaker. Add 1 g phenolphthalein, A.R., and stir to dissolve. Store in a glass dropper bottle labeled properly with concentration, date and initials. Label FLAMMABLE! on bottle.

I-13 Methyl Orange—Xylene Cyanol Transition pH 2.9 Red, pH 4.6 Green
Place 100 ml distilled water into 250 ml beaker. Add 0.1 g methyl orange and 0.14 g xylene cyanol FF. Stir to dissolve. Store in a glass dropper bottle and label properly with contents, date and initials.

I-14 0.5% Diphenylcarbazone
Weigh out 0.5 g diphenylcarbazone, A.R., transfer to a 250 ml beaker, add 100 ml isopropyl alcohol and stir to dissolve. Store in a dark glass bottle labeled properly with concentration, date and initials.

I-15 0.1% 1-(2-Pyridylazo)-2-Naphthol (PAN)
Place 100 ml isopropyl alcohol into a 250 ml beaker, add 0.1 g PAN and stir to dissolve. Store in glass bottle labeled properly with concentration, date and initials.

Analytical Procedures

(1) Cadmium, Copper, Indium, Iron, Nickel and Zinc by Atomic Absorption Micro-Method

Micropipet 0.100 ml sample into a test tube. Add 9.9 ml distilled water and mix well. Micropipet 0.100 ml aliquot into another test tube, add 0.1 ml concentrated sulfuric acid and mix by swirling. Add 9.8 ml distilled water and shake well. Use this solution for analysis, setting up metal standards of 0.2, 2.0 and 3.0 mg/L. Dilute sample as necessary when above "Reading not to exceed" value and adjust calculations by multiplying reading times factor. See example below.

Metal	Wavelength, nm, UV	Reading Not to Exceed
Cadmium	228.8	2.0
Copper	324.8	4.0
Indium	303.9	30.0
Iron	248.3	4.0
Nickel	232.2	4.0
Zinc	213.9 (Rich Flame)	1.0

Calculations:
  Reading, mg/L $\times$ 10 = g/L Metal
  Reading, mg/L $\times$ 1.33 = oz/gal Metal

Examples of Dilution Factor

  If diluted 1 + 1 water, factor = 2
  If diluted 1 + 3 water, factor = 4
  If diluted 1 + 9 water, factor = 10
  If diluted 1 + 99 water, factor = 100

(2) Tin by Atomic Absorption Micro-Method

Micropipet 0.500 ml sample into a test tube. Add 9.5 ml distilled water and mix

well. Micropipet 0.100 ml aliquot into another test tube, add 0.1 ml concentrated sulfuric acid and mix by swirling. Add 9.8 ml distilled water and shake well. Use this solution for analysis, setting up tin standards of 25, 50, 75, and 100 mg/L. Use wavelength 286.3 nm, UV for tin and nitrous oxide flame.
Calculations:
Reading, mg/L $\times$ 2 = g/L tin
Reading, mg/L $\times$ 0.27 = oz/gal tin

(3) Gold or Rhodium by Atomic Absorption Micro-Method

Micropipet 0.500 ml sample of gold or rhodium solution into a test tube. Add 9.50 ml distilled water and shake well. Micropipet 0.100 ml aliquot into another test tube and add 0.5 ml concentrated hydrochloric acid and 1 drop 10% hydrogen peroxide (reagent NS-7). Mix by swirling and add 9.35 ml distilled water and shake well. Use this solution for analysis. Set up gold standards of 1, 5, 10 and 15 mg/L or rhodium standards of 5, 10, 25 and 40 mg/L. Use gold wavelength 242.8 nm, UV or use rhodium wavelength 343.8 nm, UV. If gold reading exceeds 15 mg/L or rhodium reading exceeds 40 mg/L, then make further dilutions and adjust calculations accordingly as in (1).
Calculations:
Reading, mg/L $\times$ 2
 = g/L gold or rhodium
Reading, mg/L $\times$ 0.24
 = troy oz/gal gold or rhodium

(4) Silver by Atomic Absorption Micro-Method

Micropipet 0.100 ml sample of silver solution into a test tube. Add 9.90 ml distilled water and shake well. Micropipet 0.100 ml aliquot into another test tube and add 0.20 ml concentrated nitric acid. Mix by swirling, add 9.70 ml distilled water and shake well. Filter if necessary. Use this solution for analysis. Set up silver standards of 1, 2, 3 and 4 mg/L. Use wavelength 328.1 nm, UV for silver. If reading exceeds 4 mg/L, make further dilutions and adjust calculations accordingly as in (1).
Calculation:
Reading, mg/L $\times$ 10 = g/L silver
Reading, mg/L $\times$ 1.22 = troy oz/gal silver

(5) Aluminum by Atomic Absorption Micro-Method

Micropipet 0.100 ml anodizing solution into a test tube. Add 9.9 ml distilled water and shake well. Pipet 1.0 ml aliquot into another test tube and add 9.0 ml distilled water. Use this solution for analysis. Set up aluminum standards of 5, 15, 25 and 40 ml/L. If reading exceeds highest standard, make dilutions and adjust calculations accordingly as in (1). Use 309.3 nm, UV wavelength and nitrous oxide flame for aluminum.
Calculations:
Reading, mg/L $\times$ 1 = g/L aluminum
Reading, g/L $\times$ 0.133 = oz/gal aluminum

(6) Chromium by Atomic Absorption Micro-Method

Micropipet 0.100 ml chrome solution into a 100 ml volumetric flask, dilute to the mark with distilled water and shake well to mix. Micropipet 0.100 ml aliquot into a test tube, add 9.90 ml distilled water and shake well. Use this solution for analysis. Set up chromium standards of 0.5, 1.0, 2 and 4 mg/L. If reading exceeds highest standard make dilutions and adjust calculations accordingly as in (1). Use 357.9 nm, UV wavelength with rich reducing flame for chromium.
Calculations:
Reading, g/L $\times$ 192.3 = g/L CrO_3
Reading, g/L $\times$ 0.133 = oz/gal CrO_3

(7) Acid Tin/Lead and Tin/Lead/Copper by Atomic Absorption Micro-Method

Micropipet 0.050 ml sample into a 50 ml volumetric flask, dilute to the mark with distilled water and mix well. Set up lead standards at 5, 10 and 15 mg/L; tin standards at 10, 20 and 30 mg/L; and copper standards at 1, 2 and 4 mg/L. If it is necessary to make dilutions, make adjustments in the calculations as in (1)

Metal	Wavelength nm, UV	Reading Not to Exceed
Lead	283.3	15
Tin	286.3 (nitrous oxide flame)	200
Copper	324.8	4

Calculation:
Reading, mg/L = g/L metal
Reading, g/L $\times$ 0.133 = oz/gal metal

(8) "Free" Cyanide

Pipet 10-ml sample of solution into a 250 ml Erlenmeyer flask. Add 100 ml distilled water and 2 ml 20% KI (reagent I-3). Zero a 50 ml burette with standard 0.1N $AgNO_3$ (reagent S-4). Titrate while swirling to a pearlescent yellow endpoint.
Calculation:

ml $AgNO_3$ × N $AgNO_3$ × 0.98
$\qquad$ = g/L "free" NaCN
ml $AgNO_3$ × N $AgNO_3$ × 1.30
$\qquad$ = g/L "free" KCN
$\qquad$ g/L × 0.133 = oz/gal

(9) Total Cyanide

Pipet 10 ml sample of solution into a 250 ml Erlenmeyer flask. Add 100 ml distilled water, 10 ml concentrated ammonium hydroxide (reagent NS-8) and 2 ml 20% KI (reagent I-3). Zero the 50 ml burette with standard 0.1 N $AgNO_3$ (reagent S-4). Titrate while swirling to pearlescent yellow endpoint.
Calculation:

ml $AgNO_3$ × N $AgNO_3$ × 0.98
$\qquad$ = g/L total NaCN
ml $AgNO_3$ × N $AgNO_3$ × 1.30
$\qquad$ = g/L total KCN
$\qquad$ g/L × 0.133 = oz/gal

(10) Hydroxide, Sodium or Potassium

Pipet 10 ml sample into a 125 ml Erlenmeyer flask. Add 10 ml 10% NaCN (reagent NS-12) and 0.1 g caustic blue (reagent I-5). Zero the burette with standard 1.0 N H_2SO_4 (reagent S-2). Titrate while swirling from yellow to royal blue endpoint. (It is advisable to make up a sodium hydroxide standard at 75 g/L, follow this procedure and compare endpoint detection.)
Calculation:

ml H_2SO_4 × N H_2SO_4 × 4.0 = g/L NaOH
Ml H_2SO_4 × N H_2SO_4 × 5.6 = g/L KOH
$\qquad$ g/L × 0.133 = oz/gal

(11) Carbonate, Sodium or Potassium

Heat 300 ml distilled water to boiling. This will be used for washing the precipitate. Pipet 10 ml of plating solution into a 250 ml beaker. Add 90 ml distilled water and 10 ml 10% barium solution (reagent NS-1) using a stirring rod to stir the solution. Cover the beaker and the stirring rod with watch glass and heat to almost boiling. (Watch for bumping!) Remove from heat, allow precipitate to settle and pour supernatant while still hot through quantitative grade filter paper (Whatman No. 40 or equivalent).

Wash precipitate in beaker with hot water, allow to settle a moment and decant supernatant through filter. Repeat about three or four times to ensure removal of cyanide. Test filtrate for presence of carbonate with a few drops of barium (reagent NS-1). (If precipitate forms, it is better to start all over, using 20 ml barium reagent instead of 10 ml.) Carefully remove filter paper, place it into the beaker with any precipitate remaining in the beaker. Zero a 50-ml burette with standard 1.0 N HCl (reagent S-1). Add 100 ml boiling water to the beaker, break up filter paper with the stirring rod to release all the precipitate, add 5 to 10 drops methyl orange-xylene cyanol indicator (reagent I-13) and titrate while stirring from green to red endpoint.
Calculation:

ml HCl × N HCl × 5.3 = g/L Na_2CO_3
ml HCl × N HCl × 6.9 = g/L K_2CO_3
$\qquad$ g/L × 0.133 = oz/gal

(12) Preliminary Cyanide Destruction Process for Copper, Gold, Silver and Tin Analysis

This procedure should be performed only in a hood, preferably one with a fume scrubber in order to prevent poisonous toxic fumes from injuring anyone and from entering the atmosphere. The waste treatment systems are usually well equipped to handle the waste from the scrubber.

Pipet 10 ml sample of solution into a 250 ml beaker. While mixing, slowly add 15 ml concentrated sulfuric acid. Heat gently and carefully add 1 ml portions of concentrated nitric acid to destroy any organic matter. Heat to white fumes. Cool. If solution is not clear, then add further portions of nitric acid and heat until the brown fumes are gone and the solution clears. Cool and add 100 ml distilled water with mixing.

COPPER: While mixing, add concentrated NH_4OH dropwise until the dark blue color persists. Boil for 15 minutes. If precipitate forms, filter solution, wash filter with distilled water and transfer entire filtrate

which contains the copper to a 500 ml Erlenmeyer flask. Add 10 ml concentrated acetic acid, 2-3g KI salt (reagent NS-4) and titrate with standard 0.1 N thiosulfate (reagent S-5) until solution turns from dark brown to light brown. Add 5 ml starch indicator (reagent I-1) and continue titrating until solution becomes creamy white and the blue color does not return for 1 minute
Calculation:
ml thio $\times$ N thio $\times$ 6.36 = g/L copper
g/L $\times$ 0.133 = oz/gal copper

TIN: Add 5 g iron powder to reduce the tin and remove any copper. Heat gently for 10 minutes to reduce the tin completely. Bring to a boil, cool and filter the solution. Wash filter with distilled water, saving entire filtrate for analysis. Transfer filtrate to 500 ml Erlenmeyer flask, add 5 ml concentrated HCl and 2 ml 10% Na_2CO_3 (reagent NS-3). Add 5 ml starch indicator (reagent I-1 and titrate with 0.1 N KIO_3 (reagent PS-4) to blue endpoint.
Calculation:
ml KIO_3 $\times$ N KIO_3 $\times$ 1.19 = g/L tin
g/L $\times$ 0.133 = oz/gal tin

GOLD: Carefully add 10 ml 10% H_2O_2 (reagent NS-7) and bring to a boil to reduce gold metal. Continue boiling until sponge coagulates and solution clears. If not clear, repeat H_2O_2 treatment. Cool. Weight clean dry gooch crucible with fiberglass filter and record exact weight, (A). Filter gold solution and sponge, wash filter with distilled water and finally with methyl alcohol. Dry in oven at 110 C for 2 hr. Cool in desiccator. Weigh and record exact weight, (B).
Calculation:
(B − A) 100 = g/L gold
g/L $\times$ 0.122 = troy oz/gal gold

SILVER: Add 3 ml concentrated HNO_3 and 2 ml ferric ammonium sulfate (reagent I-9). Titrate with 0.1 N thiocyanate (reagent S-8) until solution turns slightly pink (not the precipitate).
Calculation:
ml NaSCN $\times$ N NaSCN $\times$ 10.79
= g/L silver
g/L $\times$ 0.133 = oz/gal silver
g/L $\times$ 0.122 = troy oz/gal silver

(13) Cadmium, Indium and Zinc—EDTA Method

Pipet 2.0 ml sample into a 250 ml Erlenmeyer flask and add 100 ml distilled water. Add 10 ml ammonium buffer (reagent NS-5) and about 0.1 g Eriochrome indicator (reagent I-8) so that print can just be seen through the solution. Solution will be blue at this point. Zero burette with 0.05 M EDTA (reagent PS-2). [For copper alloy solutions cool in refrigerator, add 2 ml 10% HCHO (reagent NS-11).] Add 5 ml 10% HCHO (reagent NS-11) to all other solutions. The solution changes to red color. Titrate with EDTA while swirling to blue endpoint.
Calculation:
Cadmium, g/L = ml EDTA $\times$ M EDTA $\times$ 56.2
Zinc, g/L = ml EDTA $\times$ M EDTA $\times$ 32.7
Indium, g/L = ml EDTA $\times$ M EDTA $\times$ 57.4
g/L $\times$ 0.133 = oz/gal

(14) Total Nickel or Cobalt—EDTA Method

Pipet 2.0 ml plating solution into a 250 ml Erlenmeyer flask. Add 100 ml distilled water and 10 ml concentrated NH_4OH. Zero burette with 0.05 M EDTA (reagent PS-2). Add 0.2 g murexide indicator (reagent I-7) and titrate while swirling with EDTA. Color change is from yellow to purple endpoint.
Calculation:
ml EDTA $\times$ M EDTA $\times$ 29.35 = g/L nickel
ml EDTA $\times$ M EDTA $\times$ 29.47 = g/L cobalt
g/l $\times$ 0.133 = oz/gal

Calculation for Black Nickel:
a. Record total ml EDTA used in procedure (14)
b. Total ml EDTA − ml EDTA used for zinc in procedure (26) = ml EDTA for nickel
c. ml EDTA for nickel $\times$ M EDTA $\times$ 29.35 = g/L nickel.

(15) Chloride—Silver Nitrate Method
Pipet 5.0 ml sample of solution into a 250-ml Erlenmeyer flask. Add 100 ml distilled water and 10 ml 20% ammonium acetate (reagent NS-2). Zero burette with 0.1 N silver nitrate (reagent S-4). Add 2 ml sodium chromate indicator (reagent I-2) and titrate with silver nitrate while swirling until precipitate that forms turns pink.
Calculation:
ml $AgNO_3$ × N $AgNO_3$ × 7.1
$$= g/L \text{ chloride}$$
ml $AgNO_3$ × N $AgNO_3$ × 23.75
$$= g/L \text{ } NiCl_2 \text{ } 6H_2O$$
$$g/L \times 0.133 = oz/gal$$

(16) Nickel Sulfate—By Difference
[(Total Nickel) − (0.247)
× (Nickel Chloride)] × (4.48)
= Nickel Sulfate $NiSO_4 \cdot 6H_2O$

(17) Acid Content of Chloride Nickel Solution
Pipet 1 ml plating solution into a 125 ml Erlenmeyer flask and add 25 ml distilled water. Zero a 50-ml burette with 0.1 N NaOH (reagent S-3). Add 2 to 3 drops methyl yellow (reagent I-4) and titrate with NaOH to yellow-green endpoint.
ml/L concentrated HCl
$$= \text{ml NaOH} \times \text{N NaOH} \times 83.6$$
ml/L × 0.133 = fl. oz/gal concentrated HCl

(18) Boric Acid
Pipet 2.0 ml sample into a 100 ml beaker and add Mannitol Powder (reagent NS-10) to make a thick paste using a stirring rod to mix it. Zero a 50 ml burette with 0.1 N NaOH (reagent S-3). Add 0.5 ml bromcresol purple (reagent I-11) and titrate with NaOH while mixing with stirring rod until mixture turns from yellow green to purple.
Calculation:
ml NaOH × N NaOH × 30.92 = g/L boric acid
$$g/L \times 0.133 = oz/gal$$

(19) Rochelle Salt ($KNaC_4H_4O_6 \cdot 4H_2O$)
NOTE: This procedure should be carried out in a fume hood because of poisonous toxic fumes which are evolved. Pipet a 1.0 ml sample of plating solution into a 500 ml Erlenmeyer flask, add 100 ml distilled water, and carefully add 20 ml of concentrated H_2SO_4 while swirling. Boil gently until completely dissolved, maintaining solution level by adding distilled water. Cool. Zero a 50 ml burette with 0.1 N ceric sulfate (reagent S-6). Add 2 drops ferroin (reagent I-10) and titrate with ceric sulfate until solution changes from red to colorless. Pipet an additional 25.0 ml 0.1 ceric sulfate (reagent S-6) into the flask and boil for 10 minutes. Cool to room temperature. Zero a 50-ml burette with 0.1 N F.A.S. (reagent S-10). Add 2 drops ferroin (reagent I-10) and titrate excess ceric sulfate with F.A.S. until color changes from orange through colorless to red. Record volume.
Calculation:
(ml ceric × N ceric
− ml F.A.S. × N F.A.S.)
× (28.9) = g/L rochelle salt
$$g/L \times 0.133 = oz/gal$$

(20) Chromium Analysis
(A) CrO_3 (Chromic Acid)
Pipet 10.0 ml sample into a 100 ml volumetric flask and dilute to the mark with distilled water. Pipet a 2.0 ml aliquot into a 250-ml Erlenmeyer flask. Add 100 ml distilled water and 2 ml concentrated sulfuric acid. Add 2 to 3 drops ferroin (reagent I-10). Zero a 50 ml burette with 0.1 N F.A.S. (reagent S-10). Titrate while swirling with F.A.S. Color change is from yellow to green through grey to red-rust endpoint.

g/L chromic acid = ml F.A.S.
× N F.A.S. × 16.65

(B) Trivalent Chromium
Make dilution as (A) above. Pipet 20 ml aliquot into a 250 ml Erlenmeyer flask. While swirling, add a slight excess of ammonium hydroxide (reagent NS-8). CAUTIOUSLY sniff for ammonia fumes. Bring to boil. Remove from heat. Allow to settle and filter through Whatman No. 40 or equivalent filter paper. Wash precipitate with hot distilled water. Discard filtrate. Save filter paper and its precipitate. Place a 250 ml Erlenmeyer flask below the filter funnel containing the precipitate on the filter. Add 25 ml 10% H_2SO_4 to dissolve precipitate on filter. Wash with 50 ml distilled water, saving the entire filtrate. Add NH_4OH (reagent NS-8) dropwise while swirling to make solution alkaline, then add 1 ml slight excess. Add 10 ml

H_2O_2 (reagent NS-7) and boil for 15 minutes. Cool and filter through Whatman No. 40 filter paper into a 500 ml Erlenmeyer flask. Wash filter with distilled water and save entire filtrate. Add 6 g potassium acid sulfate ($KHSO_4$) to the 500-ml Erlenmeyer and boil for 15 minutes. Cool. Add 10 ml concentrated H_2SO_4. Cool and add 2 to 3 drops ferroin (reagent I-10). Zero a 50 ml burette with standardized F.A.S. (reagent S-10). Titrate to redrust endpoint.

g/L trivalent chrome (Cr^{+3})
$$= \text{ml F.A.S.} \times \text{N F.A.S.} \times 8.67$$

(C) Alternate Trivalent Chrome
Pipet 2.0 ml sample into a 250 ml Erlenmeyer flask and add 100 ml distilled water. Dropwise add NH_4OH (reagent NS-8) until neutral to litmus and then add 1 ml excess. Add 2 ml 10% H_2O_2 (reagent NS-7) and boil for 15 minutes. Cool and filter through Whatman No. 40 or equivalent filter paper into a 500 ml Erlenmeyer flask to remove the iron. Save the entire filtrate. Add 10 ml concentrated H_2SO_4 and boil for 5 minutes. Cool, add 2 to 3 drops ferroin (reagent I-10). Zero a 50 ml burette with 0.1 N F.A.S. (reagent S-10). Titrate with F.A.S. to a red-rust endpoint.

g/L trivalent chrome (Cr^{+3})
$$= \text{ml F.A.S.} \times \text{N F.A.S.} \times 8.67$$

(21) Sulfate
(A) Gravimetric
Pipet a 25 ml sample plating solution into a 500 ml beaker. Add 100 ml distilled water, 25 ml glacial acetic acid, 25 ml concentrated HCl and 50 ml isopropyl alcohol. Boil for 30 minutes and then, while still boiling, add 20 ml barium (reagent NS-1). Boil for 2 to 3 minutes longer and allow to settle for 1 hr. Prepare a clean, dry gooch crucible with a fiberglass filter and weigh accurately to obtain TARE value. Filter the solution through the gooch crucible. Wash several times with distilled water and finally with methyl alcohol. Dry the crucible and filter in the oven at 110 C for 2 to 3 hr. Cool in a desiccator and reweigh accurately.
Calculations:
g/L sulfate = (final wt of crucible, g—TARE value, g) × 16.5

g/L sulfuric acid = (final wt of crucible, g—TARE value, g) × 16.8
ml/L sulfuric acid = (final wt of crucible g—TARE value) × 9.3
g/L × 0.133 = oz/gal

(B) Centrifugation
NOTE: Older hand-operated or electric centrifuges have exposed cups which can be very dangerous, so a great deal of care must be exercised. This procedure requires the use of special centrifuge cups for the special 100 ml pear-shaped Goetz tubes, as well as a centrifuge.
Prepare 500 ml of standard chrome solution (similar to the one in use). Add a known amount of concentrated sulfuric acid to give the same range of sulfate concentration as in the plating solution being analyzed. Treat this standard the same way as the unknown. Pipet 20.0 ml sample, 25 ml distilled water and 5 ml concentrated HCl into each centrifuge tube. Add 15 ml barium (reagent NS-1) and shake for one minute. Allow to stand for 5 minutes. Place in centrifuge for 1 to 2 minutes. Remove tubes, tap each to dislodge any precipitate and recentrifuge for 1 to 2 minutes. Remove tubes, tap gently while holding vertically in order to give precipitate a flat surface. Read volume of each tube.
Calculation:
g/L sulfuric acid
$$= \frac{\text{reading unknown}}{\text{reading standard}} \times \text{STD g/L}$$

(22) Fluoride in Chromium Solution by Specific Ion Electrode (SIE)
NOTE: This procedure requires the use of a millivolt meter (pH meter) capable of reading to 1 mv or less. The fluoride specific ion electrode is available through any chemical supply house and instructions for its use should be followed explicitly.

Make up 1 L of standard chrome solution in the range of the one(s) presently in operation; e.g., 250 g/L (33.3 oz/gal). To 1.0-ml aliquots of standard chrome solution in 100 ml volumetric flasks, add 0.5, 1.0, 1.5 and 2.0 ml of 100 mg/L (1 + 9 Reagent S-12) standard fluoride. These are equivalent to 50, 100, 150 and 200 mg/L, respectively. Dilute to the mark with

TISAB Buffer (reagent NS-15). Pipet 1 ml aliquot of the plating solution being used into a 100-ml volumetric flask, add 10 ml TISAB Buffer and dilute to the mark with distilled water. Pour each solution into marked plastic beakers. With constant stirring, place the fluoride ion electrode in each beaker and record each mv reading. Plot standard mv readings (horizontal x axis) versus concentration (vertical y axis) on two-phase semi-log paper and draw line through points. From chrome plating solution mv reading, read corresponding concentration in mg/L.

(23) Copper in Acid Copper Solutions

Pipet 5.0 ml sample into a 250 ml Erlenmeyer flask. Add 50 ml distilled water and 10 ml concentrated acetic acid. Add 2 g solid KI, swirl to mix, forming a dark brown turbid solution. Titrate immediately while swirling with 0.1 N thiosulfate (reagent S-5) until solution becomes light yellow. Add 2 ml starch solution (reagent I-1) and continue titration until blue disappears and solution becomes white.

g/L copper metal = ml thio $\times$ N thio $\times$ 12.71
g/L copper sulfate [$CuSO_4 \cdot 5H_2O$]
 = ml thio $\times$ N thio $\times$ 49.92
g/L copper fluoborate $Cu(BF_4)_2$
 = ml thio $\times$ N thio $\times$ 47.43
g/L $\times$ 0.133 = oz/gal

(24) Acid Content of Copper Plating Solutions

Pipet 5.0 ml sample into a 125-ml Erlenmeyer flask, add 50 ml distilled water and 3 to 5 drops methyl yellow (reagent I-4). Titrate while swirling with 1.0 N NaOH (reagent S-3a) until the last trace of pink disappears and before the solution turns green.

g/L sulfuric acid, H_2SO_4
 = ml NaOH $\times$ N NaOH $\times$ 98
g/L fluoboric acid, HBF_4
 = ml NaOH $\times$ N NaOH $\times$ 6.17
g/L $\times$ 0.133 = oz/gal

(25) Chloride Content in Acid Copper

NOTE: Omit Step A if there are no additives present in the copper plating solution.

(A) Place 50 ml sample into a 100 ml beaker and add 2 g granular or 1 g powdered activated charcoal which has been washed with distilled water until no trace of chloride is discernible and dried. (Use $AgNO_3$ test.) Heat to almost boiling, mixing solution from time to time. Allow to cool, filter through Whatman No. 40 or equivalent and save filtrate.

(B) Pipet 25.0 ml sample of filtrate into a 250 ml Erlenmeyer flask. Add 100 ml distilled water and add 10% NaOH dropwise until precipitate that forms just dissolves. Add 10 ml 20% TCA (reagent NS-16) and 1 ml diphenylcarbazone (reagent I-14). Titrate while swirling with 0.005N $Hg(NO_3)_2$ (reagent S-11) to the first faint purple endpoint.

mg/L chloride
 = ml $Hg(NO_3)$ $\times$ N $Hg(NO_3)_2$ $\times$ 1,420
ml/100 gal 37% hydrochloric acid
 = ml $Hg(NO_3)$ $\times$ N $Hg(NO_3)_2$ $\times$ 1,264

(26) Non-Cyanide Zinc—Metal Content

Pipet 2.0 ml sample into a 250 ml Erlenmeyer flask, add 100 ml distilled water, 10 ml buffer (reagent NS-5) and 2 ml cyanide (reagent NS-12). Add 0.1 g eriochrome indicator (reagent I-8) to turn the solution blue. Add 5 ml 10% formaldehyde (reagent NS-11) which turns the solution red. Titrate, while swirling, with 0.05 M EDTA (reagent PS-2) to purple endpoint.
Calculation:

ml EDTA $\times$ M EDTA $\times$ 32.7 = g/L zinc metal

(27) Ferrous Iron Analysis

(A) Pipet 2.0 ml sample of plating solution into a 250 ml Erlenmeyer flask, add while swirling 100 ml distilled water, 5 ml concentrated H_2SO_4, 5 ml concentrated phosphoric acid and 2 drops of 1% diphenylamine (reagent I-6). Titrate with 0.1 N dichromate (reagent PS-5) to the first permanent violet-blue color. Record ml.
Calculation:

g/L ferrous iron = ml Cr_2O_7 $\times$ N Cr_2O_7 $\times$ 27.93
g/L $\times$ 0.133 = oz/gal

(B) Alternate Method

Pipet 2.0 ml sample of plating solution into a 250 ml Erlenmeyer flask, and while swirling 100 ml distilled water, 5 ml concentrated H_2SO_4, 5 ml concentrated H_3PO_4 and 2 drops ferroin (reagent I-10). Titrate with 0.1 N ceric sulfate (reagent S-6) until color changes from red to pale blue. Record ml.
Calculation:

g/L ferrous iron = ml ceric $\times$ N ceric $\times$ 27.93
g/L $\times$ 0.133 = oz/gal

(28) Sulfuric Acid Content in Anodizing Solution

Pipet 5.0 ml sample of anodizing solution into a 250 ml Erlenmeyer flask, add 100 ml distilled water, 3 to 5 drops methyl yellow (reagent I-4) and titrate with 1.0 N NaOH (reagent S-3a) from red to yellow endpoint.

Calculation:
g/L concentrated H_2SO_4
  = ml NaOH × N NaOH × 9.8
ml/L concentrated H_2SO_4
  = ml NaOH × N NaOH × 5.4
g/L × 0.133 = wt oz/gal
ml/L × 0.133 = fl. oz/gal

(29) Aluminum Content in Anodizing Solution

Pipet 5.0 ml sample of anodizing solution into a 250 ml Erlenmeyer flask, add 100 ml distilled water, 0.5 g sodium fluoride and swirl to dissolve. Add 3 to 5 drops phenolphthalein (reagent I-12) and titrate with 1.0 N NaOH (reagent S-3a) from colorless to pink endpoint. Record volume as A ml.

Repeat above procedure except DO NOT add sodium fluoride. Record volume as B ml.

Calculation:
g/L aluminum = $(A - B)$ × N NaOH × 1.8
g/L × 0.133 = oz/gal

(30) Sodium Hypophosphite in Electroless Nickel

Pipet 5 ml sample of plating solution into a 500 ml glass stoppered iodine Erlenmeyer flask. Add 50 ml distilled water, 50 ml concentrated HCl and pipet 50.0 ml 0.1N I_2 (reagent S-7) into the flask. Swirl to mix; place in the dark for 20 minutes.

Titrate with 0.1 N thiosulfate (reagent S-5) until solution turns from brown to light yellow, add 1 ml starch (reagent I-1) and continue titrating until blue color disappears for 30 to 60 seconds. Record ml thiosulfate.

Calculation:
g/L sodium hypophosphite ($NaHPO_2 \cdot H_2O$)
  = [(ml I_2 × N I_2) − (ml thio × N thio)] (10.6)

(31) Rhodium Metal

Pipet 50.0 ml sample into a 250 ml beaker and add 50 ml distilled water. Add 20 ml hydrazine (reagent NS-17), cover with watch glass and boil solution until rhodium precipitate coagulates and supernatant is clear. Prepare gooch filter with fiberglass filter and weigh accurately on analytical balance. Filter solution and precipitate, wash several times with hot water and finally with methanol. Dry in oven at 110 C for 2 hr, cool in desiccator and weigh accurately.

Calculation:
g/L rhodium metal = (wt crucible, g
  − crucible tare, g) × 20
g/L × 0.133 = oz/gal
g/L × 0.122 = troy oz/gal

(32) Electroless Copper

(A) Copper Metal

Pipet 20.0 ml into a 500 ml Erlenmeyer flask, add 50 ml distilled water and 25 ml buffer (reagent NS-5). Heat to about 60 C, add 1 ml PAN (reagent I-15) and titrate while swirling with 0.05 M EDTA (reagent PS-2) from purple to light green. Record volume.

Calculation:
g/L copper metal
  = ml EDTA × M EDTA × 3.65
g/L × 0.133 = oz/gal

NOTE: If no reaction occurs with this procedure, use alternate procedure.

Alternate Procedure for Copper in Electroless Copper

Pipet 1.0 ml sample into a 500 ml Erlenmeyer flask, add 50 ml 0.1 N dichromate (reagent PS-5) and 5 ml concentrated sulfuric acid. Heat to boiling, boil for 5 minutes, remove from heat, add 500 ml isopropyl or methyl alcohol while still warm, mix and allow to cool.

Add 50 ml buffer (reagent NS-5) and 6 to 10 drops PAN (reagent I-15). Prepare 0.005 M EDTA by pipetting 100.0 ml 0.05 M EDTA into a 1 L volumetric flask and diluting to the mark with distilled water. Titrate with 0.005 M EDTA until the solution turns from purple to yellow-green endpoint.

Calculation:
ml EDTA × M EDTA × 63.54 = g/L copper

(B) Sodium Hydroxide and Formaldehyde

The procedure requires the use of a pH meter standardized with pH 4, pH 7 and pH 10 buffer.

Pipet 5 ml plating solution into a 250-ml beaker, add 100 ml distilled water and immerse pH electrodes. Titrate while mixing with 0.1 N H_2SO_4 (reagent S-2a) to a pH of 10.2. Record volume A. Add additional reagent S-2a to pH 10.0. Zero the burette, add 25 ml sodium sulfite (reagent NS-18) to the beaker, wait 5 minutes and titrate again with 0.1 N H_2SO_4 (reagent S-2a) to pH 10.0. Record volume B.

Calculations:

g/L sodium hydroxide = $A \times$ N $H_2SO_4 \times 8$
g/L 37% formaldehyde = $B \times$ N $H_2SO_4 \times 16.12$
g/L $\times 0.133$ = oz/gal

(33) Stannous Tin

Pipet 5.0 ml sample into a 250 ml Erlenmeyer flask. Add 50 ml distilled water and 50 ml concentrated HCl. Add 1 ml starch (reagent I-1) and titrate with 0.1 N KIO_3 (reagent PS-4) until solution turns from colorless to light blue.

g/L stannous tin = ml $KIO_3 \times$ N $KIO_3 \times 11.9$
g/L $\times 0.133$ = oz/gal

(34) Lead

(A) Pipet 10.0 ml sample into a 500 ml beaker and add 300 ml distilled water. While mixing, carefully and slowly add 10 ml concentrated H_2SO_4 to precipitate lead sulfate. Cover with watch glass. Bring to a boil for 5 minutes and cool. Prepare a clean, dry gooch crucible with fiberglass filter and weigh accurately to obtain TARE value. Filter the solution through the gooch crucible. Wash several times with distilled water and finally with methyl alcohol. Dry in oven at 110 C for 2 hr. Cool in desiccator and weigh again accurately.

g/L lead metal = (wt crucible with $PbSO_4$ − crucible tare value) $\times 68.3$

(B) Pipet 1.0 ml sample into a 250 ml Erlenmeyer flask. Add 50 ml distilled water and 1 ml 10% H_2O_2 (reagent NS-7). Swirl to mix. Add 10 ml ammonium buffer (reagent NS-5) and 5 ml triethanolamine (reagent NS-9). Pipet 25 ml 0.05 M EDTA (reagent PS-2) into the flask and mix. Let stand for 2 to 3 minutes. Add 0.1 g Eriochrome (reagent I-8) and titrate with 0.05 M Zinc (reagent PS-3) until solution turns from blue to wine red.

g/L lead metal = (ml EDTA $\times$ M EDTA) − (ml zinc $\times$ zinc) $\times 207.2$

(35) Pyrophosphate

Pipet 5.0 ml sample into a 250 ml volumetric flask. Into a second 250 ml volumetric flask, pipet 10.0 ml of standard pyrophosphate (reagent PS-8). Carry out the following procedure on each volumetric flask. Add 70 ml hot (50 to 60 C) distilled water. Add 4 g sodium hydroxide and dissolve by swirling. Add 100 ml guanylurea (reagent NS-19). Mix and allow to cool. Dilute to the mark with distilled water and mix well. Filter through 15 cm quantitative filter paper such as S&S No. 597 or Whatman No. 40 into a clean, dry 500-ml beaker. Collect at least 150 ml of filtrate. Pipet 100.0 ml aliquot into a 500 ml Erlenmeyer flask. Add 2 to 3 drops phenolphthalein (reagent I-12). Add concentrated HCl dropwise until the red color becomes light pink or colorless. Add 10 drops methyl orange-xylene cyanol (reagent I-13) and titrate immediately with 1.0 N H_2SO_4 (reagent S-2) until color changes from purple through grey to red. Add 50 ml zinc sulfate (reagent NS-20). Allow to stand for 5 minutes with occasional swirling. Titrate with standard 0.1 N NaOH (reagent S-3) until solution changes from red to purple. Record the volume of NaOH.

$2 \times \dfrac{\text{ml NaOH required for sample}}{\text{ml NaOH required for standard}}$
$\times$ concentration of standard $(P_2O_7)^{-4}$
= g/L $(P_2O_7)^{-4}$ pyrophosphate

If the 10.0 ml standard is not run, the calculations are as follows:

ml NaOH $\times$ N NaOH $\times 43.5$
= g/L $(P_2O_7)^{-4}$, pyrophosphate

(36) Thiocyanate

Pipet 10.0 ml sample into a 250 ml Erlenmeyer flask. Add 90 ml distilled water, 2 ml concentrated sulfuric acid and 15 ml 2% ferric ammonium sulfate (reagent I-9). Zero burette with 0.1 N silver nitrate (reagent S-4) and titrate the solution until the red color disappears.

ml $AgNO_3 \times$ N $AgNO_3 \times 8.1$
= g/L NaSCN, sodium thiocyanate

10. TESTING ELECTRODEPOSITED COATINGS

A. Thickness Tests

ISIDORE CROSS

Isidore Cross Associates, Metal Finishing Consultants, Waterbury, Connecticut

Tests for the determination of chemical, physical or mechanical properties of electrodeposits may be made in the course of fundamental research, or for control purposes in development or production. Although, to a certain extent, similar tests and testing techniques are used for all three purposes, most routine testing work is done in the fields of development or production. Tests which may be required in the gathering of fundamental data are often highly specialized, being designed particularly for the problem at hand. Aside from the difficulty of adequately covering this subject in a reasonably brief treatment, such tests have little or no place in a handbook devoted primarily to engineering matters. Therefore, attention will be given in this chapter only to those matters which are likely to be useful in plant development, in production, and for making calibrated thickness standards so necessary in order to use the various methods that will be described.

Chemical testing methods have as their objective one or the other of two major goals, namely, chemical analysis or determination of corrosion resistance. With the exception of alloy coatings, deposits are rarely if every analyzed chemically in production. In the case of alloy coatings, it is almost always possible to use some simple adaptation of the analytical scheme which is employed to analyze the solution from which the alloy was plated. The only problem involved may be that of obtaining a suitable sample from the basis metal in instances where the ions of the basis metal would interfere with the normal analytical procedure. In these cases it is usually most convenient to plate a special specimen on a platinum cathode from which it may be dissolved chemically.

The other aspect of chemical testing, i.e., the determination of corrosion resistance, is one which is fraught with difficulty and uncertainty. The only absolutely reliable measure of corrosion resistance is exposure of the plated work to the environment in which it is to be used. In almost every instance this procedure is too slow to be useful for production testing. Perhaps more important is the unfortunate fact that the producer of plated ware has no control over the environment in which his product will be used. In any case, so many of these environments will be encountered that testing in each of them is utterly impractical. In a search for accelerated tests which will simulate actual exposure conditions, investigators have attempted to correlate results obtained in reproducible synthetic environments with those that are obtained under conditions of practical use. The results have been for the most part confusing and unsatisfactory. Nonetheless, one such test, the salt-spray test, has gained a certain popularity and has been worked into many specifications. Attempts are constantly being made to make accelerated corrosion test results less confusing and more relevant.

The American Society for Testing Materials (ASTM) has added no less than four different modifications to the original salt-spray test specification so far.

The time required to conduct outdoor exposure tests is not as serious a matter in research work as it is in production activities. The expense and trouble involved in such tests have made it desirable to obtain as much information as possible from work of this sort. As a result, a great deal of work has been done by scientific societies wherein many sponsors can participate in mutually interesting and valuable testing. The American Society for Testing Materials has been particularly active in this respect, and the American Electroplaters' Society has cooperated in several of the programs. It is essential in outdoor exposure testing to carry out replicate work at several locations having a variety of environments, e.g., a rural atmosphere, an industrial location, a seacoast atmosphere and various combinations of these. The sites must be chosen with care to avoid misleading and atypical influences; they must be adequately protected so that the specimens are not disturbed during the period of their exposure. Inspection of the specimens must be carried out at appropriate intervals by qualified inspectors, and the specimens must be evaluated according to some rating scheme which will be comprehensible, or potentially so, for all those who wish to use the data. Records must be adequately kept, summarized and clearly presented in reports. Each of these operations presents manifold problems but perhaps none of these is more difficult than that of devising an adequate rating system. The great variety of defects which may develop makes it extremely hard to apply a single simple rating system to all sorts of specimens. Ideally, the rating system should be totally objective and independent of the judgment of the inspector. This stage has not been reached at the present time; hence, it is desirable to have one inspector or a group of inspectors evaluate all the specimens of a given series which will be exposed in several environments. Periodically, several rating systems have been proposed, and for these, reference should be made to the publications of the American Society for Testing Materials (ASTM).

The controversial salt-spray test is carried out in a closed cabinet designed to expose the specimens to a fog or mist of an aqueous solution of sodium chloride. About the only detail of testing which appears in many specifications calling for this test is the concentration of the sodium chloride solution. The design of the nozzle which produces the mist and the arrangement of the specimens in the cabinet are often not rigidly specified. Furthermore, the temperature at which the test is conducted is of considerable importance and must either be specified or agreed upon by the parties involved.

Committee B-8 of the American Society for Testing Materials, as well as the American Electroplaters' Society, has approved the tentative method of salt-spray testing[1] bearing the ASTM Designation B117-73 and B287-74. The published descriptions for these tentative methods contain minute details of apparatus and procedure, but they do not deal with types of specimens, exposure periods for specific products or interpretation of results. The latter points must be covered by individual specifications covering the product being tested or be agreed upon by parties interested in the results of the test.

The porosity of electrodeposited coatings is often intimately connected with corrosion resistance, particularly when the coating is cathodic to the basis metal. Although several methods have been used to measure porosity, recent work has cast serious doubt on the absolute value of any of these tests, but they can be employed for comparative quality tests.

The salt-spray test, referred to above, is often considered to be a porosity test and is used as such. The ferroxyl test involves a reagent containing sodium chloride and potassium ferricyanide and is used as a test for porosity of coatings on ferrous basis metals. Local attack of the steel by the reagent is revealed by spots of Prussian blue which will appear wherever the steel is attacked through the pores in the coating. Those who object to the use of this test point out that the reagent itself attacks nickel and produces pores by its own action. In spite of the objections, the ferroxyl test has enjoyed considerable popularity. Another means of detecting pores, particularly in nickel coatings on steel, has been the simple immersion of specimens in hot water. Here again the test may result in the production of pores as well as their detection. The hot water test appears to be less sensitive than the salt-spray test or the ferroxyl test and is not much used.

The basis for porosity tests is the extent of the reaction of the basis metal to specific, identifying reagents without unduly affecting the coating to be tested. Porosity of gold plated items over nickel or copper are put in sulfur dioxide or nitric acid fumes. Gold over copper or silver is tested in a hydrogen sulfide atmosphere. In electrographic testing to determine the porosity of gold over nickel, dimethyl glyoxine is the identifying agent for nickel in a conducting medium, whether in a salt solution or a gel.

Some of the physical and mechanical property tests which may at one time or another be conducted on electrodeposited coatings are thickness, appearance, porosity, adhesion, stress, hardness, wear and abrasion resistance, tensile strength and ductility. Of these, only one is commonly used as a basis for production specifications, namely thickness. It is true that the appearance of the deposit is often an extremely important property of the plated coating with respect to its acceptability, but it is equally true that only qualitative measures which are largely subjective can be applied to the evaluation of appearance. The remaining physical and mechanical tests are of interest primarily in research and development.

THICKNESS TESTS

Inasmuch as corrosion resistance has often been shown to be intimately related to the thickness of the deposit, the stipulation of minimum thickness in a product specification is obvious. However, corrosion resistance is not the only criterion that makes thickness specification necessary; at least of equal importance is the functional requirements of the deposit itself. Many products are plated to achieve definite physical and chemical properties such as conductivity on printed wiring boards and other electronic devices, wear resistance in industrial chromium plating and, in some instances, electroless nickel plating, and silver plating on bearing retainers to impart lubricity at relatively high temperatures. Plating specific coatings greatly enhances the functions of a particular item, and there are minimums and maximums in plating thickness specifications which must be adhered to in order for the item to perform as designed.

There are only a few instruments popularly used today that read thickness directly. Aside from the micrometer, microscopic cross-sectioning is the most commonly used direct reading method. Its much more accurate and sophisticated relative, the double-beam interference microscope, also referred to as interferometry, is mainly used for calibration of thickness standards and research. All of the other instruments, to varying degrees of accuracy and sophistication, take ingenious advantage of the differences in the physical and chemical properties of the coating and substrate.

The methods commonly used take advantage of the electrical, magnetic and chemical properties of the coating-substrate combination, which also includes the effect of irradiating the specimens with X-rays or beta rays. A characteristic common to these various methods is that these various properties must be calibrated against known thickness standards, and the accuracy of a measurement is dependent upon the reliability of the calibration standards. The calibration standards should correspond to the coating-substrate combination that is being measured. If the specimen to be measured is Watts nickel over low carbon steel, the calibrated standard should also be Watts nickel over low carbon steel. Calibrated standards are usually obtained from the instrument manufacturers and are referenced to standards calibrated by the National Bureau of Standards. Standards may be purchased directly from the Bureau. Often, because of varying densities of electrodeposited coatings, it is impossible to obtain calibrated standards for every conceivable coating, and the platers may have to make up their own reference standards to achieve a greater degree of accuracy. Standards used for measuring the thickness of various coating-substrate combinations are the necessary points of reference for that particular coating combination when measuring with a particular device.

Standards, besides having to be of uniform thickness, should not be measured by the same instruments being calibrated; otherwise, errors introduced by one method will go on being propagated.[2] Calibrating standards issued by the National Bureau of Standards are measured by X-ray fluorescence, profilometer, interferometrics, SEM (scanning electron microscope), microscope and coating weight. With the exception of X-ray fluorescence, none of the methods require calibration. Coating weight or mass gain, which is considered extremely accurate, requires a good uniform plate, and with care this can be accomplished; however, the density of the plate may be a problem. A simple plating cell for making uniformly coated standards whose thickness is determined by the mass gain method has been described,[2,3] and the manufacturers of proprietary processes should be able to supply the density of the metal deposited from their baths.

Because of the variety of coatings and basis material combinations used today, there is no single, universal thickness testing device. Each method takes advantage of the unique characteristics of a particular coating-substrate system. Many instruments are excellent only for a particular thickness range and fall far short of being accurate outside of their optimum ranges. Some instruments, while only having a limited range for certain combinations, may be the ideal choice because the user is concerned only for that range, and can forego using more elaborate equipment.

MICROSCOPIC-OPTICAL METHODS

A destructive test is one where the specimen must be destroyed in order to obtain a reading; this is such a test. Many people are prone to use the old cliche, "but I saw it with my own eyes!"—as if actually seeing a plating thickness attests to its reliability. For many years, microscopic cross-sections were considered the referee method when thickness tests by other means were in dispute. However, as far back as 1947, Dr. A. Brenner[4] commented that microscopic cross-sectioning was not the ultimate in accuracy for measuring coating thickness. In a report on the American Electroplaters' Society Research Project #7, evaluating various methods for determining electrodeposited coating thickness, Dr. H. J. Read wrote that while microscopic cross-sections were supposed to be a primary standard for thickness testing, there was very little material proof that indicated that the microscope had the reliability and reproducibility ascribed to it. It was important that the personnel using it be experienced and it was found that the measured area was often not truly representative of the specimen. In a series of interlab comparison studies, using the same mounts, it was shown that the microscopical method of measuring thickness was found to be less reliable than was realized.[6]

All this is not to denigrate microscopic cross-sectioning, but rather to put this method of thickness determination in its proper perspective. It's good; it's more than adequate, but it's not the ultimate as far as reliability. For thickness determinations under $2\mu m$ (80 μin.), the microscope is quite unreliable. Wilson[7] reported an error of almost 50% in viewing deposits in the order of 1 μm (40 μin.) thick. The limiting number for any degree of acceptable accuracy is 2 μm (80 μin.), technician's skill notwithstanding. With multiple plates and in instances where calibrated standards are difficult to obtain, the microscope may be the only practical means of determining thickness.

The accuracy and reliability of microscopical cross-sectioning is highly dependent upon the skill and technique of the operator, and even if this skill is there, sources of error are built into the instrument that are beyond the capabilities of the operator.[6] Possible sources of error in the use of the microscope can be traced back to the calibration of the stage micrometer and the alignment and focus of the mount. The stage micrometer used in calibrating the filar eyepiece is the ultimate basis for microscopic thickness determinations, so care must be taken that the objective used in making the calibration is the same one used in measuring.

It is worth noting that calibrating a microscope has an entirely different meaning from calibrating other thickness measuring devices. With a microscope, the vernier divisions on the filar eyepiece are calibrated against a stage micrometer which is a direct thickness measurement; with the various other methods, physical data, such as reluctance, impedance or electron backscatter intensity, have to be compared and calibrated against known thickness standards in order to convert backscatter intensity or impedance into meaningful thickness measurements.

ASTM's standard for measuring thickness using the microscope, B487-75,[8] cautions pointedly that microscopical sections shall not be used as a referee method for coatings thinner than 8 μm (300 μin.) with optimum specimen preparation. Albrecht,[9] using sophisticated equipment, found an average deviation of 0.15 to 0.25 μm (6 to 10 μin.) whether reading samples of 1.25 μm (50 μin.) or 3.75 μm (150 μin.) thick. The interesting point about that report is that no matter what the thickness, the average deviation seemed to be the same. With that being the case, the more accurate determinations were the thicker coatings, which confirms ASTM's cautionary note of not using microscopic cross-sectioning as a referee method with thicknesses under 8 μm (300 μin.).

One cannot stress too much (and this applies to

any thickness determination, no matter what method or instrument used) that the rougher the deposit is, the less accurate is the reading. Smooth deposits make for much more accurate thickness measurements.

General metallographic procedures for specimen preparation have been described by Kehl,[10] and details for polishing, etching and microscopy for various metallic and oxide coatings are found in two ASTM standards, A219-58 and B487-75.

DOUBLE-BEAM INTERFERENCE MICROSCOPE, INTERFEROMETRY

Like most optical methods for measuring coating thickness, it is a destructive test and is used mostly to measure calibration standards for other measuring devices. Interferometry is also used to make topographic measurements of surface imperfections like pits and scratches. A double-beam interferometer can be incorporated as an attachment to an ordinary metallurgical microscope.[11,12,13,14] Wild Heerbrugg of Switzerland is among several optical companies that manufacture such an attachment.

The interferometer utilizes the principle of optical interference to measure the magnitude of irregularities on reflective surfaces.[11] A double-beam interference microscope (interferometer) can determine a surface irregularity at a particular point, enabling the observer to measure a hill or a valley as well as the size and depth of a pit or a scratch. If an area is stripped, or stopped off, the difference in elevation (thickness) between the plated and unplated portion can be measured very accurately.

The interferometer is not a complicated instrument. A filtered, monochromatic light such as that from a mercury vapor lamp, green in color, is fed into a beam splitter (Fig. 1).[11] One part of the beam goes to a reference mirror, and the other part to the specimen. The split beams are reflected back to the beam splitter, where they are recombined and observed in the microscope's eyepiece. Interference lines, or fringes, are observed due to a shift in phase of the recombined reflected light (Fig. 2).[12] The split beams, having traveled two different paths, one longer than the other, are out of synchronization, resulting in an interference pattern or fringe. The difference between each fringe is equal to half the wavelength of the particular light source. The fringe also functions as a topographic map of the specimen's surface.

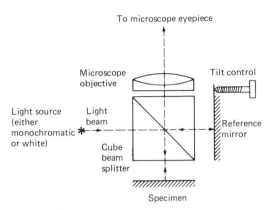

FIG. 1. Simplified diagram of an interference objective which can be substituted for a conventional microscope objective to form an interference microscope.[11]

When a flat, mirror-like specimen is illuminated with monochromatic light, numerous parallel fringes are observed. A fringe shift by an amount equal to the distance between adjacent fringes indicates a change in surface elevation which can be readily measured. Since the wavelength of a spectral monochromatic light source is invariant or constant, the fringes are self-calibrated profile lines of the specimen's

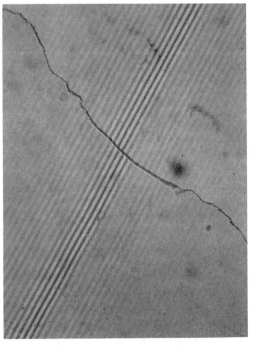

FIG. 2. Interference lines.[12]

surface. Monochromatic fringes are identical, so it is only necessary to trace a single fringe over a topographic deviation to determine its height.[12] White light is often used for better observation because there is always one intense black line which can be followed more easily in order to make a measurement (Fig. 2).[12]

A highly polished sample is prepared, and one can go about conducting the thickness evaluation in two ways. One is by masking a small square area (squares ensure a sharper profile) with lacquer or masking tape and then plating the sample. After plating, the stop-off material is removed and the step between the plated and unplated portions is observed and measured.[12,13] The other means of sample preparation is to mask after finishing, and then strip part of the coating to be measured to the layer underneath, making sure that the basis material's surface is undisturbed by the stripping operation. Since chromium and anodized aluminum can be readily stripped without disturbing the surface underneath, the second method is the preferred one for those metals. In the case of gold and other precious metals, masking prior to plating is preferred because the stripping operation is more difficult.

The test panel is then observed with the interferometer to measure the difference in height between the finished and unfinished areas.

Usually, when any light is reflected, it undergoes a slight phase shift, the magnitude of which depends upon the material reflecting the light. For instance, there is a difference in optical properties between two different metals, and a slightly greater difference between metals and plastic. This phase shift creates an uncertainty in the measurement. (The word "uncertainty" pervades all of the literature concerned with thickness testing. It actually means the amount of deviation or margin of error to expect from a measurement. Thus an uncertainty of ± 10% is greater than one of ± 5%; the lower figure means that there is more accuracy to the measurement.) This uncertainty can be overcome by evaporating a thin, but uniform, reflecting layer of metal, usually aluminum, over the whole surface so that the two reflecting surfaces are the same, and the phase shift will be the same canceling out the error from, for instance, a gold and nickel surface.[15] The presence of the evaporated surface will not change the results because what is actually being measured is the topographic difference of the surface and not whether it's aluminum oxide, gold, copper or any other metal.

The thickness range that is measurable by interferometry is from 0.12 to 10 μm (5 to 400 μin.). A highly polished surface increases the sensitivity of the measurement to within 0.05 μm (2 μin.).[11] Measurements taken with interferometry can be within 5% of the coating thickness, one of the most sensitive measuring methods in use today.

CHEMICAL METHODS

Many individual techniques fall under this heading, like gas evolution, dipping, jet, dropping and drop methods. This section will only cover those methods that are currently used because they have, to a degree, withstood the tests of time and technology and give reasonable approximate results when only reasonable approximate results can be acceptable.

Dipping methods, like the once well-known Preece Test, where zinc coated steel specimens were dipped in a neutral copper sulfate solution, are anachronistic to modern electroplating, as are the jet and gas evolution tests, because far better and more reliable methods are readily available.

The dropping test method consists of applying a corrosive solution at a constant dropping rate to an electroplated surface and with a stopwatch, measuring the time it takes to penetrate the coating and expose the substrate. The time it takes for the coating to be penetrated is proportional to the coating thickness. This test is applicable for electrodeposited zinc, cadmium, copper and tin coatings.

The test solution must be dropped on the part, held at a 45° angle, at a constant rate of 100 ± 5 drops per minute from a dropping funnel (Fig. 3).[16] Because this is a chemical reaction, and the composition of the dropping solution changes, the solutions, once applied, cannot be reused. A dropping test kit may be purchased, with the solutions, from the Kocour Company, Chicago, Illinois 60632.

Temperature is a very important factor in any chemical test, because the speed of a reaction is greatly affected by it. An example is the following. The penetration time for a part 15 μm (600 μin.) thick, tested at 21 C (70 F) is 45 seconds. At 32 C (90 F) the same thickness took only 37

TESTING ELECTRODEPOSITED COATINGS 315

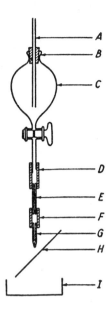

A—Glass tube to maintain constant head of solution.
B—Rubber stopper.
C—Separatory funnel.
D—Wide-bore rubber tubing (3 to 5 in. in length) or sealed glass joints.
E—Capillary tubing (5½ in. in length, 0.025 in. bore).
F—Rubber tubing or sealed glass joints.
G—Capillary glass tip (with ½ in. taper to outside diameter of 0.14 in.).
H—Specimen.
I—Collector of spent solution.

FIG. 3. Dropping funnel.[16]

TABLE 1. TIME REQUIRED TO DISSOLVE 0.00001 IN. OF CHROMIUM IN SPOT TEST

Temperature		Time (sec)
F	C	
64	17.8	13.5
66	18.9	13.0
68	20.0	12.5
70	21.1	12.0
72	22.2	11.5
74	23.4	11.0
76	24.5	10.5
78	25.5	10.0
80	26.7	9.5
82	27.8	8.5
84	28.9	8.0
86	30.0	7.5

seconds for the substrate to be exposed. Put another way, 37 seconds at 21 C (70 F) is the time it takes to test a specimen 12.5 µm (500 µin.) thick, a 20% difference. Thus if one fails to take temperature into consideration, an error of 20% can be made, in addition to the intrinsic uncertainties associated with this test under the best of conditions. Because of the nature of this test, coatings below 2.5 µm (100 µin.) in thickness cannot be tested with any degree of accuracy.

Typical solutions for this test can be found in ASTM Test Method B555-75. Brighteners and alloying materials may affect the accuracy of the test. The reproducibility of this test by an operator may be good; however, the overall accuracy of the test is not. For maximum accuracy, this test method should be standardized with specimens identical to those being tested and having known coating thicknesses determined by other means. This should compensate for any variation in the coating composition because of addition agents.

A rather primitive, "quick and dirty" test for determining coating thickness is the drop test used for very thin coatings. This test is conducted by putting a single drop of a stripping solution on the deposit and then determining the time required to expose the immediate substrate. This test is used almost exclusively for thin chromium deposits, no thicker than 1.2 µm (50 µin.).[17]

A wax ring surrounding an open area, about ¼ in. in diameter, is painted on the specimen, usually with a wax pencil. A single drop of hydrochloric acid (specific gravity 1.18) is then placed in the wax ring. The time required from the beginning of gas evolution to the first appearance of the basis metal (nickel) is determined using a stopwatch. As with all chemical tests, the temperature of the acid and the part must be the same, at room temperature. Table 1 shows the effect of temperature on the test, giving data for converting the time of the test to the thickness of chromium over nickel.

In Table 1 it can be seen that at 21.1 C (70 F), 0.25 µm (10 µin. = 0.00001 in.) of chromium is dissolved in 12 seconds; in 24 seconds at that temperature, the chromium coating would be 0.5 µm (0.00002 in.) thick.

This test has an average error, or "uncertainty" of ±20%, and if other testing facilities are available, like a coulometric instrument, it would be much preferred because of greater accuracy.[17]

Considering the great uncertainty of this method, testing chromium thickness directly over copper did not even give results that are reproducible and this method should not be considered when testing chromium plated directly on brass or copper.

MAGNETIC METHOD

Magnetic methods of measuring electrodeposit thicknesses are very popular because the instruments used are relatively cheap, the test is nondestructive, it can be used as a go/no go gage when 100% sampling is required, and can be adapted to highly localized measurements. The principle disadvantage, as this method implies, is the necessity of having either a magnetic coating, a magnetic basis metal, or both, such as electrodeposited nickel on steel.

Common measuring combinations are paint, plastic, copper, nickel, chromium, electroless nickel which is nonmagnetic with more than 7 to 8% phosphorus, zinc, cadmium, silver, tin, lead and solder on steel and kovar. Nickel can be measured on most substrates with varying results due to the permeability variations of the nickel coating. With multiple plates such as copper and silver on steel, only the combined plates can be read. The way to read a combination such as this, magnetically, is to read the copper only, first, and then subtract the copper reading from the total to determine the silver thickness.

Because the various devices actually measure the magnetic attraction between a magnet and the subject coating-substrate combination, or the reluctance of a magnetic flux path passing through the coating and substrate, with reluctance being the magnetic resistance, the instruments must be calibrated against standards of known thickness.[19,20]

Proper calibration cannot be stressed too strongly because there are many physical and structural factors of the coating and substrate which may affect the magnetic properties of either one or both. It is for this reason that calibration standards should be identical to the specimens undergoing thickness examination such as having been plated in the same bath as the specimen in the case of electrodeposited nickel, and with the unplated zeroing specimen identical to the sample.

The test can be affected by variations in the ferromagnetic characteristics of the base material. These variations are due to mechanical working, welding, heat treatment and steel composition. Low-carbon steels are least affected magnetically by the various operations mentioned. High-carbon steels, however, are greatly affected, varying in degree depending upon their composition. Reluctance measuring instruments, in addition, are affected by the electrical conductivity of the coating-substrate combination.

With nickel, the physical properties of the deposit will affect the accuracy of the test. Plating conditions, plate composition and, particularly, stress in the deposit will affect the results. With suitable heat treatment, stress can be relieved prior to making a thickness determination. A heat treatment at 400 C (750 F) for 30 minutes will equalize the magnetic permeability of a dull Watts nickel.[21] Multi-layered nickel deposits can give erroneous results because each layer is of a different composition with different magnetic properties.

The most widely used magnetic techniques are (1) measurement of the force required to detach a small permanent magnet from the coated surface of the specimen, (2) measurement of the change in reluctance caused by interposing a nonmagnetic material between an energized solenoid and a ferromagnetic basis metal and (3) an instrument which generates a magnetic field and senses magnetic flux changes from a set reference value due to the presence of the nickel to be measured.

The first method is adaptable to the thickness measurement of the following combinations: (a) a nonmagnetic coating on a magnetic base, (b) a magnetic coating on a nonmagnetic base, and (c) a magnetic coating on a magnetic base.

The best known instrument based on the principle of detaching a magnet from the specimen to be measured is the Magne-gage developed by A. Brenner of the NBS and manufactured by the American Instrument Company of Silver Spring, Maryland. This instrument is comprised of a small torsion balance at the end of whose beam is located, dangling, a small needle-like permanent magnet. Measurements are made by decreasing the tension in the torsion spring until the magnet drops down on the specimen. The spring is then wound up until the force is sufficient to pull the magnet away from the specimen, with the required force being indicated in arbitrary units on a dial. The instrument is furnished with calibration charts which relate the dial readings to the thickness of the coating being measured. In order to encompass the normal ranges of thickness and combinations usually encountered, four magnets and six calibrations are required.

This instrument does have a limiting sensitivity, and cannot be used for thin coatings because there is a constant uncertainty (error) which is independent of thickness; however, for coatings heavier than 25 µm (1 mil), this uncertainty becomes proportional to thickness. In other words, for thicker coatings, the magnetic measurements become less sensitive as they are further removed from their magnetic base, and where the coating is too thin, the magnet is insensitive to coatings of less than a certain thickness. Small parts, because of the lack of surface for the magnet to adhere to, can't be measured with any degree of accuracy. Vibrations on the measuring table will also give varying results. The effective measuring range for nickel on a magnetic substrate is much wider, going up to 50 µm (2 mils) as against 25 µm (1 mil) for nickel on a nonmagnetic base.

The instruments measuring reluctance of a magnetic flux path passing through the coating and substrate (Fig. 4) are simple to use and are made by a variety of manufacturers. Twin City Testing of Amherst, New York, makes an Autotest MS and UPA Technology and Signa Labs, both of Long Island, New York, also manufacture reluctance type instruments. These instruments measure the same coating-substrate combinations as the magne-gage, but have extended ranges to measure thicker coatings. The effective range for reluctance gages is quite high and measurements of up to 1 mm (40 mils) have been made. There is a limiting thickness in the lower ranges for this instrument in the order of about 5 µm (200 µin.).

Reluctance measuring instruments are affected by a number of factors unique to this type of a device. With thicker, highly conductive coatings, eddy current effects may interfere with the thickness tests and render improbable readings.[20] Uncertain probe pressures on the specimen can affect the consistency of the readings, but all manufacturers offer equipment with constant pressure devices to mitigate against this problem.

Since the instruments sense no difference among the various nonmagnetic coatings, whether they are metallic or nonmetallic, thickness standards are usually calibrated plastic foils. Except with nickel, the foils are placed in intimate contact with a sample of the uncoated substrate for calibration. These foils come into their own when calibrating a curved surface, because the radius of curvature is a possible source of error. When testing a round specimen, it is recommended that the calibration be done on a round specimen rather than on a flat test panel. The plastic foils may not be the ideal calibrating device when testing a heavy, conductive coating because of the above-mentioned eddy current side effects on the specimen that is being tested.

Problems common to all instruments employing the magnetic principle is that this method is sensitive to surface conditions (surface roughness) and the geometry of the test specimen, which implies the following; (a) curvature of the part, as this effect becomes more apparent as the radius diminishes, and (b) edge effect. Edge effect is the sensitivity to abrupt changes in surface contour. To lessen against this, measurements should not be made too near an edge or a corner.

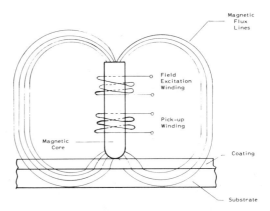

FIG. 4. A magnetic probe has a primary field excitation winding (coil) which is driven from a low-frequency AC source. This generates magnetic flux lines in a magnetic core. The magnetic core is rounded at the tip, rendering the probe, for all practical purposes, insensitive to tilt. A secondary pick-up winning is used to induce a probe output voltage which is a function of the flux density in the magnetic core. The instrument records this output voltage. The flux density is a function of the distance between the probe tip and the substrate material, that is, coating thickness. (*Courtesy Twin City Testing.*)

There are some instruments, such as the Nickelderm, manufactured by UPA Technologies, and others that measure the thickness of a nickel coating by passing a regulated DC current through a semiconductor known as a Hall Detector. The "Hall Voltage" increases in a direct linear response to increases in magnetic flux due to increased nickel thickness. The probe voltage output is translated into nickel thickness after calibration. Since no AC is used, this instrument is considered to be free of errors induced by possible eddy current effects due to the high conductivity of the coating-substrate combination. With a multi-layered coating, if the top layer of gold or rhodium is less than 2.5 μm (100 μin.), it is claimed that the thickness of the intermediate nickel plate can be accurately determined.

Properly calibrated, the magnetic method is capable of determining thickness to an accuracy of ±10%.

EDDY CURRENT

Just as magnetic type instruments measure reluctance which must be calibrated against thickness standards, eddy current devices measure impedance, which, too, must be calibrated against thickness standards in order to render meaningful thickness measurements, impedance being the resistance met by an alternating current in passing through a conductor. There is a great similarity between the magnetic and eddy current type instruments—so much so, that several companies now manufacture a dual instrument, such as Twin City Testing's Autotest DM, to measure by either method with just a flick of the switch. In this way, the best features of either method can best be exploited for a particular thickness determination.

An eddy-current probe (Fig. 5) generates a high-frequency electric field in a coil mounted in an open ferrite pot core. When the probe is placed on a specimen with a metallic substrate, eddy currents are induced in the metal. The magnitude of the eddy currents depends primarily upon the distance between the metal surface and the probe, the distance being equivalent to the insulating coating thickness, or, if a metal, the plating thickness. The strength of the induced current is indirectly measured through the interaction of its magnetic field with the probe coil. The changes in the probe coil are measured with a microammeter and are related to the coating thickness through the use of calibrated reference coating thickness standards.[22]

Eddy currents are affected by the frequency applied to the coil, the electrical conductivity of the sample, its magnetic permeability and the heat treatment history of the basis metal. The higher the frequency of the applied AC to the coil, the shallower will be the penetration of the eddy currents in any given material. Thus, lower frequencies are used for thicker coatings. For a given frequency of the inducing current, the depth of eddy current penetration varies inversely with the electrical conductivity of the specimen under examination. The difference in conductivity between coating and substrate is what actually makes this instrument a thickness gage, because it measures the degree of difference in conductivity between the two. This difference is read as impedance on the microammeter. Knowing this, we can readily see why the thickness range for copper coatings will be much smaller than for the far less conductive metals like zinc and cadmium. By far, the most widely used application for eddy current testing is when a conductor and nonconductor are involved. When the coating and substrate are both metallic, the bigger the difference in their conductivities, the more accurate the test, providing other mitigating factors are taken into account.

The magnetic permeability of ferrous metals tend to concentrate the magnetic field produced by the eddy current probe toward the surface of the specimen. This limits the depth of penetration of the eddy currents and limits the measurement by this method of nonmagnetic coatings on ferromagnetic substrates, and also the reverse

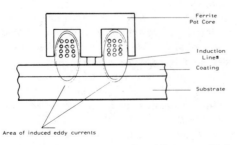

FIG. 5. Eddy current probe. (*Courtesy Twin City Testing.*)

combination, such as nickel on copper. It is recommended that these instruments not be used when either or both components are ferromagnetic. In a similar vein, ASTM's B529-70, note 2, suggests that coatings of nonconductors, less than 25 μm (1 mil) thick, on magnetic substrates not be determined by eddy current, but to use the magnetic method instead.

The conductivity of a plated metal can be influenced by solution composition, brighteners and addition agents, as well as the current density at which it was plated. An example of this is the difference in conductivity between zinc deposits from a noncyanide or low-cyanide bath and the more conductive deposits from a high-cyanide bath. Unless the calibrated samples came from the same baths in which the specimens were processed, errors in thickness readings can result. This is why the operator should be wary of calibrated samples that come with the instrument.

The eddy current principle is best applied to the following combinations: oxides,[23] paint, lacquer, teflon, rubber and photo-resist on conductive substrates such as aluminum, copper, brass, zinc, magnesium and their alloys. In some cases, this principle is also used to measure poor conductors, such as chromium, on good conductors, such as aluminum or copper. This method may also be used to measure conductive coatings, such as copper, on circuit board laminates.

While the ASTM and some instrument manufacturers don't especially recommend the use of the eddy current principle for conductive coatings on metal, other instrument manufacturers, notably UPA Technologies, claim that with its Dermitron, so long as there's a significant difference in electrical conductivity between the coating and the substrate, eddy current devices can be used, if properly calibrated.

Eddy current instruments are essentially comparators, hence they must undergo calibration. As with other instruments, they must be set at zero for the uncoated sample, and then be standardized using coated samples of known thickness as reference values. When testing nonconductive coatings, calibrated plastic foils supplied by the manufacturer are commonly used.

As with magnetic testing devices, eddy current instruments are also affected by curvature of the specimen, surface roughness, edge effects and the amount of pressure put on the probe. The probe pressure problem is easily overcome by using a constant pressure device. All eddy current probes are "tilt-sensitive," which means that their position with respect to the test specimen must remain fixed during the measurement.

Eddy current instruments are fairly accurate to within ± 10% at thickness in excess of 5 μm (200 μin.).[23]

MASS PER UNIT AREA

The measurement of mass per unit area is considered the most accurate means for determining thickness.[4] There are a number of methods used to determine mass per unit area, the most common being weight gain, coulometric, and actually counting the number of atoms in a specific area by bombarding it with either X-rays or beta rays.

However differently we "weigh" the coating in a designated area, thickness is derived by dividing mass per unit area by density, which is mass per unit volume. The flaw with this method is that sometimes density isn't always what the tables say it is. Addition agents, added to a plating process to modify or give a deposit certain properties, often intrude into the deposit and thereby change its density. In the case of gold, the difference in density may vary by as much as 10 to 15% from the standard 19.3 given in the tables, but, as we shall see, there are means to overcome this difference.

Weight Gain Method

Obtaining the weight of a coating is accomplished in several ways, with related techniques.

A plated specimen of known area is weighed, and then the plating is removed by a suitable stripper that doesn't attack the basis metal, which is then reweighed. From the difference in weight, the area and the density of the plated coating, the thickness can be readily calculated. Another method, modified slightly and used when a suitable solvent cannot be found, is to weigh a specimen of a known area before plating, and then weigh again after plating. Again, from the difference in weight, area, and density, a calculation for thickness can be made. Sometimes, the basis metal, whose area is known, is completely dissolved and the undissolved coating is dried and weighed; this is sometimes done with gold on copper or brass. Yet another method is to dissolve the coating of a specimen of known area and then analyze quantitatively the stripping solution for the coating.

All of these methods will only give an average thickness of the entire specimen. The weight gain method, however, by using proper plating technique, is capable of supplying standards for the calibration of many of our most sophisticated thickness testing instruments.[2,3]

Because of differences of deposit densities inherent in various plating processes, it has been recommended that quality control people make their own calibration standards for use with their specific instrument. If the instrument is calibrated with standards plated in a particular solution, then the instrument should be standardized for parts coming out of that particular solution.

Coulometric

Just as one is often prone to say "Xerox this" when requesting a photocopy of a document, "Kocour this" is the common generic expression used to request a test of the thickness of a part, coulometrically—even though several other companies have begun manufacturing coulometric thickness measuring devices as well.

This method is electrochemistry at its purest, being based upon an electrochemical fundamental, Faraday's Law. The law states that 96,500 coulombs (1 coulomb is equal to 1 amp/second) will, at 100% efficiency, deposit or strip 1g equivalent weight of a metal. Silver, with a valence of 1, has a molecular weight of 107.87, which also is its equivalent weight; 96,500 coulombs will deposit or strip 107.87 g silver, and any fraction of 96,500 coulombs will deposit or strip an equivalent fractional weight. Nickel, with a valence of 2 and an atomic weight of 58.70, has an equivalent weight of $\frac{58.70}{2}$ or 29.35; 96,500 coulombs will deposit or strip 29.35 g nickel. Thus, 96,500 coulombs will deposit 107.87 g silver or 29.35 g nickel.

Combine this fact with another electrochemical phenomenon: If a metal is made anodic in a certain solution, a definite potential will develop between that metal and the solution. Should there be a change in one of the metals exposed to that solution, as when a coating is completely dissolved and another metal exposed, in this case, the basis metal, a different potential will develop between the new metal and the solution.

These are the fundamental principles for coulometric or electrochemical devices used to determine plating thicknesses. If a metal can be dissolved anodically in a suitable solution at essentially 100% current efficiency, the weight of the metal dissolved can easily be calculated from the elapsed time, the current, the defined area and a known constant for the metal, which is its density.[24]

With commercial instruments, the area is usually defined by an orifice in a rubber gasket which fits on to a small metal cell or probe containing the test solution. This metal cell also serves as the cathode during the test. The specimen becomes the anode and the rubber gasket with the orifice also serves as an insulator between the anode and the cathode. When a constant current is passed through the system, the voltage in the cell remains constant while the plate is dissolving electrochemically. When the plating is all removed and a new metal, the substrate, is exposed, a sharp voltage change takes place. This is the endpoint. The voltage change, properly amplified, is enough to stop the timer, thus ending the test automatically. By correlating the test current with the area, the time readout is made numerically equal to the plating thickness, thereby giving us a direct reading instrument

All of the commercial instruments incorporate a small air agitator to stir the solution in the cell to minimize against polarization effects, thereby allowing higher current densities to be used and still maintain 100% current efficiencies. The electrolytes or test solutions used are very important inasmuch as they must not attack or dissolve the coating chemically, must permit the plating to be stripped at 100% efficiency and, finally, must not attack the substrate. This is why not every metal can be determined electrochemically, because, literally, there isn't a solution available.

The coulometric method is not affected by the electrical or magnetic properties of the coating-substrate system and is generally classified as a destructive test because the spots where the tests are made may impair the appearance of the specimen. If it isn't too objectionable, the spots may be flashed over, or, at worst, the part stripped and easily replated because the coulometric stripping solution left the basis metal finish unimpaired.

A limitation to the use of a coulometric device

is the size of the probe itself, because if the part is too irregular or too small, then a suitable test cell cannot be set up. A possible source of error may be variations in the test current, but constant mini-power sources are becoming available, and for added insurance against current variations, the instruments come equipped with a current-time integrator so that any variations in current during a test will not result in an error. To minimize any distortion of the rubber test area by too much pressure of the gasket against the part, constant pressure devices are readily available.

The coulometric device must be calibrated against standards of known coating thickness. These standards, in order to calibrate accurately, must have the same type of coating-substrate combination as the specimen. If the reading on the instrument during the calibration procedure is other than what the standard is supposed to be, then the following factor is used when conducting thickness tests.

$$\text{Factor} = \frac{\text{Standard calibration reading}}{\text{Machine reading}}$$

The machine reading for the unknown specimen is multiplied by the factor to obtain an accurate reading.

The operating range for this method is from 0.75 to 50 μm (30 to 2000 μin.); the practical operating range is from 2 to 35 μm (80 to 1500 μin.). Chromium is a notable exception because it can be measured to as low a thickness as 0.075 μm (3 μin.). The accuracy of the coulometric instrument is ±10%, and it is assumed that pure coatings are being tested because there is no calibration for alloying impurities. However, where an alloy is known, such as in the case with electroless nickel, a factor may be used to give good, acceptable results.

A unique advantage of this method is the possibility it offers for selective thickness determinations of the component members of composite electrodeposits. Just by changing the solution in the cell after an endpoint is reached, another suitable solution can be substituted and the other component of the composite plate can be checked for thickness.

The coulometric method of measuring can be further sensitized by using a chart to record voltage changes.[25] In this manner, more sensitive differences in voltages between a Watts nickel deposit and a sulfamate nickel deposit have been observed, as well as differences between a pyrophosphate copper deposit on a rolled copper substrate. This is due to detectable voltage changes between materials with different crystal structures.

Given the proper stripping solutions within the established parameters of this method, coulometric instruments are quite versatile.

Typical electrolytes may be found in the ASTM's Standard[26] covering thickness testing by coulometric devices.

X-Ray Methods

X-ray spectrometry is quite similar to beta backscatter (BBS), but is unencumbered by the margin of difference between the atomic numbers of the coating-substrate system so critical with BBS. X-ray spectrometry, however, is encumbered with an expensive price tag for the equipment and it does lack the thickness range of other methods, particularly BBS, for instance.

Properly calibrated, X-ray fluorescence measurements are very accurate within their range of operation. This means of testing is rapid, nondestructive and independent of chemical reactions, magnetic effects, differences in atomic numbers and almost every other effect except the thickness of the coating system itself.

Two techniques are generally used. One is a direct emission method which involves the measurement of secondary X-ray fluorescence emanating from the coating (Fig. 6c),[27] and the other is the indirect, or absorption, technique (Fig. 6b)[27] which measures the X-ray fluorescence of the basis material. The latter technique measures the attenuated (weakened) emissions of the substrate. As the thickness of the coating increases, the intensity of the secondary X-rays from the substrate decreases.

The absorption of the X-rays is a function of the mass of the material traversed. Thickness, mass-absorption coefficient and density are factors which determine the use of the emission technique over absorption, because these factors affect the practical operational thickness range for a particular coating-substrate system.

The X-ray emission method for measuring plate thickness is based upon irradiating a defined sample area with an intense primary beam from an X-ray tube placed close to the specimen. This primary beam stimulates from

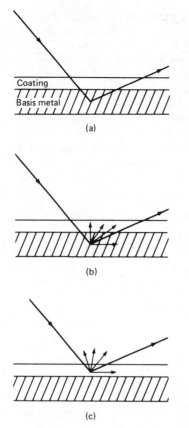

Fig. 6. Schematic diagram of X-ray methods for measurement of metal coating thickness.
  a. Diffraction from basis metal.
  b. Fluorescence from basis metal.
  c. Fluorescence from coating.

the plated material and the substrate an emission of a series of definite secondary wavelengths characteristic of the individual elements comprising the coating and the substrate. From these beams, a selected wavelength, usually the most intense, is separated and filtered by an X-ray monochromator to sort out the secondary emissions other than those from the coating.[28] The measurement of the intensity of the characteristic secondary X-rays is expressed in counts/second by a suitable detector and associated circuitry. This intensity is a function of the coating mass per unit area.

X-ray spectrometry is applicable for determining the thickness of any metallic coating, with the maximum measurable thickness for a given coating being that beyond which the intensity of the secondary radiation is no longer sensitive to small changes in thickness.[29] In other words, when the coating thickness increases to a certain value, the characteristic radiation from the substrate will disappear and the coating will assume infinite thickness. There must be a discernible beginning and end to a coating in order for it to be measured.

To convert the intensity readings of the secondary X-rays to linear thickness, the weight of coating per unit area of a specimen is compared to the weight per unit area of calibrated standards of the same coating-substrate combination. If the density of the specimen coating is not the same as that for the calibration standard, then the comparison in linear thicknesss will be in error. If the calibration is for gold, then the X-rays will "count" the gold atoms, *only*, and not the other extraneous atoms that may be present in the deposit. This is why there is a difference in density. The calibrated standard has more gold atoms per unit area than an alloyed or impure specimen.

Another basic difference between BBS and X-ray spectrometry is that BBS can be "fooled" by the use of so-called synthetic standards whose count-rates are equivalent to certain coating-substrate combinations. X-rays, on the other hand, can't be "fooled" because each element reflects back its own characteristic identifying wavelength.

Atomic number does have a bearing on X-ray spectrometry, but only so far as effective measuring thickness ranges are concerned. The higher the atomic number, the thinner is the higher limit for measuring the thickness for that material. Put another way, we can effectively measure thicker deposits of silver, with an atomic number of 47, than we can of gold, whose atomic number is 79, and thicker deposits of copper, atomic number 29, than of silver. The heavier the nucleus of the atom, the more it impedes the penetration of the primary beam and prevents it from reaching the substrate.

The lower limit for X-ray measurements is less than 0.1 μm (4 μin.); this limit is used routinely for gold on nickel, copper or kovar, nickel on steel, or brass and tin on steel. The maximum measurable thicknesses are in the order of 1 to 100 μm (40 to 4,000 μin.), depending upon the coating system. The accuracy of this method is well within ±10% and is greatly dependent upon calibration.

The precision of measurement is limited somewhat by the statistical nature of X-ray emission, (or absorption), since not all of the X-rays can be counted at the same time. There can be as many as 200,000 counts per minute, so several readings have to be made to get a better statistical average. The count rate factor makes for about a 1.3% error built into the system, aside from the errors that may be encountered from the calibrated standards.[30] Another source of error can be the impurities present in the deposit. For instance, if a nickel coating contains iron, and the substrate is steel, the X-rays cannot distinguish where the iron comes from. The same may be true with gold alloyed with nickel on a nickel substrate. These problems, however, if known, are not insurmountable, as different techniques may be employed.[30]

In measuring with X-ray fluorescence, the percentage of error increases as the thickness increases because of the attenuation from the basis metal. This problem is not only unique with this method, but is present in every other method as well.

Kriegler and Schumacher have described the use of an electron probe technique for thickness measurements using X-ray equipment.[31]

Beta Backscatter (BBS)

While beta backscatter is probably the most widely accepted method for determining coating thicknesses under 2.5 μm (100 μin.), this should not preclude its ability to accurately measure thicknesses in excess of that figure. With the exception of X-ray fluorescence, no other popularly used method, if one can refer to X-ray fluorescence as popular, comes close to it for dependability and acceptability.[32]

A drawback to BBS, when compared to X-ray fluorescence, is that the difference in atomic numbers between substrate and coating must be at least 20%. This, for all practical purposes, is no great impediment because most of the common combinations under testing very often exceed that figure. Rhodium on silver, with atomic numbers of 45 and 47, respectively, is an illustration of a system that cannot be measured with BBS. The accuracy of the thickness determinations by BBS increases as the difference between the atomic numbers of the substrate and coating increase.[33] BBS instruments, with all of the extra paraphernalia offered, are still cheaper than those used for X-ray spectrometry and are far more versatile, having a much wider operating range.

While the name, Betascope, has achieved a generic connotation for the BBS instrument, there are other manufacturers besides Twin City Testing that make such a device, UPA Technology and Sigma Laboratories among them.

Many of the problems commonly attributed to the use of BBS are self-inflicted by the user. Manufacturers' precautionary instructions are very often overlooked, in part perhaps due to overzealous advertising blurbs which tend to minimize operational shortcomings. Users have a tendency to read into the operating manuals implied warranties which are never specifically given. This instrument, unlike other thickness testers, does have "perishable" parts. Radioactive sources do have a half-life, Geiger-Mueller tubes are subject to aging, and platens, which define the subject area under examination, become worn or chipped depending upon the material used. All of the instrument manufacturers, however, remind their customers about these matters with the same diligence that a magazine reminds its readers of a lapsing subscription.

Backscattering is proportional to the number of atoms per unit area encountered by the beta ray beam. This is, in effect, weight per unit area.[34] The unit area is defined by the opening in the platen (Fig. 7). With X-ray fluorescence, the secondary emissions are characteristic of the individual elements encountered by the primary beam. With BBS, on the other hand, the intensity of the beta backscatter rate is a function of the

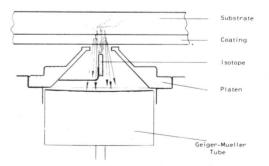

FIG. 7. Probe system with a beta ray source directs a collimated beam of beta rays toward the coated surface of the specimen. Some of the beta rays are absorbed by the sample, while others are backscattered and detected by a Geiger-Mueller tube and transmitted as electrical impulses to the basic unit. With appropriate electronic gear, these electrical impulses are converted to a thickness measurement after calibration. (*Courtesy Twin City Testing.*)

atomic number of the elements encountered. The higher the atomic number, the more intense the backscatter. This intensity is the distinguishing characteristic between one atom and another. From this, it can be deduced that the reason there has to be a 20% difference in atomic number between coating and substrate is that this is the minimum difference in backscatter radiation that can be distinguished with a high degree of certainty and precision.

The depth of penetration of the backscattered particles is dependent upon the energy of the particles from the radioactive source and the density of the material to be tested. This is why different materials have broader thickness ranges than others, and why different radioactive sources must be used for different coating-substrate systems.

Each radioactive source has an optimum range for determining the thickness of a given metal. The higher the atomic number of the material under test, the lower the penetrating power of the source. The heavier metals with larger nuclei offer the most resistance to the penetrating power of the beta rays just as with X-ray fluorescence. A weak energy source like promethium 147 has an upper operating thickness limit of 2 μm (80 μin.) for gold, but 3 μm (120 μin.) for copper or nickel. On the other hand, thallium, a much "hotter" source, has a much higher upper thickness limit, as well as a much higher low limit, because it can "burn" through a thin coating as if it were hardly there, giving erroneous results.

The measurement of BBS is based on the principle that the intensity of backscatter from a coated specimen is always between two limit values, which are the intensities of backscatter from the bare substrate and from the coating material.[35] A reading of the uncoated substrate is absolutely necessary because in order to ascertain the depth (thickness) of the coating, a piece of the substrate must be recorded by the instrument. Without a reading of the substrate intensity, the coating will appear to have infinite thickness. Thus a gold coating of 3 μm (120 μin.) will appear to be infinitely thick to a promethium 147 source, because that source lacks the penetrating power to reach and read the substrate; a thallium source must be used instead to achieve good measurable results. By comparing the backscatter rate of the specimen with those from a set of appropriate reference standards, one can very readily determine thickness.

"Calibration" and "standardization" are two terms that are often used when calibrating a BBS instrument. "Calibration" is the process of mathematically calculating the curve which defines the specific measurement application. With the late model instruments, many different calibrations can be stored in a memory for specific jobs without having to resort to recalibration.

"Standardization" is the process of updating the limit count-rate values of the substrate and the coating. "Standardization" has to be done more frequently in order to compensate for radioactive decay and drift condition, which are natural ongoing phenomena.

Some fractions of the beta rays, because of angle, can miss the Geiger-Mueller tube, so for statistical reasons, a number of readings should be taken in order to ensure a higher degree of accuracy, increasing the statistical confidence level of the thickness determination. Thus, when calibrating or standardizing, multiple readings on the standards are necessary because coated thickness standards, even if within tolerance, may vary slightly in thickness over the area defined by the aperture.

With BBS, the time of measurement is the only dimension which can be manipulated to hold the measurement uncertainty within acceptable limits, because everything else is fixed. While testing times of 10 seconds may be adequate in some instances, longer times of 30 seconds will ensure greater consistency. Count times in excess of 30 seconds do little to improve accuracy. The larger the aperture to expose more of the part to beta rays without running the risk of leaking beta rays, the more accurate the test, again for the statistical reason of measuring a greater number of backscattered particles. An opening which leaks beta-rays, however, will give a reduced count-rate and a concomitant under-estimation of the coating thickness.

In the case of gold, particularly, much has been made of the varying densities of gold deposits from sundry baths, and with good reason. Density, as was pointed out, is an important part of the equation from which thickness is derived. There are several ways to neutralize that problem.

One way is to obtain from the supplier of the proprietary bath, the density of the deposit and then use the following formula if you "standardized" with 19.3 gold.

Actual thickness = BBS reading
$$\frac{\text{Density of theoretical gold, 19.3}}{\text{Density of proprietary gold}}$$

or use the following table,[36] which is the same as the above equation.

Density (g/cc)	Multiplying Factor
17.4	1.11
17.5	1.10
18.0	1.07
18.3	1.05
19.3	1.00

If the given density of a proprietary gold is 17.5 and the BBS reading is 3 μm (120 μin.), then multiply 3 times 1.1, which equals 3.3 μm (132 μin.), the actual thickness for the 17.5 density gold deposit if it were calibrated against a gold standard with a density of 19.3.

Another way to compensate for differences in density is with the use of calibrated standards that were plated from the particular bath whose deposits are being tested. The calibration of the standards should be done by methods other than those where the instrument requires calibration. Methods for calibration have been described elsewhere in this chapter.[2,3]

BBS measures gold, solder, copper, photoresist coatings, silver, rhodium, nickel and also the percentage of tin-lead compositions and other alloys. There are countless other applications for BBS, provided the proper difference for atomic number is respected.

Measurements of thickness by transmission is another technique for obtaining thickness measurements with BBS. Transmission is used, for example, to make the measurement of copper in the plated-through holes, or to measure solid materials, such as silicon. In transmission, the specimen is interposed between the source and the Geiger-Mueller tube.

It is worth noting that the instrument's efficiency is limited if the line voltage falls below 100 volts AC. It is recommended that a constant voltage transformer be used with each instrument.

Notwithstanding the potential problems users of BBS might encounter, it is currently without peer as a thin film measurement method. It is rapid, economically sound, and supported by knowledgeable vendors. With sufficient understanding and attention to detail, thickness determinations of ±5% can be obtained.[37]

MICRORESISTANCE TECHNIQUE

This technique is uniquely suited for microresistance measurements of copper plated-through holes in printed wiring boards, (PWB). The measured resistance reflects the current-carrying capacity of the copper cylinder. (The copper plated through-hole is actually cylindrical in shape.) The microhm readings can be converted to give the average copper thickness of the hole. But, as a matter of fact, the reason there is a concern about the hole plate thickness is to determine the current-carrying capacity of the copper cylinder; with microresistance figures available, thickness readings become incidental.

The microresistance principle operates by delivering precisely controlled pulses of DC current to the four-point probe system (Fig. 8), which in turn applies these pulses to the copper plated hole. DC current is used to eliminate electromagnetic interference. One of the probes senses the microvoltage developed across the copper in the hole and feeds this minute voltage back to the basic unit to be displayed in microhms, or in a thickness which corresponds to the theoretically calculated resistance of the copper cylinder under examination. The probe system is designed to distribute the current uniformly and also to detect any voltage drop.

The microhm readings also serve as an important inspection feature because in this way plating defects such as insufficient plating, cracks and discontinuities can be detected.

Electrically parallel holes cannot be tested with this technique because they give the probe current an alternative path to travel and it is for this reason that this method cannot be used on in-

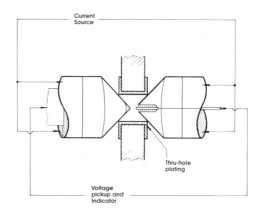

FIG. 8. Advanced hand-held probes. (*Courtesy UPA Technologies.*)

ASTM B 659

TABLE 2. Applicability of Coating Thickness Measuring Methods

NOTE—B = Beta backscatter; C = Coulometric; E = Eddy current; and M = Magnetic.

Coatings / Substrates	Copper	Nickel	Chromium	Autocatalytic Nickel	Zinc	Cadmium	Gold	Palladium	Rhodium	Silver	Tin	Lead	Tin-Lead Alloys	Non-metals	Vitreous and Porcelain Enamels
Magnetic steel (including corrosion-resisting steel)	CM	CM^A	CM	C^B M^A	CM	BCM	BM	BM	BM	BCM	BCM	BCM	B^C C^C M	BM	M
Nonmagnetic stainless steels	CE^D	CM^A	C	C^B	C	BC	B	B	B	BCE^D	BC	BC	B^C C^C	BE	E
Copper and alloys	C only on brass and Cu-Be	CM^A	C	C^B	C	BC	B	B	B	BC	BC	BC	B^C C^C	BE	E
Zinc and alloys	C	M^A	...	...	...	B	B	B	B	B	B	B	B^C	BE	...
Aluminum and alloys	BC	BCM^A	BC	BC^B E^{A,B}	BC	BC	B	B	B	BC	BC	BC	B^C C^C	E	E
Magnesium and alloys	B	BM^A	B	B	B	B	B	B	B	B	B	B	B^C	E	...
Nickel	C	...	C	...	C	BC	B	B	B	BC	BC	BC	B^C C^C	BE	...
Silver	B	BM^A	B	B	B	...	B	...	...	...	...	BC	B^C	BE	E
Glass Sealing Nickel-cobalt-iron alloys UNS No. K94610	M	CM^A	M	C^B M^A	M	BM	BM	BM	BM	BM	BM	BCM	B^A C^C M	BM	...
Nonmetals	BCE^D	BCM^A	BC	BC^B	BC	BC	B	B	B	BC	BC	BC	B^C C^C	...	...
Titanium	B	BM^A	...	BE^{A,B}	B	B	B	B	B	B	B	B	B^C	BE	...

^{A} Method is sensitive to permeability variations of the coating.
^{B} Method is sensitive to variations in the phosphorus content of the coating.
^{C} Method is sensitive to alloy composition.
^{D} Method is sensitive to conductivity variations of the coating.

The American Society for Testing and Materials takes no position respecting the validity of any patent rights asserted in connection with any item mentioned in this standard. Users of this standard are expressly advised that determination of the validity of any such patent rights, and the risk of infringement of such rights, are entirely their own responsibility.

process boards. The holes to be measured must be electrically isolated as in the finished product.

Commercially, this instrument is sold under the CAVIDERM trade name, manufactured by UPA Technologies.

MISCELLANEOUS TESTS

If the thickness of the basic metal is accurately known, the geometry of it relatively simple and the surface smooth, measurements with a micrometer are often used. A micrometer is most often used with heavy plating thicknesses, such as hard chromium deposits on shafts. The micrometer is reasonably accurate and a very convenient tool because the final dimension is usually the critical dimension for assembly.

If the plating specification is 25 μm (1 mil), the subject item is made undersize by that figure and plated to the final dimension. Quite often, with heavier plating requirements, the coating will not be uniform and the plater will have to apply a deposit in excess of the specification and then the item will have to be finish-ground to size. A micrometer is the usual thickness testing device.

Profilometry[14] has sometimes been used to trace surface roughness profiles by the use of a simple lever with the scanning tip terminating in a sharp diamond point, just like the arm of a phonograph record player. Plating thickness can be determined by tracing the surface steps from an unplated portion to the plated section. This is strictly a laboratory method and it is worth mentioning just in passing as yet another technique that is used. This method is considered very accurate and it's been applied to verify the thickness of calibration standards.[2]

Just as cutting instruments may range from cleavers to microtomes, plating thickness measuring devices and methods, as we have seen, are almost as wide-ranging. Each method has its own unique place that suits its use best. Just as one wouldn't think of cutting a side of beef with a microtome, no one should consider measuring a 5-mil plate of hard chrome on a printing role with a profilometer. Methods and instruments are dictated by the degree of accuracy necessary to meet specifications and requirements. Only the most economically expedient methods to obtain the sought-after answers need be used. On the other hand, a computerized BBS instrument with all the trappings may cost more than a good metallurgical microscope, but if a considerable amount of thickness determinations have to be made, the fancy BBS setup is by far the most economical way to go.

Having said all this, it is well to bear in mind, when thickness testing, that consistency of results using a particular method is no criterion of its accuracy. You may just be propagating the same error inherent in the method.

References

1. *Specifications and Tests for Electrodeposited Coatings*, ASTM, Philadelphia, PA (1980).
2. Ogburn, F. *Proc. 1st AES Symp. on Thickness Testing* (February 1978).
3. Baldwin, P. C. *Plating and Surface Fin.* **68**, 9:14. (1981).
4. Brenner, A. *34th Ann. Proc. AES* p. 143 (1947).
5. Read, H. J. *Ibid.*, p. 135.
6. "Interlab Comparison Report," *Plating* **59**, 4:320 (1972).
7. Wilson, G. A. *Metal Finishing*, **58**, 6 (1960).
8. Test Method B487-75, ASTM, Philadelphia, PA.
9. Albrecht, W. M. *Proc. 1st AES Symp. on Thickness Testing* (February 1978).
10. Kehl, G. L. *Principles of Metallographic Laboratory Practice.* New York: McGraw-Hill, 1949.
11. Saur, R. L. *Plating* **52**, 7:663 (1965).
12. Thomas, J. D. and S. R. Rouze. *42nd Ann. Proc. AES*, p. 49 (1955).
13. Fluhmann, W. and W. Saxer. *Publication of Wild Heerbrugg* **6**. Heerbrugg, Switzerland (January 1969).
14. Cooley, R. L. and K. E. Lemons. *Plating* **56**, 5:511 (1969).
15. Test Method B588-75, ASTM, Philadephia, PA.
16. Test Method B555-75, ASTM, Philadelphia, PA.
17. Test Method B556-71, ASTM, Philadelphia, PA.
18. Waite, C. F. *40th Ann. Proc. AES*, p. 113 (1953).
19. Brenner, A. *Journal of Research*, Natl. Bur. Stds., 18, p. 565 (1937).
20. Test Method B499-75, ASTM, Philadelphia, PA.
21. Test Method B530-75, ASTM, Philadelphia, PA.
22. Brodell, F. B. and A. Brenner. *Plating* **44**,6:591 (1957).
23. Test Method B529-70, ASTM, Philadelphia, PA.
24. Hodges, E. A. *Proc. 1st AES Symp. of Thickness Testing* (February 1978).
25. Harbulak, E. P. *Plating & Surface Fin.* **67**, 2:49 (1980).
26. Test Method B504-70, ASTM, Philadelphia, PA.
27. Sellers, W. W. and K. G. Carrol. *43rd Ann. Proc. AES*, p. 97 (1956).
28. Heller, H. A. *Plating* **56**, 3:281 (1969).

29. Test Method B568-72, ASTM, Philadelphia, PA.
30. Ogburn, F. and J. Smit. *Plating* **60**, *2*:149 (1973).
31. Kriegler, R. and B. W. Schumacher. *Plating* **47**, *4*:393 (1960).
32. Cross, I. *Plating & Surface Fin.* **66**, *11*:591 (1979).
33. Ott, A. *Plating* **63**, *1*:28 (1976).
34. Meek, R. L. *Plating & Surface Fin.* **62**, *6*:593 (1975).
35. Joffe, B. B. *Proc. 1st AES Symp. on Thickness Testing* (February 1978).
36. Lieber, S. *Metal Finishing* **72**, *12* (1975).
37. Buckley, R. R. *Proc. 1st AES Symp. on Thickness Testing* (February 1978).

B. Corrosion Tests

JAMES DOUGLAS THOMAS

Assistant Head, Electrochemistry Department, General Motors Research Laboratories, Warren, Michigan

Corrosion tests are used to evaluate how well plated coatings protect the basis material against corrosion, and how well decorative coatings retain their appearance. Ideally, it would be desirable to have a universal corrosion test that would show the type and extent of deterioration experienced in any environment to which the various coatings are exposed. Such a test is not possible because climatic conditions and corrosion processes for different metals are not alike.

Because of the differences in environmental conditions, considerable research has been done to get data on corrosion rates in many parts of the world.[1] The results of these programs are studied to obtain the cause and effect of climatic conditions on corrosion and relate these to the properties of the materials used in the exposure tests. In most of these programs, corrosion rates were expressed as weight loss with time which assumes uniform corrosion over the entire surface; both weight loss and time can be measured accurately. Thus, results of different programs or effects of different exposure locations can be assessed and compared with each other. These types of exposure tests are useful for assessing the performance of bulk metals. They can also provide guidance on relative corrosivity of various environments, or locations, for planning exposure tests for plated coatings, which do not usually corrode uniformly as do most bulk metals.

CORROSION PROCESSES

Generally, corrosion is an electrochemical process somewhat akin to a plating process. For corrosion to proceed, there must be an anode, a cathode, and an electrolyte. This is the basic reason that the corrosion in hot, dry climates is not very noticeable; there is rarely an electrolyte present to complete the electrical circuit. Under conditions where the electrolyte is present most of the time—corrosion occurs more rapidly. The nature of the metallic coating provides the anodic and cathodic areas, so the presence or absense of the electrolyte and its properties are very important in determining corrosion rates and methods to control them.

An example of the reactions involved were described by Saur[2] for the copper-nickel-chromium system:

$$Ni \rightarrow Ni^{++} + 2e \qquad \text{anodic}$$

$$2H^+ + 1/2\, O_2 + 2e \rightarrow H_2O \qquad \text{cathodic}$$

By controlling these reactions, the degradation of the coating can be minimized.

The considerable amount of research on corrosion processes and the application of the knowledge gained toward the development of accelerated tests has aided the overall effort to improve performance of deposits.

INTERPRETATION OF RESULTS

The interpretation of results of corrosion tests on plated coatings has also been improved because of the better understanding of the corrosion process. Parts or samples exposed outdoors or in the salt-spray type of accelerated test are assessed as to performance by visual examination, and the judgment of quality is very subjective. The opinion of the inspector is inherent in this procedure. As in the case of any subjective evaluation, the extremes—no failure or total failure—are readily recognized and are noncontroversial; the range in between is where opinions can differ. It is difficult to compare results of test programs that are run by different groups in different locations. Training and experience have helped to alleviate this difference. Also, ASTM Method B-537[3] is an aid to give ratings which can be compared.

Information that is necessary for interpreting results is knowledge of the coating characteristics that cause certain types of behavior. This is very important when test results are used for process control. An example of this is when the sulfur content of the different layers of nickel, in a multilayer system, have an improper ratio. The result is that protection of the basis metal is decreased—the two layers act like a single layer, surface pitting is enhanced and the bright layer becomes brittle as the result of excessive sulfur from too much brightener. Again, training and experience help in recognizing these symptoms, and in recommending proper adjustments to the plating process.

RELIABILITY/CONFIDENCE OF TEST RESULTS

Although there is a performance (corrosion test) requirement in all specifications for plated coatings intended for outdoor exposure, it is noted that complete confidence in the results is apparently lacking. This is the probable reason that specifications also contain thickness requirements. Thickness can be measured reliably, and, therefore, provides confidence in describing overall results of tests on sample parts. There is some credence to this attitude, because parts that fail the corrosion test generally fail in service. On the other hand, parts that pass the test sometimes fail in service; often this can be attributed to inadequate thickness, which can be remedied. As more knowledge is gained with regard to the influence of deposit properties on corrosion characteristics, better tests and confidence will result.

Test development can be hampered by variations in the samples used to determine effects of variables in test procedure. This can be alleviated by using a large number of samples and a statistical approach to determine the optimum use of the samples.[4] To illustrate this, the Research Project 15 Committee of the American Electroplaters' Society obtained 2000 steel and 2000 die cast parts plated consecutively in production. These parts were randomized and numbered. The statisticians suggested that a corrosion test be run with the first 50 pieces and a second set of 50 pieces. From the results, it was determined that sets of 11 pieces would be representative of the entire 2000. For the development of the Cass and Corrodkote tests, these groups of 11 pieces were invaluable in determining the effects of test variables.

Unfortunately, it is impractical to obtain a year's supply of each plated part and determine the acceptability of the entire production run and have confidence in the results. From a realistic point, samples from various lots of parts are checked and the lot, from which the samples were taken, is accepted or rejected. This has worked quite satisfactorily as the field performance has improved. The increased knowledge of deposit characteristics as related to corrosion processes has resulted in more reliable interpretation of test results. This knowledge has also been beneficial for improving process control and, therefore, has enhanced product performance.

TEST METHODS

The methods used for assessing the performance of coating systems can be classified as (1) outdoor tests, both at static sites for atmospheric effects or on vehicles for "roadesphere" effects, (2) salt spray tests for accelerating the corrosion process, and (3) electrochemical tests which can incorporate well-defined controls for inducing corrosion. As accelerated tests have been developed, many exposure programs included these tests along with outdoor exposures. Correlating results from both exposures helped the acceptance of the accelerated tests for predicting field performance.

Because the procedural details for the various test methods are available from many sources, they do not need to be repeated here. In the following sections, where necessary, comments will be made with respect to precautions to be taken and the reasons for them. Some guidance for interpreting the results will also be made, as these differ for each test.

Outdoor Exposure Tests

Racks for static-site exposure are constructed to hold test panels so that the significant surfaces of the panels are a 45° angle and face south. These conditions have become standard and are mentioned in most references to outdoor test programs. The angle of the panel surface provides for exposure to fallout (e.g., rain, snow, particulates) from the atmosphere but also provides for adequate run-off to avoid immersion effects. This is, therefore, a controlled exposure position which is adequately representative of service conditions. The southward orientation, because of the relationship of the sun and the earth, eliminates long-term exposure in shadows; this could affect the time the surface remains wet and thus provide a path for corrosion currents. (Although no literature reference has been seen regarding exposures in the southern hemisphere, one would conclude that the orientation would be northerly.) For contoured or shaped parts, some judgment is needed, as the 45° criterion cannot, obviously, be complied with.

For mobile exposure of test panels or parts, some judgment is needed for exact location on the vehicle and what is being sought in the way of results. In some cases, standard parts are replaced with test parts (e.g., door handles, bumpers). Panels have been located beneath the front bumper of cars, behind the wheels of trailers to get road splash from tires, on the front of trailers to get mist, and on the sides and rear of trailers, trucks or station wagons. This illustrates the difficulty of correlating results and the importance of a full description of the exposure location.

When static or mobile tests are performed, data regarding the exposure environments are desirable. To illustrate this, when the Cass and Corrodkote Tests were being developed, under the auspices of the Research Board of the American Electroplaters' Society, the Committee members had plastic collection bottles mounted on the front ends of their cars. The bottles had windows cut in them near the top to catch the splash from the streets of Detroit. Collectors were also located near the roof exposure sites in various locations. Analyses of the collected materials provided valuable information for test development. This information also helped to explain differences in results from the various roof locations and from the "roadesphere." Important were such things as pH; composition and concentration of anions and cations in the precipitation; and fallout of particles. A low pH, copper and iron ions and chlorides were found to be present in the severe (corrosive) environments. This information was used[5,6] in the development of the two accelerated corrosion tests.

Salt Spray (Fog) Test[7]

This method is used for corrosion testing of metallic coatings (e.g., zinc and cadmium) and organic coatings (e.g., paints). It is not recommended for decorative copper-nickel-chromium systems. The solution is 5 w/o sodium chloride (NaCl). The pH of the collected solution is 6.5-7.2; the temperature of the test is $35 + 1.1 - 1.7$ C ($95 + 2.0 - 3.0$ F). Fog collection rate, using an 80 cm^2 horizontal area collector, is 1.0 to 2.0 ml/hr. Commercial test cabinets are designed to meet these requirements.

Copper-Accelerated Acetic Acid-Salt Spray (Fog) Test (Cass Test)[8]

This test was developed for corrosion testing of copper-nickel-chromium coatings on steel- or zinc-based substrates. The basic differences between the Cass and Neutral Salt Spray Tests are: (1) the addition of 1 g of cupric chloride ($CuCl_2$) to each 3.8 L salt solution, (2) the adjustment, with acetic acid, of the pH to 3.2 ± 0.1 as measured on the collected spray and (3) that the temperature is 49 ± 1 C (120 ± 2.0 F). Commercial cabinets are designed for these test requirements.

Cleaning the surface of the test piece with magnesium oxide, as required, is critical. Without this cleaning, the onset of corrosion is delayed and parts perceived to be satisfactorily may actually be inferior in performance.

This test has become one of the specified tests, for decorative chromium finishes, in the automotive and other industries where the plated coatings are exposed to severe corrosive environments. It is also being used for testing anodized aluminum.

Corrodkote Test[9]

The Corrodkote Test was developed simultaneously with the Cass Test. In the Corrodkote

Test, a paste containing cupric nitrate, ferric chloride, ammonium chloride and kaolin (clay) is painted on the metal surface. After partial drying, the sample is exposed in a humidity cabinet maintained at $100 + 2 - 3$ C ($212 + 3.5 - 5.0$ F) and a relative humidity between 80 and 90%. After cleaning the corroded surface, the corrosion sites are redeveloped in a salt spray or humidity cabinet. Commercial equipment is available.

This test is specified, as an alternate to the Cass Test, for automotive and other applications subjected to severe environments.

Sulfur Dioxide Test[10,11]

This test finds its widest use in Germany and, to a lesser extent, in other parts of Europe. It is included here because United States manufacturers of parts and components for the automotive industry "World Cars" are finding it specified for those parts shipped to some European assembly plants. The test is used for corrosion testing of coatings subjected to SO_2-containing environments.

The test is cyclic in that the cabinet is heated to 40 ± 3 C (104 ± 5.5 F) for 8 hr and cooled for 16 hr. The test atmosphere is the same as that in a humidity cabinet, 100% relative humidity, to which SO_2 is added. A commercial cabinet* is available which meets DIN and forthcoming ASTM specifications.

Simultaneous Thickness and Electrochemical Potential (STEP) Test[12]

Although this test is not a corrosion test, it is included here because of its close relationship to methods for evaluating performance of multi-layered nickel-chromium decorative coatings. The test consists of the electrochemical dissolution of the various metallic layers and monitoring the potential changes as each layer is removed. The magnitude of the potential differences can be interpreted in terms of performance of the coating and to monitor the plating operation for corrective action when necessary. Thickness of the nickel layers is also measured.

It is currently being used internally by some automotive companies and will be a required test, along with the usual corrosion tests, in the near future.

*Harshaw Chemical Co., Cleveland, OH 44142.

Electrolytic Corrosion (EC) Test[2,13]

The EC Test is a rapid method for assessing the performance of copper-nickel-chromium coatings on steel or zinc substrates. It is a potentiostatically controlled method that corrodes the various layers of metal in a relationship found in outdoor exposure. Indicators can be used in the electrolyte to show penetration of the coating (substrate attack). Alternately, the test can be run for a specified time and corrosion assessed by a measurement of the corrosion sites by an interference microscope.[14]

This test has not gained wide acceptance because of the labor involved, but ingenuity can alleviate this problem. The features that are attractive are: (1) it is not subjective but gives numbers as to performance of the coating; (2) it tells the nature of the chromium discontinuities quickly; and (3) it gives results of one-year equivalent outdoor performance in minutes.

Other tests

A few other tests have been developed but are not used, generally because other tests have replaced them. These are mentioned briefly because they are cited in the literature and there are ASTM Methods available.

FACT (Ford Anodized Aluminum Corrosion Test).[15,16] This test is a rapid electrochemical method that is suggested for use in quality control-acceptance test for decorative anodized coatings on aluminum.

Acetic Acid-Salt Spray (Fog) Testing.[17] This method, similar to other salt spray tests, was originally developed for testing plated zinc-base die castings because it developed blistering similar to that on castings exposed outdoors. It has a pH of 3.1 to 3.3 adjusted with acetic acid. The modification of this test with cupric chloride evolved into the more rapid Cass Test.

GENERAL COMMENTS

Although the test methods described in this chapter are referenced primarily to ASTM documents, customers' specifications may vary slightly from these procedures, but will not affect the equipment required. In addition, many of these methods have been approved, or are in the reviewing stage, by the International Standards Organization. Thus, these methods will become available and may be specified by customers throughout the world. Test times and acceptance criteria, of course, are stated in the part specifications issued by the purchaser of the finished articles.

References

1. LaQue, F. L. *46th Ann. Tech. Proc. Amer. Electroplaters' Soc.*, p. 142 (1959).
2. Saur, R. L. and R. P. Basco. *Plating* **53**:35 (1966).
3. ASTM Method B-537, *1980 Annual Book of ASTM Stds.*, Part 9.
4. ASTM Method B-602, *1980 Annual Book of ASTM Stds.*, Part 9.
5. Bigge, D. M. *46th Ann. Tech. Proc. Amer. Electroplaters' Soc.*, p. 149 (1959).
6. Nixon, C. F., J. D. Thomas and D. W. Hardesty. *46th Ann. Tech. Proc. Amer. Electroplaters' Soc.*, p. 159 (1959).
7. ASTM Method B-117, *1980 Annual Book of ASTM Stds.*, Part 9.
8. ASTM Method B-368, *1980 Annual Book of ASTM Stds.*, Part 9.
9. ASTM Method B-380, *1980 Annual Book of ASTM Stds.*, Part 9.
10. German Method DIN 50018.
11. Edwards, J. *46th Ann. Tech. Proc. Amer. Electroplaters' Soc.*, p. 154 (1959).
12. Harbulak, E. P. *Plating & Surface Fin.* **67**, n2:49 (1980).
13. ASTM Method B-627, *1980 Annual Book of ASTM Stds.*, Part 9.
14. ASTM Method B-651, *1980 Annual Book of ASTM Stds.*, Part 9.
15. Stone, J., H. A. Tuttle and H. N. Bogart. *Plating* **53**:877, (1966).
16. ASTM Method B-538, *1980 Annual Book of ASTM Stds.*, Part 9.
17. ASTM Method B-287, *1980 Annual Book of ASTM Stds.*, Part 9.

C. Inspection

LELAND M. MORSE

Managing Engineer, Electrochemistry and Corrosion Department, Product Planning and Development Staff, Chrysler Corporation, Detroit, Michigan

Visual Inspection of Plated Parts

Of all the qualities of electroplated finishes, measurable and otherwise, the one most widely recognized by the ultimate consumer is appearance. The purchaser or user of a plated object, or of a product having plated components, judges the quality of the plating primarily by its finished appearance. A high degree of luster, smooth mirror-like reflectivity, and freedom from surface defects will have a strong effect on the prospective buyer's opinion of the quality and value of the product.

One may show the average consumer an article that has been heavily plated using the best available techniques to obtain good adhesion and ductility and excellent corrosion resistance, but unless the finish appearance-wise is also satisfactory, the consumer will not recognize the hidden values of the plating and will judge the product to be cheap and inferior.

This works the other way, too. Inferior plating jobs have gotten by on good appearance alone.

Adequate visual inspection methods are therefore necessary in order to maintain a salable product and to satisfy the purchaser when plated parts are purchased to go into an assembly.

Classes of Inspection

There are three classes of inspection with which we are usually concerned. Although similar in technique, they differ in their purpose.

(1) In-process inspection. This ties in closely with manufacturing quality control.

(2) Final inspection by the producer. (Often combined with packing for shipment.)

(3) Receiving inspection by the customer. This type is used when the customer is a jobber or large retailer and also when the customer purchases parts which are to be assembled into a final product.

The latter two should be closely correlated. The producer obviously should not pass any parts that the customer will not accept. On the other hand it would be poor business for the supplier to reject material that the customer's receiving inspection department would accept without question.

It is most desirable that the producer and the consumer come to a mutually satisfactory agreement regarding such things as significant surfaces, minimum luster, and the degrees of various defects acceptable. When possible, this should be done prior to the acceptance of a contract, since price is largely affected by the degree of finish required. Although the responsibility for bringing about such an agreement lies with the producer's sales department and consumer's purchasing department, the estimating and/or engineering department of the producer company must be cognizant of the quality level required.

It is to the interest of all concerned to establish criteria which are no more stringent than necessary. To demand near perfection will frequently result in excessive rejections and restrict production which, in turn, will be reflected in higher prices.

Factors in Visual Inspection

Visual inspection is employed primarily with decorative coatings or finishes. The most frequently specified decorative finishes are the bright reflective ones where deposits are either buffed to a bright finish or bright electroplating is used. These include electroplates of nickel, chromium, brass, bronze, silver, gold, and rhodium. In the case of anodic coatings on aluminum, varied degrees of brightness may be produced by buffing the basis metal and by bright dipping.

In inspecting the above types of bright finishes the primary factor is luster or reflectivity. The degree and uniformity of luster is occasionally a factor in the inspection of parts plated with cadmium, zinc or tin, primarily for corrosion resistance. This usually is true when the parts are to be displayed or are visible components of an assembly sold to the general public, and uniformity of luster and a lustrous metallic appearance lend an air of quality and good workmanship to the product.

The second factor is color, which is extremely important in the inspection of gold plating and of the alloys of brass and bronze, both of which can vary over a wide range of shades. Color inspection is especially important in the control of dyed anodic coatings on aluminum.

Texture is a factor in the inspection of satin, matté, or frosty finishes. It is also important in the inspection of anodized aluminum finishes, since they can be produced in a wide range of lusters and textures from etched to matté to pearly and bright.

There is always considerable difficulty in defining appearance requirements. The inspector is often required to set his own standards of acceptance because of an inadequate means of specifying what is desired to be expected. It may be difficult to maintain a constant acceptance level because of the influence of manufacturing problems, schedules and price. He may tighten or relax requirements, even unconsciously, depending on his state of mind and on conditions prevailing at the time. His standard of acceptance may gradually change over a period of time. From constant observation of certain defects he may become less critical, then, suddenly tighten up because of a specific complaint or criticism.

Because standards for appearance are difficult to put into written or quantitative terms, the criteria of acceptability of metal finishes are usually resolved by the use of standard samples. The preparation, selection and maintenance of standard samples are complicated by the many ways in which a plated product can fall short of a required degree of luster or reflectivity.

Arriving at a Standard of Acceptability

The first consideration in the inspection of a plated product is to define the "significant surfaces." Very few decorative parts must be highly finished all over. Significant surfaces for visual inspection are those surfaces of the finished part that must have the specified appearance or quality of "finish," e.g., those surfaces of the part that are exposed to view in the final product in normal use.

Areas of the part other than significant surfaces will usually be plated without any regard for appearance and will generally have less than the specified plate thickness. In cases where certain surfaces are to receive no deposit, it is customary to indicate such areas on the drawing of the part which serves as part of the specification. If no such areas are indicated on the drawing, it may be assumed that plating is required, or at least allowed, on all surfaces. If, however, drawings do not indicate the significant surfaces for plating, it is necessary for the inspector to determine which surfaces require the specified plating and finish. This may be done in a number of ways:

(1) Determine the use and application of the part or product and which surfaces are plainly visible.

(2) Request information from customer's engineering or design department when product is sold on specification.

(3) Consult sales or design department of producing company when plated product is a finished consumer product.

The answer to the question is not always obvious and some agreement must be reached between the parties concerned. For example, the lower one-fourth to one-eighth of the face of an automobile bumper is usually turned under and is not plainly visible to a person standing near the car. As the observer moves away however he gradually sees more and more of this lower section but the polishing wheel marks present are not noticeable at a distance. Furthermore this surface reflects only the roadway and does not pick up any highlights. For these reasons most companies do not consider the lower or turned under part of a bumper to be a significant surface of prime importance; it might be termed a secondary surface, since the specification is partially relaxed to the extent of permitting some wheel marks, although brightness and specified plate thickness may be required. By eliminating some unnecessary fine polishing a worthwhile saving is thus effected without sacrificing quality.

Whenever possible such distinctions should be made a part of the original agreement between the purchaser and supplier, either by inclusion of proper notes on the drawings of the product or by incorporation into the wording of the purchase order.

Degree of Finish

It can safely be said that an inspector who requires absolute perfection in finish will find no parts that he can accept. It follows that for a given run of a given product the quantity of pieces rejected will be directly related to the appearance quality level required. Since the cost of production increases with the percentage of rejections, the inspector can influence the cost of production by the level of quality he requires. It is important therefore that the inspector be fully aware both of the problems of production and of the requirements dictated by the use of the product.

The degree of finish required in any product is dependent on many factors, including the following:

(1) Use—whether purely decorative or purely functional, or a combination of both.

(2) Distance at which the object will be viewed under normal use.

(3) Price class of the product in which the plating is used.

(4) Comparison with competitive products.

(5) Comparison with other plated parts in the same assembly.

(6) Established standards of customary practice in the past, including accepted standards associated with a given field such as the plumbing or the builders' hardware fields.

The requirements of a satisfactory bright metal finish never have been clearly defined. The important elements are:

(1) High reflectance of incident light, combined with the following:

(a) Clear image reflection. Perhaps the basic quality of a good bright finish on a part is that it allows one to see his face in it. Who can deny that vanity is one of the basic elements of human nature? This being so, the extent to which a person can see himself may be the secret of the degree of personal satisfaction derived from plated surfaces.

(b) Absence of diffuse reflection of any sort. Diffuse reflection may be caused by haze, surface irregularity, fine or coarse scratch lines, pits, graininess, crazing or many other defects.

(2) Color of reflected light. Optical instruments are available for precise measurement of the above characteristics on perfectly flat samples. If all plating were done on perfectly flat surfaces the problem of inspection would be purely one of optical instrumentation. The flat surface, however, is the exception; practically all plating must, therefore, be rated visually.

One approach to the evaluation of brightness qualities has been comparison with standard panels. Usually, however, a curved plated surface will look better than a flat panel with equal brightness. The eye seems to correct for defects in curved surfaces. In general, the smaller the radius of curvature of the surface the greater the defects that can be tolerated.

Convex surfaces diminish the reflected image of any object and increase the apparent distinctness of the image. Also, broad, plain, unbroken surfaces will show up defects much more than narrow, interrupted surfaces.

Attempts to use photographs as standards for inspection of decorative finishes have been unsuccessful. The visual impression of a metallic surface is a binocular effect due to the differences between the reflections reaching the eyes of the observer. The difficulty of photographing any bright metallic object is well known to photographers. Color photography helps to some extent, but the only photographs which depict the natural metallic appearance of an object are those made with stereoscopic cameras and color film. The photographs must be observed using a special viewer to obtain a realistic effect. Still lacking is the opportunity to view the object at several angles as would be done if an actual sample were held in the hands.

In view of the above comments the most satis-

TABLE 6. DEFECTS DUE TO BASIS METAL OR PREPARATION

Seams and laps	Due to uneven filling of molds in casting. In wrought material, a void due to rolling or forging of an irregular surface.
Porosity	In castings, due to presence of gas or shrinkage cracks.
Slag stringers	Slag or oxide rolled into sheet metal or bars. Shows up as lines of pits.
Flash, risers, or gates incompletely removed	Common defect of castings.
Die marks	Scoring of part that occurs sometimes in deep-drawing or stamping operation.
Trim marks	Scoring and break-out of die castings during press operation to remove gates and flash.
Weld marks or spatter	Usually the result of electric arc or resistance welding.
Poor burnishing	Rough or pebbly appearance as if work were rattled around in a tumbling barrel.
Wheel marks	Due to incomplete removal of coarse polishing scratches.
Handling marks and dings	Scratches and dents picked up during shop handling and transportation.
"Orangepeel"	Graininess of metal surface giving a pebbly appearance. Often occurs in bending and drawing.
Stretcher strains	Peculiar feathery relief pattern on surface of formed steel as a result of aging of steel in storage. Also known as Lüder's Lines.
Cross breaking	Relief pattern of parallel lines occurring on press-formed material as a result of certain aging phenomena and roller leveling.

factory standards for plating inspection are samples of the actual parts in question.

Continuing our reasoning under the premise that a perfect bright finish would have perfect image reflectivity, a bright plated finish can fall short of perfection in many different ways. Most plated surfaces show a number of different defects in various degrees. Those defects which affect appearance can be divided into two classes: (1) basis metal defects; (2) defects in plating.

Some of the most familiar of these defects are listed in Tables 6 and 7. Since there are so many kinds of defects it is not probable that one could select a standard sample which exhibits the permissible maximum of each defect. For example, if we decided to accept a certain number of pits, a certain degree of rustiness, a certain amount of polishing wheel scratching, etc., all of these could not be present in one sample without the over-all finish being far below the acceptability level. It is the over-all impression, usually, that has to be considered. One then has a problem of producing or selecting standard samples of plating.

When possible at least two matching samples of the parts in question should be retained, one by the producer's inspector, and one by the customer's inspector. It is better to select and retain four samples, two of the desired finish and

TABLE 7. DEFECTS DUE TO PLATING PROCESS

Burns	Rough or dull plate due to localized high current density or arcing.
Streaks	
Striations	Relief pattern of lines of varying plate thickness.
Pits	Gas pits and pits left by dislodged nodules.
Plateaus	Areas of heavier plate surrounded by thinner areas.
Craters	Large pits with flat area at the bottom.
Stardust	Fine nodular roughness occurring in bright plate.
Roughness	Coarse nodulation or surface inclusion of dirt.
Bleeding	Evidence of foreign solution leaking out of cracks or porous areas during plating.
Cracks and crazing	Assumes many forms.
Flaking and peeling	
Milkiness	Lack of full brightness evidenced by a bluish haze on the surface.
Frostiness	
Poor coverage	Incomplete coverage, as with chromium.
Off-color	
Buffing flare	Directional haziness caused by very fine scratches from buff wheel.
Built-up edges	Heavy plate or nodulation on sharp edges.
Blisters	

two of the minimum acceptable finish. Very often, however, it is necessary to agree on a finish prior to the production of any actual parts. Then one must use other parts for samples, preferably selecting parts of similar size, design, basis metal, and method of fabrication.

When production commences it is customary to submit parts as samples for approval by the inspector or customer. At that time it is usually possible to make the final decisions as to the minimum acceptable quality of finish and the standard quality. If the samples submitted have been made on production tools and finished in production equipment they can be used in the selection of a "master sample."

In the selection of a master sample, one acceptable procedure is to pick, at random, a number of samples that cover the range of finish quality that would be expected in normal production. These should then be arranged in order of appearance and quality, from highest to lowest in the opinion of the observer. The inspector should then examine these samples under uniform lighting conditions and select the ones he considers to be representative of the finish demanded for the particular part. After doing this, he should view the parts under the conditions of actual use. If possible when they are part of an assembly, they should be installed and viewed as a part of the finished product. This is important, since under the bright but diffused light of an inspection, bench parts may look bad and prove entirely acceptable under conditions of use. Furthermore, when the part goes into an assembly with other plated items the finish of all of them should usually be as nearly alike as possible. In an assembly consisting of a number of plated parts in close proximity to each other, one that is particularly bright may make the others look dull by comparison and thus stick out like the well-known "sore thumb." An example of this is a bright nickel and chromium plated part in an assembly of chromium plated stainless steel parts. Because it is usually not practical to obtain the high brightness of bright nickel on stainless steel parts, it would be desirable in such a case to tone down the bright nickel parts in some way, such as by buffing them prior to chromium plating.

After the inspector has selected the master samples that are representative of the desired finish, he might also select a few parts which have the permissible maximum of various defects that may be encountered. They should be chosen by viewing them in their proper positions on the finished product. Such samples might include, for example, in the case of a bumper guard, one with $\frac{1}{16}$ in. of chrome burn on the edges of a secondary surface and one with a few minor projection weld dimples.

These master samples are compared with production parts in the actual process of inspection. In use, the parts should be viewed under the diffused but intense white light of an inspection lamp, which probably will not favor their appearance at all.

It is best not to continue to use the masters in actual day-to-day service. The chief inspector must carefully protect the master samples and supply the production inspectors with matching samples selected for current production. An ideal arrangement is to keep all master samples in cabinets or drawers specially reserved for that purpose.

Other defects controlled by visual inspection but not usually the subject of standardization are chrome stains, rinse water stains, finger marks, glove marks, handling marks, scuffing, and scratches on the final finish. The visual inspector is responsible also for picking up such other defects as sharp edges and roughness, which are only observed by touching the object. Roughness can be detected very easily by an inspector wearing lightweight cotton gloves if he passes a gloved hand lightly over the surface in question.

Inspection of Colored and Other than Bright Finishes

Although present-day emphasis is predominantly on bright chromium plating for decorative finishing of appliances, motor cars, plumbing and a wide variety of hardware, certain industries favor the softer effects obtainable with plating by use of special buffing methods and other metals such as silver, brass, gold, copper, and oxidized variations of some of these. The visual inspection of such finishes is equally important, and possibly more so because of the additional problems of matching color, texture and directional characteristics of light reflection.

Some finishes in this category are satin finish, butler finish, platinum finish, matté finish, frosty finish, Roman finish. Some of these terms are interchangeable, and none is clearly defined either by description of the finish or by standardization of methods for obtaining them.

It is therefore necessary to provide the operator with a standard matching sample with which to compare his work. Here again some of the actual parts in question should be used.

With finishes of the satin type, which are produced with a fine abrasive or a scratch brush, there are two things in particular to watch for—direction of scratches and treatment of recesses. The best satin finishes use dull (unbrightened) plating as a starting point. The wheel produces many fine scratch lines which give a "satiny" effect to the surface. If the operator is careless or inexperienced, the lines may cross each other at various angles and present a confused appearance which falls short of the desired effect. The lines in any finish of this sort should be predominantly parallel on any given surface. Recesses should be dull rather than bright. Sometimes it is attempted to produce satin finishes using bright nickel as a starting point because of convenience. The effect is not quite satisfactory and usually can be detected by the presence of bright recesses. The intensity and depth of finish should be substantially uniform throughout.

Gold plating covers a variety of shades from rose to greenish. Even 24-karat gold varies slightly in color with the conditions of plating. Standard color samples of plated gold are practically a necessity in inspection control of the desired shade. The same is true in the case of brass, bronze, and other alloys as well as of "oxidized" finishes, black nickel, molybdenum oxide, and the like.

The visual inspection of decorative anodizing on aluminum alloys involves a wide range of factors including depth of luster, image reflectivity, color, texture and basis metal defects not normally affecting plating. Table 7 lists some of the defects that may be cause for rejection. Anodic films are formed by electrochemical oxidation of the basis metal producing, on aluminum alloys, a thin glass-like layer of aluminum oxide. The depth of luster and image reflectivity are strongly influenced by the composition and the metallographic structure of the basis metal. The color of plain, undyed, anodic coatings is also influenced by the chemistry of the anodizing process. The color of dyed coatings is, of course, influenced by a number of factors in the processing. Texture, in the case of specially prepared basis metal surfaces, is also strongly affected by the bright dipping process and by the anodizing stage itself.

An inspector of anodized finishes, therefore, would do well to catalog the various types of visual defects encountered in his plant operation and to save a sample of each. These will prove to be of considerable assistance in the training of personnel and in expediting corrections in processing.

TABLE 8. DEFECTS IN ANODIZING

A. Defects caused by the basis metal

Streaks (in wrought alloys)	Frequently caused by smearing of magnesium oxides into the surface of the metal in the rolling operation.
Streaks (in extrusions)	Sometimes the result of improper heat treatment of the billets, or of oxidation of the surface of the ingot during storage.
Stringers	Segregation of constituents in parallel lines that etch out during bright dipping or anodizing.
Structure lines	Segregated constituents that cause parallel bands of translucent anodizing.
Grayness (in anodizing on alloys that normally produce a bright finish)	Frequently the result of improper mill practice producing grain growth or precipitation of constituents.
A dull area with a bright zone around it	May be caused by local overheating in fabrication by welding processes.

B. Defects resulting from improper processing

Over-all dull appearance	Can result from a number of causes such as a bright-dipping solution out of balance, too low a temperature of bright-dipping solution, and excessive coating weight.
Dull chalky surface	Can result from contaminated sealing solutions or a sealing solution at the wrong pH.
Dull irregular spots with bright center	May be rack contact marks.
Dull spots on surface opposite rack contact points	Result from overheating at rack contacts.
Pale streaks in a dyed coating	From bleeding chemicals, improper neutralization after anodizing.
Iridescence	Usually results from a thin anodic coating.
Crazing	Results from excessive coating weight, too low a temperature of the anodizing bath, and baking after anodizing at too high a temperature.

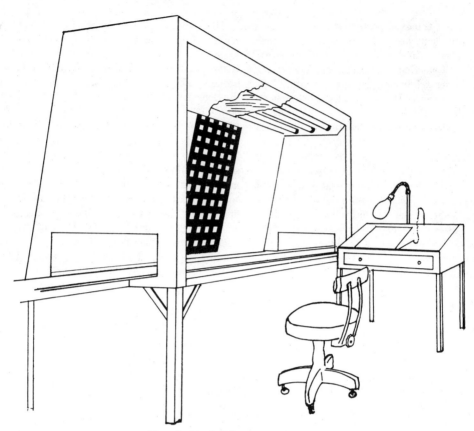

Fig. 9. Typical inspection station

In the comparison of color anodizing with preselected masters it is recommended that two or more light sources be used. A unit is manufactured for this purpose that provides "daylight" at one setting and "twilight" at another[14]. Incandescent lamps should also be used in these comparisons. It will be found that colors that match under one light source may appear different under another type of light.

Inspection Equipment

Since the whole philosophy of bright finish inspection is based on image reflectivity of the surface, an inspection station should provide the best possible means of creating image reflection. Figure 9 shows a suitable arrangement for accomplishing this.

The basic requirements are a strong source of diffused white light, and a black background screen. The light from four 40-watt fluorescent tubes, mixed daylight and white, is sufficient for an ordinary inspection booth. The fluorescent tubes should have a white reflector over them and a diffusing screen of ground or opal glass beneath. If glass of this sort is not readily available, a screen of tracing cloth or white vellum may be used, although it is a fire hazard. The black background can be black velvet, mohair, felt, or other black cloth without sheen. Flat black paint is not quite as good.

Also helpful is a black and white geometric pattern of some sort which can be placed high in the booth and next to the light source so that it is brightly illuminated. Black and white charts commercially available for testing the hiding power of paints are good for this purpose.

With lighting of this sort, the defects in plating are easily seen. The diffused light shows lack of chromium coverage on nickel or stainless steel. The black screen makes any imperfection in the bright plating show up as grayness or white lines or spots. The geometric pattern shows gross defects such as wheel marks, waviness, or orange-peel and provides a standard object for checking the image reflection quality.

There are numerous other refinements which could be added to an inspection booth, but the above are the basic requirements. A booth of the type described will show defects so plainly that at first the inspector will have to be restrained from throwing out nearly everything that comes down the line. Frequent comparison with the established standards must be the rule.

The parts may be carried through the booth on a belt conveyor, or they may be inspected while on the plating racks as in nickel inspection prior to chromium plating. In the latter case the racks should be suspended on swivel hooks so that the inspector may turn the rack around. Provision for holding a pad of paper is helpful when quality records are to be kept. Producing plants usually require 100 per cent inspection of parts. Spot checking or partial inspection of lots as used by consumers should be based on accepted sampling methods as described in the books on quality control.

Inspection Personnel

Different plants handle the problem of inspection in different ways. In small shops the responsibility sometimes rests with the plater or unracker. With reliable people who have pride in good workmanship this can be satisfactory. In many large organizations the inspectors all report to a chief inspector who in turn reports directly to management, not to the plating or polishing foreman. In other companies the inspector reports to the production supervisor. Each method has its advantages and drawbacks.

When the inspector reports to management, theoretically no bad work is shipped. The situation may however develop into a contest between production supervisor and the inspector to see which one can "out-fox" the other. When the inspectors are subordinate to production supervision the supervisor is solely responsible for the quality of his department's product. Since he is responsible also for meeting production schedules he will occasionally sacrifice quality in order to ship parts on schedule. An effective solution may be to have as chief inspector, an experienced production man who is also quality minded. With this background he is unlikely to be the type that feels he has not done his job if he goes home at night without having stopped a production line at least once that day. He will also be able to make worthwhile suggestions of corrective measures.

Final inspection is frequently combined with the packing operation. This is a logical and convenient method and can be conveniently kept under separate control from production.

In-process inspection is frequently coordinated with repair or salvage operations. This is convenient because a trained inspector can be conversant with the repair operations and route the parts back into process with the least loss of time, thereby minimizing the number of rejects in inventory.

The chief inspector or quality control supervisor may act as a roving inspector, and may assist in training inspectors, setting and maintaining standards, and maintaining liaison with the customers' inspection departments.

Use of Inspection Results

The effectiveness of uniform inspection methods is lost unless some concrete action is taken on results. A number of possible courses of action may be taken depending on whether the inspection is in the producing plant or in the customer's plant.

In the producer's plant a system of regular inspection and corrective action may be called "quality control." Since visual inspection results are not of a quantitative nature "statistical quality control" is difficult to apply. A systematic analysis of rejection may however be made and much valuable information obtained. Such analysis is valuable only if made promptly and almost continuously.

Forms used for recording inspection data vary greatly. Failures are frequently grouped according to the rework procedure necessary. The following types of information should be recorded:

Date
Shift
Hour
Inspection station
Inspector
Reasons for rejection, i.e., defects
Number of pieces rejected for each defect and each part number
Department responsible
Refinishing steps required
Number of pieces scrapped

The important thing is to obtain enough data on the report forms to permit the foremen, operators, and chemists to see the condition at a glance. Posting of the data in chart form may further the beneficial effect of creating competition between departments or between shifts within the same department. These data must also be made available for further analysis by the cost accountant and may even be used as a basis for some form of incentive plan to aid in keeping quality up and cost down.

Final inspection will further ensure that no defective work is shipped. Any consistent defects may indicate the need for changes in the process, or added operations, which may require a management decision.

When the inspection data are obtained by the customer, rejections may be followed up by one or more of the following steps:

(1) Advise supplier immediately so that he may correct defects.

(2) Arrange to have the supplier's inspector receive or see typical samples of the rejected parts.

(3) Sort out bad parts and return them to supplier or have supplier's inspector sort them.

(4) Return entire shipment for sorting and repair by supplier.

(5) Repair parts and charge supplier.

References

1. "Specifications and Tests for Electrodeposited Coatings," American Society for Testing Materials, Philadelphia, 1969.
2. Read, H. J., *Proc. Am. Electroplaters' Soc.*, **34**, 135–45 (1947).
3. Read, H. J., and Lorenz, F. R., *Plating*, **38**, 255–63 (1951); *ibid*, **38**, 945–52, 958 (1951).
4. Kehl, G. L., "Principles of Metallographic Laboratory Practice," New York, McGraw-Hill, 1949.
5. Clarke, S. G., *Proc. Am. Electroplaters' Soc.*, 24–30 (1939).
6. Read, H. J., and Thompson, J. H., *Proc. Am. Electroplaters' Soc.*, **35**, 79–86 (1948).
7. Francis, H. T., *Trans. Electrochem. Soc.*, **93**, 79–83 (1940).
8. Brenner, A., *J. Research Natl. Bur. Standards*, **18**, 565 (1937).
9. Brenner, A., and Garcia-Rivera, J., *Plating*, **40**, 1238 (1953).
9a. Brodell, F. P., and Brenner, A., *Plating*, **44**, 591 (1957).
10. Soderberg, K. G., and Graham, A. K., *Proc. Am. Electroplaters' Soc.*, **34**, 74 (1947).
10a. Goodwin, P. S., and Winchester, C. L., *Plating*, **46**, 41 (1959).
10b. Joffe, B., and Modjeska, R. S., *Metal Finishing*, **61**, No. 12, 4 (1963).
11. Brenner, A., and Senderoff, S., *Proc. Am. Electroplaters' Soc.*, **35**, 53 (1948).
11a. Kushner, J. B., *Proc. Am. Electroplaters' Soc.*, **41**, 188 (1954).
11b. Brenner, A., and Morgan, V. D., *Proc. Am. Electroplaters' Soc.*, **37**, 51 (1950).
12. Ferguson, A. L., and Stephan, L. F., *Monthly Rev. Am. Electroplaters' Soc.*, **32**, 168, 894, 1006, 1116, 1237 (1945); **33**, 45, 166, 279, 620, 1283 (1946).
13. Kellner, H. L., *Proc. Am. Electroplaters' Soc.*, **37**, 105–24 (1950).
14. LaQue, F. L., *Proc. Am. Electroplaters' Soc.*, **46**, 141–148 (1959).
15. Wesley, W. A., *Proc. Am. Soc. Testing Materials*, **47**, 803 (1947).
16. Pray, H. A., *Proc. Am. Soc. Testing Materials*, **47**, 822 (1947).
17. Soderberg, G., *Proc. Am. Soc. Testing Materials*, **47**, 826 (1947).
18. Report of Committee B-8, Subcommittee II, *Proc. Am. Soc. Testing Materials*, **49**, 226 (1949).
19. LaQue, F. L., Report of Advisory Committee on Accelerated Corrosion Tests, *Plating*, **39**, 65 (1952).
20. Pinner, W. L., *Plating*, **44**, 763 (1957).
21. Sample, C. H., Am. Soc. Testing Materials Bull., No. 123, 19, Aug. 1943.
22. Durbin, C., *Proc. Am. Electroplaters' Soc.*, **38**, 119–132 (1951).
23. LaQue, F. L., *Materials and Methods*, **35**, Feb., 77 and Mar., 77 (1952).
24. May, T. P., and Alexander, A. L., *Proc. Am. Soc. Testing Materials*, **50**, 1131 (1950).
25. Report of Section B on Porosity Tests of Subcommittee III on Conformance Tests, Proc. Am. Soc. Testing Materials, **54**, 299 (1954).
26. Pinner, W. L., *Plating*, **44**, 763–766 (1957); *Proc. Am. Electroplaters' Soc.*, **43**, 50–55 (1959).
27. Nixon, C. L., Thomas, J. D., and Hardesty, D. W., *Proc. Am. Electroplaters' Soc.*, **46**, 156–163 (1959).
28. Sukes, G. L., *Metal Finishing*, **57**, 56–65 (1959).
29. Pinner, W. L., *Proc. Am. Electroplaters' Soc.*, **43**, 50–55 (1956).
30. Bigge, D. M., *Proc. Am. Electroplaters' Soc.*, **46**, 149–153 (1959).
31. Edwards, J., *Trans. Inst. Metal Finishing*, **35**, 277–294 (1958); *Proc. Am. Electroplaters' Soc.*, **46**, 154–158 (1959).
32. Brace, A. W., and Pocock, K., *Trans. Inst. Metal Finishing*, **35**, 277–294 (1958).
33. Hooper, J. H., *Trans. Inst. Metal Finishing*, **35**, 79–90 (1958).
34. The Macbeth Daylighting Corporation, Newburgh, N. Y.
35. "Report of Section A, Subcommittee II of A.S.T.M. Committee B-8," *Am. Soc. Testing Materials Proc.*, **65**, 270 (1966).
36. Snyder, W., and Saltonstall, R. B., *Plating*, **54**, 270–271 (1967).
37. Saur, R. L., and Basco, R. P., *Plating*, **53**, 35–38 (1966); *ibid*, **53**, 320–325 (1966); *ibid*, **53**, 981–985 (1966); *ibid*, **53**, 1124 (1966); *ibid*, **54**, 393–394 (1967).

11. INDUSTRIAL HYGIENE AND SAFETY

Bryan E. Winter

Manager, Safety and Environmental Control, Alexandria Metal Finishers, Inc., Alexandria, Va.

AND

Anthony O. Facciolo, Jr.

Safety Engineer, Alexandria Metal Finishers, Inc., Alexandria, Va.

This chapter is included so that the reader may be aware of the different and often serious problems in industrial hygiene and safety that occur in the metal finishing industry, particularly in the area of electroplating. This chapter is intended to serve as an introduction only. Readers charged with the responsibility of engineering a safe environment in a finishing operation should consult the references listed at the end of this chapter, and thoroughly review the applicable federal, state and local statutes.

CONTROLLING LEGISLATION

Federal

The controlling legislation is the Williams-Stieger "Occupational Safety and Health Act of 1970" (PL-91-596) which was passed into law "to assure safe and healthful working conditions for working men and women. . . ." This act established the National Institute for Occupational Safety and Health (NIOSH) under the Department of Health, Education and Welfare (DHEW); and the Occupational Safety and Health Administration (OSHA) under the Department of Labor (DOL). The act provides for research, information, education and training in the field of safety and health, and *authorizes enforcement of the Standards.* Of particular interest to the safety engineer in the field of metal finishing is Title 29, Labor, Chapter XVII "Occupational Safety and Health Administration," Department of Labor, Part 1910. Under Title 29, the safety and health of employees is covered by the standards of the act.

In more recent years, the increasing concern for the protection of the environment has created a new act which has had a direct effect with respect to environmental safety in the metal finishing industry. This act is known as the "Resource Conservation and Recovery Act of 1976" (RCRA) which was passed into law to assure the safe handling and proper disposal of all hazardous wastes (almost all electroplating wastes fit the definition of hazardous waste). The Environmental Protection Agency (EPA) is responsible for the enforcement of the standards set forth by this act. RCRA stipulates that all generators of hazardous wastes are responsible for assuring proper storage, handling and disposal of these designated hazardous wastes, and that they further assure the safe storage and handling of these wastes by developing emergency spill control procedures and emergency contingency plans for the control of accidental spills.

All regulations referring to these acts should be observed, particularly those dealing with toxic and hazardous materials or waste as they relate to electroplating, occupational health and environmental control, personal protective equipment, medical and first aid, walking and working surfaces, fire protection, materials handling and storage, machinery and machine guarding and electrical hazards.

State and Local

Most state and local jurisdictions have regulations dealing with industrial hygiene and safety which copy the federal statutes. Occasionally, however, state and local regulations may vary significantly from federal requirements. It is advisable, therefore, for the safety engineer to become familiar with local requirements and to comply with them as closely as possible. The authors' experience in dealing with local health departments is that they are extremely knowledgeable and helpful. It is advisable to make contact and consult with your local people before initiating any safety engineering or safety programs.

TOXICOLOGY AND ELECTROPLATING

Toxicology is the science which deals with determining the effects of short-term (acute) and long-term (chronic) exposure to various chemical environments. Exposure can take place four different ways: 1) dermal (skin) exposure to substance; 2) inhalation of substance; 3) ingestion of substance; or 4) absorption of substance through the skin.

Toxicology not only determines the effects but it determines the amount of exposure to various chemical environments that the human body can tolerate in an eight-hour day before possible injury may result. These amounts are commonly known as "threshold limit values" or "TLVs". (Some state health regulatory agencies prefer to use "PEL"—permissible exposure limits.) The TLVs for plating shop substances vary greatly, as does the toxicity of the substances. TLVs may be exceeded from time to time for short periods as long as the daily average concentration does not exceed the TLV. Some substances, in addition to an assigned TLV, also have a "C" rating. These substances may never exceed the TLV even for a brief period of time.

The NIOSH "Registry of Toxic Effects of Chemical Substances," Volumes I and II (listed under references at the end of this chapter), can be a valuable reference for locating various TLVs for specific substances. Many electroplating chemicals have proprietary names, making it difficult to identify their primary constituents.

In these instances it will be necessary to request OSHA form 20's, otherwise known as "Material Safety Data Sheets," from the chemical suppliers. The Material Safety Data Sheets will give a threshold limit value for the material in question based on its constituents.

Exposure to hazardous materials and toxic chemicals is considered to be the most important health and safety problem in electroplating, despite recent studies that indicate that systematic occupational illness among electroplating operators rarely occurs.

What constitutes toxic or hazardous materials is now a matter of law as defined by the various federal and state statutes. Since experts can and do differ on the toxicity of materials, only those materials which have been proven historically to cause serious health or safety problems will be discussed in this chapter.

Dermatitis

Dermatitis is the general term given to diseases or injuries of the skin. Approximately one half of all compensational claims are due to occupational skin dermatoses.

There are basically two classes of dermatitis: 1) primary irritation dermatitis, which is caused by physical, mechanical or chemical agents, and 2) sensitization dermatitis, which is an allergic reaction to a given substance. This sensitivity becomes established over a period of time. Once the sensitivity is fully established, short exposure to a small amount of the material can cause a severe reaction. This type of dermatitis is somewhat dependent on the individual's own tolerance levels.

Some substances can cause both primary irritation dermatitis and sensitization dermatitis. The most common plating shop materials that cause both types of dermatitis are chromic acid, nickel plating baths and trichloroethylene or perchloroethylene (tetrachloroethylene) used in vapor degreasing. Table 1 lists commonly found materials in electroplating operations that can cause dermatoses.

Many of the dermatological problems brought on by electroplating chemicals, such as ulcerations, blisters and cracking of the skin, can often result in painful secondary infections if left unattended.

Cyanide

Although strong arguments can be made for the equal toxicity of other chemicals, cyanide is generally accepted as the most dangerous mate-

TABLE 1. COMMONLY FOUND CHEMICALS IN ELECTROPLATING OPERATIONS THAT AFFECT THE SKIN

Chemical	Use	Irritating Action
A) Acids		
1) Chromic	Chrome plating baths	Ulceration on the skin—"chrome holes."
	Chromic anodizing	
	Chromate conversion coatings	Breathing exposure to chromic acid mists can cause a perforation of nasal septum.
2) Hydrochloric	Metal pickling (derusting)	Irritation and possible ulceration of skin.
3) Nitric	Stripping of metals	Severe burns and ulcers.
4) Sulfuric	Stripping and metal pickling	Severe inflammation of mucous membranes. Corrosive action on skin.
B) Alkalies (caustics)		
1) Potassium or sodium hydroxide	Stripping of metals PH adjustments in plating baths	Severe burns, persistent ulcers and loss of fingernails.
2) Potassium or sodium cyanide	Plating baths	Blisters and ulcers.
C) Elements		
1) Nickel salts	Nickel plating baths	"Nickel Itch"—a rash causing severe itching. Some people are more sensitive than others.
D) Solvents		
1) Trichloroethylene and perchloroethylene (tetrachloroethylene)	Degreasing of metals	Dry, cracked skin due to removal of skin oils.

rial in electroplating. The metallic cyanide salts utilized in the electroplating process are all extremely toxic if swallowed or absorbed through the skin. Concentrations as low as 50 to 100 mg absorbed in the body will produce death caused by the ability of the cyanides to block the exchange of oxygen between blood and body tissues. A lethal dose will produce an almost instantaneous death and the only treatment is immediate first aid. Despite this, there are very few deaths or illnesses from exposure to cyanide. Those that do occur are usually the result of the inhalation of hydrogen cyanide gas, which is liberated from a mixture of cyanide salts with an acid. This gas is extremely lethal if inhaled and is most often generated by accidental spills of acid and cyanide which combine on the shop floor or in floor drains. This condition can be further aggravated by poor ventilation, resulting in serious injury or death to the worker. It should be emphasized that cyanide solutions have been in constant and widespread use with few reported injuries or deaths, which is the best evidence of the careful use and control of this material in electroplating establishments.

Acids

Acids are used frequently in most electroplating processes to remove metal oxides or to properly prepare a surface to accept the subsequent finish.

Generally, cold dilute acids such as sulfuric, hydrochloric, nitric and chromic are used; however, you may find any type of acid involved in either a cleaning or electroplating process, and the concentration of acid and the temperature at which it is being used will vary according to the particular requirement.

The health hazards most frequently arising out of the use of acids are dermatitis (skin disease, as mentioned), burns and fires.

While all acids present special health and safety hazards when handled carelessly or used in poorly ventilated areas, two in particular present special problems. Sulfuric acid must be handled and mixed very carefully because it is an extremely reactive acid when mixed with other acids or water; chromic acid, chromates and dichromates are strong oxidizing agents, and while the acid itself is not a fire hazard, it can cause fires and minor explosions if allowed to come in contact with acetone, alcohol, glycerine and lacquer thinners. Chromic acid must be kept dry and should be stored away from wood, paper and other reducing agents. Empty chromic acid containers can be a fire hazard if used to store trash and should not be used for that purpose. As with all chemicals, chromic acid drums should be kept tightly closed when not in use.

Dilute chromic acid and chromic acid mist generated by the chromium plating process, which comes in contact with the exposed area of the workers, can cause skin infections which usually appear as wet or dry eruptions and usually take a long time to heal. These eruptions or ulcers can form at breaks in the skin, and the ulcers produce what are commonly referred to as "chrome holes" and other sores on skin surfaces. The mist generated in the electroplating process can cause nasal irritation and perforation of the nasal septum (hole in the nose).

Alkalies and Caustics

Alkaline solutions, usually very hot, are used to clean metals or plastics prior to electroplating. Severe, diffuse burns can occur from extensive contact with these materials, and the possibility for splash burns involving the eyes is present whenever caustics are handled.

Most of the accidents involving alkaline cleaners are caused by carelessness in making additions to the tank. Additions should be made slowly, using mild agitation to ensure a thorough mixing. It is advisable to preheat the water to 120 F to facilitate rapid mixing. Cleaner or caustic which is added to cold water in a rapid manner can collect at the bottom of a tank and cause a flash steam explosion. A large amount of heat is generated when this occurs, and in many cases it will cause a violent eruption of the solution.

Electrocleaning involves the use of alkaline solutions but differs from soak cleaning in that the use of an electric current is employed to generate bubbles at the surface to be cleaned. These bubbles produce a scrubbing effect which in turn creates a cleaner plating surface than the non-electrolytic soak cleaner. The alkaline solution is basically the same for the electrocleaner as it is for the alkaline soak; therefore, the same dangers may be encountered during tank recharging as previously mentioned. In addition, electrocleaning solutions produce hydrogen, which is a highly explosive gas that can be ignited by a mere electrical spark. Therefore, it is of extreme importance that the electrical current to the cleaning tank be shut off when entering or removing work from the tank. If the current cannot be turned off, extreme care must be used to prevent sparking as the work is removed from the tank.

Solvents

Solvents are used in the electroplating industry to remove oil, grease and loose dirt from the part to be plated. These solvents have the same effect on human skin, and while the removal of loose dirt is sometimes desirable, the removal of natural oils from your skin in this manner is not a pleasant experience. The two most commonly used degreasing solvents in the electroplating shop are trichloroethylene and perchloroethylene (tetrachloroethylene), two "chlorinated hydrocarbons." Prolonged exposure to these solvents can cause headaches, dizziness, nausea, fainting and even death. Fortunately, solvents of this type can be readily expelled from the body by getting to fresh air as soon as possible. The most serious accidents involving these degreasing solvents occur during tank clean-out operations. Obviously, it is important to utilize all safety guidelines and precautionary measures during a tank clean-out operation.

SPECIFIC HAZARDS

In addition to the chemical-related health hazards in plating operations, there are many specific hazards which can occur in the day-to-day operation of the shop.

Handling of materials accounts for 20 to 25% of all occupational injuries. Strains, sprains, fractures, cuts and bruises are the common injuries. They are caused primarily by unsafe work practices—improper lifting, carrying too heavy of a load, incorrect gripping, failing to observe proper foot or hand clearances, and failing to wear personal protective equipment.

Back injuries have the most significant effect on a shop's "workmen's compensation rate." Lifting is so much a part of everyday jobs that most workers don't think about it. But it is often done incorrectly, resulting in pulled muscles, disk lesions or even painful hernias.

Many employee injuries in plating shops can be attributed to slippery walk surfaces resulting from tank drag-outs between tanks and increased humidity in plating operations. To help remedy this problem, many shops install walkboards which keep employees off of the slippery floor surface. In time, many of these walkboards deteriorate due to the corrosive environment of the shop, and, if not properly maintained, can result in serious injury to the employees.

Eye injuries can result by carelessly working around the process tanks without protective goggles. The dangers of corrosive plating chemicals entering the eye cannot be overemphasized.

Since the plating process almost always utilizes electric current, there exists the danger of electrical shock. Not only is electric power used during the plating process, but it is used to power much of the equipment, such as air compressors, rotary air blowers, rectifiers and even tank heaters. Therefore, it is important that this equipment be wired and grounded in accordance with the national electric code to guard employees against electrical shock.

Fires within a plating shop are rarely given much consideration. But plating plant fires do occur and usually with devastating results. Fires can occur from improper storage or handling of stock chemicals or from equipment which is used improperly. For example: Soluble cyanides are reducing agents, and mixtures with strong oxidizers such as ammonium nitrate or chromic acid can result in fiery explosions. Electrical immersion heaters used to heat corrosive solutions in plastic-lined tanks can cause highly destructive fires if the liquid level of the tank is allowed to evaporate to a low level. As the liquid level drops, the heating element is exposed and can ignite the plastic side wall of the tank. If the destruction of a single tank and its contents does not seem serious enough, further complication could arise if the fire progresses into the plastic duct work that is often present in the ventilation systems around these tanks.

Many of these specific hazards arise from frequently violated regulations under Title 29, Part 1910, Code of Federal Regulations. An extremely useful publication put out by NIOSH, entitled "Health and Safety Guide for Electroplating Shops," contains a safety checklist on these most frequently violated regulations, and, if used properly, it can be an invaluable tool in attaining safe working conditions in an electroplating operation.

CONTROLS

Control technology pertaining to the plating and finishing industry can be categorized as: 1) environmental engineering concepts; 2) employee education; 3) personal protective and first aid equipment; and 4) ventilation.

Environmental Engineering Concepts

Engineering controls are the preferred method of reducing employee exposure to hazardous substances. They include a wide range of control concepts which can be used independently or together. Some basic principles for controlling the workers' environment utilizing engineering controls are substitution, isolation or process modification and ventilation.

Substitution

Substituting less hazardous materials, equipment or even processes is probably the least expensive and most positive method of controlling a certain occupational health hazard. This is the best method of control since it greatly reduces, or, in some cases, eliminates entirely the health hazard associated with the process. Of course, this type of control is limited by the availability of substitute materials. Some forms of substitute control which have been utilized in the plating shop are: 1) replacing cyanide plating solutions with non-cyanide baths; 2) the use of trivalent chrome plating instead of the more common hexavalent bath; and 3) replacing liquid acids with powdered acids for storage safety purposes.

Isolation or Process Modification

Isolation or process modification can be achieved by using a physical barrier-enclosure or by separating the employee from the hazard by time or space. For example, the use of an automatic plating line will reduce the health hazards by isolating the employee from the hazardous area and/or reducing the time spent in the area. Isolation or process modification techniques are best implemented during the design and construction of new plants, the expansion of existing plants or the replacement of obsolete equipment.

Employee Education

Employee training and awareness is the single most useful control technology to reduce or eliminate worker exposure to the hazardous chemical environment of a plating operation. The following suggestions have been found to dramatically reduce accidents and increase safety:

1. Have regular safety meetings with the employees and keep records to document the fact that these meetings have been held.
2. Develop a safety manual for your specific operation and give a copy to each worker.
3. Familiarize local authorities such as police, fire departments and emergency response teams with the layout of your plant, the types of chemicals and waste chemicals that are handled and their associated hazards, as well as possible evacuation routes.
4. Make arrangements with all local hospitals to familiarize them with the types of chemical exposures, injuries and illnesses which could result from fires, explosions or hazardous chemical spills.
5. Develop an emergency contingency plan describing action to be taken in case of an emergency resulting from fires, explosions or any unplanned release of a hazardous chemical to the environment. The plan should include the following:
   a. A description of the arrangements with local authorities that you have made, as mentioned in 3.
   b. A list of names, addresses and phone numbers of company personnel qualified to act as "emergency coordinators." These personnel should have a copy of the contingency plan at home as well as at work.
   c. An evacuation procedure describing a method of signaling personnel that evacuation is necessary, and designating both primary and secondary escape routes from the facility.
   d. An orderly list of procedures to be implemented by the acting emergency coordinator. (A sample "contingency plan" outline developed by the National Association of Metal Finishers' Solid Waste Disposal Committee has been reprinted at the end of this chapter for your use.)
6. Make employees aware of the location and use of various safety equipment such as eyewash fountains, emergency deluge showers and respirators.
7. Familiarize employees with basic first aid techniques.

All employees should be reminded repeatedly of the hazards which exist. Warning signs or posters should be located in conspicuous places to act as reminders of the potential dangers associated with certain operations or areas. All tanks should be labeled giving appropriate warnings where applicable, and a description of their contents. In addition to this, you may find that color-coding the plating tanks (for example, blue for rinse tanks, black for tanks containing acidic solutions, yellow for tanks containing alkaline solutions and red for tanks containing cyanide solutions) can help significantly by quickly identifying a tank's contents.

Incentives for employees to develop good work practices can be accomplished by giving yearly safety awards, mentioning accomplishments during an employee yearly review and even considering safety habits as a factor in wage increases.

Personal hygiene and general housekeeping measures aid in establishing and maintaining a low employee injury or illness rate. Some basic hygiene and housekeeping requirements are given below.

1. Each worker should have a clean change of clothing at the beginning and end of each shift, and the worker should be provided with two lockers, if possible; one for street clothing, the other for soiled work clothing so that there will be little possibility of cross-contamination.
2. Employees who work with toxic plating substances should be required to remove contaminated clothing before eating, drinking or smoking.
3. A chemical storage area should be set up and properly maintained. The different classes of chemicals should be separated and stored as follows:
   a. All cyanide salts should have a separate storage room incorporating a dike around its perimeter, especially if you are storing cyanides in liquid form. The storage room should be kept locked, and appropriate warning signs should be hung on exterior and interior walls.
   b. All acids, including chromic acid, should be stored together at a substantial distance from the cyanide storage room. Diking, or a partition, is recommended to contain any spills.
   c. All strong oxidizers should be stored together.
   d. All solvents, chlorinated and nonchlorinated, should be stored in a separate storage room incorporating a B-rated

fire door; diking and explosion-proof or vapor-proof lighting fixtures (check local fire ordinances). All metal containers of flammable solvents should be grounded.

All storage areas should be kept clean, orderly and dry to the greatest extent possible.
4. No employee should be allowed to eat or drink in an area that is exposed to toxic plating materials. A separate lunch and break room should be provided for this purpose.
5. Food or beverages should not be stored in an area exposed to toxic materials.
6. The work area should be kept clean, orderly, sanitary and dry to the greatest extent possible. All spills should be attended to immediately. Spill control equipment such as absorbent pillows or pads, neutralizing agents (J. T. Baker Chemical Company has an excellent line of neutralizing agents for accidental spills), scoops, shovels and drums (to dispose of absorbed material) should be kept available at all times.
7. Floor areas which are constantly wet should have a non-slip surface where employees walk or work.
8. Walkboards, if used, must be kept free from protruding nails, splinters, holes and loose boards.

As stated previously, employee training, along with good personal hygiene and housekeeping practices, will aid significantly in attaining a low injury or illness rate in an electroplating operation.

Personal Protective and First Aid Equipment

In most cases, it is essential that personal protective equipment be used during the plating processes to prevent worker exposure to the hazardous substances. Available personal protective equipment should include 1) chemical safety glasses with side shields or safety goggles (preferably of the antifog type); 2) full-face shields; 3) rubber gloves (barrier creams can be applied to the skin for added protection, but barrier creams alone should not be used in place of gloves), bibs, aprons, boots and leggings; 4) rubber capes or hoods; and 5) respirators which are specific for hazardous substances such as acid, gas, caustic fumes and solvent fumes (these should be kept on hand where in-plant exposure levels exceed the appropriate TLV for that substance).

No matter how safe the working environment, there is always the possibility of an on-site injury occurring, and in most cases immediate action utilizing the proper first aid equipment is necessary. The following equipment and supplies are recommended as part of your safety program: 1) rapid deluge showers; 2) eye-washes, located near the hazard area (because the time factor involving the eye is very critical); 3) portable oxygen units; 4) a cyanide antidote kit (this item is not usually available at hospitals but can be purchased through your physician and kept at the plant to accompany the victim being transported to a hospital under emergency circumstances); 5) emetics to induce vomiting in the case of some poisons; 6) universal antidotes for other poisons; and 7) a stretcher (may be necessary) and various types of splints. First aid supplies should include angular bandages, band aids, eye patches, adhesive tape, sterile antiseptic eye-wash solution, boric acid, heat tablets, aspirin, gauze, antiseptic ointment, petroleum jelly and iodine.

The old adage that "an ounce of prevention is worth a pound of cure" is really the first commandment of all safety programs. The second commandment is that employees must be properly trained in the use of safety equipment and supplies. Without proper training and motivation, your safety program cannot succeed.

Ventilation

Local exhaust ventilation consisting of hood, ducting and fans is the primary method for controlling emissions from open surface tanks in the plating and finishing industry. The local exhaust system design depends on 1) the health hazard or TLV of the plating or cleaning solution; 2) the evaporation or gassing rate; 3) the temperature and liquid level of the solution; 4) the degree of air agitation; and 5) the tank's width to length ratio.

Hoods for open tanks are usually lateral, consisting of two slots or an open face. Canopy type hoods are not recommended for use in local exhaust systems since the plater frequently bends over the tank. This would allow direct exposure of the worker to the toxic fumes being exhausted. Local ventilation systems should include the proper hood size, plenum size, duct sizes and airflow rate for the specific hazard. When designing a local exhaust ventilation system, it is further recommended that the system be designed so as not to mix different incompatible exhaust fumes. Mixing the exhaust from a cyanide plating bath along with a chromic acid exhaust could result in an extremely dangerous explosion or the production of lethal hydrogen cyanide gas.

General exhaust ventilation is the venting of air from a building with no specific connection to the operation. General ventilation alone can be used to control in-plant emission when the health hazard potential is relatively low. It is recommended that both general and local exhaust ventilation be employed in an electroplating operation.

Chemical inhibitors and plastic floating balls

CONTINGENCY PLAN

1. FACILITY DESCRIPTION

   Location

   NAME: _____

   ADDRESS: _____

   CITY _____ STATE _____ ZIP _____ PHONE _____

2. GENERAL

This plan describes the actions the facility personnel must take in response to fires, explosions, or any unplanned sudden or non-sudden release of hazardous waste or hazardous waste constituents to air, soil or surface water at the facility.

All personnel should be familiar with the procedures and equipment described in this contingency plan. In the event of a fire, explosion or any unplanned release of hazardous waste or hazardous constituents into the air, soil or surface water at the facility, facility personnel should immediately notify the following emergency coordinators:

3. INDIVIDUALS QUALIFIED AS EMERGENCY COORDINATORS

Name	Responsibility	Telephone & Address (Home and Office)
_____	Primary Emergency Coordinator	_____ _____ _____
_____	Alternate	_____ _____ _____
_____	Alternate	_____ _____ _____
_____	Alternate	_____ _____ _____

If none of these people are available, facility personnel should, where necessary, activate the alarm, evacuate the premises and notify the listed authorities.

INDUSTRIAL HYGIENE AND SAFETY 349

4. ARRANGEMENTS AGREED TO BY LOCAL AND STATE FIRE, POLICE, HOSPITAL AND RESPONSE OFFICIALS TO COORDINATE EMERGENCY SERVICES

_____
_____
_____
_____
_____
_____
_____
_____

5. AGENCIES TO BE NOTIFIED   Telephone Number

Police: _____

Fire/Emergency: _____

Hospital: _____

Emergency response teams: _____

Local regulatory agency: _____

State regulatory agency: _____

Federal regulatory agency: _____

National response center: _____

6. EMERGENCY EQUIPMENT

Physical Description	Location	Capabilities
Fire extinguishers	_____	_____
Water hoses	_____	_____
Telephones, P.A. systems, etc.	_____	_____

7. EVACUATION PLAN

Signal used to begin evacuation: _____

Evacuation route: _____

Alternate evacuation route: _____

8. RESPONSIBILITIES OF EMERGENCY COORDINATOR

A. Activate alarm system or other internal notification device to initiate evacuation.
B. Notify local authorities: police, fire department, etc. (as in "arrangements" section).
C. Identify the type, source, amount and degree of release of any hazardous materials due to the emergency. This includes materials which become a waste as a result of the incident, as by being spilled, discharge from a ruptured tank, etc. When necessary, you are required to do a chemical analysis.

D. Assess the potential hazards to human health or the environment. The assessment must include direct and indirect effects from fires, explosions and spills (e.g., toxic, irritating and asphyxiating gases, surface water run-off, etc.).
E. If the emergency could cause a threat to human health or the environment outside the facility:
   (1) Notify the local authorities that evacuation of local areas may be necessary.
   (2) Notify the government official in your area responsible for emergency coordination. If unknown, you must call the National Response Center at (800) 424-8802 and report your name and phone number, name and address of your facility, time and type of incident, name and quantity of materials involved, the extent of any injuries and possible hazards to health or environment (if known).
F. During any emergency, the coordinator must take all reasonable steps necessary to assure that any *other* fires, explosions or releases of hazardous substances do not occur or spread. These steps may include halting production, removing or isolating all hazardous waste containers and cleaning up any spills.
G. The emergency coordinator is obligated to monitor for all other possible problems (e.g., pressure buildup, leaks, etc.) caused by the emergency.
H. Immediately following any emergency, you must provide for complete cleanup measures as necessary, including any treatment, storage or disposal of recovered wastes, contaminated soil or waters. No incompatible wastes may be treated, stored or disposed of until cleanup is complete. All emergency equipment must be cleaned up.
I. Reporting Requirements
   (1) Following cleanup and before resuming production you must notify the EPA Regional Administrator and your local authorities that your facility is completely cleaned up in accordance with your "contingency plan" and ready to resume operations.
   (2) Within 15 days, you must report details of the incident in writing to your EPA Regional Administrator. The report must include your name, address and telephone number; name, address and telephone number of your facility; date, time and type of incident; name and quantity of hazardous materials involved; any injuries; any actual or potential hazards to health or the environment resulting from the incident; and the estimated quantity and disposition of any materials recovered after the incident.

9. COPIES OF CONTINGENCY PLAN

Copies of this contingency plan shall be maintained at this facility and shall be sent to local and state officials that may be called upon to provide emergency services.

can be utilized in a plating bath to slow the rate of contaminant escape. Chemical inhibitors are surface tension reducers or foam-producing compounds that form a partial barrier at the surface of the plating solution. Plastic floating balls are used to create a physical barrier by reducing the surface area of the solution. Although both chemical inhibitors and plastic floating balls are effective in reducing emissions from open plating tanks, they should not be considered a substitute for ventilation control.

The observations and recommendations of the authors are the result of many years of "hands on" experience in a metal finishing job shop. The safety principals and guidelines discussed in this chapter have been successfully implemented in many electroplating environments, but no claim is made that they constitute the answer in every case.

References

1. Clayton, George D. and Florence E. Clayton. *Patty's Industrial Hygiene and Toxicology*, Vols. I, 2C and III. New York: John Wiley & Sons, 1977.
2. *Accident Prevention Manual for Industrial Operations*, Seventh Edition, National Safety Council, Chicago, 1974.
3. "Good Work Practices for Electroplaters." DHEW (NIOSH) Publication No. 77-201, Cincinnati, U.S. Department of Health, Education and Welfare, 1977.
4. "Health and Safety Guide for Electroplating

Shops." DHEW (NIOSH) Publication No. 75-145, Cincinnati, U.S. Department of Health, Education and Welfare, 1975.
5. "Respiratory Protection...A Guide for the Employee." DHEW (NIOSH) Publication No. 78-193B, Cincinnati, U.S. Department of Health, Education, and Welfare, 1978.
6. Zopic, J. E. and W. A. Himmelmen. *Highly Hazardous Materials, Spills and Emergency Planning.* New York: Marcel Dekker, 1978.
7. *Hazardous Chemical Safety,* Volumes I and II. New Jersey: J. T. Baker Chemical Co., 1979.
8. *Fire Protection Guide on Hazardous Materials,* Seventh Edition, National Fire Protection Association, NFPA No. SPP-1D, Boston, 1978.
9. *National Fire Codes,* Vols. 1-16 NFPA, Boston, 1981.
10. "Code of Federal Regulations 29, Parts 1911 to 1919 and Part 1920 to End, Labor," Washington, D. C., Office of the Federal Register, National Archives and Records Service, General Services Administration, 1980.
11. "The Resource Conservation and Recovery Act," United States Environmental Protection Agency, Washington, D. C., 1981 (SW-171).
12. Sax. *Dangerous Properties of Industrial Materials,* Third Edition. New Irvsey, New York: Van Nostrand Reinhold, 1968.
13. Robinson, J. S. *Hazardous Chemical Spill Cleanup.* New Jersey: Noyes Data Corp., 1979.
14. Sittig, Marshall. *Handbook of Toxic and Hazardous Chemicals.* New Jersey: Noyes Data Corp., 1981.
15. Weiss, G. *Hazardous Chemicals Data Book* New Jersey: Noyes Data Corp., 1980.
16. "Registry of Toxic Effects of Chemical Substances," Vols. I and II, National Institute for Occupational Safety and Health, Cincinnati, 1980.
17. Hurley, Donald E., John W. Sheehy and Alfred A. Amendolo. *Study Plan for an Evaluation of Engineering Control Technology for Plating and Cleaning Operations.* National Institute for Occupational Safety and Health, Division of Physical Sciences and Engineering, Cincinnati, Ohio, 1980.
18. Meyer, W. R. "Hazards in the Electroplating Room," *Technical Proceedings of the American Electroplaters Society* **47**:55–58 (1960).
19. Young, M. A., "Health Hazards of Electroplating," *Journal of Occupational Medicine* **7**:348–352 (1965).
20. Joger, L. E. "Hazards in the Plating Industry," *Occupational Health Review* **18**:3–10 (1966).
21. Flanigan, L. S. "Development of Design for Exhaust Systems for Open Surface Tanks." DHEW (NIOSH) Publication No. 75-108, U.S. Department of Health, Education and Welfare, Cincinnati, 1974.
22. Schuman, M. "Designing Ventilation for Electroplating Plants," *Air Engineering* **10**:19-20 (July 1968).
23. Scheldon, A. W. "Plating Plant Preventive Precautions," *National Safety News,* **100**:54-58 (November 1965).
24. Craig A. and R. W. L. Jones, "Occupational Health in the Metal Finishing Industry," *Occupational Health* **23** No. 7:215-223 (July 1971).

12. RELEVANT MATERIALS SCIENCE FOR ELECTROPLATERS

Rolf Weil

*Professor of Materials and Metallurgical Engineering,
Stevens Institute of Technology, Hoboken, N.J.*

The products of electrodeposition, namely, coated or electroformed articles, are solid materials. The topics of materials science which deal with the nature and properties of solids are therefore relevant for the electroplater. *Materials science* deals with how properties depend not only on chemical composition, but also on structure, which is the arrangement of the various atomic species.

The main groups of engineering materials are metals, semiconductors, ceramics and polymers. Electrodeposits are almost always metals. The starting work pieces, the substrates which are plated, can belong to any of the four groups, but again they are mostly metals. The important properties of metals are ductility, which is the ability to undergo plastic deformation without breaking, good thermal and electrical conductivity and lustre. Some materials which do not possess all these properties are still considered to be metals. For example, manganese and chromium, which fracture without any appreciable prior deformation are still metals. Most practically used metals are *alloys*, which are mixtures of elements of which at least the one present in the greatest relative quantity is a metal. For example, brass is an alloy of two metals, namely copper and zinc. Electroless nickel is an alloy containing appreciable amounts of phosphorus or boron, which are not metals.

Semiconducting materials are of great importance in electronic applications. Silicon and to a lesser degree, germanium are the semiconductor materials used in electronic devices. They contain a very small amount of another element called a *dopant*. Dopants of lower valence provide the positive (p) charge and those of higher valence the negative (n) charge necessary for device functioning. Certain compounds such as lead sulfide, zinc oxide and gallium arsenide are also semiconductors. In these materials, slight deviations from stoichiometry can provide the p or n characteristics. When semiconducting materials are substrates for electroplating, special prior surface treatments are necessary.

Ceramics are nonmetallic, inorganic materials. They may be composed of oxides of nonmetals such as magnesium oxide or oxides of nonmetals such as silicates or, as is frequently the case, a combination of metallic and nonmetallic oxides. Ceramics find applications as refractory materials because of their load-bearing capabilities at elevated temperature. Glasses, which are also generally oxide mixtures are also ceramic materials.

Most polymeric materials are based on carbon-to-carbon bonds. In many polymers the carbon atoms of small molecular groups, which are called monomers or simply mers, are joined to form long chains. On the sides of the chains are either hydrogen atoms or molecular groups. On heating, these polymers become soft and are therefore called *thermoplastics*. If the side groups are linked together into a tight network arrangement, the materials are rigid and called *thermosetting plastics*. Ceramics and polymer substrates must also be subjected to special surface treatments prior to electroplating.

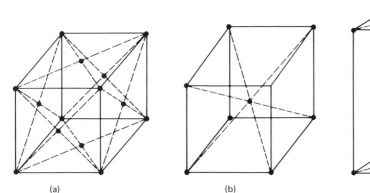
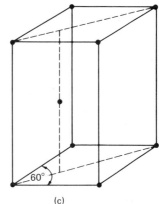

FIG. 1. Unit cells. a. Face-centered cubic. b. Body-centered cubic. c. Hexagonal close-packed. Positions of atom centers are shown.

Materials can also be classified as *crystalline* or *amorphous*. In crystalline materials the same three-dimensional arrangement of atoms or atom groups extends over comparatively large distances. Crystalline materials melt at a definite temperature or over a range of temperatures. In amorphous materials the atom arrangements change over comparatively very short distances. Amorphous materials soften and assume more liquid-like properties as the temperature is increased rather than having a definite melting point.

STRUCTURE OF MATERIALS

The three-dimensional arrangement of crystalline materials is called the *lattice*. The lattice can be thought of as being composed of building blocks called *unit cells*. The three most important lattices are face-centered cubic, body-centered cubic and hexagonal. The unit cells of these lattices are shown in Fig. 1. Of the commonly plated metals, copper, nickel, gold, silver and the platinum group have a face-centered-cubic lattice, the unit cell of which is depicted in Fig. 1a. Iron and chromium possess the body-centered-cubic lattice represented by the unit cell in Fig. 1b. Zinc and cadmium possess the hexagonal unit cell shown in Fig. 1c.

It is convenient to be able to specify certain planes and directions in crystal lattices. The three most important planes are {100}, {110} and {111} shown for cubic lattices by the hatched lines in Fig. 2. Also depicted in Fig. 2 are the three most important directions, namely, <100>, <110> and <111>. As can be seen here, the directions and planes designated by the same set of three digits are perpendicular to each other. The method of naming the planes and directions is described in any materials science textbook.

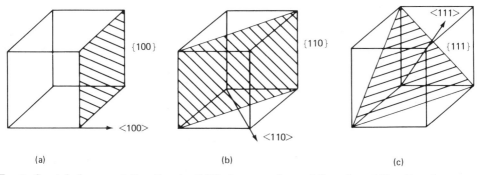

FIG. 2. Crystal planes and directions. a. {100} planes are faces of the cube. <100> directions are cube edges. b. The sides of {110} planes are two face diagonals and two cube edges. <110> directions are face diagonals. c. The sides of {111} planes are face diagonals. <111> directions are cube diagonals. One plane of each type is shown hatched. One direction of each type is shown by an arrow.

The volume of materials in which all the unit cells are aligned in a three-dimensional arrangement is a *crystal*. When a solid is composed of many crystals, they are also called *grains*. The term *crystallite* is frequently used in describing the structure of electrodeposits. In the literature the term is sometimes used to mean very small grains. However, it is defined most often and for purposes of subsequent discussion here to mean a small group of unit cells which may be slightly misoriented with respect to their neighbors. In electrodeposits they develop initially as discrete particles. On joining they can form one crystal.

In practical materials the atomic arrangement is not perfect; there are many defects. There are point defects, such as a missing atom called a *vacancy* or atoms called *interstitials* which occupy sites where they should not be. A foreign atom in the lattice of a pure element is actually also a point defect. In electrodeposits atomic groups or molecules are often included in the crystal lattice, as will be discussed later.

The most frequently encountered line defects are *dislocations*. Figure 3 is a schematic representation of a dislocation. While there are several causes for the formation of dislocations, their properties are independent of their origins. The dislocation in Fig. 3 can be considered to be caused by two shear forces, labelled F which resulted in a partial displacement of the top portion of the block with respect to the bottom. The dislocation is the boundary between the displaced and original portions which in Fig. 3 is a curved line from A to B marked D. The arrows show the direction of motion under the force F. At A a ledge has been created. When the dislocation line is parallel to the force F as it is at A, it is called a *screw dislocation*. At B the line is perpendicular to the force F and is called an *edge dislocation*. The edge dislocation is seen to be the end of an extra plane of atoms. Between A and B the dislocation line is inclined to the force F and therefore is a mixed dislocation. Most dislocations in electrodeposits are of the mixed type.

Grain boundaries are planar defects. A special type of boundary is that of a *twin*. Twins, which are frequently present in electrodeposits, are volumes of material bounded by two parallel planes as represented in Fig. 4. The crystal lattice in the twin is the mirror image of that in the matrix. In face-centered-cubic materials the boundaries or twinning planes of the {111} type are shown in Fig. 4. Some plated materials, such as electroless Ni-P and Ni-B, have such small grains, and even these contain defects, that they can be considered to be amorphous.

ATOMIC DISTRIBUTION IN MATERIALS

Most engineering materials are composed of several different atomic species. The atomic species are called *components* and their relative amounts give the *composition* of the material. For purpose of simplicity, only *binary systems*, composition ranges of two atomic species will be discussed here. If two components mix in the unit cells in the same average atomic ratio as in the material as a whole, there is only one solid phase. A *phase* is a homogeneous portion of matter. For example, in brass containing 25% zinc by weight three out of four atoms are copper and one out of four is zinc because the atomic weights of the two elements are nearly the same. Brass of this composition has a face-centered-cubic unit cell such as is shown in Fig. 1a. One such unit cell contains four atoms because one-eighth of the eight corner atoms is in the unit cell (eight unit cells share a corner) and half of the six face atoms are in the unit cell (a face is shared by two unit cells). If in the unit cell of brass, on the average one of the four atoms is zinc and the other three copper, the atomic ratio there is the same as in the material as a whole. Furthermore,

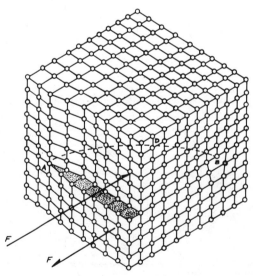

FIG. 3. Dislocation formation by shear. (a) The *dislocation line*, D, expands through the crystal until displacement is complete. (b) This defect forms a screw dislocation where the line is parallel to the shear direction. (c) The linear defect is an edge dislocation where the line is perpendicular to the shear direction.

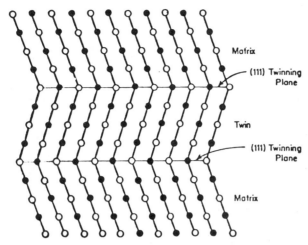

Fig. 4. Atomic arrangement at the twinning plane in a face-centered cubic metal. Black and white circles represent atoms on different levels (planes). (After Barrett, C. S., ASM Seminar, *Cold Working of Metals*, 1949, Cleveland, Ohio, p. 65.)

as the material must then be homogeneous, there is therefore only one phase. If materials of all the various possible compositions are the same solid phase, there is complete solid solubility. A component is soluble in another if their mixing does not result in new phases. A graphical representation showing the existence of various phases under equilibrium conditions as a function of temperature and composition is called a *phase diagram.*

The phase diagrams of most binary systems exhibit more than one solid phase because the solubility of the components in each other is limited. For example, there is limited solubility in the copper-silver system. Under equilibrium conditions the maximum solubility of copper in silver is about 9% by weight. By a simple calculation using the atomic weights, 9% copper by weight translates to about one-seventh of the atoms. Thus there would be four copper atoms out of a total of 28. In a face-centered-cubic material 28 atoms constitute seven unit cells. Thus a maximum of four copper atoms would be distributed among these seven unit cells. A similar limited solubility exists for silver in copper. There are therefore two solid phases each having the same crystal structure as the pure components. These two phases are the *terminal phases* as they are found at the two ends of the composition coordinate on the phase diagram. Most phase diagrams also exhibit phases which are different from the pure components. They are called *intermediate phases.* Some intermediate phases are true chemical compounds; most are not. An example of a compound is Cu_2O, which occurs at 67 atomic % copper in the copper-oxygen phase diagram. Different types of phase changes occur as a function of temperature. The gas phase is usually not shown on phase diagrams, but the phase change from liquid to solid is. Some components exist in different crystal structures. For example, iron changes from body-centered cubic to face-centered cubic and back to body-centered cubic with decreasing temperature. Of course, each different crystal lattice is a different phase. The phase changes are called *polymorphic transformations.*

Three types of phase changes can occur during solidification. In the composition range in which there is solid solubility, the liquid freezes to become the one solid phase. In pure components, freezing occurs at one temperature, the melting point, while in mixtures it occurs over a temperature range. In this temperature range, the solid and liquid have different compositions which change continuously with temperature. Some intermediate phases, which are said to be *congruently* melting, behave like pure components. In composition ranges in which two solid phases exist in equilibrium, solidification occurs in most binary systems by the eutectic reaction. With decreasing temperature, first one of the two phases solidifies over a range of temperatures. Which of the two phases solidifies first depends on the composition. There is also one composition where neither phase forms first. At one temperature, called the eutectic temperature, a mixture of the two phases forms from the remaining liquid. The material having the one composition where neither phase forms first, freezes to become only the mixture of the two phases. There is another reaction in which one phase again solidifies first over a temperature range. But then at one temperature all or part of this phase redissolves in the liquid to form the other phase. This reaction is termed *peritectic*.

Four types of reactions can occur in the solid state when the temperature is decreased. The polymorphic transformation from one solid state to another is analogous to solidification when there is only one solid phase. A second transformation occurs when the solid solubility decreases with decreasing temperature. Then a second solid phase forms. This reaction is called *precipitation*. The other two reactions are analogous to the eutectic and peritectic ones except that a solid takes the place of the liquid. These reactions are called *eutectoid* and *peritectoid*, respectively.

Reactions in real materials do not generally occur under equilibrium conditions; therefore, the temperature and composition stability ranges of the phases differ somewhat from those shown on phase diagrams. For example, solidification generally occurs at lower temperatures than the equilibrium ones. Phases sometimes form which are different from those shown on phase diagrams. They are called *metastable* phases.

MECHANICAL PROPERTIES

The engineering applications of materials often depend on their mechanical properties. The mechanical properties are the responses of a material to stresses; i.e., the loads per unit area. The pertinent properties are the modulus of elasticity, yield strength, tensile strength, elongation and reduction in area. These properties are usually determined by means of a tensile test. One end of a specimen is fixed. By applying increasing loads to the other end, the specimen is stretched. The length change between two preselected points on the specimen is recorded as a function of the applied load. The separation of the two points is called the *gage length*. From the recorded data, an engineering stress-strain curve is usually constructed, as shown in Fig. 5. The *engineering stresses* are the applied loads divided by the original cross-sectional areas. The *engineering strains* are the length changes divided by the original gage length. The *modulus of elasticity* is the slope of the initial portion of the engineering stress-strain curves where the material behaves elastically, i.e., there is no remaining length change if the applied load is removed. *Plastic* behavior occurs if there is a remaining length change after the applied load is released. The *yield strength* is defined as the engineering stress which results in a specified plastic strain, called the *offset*. The yield strength is found, as illustrated in Fig. 5, by drawing a straight line with the same slope as the modulus of elasticity from the offset laid off on the strain coordinate. The yield strength is the intercept of the straight line and the stress-strain curve. The *tensile strength* is the maximum engineering stress. On the graph illustrated in Fig. 5, the maximum stress occurs at a smaller strain than fracture because of a phenomenon called *necking*. Necking occurs where there is greater reduction in the cross-sectional area over only a part of the gage length. The elongation is the plastic engineering strain after fracture. It is determined by placing the two fractured parts of the specimen together and measuring the distance between the two preselected points. This length change, i.e., the measured distance minus the original gage length, divided by the original gage length is the elongation expressed as a percentage. As the elastic strain is usually very small compared to the plastic one, the elongation is essentially the same as the engineering strain at fracture. The reduction in area is the quantity, the original cross-sectional area minus the one at the fracture, divided by the original one expressed as a percentage. A true stress-strain diagram, which takes into account the changes in the cross-sectional area and in the gage length during stretching and is therefore a better indication of the mechanical behavior of the material, can also be constructed from the data obtained in a tensile test. The *true stress* is the applied load

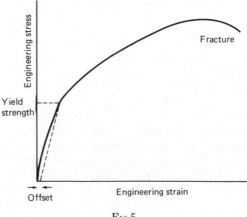

Fig 5.

divided by the instantaneous cross-sectional area. The *true strain* is the natural logarithm of the ratio of the instantaneous gage length divided by the original one.

Tensile testing of electrodeposits and electroforms has to be performed with great care. Scratches, porosity and serrated edges, which could result from cutting, can significantly reduce both the fracture stress and the ductility as indicated by the elongation and reduction of area. In a tensile test it is also important that the direction of load application be parallel to the sides of the specimen. Some of the problems involved in tensile testing of electrodeposits can be mitigated by bulge testing. A sample in the shape of a circular disk is clamped around its circumference. The pressure of a fluid is increased under the clamped disk to cause it to bulge upward. The height of the bulge is recorded. The stress and strains can be calculated from the fluid pressures, the bulge heights and the specimen dimensions as described by Prater and Read.*

The hardness of electrodeposits is frequently measured because it is easy to do so. However, hardness is not a real mechanical property. It may serve as an indication of strength or wear resistance. Hardness is determined from the depth of penetration of the material by an indentor. Great care has to be exercised in the interpretation of the results as factors other than the properties of the material affect the measured values.

Different structural changes occur when materials are subjected to stresses. The elastic behavior consists of changes in the interatomic spacings. The atoms move only fractions of the spacing so that they can readily return to their original positions when the stresses are released. In crystalline materials, plastic deformation most often occurs by slip. *Slip* is the relative displacement of planes of atoms analogous to the shearing of a deck of cards. The planes which slip are usually those on which the atoms are packed the closest. When slip is difficult or impossible such as in very thin electrodeposits or under very high loading rates or if the slip planes are unfavorably oriented with respect to the applied stresses, plastic deformation can occur by twinning. It is evident from Fig. 4 how twinning can lead to the distortions. During plastic deformation whole planes of atoms do not move at one time. Thus there is a boundary between the portion which moved and the one which has not yet done so. As already discussed and illustrated in Fig. 3, this boundary is a dislocation. Dislocations are already present in all practical crystalline materials. Their movement then results in plastic deformation. For example, if the dislocation in Fig. 3 moves to the left and the back edges of the block, the top portion will have been permanently displaced by one interatomic distance with respect to the bottom part.

The yield strength of crystalline materials which is essentially the stress required to start plastic deformation is proportional to the difficulty of moving dislocations. The more obstacles to dislocation motion exist, the higher is the yield strength. Grain and twin boundaries are obstacles to dislocation motion because they cannot be crossed. Therefore, the smaller the grain size, the greater is the yield strength. During plastic deformation which occurs during such forming operations as rolling, forging, swaging or wire drawing, new dislocations are created. The greater the number of dislocations, the less the individual ones can move. Therefore the yield strength increases after forming operations. This phenomenon is called *work hardening*. Foreign atoms in solid solutions or particles of a second phase also interfere with dislocation motion. Therefore alloys are generally stronger than pure metals.

Fracture occurs by the spreading of small cracks, pores or similar flaws in materials. The stress at such flaws exceeds the magnitude of the applied ones. The sharper the tip of a crack, the greater the stress increase. In such brittle materials as ceramics, thermosetting plastics and some metals and alloys, flaws are already present before loads are applied. In ductile materials, cracks form after plastic deformation as it becomes increasingly more difficult for dislocations to move. For these materials, therefore, the higher the strength (i.e., the more obstacles to dislocation motion), the less ductile they are. Ductility is equivalent to how much plastic deformation (dislocation motion) is possible before cracks form which lead to fracture. When the cracks are already present in brittle materials, the stress to propagate the crack to fracture may be less than that to move dislocations. Thus there may be essentially no plastic deformation before fracture which constitutes brittle behavior. For the same reason, if cracks or such other defects as deep scratches, pores, etc. are

---

*Prater, T. A. and H. J. Read, *Plating* **36**:1221 (1949).

present on materials which without them would be ductile, they behave in a more brittle fashion.

Most partially or wholly non-crystalline materials, which includes the polymeric substrates frequently used for electroless plating, do not exhibit distinctive elastic and plastic behavior. Rather their strength depends strongly on the rate of straining. The faster they are deformed, the stronger they tend to be. Most materials which behave in this way soften with increasing temperature. There is usually a large change in their behavior over a narrow temperature range called the *glass-transition temperature*. The familiar behavior of a glass rod when it is pulled apart while being heated is an example. When a certain temperature is reached, the rod begins to flow and can be drawn to a fine point. The behavior of elastomers such as rubber is due to long carbon-to-carbon side chains which link the main chains. When not under stress, the side chains are coiled. Under stress, they uncoil. As they revert to the uncoiled state after unloading, there is no permanent deformation.

The stresses to which materials are subjected need not be externally applied. *Internal stresses* can originate from non-uniform forming processes. For example, when a strip is rolled, the near-surface layer is lengthened more than the interior because of friction. The interior tries to prevent this extra lengthening of the near-surface layer, while the near-surface layer tries to stretch the interior to its length. In this way, the interior exerts a compressive stress on the surface layer while the surface layer exerts a tensile stress on the interior. Such internal stresses can also develop when only the near-surface region is deformed by polishing or machining. The internal stresses which develop during electroplating are discussed later.

HEAT TREATMENT

Work hardening, the increase in the yield strength and hardness with plastic deformation such as occurs during forming processes can be relieved by a heat treatment called *annealing*. Annealing is usually necessary not only because continued plastic deformation without it results in fracture, but also as the yield strength increases, more energy is needed and therefore there are higher costs. Annealing at relatively low temperatures results in the relief of internal stresses and a restoration of the electrical conductivity which is reduced during plastic deformation. The yield strength measured after the material is again at room temperature does not change significantly after the low-temperature annealing process, which is called *recovery*. To reduce the ambient-temperature yield strength and hardness it is necessary to anneal at a higher temperature. This temperature is roughly half the melting point on the Kelvin scale. Also the greater the prior plastic deformation and the purer the material, the lower is this temperature. A process called *recrystallization* occurs in which new grains form at the expense of the deformed ones. The grain size after recrystallization is determined by the amount of prior plastic deformation. The more deformation, the smaller are the grains. Plastic deformation and recrystallization are therefore used to control the grain size. If annealing is continued after recrystallization is complete, the grain size increases.

Precipitation or *age hardening* is a heat treatment to increase the yield strength and hardness of materials. Aluminum alloys, beryllium-copper and Ni-P and Ni-B alloys of the type obtained by electroless plating can be age hardened. To age harden an alloy, it is first necessary to produce a supersaturated solid solution. It is produced by heating to a temperature at which the minor constituents (the elements present in the smaller concentrations) are soluble and then rapidly cooling to room temperature. The Ni-P and Ni-B alloys are plated as supersaturated solid solutions. Therefore, they need not be heated to dissolve the minor constituents or to be rapidly cooled. The age-hardening treatment usually involves heating to a temperature which is well below that at which the secondary components are soluble. Atoms diffuse to attain the composition changes to form the new phase, which should precipitate. For example, an alloy of about 2 atomic % copper, balance aluminum can be precipitation hardened. Above a temperature of about 530 C, this amount of copper is soluble in aluminum. Thus there would be on the average two copper atoms in twenty-five face-centered-cubic unit cells. According to the phase diagram, the alloy should consist of two phases, the terminal solid solution and an intermediate phase having the formula $CuAl_2$. To form this intermediate phase, the copper atoms must become much more concentrated in certain locations. It is necessary that there be one copper atom for every two of aluminum. Such a local increase in the concentration of the minor constituents before the new phase can actually form, results in substantial alterations of the interatomic spacing which causes severe distortions of the crystal planes of the alloys. The distortion impedes dislocation motion and thereby hardens the materials. If the alloys are heated so that crystals of the second phase actually form, the distortion and consequently the strength and hardness decrease and the material is said to be *overaged*. However, small well-distributed particles of a second phase can also have considerable hardening effects.

The heat treatment to harden steels, which are alloys of iron and carbon, also involves the creation of a supersaturated solid solution. Carbon is considerably more soluble in the face-centered-cubic lattice of iron which exists at elevated temperatures than in the body-centered-cubic one. If face-centered-cubic iron containing considerable amounts, i.e., 0.3 to 1.5% carbon in solid solution is rapidly cooled or quenched, a material called *martensite* can be formed. The carbon which is trapped in solid solution causes distortions and the hardening. As-quenched martensite is very brittle. Heating at relatively low temperatures, which is called *tempering*, results in increased ductility with a small reduction in hardness.

STRUCTURE AND PROPERTIES OF SURFACES

In plating operations, the surface properties of the substrate as well as those of the deposit are of practical importance. Surfaces differ from the interior of a material because atoms have fewer neighbors. To create a surface, it is necessary to break the bonds between planes of atoms which had been neighbors. Atoms with fewer neighbors are more unstable and therefore have a higher energy. This extra energy, the *surface energy*, is defined as the energy per unit area which is added when a surface is created. Usually other atoms or groups of atoms take the place of the missing neighbors, the atom layer which would have been removed when a new surface is created. On most metal surfaces there are oxygen ions which have chemically reacted to form oxides. Gas molecules or atom groups such as the organic molecules of greases can also take the place of the missing neighbors. Their bond to the material surface is generally weaker than a chemical one. Such attachment, which is called *adsorption*, is dependent on weak attraction of opposite charges. Adsorption occurs preferentially at sites of such surface irregularities as protrusions, ledges and grain boundaries.

The *adhesion* of one material to another such as an electrodeposit to its substrate depends on the condition of the surface. If the substrate surface is free of oxides and of adsorbed substances other than the water molecules of the plating solution, the atoms of the deposited metal can assume the positions of the missing neighbors. Strong metallic bonding can then result. Bonding to oxides is generally not strong. Therefore, it is difficult to obtain good adhesion between electrodeposits and such substrates as stainless steel, nickel and aluminum because of the affinity of these materials for oxygen. Bonding of metals to ceramics and polymers is also generally weak. In such cases, adhesion is attained principally by mechanical means. Voids are created on the substrate surface. Metal can be deposited usually by electroless plating into the voids and held there by a keying action. The rest of the deposit adheres to the metal in the voids of the substrate.

The treatments given to substrate surfaces can affect the adhesion of a deposit. Buffing, polishing and lapping can result in severe plastic deformation of surface layers. Such plastic deformation can cause the surface layer to be brittle and introduce internal stresses. Both of these conditions can lead to cracking at or near the deposit-substrate interface or peeling especially when the plated parts are subsequently subjected to loads. If plated parts are subjected to elevated temperatures, intermediate phases which are often brittle, can form and result in a loss of adhesion of the deposit. Such a condition can exist, e.g., when zinc is copper plated or steel coated with zinc. If particles of a second phase which is not electrically conductive are present on substrate surfaces or if such particles become embedded from surface treatments, metal cannot be plated on them and porosity results in the deposit. As the deposits become thicker, they also tend to grow sideways to cover nonconducting areas. While the pores can close in this way, the deposit still does not adhere to the substrate there.

Wear resistance is an important surface property in many applications of electrodeposits. The two most important types in such applications are adhesive and abrasive wear. Wear results when surfaces or particles rub against each other so that there is a loss of material. Practical surfaces are composed of protruding and recessed areas. The protrusions are the areas which are generally in contact. Even under light applied loads, the stresses on protrusions are large because of the small contact area. The rubbing also results in heating of the contact area and the removal of oxides and other foreign materials there. These three factors, the high stress, the temperature rise and the removal of foreign materials can result in the bonding of the opposing contacted areas. The rubbing motion results in the breaking of the bonds again. If the fracture does not occur at the originally bonded interface, one of the parts suffers a weight loss. It is also possible for particles to break off when bonds are sheared so that both parts which rub against each other suffer weight loss. A weight loss due to the fracture of bonded areas away from their original interface is *adhesive wear*. When harder protrusions or particles plow furrows into softer surfaces and remove material, *abrasive wear* is taking place. Grinding and polishing operations are therefore a form of abrasive wear. In such surface treatments, furrows are always created. In general, the depth of the furrows is half the size of the particles used in the last operation. Adhesive wear can be minimized by the use of lubricants which prevent bonding. In general, harder materials are less susceptible to abrasive wear.

The processes which occur during machining operations are very complicated. The cutting processes usually result in severe plastic deformation and work hardening not only of the chips which are removed, but also of the remaining surface. There is also generation of heat. Adhesive wear of tools is possible and can be avoided to some degree by lubricants. Abrasive wear of the materials is, of course, one of the main processes occurring during machining. However, abrasive tool wear also occurs especially if hard materials are being machined. Substances are sometimes present in materials to improve their machinability. These substances may provide sites for fracture and thereby facilitate chip formation. Some substances (e.g., lead in brass) can also act as lubricants. In such cases and also generally when there are several phases of different hardnesses present at a surface, the softer one tends to smear over the harder one. There is poor adhesion between the smeared layer and the actual surface of the part. This layer has to be removed therefore before plating.

When metal surfaces are in contact with aqueous media, electrochemical reactions are likely to occur. The surface metal atoms have such ligands as water molecules or chloride ions as neighbors. It is thus often relatively easy for a metal atom to leave the surface and become an ion surrounded by ligands. This process is *corrosion*. When a metal atom becomes an ion, it leaves electrons behind. The sites where the atoms left thus become the *anode*. The electrons (i.e., the current) flow to another site where a reduction reaction, one which uses electrons, occurs. Such sites are the *cathode*. In acidic media, the cathodic reaction generally is the reduction of hydrogen ions to atoms. In neutral or basic media, the cathodic reaction is usually the conversion of dissolved oxygen to hydrated oxide ions.

The fewer neighbors metal atoms have on a surface, the easier it is for them to become ions. At such surface sites as edges of atom rows, grain boundaries, cracks and dislocations, atoms have fewer neighbors. Consequently, these sites preferentially corrode. Plastically deformed metals, because of their greater dislocation content, also corrode more readily. On surfaces of non-homogeneous composition such as those of multiphase alloys or made up of different metals, it is relatively easy to form the anodic and cathodic sites, which are necessary for corrosion. Therefore, such materials are more susceptible to corrosion. When the concentration of dissolved oxygen is lower because of lack of convection in such places as under a washer, in a crevice or under non-adherent soils, corrosion is enhanced. The sites where more oxygen is available become the cathode. The surface areas in contact with the aqueous medium with little dissolved oxygen then become the anode and corrode. Insoluble, adherent corrosion products can limit the opportunity for metal atoms to become ions and reduce corrosion. A significant reduction of the corrosion rates, such as produced by the thin adherent oxide on stainless steel, is called *passivity*. If the anodic areas are very small, corrosion is concentrated there and pitting occurs.

Corrosion can lead to fracture of components under relatively low stresses. It can result in the formation of pits, pores or crevices where fracture can start. Such defects will generally have sharp tips so that the stress there is increased over that which is applied. Corrosive media can lower the surface energy which facilitates the creation of the new surfaces which constitutes the fracture process. Externally applied or internal stresses can also enhance corrosion. They can cause brittle passive films to fracture and they increase the number of dislocations. These phenomena of stress enhancing corrosion and corrosion increasing the stress and facilitating fracture are called *stress corrosion*. Examples of stress corrosion are brasses in ammonia or certain amines, stainless steels in media containing chloride ions and aluminum alloys in liquid metal such as mercury or gallium.

Cathodic reactions involving the reduction of hydrogen ions to atoms can lead to *hydrogen embrittlement*, which is also a phenomenon in which components fracture under relatively low stresses. Embrittlement can occur when atomic hydrogen is produced during pickling, cathodic cleaning or electroplating at less than 100% efficiency. Certain corrosion-preventive measures can also lead to hydrogen embrittlement. A piece of a less noble metal such as zinc, which corrodes preferentially, can protect a steel component when it is in electrical contact with it. However, in acidic media hydrogen atoms are produced on the protected component which in this case is the cathode. The atomic hydrogen produced by cathodic reactions usually combines to form bubbles of molecular hydrogen which leave the surface. But the tiny hydrogen atom can also move (i.e., diffuse readily into the material between the larger atoms of a metal lattice). Once in the metal, hydrogen atoms then diffuse into voids, cracks or pores and there combine into the molecular form. The gas pressure in the voids thus increases causing them to open and propagate to fracture under relatively low applied stresses. If the tips of cracks can be blunted by plastic deformation in ductile materials, hydrogen embrittlement is not a likely problem. However, such materials as martensitic steels, which do not plastically deform readily, are susceptible to hydrogen embrittlement. The presence of hydrogen embrittlement is usually determined by a bend test. A strip is bent and the radius of curvature when fracture occurs is noted.

The removal of hydrogen by heating components eliminates embrittlement. However, as hydrogen embrittlement is often a problem in heat-treated parts, care must be taken that heating to remove hydrogen does not adversely affect the mechanical properties. Minimizing the time of pickling in acids and using inhibitors is advisable for materials susceptible to hydrogen embrittlement. Other means of reducing the problem are restrictions on cathodic cleaning and the use of plating solutions designed to reduce hydrogen codeposition.

The light-reflecting property of a surface is generally called *brightness*. Brightness depends on the degree to which its protrusions and depressions greater than the wavelength of visible light deviate from a given plane. Polishing and buffing is the removal of such protrusions and sometimes the filling of depressions. Electropolishing is also based on the preferential dissolution of protrusions. Protrusions are closer to the counter-electrode so the current density is higher and they dissolve preferentially. The brightness of electrodeposits will be discussed later.

STRUCTURE OF ELECTRODEPOSITS

There are some aspects of structure which are specific to electroplated metals. When a metal is deposited it will try, if possible, to copy the structure of the substrate. Such copying generally involves *epitaxy*, which occurs when definite crystal planes and directions are parallel in the deposit and substrate, respectively. If the deposit and substrate have the same crystal structure and the difference in the lattice parameters (i.e., the dimensions of the unit cells) is not great, the same plane and direction are parallel. An example is nickel plated on a copper substrate. Nickel and copper have a face-centered cubic crystal structure and their lattice parameters are 0.352 and 0.361 nm, respectively. Thus, e.g, when a {100} plane is parallel to the surface of a copper-substrate grain, the same crystal plane of the nickel deposit is parallel to the surface. Also corresponding crystallographic directions in the deposit and substrate are parallel. Even if the deposit and substrate do not have the same crystal structure or if their interatomic spacings differ considerably, there can still be epitaxy. Epitaxy is possible then if the atomic arrangement in a certain crystal direction of the deposit matches that in the substrate. The direction in the substrate is then probably different from that in the deposit.

The first atom layers of an epitaxial deposit are usually stretched or compressed to the same interatomic spacing as that of the much more massive substrate, which is usually a different metal. In the example previously cited, the nickel deposit, having the smaller lattice parameter, would be stretched to fit on the atoms of the copper substrate, which have the larger spacing. A deposit with a larger interatomic spacing as, e.g., gold plated on a copper substrate, would be compressed. A deposit stressed to fit on the substrate is said to be *pseudomorphic*. The stresses in a pseudomorphic deposit can be quite large in magnitude. As the deposit becomes thicker, it can assume its own equilibrium interatomic spacing and relieve the stresses by the introduction of edge dislocations. As was illustrated at B in Fig. 3, an edge dislocation is the end of an extra row of atoms. These extra atom rows in either the deposit or substrate, whichever has the smaller interatomic spacing, can compensate for the dimensional changes.

The mechanism of epitaxial deposition involves atoms of the electroplated metal attaching themselves at suitable sites on the substrate surface. Such suitable sites, called *kink sites*, are the ends of as-yet incomplete rows of atoms. The atoms of the metal being deposited can be discharged (i.e., converted from ions in the plating solution) on the substrate surface and then diffuse to kink sites. An alternative is for the ions to diffuse in the solution until they are situated opposite a kink site and there undergo discharge. When an atom row is completed at a grain boundary, a new one can start when a few depositing atoms meet at the edge of the completed row while they are diffusing on the surface seeking kink sites. If screw dislocations intersect the surface, atoms attaching themselves at kink sites can form spirals which do not require the formation of new rows. The next layer of atoms can also start from such a spiral or it can begin by the formation of a new row of atoms as just described. Epitaxial growth can thus be visualized as the lateral spreading of one-atom-high layers (i.e., monolayers) over the surface.

In practical plating solutions there are foreign species present which can limit the lateral spreading of monolayers. These foreign species include addition agents, products of reactions occurring in the plating solution, impurities from chemicals and water and in some cases even the negative ions. These foreign species can be adsorbed at kink sites and thereby prevent the attachment of the atoms of the depositing metal. In this way the lateral spreading of the monolayer is either stopped or slowed while the deposit grows around the blocked site. The monolayer on top of a blocked or slowed one cannot overtake it. When monolayers can only spread laterally to the edge of a blocked one, a step several atoms high is formed. The process is called *bunching*. The greater the number of growth-impeding foreign substances, the smaller will be the lateral dimensions of the three-dimensional structures which form by bunching. Epitaxial deposition therefore occurs in practical plating solutions by the formation of relatively small three-dimensional crystallites. The structures are crystallites as previously defined because all of them which formed on one substrate grain have the same crystal plane essentially parallel to the surface. The crystallites join or coalesce to form a multi-atom-high layer of a grain of the deposit. The coalescence of crystallites which are slightly misoriented with respect to their neighbors is equivalent to the conditions shown in Fig. 3. In Fig. 3, the deformed portion was slightly misoriented with respect to the undeformed one. Thus crystallite coalescence can lead to the formation of dislocations in electrodeposits. The codeposition of foreign species can also cause dislocation formation. Consequently, electroplated metals generally contain large numbers of dislocations.

In plating solutions with relatively low concentrations of growth-impeding foreign substances, such as, e.g., acid-copper baths without addition agents, epitaxial deposition can continue throughout the thickness. The grains then have a columnar shape which is seen when the cross-section of the deposit is examined microscopically. When there are more foreign species present, the epitaxial growth is limited to relatively thin deposits. Such is the case, e.g., in nickel plating solutions in which the basic substances, which form at the cathode due to the pH rise when hydrogen is reduced, have a growth-impeding role. All layers which constitute a grain of a deposit are epitaxial, because each layer which is the substrate for the next one must have the same crystal plane parallel to the instantaneous surface.

The change in orientation when a deposit ceases to be epitaxial with the substrate often occurs by the formation of twins like the one illustrated in Fig. 4. Foreign species can cause atoms to become misplaced in a way to develop the mirror images of twinning. If twinning is repeated several times, different crystal planes can become parallel to the surface so that grains of new orientations form. The thickness of the deposit which is epitaxial with the original work piece depends on the composition of the plating solution and also on the grain orientation. The thickness of the epitaxial deposit is greater when a certain crystal plane is parallel to the surface. For example, nickel deposits from an additive-free Watts bath grow epitaxially on their substrate to greater thicknesses when a {100} plane is parallel to the surface. For copper plated in acid sulfate or gold in sulfite solution, such is the case when a {110} is parallel to the surface.

If the substrate surface is covered with foreign material so that epitaxial deposition is not possible, new grains have to be formed, i.e., nucleated. New grains also have to be nucleated if the number of kink sites available for deposited atoms is severely limited. Such a condition can occur if foreign species block the sites or if the atoms arrive in very large numbers, as is the case at high current densities. New grains are probably nucleated by atoms meeting to form three-dimensional arrays consisting of a few unit cells.

When new grains are nucleated, under some conditions a relatively large fraction of them may have a certain crystal plane parallel to the surface. Under other conditions, the orientation may be initially random; i.e., there are about the same number of grains having all the possible crystal planes parallel to the surface. However, grains with a certain crystal plane parallel to the surface may grow faster in the direction perpendicular to the substrate. Such faster-growing grains can also spread laterally until they meet other faster-growing grains of the same orientation. In this way, the grains with other orientations become covered and cease to grow. Under both conditions either when a greater fraction is nucleated with a certain orientation or if there are faster-growing grains, there is a certain crystal plane preferentially parallel to the deposit surface. This preferred orientation is called a *fiber texture*. The crystal direction which is then preferentially perpendicular to the deposit surface is the *fiber axis*. The fiber texture depends on such factors as the composition of the plating solution, the deposition potential and the type of agitation. Certain addition agents in the plating solution can alter the fiber axis or cause the deposit to become randomly oriented probably because of the way they are adsorbed on certain crystal planes. When there is a fiber texture the grains, even if very small, tend to be columnar in shape.

When the deposit is not epitaxial with the substrate, the grain size is affected by the composition of the plating solution and the plating condition. Growth-inhibiting substances in the plating solution tend to result in finer-grained deposits. The higher the current density, in general the smaller the grain size. However, very low-current densities can also result in smaller grains because the relatively long time interval before metal atoms are incorporated at kink sites favors their becoming blocked by foreign substances. As already pointed out, when kink sites are blocked new grains usually have to be nucleated. Elevated solution temperatures favor continued incorporation of metal atoms at kink sites and therefore a larger grain size. When an increase in the pH at the cathode results in the formation of growth-inhibiting basic substances, the grain size tends to decrease. Pulse plating also tends to result in grain nucleation and therefore a smaller size because while the current is off, foreign materials can diffuse to kink sites and block them.

The measurement of the grain size of electrodeposits must be done very carefully. Very often there are crevices around clusters of small grains. Such crevices can be easily mistaken for grain boundaries. It is necessary to use X-ray or electron diffraction in addition to microscopic methods to determine the grain size of electrodeposits. A particle size, which in the case of very small-grained deposits is essentially the same as the grain size, can be determined from the broadening of X-ray diffraction lines as described in any textbook dealing with crystallography. The true grain size can also be ascertained by obtaining an electron-diffraction pattern from a given area. If the pattern is that of a single crystal, the area is one grain. Such patterns can be obtained by selected-area diffraction as performed in a transmission-electron microscope. The fiber axis is also sometimes determined incorrectly. The intensity of X-rays reflected from a particular crystal plane is, among other factors, proportional to the number of grains with that plane parallel to the surface. It is often assumed that the crystal plane with the highest intensity of reflected X-rays as compared to a randomly oriented sample to account for the other factors constitutes the fiber texture. It is sometimes forgotten that the reflected X-ray waves from certain crystal planes cancel each other so that their intensity is zero. Such planes may be preferentially parallel to the deposit surface and therefore constitute the true fiber texture. But it would not be evident from the relative intensities of reflected X-rays. If a beam of X-rays is directed perpendicular to the fiber axis, which for electrodeposits would be parallel to the surface, and the reflected rays are recorded on film, there is a method also described in crystallography textbooks which permits a determination of the real texture. This method should be used first to ascertain which crystal phase is preferentially parallel to the deposit surface. Then, a more quantitative determination can be made from the intensity of the

X-rays reflected by this plane or planes parallel to it.

Alloy deposits and those plated in solutions containing relatively high concentrations of addition agents often exhibit a structure consisting of alternate lighter and darker bands when their cross-sections are examined microscopically. This *banded structure* is only visible after the proper etching, i.e., intentional attack by corrosive solutions which are usually acids. The bands are therefore a result of different severities of corrosive attack by the etchants. In alloy deposits there is often a variation in composition with thickness which corresponds to the bands and is the cause of the different etching attack. A variation in the grain size in which smaller grains are preferentially etched can also be the cause of the banded structure. The variation of the grain size has been attributed to periodic blocking of growth by addition agents which requires nucleation of a whole new layer of grains.

PROPERTIES OF ELECTRODEPOSITS

The mechanical properties of electrodeposits depend to a considerable extent, though indirectly, on the types and amounts of growth-impeding substances at the cathode surfaces at which deposition occurs. Deposits plated in solutions containing growth-inhibiting species so as to result in very small grain sizes tend to have high yield strengths and to be very hard. The grain boundaries in such deposits are the main obstacles to dislocation motion. Even relatively large-grained deposits contain many dislocations and are also frequently heavily twinned. Such deposits are therefore also generally stronger and harder than the same metals produced by means other than electroplating. The same factors which raise the strength of electrodeposits generally result in lower ductility.

The electrodeposition processes frequently result in the development of internal stresses. The stresses due to pseudomorphism have already been discussed. The coalescence of three-dimensional, epitaxial crystallites can be a cause of internal stresses. If the crystallites are pulled together by surface tension before the spaces between them are completely filled in, but are restrained from bending by the substrate, stresses develop. Stresses can also develop in a similar way when small grains coalesce. Certain dislocation configurations such as, e.g., a series of edge dislocations so that all the extra atom planes face in the same direction, have been postulated as a cause of internal stress. Hydrogen is also a probable cause of internal stresses in electrodeposits. If hydrogen atoms are included in the crystal lattice and subsequently diffuse out again or towards the substrate, the new surface layer shrinks. The material below either expands or stays the same and by trying to prevent the surface layer from shrinking causes it to be in tension. If hydrogen atoms diffuse laterally into voids and there combine to form gas molecules, there is an expansion so that a compressive stress can develop. There may also be other causes of internal stress.

Tensile stresses are the more detrimental. They can cause a deposit to crack. They can also result in loss of adhesion to the substrate and reduce the fracture strength and the ductility. They are especially detrimental if the plated part is subjected to external forces which vary cyclically with time as, e.g., a rotating shaft. The presence of certain addition agents in the plating solution has been found to reduce tensile stresses or to result in the development of compressive ones, which are desirable in some cases. Annealing also relieves internal stresses. However, some deposits, e.g., nickel containing sulfur on annealing form a brittle sulfide usually in the grain boundaries. Such a sulfide renders the deposit very brittle and is therefore not an effective means of improving the properties.

Brightness is the most important property of decorative deposits. The brightness of thin deposits depends on that of the substrate. Thicker bright deposits are produced by addition agents in the plating solution which result in the elimination of protrusions or crevices which deviate from the surface plane by about the wavelength of visible light. Deposits plated in solutions containing relatively low concentrations of growth-inhibiting, foreign substances and which therefore tend to consist of large, columnar grains are usually dull. The tops of large, columnar grains are generally dome shaped and their boundaries are deep crevices. Thus there are essentially no light-reflecting flat areas on the surface of the deposits. Certain addition agents cause the tops of columnar grains to become flatter; other additives result in smaller grains so that the height differences between their tops and boundaries is reduced. In both instances, the deposits are semibright. For thick deposits to be specularly bright, the grain diameter has to be less than the wavelength of light. However, a small grain size is not a sufficient condition for brightness. The growth-inhibiting foreign substances, which result in the small grains, must be uniformly distributed over the cathode surface. If there is a locally greater growth impediment, a crevice results and the deposit has a hazy appearance.

Levelling is the preferential deposition into crevices on the surface of the work piece. When the deposit is thicker in such crevices as, e.g., the tool marks from machining, the deposit becomes smoother than the substrate. Levelling is again achieved by means of certain growth-inhibiting, foreign substances in the plating solution. These substances are adsorbed preferentially on the protrusions hindering deposition there and thus result in heavier plating into the crevices. There are two probable reasons for the preferential adsorption of the growth-inhibiting substances on the protrusions. The current density is greater on the protrusions because of the higher electric field. Some growth-inhibiting substances adsorb preferentially at sites of higher current density. Secondly, the growth-inhibiting substances become incorporated in the deposit and have to be replenished at the cathode by diffusion from the bulk of the plating solution. As the distance from the bulk of the solution to protrusions on the work piece is somewhat shorter, the substances are replenished there more rapidly and their concentration is consequently greater. Therefore, there is more inhibition to deposition at the protrusions.

In certain applications of plated components the electrical contact resistance is important. As the electrical conductivity of metals is much greater than of other materials, the most important factor in contact resistance is the absence of non-metallic substances where the parts touch. The presence of oxides is obviously detrimental, which is the reason plated noble metals are used. In some cases making contact can result in the wearing away of undesirable surface materials. The electrical resistivity of electroplated metals is generally somewhat greater than that of the same metal produced by other means. The reason is that such point defects as foreign species in solid solutions, vacancies and interstitials, which are the main cause of a higher electrical resistivity in metals, are more prevalent in electrodeposits.

Some electrodeposits are used for computer memory elements and similar applications in which the magnetic properties are important. There are crystallographic directions in which ferromagnetic materials are easiest or hardest to magnetize, respectively. As was already discussed, it is possible to control to some degree the fiber axes of electrodeposits. In electrodeposits, it is therefore possible to control the directions of easy or hard magnetization. The imposition of an external magnetic field during plating of magnetic metals can also change the fiber axis. The generally small grain size of plated metals is advantageous for providing greater specific information storage.

13. SURFACE PROTECTION AND FINISHING TREATMENTS

A. Phosphate Coating Processes

ALFRED DOUTY

Senior Technical Advisor, Amchem Products, Inc., Ambler, Pa.

AND

ELLSWORTH A. STOCKBOWER

Industry Manager, Metal Working Chemical Division, Amchem Products, Inc., Ambler, Pa.

Revised by

WILLIAM C. JONES

Research Director, Heatbath Corp., Springfield, Mass.

Phosphate coatings are specified by the metal finishing engineer for a variety of end uses and may be formed on the surfaces of iron, steel, galvanized steel, aluminum and electrodeposited zinc and cadmium. Phosphate coatings are used to promote the adhesion of organic coatings to metal substrates and to retard the rate of interfacial corrosion. Phosphate coatings are used to retain and enhance the performance of corrosion resistant oils and waxes on metallic surfaces. Phosphate coatings, with supplementary lubrication, are used to assist in cold deformation processes such as wire drawing, cold heading and impact forming.

Phosphate coatings are able to provide these valuable functions because of their particular composition and structure. They are formed by the controlled corrosion of the metal surface being treated. The resulting coating is tightly bonded to the surface and is mineral in nature. The coating has a definite chemical composition and crystal structure. When fully developed, the coating will cover the surface completely and can be shown to exhibit electrical insulating properties.

The selection of the best coating for a particular application can be facilitated by an understanding of the characteristics of various coatings now being applied in commercial practice. Of particular importance is the physical structure. A series of scanning electron micrographs ilustrates the difference in structure. A typical manganese phosphate coating on steel (Fig. 1) is characterized by a dense block-like deposit of angular crystals densely packed in a random orientation. This heavy coating provides a cor-

FIG. 1. Manganese phosphate on steel.

rosion resistant or lubricating surface when the proper oil is applied in a supplementary treatment. The composite system of manganese phosphate and oil is used to reduce wear and prevent galling of moving parts such as piston rings, shafts, gears, cylinders, and pistons. In some cases, the etch pattern formed in the metal surface by the formation of the manganese phosphate coating is also of importance since it promotes oil retention on the sliding surface after break-in.

Heavy zinc phosphate coatings on steel (Fig. 2) exhibit a more platelike structure than manganese phosphate. Two varieties are normally encountered. When maximum corrosion protection is desired, an iron content is maintained in the phosphating solution and a crystal structure as shown in Fig. 2 is produced. The orientation is principally parallel to the metal surface and the layered structure offers an excellent surface for the retention of corrosion resistant oils and waxes.

When a heavy zinc phosphate is used to facilitate cold deformation such as encountered in the drawing of wire and tubing, cold shaping, cold extrusion, etc., the phosphating solution is operated under iron-free conditions resulting in a crystal structure showing a high percentage of crystals oriented vertically to the metal surface. This is illustrated in Fig. 3. These thin plates may be lubricated by one of several methods, including lime, borax, specialized soaps and oils. In some cases, the surfaces of the individual crystals may be converted to a zinc soap. These lubricated plates then provide a multitude of slipping surfaces that promote the ease of cold deformation.

When a phosphate coating is used to promote paint bonding on steel or zinc coated steel, a number of options are available to the finishing engineer. A zinc phosphate coating may be desired for superior performance in retarding the lateral creep of corrosion between the organic coating and the metal originating at a scratch through the coating to the metal. Typical crystal structures for a zinc phosphate formed from the same solution on steel and galvanized steel are shown in Figs. 4 and 5. The coating

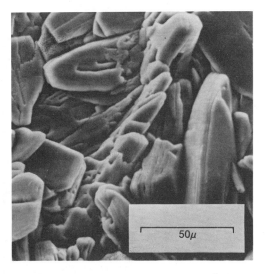

FIG. 2. Heavy zinc phosphate on steel.

FIG. 3. Heavy zinc phosphate on steel (iron free).

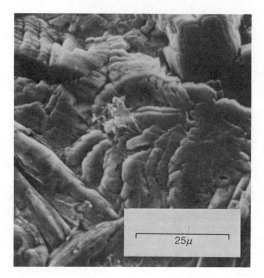

FIG. 4. Zinc phosphate for paint bonding on steel.

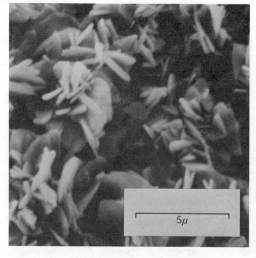

FIG. 6. Iron phosphate for paint bonding on steel.

on steel is tightly bonded to the surface and the effective surface area of the phosphate coating is considerably greater than the underlying metal. The phenomenon assists in increasing the adhesion of the organic coating which is further promoted by increasing the number of sites for mechanical locking of the coating to the phosphate crystals.

A zinc phosphate coating on a galvanized surface (Fig. 5) shows a different structure. In this case, the individual crystals are well-defined and are distributed in a random orientation on the surface. Note that these photomicrographs are at a higher magnification than

that used for the corrosion resistant or drawing phosphates. A fine grained phosphate is desirable for a paint base in order to provide acceptable gloss with minimum organic coating thickness.

A less expensive but still effective phosphate treatment to promote paint bonding utilizes a so-called iron phosphate treatment on steel or zinc rich surfaces. The structure formed on a steel surface is illustrated in Fig. 6. It consists of an extremely fine grained deposit of mixed oxides and phosphates of iron. The crystal structure is so fine that the coating is often classified as amorphous. When the same solution used to

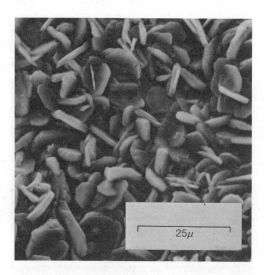

FIG. 5. Zinc phosphate for paint bonding on galvanized steel.

FIG. 7. Iron phosphate for paint bonding on galvanized steel.

FIG. 8. Calcium-modified zinc phosphate on steel.

produce this coating on steel is used to treat a galvanized surface, the deposit shown in Fig. 7 is formed. This coating differs in structure and composition from that produced on steel and has the characteristics of a fine grained zinc phosphate.

Another popular prepaint treatment is calcium modified zinc phosphate. The coating as shown in Fig. 8 consists of densely packed granular crystals showing a high degree of uniformity that create a surface with greatly increased surface area and an intricate void pattern well-suited to provide a high degree of mechanical lock of the organic coating to the treated surface.

AMORPHOUS PHOSPHATE COATINGS ON ALUMINUM SURFACES

Coatings consisting mainly of aluminum and chromium (trivalent) phosphates with traces of other elements are formed on aluminum alloy surfaces by treating them with acid aqueous solutions containing phosphorus (pentavalent), chromium (hexavalent) and fluorine. Other elements may be present in the solutions including boron, silicon, zirconium, titanium, etc. These coatings are characterized by a degree of crystallinity too low to be readily detected by electron diffraction or X-ray techniques. They are accordingly designated as amorphous.

The chemistry of formation of these coatings is not entirely understood. Apparently direct oxidation of the aluminum surface by the hexavalent chromium and hydrogen ion couple in the presence of the phosphate radical leads to the formation of a trivalent chromium and aluminum phosphate layer. The role of either the fluoride or complex fluoride seems to be one of depassivation. Suffice it to say that very little fluorine appears in the coating produced.

While the coatings described may be considered to be typical of several of the types available, their structure and performance may vary with the specific chemical composition and operating conditions of the phosphating solution as well as a variety of other factors, including the composition of the metal being treated, its physical form and surface history, its degree of cleanliness and surface activation, the effectiveness of rinsing between stages, and the nature of any seal or other post-treatment that may be applied. The chemical compositions used to clean, activate, phosphate, seal or post-treat are usually proprietary products and the specific recommendations of the supplier should be followed. However certain general comments may be beneficial.

Because the metal surface being phosphated participates in the chemical reaction to form the coating, the ease and uniformity of its ability to corrode in the reaction plays an important part in proper coating formation. The chemical and physical nature of the extreme outer surface is much more important than the same properties for the article as a whole. The presence of stressed areas due to cold working, the presence of previously applied corrosion resisting films, the possibility of decarburization, and the existence of pre-established etching patterns, all have an influence on the proper development of the phosphate coating.

In contrast to the stringent cleaning requirements for electroplating, the surface to be phosphated should be cleaned and conditioned by techniques designed to form a large number of active sites on the surface in order to form a fine grained phosphate coating with even coverage. In general, minimal chemical cleaning is indicated. Mechanical methods for rust removal are preferred. Grit blasting and tumbling procedures are effective and are preferred over acid types of pretreatment. Organic contamination may be removed by solvent degreasing procedures, emulsion cleaning or alkaline cleaning, providing that the cleaner will not inhibit the surface and interfere with phosphate coating formation. In the case of iron phosphates used as a pre-paint treatment, the cleaning and phosphating operations may be performed by the same solution.

When the maximum in grain refinement is required, a conditioning process is included. The active materials tend to be finely divided to colloidal in nature and their application increases the number of active sites for the development of the phosphate coating. Sometimes this procedure may be incorporated in the cleaning steps but, preferably, it is applied as a separate treatment after cleaning. The effective life of these solutions is limited and they should be changed frequently for maximum effectiveness.

In the case of manganese and zinc phosphating solutions, the chemical composition is such that it provides a source of these elements in solution together with a source of phosphate. The function of these essential ingredients is often enhanced by the presence of oxidizing materials such as nitrate, nitrite, or chlorate. These materials accelerate the formation of the phosphate coating by increasing the aggressiveness of the acid solution. In addition, the solution may also contain catalytic amounts of other metals, complexing agents, and depolarizing agents.

Conventional chemical control of these solutions may include the determination of free acid, total acid and ferrous iron. The free acid is expressed as the number of milliliters of tenth normal sodium hydroxide, often referred to as "points," necessary to titrate a known volume, usually ten milliliters, of phosphate solution to a pH of 4.0. Total acid is the number of milliliters of tenth normal sodium hydroxide required to titrate the same volume of sample to a pH of 8.2. Ferrous iron is usually determined by titration with potassium permanganate.

The free acid maintains the solubility of the zinc or manganese phosphate and provides a measure of degree of aggressiveness of the phosphating solution toward the metal to be coated. Usually maximum coating rates will be obtained when the free acid is only slightly higher than that necessary to maintain solubility of the metal phosphates in the working solution.

The total acid may be considered to be a measure of the amount of coating materials in solution and its ratio to the free acid is more significant than either individual value. The ratio of total to free acid in a liquid concentrate is usually lower than that of the operating bath and either partial neutralization or processing of scrap work may be necessary to establish the proper balance between the free and total acid. The latter procedure also establishes an iron level in the solution which is generally desirable when the phosphate coating is applied to assist in corrosion protection. When the phosphate is to be used in a cold deformation process, soluble iron is excluded from the solution by the use of an oxidizing agent such as sodium nitrite.

Whenever ferrous metals are phosphated, a stock loss is generated and iron is introduced into the phosphate solution. Ultimately this results in the formation of ferric phosphate, which has very limited solubility. This material is the principal ingredient in the sludge which is generated as an inevitable by-product of the phosphating process. It may be removed by filtration or settling and decantation. Both high-temperature and low-temperature phosphating solutions will produce sludge.

The production of another insoluble material, scale, is characteristic of phosphate solutions operated at high temperature. A general characteristic of zinc and manganese phosphating solutions is that the metal phosphates become less soluble as the temperature is raised. This leads to selective deposition of scale on those surfaces that have the highest temperature; i.e., the heating surfaces used to maintain the operating temperature. When possible, the development of this type of scale may be limited by the use of a tank within a tank type of construction wherein the annular space between the tanks contains a suitable heat transfer fluid, or by the use of a properly designed external heat exchanger.

Phosphating solutions are sensitive to metallic contamination and materials such as lead and aluminum should be excluded in those cases where the solution is not designed to compensate for their introduction. Obviously, the introduction of cleaner residues and organic contamination will have a detrimental effect on the proper operation of a phosphating solution.

Reference to Tables 1 to 3 will immediately call attention to the wide variety of objects which are regularly phosphate-coated. Wherever the type of coating to be applied offers a choice as to method of application, the method selected must be governed by the usual engineering principles. However, within the allowable process limitations (as given in the tables for particular types of coatings) emphasis in making a choice of equipment will depend on the following factors:

(a) Size of product, including ease of transport.
(b) Uniformity of pieces or unit groups of pieces.
(c) Scale of production.

TABLE 1. PHOSPHATE COATINGS FOR STEEL AND IRON

Type of Surface	Cold-rolled Stampings	Cast, Forged, Hot-rolled	Cast Iron, Cast Steel	Cold-rolled Stampings
General form and examples of products	Automobile bodies, refrigerators	Hardware, nuts, bolts, guns, cartridge clips, etc.	Small arms components, nuts, bolts, etc.	Stampings for cabinets, metal furniture
Supplementary treatment	Painting	Protective oils, waxes, etc.[1]	Protective oils, waxes, etc.[1]	Painting
Type of coating	Zinc phosphate[2,3]	Zinc phosphate[2,3]	Manganese phosphate[4]	Iron phosphate
Available methods of application	Power spray, washer, dipping, hand brush or spray	Dipping, barrel-tumbling	Dipping, barrel-tumbling	Power spray washer
Coating weight mg/sq ft	100 to 600	1000 to 4000	1000 to 4000	30 to 100
Usual coating time, min	Power spray 1 to 2; dipping 2 to 5; hand brush 2 to 5	20 to 30	15 to 30	2 to 3

[1] These coatings, especially the zinc phosphate, are occasionally relied on for corrosion resistance unprotected by oil, etc.

[2] Zinc phosphate coatings have also been used as insulation between laminations in transformers and other electromagnetic equipment. Core powder for similar use is also so coated.

[3] Zinc phosphate coatings of coating weights 200 to 800 mg/sq ft, applied by means of power spray washer as well as by dipping have been successfully used as *aids to drawing*. They have been successfully used on wire, tubing, partly formed pressed objects such as shells, etc. in the forming of flat-polished auto bumper stock, etc. Wire has been drawn "wet" or dry, and dried soap-type lubricants have been used over the phosphate coating.

[4] This type of coating is extensively used to impart increased *wear resistance* to ferrous metal bearing surfaces. It is used regularly on piston rings, cam shafts, valve tappets, gears, universal joint parts, gun lock parts, etc. It is the most extensively used type of phosphate coating for this purpose.

(d) Material handling methods preferred for use in the particular plant.

Treatment of a specific type of process will include examples of the products made by that process, some light may thus be shed on the choice of process type.

Process Cycles

Requirements. Starting with sheets, continuous strip, completed objects, or partial assemblies, the complete phosphate coating process includes at least the following operations:
(1) Preparation for application of the coating which may include:
  (a) Cleaning in its usual sense
  (b) Removal of oxides or corrosion products
  (c) Conditioning for the reception of coatings of finer grain structure
(2) Application of the coating proper
(3) Rinsing of the coated surfaces:
  (a) Water rinsing
  (b) Conditioning rinsing
(4) Drying

(5) Application of supplementary coatings.

Each of these operations may require one or more steps. Occasionally some combination of (1) and (2) above is possible.

Following is a classification of the steps that may be required in any complete phosphate-coating cycle.

Classification of Process Steps

A. Cleaning. This term has the usual meaning in electroplating and allied techniques. It may consist of one of the following:

(a) Hand wiping with or without organic solvent.
(b) Vapor degreasing.
(c) Alkali cleaning, with or without completely or partially emulsified solvent.
(d) Treatment with self-emulsifying solvent.
(e) Application of phosphoric acid-solvent-detergent cleaners, by hand or otherwise (see C(b) below).
(f) Special procedures such as steam jet cleaning, blasting with shells, or oat hulls.

B. Rinsing after cleaning.

TABLE 2. PHOSPHATE COATINGS FOR ZINC AND CADMIUM

Type of surface, general form and examples	Galvanized steel or cadmium plated steel. Heavy plated coatings, erected structures, signs, buildings, truck bodies	Zinc or cadmium electroplate especially when thin-base die castings, zinc surfaces of any form. Plated small parts, continuous strip
Supplementary treatments	Painting	Painting
Type of coating	Zinc phosphate[2,3]	Zinc phosphate[1,3]
Available methods of application	Hand brushing or spraying, brief dipping, followed by delay before rinsing	Power spray washer, dipping
Coating weight, mg/sq ft	200 to 600	75 to 500
Usual coating time, min	2 to 15 including delay before rinsing	0.5 to 2.0

[1] This type of coating has been also used as an aid in the drawing of galvanized wire.

[2] This type of coating is formed by reaction of a film of solution with the surface during the delay before rinsing. It is designed particularly for large or relatively immovable pieces which are not readily dipped in a tank or carried on a conveyor.

[3] Both the coating types described here are specified for use under paint. The life of paint over zinc exposed to the weather is greatly improved by these coatings for two reasons:
(a) Prevention of spread of underfilm corrosion.
(b) Prevention of impairment of paint bond because of reaction of alkaline zinc corrosion products with the finish which otherwise frequently results in paint flaking off to zinc surfaces.

TABLE 3. PHOSPHATE COATINGS FOR ALUMINUM AND ITS ALLOYS

Type of coating	Zinc phosphate crystalline[1]	Aluminum-chromium phosphate, amorphous[2]
Type of surface general form, and examples of products	Aluminum alloys, wrought, sheet metal stampings, truck bodies, aluminum tiles	Aluminum alloys, wrought or cast. Any continuous strip, aircraft parts, automotive parts
Supplementary treatment	Painting	Painting or none
Method of application	Power spray washer dipping tanks	Power spray washer, dipping tanks manual brush or spray
Coating weight, mg/sq ft	100 to 300	100 to 250
Usual coating time, min	1 to 3	20 sec to 2 min

[1] This type of process is designed to furnish crystalline zinc phosphate-coatings somewhat similar to those used on sheet steel, for paint bonding purposes. It is possible to process work of several different kinds of metal in the same solution.

[2] This type of coating affords a superlative paint-bonding and enormously increased organic finish life. It is also suited to resist corrosion without supplementary coatings. There are several modifications of this process for use by different methods of application and particularly adapted for different services.

C. *Oxide removal.* This may require one of the following operations:
(a) Pickling in phosphoric or other acids.
(b) Rust removal with phosphoric acid-solvent-detergent type cleaners.
(c) Sand or grit blasting with air pressure or centrifugal application.
(d) Grinding of welding flash.
(e) Sand or grit tumbling.

D. *Rinsing after oxide removal.*

E. *Grain refining pretreatment.* This consists of treating the surface with a special reagent which may be alkaline or acid.

F. *Rinsing after oxide removal.*

G. *Phosphate coating proper.*

H. *Rinsing after phosphating.* Use clean flowing water.

I. *Conditioning rinsing.* This may involve either:
(a) Dilute chromic and/or chromic phosphoric acid treatment for use on surfaces which are to be protected against corrosion.
(b) Alkalining rust-inhibiting treatment for objects to be cold-deformed. Not used under paint, etc.

J. *Drying.* Use any of the following methods:
(a) Air oven drying.
(b) Air drying for heavier objects.
(c) Water displacing oil treatment.

K. *Supplementary coating.* This may include applying the following and may also require subsequent drying:
(a) An oily or waxy corrosion retardant.
(b) A lubricant to retard wear or assist cold deformation.

L. *Painting, lacquering, or the like.* Although this, too, may be considered a "supplemental coating" it is usually a more elaborate procedure than (K) and is considered separately.

Discussion of Process Steps in Practical Procedures

A. *Cleaning.* While all the above methods are used under special circumstances depending on the nature of the surface and its contaminants, a few details will be discussed for later reference.

(a) Alkali cleaning is preferable wherever applicable for all mass production because of its simplicity and economy. In immersion processes alkali cleaning may require 3 to 10 min in solutions maintained at 80–98 C (180–210 F). In recirculating spray applications, 1 to 2 min at

80 C (180 F) maximum temperature are usual conditions.

(b) Self-emulsifying solvent cleaning involves brief dipping into the cleaner, followed by a drainage period of 0.5 to 1.0 min. Subsequent rinsing usually requires forcible spraying, or at least two hot 80–90 C (180–200 F) immersion rinses of 0.5–1.0 min each. Agitation in these rinses is very important. Cleaning of this kind alone is adequate only when the phosphate coating operation employs high temperature solutions for long times.

(c) Vapor degreasing is generally applicable for grease removal. It requires no after-rinsing. However, it is quite expensive compared to alkali cleaning. In certain cases objectionable finely divided solid dirt will not be removed by this process.

(d) Acid cleaning is carried out in various ways. Such cleaners are used for example:

   (i) To derust large structures in a hand operation by brush, "paint-gun" spray, or flowing application.

   (ii) As a "rust-spotting" operation to large subassemblies such as automobile and refrigerator parts before they enter the phosphate-coating cycle proper.

   (iii) As a pickling cleaner, in which hot solutions are used as described under pickling. This requires the use of large acid-resistant tanks.

B. Rinsing after cleaning. Generally this is done in overflowing water. Temperature is not critical, but hot water rinsing assists cleaning and may make a cleaning procedure passable under certain conditions. Immersion rinses are preferably from 0.5 to 1.0 min long; spray rinses of 0.5 min are generally sufficient.

The overflow rate of rinses should be sufficient to ensure a degree of contamination so low that drag-in of rinse solution into the phosphate-coating stage does not result in coating difficulties or wasteful precipitation of phosphating chemicals.

Until practical operating limits are established it is safe to maintain overflow amounts or replacement schedules for rinse water adequate to keep the contamination of the rinse water down to 1.0 per cent or less of the preceding tank concentration. If two rinse stages are used in series the contamination in the second stage is, of course, the critical one.

Rinse tanks need to be acid resistant only when they immediately follow acid cleaning.

C. Pickling. Pickling removes scale, heavy oxides, rust, or corrosion products if present.

Whenever possible it is economical to separate mill scale removal or analogous operations completely from the phosphate-coating apparatus. Strong dilute mineral acids such as sulfuric acid, hydrochloric acid, etc., which may be hot, can be used for these operations in the ordinary "pickle house." Thus the corrosion stimulative fumes are kept out of the phosphate-coating area which is often in or adjacent to the paint shop.

If the presence of rust or scale from brazing or welding operations requires it, the phosphating process must include a pickling stage. Phosphoric acid is generally used since fumes of this acid are not corrosion-stimulative to finished work as are those of other mineral acids.

For immersion processes with their large open tanks, pickling is in phosphoric acid which may also contain a detergent to assist cleaning and smut removal. In immersion processes the pickling solution is usually at 60–80 C (140–180 F); pickling time may range from 2 to 10 min, most usually 3 to 5 min. Actual tests on the work to be treated are suggested before the cycle is decided upon. Tanks for phosphoric acid pickling may be of stainless steel (Grade 301 or 316) or of other acid-resistant construction. Lead is not recommended for phosphoric acid pickling.

In continuous spray processes pickling with sulfuric acid has been used. In this case the pickling section must be flanked by "water curtains," and ventilation of the conveyor tunnel must be so arranged that all air currents travel toward the pickling section. The following precautions are absolutely essential in conveyorized continuous spray machines which employ sulfuric acid pickling:

   (i) Rubber-lined or other acid-resistant construction for pickling stage, including tanks, pumps, piping, nozzles, and housing.

   (ii) Acid-resisting hangers or conveyor parts which enter the pickling section.

   (iii) Acid-resisting construction for the "water curtain" section at each end of the pickling section, as well as for at least one following rinsing stage.

   (iv) Acid-resisting ventilating fan and duct operating from the top of the tunnel in the pickling section.

   (v) If an overhead chain conveyor is used, all hooks must pass through a water-seal trough containing running water to prevent acid corrosion of chain, rollers, and supporting beams.

(vi) In sulfuric acid pickling concentrations are from 5 to 15 per cent by volume of 66 Be′ sulfuric acid; temperatures are from 50 to 70 C (120 to 160 F), generally limited by the properties of rubber, plastic piping, pump impeller alloy or other material of construction.

D. Rinsing after pickling. If sulfuric acid or other than phosphoric acid is used in pickling, see section C.

In immersion processes using phosphoric acid pickling, rinsing requires one unheated, overflowing rinse in clean water. Rinsing time 1 min. In spray processes 30 sec may suffice. However, unless a grain refining pretreatment follows, rinsing must be very thorough. Contamination must be kept very low. After phosphoric acid pickling, low rinse contamination will usually make unnecessary the use of acid-resistant equipment in this rinsing stage.

E. Grain refining pretreatment. This treatment may be required if an especially fine grain structure is desired in the phosphate coating. It is usually not required on steel surfaces unless pickling or acid cleaning has been employed. The treatment may result in slightly more rapid and uniform coating of galvanized steel or pickled steel surfaces.

Two kinds of grain refining pretreatment are available, namely, one using an alkaline solution, another an acid solution. Both are used at room temperature, and application time need not exceed 30 sec. The acid type of pretreating solution requires equipment resistant to cold dilute oxalic acid. The alkaline type may be handled in mild steel.

F. Rinsing after grain refining treatment. Cold overflowing water rinsing is required. 0.5 to 1.0 min is required in immersion processes— 0.5 min is sufficient in spray processes.

G. Phosphate coating proper. For details see the following tables for different kinds of coating.

H. Rinsing immediately after phosphate coating. A cold overflowing water rinse lasting 0.5 to 1.0 min is satisfactory for immersion processes. In continuous spraying processes 0.5 min is sufficient.

Note that by keeping this rinse low in contamination, the life of the final conditioning rinse solution is prolonged.

I. Conditioning rinse. Surfaces which are phosphated as part of a corrosion-retarding finish, whether they are to remain bare, or are to receive paint or other supplementary finish, invariably have a conditioning rinse in very dilute solutions usually containing chromic acid or chromic acid and phosphoric acid to furnish an additional anti-corrosive effect and to minimize the deleterious effect of residual soluble matter, which can stimulate blistering of paint, rusting, etc.

This treatment may be omitted on surfaces which have been phosphate coated to assist drawing or forming. Since such surfaces are often dipped into soap or soap emulsions of grease as lubricants, the final conditioning rinse may consist of very dilute alkali plus a corrosion retardant salt. In this case, of course, the work enters the lubricating bath without previous drying, directly from the conditioning rinse.

The conditioning rinse is maintained sufficiently warm to aid subsequent drying of the metal usually in the range of 60 to 80 C (140 to 180 F). In immersion processes the time of treatment is 0.5 to 1.0 min; in spray processes the time is usually about 0.5 min.

J. Drying the coated and rinsed surfaces. In mechanized processes this operation is generally accomplished by means of a hot-air oven. A steam heated or indirect fired oven is best as it is safer not to allow coated work to come into contact with products of combustion. Oven temperatures may be as high as 175 C (350 F), but there is no advantage in temperatures higher than needed to dry the surfaces completely within about 3 to 5 min. For amorphous phosphate coatings the metal temperature during drying should not exceed 80 C (180 F).

For objects having pockets or areas which do not drain freely, a blow-off with compressed air before or after drying may be required.

When "water-displacing" rust preventives are satisfactory as supplementary coatings, the drying step may be omitted altogether. In this case, accumulated water must be withdrawn from beneath the water displacing material.

K. Supplementary coating. Oils, so-called "cutback" petrolatum, soluble oils, and waxes are among the materials often applied as supplementary coatings for corrosion protection. For lubrication, oils, drawing compounds, soap solutions, and emulsions are common. Application is usually by dipping; roller coating may be used for flat sheets or continuous strips. Nonaqueous materials are generally applied cold; soluble oils, soap solutions and emulsions may be heated to accelerate later drying. In dipping processes immersion need rarely exceed 0.5 min. Depending on viscosity, drying rate, etc., draining time over the tank or drip-pan varies between 30 sec and 2 min.

Soap and similar "dry film" lubricants are dried onto the surface by means of a drying oven or by radiant heating lamps.

L. *Painting.* Paint, lacquer, and the like may be applied to phosphate-coated surfaces by any of the usual painting methods. Dipping, flow coating, brushing, roller coating, and spraying have all been used on phosphate-coated work.

In a mechanized phosphating system, whether immersion or spray coating is practiced, it is generally undesirable to perform the painting operation while the work is being transported on the same conveyor, hooks, baskets, or racks as those on which it was carried through the phosphating system.

The apparent advantage of carrying work through painting without transferring to new carriers is that contamination of the surface after paint-bonding and conditioning treatment is less likely if the work is painted immediately and without handling. This advantage is overshadowed, however, by the following considerations:

(a) An interruption in the painting operation will cause a stoppage of work in the phosphate-coating operation with consequent misprocessing of all the pieces in process. Such a stoppage will occur not only in case of difficulties but also at lunch time, etc., unless work schedules of painters and of phosphating machine staff are staggered.

(b) Hooks, racks, baskets, etc., gradually accumulate heavy paint layers which must be removed from time to time. This is difficult and expensive.

Immersion Processes*

An immersion process is one in which the work to be coated is dipped successively for appropriate intervals into a series of processing tanks.

Applications. Immersion processes are applied:

(a) To produce heavy phosphate coatings

(b) To produce light to medium paint bonding phosphate coatings when:

(i) the scale of production is relatively small

(ii) the shape of parts makes significant surfaces inaccessible to a spray

*See Table 4.

(iii) parts are too small to allow individual racking or handling

(a) Heavy phosphate coatings are usually applied in solutions at 90 to 100 C (195 to 210 F), coating times range from 10 to 40 min thus making spraying impractical.

(b) The lighter paint bonding coatings may be applied by immersion or spraying. Application by immersion (rather than spraying) is preferable under the above conditions. For the purpose of selection, the coating of less than approximately 2,000 ft^2/hr can be considered small production, although no definite limit can be established which will meet all conditions encountered in industry.

Equipment. The equipment for immersion phosphate coating may be classified as follows:

(1) Conveying equipment
(2) Work-supporting equipment, such as
   (a) Hooks
   (b) Racks
   (c) Baskets
   (d) Tumbling barrels
(3) Tanks with associated services
   (a) Water supply
   (b) Heating supply—steam, electricity, gas, or oil
   (c) Drain and overflow lines to sewer
   (d) Ventilating equipment
(4) Drying equipment
   (a) Ovens
   (b) Air heaters and fans
   (c) Compressed air blow-off

(1) *Conveying equipment.* Conveying equipment for immersion phosphate-coating may be of any type suitable for transporting the work from loading to unloading stage. It must also provide for lowering the work into and raising it out of the various tanks in proper sequence and rhythm.

Commonly used equipment includes:

(a) Overhead monorail conveyors with manual, air or electrically operated hoists. For very small production, manual progression of the work carrier is suitable.

(b) Continuously moving rising and falling chain driven conveyor in association with boat-shaped ("gondola") tanks (see Fig. 9).

(c) Specialized automatic equipment strictly analogous to automatic plating equipment. Usually, however, without equipment for supplying electric current to the tanks.

Machinery of this kind may use continuously or intermittently moving conveyor elements. Its

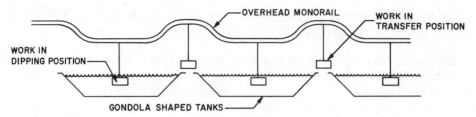

Fig. 9. Conveyor with boat-shaped gondola tanks.

distinguishing feature is usually that it permits rapid transfer from tank to tank. Because vertical ended tanks may be used, such equipment, though generally more expensive than (b), occupies less floor space (see Fig. 10).

Design features of conveying equipment are not peculiar to the immersion phosphate-coating art. The following constants are recommended:

(a) *Draining time* during which solution running off the work, hooks, baskets, etc., falls into the tank from which it has just been raised (or into an interstage drain pan discharging thereto) should be 5 to 15 sec depending on the ease of draining of the work. Hollow pieces, closed at one end, may require special handling.

(b) *Transfer time* is preferably held at below 30 sec from emergence to re-immersion for any surface of the object in process. If this is difficult (as is sometimes the case with conveyors of the rising and falling monorail type) discoloration of the work or unevenness of appearance may result from partial drying-up of the surfaces between stages. Wherever this is possible, outlets should be provided for misting nozzles to keep the work wet during the transfer. These nozzles are generally fed with fresh tap water.

(c) *Construction* of the conveyor need not, in general, be acid-resistant. Due attention should be paid to possible condensation of moisture from heated solutions upon overhead conveyors. This may result in dripping of grease, etc., upon the work with consequent imperfect cleaning or coating, or contamination of the coated surface, unless suitable splash guards are provided.

(2) *Work supporting equipment*. The supporting equipment such as hooks, racks, baskets, and tumbling barrels is similar to that used in electroplating, with the exceptions noted below:

(a) Hooks (and baskets) need not be electrically insulated. Construction materials resistant to alkali cleaners, acid phosphating solutions, etc., are desirable. Stainless steel is the best; however, mild steel has been used, the life of the carriers depending on the time and temperature in the phosphating stage. Of course, if the cycle includes pickling, mild steel is precluded.

(b) Tight contact of significant surfaces of the work with hooks and racks is undesirable. Tight contact with more cathodic metal such as stainless steel may cause lightly coated or bare areas on aluminum treated to produce amorphous coatings. For coating aluminum in this case, racks, hooks, etc., covered with acid-resistant plastic may be necessary unless the pieces can be supported from nonsignificant surfaces.

Light contact or vibrating contact with baskets, hooks or other parts is permissible in phosphate coating. The coating, being insulating, tends to intrude itself between lightly or intermittently touching surfaces. However, for work to be piled loosely, baskets should be so designed that by tipping, turning over, etc., the points of contact between pieces or pieces and baskets are shifted several times during coating and cleaning operations.

(c) Tumbling barrels require no insulation and no chains, studs, or other electrical contact points. Slowly turning barrels (2 to 5 fpm) with

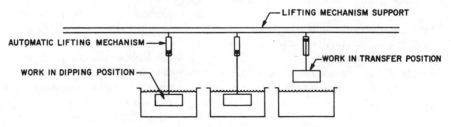

Fig. 10. Automatic conveying equipment with vertical-ended tanks.

low flights are desirable to avoid abrading off the freshly formed coating. Work depth over 6 in. in the barrel is undesirable.

Barrels may be constructed of stainless steel (Type 301 or 316) or, if shorter life is permissible, of ordinary steel, in the production of crystalline phosphate coatings. Amorphous coatings on aluminum require stainless steel or heavily plastic coated barrels.

Usually, drive mechanisms involving a pair of half bearings and oscillating ratchet arm are installed on each tank. The barrel with supporting trunnions and ratchet wheel is moved from tank to tank by means of hooks passing through lugs on the barrel (see Fig. 11).

(3) *Tanks with associated services.* Tank types used in immersion phosphate coating are illustrated in Fig. 12. Except for obvious features these differ little from tanks used for other processes.

Cleaning and rinsing tank dimensions are frequently fixed by the size and quantity of work which must be in process at one time. The size of a phosphate coating tank proper, however, is affected by other considerations.

(i) Since phosphate coating processes normally are attended by the formation of precipitates known as "sludge," it is necessary to allow a tank *depth* such that the work need not pass through precipitated sludge. Good practice dictates leaving 6 in. to 1 ft clearance between work and tank bottom.

(ii) Although in the ordinary immersion coating process no consideration need be given to electric current flow paths, *bath loading* in terms of surface to be coated in square feet per hour per gallon of solution is important. Necessary frequency of control analysis of coating solution and frequency of required sludge removal are reduced at low bath loading. While no absolute figures can be given, experience indicates the minimum working volumes, as shown in Table 4A, should be given.

Services indicated in Fig. 12 are conventional except that the *water supply* should be adequate to fill any tank within 30 min.

Heating may be by steam coils or flat "plate coils," by electricity, gas, or oil. The heating of cleaning, pickling, or rinsing provides no special problems. The same is true for the phosphate-coating tank when producing amorphous coatings on aluminum, except that all heating surfaces should be kept at least 4 in. above the bottom of the tanks. Heating elements in all tanks, and especially in phosphating tanks, must be mounted along the sides of the tanks. Bottom heating is not recommended because it causes undue circulation of sludge which will probably result in a loose deposit on the coated work.

In tanks containing solutions for zinc or manganese phosphate coating the heating surfaces become more or less rapidly coated with a scale of insoluble phosphate. In such tanks, especially when operating above 80 C (180 F), it is desirable to provide a set of spare heating coils and to arrange the coil installation to facilitate ready removal and replacement of scaled coils. The coils may be descaled mechanically or by immersion in inhibited 5 per cent (by vol.) sulfuric acid in a tank especially provided for this purpose. Manganese phosphate solutions are particularly rapid scale-formers. It is not uncommon that coil replacement for scale removal is required weekly.

Accurate thermostatic controllers are highly desirable, especially in phosphate-coating tanks, where fluctuations of solution temperature are often deleterious to solutions and to work quality.

It is usually economical to provide thermal insulation for heated tanks.

Calculation of necessary heating surface for tanks other than phosphate coating tanks proper may be made as though the tanks contained water only. For phosphate-coating tanks in immersion processes, the heating surface provided should be from two to three times as large as would be required to heat water only under

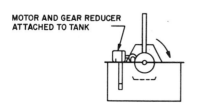
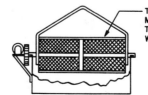

Fig. 11. Tumbling barrel arrangement.

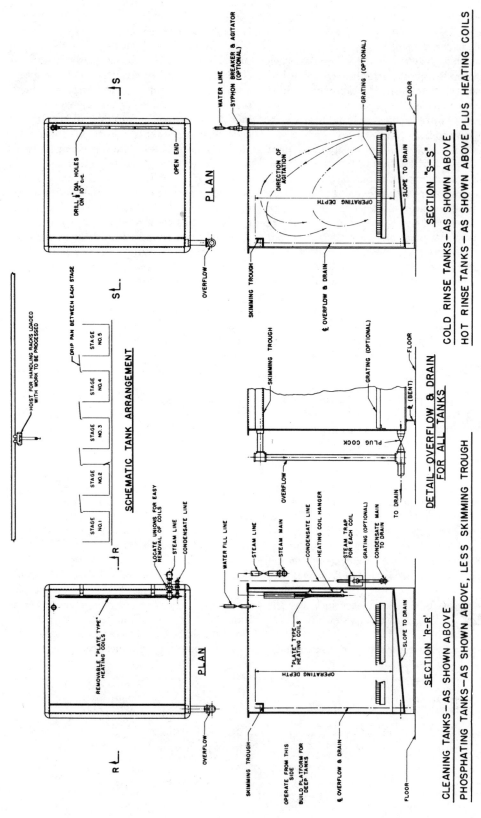

FIG. 12. Typical tank construction for immersion processes.

TABLE 4. IMMERSION COATINGS
(Type of Coatings and Process Details)

Type of Coating	Metal Treated	Purpose of Treatment	Coating Time (min)	Range of Operating Temp. (F)	Range of Operating Temp. (C)	Typical Operation Sequence (see Classification of Process Steps)	Recommended Material of Construction for Phosphating Stage (see note 7)
Heavy zinc phosphate	Steel	Rustproofing Corrosion resistance	15–40	190–205	90–95	A(c),B,G,H,I,J,K See Note 1	Tank: acid-resistant or ⅜ in. mild steel. Heating elements: acid-resistant
Heavy manganese phosphate	Steel	Wear-resistance	15–40	200–210	95–98	A(c),B,G,H,I,J,K See Notes 1 and 2	Tank: acid-resistant. Heat-elements: acid-resistant
Light to medium zinc phosphate	Steel	Paint bonding or assisting cold deformation	2–5	80–180	27–82	A(c),B,G,H,I,J,L See Note 1	Tank: acid-resistant or ⅜ in. mild steel. Heating elements: acid-resistant
Light iron phosphate	Steel	Paint bonding	2–5	140–180	60–82	A(c),B,G,H,I,J,L See Note 3	Tank: M.S. or acid-resistant. Heating elements: mild or acid-resistant steel. See Note 4.

TABLE 4. (*Continued*)

Type of Coating	Metal Treated	Purpose of Treatment	Coating Time (min)	Range of Operating Temp. (F)	Range of Operating Temp. (C)	Typical Operation Sequence (see Classification of Process Steps)	Recommended Material of Construction for Phosphating Stage (see note 7)
Light to medium zinc phosphate	Zinc or cadmium	Paint bonding	1-3	140-180	60-82	A(c),B,G,H,I,J,L See Notes 5 and 6	Tank: acid-resistant or ⅜ in. mild steel. Heating elements: acid-resistant
Light to medium amorphous phosphate	Aluminum	Paint bonding or corrosion prevention unpainted	1-3	110-120	43-49	A(c),B,G,H,I,J,L	Tank: acid-resistant. Heating elements: acid-resistant
Light to medium zinc phosphate	Aluminum	Paint bonding	1-3	130-135	55-57	A(c),B,G,H,I,J,L	Tank: acid-resistant. Heating elements: acid-resistant

[1] Operations C and D can be inserted in sequence shown if rust or scale must be removed. If necessary, Operations E and F can also be inserted.
[2] Often Operations I, J, and K of typical sequence shown are omitted and a soluble oil immersion step substituted.
[3] Operations C and D can be inserted in sequence shown if rust or scale must be removed.
[4] Proprietary processes available operate at various degrees of acidity; accordingly, process manufacturer should be consulted before employing either mild steel or acid-resistant equipment.
[5] Inhibited alkaline cleaners normally are used in Operation A when cleaning zinc or cadmium.
[6] Occasionally Operations C and D are inserted in above operational sequence, but immersion time in Operation C is usually limited to 10 to 30 sec and is used primarily as "activation" dip. Operations E and F are also often inserted in above typical cycle.
[7] In order of decreasing corrosion-resistance in acid phosphate solutions, the following materials of tank construction are listed: Stainless Steels AISI Types 316, 304, 302, 430, 410. All of these are considerably more resistant than mild steel or low-alloy steels.

TABLE 4A. IMMERSION PHOSPHATE COATING, RECOMMENDED MINIMUM TANK VOLUMES

Type of Coating	Metal Treated	Min Vol,* gal/sq ft surface coated per hr
Manganese phosphate, heavy	Steel or cast iron	2.5
Zinc phosphate, heavy	Steel	2.0
Zinc phosphate, light to medium	Steel, Zn, Al, Cd	1.0
Amorphous phosphate	Al	0.1

* To get minimum tank volume multiply square feet of surface coated per hour by factor.

the same conditions to provide for accumulation of scale on the heating coils.

Heating requirements for tanks should be calculated both on the basis of (a) the heat necessary to maintain temperature and (b) on the basis of heat necessary to raise the solution from room or water supply temperature to operating temperature within the time available after shutdown periods. Often it is wise to provide for sufficient heat input to heat the solutions to operating temperature in 1 hr. Whichever figure is the higher should be chosen as the basis of design.

Ventilation of immersion phosphate-coating equipment is sometimes required to lower the concentration in building or working space of:

(1) Steam or condensed vapor
(2) Pickling bath spray, especially if acids other than phosphoric are employed
(3) Spray or vapors from phosphate coating tanks.

In most instances reasonable rates of air change in the working space are adequate to eliminate discomfort to personnel and dripping of condensed moisture from pipes and structures.

However, many phosphate coating solutions emit small amounts of oxides of nitrogen as well as hydrogen. Solutions for producing amorphous coatings on aluminum give off small amounts of hydrofluoric acid and spray from slight gas evolution contains fluorides, chromic acid, etc. Pickling baths always evolve acid spray. Phosphoric acid spray is not very irritating to personnel or damaging to structures; the contrary is true of sulfuric or hydrochloric acid spray which is irritating to personnel and both corrosive and rust stimulative to structures and to work in process.

Therefore, when phosphating is carried out on a large scale in confined quarters, ventilation should be designed to take care of these air contaminants. In addition to providing a general air change in the working space, however, it will usually be found desirable to prevent the fumes from pickling tanks and amorphous phosphate-coating tanks from entering the working space. In order not to interfere with overhead conveyors or structures this is best done by providing these tanks with slotted suction boxes exhausting air across the solution surface at a high enough velocity to prevent fumes from entering the working space. Equipment similar to that used in connection with chromium plating is recommended.

(4) *Drying equipment.* This may be of any type suitable for drying the coatings within about 3 min. Oven temperatures may be as high as 175 C (350 F) for crystalline phosphate coatings. Amorphous coatings on aluminum, however, must not be heated above 80 C (180 F) during drying.

The designer will do well to make drying equipment controllable as to temperature and air circulation so as to accommodate a variety of coated objects of different mass and cross section.

Spray Processes, with Solution Recirculation

Probably the largest area of metal surface is phosphate-coated by passing the work, supported on a suitable conveyor, through a succession of stages in each of which it is sprayed for an appropriate time with a different solution. Each solution, contained in a tank, is pumped through suitable piping to the nozzles which discharge it against the work surface. The solution which runs off of the work returns to the tank from which it came.

Applications. Spray processes (as defined above)— are applied:

(a) To produce light to medium weight coatings for paint bonding or to aid mechanical deformation
(b) On iron, steel, zinc, cadmium, or aluminum surfaces
(c) When the scale of production is relatively large (at least 2000 ft^2 of surface treated per hour)
(d) When all significant surfaces can readily be reached by solution projected by the nozzles
(e) When the work items lend themselves readily on individual hanging on appropriate racks or hooks.

(f) When continuous strip of steel, aluminum, or zinc electroplate is to be coated.

Tyes of Metal Coated and Process Details

Data on the types of metal coated and related process details are given in condensed form in Table 5.

In addition to the above, the following observations concerning Table 5 are in order for spray processes in general.

(1) Operations C and D may be inserted wherever rust and scale must be removed.

(2) Any operating sequence resulting in coating requiring grain refinement will need the interpolation of steps E and F before step A.

(3) Pump impellers and shafts in the acid final rinse, step I(a) should be made of stainless steel (301 or 316) if long life is required.

Equipment. The equipment for spray coating processes may be classified as follows:
(1) Conveying equipment
(2) Work supporting equipment
   (a) Hooks
   (b) Racks
   (c) Baskets
(3) Tanks with associated services
   (a) Water supply
   (b) Drain and overflow lines to sewer
   (c) Heating supply—steam, electricity, gas, or oil
   (d) Screens
(4) Pumps, piping, and nozzles
(5) Tunnels and associated equipment
   (a) Baffles, grating, access doors and lighting
   (b) Drain areas
   (c) Vestibules
   (d) Ventilating equipment
(6) Drying equipment
   (a) Ovens
   (b) Air blow-off stations
(7) Auxiliary equipment for
   (a) Sludge removal
   (b) Chemical feeding and metering

(1) *Conveying Equipment.* Conveying equipment for spray phosphate coating may be of any kind suitable for transporting the work through the various processing and draining stages. The most commonly used conveyors are:

(a) Overhead continuously moving chain-driven conveyors have the advantage of being relatively easily shielded from water or chemical solutions by suitable spray-guards. In especially corrosive areas water seal troughs are required (see process step C Pickling). The machine may be designed to pass work around turns as well as in a straight line.

(b) Bar conveyors, in which the work is supported on parallel bars carried by a pair of parallel chains. These have an advantage in that they permit lower, wider, and, therefore, shorter machines to be constructed. They have the disadvantage that it is practically impossible to prevent not only the cross bars but even the chains and sprockets from being reached by the chemical sprays. Accordingly these must be made chemically resistant. Moreover, the work may not be made to turn in a horizontal plane; thus the machine must be designed in a straight line.

(c) Continuous strip is handled by means of a series of drive rollers, supporting rollers, and "squeegee" rollers.

(d) Various special equipment has been designed for treating partially fabricated articles. Some of these machines embody intermittently moving conveyors with doors between the stages and intermittently operating pumps. A large German automobile manufacturer uses a number of machines of this type. It is true that a machine of this kind, requiring as it does no draining sections, (drainage being effected over the tanks while the pumps are shut off), and requiring no spacing to prevent overspray from stage to stage, can be made shorter in length than a design embodying a continuously running conveyor and pumps. The disadvantages, however, seem generally to outweigh the advantages. The complications involved in the door operating mechanism and the timing controls increase the cost of the machine and decrease its flexibility with respect to changes in line speed, etc.

Design features for conveying equipment are not peculiar to this art except in the special details mentioned. A few critical features of machine design with respect to speed, length of drain section, etc., are:

(a) Transfer time should be kept as short as possible. Drainage requirements rarely exceed 20 sec. Excessive transfer time is likely to cause great difficulty. Not only is partial drying of work likely to occur during transfer, with consequent irregularity of work appearance, but the fact that there is an enclosed tunnel gives rise to a "humidity cabinet" effect which is almost certain to lead to corrosion or "rust blushing" especially after the acid cleaning, pickling, or phosphate-coating stages.

TABLE 5. SPRAY PROCESSES
Types of Metal and Coating: Process Details

Type of Metal Treated	Type of Phosphate-Coating	Purpose of Treatment	Coating Time*	Range of Operating Temperature (F)	Range of Operating Temp. (C)	Typical Operation Sequence (see classification of process steps)	Recommended Material of Construction for Phosphating Stage
Cold rolled sheet steel partly fabricated	Light zinc	Paint bonding	0.75–1.0 min	110–135 Note (7)	43–57	A(c), B, G, H, I(a), J(a), L. Note 1	Pumps, piping, nozzles, stainless steel tanks; Note 2
Hot rolled sheet steel partly fabricated	Light zinc	Paint bonding	0.75–1.0 min	120–135	49–57	A(c), B, C(a), D, E, F, G, H, I(a), J(a), K, L	As above; Notes 2 and 3
Steel, partly fabricated	Medium zinc	Cold deformation	0.75–1.5 min	135–165	57–74	A(f), A(c), B, G, H, I(b), K(b), J	As above; Notes 2 and 4
Galvanized, cadmium plated partly fabricated	Medium zinc	Paint bonding	0.75–1.0 min	125–140	52–60	A(c), B, G, H, I(a), J(a), L	As above; Notes 2 and 5
Continuous electro-galvanized strip	Light zinc	Paint bonding	5–30 sec	125–130	52–54	G, H, I(a), J(a), L	As above; Notes 2 and 6
Continuous aluminum strip	Amorphous Al-Cr Phos.	Paint bonding	10–20 sec	110–120	43–49	A(c), B, G, H, I(a), K	Pumps, piping, nozzles, stainless steel; tank stainless or plastic lined
Aluminum, partly fabricated	Amorphous Al-Cr Phos.	Paint bonding	30–45 sec	110–120	43–49	A(c), B, G, H, I(a), K	Stainless throughout; tank may be plastic lined

*All times and stage lengths referred to in this section extend from the center lines of the first spray risers to the center lines of the last risers.
(1) Occasionally processes are used in which the coating solution is not heated. In these cases coating time may be as high as 2 min.
(2) Double rinsing after cleaning is recommended. First rinse, hot 60 to 77 C (140 to 170 F). Second rinse unheated.
(3) Stainless steel tanks are preferred. ⅛ in. mild steel is usable for tanks. Housing may be of mild steel.
(4) Entire pickling stage and first rinsing stage must be acid resistant. See all details given under discussion of process steps, section C Pickling. Note particularly precautions with respect to conveyor water seals, acid resisting hangers, and ventilation.
(5) Certain work which is to be plated after cold deformation first ground and polished as on a belt sander (Step A(f)). After cleaning and phosphating it is dried only after receiving a roller coat of lubricant. See coating steps H, I(b), J. The drying of such objects (as, for example, blanks for automobile bumpers) is accomplished by infrared lamps while the blanks are on a roller conveyor.
(6) Galvanized objects are very variable in character. In some instances cleaning may be omitted. Some metal coats irregularly unless grain refining pretreatment and rinse (Steps E and F) are inserted before G.
(7) Freshly electroplated strip requires no cleaning. If old, alkali cleaning may be necessary.

(b) *The length of the interstage drainage area* is closely related to conveyor speed. With the usual continuously moving conveyor, even the best type of profile baffling to limit overspray, directly from nozzles or deflected from workpieces, will still hardly permit making the horizontal distance from the last spray nozzle to the top of the slope of drain pan less than the height of the tunnel profile. This means that the horizontal distance from the last spray nozzle riser of one stage to the first spray riser of the succeeding stage is usually from 1.5 to 2.0 times the height of the tunnel profile. This distance is likewise limited by the horizontal length of the workpieces. No surfaces or channels along which solutions may travel should span the distance from the highest point of the drain pan to a nozzle in preceding or succeeding stage. Otherwise solutions may pass from stage to stage, resulting in unwanted mixing of different liquids.

In general it will be found that conveyor speeds less than about 5 fpm will lead to operating complications in machines with continuously running conveyors. In any case, transfer times of over 1 min indicate the desirability of installing interstage fresh-water misting nozzles, especially in the drain areas following stages in which heated solutions are applied.

(c) Other design features may be noted in Fig. 13.

(2) *Work Supporting Equipment.* Work supporting equipment for spray processes needs no comment in addition to that made in connection with immersion processes.

(3) *Tanks and Associated Services.* In addition to the information given for the general features of immersion tanks in the section on immersion processes, the major points of interest in connection with tanks used in spray processes are the following:

Tanks are divided by suitable partitions into a *main section* separated from the *pump sump* section. A baffle 12 to 15 in. high should be installed between these sections to prevent sludge, accumulated in the main section, from entering the pump. Solution entering the pump sump section should pass through a set of double *screens* of about 4 meshes per inch for cleaning and rinsing stages and 6 meshes per inch for the phosphate coating stage. These are to protect the pumps and nozzles from circulating solid contaminants, etc. The screens should be ruggedly constructed of 16 gage or heavier wire, in substantial frames, and should be provided with handles. Accurately constructed slides should permit independent removal and replacement of each screen in a set. For high production lines, an additional set of screens is often provided (see Fig. 13).

Screens in the phosphate-coating section should be of 18:8 stainless steel. In a pickling section using sulfuric acid, rubber-covered construction is recommended. For each 100 gpm pumped, at least 2 ft^2 of screen area should be provided.

Spray process tanks should be designed to give a solution retention time of from 2.5 to 5.0 min. In general the tank size should be related to the pumping rate to give the following retention times:

Process stage	Recommended solution retention time (min)
Cleaning	2.5–3.0
Rinse after cleaning	2.0–2.5
Phosphate-coating	3.0–5.0
Conditioning rinses	2.5–3.0

A small retention time may result in rapid fouling or deterioration of solutions or in over criticality of control.

Solution depth in tank should be at least 30 in. to permit locating the pump suction line:

(a) Well off the bottom to stay clear of trash falling into the suction section.

(b) Well below the surface to avoid sucking air, even when the solution level is accidentally allowed to fall a little too low, or when partly plugged screens cause a fall of level in the pump suction section compared to the main section.

Services indicated in Fig. 13 are conventional except that the water supply should be adequate to fill any tank in 30 min.

Drains and overflow lines need no special comment. Discharge of solutions from tanks must be considered from the point of view of waste disposal as affected by local conditions and regulations.

Accurate calculation of heating requirements to maintain the temperature of a solution in a tank while the solution is being sprayed in an enclosed space is very difficult. The heat input necessary to maintain solution temperature depends on the following heat losses:

(a) Radiation and convection from tank wall and piping.

(b) Radiation and convection from solution surface.

(c) Increment of heat content of the metal objects passing through the solution.

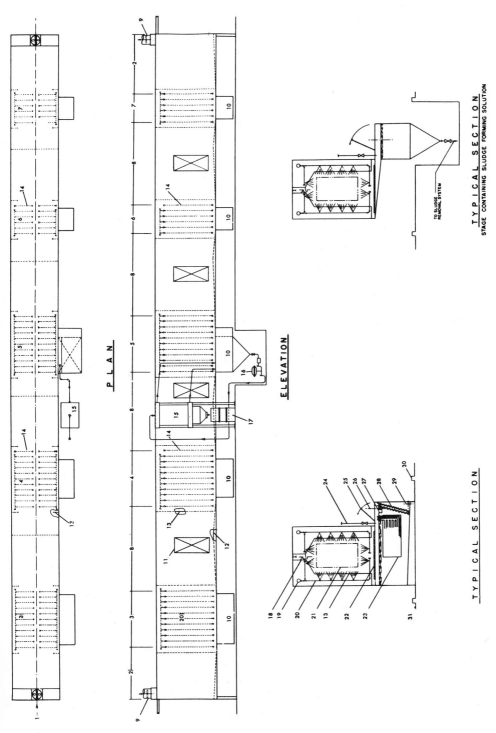

Fig. 13. Schematic layout of five-stage spray phosphating machine. (*See Reference Table for Fig. 13.*)

Reference Table for Fig. 13

Ref. No.	Machine Feature	Ref. No.	Machine Feature
1	Conveyor center line and direction of travel	17	Sludge drum
2	Vestibules, entrance and exit.	18	Conveyor guard
3	Cleaning stage	19	Header pipes from pumps
4	Water rinsing stage	20	Riser pipes bearing nozzles
5	Phosphating stage	21	Outline of work in process
6	Water rinsing stage	22	Grating forming floor of tunnel
7	Conditioning rinsing stage	23	Heating coils
8	Drain areas	24	Water fill line
9	Ventilating ducts with fans	25	Access lid to solution tank
10	Solution tanks	26	Pump center line
11	Access doors	27	Skimming trough
12	Sloping drain pans	28	Double removable screens
13	Baffles against overspray	29	Tank drain center line
14	Riser connected to fresh water line	29	Tank drain center line
15	Sludge filtering unit	30	Curbing around tanks, with drains within
16	Transfer pump—sludge laden solution	31	Floor

(d) Physical loss of solution with its heat content above cold water (and replenishing chemical) supply temperature (1) by dragout due to incomplete drainage, and (2) by entrainment of drops in the ventilating air.

(e) Total heat content of water evaporated from tank and spray.

(f) Heat gain of ventilating air.

Losses under a, b, c, and d(1) above are readily calculable by well-known methods, making reasonable assumptions as to entering and leaving temperature of the work, and as to rate of solution drag-out. It should be noted, however, that radiation from the solution surface is to surfaces at a temperature near that of the solution. These losses, however, are usually small compared to these under d(2), e, and f. The magnitude of these larger losses is dependent in a complicated manner on several factors in the operation of the apparatus, particularly on the pumping rate of the solution, the fineness of the spray, and the volume and direction of flow of ventilating air.

Empirical data taken from the operation of successful machines indicate that the entire heat requirement for a stage may be expressed in terms of a reasonable assumed temperature drop experienced by the solution in its circuit through pump and nozzles, over the work and back to the tank. While this procedure is not entirely justified on a scientific basis, the results are usually quite accurate enough for design purposes. See Table 6.

For tanks which are preceded by *unheated* rinse solutions or for first-system tanks, the following data apply.

B.T.U. per hour = gallons per minute $\times 500 \times$ temperature drop (F)

Metal treated	Steel
Type of phosphate coating	Medium weight zinc phosphate
Purpose of coating	Rustproofing, paint bonding
Coating time, min	2 to 4
Temperature	150 to 170 F
Type of current	60 cycles/sec, AC
Operating voltage	6 to 12 rms
Current density on work	25 to 50 amp/sq ft
Recommended material of construction	Tank, heating coils stainless steel 18:8

Notes:

(1) *Example:* A cleaning stage receives cold work entering a machine. The pumping rate is

TABLE 6. HEAT REQUIREMENTS FOR SPRAY STAGES
Estimated Temp. Drop Selection
(May Vary ± 25% Due to Conditions)

	Temperature Drop (F)	
Operating Temperature (F)	First and Last Stages and Any Following Cold Ones	Intermediate Stages
70–100	4	2
101–120	5	3
121–130	6	4
131–140	7	5
141–150	8	6
151–160	9	7
161–170	10	8
171–180	11	9
181–190	12	10

550 gpm and the solution operating temperature is 160 F. What is the heat requirement for this stage?

Answer: $550 \times 500 \times 9 = 2{,}475{,}000$ Btu/hr

(2) The above Btu requirements may be decreased by about 20 per cent when the preceding stage operates at a temperature within 25° of the one under calculation.

(3) Heat requirements calculated in the above manner should be sufficient to heat a cold solution to operating temperature within about 2 hr.

(4) If any unusually high losses are contemplated under items a, b, c, and d(1) above, these factors must be considered separately.

(5) The heat transfer area required to supply the calculated Btu input to the solution may be computed as though the tanks contained water only, except in the case of the phosphate-coating stages. In this latter case, the calculated heating areas should be substantially increased to provide for possible accumulation of scale.

(4) *Pumps, Piping, and Nozzles.* Any suitable arrangement of pumps, piping, and nozzles may be used which will give an adequate density of solution evenly distributed over the work. A common arrangement for coating partly fabricated objects is shown in Fig. 13.

Obviously not all possible arrangements can be discussed. However, certain recommendations will be made with respect to the design of such equipment for use in producing low to medium weight phosphate coatings on steel and zinc surfaces, together with some comments as to how these recommendations may be modified for other cases.

Design Features

In Table 7 are tabulated certain design constants compiled from many successfully operating power-spray washing machines.

(5) *Tunnels and Associated Equipment.* The spray zones and drain areas between spray zones are normally enclosed by a sheet-metal housing, often referred to as the tunnel. The tunnel (a) confines the sprayed solutions and (b) provides a channel for exhausting vapors to exhaust stacks (see Fig. 13 for a typical cross-section).

Normally baffles are mounted to the tunnel at the entrance and exit ends of each spray zone to minimize over-spray from stage to stage (see Fig. 13 for typical example).

A heavy grating is often placed at the bottom of the spray tunnel to provide a cat-walk through the machine and to prevent any work dislodged from conveyor from dropping into tanks (see Fig. 13).

It is desirable to have access doors in each drain area for entry to the tunnel, as well as for viewing treated work, carry-over of solution by the work, over-spray, etc. (see Fig. 13).

It is also desirable (especially for large units) to have lights on hinges mounted outside the tunnel adjacent to the access doors. These lights are adjustable and can be swung into the tunnel when needed. These fixtures can be purchased from any supply house dealing in lighting equipment for use on truck loading docks or they can be fabricated by the equipment manufacturer.

The area between two spray zones within the tunnel is the drain area, and the design of the length of these areas is fairly critical. The length of drain areas is based on the two following functions which this area must perform:

(a) Prevent, or maintain at an absolute minimum, over-spray of solution from stage to adjacent stage.

(b) Prevent or minimize physical carry-over or channeling of the solution (by work being treated) from stage to adjacent stages.

Intermixing of chemicals or solutions between any two adjacent stages is costly and very possibly, deleterious to quality. In general, drain area lengths should be only long enough to accomplish the functions listed in (a) and (b) above.

Usually the drain areas are sloped as shown in Fig. 13 with a 3-in. rise for every 10 ft of lineal drain.

The tunnel is normally extended beyond the first and last stages, to provide space for vestibules and exhaust "stacks." Also, the floor of the vestibule is sloped so as to serve as a drain pan for over-spray. Normally, the lengths of the vestibules are equal to about the height of the tunnel but usually not less than 3 ft (see Fig. 13 for typical section).

Exhaust fans are normally provided at the top of the vestibules at entrance and exit end of washer, as shown in Fig. 13. The fans are designed to prevent mists and vapors from escaping from the ends of the tunnel. As a general rule, mists will not escape if the velocity of the air into the ends of the tunnel is at least 125 fpm. Excessive heat loss is encountered if exhausts are overdesigned. It is also desirable to have drip baffles below the exhaust stacks to prevent condensed vapors from dripping onto work in process.

(6) *Drying Equipment.* A circulating hot-air oven, preferably indirect fired, is normally pro-

TABLE 7. TYPICAL DESIGN CONSTANTS FOR SPRAY COATING MACHINES

Purpose of Stage	Retention Time (min)	Spray Density (gpm/sq ft)	Horizontal Spray, Riser Spacing, Center to Center (in.)*	Average Vertical Spray Nozzle Spacing (in.)	Type of Nozzle	Nozzle Pressure (psi)
Cleaning	2.5–3.0	2.5–3.0	12–20	10–14	Flat spray	15–25
Rinsing	2.0–2.5	2.0–2.5	12–20	10–14	Flat spray	15–25
Pickling	1.5–2.5	1.5–2.0	12–20	10–14	Flat spray	10–15
Rinsing	2.0–2.5	2.0–2.5	12–20	10–14	Flat spray	10–15
Coating	3.0–5.0	1.6–3.0	12–20	10–14	Hollow-cone spray	8–12
Rinsing	2.0–2.5	2.0–2.5	12–20	10–14	Hollow-cone spray	10–15
Conditioning rinsing	2.0–3.0	1.5–2.5	12–20	10–14	Hollow-cone spray	

* Spray riser spacing may increase with increasing conveyor speed from 5 ft/min to 25 ft/min within the limits shown.

Notes for Table 7:

(a) Retention time means volume of solution in tank at working level in gallons, divided by pumping rate in gallons per minute.

(b) Spray density refers to the rate of solution delivery into the working zone per unit of maximum exterior work area which can be fitted into the clear working space. If a spray tunnel for partly fabricated work is in question, the "work area" is defined as the perimeter of the tunnel cross section inside of the baffles defining the silhouette, multiplied by the length of the stage between centers of the extreme spray risers.

If continuous strip is being treated, the "work area" is defined as twice the width of the strip by the length of the stage between extreme nozzle centers. If multiple strips are being treated, the total strip width is used in the formula.

Example 1: The cross section of a phosphating stage for refrigerators has a periphery within the baffles of 27 ft. It is 10 ft long between centers of first and last spray risers. The pump delivers 550 gpm. What is the spray density?

$$\text{Answer: } \frac{550}{27 \times 10} = 2.04 \text{ gpm/sq ft}$$

Example 2: A phosphating machine stage for flat strip coats four bands each 3 in. wide in 30 sec. The web speed is 100 ft/min, and the spray density required is 1.5 gpm/sq ft. What must be the pumping rate of the solution?

$$\text{Answer: } 2 \times 4 \times \frac{3}{12} \times \frac{100}{2} \times 1.5 = 150 \text{ gpm}$$

(c) Horizontal spray riser spacing recommended varies with conveyor speed over the indicated range between conveyor speeds of 5 to 25 ft/min. Higher conveyor (or continuous web) speeds permit even larger spacing.

If nozzles are mounted on pipes parallel to the conveyor travel (a possible but unusual arrangement), the tabulated values then mean nozzle spacing along the conveyor.

(d) "Vertical" spray nozzle spacing means nozzle spacing at right angles to the line of conveyor travel. This distance is chosen so that the work is evenly covered by the nozzle discharge. Stagger nozzles on successive risers.

(e) Nozzle type refers to the kind of spray pattern delivered. "Flat spray" means a nozzle delivering a pattern which is elongated and not finely atomized. Angle subtended by spray pattern at nozzle is about 30 to 50°. Hollow-cone spray means a nozzle delivering a hollow-cone pattern-automization is quite fine.

(f) Nozzle pressures are to be interpreted literally as actual pressure at the nozzle, not at the pump.

Further notes for Table 7:

(1) The data given above are for light to medium zinc phosphate coating of steel or zinc. For amorphous coating of aluminum, the use of flat spray nozzles is recommended throughout all the sections.

(2) All piping should be of ample size to feed the nozzles at full indicated pressure and volume, with a head drop from pump outlet to nozzle of not more than about 15 ft. Piping for phosphate-coating solution should be made one pipe size larger than the computations indicate to allow for accumulation of interior scale, etc. The highest point of pump section intakes should be located at least one foot below the solution level to eliminate the possibility of sucking air into the piping system—the lowest points of intakes should be at least 6 in. off the bottom of the pump section of the tank.

vided. Oven should be of such capacity as to dry the coated work in from 3 to 5 min with air at 120 to 177 C (250 to 350 F). Vapor and spray mist from the final stage of the washer should not be allowed to enter the oven. This is accomplished by removing the oven from the spray washer a distance of 3 ft.

If drain holes cannot be located in the work so as to remove all "pocketed" or entrapped final rinse solution, suitable compressed air jets can be directed to these areas to disperse this solution before work enters the drying oven. Other methods of solving this drainage problem may be employed such as tilting the racked work.

(7) *Auxiliary Equipment.* It will be necessary in most spray zinc phosphate-coating systems to remove sludge from the coating stage tank from time to time. This removal can be accomplished by (a) settling and decantation, or (b) filtration.

For decantation a conical bottomed tank of capacity sufficient to hold about 1/3 to 1/2 the volume of the coating solution is provided. This tank has a bottom valve of large diameter for removal of settled sludge, and a side opening with valve (about 1/3 of the volume up the tank) for returning the clarified solution to the phosphating tank by gravity. In order to transfer settled sludge from the phosphating tank to the settling tank periodically, a pump of 75 to 150 gpm capacity equipped with flexible hose on the suction side is normally employed.

Continuously or intermittently operated filtration equipment has also been used advantageously. Such equipment is a little more expensive initially, but is somewhat more efficient in the recovery of clear solution than is the decantation equipment.

Phosphating processes normally operate more efficiently if replenishment of coating chemicals to the phosphating tank is continuous. This is best accomplished by feeding the chemicals in solution by means of metering pumps adjustable for deliveries in range of 0 to 5 gal/hr. Of course, pumps must be chemically resistant to the solutions being fed. Translucent plastic tubing is recommended for use on suction and discharge lines.

Simplified and Specialized Processes

A simplified process is one in which one or more of the process steps is accomplished manually instead of mechanically.

A specialized process is one in which unusual properties are imparted to the coating being applied.

Simplified Processes. Quite often simplified processes are employed when the items to be treated are too large for convenient application of immersion or conventional spray technique, or when the production is too small to warrant the installation of fully mechanized equipment.

Purely manual cleaning and application of paint bonding phosphate coatings is sometimes employed in the treatment of aluminum, steel, zinc, or cadmium surfaces. Examples of such applications are: large structures, such as buildings, bridges, gas holders, ships, etc., as well the refinishing of automobile bodies.

An example of a semimechanized process is illustrated in Fig. 14.

Specialized Processes. Specialized processes involve additional steps or equipment to those which have been described above. Two kinds of specialized processes are of sufficient interest to require mention: (1) Production of colored phosphate coatings. (2) Electrophosphating.

(1) *Production of Colored Phosphate Coatings.* Such coatings are produced by adding one or more steps to the usual coating process, by which colored compounds are formed in or on the phosphate coating. These steps, however, consist merely in treating the coated surface with one or more aqueous solutions, with or without intermediate rinsing. From the engineering standpoint, the fitting of such additional steps into a phosphate coating process involves no principles which have not been discussed. Cycle time, temperature, etc., as well as materials of con-

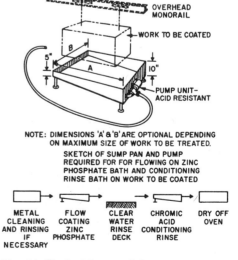

Fig. 14. Typical layout of flow zinc phosphate process.

struction for the coloring chemical stages of such proprietary process will be specified by their proprietors. Usually short treatment times, low temperatures and dilute metal salt solutions are involved.

The requirements of waste disposal and fuel economy have resulted in the development of very low-temperature 20 to 30 C (70 to 90 F) iron phosphating systems, non chromated sealing solutions and solvent based systems. The basic principles of equipment design are, however, similar.

Cathodic electrodeposition systems for applying paint are becoming increasingly popular (see Chapter 38). These systems require special characteristics in the phosphate coatings used in the pretreatments. It is important, therefore, that these special factors be considered when selecting and designing a pretreatment system for such a coating. Equipment design, however, will not usually be different from that described above.

Acknowledgments

Sincere thanks are due to George A. Russell, P.E., Director of Engineering, and to other members of the Engineering Department, Amchem Products, Inc., for their helpfulness in the preparation of this section.

Bibliography

Only a sufficient bibliography is listed here to give the reader a general approach to the subject of phosphating. Little or no literature exists on the engineering phases of the art.

Books:

Machu, W., Die Phosphatierung, Verlag Chemie, Weinheim/Bergstrasse, 1950.

Kohler, W., Fortschritte auf dem Gebiete der Phosphatierung, Verlag Chemie, Weinheim/Bergstrasse, 1950.

Articles, Pamphlets, etc.:

Burbank, J. B., Phosphate Coatings on Steel, Parts 1 and 2, Naval Res. Lab. Reports C-3481, June 1, 1949, and C-3510, July 28, 1949.

Holden, H. A., Latest Developments in Phosphate Coating Methods and Techniques, *J. Electrodepositors' Tech. Soc.*, **24**, 111 (1949).

Machu, W., Phosphate Coating and Its Scientific Fundamentals, *Metallwirtschaft*, **22**, 33-35, 481-487 (1943).

Machu, W., Porosity and Rate of Formation of Phosphate Coatings, *Arch. f. Metallkunde*, **8**, 278 (1949).

Machu, W., Porosity of Nonconducting Coatings, etc., *Korr. u. Metallschutz*, **20**, Nos. 1 and 6 (1944).

Machu, W., Kinetics of Coating Formation, *Arch. f. Metallkunde*, **3**, 203 (1949).

Streicher, M. A., Phosphatization of Metallic Surfaces, *Metal Finishing*, **46**, 61-69 (1948).

Reviews:

Clarke, S. G., Phosphating Processes in Germany PB 79180, Jan. 1946, BIOS Trip 1670, British Intelligence Objectives Sub-committee, London, England.

Macchia, O., Researches into Phosphate, etc., Part 3, Critical Review of Patents, *L'Industria Meccanica*, **18**, 566, 620, 694 (1936).

B. Chromate Conversion Coatings

WILLIAM P. INNES

Vice President and Technical Director, MacDermid Inc., Waterbury, Conn.

Chromate conversion coatings were developed to minimize the formation of "white rust" on zinc electroplated coatings and to guarantee paint adhesion. Success with zinc led to treating cadmium in the same manner. Later, a broad range of conversion coatings[1-7,11,12,13,16,17,18,19] were developed for silver, copper, beryllium, tin, copper alloys, aluminum, magnesium, zinc base die castings and electroplated chromium.

Chromate conversion coatings are formed because the metal surface dissolves to a small extent, causing a pH rise at the surface-liquid interface. This results in the precipitation of a thin complex chromium-metal gel on the surface, composed of hexavalent and trivalent chromium and the coated metal itself. This gel is normally soft when formed and therefore the coated work must be carefully handled. After drying, the coating becomes hard and relatively abrasion-resistant. If heat treating for relief of hydrogen embrittlement is required, this should be done before chromating because the chromate film will not withstand high temperature > 65 C (150 F) treatment unless protected from dehydration with a supplemental protective organic coating.

Good processing of electroplated zinc and cadmium is aided by keeping in mind the following.

1. Smooth, fine grained deposits with uniform thickness respond best to chromating.
2. The codeposition of other metallic impurities, especially copper and lead, can result in defective conversion coatings.
3. Good plating practice dictates the removal and neutralizing of plating solutions prior to chromating. An activating solution consisting of dilute nitric acid (½%) or a proprietary solution is used to neutralize and remove the plating solution.
4. Maintenance of the chromate solution is controlled by vendor methods and the experience of the operator.
5. Good water rinsing after chromate is essential. Bleaching (NaOH or Na_2CO_3) may be required to provide a clear noniridescent coating.
6. The final rinse should be agitated and warm 50 to 60 C (120 to 140 F) with sufficient flow to avoid the buildup of acid or alkali to prevent the removal of the coating.
7. The drying temperature should not exceed 70 C (160 F) to prevent destroying the protective quality of the coating.
8. If chromating is not done immediately after plating, the zinc or cadmium should be activated by suitable alkaline cleaning and acid dipping before conversion coating.

MECHANISM OF CORROSION PROTECTION

Chromate coatings are adherent to the metal and provide self-healing protection from atmospheric conditions in rural, industrial, inland, and seacoast locations. Protection is provided by the leaching of the soluble chromate content in the film. The degree of protection is improved by increasing both the chromate content in the film and the thickness of the metal deposit.

The iridescent and olive drab coatings have a higher chromate content than the clear coatings and provide the maximum corrosion resistance. A typical analysis of a chromate conversion coating is:

	%
Cr^{+6}	8.68
Cr^{+3}	20.22
S (as SO_4)	3.27
Zn^{+2}	2.12
Na^+	.32
H_2O	19.3
Oxygen	Balance

TYPE OF COATINGS

Coatings fall into one of two categories:

1. Clear
   a. Single dip blue bright or white
   b. Bleached white or blue bright
2. Colored
   a. Dyed
   b. Iridescent bronze
   c. Olive drab
   d. Black

Zinc and cadmium are widely used primarily because of their unique property of sacrificially protecting carbon steel from corroding. This is due to the electrochemical cathode protection provided by these metals.

The decorative appearance of these coatings is improved by the chromate coating. The solution polishes or smoothens the surface, which improves the specular reflectivity or brightness. Surface film and nonuniform haze, which would interfere with specular reflection, are removed.

Greater resistance to finger staining during handling is provided by the coatings. The conversion coating provides improvement in corrosion resistance and the treated work retains its brilliance longer in mild exposure conditions. Shelf life is significantly improved.

Generally, chromated zinc is more resistant in industrial atmospheres. Cadmium corrodes less in rural and seacoast environments.

Dyed finishes are produced in a supplemental processing step after the chromate coating. Once the chromate coating has been dried, it cannot be dyed. Applications for dyed coatings are:

1. Color coding; e.g., blue for metric and red for other than metric. This is widely used in the fastener industry to provide identification for different sizes.
2. Black for camera parts and where optical properties are required.
3. Simulated brass for lamps, wire goods and furniture. This type of work is usually painted to provide added durability.

METALS COMMONLY CHROMATED

Commonly chromated metals are:

1. Electroplated metals
   a. Zinc—decorative, corrosion protection and paint base.
   b. Cadmium—decorative, corrosion protection and paint base.
   c. Copper—decorative, tarnish resistance and paint base.
   d. Tin—tarnish and corrosion resistance.
   e. Chromium—corrosion resistance and paint base.
   f. Brass—tarnish resistance and prior to painting.
   g. Silver—tarnish resistance due to sulfide.
2. Zinc base die castings for deburring, decorative, corrosion protection and paint base.[17]
3. Copper and copper alloys, aluminum and magnesium—oxide removal, decorative, tarnish resistance and paint base.
4. 316 stainless steel—black decorative coating.[15]
5. Hot dipped galvanized and mechanically plated zinc coatings.

ELECTRICAL RESISTIVITY OF CHROMATE COATINGS

Metal	Type of Film	microhm-cm
Zinc	None	20–50
Zinc	Clear	50–100
Zinc	Yellow	100–1,000
Cadmium	Clear	70–120
Aluminum	None	200–500
Aluminum	Clear	500–900
Aluminum	Yellow	800–2,000
Silver on steel	Clear	Increase of 27

The above chart shows the resistivity of chromate coating on zinc, cadmium, aluminum and silver. The resistivity of the chromate film is low and can be classified as an electrical conductor. This enables electrical connectors to be protected without increasing the surface resistance. The major advantage is to provide constant contact resistance in various environmental exposures. Silver is an excellent conductor, but the contact resistance of uncoated silver increases after aging due to the formation of silver sulfide. A chromate coating solves this problem.

Chromate coatings may be soldered with a suitable rosin flux. Spot welding or fusion welding also works well.

THE CHROMATE COATING SOLUTION

Chromate solutions are generally preferred with proprietary liquid or dry components and made up to the vendor's recommendations. There are three basic types of solutions:

1. *Hexavalent chromium*—provides maximum corrosion protection
   a. Immersion
   b. Electrolytic
2. *Trivalent chromium*—provides less protection[8,9]
3. *Chromium-free coatings*—provide very little protection[10,20]

Control Factors for Hexavalent Chromate Solution

1. Concentration—by titration
   Cr^{+6}
   Cr^{+3}
2. pH—by pH meter or conductivity cell
   pH—0 to about 2.8
3. Temperature, 24 to 32 C (75 to 90 F)
4. Immersion time
   10 to 30 seconds (less for continuous strip lines)
5. Agitation
   Air
   Mechanical
   Solution flow
6. Transfer time
   Should be ≤ 25 seconds

TYPE OF PROCESSING

1. Spray, transfer roller coat or flooding
   a. With subsequent water rinsing

b. Without water rinsing for some coatings on strip lines
2. Immersion
   a. Barrel
   b. Basket
   c. Rack

EQUIPMENT

Tanks
316 stainless steel with 316 ELC welds
Polypropylene, polyethylene, PVC, Teflon
Glass for F^- (fluoride) free solutions

Heaters
Type 316 stainless steel
Teflon
Titanium or tantalum for F^- free solutions
Glass or quartz for F^- free solutions

Pumps
CPVC
Polypropylene
Stainless steel type 316

Air coils
PVC

Ventilation
PVC

Baskets
PVC
Stainless steel type 316

Racks
Plastisol with 302 stainless steel rack tips

AUTOMATIC CONTROL

1. A control device is used to maintain the concentration by conductivity which feeds a signal to pump fresh concentrate.
2. Automatic temperature control.
3. Ion exchange to remove contaminants with or without evaporation for recovery and reuse.

It is good practice to maintain the chromate processing solution by chemical analysis and to use automatic control. This provides maximum utilization of the chromate and reduces frequency of waste treatment.

Ion exchange in a reciprocating (cation/anionic bed) system is in use to remove decomposition cations and anions. It is coupled with evaporation for recovery and reuse of chromate dragout, bailout and spent solutions. This can be used in some chromate solutions to recover and reuse solution which lowers operating cost and reduces the quantity of spent solution that requires treatment.

PROCESS SEQUENCE

1. Electroplated metals
2. CW (cold water rinse)
3. CW (indicates counterflow rinsing)
4. ½% HNO_3 or proprietary activator
5. CW
6. Chromate
7. CW
8.* CW
9. HW, maximum 50 to 60 C (120 to 140 F) (hot water rinse)
10. Air dry, maximum 70 C (160 F)

Bulk plating procedure for iridescent bronze, olive drab or black:

1. Do not rotate the barrel in the chromate or during subsequent rinsing and drying to avoid rubbing off or scratching the coating.
2. Or transfer to a basket for chromating, rinsing or drying.

Processing copper, copper alloys, aluminum and magnesium:

1. Clean
2. CW
   CW
3. Oxide removal
4. CW
5. Chromate
6. CW
   CW
   HW maximum 60 C (140 F)
7. Dry, maximum 70 C (160 F)

CONTROL OF ELECTROPLATING SOLUTIONS

1. Control of contaminants:
   a. Zinc: cyanide, alkaline (CN-free) and chloride
   Remove:

Cu	Zinc dust
Pb	Sulfide (for cyanide) and zinc dust for alkaline and chloride
Cr	H_2O_2—chloride $Na_2S_2O_4$—cyanide and alkaline
Fe	H_2O_2 or $KMnO_4$—chloride

*Bleached coating requires a NaOH or Na_2CO_3 leach.

b. Cadmium: cyanide and acid
Remove:

Cu	Zinc dust
Cu	Zinc dust
Pb	Zinc dust
Cr	$Na_2S_2O_4$—cyanide
	H_2O_2 acid

2. Chemistry:
Control factors within recommended range including brightener and other additions to produce smooth, fine grained deposits free of impurities and, when needed, uniform brightness without burn and uniform thickness.
3. Thickness of zinc and cadmium:
The thickness should be not less than 3.75 μm average. Chromating generally removes 0.25 to 0.5 μm of the deposit.

Troubleshooting

Problem with Chromate Coating	Cause
No coating	Chromate depleted or pH too low
Leaching in teardrop	Transfer time too long (> 25 seconds)
Blue bright to white	Immersion time too long or chromate concentration too high
Bronze rub-off	Immersion time too long or pH too low
Iridescent blue bright	Immersion time too long or lack of air agitation
Stripped zinc or cadmium	Thin electroplate; immersion in chromate too long; temperature too high
Cloudy blue bright	Temperature too high
Dull blue bright	Temperature too low
Loose bronze coating	Temperature too high
Dehydrated coating	Drying temperature > 70 C
Poor corrosion resistance	Drying temperature > 70 C

SALVAGING REJECTS

Electroplated zinc, chromated and unchromated, can be stripped in mineral acid (HCl or H_2SO_4) and reprocessed. Cadmium can be stripped in other chemical solutions. Chromate coatings on zinc die castings, copper, copper alloys, aluminum and magnesium can be stripped in alkaline and/or acid and reprocessed.

COATINGS FOR CONVERSION COATINGS

Wax, water base polymers, solvent based polymers and powder coated polymer coatings are used on conversion coatings for the following reasons:

1. To improve the durability of the coated zinc:
   a. Water and solvent based polymer coatings
   b. Powder polymer coatings[14]
2. To provide better abrasion resistance (e.g., for refrigerator shelving):
   a. Water and solvent based polymer coatings
   b. Powder polymer coatings
3. To provide lubricity and improved durability:
   a. Water emulsified wax.

ACCELERATED CORROSION TESTS

Accelerated corrosion tests are used to test durability of the chromated work after processing both by the plater and the buyer. It is difficult to correlate the test to actual durability in service. Nevertheless, these tests, coupled with field service experience, serve as a valuable tool in the quality control of chromate coatings.

The 5% lead acetate test is simple to run and interpret. A drop of solution (5 g $Pb(C_2H_3O_2)_2$ per 100 ml water) is placed on the surface. The time required to obtain a black spot in the drop of liquid is recorded. A clear blue bright single dip (no bleach) chromate on zinc plate should withstand the lead acetate for $\geq$ 5 seconds.

The time for first white corrosion on chromated zinc in 5% neutral salt spray is given below:

Coating	Hours to First White Corrosion
Zinc plate, no chromate	1 or less
Single dip clear	8 to 24
Bleached or leached clear	32 to 48
Iridescent bronze or yellow	72 to 96
Dark bronze	96 to 144
Drab	144 to 250

Painted chromate coatings are checked in 100% relative humidity at 95 F (36 C) and in 5% neutral salt spray. The painted test panels are cross-hatched with an X. The X is cut through the paint. First failure occurs by blistering or poor paint adhesion (peeling) in the X area.

Sodium sulfide test on chromated zinc plate is used to check the quality of the conversion coating.

SPECIFICATIONS

Military, Federal, SAE, ASTM, ASI and other specifications are used as a standard agreement between the processor, buyer and user to measure:

1. Type of coating.
2. Thickness: where and how to check.
3. Corrosion resistance and specifically how to check.
4. Weight, color.
5. Method of sampling for testing.
6. Qualified product lists, certification of processor by procurement and other requirements.

Government Specification	Coating for
USA 57-0-2C	Zinc
MIL-F-14072	Zinc die castings
QQ-Z-325 A	Zinc
MIL-C-17711	Zinc die castings and hot dip galvanized surfaces
MIL-T-12879	Zinc die castings
QQ-P-416C	Cadmium
MIL-M-3171C	Magnesium
QQ-S-365B	Silver
MIL-C-5541C	Aluminum
MIL-C-81706	Aluminum, also QPL-81706-1

AMS Specification	Coating for
2400B	Cadmium
2400K	Cadmium
2402D	Zinc
2473C	Aluminum
2474C	Aluminum
2475B	Magnesium
B449	Aluminum
B201	Zinc

SAFETY

Chromate dry and liquid chemicals are acid and strong oxidizing agents. Storage should be segregated from any combustible (organic) chemicals or materials (rags, paper, etc.). Spills should be confined, picked up with suitable safety equipment and the area flushed thoroughly with cold water. The spilled material should be treated prior to disposal by reducing any Cr^{+6} to Cr^{+3}, neutralizing and precipitating.

Personnel handling chromating chemicals should wear eye protection (goggles or a face shield). Protective clothing and gloves should be worn. Exposed skin or eyes should be flushed immediately with clear, cold water. Eye exposure must be flushed continuously for more than 20 minutes with cold water. Contact a doctor immediately for the treatment of any injury caused by chemical contact.

WASTE TREATMENT

Chromate containing conversion coatings which require treatment should be handled as follows:

1. Reduce Cr^{+6} to Cr^{+3} with SO_2, $NaHSO_3$ or $Na_2S_2O_5$ in solution below pH 6.0. Above pH 6.5, use sodium hydrosulfite.
2. Raise pH to 8.5 to 9.5 with lime or sodium hydroxide.
3. Remove $Cr(OH)_3$ by settling or filtration.

References

1. Wilhem, E. U.S. Patent 2,035,380 (1936).
2. Meyer, W. and M. Dunlevy. U.S. Patents 2,088,429 (1937); 2,172,171 (1939).
3. King, V. U.S. Patent 2,393,640 (1946).
4. Anderson, E. *Proc. Am. Electroplat. Soc.* **31**:6 (1943).
5. Thomas, R. and C. Ostrander. U.S. Patents 2,393,663-4; 2,393,943 (1946).
6. Ostrander, C. In *Modern Electroplating*, 2nd Ed. F. A. Lowenheim, Ed. New York: Wiley, 1963, pp. 624-626, 630.
7. Halls, E. E. *Metallurgia* **38**:75 (1948).
8. Bishop, C. and T. Foley. U.S. Patent 4,171,231.
9. Guhde, D. and E. Burdt. U.S. Patent 4,263,059.
10. Zuenot, R. and A. B. Bengali. U.S. Patent 4,225,351.
11. DeLong, H. and D. Ritzema. U.S. Patent 3,447,973.
12. Pearlstein, F. *et al.* U.S. Patent, 3,485,682.
13. Shackleford, R. *et al.* U.S. Patent 3,419,001.
14. Hamilton, D. *et al.* U.S. Patent 3,535,166.
15. Hart, A. U.S. Patent 3,755,117.
16. Caule, F. U.S. Patent 3,833,433.
17. Kinder, D. U.S. Patent 3,060,071.
18. Bellinger, K. *et al.* U.S. Patent 3,171,765.
19. Giesker, W. *et al.* U.S. Patent 2,746,915.
20. Kinder, D. U.S. Patent 3,843,430.

C. Sulfuric and Chromic Acid Anodizing of Aluminum

W. C. Cochran

Recently retired from Alcoa Laboratories, Alcoa Center, Pa.

Anodizing is an electrolytic oxidation process in which an aluminum component is made anodic in a cell containing an aqueous, acidic electrolyte and a metal cathode. When electric current is passed through the cell, the aluminum surface is converted to an adherent aluminum oxide coating which is integral with the aluminum substrate. Anodizing may be regarded as the artificial thickening of the thin (1 to 5 nm) native oxide film that is always present on aluminum exposed to the atmosphere. During anodizing, oxide and hydroxyl ions from the electrolyte are driven to the aluminum surface where they penetrate the existing oxide film and combine with aluminum ions from the metal to build up aluminum oxide/hydroxide coating near the metal/oxide interface. A nonporous barrier oxide layer whose thickness is proportional to the applied voltage (about 1 nm/volt) forms initially and persists as an advancing front into the metal during the anodizing process. Porous oxide coating soon develops above the barrier zone due to the dissolving action of the acidic electrolyte. The porous structure permits continued growth in thickness of the coating until equilibrium is established between formation and dissolution of coating. Anions from the electrolyte are incorporated in the coating, typically about 15 weight % sulfate for sulfuric acid anodic coatings, but only trace amounts of chromate (< 0.1 weight %) for chromic acid anodic coatings. The porous nature of anodic coatings is advantageously used to incorporate dyes, pigments, corrosion inhibitors or lubricants. Normally, the porosity of the coatings must be closed by a sealing process for best performance in service.

Sulfuric acid is the most widely used anodizing process for aluminum due to low electrolyte and operating costs and versatility in coating thicknesses possible, up to 25 µm (1 mil) for conventional coatings and to 100 µm (4 mils) for hard coatings. Chromic acid anodic coatings are usually limited to thicknesses of from 0.5 to 3 µm (0.02 to 0.12 mil) and owe their use to the fact that chromic acid does not appreciably attack aluminum alloys. Thus, there is no danger of corrosion from inadvertent entrapment of chromic acid electrolyte in joints and crevices of components. To avoid pollution hazards and waste treatment costs, there has been a trend away from chromate-based processes. In spite of this, chromic acid anodizing continues to be used in aircraft and defense-related applications.

Classes of Coatings

Table 1 classifies sulfuric and chromic acid anodic coating types by thickness, service conditions and typical applications.[1] The classification and minimum requirements for anodic coatings, as used by agencies of the Department of Defense, are summarized in Table 2.[2] Weight of anodic coating and its protective ability in neutral salt spray are principal criteria for Types I and II coatings. Type III coatings have thickness and abrasion test requirements. Coatings are further classified as dyed or not dyed.

Coating Functions

The principal functions of anodic coatings are to provide hard, wear- and scratch-resistant surfaces on aluminum parts, to protect the metal surface from corrosion and staining and to achieve and maintain specific decorative texture, color and appearance effects. Additional functions include maintaining high reflectivity, controlling thermal properties (heat absorption and emittance), and providing electrical insulation. Anodic coatings also find use as a basis for organic coatings, adhesives, electroplated metallic coatings or lubricants.

SULFURIC ACID ANODIZING

Sulfuric acid, the most widely used anodizing process for aluminum, was patented in 1927 by Gower and O'Brien in England.[3] Various concen-

TABLE 1. Anodic Coating Types and Service Conditions (Adapted from ASTM B580)

Type	Minimum Coating Thickness, μm (mils)	Electrolyte	Service Condition	Typical Applications
A—Engineering, Hard Coat	50 (2.0)	H_2SO_4 or H_2SO_4 + oxalic acid	5—*Very severe.* Prolonged atmospheric weathering, high wear conditions.	A, B—Unmaintained exterior architectural components. Machinery or marine parts.
B—Architectural Class I	18 (0.7)	H_2SO_4		
C—Architectural Class II	10 (0.4)	H_2SO_4	4—*Severe.* Resist scratching, abrasion, weathering and corrosion.	C—Maintained exterior architectural parts.
D—Automotive Exterior	8 (0.3)	H_2SO_4		D—Exterior automotive trim.
E—Interior, Moderate Abrasion	5 (0.2)	H_2SO_4	3—*Moderate* abrasion, occasional wetting.	E—Appliances, furniture, nameplates, reflectors.
F—Interior, Limited Abrasion	3 (0.1)	H_2SO_4	2—*Mild* indoors, minimum wear.	F—Automotive—interior, housewares, enclosed reflectors.
G—Chromic Acid	1 (0.04)	CrO_3	1—*Crevice condition*, humid, little or no abrasion.	G—Aircraft assemblies with lap joints. Base for paint.

TABLE 2. Miltary Specification MIL-A-8625C Requirements: Anodic Coatings for Aluminuim Alloys

Coating Type	Minimum Coating Weight, g/m² (mg/ft²)	Salt Spray Resistance, 336 hours	Maximum Color Change, Class 2 Coatings, 200 hours— Carbon Arc Light	Alloy Limitations*
I. Chromic acid, sealed		Maximum of 15 isolated pits in 10 dm² (150 in²), or 5 in 2 dm² (30 in²), none >0.8 mm (1/32 in.) in diameter		Maximum Cu = 5% Maximum Si = 7%
Class 1 (not dyed)	2.15 (200)		3 E units	
Class 2 (dyed)	5.38 (500)			
II. Conventional sulfuric acid, sealed		Same		—
Class 1 (not dyed)	6.46 (600)			
Class 2 (dyed)	26.91 (2500)**		3 E units	
III. Hard coatings,+ not sealed*	— —	—	3 E units	Maximum Cu = 5% Maximum Si = 8%

*Unless otherwise specified in contract.
**15.07 g/dm² (1400 mg/ft²) for 2XXX alloys and certain casting alloys.
+Nominal thickness = 51 μm ± 10% (2 mils ± 10%), unless otherwise specified. There is an abrasive wear test requirement.

trations of sulfuric acid in water have been used as the electrolyte, ranging from 1 to 65 weight %, but 12 to 22% concentrations are most often employed. Two general types of coating are produced, conventional and hardcoat. Conventional coatings are used for protective and decorative purposes in thicknesses from 2 to 25 μm (0.08 to 1.0 mil).

Hard coatings are mainly specified for engineering purposes. They are produced with lower solvent action of the electrolyte on the oxide, and hence are denser and can be made thicker than conventional anodic coatings. Thicknesses of 25 to 100 μm (1 to 4 mils) are employed. Hard coatings are processed at higher voltages (25 to 100 V) and current densities (2.6 to 3.9 A/dm^2) (24 to 36 A/ft^2) than conventional coatings which require only 12 to 20 V and 1.3 A/dm^2 (12 A/ft^2).

Figure 1 shows the weights of anodic coatings produced on various wrought aluminum alloys by the conventional sulfuric acid process and by a hard coating process.[4]

Processing Steps

The following is a general description of the processing steps for producing conventional and hard coatings.

1. *Cleaning.* Organic soils such as oils and greases are removed by vapor degreasing, solvent washing, inhibited alkaline soak cleaners, emulsion cleaners, ultrasonic cleaning or cathodic cleaning. (See Chapter 3, Part F.)
2. *Rinsing.* Flowing cold water.
3. *Deoxidizing.* A deoxidizing type etch should be used on alloys that have a heat treat oxide film; i.e., alloys in a "T" temper. Deoxidizing is unnecessary for nonheat-treated alloys ("H" tempers). Hot sulfuric-chromic acid solution is an excellent deoxidizer. A solution containing 16 weight % H_2SO_4 + 3 weight % CrO_3 is used for 2 to 5 min. at 70 to 80 C (158 to 176 F). Suitable proprietary deoxidizing etches

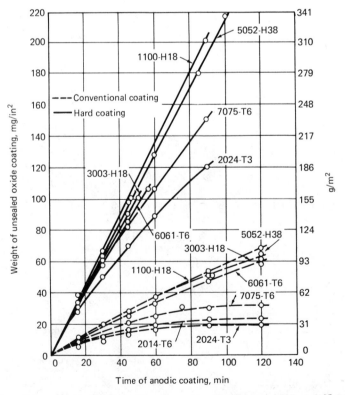

FIG. 1. Relation of coating weight to anodizing time for conventional process, 15 wt. % H_2SO_4, 21 C (70 F), 1.3 A/dm^2 (12 A/ft^2), and for a hard coating process, 12 wt. % H_2SO_4 + 1 wt. % oxalic acid, 10 C (50 F), 3.9 A/dm^2 (36 A/ft^2).[4]

that contain no chromates are available. (See Chapter 3, Part G.)
4. *Rinsing.*
5. *Etching.* Etching is most commonly done in warm 5 weight % caustic soda solution. Temperatures are from 43 to 66 C (110 to 150 F) and contact times vary from 0.5 to 10 min, depending upon the amount of metal to be removed or the textural effect desired. Acid etches are sometimes used in place of caustic etching.
6. *Brightening.* In bright anodizing, chemical or electrolytic brightening is used in place of etching. The more common procedure is chemical brightening by immersion of the work in a bath comprising 81 weight % H_3PO_4 + 3 weight % HNO_3 + 16 weight % H_2O + 0.5 g/L copper nitrate crystals for from ½ to 5 min at 85 to 105 C (185 to 220 F). Another composition, sometimes used in small installations, contains 70 weight % H_3PO_4 + 2 weight % HNO_3 + 15 weight % acetic acid + 13 weight % H_2O.[5] Acetic acid reduces dragout loss by lowering the viscosity of the bath. Electrolytic brightening in special electrolytes is sometimes used, but is more costly than chemical brightening.
7. *Rinsing.*
8. *Desmutting.* A room temperature 50 volume % solution of nitric acid is used to dissolve the dark smut present on the surfaces of items that have been etched or chemically brightened. The deoxidizing etch (Step 3) may also be used for desmutting etched surfaces. Treatment time is usually from 5 to 30 sec.
9. *Rinsing.*
10. *Anodizing.*
    a. *Conventional coatings* are produced typically in a 15 to 18 weight % H_2SO_4 electrolyte with a current density of 1.3 A/dm^2 (12 A/ft^2) at a temperature of 21 C (70 F). Time of anodizing is from 10 to 60 min, depending on the coating thickness desired. Thickness development for common alloys is at a rate of 0.36 $\mu m/min$ (0.014 mil/min). Aluminum-copper alloys (2XXX series) have a lower rate. For coatings that are to be colored by dyeing, an electrolyte temperature of 27 to 30 C (80 to 86 F) is employed.
    b. *Hard coatings* are produced within a range of conditions that include the use of electrolytes of 12 to 22 weight % H_2SO_4 with current densities of from 2.6 to 3.9 A/dm^2 (24 to 36 A/ft^2) and temperatures of 0 to 10 C (32 to 50 F). With a current density of 3.9 A/dm^2 (36 A/ft^2), coating thickness growth is at a rate of about 1.3 $\mu m/min$ (0.05 mil/min).

    Several proprietary hardcoat processes are used which include minor additions of chemicals to the H_2SO_4 electrolyte, or the use of different wave forms of electric current such as pulsed dc, or ac superimposed on dc.
11. *Rinsing.*
12. *Coloring.* When desired, color is imparted to the anodic coating by impregnating its pores with dyes or mineral pigments, or by electrodeposition of metallic pigments (see the section on coloring).
13. *Rinsing.*
14. *Sealing.* The porosity in anodic coatings is closed by sealing for 15 to 25 min in boiling deionized water (pH 5.6 to 6.5), or for 5 min in 0.5 weight % nickel acetate solution (pH 5.5 to 5.8). Proprietary chemicals that minimize sealing smudge are often added to sealing baths. These are polyhydroxyl or polycarboxylate organic compounds having molecular weights and geometries that prevent the formation of surface smut without interfering with the sealing reaction in the pores.[6]

    For improved corrosion protection in salt environments, sealing is done in 5 weight % sodium or potassium dichromate solution, for 15 min at 95 to 100 C (203 to 212 F) at a pH of 5 to 6.5.

    Hard coatings for engineering uses are not sealed because sealing reduces the coatings' abrasion resistance. However, when intended for outdoor use without maintenance, hard coatings are sealed for optimum resistance to weathering.
15. *Rinsing.* (Optional when boiling water seal is used.)
16. *Drying.* The parts are permitted to dry in ambient air, or may be force-dried in a stream of oil-free compressed air, or in heated air at a maximum part temperature of 105 C (221 F). If higher tempera-

tures are used, fine, hairline craze cracks may form in the anodic coating.

Coloring

Coloring is useful for decoration or for coding of small components such as rivets and other fasteners. Colored anodic coatings are also used to control the optical and thermal properties of aluminum components. Color is produced by different methods, the most common of which is *dyeing*. The porous surface of an unsealed anodic coating is an excellent base for adsorbing and retaining dyes. Water-soluble dyes of the acid (anionic) type are most commonly used, typically at concentrations of about 0.1 to 1 weight %, with contact times of 5 to 15 min at temperatures of 49 to 66 C (120 to 150 F), in a pH range of 5 to 7. Dyeing conditions recommended by the dye manufacturer should be used. Most dyed anodic coatings are not sufficiently colorfast for prolonged outdoor use, but selected dyes are available that provide weather-resistant color finishes for anodized architectural components. Exemplary of such dyes are the "Sanodal" series, Deep Black MLW, Black GL Paste, Turquoise PLW, Blue G, Red B3LW and Yellow 3GL.[7]

Inorganic pigments can be precipitated in the pores of anodic coatings to produce stable colors. The only two that have had significant commercial use are iron oxide for golden hues, and cobalt oxide for bronze shades. Iron oxide is precipitated in the anodic coating by contacting it with a warm, dilute solution of ferric ammonium oxalate. Cobalt oxide is formed in the coating by a two-step procedure—immersion of the work in a cobalt acetate solution followed by rinsing and immersion in a potassium permanganate solution.

Integrally colored anodic coatings are produced on aluminum alloys containing silicon, chromium or manganese. The colors are developed during the anodizing process and do not require a separate coloring step, though they may be overdyed to modify the color. Aluminum-silicon alloys (4XXX series) and aluminum casting alloys containing silicon yield light to dark gray colors when anodized, due to occlusion of microparticles of silicon in the coating matrix. Aluminum alloys containing chromium in solid solution produce golden shades and those containing manganese yield light beige hues when anodized by the conventional sulfuric acid process. Aluminum alloys anodized by a hardcoat process are inherently colored bronze, or from gray to near-black, depending on the alloy composition and anodizing conditions. The colors arise from light-absorbing microparticles entrapped in the hard coating. Integral colors are light- and heat-fast and are used as architectural finishes on building facades.

Electrolytic coloring is the most recently developed method for coloring anodized aluminum. It involves first applying a clear anodic coating of the desired thickness to the aluminum parts, using conventional sulfuric acid anodizing. The second step is electrodeposition of metallic pigment (tin, cobalt, or nickel) in the pores of the anodic coating, using alternating current, from an aqueous, acidic electrolyte containing the dissolved metal salt. The colors produced can be controlled to give light to dark bronze shades, and black. Many proprietary processes have been developed.[8] Electrolytically deposited color finishes are both heat- and colorfast and are widely used in architectural applications.

Equipment

The equipment for sulfuric acid anodizing includes a series of tanks for holding the various processing solutions, a refrigeration and heat exchange system to remove heat from the anodizing bath, a dc power supply for anodizing, a heat source (steam, natural gas or electrical) for heating the etching, dyeing and sealing solutions, a compressor to furnish compressed air for agitation or stirring of processing baths and an exhaust system for the removal of fumes from processing tanks.

Tanks

The tanks for holding alkaline cleaning and etching solutions are usually mild steel of welded construction, equipped with heating coils. Proprietary deoxidizing etches usually require tanks of type 316 stainless steel; however, those containing fluoride may require plastic or plastic-lined tanks. The hot sulfuric/chromic acid deoxodizing etch requires a lead-lined steel tank with lead heating coils. The nitric acid desmutting bath is held in a type 316 stainless steel tank. A stainless steel tank with S.S. heating coils is used to contain the phosphoric/nitric acid chemical brightening solution.

The tank for holding the sulfuric acid anodizing electrolyte is usually steel with a suitable plastic lining. Lead-lined, or rubber-lined steel tanks may also be used. Lead cooling coils or plate-coils are mounted in the tank for controlling temperature of the electrolyte. In large installations, the electrolyte is cooled by circulating it through external heat exchangers. The lead lining of an anodizing tank serves as the cathode. In plastic or rubber-lined tanks, the lead cooling coils or plates may be used as cathodes, or additional cathodes may be added in the form of lead strips or grids.

Dye solutions are contained in stainless steel tanks with S.S. coils for heating. When a proprietary electrolytic coloring process is used, the tank for holding the electrolyte should be of a construction recommended by the purveyor of the process. For the boiling water seal or nickel acetate seal, tanks and heating coils of type 347 S.S. are used to hold the solutions. The dichromate sealer can be held in a plain welded steel tank with heating coils. Water-rinse tanks are usually made of mild steel. The flow of water through the rinse tanks should be sufficient to prevent buildup of chemicals corrosive to the tanks.

Power Supply

Silicon type rectifiers are most often used to supply direct current for anodizing. The voltage should be continuously or stepwise variable from "off" to the desired level during anodizing. Conventional coating requires a power supply capable of supplying up to 24 volts. Hard coating requires up to 100 volts. Both anodizing processes are carried out at constant current which necessitates gradually increasing the voltage as necessary during the anodizing cycle. Programmable power supplies that provide automatic voltage and current control are available (see chapter on rectifiers). The amperage capacity of the power supply needed for an installation can be calculated by multiplying the surface area of the maximum size load of aluminum components envisaged by the current density to be employed. Power supplies vary from a few hundred amperes for small anodizing tanks to as much as 10,000 amps for very large installations. Ammeters and voltmeters are essential for monitoring the anodizing processes.

If an electrolytic coloring process is to be used, an alternating current power source of 25 volts capability is required. Coloring current densities used vary from 0.2 to 0.8 A/dm^2 (2 to 8 A/ft^2). Some processes employ more complex power sources to supply different wave forms of current/voltage. The licensors of those processes should be consulted for recommendations.

Racks

Racks are used to support the articles to be anodized and to connect them to the anode bar of the anodizing tank. Racks should be of sturdy construction to ensure long life and must be designed to make good electrical contact with the articles and maintain this contact throughout the anodizing cycle. Because the labor involved in racking and unracking the articles is the highest cost item in anodizing, it is important that the racks be designed so that the articles can be readily inserted and removed from them.

Aluminum alloys are the most common rack materials, but titanium is also used, especially for exposed contacts. Titanium, though more expensive, has the advantage that it polarizes in the electrolyte but does not form a thick, hard oxide coating, hence does not require stripping of the contact points after each anodizing cycle. Aluminum rack contacts must be stripped of the insulating oxide coating after each processing load so that good electrical contact is assured for the next load of parts. Stripping is usually done by immersion in the caustic etch bath. This results in gradual dissolving away of the aluminum racks until they become too thin to support the parts. Titanium racks or contacts are little affected by the processing solutions, hence last much longer than aluminum racks. To keep aluminum racks from dissolving away during processing, they are often coated with a plastisol or other suitable organic resist material, except at the electrical contact areas. Contact between the racks and the parts being anodized can be maintained by spring pressure (Fig. 2), or by means of bolted or clamped connections.

Aluminum articles having cavities should be racked in a manner so that gas liberated from their surfaces can escape upwards and not be held in pockets during processing. In a given tank load, all items being anodized should be of the same aluminum alloy and temper. This is because different alloys or tempers draw different amounts of current at a given voltage and consequently would end up with unequal thicknesses of anodic coating.

Bulk anodizing is employed for small aluminum parts such as rivets, bolts and nuts. In place of racks, a perforated cylindrical container (Fig. 2) or basket is tightly packed with the small parts so that they make electrical contact with one another. Electrolyte is caused to flow through the container of small parts by strong air or mechanical agitation during anodizing. Bulk anodized parts will be bare of anodic

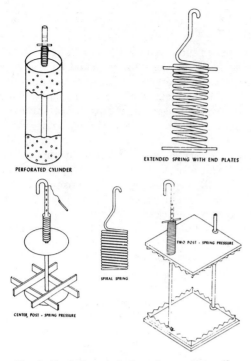

FIG. 2. Typical rack designs for anodizing.[13]

coating wherever they contacted one another during anodizing.

Electrolyte Maintenance

The sulfuric acid electrolyte is relatively inexpensive and easy to maintain compared to other electrolytes. It is stable under anodizing conditions, so the main loss of electrolyte, which must be made up by periodic additions, is due to dragout. During use, dissolved aluminum slowly builds up in the electrolyte. The limiting concentration is about 20 g/L. Above this concentration, aluminum sulfate can precipitate out on tank walls at the solution level. For process consistency, it is common practice to maintain the electrolyte at constant levels of aluminum and sulfuric acid; e.g., 10 g/L aluminum and 175 ± 10 g/L sulfuric acid. Frequently, depending on the type of work load, dragout of electrolyte tends to keep the aluminum content from building up beyond a certain level so that an equilibrium is reached. Decanting and disposal of a portion of the electrolyte with additions of acid and water, based upon analytical determinations of total and free sulfuric acid, is the usual maintenance practice. It was established that 20% of the combined acid (total minus free acid) behaves as free acid; i.e., is available for anodizing. Chlorides, when present as contaminants, can cause "burning" or localized pitting during anodizing. It is good practice to keep the chloride level below 200 ppm. Generally, the higher the anodizing voltage, the lower the tolerance for chloride contamination.

Because sulfuric acid is relatively inexpensive, it has been the practice to dump the bath, or portions of it, periodically, as necessary. However, in view of waste disposal regulations it is sometimes economical to reclaim and recycle the electrolyte. Proprietary ion-exchange methods have been developed for regenerating the electrolyte to remove dissolved aluminum and heavy metal contaminants. In other processes, dissolved aluminum is precipitated and recovered as an alum salt.

Cast Alloys

The greatest volume of anodizing is performed on wrought aluminum alloys; e.g., extrusions, sheet and forgings. Cast aluminum alloy products are sometimes anodized to provide increased resistance to corrosion and abrasion, or decorative appearance. Anodic finishes may be applied to all three types of cast products: die, permanent mold and sand castings. The best-appearing finishes are produced on Al-Mg casting alloys—e.g., 511.0 and 514.0—which yield clear, colorless coatings having some metallic luster when anodized by the conventional sulfuric acid process. The Al-Mg-Zn alloys (7XX.0 series) also yield light-colored conventional coatings. Casting alloys containing 1 weight % or more of copper give coatings that are less dense and less protective than coatings on alloys having little or no copper. This is because the copper dissolves out during anodizing. Aluminum alloys containing more than about 0.5 weight % silicon yield coatings having a gray color.

It is usually more difficult to produce a uniform anodized appearance on cast alloys than on wrought alloys. Large grains, surface porosity, segregation, dross inclusions, or flow lines lead to nonuniform coating appearance. These factors can be minimized through foundry practices when it is known in advance that the castings are to be anodized. Mechanical pretreatment by abrasive blasting, shot peening, satin finishing or polishing the surface of the casting before anodizing can also help in achieving a uniform appearance.

Porosity in the surface of a casting can entrap

acid electrolyte which is not completely removed in the water rinse. In subsequent dyeing or sealing treatments, the acid can leach out and prevent dye uptake in areas of porosity, and can also prevent sealing of those areas. This problem is avoided by immersing the anodized casting in dilute ammonia or bicarbonate solution, or in a 50 volume % nitric acid solution, at room temperature, followed by rinsing in water. These solutions remove the acid electrolyte before proceeding to the dyeing and sealing treatments.

Coating Properties

Sulfuric acid anodizing has good *covering power;* i.e., the ability to develop coating over all surfaces of an aluminum part exposed to the electrolyte. This is due to the good conductance of the electrolyte (0.5 siemens/cm) and to the nature of the anodizing process. As anodic coating forms on one area, it partially impedes the current flowing there, forcing it to flow to uncoated or more thinly coated surfaces where the film electrical resistance is lower. Thus, recessed areas and the inside surfaces of holes on aluminum parts exhibit coating coverage. When necessary, auxiliary cathodes are used to carry current to extreme recessed areas; e.g., a cathode connected aluminum rod is inserted through the length of a long pipe when it is desired to anodize the internal surface uniformly.

The *adhesion* of the anodic coating is excellent since the oxide is integral with the metal substrate from which it was formed. The coating will not spall off upon bending or impact, nor when subjected to thermal shock by immersion in liquid nitrogen or oxygen, nor when exposed to temperatures up to the melting point of the metal substrate. An exception is that thick, hard coatings may exhibit spalling or flaking on the compression side of a severely bent part. The anodic coating is as hard as corundum, but is brittle, having an elongation value of only about 0.4%. The coating will crack from deformation of the anodized part by bending, stretching or thermal expansion, because the linear coefficient of expansion is much lower than that of aluminum. The thicker the coating, the greater the tendency for *craze-cracking*. These fine, hairline cracks can act as stress risers and lower the fatigue strength of anodized structural parts, depending upon factors such as thickness and sealing of the coating. In many service environments, craze cracks do not reduce the protective value of the coating (e.g., in most architectural applications). In severe corrosive environments such as acidified salt spray, craze cracks can act as sites for corrosive attack of the basis metal. Anodic coatings alone are not reliable for the protection of susceptible alloys against stress corrosion cracking.[9]

The resistance of anodic coatings to *chemical attack* by dilute aqueous solutions is good in the pH range of 4 to 8.5. The anodic oxide is amphoteric and dissolves readily in strong acids and bases. Sealing the porosity of the coating lowers its rate of dissolution significantly. Conventional sulfuric acid anodic coatings show excellent resistance to erosion by weathering in an industrial atmosphere. In a series of tests of up to 18 years exposure, it was shown that the coatings thin down gradually and uniformly at an average rate of only 0.33 μm/year (0.013 mil/year).[10] The loss in coating thickness appears to be linear with exposure time. Thus, theoretically, a 25 μm (1 mil) thick coating should last for 76 years before bare metal becomes exposed.

The resistance to *abrasive wear* of anodic coatings is good and accounts for their use in many engineering applications. Sealing the coating reduces the abrasion resistance significantly. For applications requiring optimum wear resistance, anodic coatings are used in the unsealed condition. Hard coatings have about twice the resistance to abrasive wear as conventional sulfuric acid coatings.

Anodic oxide is a good insulator and some use is made of anodic coatings as *electrical insulation* on aluminum electrical conductors such as wire or strip for coil windings. Anodized strip conductor has a space saving advantage over insulated copper wire, and the anodic coating withstands higher temperatures than organic insulation. The average voltage breakdown value for anodic coatings is about 40 V/μm (1000 V/mil). However, the minimum breakdown values are usually much lower than the average owing to microflaws in the anodic coating. Thus, a 25-μm (1-mil) thick coating may show spot-to-spot values ranging from 250 to 1750 volts.

Testing of Coatings

Table 3 summarizes the principal ASTM test methods, with their ISO (International Organization for Standardization) equivalents, used for evaluating sulfuric acid anodic coatings.

TABLE 3. STANDARD TESTS FOR
ANODIC COATING QUALITY

Property	Principle	Standard Test ASTM	Standard Test ISO
Thickness	Eddy Current	B244*	2360
	Light Section Microscope	B681	2128
	Interference Microscope	B588	—
	Microscopic Measurement of a Cross-Section	B487*	—
Stain Resistance	Dye Stain	B136*	2143
Weight	Weigh/Strip/Reweigh	B137*	2106
Seal Quality	Acid Dissolution	B680	3210 2932
	Impedance/Admittance	B457	2931
Protective Value	Neutral Salt Spray	B117*	—
	CASS Test	B386*	—
Insulation	Voltage Breakdown	B110	2376

*Approved for use by agencies of the Department of Defense.

Many of these methods are called for in industry and government specifications.

Determination of coating thickness by the nondestructive eddy current method, ASTM B244, is the most frequently applied test and is used in nearly all anodizing shops. Microscopical measurement of polished cross-sections of the anodized part, ASTM B487, is a useful though time-consuming means to check coating thickness and its uniformity at specific locations. Less used coating thickness methods employ the light section and interference microscopes.

The second most used test is the dye stain test, ASTM B136, which serves to check the resistance to staining of anodic coatings that have been given a sealing treatment. The quality of seal of sealed sulfuric acid anodic coatings can be determined by the acid dissolution test, ASTM B680. In this test, the weight loss of the coating is determined after immersion in a solution containing 5 weight % H_3PO_4 + 2 weight % CrO_3 for 15 min at 38 C (100 F). The lower the weight loss, the better the quality of seal.

Determination of anodic coating weight per unit area of substrate, ASTM B137, is required in some specifications. This is a destructive test in which a coated part of known area is weighed before and after dissolving the coating in near-boiling 5 weight % H_3PO_4 + 2 weight % CrO_3 solution, known as "stripping solution." It does not appreciably attack the basis metal, but is a good solvent for anodic oxide. The difference in weighings is the coating weight, usually expressed in g/m^2 (mg/ft^2).

A coating quality test used more in Europe than in the U.S. consists of measuring the admittance (ac conductance) or its reciprocal, the impedance, of coated parts in contact with an electrolyte. The numerical value obtained is influenced by both coating thickness and unsealed porosity.

How well the anodic coating is expected to protect the metal substrate in a chloride environment is gaged in the neutral salt spray test, ASTM B117, which can be called for in the Military Specification, MIL-A-8625C. The copper-accelerated, acetic acid salt-spray (CASS) test, ASTM B368, is often required by the automobile industry for anodized components such as trim and bumpers. The ASTM B110 ac voltage breakdown test provides a measure of the anodic coating's electrical insulation value in air.

CHROMIC ACID ANODIZING

Chromic acid was the first anodizing process to be developed industrially and was patented in 1923 by Bengough and Stuart. Their process, still used by some, involves a rather complex voltage control procedure of slowly raising the voltage to 40 V over the first 10 to 15 min, holding it constant for 20 to 40 min, then gradually raising it to 50 V and maintaining it for a final 5 min. The total time required varies from 40 to 60 min. A 3 weight % chromic acid electrolyte is used at 40 C (104 F). A constant voltage procedure was later developed (see under Processing Steps) and is more widely used than the Bengough-Stuart procedure. Additional processes were developed to produce thicker (up to 8 μm, or 0.3 mil) coatings having improved dye uptake properties for decorative purposes. These processes, applicable to the alloy series 1XXX, 5XXX and 6XXX, yield an opaque, enamel-like appearance. An electrolyte containing about 12 weight % chromic acid is employed at 50 to 57 C (122 to 135 F) for 25 min at a current density of about 1.3 A/dm^2 (12

A/ft²) and voltages of 20 to 25. The decorative chromic acid anodic coatings are little used because they are softer and do not resist abrasion and scratching as well as sulfuric acid coatings.

Though not as widely used as sulfuric acid anodizing, chromic acid anodizing is highly desirable for protecting critical items that have lap joints, crevices, recesses or blind holes that can entrap electrolyte. Chromic acid does not appreciably attack aluminum and its alloys. If acid is entrapped, it does not cause corrosion of anodized components in service. This accounts for the continuing use of chromic acid anodizing for protective coatings in the aircraft and defense-related industries. The coating is applied to spot-welded or riveted aircraft assemblies because it is a good base for paint or adhesive bonding and provides good corrosion resistance despite its thinness.[11] Additional advantages of chromic acid anodizing are minimal dimensional changes in parts, and no harmful effect of the coatings on the fatigue life of components.

Another use for chromic acid anodizing is to detect cracks or other flaws in critical, finished aluminum alloy parts. After anodizing, rinsing and drying, cracks are revealed when chromic acid bleeds out to produce a brown stain.

Processing Steps

1. *Cleaning.* When organic soils such as oils and greases are present on the parts, they should be cleaned in a non-etching, alkaline soak cleaner. An emulsion cleaner, or vapor degreasing or solvent washing may also be used. It is important to remove organic contaminants before anodizing, otherwise they will react with (be oxidized by) the hexavalent chromium in the anodizing electrolyte. Reduction of hexavalent to trivalent chromium reduces the amount of chromic acid available for anodizing. For aluminum parts that have been machined, cleaning is the only surface preparation necessary before anodizing. (See also Chapter 3, Part F.)
2. *Rinsing.* Cold flowing water.
3. *Deoxidizing Etch.* Immersion of the parts in a solution containing 16 weight % H_2SO_4 + 3 weight % CrO_3 at 70 to 80 C (158 to 176 F) for 2 to 5 min. Proprietary deoxidizing etches that do not contain chromates are available. A deoxidizing etch is useful for removing the heat treat oxide film present on alloys in a "T" temper. Removal of heat treat oxide film permits the anodizing process to start uniformly. (See also Chapter 3, Part G.)
4. *Rinsing.*
5. *Etching.* Frequently, parts that are to be chromic acid anodized do not require an alkaline or acid etching treatment. This is the case for machined parts that have been cleaned, or for parts that have received a deoxidizing etch treatment (Step 3). When an etching treatment is employed, it is usually done with a minimum amount of metal removal to maintain close dimensional tolerances on critical parts, and in the case of alclad sheet products, to avoid dissolving off too much of the cladding. When necessary, a light etch in a conventional caustic etch bath may be used.
6. *Rinsing.*
7. *Desmutting.* If an alkaline or inhibited alkaline etch has been used, the etch smut, or any siliceous material from the inhibitor, should be removed by immersion in 50 volume % nitric acid solution for 5 to 30 sec at room temperature. (See Chapter 3, Part G.)
8. *Rinsing.*
9. *Anodizing.* A constant voltage procedure typical of commercial practice is as follows: The electrolyte contains from 3 to 10 weight % chromic acid, expressed as CrO_3, chromic acid anhydride. Electrolyte temperature is 35 C (95 F). A low starting voltage of about 5 V minimizes a current surge and possible "burning" of parts. The voltage is then raised gradually during the first 5 min to 40 V and is then held constant for 30 min. Figure 3 shows the anodic

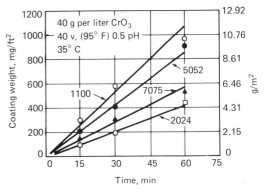

FIG. 3. Relation of coating weight to anodizing time, using a constant-voltage chromic acid anodizing procedure.[12]

coating weights produced by such a procedure on four wrought aluminum alloys.[12] Experience has indicated that coatings of less than 1.55 g/m² (144 mg/ft²) weight will not meet salt-spray test requirements.
10. *Rinsing.*
11. *Coloring.* When color is desired for identification or decorative purposes, this is achieved by immersion of the rinsed, still-wet parts in a dye solution. Acid type, water-soluble dyes are employed under conditions recommended by the dye manufacturer. To obtain improved dye uptake, chromic acid coatings are sometimes anodized at elevated temperatures of 50 to 57 C (122 to 135 F).
12. *Rinsing.*
13. *Sealing.* Nondyed chromic acid coatings are sealed for 10 min in boiling deionized water. Improved resistance to salt water corrosion is obtained when sufficient CrO_3 is added to the sealing water to lower the pH to 5. Sealing may also be performed in 5 weight % sodium or potassium dichromate solution for 10 to 15 min at 95 to 100 C (203 to 212 F), at pH 5 to 6.5. Dyed chromic acid coatings are usually sealed in 0.5 weight % nickel acetate solution, or as recommended by the dye manufacturer.
14. *Rinsing.* (Unnecessary when boiling water seal is used.)
15. *Drying.* The parts may be dried in ambient air, in a stream of oil-free compressed air, or in heated air. Because they are less dense and thinner than sulfuric acid anodic coatings, chromic acid coatings are much less susceptible to heat crazing.

Equipment

A chromic acid anodizing installation requires a series of tanks to hold the different processing solutions, an anodizing power supply, a heat source for heating processing solutions, sources of deionized water and compressed air and an exhaust system for fume control. A cation exchanger for regeneration of spent chromic acid electrolyte is highly desirable.

Tanks

Suitable tank materials for holding the deoxidizing etch, caustic etches, nitric acid desmutter, dye solutions and sealing baths are as described in the section on sulfuric acid anodizing (see under "Tanks").

The chromic acid electrolyte is held in a mild steel tank of welded construction. An iron pipe coil installed in the tank can serve for both heating and cooling the electrolyte by passing steam or cold water through the coil intermittently. Steel or stainless steel "plate coils" can be used in place of a pipe coil. Optionally, heating may be accomplished by means of electric immersion heaters. Maintenance of a constant uniform temperature throughout the electrolyte during anodizing is essential for consistently good results. Thermostatic control of the bath to about ±2 C (±4 F) is recommended. Circulation of the bath aids in maintaining a uniform temperature and is accomplished by means of motor stirrers, or by compressed air bubbling upwards from a perforated pipe lying on the tank bottom. The steel tank and its heat exchanger coils are connected as the cathode in the electric circuit. Because large cathode area (low cathode current density) favors the electrolytic reduction of hexavalent chromium to the trivalent form, the total cathode surface should be limited to between one-fifth and one-tenth of the total anode area. This is accomplished by shielding part of the cathode area by means of wire-reinforced glass plates. Laying acid-proof bricks without mortar on the tank bottom is another shielding method. This has a further advantage of avoiding accidental short circuiting of the work with the tank.

Because human contact with chromic acid is hazardous, it is essential to have adequate exhaust ventilation at the anodizing tank. In small installations, the entire tank may have an exhaust hood over it. For larger tanks, where overhead cranes or hoists are used to manipulate the work, the usual practice is to employ slotted ducts mounted along the top edge of the tank sidewalls. The ventilating ducts should be designed to remove 1 cubic meter of air per second per square meter of electrolyte surface (200 ft³/min/ft²). (See chapter on ventilation.)

Electrical

Silicon type rectifiers are usually employed to furnish direct current for the anodizing process. They should be capable of supplying up to 40 V and current sufficient for a current density of 0.54 A/dm² (5 A/ft²). When thick, decorative type coatings are to be produced, the voltage and current density requirements are 25 V and 1.3

A/dm² (12 A/ft²). The power supply must be regulatable from "zero" voltage up to the maximum voltage, either continuously, or in steps of 5 V or less. Ammeters and voltmeters are necessary for proper control of the process.

Racks

The racks for holding the work must insure firm mechanical contact to prevent the build-up of insulating oxide from breaking the electrical circuit during anodizing. Mechanical contact may be by spring pressure, bolted connections or adjustable clamps. Racks are usually constructed of aluminum alloys 6061, 6063, 2024 or 3003. Plastic coatings may be applied to the racks, except at electrical contact areas, to give them a longer life than bare aluminum racks. The plastic coating must be resistant to the highly oxidizing, warm chromic acid electrolyte. Coated racks have the further advantage of not consuming as much of the anodizing current as do bare racks. Small parts such as fasteners can be bulk-anodized by packing them tightly in a perforated cylinder through which electrolyte can flow (Fig. 2).

Electrolyte Maintenance

Chromic acid is lost from the electrolyte as mist, by solution dragout on the parts, and by reduction of hexavalent chromium to the trivalent state at the cathode during anodizing. A further reduction in the acid available for anodizing occurs because dissolved aluminum and trivalent chromium combine with hexavalent chromium as aluminum dichromate and chromium dichromate complexes which are ineffective for anodizing. As these materials accumulate in the electrolyte, it is necessary to make additions of chromic acid or, preferably, to remove the aluminum and trivalent chromium by ion exchange. Stable cation exchange resins are available for regenerating the electrolyte to purified chromic acid solution for reuse. Ion exchange regeneration has the further advantage of eliminating the need for waste treatment of spent bath. A convenient way to check the effective, or free, chromic acid in the electrolyte is by pH measurement. The relationship between pH value and free chromic acid is shown in Fig. 4.[13] The pH of the electrolyte, measured at room temperature, is usually maintained at a value below 0.6.

The aluminum content of the electrolyte is usually kept below 0.3 weight %. The most

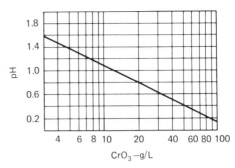

FIG. 4. Variation of pH with free chromic acid concentration.[13]

harmful contaminant that is sometimes encountered is chlorides, which should not be permitted to exceed 200 ppm. Excessive chloride may cause burning or pitting of the parts. Sulfates are normally present as an impurity in technical grade chromic acid anhydride and it is not unusual to operate the anodizing electrolyte at levels of up to 300 ppm sulfate. It is the presence of sulfate in the 100 to 300 ppm range that gives rise to the opaque appearance of chromic acid anodic coatings. The sulfate causes a fine microroughening of the aluminum surface, giving it a diffuse, whitish aspect. Excessive sulfate contamination will make the electrolyte corrosive to aluminum and to the steel tank. It is good practice to keep the sulfate level below 500 ppm. If it is necessary to reduce the sulfate content of the electrolyte, this can be accomplished by the addition of a calculated amount of barium carbonate or barium hydroxide to form insoluble barium sulfate.

Testing of Coatings

The main requirements of chromic acid anodic coatings are coating weight per unit area of basis metal and protective value in the neutral salt spray test (Table 2). Of the various standard tests for coating quality listed in Table 3, only ASTM Methods B117 (neutral salt spray test) and B137 (coating weight determination) are widely used for chromic acid anodic coatings. The other listed tests are generally unsuitable because of the thinness of the coatings.

Designation System for Anodic Coatings

The "Aluminum Association Designation System for Aluminum Finishes"[14] was established to assist users in designating the kind of finish desired so that suppliers would know what to furnish. Three major categories are covered;

TABLE 4. Anodic Coatings (A)[14]

Type of Finish	Designation*	Description	Examples of Methods of Finishing**
General	A10	Unspecified	
	A11	Preparation for other applied coatings	3 μm (0.1 mil) anodic coating produced in 15% H_2SO_4 at 21 C ± 1 C (70 F ± 2 F) at 129 A/m^2 (12 A/ft^2) for 7 min, or equivalent.
	A12	Chromic acid anodic coatings	To be specified.
	A13	Hard, wear and abrasion resistant coatings	To be specified.
	A1X	Other	To be specified.
Protective and Decorative Coatings less than 10 μm (0.4 mil) thick	A21	Clear coating	Coating thickness to be specified. 15% H_2SO_4 used at 21 C± 1 C (70 F ± 2 F) at 129 A/m^2 (12 A/ft^2).
	A211	Clear coating	Coating thickness—3μm (0.1 mil) minimum. Coating weight—6.2 g/m^2 (4 mg/in^2) minimum.
	A212	Clear coating	Coating thickness—5μm (0.2 mil) minimum. Coating weight—12.4 g/m^2 (8 mg/in^2) minimum.
	A213	Clear coating	Coating thickness—8 μm (0.3 mil) minimum. Coating weight—18.6 g/m^2 (12 mg/in^2) minimum.
	A22	Coating with integral color	Coating thickness to be specified. Color dependent on alloy and process methods.
	A221	Coating with integral color	Coating thickness—3μm (0.1 mil) minimum. Coating weight—6.2 g/m^2 (4 mg/in^2) minimum.
	A222	Coating with integral color	Coating thickness—5 μm (0.2 mil) minimum. Coating weight—12.4 g/m^2 (8 mg/in^2) minimum.
	A223	Coating with integral color	Coating thickness—8 μm (0.3 mil) minimum. Coating weight—18.6 g/m^2 (12 mg/in^2) minimum.
	A23	Coating with impregnated color	Coating thickness to be specified. 15% H_2SO_4 used at 27 C ± 1 C (80 F ± 2 F) at 129 A/m^2 (12 A/ft^2) followed by dyeing with organic or inorganic colors.
	A231	Coating with impregnated color	Coating thickness—3 μm (0.1 mil) minimum. Coating weight—6.2 g/m^2 (4 mg/in^2) minimum.
	A232	Coating with impregnated color	Coating thickness—5 μm (0.2 mil) minimum. Coating weight—12.4 g/m^2 (8 mg/in^2) minimum.
	A233	Coating with impregnated color	Coating thickness—8 μm (0.3 mil) minimum. Coating weight—18.6 g/m^2 (12 mg/in^2) minimum.
	A24	Coating with electrolytically deposited color	Coating thickness to be specified. Application of the anodic coating, followed by electrolytic deposition of inorganic pigment in the coating.
	A2X	Other	To be specified.

TABLE 4. (continued)

Type of Finish	Designation*	Description	Examples of Methods of Finishing**
Architectural Class II† 10 to 18 μm (0.4 to 0.7 mil) coating	A31	Clear coating	15% H_2SO_4 used at 21 C ± 1 C (70 F ± 2 F) at 129 A/m^2 (12 A/ft^2) for 30 min, or equivalent.
	A32	Coating with integral color	Color dependent on alloy and anodic process.
	A33	Coating with impregnated color	15% H_2SO_4 used at 21 C ± 1 C (70 F ± 2 F) at 129 A/m^2 (12 A/ft^2) for 30 min, followed by dyeing with organic or inorganic colors.
	A34	Coating with electrolytically deposited color	Application of the anodic coating followed by electrolytic deposition of inorganic pigment in the coating.
	A3X	Other	To be specified.
Architectural Class I† 18 μm (0.7 mil) and thicker coatings	A41	Clear coating	15% H_2SO_4 used at 21 C ± 1 C (70 F ± 2 F) at 129 A/m^2 (12 A/ft^2) for 60 min, or equivalent.
	A42	Coating with integral color	Color dependent on alloy and anodic process.
	A43	Coating with impregnated color	15% H_2SO_4 used at 21 C ± 1 C (70 F ± 2 F) at 129 A/m^2 (12 A/ft^2) for 60 min. followed by dyeing with organic or inorganic colors, or equivalent.
	A44	Coating with electrolytically deposited color	Application of the anodic coating followed by electrolytic deposition of inorganic pigment in the coating.
	A4X	Other	To be specified.

*The complete designation must be preceded by AA—signifying Aluminum Association.
**Examples of methods of finishing are intended for illustrative purposes only.
†Aluminum Association Standards for Anodized Architectural Aluminum.

mechanical finishes, chemical finishes and coatings. The anodic coatings part of this system is reproduced in Table 4. Anodic coatings are designated by the letter "A," followed by digits defining specific finishes of the various types. Brief descriptions with examples of methods of finishing are given for each designation.

References

1. ASTM B580, Standard Specifications for Anodic Oxide Coatings on Aluminum.
2. MIL-A-8625C, Military Specification, Anodic Coatings, for Aluminum and Aluminum Alloys, June 15, 1968, and Amendment 1, March 13, 1969.
3. British Patent 290,901.
4. Wodehouse, R. A. "Sulfuric Acid Anodizing and Oxide Treatments for Aluminum," *Electroplating Engineering Handbook*, 3rd Edition. New York: Van Nostrand Reinhold, 1971, Chapter 14, p. 462.
5. U.S. Patent 2,650,157.
6. Gohausen, H. J. "Consideration for the Sealing of Anodized Aluminum to a Smut-free Condition," *Aluminum Association Finishing Seminar Technical Papers* I:161-175. The Aluminum Association, Washington, D.C. (March 30-April 1, 1982).
7. Muller, E. "Weather Resistant Colored Anodic Finishes on Aluminum," *Prod. Finishing*, 20-21 (Feb. 1977).
8. Sheasby, P. G. "Electrolytic Coloring—Processes and Properties." *Ibid*, Ref. 6, 145-160.
9. Cochran, W. C. and D. O. Sprowls. "Anodic Coatings for Aluminum," *Corrosion Control by Coatings*. H. Leidheiser, Jr., Editor. Princeton: Science Press, 1979, p. 179.
10. Mader, O. M. "A Critical Review of Aluminum in the Building Industry," *Metals and Materials*, 303-307 (July, 1972).
11. Pearlstein, F. "Selection and Application of Inorganic Finishes, Part III—Anodic Coatings for Aluminum," *Plating and Surf. Fin.*, 38 (Feb. 1979).

12. Best, G. E. *et al.* "Chromic Acid Anodizing Characteristics of Wrought Aluminum Alloys," *Proc. Am. Soc. for Testing Materials* **59** (1959).
13. Best, G. E. "Chromic Acid Anodizing of Aluminum," *Electroplating Engineering Handbook*, 3rd Edition. New York: Van Nostrand Reinhold, 1971, Chapter 14, p. 451.
14. The Aluminum Association, 818 Connecticut Ave. N.W., Washington, D.C. 20006.

Bibliography (*Sulfuric and Chromic Acid Anodizing of Aluminum*)

1. Wernick, S. and R. Pinner. *The Surface Treatment and Finishing of Aluminum and its Alloys.* Two volumes, Fourth Edition. Teddington: Robert Draper, 1972.
2. Brace, A. W. and P. G. Sheasby. *The Technology of Anodizing Aluminum*, Second Edition. Teddington: Robert Draper, 1979.
3. Van Horn, K. R. (Ed.). *Aluminum Vol. III, Fabrication and Finishing.* Metals Park, Ohio: American Society for Metals, 1967, Chapters 16–19.
4. Henley, V. F. H. *Anodic Oxidation of Aluminum and its Alloys.* Oxford: Pergamon Press, 1982.
5. Tajima, S. "Anodic Oxidation of Aluminum," *Advances in Corrosion Sciences and Technology, Vol. I.* New York: Plenum Press, 1970, pp. 229–362.
6. Wood, G. C. "Porous Anodic Films on Aluminum," *Oxides and Oxide Films, Vol. 2.* J. W. Diggle, Editor. New York: Marcel Dekker, 1973, 167–279.
7. The Aluminum Finishing Seminar, *Anodizing Session Technical Papers*, March 1973, Chicago. The Aluminum Association, Washington, D.C.
8. *Aluminum Finishing Seminar Technical Papers* I, 297 pp., March 30–April 1, 1982, St. Louis. The Aluminum Association, Washington, D.C.

D. Anodizing and Surface Conversion Treatments for Magnesium

H. K. DeLong

Associate Scientist, Metallurgical Laboratory, Dow Chemical Company, Midland, Mich.

The increasing use of magnesium alloys in numerous industrial fields, especially the aircraft, automotive, materials handling, and portable tool industries, has focused attention on this metal due to its inherent lightness, and good strength to weight ratio. The present increased confidence can largely be traced to the development of more corrosion-resistant alloys and improved methods of surface protection. Unlike bare mild steel, inland and moderate marine atmospheric exposures have no serious corrosive effects other than to form a gray discoloration due to oxidation products after several years of exposure. However, in contrast unprotected magnesium may be severely corroded when immersed in sea water or subjected to sea spray or chloride-laden road splash.

The oxide-carbonate film formed on magnesium is not entirely self-healing, is alkaline in nature, and presents conditions analogous to those experienced on a zinc surface. Many methods have been investigated for treating magnesium by chemical immersion or anodizing processes in order to produce a more protective surface film or coating. These surface films require supplementary protective organic coatings for the majority of applications. They are designed to replace the naturally alkaline but less protective oxide-carbonate film with a more corrosion inhibiting and less alkaline or slightly acid coating. Such a surface is more compatible with organic coatings serving as a base for paints and it provides better protection when used alone. A number of protective treatments for magnesium are now being applied in the field, each of which is useful, depending on the application and severity of exposure for which the magnesium alloy is intended.

Cleaning

As with all other metals being prepared for finishing treatments, a chemically clean surface is necessary if magnesium parts are to be treated by chemical or electrochemical processes to ensure adherent and continuous protective films. Therefore, all surface contaminants such as oil, grease, dirt, oxides, die-forming compounds, etc., must be removed. This requires the establishment of a suitable cleaning cycle depending on the condition of the metal and the nature of the contaminant to be removed. In some cases it may also be necessary to remove previously applied surface treatments to ensure proper coverage of the final treatment.

Mechanical Cleaning. Methods of mechanically cleaning magnesium are very similar in detail to those used on zinc and aluminum alloys. These include blasting, sanding and wire brushing. The operational details for these metals are described in Chapter 3. When blasting methods of any type are employed there is danger of embedding surface contamination which will greatly increase the basic surface corrosion rate. In such cases subsequent acid pickling must be used to etch up to 0.002 in. from the surface prior to application of a protective coating where maximum corrosion resistance is a requisite.

Solvent Cleaning. Solvent cleaners are employed to remove abnormal amounts of grease and oil and as a precleaning step prior to painting when the treated work has been soiled in handling. Initial removal in this manner prevents a rapid build up of oil and grease in the alkaline cleaner and reduces the time in this operation. Chlorinated solvent vapor degreasing, hydrocarbon solvents, paint thinners, and similar solvents may be used. Emulsion cleaners may also be used in the precleaning cycle instead of straight solvents. These cleaners are essentially a mixture of emulsifying agents and hydrocarbon solvents in which it is possible to water-rinse the solvent and contaminant from the surface after cleaning by dip or spray methods (see Chapter 3).

Alkaline Cleaning. Magnesium, unlike aluminum, is not appreciably attacked by caustic solutions. Therefore, heavy-duty cleaners high in caustic soda similar to those used on steel are quite satisfactory. These types are particularly beneficial when old chromate type coatings as well as oil and grease are to be removed.

Most proprietary cleaning solutions with a pH preferably above 11 have been satisfactory. The following mixture is an example of a suitable cleaner for magnesium.

Sodium hydroxide (NaOH)	8 oz/gal
Trisodium phosphate ($Na_3PO_4 \cdot 12H_2O$)	1⅓ oz/gal

The addition of 0.05 to 0.1 oz/gal of soap or suitable commercial wetting agent may be added if the bath is to be used as a soak cleaner. The sodium hydroxide can be varied from 2 to 12 oz/gal with the higher concentration being preferred for the removal of "burned on" die lubricants and the lower concentration for special alloys high in zinc and zirconium that are less resistant to strongly alkaline solutions. The temperature of the soak cleaner is maintained at 180 to 210 F and the time of immersion is 3 to 20 min. It is beneficial to agitate the work or solution in the interest of shortening the time requirement.

Electrocleaning also shortens the cleaning time considerably. When using this method the work should be made the cathode (anodic cleaning usually produces an undesirable coating of magnesium hydroxide or can result in local pitting depending on the bath composition, etc.). The cleaner composition is basically the same as for soak cleaning except that the presence of soaps or wetting agents generally is undesirable. A current density of 20 to 40 asf is employed for 1 to 5 min and no agitation of the work or solution is required.

Rinsing after the alkaline cleaning must be very thorough to avoid neutralizing or contaminating the pickling or treating baths which usually follow this operation.

Pickling

Oxide layers, certain old chemical coatings, burned-on drawing and forming lubricants or other water-insoluble or unemulsifiable substances are not removed entirely by solvent or alkaline cleaning. To remove these residual materials acid cleaning by pickling is required. Numerous pickling solutions have been used in the past for magnesium but only those that are now commonly used will be described. *In all cases these pickling solutions should be preceded by appropriate alkaline and/or solvent plus alkaline cleaning and thorough water rinsing.*

Chromic Acid-Nitrate. Solutions containing chromic acid and a nitrate are commonly used for cleaning various forms of magnesium and, in particular, wrought magnesium alloys that are low in aluminum content. Some of the more important uses for this solution are for the removal of "burned on" graphite lubricants from hot-formed sheet parts, for cleaning prior to arc or gas welding assemblies, and for removal of mill scale from sheet. The parts are immersed at a temperature of 60 to 100 F for 1 to 3 min or longer as required, in a solution of the following composition.

Chromic acid (CrO_3)	24 oz/gal
Sodium nitrate ($NaNO_3$)	4 oz/gal

The bath constituents can be varied to reduce

the reaction rate. Increasing the chromic acid to 32 oz/gal and decreasing the sodium nitrate to 2 oz/gal is best for basket pickling of small parts. For removing contamination from sheet or stampings it is important to remove a minimum of 0.5 mil of surface in this pickle. Depletion is indicated by lack of chemical action and a pH 1.7 or higher indicates a need for additions. The bath may be controlled by the addition of chromic acid to bring the solution back to the original pH range of 0.5 to 0.7. The solution may be revivified about four times in this manner.

Ferric Nitrate (Treatment No. 21). This pickling process is gradually replacing the chromic-nitrate listed above and the chromic-nitrate-hydrofluoric acid pickle required for high aluminum content magnesium alloy castings. It produces a chemical polishing effect and provides a smut-free surface on castings. It is, therefore, applicable to all the common alloys and forms. Although developed primarily as a means of economically applying an attractive shelf-life finish, it also is being utilized effectively for a welding precleaner, for the removal of hot-forming lubricants, oxidation or corrosion products, mill scale, etc.

The cleaned parts are dipped in or sprayed with a solution of the following composition:

Chromic acid (CrO_3)	24 oz/gal
Iron nitrate ($Fe(NO_3)_3 \cdot 9H_2O$)	5.3 oz/gal
Potassium fluoride (KF)	0.5 oz/gal

This bath is operated at a temperature of 60 to 100 F and the treatment time is from 15 sec to 3 min or longer depending on the degree of brightness desired or the nature of the surface contamination to be removed. In this solution approximately 0.15 mil of surface is removed per minute. Increasing the fluoride tends to increase the reactivity and to a degree the brightening effect. The operational range of the potassium fluoride is from 0.27 to 0.94 oz/gal. For castings the higher fluoride concentration is preferred.

Chromic-Sulfuric Acid. The chromic-sulfuric acid pickle is used primarily as a chemical spotwelding surface precleaner. This solution gives consistently low surface resistance for spotwelding. The work is immersed for 2 to 3 min at a temperature of 70 to 90 F in a bath of the following composition:

Chromic acid (CrO_3)	24 oz/gal
Conc. sulfuric acid (H_2SO_4)	0.065 fl oz/gal

In high production operations it is advantageous to pickle for 10 to 30 sec in 0.5 to 1.0 per cent by volume of sulfuric acid prior to treatment in this bath. This assures neutralization of any alkali carried over from the alkaline cleaner and prolongs the life of the spotweld cleaning bath.

Chromic Acid. Chromic acid pickling is used primarily for removing corrosion product and old chemical treatments from all forms of magnesium and flux from castings. It is also used on machined or close tolerance parts to remove superficial oxide, etc., where no significant dimensional change can be tolerated. The magnesium parts to be cleaned are immersed in a solution of the following composition:

Chromic acid (CrO_3)	24 oz/gal

Immersion time varies from 1 to 15 min depending on the condition of the parts and the bath temperature. The temperature of operation can be varied from room to that of boiling.

Phosphoric Acid. This pickle bath is most commonly used for preparing castings for further surface finishing and for sheet products with deeply embedded mill scale. It removes surface alloy segregation from as-cast surfaces and allows more uniform appearing surface conversion coatings to be applied. The parts are pickled for 30 sec to 1 min or longer in a bath of the following composition:

Phosphoric acid (85% H_3PO_4)	0.9 oz/gal

The bath is maintained at a temperature of 70 to 100 F. Metal loss due to pickling is 0.5 mil per minute of immersion.

Acetic-Nitrate. The acetic-nitrate pickle is primarily a producers in-process pickle for the removal of mill scale and oxide from sheet products. The pickle can be used on other wrought forms and on non-aluminum alloy content magnesium castings. The following solution operated at 70 to 100 F is used:

Glacial acetic acid (CH_3COOH)	25.5 fl oz
Sodium nitrate ($NaNO_3$)	6.7 oz

The usual pickling time is 30 sec to 1 min. A metal removal of approximately 1.0 mil/min is produced by this solution.

Other Acid Pickling Solutions. Other acid pickling solutions are available but are used primarily on rough castings by the producer or for special purposes. These are as follows:

(1) Concentrated sulfuric acid (H_2SO_4)	4 fl oz/gal
(2) Nitric acid (70% HNO_3)	8 fl oz/gal
Concentrated sulfuric acid (H_2SO_4)	2 fl oz/gal
(3) Nitric acid (70% HNO_3)	9 fl oz/gal

Tank Equipment For Cleaning And Pickling

Chromic acid: lead, vinyl-lined steel, stainless steel or 1100 (2S) aluminum.

Chromic acid nitrate: vinyl-lined steel, stainless steel (Type 316), 1100 (2S) aluminum.

Ferric nitrate: vinyl-lined steel, stainless steel (Type 316).

Chromic acid-sulfuric: vinyl-lined steel, stainless steel (Type 316).

Phosphoric acid: lead, ceramic, rubber, vinyl-lined steel.

Acetic-nitrate: 1100 (2S) aluminum, ceramic, stainless steel, rubber-lined steel.

Paint-Base Treatments

Numerous chemical dip-type or surface-conversion treatments are available for magnesium alloys. These are primarily used as a base for improving the corrosion resistance and paint base properties. Only those in current production use that are economical or perform satisfactorily will be considered. Others may be competitive in some specific quality but are no longer used because they were not economically feasible or were lacking in other qualities. Most of these treatments are covered or are allowed by Military Specification Mil-M-3171A (AER) under Types I, II, III, IV, and V treatments. In some special cases these treatments provide adequate protection in themselves, e.g., interior parts and automotive engine parts that normally are covered with oil which adds to the protection.

Chrome Pickle (Mil-M-3171A, Type I, or Treatment No. 1 and 20). The chrome-pickle treatment is the most commonly used of all chemical treatments that have been developed for magnesium. It is used for many commercial applications but is limited to touch-up repair work for most military applications. The coating is used to protect magnesium during shipment and storage as well as a final treatment. Good paint base properties are exhibited by this coating.

The treatment is applied by dipping, spraying or brush methods. The etching action of the Chrome Pickle removes up to 0.6 mil of surface thus limiting its use on close tolerance machined parts unless allowances are made or the amount of surface removal can be tolerated.

In applying the chrome-pickle treatment a solution of the following composition is generally used for wrought products:

Sodium dichromate ($Na_2Cr_2O_7 \cdot 2H_2O$)	1.5 lb/gal
Concentrated nitric acid (70% HNO_3)	1.5 pt/gal

Magnesium sand, permanent mold, and die castings are preferably treated in a modified chrome pickle (Treatment No. 20) of the following composition:

Sodium acid fluoride ($NaHF_2$)	2 oz/gal
Sodium dichromate ($Na_2Cr_2O_7 \cdot 2H_2O$)	1.5 lb/gal
Aluminum sulfate ($Al_2(SO_4)_3 \cdot 14H_2O$)	1.3 oz/gal
Concentrated nitric acid (70% HNO_3)	1 pt/gal

The above baths are operated at a temperature of 70 to 90 F and the treatment time is 20 sec to 2 min. A 1 min standard treatment time is used for all alloys and forms except die castings where the time is reduced to 20 to 30 sec. The transfer time between pickling and initial cold-water rinsing should be kept to a minimum (5 to 30 sec) to avoid the formation of loose powdery coatings.

Articles too large to immerse and repair of abraded areas of previously applied coatings are processed by brush application using a generous amount of fresh solution which must be allowed to remain on the surface for at least 1 min while brushing and then be washed off immediately with cold running water. The coating thus produced is less uniform in appearance than that produced by the dip or spray processes but is equally good as a paint base.

The color of the coating produced by this treatment is matte gray to yellow-red iridescent with a degree of fine surface etching for best paint adhesion. Smooth, bright brassy, coated surfaces are somewhat inferior for painting but are equally protective. The freshness of the solution, magnesium alloy being processed, and number of times the bath has been revivified influence the color of the coating.

Sealed - Chrome Pickle (Mil - M - 3171A, Type II, Treatment No. 10). The sealed-chrome pickle is essentially a two-step treatment of first applying the chrome pickle (Mil-M-3171A, Type I) and then sealing this coating by a subsequent step of boiling in the dichromate-fluoride bath of the dichromate process mentioned below. The sealing treatment which improves the protection of the chrome-pickle coating is most commonly applied to wrought products but also can be applied to castings if the amount of metal removed by the chrome-pickle dip is not a limitation or is allowed for in machining. The sealed-chrome-pickle provides corrosion resistance equal to the dichromate process. The coating formed is matte gray to yellow iridescent through to brown depending on the particular magnesium alloy being processed.

The steps in this process include chrome pickling as described, then cold-water rinsing. Immediately after chrome pickling the parts are boiled for 30 min in the dichromate-fluoride solution shown below under "Dichromate" treatment. This operation is followed by cold-water rinsing and a dip in hot water to facilitate drying.

It is important to apply a fresh chrome pickle with this process for best results. Sealing an old chrome-pickle film as supplied by the metal producer as a protective coating for shipping is not as effective since ageing of the film prevents proper sealing.

Dichromate (Mil-M-3171A, Type III, Treatment No. 7). The dichromate process is the most common of the dip-type chemical treatments in use today for magnesium applications wherein maximum corrosion protection is desired. The treatment effects no significant dimensional change and normally is applied after machining and prior to painting. The coating varies from light to dark brown in color depending on the alloy to which it is applied.

The following steps are taken in applying the dichromate treatment after proper degreasing and prepickling, if required, as previously described.

Immerse for 30 sec to 5 min at a temperature of 70 to 90 F in the following solution:

Hydrofluoric acid (60% HF) 24 fl oz/gal

All magnesium alloys except AZ31B are immersed for 5 min. A 30 sec dip, or prolonged cold-water rinsing for 5 min, is used on AZ31B alloy, otherwise the passive fluoride film that is produced will tend to retard the formation of the subsequent chromate film on this particular alloy. Careful rinsing is important to minimize the carry-over of excess fluoride into the dichromate bath rendering it inoperative.

An alternative fluoride treatment may be used on wrought products and on castings which have been acid pickled after blasting or casting to remove surface contamination, surface alloy segregation, casting skin, etc. This bath is an aqueous solution of sodium, potassium, or ammonium acid fluoride or mixtures of these salts. The acid does not rapidly attack aluminum inserts, rivets, etc., and is more economical to use. The solution consists of the following:

Sodium, potassium, or ammonium acid   6.7 oz/gal
fluoride ($NaHF_2$, KHF_2, or NH_4HF_2)

The parts are immersed for 15 min at a temperature of 70 to 90 F and are then rinsed in cold-running water after the fluoride dip and then the parts are immersed for 30 min in the following aqueous solution operated at or near boiling (210 to 212 F):

Sodium dichromate   16–24 oz/gal
  ($Na_2Cr_2O_7 \cdot 2H_2O$)
Calcium or magnesium fluoride   0.33 oz/gal
  (CaF_2 or MgF_2)

The presence of calcium or magnesium fluoride assists in chromate film formation in the dichromate bath. They are only slightly soluble and thus control the proper concentration in solution. An excess of these fluorides is added above that which is not completely soluble in a fresh unused solution. They can be conveniently suspended in the solution in cloth bags or the excess can be added to the tank so that the bath remains saturated with them. However, it should be noted that hydrofluoric acid or acid fluoride carried over into this bath can cause an excess of free fluoride because of their high solubility, thus rendering the bath inoperable. When the free fluoride raises above about 0.2 per cent a coating will not readily form.

After treating, the parts are rinsed thoroughly in cold water and then dipped in hot water to facilitate drying.

Galvanic Anodize (Mil-M-3171A, Type IV; Treatment No. 9). This treatment was designed specifically as an alternative process for those alloys which do not react to give a comparative higher performance protective film in the "Dichromate" process. These magnesium alloys include EK30A, EK41A, HK31A, and M1A. However, treatment is not limited to these alloys but is applicable to all forms and alloys. The process makes use of the relatively high potential difference existing between the magnesium and the steel tank or steel cathodes hung in the tank. It causes no appreciable dimensional change and is normally applied after machining operations. The coating formed is dark brown to black in color. Parts must be racked and electrically connected to the dissimilar metal tank or steel cathode plates for galvanic action to take place.

The following steps are used in applying this treatment after cleaning as previously described. They are then treated in the hydrofluoric acid or acid fluoride dip exactly as shown for the "Dichromate" treatment. Next, the work is galvanically anodized for at least 10 min and for as long as 30 min (depending on the alloy and bath condition) at 120 to 140 F until uniformly covered with a dark brown to black coating. Prolonging the treatment beyond the time for complete coverage can result in the production of nonadherent powdery coatings. The following aqueous solution is employed:

Ammonium sulfate $(NH_4)_2SO_4$	4 oz/gal
Sodium dichromate $(Na_2Cr_2O_7 \cdot 2H_2O)$	4 oz/gal
Ammonium hydroxide (Sp gr 0.880-NH_4OH)	0.33 fl oz/gal

The steel tank usually provides the cathode in this process. If the tank is lined using a nonmetallic coating or lead, then steel cathode plates should be used. After the treatment the work is rinsed in cold water, followed by a hot water dip to facilitate drying.

NOTE: Where the object is to produce a black color the solution can be applied by dipping at or near boiling temperature (210 to 212 F) instead of galvanic anodizing. However, this method is limited in application to those alloys that are receptive to the "Dichromate" treatment. No racking is required and the treating time is increased to 30 to 45 min in the sulfate-dichromate-ammonia bath. When used as a boiling treatment the process is known as the "Alkaline-Dichromate" or Treatment No. 8 (see Table 2).

A more pronounced black coating is usually obtained by adding one of the following organic dyes to the treating solution at a concentration of 0.25 oz/gal.

Indulin Blue W.S. (Color index #861).
Nigrosine Black W.S. (Color index #865).

Other Treatments. *Phosphate (Treatment No. 18).* This treatment has been used primarily for brush or spray touch-up on small areas of previously chemically treated or anodized work surfaces that have been damaged by scratching or bared for other causes prior to painting. It has replaced the chrome-pickle as a repair treatment to some extent primarily because of better reproducibility under production shop conditions and the fact that it is less corrosive if allowed to become entrapped between faying surfaces or in pocketed areas of an assembly. A solution of the following composition is used:

Ammonium acid phosphate $(NH_4)_2H_2PO_4$	16 oz/gal
Ammonium sulfite $(NH_4)_2SO_3 \cdot H_2O$	4 oz/gal
Ethyl alcohol (denatured)	8–20 fl oz/gal

The solution is brushed or sprayed on for 1 min or until a continuous gray coating is formed. The parts may also be dipped for 1.5 to 2 min or until gassing stops. For best results a cold-water rinse is required after this treatment. Hot-water rinsing is to be avoided.

Dilute Chromic Acid (Treatment No. 19). Of the many suitable processes for magnesium this treatment is the least expensive. It can be applied by brush, spray or dip methods. Better corrosion protection is afforded than with the "Phosphate" treatment although the paint base properties are similar. It is less critical to apply than the "Chrome-Pickle" for touch-up work or repair work on previously coated surfaces and is not harmful if entrapped between faying surfaces, etc.

A solution of the following composition operated at room temperature is required:

Chromic acid (CrO_3)	1.3 oz/gal
Calcium sulfate $(CaSO_4 \cdot 2H_2O)$	1 oz/gal

The chemicals are added to water in the order shown, and stirred vigorously for about 15 min to ensure saturation of the solution with the calcium sulfate.

Proper application of the coating requires that the surface be kept wet with solution until a brassy iridescent film is formed. For brush or spray application this is for 1 to 3 min. Dip processing requires 30 to 60 sec.

Unlike the "Chrome-Pickle," the time between pickling and cold-water rinsing is not critical. In fact, where rinsing is not feasible, it can be eliminated without materially altering the effectiveness of the coating; it is necessary only to sponge dry the drain-off liquid at the edges of the work.

Anodizing Processes

Maximum corrosion protection and paint-base properties for magnesium alloys are provided by anodic treatments, particularly by certain processes that have been developed in recent years. Many of the older electrochemical treatments have not survived, having failed to offer any major benefits over the chemical dip treatments as the latter were improved from time to time. Currently available anodic treatments produce relatively thick, dense, adherent, abrasion-resistant protective coatings. The coatings in general are hard and possess a high degree of electrical resistance.

Several anodic treatments have been used for the finishing of magnesium. Only those that provided greater protection, now currently in production use, will be discussed in detail. These are based on the use of modified acid fluoride and caustic type electrolytes.

Modified Acid Fluoride Anodize (Mil-M-45202 (ORD), Type I and II, Class C, D and E); No. 17 Anodize Treatment*. The No. 17 anodize coating is applicable to all forms and alloys of magnesium. The process is based on an aqueous acidic electrolyte containing a combination of fluoride, phosphate, chromate, and sodium ions. Either alternating or direct current is suitable, but alternating current is most commonly used because of lower cost of electrical equipment. However, approximately 30 per cent less time is required for anodizing at a given current density when direct current is used.

The No. 17 coating is applied in three specific types depending on the end or terminal voltage employed and the requirement of the coating. These types are as follows:

Clear. A very thin 40 v coating used as a base for subsequent clear lacquers or paints to produce a final appearance similar to clear anodizing on aluminum. Anodizing time is 1 to 2 min for this procedure.

Low-Voltage Thin Coating. End voltage of 60 to 75 for the best combination of protective value and paint-base properties. The anodizing time is from 2.5 to 5 min depending on alloy and current density. Coating is light gray to pale green in color. The thickness is 0.2 to 0.3 mil.

Regular Full Coating. This type is applied in the range of 75 to 95 v. It offers the best combination of abrasion resistance, protective value, and paint-base characteristics. The coating is a full medium green in color. The thickness is 0.9 1.2 mil.

Parts to be processed are first degreased, pickled if necessary to remove gross surface contamination and then anodized in either of the following solutions:

Solution A

	For A-C Use	For D-C Use
Ammonium acid fluoride (NH_4HF_2)	32 oz/gal	48 oz/gal
Sodium dichromate ($Na_2Cr_2O_7 \cdot 2H_2O$)	13.3 oz/gal	13.3 oz/gal
Phosphoric acid (85% H_3PO_4)	11.5 fl oz/gal	11.5 fl oz/gal

Solution B

	For A-C Use	For D-C Use
Ammonium acid fluoride (NH_4HF_2)	27 oz/gal	36 oz/gal
Ammonium acid phosphate ($NH_4H_2PO_4$)	13.3 oz/gal	13.3 oz/gal
Sodium dichromate ($Na_2Cr_2O_7 \cdot 2H_2O$)	10.6 oz/gal	10.6 oz/gal

* Patented and subject to special licensing arrangement with The Dow Chemical Company.

In preparing the bath the chemicals are added in the order shown while stirring. After all chemicals have been added the solution is heated to 180 F with stirring continued to ensure thorough mixing. Reheating to 180 F while stirring is required each time before using if the bath is allowed to cool to room temperature when not in operation.

The operating conditions for anodizing may briefly be stated as follows:

Current density: 5 to 50 asf or more either a-c or d-c.

Time: $\dfrac{20 \text{ to } 500 \text{ amp min/sq ft}}{\text{Current density, asf}} = $ minutes

Operating temperature: 160 asf to 180 F (Bath will not operate below 140 F but can be operated up to boiling)

Voltage: a-c or d-c up to 95 v (110 v where extremely heavy coatings are desired)

Anodizing is carried out preferably using a reasonably constant applied current. Initially 1 to 30 v is applied and the voltage is continuously raised to maintain the desired current density as the coating thickness and resistance increases. The terminating voltage varies with magnesium alloys and current density but the number of ampere minutes per unit of area remains constant for any one alloy. The data in Table 1 are a guide to indicate the time required.

After anodizing the parts are rinsed in cold and then hot water to facilitate drying. Where the anodized parts are to be left unpainted or painted only in certain areas they can be sealed in an aqueous solution containing 7 oz/gal of sodium silicate (waterglass—Na_2Si_4O) by dipping for 15 min at 200 to 212 F. After sealing, the parts are again water rinsed and then dried.

Equipment. A steel tank and heating coil is normally used for this solution. Copper, nickel, stainless steel, lead, and aluminum are attacked. The tank can be lined with synthetic rubber or a vinyl-type lining which is helpful in preventing electrical grounding of the work through contact with the tank during anodizing. Either an unlined tank or mild steel electrodes may be used as cathodes with d-c but the work itself is used as electrodes with a-c.

Electrical equipment for both manual and automatic operation of current control is readily available. Rectifiers with saturable core reactors or motor generators have all been used successfully as a source of d-c. Racks can be made of any suitable material. If they are other than magnesium or high magnesium content alumi-

TABLE 1. SUGGESTED FINAL VOLTAGE FOR APPLYING NO. 17 ANODIZE TREATMENT TO MAGNESIUM AT 20 ASF

Magnesium Alloy	Form	Low Voltage Coating			High Voltage Coating		
		Current	Final voltage	Time (min)	Current	Final voltage	Time (min)
AZ31B AZ61A	Extrusions	a-c	70	3.5	a-c	90	15
M1A ZK60A	Extrusions	d-c	75	2.5	d-c	95	11
AZ80A	Extrusions	a-c d-c	60 65	5.0 4.0	a-c d-c	75 80	24 16
AZ63A AZ91C AZ92A	Castings Castings Castings	a-c d-c	60 65	5.0 4.0	a-c d-c	75 80	24 16
EK30A EK41A EZ33A	Castings Castings Castings	a-c d-c	70 75	5.0 4.0	a-c d-c	90 95	24 16
HK31A	Sheet	a-c d-c	70 75	3.5 2.5	a-c d-c	90 95	15 11
AZ31B HK31A	Sheet Sheet	a-c d-c	70 75	3.5 2.5	a-c d-c	90 95	15 11

num alloys, they must be coated below the solution level with a vinyl plastisol type coating. (Magnesium or high magnesium alloy content aluminum such as 5056, 5083, 5456, etc. must be used for the exposed contact areas.)

Modified Caustic Anodize (Mil-M-45202 (ORD), Type I and II, Grades 1, 2, 3, 4, 5); H.A.E. Anodize Treatment. The H.A.E. anodize for magnesium alloys, developed at Frankford Arsenal, is another modification of the many "caustic"-type anodic treatments that have been suggested for use on magnesium in the past. Coatings produced by this process are equivalent in corrosion protection to the No. 17 anodize. The highest voltage type of coating is the hardest now available for magnesium. The H.A.E. anodize can be sealed and also neutralized to improve its corrosion resistance and paint base properties.

There are three types of coatings that can be produced from the same electrolyte with this process, with some variation in the operating conditions. These are as follows:

Low Voltage. This coating is of the softer type, smooth, and olive-drab in color. It is applied in 15 to 20 min using approximately 9 v a-c at a current density of 40 asf and a bath temperature of 140 to 150 F.

High Voltage (Light Coating). This coating is also of the softer type, smooth, and tan in color. The bath is operated at a temperature of 70 to 80 F and cooling is usually required. The end or terminating voltage is 60 v at a current density of 18 to 20 asf and a processing time of approximately 8 min.

High Voltage (Hard Coating). The "hard coating" is produced by anodizing to a terminating voltage of 85 v. The coating is brown in color and somewhat rougher than the lower voltage coatings. A current density of 18 to 20 asf is used and the time of treatment is 60 to 75 min. The bath is operated at a temperature of 70 to 80 F and a means of cooling the solution during operation is required.

The solution for anodizing has the following composition:

Potassium hydroxide (KOH)	22 oz/gal
Aluminum hydroxide (Al(OH)$_3$)	4.5 oz/gal
Potassium fluoride (KF)	4.5 oz/gal
Trisodium phosphate (Na$_3$PO$_4$·12H$_2$O)	4.5 oz/gal
Potassium manganate (K$_2$MnO$_4$) or potassium permanganate (KMnO$_4$)	2.5 oz/gal

The ingredients are dissolved in water in the order given. Instead of aluminum hydroxide,

Table 2. Chemical and Anodic Treatments for Magnesium

Dow No.	Name and Type of Treatment[1]	Specifications[2]	Tanks[3]	Alloys which can be treated	Appearance	Time in Bath (min)	Uses	Remarks
1	Chrome-pickle, Type C	Mil-M-3171A, Type 1, AMS-2475	A	All	Matte gray to yellow iridescent	0.25–2	General purpose. Applied for protection during shipment and storage. Good paint base.	Simple dip treatment. Etching action causes up to 0.6 mil surface dimensional loss.
7	Dichromate, Type C	Mil-M-3171A, AMS-2475	B,C	All[6]	Light to dark brown	30	Provides good combination of paint base and protective qualities.	Does not materially affect dimensions. Requires 0.5 hr or more in boiling solution.
8	Alkaline dichromate, Type C		B,C	All[6]	Dark brown to black	30–45	For black finish.	Has protective and paint base properties plus provides black color.
9	Galvanic anodize, Type EC	Mil-M-3171A, Type IV	B,C	All	Dark brown to black	10–30	For alloys that are not receptive to the dichromate treatment.	Requires galvanic couple between work and tank. Work needs to be racked.
10	Sealed chrome-pickle, Type C	Mil-M-3171A, Type II	A,C	All	Similar to No. 1, light to dark brown	0.25–30	Alternate for the dichromate when dimensional loss can be allowed.	Improved protection over regular chrome-pickle.
17	No. 17 Anodize[4], Type EC	Mil-M-45202 (ORD), AMS-2478	D	All	Clear, gray light to dark green	1–25	Excellent corrosion and abrasion resistance. Good paint base.	Either a-c or d-c may be used. Available through licensing arrangement.
18	Phosphate, Type C		D	All	Medium to dark gray	1–3	For touch-up or repair of damaged coatings.	Used as alternative for chrome-pickle as a touch-up for damaged coatings.
19	Dilute chromic acid, Type C		A	All	Yellow iridescent to dark brown	0.5–2	For touch-up or repair of damaged coatings. Cheapest of all treatments.	Used as alternative for chrome-pickle as a touch-up for damaged coatings. Also used as a tank treatment where cost is important and protection secondary.
20	Modified chrome-pickle, Type C	Mil-M-3171A, Type I	A (316)	All[7]	Yellow iridescent	0.25–2	Used primarily on castings instead of regular chrome-pickle.	Provides better appearing, powder-free coating on castings.
21	Ferric-nitrate bright pickle, Type C		A (316)	All	Silvery	0.25–5	Decorative finish. Good clean-up pickle.	Provides better appearing, powder-free coating on castings. Good shelf life finish. Also used to remove lubricants, oxide, corrosion, etc.
H.A.E.[5]	H.A.E.[5], Type EC	Mil-M-45202 (ORD), AMS-2476	D	All	Tan, brown, or olive drab	8–90	High corrosion and excellent abrasion resistance.	Only a-c current. Hardest of all coatings. Available with permission.

[1] Type of treatment: C = chemical; EC = electrochemical. [2] Mil = Military Aeronautical or Ordnance Specification; AMS = Aircraft Material Specification of the SAE. [3] Tank materials: A = stainless steel; B = rubber-lined steel for HF; C = mild steel for dichromate; D = mild steel. [4] Further information on No. 17 anodize treatment may be obtained from Dow Metal Products Co., Division of Dow Chemical Co., Midland, Mich. [5] Further information about H.A.E. may be obtained from Western Sealant Metal Finishing Co., 1719-21 N. Front St., Philadelphia, Pa. [6] Except M1A, EK30A, EK41A, HK31 and HM21. [7] Normally used only on castings.

scrap aluminum may be used at the rate of 1.5 oz/gal. The scrap should contain at least 99 per cent aluminum to avoid contamination. If potassium manganate is not available potassium permanganate may be used. The permanganate should be completely dissolved in hot water before adding to the anodizing tank.

After anodizing the parts are thoroughly rinsed in cold water and then given a post-treatment dip for 1 to 2 min at room temperature in a solution containing 13.3 oz/gal ammonium acid fluoride and 2.7 oz/gal of sodium dichromate. The parts are again rinsed in cold and hot water, then dried. This post-treatment is necessary to neutralize the alkali retained in the coating to allow good paint adhesion and it also improves the protective value of the coating, particularly in the unpainted condition. (*Note:* Several other post-treatments other than this have been suggested for this process. It has been the writer's experience that these add little improvement over the dichromate-fluoride post-treatment and that they are too complex to be practical; therefore, these are not being used currently to any extent and, in cases where surfaces are not to be painted, no post-treatment has been found necessary.)

Equipment. The H.A.E. anodize treatment is used with alternating current since direct current is not satisfactory. Equipment similar to that used with the No. 17 anodize is satisfactory. The racks for holding the work are also the same. Aluminum alloys are not as suitable for making the racks because of the strongly alkaline nature of the electrolyte.

Chemical Treatment No. 23. This "Stannate Immersion Treatment", for which a royalty free license is required from The Dow Chemical Company, was developed under an Ordnance Contract by Dow. It is applicable to all magnesium alloys, even those which contain dissimilar metal fasteners or inserts (other than aluminum). It is comparable to treatments No. 7 and No. 17. It is also used as a paint base, especially (because of its low electrical resistance) where RF grounding is required.

The preferred cleaning procedure includes solvent degreasing, when necessary, followed by alkali soak cleaning, hydrofluoric acid pickling (24 fl oz/gal 60% HF, 5 min at 70 F), Stannate Immersion Treating, neutralizing (if parts are to be painted), and suitable rinsing steps. Any heavy duty alkali soak cleaner may be used at 170 to 180 F for 2 to 5 min.

The operational limits of the stannate bath are as follows:

Potassium Stannate ($K_2SnO_3 \cdot 3 H_2O$)	5–7.3 oz/gal
Tetrasodium pyrophosphate ($Na_4P_2O_7$)	4–7.3 oz/gal
Sodium hydroxide (NaOH)	1–1.7 oz/gal
Sodium acetate ($NaC_2H_3O_2 \cdot 3 H_2O$)	1–4 oz/gal

The bath is prepared by dissolving the NaOH first followed by the $K_2SnO_3 \cdot 3H_2O$ and then the $NaC_2H_3O_2 \cdot 3H_2O$. The solution is then heated to 140–150 F before dissolving the $Na_4P_2O_7$, using good agitation. Approximately 8 to 10 sq ft may be treated per gallon of solution before chemical additions, based on analyses*, become necessary.

Parts should be stannate treated at 180 F until a minimum of 0.15 mil coating is formed, or until no bare untreated metal (including inserts) is visible. A recommended 20-min immersion period normally produces a 0.2 mil thick coating on AZ31B alloy and a slightly heavier coating on steel surfaces.

If parts are to be painted, they should be neutralized in a sodium acid fluoride ($NaHF_2$) solution of 6.3 oz/gal. Furthermore, use of wash primers is mandatory when painting over the stannate treatment film.

Equipment. Steel tanks are used for the alkali cleaner, immersion stannate solution and all rinses, except those following the HF pickling and the $NaHF_2$ neutralizing solutions. The latter rinses and fluoride treatment solutions require steel tanks suitably protected with chemically resistant coatings. (See Chapter 21.)

Steel steam coils or plate coils may be used in the heated solutions.

---

* Analytical methods for the No. 33 bath may be obtained from The Dow Chemical Company, Midland, Michigan, USA.

14. NONELECTROLYTIC METAL COATING PROCESSES

A. Non-Catalytic Chemical Methods

H. L. PINKERTON

Fellow Engineer, Process Engineering Department, Aerospace Division, Westinghouse Electric Corporation, Baltimore, Md.

Methods for coating an object with metal without the use of an applied electric current may be classified broadly as chemical or physical. The chemical processes may be further subdivided into chemical displacement or immersion methods, contact plating, and chemical reduction methods. The chemical reduction methods include a few reactions characterized by the unusual fact that the reduction reaction is catalyzed by the very metal that is being deposited. Such reactions can therefore continue as long as the supply of metal ions and reductant is maintained; in contradistinction to displacement reactions, this means that the thickness of deposit which is obtainable is unlimited. These reactions will be discussed later in this chapter.

Aside from the catalytic reactions, all other chemical methods have, for various reasons, a practical upper limit of deposit thickness that can be obtained. Generally, this upper limit is of the order of 10 millionths of an in., or less frequently 100 to 200 millionths. This thickness limits the practical usefulness of these processes to applications where the coating serves for other than corrosion protection, as for decoration or preparation for some other processing such as painting or further electroplating.

To avoid confusion, it will be well to review a few fundamental facts and to define some terms that will be used. The electromotive force (emf) series of elements (Table 27 p. 49 is as well-known as is the fact that metallic iron, for instance, will dissolve in and displace copper from a copper sulfate solution. This is said because iron is "above" copper in the emf series, but every plater knows that iron will not displace copper from a cyanide copper-plating bath. This illustrates that the usual emf series applies only to the electrode potentials of the metals in solutions of their simple salts at unit activity at 77 F (25 C). At other ion concentrations, temperatures, and especially in the presence of complexing agents, the electrode potentials will differ greatly; furthermore the order of the metals may be changed. Thus, in the copper cyanide bath, the potential of the copper has been shifted toward the electropositive end of the series at least as far as iron (in the same solution) chiefly by its complexing with cyanide.

In this country the electropositive end of the emf series refers to the alkali metal (or "base" metal) end; whereas English practice, for instance, is the reverse. Also, in this section the term "base metal" will be used in contradistinction to "noble" (or "electronegative") metal and never to connote the substrate or basis metal.

Coating by Displacement (Simple Immersion)

The formation of a coating by chemical displacement is not as simple a process as may appear, and is in many respects related to electrodeposition. Several different mechanisms may be involved. First, there may be only a simple interchange of electrons between the atoms of the solid base metal and the ions of the nobler metal in solution; however if this were all, deposition would cease when the noble metal layer were a few atoms thick. Second, due to some inhomogeneities in the base-metal surface, local galvanic couples may be set up, whereby the nobler metal is deposited upon the cathodic areas, rendering these areas even more electronegative. Thus the deposition continues until by sidewise growth of the deposit the anodic areas are either covered or reduced to such a size that the current density becomes excessive, thereby polarizing the "anode" and stopping the action. The nature of the deposit on the cathodic areas will be determined by the usual factors that control the structure of an electrodeposit. Thus, if the emf of the local couples, including the base-noble metal couple, is sufficiently high, the current density may exceed the limiting cathode current density of the bath, resulting in the formation of spongy deposits. Therefore, the greater the distance between the two metals in the emf series, the more difficult will it be to obtain dense, coherent deposits. The usual measures, however, will be effective in improving the character of the deposit: increase in noble metal concentration, increase in temperature, agitation, and proper buffering of the solution. Also, complexing the noble metal ion in solution will raise its deposition potential, which may reduce the effective emf and thus improve the quality of the deposit. If the complexing agent is too strong, the deposition potential may be raised so high that the reaction cannot proceed at all.

Several authors[1-3] have studied immersion coatings produced on rotating specimens in simple salt solutions, and it has been shown that for most of the systems studied, there was a threshold speed of rotation, i.e., degree of agitation, below which the deposit was spongy, and above which it was sound, dense, and adherent. Furthermore, the greater the separation of the two metals in the emf series, the higher was this speed threshold. For Ag on Zn, no speed was sufficient to prevent spongy deposits, up to the point where centrifugal force detached the deposit as fast as it formed.

Theoretically, at least, any of the metals of interest to electroplaters can be made to form immersion deposits on a less noble metal, but the thickness limitation restricts their utility so that commercial applications of simple immersion coatings are not numerous.

Alloys. Alloys may also be deposited by immersion, the oldest example of which is the so-called "liquor finish," "straw color" or "brass" on steel wire[4]. This is actually a bronze deposited from an acidified solution of the sulfates of copper and tin. A bath has been patented[5] for depositing a true brass on aluminum or its alloys from a mixture of zincate and copper cyanide containing also a small amount of lead. The lead apparently raises the deposition potential of the other metals; otherwise it operates to inhibit the speed of reaction so that the quality of the deposit and its adhesion are allegedly improved.

Copper. Coppering of steel by immersion is used on steel wire prior to drawing, to minimize wear on the dies, and on sheet to facilitate deep-drawing operations. The familiar slightly acid copper sulfate solution is used, or similar proprietary solutions containing appropriate complexing compounds, to improve the properties of the deposit.

Gold. Since before the days of electroplating articles of inexpensive costume jewelry have been gilded by immersion, and the older literature is replete with formulas, typical of which are[6]:

(1) Gold, as chloride   1¼ oz Troy/gal
     Potassium bicarbonate   4 lb/gal
     Boil for 2 hr. Dip brass or copper articles in warm solution.

(2) Potassium pyrophosphate   10.7 oz/gal
     Potassium cyanide   0.033 oz/gal
     Gold, as chloride   0.27 oz/gal
     Used hot, for copper and its alloys.

This bath is attributed to Roseleur, who recommends a mercury "quicking" dip prior to gilding, especially for soldered articles. He further states that repetitive cycles of quicking (in weak mercuric nitrate) and gilding can build up thick enough deposits "to resist the action of nitric acid for several hours." If true, this is interesting, for it is an example of the deposition of a baser metal, i.e., mercury upon one apparently more noble, from a simple salt solution. The fact that mercury will instantly amalgamate with gold probably alters its deposition potential favorably; also the thin gold layer is, very likely, quickly alloyed with mercury, copper, or both, from the substrate, and does not exhibit as noble a potential as pure gold.

Tin. In all probability more area of metal surface, and certainly more articles, have been immersion-coated with tin than with any other metal. Literally countless pins, hooks, eyelets, screws, buttons, etc., of brass, copper or steel have been "whitened" by immersion tinning. Approximately all aluminum alloy pistons for internal combustion engines are coated with an immersion tin deposit to assist in proper running-in of the piston.

Numerous baths have been used for tinning steel and copper alloys, typical of which are:

(3) A saturated solution of potassium bitartrate (cream of tartar) containing 0.5 to 4 oz/gal $SnCl_2 \cdot 2H_2O$ and operated boiling.

(4) $NH_4Al(SO_4)_2 \cdot 12H_2O$ 2.7–4.2 oz/gal
    $SnCl_2 \cdot 2H_2O$ 0.1–0.33 oz/gal
    Operated boiling.

(5) $SnCl_2 \cdot 2H_2O$ 2.0 oz/gal
    NaOH   2.5 oz/gal
    NaCN   1.0 oz/gal
    Operated boiling.

A variation of bath (5) may be operated cold[7].

(6) $SnCl_2 \cdot 2H_2O$ 0.66 oz/gal
    NaOH   0.7 oz/gal
    NaCN   6.7 oz/gal

The deposition of tin on copper is another example of an apparent contradiction to the emf series; in fact, in simple salt solutions tin will dissolve, thus replacing copper which plates out. In these solutions containing complexing ions, the deposition potentials of both metals must be raised, but copper more than tin, until tin becomes the nobler metal. All the solutions contain radicals (NH_4 or CN) with which copper complexes strongly; however tin complexes weakly if at all.

The bath used for tinning aluminum pistons[8, 9, 10] is a 6 oz/gal solution of sodium stannate at 175 to 180 F for 3 to 5 min. In this time, an exceptionally thick deposit of 150 to 200 millionths of an in. is obtained. Although many other tinning baths have been operated[6, 11] in the past, they are too numerous to detail.

Nickel. The only large-scale operation employing the immersion nickel process is the well-known "nickel dip" for steel in the ceramic enameling industry, which is a dilute (1 to 2 oz/gal) solution of nickel sulfate, sometimes with ¼ oz/gal of boric acid, maintained at a pH of 3 to 4 and used at a temperature of 160 F. It has been shown[12] that a more concentrated chloride bath (60 oz/gal nickel chloride, 4 oz/gal boric acid) at a pH of 3.5 to 4.5, and 160 F will produce a coating on steel about 20 millionths of an in. thick in 10 min., almost 30 millionths in 1 hr, and approximately 90 millionths in a day. A patent[13] describing a similar bath claims advantages in its use in securing adhesion of silver to steel in making aircraft engine bearings.

Zinc. Zinc is so high in the emf series that only aluminum and magnesium are more electropositive, and it is only with these metals that zinc-immersion processes have found any application. The bath is always a highly concentrated, highly alkaline zincate solution, and the various modifications of it and conditions of use are outlined in Chapter 4.

Contact Plating

Contact plating was discovered[14] by Henry Bessemer, subsequently famous in steelmaking, in 1831, who found that a casting of type metal could be coppered more quickly and completely if it were placed in a zinc tray and the coppering solution poured over it.

In simple immersion processes, the basis metal dissolves and is stoichiometrically replaced by the depositing metal:

$Zn^0 + Cu^{++} \rightarrow Zn^{++} + Cu^0$ (deposits on Zn)

In contact plating, the basis metal does not dissolve; instead, a piece of more electropositive metal is placed in contact with the basis metal. The more electropositive metal dissolves, while both it and the basis metal are covered with depositing metal, as illustrated below:

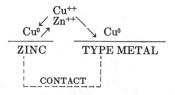

This is truly an electroplating process, evidenced by the fact that the areas of basis metal far removed from the point of contact, i.e., the anode, receive less plate than those nearer. The contact process is generally capable of producing heavier deposits at a faster rate than immersion processes, but its chief advantage lies in the fact that deposition will occur in systems that are incapable of producing any deposit by immersion. Thus, if steel is placed in a hot standard sodium stannate plating bath, nothing happens; however if the steel is touched with a piece of zinc or aluminum, it will plate, as will be described.

The positioning and relative area of the sacrificial contact metal is important in determining the distribution and thickness of the deposit;

furthermore care must be taken to avoid lack of deposit at the point of contact. Since the action ceases if and when the contact metal becomes completely covered with depositing metal, the nature of the coating on the contact is important. If it is spongy and nonadherent, the plating action may be expected to last longer, but such loose particles detached from the "anode," i.e., contact metal, may cause roughness of the deposit on the main article.

Copper. Copper contact plating has already been mentioned. Another more unusual example is the old practice of "oxidizing" the graphited surface of wax masters in electrotyping. In one version of the process, iron powder is sprinkled on the graphite and a copper sulfate solution is poured over it, whereby the graphite becomes coated with copper, which improves its conductivity. There are no large-scale applications of contact copper plating.

Gold. Although entirely possible, contact plating with gold has not been much practiced since gold is also deposited on the contact where it is not wanted and must be recovered.

Tin. Although contact tinning has been known for many years[6, 11], it never received much attention until recently, probably because immersion tinning is so easy and simple. Wilson and Wright reported[15] that in a solution with the usual stannate plating bath composition, an aluminum contact made it possible to produce contact deposits up to 0.002 in. thick, and that the process could be employed concurrently with electrodeposition to enable blind holes and other places inaccessible to the current to be plated. Weimer and Price[10], reporting on this process, stated that they found deposits above 0.0002 in. to be dark and nonadherent. They also reported much faster deposition rates from other baths, for example, from the standard acid sulfate plating solution with a magnesium contact, where deposit thicknesses were about five times those obtained in the stannate bath under the same conditions. Good results were obtained until the bath had been worked to a magnesium concentration of 5.3 oz/gal.

Zinc. There are no practical examples of contact zinc plating, but Raoult[16] showed that gold or copper could be zinc-coated in a solution of $ZnSO_4$ using a zinc wire. This is only of interest because in this case there is merely a transfer of zinc from the contact (now truly an anode) through the solution to the work, but the action will cease when the work is coated.

Maintainence of Immersion and Contact Baths

Most baths for immersion and contact plating are doomed to rather short lives because (a) the coating metal is quickly depleted, (b) the bath becomes intolerably contaminated with the *base* metal whether from the contact or basis metal, and (c) side reactions often occur, usually involving pH changes, which usually cannot be corrected without upsetting the bath balance in some unavoidable way. A notable exception is the last-cited process of Raoult, in which the bath should last indefinitely.

In most cases the operators of these processes are satisfied to take such steps as are relatively easy to extend the life of the bath and then discard it or a portion of it when something is intolerably out of control. In the "liquor finishing" of steel wire, for instance, additions of tin and copper sulfate are made under analytical control, while the normal drag-out of solution is sufficient to keep the iron concentration from building up above some tolerable, equilibrium concentration. If there is further buildup, a portion of the bath may be discarded and replaced with fresh bath.

Non-Catalytic Chemical Reduction

The non-catalytic chemical reduction processes in general suffer from several disadvantages, as compared with catalytic processes: (a) the deposition ceases about as soon as the object is covered, (b) the bath continues to complete decomposition after the object is coated, and (c) the coating is formed indiscriminately on all objects in the bath, including the walls of the vessel. While the catalytic processes are not perfect in all these respects, they are greatly superior.

The form in which a metal appears when it is reduced in solution depends on the conditions of reduction. If the reaction is very fast, as when strong reducing agents are added to simple metal salt solutions, the metal precipitates as an unattractive sponge, which is gray, brown or black, depending on the metal and its particle size. When the reaction is substantially slower, the metal may deposit in the form of a continuous film on any solid in contact with the solution. When the film has reached a certain thickness (in the absence of any catalytic effect) film formation ceases, and any metal remaining in solution will precipitate as a sponge.

The oldest and still most widely used chemical reduction process is the reduction of silver to form mirrors, ornamental films, or conducting films on nonconductors for further electrodeposition. Some copper has been used for the same purposes. The ability to form metallic films on nonconductors is about the only justification for choosing the rather ticklish reduction processes over the simpler immersion or contact processes since the maximum film thickness that can be achieved is similar.

Silver. The old familiar reduction processes (Brashear or cane sugar, rochelle salt and formaldehyde processes)[17-19] are no longer widely favored, because of their inefficiency and associated hazard of forming the dangerously explosive fulminating silver. All these processes used a solution of silver ammonio-nitrate in which an excess of ammonium hydroxide was deliberately avoided. In modern processes, reducing agents are used which are capable of precipitating silver films from similar solutions in the presence of excess ammonia, which eliminates the danger of forming fulminating silver. Such solutions may be safely stored, whereas the older, carefully balanced solution had to be used immediately or destroyed. These solutions can be formulated so that the silver film is formed in a matter of seconds, making possible the application of the film by means of a two-solution spray gun. Spray-silvering methods offer the advantages of a high production rate, simplicity of operation and much higher efficiency of silver utilization than tray-silvering methods. One recommended formula[20] uses two solutions which are mixed at the gun in equal proportions:

	oz/gal	Molarity
(7) Solution A		
$AgNO_3$	1.3	0.0566
NH_3 (added as 28% ammonium hydroxide)	0.59	0.258
Solution B		
Hydrazine sulfate (or nitrate or hydrate)	2.56	
NaOH, or equiv. KOH (optional)	0.64	

The speed of reaction may be reduced by omitting the alkali hydroxide from solution B and still further reduced by a substantial increase in the ammonia concentration in solution A. Glyoxal may also be used as a reducing agent instead of the hydrazine compound.

From the spent solutions, any remaining silver may be recovered in the usual way, i.e., by precipitating it as chloride with hydrochloric acid and later collecting it for sale to a refiner.

Copper. Most of the early solutions proposed for copper reduction produced films that tended to be nonadherent and spongy owing to incomplete reduction and co-deposition of cuprous compounds in the film. A process[21] allegedly free of this difficulty utilizes a 4-per cent solution of zinc sulfoxylate formaldehyde to reduce Fehling's solution, whereby a copper film of 12 to 13 millionths of an in. can be deposited in 35 to 45 min.

Saubestre[22] made an extensive study of the formation of copper films. While this work was concerned primarily with developing a catalytic process (which will not be considered in this section) in the course of his work, he developed non-catalytic formulas for tray-coppering and for spray application:

(8) Solution A—23.7 oz/gal $NaKC_4H_4O_6 \cdot 4H_2O$ (rochelle salt)
6.7 oz/gal NaOH
4.7 oz/gal $CuSO_4 \cdot 5H_2O$
B—2.7 oz/gal $(NH_2)_2 \cdot H_2SO_4$ (hydrazine sulfate)
Mix 10 vol A and 3 vol B at 176 F before use.

(9) Solution A—4.7 oz/gal $CuSO_4 \cdot 5H_2O$
6.7 oz/gal conc. H_2SO_4 (66° Be)
B—67 oz/gal NaH_2PO_2 (sodium hypophosphite)
Mix on work with two-gun spray 2 vol A and 1 vol B at 176 F

(10) Solution A—Same as 9A
B—13.6 oz/gal $Na_2S_2O_4 \cdot 2H_2O$ sodium hydrosulfite)
Operate same as 3

Formula 8 has a tendency toward instability, which makes its value questionable; nevertheless its action is very fast. The last two formulas are so fast that they cannot be used except in a two-gun spray.

Other Metals. No commercial importance is attached to the production of films of other metals by non-catalytic chemical reduction, but two observations in the author's experience may be of interest. The reduction of gold chloride by ferrous sulfate in one experiment gave a massive coherent gold deposit on the sides of the beaker. The deposit, which could be lifted out easily and handled was about 10 to 12 mil thick; consequently this reaction must be catalytic.

A solution of iodoplatinic acid in lithium iodide precipitated its platinum quantitatively on the sides of the beaker as a platinum mirror. The introduction of a small quantity of iodide ions into a stable solution of lithium bromoplatinate or of lithium ions into any iodoplatinate solution had the same effect.

Sensitizing for Chemical Reduction

In the production of metallic films by chemical reduction, the cleanliness of the substrate is important; also it is helpful, and in many cases necessary, to "sensitize" the surface by special means. The sensitizing procedure enhances the uniformity and adhesion of the deposit[17] and may influence the speed of deposition and the thickness obtainable. Usually, sensitizing consists of wetting the cleaned surface of the substrate with a solution containing ions of another metal, that is, tin or a precious metal.

Of a great number of possible sensitizers for silver reduction, a tin salt has been shown[23] to be the most satisfactory. This is usually stannous chloride; the concentration and pH do not appear to be significant since an alkaline stannite solution will serve equally well. Saubestre's paper[22] included a study of sensitizers for coppering, in which it is shown that stannous chloride is ineffective as a sensitizer for copper, but stannite has some value, although it is less effective than the precious metals. A weak (>0.001 wt per cent) solution of gold as chloride or aurate was preferred.

The sensitized surface is always rinsed well before plating; distilled or deionized water should be used for this purpose. When sensitizing intensely hydrophobic surfaces such as wax and certain plastics, it is helpful if the solution be made up in alcoholic rather than aqueous medium[24]. In such cases also both the silver and reducing solutions should contain a wetting agent, which may be of the anionic type such as an alkyl aryl sulfonate.

References

1. Centnewszwer, M., and Heller, W., *Z. physik. Chem.* (A), **161,** 113 (1932).
2. King, C. V., and Burger, M. M., *Trans. Electrochem. Soc.*, **65,** 403–411 (1934).
3. Piontelli, R., *Trans. Electrochem. Soc.*, **77,** 267–277 (1940).
4. Richter, H. W., *Metal Finishing*, **53,** No. 5, 66–67, 76 (1953).
5. Balden, A. R., and Morse, L. M., U.S. Patent 2,496,845 (Feb. 7, 1950).
6. Watt, A., and Philip, A., "The Electro-Plating and Electro-Refining of Metals," D. Van Nostrand and Co., 2nd Ed., New York, 1911.
7. Sullivan, J. D., and Pavlish, A. E., U.S. Patent 2,159,510 (May 23, 1939).
8. Brown, D., Shulnberg, V., and Dell, G., *Metal Ind.*, **36,** 11–19 (1938).
9. Finnie, E., *Products Finishing*, **6,** 42–44 (1942).
10. Weimer, D. E., and Price, J. W., *Trans. Inst. Met. Finishing*, **30,** 95–111 (1954).
11. Langbein, G., and Brannt, W. T., "Electro-Deposition of Metals," 9th Ed., New York, 1924.
12. Wesley, W. A., and Copson, H. R., *Trans. Electrochem. Soc.*, **94,** 30–31 (1948).
13. Schaefer, R. A., U.S. Patent 2,391,039 (Dec. 18, 1945).
14. Lovett, M., *Electroplating and Metal Finishing*, **4,** 152–54, 162, 221–224, 231 (1951).
15. Wilson, J. K., and Wright, O., *Electroplating and Metal Finishing*, **4,** 274–276 (1951).
16. Raoult, *Compt. rend.*, **76,** 156 (1883).
17. Hepburn, J. R. I., *J. Electrodepositors' Tech. Soc.*, **17,** 1–13 (1941).
18. Wein, S., *Metal Finishing*, **39,** 666–672 (1941); **40,** 24–25 (1942).
19. Upton, R. B., *J. Electrodepositors' Tech. Soc.*, **22,** 45–72 (1947).
20. Peacock, W., U.S. Patent 2,214,476 (Sept. 10, 1940).
21. Narcus, H., *Proc. Am. Electroplaters' Soc.*, **35,** 157–168 (1948).
22. Saubestre, E. B., *Proc. Am. Electroplaters' Soc.*, **46,** 264–276 (1959).
23. Macchia, O., *Chem. News*, **135,** 197 (1927).
24. Milton, C. L. Jr., Cort. I., Nielson, C. A., and Cowan, I., *Metal Finishing*, **45,** No. 8, 61–62, 64 (1947).

B. Vapor-Phase Methods

JOHN A. THORNTON

*Department of Metallurgy and Mining Engineering,
and The Coordinated Science Laboratory
University of Illinois, Urbana, Ill.*

Any coating process can be divided into three steps: (1) passage of coating material from the solid phase into a suitable transport phase; (2) transport of coating material from the source to the vicinity of the substrate; (3) passage of coating material from the transport phase back into the solid phase on the substrate surface. This chapter treats those methods where the transport phase is a vapor. Six of the most prominent vapor phase methods are described: (1) vacuum evaporation, (2) sputtering, (3) ion plating, (4) chemical vapor deposition, (5) plasma-assisted chemical vapor deposition and (6) plasma spraying. As a group they permit virtually any inorganic coating material to be deposited and are therefore becoming increasingly important for a wide range of applications.

The vapor phase coating methods differ from one another primarily in the mechanism by which the coating material is passed into the vapor phase. This mechanism determines the range of coating materials for which the methods are applicable, and to a large measure the deposition rates that can be achieved.

The transport step determines the size and shape of the substrates that can be coated. The transport also influences the physical properties of the coatings because of the influence of the coating flux arrival directions on the growth morphologies of the deposits. In contrast to electroplating, vapor phase coating methods do not require that the substrates be electrically conducting.

The coating growth step involves a vapor-phase to solid-phase transition that will, in general, be far from equilibrium. Thus coatings with unusual microstructures, and with physical properties that are vastly different from those produced by conventional bulk-material preparation methods, can be deposited. The fact that the substrates are placed in a vapor rather than liquid ambient permits the substrate temperature to be used to influence coating properties.

VACUUM EVAPORATION[1-5]

Basic Principles

The coating material is passed into the transport phase by sublimation or evaporation. The source material may be heated directly or placed in contact with a heated surface. Heating is typically done by electrical resistance, eddy current, thermal radiation, electron beam bombardment, laser beam or electric discharge methods. The vaporized material leaves the surface of the source as atoms or molecules traveling in directions defined by a cosine distribution with thermal energies (0.02 to 0.05 eV). The process is conducted under vacuum (typically 10^{-3} to 10^{-4} Pa). Consequently, the evaporated flux undergoes an essentially collisionless line-of-sight transport to the substrates, which are placed to intercept the flux. The only fundamental requirement on the substrates is that they be maintained at sufficiently low temperature so that the coating flux constitutes a supersaturated vapor over their surfaces.

The vacuum evaporation process itself is virtually non-polluting. Of course, the evaporation of materials containing hazardous constituents will require special handling methods.

Range of Applicability

Any material can be passed into the vapor phase if sufficient energy is delivered to its surface. Thus vacuum evaporation is an extremely versatile process. However, the method is easiest to apply to, and is most generally used for, elemental materials with low melting points and high vapor pressures. Thus aluminum constitutes 90% of all the material evaporated for industrial applications. Deposition rates of as high as 1000 nm/sec (2 mils/min) can be achieved for such materials. Refractory materials such as tungsten can be effectively evaporated by electron beam methods, but the rates are considerably lower (10 to 50 nm/sec).

Alloys having constituents with similar vapor

pressures can be passed into the vapor phase with little change in composition, particularly when rod feed electron beam sources are used. Thus alloys such as Ni-Cr, Ag-Cu, Ti-Al-V, and Ni-Cr-Al-Y have been effectively deposited by evaporation. However, difficulty can be encountered in evaporating alloys or compounds which contain constituents having widely different vapor pressures (factor of 100 or more).

Flash evaporation, where powder or chips of the material of interest are sprinkled onto a superheated surface, is sometimes used to produce complete evaporation of all the constituents. Multi-source evaporation, in which a number of independently controlled sources are operated simultaneously, with each source supplying one or more of the required constituents, can also be used to control the stoichiometry. However, multi-source methods are tedious to apply and are used only for very special applications. An example is the process of molecular beam epitaxy (MBE), where coatings of materials such as compound semiconductors are deposited in precisely controlled structures onto relatively small area substrates through the operation of multiple sources at low rates in an ultra high vacuum (10^{-11} Torr) chamber.[6]

In some cases compounds will evaporate primarily as molecules without dissociation. Examples are SiO, MgF_2, B_2O_3 and CaF_2. These materials can be effectively evaporated directly to yield deposits with the same composition as the starting material. Thus, for example, evaporated SiO and MgF_2 coatings are widely used in the optics industry. Compounds that dissociate on evaporation will often yield deposits which are deficient in the nonmetallic species. This difficulty can be overcome by evaporating in an atmosphere of the deficient species. An extension of this approach is the process of reactive evaporation, where metal atoms are evaporated with a concurrent flow of reactive gas injected into the chamber. Reactions between the metal atoms and the gas molecules occur at the substrate to form a compound. This reaction can be promoted by maintaining a plasma discharge in the atmosphere of reactive gas adjacent to the substrate. This latter process is known as activated reactive evaporation (ARE). It has been effectively used, for example, to deposit wear-resistant TiC coatings and transparent conducting Sn doped In_2O_3 coatings.

When coatings are deposited by evaporation, the substrates are subjected to thermal radiation from the evaporation source. This radiation can be significant, particularly when refractory materials are evaporated. Therefore, caution should be exercised when one is coating substrates that are subject to damage or outgassing when heated.

Apparatus Configuration

An evaporation coating apparatus consists typically of 1) a vacuum chamber, 2) a pumping system, 3) pressure measuring instrumentation, 4) evaporation source(s) and power supplies, 5) substrate fixturing and 6) deposition rate detectors (See Fig. 1). Pumping systems combine high-vacuum pumps with mechanical backing and roughing pumps. The high-vacuum pumps are generally of the oil diffusion type, although getter-ion, turbomolecular or cryogenic pumps are sometimes used.

The coating flux from an evaporation source is in general nonuniform over most substrate surface shapes. The problem is eased for many applications by arranging an array of low-cost resistant-heated wire sources surrounding the objects to be coated. For example, tungsten wire sources are commonly used to evaporate metals such as Al, Ni, Fe, or Pt. Nevertheless, fixturing capable of imparting motion to the substrates is needed if uniform coating thicknesses are required. Suitable fixturing can provide thickness uniformities of about ±10% on many substrate shapes.

The rate of evaporation is an exponential function of the source temperature. Coating thicknesses deposited from simple resistive heated sources are generally established by placing a fixed amount of coating material in the sources and evaporating all of it. However, deposition rate monitors must in general be used

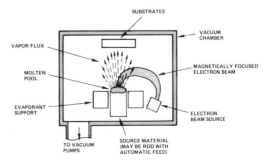

FIG. 1. Schematic illustration of evaporation system with electron beam source. (*Courtesy Telic Company.*)

to control the rate of evaporation and to determine accumulated coating thickness.

Electron beams (Fig. 1) are the most versatile of the vacuum evaporation sources for depositing a wide range of materials at high rates. Beam powers are typically 3 to 50 KW with electron energies of 10 keV.

Coating Properties

The properties of evaporated coatings are strongly dependent on (1) the substrate temperature T relative to the melting point T_M of the coating material, and (2) the arrival directions of the coating flux relative to the substrate surface (See Fig. 2). The coating atoms in vacuum evaporation tend to bond strongly with the substrate (or the surface of the growing coating) almost immediately upon their arrival. This is in contrast to electroplating, where the condensation process involves a successive diminution in the number of water molecules as hydrated ions pass to lattice sites. Thus the adatom mobility at a given T/T_M is much lower in vacuum evaporation than in electroplating.[8]

At low T/T_M the adatom mobility is incapable of overcoming the atomic scale shadowing effects that are caused by an oblique component to the coating flux or by a rough substrate or coating surface.[8] This shadowing results in the Zone 1 structure (Fig. 2) which is characterized by columnar voids. Care must be exercised when coating complex shaped substrates to avoid oblique deposition at low T/T_M. When the flux is normal to the surface of a smooth substrate, a fine-grained fibrous structure results (Zone T). At $T/T_M \sim 0.5$ dense columnar grains similar to those in electroplated coatings are formed (Zone 2). At high T/T_M recrystallization can result in large columnar or equiaxed grains (Zone 3).

Coatings deposited at high T/T_M tend to have the properties of fully annealed bulk materials. Metal coatings deposited at low T/T_M have high dislocation densities and properties typical of highly cold-worked materials. Coatings of compounds deposited at low temperatures generally have properties that are inferior to their bulk counterparts. Refractory materials deposited at low temperatures are typically in a state of high intrinsic stress, which is usually tensile for metals and may be tensile or compressive for compounds. The stress state in coatings deposited at high T/T_M is generally dominated by the thermal stresses.

Application

Vacuum evaporation equipment is relatively expensive. Therefore, the process is most effective when large numbers of substrates are to be coated. Under these conditions coating costs can be low, as is seen by the wide use of evaporated aluminum on everything from bottle caps and trophies to polymer sheet for capacitors. Other applications include (1) aluminum corrosion-protective coatings on steel strip; (2) Al or Al-Si conductor coatings for microelectronic devices; (3) M-Cr-Al-Y coatings, where M is Co, Ni or Fe, on nickel-base jet-engine blades used in the hot end of aircraft jet engine turbines; (4) MgF_2 and SiO coatings for optical filters; and (5) wear-resistant coatings such as TiC and TiN for high-speed cutting tools.

SPUTTERING[1-4,9,10]

Basic Principles

Sputtering is a vacuum coating method wherein coating material is passed into a vapor transport phase by the momentum exchange associated with particle bombardment of a source (target) composed of the coating material in question. The bombarding species are generally ions of a heavy inert gas. Argon is most commonly used. The source of energetic ions may be an ion beam source or a plasma discharge within which the target is immersed. The most common method makes the target the cathode electrode of a low-pressure glow discharge which is sustained in the argon working gas. The current passing into the target is composed largely of ions from the plasma which are accelerated to energies of several hundred eV by the electric field which develops in the cathode sheath region adjacent to the target.

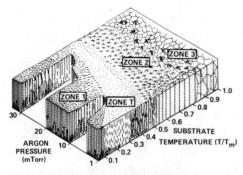

FIG. 2. Zone diagram showing microstructure of coatings deposited by evaporation and sputtering in the absence of ion bombardment.[7] (*Courtesy American Institute of Physics.*)

Such an apparatus is shown schematically in Fig. 3.

The most striking characteristic of the sputtering process is its universality. Since the coating material is passed into the vapor phase by a mechanical (momentum exchange) rather than a chemical or thermal process, virtually any material is a candidate coating. Direct current discharges are generally used for sputtering metals. An rf potential must be applied to the target to sputter a nonconducting material.

The sputtered species are primarily neutral atoms. They are ejected from polycrystalline type targets in a near-cosine distribution with relatively high energies (average of 10 to 40 eV). The sputtering process is quantified in terms of the sputtering yield, defined as the number of atoms ejected per incident ion. The sputtering yields of most materials are about unity and within an order of magnitude of one another. The sputtering rates for all materials at a given target power input are therefore similar. By contrast, in evaporation, the rates for different materials at a given power input to the source can differ by several orders of magnitude.

Sputter coating chambers are typically evacuated to pressures in the 10^{-3} to 10^{-5} Pa range before backing-filling with Ar to pressures in the 0.1 to 10 Pa range. The intensity of the plasma discharge, and thus the ion flux and sputtering rate that can be achieved, depend on the configuration of the cathode electrode. The deposition rate depends on the working gas pressure, as high pressures limit the passage of the sputtered flux to the substrates. For many years most sputtering was done using simple planar electrode systems of the type shown in Fig. 3. Even when operated under the most favorable conditions (pressures in the 1 to 10 Pa range), these devices yield relatively weak discharges and deposition rates that are typically less than 2 nm/s. Consequently, sputtering developed a reputation for very low deposition rates.

The development during the past few years of a class of sputtering sources with magnetic plasma confinement, called magnetrons, has greatly enhanced the capabilities of the sputtering process.[9] These devices can provide order-of-magnitude increases in sputtering rates. They can be operated at such low pressures (0.1 to 0.5 Pa) that the sputtered atoms pass line-of-sight to the substrates. They can be scaled to large sizes that provide uniform deposition over very large areas (many m²). The advent of the magnetron technology has greatly increased the range of applications for sputtering.

There are many variations of the basic sputtering process. The substrates may be biased as electrodes prior to coating, so that contamination is removed by sputtering, and coating nucleation sites are generated on the surface. This is known as sputter cleaning. They may also be biased to cause ion bombardment during deposition for the purpose of removing loosely bonded contamination or modifying the structure of the resulting coating. This is known as bias sputtering. A gas may be used to introduce one or more of the coating constituents into the chamber. This process is known as reactive sputtering.

Sputtering, like evaporation, is by its basic nature a nonpolluting process.

Range of Applicability

One of the most important features of the sputtering process is that, after a short equilibration period, the flux of sputtered material leaving a target will be identical in composition to the target, provided that the target (1) is maintained sufficiently cool to avoid interdiffusion of the constituents, and (2) does not decompose. Thus, if the gas phase transport and substrate sticking coefficients are the same for all the constituents, the coating composition will be identical to that of the target. Therefore, coatings of alloys such as stainless steels, compounds such as Al_2O_3 and even bone[11] and a polymer (PTFE)[12] have been successfully deposited by sputtering.

Often, application of the sputtering process is limited by target considerations. Metal targets are in general relatively easy to fabricate and to cool. However, magnetron sources can accommodate only relatively thin targets of magnetic

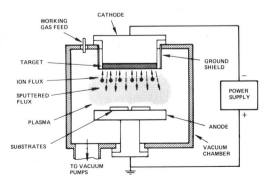

FIG. 3. Schematic illustration of sputtering process in simple planar diode apparatus. (*Courtesy Telic Company.*)

materials because of the influence of the target in distorting the magnetic field used for plasma confinement. Targets of compounds such as oxides, carbides, nitrides, etc., can be formed by hot-pressing powders. However, poor heat transfer, coupled with vulnerability to target cracking or the loss of volatile constituents, generally limits the power levels that can be delivered to nonmetallic targets.

Metal deposition rates with dc-driven magnetron sources are typically in the range from 5 to 30 nm/s.[13] Target cracking and decomposition effects, along with the basic character of rf discharges, typically limit the deposition rates for directly sputtered compounds to the range 1 to 10 nm/s, even when magnetrons are used.[13] Reactive sputtering with metal targets greatly simplifies the target fabrication problem. However, deposition rates are still generally limited to the 1 to 10 nm/s range because of surface layers of modified composition which tend to form on the target and effectively reduce the sputtering yields.

Substrate heating depends on the deposition rate and on whether the substrates are in contact with the plasma. Substrates in contact with the plasma can reach temperatures in the 300 to 500 C (572 to 932 F) range, even at moderate deposition rates (~1 nm/s). The magnetic plasma confinement permits the substrates in magnetron sources to be removed from the plasma. Thus dc magnetrons can be used to deposit metal coatings onto heat-sensitive substrates such as plastics.

Apparatus Configuration

The vacuum chambers and pumping systems required for sputtering are identical in their essential elements to those used for evaporation (see previous discussion). In addition, one has the sputtering sources and power supplies, and a feed and control system for the working gas. Chambers with vacuum interlocks which permit the substrates to be introduced without exposing the chamber walls and sputtering sources to the atmosphere are coming into increased usage. The rate of sputtering from a given target is proportional to the power delivered to the target and can be controlled relatively accurately. Thus deposit thicknesses are often established simply by the time of deposition rather than by using a rate monitor.

Electrode configurations of the type shown schematically in Fig. 3 are known as planar diodes. Although limited to moderate deposition rates, as noted previously, they have been widely used in the electronics industry to deposit a range of coating materials onto flat wafer substrates using dc or rf power. Rf sputtering is generally done at the industrially allowed frequency of 13.56 MHz. Power levels are typically in the 500 W to 5 KW range. Voltages (dc, or rf peak-to-peak) are typically 2 to 4 kV. In the case of the magnetron sources,[9,13] magnetic fields are used to confine intense plasma rings or sheets, containing circulating currents, over planar or cylindrical targets.

The planar magnetrons are often configured with race-track shaped plasma rings confined on rectangular targets. Sputtered atoms are emitted from beneath the plasma ring and deposited onto substrates which are transported in a direction perpendicular to the long axis of the target. The cylindrical targets may be hollow cathodes, where the sputtered atoms pass inward to axially mounted substrates. They may also be in the form of central posts from which the sputtered flux passes radially outward to circumferentially mounted substrates. Cylindrical magnetrons are particularly effective for batch process applications. Magnetrons are typically driven by constant current power supplies at current levels in the 5 to 50 A range with voltages of 0.5 to 1 kV.

Sputtering is an inefficient process from an energy point of view, with most of the power input appearing as target heating. Accordingly, sputtering targets are generally water-cooled. Substrates may be heated or cooled depending on the application.

Coating Properties

A schematic representation showing the dependence of the microstructure of sputtered coatings on the Ar pressure and substrate temperature was given in Fig. 2. The same basic diagram (at zero Ar pressure) also applies to evaporated coatings, as discussed previously. There are three important ways in which the structures of sputtered coatings may differ from those of evaporated coatings. First, the sputtered atoms can become scattered by the working gas so that they approach the substrates at very oblique angles. Thus elevated working pressures at low T/T_M tend to promote the Zone 1 type structure. Second, significant numbers of ions can be neutralized and reflected at the target surface and then pass back in the direction of the substrates with energies as high as several hundred eV. At low working pressures these atoms will reach the substrates and subject them to a bombardment which tends to promote the dense, fine-grained Zone T structure. Third, the substrates may be in contact with the plasma. Plasma ion bombardment tends to suppress the Zone 1 structure, even at elevated Ar pressures, and to promote a dense, fine-grained structure of the Zone T type. Ion bombardment on uncooled substrates will also promote higher temperature structures. These bombardment effects have given sputtering a reputation for producing dense, high-quality deposits. Coatings which are subjected to energetic reflected atom or ion bombardment during growth often exhibit high compression type internal stresses. Coatings deposited under other conditions are often in tension.

Applications

The ability to control coating composition has caused sputtering to become used in the electronics industry for applications such as aluminum alloy and refractory metal microcircuit metallization layers, oxide microcircuit insulation layers, transparent conducting electrodes, amorphous optical films for integrated optics devices, piezoelectric transducers, photoconductors and luminescent films for display devices, optically addressed memory devices, amorphous bubble memory devices, thin film resisters and capacitors, videodiscs, solid electrolytes, thin film lasers and microcircuit photolithographic mask blanks. Other applications include wear- and corrosion-resistant coatings on substrates ranging from razor blades to machine tools.

Magnetron sources have opened up new applications because of their large area capability and reduced substrate heating. Thus large in-line systems with vacuum interlocks are used to coat 2 × 3.5 m architectural glass plates. Magnetrons are also used on a production basis to deposit chromium decorative coatings on plastic automobile grilles and other exterior trim.

ION PLATING[1, 14, 15]

Basic Principles

Ion plating is a generic term applied to those deposition processes in which the substrate surface and/or the depositing film is subjected to bombardment by a flux of high-energy particles sufficient to cause changes in the properties of the interfacial region and the coating.[1] When the source of coating atoms is via sputtering, these bombardment processes are usually called sputter cleaning and bias sputtering, as discussed previously. Thus the term ion plating is generally used to refer to the case where evaporation is the source of coating flux. The process is typically executed by maintaining an argon pressure of 1 to 7 Pa in an evaporation chamber and striking a plasma discharge between the substrate(s), connected as a cathode, and the evaporation source and chamber, which are at ground potential. Such an apparatus is shown schematically in Fig. 4.

A functional difference between most ion plating and bias sputtering systems is that the substrates in ion plating are generally the cathodes that sustain the plasma discharge and consequently operate at voltages of 2 to 5 kV. In bias sputtering the substrates are usually immersed in the plasma discharge created by the target electrode, and the substrate voltages are typically in the 50 to 500 V range. Substrate current densities are typically about 0.5 mA/cm^2 in both cases. The particle bombardment is by ions and atoms of the working gas.

The principal benefits obtained from the ion plating process are (1) improved adhesion; (2) substrate heating which enhances diffusion and chemical reactions; (3) particle bombardment modifications to the morphology, crystallography, composition and physical properties of the coating; and (4) improved coating coverage on substrates with complex shapes because of a combination of coating flux gas scattering and resputtering of the deposited coating.

In a somewhat similar process called cluster ion beam deposition,[16] the evaporated flux is

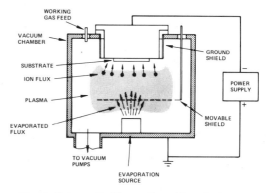

FIG. 4. Schematic illustration of ion plating system with evaporation source of coating material. (*Courtesy Telic Company.*)

caused to pass through a restricting orifice which raises the vapor flux density to the point where gas phase nucleation occurs. The nucleated droplets typically contain 10^2 to 10^3 atoms. The droplets are passed through an electron beam, which induces a surface charge, and a grid system which accelerates them through a several thousand volt potential prior to their impact on the substrates. The method is reported to yield dense coatings with excellent physical properties. The substrate heating is sufficiently low so that polymeric materials can be coated.

Range of Applicability

Since ion plating is generally practiced with evaporation as the method of providing coating flux, its range of coating material applicability is the same as was discussed previously with respect to evaporation. The use of a reactive working gas allows the deposition of compounds by reactive ion plating.

If the substrate is not electrically conducting, an rf potential, or a screen electrode in front of the substrate, must be used to induce the ion bombardment. Difficulties are generally encountered in producing uniform particle bombardment on substrates having irregular shapes.[14] Blind holes and re-entrant corners are particularly troublesome. Care must also be exercised when depositing alloys and compounds, because the particle bombardment may change the coating composition by preferential sputter-removal of one or more of the constituents.

The use of particle bombardment to modify the structure of a growing coating is limited by the requirement that the bombardment flux be adequate to produce resputtering at a rate which is a significant fraction of the deposition rate for the arriving coating flux. The high deposition rates that can be achieved with evaporation therefore require bombardment fluxes that are difficult to achieve for many substrate shapes. Net deposition rates in ion plating are typically those achieved for evaporation, modified by the resputtering effects.

Apparatus Configuration

Ion plating apparatuses are essentially identical to the evaporation systems discussed previously, but with provisions for producing ion bombardment of the substrates. They therefore include working gas feed and control systems and biasing power supplies essentially identical to those used in sputtering. The substrate bombardment may be produced by (1) making the substrates the cathode of a glow discharge, as shown in Fig. 4, (2) applying a negative bias to the substrate and immersing it in a plasma generated by independent electrodes, (3) placing the substrate(s) behind a grid electrode or within a screen barrel that induces a bombarding ion flux, or (4) incorporating an ion beam source within the deposition chamber.

Coating Properties

Ion plated coatings have an excellent reputation for adhesion. This is because of the effect of the substrate sputter cleaning which is done prior to and at least through the initial period of deposition. The bombardment removes contamination, creates nucleation sites, causes surface disruptions that enhance interface diffusion when the coatings are deposited, causes physical mixing of the near-surface material, and raises the substrate temperature. Particle bombardment during coating growth produces the same beneficial effects that were discussed previously with reference to sputtering. Thus dense structures of the Zone T and Zone 2 type are promoted and the porous Zone 1 structure is suppressed, even though ion plating may be conducted in the presence of a working gas pressure in the 1 to 7 Pa range. Coatings grown at modest T/T_M under conditions of significant particle bombardment may contain several atomic percent of entrapped working gas. Ion plating coatings are often in a state of high compressive stress.

Applications

The improved adhesion obtained with the ion plating process has been the basis for most of the applications. It has generally been used where no simpler technique was found to be adequate. An exception has been in Japan, where ion plating has come into wide use to replace electrodeposition because of concerns over pollution.

Ion plating is generally used to deposit metal coatings on metal substrates. Some work with plastic substrates has been reported. Most ion plating applications fall into three categories: (1) wear and erosion—typical coatings are metals, carbides and nitrides of materials such as Cr, Ti and Zr; (2) corrosion inhibition—examples are Al, Cd, Cr, Ti and Al_2O_3 on steel; (3) joining—examples are Ag on U and Be, Cu on Ta, and Au on Mo. Because of the wide range of applicable coating substrate combinations, ion plated and sputtered coatings are also used as a "strike" for subsequent deposition by electroplating.

CHEMICAL VAPOR DEPOSITION (CVD)[1, 2, 9, 17-19]

Basic Principles

The CVD process is based on the fact that a large number of desirable coating material constituents can exist in a chemical form that is a suitable transport phase. Examples are fluorides, chlorides, bromides, iodides, hydrides, organometallics and hydrocarbons. The solid-to-gas-state reaction can be made to occur as part of the coating apparatus. This is called the closed tube process. However, the most common procedure is to procure the coating constituents as gas or liquid phase starting materials. The process then consists of passing these vapors (liquids must be boiled) fluid-mechanically, often with the assistance of a carrier such as Ar, He or N_2, to the vicinity of the substrates, where a chemical reaction is initiated which decomposes the reactant gases to give the desired coating material as a reaction product on the substrate. This is called the open tube process.

Most commonly, activation for the deposition reaction is provided by heating the substrates. It is this type of processss that is discussed here. An alternate method uses a plasma discharge to assist the reaction. That process will be discussed in the next section. Other methods include the use of rf fields, light or X-ray radiation, electron bombardment or catalytic action of the substrate surface.

Thermal reduction may be just simple pyrolysis. However, the more common approach is to use a second reactive species, usually hydrogen, to induce the reaction at lower temperatures. Typical required substrate temperatures are in the range of 500 to 2000 C (932 to 3632 F). Reactant flows can be combined to give various compounds in addition to elemental metals and semiconductors. Some examples are given in Table 1.

The advantages of the CVD process are that complex shapes can be coated with refractory materials at relatively high deposition rates (10 to 100 nm/sec) using relatively simple equipment. The process is generally carried out in the pressure range 100 Pa to 1 Atm, so that high vacuum equipment is not required. Compound stoichiometry can be controlled by controlling the constituent gas injection rates. Disadvantages are the elevated substrate temperatures and the toxic and corrosive nature of many of the working gases. Sometimes the latter difficulty can be overcome by precoating the substrates with a protective layer, using an alternative process such as sputtering.

Range of Applicability

Basically any chemical reaction between reactive vapors that leads to a solid reaction product can be used for CVD.[18] CVD is very effective for depositing the group IV, V and VI refractory metals, Ti, V, Cr, Zr, Nb, Mo, Hf, Ta, W. Adding suitable reactive gases (see Table 1) to the metallic transport vapors permits refractory compounds such as metal carbides, nitrides, silicides, borides and oxides to be formed. Other metals which are less commonly deposited by CVD are Al, Ni, Fe, Cu, Th, U, Rh, Ru, Os, Ir, Pt and Re. Carbon deposition is important for several applications. CVD plays an important role in microelectronics technology. Elemental semiconductors (Si and Ge), compound semiconductors (GaAs, GaP, ZnSe and SiC) and dielectric compounds (SiO_2, Si_3N_4, BN, Al_2O_3 and TiO_2) have been deposited by CVD.

CVD is very effective in providing a coating over all surfaces of a complex shaped substrate. However, it is often difficult to achieve uniform

TABLE 1

Coating	Starting Constituents	Decomposition Temperature
Si	SiH_4	800-1300 C (1472-2372 F)
Ni	$Ni(CO)_4$	200-300 C (392-572 F)
W	$WCl_6 + H_2$	850-1400 C (1562-2552 F)
TiC	$TiCl_4 + CH_4$	800-1100 C (1472-2012 F)
GaAs	$GaCl_3 + As_4 + H_2$	~700 C (~1292 F)
Al_2O_3	$Al_2Cl_6 + CO_2 + H_2$	800-1400 C (1472-2552 F)

coating thicknesses over complex shapes or large areas. Allowable substrate materials may be limited by the high temperatures required to induce many of the reactions and by the corrosive nature of product gases such as HCl and HF. Special attention must be given to thermal stresses.

Apparatus Configuration

A CVD deposition apparatus for the open tube process consists typically of (1) a reaction chamber with provisions for substrate mounting and for controlling the fluid mechanics of the reactant gases, (2) equipment for heating the substrates, (3) a reactant gas supply system and (4) an exhaust gas disposal system. The achievement of uniform coatings with controlled properties depends on the uniformity of the substrate temperature and the flow of reactants over the surface. Therefore, apparatus configurations are generally dictated by these heating and fluid mechanics considerations and depend on the substrate shape. In the case of the microelectronics applications the wafers have a common shape, and highly automated commercial coating systems have been developed in which, for example, a controlled reactant flow passes over a heated table on which the wafers are placed. For mechanical parts special apparatus configurations are often assembled for each application. Such an apparatus is indicated schematically in Fig. 5, where it is assumed that the substrate is of a material that can be inductively heated. Another example of a specialized apparatus is that of carrying out the reaction in a fluidized bed to obtain uniform coatings of small particles in the 5 μm to 1000 μm range.

In the closed tube process a thermal or chemical driving force causes coating material to pass into the vapor transport phase at one position in the reactor and out of the vapor phase onto the substrates at another position. The pack cementation process is an example of chemical transport in an isothermal chamber.[1,17] For example, in the chromizing process iron objects such as turbine blades are packed in a mixture of chromium powder, ammonium iodide (NH_4I) as an activator, and aluminum oxide powder as a porous mass. Chromium is deposited onto the turbine blades and diffuses into the iron. The process is driven by the difference in activity between the chromium metal in its free state and in solution with the iron as an alloy coating.

Coating Properties

The CVD deposition rate is a function of the substrate temperature and the partial pressures of the plating vapor and reducing gas. When these parameters are optimized to give a maximum deposition rate, the process becomes limited by the fluid transport. Under these conditions it is difficult to obtain uniform coatings over large areas or in recesses. Lowering the substrate temperature causes the substrate reactions to be rate-limiting. This improves coating uniformity and permits transport into crevices.

The coating structure is complicated by the fact that the substrate temperature influences both the reaction and deposition rates as well as the adatom mobility. Conditions which optimize the deposition rate tend to give a Zone 1 structure (see Fig. 2). As a general rule, low temperatures and high vapor concentrations promote amorphous or finely crystalline deposits of the fibrous Zone T type, while high-substrate temperatures and low vapor concentrations promote the more coarsely crystalline columnar grains representative of the Zone 2 structure. Sometimes a co-deposited grain growth inhibitor is used to promote a fine grain deposit. In general, if conditions and techniques can be adjusted to give dense, fine-grained material, its mechanical properties approach those of its wrought counterpart. CVD deposits have an excellent reputation for good adhesion.

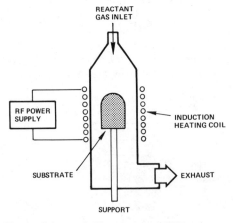

FIG. 5. Schematic illustration of CVD apparatus designed to coat a part that can be heated inductively.

Applications

CVD is widely used in the electronics industry to deposit semiconductor, insulating and resistive coatings. All device grade silicon used by that industry is made by H_2 reduction of $SiHCl_3$ or thermal decomposition of SiH_4. Films of compound semiconductors such as GaAs are used for light-emitting diodes.

Fluidized beds are used to deposit pyrolytic carbon onto nuclear fuel particles. The high deposition rates make CVD effective for forming free-standing shapes. Thus CVD-formed crucibles and tubing of W and Si are commercially available. CVD is used to form boron fibers for use in composite materials. CVD coatings of TiC are widely used on cemented carbide tool inserts. Ta coatings on steel are being examined for use in the chemical process industry. A recently developed process which allows WC to be deposited at low temperatures [350 to 500 C (662 to 932 F)] shows promise for applying wear-resistant coating onto heat-treated steel or other substrates with thermal limitations.[20]

PLASMA ASSISTED CVD[1, 9, 21]

Basic Principles

In the plasma assisted CVD process the reactant gases are passed through a plasma discharge. Gas phase reactions reduce the activation energies for the substrate reactions, with the consequence that deposition temperatures can be low, typically less than 300 C (572 F). Consider, for example, the pyrolytic decomposition of monosilane (SiH_4). Decomposition starts at about 400 C (752 F), but temperatures in the 800 to 1300 C (1472 to 2372 F) range are required (Table 1) to achieve reasonable deposition rates. When the SiH_4 is passed through a glow discharge, practical growth rates (2 nm/s) can be achieved at temperatures of 200 C (392 F) or less. The plasmas are low-pressure (10 to 10^3 Pa) glow discharges in which the gas temperatures are low but the electron energies are relatively high. The primary role of the plasma is believed to be electron-collision-induced dissociation and ionization of the reactant gases. Radiation from the plasma, and, in cases where the substrates are in contact with the plasma, electron and ion bombardment, may play a role in the coating growth.

Range of Applicability

A plasma can probably be used to assist any CVD reaction. However, intermediate products of the plasma decomposition may cause etching of solid surfaces. (This is the basis for the plasma etching processes that are becoming used in the electronics industry.) Most of the plasma deposition reactions that have been refined are for wafer coating applications in the electronics industry. Deposition rates are typically in the range of 0.5 to 10 nm/s. Example coatings are Si, Ge, silicon nitride, silicon oxide, silicon carbide, carbon, aluminum oxide, tin oxide and boron nitride.

The uniformity of the deposit depends on the uniformity of the plasma discharge. Thus, plasma-assisted CVD cannot be expected to offer the flexibility for coating complex substrate shapes that is provided by conventional CVD.

The plasma assisted CVD process is applicable to a broad range of substrate materials. The primary limitation is possible attack from intermediate products of the plasma discharge.

Apparatus Configuration

Plasma-assisted CVD apparatuses consist typically of a reaction chamber and pumping system, an electrode system and power supplies for sustaining the plasma discharge, a reactive gas feed and control system and an exhaust gas disposal system.

A wide range of apparatus configurations are used. The substrates may or may not be in actual contact with the plasma. Many applications result in the formation of poorly conducting layers on electrodes. Therefore, the discharges are generally driven at frequencies in the 50 kHz to 13.56 MHz range, although dc is sometimes used. Electrodes can vary from solenoidal coils surrounding a reactor tube, to the parallel plate type shown in Fig. 6. Power densities are low, typically in the range of 10 to 50 mW/cm². Radial flow parallel plate reactors of the type shown in Fig. 6 have become widely used. Parallel plate

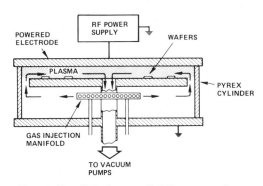

Fig. 6. Parallel-plate, radial-flow type plasma CVD reactor designed for coating wafers in the electronics industry. (*Courtesy Telic Company.*)

reactors are being scaled to diameters of about 1 m for wafer processing.

A high-vacuum pumping system may be used to initially evacuate the chamber. However, the plasma discharges are generally operated at relatively high working gas flow rates so that oil sealed type rotary vacuum pumps are used during deposition. The corrosive and toxic character of the exhaust gases may require that special traps be used to protect the pumps and the environment.

Coating Properties

Plasma-assisted CVD tends to produce a generic class of materials that in some ways have more in common with one another than they do with their bulk counterparts. When deposited at low temperatures, these coatings exhibit less than bulk densities, have amorphous or polymeric type structures, and typically contain large amounts (often 10% or more) of hydrogen. They can have a wide range of stoichiometric compositions, which can be controlled by the working gas composition. The electrical and optical properties are strongly dependent on the stoichiometry and deposition conditions.

Much is still to be learned about the properties of the coatings. Recent studies of amorphous Si films indicate that they can have the general Zone 1 and Zone T forms shown in Fig. 2. The Zone 1 coatings appear to consist of relatively dense amorphous grains, with hydrogen incorporated in Si-H type bonding at "point type" defects. The "voids" appear to consist of a lower-density polymeric material of the $(Si-H_2)_n$ type.

Applications

The applications are a consequence of the unique properties of particular plasma CVD coatings. Thus amorphous Si coatings can be deposited under conditions in which the incorporated hydrogen passivates the electrical defects generated by the amorphous structure; consequently these hydrogenated amorphous silicon films (termed α-Si:H) can be impurity-doped to yield both n- and p-type materials. Accordingly, α-Si:H films are currently being used to make low-cost terrestrial photovoltaic devices.

Si_3N_4 films deposited by conventional CVD require a substrate temperature of 750 C (1382 F). Silicon nitride films produced by plasma-assisted CVD at substrate temperatures in the 200 to 400 C (392 to 752 F) range are used as passivating layers for electronic devices. In spite of their lower densities, these films have very low diffusion rates for alkali ion contaminants and other species such as water vapor that are important in electronic devices.

Diamond-like carbon films with extreme hardness and chemical inertness have been produced by cracking hydrocarbon gases such as butane in rf discharges and condensing the products on substrates which are subjected to ion bombardment.[22]

PLASMA SPRAYING[1, 23, 24]

Basic Principles

In this category we include plasma torch and detonation gun (d-gun) coatings.[1] Both are line-of-sight processes in which the coating material in powder form is heated to near or above its melting point and accelerated by either a plasma gas stream or a detonation wave. The spray of molten powder is directed at the surface to be coated and on impact forms a coating consisting of many layers of overlapping thin lenticular particles or splats. The plasma torch process is generally carried out in an inert gas to minimize oxidation. Recent work in which the spraying is carried out in vacuum chambers has yielded superior coatings. This process is known as low-pressure plasma spraying (LPPS). LPPS has, for example, been used to deposit very reactive metals such as Ti and Ta.[25]

Range of Applicability

Almost any material that can be melted without decomposition can be used to form the coating. Therefore, coating materials include a wide range of metals, alloys, oxides, carbides, nitrides and borides. Powder particle sizes fall typically in the 5 to 60 μm range. The substrate is seldom heated above about 150 C (302 F). Coating thicknesses are typically in the range from 0.05 to 0.5 mm, but in a few applications may exceed 5 mm.

Normally, the highest quality coatings are achieved when the spray arrives perpendicular to the substrate surface. Thus some difficulties may be encountered in coating complex parts, particularly those with narrow grooves or sharp angles. Small right angle torches permit the coating of tubes with inside diameters down to about 4 cm. D-guns can be used only to coat external surfaces.

Apparatus Configurations

A plasma torch or d-gun apparatus consists of the torch or gun, gas controls, power supplies and powder feeders. Figure 7 shows a schematic illustration of a plasma torch. A carrier gas, usually Ar or N_2 or a mixture of these with H_2 or He, flows around the cathode and through the anode which serves as a constricting nozzle. A direct current arc is maintained between the electrodes. Arc power levels vary between 5 and 80 kW. In contrast to the low-pressure glow discharges, the electrons and heavy particles in the arc are at temperatures which are comparable, and equal to about 30,000 K. The plasma gas velocities exiting from most conventional torches are subsonic (100 to 300 m/sec), but supersonic velocities can be generated by using converging-diverging nozzles and spraying into a low-pressure ambient. The powders are generally injected just downstream of the nozzle throat, as shown in Fig. 7.

The d-gun process was developed by Union Carbide Corporation, and d-gun coatings are available only through that corporation. A d-gun is typically about 2.5 cm in inside diameter by about 1 m long. A mixture of O_2 and C_2H_2 is fed into the barrel along with a charge of coating powder. The gas is ignited and the detonation wave heats the powder to close to its melting point and accelerates it to a velocity of about 750 m/sec. The gun is automatically cycled about four to eight times per second. Each pulse results in the deposition of a circle of coating about 25 mm in diameter and a few microns thick.

For most industrial applications involving the use of torches or d-guns, the parts and/or the torch are translated automatically or semiautomatically to provide uniform coatings.

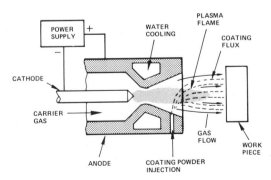

FIG. 7. Schematic illustration of plasma torch deposition system.

Coating Properties

Virtually all plasma sprayed coatings require a roughened surface to obtain good adhesion. Grit blasting is the method most frequently used. Mechanical interlocking is the primary bonding mechanism. Consequently, the coating surfaces are also rough. Therefore, in the majority of applications the coating surfaces are finished by machining or grinding.

Both plasma sprayed and d-gun coatings consist of many layers of thin lenticular particles. The coating grain sizes are generally small, and non-equilibrium phases and stresses may be present because of the rapid cooling rate. Coating properties are influenced by the powder composition and size distribution, the carrier gas, and the ambient at the substrate location. For example, compositional changes can result because of surface evaporation losses from very small particles. Coating hardnesses generally increase with the use of finer powder. For a given material, d-gun coatings are usually harder than torch sprayed coatings. The d-gun coatings also have a higher density (> 95% of bulk compared to 80 to 95% of bulk for torch sprayed coatings) and better adhesion because of the higher impacting velocity of the droplets.

Applications

A common application is to decrease wear. D-gun WC-Co coatings are used on gas turbine compressor blades for erosion resistance and on roller guides for steel mill pickle lines. Co-spraying of aluminum bronze with Al_2O_3 or carbide particles has yielded coatings which combine good wear resistance with lubricity. These coatings have been used for bearing applications.

Coatings consisting of a metallic undercoat such as Ni-Cr or Ni-Al and with an oxide outer layer such as zirconia or magnesium zirconate are used as thermal barrier layers for applications such as combustion chambers, thrust reversers and piston heads.

The usefulness of the coatings for corrosion protection depends on the degree to which interconnected porosity provides a path to the substrate. One approach is to use the plasma sprayed coatings to provide abrasion resistance and to seal the coating with epoxy to prevent substrate corrosion. Recent work indicates that high quality M-Cr-Al-Y type coatings, which provide protection against hot corrosion, can be deposited using the low-pressure plasma spray method.

References

1. Bunshah, R. F. (Ed.). *Films and Coatings for Technology*. Park Ridge, N.J.: Noyes, 1982. (This book includes chapters on vacuum evaporation by R. F. Bunshah; sputtering by J. A. Thornton; ion plating by D. M. Mattox; chemical vapor deposition by J. M. Blocher, Jr.; plasma assisted CVD by A. R. Reinberg; and plasma spraying by R. C. Tucker, Jr.)
2. Powell, C. F., J. H. Oxley and J. M. Blocher, Jr. *Vapor Deposition*. New York: Wiley, 1966. (This book contains numerous chapters on various aspects of chemical vapor deposition and vacuum deposition.)
3. Holland, L. *Vacuum Deposition of Thin Films*. London: Chapman and Hall, 1966.
4. Chopra, K. *Thin Film Phenomena*. New York: McGraw Hill, 1969.
5. Glang, R. "Vacuum Evaporation," *Handbook of Thin Film Technology*. L. I. Maissel and R. Glang (Eds.). New York: McGraw Hill, 1970, pp. 1-3 to 1-130.
6. Bachrach, R. Z. in *Crystal Growth*, 2nd Edition. B. R. Pamplin (Ed.). New York: Pergamon, 1980, Chapter 6.
7. Thornton, J. A. *J. Vac. Sci. Technol.* **11**:666 (1974).
8. Thornton, J. A. *Ann. Rev. Mater. Sci.* **7**:239-260 (1977).
9. Vossen, J. L. and W. Kern (Ed.). *Thin Film Processes*. New York: Academic Press, 1978. (This book includes chapters on glow discharge sputtering by J. L. Vossen and J. J. Cuomo; cylindrical magnetron sputtering by J. A. Thornton and A. S. Penfold; planar magnetron sputtering by R. K. Waits; ion beam deposition by J. M. E. Harper; chemical vapor deposition by W. Kern and V. S. Ban; and plasma deposition of inorganic films by J. R. Hollahan and R. S. Rosler.
10. Chapman, B. *Glow Discharge Processes-Sputtering and Plasma Etching*. New York; Wiley, 1980.
11. Shaw, B. J. and R. P. Miller. U.S. Patent 3,918,100 (Nov. 11, 1975).
12. Morrison, D. T. and T. Robertson. *Thin Solid Films* **15**:87 (1973).
13. Thornton, J. A. *Thin Solid Films* **80**:1 (1981).
14. Mattox, D. M. *J. Vac. Sci. Technol.* **10**:47 (1973).
15. Spalvins, T. *J. Vac. Sci. Technol.* **17**:315 (1980).
16. Takagi, T., I. Yamada and A. Sasaki. *J. Vac. Sci. Technol.* **12**:1128 (1975).
17. Schäfer, H. *Chemical Transport Reactions*. New York: Academic Press, 1964.
18. Feist, W. M., S. R. Steele and D. W. Readey. *Physics of Thin Films* **5**:237 (1969).
19. Holzl, R. A. "Chemical Vapor Deposition Techniques," *Techniques of Metals Research*, Vol. 1. R. F. Bunshah (Ed.). New York: Interscience, 1968, Part 5, Chapter 33, pp. 1377-1405.
20. Bhat, D. G. San Fernando Laboratories, Pacoima, Calif. private communication, December 1981.
21. Reinberg, A. R. *Ann. Rev. Mater. Sci.* **9**:341 (1979).
22. Holland, L. *J. Vac. Sci. Technol.* **14**:5 (1977).
23. Gerdeman, D. A. and N. L. Hecht. *Arc Plasma Technology in Materials Science*. New York: Wien, Springer-Verlag, 1972.
24. Smart, R. F. and J. A. Catherall. *Plasma Spraying*, London: Mills and Boon, 1972.
25. Steffens, H. D., H. M. Höhle and E. Ertürk. *Thin Solid Films* **73**:19 (1980).

C. Catalytic Methods

Ian McDonald

Frederick Gumm Chemical Company, Lyndhurst, N. J.

The term *electroless plating* originated with the work of Brenner and Riddell[1] in the development of a method of covering metallic bodies with ultramicrocrystalline alloys of nickel or cobalt phosphorus. This process was unique in that it produced metal coatings by the controlled autocatalytic reduction of metallic cations from aqueous solution, without the use of an externally applied potential. A chemical reducing agent in the solution supplies the electrons for the reduction of the metallic salts to their elemental form:

$$M^{2+} + 2e \begin{pmatrix} \text{supplied by the} \\ \text{reducing agent} \end{pmatrix} \xrightarrow{\text{catalytic surface}} M^0.$$

The reduction reaction can only occur at a catalytic surface. This demands that the metal being deposited is itself catalytic in order that deposition may be maintained.

Electroless deposition is usually selected because it provides one or more of the following advantages over electrodeposition:

1. Uniform deposits are produced on plated components with no excessive build-up on projections or edges, even on complex shapes.
2. Electroless deposits are usually less porous than their electrolytic counterparts.
3. Externally applied current is not needed so that anodes or sophisticated racking is not required. (Provision for electrical contact may have to be made so that parts may be cleaned satisfactorily).
4. Deposits may be produced directly on nonconductors (with suitable pretreatment).
5. Coatings produced by electroless means often have unique chemical or physical properties.

CATALYTIC ARSENIC PLATING

To date, no major commercial applications have appeared. The process is technically valid, and the following is a typical formulation.

Zinc sulfate, g/L (oz/gal)	30 (4.0)
Arsenic trioxide, g/L (oz/gal)	30 (4.0)
Sodium citrate, g/L (oz/gal)	60 (8.0)
Sodium cyanide, g/L (oz/gal)	60 (8.0)
Sodium hydroxide, g/L (oz/gal)	7.5 (1.0)
Ammonium hydroxide, g/L (oz/gal)	60 (8.0)
Sodium hypophosphite, g/L (oz/gal)	60 (8.0)
ph (adjust to this value)	11.5
Temperature, C (F)	80–90 (115–195)

Ammonia must be added regularly to maintain pH. The initial deposition rate is not likely to exceed 2.5 μm/hr, and immersion times of more than 15 min are not recommended. Due to the relatively high ammonium hydroxide concentration and operational temperature this bath could present a health hazard if used on a commercial scale.

CATALYTIC CHROMIUM PLATING

Many articles and patents have been published on this subject. To date these have not been true autocatalytic processes, but essentially rely on high-temperature diffusion of chromium into a suitable substrate. Since no solid evidence exists that a commercially viable bath is available, no formulation will be suggested for further investigation, but the work of Holker and Wells[96] would be worth examination by those interested.

CATALYTIC COBALT PLATING

Electroless cobalt is used primarily in the production of magnetic memory surfaces. It is possible in the future that the process may have applications for use in high-temperature bearing surfaces. The method was originally patented by Brenner and Riddell in 1950. To date no commercial processes exist as most of the work in this field has been directed towards the development of solutions to produce proprietary magnetic films.[2]

The reaction is similar (except for operating pH) to that which occurs in chemical nickel plating. To date all of the published work has indicated that the reaction will not occur in acid media. Essentially, cobalt ions are catalytically reduced to the metal in alkaline solution by hypophosphite ions, which are thereby oxidized to orthophosphite ions:

$$Co^{++} + (H_2PO_2)^- + OH^- \xrightarrow{\text{catalytic surface}} Co^0 + 2H^+ + (HPO_3)^{--}.$$

At the same time, elemental phosphorus, which immediately alloys with the cobalt, is produced through reduction-hydrolysis of a fraction of the hypophosphite ions while, simultaneously, more hypophosphite anions are oxidized to phosphite with evolution of hydrogen gas:

$$(H_2PO_2)^- + OH^- \rightarrow (HPO_3)^{--} + H_2.$$

To avoid random reduction of the cobalt ions, the concentration of all reagents has to be main-

TABLE 1. ELECTROLESS COBALT BATH FORMULATIONS

	A		B	
Cobalt chloride (deca hydrate), g/L (oz/gal)	30	(4.0)	30	(4.0)
Sodium hypophosphite, g/L (oz/gal)	20	(2.7)	10	(1.3)
Sodium citrate, g/L (oz/gal)	35	(4.7)	100	(13.4)
Ammonium chloride, g/L (oz/gal)	50	(6.7)	50	(6.7)
pH (adjust with NH_4OH)	9–10		9–10	
Temperature, C (F)	95–100	(200–212)	95–100	(200–212)
Deposition rate μ/hr (mil/hr)	15	(0.6)	5	(0.6)
Deposit appearance	Dull		Dull	

tained at a controlled, low level. The bath must contain a complexing (chelating) agent to prevent precipitation of cobalt hydroxide or basic salts. The presence of certain selected hydroxycarboxylic anions, particularly citrates, also results in better physical appearance. Tartrates can be used, although their presence causes the formation of grey, dull coatings; however, hydroxyacetate, gluconate or salicylate are not recommended because they substantially slow down the deposition rate.

Two typical bath compositions according to Brenner and Riddell[1] and U.S. Patent 2,532,284 are given in Table 1. Bath *B* produces deposits of superior physical properties; however, initiation may be slow. In such cases, immerse first into Bath A (Bath A is used primarily in memory disc application).

Electroless cobalt plating is considerably improved by selecting chelating agents other than hydroxycarboxylic acids, and using some of the additives ("exalting" compounds, stabilizers) developed for electroless nickel deposition.[18, 22, 23]

Of particular interest is the fact that the efficiency of the cobalt reaction (hypophosphite utilization) has been found to be 66% under proper conditions, as compared to about 33% for nickel.

Cobalt alloys are also known and are covered in the nickel section.

ELECTROLESS COPPER PLATING

Electroless copper deposition is a relatively new science as its feasibility as a technical process was only proved at the end of the 1950s.[61] The major application for the deposit has been the production of printed circuit boards, and in particular for through-hole interconnection of double-sided boards.

In the production of a printed circuit board the electroless copper deposit provides a conductive surface coating, which allows subsequent deposition of electrolytic copper to develop the required thickness of track. The actual thickness of the electroless deposit is determined by the number of electrical contact points for the electrolytic build-up stage, and also by the current density to be used. If too thin an electroless deposit is used for the contact spacing, or for the applied current density, the electroless deposit will "burn off." Contact "burn off" results in unsatisfactory deposits and the rejection of plated parts.

The other major industrial application for electroless copper is the decorative plating of plastic moldings, and sometimes the selective plating of other nonconductors. With all of these processes the techniques are the same.

A nonmetallic substrate is cleaned and etched, providing a mechanical key for the subsequent metallizing stages. The components are then activated and sensitized in a mixed tin and palladium process which provides a catalytic activator for the electroless bath. The exposed palladium surface acts as a catalytic trigger for the initial deposition of copper. The deposition process becomes autocatalytic at this stage and the small islands of deposit that develop around the adsorbed tin/palladium sites spread laterally across the surface. After a short period of time the isolated islands join to form a continuous, conductive deposit.

Electroless nickel has been proposed as an alternative to electroless copper. The electroless copper systems do offer some advantages over electroless nickel in the rate of lateral coverage and bridging. At the present stage in the development of electroless copper technology the processes may still be a little less stable than the electroless nickel alternatives.

Plating Bath Formulation

The first application for chemical copper reduction was in the qualitative analysis of aldehydes. A solution of copper sulfate in an alkaline solution of a soluble salt of tartaric acid is known as Fehlings solution. When aldehydes are added to Fehlings solution, copper is reduced and a red precipitate of cuprous oxide forms. Alcohols or ketones do not produce the same reaction. This test for aldehydes served as the basis for the development of a practical catalytic copper plating process.

Cahill[61] first showed that the reduction of cupric ions with formaldehyde in alkaline solution could be autocatalytic. Okinaka[62] established that the overall reaction of the catalytic copper plating process could be represented as:

$$Cu^{++} + 2HCHO + 4OH^-$$
$$\downarrow$$
$$Cu + H_2 + 2H_2O + 2HCOO^-.$$

Bath Constituents

The major additive groups for the successful deposition of electroless copper are similar to those found in electroless nickel solutions. They may be listed as:

a. Source of copper ions.
b. Complexing agent that will maintain the metal ions in solution at operating pH.
c. Compatible reducing agent for the copper ions.
d. A material to adjust the pH of the bath.
e. An additive to stabilize the solution.
f. Other groups such as wetters and stress relievers.

Each of the additive groups deserve further examination and explanation of their effects.

Source of Copper Ions

Electroless copper plating involves the reduction of cupric ions to their metallic state at a controlled rate. The source of the ions in the solution is not important as long as the associated anion does not have an adverse effect on the process. The cupric ion source that is most usually used is cupric sulfate. The commercial alternatives may be to use the cupric salt of one of the complexing agents in the bath formulation. When using cupric sulfate the concentrations that may be used are between 3 g/L to 15 g/L, with a preferred concentration range of 8 g/L to 12 g/L.

Complexing Agents

Electroless copper baths must be alkaline in order for reduction (by the commercially viable reducing agents) to take place. This means that a complexing system in the formulation must be capable of maintaining the cupric ions in solution. In general, the more stable the complex formed (i.e., the lower the free cupric ion concentration), the greater the reducing ability of the reducing agent that is required to initiate deposition.

Examples of some suitable complexing agents include Rochelle salts,* ethylenediaminetetracetic acid, the sodium (mono-, di, tri- and tetrasodium) salts of EDTA acid, nitrilotriacetic acid and its alkaline salts, gluconic acid, gluconates, triethanolamine, glucono- lactone and modified EDTA salts such as N-hydroxyethylenediaminetetracetate. The actual concentration of complexants will be controlled by the cupric ion concentration in the solution. Usually this is in the range of 20 to 50 g/L or three to four times the concentration required to complex the cupric ion content of the bath.

Rochelle salt is probably the complexing agent used most frequently in catalytic copper plating solutions. EDTA can be substituted for the Rochelle salt in equal molar quantities. The copper complex formed with EDTA is considered to be more stable than that formed with tartrate. Combinations of the two complexing agents have an interesting effect, as demonstrated by Saito and Saubestre. The copper deposition rate increases when EDTA is added to a catalytic copper plating solution containing Rochelle salt complexing agents. The copper plating rate reaches a limiting value of about double the initial rate with 15 g/L EDTA in 28 g/L Rochelle salt solution.

Reducing Agents

Formaldehyde is the most widely used reducing agent for electroless copper plating. The reducing ability of this compound is related to the pH of the bath. As the solution pH rises, so does the rate at which the formaldehyde will reduce the cupric ions to their metallic state. The effect of increasing the deposition rate is a reduction in the bath stability, as would be expected.

Paraformaldehyde, a solid polymer of formaldehyde, is often used as the reducing agent. The chemical composition of paraformaldehyde is best expressed by the formula $HO(CH_2O)_nH$. Staudinger and co-workers[67] have proposed that n is in the area of 8 to 100 and that paraformalde-

*Sodium potassium tartrate.

hyde may be described as a mixture of polyoxymethylene glycols. Paraformaldehyde is available commercially in higher purities than the 40% aqueous solutions of formaldehyde known as Formaline. Commercial Formaline may also contain antipolymerization compounds that might have some poisoning effect on the catalytic surface.

Examples of some other reducing agents include formaldehyde precursors or derivatives such as trioxane, dimethylhydantoin, glyoxal, borohydrides such as alkali metal borohydrides (sodium and potassium borohydride), substituted borohydrides such as sodium trimethoxyborohydride and boranes such as amine borane (isopropylamine borane and monopholine borane). Hypophosphite and hydrazine have been used as reducing agents for copper deposition, but they do not appear to support fully autocatalytic reduction. The original use for these compounds was in the reduction of copper for mirror backing purposes.

pH Adjustment

The majority of electroless copper baths operate in alkaline pH ranges, since very few of the reducing agents are effective in acidic solutions. As the bath is operated the pH will drop and adjustment is required to maintain the solution within the operational range. The most commonly used materials for this purpose are sodium or potassium hydroxide.

Bath Stabilization

As a general rule, catalytic copper plating conditions which favor rapid plating also reduce bath stability. A high plating rate bath, such as the one developed by Saubestre, will last only a few hours. This is the result of catalytic particles forming in the bath, which soon deplete the bath of copper. Periodic filtration will extend the bath life.[64]

The recent patent literature reveals that a concerted effort is underway to find additives which will improve bath stability. Ideally, one would like to improve bath stability without reducing plating rate. Some proprietary baths are stabilized with organic polymer materials which adsorb on metallic surfaces. These additives can markedly reduce the copper deposition rate. 2-Mercaptobenzothiazole is reported to be a successful stabilizer, but too high a concentration increases stability with a subsequent lowering of the deposition rate. The useful concentration of 2-mercaptobenzothiazole is approximately 0.003%.

Saubestre states that bath stability is obtained with increasing carbonate content, but only at the expense of decreasing deposition rates. Divalent sulfur compounds, alkyl mercaptans and thiourea have also been used as stabilizers.

Soluble alkaline alcohols or large amounts of methanol (10 to 70%) are effective in stabilizing catalytic copper plating baths. The stabilizing effect of methanol is unique in that plating rate is decreased only to a small extent with the increase in methanol concentration and bath stability. Saito found that 0.1 mg/L thiourea added to the bath distinctly improves bath stability, without a significant reduction in the plating rate. At 1 mg/L thiourea, bath stability was further improved, but the plating rate was reduced by one-half. At 10 mg/L thiourea, the plating reaction stopped. A thiourea concentration of between 0.05 and 0.1 mg/L is recommended. Organic compounds similar in structure to thiourea have also been patented.[66,71]

Effects of Other Additives

The origin of stress in catalytic copper deposits in not known. However, it is believed to be associated with hydrogen co-deposited with the copper and possibly included in the deposit. Without a suitable additive, which presumably eliminates incorporation of hydrogen into the deposits, the catalytically plated copper is brittle and can withstand only limited bending or thermal stress without fracture. This is not a problem when the deposit thickness is in the order of a millionth of an inch in thickness and is overplated with ductile electrolytic copper. However, where the entire desired catalytically plated copper thickness is in the order of mils, it is necessary to provide an additive which will eliminate or minimize the internal stress in the deposit. For example, the use of small amounts of cyanide in the plating solution is suggested to relieve stress. Water soluble cyanide compounds of various metals have been claimed as stress relievers for catalytic copper plating. Another recent patent claims polysiloxanes, such as G.E. Silicon fluid SF-96, as effective stress relieving agents.[69]

It is believed that the silicon in the compound is the important constituent. Polysiloxanes are the least soluble silicon compounds in basic copper solutions, but are preferred since they provide the greatest increase in ductility and also enhance appearance by yielding a finer grain deposit. Substituted silane compounds such as methyldichlorosilane are also effective. The silanes must be handled with caution, as some are violently reactive with water. Silicon compound concentrations in the order of 250 ppm are recommended.

THE OPERATION OF ELECTROLESS COPPER BATHS

The majority of the solutions described in Table 2 will operate at room temperature. The rates of deposition will increase rapidly as the solution temperature is increased. As will be expected, the stability of the bath decreases with increasing temperature.

Suitable stabilizing agents may be used to counteract the tendency of the bath to break down and plate in an uncontrolled manner. The agitation rate will accentuate the effect of any stabilizing compounds used in the solutions and excessive agitation may lead to misplating on exposed edges of parts.

One important factor about electroless copper plating solutions is the manner in which the solutions are made up and stored. It is usual to make up the baths as a two-part solution which may be mixed when the bath is needed. If the reducing agent used is formaldehyde and this is mixed and stored with the hydroxide, the compounds may react together on prolonged storage. This is known as the Cannizzaro reaction, in which formaldehyde breaks down into formic acid and methyl alcohol. As a result of this it is usual to dissolve the copper sulfate and formaldehyde in one solution and the complexing agent as well as the alkali in the other additive. In this form the solutions will be stable for extended periods of storage. On occasions when the additives are mixed for use, there may be some copper hydroxide formed; this precipitate will redissolve as the water soluble copper complex forms fully.

It has proved beneficial to add small amounts of cyanide ions to electroless copper baths at the rate of 10 to 25 mg/L. One of the effects this has is to reduce the level of internal stress within the deposits. Wetting agents may also be used to aid in the release of gas bubbles from the depositing surface and to help with wetting of the substrate.

ELECTROLESS COPPER TREATMENT SEQUENCE

The majority of electroless copper deposited throughout the world is used in the plating of nonconductors. This field is divided between plating of printed circuit boards and the decorative metallizing of plastic parts. The early stages of the process are common to both fields.[28,29,30] They may be listed as:

1. Chromic/sulfuric etch with agitation for 3 to 10 minutes.
2. Drag-out.

TABLE 2. ELECTROLESS COPPER SOLUTION FORMULATIONS

	1		2		3		4		5		6	
	g/L	oz/gal	g/L	oz/gal	g/L	oz/gal	g/L	oz/gal	g/L	oz/gal	g/L	oz/gal
Copper sulfate ($CuSO_4 \cdot 5H_2O$)	30	4.0	10	1.3	10	1.3	28	3.8	7.5	1.00	7.5	1.0
Rochelle salt ($NaKC_4H_4O_6 \cdot 4H_2O$)	135	18			40	5.4			85	11.3		
EDTA (sodium salt)			20	2.6			50	6.6			21.0	2.8
Formaldehyde (37% by weight)	150	20	7.5	1.0	32	4.3	12	1.6	36.0	4.8	7.0	0.9
Sodium carbonate									15	2.0		
Paraformaldehyde					12	1.7						
Sodium hydroxide (NaOH)	40	5.4	10	1.3	10	1.3	20	2.7	12	1.6	3	0.4
Temperature	25 C	77 F	25 C	77 F	25 C	77 F	25 C	77 F	25 C	77 F	70 C	160 F
	May be used for high-speed deposition of up to 25 μ/hr.		Use polysilixane stress reliever.		Use 1 ppm thiourea as stabilizer.		Use 35% methanol as stabilizer.		Solutions stabilized with nitrobenzene derivatives such as p-nitrobenzaldehyde.			

3. Rinse in cold water.
4. Rinse in cold water.
5. Neutralize in alkaline solution.
6. Rinse in cold water.
7. Rinse in cold water.
8. Activate in stannous chloride solution for 3 to 5 minutes.
9. Rinse in cold water.
10. Rinse in cold water.
11. Accelerator solution (palladium based), 30 to 60 seconds.
12. Rinse in cold water.
13. Electroless copper plate to required thickness.
14. Rinse in cold water.
15. Activate in copper sulfate/sulfuric acid solution.
16. Rinse.
17. Acid copper strike for 5 minutes.
18. Bright acid copper plate for 20 minutes at 30 A/ft^2 (3 A/dm^2).
19. Rinse in cold water.

From this stage, printed circuit boards may go on for subsequent specialist treatment and decorative finished plastic parts would be nickel plated for possible chromium or brass over plating.

The activation and acceleration steps are required to make the nonconducting polymer catalytic so that electroless deposition may be initiated.

CATALYTIC GOLD PLATING

The following baths are subject to United States Patents:

No. 1

Potassium gold cyanide, g/L (oz/gal)	2 (0.25)
Ammonium chloride, g/L (oz/gal)	75 (10.0)
Sodium citrate, g/L (oz/gal)	50 (6.7)
Sodium hypophosphite, g/L (oz/gal)	10 (1.3)
pH (adjusted with NH$_4$OH)	7–7.5
Temperature C (F)	90 (195)

This bath will deposit up to 7.5 μm/hr when freshly prepared, but has a limited life.

No. 2

Potassium gold cyanide, g/L (oz/gal)	28 (3.75)
Citric acid, g/L (oz/gal)	60 (8.0)
Monopotassium acid phthallate, g/L (oz/gal)	25 (3.4)
Tungstic acid, g/L (oz/gal)	45 (6.0)
Sodium hydroxide, g/L (oz/gal)	16 (2.1)
Sodium salt of N, N, diethylglycine, g/L (oz/gal)	3.75 (0.5)
pH (adjusted with citric acid)	5.6
Temperature C (F)	85–95 (185–200)

This bath will deposit gold over copper, nickel or Kovar substrates, the deposition rate being approximately 0.04 μm/hr.

Both of these solutions exhibit difficulties, as there is some doubt if they are truly autocatalytic; there may also be questions about the useful solution life.

CATALYTIC IRON PLATING

The first in-depth investigation of the conditions of deposition of iron using hypophosphite was carried out by Ruscior and Croila[115] in 1971. At temperatures of 80 C and at a pH of 10.5 the deposit contained 3% phosphorus, but at a pH of 11.4 the phosphorus content was only 1%. The rate of deposition was very low at only 2 μm/hr and the coating was shown to be ferromagnetic.

Ruscior and Croila bath:

Ammonium ferrous sulfate	30 g/L (4 oz/gal)
Sodium citrate	45 g/L (6 oz/gal)
Sodium hypophosphite	30 g/L (4 oz/gal)
pH (adjusted with NaOH)	Above 9.5
Temperature	60–90 C (140–195 F)

Another formulation that would be worth examination is listed in U.S. Patent 2,532,283:

Ferrous sulfate	30 g/L (4 oz/gal)
Sodium potassium tartrate	50 g/L (6.7 oz/gal)
Sodium hypophosphite	10 g/L (1.3 oz/gal)

pH (adjusted with
NaOH) 8-10
Temperature 75-90 C (167-195 F)

Other valid published patents are U.S. 3,597,267 and U.S. 3,627,545.

CATALYTIC NICKEL DEPOSITION

In terms of pure volume, acidic nickel/phosphorus solutions occupy the greatest share of the commercial market today, mainly due to their combinations of wear and corrosion resistance.

In order to study the electroless deposition of nickel, a Pourbaix diagram (Fig. 1) has been constructed for the system. This was achieved by superimposing the equilibrium diagram for nickel/water onto the equilibrium diagram for phosphorus/water. It can be seen that in the hatched areas it is possible that hypophosphites can be oxidized to phosphites and phosphates, this oxidation giving rise to a reduction of divalent nickel ions (Ni^{++}) to metallic nickel. Therefore, if a hypophosphite is added to a solution of a nickel salt (in the presence of a suitable catalyst) it is at least theoretically possible for metallic nickel to be deposited from this solution.

In practice, if a solution of a nickel salt and sodium hypophosphite is allowed to react, the nickel is precipitated as a fine sludge which is of no commercial use. It is therefore necessary to add another salt (usually of an organic acid) which acts both as a buffer and as a mild complexing agent for nickel. This leads to a stable solution which deposits metal at controlled rates and increases the pH range over which the bath may be operated.

Scholder and Heckel[8] showed in 1931 that the deposit formed during the reduction of nickel by hypophosphite consists of a mixture of phosphide with metallic nickel. According to these authors, the composition of the phosphides is represented by the formulas Ni_5P_2 and Ni_2P. The compounds were separated by the successive action of acetic acid and hydrochloric acid on the deposit. It is difficult to decide from the data given by these authors whether these phosphides are really formed in the process of reduction by the hypophosphite, or whether they are formed during the subsequent treatment of the deposit with the acids.

According to the concepts of Scholder and Heckel, the deposit consists of a mixture of pure metal and metallic phosphides. The reactions they suggested were:

$$NiO + H_3PO_2 \rightarrow Ni + H_3PO_3 \quad (a)$$

$$3H_3PO_2 \rightarrow H_3PO_3 + 2P + 3H_2O \quad (b)$$

$$2\,Ni + P \rightarrow Ni_2P_6 \quad (c)$$

The most interesting fact about this theory is that the authors did not connect the reduction of nickel with the evolution of hydrogen. They made the assumption that hydrogen is formed as a consequence of the presence of an active metallic surface which catalyzes the decomposition of hypophosphite.

Brenner and Riddell[1,2] proposed that the initial stage of the deposition process in an alkaline medium is the transfer of electrons from the metal catalyzing the reaction to the nickel ions, with the reduction of the latter to the monovalent state:

$$Ni^{2+} + M \rightarrow Ni^+ + M^+.$$

A further reaction results in the reduction of nickel to the metallic state:

$$2\,Ni^+ \rightarrow Ni^{2+} + Ni.$$

The transfer of electrons back to the catalyzing metal is accomplished by hydroxyl ions, which are first converted to the radical (a) and then

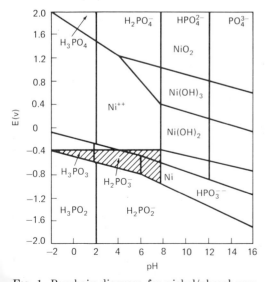

FIG. 1. Pourbaix diagram for nickel/phosphorus.

reduced by the hypophosphite (b):

(a) $OH^- + M^+ \rightarrow OH + M$

(b) $H_2PO_2^- + 2OH \rightarrow H_2PO_3^- + H_2O$.

An analogy with the reduction of nickel salts by boron hydride has been proposed by Lukes.[105] According to the theory, the interaction between the hypophosphite ion and water leads to the formation of a hydride ion, H^-, which then acts to reduce the nickel ions in solution. The hydride ion does not remain as a proton, but it is reduced to its atomic state. The atomic hydrogen forms molecules which are subsequently liberated as hydrogen gas. According to the theory for the hydride mechanism, the reactions may be represented by the following equations:

$$H_2PO_2^- + H_2O \rightarrow HPO_3^{2-} + 2H^+ + H^-$$
(in acid solutions) (1)

$$H_2PO_2^- + 2OH^- \rightarrow HPO_3^{2-} + H_2O + H^-$$
(in alkaline solutions) (2)

$$Ni^{2+} + 2H^- \rightarrow Ni^0 + 2H$$

$$H^+ + M^- \rightarrow H_2$$

Lukes' theory gives a satisfactory explanation of the simultaneous reduction of nickel and hydrogen.

Many other mechanisms for the reduction process have been proposed. The possibility of a simultaneous catalytic and electrochemical oxidation of hypophosphite was proposed by Hickling and Johnson,[106] who were investigating the anodic behavior of hypophosphite. They established that oxidation is brought about at static potential by a catalytic reaction as well as by an electrochemical one. There is also a colloid theory due to the work of a number of researchers.[113, 119, 120]

The wide range of theoretical explanations for the mechanism of autocatalytic deposition demonstrates the complicated nature of the process.

SOLUTION FORMULATIONS

As chemical nickel solutions started to become established as commercial processes during the mid-1950s, the advantages of the acid formulations became apparent. The initial benefits that were noted were the higher speeds of deposition, the greater stability of the baths and the improved physical characteristics of the deposits. Earlier ammoniacal alkaline baths had shown problems in all of these areas.

A modern electroless nickel solution may contain a wide range of organic and inorganic acid salts, giving each formulation its individual deposit characteristics. The general mechanism of solution operation may be written as follows.

Ni^{2+} + Sodium hypophosphite +

Buffers, complexors, accelerators,

stabilizers, wetters, moderators. $\xrightarrow{\text{Catalytic surface}}$

(Ni + P) deposit + H_2
 + Sodium orthophosphite

The principles of the formulation of electroless nickel bath are best demonstrated by examining each additive group individually. A range of solution formulations may then be examined in Table 3.

Source of Nickel Ions

The most common forms of nickel salts used are nickel chloride and nickel sulfate. There have been other alternatives used in the form of nickel sulfamate and nickel hypophosphite (produced by the action of hypophosphorous acid on nickel carbonate). The governing technical criteria for the choice of metal salt must be high solubility and purity. These requirements are necessitated by the need to produce bath addition solutions at high concentration and with low levels of detrimental contamination.

The concentration of nickel in operational baths, particularly in acid baths, has effects on both the rate of deposition and the appearance of the deposit.[100] The normal range of Ni^{2+} concentration in acid solutions is 4 to 10 g/L (0.5 to 1.3 oz/gal). At lower concentrations, the reduction rate of the solutions is impaired. As the optimum concentration is reached, the maximum deposition rate is achieved, usually combined with the best deposit appearance.

Sodium Hypophosphite

This material reduces the nickel ions in solution to their elemental form. During this reduction process, the hypophosphite anion decomposes to orthophosphite, with the production of gaseous hydrogen. A small percentage of the breakdown products are included in the deposit, resulting in phosphorous contents of up to 15% by weight. As the electroless baths are operated for extended periods of time, the orthophosphite content of the solution increases to the stage where it may precipitate as sodium phosphate.

The reduction reaction is not 100% efficient in that it requires approximately three moles of hypophosphite to reduce one mole of nickel on a practical scale, the additional reducer possibly being catalytically oxidized on the surface of the active nickel or being lost by spontaneous thermal decomposition.

At the concentrations below optimum, the deposition rate of the solution is slow, and blistered deposits may result, as the bath does not initiate deposition immediately. As optimum concentration is achieved, the deposition rate of nickel reaches its maximum. Curiously as this hypophosphite level is exceeded the deposition rate of the solution drops until a concentration at which the bath spontaneously decomposes is attained.[15, 16, 101]

Buffers

This group is closely linked with the organic complexors in structure and empirical formula. Their main function is to maintain the bath pH and prevent a rapid increase in the acidity of the solution. Even with the use of strong buffers, chemical plating systems require additions of an alkaline compound such as ammonium hydroxide, or sodium hydroxide, to maintain the bath within operational parameters.

Rapid changes in the pH of electroless nickel solutions have a number of interconnected effects. As pH drops so does the plating rate until a point is reached where the solution will not initiate or sustain deposition. There are also changes in the physical properties of the deposits with pH. Randin and Hinterminn[97] noted that in acid chloride solutions with a glycollic/succinic complexant system, the percentage of phosphorus increased with decreasing pH.

Complexors

These compounds are almost all organic acids and usually dicarboxylates. They have a dual function in that the salt may be selected primarily for its ability to chelate nickel ions, but will usually have a secondary buffering effect.

Primarily this group forms a variety of complexes with nickel thereby reducing the concentration of free nickel ions. This stabilizes the solution and tends to inhibit the production of nickel phosphite precipitates. The efficiency of a given complexing agent can be determined by the limit of the concentration of phosphite at which a precipitate forms. By the very nature of complexors they will tend to slow down the plating rate of a bath. As a general statement, the higher the concentration of any one complexor the greater the loss in plating speed.[18, 19, 21] The only exception to that statement is lactic acid, which has the unusual capability of increasing the deposition rate in some formulations while still acting as a very effective buffer and complexing agent.

Accelerators

As the concentration of complexors is increased to a point at which the bath is stable and does not precipitate phosphites, the deposition rate drops substantially. Improvements in this rate may be obtained by using soluble fluorides[97, 98, 99] or sulfur-bearing compounds such as thiourea.[100] The concentration of these materials is usually 2 to 20 ppm. If this level is exceeded, the additives then start to poison the reduction reaction and will inhibit the bath to the point of stopping all deposition.[19]

Stabilizers

The disadvantage of any truly electroless process is that it must display some degree of limited instability to operate at all. Under the most unfavorable conditions the solution may actually spontaneously decompose. This is a violent reaction in which, in a relatively short period of time, the bath reduces all of the nickel salt content in solution to finely divided metallic nickel powder. Such reactions are exceptionally rare in commercial electroless nickel solutions, but more localized solution plate-out may still be triggered by a number of circumstances:

1. If localized overheating of the bath occurs.
2. When additions of reducer are made to a solution (high concentrations of hypophosphite may develop in isolated areas).
3. If solids contaminating the bath act as catalytic bodies. (Due to the high surface area to mass ratio of powdered contaminants such as dust or other finely divided materials, a small amount of contaminant may prove a serious problem.)
4. If the introduction of trace elements into the solution leads to unwanted reduction of

metal. (This may be caused by drag in of chemicals from pretreatment processes or by the use of a material such as palladium to initiate deposition on noncatalytic copper alloys.)

There are two main answers to these problems, the first being the circulation and filtration of the bath. This produces a homogeneous solution devoid of particulate matter that may cause plate-out; the second is a chemical modification of the bath.

During the 1950s ions such as Pb^{++}, Cd^{++}, Zn^{++}, SCN^- and CN^- were regarded as catalytic poisons, as at moderate concentrations they stopped reduction totally. The work of Talmey and Gutzeit[13, 22, 23, 57] established that at very low concentrations of 1 to 10 ppm these ions act as stabilizers. Gendrzynski followed this work by patenting[104] the use of molybdic acid anhydride, arsenious acid, hydroxyl amino sulfate and hydrazine. Subsequently a number of patents were granted covering the use of selenium and tellurium compounds as well as a large number of sulfur-containing organic compounds. Examples of these are mercaptanes, thiourea, xanthogenates and sulfonates.

The exact mechanism of stabilization by these materials has not been explained fully. Rozenblyum[121] proposed a theory for the effect of "poisonous additives" by suggesting that two reactions are competing on a catalytic surface, the first being the deposition of nickel, and the second being the diffusion-controlled adsorption of stabilizers. The first reaction is kinetically controlled, and at normal deposition rates the renewal of active nickel surfaces with fresh deposits takes place at a faster rate than the adsorption of poisonous ions onto the surface. Poisoning begins when the rate of adsorption of catalytic poisons exceeds the rate of nickel deposition.

Some stabilizers have secondary brightening effects in that they are capable of producing deposits with quite definite improvements in surface luster.

Wetting Agents

These materials are used to reduce the surface tension of the plating solutions in an attempt to control the formation of gas bubbles. As with electrolytic processes, if a gas bubble adheres to the depositing surface a pit may grow around it, leaving a void in the deposit. The most commonly used materials are alcohol sulfonates, fatty acid sulfonates and non-ionic products such as ethylene oxide derivatives.

Moderators

During the operation of a bath a number of materials build up in the solution as byproducts of the reduction reaction, or from maintenance of the system. These compounds do not directly take part in the reduction of nickel but they influence the rate of deposition of metal.

There are two major groups in this field. The first is the neutral salt of the anion of the nickel compound used to regenerate the bath. In the case of nickel sulfate used in a bath that is adjusted with sodium hydroxide, this would be sodium sulfate. The second group is the byproduct of the reducer oxidation, usually sodium orthophosphite.

Moderators will eventually build up in significant concentrations, and can stabilize the reduction process to a great extent. Eventually, these materials result in the bath being dumped due to their effects of reduction in plating rate, and their eventual saturation and precipitation from solution.

Many attempts have been made to extend the operational life of chemical plating solutions. The removal of materials such as sodium sulfate is relatively easy. If the bath temperature is allowed to fall to 5 to 10 C (41 to 50 F) the majority of the sulfate salts will crystallize out, and may be removed by filtration.

The reduction in the concentration of sodium phosphite can be achieved using ion exchange techniques. If a suitable exchange resin is activated with hypophosphorous acid, and the bath is passed through this, the orthophosphite may be replaced with hypophosphite.

These techniques, unfortunately, do not deal with other contaminants such as materials from pretreatment processes that build up during the life of a solution. These other contaminants have made solution reclaim on an industrial scale unworkable up to the present time.[40, 41, 42, 43, 44, 45]

THE FACTORS THAT INFLUENCE THE DEPOSITION RATE

The factors that influence the rate of deposition of a chemical nickel bath are wide ranging. There is no simple connection between the effect of one material in any electroless nickel bath. The effects of any one compound will vary with the type of solution and the concentration of other components. Gawrilov[114] defined the factors affecting the speed of deposition in this way:

$$S.D. = f(T, \text{pH}, C_{Ni^{2+}}, C_{H_2PO_2^-}, C_{HPO_3^{2-}}, S/V, C, A, S, n_1, n_2 \ldots)$$

where:
- T = temperature of deposition
- pH = pH value
- $C_{Ni^{2+}}$ = concentration of nickel
- $C_{H_2PO_2^-}$ = concentration of hypophosphite
- $C_{HPO_3^{2-}}$ = concentration of phosphite
- S/V = relationship of plated surface to volume of plating solution
- C = nature and concentration of complexing agent
- A = nature and concentration of accelerator
- S = nature and concentration of stabilizer
- n_1, n_2 = additional factors such as pressure, speed of agitation, etc.

The Effects of Temperature

The rate-controlling factor of most chemical nickel solutions is the bath temperature. No reaction at all is to be expected from an acid bath at temperatures below 50 C (122 F). The majority of acid type formulations operate within an 80 to 100 C (176 to 212 F) temperature range, with specific baths normally having operational temperature bands of ±3 C (5.4 F) from their optimum.[14]

As temperature increases so will the rate of deposition of nickel, and the rate at which initiation of the substrate occurs increases accordingly.[33]

It has been demonstrated by Baldwin and Such[108] that the major effect of temperature on deposit analysis is a decrease in phosphorus content with increasing temperature.

The Effects of pH

As a general statement the rate of deposition tends to increase with pH. The minimum value for any deposition being 3.0, as the pH value reaches 7.0 the hypophosphite spontaneously oxidizes by the reaction:

$$(H_2PO_2)^- + OH^- \rightarrow (HPO_3)^{2-} + H_2.$$

The optimum pH value for acidic baths will always be a compromise between commercially acceptable plating rates and technically viable stability. In practice the preferable pH range is between 4.3 and 4.9. There has, however, been some use made of alkaline solutions, but these baths exhibit poor physical properties.[11, 12, 16, 17, 34, 61]

ANALYSIS OF DEPOSIT

The analysis of the alloy deposit is dependent on the bath formulation used and the kinetics of each individual reaction in the reduction process. Each of these reactions may be effected by bath pH and temperature. Electroless nickel coatings are alloys primarily of nickel and 2 to 14% phosphorus by weight. There may also be traces of other elements introduced into the deposit. These can result from additives in the bath or contaminants that have been introduced being co-deposited.

Some of these elements, such as antimony, bismuth, chromium, lead, mercury and selenium, may have detrimental effects on the corrosion resistance of the deposits. These phenomena are being investigated by a number of researchers at the moment, but conclusive evidence is yet to be published.

STRUCTURE

In the as-plated state, chemical nickel deposits have been shown by X-ray analytical techniques to have a microcrystalline structure.[16, 26] Metallographic investigation shows a laminar structure, after anodic etching in chromic acid solutions. It has been proposed that the visible laminations are due to variation of the phosphorus content, or changes in the structure of the nickel phase itself. The effect is most likely due to the nature of the way in which the chemical reduction reaction occurs.

Research has been conducted by a number of laboratories which have sometimes produced conflicting results. Gutzeit and his fellow researchers reported on the structures of 7 to 10% phosphorous alloys in 1959. They concluded that the deposit was a super-saturated solution of phosphorus in a noncrystalline nickel matrix, with a level of disorder close to that of a liquid.[110] This theme was supported by Moissejer[111] in 1962, who further proposed the existence of a three-phase system. This was to consist of a beta phase of less than 5% phosphorus. At concentra-

TABLE 3. ELECTROLESS NICKEL SOLUTION FORMULATIONS.

	Boric	Fluoride	Fluoboric	Citric	Ammonium/citrate	Lactic/acetic	Lactic/propionic	Lactic/succinic	Glycollic	Acetic/malic
Nickel chloride, g/L (oz/gal)	12 (1.6)				44 (5.9)		26 (3.5)		30 (4.0)	20 (2.7)
Nickel Sulfate, g/L (oz/gal)		30 (4.0)	14 (1.9)					20 (2.7)		
Sodium hypophosphite, g/L (oz/gal)		30 (4.0)		80 (10.7)		20 (2.7)	24 (3.2)	25 (3.3)		25 (3.3)
Acetic acid (80%), g/L (oz/gal)	10 (1.3)		25 (3.3)	25 (3.3)	11 (1.5)	25 (3.3)			10 (1.3)	20 (2.7)
Glycollic acid, g/L (oz/gal)									35 (4.7)	
Lactic acid (80%), g/L (oz/gal)						35 (4.7)	27 (3.6)	34 (4.5)		
Propionic acid, g/L (oz/gal)						2 (0.26)	2.2 (0.29)			
Fluoboric acid (40%), g/L (oz/gal)			28 (3.8)							
Sodium fluoride, g/L (oz/gal)										
Ammonium fluoride, g/L (oz/gal)	5 (0.7)		4 (0.53)			2 (0.26)		2.5 (0.33)		
Citric acid, g/L (oz/gal)				95 (12.7)	95 (12.7)					
Succinic acid, g/L (oz/gal)								7 (0.9)		
Malic acid, g/L (oz/gal)										20 (2.7)
Amino acetic acid, g/L (oz/gal)										
Ammonium chloride, g/L (oz/gal)			6 (0.8)	50 (6.7)						
Boric acid, g/L (oz/gal)	15 (2.0)	15 (2.0)	8 (1.0)							5 (0.7)
Lead (ppm)						0.7-3		0.7-3		0.7-2
Tellurium (ppm)										1.0
Organic nitrogen compound (ppm)			4				4			
Operational pH value	6.0	6-7	6-6.5	4.8-5.8	8.5-9.0	4.5-4.7	4.6	4.5-4.7	4-6	6.5-6.7
Used to adjust pH	NaOH	NaOH	NaOH	NaOH	NH$_4$OH	NaOH	NaOH	NaOH	NaOH	NaOH
Temperature, C (F)	60 (140)	88 (190)	90 (195)	93 (200)	95 (203)	93 (200)	93 (200)	95 (203)	90 (195)	90 (195)
Deposition rate, u/hr	9	30	32	15-20	15	22-27	25	32-35	15	55

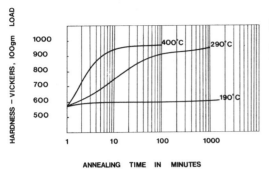

FIG. 2. Time at temperature hardening curve.

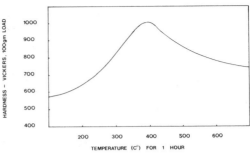

FIG. 4. Phase diagram, nickel/phosphorus system.

tions of 5 to 8.5% this would be a mixture of alpha and beta nickel, and above 8.5% the structure would be pure alpha.

The finding of Graham, Lindsay and Read[112] contested the earlier investigations by reporting that a supersaturated solid solution of phosphorus (7 to 9%) and crystalline nickel was deposited onto copper targets. This deposit was crystalline and fibrous with a preferred nickel orientation of (1,1,1,). By 1970 Randin and Mintermann had published a paper describing the existence of Ni_2P in as-plated matricies with very little free phosphorus.[113]

Most of the researchers agreed that the grain size of the nickel phosphide crystals ranged from 15 to 100Å with an average size of 20 to 60Å.[114] As the deposit is heated above 200 C (392 F), precipitation of the Ni_3P begins to occur from the super-cooled liquid structure. The deposit then develops into a super-saturated crystalline solid solution. This subsequently decomposes as the intermetallic compound Ni_3P segregates.[4, 16, 26, 46]

HEAT TREATMENT

The effect of thermal precipitation is to produce a coating with higher hardness values and greater wear resistance. This may also be accompanied by a slight shrinkage of the deposit and a loss of ductility.

As the temperature of treatment exceeds 425 C (797 F) or is held at temperatures close to this value for extended periods, the deposit will soften again. With this reduction in hardness, the ductility of the deposit reverts to its original levels.[4, 26, 32, 49]

Another effect of heat treatment is an improvement in the levels of substrate adhesion. This change is most marked in the case of nonferrous

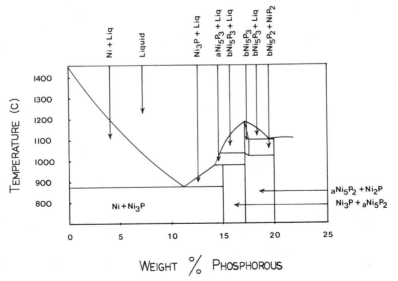

FIG. 3. Hardness curve for electroless nickel.

substrates. The mechanism for this reaction has been investigated by several workers.[116,117] Their results show a limited diffusion of nickel into the substrate. In the case of ferrous substrates an iron-rich layer may develop in the nickel deposit as well.

The diffusion phenomenon may also be found when attempts are made to weld electroless nickel coated substrates. The phosphorus in the deposit diffuses into the weld area and is very likely to cause weld failure due to localized cracking.

CORROSION RESISTANCE OF DEPOSITS

Electroless nickel deposits are more electronegative than the majority of the ferrous and aluminum alloys that they are called on to protect. This means that the deposits can only protect these substrates by providing a mechanical barrier to corrosive media, and they cannot offer any sacrificial electrochemical defense. The mechanism of this barrier protection effect would appear to be the development of a passive surface oxide coating reducing the galvanic currents between the grain boundaries thereby restricting the total corrosion current. If this film is disrupted it may adversely effect the rate of corrosion of the deposit.

Due to the mechanism of protection any voids or pits in the deposit will have very serious consequences for the overall protection of the component.

The greatest care must be paid to pretreatment systems, in both the mechanical and chemical preparation of the substrate. Steps must also be taken to limit the drag-in of contaminants to the plating solution itself.

The corrosion resistance of electroless deposits will vary from one bath formulation to another, and it may become necessary to test individual solution types to ensure that they meet the end user's needs. Porosity tests can provide a great deal of information about the overall corrosion resistance of a specific bath, and can also become a useful examination for quality control on a day-to-day basis.[17,47,48]

PHYSICAL PROPERTIES OF NICKEL PHOSPHOROUS ALLOYS

The physical properties of each individual alloy produced from different bath formulations will vary slightly. As a general guide they should be close to the range of attributes listed below.

Phosphorus content	8–12%
Hardness: as plated	550–650 VPN (100 g load)
heat treatable to	950–1050 VPN (100 g load)
Elongation before fracture	0.5–2%*
Tensile strength	over 700 MPa
Coefficient of thermal expansion	$10\text{–}13 \times 10^{-6}\ \alpha/k^{-1}$
Melting point	880–890 C
Electrical resistivity (λ)	$520\text{–}680\ \rho/m \times 10^{-8}$
Magnetic coercivity	above 9%, nonmagnetic as plated
Density	$7850\text{–}8050\ \rho/kg\ m^{-3}$

Ternary Alloys

Research into the production of ternary and quaternary alloys of nickel or cobalt with a variety of other alloying metals and reducers has been carried out since 1950. The greater part of this work has been concentrated on the ferromagnetic properties of the deposits, for audio and video tape production.

The most successful commercial alloy process to date has been the "Niculoy 22" bath,** which is a Ni-Cu-P containing approximately ½ to 1% copper. (The most comprehensive survey of these solutions has been carried out by G. G. Gawrilov.[114])

MECHANICAL WEAR

The abrasion resistance of electroless nickel deposits is very good in lubricated wear situations as long as the temperature is not too high. The only comparative data available are the wear index numbers from the Taber Abrader. An example of the results to be expected are listed below.

Electrolytic nickel plate (Watts bath)	14.7
Electroless nickel (as deposited)	9.6
Electroless nickel, heat treated at 300 C (572 F)	4.4
Electroless nickel, heat treated at 500 C (932 F)	2.7
Electroless nickel, heat treated at 650 C (1202 F)	1.3
Hard chromium	2.0

*This figure is for the as plated form and may rise to 6% after heat treating at 625 C (1157 F).

**Niculoy 22 is a trade name of the Shipley Company (Europe), Coventry, England.

These tests were performed with a CS-10 wheel at 1000-g loading, for 5000 cycles. The Tabor Wear Index quoted is an empirical unit, and the higher the number the greater the wear rate.

Electroless nickel may be used as a wear surface when difficulty has been experienced, particularly with galling. An excellent example of this may be when two stainless steel surfaces bear on each other. In the untreated state, immediate galling will occur. If one surface is chemically nickel plated then an excellent bearing system develops.

Examinations involving two metal systems under chlorinated paraffin boundary lubrication gave the following coefficients of friction.

Chromium vs. nickel	0.20
Chromium vs. electroless nickel	0.30
Nickel vs. steel	0.20
Electroless nickel vs. steel	0.26
Nickel vs. nickel	Galling
Electroless nickel vs. nickel	0.25

CHEMICAL PRETREATMENT

The systems for the preparation of substrates that are to be electroless nickel plated demand very careful control. Autocatalytic nickel deposits are very sensitive to the physical and chemical condition of the underlying metal. It is possible to cause substantial variation in properties such as corrosion resistance by cleaning using an incompatible system.

The primary requirement in any pretreatment cycle must be the removal of insoluble soil, followed by the displacement of attached chemical debris. In cleaning for plating it is not possible or practical to produce a physically and chemically clean surface. What is really required is that all detrimental contamination of the metal substrate is detached and is replaced with a chemically compatible surface. Examinations such as water break tests indicate physical cleanliness. These tests do not give any guide to the chemical activity of the metal that will form the bond at the deposit interface.

It has been suggested that the use of any form of inhibitor in acid activators may be prejudicial to the corrosion performance of articles[109] and that the last electroclean stage before plating should be free of surfactants or cyanides. These measures are intended to limit the level of adsorbed materials on the bonding surface and to leave the substrate as active as possible.

Some suggested pretreatment cycles have been listed below; these are by no means hard and fast sequences. A wide range of special pretreatments which are specific to alloy type or geometry of component have been developed, and each one of these is valid in its own right.[24, 25, 60]

Low-Alloy Mild Steels

1. Vapor degrease or emulsion soak cleaner.
2. Alkaline soak cleaner.
3. Anodic electrolytic cleaner.
4. Rinse, cold water.
5. Acid activate 10–15% sulfuric acid.
6. Rinse, cold water.
7. Electroless nickel plate.

Aluminum and Aluminum Alloys

Aluminum may be electroless nickel plated very successfully, but it is essential to know which alloy is being treated. Different alloys will require variations in processing, specifically in acid etching techniques (see Chapters 3 and 4).

Sequence:

1. Solvent clean, if required.
2. No-etch soak clean.
3. Cold water rinse.
4. Deoxidize surface to remove any heavy oxide films, using 50% v/v nitric acid or proprietary deoxidizer.
5. Cold water rinse.
6. Alkaline etch if required by the condition of the substrate or if requested by customer.
7. Cold water rinse.
8. Acid pickle (the formulation of the acid pickle is determined by the alloy to be plated).
9. Cold water rinse.
10. Zincate following supplier's recommendations.
11. Cold water rinse.
12. Strip zincate in 50% v/v nitric acid.
13. Cold water rinse.
14. Zincate.
15. Cold water rinse.
16. Cold water rinse.
17. Electroless nickel plate to required thickness.

Copper, Brass, and Beryllium Copper Alloys

A valid treatment sequence would be:

1. Solvent degrease if required.

2. Soak clean.
3. Cold water rinse.
4. Cathodic electroclean.
5. Cold water rinse.
6. Acid dip in proprietary dry acid salt.
7. Cold water rinse.
8. Optional-strike in electrolytic nickel bath.*
9. Cold water rinse.
10. Electroless nickel plate.

APPLICATIONS FOR ELECTROLESS NICKEL

Electroless nickel should not be considered as a decorative surface finish, or as a direct replacement for any single electrolytic process. The electroless nickel coatings by comparison with alternative finishes are relatively expensive but they display unique physical properties.[17,48]

It is worth considering the range of uses that nickel/phosphorus alloy coatings display:

a. Electroless nickel plated low-alloy steel components can often be used to replace more expensive exotic alloy fabrications. The impervious alloy deposit will give the same component life as the much more expensive metal alloys. Even in the case of environments that are not particularly corrosive, electroless nickel acts as a barrier, preventing contamination of manufactured commodities with small amounts of the substrate metal.

b. Chemical nickel coatings exhibit excellent wear resistance which is due to a combination of factors. The coatings may be heat treated to produce very high hardness values while still retaining the natural lubricity of the as plated finish. When nonpolar or long-chain polar aliphatic compounds are used in conjunction with the coatings, electroless nickel exhibits excellent wettability. This further decreases mechanical friction.

c. In many bearing applications electroless nickel may be used to prevent galling of aluminum and stainless steel wear surfaces.

d. Relatively thin coatings will allow the soldering of aluminum, beryllium and magnesium, for electrical engineering uses.

e. Chemical nickel deposits may be used as a bonding material. If two metal surfaces are coated with electroless nickel and then furnace brased in a reducing atmosphere an extremely effective braised joint is produced.

*The parts may be initiated in the electroless bath.

f. Due to the uniformity of the electroless deposits they are particularly suited for recovery of worn components, or mismachined parts. The limit of the deposit thickness is really in the order of 150 μ (0.006 in.), because of the financial constraints of very heavy build-up.

The limitations that apply to electroless nickel are the same as those that apply to other nickel deposits. Specifically, almost all inorganic acids and short chain carboxylic acids exhibit penetration rates that are too large for the practical use of electroless nickel. This is also true of materials that form soluble nickel complexes such as cyanides, ammonia, short chain organic amines and many mono- and polyhydroxycarboxylic anions.

THE USE OF ULTRASONICS WITH CHEMICAL NICKEL SOLUTIONS

There have been widely diverging reports on the effects of ultrasonic radiation in electroless plating baths. The majority of beneficial claims have been for increases in deposition rate. Kosub and his fellow workers[118] described rates of up to 60μ/hr. Conversely, other researchers have found very little benefit from the use of ultrasonic transducers, and have even reported increases in porosity in the deposit.

The reported work to date has mostly been carried out in very small-scale units and no industrial experience with the process appears to have been gained. It may be likely that future research will lead to limited specialized applications.

STRIPPING OF DEPOSITS

Unsatisfactory deposits may result from a wide variety of sources. It is not good practice to replate a component over faulty plating, as lamination may result. The only action that may be taken is to totally strip the part. This is best carried out before heat treatment of the coating, as the precipitation of phosphides in the deposit makes its removal much more difficult.

Nitric acid has been used as a stripping media, but it is possible that the use of such a strongly oxidizing acid may result in porosity in the substrate on ferrous alloys. The safest method may be to use a proprietary nitroaromatic based stripper with an effective inhibitor system.

MECHANICAL PRETREATMENT

Some components may be contaminated with soils that do not easily respond to chemical methods of removal. These may be heavy rusting heat treatment scale, or weld slag. Even machining methods such as lapping may leave abrasive materials imbedded in critical surfaces.

The best method of treating these problems is by mechanical means. The components may require further machining such as honing or grinding. Blasting methods such as shot blasting or bead blasting with a wide variety of media are very effective in both severity of cutting action and removal rate of metal. Vibratory finishing is also a method of mechanical finishing that should not be discounted.

Whichever method is used, the final aims must be the same. The parts must be free of pinholes, burrs or segregations of foreign materials and must have the smoothest possible surface finish before plating.

OPERATIONAL PLATING SYSTEMS

In electroless deposition the equipment used to contain the solution plays a very important role. The tank system is an integral part of the deposition process. A well-designed tank can make the difference between a successful, profitable operation or one with continued problems.

An obvious difficulty with electroless deposition is that the solution is inclined to deposit onto any surface with which it comes in contact. This includes the plating vessel and its auxiliary apparatus. Therefore, the equipment must be inherently inert or must be capable of being rendered passive.

A minimum number of basic services are necessary for the operation of chemical plating baths on an industrial scale. These break down into four areas—the tank, a circulation pump, a filter unit and a heating system.

The Tank

As electroless nickels have developed in sophistication, the materials of construction and design of the plating tank have changed to keep pace. The earliest satisfactory industrial scale plating units were designed for Kanigen[53] operations and were all stainless steel fabrications. The tanks were treated with 30 to 50% nitric acid, to remove any build-up of nickel and to repassivate the vulnerable metal surfaces. During the late 1950s, titanium also became available for smaller tank units. With either type of metal tank, the oxide film may be maintained by making the tank anodic. A small cathode rod is used at an anodic current density of 50 mA/ft^2 (5mA/dm^2).

Tanks may also be built with a water jacket which may be filled with a glycol solution. The "indirect" system allows the use of any type of available heating method to raise the temperature of the outer jacket. The inner plating tank walls are used as a heat exchange surface to heat the plating solution.

The alternative for larger systems of over 500 gal is to use an external heat exchanger. As polymer technology developed it became fashionable to use loose fitted plastic liners inside metal tanks. Although these disposable liners were effective for isolation of the solution from the walls of the vessel, they made heat transfer from indirect heating more difficult as they insulated the plating solution from the heating media.

The latest development in electroless plating systems has been the use of polypropylene. The only acceptable grades of the material are the unpigmented, natural, stress-relieved types. Polypropylene may be welded to form rigid freestanding units or as a heavy liner for larger, steel-supporting tanks. This material has one great advantage over the alternative metal tanks in that it is relatively passive, but it limits the types of heating that may be used.

Heating

Increasing and maintaining the temperature of chemical plating systems is a critical point in their operation. It is obvious that any localized overheating may give rise to a situation in which the bath may become locally unstable, and either plate onto the heater or initiate a total solution breakdown.

The choice of heating method must be closely linked to the construction material of the tank. With metal tank units of less than 500 gal, indirect heating through the vessel walls is quite effective. The limitations of this method are that the wall thickness of the inner tank must be kept to the minimum possible to allow effective, efficient heat transfer. There is also a limit to the size of this type of unit. If the tanks are too big, the area of available heating surface in relation to volume of solution becomes too small and the maintenance of operational temperature becomes difficult. Heating of the outer jacket may be by a wide range of methods. As long as localized overheating due to poor circulation of the intermediate solution is avoided, few problems develop. It may be worth noting that the inner tank should be electrically isolated from the heater coils. This is required, as stray currents from other areas of the plating shop may be carried through to the walls of the plating tank and may initiate deposition. The easiest way to isolate the tank is to paint its outer surfaces (which are immersed in the heat transfer solution) with an epoxy based paint.

With direct heating systems the heater is immersed in the plating solution itself. The limitations on this type of system are that the surface temperature of the heater, particularly with electrical resistance heaters, must not be allowed to become too high. The solution must also be circulated around the heaters. If these points are not treated with great care the bath will be pushed into an unstable state. Derating of electrical systems helps a great deal in reducing the energy output per unit area of heater used. If steam heating is to be considered there are a range of Teflon coils available that have been proven to be very effective. Irrespective of the type of direct immersion heater used, it is good practice to circulate the bath around the area of the heaters themselves. This is to maintain an even bath temperature and therefore an even rate of deposition over the entire tank area.

External heat exchangers have been used for a number of large plating systems where the tanks are too large for indirect heating, and where the cost of direct immersion heaters may be too high. The exchanger may be stainless steel, titanium or glass lined. If the flow rate of the solution is sufficiently high, few problems develop with this system.[13, 27, 35, 36, 37]

Circulation Pumps and Filter Units

In smaller plating systems, the pump and filter may be combined into a single unit, particularly in the case of cartridge filters. The most satisfactory design of pump has proved to be the mechanically coupled rotary design with water-cooled seals, constructed from polysulfone, polypropylene or stainless steel. For tank volumes of under 100 gal, submersible sealless pumps are an acceptable substitute.

The filter medium may be wound cartridge or bag type polypropylene felt construction with a pore size of 2 to 5μ. It is more important to improve on the throughput of the filter than to try to reduce the porosity of the filter media. The ideal filtration rate for units of under 500 gal is 10 turnovers of the tank volume per hour. For example, a 100-gal tank should be filtered at 1000 gal/hr. As the size of the tank increases, this ratio may be reduced a little.[13, 35, 36]

Selective Plating

In some cases it may be desirable to plate only selected areas of certain parts. With chemical deposition of nickel, stop-off lacquers must fulfill a wide range of requirements. The initial stages of pretreatment expose the stop-off to highly alkaline cleaners; this is then followed by acid activation and an electroless plating solution operating at 90 C.

One of the few groups of brushable or sprayable paints that are resistant to these service conditions is epoxy based. The drawbacks to these materials are the necessity to bake them to effect a good cure and the need to use paint strippers for their removal.

The alternatives are a range of mechanical devices. For internally or externally threaded areas, tapped polypropylene plugs may be screwed into place. Shrink fitted sleeves can be applied to the external diameter of shafts. Plastic tapes may also be used to isolate some accessible faces of parts.

The greatest difficulty with selective electroless nickel deposition is that the process is very labor-intensive and time-consuming. It is usually less expensive to plate a component all over than to stop-off certain areas.

Racking

As with any other electroless processes, the work racks must fulfill two main needs, the first being to suspend the components and to locate them positively, and the second to make definite electrical contact with jobs if electrolytic treatments are required.

In the case of parts with complicated shapes or blind holes, attention must be paid to the entrapment of gas evolved during plating. If a gas pocket is allowed to form, then a void in the deposit will result. Due to the nature of electroless deposition, agitation of the solution or the parts is very desirable. This may be achieved by mounting the rack on a reciprocating work bar moving in either the horizontal or vertical plane.

Barrel Plating

Chemical nickel plating of small mass-produced components is often required. It may be possible to carry this out in polypropylene baskets with intermittent agitation, to prevent contact marking, or by using barrel plating techniques.

When barrel plating with electroless solutions the rotational rate of the barrels must be reduced greatly over that which would be used for electrolytic techniques. Rotational speeds of one to five revolutions per minute are not unusual. Allowance must be made for flow of solution through the barrel. If the perforations or the barrel configuration do not allow free movement of plating solution through the barrel, and hence to the work load, reduced rates of deposition may result.

Galvanic Initiation

A wide range of ferrous alloys and zincated aluminum will activate directly in electroless nickel solutions. For noncatalytic metals which are more noble than nickel, such as copper, manganese and some high-strength ferrous alloys, some form of initiation is required to start deposition. The easiest method is to make electrical contact with a part which is already plating, or to impose an external current by making the component cathodic for 10 to 15 seconds. The tank wall (if it is metal) or a small nickel anode may be used as the positive electrode.

BORON NICKEL ALLOYS

Nickel alloys reduced by the action of boron salts have not achieved the same level of industrial acceptance as their phosphorous alloy counterparts. The nickel boron deposits exhibit high as-plated hardness and have a melting point some 200 to 400 C (360 to 720 F) higher than comparable phosphorous alloys. Applications for the boron based baths are found mainly in the electronics field, and in areas where higher-temperature wear resistance is required.

Interest in this field was first aroused when Schlesinger produced pure sodium borohydride in 1942. He noted the formation of Ni_2B and Ni_3B borides, but it was not until 1954 that the first usable coatings were deposited. By the end of the 1950s technically effective processes had been developed. These solutions consisted of an aqueous, highly alkaline solution of nickel salts chelated with organic complexing agents, and reduced by a borohydride salt. The bath required careful control of its heavy metal stabilization system to prevent the solution decomposing. In using sodium borohydride the bath pH must be maintained above 10.0, and preferably in the range of 12 to 14. This is to limit the hydrolysis of the BH_4 species.

More recent solution formulations have moved toward the use of dimethylaminoborane (DMAB) as a reducer. These boron-nitrogen compounds have a number of distinct advantages. It is possible to operate DMAB solutions over a wider pH range than the sodium borohydride formulations. The nickel aminoborane baths have almost unlimited life, as the breakdown products have very little effect on the operation of the solution. Deposition rates of 6 to 9 μm/hr are maintained, to a large extent independently of the nickel concentration in the tank, and are really only controlled by the reducer concentration.

The reduction reactions of these formulations can be described by the following equations:

$$R_2NBH_3 + 3Ni^{2+} + 3H_2O$$
$$\rightarrow 3Ni + (CH_3)H_2N^+ + H_3BO_3 + 5H^+$$

TABLE 4. FORMULATIONS FOR BORON COMPOUND REDUCED BATHS

COMPONENTS	1		2		3		4	
Nickel sulfate, g/L (oz/gal)					50	(6.7)		
Nickel chloride, g/L (oz/gal)	93	(12.4)	24–48	(3.2–6.4)			20	(2.7)
DMAB, g/L (oz/gal)	37	(5.0)	3–4.8	(0.4–0.64)	3	(0.4)		
Hydrazine borane, g/L (oz/gal)							0.5–1.5	(0.06–0.18)
Ethylene diamine, 50%, g/L (oz/gal)			18–37	(5–10)			20	(2.7)
Potassium acetate, g/L (oz/gal)			18–37	(5–10)			40	(5.4)
Sodium pyrophosphate, g/L (oz/gal)					100	(13.3)		
Boric acid, g/L (oz/gal)	25	(3.3)						
pH	4.3		5.5		10		8.5–10	
Temperature, C (F)	27	(80)	70	(158)	25	(77)	68	(122)

$$2R_2\text{NHBH}_3 + 4\text{Ni}^{2+} + 3\text{H}_2\text{O} \rightarrow \text{Ni}_2\text{B} + 2\text{Ni} + 2R_2\text{H}_2\text{N}^+ + \text{H}_3\text{BO}_3 + \tfrac{1}{2}\text{H}_2 + 6\text{H}^+$$

where R is a $(\text{CH}_3)_n$ species.

References

1. a. Brenner, A. and G. E. Riddell. *J. Research Natl. Bur. Standards* **37**:1 (1946); *Proc. Am. Electroplaters' Soc.* **33**:16 (1946).
   b. Brenner, A. and G. E. Riddell. *J. Research Natl. Bur. Standards* **33**:385 (1947); *Proc. Am. Electroplaters' Soc.* **34**:156 (1947).
2. Brenner, A. and G. E. Riddell. U.S. Patent 2,532,284 (Dec. 5, 1950).
3. Saubestre, E. B. *Sylvania Technologist* **12**, Nos. *1, 5* (1959); *Am. Electroplaters' Soc.* **46**:264 (1959).
4. "Symposium on Electroless Nickel Plating," Am. Soc. Testing Materials, Technical Publication No. 265 (1959).
5. Wurtz. A. *Ann. chim et phys.* **3**, Ser. *16*:198 (1846); Wurtz, A. *C. r. de l'Academie des Sciences* **18**:702 (1844); **21**:149 (1845).
6. Breteau, P. *Bull. soc. chim.* **9**:515–518 (1911).
7. Paal, C. and L. Frederici. *Ber. deut. chem. Ges.* **64B**:1766 (1931).
8. Scholder, R. and H. Hecket. *Z. anorg u. allgem Chem.* **198**:329 (1931).
9. Scholder, R. and H. L. Kaken. *Ber. deut. chem Ges.* **64B**:2870 (1931).
10. Roux, F. G. U.S. Patent, 1,207,218 (Dec. 5, 1916).
11. Brenner, A. and G. E. Riddell. U.S. Patent 2,532,283 (Dec. 5, 1950).
12. Gutzeit, G. *Plating* **46**, Nos. *10, 11, 12* (1959); **47**, No. *1* (1960).
13. Metheny, D. and E. Browar. U.S. Patent 2,874,073 (Feb. 17, 1959).
14. Brenner, A. *Metal Finishing.* **52**, *11*:68 (1954); **52**, *12*:61 (1954).
15. Gutzeit, G. and A. Krieg. U.S. Patent 2,658,841 (Nov. 10, 1953).
16. Gutzeit, G. *Metal Progr.* **65**,*1*:113 (1954); Gutzeit, G. *Trans. Inst. Metal Finishing* **33**:1–29 (1956).
17. de Minjer, C. H. and A. Brenner. *Plating* **44**:1294 (1957).
18. Gutzeit, G. and E. Ramirez. U.S. Patent 2,658,842 (Nov. 10, 1953).
19. Gutzeit, G. U.S. Patent, 2,694,019 (Nov. 9, 1954).
20. Gutzeit, G., P. Talmey and W. Lee. U.S. Patent 2,822,294 (Feb. 4, 1958).
21. Gutzeit, G., P. Talmey and W. Lee. U.S. Patent 2,935,425 (May 3, 1960).
22. Talmey, P. and G. Gutzeit. U.S. Patent 2,762,723 (Sept. 11, 1956).
23. Talmey, P. and G. Gutzeit. U.S. Patent 2,846,327 (Aug. 12, 1958).
24. Maclean, J. D. and S. Karten. *Plating* **41**:1284 (1954).
25. Lee, W. G. and E. Browar. U.S. Patent 2,928,757 (Mar. 15, 1960).
26. Goldenstein, A. W., F. Schoosberger and G. Gutzeit. *J. Electrochem. Soc.* **104**:104 (1957).
27. Talmey, P. and W. J. Crehan. U.S. Patent 2,717,218 (Sept. 6, 1955).
28. Brenner, A. *Metal Finishing* **52**, Nos. *11, 12* (1954).
29. Crehan, W. J. U.S. Patent 2,690,402 (Sept. 28, 1954).
30. Lee, W. U.S. Patent 2,949,018 (June 7, 1960).
31. Gutzeit, G., W. J. Crehan and A. Krieg. U.S. Patents 2,690,401 and 2,690,402 (Sept. 28, 1954).
32. Ziehlke, K. T., W. S. Dritt and C. H. Mahoney. *Am. Electroplaters' Soc.*, Research Report K-1308 (Feb. 2, 1959).
33. Gostin, E. L. *Iron Age* **171**:115 (1953); Jendryznski, H. J. and T. F. Stapleton. U.S. Patent 2,721,814 (Oct. 25, 1955).
34. Metheny, D. D. U.S. Patent 2,872,353 (Feb. 3, 1959); Talmey, P. and D. E. Metheny. U.S. Patent 2,837,445 (June 3, 1958).
35. Talmey, P. and W. J. Crehan. U.S. Patent 2,658,839 (Nov. 10, 1953).
36. Talmey, P. and W. J. Crehan. U.S. Patent 2,941,902 (June 21, 1960).
37. Gutzeit, G. and R. W. Landon. *Am. Electroplaters' Soc.* **41**:1416 (1954).
38. Talmey, P. and W. J. Crehan. U.S. Patent 2,816,846 (Dec. 17, 1957).
39. Talmey, P. and W. J. Crehan. U.S. Patent 2,908,419 (Oct. 13, 1959).
40. Spaulding, R. A. U.S. Patent 2,726,968 (Dec. 13, 1955).
41. Spaulding, R. A. U.S. Patent 2,726,969 (Dec. 13, 1955).
42. Chambers, C. G. and R. A. Spaulding. U.S. Patent 2,791,516 (May 7, 1957).
43. Duvall, N. U.S. Patent 2,886,452 (May 12, 1959).
44. Budininkas, P. U.S. Patent 2,886,451 (May 12, 1959).
45. Lee, W. U.S. Patent 2,872,354 (Feb. 3, 1959).
46. Wolff, R. H., M. A. Henderson and S. L. Eisler. *Plating* **42**:537–544 (1955).
47. Gutzeit, G. and R. T. Mapp. *Corrosion Technol.* **3**, No. *10*:331 (1956).
48. Spraul, J. R. *Plating* **46**:1363 (1959).
49. Lee, W. G. *Plating* **47**:288–290 (1960).
50. Gutzeit, G. U.S. Patent 2,697,651 (Dec. 21, 1954).
51. Crehan, W. J. U.S. Patent 2,908,568 (Oct. 13, 1959); Crehan, W. J., W. Klouse and P. Talmey. U.S. Patent 2,908,568 (Oct. 13, 1959).

52. Talmey, P. U.S. Patent 2,848,359 (Aug. 19, 1958).
53. "Kanigen" is a General American's registered trademark for its process of chemical nickel alloy deposition and the coating produced in its plants and in those of its licensees.
54. Talmey, P. U.S. Patent 2,766,138 (Oct. 9, 1956).
55. Girard, R. J. U.S. Patent 2,774,688 (Dec. 18, 1956).
56. Gutzeit, G., P. Talmey and W. G. Lee. U.S. Patent 2,822,293 (Feb. 4, 1958).
57. Talmey, P. and G. Gutzeit. U.S. Patent 2,767,723 (Sept. 11, 1956).
58. Gutzeit, G., P. Talmey and W. G. Lee. U.S. Patent 2,935,425 (May 3, 1960).
59. Garland, D. L. and R. D. Gray, Jr. U.S. Patent 2,575,214 (Nov. 13, 1951).
60. Schwartz, M. *Proc. Am. Electroplaters' Soc.* **46**:176-183, 249 (1960).
61. Cahill, A. E. *Proc. Am. Electroplaters' Soc.* **46**:139 (1957).
62. Okinaka, Y. Paper presented at Electrochem. Soc. Meeting, Detroit, Oct. 1969. *Extended Abstracts*, p. 383.
63. Chattaway, F. D. *Proc. Royal Soc.* **80**:88 (1947).
64. Saubestre, E. B. *Proc. Am. Electroplaters' Soc.* **46**:264 (1959).
65. Goldie, W. *Metallic Coating of Plastics, Vol. 1.* Middlesex, England: Electrochemical Publications, 1968, p. 66.
66. Saito, M. *J. Metal Finishing Soc. of Japan,* **16**, *No. 7*:16 (1965); **17**, *No. 1*:14 (1966); **17**, *No. 7*:10 (1966).
67. Staudinger, H. *et. al. Ann.* **474**:241 (1929).
68. French, E. A. *Trans. Opt. Soc.* **25**:229 (1923-24).
69. Sharp, D. J. Private communication.
70. Pearlstein, F., K. T. Fujimoto and R. Wick. *Electron. Inds.* **21**:117 (1962).
71. Saito, M. *Products Finishing,* July, p. 57 (1959).
72. Feldstein, N. and P. R. Amodio. *Proc. Am. Electroplaters' Soc.*, 2nd Plating in the Electronics Industry Symposium, Feb. 1969, p. 219.
73. Sharp, D. J. Paper presented at Electrochem. Soc., New York Meeting, May 1969. *Extended Abstracts*, p. 370.
74. Turner, D. R. Paper presented at Electrochem. Soc., Detroit Meeting, Oct. 1969. *Extended Abstracts*, p. 368.
75. Lukes, R. M. U.S. Patent 2,996,408 (Aug. 15, 1961).
76. Lukes, R. M. U.S. Patent 3,075,856 (Jan. 29, 1963).
77. Lukes, R. M. U.S. Patent 3,075,855 (Jan. 29, 1963).
78. Dutkewyek, O. B. U.S. Patent 3,383,224 (May 14, 1968).
79. Cahill, A. E. and V. P. McConnell. U.S. Patent 2,874,072 (Feb. 17, 1959).
80. Agens, M. C. U.S. Patent 3,975,855 (Jan. 29, 1963).
81. Brookshire, R. R. U.S. Patent 3,046,159 (July 24, 1962).
82. McCormack, J. F. British Patent 1,058,915 (Feb. 15, 1967).
83. Weisenberger, L. M. U.S. Patent 3,431,120 (March 4, 1969).
84. Jackson, H. U.S. Patent 3,436,233 (April 1, 1969).
85. Shipley, C. R., Jr. and M. Gulla. U.S. Patent 3,329,512 (July 4, 1967).
86. Schneble, F. W. U.S. Patent 3,361,580 (Jan. 2, 1968).
87. Torigai, E. *et. al.* U.S. Patent 3,392,035 (July 9, 1968).
88. Shipley, C. R. and M. Gulla. U.S. Patent 3,457,089 (July 22, 1969).
89. Merker, R. and S. Lucca. U.S. Patent 3,463,123 (July 1, 1969).
90. Heymann, K. and H. Grundel. U.S. Patent 3,454,516 (July 3, 1969).
91. Schneble, F. W. *et al.* U.S. Patent 3,257,215 (June 21, 1966).
92. Schneble, F. W. *et al.* U.S. Patent 3,310,430 (Mar. 21, 1967).
93. Dutkewyek, O. B. U.S. Patent 3,475,186 (Oct. 28, 1969).
94. Argens, M. C. U.S. Patent 2,938,805 (May 31, 1960).
95. Holker, K. U. and C. Wells. U.S. Patent 3,589,927 (June 29, 1971), assigned to Albright & Wilson, England.
96. Randin and Hintermann. *Plating* **54**, *5*:523 (1967).
97. Jungslager, E. F. *Metalloberfl.* **22**, *6*:170 (1968).
98. Gawrilov, G. *Metalloberfl. Agnew. Electrochemie.* **25**, *8*:277 (1971); Bulgarian Patent 16114 (1970).
99. U.S. Patent 2,694,019 (1954).
100. Brenner A. *Handbuch der Galvanotechnik, Bd II.* Munich: Hanser Verlag, 1966, Kap 21.
101. U.S. Patent 2,762,723 (1956).
102. Elze, J. *Metall.* **14**, *2*:104 (1960).
103. U.S. Patent 2,876,116 (1959).
104. Lukes, R. M. *Plating* **51**, *10*:969 (1964).
105. Hickling and Johnson. *J. Electroanal. Chem.* **13**:100 (1967).
106. Gorbunova, K. M. and A. A. Nikiforova. *Physico-Chemical Principles of Nickel Plating,* 63-11003, U.S. Dept. Commerce (1963).

107. Baldwin, C. U. and T. E. Such. *Trans. Inst. Metal Fin.* **46**, *2*:73 (1968).
108. Beer, C. F. *The Importance of Metallurgical State.* Paper presented at Electroless Nickel II Conference, Cincinnati (1980).
109. Goldenstein, Rostoker, Schossburg and Gutzeit. *J. Electrochem. Soc.* **104** (1959).
110. Moissejev, V. P. *Issvestia An U.S.S.R., Phys.* **26**, 3:378 (1962).
111. Graham, Lindsay and Read. *J. Electrochem. Soc.* **12,***1200*:112 (1963); 4:401 (1965).
112. Randin and Hintermann. *J. Electrochem. Soc.* **117**, 2:160 (1970).
113. Gawrilov, G. G. *Chemical Nickel Plating.* Pertcullis Press, 1979.
114. Ruscior, C. and E. Croiala. *J. Electrochem. Soc.* **118**, 5:696 (1971).
115. Hirth, F. W. and H. Speckhart. *Metall.* **26**, 10:1012 (1972).
116. Rjabchenkov, A. V., V. V. Ovsjankin and J. A. Sotev. *Metalloveo i Term. Obrb. Metallov* **6**:30 (1972).
117. Kosub, V. S. et al. *Jurnal Prikl. Chimij* **36**, *12*:2762 (1963).
118. Cavallotti, P. and G. Salvago. *Plating* **59**, 7:665 (1972).
119. Kovac, Z. and I. M. Croll. *Electrochemical Meeting* (Oct. 1968), Montreal Abstr. 448.
120. Rosenblyum, R. and A. Dyakov *Jurnal Prikl. Chimij* **39**, *9*:1987 (1966).

15. CURRENT AND METAL DISTRIBUTION

H. L. Pinkerton

Fellow Engineer, Process Engineering Department, Aerospace Division, Westinghouse Electric Corporation, Baltimore, Md.

Revised by Lawrence J. Durney

It is scarcely necessary to emphasize the importance of metal distribution in electroplating, regardless of the purpose of the plating operation. The ability to control the distribution of metal over the surface of the workpiece is the key to successful plating for corrosion resistance, and more especially to meeting the very exacting requirements of electroforming and other engineering uses of plating.

The distribution of an electrodeposit over an object is determined by the local current density at each point as well as by the cathode efficiency of the bath at that current density. The local current density in turn is determined by the *primary current distribution* and the local polarization. The primary current distribution, as will be explained later, is determined completely by the geometry of the plating cell. It is independent of the properties of the solution provided they are uniform throughout the solution, right up to the electrode surface. The term "polarization" combines all those phenomena, both physical and electrochemical, which operate to change the properties of the solution in the vicinity of the electrode. These changes result in an actual local or "secondary" current distribution which differs from the primary distribution.

Many subtle variables influence the distribution of electrodeposited metal, however each may be put into one of four classes:

1. Those governing the primary current distribution.
2. Those related to the plating bath and conditions of operation, which function to alter the primary current distribution.
3. Those aspects of the shape and design of the workpiece which do not *per se* control the primary current distribution, yet operate to alter it.
4. The relation of plating bath efficiency to the current density.

In most practical cases, the primary current distribution is the controlling factor in determining the metal distribution. When the primary current distribution is uniform or very nearly so, the cathode polarization will usually be equally uniform, and so will the metal distribution. When the primary current distribution is markedly non-uniform, the other classes of factors may operate either to improve or worsen the uniformity of metal distribution over what would be expected in their absence, but in general, they cannot completely overcome the overriding influence of the primary current distribution.

PRIMARY CURRENT DISTRIBUTION

The current distribution over an electrode in the absence of polarization and other disturbing factors at the electrode is called the primary current distribution. It is determined solely and completely by the geometry of the system which includes the shape and size of the electrodes, their conductivity, in the rare case when they are not equipotential surfaces, the spatial relationships of the electrodes to each other *and to the electrolyte boundaries*, the conductive nature of these boundaries, and the shape and location of any other conductive or nonconductive bodies in the electrolyte. In practice, such bodies would be either thieves or shields.

The general theory of the potential field is of such universal scientific interest that it attracted the attention of mathematical physicists as soon as the concept of a potential had appeared. The first application to electrolytes was made by Riemann[1] in 1855 and later by Weber[2], but it was not until the work of Kasper[3] in 1939–42 and later of Kronsbein[4] that any serious application of the theory was made to the art of electrodeposition. The subject is mathematically so complex that at the present time only the simplest cases of geometry can be considered, and even then, cathode polarization must be assumed to be either absent or uniform. Therefore it is not now of any practical use to discuss in detail the results of these theoretical investigations. However, a general grasp of the theory of the potential field and a few general rules arising from the mathematical work will be of immense help to the practical plater and will prevent many misconceptions and misinterpretations of experiments.

Whenever an electric potential, i.e., voltage is applied between two electrodes in an electrolyte, *every* point in the electrolyte assumes some potential intermediate between that of the two electrodes. Since the conductivity of the metal electrodes is usually several million times that of the electrolyte, it may be assumed with a slight margin of error that every point on the electrode surface is at the same potential, i.e., it is an equipotential surface. Similarly, equipotential surfaces will be found in the electrolyte surrounding each electrode. Close to one electrode, the equipotential surface will resemble that electrode in shape; however the shape will change as one moves farther away until at last the equipotential surface close to the second electrode assumes a shape more or less conforming to that electrode (Fig. 1). The shape of these equipotential surfaces in this and similar simple cases may be computed mathematically[3,5] and has been verified experimentally[6,7,29]. It is also possible to calculate or map experimentally the

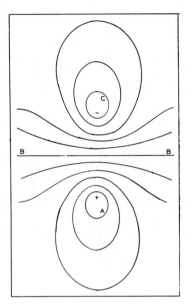

FIG. 1. Trace of equipotential surfaces in a tank of electrolyte with two line electrodes (+ and −).

path that the current follows (Fig. 2). In this plot, an equal amount of current, e.g., one ampere, is flowing along each of the electrolyte paths bounded by the solid lines, and some undefined current between the outer lines and the insulating boundaries.

In understanding the significance of these

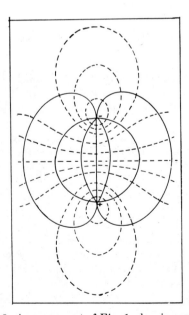

FIG. 2. Arrangement of Fig. 1, showing equipotential surface traces (dotted) and traces of surfaces of force (full lines).

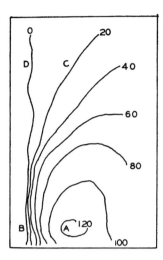

FIG. 3. Hypothetical contour map, illustrating analogy between contour lines and equipotential lines.

diagrams of equipotential lines, it may be helpful to compare them to the more familiar contour map used by geographers (Fig. 3), in which points of equal elevation are connected by lines. Figure 3 represents a hill and cliff along a shore (elevation zero), and each successive contour line is 20 ft higher. It is easy to see that over the equal distances AB and CD there is a considerable difference in the slope of the ground, and rainfall landing on the hill at A will race to the sea; whereas from C its movement will be much slower. In a very real sense, this is also a diagram of equipotential lines, and the closer they are crowded together, the steeper is the slope. In our diagrams of electric potentials, the current density, similar to the water flow, will be greatest where the equipotential lines are crowded most closely together.

Figure 1, in mapping terms, would be interpreted as a hill with a steep cliff in front and a gradual approach at the rear. A little distance away across rather level land is a spoonshaped quarry hole with a very steep declivity on the hill side of the hole, less steep on the other side.

It is a basic law of the potential field theory that the traces of the equipotential surfaces and surfaces of force in any plane form an orthogonal net, that is, they everywhere cross each other at right angles. Since the electrodes themselves are also equipotential surfaces, this means that the current enters or leaves the electrode at any point in a direction perpendicular to the surface at that point.

It is also fundamental in field theory that any equipotential surface may be replaced by a (perfect) conductor without disturbing the field. Similarly, the substitution of an insulating boundary for any line of force does not disturb the field. It is important to recognize that this is true only if the substitute surface *completely* replaces the original. Conversely, when *anything* cuts one of the lines of the net, the field is disturbed; moreover the effect of moving one insulating tank wall closer to the electrodes is shown in Fig. 4[6]. It follows from what has been said that the current distribution over a cylindrical (A of Fig. 1) and plane (B-B of Fig. 1) electrode is defined by the same plot, as is the distribution between wire (+, Fig. 1) and outside eccentric cylinder (C of Fig. 1)* Gilmont and Walton[8] utilized these concepts in the design of their modified Hull cell, which gives a linear current distribution over a flat electrode, the other electrode being curved, and placed along an equipotential surface.

It should be observed that in these diagrams we are actually representing a three-dimensional system in two dimensions, which means that to represent the actual system, the diagrams must be moved in a direction perpendicular to the paper between two insulating bound-

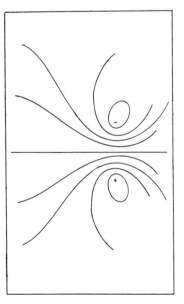

FIG. 4. Arrangement of Fig. 1, with electrodes displaced from tank center line.

*Mathematically this is not rigorously true in the practical case, because then the "cylinders" are not true circles, which occurs only in an unbounded (i.e., infinite) field.

aries. Practically, this means only that the electrodes extend from the surface of the solution to the bottom of the tank. What happens when these conditions are not met, as when the plane electrode stops short of the bank walls, is discussed by Hine[5]. Another interesting observation may be made from these diagrams. A single anode, if it is significantly separated from the cathode, will give the same current distribution as a row of anodes rather closer to the cathode, along plant *B-B* of Fig. 1, for instance.

The substitution of conducting for insulating walls has been well discussed by Kasper[3] and illustrated by Lukens[4b]. As would be expected, the effect is profound, and in the case of two isolated electrodes well away from the walls, as Lukens illustrated, results in a much more uniform distribution, front to back, over the electrodes. However, this is not a practical plating tank arrangement, and in most cases the benefit derived would be slight, if any, in the more conventional plating tank setup. The preference for insulated tanks is well justified on many other grounds.

THE CONTROL OF PRIMARY CURRENT DISTRIBUTION

Not much more can be said about the theory of current distribution without getting more involved in detail than is the intent of this chapter. Interested readers should consult the references cited. There are also numerous references[9,10,14-19] having a more immediately practical value, but unfortunately some authors[20,21,23-26] who are unaware of the basic theory become frequently confused and puzzled by their results and offer untenable explanations. Even a reader with only a fair grasp of the theory can, however, extract much of value from the data they present. It is the purpose of this section to present briefly a set of generalizations about the practical control of primary current distribution. Much that will be said is intuitively recognized by the experienced plater, but some ideas may be new to him.

Perfectly Uniform Current Distribution

There are basically only three possible arrangements giving perfectly uniform current distribution[3]: infinite parallel planes, infinite concentric cylinders, and concentric spheres. None of these arrangements has as yet been offered to any shop to be plated, but under the rules we have certain freedoms: (a) we may replace any line or surface of current flow with an insulating boundary, or (b) we may replace any equipotential surface with an electrode. Thus, we may box in our infinite parallel planes with three perpendicular insulating walls and an air surface to form the well-known Haring cell[22]. To work properly, the walls must be *exactly* perpendicular to the electrodes[3], consequently the Haring cell should always be placed on a truly horizontal surface (a precaution that has not been found mentioned in the literature). A plating rack full of closely spaced flat parts, equidistant between two rows of anodes, in a tank just big enough to hold the rack, is an arrangement about as similar to the Haring cell as the practical plater will probably ever see.

In the same way, we may eliminate a section of our infinite cylinders by means of insulating planes at right angles to the axis and obtain an arrangement for uniform primary current distribution. The central cylinder, often the anode, must be perfectly positioned except when its diameter is very small with respect to the outer cylinder. Thus it is best to use the smallest anode that will adequately carry the current, but in most cases this poses a problem because the cathode current density is perforce smaller than the anode current density, yet usually must be above some minimum value. Hammond[27] discusses some of the problems of plating gun barrels, a modified example of concentric cylinders. An interesting property of the concentric cylinder system is that the voltage drop between the electrodes at a given current density is dependent only on the *ratio* of their radii, not their absolute value, which may be mils or miles. As Kasper[3] points out, radial sections of concentric cylinders (the radii being insulating walls) also give perfect primary current distribution.

The case of concentric spheres, or radial sections thereof, which are equivalent, is very rare and will not be discussed.

Control of Nonuniform Current Distribution

It is well recognized that almost every practical plating arrangement results in a nonuniform current distribution. This is tolerable to a degree but often requires improvement to produce acceptable work. The ability to visualize, even if imperfectly, the potential field around the workpiece is a great aid in making favorable adjustments in the geometry of the setup. In doing this it is most important not to neglect the effect of the insulating boundaries (the tank members and the solution surface). Thus, a rack which may give good distribution over the workload when operated in a tank full of other racks will give a very different and less uniform distribution if plated alone or in a partially loaded tank.

An important problem in metal distribution, with which present theory is unable to deal quantitatively, is deposition into recesses, typified by an angle (Fig. 5). Theory does tell us that in a perfect angle, with perfectly sharp corners, the primary current density is infinite on the sharp external corners of the angle and zero on the internal corner. Thus it is theoretically impossible to deposit plate on the internal corner of a true (zero-fillet) angle (even one of 179 degrees) while the deposit on the external corners of zero radius must inevitably be burnt, statements which any practical plater will find easy to believe. Many authors have discussed this problem[3, 9, 13]. The primary current distribution is not influenced by the size of the angle, but only its shape, and the last two references cited deal with micro-angles (one outstanding case in which factors other than primary current distribution are controlling, as is ably discussed by Kardos[13]. A common misconception with respect to angles and recesses in general is that the farther away the anode, the better is the current distribution. This is based on the false premise that the current flow is in straight lines, and deducing from this, that when the anode is farther away, the distances from the anode to near and far points of the angle (or recess) are more nearly equal. However, as can be seen from Fig. 5, the anode AB could be moved up as close as position CD without affecting the current distribution.

The cathode or cathodes should be arranged so as to present as simple a surface over-all as possible, whether plane or curved. The anodes should be arranged to conform, not necessarily to the shape of the cathode assembly, but to the estimated form of that equipotential surface which in conjunction with the boundaries will tend to give the nearest approximation to straight-line current flow. For instance, in plating into a large angle (Fig. 6), the best anode arrangement is not a conforming angle (Fig. 6a). A better arrangement is a wedge (Fig. 6b); still better may be a single wire (Fig. 6c) or its equivalent, a distant anode and insulating masks A-B forming a slot (Fig. 6d).

The potential surfaces in Fig. 5 have been drawn from pure conjecture. In such a problem, where the cost was justified, they could be actually mapped, probably most easily by the technique of Lukens[6], or with more difficulty

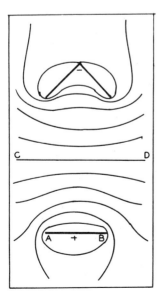

FIG. 5. Equipotential lines between plane anode A-B and angle cathode in tank of electrolyte.

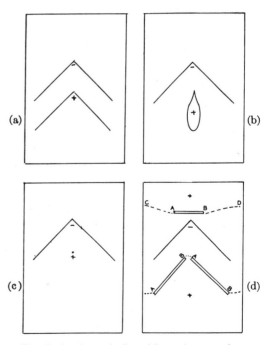

FIG. 6. Angle cathode with various anode arrangements: (a) conforming angle (b) tapered anode (c) point (wire) anode (d) virtual anodes formed by insulating marks A-B.

by the analog technique[7,39], with conductive paper, using various shapes and sizes of insulating shields. The use of shields is in this case preferable to a "thief" or "robber" at the outer angle corners because although the latter will protect the corners from overplating, they will still further decrease the plate deposited in the interior of the angle. Figure 6d also shows the use of an insulating shield AB to protect the outside corner. In this case, a thief could do as well, but is more complicated and is wasteful of metal. In visualizing the probable effect of shields, it is convenient to remember that the equipotential surface formed by the slot or openings left by the shield becomes the effective anode as AC, BD and the outer broken lines of Fig. 6d.

Rousselot[42] has recently published the results of an experimental evaluation of the effects of shields, thieves and bipolar conductors on the field, using conductive paper for a two-dimensional analysis of simple configurations. He shows that a properly placed shield is as effective as a thief, whereas a thief may rob 25 to 50 per cent of the current, or more.

Kronsbein[11] made quantitative measurements of the effect of fillet radius in improving deposit distribution on the interior of an angle and concluded that as a general rule no further improvement was afforded by increasing the fillet radius beyond 0.010 in./in. of angle shank. No similar rule is yet available for external sharp edges or corners, but here it is obvious that the only upper limit of desirable radius is that set by other functional considerations.

Hine[5] and co-workers have shown that when two parallel electrodes of equal lengths are centered between and perpendicular to two insulating side walls at varying distances from them, with distant back walls, the proportioning of current between the face and back of the electrodes can be calculated. It is, as would be expected, a function of the degree to which the electrodes fill the distance between the walls. Their rounded off figures show:

Cathode Width, as Per Cent of Wall-to-Wall Distance	Primary Current Distribution Per Cent of Total Current on Face
95	97.5
84	92
72	86.5
58	79
50	75

Each automobile bumper in a rack of bumpers being plated back-to-back is an approximation of this arrangement, since between each two bumpers is an equipotential plane equivalent to an insulating boundary in the sense of the above example. The spacing of the bumpers than determines how much plate is deposited on the back.

Sometimes irregularities in current distribution may be overcome by rotating or oscillating the anode or cathode; thus in any problem of plating a cylinder, if it can be rotated about its axis, inequalities of anode spacing and hence of distribution *around the circumference* can be overcome. Inequalities along the axis due to tapering sections or high resistance in the smaller electrode[28] have to be dealt with by other means. This may require purposeful tapering of one of the electrodes. In the case where the larger electrode, because of thin wall section or material of construction, also has appreciable voltage drop along its length, then by making the anode connection at one end and the cathode connection at the opposite end, and choosing the anode cross section to give the same voltage drop per foot as the cathode, it is sometimes possible to cancel out the effect of the electrode resistance. This has been done in at least one instance[29]. In any such arrangement the anode must be insoluble or must not change in cross section enough to upset the calculations.

With regard to anode-cathode spacing, there is usually little to be gained by increasing this beyond the 8 to 16 in. normally found in most commercial installations. There is, however, one consideration, frequently overlooked, from which arises the often-heard complaint that work on the bottom or top of a rack is burnt. This condition can arise from having too much solution above and below the rack. The top of the work should usually not be more than 2 to 3 in. below the surface *unless* there is good reason to want a concentration of current at this point. The bottom of the work presents a more ticklish problem because many solutions are operated with a layer of sludge at the bottom which the plater does not wish to disturb; hence work is usually kept 8 to 12 in., and frequently more, off the tank bottom. A fairly simple way out of this difficulty is to erect insulating walls designed into the tank, or built up with plastic or loose brick, as shown in Fig. 7. The walls should be as close as practicable to the plane of the cathode. This will effectively reduce the amount of current received by the bottom of the rack without aggravating the sludge problem. The common remedy of using anodes shorter than the work is

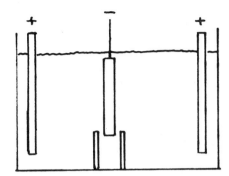

FIG. 7. Shielding bottom of workpiece from excessive current.

less effective and will not work at all in many cases where the anode-cathode distance may be great enough to allow the formation of an equipotential field in the shape of a plane, which is equivalent to using full-length anodes much closer to the work.

The number of anodes per foot of tank is not usually significant, since the equipotential surface a little distance away becomes practically a plane and is equivalent to a sheet anode. However, if the cathodes do not cover the entire length of the tank as they should, the situation can be improved to a certain extent by making the effective width of anodes somewhat shorter than the cathode width. How much shorter can only be determined by experiment, and the anode-cathode distance is important as explained above.

The arrangement of work on a rack, i.e., closeness of approach of the parts and of racks in the tank will be discussed later because considerations other than primary current distribution may apply.

SECONDARY CURRENT DISTRIBUTION

The distortion of the primary current distribution by the second and third classes of factors mentioned in the opening of this chapter results in a secondary current distribution, which is really what does the work of depositing metal. It is the intent of this chapter merely to show how metal distribution may be controlled; hence many of these factors will be mentioned only briefly since many are not completely within the control of the plater, although inherent in the selection of the bath and operating conditions.

The effects of the second class are included under the term "polarization," which is any effect of the passage of current that results in increased resistance or in phenomena which have the same effect, insofar as current distribution is concerned. Thus, for example, the composition of the cathode film is changed by the action of the current, reducing the concentration of conducting ions and causing a rise in resistance (concentration polarization). An anode may become coated with a more resistant film (chemical polarization) or gas discharge may decrease the effective cross section of the electrical path (gas polarization). Provided the anodes are uniformly polarized, there is no effect on the cathode current distribution; however nonuniform polarization will have an effect which, if desired, may occasionally be overcome by altering the anode arrangement to provide more uniform anode current distribution; by altering the bath composition or conditions, (including agitation) to reduce anode polarization, or to make it more uniform. Gas polarization may be reduced by agitation or reorientation of the workpiece. Agitation will also reduce concentration polarization, but this is not always desirable since it may adversely affect throwing power (q.v.).

With reference to the cathode, the electrochemical properties which determine the secondary current distribution are the cathode polarization and the conductivity of the solution. The primary current distribution between two points, 1 and 2, on a cathode (Fig. 8) is given by the equation

$$\frac{i_1}{i_2} = \frac{d_2}{d_1}$$

where i_1 and i_2 are the current densities at points 1 and 2, respectively, at distances d_1 and d_2 from the anode *measured along the lines of force traversed by the current.* The secondary current distribution is generally accepted[35,39,40] to be

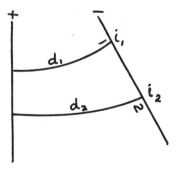

FIG. 8. Primary current distribution on a cathode.

given by the equation

$$\frac{i_1'}{i_2'} = \frac{d_2 + k\,\dfrac{dE}{dI}}{d_1 + k\,\dfrac{dE}{dI}}$$

where k is the solution conductivity and dE/dI is the slope of the cathode polarization curve between the current densities i_1 and i_2. The second terms in the numerator and denominator are identical and have the dimensions of a length. The combined effect of polarization and conductivity is therefore equivalent to *adding* an equal solution length to the path lengths d_1 and d_2. It will thus be seen that the secondary current distribution is *always* more or less of an improvement over the primary distribution.* It is also to be noted that the polarization and conductivity are not inherent properties of the solution alone, but are dependent as well on other outside factors such as temperature, agitation and current density. Furthermore, concentration polarization can alter the conductivity of the solution in the vicinity of the cathode. The degree of improvement of secondary over primary current distribution becomes greater as the value of the second numerator and denominator terms increases, that is, there is relatively more improvement in a solution of high conductivity, operating under conditions where the cathode polarization is increasing rapidly with curent density. The latter condition is usually obtained only at rather low current densities.

The third group of factors affecting metal distribution comprises unusual geometry or metallic nature of the workpiece; thus cup-shaped workpieces or the presence of blind holes may require special positioning or even reorientation during plating. Chamfering or rounding the sharp edges of blind holes will improve the primary current distribution to a certain extent. When the nature of the cathode metal electrolyte system is such that the depositing metal has a high overvoltage, recesses may not plate at all without special procedures such as striking initially at high current density until the recess is covered, or striking with another metal having a lower overvoltage, or using auxiliary or bipolar electrodes (Chapter 21).

It is thus seen that the secondary current distribution is controlled by a very complicated set of variables, and the effect of any one of these cannot be predicted in any general way.

METAL DISTRIBUTION

The rate of metal deposition at any point on a cathode is determined by the current density existing at that point in the secondary current distribution, and by the cathode efficiency of the bath at that current density. Since the secondary current distribution is unknown, the metal distribution also cannot be predicted. With a sufficient knowledge of a given system, however, the *direction* of the effect of a certain change in variables can often be stated, so that an improvement in metal distribution can be effected. It should still be remembered that the primary current distribution, which can be manipulated in an empirical sort of way, usually is the controlling influence on metal distribution. Control of the secondary current distribution is less effective and is complicated by the fact that the controlling variables are completely interrelated, e.g., a temperature change affects the resistivity and viscosity of the cathode film, the ionization constant of complex compounds, the hydration and mobility of ions, and the cathode efficiency, which variables may have conflicting effects on metal distribution and the net effect of their interaction is unpredictable.

Throwing Power

"Throwing power" is a term invented to give a name to a phenomenon. It was observed that there was usually a disparity between the primary current distribution and the metal distribution, and that the degree of disparity appeared to vary from solution to solution. In their classic paper on the subject, Haring and Blum[22] defined throwing power as a number derived from the deviation of the actual metal distribution from the primary current distribution. They considered it a property of the solution, which it is not,* and measured it in the now familiar throwing power box. Numerous researches[23,25,30,31,32,34] followed along the same

---

*This is true so long as the slope of the polarization-current density curve is not negative. No such case has yet been reported.

*The *concept* of throwing power may properly be considered as describing a property of a solution, but any numerical value assigned to it has no meaning outside the apparatus in which it was measured. The *throwing power number*, however defined, is not a useful property as is conductivity, for example, in that no calculations can be based upon it in any practical case.

lines, and many different ways of expressing throwing power as a number were developed. As practical tools these numbers, however defined, are useless since they apply only to the particular shape and size of box used in the investigation, and there are other limitations and objections[3b,33,35,36]. This should now be obvious from what has been said in the preceding section, because throwing power in the case of a practical plating job actually combines the effects of the distorting variables producing secondary current distribution from the primary with the current density-cathode efficiency relationship. In so doing, it absorbs all the unpredictable relationships of all these variables and their effect on the particular electric field under consideration.

It has been previously stated that the secondary current distribution is always somewhat better than the primary distribution. Whether the metal distribution is also better, i.e., positive throwing power as defined by Haring and Blum, depends entirely on the slope of the cathode efficiency-current density curve between the current densities in question. If the slope is negative (efficiency decreasing with increasing current density), the metal distribution will be even better than the secondary current distribution. If the slope is small and positive, the matal distribution will not be as good as the secondary current distribution, and may or may not be better than the primary current distribution. When the slope is strongly positive, e.g., a chromium bath, the metal distribution is virtually certain to be poorer than the primary current distribution.

About all that is practically useful in the throwing power concept is what every plater knows, namely, that the chromic acid bath has miserable "throw"; other acid baths can be expected to throw somewhat better but still not too well; a cyanide bath usually throws quite well, alkaline tin and tin alloy baths are about the best.

Illustrating the effect of geometry on throwing power, Graham[37] and co-workers first made the surprising observation that in micro-pores, acid copper, nickel and even chromium could fill the pores completely, but that they were never filled by deposits from cyanide baths. This work was confirmed and extended by Reinhard[38] and Foulke and Kardos[12] and elegantly explained by Kardos[13]. The latter showed that when the dimensions of a recess approached the order of magnitude of the thickness of the cathode film, some unusual effects will be obtained, including even a complete reversal of the usual order of throwing power of various types of bath.

Control of Throwing Power

After all that has been said, it may seem strange to suggest control of throwing power. Although no rules are generally applicable, it is still possible to exert a degree of control over throwing power, but in an empirical fashion only. Such measures should be undertaken only after all obvious and practical methods of improving the primary current distribution have been exhausted to no avail. In spite of many theoretical objections to it, the cavity scale of Pan[31] is probably as practical a tool as any. Using a sample of the process bath in an experimental tank, the effect of agitation, temperature and bath composition variables can be studied. A change that produces a distinct improvement, as measured on the cavity scale, will *probably* produce a somewhat similar improvement in production. The cavity scale employed should be one of the coarser scales suggested by Pan, as the finer scales tend to show capricious differences that are misleading. To obtain a meaningful improvement in production runs, the improvement in throwing power must be very marked. Also, it is important to note that for this purpose, visual observation of the throw into the cavities at varying depths is meaningless. The actual thickness must be measured; otherwise the scale is only a measure of covering power (see below).

The following example is actually a matter of controlling the primary current distribution rather than the throwing power, but it well illustrates how dependent throwing power is on the geometry, if the throwing power is measured on a particular workpiece in a tank instead of in a box. Shaefer and Pochapsky[15] have shown the effect of the spacing of parts over the rack on the current distribution in different solutions. If parts with central recesses are widely spaced on the rack in a solution of poor throwing power such as nickel, the edges will draw excessive current, thus distorting the electric field somewhat as shown in Fig. 9a. By moving the pieces closer together (Fig. 9b), the equipotential lines become more nearly planar, favoring deposition of more metal in the cavities than the arrangement of Fig. 9a.

An interesting illustration of the effect of these factors in a zinc plating solution was developed

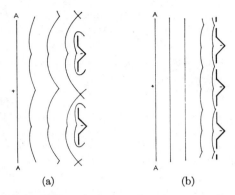

Fig. 9. Effect of spacing of parts on a rack on the equipotential lines.

by Swalheim and Jackson[43]. Using panels plated in a Hull cell, the expected thickness was calculated, and the actual thickness measured. The results were expressed in terms of the apparent solution efficiency, although they actually combine the effects of the current density/efficiency curve, and the current density/polarization curve. The figures obtained can be summarized:

Theoretical C.D.		Apparent Efficiency
8 A/dm^2	(80 A/ft^2)	48%
4 A/dm^2	(40 A/ft^2)	70%
2 A/dm^2	(20 A/ft^2)	100%
0.5 A/dm^2	(5 A/ft^2)	137%

It is important to point out that this effect can be highly variable. Aside from the effects of variation in solution composition, temperature and agitation, the final steady state of the system, and therefore the magnitude of the change in plate distribution, is a function of the average current density. Had the panels been run at different applied currents, the results would have been somewhat different in magnitude, although, in general, the results would be the same.

Experienced platers often manipulate solution compositions to alter either the polarization or efficiency relationships to improve throwing power. Thus in cyanide zincs, it is quite possible to improve throwing power at the expense of efficiency, or improve efficiency at the expense of throwing power.

Even in baths with high efficiencies, and a relatively flat current density/efficiency curve, control of the current density/polarization curve can result in an improvement of throwing power. Many of the newer bright plating systems take advantage of this approach. Some sacrifice in the maximum allowable current density generally will have to be made in order to accomplish this, however.

Barrel plating represents a special case of controlled geometry. Rather than measuring throwing power, the spread of observed thicknesses around the nominal thickness is considered. Variables include the dimension of the barrel, the speed of rotation of the barrel, the method of contacting the parts, and the composition characteristics of the plating solution. In work done for the American Electroplaters Society research program, Craig and Harr established the effect of some of these variables.* The spread in general follows the general bell curve for distribution. However, the spread factors for a 6 × 6 in. barrel average 55% larger than those for a 4 × 6 in. barrel at the same rotational speed. Thus it appears that the spread factor is almost directly proportional to the barrel diameter.

The effect of rotational speed is also of course related to the barrel geometry, since it is the combination of barrel geometry and rotational speed that determines the mixing of the load, and therefore the spread factor. For loads that slide easily and mix properly, the spread is approximately inversely proportional to the square root of the rotational speed. Parts which tangle or nest can produce gross deviations in the spread factor, since mixing will be nonuniform. This problem often can be corrected by changing the method of contacting the load to produce better mixing; e.g., changing from a dangler contact to a bar contact.

Covering Power

One aspect of metal distribution has to do with the ability of a bath to deposit metal at very low current densities, as in deep cavities or holes. This is termed "covering power" and is best demonstrated by the cavity scale of Pan[31]. The covering power is governed first by the bath composition, which determines the decomposition potential, i.e., the lowest voltage at which metal can be deposited from the solution, and secondly, by the cathode metal and its surface condition, which determines the overvoltage of

*Craig and Harr, AES Research Projects #34 and #44. Reports available from AES, 1201 Louisiana Ave., Winter Park, Fla. 32789.

metal deposition, i.e., the voltage in excess of the decomposition voltage necessary to produce discernible deposit.

This is quite different from throwing power, which determines *how thick* the deposit in a recess will be; whereas covering power is concerned simply with whether any deposit at all can be obtained in the recess. In using the cavity scale to measure relative covering power under different conditions, a simple visual observation suffices to determine the deepest cavity in which a deposit is obtained.

Many metals can be deposited at extremely low current densities, which means that in these cases, the decomposition potential is readily attained and the metal overvoltage is insignificant.

Nickel, and more especially chromium, have appreciable decomposition potentials; furthermore chromium requires a rather high minimum current density to produce any deposit, under the most favorable conditions. At least 10 to 12 asf are required to deposit chromium, and the threshold current density is usually much higher. Nickel and chromium are also more apt to be sensitive to the basis metal. Chromium usually covers nickel better than it does brass, although the preparation of the brass surface is an important factor; the coverage on buffed and conventionally cleaned brass is much poorer than over buffed brass that has simply been degreased. As another example in the author's experience, lead covers tin much better than it does copper. A cavity in a copper casting that could not be covered at all if plated directly in lead readily accepted a deposit after the piece had been plated in alkaline tin, which completely coated the cavity. The lead "followed" the tin, and a deposit of reasonable thickness could be obtained. A well-known case illustrating this phenomenon is the reluctance of a cast-iron surface to accept a deposit of zinc from the usual cyanide bath; whereas the same casting can readily be plated in other cyanide baths or in an acid zinc bath. This is because the hydrogen overvoltage on cast iron is lower than the deposition potential of zinc from a cyanide bath; thus under normal conditions only hydrogen is obtained. Special measures[41] can be taken to enable a zinc deposit to be obtained under such conditions. These involve either raising the hydrogen overvoltage, as by omitting the acid-pickling operation in the preplating treatment, or altering the bath slightly so that a good initial deposit of zinc can be obtained by striking at high voltage, i.e., above the deposition potential of zinc, after which plating can proceed normally.

The discussion thus far has been based on theory, which presupposes that all conditions are ideal. Practically, this means that all contacts in the electrical system are perfect; that the tank bus and rack members are adequate to carry the current without appreciable voltage drop; and that the electrolyte is electrically isotropic, that is, the conductivity is equal everywhere in the solution, which requires that there be no stratification with respect either to temperature or concentration. A final and very important condition is that all the current must pass through the solution directly from one electrode to the other without following other accidental haphazard paths, or "leaking" out of the system altogether, i.e., there must be no stray currents. Since these conditions will normally be met in good plating practice, one may justifiably apply the precepts of this chapter to a practical problem with some confidence.

Stray Currents

It is appropriate at this point to consider the effects of one form of deviation from the ideal conditions set forth above, namely, the presence of stray current.

Two types of stray current should be recognized. In the first, a portion of the current takes some path other than the intended one through the solution, and in the most common case this is through the tank walls or bottom when an unlined tank is used, or when the lining is perforated in several places. Other than the obvious fact that the desired current distribution is upset, there are side effects which may sometimes be more serious. Thus, the tank becomes in effect a series or intermediate electrode and will tend to plate up on areas near the anode, whereas the areas near the work act as anodes. The plating build-up may eventually short the tank out solidly to one or more anodes and greatly increase the magnitude of the stray current. The seriousness of the anodic tank areas, aside from the effect on current distribution, which will depend on the nature of the solution, will usually be detrimental. Another stray current phenomenon of this class is associated with the placing of metallic elements such as heating coils or thermostat bulbs, etc., in the path of the current where they may act as intermediate anodes with similar results.

Parts which fall off racks and are allowed to remain in the bottom of the tank represent a very serious example of this effect. They not only alter the current distribution, but may also be a massive source of metallic contamination. Such parts become bipolar anodes, and the portion of the part nearest the work becomes anodic. This portion may rapidly dissolve, introducing damaging contamination. Such is the case for zinc based die castings or brass parts in a nickel tank, lead parts in a copper tank, etc. Tanks should be checked at least daily and all such parts removed.

The second class of stray currents involves the leakage of current from the tank to the floor, adjacent tanks or to auxiliary equipment installed in or on the tank such as ventilating duct, piping to coils, electrical conduit, etc. Once again, aside from the effect of this lost current on the current distribution to the work, there may be very serious consequences. Although the voltage driving such stray currents may be low, they have an insidious way of meandering through the shop and may cause serious accelerated corrosion in the most unexpected places, often remote from the source of the current. Consequently, all electrified tanks should be installed so as to be completely insulated from ground, and the integrity of this insulation should be checked regularly. A stray current alarm is described in Chapter 37, Auxiliaries, capable of detecting stray currents of this class, will not indicate internal stray currents of the first class. AC shorts can be particularly damaging since AC acts as a depolarizer and permits or accelerates corrosion under conditions normally safe and acceptable. They are extremely difficult to trace, since, in some cases, they may be induced simply by close proximity to a power supply line. Typical examples which illustrate the problem include the following.

1. Two titanium coils, used in alkaline solution, were attacked and perforated as a result of an AC leak from a defective solenoid on a temperature control valve. Time to failure—4 weeks.
2. A titanium coil in a sulfuric acid solution was destroyed by an AC short induced by laying the steam line immediately next to a 440 V cable supplying a 6000 A rectifier. Time to failure—6 weeks.
3. A stainless steel tank holding a chromating solution was twice perforated by an AC short from a hoist motor. Time to first failure—unknown. After repair by welding, time to failure the second time—12 days.

Stray currents of either kind may also cause plating defects that can be very baffling to trace. Sometimes these currents have such a direction and strength that the work, or some portion of it becomes anodic during part of the plating cycle. This can cause these areas to have an etched, dark or rough appearance, or in the case of bright plating, to be cloudy or dull. It is a particularly common problem in nickel and chromium plating. Hazing, failure to cover with chromium and even peeling can result. Careful attention to the placement of the anodes at the entry and exit stations is important in avoiding this effect. In extreme cases, live entry and/or exit may be necessary.

References

1. Riemann, B., *Pogg. Ann.*, **95**, (1855).
2. Weber, H., *Crelles Journal fur die reine und angew. Mathemetik*, **75**, 95–103 (1873); **76**, 7–20 (1873).
3. Kasper, C., (a) *Monthly Rev. Am. Electroplaters' Soc.*, **26**, 11–26, 91–109 (1939); (b) *Trans. Electrochem. Soc.*, **77**, 353–383 (1940); (c) *Ibid* **78**, 131–161 (1941); (d) *Ibid*. **82**, 153–185 (1942).
4. Kronsbein, J., *J. London Math. Soc.*, **17**, 152–157 (1942); *Proc. London Math. Soc.*, **49**, (2) 260–281 (1947); *Plating*, **36**, 851–854 (1950); **39**, 165–170 (1952).
5. Hine, F., Yoshizawa, S., and Okada, S., *Trans. Electrochem. Soc.*, **103**, 186–193 (1956).
6. Lukens, H. S., *Trans. Electrochem. Soc.*, **67**, 29–36 (1935); *Proc. Am. Electroplaters' Soc.*, **23**, 186–197 (1935).
7. Kinney, G. F., and Festa, J. V., *Plating*, **41**, 380–384 (1954).
8. Gilmont, R., and Walton, R. F., *J. Electrochem. Soc.*, **103**, 549–552 (1956); U. S. Pat. 2,913,375 (Nov. 17, 1959); and *Proc. Am. Electroplaters' Soc.*, **43**, 239–245 (1956).
9. Kronsbein, J., and Morton, L. C., *Proc. Am. Electroplaters' Soc.*, **36**, 229–239 (1949).
10. Kronsbein, J., *Proc. Am. Electroplaters' Soc.*, **37**, 279–284 (1950).
11. Kronsbein, J., *Plating*, **40**, 898–901 (1953).
12. Foulke, D. G., and Kardos, O., *Proc Am. Electroplaters' Soc.*, **43**, 171–180 (1956).
13. Kardos, O., *Proc. Am. Electroplaters' Soc.*, **43**, 181–194 (1956).

14. Schaefer, R. A., and Mohler, J. B., *Trans. Electrochem. Soc.*, **85**, 431–440 (1944).
15. Schaefer, R. A., and Pochapsky, H., *Proc. Am. Electroplaters' Soc.*, **38**, 155–166 (1951).
16. Krosbein, J., *J. Electrodepositors' Tech. Soc.*, **17**, 83–104 (1942).
17. Gardam, G. E., *Proc. Am. Electroplaters' Soc.*, **27**, 16–23 (1939).
18. Watson, S. A., and Edwards, J., *Trans. Inst. Met. Fin.*, **34**, 167–198 (1957).
19. Barnartt, S., *Trans. Electrochem. Soc*, **98**, 311–317 (1951).
20. Tope, N. A., *J. Electrodepositors' Tech. Soc.*, **7**, 83–90 (1932).
21. Burt-Gerrans, J. T., *Proc. Am. Electroplaters' Soc.*, **27**, 39–43 (1939).
22. Haring, H. E., and Blum, W., *Trans. Electrochem. Soc.*, **44**, 313–345 (1923).
23. Heatley, A. H., *Trans. Electrochem. Soc.*, **44**, 283–303 (1923).
24. Engel, A. V., *Nature*, **146**, 66 (1940).
25. Field, S., *J. Electrodepositors' Tech. Soc.*, **7**, 83–90 (1932).
26. Field, S., and Weill, A. D., "Electroplating," 4th Ed., 1943.
27. Hammond, R. A. F., *Trans. Inst. Met. Fin.*, **34**, 83–105 (1957).
28. Weisselberg, A., and Staff, *Trans. Electrochem. Soc.*, **90**, 235–245 (1946).
29. Unpublished data of author's laboratory.
30. Pan, L. C., *Trans. Electrochem. Soc.*, **58**, 423–434 (1930).
31. Pan, L. C., *Metal Ind. (N.Y.)*, **28**, 271–274 (1930).
32. Pinner, R., *Electroplating*, **7**, 9–15, 49–53, 59 (1954).
33. Gardam, G. E., *J. Electrodepositors' Tech. Soc.*, **25**, 77–81 (1950).
34. Jelinek, R. V., and David, H. F., *J. Electrochem. Soc.*, **104**, 279–281 (1957).
35. Hoar, T. P., and Agar, J. N., *Disc. Faraday Soc.*, **1**, 159–162 (1947).
36. Schlotter, M., and Korpiun, J., *Trans. Electrochem. Soc.*, **62**, 129 (1932).
37. Graham, A. K., Anderson, E. A., Pinkerton, H. L., and Reinhard, C. E., *Plating*, **36**, 702–709 (1949).
38. Reinhard, C. E., *Proc. Am. Electroplaters' Soc.*, **37**, 171–183 (1950).
39. Rousselot, R. H., *Metal Finishing*, **57**, No. 10, 56–61 (1959).
40. Tucker, W. M., and Flint, R. L., *Trans. Electrochem. Soc.*, **88**, 335–358 (1945).
41. Diggin, M. D., *Metal Finishing*, **41**, 277–281 (1943).
42. Rousselot, R. H., *Metal Finishing*, **59**, No. 3, 57–63 (1961).
43. Jackson, J. L. and Swalheim, D. A. *Hull Cell Tests for Quality Plating* Illustrated Lecture, American Electroplaters Society.

16. ELECTROFORMING

Dr. George A. DiBari

Senior Program Manager, The International Nickel Company, Inc., New York, N.Y.

ELECTROFORMING DEFINED

Electroforming is the fabrication of simple and complex shaped articles by means of the electroplating process. The basic steps in electroforming are the following: a suitable mandrel is obtained and prepared for electroplating; the mandrel is placed in the appropriate electroplating solution and metal is deposited upon the mandrel by electrolysis; when the required thickness of metal has been applied, the metal-covered mandrel is removed from the solution; the mandrel is separated from the electrodeposited metal. The electroform is a separate, free-standing entity composed essentially of electrodeposited metal. These basic steps are implicit in the definition adopted by ASTM Committee B8, namely "electroforming is the production or reproduction of articles by electrodeposition upon a mandrel or mold that is subsequently separated from the deposit."[1]

The definition stresses that the mandrel or mold is removed from the electroformed article, thus excluding decorative and engineering applications of electroplating. In decorative applications, an adherent metallic coating is applied to an existing article and becomes a permanent part of it to improve its appearance and corrosion resistance. In electroplating for engineering purposes, adherent metallic coatings are applied to an existing article to improve corrosion resistance, to reduce surface wear, to render a surface solderable, to build up worn or undersized parts for salvage, to modify magnetic properties and for other purposes. Electroforming is concerned with the production of parts, the electrodeposited metal must be sufficiently non-adherent to permit separation from the mandrel and, in the most successful examples, the articles produced by electroforming are unique.

ORIGINS

Electroforming originated amidst the excitement generated by the discovery of electricity in the early 19th century. The discovery led to scientific studies of electrochemical phenomena throughout the world. One critical development was the construction of a cell for producing direct current from chemical sources, by Alexander Volta in March 1800. Before the end of that year, Carlisle and Nicholson described the first copper deposits obtained by electrolysis of an aqueous solution using the voltaic cell as the current source. Before the middle of the century, the metals—copper, lead, silver, tin, mercury, cobalt, arsenic, platinum, gold, nickel, iron and mixtures of iron and copper—had reportedly been produced electrolytically in the form of powders and coarse crystals of unknown purity.

Professor B. S. Jacobi of the Academy of Sciences, St. Petersburg, Russia, is credited with discovering electroforming during his investigations on galvanic cells. One of these cells consisted of a copper cylinder filled with copper sulfate, a semipermeable membrane and a zinc electrode immersed in a sodium chloride solution. He observed that the copper cylinder became covered with a dense, non-adherent deposit of copper. When he removed the deposit, he saw that the scratches on the cylinder were reproduced on the surface of the copper deposit. Early in 1837, Jacobi used an engraved copper printing plate as the copper cathode in a copper sulfate solution and electrodeposited copper upon it. This experiment was only partially successful, because he had difficulty separating the electrodeposit from the engraved printing plate. Where it could be separated, the deposit had accurately reproduced the details engraved on the original printing plate. Jacobi continued his work and officially described his results to the Academy of Sciences, October 4, 1838, in the following way: "In the experiment with an engraved plate covered by a very thin layer of vegetable oil, it was possible to produce another copper plate on which even the slightest imprints on the original plate were reproduced with the highest accuracy."[5, 6]

In 1842, Professor R. Bottger of Germany reported that he had successfully electroformed articles of nickel using the double-salt solution of nickel ammonium sulfate which he had formulated and which he believed was the most suitable one for electrodepositing nickel. During the latter half of the 19th century electroforming with iron was also investigated. In 1869, iron replaced copper electroforming at the Russian Treasury Printing Press, where it was used for making engraved plates for printing banknotes. The three metals most widely used today—copper, nickel and iron—were, thus, electrodeposited and applied to the production of electroformed articles before 1870.

Following its discovery, electroforming was immediately applied to the reproduction of art objects, such as bas-reliefs, sculptures and statues, and to the duplication of engraved plates for the printing of money. The capability of the electroforming process to reproduce minute surface detail with great fidelity was the main reason these particular applications were the first to arise. Modern applications of electroforming are diverse and will be considered in detail later.

WHEN TO SPECIFY ELECTROFORMING

The question of when to specify electroforming as the method of producing a part is an important one for the designer or manufacturing engineer.[16] The answer will depend on the extent to which the designer or engineer can take advantage of the capabilities of the process. Consider the following capabilities:

1. The process can reproduce fine surface detail with great accuracy, as mentioned. An excellent example of this is in the production of electroformed nickel stampers (molds) for pressing phonograph records, where the accuracy of detail must be maintained through several generations of positive and negative replications of the original sound track. The accuracy of reproduction is within a fraction of a micrometer and this degree of accuracy makes possible the high quality stereophonic recordings available today. An extension of this capability is in the duplication of complex surface finishes. Bright, semi-bright, matte, smooth, brushed, roughened surfaces and combinations of these can be reproduced without the need for machining or polishing individual components after fabrication. Wood-grains, leather patterns and other textures can be incorporated into molds and dies for plastics, zinc and other materials—natural textures impossible to duplicate as faithfully as by electroforming. The combination of modern photolithographic methods of generating patterns with the electroforming process makes it possible to reproduce flat parts with extreme precision and fineness of detail, including giving edges and walls of holes predetermined angles.[113] The ability to reproduce fine detail makes the production of flexible printed circuitry entirely feasible.[76]

2. Parts can be reproduced in quantity with a very high order of dimensional accuracy. The accuracy attainable will depend on the nature of the mandrel material and the accuracy with which it can be machined—usually within about 0.004 μm. Once the mandrel is prepared, all parts produced from it will be dimensionally accurate.

3. The mechanical and physical properties of the electroform can be closely controlled by selecting the metal to be used, and by adjusting the composition of the electroplating solution and the conditions of deposition. Given the large number of metals, alloys and materials that can be electrodeposited, the possibilities are broad enough to satisfy most property requirements. Composites of various metals can be created; for example, radar wave guides of copper and nickel with inner linings of silver or gold have been fabricated by electroforming.

4. There is virtually no limit to the size of the object that can be electroformed. Nickel foil 4 μm thick is produced on a continuous basis, as well as textile printing screens up to 6 m long.

5. Shapes can be made and reproduced that are not possible by any other method of fabrication; for example, a seamless radar wave guide with two right-angle bends and with the interior made to close dimensional tolerances and with

high surface finish. Seamless cylindrical belts, also, fall into this category.

6. Electroforming is applicable to the making of single pieces or large production runs. In the latter case, it is possible to start with one master and build up in several generations a number of successively negative and positive electroforms. As the number of positives produced increases, these are used to make more negatives, thus increasing the production rate until the desired capacity is reached, after which nothing but negatives are produced. The practical limit to the maximum number of pieces that can be produced by electroforming depends on a number of factors and is difficult to define with certainty.

7. In the specific case of molds and dies for making plastics, zinc and glass parts, electroforming can provide tooling with resistance to corrosion, erosion and abrasion; with good heat conductivity and precise parting lines to minimize and eliminate flashing; and with high wear-resistance over long production runs.[27]

The engineer should also be familiar with the limitations of the electroforming process. The cost of an electroformed article may be relatively high if the same article can be mass-produced by a completely automated process. In the case of shapes, surface finishes, and fineness of detail that cannot be reproduced by any other means, cost becomes a secondary consideration. In the case of tooling, electroforming can be more versatile than, and cost-competitive with, pantograph machining, hobbing, chemical and electrochemical machining and other die-making techniques. The time to produce an individual piece by electroforming may be relatively long if wall thickness is great. There are limitations in design; for example, sharp angles, corners and very deep, narrow recesses may cause problems.[45] It is difficult to achieve great or sudden changes in wall thickness by electroforming. Mandrels must be handled with care since defects caused by accidental damage will be reproduced in the electroform.

The engineer should, therefore, specify electroforming whenever:

1. The difficulty and, hence, the cost of producing the object by mechanical means is unusually high.
2. Unusual mechanical and physical properties are required in the finished piece.
3. Extremely close dimensional tolerances must be held on internal dimensions and on surfaces of irregular contour.
4. Very fine reproduction of detail and complex combinations of surface finish are required.
5. The part cannot be made by other available methods.

MANDREL TYPES AND MATERIALS

The success of the electroforming process will depend on the type of mandrel chosen and the material used in its fabrication.

Mandrels may be classified as conductors or nonconductors of electricity, and each of these may be permanent, semi-permanent or expendable (Table 1). Whether or not a mandrel is a conductor will determine the procedures required to prepare it for electroforming. Conductive mandrels are usually pure metals or alloys of metals and are prepared by the usual procedures, but may require application of a thin parting film to facilitate separation of the electroform from the mandrel (unless the mandrel is removed by melting or chemical dissolution). Nonconductors must be made conductive by spraying the surface with a thin metallic film, usually silver. The thin film of silver, also, serves

TABLE 1. TYPES OF MANDREL MATERIALS

Types	Typical Materials
Conductors	
Expendable	Low-melting alloys; e.g., bismuth-free 92% tin and 8% zinc
	Aluminum alloys
	Zinc alloys
Permanent	Nickel
	Austenitic stainless
	Invar, Kovar
	Copper and brass
	Nickel plated steel
Nonconductors	
Expendable	Wax
	Glass
Permanent (or semi-permanent)	Rigid and collapsible plastics; e.g., epoxy resins and polyvinyl chloride, respectively
	Wood

to facilitate separation of the electroform from the mandrel.

Whether or not a permanent or expendable mandrel should be used is largely dependent on the particular article that is to be electroformed. If no re-entrant shapes or angles are involved, it is possible to use permanent, rigid mandrels that can be separated from the finished electroform mechanically and re-used. If re-entrant angles and shapes are involved, it is necessary to use mandrel materials that can be removed by melting or by chemical dissolution, or materials that are collapsible, such as polyvinyl chloride and other plastics.[21, 22, 25]

The various types of mandrel materials each have their own advantages and disadvantages.

Austenitic stainless steel is an excellent material to use as a permanent, conductive mandrel. Although machining may be relatively difficult, it can be machined to close tolerances and given a high finish. Resistance to damage in use is good and, as it is conductive, there is no need to metallize the surface before beginning the electroforming process. Austenitic grades of stainless steel are preferred because they readily form stable, passive films that prevent adhesion between electroform and mandrel.

Copper and brass are also used as permanent, conductive mandrels and are less costly than stainless steel. They are relatively easy to machine and are ideal when intricately engraved or textured surfaces are required. Separating the electroform from a copper or brass mandrel is easy provided the proper parting film is used (see below). Alternatively, copper and brass mandrels can be electroplated with thin deposits of nickel and/or chromium to facilitate parting and to prevent attack of the mandrel by the processing solutions.

Conventional steel can be used as a mandrel material; it is inexpensive and easily machined. It is usually necessary to nickel plate the entire mandrel and use a parting film.

"Invar" or "Kovar" are alloys that have low coefficients of thermal expansion and are relatively expensive. They are used in critical cases where high dimensional accuracy is required. The electroform and mandrel are heated, and the greater thermal expansion of the electroform permits withdrawal of the mandrel. They are conductive and resistant to damage in handling.

Pure nickel, although difficult to machine, is used as a permanent mandrel. It requires simple treatment to ensure release of the electroform from the mandrel.

Expendable conductive mandrel materials include aluminum and zinc alloys, and low-melting materials, such as bismuth-free tin-zinc alloys.

Aluminum or one of its alloys is used because it is easy to machine to close dimensional tolerances and can be given a high surface finish. It is relatively high in cost, easily damaged and does not always require metallization prior to electroforming. When used as an expendable mandrel, aluminum or its alloys can be dissolved away from a nickel electroform in a 10% solution of sodium or potassium hydroxide solution operating at room temperature.

Zinc-based alloys have similar characteristics and are also used as expendable materials. The mandrel can be dissolved in a cold 10% hydrochloric acid solution without attacking a nickel electroform. Superplastic zinc alloys (zinc-aluminum-magnesium and zinc-aluminum alloys) can be blow-molded into complex shapes at relatively low temperatures, 250 C(480 F) and have been used to produce thin-walled expendable mandrels.

Tin-zinc alloys, with and without bismuth, have low melting points and thus can be used to make mandrels that can be removed from the electroform by melting the alloy. This can usually be accomplished without damaging or distorting the electroform provided the temperature is less than 250 C(480 F). Bismuth can embrittle copper and nickel electroforms if it is not completely removed from the inside of the electroform, in which case the use of bismuth-free tin-zinc alloys are required

Nonconductive mandrel materials include wax, glass and plastics and these materials are expendable or, at best, semi-permanent.

Wax is useful because it can be readily cast and easily melted out of the finished electroform, but it has many disadvantages. It may be brittle, it may have a poor surface finish and it is easily damaged. It is inexpensive, requires metallization and cannot be used to maintain high tolerances. Cheap grades of wax melt or soften when exposed to electroplating solutions operated at normal temperatures. There are only a few grades of wax, such as beeswax, that can be successfully silver-sprayed. Graphite-loaded wax which is conductive has been used successfully as an expendable mandrel.

Plastics materials can be used to fabricate

permanent or semi-permanent mandrels, but, being nonconductive, require the use of silver-spraying or other metallization techniques. In general, plastics are relatively cheaper and have inferior mechanical properties than metal mandrels, and cannot be made to close dimensional tolerances and with very high surface finish. Rigid plastic mandrels are often made from epoxy resins in the case of electroforms without reentrant angles. For electroforms with reentrant angles or shapes, collapsible plastic materials, such as polyvinyl chloride (PVC), are used. Special care is required in using these materials if an acceptable mandrel is to result.

Glass is very difficult to machine, is easily damaged, is relatively expensive and requires metallization, but surface finish can be excellent and close tolerances can be maintained. It is, therefore, sometimes used as a mandrel material.

Wood, leather and *fabrics* can be used as mandrel materials when it is desired to reproduce their surface textures and patterns. These materials are porous and must be sealed and then made conductive. It is often easier to duplicate the desired surface finish by casting with polyvinyl chloride and then electroforming onto the plastic replica of the surface.

Any solid material can be used to fabricate a mandrel for electroforming, but the following generalizations may help in selecting a suitable material:

1. Permanent mandrels are preferred for accuracy and for large production runs.
2. Expendable mandrels must be used whenever the part is so designed that a permanent mandrel cannot be withdrawn. Expendable mandrels of low-melting point alloys may be used for low-cost items not requiring close tolerances. Collapsible plastic mandrels have been used to fabricate parts with reentrant shapes or angles.
3. It is important that the mandrel retain its dimensional stability in warm plating baths. Wax and most plastics expand when exposed to electroplating solutions operated at elevated temperatures. In such cases, it may be necessary to use acid copper, nickel sulfamate and other electroplating solutions that function at room temperature.

MANDREL DESIGN AND FABRICATION

The ability to produce an electroform will also depend on the design of the mandrel. The electroforming operation can often be simplified by a few design changes which do not impair the functioning of the piece. ASTM Standard B450, "Engineering Design of Electroformed Articles," contains useful information on this subject.[3] Some of the design considerations include the following:

1. Exterior angles should be provided with as generous a radius as possible to avoid excessive buildup and treeing of the deposit during electroforming. Interior angles on the mandrel should be provided with a fillet radius of at least 0.05 cm/5 cm of length of a side of the angle.
2. Wherever possible, permanent mandrels should be tapered at least 0.08 mm/m (0.001 in./ft) to facilitate removal from the electroform. (Where this is not permissible, the mandrel may be made of a material with a high or low coefficient of thermal expansion so that separation can be effected by heating or cooling).
3. A fine surface finish on the mandrel, achieved by lapping or electropolishing, will generally facilitate separation of mandrel and electroform. A finish of 2 μin. RMS is frequently specified.
4. Flat bottom grooves, sharp angle indentations, blind holes, fins, v-shaped projections, v-bottom grooves, deep scoops, slots, concave recesses, rings and ribs can cause problems with metal distribution during electroforming, unless inside and outside angles and corners are rounded.

The method of fabrication of the mandrel will depend on the type selected, the material chosen and the object to be electroformed. Mandrels may be manufactured by casting, machining, electroforming and other techniques. Permanent mandrels may be made by any of the conventional pattern-making processes.

PREPARATION OF MANDREL SURFACES

Nonconducting mandrels must be made impervious to water and to the processing solutions, and then rendered conductive. Porous materials—for example, leather and plastic—may be impregnated with wax, shellac, lacquer or a synthetic resin formulation to make them impervious to water. A molten mixture of wax containing 4 parts beeswax, 12 parts paraffin and 2 parts resin can be used for this purpose. The nonconducting mandrel is immersed in the wax mixture for at least 30 minutes or until bubbling ceases, when it is removed and cooled. This will result in a considerable loss of detail and it may be necessary to use thin films of lacquer to seal porous, nonmetallic mandrels.

Nonconducting materials may be rendered conductive in a number of ways, but perhaps the most common is to apply a chemically reduced film of silver to the surface. The surface of the mandrel must be cleaned before the silver film can be applied. It may need to be cleaned with an organic solvent that does not attack the mandrel surface and/or the surface may need to be scrubbed with a slurry of magnesium oxide using a cotton-wool pad and then thoroughly rinsed to remove all traces of the magnesium oxide.

The silver-spraying process consists of essentially two steps: sensitization and silver reduction. Sensitization is achieved by spraying the mandrel with a solution of the following composition:

Stannous chloride	30 g/L
Stannic chloride	3 g/L
Hydrochloric acid (35% w/w)	40 ml/L
Sodium chloride	175 g/L

The solution is prepared by heating 40 ml of concentrated hydrochloric acid and dissolving first the stannous chloride and then the stannic chloride in the hot acid. This solution is diluted, sodium chloride is dissolved in it and water is added to adjust the volume to 1 L. The sensitizing solution is sprayed on the mandrel by means of a single-nozzle spray gun until the surface is uniformly and completely wet.

A double-nozzle spray gun is used to produce the silver film. The silver solution is sprayed from one nozzle, the reducing solution from the other, simultaneously.

The silver solution is prepared as follows:

1. Dissolve 20 g silver nitrate in 1 L deionized water.
2. Add 30 ml 0.88 molar ammonium hydroxide.
3. When the solution clears, add 2 g sodium hydroxide.
4. Adjust the volume of the solution to 1.25 L with deionized water.

To avoid the formation of explosive silver compounds, this formula should never be mixed in the form of a concentrate. The instructions for mixing must be followed exactly. Containers used for mixing and storing this solution should be washed carefully after use. The solution should not be allowed to evaporate to a concentrated form or to form dry crystals. Dry, the crystals are explosive.

The reducing solution is prepared as follows:

1. Add 22 ml of formaldehyde (40% solution) to 1 L of deionized water.
2. Dissolve 0.5 sodium hydroxide in 5 ml deionized water and add 0.5 g lead acetate.
3. Add the lead acetate solution to the formaldehyde solution.
4. Adjust the volume to 1.25 L with deionized water.

There are other ways of making nonconducting mandrels conductive: 1) by use of finely divided metal powders dispersed in a binder ("bronzing"); 2) by application of finely divided graphite to wax, natural and synthetic rubbers and similar materials that have an affinity for graphite; and 3) application of graphite with a binder. The chemical precipitation of films, as exemplified by silver, however, is often preferred because dimensional accuracy is not affected, the film has little adhesion and parting is not difficult. If necessary, the silver film can be stripped from the electroform with either nitric acid, warm sulfuric acid or a cyanide solution.

It is also possible to metallize nonconducting mandrels using electroless nickel and electroless copper processes available from most plating supply houses. Specific preparation techniques may be required to use these processes for electroforming.

With most metallic mandrels, a parting film on the surface helps to assure that the electroform can be separated from the mandrel. After removing all traces of grease or oil by means of chlorinated hydrocarbon solvents, various metallic mandrels are given different treatments for this purpose.

Stainless steel, nickel, nickel- or chromium-plated steel should be scrubbed with magnesium oxide, rinsed and passivated by immersion in a 2% solution of sodium dichromate for 30 to 60 seconds at room temperature. After this, the mandrel must be rinsed several times to remove all traces of the dichromate solution.

Copper and brass mandrels that have been nickel and/or chromium-plated may be passivated in the dilute sodium dichromate solution, as described in the preceding paragraph. If not electroplated, the surface can be made passive by immersion in a solution containing 8 g/L of sodium sulfide. Trial and error may be required to obtain the proper thickness of film to prevent excessive adhesion while avoiding premature separation of the electroform.

Aluminum alloys may require special treatments even when they are used as expendable mandrels to be separated from the electroform by chemical dissolution. The reason for this is the difficulty of obtaining sound electrodeposits directly on aluminum from most electroforming solutions, especially if the deposits are highly stressed. In this case, it may be necessary to use proprietary zincate or stannate treatments to achieve a degree of adhesion that will prevent lifting and curling of the deposit from the mandrel. These special treatments are described elsewhere in this book. When low-stressed deposits (near zero) are being produced, treatment of the aluminum alloy by degreasing, cathodic alkaline cleaning and immersion in a 50% solution of nitric acid, by volume, may be sufficient.

Zinc and its alloys may require no other preparation than conventional cleaning if they are used for expendable mandrels and are to be parted by chemical dissolution. But, depending on what metal is being used for electroforming, it may be necessary to electroplate the zinc alloy metal with copper from a neutral or alkaline cyanide solution to prevent attack of the mandrel.

The *fusible alloys* used to make expendable conductive mandrels that can be melted out have a tendency to leave a residue of tin on the surface of the electroform. This can be prevented by electroplating the fusible mandrel with a layer of copper that is electropolished prior to beginning the electroforming process.[51]

The processing of mandrels for electroforming is discussed in an ASTM Standard where details of some of the above preparation procedures are given.[4]

ELECTROFORMING SOLUTIONS AND DEPOSIT PROPERTIES

The electroplating solution and, hence, the metal to be electrodeposited onto the mandrel surface (the cathode) are chosen to satisfy the functional requirements of a particular application. To the extent possible, the properties of the electroform are specified with emphasis on end-use requirements. This approach provides a rational basis for selecting the electroplating solution and, if necessary, modifying it to control the properties of the electrodeposited metal within specified limits.

The electroplating solutions used for electroforming are basically the same as those used for other applications. In general, any practical plating solution can be applied to electroforming. In practice, nickel, copper and iron electroplating solutions are the ones most widely used. Electroforming of gold, silver, platinum metals, alloys, combinations of these and composites of various kinds is possible, and this versatility has extended the range of properties that can be obtained by electroforming.

The composition and operating conditions of nickel, copper and iron electroforming solutions are given in Table 2, along with typical values of tensile strength, percent elongation, hardness and internal stress of deposits from these solutions. The properties given in the table are those obtained from additive-free baths operated within the specified conditions.

The properties of electrodeposits are interrelated and are influenced by operating conditions. For example, steps taken to increase the hardness of the deposit usually increase its strength and reduce its ductility. Operating variables that influence the properties of deposits include temperature, pH and cathode current density. The constituents of the solution, if their concentrations are not kept within certain limits, can, also, affect the properties of deposits. The influence of these variables on some of the properties of nickel deposited from Watts and sulfamate solutions is shown in a qualitative fashion in Fig. 1. Deposits from these two nickel solutions are affected differently by the same variables, in many cases. Deposits from copper

TABLE 2. NICKEL, COPPER AND IRON ELECTROFORMING SOLUTIONS AND SOME PROPERTIES OF THE DEPOSITS

Nickel Electroforming Solutions*

	Electrolyte Composition, g/L	
	Watts Nickel	Nickel Sulfamate
	$NiSO_4 \cdot 6H_2O$ 225–300	$Ni(SO_3NH_2)_2$ 315–450
	$NiCl_2 \cdot 6H_2O$ 37.5–52.5	$NiCl_2 \cdot 6H_2O$ 0–22.5
	H_3BO_3 30–45	H_3BO_3 30–45
	Operating Conditions	
Temperature	44–66 C	32–60 C
Agitation	Air or mechanical	Air or mechanical
Cathode current density	3–11 A/dm²	0.5–32 A/dm²
Anodes	Nickel	Nickel
pH	3.0–4.2	3.5–4.5
	Mechanical Properties	
Tensile strength, MPa	345–485	415–620
Elongation, %	15–25	10–25
Hardness (Vickers, 100 g load)	130–200	170–230
Internal stress, MPa	125–185 (tensile)	0–55 (tensile)

Copper Electroforming Solutions*

	Electrolyte Composition, g/L	
	Copper Sulfate	Copper Fluoborate
	$CuSO_4 \cdot 5H_2O$ 210–240	$Cu(BF_4)_2$ 225–450
	H_2SO_4 52–75	HBF_4 maintain pH at 0.15–1.5
	Operating Conditions	
Temperature	21–32 C	21–54 C
Agitation	Air or mechanical	Air or mechanical
Cathode current density	1–10 A/dm²	8–44 A/dm²
Anodes	Soluble copper	Soluble copper
	Mechanical Properties	
Tensile strength, MPa	205–380	140–345
Elongation, %	15–25	5–25
Hardness (Vickers, 100 g/l)	45–70	40–80
Internal stress, MPa	0–10 tensile	0–105 tensile

Iron Electroforming Solutions

	Electrolyte Composition, g/L	
	Ferrous Chloride	Fluoborate Bath
	$FeCl_2 \cdot 4H_2O$ 300–450	$Fe(BF_4)_2$ 225
	$CaCl_2$ 125–335	NaCl 10
	(Iron content 160 g/L)	
	Operating Conditions	
Temperature	90 C	55–60 C
Agitation	Cathode and/or solution	Cathode and/or solution
Cathode current density	4–6 A/dm²	2–10 A/dm²
Anodes	Low-carbon steel (bagged)	Low-carbon steel (bagged)
pH	0.5–1.5	2.0–3.0
	Mechanical Properties**	
Tensile strength, MPa	345–450	
Elongation, %	20–40	
Hardness, Brinnel	215–270	
Internal stress, MPa	90–140 (tensile)	

*Based on ASTM Standard B503.
**Approximate values.

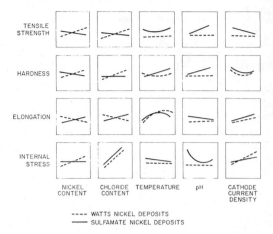

Fig. 1. Influence of variables on properties of nickel deposits.

sulfate and copper fluoborate solutions also respond differently when changes are made in the operating variables.[2]

Nickel can be electrodeposited over a wider controlled range of physical properties than any other metal, and more is known about the effects of bath composition, operating conditions and addition agents on the mechanical, physical and corrosion-resistant properties of the deposit.[30, 32, 36, 42, 43, 44, 47, 50]

Copper is used for electroforming because it is relatively inexpensive and strong enough for many applications. It can be deposited at high rates from the acid sulfate bath and in a reasonably low state of stress. Information on the properties of electrodeposited copper is available.[30, 31, 33, 42]

Iron is inexpensive and strong, and electrodeposits from the ferrous chloride solution are surprisingly ductile and stress-free when the solution is operated very hot (near boiling). The ferrous chloride solution is very corrosive and equipment for containing and handling the solution must be designed accordingly. Information on the properties of electrodeposited iron and its use for electroforming is available in the technical literature.[114]

OTHER METALS AND PROCESSES

The electroforming of gold, silver, platinum metals and various alloys has been described in the literature. Gold and silver objects have been electroformed from cyanide solutions.[63] Platinum, palladium and iridium have been electrodeposited from molten cyanides, and these processes have been used to electroform platinum and iridium crucibles.[61] The electroforming of nickel-cobalt molds and dies has been explored because of the improved high-temperature properties of the alloy deposits.[40,41] Aluminum, deposited from an organic, non-aqueous electrolyte, has been successfully electroformed.[115] Refractory metals have been electroformed from fused salt baths.[116]

Electroplating processes that result in the incorporation of particles uniformly dispersed throughout an electrodeposited metal matrix have been investigated in connection with electroforming. By judiciously selecting the particle or particles to be occluded, it should be possible to obtain electroforms with unique physical and mechanical properties, including high resistance to wear and to elevated temperatures, and with unusual anti-friction and lubricious characteristics. Oxides, carbides, nitrides and borides have been incorporated in nickel, iron, cobalt and tungsten-cobalt electrodeposited matrices to produce high-temperature, oxidation-resistant materials. Mica has been codeposited with nickel and the deposit has interesting anti-friction properties. Although these processes have been investigated for electroforming, the possibilities for creating articles with new and novel properties have not been exhausted.[65-72]

CONTROL OF ELECTROFORMING PROCESSES

Successful electroforming requires careful control of the electrolyte and of the operating variables—pH, temperature, agitation and current density. In this respect, control of electroforming is similar to that of decorative electroplating. Control is more difficult and, perhaps, more critical in the case of electroforming because processing may take hours or days to complete. The most difficult problems encountered in electroforming are usually associated with controlling metal distribution, internal stress, and roughness of the deposits. Addition agents may overcome some of the difficulties, but additives must be controlled and applied judiciously.

Metal distribution, that is, the variation of the thickness of metal deposited at various points on the surface of a mandrel, is related to current distribution. Recessed areas will receive less current; areas that project from the surface will receive higher current. The current density and, consequently, the rate of metal deposition will be lower in recessed areas than at areas which project from the surface. The result is that metal distribution will be non-uniform in most cases. The deposit will be relatively thin in recessed areas and relatively thick on projections.

Some electroplating processes are less sensitive than others to the variations in current density that are characteristic of the electroforming process. The capability of a plating solution to deposit uniformly thick coatings despite variations in current density on the article being electroformed is measured by its throwing power. Throwing power describes the complex relationship between metal distribution and the variables—conductivity, polarization, current efficiency and geometry.[52] The throwing power of a plating solution can be measured directly in a Haring cell or in an open Hull cell.[51] Studies of the throwing power of plating solutions suggest that throwing power can be improved by reducing the current density and increasing the pH, temperature and metal content of the bath. All-chloride nickel plating solutions have better throwing power than Watts and nickel sulfamate baths. Copper cyanide plating solutions have significantly better throwing power than copper sulfate baths.

Although throwing power is dependent on the composition of the plating bath and can be controlled to some extent by adjusting operating variables, the electroformer will usually select a process based on the mechanical and physical properties of the deposit, rather than the throwing power of the solution. Metal distribution is then improved by proper racking, and by the use of thieves, shields and/or conforming or auxiliary anodes. (This subject is discussed in detail in another chapter of this book.) The use of these processing aids makes it possible to improve metal distribution significantly. Although the application of these techniques can be optimized by trial and error, simulation of the electroforming operation combined with the determination of cathode potential profiles can facilitate the design and placement of shields and auxiliary anodes.[55-57]

The control of *internal stress* of electrodeposited metals is especially important in electroforming. Internal stress refers to forces created within an electrodeposit as a result of the crystallization process itself and/or the codeposition of impurities such as hydrogen, sulfur and other elements. The forces are either tensile (contractile) or compressive (expansive); rarely are electrodeposits free of some degree of internal stress. Excessive tensile or compressive stress can cause the following problems: 1) distortion of the electroform when it is separated from the mandrel; 2) difficulty in separating the electroform from the mandrel; 3) curling, peeling or separation of the electrodeposit from the mandrel prematurely and 4) buckling and blistering of the deposit (usually associated with high compressive stresses). Control of internal stress is, therefore, critical for successful electroforming.

Typical values of internal stress are given in Table 3. Copper, nickel, cobalt, iron, silver, rhodium, palladium and chromium are generally deposited in a state of tensile stress. Zinc, cadmium and lead, although not widely used for electroforming, usually are deposited in the compressive state.

Internal stress can vary widely depending on the metal and the composition of the electrolyte. Copper and silver are normally deposited with quite low stress. Internal stress is greater for copper deposited from a cyanide bath than from an acid copper sulfate bath.

Nickel is a good example of how internal stress

TABLE 3. TYPICAL VALUES OF INTERNAL STRESS OF ELECTRODEPOSITED METALS

Metal	Bath	Stress, MPa*
Copper	Cyanide	34–85
	Acid	0–14
	Fluoborate	0–20
Nickel	Watts	110–210
	Watts, with H_2O_2	275 or higher
	All chloride	205–310
	Fluoborate	100–175
	Fluoborate, with H_2O_2	100–175
	Sulfamate, no chloride	0–55
	Sulfamate, with chloride	55–110
	Sulfate	110–138
Iron	Chloride	90–138
Chromium	Chromic acid	−115 to 345 (depends on extent of cracking)
Silver	Cyanide	12 to 25
Gold	Cyanide	Compressive

*Positive values are tensile.

is influenced by the nature and composition of the electroplating solution. The all-chloride solution produces deposits with the highest internal tensile stresses. Nickel sulfamate solutions, without chlorides and addition agents, produce deposits with the lowest internal tensile stresses. Increasing the chloride content of a nickel bath tends to increase the tensile stress of the deposits.[34, 35, 47]

Organic additives can be used to control the internal stress of nickel, copper and other electrodeposited metals. In the case of nickel, sulfur-bearing materials, such as saccharin or p-toluene sulfonamide, can be used to modify internal tensile stress; that is, reduce it to zero or cause it to become compressive. The additives generally increase the hardness and strength of the nickel deposits, and affect other properties and so must be used judiciously. Effective stress-controlling agents result in the codeposition of small amounts of sulfur with nickel. The presence of sulfur affects the high-temperature properties adversely, and nickel deposits containing sulfur cannot be heated above 260 C(500 F) without severe embrittlement. The codeposition of small amounts of manganese prevents embrittlement of sulfur-containing nickel deposits to some extent and allows heating at or slightly above that temperature.[38] The concentrated nickel sulfamate bath can be operated at high current densities to produce deposits with zero or very low internal stresses by using continuous conditioning of the electrolyte in an auxiliary tank, the technique being shown to be effective with nickel as well as nickel-cobalt alloy plating.[26, 117] The internal stress of copper deposits can also be controlled by the use of organic addition agents; at low current densities the stress in copper from acid baths is reportedly compressive.[118]

Internal stress is thus controlled by selecting the metal, by specifying a specific electrolyte and by using organic addition agents. The control of current density, as well as other operating variables, is also essential.

Roughness can also be a source of difficulty in electroforming. Any condition which would tend to cause roughness in decorative plating will have much more serious results in electroforming. Nodules, nuggets and trees will form. These become high-current density areas, and the larger they get the faster they grow, and the more they "rob" surrounding areas. An unsound electroform results, which will usually be a reject. As a consequence, the filtration rates used in electroforming are very high and may amount to passing the bath through the filter one-half to one or more times per hour.

Airborne dirt is a serious source of roughness; it is good practice to keep the plating area clean (by vacuum cleaning, not sweeping) and to supply it with filtered air keeping the room under an inch or so of positive pressure. These precautions are particularly necessary when the plating tank has an exhaust system that draws air across the bath.

Anode particles will also cause roughness and must be controlled by the usual means of anode bags or diaphragms; where even this is inadequate, higher filtration rates and cathode agitation may serve to overcome the difficulty.

A frequently overlooked source of roughness is the crystallization of bath chemicals on anode bags, tank walls and the super-structure. When these particles fall or are brushed into the solution, they are usually slow to dissolve and may become attached to the cathode. Such accumulated salts should be removed carefully, not allowing them to fall into the bath, or they should be flushed off with a stream of water at a time when the tank is not working. Solutions in which all components are highly soluble, such as sulfamate or fluoborate baths, do not have this tendency and for this reason may be preferred when other considerations permit.

To avoid particles from precipitation of hard water constituents, deionized water should be used to make up the plating bath and in the rinse prior to it.

Treeing at edges and corners is frequently troublesome and may be minimized by the use of shields or by the choice of bath. For instance, where the deposit from a high-chloride or all-chloride nickel bath can be accepted, these baths are much better than the Watts bath in this respect. Certain addition agents may also suppress the treeing tendency, but should only be used after careful consideration of possible effect on the physical properties of the deposit. Another expedient, applicable to certain shapes of mandrels, is to extend the mandrel beyond the dimensions actually desired, so that the treeing will occur on a part of the electroform that will be machined away.

Should it become necessary to interrupt processing of the electroform, to remove excessive nodular growths and trees by machining, extreme care should be exercised to insure that the machined surface is properly treated to accept an adherent deposit. Preferred treatments to accomplish this on nickel surfaces have been described.[119]

To be able to use higher current densities, baths such as sulfamate[120] or fluoborate[121] may be chosen (other considerations permitting) that are known to have high limiting current densities, especially when operated at high metal content, high temperature and with adequate agitation.

Agitation of every kind, singly or in combination, should be employed whenever possible, to minimize burning and pitting at high-current density sites. Solution agitation, either air or mechanical, may induce roughness, unless the solution is kept clean by using a very high filtering rate. Cathode agitation is very effective, and rotation of the cathode is desirable when feasible, especially if it is a surface of revolution. In this case, besides the advantages of agitation, rotation ensures the uniformity of deposit thickness from point to point around the circumference. (Inequalities *along* the axis have to be overcome otherwise). When rotation is used, brushes or other commutating devices must be provided to carry the current to the cathode. Care must be taken that the rotating and commutating mechanisms are so designed that no lubricating grease and metal particles resulting from wear of bearing surfaces are permitted to fall into the plating solution.

MACHINING AND FINAL FINISHING OF THE ELECTROFORM

To avoid deformation, any necessary machining or other mechanical finishing operations are usually performed before the electroform is parted from the mandrel. There are no particular problems associated with the finishing of copper electroforms, but nickel does not machine readily. Even grinding, which is preferable, may present some difficulty, especially with soft deposits which quickly load up the grinding wheels and necessitate frequent dressing with a diamond tool. Carr[122] gives explicit directions for machining and grinding various deposits.

PARTING

Electroforms are removed from permanent mandrels mechanically, by the use of one or a combination of several of the following techniques:

Impact, by a sudden hammer blow or pull.

Gradual force, applied by a hydraulic ram to push, or a jack-screw or wheel-puller to pull the pieces apart.

Heating with a torch or hot oil bath, either to melt or soften a parting compound or to take advantage of a favorable differential in the coefficients of expansion of mandrel and deposit.

Cooling, as in a mixture of dry ice (solid CO_2) and acetone. In this case, the mandrel must have a considerably lower coefficient of thermal expansion, and then on withdrawal from the cold bath, the electroform will expend faster than the mandrel, permitting withdrawal.

Prying with a sharp tool may be used with care to separate relatively flat pieces, such as phonograph record stampers or engraving plates from their mandrels.

Expendable molds are melted or dissolved out:

Zinc alloys are dissolved out in a hydrochloric acid solution.

Aluminum alloys are readily soluble in a strong, hot, sodium hydroxide solution.

Low-melting alloys are melted and shaken out. The alloy may be used over and over. If "tinning" occurs, the electroform may (if it is nickel) be cleaned up with strong nitric acid, but it is preferable to avoid "tinning" by other means, already discussed.

Plastics of the thermoplastic type may be softened by heat so that the bulk of the mandrel may be withdrawn, after which the electroform is cleaned with a suitable solvent. An expensive alternative is to dissolve the entire mandrel with a solvent. Plastics are more often suitable for permanent than expendable molds.

Wax may be melted out and any residues dissolved by a suitable organic solvent.

BACKING UP THE ELECTROFORM

For certain applications, as in the electroforming of molds and printing plates, it is necessary that the electroform be backed with some other material, which is then finished to specified dimensions to fit into a bolster or onto a printing press. The materials used for backing will vary widely depending on the application. The most important backing methods involve the following: 1) casting with low-melting temperature alloys; 2) spraying with various metals and alloys; 3) electroplating with other metals; 4) use of thermosetting plastics and 5) spark-eroded steel back-ups. Advancements in backing techniques have been made in recent years and are described in the technical literature.[18, 19, 20, 23]

ELECTROFORMING APPLICATIONS

Modern applications of electroforming are diverse and may be placed in the following categories:[73]

A. *Tools*
   1. *Molds and dies*—for producing leather-grained textures and lustrous finishes on the surfaces of plastics and other molded materials; for example, electroformed molds have been used in making auto armrests; taillight reflectors; ballpoint pen bodies; plastic caps; switch plates; chocolate and rubber products; ice cream and ice pops; plastic dolls; plastic fresnel lenses.[73, 78, 92, 98, 108] Applications where the ability to reproduce detail on an extremely fine scale is critical include: phonograph record and video-disc stampers; embossing plates; printing plates for stamps and currency.[86]
   2. *Press tools*[73]
   3. *Foundry patterns*[73]
   4. *Diamond cutting bands*[73, 77]
   5. *Abrasive sheets*[73]
B. *Mesh products*
   1. Textile printing screens[73]
   2. Battery mesh and other porous structures[93]
   3. Filters[73]
   4. Razor screens[73]
C. *Other products*
   1. Foil for printed circuit and other purposes[73, 111]
   2. Space mirrors and metal optical parts[87, 88]
   3. Bellows[13, 89]
   4. Radar wave guides[96, 100, 110]
   5. Heat exchangers[73, 112]
   6. Seamless belts of various kinds
   7. Venturi meters[123]
   8. Pitot tubes[124]
   9. Rocket thrust chambers, nozzles and motor cases[15, 83, 90, 91, 95]
   10. Consumer products[99]

These applications illustrate the scope and versatility of modern electroforming as a method of fabrication. The method is not universally applicable, but when it is used to best advantage, it results in unique products, components and tools.

References

ASTM Standards

1. ASTM Standard B374. "Definitions of Terms Relating to Electroplating," *1981 Annual Book of ASTM Standards*, p. 169.
2. ASTM Standard B503. "Use of Copper and Nickel Electroplating Solutions for Electroforming," *1981 Annual Book of ASTM Standards*, p. 307.
3. ASTM Standard B450. "Engineering Design of Electroformed Articles," *1981 Annual Book of ASTM Standards*, p. 236.
4. ASTM Standard B431. "Processing of Mandrels for Electroforming," *1981 Annual Book of ASTM Standards*, p. 207.

Historical

5. Pavlova, O. I. "Electrodeposition of Metals—A Historical Survey," U.S. Department of Commerce, Springfield, VA 22151.
6. Smith, C. A. "Early Electroplating, Part 2, Commencement of Industrial Applications (1836-1852)," *Finish. Ind.* 1(3):24-28 (1977).

Reviews

7. Dambal, R. P. "Industrial Applications of Electroforming—a Review," *J. Electrochem. Soc. India* 24(4):179 (1975).
8. Squitero, A. D. "Take Advantage of Modern Electroforming," *Mach. Tool Blue Book* 69(5):78-86 (1974).
9. Faruq, Marikar Y. M. and K. I. Vasu. "Heavy Deposition—Electroforming and Electrosalvaging," *J. Electrochem. Soc. India* 23(3):171 (1974).
10. Hartley, John. "Electroforming," *Eng. Mat. Design* 17(7):13-14 (1973).
11. Spencer, Lester F. "Modern Electroforming, Part 2, Electroforming Solutions," *Met. Finish.* 71(4):53-59 (1973); 71(3):54-57 (1973).
12. Spencer, Lester F. "Modern Electroforming, Part 1, Requirements and Mandrels," *Met. Finish.* 71(2):64-72 (1973).
13. Sanborn, C. B. "Nickel Electroforming Applications—Why They Exist," Symposium on Electrodeposited Metals as Materials for Selected Applications, AD-738272, NTIS, Springfield, Va., 1972, pp. 65-81.
14. Spiro, Peter. *Electroforming*. Teddington, England: Robert Draper, 1971, 335 pages.
15. Suchentrunk, R. and O. Tuscher. "Electroforming in Aerospace Engineering," *Galvanotechnik* 70(12):1178-1184 (1979) (in German).

16. Hartley, Peter. "When to Specify Electroforming," *Des. Eng.* (London) (Feb. 1975) pp. 47, 49, 51.
17. Weiler, Gerd G. "Electroforming, Definition and Literature Survey," *Metalloberflaeche* **26***(10)*:381 (1972) (in German).

Mandrels and Backing Techniques

18. Wearmouth, W. R. "Application of New Developments in Electroforming Technology in the Toolmaking Industry," Interfinish 76—Proceedings of the Ninth World Congress on Metal Finishing, Amsterdam, 1976, 21 pages.
19. Watson, S. A. "Recent Developments in (Nickel) Electroforming and Backing Mould Cavities," *Het. Ingenieursblad* **45***(9)*:279-287 (1976).
20. Dean, A. V. and W. R. Wearmouth. "New Backing Techniques for Electroformed Moulds and Dies," *Electroplat. Met. Finish.* **28***(12)*:18-23, 29 (1975).
21. Spencer, Lester F. "Modern Electroforming, Part 1, Requirements and Mandrels," *Met. Finish.* **71***(2)*:64-72 (1973).
22. Stekelbach, W. "Electroforming," *Galvanotechnik* **66***(4)*:312-316 (1975).
23. Dean, A. V. "Further Developments in the Use of Cast and Sprayed Backings on Electroformed Moulds and Dies," *Metallurgia* **45***(5)*:243-248 (1978).
24. Hartley, John. "Electroforming," *Eng. Mater. Des.* **17***(7)*:13-14 (1973).
25. Bertucio, E. C. "Electroforming with Collapsible Mandrels, *Metal Finish.* **64**:61-66 (1966).

Electroforming with Nickel

26. DiBari, G. A. "Electroforming," *Metal Finishing Guidebook and Directory Issue '82*. Hackensack, N.J.: Metals and Plastics Publications, p. 456.
27. "Inco Guide to Nickel Electroforming—Moulds and Dies for Plastics, Zinc, and Glass," Inco Europe, London, 1978, 83 pages (booklet).
28. "Electroforming with Nickel," International Nickel Company, New York (1976), 19 pages (booklet).
29. "Electroforming with Nickel," American Electroplaters' Society, Winter Park, Fla. (1965), 27 pages (booklet).

Properties of Electroformed Materials

30. Safranek, W. H. *The Properties of Electrodeposited Metals and Alloys—A Handbook.* New York: American Elsevier Publishing, 1974.
31. Safranek, W. H. "Physical and Mechanical Properties of Electroformed Copper," ASTM STP 318, Am. Soc. Testing Mats., 1962, pp. 44-53.
32. Zenter, V., A. Brenner and C. W. Jennings. "Physical Properties of Electrodeposited Metals, I—Nickel," *Plating* **39***(8)*:865-927 (1952).
33. Lamb, V. A. and D. R. Valentine. "Physical and Mechanical Properties of Electrodeposited Copper," *Plating* **52***(12)*:1289-1311 (1965); **53***(1)*:86-95 (1966).
34. Marti, J. L. "The Effect of Some Variables Upon Internal Stress of Nickel as Deposited from Sulfamate Electrolytes," *Plating* **53***(1)*:61-71 (1966).
35. Yamamoto, M. and T. Sato. "The Effect of Superimposed Sine and Square Current to Direct Current on the Internal Stress in High Speed Electroformed Nickel," *Bull. Jpn. Soc. Precis. Eng.* **14***(2)*:119-120 (1980).
36. Brooks, J. A., J. W. Dini and H. R. Johnson. "Effects of Impurities on the Weldability of Electroformed Nickel," Sandia Laboratories, Report Number N80-11214/7, Dec. 1978, p. 36.
37. Lai, S. H. F. and J. A. McGeough. "Electroforming and Mechanical Properties of Iron-Nickel Alloy Foil," *J. Mech. Eng. Sci.* **21***(6)*:411-417 (1979).
38. Wearmouth, W. R. and K. C. Belt. "Electroforming with Heat-Resistant, Sulfur-Hardened Nickel," *Plat. Surf. Finish.* **66***(10)*:53-57 (1979).
39. Dini, J. W., H. R. Johnson and J. A. Brooks. "Zinc in Sulfamate Nickel Deposits: Influence on Weldability of Electroforms," *Met. Finish.* **77***(2)*:99-101 (1979).
40. Dini, J. W. and H. R. Johnson. "Some Property Data for Nickel-Cobalt Electrodeposits," *J. Mater. Sci.* **11***(9)*:1779-1780 (1976).
41. Wearmouth, W. R. and K. C. Belt. "Mechanical Properties and Electroforming Applications of Nickel-Cobalt Electrodeposits," *Trans. Inst. Metal Finish.* (Summer) **52***(3)*:114-118 (1974).
42. Safranek, W. H. and C. H. Layer. "Property Data for Electrodeposits Produced at Rates Above One Mil Per Minute," *Proceedings of the 1973 Symposium on Electrodeposited Metals for Selected Applications*, 1974, 103-110.
43. Dini, J. W., H. R. Johnson and H. J. Saxton. "Influence of Sulfur Content on the Impact Strength of Electroformed Nickel," *Electrodeposition Surface Treatment* **2***(3)*:165-176 (1974).
44. Whitehurst, M. L. "Strength and Ductility of Electroformed Nickel," Symposium on Electrodeposited Metals as Materials for Selected Applications, AD-738272, NTIS, Springfield, Va., 1972, pp. 53-64.

45. Notley, J. M. "Corner Weakness in Nickel Electroforms," *Trans. Inst. Metal Finish.* **50**(1):6-10 (1972).
46. MacInnis, R. D. and K. V. Gow. "Tensile Strength and Hardness of Electrodeposited Nickel-Iron Alloy Foil," *Plating* **58**(2):135-136 (1971).
47. Hammond, R. A. F. "Nickel Plating from Sulphamate Solutions, Part 3, Structure and Properties of Deposits from Conventional Solutions," *Metal Finish.* **16**(188):234-243 (1970).
48. Dini, J. W. and J. R. Helms. "Properties of Electroformed Lead," *Metal Finish.* **67**(8):53-55 (1969).
49. Dini, J. W., H. R. Johnson and M. I. Baskes. "Fracture Toughness and Reduction in Area Data for Some Electrodeposits," *Met. Technol.* **1**(8):391 (1974).
50. Sample, C. H. and B. B. Knapp. "Physical and Mechanical Properties of Electroformed Nickel at Elevated and Subzero Temperatures," *Special Technical Publication No. 318*, American Society for Testing and Materials, Philadelphia, Penn. (1962).

Metal Distribution

51. Watson, S. A. "The Throwing Power of Nickel and Other Plating Solutions," *Trans. Inst. Metal Finish.* **37**:28 (1960).
52. Lowenheim, F. A. (Ed.) *Modern Electroplating.* New York: John Wiley 1974.
53. Dalby, S., J. Nickelsen and L. Alting. "Metal Distribution in Electroplating," *Electroplat. Metal Finish.* **28**(10):18-23 (1975).
54. Chisholm, C. U. and K. Clements-Jewery. "Controlled Thickness Electroforming: Initial Studies," *Proc. of the Int. Mach. Tool Des. and Res. Conf.*, published by Imperial College of Science and Technology, and Macmillan Press, London, 1978, pp. 217-223.
55. McGeough, J. A. and H. Rasmussen. "Analysis of Electroforming With Direct Current," *J. Mech. Eng. Sci.* **19**(4):163-166 (1977).
56. McGeough, J. A. and H. Rasmussen. "Perturbation Analysis of the Electroforming Process," *J. Mech. Eng. Sci.* **18**(6):271-278 (1976).
57. Kelly, W. K., J. W. Dini and C. M. Logan. "Electroforming Copper Targets for a Neutron Source Facility," *Plating and Surface Finishing,* **69**(2):54-59 (1982).
58. Saubestre, E. B. *Metal Finish.* **56**(11):62, 67 (1958).
59. Hothersall, A. W. and G. E. Gardam. *J. Electrodepositors' Tech. Soc.* **27**:181-199 (1951).

Gold and Platinum Metals

60. Hydes, P. C. and H. Middleton. "The Sulphite Complexes of Gold—Their Chemistry and Applications in Gold Electrodeposition," *Gold Bull.* **12**(3):90-95 (1979).
61. Schlain, David, Frank X. McCawley and Gerald R. Smith. "Electrodeposition of Platinum Metals from Molten Cyanides—Technique Applicable to Electroforming," *Platinum Metal Rev.* **21**(2):38-42 (1977).
62. Mason, D. "Electroforming Crown," *Electroplating and Metal Finish.* **22**(7):26-28 (1969).
63. Rubenstein, M. *Metal Finish.* **54**(2):52-56; **54**(3):56-59; **54**(4):58-60 (1956).
64. Mohan, A. "Electroforming of Gold," *Gold Bull.* **8**(3):66-69 (1975).

Composite Materials

65. Malone, G. A. "Electrodeposition of Dispersion Strengthened Alloys," Symposium on Electrodeposited Metals for Selected Applications, Battelle Memorial Institute, Columbus, Ohio, November 1973.
66. Sykes, J. M. and D. J. Allner. "Mechanisms for the Formation of Electrodeposited Composite Coatings," *Trans. Inst. of Metal Finish.* **52**:28 (1974).
67. Harris, S. J. and P. J. Boden. "Electroforming with Composite Materials," *Electroplat. Metal Finish.* **26**(5):9-13 (1973).
68. Bazzard, Rosemary and P. J. Boden. "Codeposition of Chromium Particles in a Nickel Matrix," *Trans. Inst. Metal Finish.* **50**(2):63-69 (1972).
69. Wallace, W. A. and V. P. Greco. "Electroforming High-Strength, Continuous, Fiber-Reinforced Composites," *Plating* **57**(4):342-348 (1970).
70. Cooper, G. "Forming Processes for Metal Matrix Composites," *Composites* **1**(3):153-159 (1970).
71. Harris, S. J., A. A. Baker, A. F. Hall and R. J. Bache. "Electroforming—Filament Winding Process—Method of Producing Metal Matrix Composites," *Trans. Inst. of Metal Finish.* (Winter) **49**(5):205-213 (1971).
72. Snaith, D. W. and P. D. Groves. "Some Further Studies of the Mechanism of Cermet Electrodeposition," *Trans. Inst. Metal Finish.* (Autumn) **55**(3):136-140 (1977).

Applications

73. Watson, S. A. "Applications of Nickel Electroforming in Europe," *Plating and Surface Finish.* **62**(9):851-861 (1975).
74. Schaer, G. R. and T. Wada. "Novel Equipment for Electroforming Flexible Circuits," *Prod. Finish.* **45**(6):90-92 (1981).
75. Pellegrino, P. P. "Electroforming Circuit Board Process," *Proceedings of the Symposium on Plating in the Electronics Industry,*

published by American Electroplaters' Society, Winter Park, Fla., 1981, p. 9.
76. Schaer, G. R. and T. Wada. "Electroformed Flexible Printed Circuit," *Proceedings of the Symposium on Plating in the Electronics Industry*, published by American Electroplaters' Society, Winter Park, Fla., 1981, p. 16.
77. Grazen, A. E. "New Process for Making Diamond Rotary Dressers," *Cutting Tool Eng.* **33**(1-2):19-20 (1981).
78. Wearmouth, W. R. "Applications and Developments in Nickel Electroforming in Toolmaking," *Met. Finish.* **78**(11):35-39 (1980).
79. Lai, S. H. F. and J. A. McGeough. "Electroforming and Mechanical Properties of Iron-Nickel Alloy Foil," *J. Mech. Eng. Sci.* **21**(6):411-417 (1979).
80. Young, J., F. Ogburn and D. Ballard. "Electroforming a Micrometer Scale of 50 Microns Over-All Length," *Met. Finish.* **78**(8):27-29 (1980).
81. Lai, S. H. F. and J. A. McGeough. "Electroforming of Iron Foil," *J. Chem. Technol. Biotechnol.* **30**(1):7-13 (1980).
82. Berdan, B. L. and B. M. Luce. "Process for Electroforming Copper Foil," Gould, Inc., U.S. Pat. No. 4,169,018, Jan. 16, 1978, Off. Gaz., Sept. 25, 1979.
83. O'Tousa, J. E. "Electrodeposition Applications for the Space Shuttle Main Engine," *The Enigma of the Eighties—Environment, Economics, Energy*, Vol. 2. San Francisco, Calif., May 8-10, 1979. publisher Society for the Advancement of Material and Process Engineering. Azusa, Calif.
84. Lai, S. H. F. and J. A. McGeough. "Electroforming of Iron Components," *Trans. Inst. Met. Finish.* **57**(2):70-72 (1979).
85. Wilson, A. R. and W. Quartz. "RF Cavities by Electrodeposition," *Engineering* **218**(11):1214-1215 (1978).
86. Stagg, K. J. "Development of Electroforming with Particular Reference to Record Industry Requirements," *Prod. Finish.* (London) **31**(9):27-28, 30-31, 65, 67, 69 (1978).
87. Anon. "Precision Reflective Optics by Electroforming Technique," *Finish. Ind.* **2**(11):40, 42 (1978).
88. Anon. "Mass Production of Reflective Optics Given Major Assist by Electroformed Nickel," *Nickel Topics* **31**(2):11-12 (1978).
89. Campbell, Paul. "Guide to Metal Bellows," *Des. News* **34**(2):46-48 (1978).
90. Anon. "Advanced Rocket Engine Thrust Chambers Made by Nickel Electroforming," *Nickel Topics* **30**(4):14-15 (1977).
91. Dini, J. W. and H. R. Johnson. "Electroforming of a Throat Nozzle for a Combustion Facility," *Plat. Surf. Finish.* **64**(8):44-51 (1977).
92. Wearmouth, W. R. "Application of New Developments in Electroforming Technology in the Toolmaking Industry," *Interfinish 76—Proc. 9th World Congress on Metal Finish.*, Amsterdam, 1976, 21 p.
93. Bhagwat, M. J., R. Raman and D. L. Roy. "Electroforming of Porous Ni Tubes," *Metall. Eng.* IIT Bombay, **7**:9-13 (1975-76).
94. Johnson, H. R. and J. W. Dini. "How to Fabricate Thin Electroforms," *Prod. Finish.* **40**(7):46-52 (1976).
95. Johnson, H. R. and J. W. Dini. "Fabricating Closed Channels by Electroforming," *Plating Surf. Finish.* **62**(5):456-461 (1975).
96. Redy, A. L. "Behavior of Electroformed Copper and Nickel Waveguides Regarding Dimensional Stability During Heat Process," *Proc. 13th Sem. Electrochem.*, Central Electrochem. Res. Inst. India, 1974, pp. 364-372.
97. Reiner, Herbert W. "Electroforming and Electroplating as Art," *Plating* **62**(3):211-213, 217 (1975).
98. Wearmouth, W. R. and K. C. Belt. "Mechanical Properties and Electroforming Applications of Nickel-Cobalt Electrodeposits," *Trans. Inst. Metal Finish.* **52**(3):114-118 1974).
99. Hart, A. C. "Electroforming of Holloware," *Metals Australia* **6**(8):261-265 (1974).
100. Dugan, W. P. "Miniature Electroformed Waveguide With Flanges Attached," *Plating* **61**(11):1019-1021 (1974).
101. Dugan, W. P. "Tube Circuits Produced by Electroforming and Chemical Milling," *Plating* **61**(6):569-570 (1974).
102. Anon. "Nickel Electroforming for Die Production," *Mach. Prod. Eng.* **27**(2):244-246 (1974).
103. Stauffis, C. R., G. A. Malone and Rudolf A. Duscha. "Nondestructive Evaluation of Regeneratively Cooled Thrust Chambers for Rocket Engines," *Proceedings of the 1979 Symposium on Electrodeposited Metal for Selected Applications*, April 1974, pp. 65-79.
104. Anon. "Electroformed Electrodes for EDM," *Tool Prod.* **40**(1):72-73 (1974).
105. Vijayasemha, C. R. and M. A. Tirunarayanan. "Electroforming Metal Bellows," *J. Electrochem. Soc. India*, **22**(2):96-99 (1973).
106. Guha, B. R., V. Venkataraman, D. C. A. Narayan and P. C. Debnath. "Surface Roughness Comparison Specimen by Electroforming Process," *Trans. Soc. Advancement Electrochem. Sci. Technol.* **8**(2):54-56 (1973).
107. Watson, S. A. "Use of Electroforming as a Production Process," *Metals Australia* **5**(4):107-113 (1973).
108. Wearmouth, W. R. and K. C. Belt. "Thick Nickel-Cobalt Alloy Electroforms Suitable for Zn Base Die-Cast Molds," *Electroplating Met. Finish.* (July) **26**(7):9-11, 13-14 (1973).

109. Hart, A. C. "Electroforming of Holloware," *Indust. Finish.* **25**(299):14, 16, 18, 20 (1973).
110. Wills, K. G. "Electroforming Waveguides," *Electroplating* **26**(6):9-10, 12-13, 26 (1973).
111. Dugan, W. P. "Vacuum-Formed Anode Shields for Electroforming Tube Circuits," *Plating* **60**(4):362-364 (1973).
112. Wilson, H. G. E. "Electroforming Moulds, Heat Exchangers, Electrodes," *Electroplat. Met. Finish.* **26**(2):29, 31 (1973).
113. Trausch, G. "New Photolithographic Pattern Generation Process for Producing Precision Flat Parts by the Electroforming Method," *Siemens, Erlangen* **8**(6):347-351 (1979) (In German).
114. Squitero, A. D. "Electrodeposition of Iron," *Products Finish.* (March 1951). Also see: Max, A. M. "Iron Plating," *Metal Finishing Guidebook and Directory Issue 82*. Metals and Plastics Publications, 1982, pp. 282-284.
115. Schmidt, F. J. and I. J. Hess. "Electroforming Aluminum Mirrors," *Plating* **53**:183-192 (1966). Hanson, R. et al. *Plating* **55**:347-355 (1968).
116. Mellors, G. W. and S. Senderoff. *Plating* **51**:972-975 (1964).
117. Ponnuthurai, M. and S. Chakrapani. "Electroforming from Sulfamate Solutions," *Finish. Ind.* **2**(10):20, 22, 25 (1978).
118. Popereka, M. Ya. "Internal Stresses in Electrolytically Deposited Metals," published for the National Bureau of Standards and the National Science Foundation, Washington, D.C. by the Indian National Scientific Documentation Center, New Delhi, 1970. Document Number TT 68-50634.
119. Carlin, F. X. *Plating* **55**:148-151 (1968).
120. Kendrik, R. J. "High Speed Nickel Plating from Sulfamate Solutions," *Trans. Inst. of Metal Finish.* **42**:235 (1964).
121. Wesley, W. A. and E. J. Roehl. *Plating* **37**:142-146, 171 (1950).
122. Carr, D. S. *Plating* **43**:1422-1429 (1956).
123. Bull, A. W., J. W. Bishop, M. H. Orbaugh and E. H. Wallace. *Metal Finish.* **37**:461-464 (1939).
124. Kasdan, A. S. *34th Proc. Amer. Electroplaters' Soc.*, pp. 115-121 (1947).
125. Schaer, G. and P. Krasley. "Electroforming Accelerated by Forced Solution Flow," *Plat. Surf. Finish.* **66**(12):36-38 (1979).
126. Evans, C. J. "Tin Alloys Aid in Electroforming Industrial Components," *Australas. Corros. Eng.* **22**(9-10):22 (1978).
127. Missel, L. "Functional Copper-Tin Alloy Plating," *Plat. Surf. Finish.* **65**(10):36-40 (1978).
128. Yamamoto, M., Y. Shimizaki and T. Sato. "High Speed Nickel-Cobalt Alloy Electroforming by High Current Density with Flowing Electrolytes," *Mem. Fac. Eng. Hokkaido Univ.* **14**(3):45-54 (1976).
129. Ormerod, R. A. and R. W. A. Gill. "Electroforming for Electronics," *Trans. Inst. Metal Finish.* **51**(1):23-26 (1973).
130. Watson, P. "Applications of Alloy Electroforming," *Electroplat. Met. Finish.* **26**(3):7-8, 12-14 (1973).
131. Kendrick, R. J. "Production and Uses of Electroformed Strip and Electroformed Components," *Metal Finishing J.* **18**:71-72, 75 (1972).
132. Herridge, F. W. "Electroformed Fabrications Incorporating Auxiliary Components," *Machinery (London)* **120**:81-82 (1972).
133. Anon. "Electroforming on Tin Alloy Mandrels," *Tin Its Uses* **90**:9-11 (1971).
134. Johnson, H. R. "Electroforming Mates Ultra-Precision Punch and Die," *Mod. Mach. Shop* **43**(10):52-56 (1971).
135. Horschemeyer, W. "Seamless Copper Crucibles for Vacuum-Arc-Melting Furnaces," *Copper* **2**:13-15 (1970).
136. Lowry, Lloyd. "Impossible Parts—Photo Electroforming May be the Answer," *Plating* **58**(3):199-200, 202 (1971).
137. Dini, J. W. and H. R. Johnson. "Separating Large Thin Electroformed Parts from Mandrels," *Metal Finish.* **68**(9):52-55 (1970).
138. Davis, Leroy W. "Methods of Making Metal-Matrix Composites," *Society of Manufacturing Engineers Journal* **70**(4), 17 page pamphlet, Dearborn, Mich. (1970).
139. Rutter, J. "Electroforming Small Components for Laboratory Studies," *J. Physics E-Sci. Instrum.* **3**(7):568-569 (1970).
140. Passin, George E. "Method for Forming Very Small Diameter Wires," *Rev. Sci. Instrum.* **41**(5):772-774 (1970).
141. Bailin, P. "Electroforming Automation at Columbia Records," *Plating* **56**(6):658 (1969).
142. Schneck, R. and R. Edwards. "Electroformed Tooling for Reinforced Plastics and Advanced Composites," *Plast. Technol. Adv. 1980 Update*. Published by the Society of Plastic Engineers, Brookfield, Conn., 1980, pp. 39-46.
143. Crichton, W. A., J. A. McGeough and J. R. Thomson. "Effects of Cathode-Mandrel Surface Irregularities on the Mechanical Properties and Structure of Electroformed Iron Foil," *J. Mech. Eng. Sci.* **22**(2):49-54 (1980).
144. van Delft, J. Ph., J. vander Waals and A. Mohan. "Electroforming of Perforated Products," *Trans. Inst. Met. Finish.* **53**(4):178-183 (1975).
145. Omerod, R. A. and R. W. A. Gill. "Electroforming for Electronics," *Trans. Inst. Met. Finish.* **51**(1):23-26 (1973).

2 / ENGINEERING FUNDAMENTALS AND PRACTICE

17. PLANT LOCATION AND LAYOUT

Simon P. Gary

President, Scientific Control Laboratories, Chicago, Illinois

The location and layout of a plating installation will have a tremendous effect upon a facility's

- Quality of plating
- Labor efficiency
- Energy efficiency
- Ease of environmental compliance
- Quality of working environment
- Transportation costs
- Labor costs
- Labor availability
- Equipment life
- Ability to accommodate changes in product mix
- Flexibility to allow changes in processes.

Once a plant is constructed and the equipment installed, it is extremely expensive to change the location or plant layout. For this reason it is imperative that a great deal of advance planning be done so that these decisions will be the wisest possible decisions for the present and the future.

There is no clear-cut, scientific formula for evaluating the best site selection or equipment layout—nor is there any one best way. The final decisions will be reached through a series of trade-offs and judgmental decisions. These decisions can be reached based on the wisdom and preferences of a single individual, or they can be team decisions. In either case, a chart similar to the chart at the end of this chapter should be used as a check list to make sure that all pertinent factors have been considered. "Importance value" on a 1 to 10 basis should be assigned to each item. The ratings will be entirely arbitrary and will vary for different plants; and yet, there will be some universality to them. By assigning numerical value ratings one can minimize the likelihood of placing too much importance on something relatively unimportant or failing to give primary consideration to something very important.

Once a plant is erected and the equipment installed the plant will either prosper or flounder as erected. Proper advance planning will virtually assure achieving the goal of any change in plant location and plant layout; that is, greatly improved productivity and profitability. Before discussing the specifics of location and layout, let's look at the more obvious effects they have upon a company.

Quality of Plating

One's first reaction might be that quality of plating is dependent upon the processes selected and the proficiency of personnel. This is only partially true. Plant location and, especially, plant layout can have a tremendous positive or negative effect on the quality of plating. For example: If equipment is laid out with adequate, proper sized rinsing, good aisle space and easy access to controls and dials, one increases the likelihood of high-quality plating. If, however, the plant is laid out so that finished work is stored near an acid tank giving off corrosive fumes which are inadequately vented, one is asking for trouble.

Labor Efficiency

Equipment selection and plant layout have a very great impact on labor efficiency. For instance, a highly efficient, fully automatic plating machine will be severely hobbled if one has not provided sufficient staging area for unplated work, a convenient location for proper racking, or adequate loading area. On the other hand, production facilities may be laid out in such a way that the work may be processed almost untouched by human hands and labor productivity will be extremely high.

Energy Efficiency

Heating, electrical energy and water consumption costs are obviously reducible through wise location and equipment layout.

Ease of Environmental Compliance

It is almost impossible to achieve environmental compliance (water, air and solid waste) in a plant that is poorly located or poorly laid out. For many plants, the sole purpose in relocating or rearranging the plant is to be able to economically and reliably achieve environmental compliance. There is also a wide variation in environmental regulations depending on plant location. Sometimes it's a question of which side of a state border one is on; more often, it's whether one is within or outside of a metropolitan area and whether one discharges to a sewer system having high treatment capabilities or to a heavily-fished stream, to cite two extremes. Also, some states offer economic incentives (no sales tax on pollution control equipment; pollution control bonds, etc.) to encourage efforts to achieve environmental compliance; others do not.

Quality of Working Environment

With proper layout, a plant can comply with all OSHA regulations and provide a pleasing working environment at nominal additional cost. As a matter of fact, it is usually more expensive to operate a slop house than a showplace. The working environment also affects labor availability, quality and efficiency.

Transportation Costs

Locate your plant loading dock so it's only access is from a busy street in a major metropolitan area or build an undersized dock, and you will see one example of the impact of location and layout on transportation costs. Access to expressways, distance to major customers, loading and unloading facilities severely affect transportation costs.

Labor Cost and Labor Availability

The effect of location on labor costs and labor availability are direct and indirect. The direct effects are items with which management is well familiar and upon which they can easily reach decisions. However, the indirect effects are often less apparent but equally important. Indirect factors such as parking space, lunch room, locker room, public transportation, lighting and ventilation affect labor costs and labor availability and are largely determined by location and layout.

Equipment Life

Equipment life is primarily determined by the sturdiness of the equipment and the quality of maintenance provided. However, the plant layout may have a great affect on the degree of maintenance required and the likelihood of it being provided. For example, placing an air cooled rectifier so that it will draw its air from a dust-laden area or placing a motor requiring frequent oiling in a difficult spot will invite early equipment failure.

Flexibility to Adapt to Change in Product Mix or Processes

Many shops are now operating at less than peak efficiency because they have had to adapt to a new process or a new product and the old installation just couldn't be successfully changed. Whether you assign flexibility an "importance value" rating of 1 or 10 is your decision—but do give it consideration.

PLANT LOCATION

Regarding location, there are considerations that are related to what part of the country a plant shall be located in. We call these geographical considerations. There are also considerations of where a plating department shall be located within a manufacturing complex. We have identified these as internal considerations.

The importance, and freedom, of choice regarding geographical location will be different for each plant. For one job shop, proximity to a major customer may pinpoint the ideal location, while another job shop with a large customer list may be free to locate anywhere within a 100-mile radius. Some captive plating shops must locate at the site selected for the primary manufacturing operation; other captive shops have such an impact on the entire facility that the entire plant will be located at the site most advantageous for the plating operation. The order of importance of the factors to be considered when deciding upon plant location will vary for each plant; but careful evaluation will help sort them into "absolute musts," "nice to have" or "somewhere in between." Items which should be considered include the following.

Zoning

Plating companies are usually considered undesirable neighbors. They are therefore usually discriminated against in local zoning and land use ordinances and are restricted to areas which are zoned "heavy manufacturing" or "limited use." Be sure that the site you have selected will allow you to operate a plating department in full compliance with all zoning regulations. A lawyer is almost a must for this verification.

Neighbors

Plating companies are usually noisy and their exhaust air usually has a distinct odor. Even if the site you select is zoned for plating, you may encounter resentment from your neighbors if they are in a different type of business and don't understand the problems that go with some manufacturing operations.

All exhaust fans should exhaust 20 ft higher than any building within 100 ft. For this reason, you would not want to put a one-story plating shop right up against a three-story or four-story office or manufacturing facility. Some local regulations preclude a plating company within 1000 ft of a school or hospital.

Try to have neighbors you can get along with and then be a good neighbor.

Labor Cost and Availability

The factors to be considered here are economic and esoteric and there are usually a lot of trade-offs. Frequently, a location that provides access to low-cost, unskilled labor is unappealing to semi-skilled and skilled labor. Public transportation is usually more important to unskilled than skilled labor. Some states have more friendly labor laws than others. Labor attitudes sometimes change in as short a distance as 30 or 40 miles. Labor taxes including head taxes and city income taxes will have an impact on labor costs and availability. Hospitalization costs will sometimes change with location. These factors are common to all businesses, and local business development groups, chambers of commerce and manufacturers' associations are good sources of information about them.

Utility Costs

Thirty years ago utility costs were a small factor in the plating industry, while labor costs were a major factor. The gap has closed considerably, not because labor costs have gone down, but because utility costs have escalated. Once a plant is located in a specific site there is not much one can do about utilities costs except economize, but there's quite a bit one can do about these costs when making a site selection. Gas, electricity, telephone, water and sewage costs are not uniform across the country. Sometimes they vary widely within a 2-mile span. The same can be said for taxes and insurance. It is important that one not only determine these costs for various optional site locations, but it is equally important that one make an appraisal of what the future trend will be. Some plants have walked into a trap of cheap tax rates and economic incentives only to find that, after 3 or 5 years, the "honeymoon" was over and agencies and utilities began to "pluck the golden goose" at an alarming rate. We are considering taxes as a utility cost and would suggest they be evaluated in relation to the services provided, such as police protection, fire protection, sewage facilities, etc.

Environmental Regulations

In spite of efforts of the U.S. EPA to promulgate National Standards, environmental regulations still vary widely across the country. We predict that this divergency will continue far into the future. Later on, we will discuss how one can lay out a plant to more easily achieve environmental compliance, but in this section we want to more thoroughly discuss the fact that plant location will have a great effect on the specific requirements one must meet in order to obtain environmental compliance.

Probably the greatest variation in environmental regulation for a plating shop will be between the extremes of a rurally located shop discharging moderately large quantities to a small fishing stream (a direct discharger) as compared to a plating shop located in a large metropolitan city having an efficient, well-equipped, enlightened, sanitary district. The direct discharger will be completely on his own, regarding waste water treatment. He will have to run his shop in such a way that he will never discharge a waste water that would be harmful to aquatic life. Note that effect on aquatic life is the criteria for stream discharges. Animals and humans can tolerate more pollution than can fish, so the regulatory approach is to set the limits so strict that fish and other aquatic biota will be unaffected and, therefore, humans and other animals will be more than adequately protected. I don't think anyone really quarrels with this, but it does mean that the direct discharge plater will have to run a much tighter ship than his city competitor.

When a plater in a large metropolitan city discharges to a sewer, he is part of a waste treating team. He and the sanitary district, together, must produce a discharge from the sanitary district that will be harmless to fish or ground water supplies. Together they must achieve the same degree of treatment as the direct discharge plater. This usually means that the plater only need partially treat (this is called pretreatment) to a lesser degree than would be required for a direct discharger. Furthermore, if the plater commits a minor error, there is a high likelihood that the treatment afforded by the sanitary district will serve as a backup and will correct for the plater's error, the net result being that the environment will not be adversely impacted.

A middle of the road approach to selecting a plant location based on environmental regulation considerations is probably the best. One might be tempted to choose to locate in an area where almost anything goes environmentally, and then regret the decision when a very offensive company locates next door for the same "who cares" attraction. It is also highly likely that the area which has extremely loose environmental control will become enlightened at a later date, with resultant backlash and very harsh regulations.

Internal Considerations

When a plating plant is part of a larger manufacturing complex the question of location is where it should be located in relation to the rest of the manufacturing facility. Twenty or thirty years ago, plating departments were located like outhouses—"keep it away from me, but close enough to be convenient." We have come a long way since then. It is entirely feasible to now put a plating department at any location within a manufacturing complex.

Fifty years ago, there was a strong tendency to locate plating departments on the top floor of multi-storied buildings or in the far corner of single-story complexes. The reason was that the plating departments usually gave off obnoxious, corrosive fumes. These fumes tended to float upwards and out the open window and therefore the desirability of these locations. However, platers use a lot of water, and sometimes they encounter floor spills; water tends to run downhill and the upper story or off-in-the-corner locations sometimes resulted in bad leaks into other departments. These mishaps resulted in a trend to put the plating department in the basement or at least a foot or two below the rest of the manufacturing complex. Location alone will not solve either of these problems, but a wise choice of location will simplify the needed requisite plant engineering.

The factors which should be considered when deciding on the in-plant location of a plating facility are given below.

Ventilation. Plating departments will exhaust a relatively high amount of exhaust air. This air must come from somewhere. In northern climates it will come from make-up air units or other departments; in southern climates it will come from open windows or doors, and from other departments. If the plating department is located immediately adjacent to a painting operation that is exhausting more air than the plating department, the painting department will be under relative negative pressure compared to the plating department, and fumes may reach the painting department and seriously impact the quality of painting. Conversely, if the plating department is under negative pressure compared to an adjoining screw machine installation that has fine oil mists in the air, the plating department is going to have a lot of unnecessary headaches.

The point is, when locating a plating department, one must consider not only the ventilation requirements of the plating department but their impact on the ventilation system of the entire plant.

Floor Loading Capacity/Ceiling Height. Plating solutions and plating equipment are heavy and one must be sure to provide adequate floor loading capacity. Heavy floor load capacity is less apt to be available on upper floors and this often affects department location.

It's nice to have 12-ft ceiling heights even for

jewelry or laboratory plating, while almost all other plating requires 16 or 18 ft minimum of clear ceiling height. The present trend is to provide 20 ft of clear ceiling height. Added ceiling height gives greater flexibility regarding selection of automatic equipment, conveyorizing, etc.

Work Flow. We have mentioned the previous considerations regarding plating department location first so they will not be forgotten, but we feel the overwhelming criterion for plating department location should be a location which will most favorably affect the work flow throughout the manufacturing complex. Locate the plating department where it will be most convenient considering the entire manufacturing operation and then design the plating department so that location will be satisfactory. If location makes absolutely no difference to work flow, then give credence to those peripheral considerations, such as proximity to sewer connections, outside fresh air and incoming utilities.

PLANT LAYOUT

The plant layout (and by this we mean positioning of the major pieces of plating equipment and support facilities) will have a major effect on how smoothly and efficiently work moves through a plating installation. Consideration should be given to plant layout when making a new installation, when adding a major piece of new equipment and sometimes just for the sake of rearranging and improving efficiency.

When considering how to efficiently lay out a plating facility, one must first gather all the pertinent information about the "what" and "how" for that plant. These should be tabulated in writing and discussed with everyone concerned until there is complete agreement that nothing has been overlooked.

Under "what" one should consider not only what finishes are to be applied (zinc, nickel, anodize, etc.) but each and every operation and facility associated with applying these finishes, including:

- Receiving of material to be plated
- Moving material to the plating department
- The entire plating process including racking, plating, unracking and packing
- Plating related procedures including stripping, solution filtration, solution maintenance and waste treatment
- Supporting functions and facilities such as laboratory, maintenance, inspection, toilets, locker rooms, lunch rooms, office facilities and employee parking.

One should also prepare a "how" list. This list can be a side-by-side companion to the "what" list or a separate free-standing list. It should detail "how" each operation will be accomplished including production rates, personnel required, equipment required, floor space required and utilities required.

When the "what" and "how" lists are complete, a large plat of the facility should be prepared and placed on cardboard or other soft backing. Cut-outs should be prepared to scale depicting the floor space required for each function. These cut-outs can then be pinned on the plat and moved around at will to experiment with various plant layouts. Through a cut and try process an efficient layout will evolve.

When preparing the plat and cut-outs, remember that many people have trouble visualizing concepts and space relationships. An oversized plat will tend to make aisles and staging areas appear larger than actual while too small a plat will make everything seem convenient, when in actuality it may be too spread out. Too small a plat may also be inconvenient for group consideration. For most installations a plat scale of ½ in. equals 1 ft will be satisfactory.

The process of preparing a plant layout is usually rather straightforward. The pitfalls to avoid are:

- Superficiality resulting from failure to pay attention to details
- A lack of imagination coupled with a lack of a venturesome spirit.

Don't gamble too heavily on the untried, but to do something new. You will never have a better opportunity to innovate and make major improvements.

Some of the pertinent data which must be considered when planning a plant layout are as follows.

Desired Production. A careful appraisal must be made of the desired production capacity for present needs, for near-term goals and for long-term goals. Underestimate and you are a dead duck; overestimate and you'll have to pray for growth.

What Processes are to be Used—Now and Later. In one sense, this is an aspect of productive capacity, but, in another sense, it should be considered separately. Processes do change and job shops and captive shops must decide if there is a likelihood of change and plan accordingly. For example, a furniture manufacturer who senses a trend from bright chrome to brass or black—or an electronics plater who senses a trend from alkaline tin to bright solder or a hard chrome plater who knows that some of his plating will shift to electroless nickel—may want to design to be able to accommodate such changes. On the other hand, a barrel zinc plater may feel comfortable putting in a highly efficient, but almost unchangeable, plating system. Somebody has to gather the facts and make this decision.

Equipment Selection. The selection of plating equipment is a subject all its own, but it intertwines with plant location and plant layout. Sometimes ceiling heights or available floor space will limit the options of equipment selection, and some types of equipment lend themselves more readily to a specific plant layout than do others. Sometimes it's desirable to evaluate two or three types of plating equipment and two or three different plant layouts in order to plan the best plant layout/equipment combination.

Support Facilities. When planning a plating plant layout, one usually properly considers the major items of production equipment, but the support equipment and facilities are sometimes given insufficient consideration. The major characteristic of an excellent plant layout is how well the support facilities mesh into the total system. The little details are usually what spell the difference between success and failure. Some of the more obvious and, hopefully, some of the less-than-obvious, are:

1. *Storage and staging areas.* The need for these is so obvious that one is shocked to find that they are frequently undersized or poorly located. It costs money to move material and it costs a lot of money to move it inefficiently or unnecessarily.

2. *Shipping and receiving areas.* The location and design of dock facilities and other shipping and receiving facilities will determine how efficiently you handle materials. They will also determine whether truckers are glad or reluctant to service you, and the degree of service you are able to give your customer.

3. *Maintenance, rack storage, etc.* Don't be stingy! Provide good facilities with a little extra space. Regarding plating racks—if a variety of racks are being used, serious consideration must be given to how the racks not in use will be stored. Some companies store unused racks on movable carts; others have overhead conveyor storage facilities. Rack life will be greatly lengthened if good storage is available. Also, the time required to switch from one job to another will depend upon rack storage. Give it serious thought.

4. *Amenities.* Office facilities, wash room facilities, locker rooms, recreational area, etc., should all be considered when preparing a plant layout.

5. *Storage of Supplies.* When providing storage for chemicals and supplies, one should consider OSHA requirements, security and convenience. It sounds cut and dried until one asks whether money could be saved by buying and storing bulk acids, should one provide secure storage in order to play the market and buy cadmium or nickel during dips in the market, or should one minimize storage space to force reducing inventory of supplies to a bare minimum to reduce working capital and reduce exposure to theft?

Waste Treatment Consideration. While not strictly related to plant layout, two factors are worth emphasizing which will greatly facilitate compliance to waste water regulations. They are rinsing facilities and waste flow.

Rinsing Facilities. For hoist lines or automated lines, every plating operation should be followed by two rinsing processes and every major source of pollution (such as a plating tank) should be followed by a minimum of three, and preferably four, rinses. Three or four rinses may be used in different modes such as one drag-out and three counter-flow rinses or two recirculating recovery rinses and two counter-current rinses or a closed-loop, four-rinse system. The mode will depend on the waste treatment technology and the process involved; but, if one provides three or four rinses, the rinsing mode can easily be changed any time to accommodate present and future technology.

For hand lines and low-volume lines, one may wish to provide only one rinse after cleaning and pickling and only two or three after plating, etc., in order to reduce operation labor. This will be a judgmental decision balancing labor costs against water cost, ease of waste treatment and floor space.

Waste Flow. A sloppy, wet, plating room floor should be a thing of the past. (There are some installations that are well-designed where the entire floor is washed into a sump and the resulting flow of water properly treated for discharge, but even in these installations there is no haphazard, let-it-run-on-the-floor approach.) Every drop of discharge water should be discharged, as per a plan, to its proper stream for treatment. Make maximum use of gravity flow. Proper layout may require slightly elevating some rinse tanks and process tanks in relation to other tanks but it is worth it compared to having a multitude of sump pumps and transfer pumps.

To summarize, in laying out a plating plant, the primary consideration should be to move the work into and out of the department with maximum efficiency and to provide convenient support facilities.

EXAMPLE OF PLANT LAYOUTS

Just to stir your imagination, we present plating tank layouts of a few plating installations. In two cases we do not show shipping and receiving areas because they are not germane to what we are illustrating; in the other, we show them because the way they are integrated into the plating process is very significant.

The Common Well

We have assigned this terminology to those installations where all of the cleaning, acid treating and post-plating rinsing and drying are done in a "common well" or "kitchen sink" and the plating processing is done in surrounding process tanks. This configuration is especially advantageous when a wide variety of finishes must be plated but the volume of any one finish is not large. It is also advantageous when the plating times are long in relation to the time required for cleaning and rinsing. Two situations that are served well by such a line would be jewelry finishing or hard chrome plating. The disadvantages of this tank configuration are:

- *Capacity.* It's almost certain the cleaning and rinsing line will not be adequate to service all of the surrounding tanks if they are to be kept full of work at all times.
- *Uncontrolled floor spills* that may make waste treatment difficult are highly likely with this configuration.

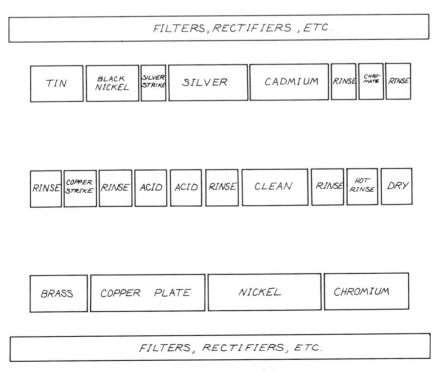

FIG. 1. A common well tank layout.

- *Lack of automation.* It is virtually impossible to automate this configuration.

Common Well Plus Transfer

When the common well configuration becomes too large or complex, it loses its inherent advantages and its limitations become serious. When this happens, one can modify the common well by installing a transfer station to move parts from the common well to the process tanks and back. One can easily partially automate such a configuration, and microprocessors are available which will fully automate such a line. This configuration maintains the advantages of minimum walking, ease of control and flexibility. It also overcomes the disadvantages of the previously illustrated common well because it also offers high production capacity, virtual elimination of floor spills and adaptability to water conservation and waste treatment. The major disadvantage is that the process solutions will probably not be operable at full capacity because the cleaning and rinsing line capacity will be the governing factor. Another disadvantage is that the person loading the work into the line must be able to look ahead and schedule his loading so that some processes are not overloaded while others starve for work.

The configuration shown is an actual installation. The thing that makes this installation significantly efficient is that more than 60% of the plating done by this company is zinc plating. The cleaning line, the post treating line and the zinc plating line are controlled by a microprocessor controlling two hoists. When plating zinc, the combined unit operates as a full, automatic, straight line hoist machine, operated by one man. This is usually done for the last hour of the day shift and all of the second shift. When zinc is not being plated, the microprocessor controls the hoist, servicing only cleaning, pickling, rinsing, etc., and two operators control the movement of parts from the transfer machine through the various plating processes and back to the transfer machine. When operating in this mode, it's a three-man operation; one loading and unloading and two doing the plating processing.

The Big Circle

For really big production of essentially the same finish, one will surely want to consider placing the plating process tanks in a big loop. The loop may be a circle, an oval, a rectangle or a back and forth loop as we have chosen for our illustration. All return type full automatic machines are of this configuration. It's just loaded with advantages. The major disadvantage is a lack of flexibility. Once the big loop has been installed, it's not easy to make changes.

The company we have chosen to illustrate this configuration plates wire goods with nickel/chrome or with bright nickel/brass and lacquer. The parts are conveyed through the various processing tanks on a conveyor system that dips down and rises up to transport the parts into, through and out of the various tanks. This conveyor mechanism is sometimes called a "dipsy-do" or "roller coaster" mechanism. Everything that's plated is hung from a conveyed hook but small "s" hook adaptors are used if required to string two or three small parts together to load the tank well. In effect then, different racks ("s" hook configurations) are used but the basic rack (primary hook) is never taken off the conveyor or changed. As can be seen on the drawing, the chromium, and brass tanks are moved into or out of the line depending on whether the parts are to be chromium plated,

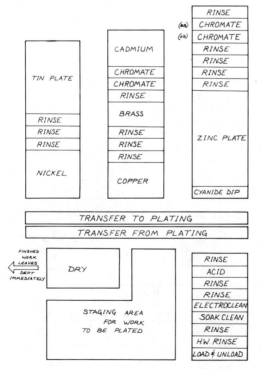

FIG. 2. Common well plus transfer tank layout.

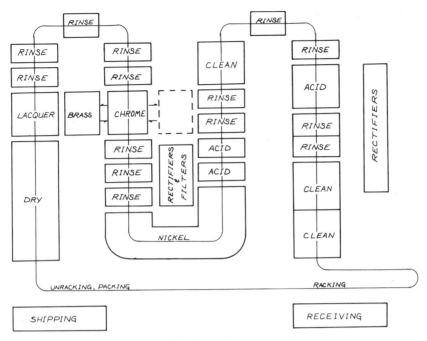

Fig. 3. The big circle tank layout.

nickel plated, or nickel/brass plated. Skip mechanism on the conveyor hoist assures rinsing of the chromium or brass in the proper rinses and skipping lacquering of chromium plated parts. It should be noted that by using double drag-out rinses after nickel and chromium, and a small evaporative recovery unit after brass plating, this company has achieved full compliance with a moderately stringent sewer pretreatment requirement. The really startling thing about this installation is the efficiency achieved in loading, unloading, etc. The company is a job shop servicing five or six major accounts. All work comes to the plant in semi-trailers. These trailers are parked at the "incoming" docks and the trailers sit there until it's time to plate the parts within them. When it's time to plate, the skids or packing cartons are pulled out of the trailers sufficiently to allow for racking of the parts right at the end of the trailers. The empty containers are moved over to the shipping area after plating. The parts are unracked in the shipping area, placed back into skids or cartons and slid into a trailer. Ninety percent of the work done by this plant is loaded from the trailers, onto a plating rack, and loaded back into the trailers. Plating is almost incidental. The total operation is highly efficient.

Straight Line

There are times when it's highly effective to lay all the tanks out in one continuous line and then move the parts through whatever sequence is desired. This is very effective for intermediate production rates. If the line becomes extremely long and a lot of different finishes or processes are involved it is usually necessary to service the line with two or three hoists; or, if it is manually operated, the operator soon feels he's on a cross-country hike each day. If multiple hoists are required, the operators sometimes get into each other's way. Not that these problems are not insurmountable, but they do place production limitations on this configuration.

Sometimes a solution to production limitations for a straight line layout serviced by hoists is to increase the width of each process tank so that the hoist can pick up more work with each movement. This works to a point, but when one goes beyond 10- or 12-ft wide tanks, other factors become a problem, such as:

- The mechanical problem of lifting long bars of work may be troublesome.

- Variations of plating thickness will be encountered between parts at the ends and in the center of the tanks.

- Long narrow tanks are more difficult to control regarding filtration, agitation and temperature.

FACTORS TO CONSIDER IN PLANT SITE SELECTION

	Importance Rating	Site I	Site II	Site III
Zoning				
Neighbors				
Local labor rates				
Labor availability				
General workers				
Skilled workers				
Office workers				
Taxes				
Tax incentives,				
Municipal services, etc.				
Utility rates				
Water and sewer				
Electricity				
Gal/oil/coal				
Environmental regulations				
Transportation				
To and from customer				
Employees				
Adequacy of space				
Now				
Future				

18. FLOORS—PLANS AND CONSTRUCTION

Robert R. Pierce

Manager, Corrosion Engineering Products Department, Pennsalt Chemical Corporation Philadelphia, Pa.

In this chapter specifications, design aids and general information are presented to make possible economical and efficient design and construction of floors, drains, sumps and equipment supports. The effect of chemicals on conventional materials and design is also described.

The importance of careful design cannot be overemphasized. Corrosion plagues those who did not plan for corrosion prevention and control before building or expansion of their finishing plant. It is possible for the maintenance cost of a finishing plant to be 300 per cent higher than the minimum for a well-planned and adequately maintained plant. It is extremely difficult to fight corrosion once conditions get out of control in the plant. Not only is maintenance cost much higher, but depreciation of buildings and equipment is accelerated.

In plants where chemicals are manufactured, design is relatively simple, since generally only one chemical is produced in a building; however, in the finishing plant many different chemicals are used in the presence of much moisture and humidity, so that corrosive conditions are more aggressive than in chemical plants.

The attitudes of management and supervisory personnel are extremely important. These people must take the attitude that excessive maintenance can be avoided and prevented. Here the secret of success is not to know success or failure alone, but rather to know why. In this section the author will attempt to show how to design for economical maintenance and why the design functions are required.

It is never economical to let corrosion run rampant as depreciation will be excessive, maintenance costs will run as much as 300 per cent higher than they should, the morale of personnel will be low, and the quality of the product being processed will be adversely affected.

Floors and Drains

Floors are the supporting structure for finishing equipment; as such, they must be protected against the spillage, splash, drainage and rinsing which inevitably occur around solution tanks in most plating rooms. They must also be constructed to withstand fork truck, dolly, hand truck, and foot traffic without dusting or abrading. Since concrete is the basis for almost all floor structures, the effects of plating and metal finishing chemicals on concrete and protective methods are shown in Table 1.

Many of the acids and salt solutions used in plating and metal finishing operations attack the concrete, forming soluble salts which dissolve, thus leaving the concrete brittle and weak. In some cases insoluble salts are formed which acquire waters of hydration and "grow" or swell, causing pressures within the concrete which disintegrate the structure.

Since the chemicals and acids used in finishing operations are so destructive to concrete, it is imperative to protect concrete from drip, splash and spillage. Practical platers know that it is virtually impossible to eliminate this contact of corrosive solutions with concrete; therefore, let us

TABLE 1. CHEMICAL RESISTANCE OF VARIOUS FLOORS AND SURFACINGS CURRENTLY AVAILABLE.

Type	pH Range	Water	Sulfuric Acid	Hydrochloric Acid	Nitric Acid	Organic Acids	Solvents	Mineral Oils	Animal Oils	Chloride Salts	Sulfate Salts	Alkalies
Good Concrete	7-12	E	NR	NR	NR	P	G	G	P	F	P	G
Good calcium aluminate cement concretes	4.5-0	E	NR	NR	NR	F	G	G	F	G	G	F
Light Duty Methods Quarry tile (portland cement joints)	7-12	E	NR	NR	NR	P	G	G	P	F	P	G
Coatings 10 to 125 mils	6-14	E	E	F	F	F	E	E	F	G	F	G
Latex surfacing (1/4 in. & up)	2-12	G	F	F	F	F	P	F	F	G	G	G
Asphalt surfacing (1/4 in. & up)	1-11	E	F	G	F	F	NR	NR	P	E	E	F
Moderate Duty Methods Epoxy surfacing (polyamide-1/8 in. & up)	5-14	E	F	F	P	F	F	G	F	F	F	E
Quarry tile (epoxy-polyamide side joints)	5-14	E	F	F	P	F	G	G	G	G	G	E
Severe Duty Methods Paver tile (furan side & epoxy or polyester)	1-13	E	G	G	F	G	E	E	E	E	E	E
Polyester surfacing (1/4 in. & up)	1-12	E	G	E	E	G	NR	P	G	E	E	F
Epoxy surfacing polyamine (1/4 in. & up)	1-14	E	G	E	F	G	E	E	E	E	E	E
Brick with furan, phenolic epoxy & polyester, side & bed joints over 1/4 in. asphalt membrane	0-14	E	E	E	G	E	G	E	E	E	E	E

E — excellent G — good F — fair P — poor NR — not recommended

consider how we can best protect the base floor (see Fig. 1).

Generally the concrete is protected by an impervious membrane such as special hot-applied asphalts about ¼ to ⅜ in. thick. Because such membranes are quickly damaged by mechanical abuse, temperature, traffic and chemical spillage, they are protected with a layer of corrosion-resistant brick. This layer of brick bonded with chemical-resistant mortar is commonly called the masonry protection for the asphalt layer or other types of membranes used on concrete. Thus, this method of protecting floors, trenches, walls and pits from chemicals is based on three construction components, which should meet the following engineering requirements.

(1) *Supporting Structure.* The supporting structure must be designed for full operating loads, including the weight of all tanks full of solution, equipment, the brick layer, stored supplies, finished parts, and moving equipment such as fork trucks, etc. The supporting structure must remain relatively rigid under these loads; consequently, it can be deduced that it is not practical to use steel plates or other flexible supporting structures such as wood. The general specifications for concrete that is to be later protected with a chemical protection method follow:

(a) The concrete base shall be finished by screeding to a wood-float finish with high peaks removed by one light pass with a steel trowel. It shall have a pitch toward drains of ¼ in/ft. Newly poured concrete shall be cured at least 10 days before any acid-proof construction is installed.

(b) Old concrete surfaces shall be free of grease and oil, and shall be thoroughly clean and dry. Uneven, eroded, broken or cracked concrete shall be repaired and cured before proceeding with acid-proofing work.

(c) 'Josam' or similar flanged-type floor drains are to be imbedded in the concrete and

FIG. 1. Illustrations of floor protection under vessels holding corrosive chemicals in a plating facility.

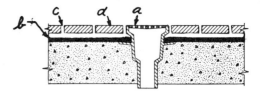

FIG. 2. Typical drain construction.
a. "Josam" or similar type floor drain
b. ¼ in. asphalt membrane primed with asphalt primer
c. Acid-proof mortar joints
d. Acid-proof brick with resin mortar bottom and side joints

made ready to receive the acid-proof floor (see Fig. 2). Particular attention shall be paid to assure the proper elevation of the drain in respect to the finished floor surface. Pipes shall be brought through floors in sleeves as shown in Fig. 3. Piers, foundations, pedestals and stairwells shall be designed so that drainage is not impeded as shown in Figs. 4, 5 and 6.

(2) *Impervious Interliner.* The impervious membrane or interliner between the brickwork and the supporting structure must, by engineering definition, be impervious to and fairly resistant to chemical attack at the anticipated service temperatures. In general practice many generic types of membranes can be considered, but practically all of those used in the finished room are based on formulations of special asphalts. These special asphalt membranes generally range in thickness between ¼ and ⅜ in. and should be reinforced with coated glass cloth at corners, columns, and positions where movement is possible. It is good practice to use reinforcing in the interliner and where the interliner is relatively thick, i.e., ⅜ in., in trenches, pits, sumps and other places where liquids are held.

To date, the best membranes of the special asphalt class are those which are applied hot and squeegeed on the surface. Membranes based on water emulsions or containing solvents are generally too permeable for this use. A general specification for an impervious membrane which is to be protected with a brick or tile sheath follows:

(a) Over the smooth, cured concrete brush one coat of an aslphalt primer designed for penetrating concrete surfaces. Brush over the entire surface and up the walls, columns, and vertical

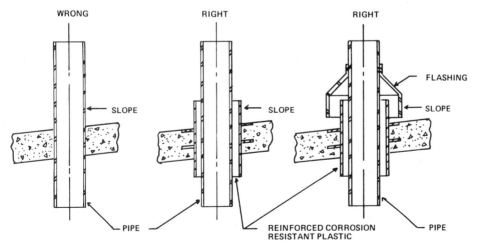

FIG. 3. Pipes through chemical floors.

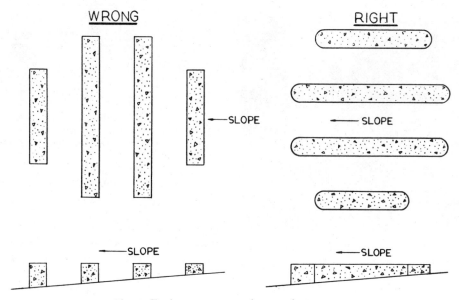

Fig. 4. Equipment supports in corrosive areas.

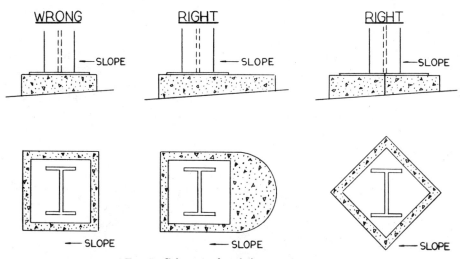

Fig. 5. Column pedestals in corrosive areas.

surfaces for a distance of 8 in. The primed area should be kept closed to other tradesmen and traffic for at least 8 hr or overnight.

(b) After the primer has dried, apply special asphalt (acid-resisting type) by squeegeeing hot molten material until a constant thickness of $\frac{1}{4}$ in. is obtained on horizontal and vertical surfaces (see Fig. 7).

(c) Where strong or oxidizing acids will be used in the area, a layer or reinforcing such as coated glass cloth should be imbedded between the layers of asphalt. In all cases, two layers of reinforcing should be used in trenches, sumps and one layer at corners of walls, curbs, drains, piers, and pipe sleeves.

(d) The coated area should be kept closed to other tradesmen and traffic until the brick floor is laid.

(3) *Masonry Protection.* The masonry protection must be completely resistant to all of the chemicals at the temperature used in the finishing operations. It serves to provide protection for the impervious membrane against mechanical abuse, to provide thermal insulation for the impervious membrane, and to provide a permeable barrier to prevent continuous flow of the chemicals over

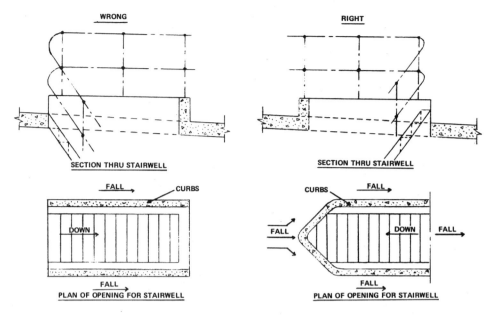

Fig. 6. Design of stairwell.

the surface of the impervious membrane, thus preventing excessive oxidation and embrittlement of the membrane. The masonry protection does not give additional strength to the supporting structure; moreover, it should not be considered impervious in itself (see Fig. 7).

Seven general types of chemical-resistant mortars are used to bond chemical-resistant brick or tile; these are classified by their resistance to plating chemicals as shown in Table 2. A selection of the proper acid-proof mortar for specific service conditions may be made from these data. The mortar should be applied on the bottom of the brick as well as the sides (see Fig. 7) and cured according to instructions of the manufacturer before being placed in service.

General specifications for brick are given in Table 3. The brick should meet the requirement given in ASTM Specification C 279, Type H or L. They are laid over the impervious membrane or interliner (as shown in Figs. 7 and 8) in such a manner that the joints should be as narrow as good workmanship will permit and will generally average $\frac{1}{8}$ in. All extruded mortar should be carefully removed.

(4) *Drains and Sumps.* Orthodox construction consists of the three-component construction using: (1) concrete as a supporting structure; (2) impervious membrane or interliner; and (3) brick sheathing. The pitch of floors to drains

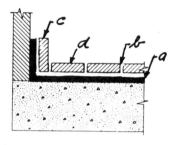

Fig. 7. Acid-proof brick floor construction with impervious interliner.

a. $\frac{1}{4}$ in. asphalt interliner with glass cloth reinforcement primed with asphalt primer
b. Properly selected acid-alkali resisting cement side and bottom joints
c. "Tie-in" wall covering
d. Acid-proof brick

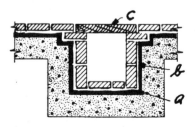

Fig. 8. Cross section of drainage trench and cover.
a. $\frac{1}{4}$ in. asphalt interliner with glass cloth and primer
b. Acid-proof joints
c. Planking (or grating)

TABLE 2. RESISTANCE TO CHEMICALS OF AVAILABLE CHEMICAL MORTARS

	Furfuryl Alcohol Types	Modified Phenolic Types	Phenolic Types	Silicate Types	Sulfur Types	Amine Epoxy Types	Polyester Types
Perchlorid acid	NR	NR	NR	E	NR	NR	NR
Nitric acid	NR	NR	NR	E	G	NR	E
Chromic acid	NR	NR	NR	E	E	NR	NR
Nitric acid plus hydrofluoric acid	NR	NR	NR	NR	G	NR	G
Sulfuric acid—strong	NR	G	F	E	G	NR	NR
Sulfuric acid—medium strength	F	E	G	E	E	F	F
Hydrochloric acid	E	E	E	E	E	E	E
Hydrofluoric acid	E*	E*	E*	NR	E*	E*	E*
Phosphoric acid	E	E	E	E	E	E	E
Steam	E	E	E	P	P	G	G
Salts—unsaturated—acid side	E	E	E	E	E	E	E
Salts—unsaturated—alkaline side	E	G	P	NR	NR	E	F
Alkalies—strong	G	F	P	NR	NR	E	F
Alkalies—medium strength	E	G	P	NR	NR	E	G
Temperature—top limit	350	375	375	750–2500	190	225	225

E—Excellent   G—Good   F—Fair   P—Poor   NR—Not recommended

* For hydrofluoric acid service, use carbon fillers.

TABLE 3. STANDARD ACID-PROOF BRICK DIMENSIONS, MATERIALS, AND COMMON USES

Length (in.)	Width (in.)	Thickness (in.)	Shale	Silica	Carbon
8	3¾	2¼	MF, SF		
8	2¼	3¾	SW		
8	4½	3¾	SW		
8	4	1⅜	LF		
9	4½	1¼		LF	LF
9	4½	2½		MF, SF	MF, SF
9	2½	4½		SW	SW
12	6	1½		LF	

LF—Light flooring (for foot traffic and hand trucks).
MF—Medium flooring (Fork lift trucks, etc.).
SW—Sump and trench walls.
SF—Sump and trench floors.

or trenches is essential to secure good drainage and prevent pockets and general wetness.

In constructing sumps or pits which are over 20 ft long, provision should be made to bow the walls. The long walls should be bowed outward a minimum of 1 per cent and the short walls a minimum of 2 per cent. In smaller tanks, straight walls can be used but walls should never bow inward.

Two types of acid brick can be used for floors, trenches, pits, and sump areas. The hard-burned fire clay type is buff in color and has the best thermal shock resistance. The red shale clay type has less porosity and is generally less expensive. In areas where there will be heavy spillage of hydrofluoric acid, carbon brick should be specified.

Floors should be sloped ¼ in/ft to drains and trenches. Three sloping techniques are illustrated in Figs. 9, 10 and 11.

A typical method of construction of a sump is illustrated in Fig. 12.

Trenches and bell-and-spigot clay pipe drains should be sloped ⅛–¼ in/ft and should discharge freely at the end. Figures 13 and 14 give flow rate

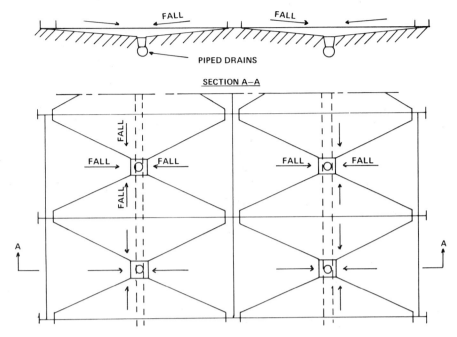

FIG. 9. Saucer slope on floors.

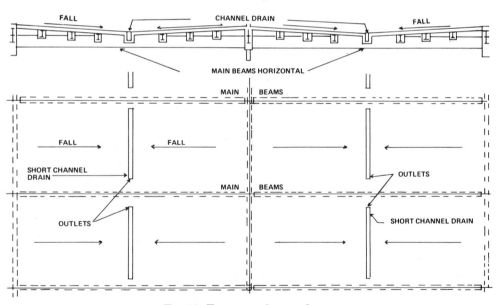

FIG. 10. Transverse slope on floors.

data in brick-lined trenches and clay bell-and-spigot pipe when a pitch of $\frac{1}{4}$ in/ft is used.

In designing drains and sumps, careful consideration must be given to peak loads when draining. During downtime, a production department can allow for draining and refilling tanks. Generous over capacity should be designed into the drainage system to take care of emergency or improper scheduling of dumpings and future additions of tank capacity, as well as routine draining requirements.

Tank and Pump Foundations

Concrete pump-and-tank foundations should be protected as shown in Fig. 15. Tanks should be placed on foundations with adequate clearance

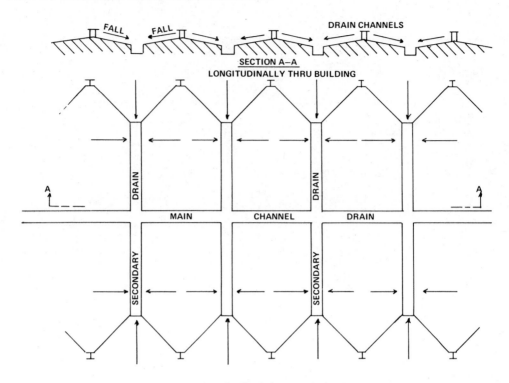

Fig. 11. Longitudinal slope on floors.

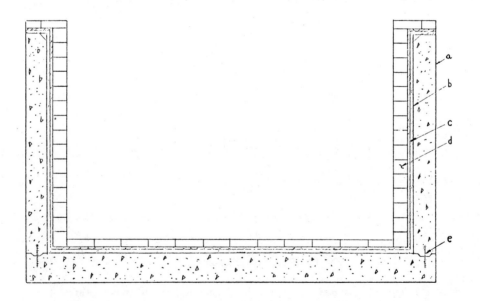

Fig. 12. Sump tank for holding chemical waste and/or neutralizing.

a. Properly reinforced liquid tight concrete
b. impervious elastic membrane ⅛ to ⅜ in. thick
c. Resin mortar bed joint
d. Acid-resisting brick wall 4 in. thick bonded with acid-alkaline resin mortar
e. Elastomeric or plastic water stops

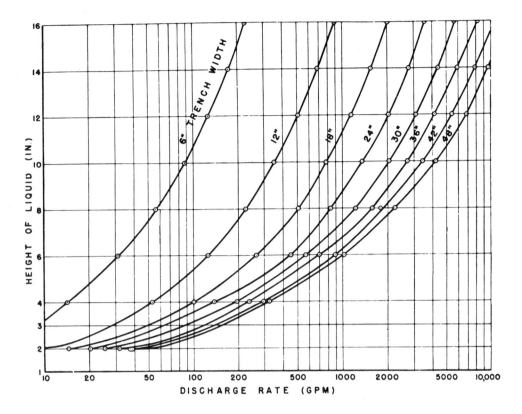

Fig. 13. Discharge rate (gpm) vs height of liquid (in.) for acid-brick trenches based on ¼ in./ft slope.

between the tank bottom and the floor to allow room for paint repairs, maintenance and dispersal of heat.

Bell-and-Spigot Pipe and Manholes

The bell-and-spigot joints of chemical stoneware pipe must be caulked and filled with suitable materials, which must be all-purpose when used to carry the variety to acids, alkalies, salts and organics from plating operations. The filled joint is illustrated in Fig. 16. In such work, the bell-and-spigot pipe are specified without glaze or ends are sandblasted to remove the glaze. All grease and foreign matter are removed and the pipe is thoroughly dried. The correct resin cement is troweled into the bell of the pipe and onto the end of the spigot, which is shoved into the resin mortar until it is centrally seated. Next, a ring of oversized asbestos rope is tamped tightly into the back of the joint. The balance of the annular space is filled with resin motar.

It is important to support the pipe so as to maintain complete rigidity.

General Information on Chemical-Resistant Floors and Drains

In some cases it is desirable to isolate certain areas where chemicals such as hydrofluoric acid or cyanides are utilized. Unless the waste solutions are extremely dilute (below 50 ppm hydrofluoric acid), stoneware or clay pipes should never be used to carry off hydrofluoric acid wastes. Cyanide wastes must be kept separate from other wastes to prevent hazards due to accidental mixing with acids; furthermore, they require special disposal techniques.

It is now common practice to provide a separate closed-pipe drainage system for safe handling of cyanide wastes.

Heavy-Filled Resin-Topped Floors. Heavy-filled resins for trowling or spraying floors (see Fig. 17) are composed of resins generally of an epoxy or polyester class, room temperature airing agents, and siliceous or carbonaceous fillers. The cost of this type of flooring is one-half to two-thirds that for acid-proof masonry floors, and can be considered for protection of the finishing room

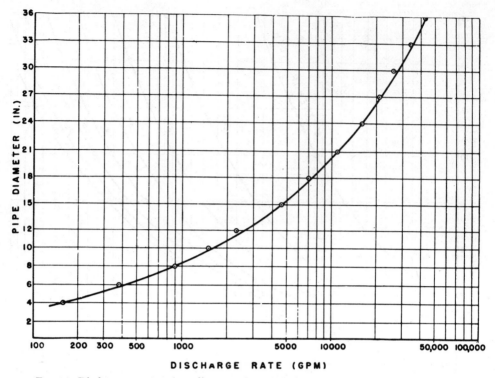

Fig. 14. Discharge rate (gpm) vs diameter (in.) of vitrified clay pipe with ¼ in./ft slope.

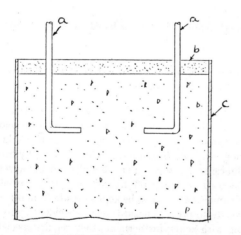

Fig. 15. Foundations and piers.
a. Anchor bolts
b. Chemical resisting grout—minimum thickness 1 in.
c. Chemical-resisting resin surfacing—⅛ in. thick

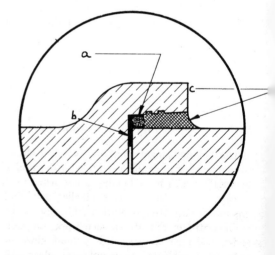

Fig. 16. Acid-proof joints for terra-cotta pipe.
a. Asbestos rope
b. Furan mortar
c. Sulfur cement

floors. These materials will withstand temperatures up to 140 to 220 F, depending on the type of resin. The floors have good wear resistance and do not dust; these toppings are applied ¼ in. in thickness. Good, strong and uncoated concrete should be designed and installed to provide a slope to drains or trenches of ¼ in/ft.

The resins are resistant to most acids, alkalies and salts used in the finishing room. Resin with carbonaceous filler is employed in areas where there will be frequent spillage of hydrofluoric acid. Load is not a factor because these materials do not flow and are stronger than concrete.

Comparative Floor Costs. In Table 4 the

Fig. 17. Heavy-filled resin-topped floors.

average initial cost of various floors are compared to slab concrete alone as a ratio of the former to the latter being equal to 1.0.

A recent development by the Bureau of Mines is a specialized sulfur concrete in which a modified sulfur cement replaces portland cement as the aggregate binder. The specially prepared sulfur is melted and mixed with the preheated aggregate at a temperature of 125 to 150 C (257 to 302 F). The resulting product is applied in a manner similar to hot asphalt mixes and on cooling forms the sulfur concrete. In extensive field testing, excellent resistance has been demonstrated to many acid and salt solutions. Additional information is available from the Bureau of Mines (500 Date Street, Boulder City, Nevada 89005) and the Sulfur Institute (1725 K Street N.W., Washington, D.C. 20006).

Dust Free Floors. Floors in the balance of the finishing plant can dust and affect the finishing work. All concrete floors in the finishing plant and areas outside of chemical exposures should be correctly designed and installed. The best mix obtainable with local aggregates and sands should be designed by a local concrete testing laboratory. The slump should ideally be 2½ in. for floors and 3½ in. for walls. No extra water should be added at the job site; the water/cement ratio should never be more than 42.3 lb of water for a 94 lb bag of cement.

Minimizing of dust is accomplished by a chemical treatment or with coatings. The chemical treatments are based on application of sodium silicate or zinc and magnesium silico fluoride solutions to concrete. Many different types of coatings can be used. The principal types for floors are based on chlorinated rubber, vinyl, urethane and catalyzed epoxy.

All such materials aid only the immediate surface and need periodic renewal.

Recently internally modified concretes have become available through development of water curing plastic resins. These materials show promise and can provide moderate chemical resistance

TABLE 4. COMPARATIVE COSTS FOR FLOORS*

(Based on about 5000 sq ft, including surface preparation and primers)

Coatings**	Thickness, in.	Comparative Floor Cost***
Chlorinated rubber	5 (mil)	1.29
Epoxy	20 (mil)	1.43
Urethane	⅛	1.90
Surfacing		
Asphaltic	1¾	1.90
Latex	¼	1.79
Epoxy (polyamide)	¼	2.32
Epoxy (polyamine)	¼	2.32
Polyester	¼	2.43
Epoxy (polyamine-glass reinforced)	⅛	2.57
Polyester-glass reinforced	⅛	2.57
Tile		
Quarry tile (portland cement joints)	¾	2.29
Quarry tile (epoxy-polyamine joints)	¾	2.61
Pavers—furan side and epoxy or polyester bed joints	1½	3.14
Brick		
1⅜ in pavers—furan side and bed joints over ¼ in. asphaltic membrane	1¾	3.32
2¼ in. brick—furan side and bed joints over ¼ in. asphaltic membrane	2⅝	3.32

* Excerpted by special permission from Chemical Engineering (December 16, 1968) Copyright © 1968, by McGraw-Hill, Inc., New York, N. Y. 10036 (see Ref. 5).

** Applied over concrete slabs.

*** Ratio = $\dfrac{\text{Floor Cost}}{\text{Concrete Slab Cost}} = 1.0$

and minimize dust with several inches of protection, in lieu of surface treatments or coatings.

Foundations and Piers

In chemical areas concrete foundations, piers and pedestals and anchor bolts should be protected. In past years, such protection was provided using acid brickwork and membranes. In recent years, an improved method has proven itself—this method is shown in Fig. 15. The grout should have chemical resistance, low shrinkage, a good bond to concrete, steel, and its own aggregate. It's coefficient of expansion should be close to that of the concrete.

Equipment Elevation

Many finishing lines are installed in pits to bring tank heights to working levels and to compensate for a lack of headroom. Quite often, the equipment receives the primary consideration and is simply installed in a concrete pit. The concrete in the bottom of these pits is attacked, and eventually the entire building may be endangered. The pit should be protected as required. Generally, the bottom of the pit, as well as any troughs in it, should be protected with the membrane and acid brickwork method or the heavyfilled resin topping compounds. The side walls of the pit should be protected in a similar manner to a height of at least 8 in. as long as the drainage trenches are designed with sufficient over-capacity to preclude overflow conditions. The side walls of the pit should also be protected from splash, spillage and solution drag-out as required. The concrete above the heavy duty protection should be coated with at least three coats of a good acid- and alkali-resistant coating, such as epoxies, polyesters, neoprenes, vinyls or chlorinated rubbers over a suitable primer. If a separation of overflows or wastes is necessary, it is occasionally desirable to design two trenches in the bottom of the pit.

Painting the Finishing Plant

Careful planning of original design and materials can greatly reduce maintenance painting costs, for example, maintenance of protective coatings over steel can cost as high as $1.00/sq ft/yr; whereas careful design, specification, application and planned maintenance can result in annual costs of 6 to 12¢/yr. If good design and planned maintenance are not executed, air contamination can affect finishing results; also, maintenance will become extremely difficult, since much dust and foreign matter will be present during corrective measures.

New Plants. Currently, the best procedure is to sandblast all steel, before erection at the plant location, and subsequently apply the proper inhibitive primer and one topcoat. After the steel is erected, all damage to the primer and coating should be repaired. At this time, any areas which will later become inaccessible for maintenance should be built up to a minimum of 7.5 mil system thickness. Construction should be finished, and finally coating thickness should be built up in each area to a minimum of 5 mil. Cinder block, concrete and asbestos boards should be coated to prevent dust in the finishing plant.

The coating system thickness over all steel should never be less than 5 mil and, in many cases of more aggressive locations, thicknesses as high as 20 mil are required. We will cover individual specifications for specific areas later. Figure 18 shows the probability of early paint failure in the average chemical exposure.

Each area is discussed individually below. In all cases sandblasting steel surfaces will give far better performance and a return on investment of over 200 per cent. If mill scale is covered by some coating and exposed in a chemical atmosphere, the bond starts to loosen in about 2 months and the coating installation fails within a year.

Structural Steel. Various coating types can be used with the neoprene, vinyl, chlorinated rubber or catalyzed epoxy types suggested. A minimum coating system thickness of 5 mil should be specified.

Window Sash. Same as Structural Steel.

Cinder Block. Fill surface with a water-based latex coating. Next apply chlorinated rubber, vinyl, neoprene or catalyzed epoxy systems.

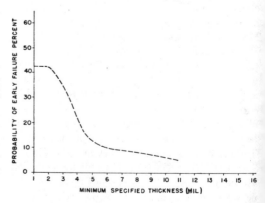

Fig. 18. Probability of early paint failure vs minimum specified thickness (mil).

Concrete and Concrete Block. Same as Cinder Block. Latex coating may be omitted if surface is smooth.

Asbestos Boards. Same as Cinder Block, except surface should be etched with muriatic acid in place of the latex coating.

Alkaline Cleaning and Etching Tanks. For protecting the outside, coatings that are resistant to strong alkaline conditions at high temperatures should be used. Such coatings include the rubber latexs, neoprenes, and catalyzed epoxies.

Sulfuric Acid Tanks. If spillage or splashing is a factor, use sheet rubber $1/32$ in. thick or a heavy neoprene applied $1/32$ in. thick to sandblasted surfaces. If no spillage or splashing is involved, a 10 mil thickness of the chlorinated rubbers, neoprenes or vinyls can be specified.

Hydrochloric Acid. Same as Sulfuric Acid.

Chromic Acid. A 5-mil thickness of the chlorinated, rubber vinyls will give good performance.

Nitric Acid. Same as Chromic Acid.

Hydrofluoric Acid. Same as Sulfuric Acid.

Plating Solutions Based on Acid, Neutral or Alkaline Salts. Five mil of any good synthetic paint can be used; however, splash or atmospheric corrosion from the preceding tanks must be considered. Also, any solutions that are strongly acid, oxidizing, or alkaline should take the recommendations of comparable acids or alkalines.

Old Plants and Maintenance. Each plant area and individual vessel should be coded, the condition of the coatings in the area should be inspected at least once per year and planned maintenance should be carried out. In most cases, deterioration will first occur in localized areas. As soon as such areas are well-defined, rust and loose paint should be removed by wire brushing or with mechanical tools. Next, these spots should be washed with a solution of 90 per cent water, 5 per cent methanol, and 5 per cent soda ash and left to dry. After drying, dust the repair areas, spot prime and apply two spot coats with alternating colors. If desired for appearance, the last topcoat can be applied over the whole surface—apply two top spot coats with alternating colors. The periodic repair method generally cost about 20 per cent as much as letting a surface go to the point where complete repainting is necessary.

Summary and Supplementary Points on Painting

(1) If a plant is in bad condition and no program for repair and maintenance painting has been studied, call in a well-qualified consultant to make recommendations for repair and to help set up a maintenance program.

(2) Set up a definite program for original design and maintenance.

(3) If studies of coating systems on panels are desired, use the KTA panel and evaluation method.

(4) Always give metal surfaces the best surface preparation possible and stop corrosion on surfaces before priming.

(5) Provide a thickness gauge so that the individual responsible for coatings will be able to measure thickness.

(6) Do not give colors too much consideration, unless it is realized that maintenance costs will be increased because of the lower resistance of colored pigments. Generally, black will give the best performance with gray second and white third and chrome green fourth.

(7) Always specify the correct type of coating for individual problems. Always apply the thickness required for each individual chemical exposure. Never use less than 5 mil dry-film thickness on metal surfaces exposed to atmospheric corrosion conditions.

Conclusion

The finishing plant can be designed for economical maintenance and cleanliness. A planned design and maintenance program requires a management decision and support to make it successful. Such programs, when well-directed, lead to lower yearly costs and will save additional money by safe-guarding finishing operations from contamination and dust.

General References

1. Kleinlogel, A., "Influence on Concrete," New York, Frederic Ungar Publishing Co., 1950.
2. Pierce, R. R., *Chem. Eng.*, 149–153 (May 1952).
3. Tator, Kenneth, "Protective Coatings," *Chem. Eng.*, 143–190 (Dec. 1952).
4. Pierce, R. R., "Why Go to Thick Protective Coatings," *Chem. Eng.*, (April 1954).
5. Pierce, R. R., "Design and Materials for Protecting Concrete Floors from Chemicals," *Chem. Eng.*, (Dec. 16, 1968).

19. TANKS—DESIGN, CONSTRUCTION INSTALLATION AND MAINTENANCE

JOHN V. DAVIS

Engineering Consultant, The Udylite Corporation, Detroit, Mich.

AND

A. KENNETH GRAHAM

Consulting Engineer, Jenkintown, Pa.

Revised by Lawrence J. Durney

Vice President for Technology, Frederick Gumm Chemical Co. Inc., Lyndhurst, N.J.

The primary requirement of a tank for an electroplating installation is that it will hold its prescribed liquid without leakage or solution contamination over a desired period of time. The specific liquid and temperature to be used will determine the selection of the material of construction or lining. The tank size, shape and construction are dependent on its use, the material of construction, and the type of installation in which it will be used. In a manually operated line, for example, a rinse tank may be as large as a plating tank in order to be able to process an occasional large piece. In a fully automatic machine, on the other hand, a rinse tank is usually made of a size sufficient to hold only one plating rack or carrier, thus limiting the volume of rinsing water required for a complete change and minimizing the length of the machine as an economic consideration.

MATERIALS OF CONSTRUCTION

The material used for constructing a tank must be chemically resistant to the liquid it is to contain unless a lining is to be used, in which case the lining must meet this requirement. In either case the liquid must not be contaminated. (See Chapters 6 and 20.)

Both the material of construction and the use of a lining influence in part the design of a tank as will be shown in the following discussion.

At one time wood was extensively used for tanks in the plating industry. Due to increases in the cost of materials and construction, and the development of improved plastic materials, it has essentially been eliminated. The most common materials now used are resin bonded fiber glass, plastic and lined or unlined steel and/or stainless steel.

Plastic Tanks

Various plastic materials are becoming more and more popular, a few of which are discussed below.

Resin Bonded Fiber Glass Tanks. Glass fiber cloth can be impregnated with a number of different chemically resistant self-curing resins to form structurally stable tanks. In one method of construction, alternate layers of cloth and resin are built up on a form in such a way that the resulting tank is essentially of one-piece molded construction.

Overflow dams and drains can be built in. Small tanks up to about $4 \times 4 \times 3$ ft deep usually are made with $\frac{1}{4}$ in. walls and these may be reinforced with ribs for extra strength if needed. Larger tanks have been made with thicker walls and relatively heavy reinforcing ribs. Repairs can be made by the user, and kits and instructions are available for this purpose.

Tanks made of this material are relatively light in weight. They are also quite strong, although care should be taken in handling, especially to avoid dropping. They should also rest directly on the floor rather than on insulating blocks.

No special corrosion-resistant paint is required for maintenance or protection of the outside of these tanks.

Acrylic Resin Tanks. High-temperature acrylic resin materials are available for tank fabrication.*† The transparency of these materials is an advantage in some cases although some solutions will stain the material. Because of the availability of other plastic materials, the use of these resins is now usually limited to the construction of small tanks for use in pilot laboratories, jewelry processing or electronic work.

Ceramic Tanks and Crocks. The small ceramic tanks and crocks formerly used extensively have now been replaced almost completely by polyethylene and polypropylene units. Many types are available as standard stock items at relatively low cost. They have the advantage over ceramics of being resistant to thermal and mechanical shock and resistant to hydrofluoric acid and fluorides. Since, in some cases, temperature resistance may be limited, care must be exercised in selecting the proper grade of material to provide the required temperature resistance at the lowest cost.

Concrete Tanks. Steel-reinforced concrete tanks are commonly used in waste treatment installations. They are asphalt-lined in many cases, especially for use with acid media. Plywood forms are favored since the concrete surface should be smooth. Too much sand in the concrete mixture will give a porous surface.

---

*E. I. duPont de Nemours & Co., Inc., Wilmington, Del.

†Rohm & Haas Co., Philadelphia, Pa.

These tanks may be lined with asphalt, or chemically resistant mastic materials.

Custom-made polyvinyl liners may occasionally be installed after the concrete has been asphalt coated. Where these liners are used, the tank must be carefully prepared to avoid any possibility of puncturing the liner by roughness or protrusions in the concrete surface.

Steel Tanks. Steel is the most commonly used material for tank construction because of its low cost, high strength and ease of fabrication. It is suitable for all the alkaline plating baths, but linings are commonly applied to prevent iron contamination, and especially to insulate against stray currents. Linings are necessary for use with all acids or acid-type plating baths.

The high strength of steel is fully utilized in designing tanks to support various items of equipment and accessories. In still tank applications the weight of the loaded racks is also supported by the tank walls. To prevent bulging, various means of stiffening or reinforcing are used as described later.

The ease of fabrication with steel enables one to maintain close dimensional tolerances. The over-all tolerance for a still tank may be $\frac{1}{4}$ in. Tanks for most fully automatic machines must be held to plus or minus $\frac{1}{8}$-in. tolerance since they must fit tightly against each other in order to control clearances at all transfer locations. In the latter case the tanks may be mutually supported by each other and/or by components of the machine. It is also relatively simple with steel construction to install inlets, outlets, overflow dams, and other special features as required.

STEEL TANK CONSTRUCTION

Many aspects of construction are applicable to all types of plating tanks, whether for still, fully automatic, semiautomatic, or barrel plating. For purposes of illustration therefore only still tank construction will be discussed here.

Material

Tanks should be constructed of mild steel plate with a minimum thickness of $\frac{3}{16}$ in. using greater thickness as required for heavier tanks. All plates should be flat and free of buckling. Where possible the sides should be made of one-piece construction, but when this is not possible all edges should be sheared square and true, and then should be accurately beveled for proper welding.

Welds

Tanks should be constructed with continuous, double electric welds. The direction of the weld should be horizontal or vertically downward. When necessary to break a weld, it should be tapered off and overlapped to give a smooth continuous weld. Any break-off points in the weld inside a tank must be out-of-line with similar break-off points outside the tank.

Welding electrodes must conform to American Welding Society Specifications, as specified for the parent metal to be welded. For hot-rolled steel, for example, the welding electrodes recommended are AWS-6013, AWS-6010 or AWS-6012.

Care must be taken to obtain clean welds with a satisfactory surface appearance. The ripples and width of the head should be uniform. Butt welds should be flush or slightly above the plate surface and without excessive build-up. Fillet welds should have equal laps on each plate. In all cases the welds must be free from splatter and flux, especially if the tank is to be lined. Transverse, lineal, or crater cracks or porosity must be avoided. If present they should be completely ground out, tapering the ground area with respect to the weld and refilled to give a smoothly welded surface. Similarly, all craters should be filled with weld metal to give a level surface. The inside welds should be flat or convex.

All welds which are to be covered with a lining must be ground if the surfaces are rough or defective.

TANK FEATURES

The general shape of a conventional plating tank is that of a lid-less box such as is shown in Fig. 1. However, certain features are added to this box for strength, drainage, liquid circulation, and efficient plating operation.

Tank Base Supports

The bottom of a tank in any plating room should be at least 4 in. above the level of the floor and supported by a nonporous material, with as much surface area exposed to the air as possible to minimize corrosion and facilitate maintenance. Supporting members should be placed crosswise on long tanks and spaced close enough

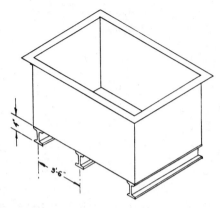

Fig. 1. Conventional small plating tank showing steel I-beam support.

to prevent the bottom of the tank from bulging due to the weight of the solution such as shown in Fig. 1. When buttress bracing is required, the supports should be I-beams and should extend under the buttress as shown in Fig. 2. Plating or electrolytic treatment tanks must be insulated to avoid stray currents. A nonporous material such as shatterproof glass or "Bakelite" can be used for this purpose. The insulating material usually is placed under the supporting members.

Tank Rims

The tank rim serves several purposes: as a stiffener to the upper edge of the tank, and as a supporting device for anode and cathode rods, ventilation boxes, coils, controllers and other items. Rims may be of any structural shape that provides adequate reinforcement. Angle iron is perhaps the most satisfactory rim material.

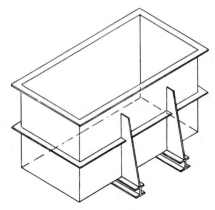

Fig. 2. Conventional tank showing the use of buttresses for bracing the sides and an additional horizontal angle.

Not only does the vertical flange provide sufficient welding area, but the horizontal flange provides the necessary mounting area for accessories. Also, the sizes of angle iron available vary widely, both in web thickness and flange width, providing an almost unlimited variety to meet requirements.

The upper edge of the tank and angle rim should be joined with a continuous arc weld, while the lower edge of the angle can be spotted with welds from 1½ to 2 in. long, placed every 6 to 8 in., as shown in Fig. 3. The top continuous weld serves a threefold purpose:

1. The horizontal leg of the angle provides the necessary stiffness and an integral part of the tank wall.
2. The continuous weld "seals-off" any gap between the outside of the tank wall and the angle, thereby preventing solution from being trapped in this space and increasing the life of the tank.
3. The continuous weld provides a solid backing for a lining if required.

For small tank construction (limited to the use of ³⁄₁₆ to ¼ in. plate and a maximum dimension of 3 to 4 ft) a broken flange of about 2 to 3 in. is satisfactory as shown in Fig. 4. Larger tanks should have the additional support provided by the welded angle construction. The thickness of the angle should be equal to or greater than the tank wall. For greater strength a channel (3 in. or larger) can be used as shown in Fig. 5.

A tank lining should preferably extend over the rim and down to the lower edge of the angle as shown in Fig. 6.

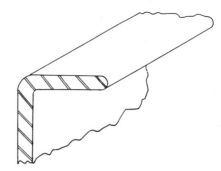

FIG. 4. Broken flange for small tank construction.

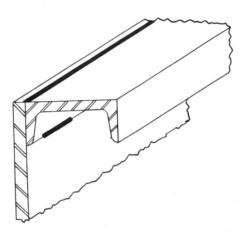

FIG. 5. Channel flange for greater strength. Note continuous dark weld line along the top with spot-welding on the bottom.

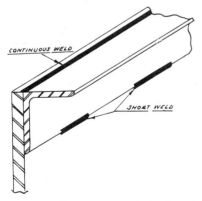

FIG. 3. Tank flange constructed of angle iron continuously welded along the top and spot-welded along the bottom.

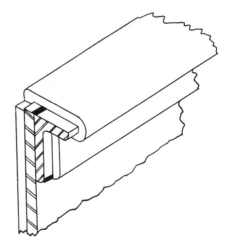

FIG. 6. Angle iron tank flange showing recommended lining down to the bottom outside edge.

Tank Reinforcing

Reinforcing Belts. These may be made of any structural shape which will provide sufficient strength to prevent any appreciable bulging. Angle iron is most desirable for reinforcing belts because of its aforementioned versatility and the added advantage that it may serve as a support for duck board flooring.

When necessary, tanks may be vertically gussetted with flat, triangular shaped plates, welded at right angles to the side of the tank. The width of the gusset at the base will vary according to the depth of the tank and the spacing of the gusset itself.

Side Braces. Long, deep tanks require more support than that afforded by the angle iron rim. A tank over 6 ft long and 3½ ft deep will require an additional horizontal angle about one-third of the way up from the bottom. A tank over 10 ft long and 4½ ft deep may require vertical buttresses on approximately 3½-ft centers, in addition to the reinforcing belt (see Fig. 2).

In some cases a tank with sufficient bracing will still bulge. This is almost always attributed to an uneven floor causing undue stress at the ends or the middle of the tank.

Bottom Designs. The design and pitch of the bottom of a tank are important because they influence the ability to drain without liquid hold-up and to flush out solids such as dirt or sludge. Many solids will settle and cling to surfaces inclined at a much steeper angle than the practical incline for a sloping bottom. However, a slightly sloping bottom is an aid to complete draining and hosing down when cleaning. These advantages should be weighed against the increased cost of a sloping bottom and its supporting members. Typical bottom designs are illustrated in Figs. 7 to 11.

As a special case, a small flat-bottomed tank may sometimes be shimmed to pitch slightly toward a bottom drain corner. A built-in sloping bottom is shown in Figs. 7 and 8.

A sump across one end (Fig. 9) or across the middle (Fig. 10) of a tank costs much less to construct than a bottom with a continuous slope, and largely serves the same purpose. After a tank is drained the solids can be collected at the sump and flushed out the drain.

The saw-tooth sloping bottom such as shown in Fig. 11 has many advantages where excessive sludging requires constant motion of the solution. The solution may be kept in motion by air

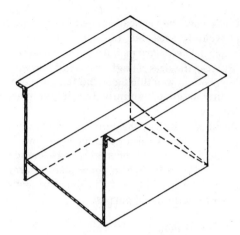

FIG. 7. A sloping tank bottom.

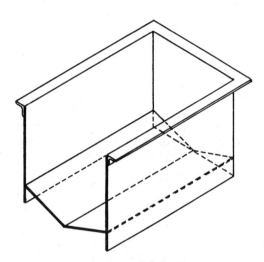

FIG. 8. A double sloping tank bottom.

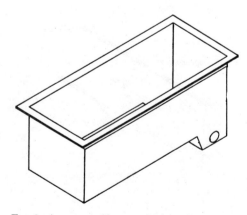

FIG. 9. A sump and bottom drain outlet at the end of a tank.

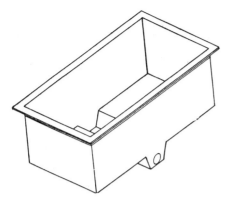

Fig. 10. A center sump and bottom drain outlet.

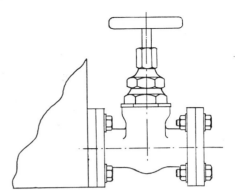

Fig. 12. Bottom flanged outlet with valve and blind flange.

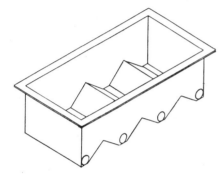

Fig. 11. A saw-tooth sloping tank bottom with multiple outets.

jets or by pumping through orifices near the bottom.

Drain Valve Mountings

Rinse tanks, and in fact any unlined tank which does not contain a dangerous solution, may be equipped merely with a nipple and cap on the bottom drain. If the tank has a sump the drain is located at the bottom of the sump. It may be on the side of the sump or at the end. In the case of a plain tank without a sump the drain should be located at the bottom and near one corner so that when cleaning, all the sludge may be washed to that corner and out the drain. In a multiple tank set-up the drains should be located so that they are all on one side for ease in installation and maintenance.

It is permissible in most cases to equip the drains with valves at the time of installation.

Many companies insist on having all tanks containing hazardous or valuable solutions provided with bottom outlets covered with blind flanges so that the solutions be moved by pumping over the side only. Others permit a bottom outlet consisting of a valve covered by a blind flange (Fig. 12). If a valve only is used one not only depends on the operator's judgment for the safety of the solution but must accept the fact that solution will be lost as the valve seat wears.

Dam-Type Overflows

A dam-type overflow is illustrated in Fig. 13. To assure good drainage the lip of the weir should be at least 2 in. above the top of the drain connection. Since most drains are of 2-in. pipe, this requires a minimum of 4 in. from the top of the weir to the bottom. A depth of 6 in. is desirable, although a 2×6-in. overflow dam cannot easily be lined. In this case a 3×4 in. overflow may be used.

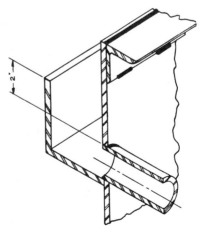

Fig. 13. A dam-type overflow with welded pipe outlet.

Flanged Outlets

When a tank is to be lined with lead, rubber, or sheet plastic material applied in a flexible state, a flanged-type fitting should be used to connect to drain lines. The flange is screwed securely to a threaded pipe, and this is then welded to a tank opening. The pipe is welded with a bead on both the inside and outside of the tank. A securing weld should be placed between the flange and the pipe itself. The pipe should be at least 2 in. in diameter and as short as possible to allow sufficient room for the lining operation (see Figs. 14 and 15). The lining then acts as a gasket for the mating flange and protects the flange surface from the solution in the tank.

Many linings fail at the flanged connections through movement of the flange during installation (because the flange was not sufficiently tightened before lining or because the flange was not also welded to the pipe).

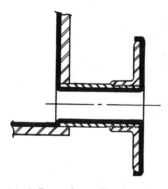

Fig. 14. A flanged-type lined tank outlet.

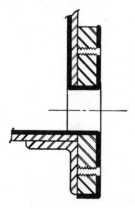

Fig. 15. A flanged-type lined tank outlet for close fitting.

Tank Accessories

Various accessory equipment and mechanisms may be mounted on a still tank. These may include anode and cathode rods, which serve as supports for anodes and loaded racks during plating, units or coils for heating and cooling, rod agitators, lines for air agitation, sprays, control devices, lining and diaphragms. Most of these accessories are discussed elsewhere under appropriate sections of this book, but brief reference will be made to several items here.

Linings

Steel tanks which require some type of lining also require special construction details. They must be double welded, the same as for bare steel tanks, but extra care must be taken to obtain smooth, clean, splatter-free welds. It may be necessary to grind the welds. Special attention must be given to construction and welding of tank outlets and overflow dams as previously discussed. (For information on linings see Chapter 20.)

Diaphragms. The segregation of a plating solution into anolyte and catholyte by means of a porous nonconducting medium is called diaphragming, and the porous medium is called the diaphragm.

The diaphragm must be held in place to avoid tearing by racks, contact hooks, or workpieces as they are moved in and out or through the plating bath. The supporting structure, especially when built into a tank as shown in Fig. 16, is sometimes loosely referred to as a diaphragm.

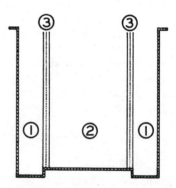

Fig. 16. Cross-sectional view of a plating tank with a built-in diaphragm showing (1) anode compartments, (2) cathode compartment, and (3) the diaphragms.

The built-in type of diaphragm consists of rubber-covered frames over which a cotton duck, nylon or a suitable cloth is stretched. The frames are fastened between rigid rubber-covered-steel supports which are a part of the tank construction.

The use of a diaphragm usually requires greater anode to cathode spacing to allow room for the diaphragm and sufficient clearance to lessen the tendency to tear the cloth when moving anodes or plating racks in and out of the tank. Because of the greater spacing the tank size must be larger and the d-c current source must have higher voltage to overcome the increased voltage drop through the plating bath and the diaphragm. The solution is filtered from anode compartment. For best results the pump suction line is split with valves in both lines so that the catholyte is filtered simultaneously.

Advantages claimed for the use of a diaphragm include:

1. Elimination of use of individual anode bags.
2. Reduction of roughness.
3. Ability to use air agitation and higher current densities without incurring roughness.
4. Freedom to inspect anodes and to replace same at any time without incurring roughness.
5. Improvement in throwing power and metal distribution due to increased anode to cathode spacing.

Other Features

Rinse tanks and many alkali cleaner and acid treatment tanks are frequently equipped with an overflow dam extending the full length of one of the sides. This enables one to more effectively skim the surface than with a corner dam. This is usually accomplished by overflowing with water in the case of a rinse tank or by pumping as in the case of an acid dip or plating solution that is being continuously filtered. The location of the water-feed line in a rinse may be used to control the direction of flow for surface skimming.

MANUALLY OPERATED TANKS

The depth of a tank and its width are limited by the length of the average man's arm. The tank length is governed either by the largest part to be treated or by the number of rack loads that can conveniently be handled with other facilities available.

Heating coils and ventilation ducts are best located on the side of the tank opposite that used by the operator. When this is not possible, the size and shape of parts or loaded racks that can be comfortably handled will be further restricted since the operator must reach further when processing workpieces.

TANKS FOR AUTOMATICS

As stated previously tanks for automatic plating machines must meet more rigid dimensional specifications because their ends must be precisely located with respect to a transfer mechanism which on a chain-driven machine must be synchronized with a definite chain pitch and rack spacing. The width is more flexible. Anode spacing or the anode to cathode distance may be designed to accommodate a variety of parts.

When ordering tanks separately for a new machine, or for a new cycle for an existing machine, the exact rack spacing around the entire machine should be determined and every precaution taken to make sure the tank ends fall into the exact position and that all tank dimensions are inside dimensions. Proper allowances should be made for wall thicknesses including the lining over the edge because error in dimensions can accumulate, and an error of $1/4$ in. at each tank end can reduce the usable rack size by several inches.

HEAD ROOM

The head room required in a plating installation may be governed by the depth of the plating rack and the means of transferring it from one tank to another. Generally a manually operated system does not encounter head room difficulties because the rack depth is limited to the length of a man's arm and even with an elevated platform around the tanks an average room of 8 or 9 ft in height is sufficient. However, with mechanical transferring devices such as an overhead hoist or fully automatic plating machine deeper racks may be employed to advantage and head room must be considered as a limiting factor.

In determining the required head room for an installation involving the mechanical transfer of racks, the following dimensions must be considered:

1. Distance from floor or pit level to the tank bottom.
2. Thickness of the tank bottom, including the lining.
3. Distance from inside tank bottom to the bottom of the rack when immersed.
4. Depth of the rack below the top of the tank.
5. Required clearance from the bottom of the rack above the top rim of the tank during transfer (not less than 2 in.).
6. Total vertical length of the rack.
7. Depth of the hook supporting the rack.
8. Vertical distance from the hook to the carrier means.
9. Total height of the carrier means and its supporting structure.

Any increase in the depth of immersion of a rack must be doubled to allow for clearance when figuring the required head room, while all other dimensions remain constant.

A pit is frequently used to compensate for limiting head room. Fig. 17 illustrates a pit installation with a floor curb, duck boards and drains on both sides. Less elaborate designs are acceptable, especially when operating from one side only. In such cases one drain may be adequate (see Chapter 18), but the floor curb is recommended to prevent dirt and debris from falling into the pit and clogging the drain.

MAINTENANCE

Preventive maintenance of tanks in a plating installation may consist of periodically washing corrosive liquids and salts from the outside of the tanks and from their accessories. When needed, a coat of protective paint should be applied.

The most likely point of failure on a tank is the bottom. Care should be taken when installing the tank to renew the bottom protection wherever it may have been damaged while moving it in position. Space for air circulation should be provided by the use of supports to help keep the bottom dry.

Five-in. steel beams placed on about 30-in. centers are recommended for tank supports. These of course should be coated with the same protecting paint as the tank.

Many tanks are damaged during installation by careless welding. A small piece of red-hot metal can melt its way through a lead, rubber, or plastic lining. A few inches of water, but preferably a tank full of water, will prevent this occurrence.

ACKNOWLEDGMENTS

The authors wish to thank Gunnar C. T. Lindh, Chester G. Clark, and Leo J. Lewandowski of The Udylite Corporation for assistance in preparing material for this chapter.

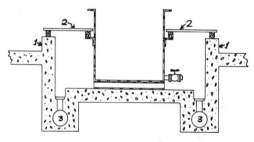

FIG. 17. Cross-sectional view of a tank installed within a pit showing (1) floor curbs, (2) duck boards, and (3) drains.

20. LININGS

Kenneth Tator*

Formerly Kenneth Tator Associates, Coraopolis, Pa.

In this chapter there is no intent to make the reader an expert in the formulation and application of linings. Space does not permit and the electroplater has no more need for such detailed information than the lining applicator has of the specialized knowledge of how to buff and polish a plated article. Information presented herein is for the use of the electroplater as a purchaser, user, and caretaker of lined equipment.

Details of lining application should be left to the more knowledgeable applicator. The lining materials manufacturer should be consulted for his best recommendation, selection and formulation of the lining required. His experience of good and bad performance goes far beyond the boundaries of any shop. If guarantees are involved, keep your hands off—one may invalidate the guarantee.

Linings are familiar to and a necessary part of the equipment of any electroplater. They protect the baths from contamination by the material of which the tank is constructed. They also protect the tanks and piping from corrosion by these baths. In addition, they are electrical insulators which restrict the plating current to productive uses.

LINING FORMS

Linings are used in a variety of forms (see Fig. 1):
(1) Sheet linings.
(2) Tank liners.

* Deceased.

(3) Liquid applied linings.
 (a) Coatings for plating racks and fixtures.
(4) Tank sheathings.

Some lining materials are available in several of these forms.

Sheet Linings

Organic Linings. These are the traditional linings. They consist of cementing sheets of lining material to the tank surfaces and lapping, weld butting, or strip sealing at the sheet junctions. The sheets are factory calendered and inspected; their thicknesses are uniform, except where they have been distorted over compound curvatures in application. They are normally inspected so as to be free from pinholes and other barrier discontinuities. Thus, their performance, assuming proper selection of material, is dependent upon adequacy of application, adhesion of the sheets to the equipment surface (without air occlusion), and the effectiveness in making the laps.

For this type of application, it is essential that the sheet possess flexibility and a reasonable amount of distensibility. Application is generally confined to rubbers which are flexible and elastic at the time of application, or thermoplastic sheetings such as the vinyls and vinylidene chlorides which become so when warmed. Most of the rubber linings require vulcanization after application to develop their full physical and chemical properties. The thermoplastics require no such heat treatment, except possibly for strain relief.

These sheet linings are normally applied in thicknesses from $1/16$ to $3/8$ in. and are available in

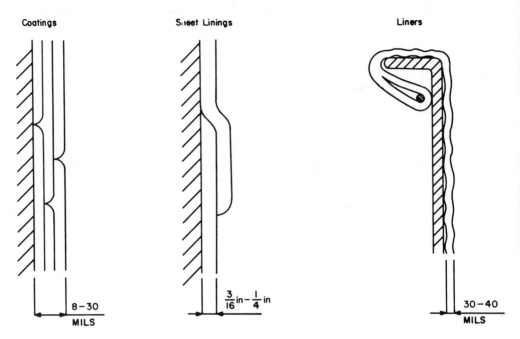

Fig. 1. Lining forms.

a wide range of hardnesses and physical properties.

Lead Linings. Lead linings are a specialized application of sheet linings (in that lead sheets are attached), but usually not homogeneously bonded to the steel tank wall. Homogeneous bonding is available, but usually too expensive to justify its use in plating services.

Sheet lead is available in various thicknesses, but it is usually applied at ⅛ in. Application is made by "lead burning" companies specializing in lead lining applications.

Tank Liners

Tank liners are a logical development of the defensible concept that a lined tank is nothing but a supported corrosion barrier. Liners are thus corrosion resistant barriers, prefabricated to fit within a structural support. This structural support may be an old or a new steel tank, a plywood or fiberglass box, or a structural framework.

Liners are bags of flexible material in thicknesses ranging between 8 mil and 3/32 in. They are custom fabricated to fit closely within tank liner's support structure.

While they can be made of rigid material, they are more practically and economically made of flexible sheetings of polyvinyl chloride, polyethylene, and natural and synthetic rubbers. They have the advantage of being relatively inexpensive, they can be installed without tank removal by plant personnel, they are readily replaceable and can be used for inexpensive renovation of old tanks and the lining of new tanks.

Liquid Applied Linings

These are linings that, in at least one stage of their application, are a liquid. Owing to their mobility in this liquid stage, they are more prone to nonuniformity of lining thickness than other lining types. Thus, they must be applied with a permissible variation in thickness. Applicators of these linings are alert to see that no thicknesses fall below protective minimums.

The spray-applied, paint-like protective coatings are recognizable as belonging to this form. However, the so-called "100 per cent solids" coatings, plastisols, flame-sprayed and fluidized-bed applications also are variants of this form.

Plating racks and fixtures are discussed fully in Chapter 23. The materials used for the insulation of these are almost exclusively the "liquid applied" materials described in this section. Rack coatings, stop-off lacquers, and masking compounds are usually of the solution coating type. Heavy duty coatings of vinyl plastisols have been in long successful use. Equally durable coatings from 100 per cent solids, flame-sprayed or fluidized-bed applications are in use—these application methods permit a much wider selection of coating materials for plating racks and fixtures than have heretofore been available.

Protective Coatings. These are normally applied in thicknesses of 7 to 30 mil. Vinyl coatings and lacquers of this type have been used by electroplaters for many years as maintenance coatings and linings, stop-off lacquers, and rack coatings.

Even though the materials may be resistant and insulating, their thin and variable thickness precludes any use with corrosive baths and cleaners. Susceptibility to impact damage makes them unreliable as insulating barriers.

These coatings are used for protection against bath and product contamination in non-corrosive baths and rinse tanks; and for protection of exterior surfaces of all tanks.

It is normal to apply these in multiple coats to insure that any discontinuities, thin areas, or other imperfections in one coat will be sealed in by succeeding coats.

Liquid applied protective coatings in air-drying or curing application are available with the epoxies, acrylates, polyesters, neoprenes, urethanes, vinyls, and vinylidene chlorides among others. The adhesion and coating resistances of all of these will be improved by baking. Baking at temperatures above 500 F is mandatory for the flurocarbon coatings and chlorinated polyesters.

"100 Per Cent Solids" Linings. These are organic coatings of which the resin is a liquid at the time of application; but by prior admixture with catalyst or reactant, they will convert to a resistant, solid form within a short time after application. These materials build linings from 10 to 30 mil readily by spray, and up to $\frac{1}{4}$ in. by trowel application when filled with fine inorganic material such as silica. They are often reinforced with glass fabric laid into the coating during application.

These materials have the advantage of building high thickness in a minimum number of coats. They are free from the shrinkage cracking, which may occur in solution coatings during evaporation of the volatile solvents.

The polyamide epoxies, polyesters, and certain formulations of neoprenes are available in these forms.

Plastisols. Plastisols are well known to the electroplaters through their extensive use as rack coatings, as well as equipment linings. Plastisols are a unique material within this classification.

Plastisols are a polyvinyl chloride resin dispersed in its liquid plasticizer; thus they are liquids in spray, dipping, or trowel consistencies. After application by one of these methods, they are baked in an oven to temperatures approximating 350 F. As the temperature increases, the polyvinyl chloride resin will absorb the plasticizer, becoming a solid of low cohesive and adhesive strength. At the temperature of approximately 350 F, it will fuse into a coherent, adhesive coating or lining, which on cooling possesses its maximum physical and chemical resistances.

The spraying formulations are currently inferior to other types of plastisols for hot aqueous services.

Flame-Sprayed and Fluidized-Bed Coatings. Coating materials which at the time of application are heat fusible, such as all thermoplastic resins and some thermosetting resins, may be applied by flame-spray or fluidized-bed application methods.

In the *flame-spray* method, the powdered resin is blown through the intense flame of the flame-spraying torch. This flame simultaneously melts the powder and preheats the steel surface on which the molten resin particles impinge. Application is continued until the desired thickness is built up. After the required build up, the torch (without powder flow) may be used to fuse the deposit into a continuous barrier coating.

The principle of flame-spray coating is to deposit and distribute a molten material upon a tank surface heated at above the melting point of the resin. This same principle can be achieved by the fluidized bed application method.

The *fluidized-bed* method consists of heating the tank or article to be lined or coated to a temperature above the melting point of the protective resin. While at this temperature, the resin (in powder form) is blown, fluffed, or tumbled in contact with the hot surface, where it melts. After the excess powder is poured or blown out, the tank is then baked at a temperature which is the fusion point of the resin and which sinters or flows the resin into a continuous barrier coating.

The thermoplastic vinyl, fluorocarbons, chlorinated polyesters, and certain formulations of the thermosetting epoxies are being applied by both the flame spray and fluidized bed techniques.

TANK SHEATHINGS (See Table 1)

Sheathings of wood or brick may be used within tanks, over the protective lining, where that lining must be protected from impact damage from

TABLE 1. SHEATHING SUITABILITIES*

Material	Nonoxidizing	Acidic Baths Oxidizing	Fluoride	Alkaline Baths pH < 12	Alkaline Baths pH > 11	Organics	Max. Temp., F.
Wood	S	X	S	X	X	X	160
Acid Brick	S	S	X	S	S	X	2200
Carbon Brick	S	X	S	S	S	S	930
Mortars:							
Epoxy (amine)	S	X	S	S	S	S	225
Furane	S	Ex. HNO_3	S	S	S	S	350
Phenolic	S	Ex. HNO_3	S	S	X	S	375
Polyester	S	S	S	S	X	X	225
Silicate	S	S	X	X	X	S	1850
Sulphur	S	Ex. CrO_3	S	X	X	X	200

* S = Suitable; X = Unsuitable; Ex. = S, except for.

cleaning, operating tools, or falling objects; they should also be used where the operating bath temperature is above the maximum operating limits for the lining.

Where impact is solely from falling objects, the sheathing is normally applied only over the lining on the bottom of the tank. Wherever there may be impact or scraping damage to the sidewalls, the entire interior surface may require sheathing.

For protection of lining from impact injury, it is not essential that brick be laid in cement; loose brick on the bottom usually will suffice.

Cemented brick sheathings over the lining will permit the use of linings for bath temperatures above the maximum limits for that lining. The brick will insulate the lining from the high temperature bath; heat conducted through the brick to the lining will be radiated from the steel tank surface to the surrounding atmosphere, thus maintaining a temperature gradient through the sheathing-lining-tank cross section. It is essential that the exterior tank surfaces be uninsulated and that there be no interference with heat dissipation from the exterior surfaces of the tank.

The following is an example of the effectiveness of brick sheathing: with a bath operating at 212 F., 4 in. of brick will bring the lining temperature down to 150 F.; 8 in. of brick will reduce this to approximately 130 F.

Mortar bonding of the brick must be used to prevent the liquid of the bath from direct contact with the lining.

LINING MATERIALS

Many different types of materials are suitable for and are used for linings. These range from the now almost discontinued asphalt, through the rubbers and plastics, to metallic linings of lead. Many of these may be used interchangeably for plating; some are available in a variety of lining forms.

Certain types of plating, especially bright nickel, are very sensitive to trace extractions from some lining formulations, producing a cloudy deposit. It is well to verify lining compatibilities with the supplier of proprietary plating baths.

Lining materials may be classified according to their type, which reflects their physical properties and resistances:

(1) Rubbers.
(2) Thermoplastics.
(3) Thermosetting resins.
(4) Metallics.

Rubbers

Except for the hard rubber variation, rubbers are recognized by their resilient, elastic characteristics. They may be natural, or synthetic rubbers such as the neoprenes, "Hypalon," nitrile, or butyl rubber.

They are normally used as sheet linings, but all can be made available as tank liners. Some formulations of neoprene and "Hypalon" are suitable as liquid-applied linings of the coatings or "100 per cent solids" type.

As sheet linings, they are normally applied in an unvulcanized condition that permits the hand forming required to follow the contours of the steel tank being lined. They are then vulcanized by heat to develop their full physical and chemical resistances.

Natural rubber linings are outstanding in their resistance to hydrochloric acid baths and cleaners; but are rapidly deteriorated by oxidizing baths such as chromic and nitric. They are softened by oils, greases, and hydrocarbon organic solvent, and may be affected or permeated by other organic materials. It is for this reason that they should not be used for acetic acid containing baths.

Natural rubber may be compounded to produce any degree of hardness including rigid hard rubber. As hardness increases, moisture permeation of the material decreases and the allowable maximum operating temperatures increase. The temperature for hard rubber is 200 F. as compared with 150 F. for soft rubbers. Laminar combinations of hardened, soft-rubber sheet linings, such as "Triflex," are available; they provide the resistances of hard rubber with the impact and strain absorbing capability of soft rubber.

Synthetic soft rubbers are available which will overcome some of the disadvantages of natural rubber, but in some cases introducing others. *Neoprene* has improved resistance to aging and to softening under mineral oils and greases. "Hypalon" is satisfactorily used for chromic acid and other acidic and alkaline baths. Resistance to organic solvents is poor. "Hycar" rubber will resist higher temperatures and is more inert to petroleum solvents than natural rubber. *Butyl* rubber is more impervious to gases, strong oxidizing acids and aging. None of these synthetics possess the outstanding resistance of natural rubber to hydrochloric acid.

Thermoplastics

Asphalt must be given recognition, as it was one of the earliest linings used by platers. Asphalt is a thermoplastic that is softened with heat and softened or dissolved by lubricating oils and greases and other organic solvents.

The use of asphalt as a lining material has been almost entirely displaced by synthetically produced thermoplastics, of which the vinyls represent a notable example. These synthetics are available in higher controlled purity and a wider range of resistance to plating solutions and cleaning baths. Depending upon their nature and degree of plasticization, they can be made available from pliable sheeting to rigid, self-supporting materials of construction for small tanks and plating barrels.

They remain susceptible to heat, i.e., they soften as the temperature increases and harden as the temperature decreases. They remain soluble in or softened by many organic materials or solvents.

Vinyls, PVC. These vinyls are available principally in two chemical forms: the vinyl chlor-acetate and the polyvinyl chlorides, PVC. In both forms they have an excellent range of utility with all plating baths and cleaners, with the exception of those containing acetic acid. As they are thermoplastic, they are sensitive to elevated temperature and can be dissolved or softened by many oils and organic solvents.

The chlor-acetate types are primarily used as protective coatings. These are used as linings of rinse tanks and for the exterior protection of other tanks. "Tygon" paint is an example of this type of coating.

The polyvinyl chlorides (PVC) are in wide use as sheet linings, fabricated tanks and equipment, and tank liners. These are available in forms ranging from flexible, extensible sheets to hard and rigid fabrications. "Koroseal" and "Tygon" sheet linings are examples.

Vinylidene Chloride. This material, "Saran", is similar in its properties and resistances to that of vinyls, except that it should not be used in strong oxidizing chromic or nitric baths nor in strong caustics.

It is available in protective coating forms useful for the protection of interior surfaces of rinse tanks and exterior surfaces of all other tankage, and in sheet linings for tanks and piping.

Polyethylene. Polyethylene is available in two states of polymerization: the standard grade with a maximum operation temperature of 140 F, and high density polyethylene with a maximum operating temperature of 180 F. Their chemical and resistant properties are substantially identical.

They may be used for all acidic plating and cleaning baths, but should not be used for strong caustic baths or cleaners, nor for organics, including acetic acid baths.

These are available as protective coatings applied to steel by flame spray or fluidized bed procedures, as fabricated, small non-supporting tanks, ducting or hoods, and as liners.

Polypropylene. Polypropylene, being a homologue of polyethylene, has most of the properties of polyethylene. As might be expected of a higher homologue, it has a somewhat higher maximum average operating temperature: 210 F.

It is available in the form of supported tanks, ducting and hoods.

Flurocarbons. Representative of the flurocarbons are "Kynar", "Kel-F", and similar organic fluorides. These are relatively inert, tough thermoplastic resins of high softening point: above 350 F.

They may be applied as protective coatings from dispersions, flame-spray, or fluidized bed application. These must be "sintered" or fused at approximately 500 F. to establish film continuity and adhesion.

These may be fabricated from sheet, or machined or molded to produce small tanks, valve parts and pipe fittings. They are resistant to all plating baths and cleaners.

Chlorinated Polyether. "Denton" is available as sheet linings, liquid-applied linings and coatings, and fabricated forms. It is resistant to all plating and cleaning baths, except chromic acid and other oxidizing baths, up to 210 F.

Thermosetting Resins

These are those materials, usually synthetic, which are applied, formed or molded in a state of incomplete polymerization; they then react or are caused to react to develop their characteristic physical and chemical properties. The vulcanizable rubbers are also representative thermosetting materials, but differentiated from these resins by their rubbery nature.

Thermosetting resins are seldom used as sheet linings or liners, but find their greatest use in protective coatings of epoxy or urethanes, as piping and fittings, plating barrels, and small self-supporting tanks of asbestos or glass reinforced phenolic, epoxy, furane, or polyester. The furanes, phenolics and silicate mortar cements for brick sheathing fall in this class of thermosetting materials.

Epoxy. Epoxies form tough resins that possess outstanding resistance to caustics and most acidic baths or cleaners of less than 20 per cent acid, except for the oxidizing baths of chromic and nitric acids. While they are unaffected by most organic solvents and oils, their use is not recommended with baths containing acetic acid or acid acetates.

They may be used in a protective coating form that is readily applied in thicknesses from 5 to 30 mil. Protective coating "100 per cent solids" formulations are available and permit attainment of high barrier thicknesses in a minimum of coats. These latter formulations are often reinforced by laying glass cloth into an intermediate wet coat.

By lay-up of epoxy-saturated glass cloth over forms, self-supporting tanks, piping, ducting and hoods are produced.

Properties of the coatings or fabricated equipment may be varied by choice of resin, and more particularly by type of curing agent used. For protective coatings in neutral and high pH immersion, the polyamide epoxies are preferred; whereas the amine cured epoxies possess better acid resistance. For epoxy-glass fabrication, the equipment fabricator will use that formulation or combination best suited for the operating conditions.

Urethanes. These copolymers are available either as protective coatings or in sheet form, either as sheet linings or tank liners. A popular example of the sheet type is "Neothane."

They are resistant to all normal plating and cleaning baths, with the exception of the strongly oxidizing chromic acid baths or nitric acid cleaners. They possess outstanding abrasion resistance.

Polyesters. These resistant, tough resins may be used in protective coating linings reinforced with glass fiber or fabric. Their greatest current use is as the binder and corrosion resistant vehicle in the lay-up construction of self-supporting, glass fabric, reinforced tanks, ducting, hoods, and piping.

Their resistance to most plating and cleaning baths is good, but the lack of resistance of any incorporated glass fiber to hydrofluoric acid containing cleaners, fluoroborate plating solutions and strong caustic plating baths or cleaners, precludes the use of glass-reinforced constructions for these uses. Resistance to organic solvents is low.

Acrylates. These are best known under the trade names of "Lucite" or "Plexiglas."

While the acrylates can be formulated into protective coatings, they lose some of their characteristic chemical resistance by such formulation. They are particularly useful to the plater in fabrications where their chemical resistance to most plating solutions and cleaners and their optical clarity permits construction in small plating tanks, plating barrels, and environmental test chambers, which permit observation of the operations through the equipment walls.

Metallics

Corrosion resistant alloys and metals are available for tank claddings and liners. Of these, only lead linings have attained stature in the plating industry.

Lead. Lead sheeting, usually alloyed with 5 per cent antimony, is formed inside of a tank as a rigid lead tank liner with adhesion to the support tank only at attachment points.

Lead linings are particularly suited for sulphuric acid and acid sulphate baths and cleaners and give adequate service in chromic acid plating baths. They should not be used for hydrofluoric acid or in fluoroborate plating baths, for nitric acid, in baths containing acetic acid, or in alkali plating baths or cleaners.

Maximum allowable temperature for lead linings is lower than might be expected for a metal: 165 F., due to the tendency of lead to creep by expansion at higher temperatures. Lead linings are often brick sheathed to permit higher temperature sue.

Lead has the disadvantage in some uses of being a conductor of electricity. Precautions need be made in plating operations to prevent current and anode metal loss.

LINING REQUIREMENTS

Platers have the choice of a wide variety of pickling, cleaning and plating solutions. At first glance, the selection of the optimum lining for each of these baths would seem to be insurmountable. Fortunately, however, lining performances depend not on specific bath compositions, but on basic characteristics of the bath.

For example, a lining which is resistant to the alkalinity produced by sodium hydroxide would be equally resistant to alkalinities produced by potassium hydroxide, sodium carbonate, trisodium phosphate, or other fixed alkali. Thermoplastic linings which are softened at a certain temperature do not know or care whether this limiting temperature is produced by an alkaline or acidic bath, or just plain rinse water. Thus, it is not required to design a lining for each specific variant in bath composition. Linings can be selected on the basis of certain general characteristics of the plating process which affect lining performance. These are:

(1) *Acidic Baths*
  (a) oxidizing
  (b) acid fluorides
  (c) others
(2) *Alkaline Baths*
(3) *Organic Solvent Action*
(4) *Temperature*
(5) *Electrical Requirements*

Lining performances are summarized in Table 2. As essential ingredients of some plating solutions can be "poisoned" or adsorbed by the constituents of some linings, it is wise to verify suitability for proprietary plating chemicals with your supplier.

TABLE 2. LINING MATERIAL SUITABILITIES*

Material	Forms**	Acidic Baths Nonoxidizing	Oxidizing	Fluoride	Alkaline Baths pH < 11	pH > 11	Organics	Max. Temp., F.
Acrylates	F	S	X	X	S	S	X	180
Chlorinated polyether	C,S,F	S	X	S	S	S	S	210
Epoxy (amine)	C	S	X	S	S	S	S	170
Epoxy-glass fibre	F	S	X	X	S	X	S	170
Fluorocarbons	C,F	S	S	S	S	S	S	275
"Hypalon"	C,S,F,L	S	S	S	S	S	X	150
Lead	S	Ex. HCl, HBF$_4$	Ex. HNO$_3$	X	X	X	X	165
Neoprene	C,S	"	X	S	S	S	X	200
Polyester	C	S	S	S	S	X	X	200
Polyester, glass fibre	F	S	S	X	S	X	X	200
Polyethylene, Std.	C,F,L	S	S	S	S	X	X	140
Polyethylene, HD	F,L	S	S	S	S	S	X	180
Polypropylene	F	S	X	S	S	S	X	210
Rubber, soft	S,L	S	X	S	S	S	X	150
Rubber, hard	S,F	S	X	S	S	S	S	200
Urethane	C,S,L	S	X	S	S	S	S	150
Vinyl	C,S,F,L	S	S	S	S	S	X	150
Vinylidene chloride	C,S	S	X	S	X	X	X	140

* S = Suitable; X = Unsuitable; Ex. = S, except for.
** Forms: C = Protective coating; S = Sheet lining; L = Liner; F = Fabricated, or molded tanks, piping, or fittings.

Acidic Baths

All lining materials, within their temperature limitations (shown in Table 2), possess resistance to acidity, but not necessarily to secondary reactions such as oxidation, solvent action, or attack by silica destroying reactants.

Thus, most lining materials, with the notable exception of neoprene and lead, will be resistant to the muriatic solutions commonly used for metal cleaning and acidic chloride plating baths. All will be resistant to sulphuric acid in concentrations below 50 per cent and acidic sulphate baths. All will be resistant to phosphoric acid cleaners and phosphating processes.

As acids are corrosive to the steel of tanks and equipment, heavy duty linings always should be used. Never use protective coatings of thickness less than 30 mil, regardless of how acid resistant the coating material might be. Sheet linings of thicknesses of $3/32$ in. or greater are indicated, or tank liners. Protection of these linings against inadvertant tears or breaks by operational impacts is desirable (by means of protective sheathings of wood or brick, or by increased lining thickness).

Oxidizing Acids. Chromic acid and nitric acid, their salt, and sulphuric acid in concentrations above 50 per cent are strong oxidizing agents. Hydrogen peroxide used as an anti-pitting agent (as in nickel baths) is oxidizing.

Oxidizable linings such as natural and many synthetic rubbers and certain thermosetting coatings (such as the epoxies) will not prove sufficiently resistant. Embrittlement and cracking of the linings will result. Oxidation of sulphur from rubber linings may also alter the sulphate-chromate balance for bright chrome plating.

Vinyl, fluorocarbon, and lead sheet linings are used. Lead is not, however, suitable for nitric acid. Polyethylene and vinyls are used for tank liners. Glass reinforced polyesters and polypropylene have been used for small tanks, plating barrels, and piping.

Acidic Fluorides. Hydrofluoric acid and acidic fluorides have the property of dissolving silica. No glass or glass fiber reinforced tanks or tank lining material should be used in such baths. For brick sheathing, carbon brick with carbon filled mortar is required, as these fluorides will attack the siliceous content of acid brick.

By product, hydrochloric acid containing hydrofluoric acid in concentrations over 50 ppm will attack ceramic acid brick. Carbon brick with carbon-filled mortars should be used as sheathings

Fluoroborate baths have been used with all linings, but prolonged use will deteriorate glass or silica containing linings.

The natural rubbers, vinyls, vinylidene chlorides, urethanes, fluorocarbons, and chlorinated polyesters are suitable for sheet linings. For tank liners, the vinyls and polyethylene are used.

Alkaline Baths

As a general rule, alkalies are not corrosive to steel and hence, no lining need be used for corrosion protection. Linings may be required to protect the bath or product from iron contamination or from rust deposits forming on the tankage above the liquid level and falling into the bath. For alkaline plating tanks, lining may be required to insulate the tank electrically, to prevent current diversion to and attack of the tank surfaces, and for plating onto these surfaces.

The vinyl and epoxy protective coatings may be used as linings for caustic cleaning and rinse tanks.

The alkaline baths may be divided into two categories: (1) those with strong caustic reactions above a pH of 11; and (2) those at a pH of 11 or below.

Baths above pH 11. This category is represented by the strong caustic baths such as caustic cleaners, stannate plating baths, and all other baths and cleaners with strong caustic reactions.

Any lining material which is either saponifiable or necessarily contains saponifiable compounding ingredients or siliceous material will be unsuitable. In the use of masonry sheathing, carbon brick with carbon-filled mortar should be used in place of the siliceous acid brick. Glass or glass fiber reinforced resins should not be used.

Baths at or Below pH 11. This category includes ammonical and cyanide baths. Linings are not destroyed by alkalinity, but by reaction with bath constituents.

Thermoplastics such as vinyls or vinylidene chlorides are satisfactory in use, and the thermosetting coatings such as the epoxies or neoprenes have special application.

Organic Solvent Action

Some lining materials, notably the thermoplastics and the natural rubbers, will be dissolved or softened by organic solvents.

Protection of these from spillage of lubricating oil or grease is required. These organic susceptible linings will not prove suitable for baths containing organic acids, such as acetic acid.

Thermosetting resins such as the polyesters, urethanes, hard rubber and the relatively inert fluorocarbons or chlorinated polyether will be suitable for acetic acid baths.

Temperature Limitations

All lining materials have maximum-operating-temperature limitations. Higher temperatures will soften and increase the permeability of thermoplastics. Thermosetting linings will be hardened to the point of cracking and loss of adhesion.

"Maximum Temperature" values in Table 2 are the normal maximum temperature limits for most baths and usual lining formulations. With some chemical baths, these permissible limits may be greater or lower. In general, the greater the acid concentration, the lower will be the allowable temperature. The lining applicator and the supplier of plating chemicals will advise if any deviations are allowable from those given in this table.

Electrical Requirements

It is usually desirable that interiors of plating tanks be electrically insulated. Without such insulation, plating losses will occur by diversion to tank surfaces and possible leakage to ground. Anode metal may be lost by being plated out on these surfaces. If any exposed tank steel becomes anodic, bath contamination will result.

All of the currently used lining materials, with the exception of lead, carbon brick, and any organic linings which are highly filled with carbon, will be non-conductive and sufficiently insulating to accomplish this function of electrical isolation.

Design for Lining (see Fig. 2)

Equipment design and preparation for lining do not particularly concern the electroplater; these provisions normally are handled between the equipment fabricator and the liner, which are often the same company. However, when the plater orders a tank lined, which had previously been in other uses or separately fabricated, he must appreciate these design principles. Either way, he will pay their cost, regardless of whether he arranged for the tank to be initially prepared, or whether the lining applicator had to perform this work before lining.

Tank design and preparation for lining is for the purpose of preparing the surface which is clean, continuously smooth and rounded, without any abrupt discontinuities and irregularities.

Sharp Edges, Prominences and Projections. Sheet linings cannot follow the curvature

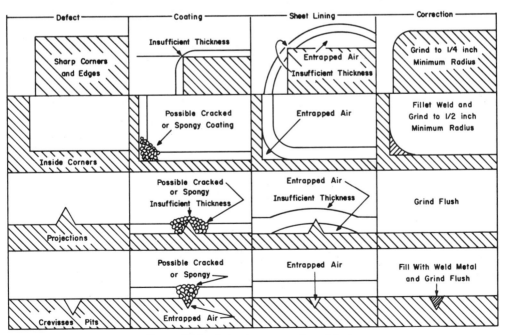

FIG. 2. Lining behavior over surface defects.

of sharp edges, prominences and projections. These will either cut into the lining, reducing the barrier thickness, or the lining will bridge over these, thus entrapping air at these points. Under fluctuating bath temperatures, this entrapped air will expand and contract, causing flex fatigue of the lining and destroying its adhesion to the tank surface.

The surface tension of liquid-applied linings will cause them to pull thin from edges and peaks, resulting in deficient thicknesses at these irregularities.

Tank liners are liable to be cut or punctured by such edges or projections.

All sharp edges should be ground to a radius sufficient that the lining material can follow the curvature without bridging. Prominences and projections should be similarly rounded, or if possible, ground flush.

Inside Corners, Crevices, and Pits. Lining materials cannot flow into and seal deep pits and crevices. Air will be entrapped in or under the lining, resulting in blow holes, through liquid applied linings, and fatigue cracking and loss of adhesion of sheet linings. Sheet linings cannot follow the contour of squared inside corners. Air entrapment will result.

All crevices should be continuously seal-welded. Pits and other abrupt surface depressions should be either fillet-welded or filled with a caulking material of the same type, or one that is compatible with the lining material. Inside corners should be given a radius by similar means.

Surface Preparation and Priming. Sheet and liquid applied linings require application over cleaned, roughened metal. Sheet and most liquid-applied linings require application of an adhesive primer over this surface before lining application.

Any residual rust or soilage on the tank surface before lining application will not only impair initial adhesion, but may cause subsequent blister development in service, through creation of osmotic pressure cells. Surface preparation is preferably by sand blasting: to a white metal finish for linings in immersion service; and to at least a near-white finish for chemical resistant coating on the exterior of tanks.

As tank liners are not in adhesive contact with the tank walls, rigid surface preparation and primer is not required. Nevertheless, it is advisable to remove any rust deposits which may abrade the liner, and to preserve this rust-free condition by coating the surface of the tank with a coat of a good moisture-resistant metal paint before liner insertion.

Design for Various Lining Forms. *Sheet linings* are used, and their higher cost is justified only for the holding of the most corrosive solutions. No equipment metal exposure can be tolerated. Thus, it is required that all design and surface preparation details be rigorously observed.

Liquid applied linings should never be used for the protection of tankage and equipment against highly corrosive baths. These are normally used for prevention of contamination, loss of electric current, and anode metal. A small frequency of minute discontinuities is permissible. Only in the fortuitous conjunction of perfect tank design, preparation, and lining application can a completely pinhole-free lining be produced. The better the design and preparation, the lower will be the frequency of such discontinuities.

Tank liners can tolerate no sharp edges or points which might puncture or abrade the liner. It is advisable to round all outside corners or flanges on which the liner will bear. To prevent air entrapment between the liner and the supporting tank or container, it is advisable to drill small vent holes through the structural shell, particularly at lower inside corners. These will also serve to warn of any breaks in the lining, by drippage from these holes.

Preconditioning for New Linings

New linings often will contain within the lining, or on its surface, foreign materials that are from or are used in the lining application. Also, the lining materials may contain small amounts of soluble materials which will be leached into the bath during initial operation of the lined tank or equipment.

These foreign or leachable materials should be removed before the lined equipment is used for production purposes, as these constituents may adversely affect bath compositions or performances. Once this is done, full resistance and inertness of the linings will result.

For such removal, it is advised to fill the equipment with water and hold at a temperature just below the top allowable operating temperature for the lining for a period of at least 24 hr. If the lining is to be used in acid service, it is advisable to acidulate the water. Add to the water such wetting agents or detergents which might be used in the production bath.

Lead linings are preconditioned with diluted

sulphuric acid to establish the protective passive surface.

After such a pre-conditioning soak, the linings should be rinsed with fresh water; they will then be ready for service.

Care of Linings

For proper maintenance of tank linings, it is essential to know at any future time the type of lining, and its limitations. Toward these ends, it is advisable to stencil each tank, or post information on or near the tank, giving the type of lining, date of installation, maximum operating temperatures permitted, and types of use or baths for which the lining should never be used.

During installation and the service life of the tank, the following precautions should be observed.

Exterior Protection. It can be anticipated that there will be spillage, condensation, and fumes from the bath contained within the tank, or from adjacent tanks. Exterior surfaces of the tank should be protected with coatings selected to resist such spillage. Except in anticipated spillage of strong acids, these may be thin film, protective coatings of total protective thicknesses between 6 and 10 mil, applied over properly primed and mill scale freed, clean surfaces.

The underside of the tank deserves special consideration. The tank preferably should be raised off the floor on insulated supports to prevent spillage from accumulating between tank bottom and floor, thus corroding the tank from the bottom in, and to facilitate maintenance attention.

Where spillage might otherwise run under the tank, deflector flanges should be affixed to the bottom edges to divert the spillage toward the floor drainage system. Such deflector flanges or drip pans should also be placed between tanks, to minimize drippage from overhead work onto the tank exteriors.

Protection Against Mechanical Injury. Linings should be protected from damage of heavy work, from fittings dropped on the bottom of the tank or from scraping the sides.

With sheet linings, it is advisable to specify greater thicknesses on the bottom. For all linings, it is advisable to sheath the bottom with loosely laid brick to take such impacts. Where injury may also be expected on the side of the tank, it may be advisable to brick sheath all lining surfaces, with wall brick mortared back and side faces. Wood sheathing may be used for acid and rinse tanks.

In the recovery of fallen objects, do not rake for these with pointed hooks or grapples. Use a tank magnet instead.

Protection From Excessive Heat. Baths should always be heated by heaters immersed in the bath or by external heat exchangers circulating the bath. Except in the case of thin thermosetting linings such as the phenolics, epoxies or furanes, it is inadvisable to heat through the lined wall as by jacketed vessels.

Where immersion heaters or coils are used, it is well to keep these coils at least 1 in. away from the lined surface, to prevent overheating of that surface and to allow free circulation of the bath around the heater surfaces. When a bath is being rapidly brought up to operating temperature, it is advisable to agitate the bath to prevent local "hot spots" adjacent to the heater area.

Where bath temperatures may approach or exceed the safe operating temperature for the lining, mortared brick sheathing should be installed over the lining.

Tank Fittings and Attachments. When attachments are made to the lined tank or flange, be sure that those fittings and their point of attachment are as well protected as the tank itself. The point of attachment should be properly gasketed and sealed, and if necessary, electrically insulated. Such consideration is normal for attachments or fittings below the bath level, but often overlooked are the attachments, rod and anode supports, hangers, exhaust ducting, and other attachments to the lined top flange of the tank.

Electrical Isolation. Plating tanks should not be grounded; they should be insulated from the floor and adjacent equipment.

At all emptyings of the tanks, the entire lining should be examined for build up of metallic electro-deposits, or treeing, growing over discontinuities in, or foreign sources of electrical attraction on the surface of the lining. These should be removed. If not removed, they will continue to spread over the tank area, thus creating increasing diversionary cathode surfaces, with resulting plating-current and anode-metal loss. In addition to cleaning, any discontinuities should be repaired.

Lining Repair

The ease of repair of linings is dependent upon the lining type.

Thermoplastic linings: small breaks in the vinyls vinylidene chlorides, polyethylenes, and polypropylenes can be repaired by melting lining material by means of a soldering iron. Larger breaks can be repaired by cementing a patch over the break, under the action of heat or solvent.

The solvent-applied, thin coatings are readily repaired by cleaning off the failing area and recoating with the same system originally applied.

Repair of vulcanized linings and thermosetting liners or self-supporting tanks usually requires the services of trained applicators.

Acknowledgments

The author gratefully acknowledges review and contributions by: A. K. Graham; L. E. Lancy, Lancy Laboratories; H. C. Klein, B. F. Goodrich Industrial Products Co.; W. J. McDowell, Fabrico Manufacturing Co.; S. J. Oechsle, Jr., Metalweld, Inc.; R. R. Pierce, Pennsalt Chemicals Corp.; R. B. Saltonstall, Udylite Corp.

21. HEATING AND COOLING EQUIPMENT

H. L. Pinkerton

Fellow Engineer, Process Engineering Department, Aerospace Division, Westinghouse Electric Corporation, Baltimore, Md.

AND

A. Kenneth Graham

Consulting Engineer, Jenkintown, Pa.

Introduction

It is the purpose of this chapter to provide the reader with the information necessary to make an intelligent selection of heat transfer equipment and to install it in conformance with approved plating practice. Those familiar with heat transfer theory may find the treatment oversimplified; the author justifies this approach on the basis that precise designing is unnecessary. It is only required that the equipment be adequate and yet not grossly oversized. One simply provides a little more capacity than is necessary and lets the automatic control do the rest. The use of theory has been confined to calculating the heat to be transferred, and to the concept of the overall heat transfer coefficient of a piece of equipment for the purpose of calculating the required area. In some instances, nomographs are given which eliminate all calculations. Those desiring a more detailed treatment of theory are referred to standard works on heat transfer.

Heat Exchanger Selection

The factors which must be considered in arriving at the most satisfactory heat exchanger are as follows:
(A) Heating or Cooling Medium
  Heating:
  (1) Steam
  (2) Hot water
  (3) Electricity
  (4) Gas

  Cooling:
  (5) Cold water
  (6) Refrigerated brine
(B) Heat-Exchanger Type
  (1) Internal (immersed)
  (2) External
(C) Materials of Construction
  (1) Corrosion resistance
  (2) Compatibility with solution
  (3) Strength and durability
  (4) Cost

Choice of *medium* is usually dictated by local considerations of availability and cost.

Choice of *exchanger* type is to a certain extent dependent on the indicated size of exchanger required. When the required transfer area is less than about 20 to 50 sq ft, an immersion exchanger is applicable, although larger immersion heaters can be used; moreover the advantages of external exchangers may make them preferable even in smaller installations. The capacity of the immersed type is limited not only by space considerations in the tank, but also because the external type can be designed for more efficient heat exchange, especially in cooling. An internal exchanger is simpler to install and will usually be less costly. It occupies valuable tank space and may be subject to stray currents unless properly installed. Immersion exchangers such as glass or graphite may be damaged by dropped or mishandled work or anodes. The external exchanger is less subject to these limitations and may, in addition, provide solution agitation promoting uniformity of solution temperature and composition.

Materials of Construction

Suitable materials of construction for use with the various plating baths are indicated in Chapter 6. For other processes such as anodizing, electropolishing, etc., the usual considerations of corrosion resistance will apply.

Basic Relations

A heat exchanger is a device for transferring heat from a hot fluid through a wall to a colder fluid. There may be five barriers to this transfer:

1 and 5—the two stagnant fluid layers adjacent to the wall, through which heat is transferred by convection and conduction.

2 and 4—scale or corrosion products on both sides of the wall, of poor thermal conductivity, and

3—the wall itself, which is usually highly conductive, and hence of little relative importance.

Electric immersion heaters are an exception to this generalization as will be seen.

The capacity of any exchanger is the number of Btu/hr which is transferred in operation. The heat transferred is proportional to the wall area in contact with the two fluids and to the temperature differential between them according to the familiar equation:

$$Q = UA\Delta T \quad \text{(Eq. 1)}$$

Here Q is the Btu/hr of total transfer, A is the transfer area, sq ft, and ΔT is the temperature differential, degrees F. The constant of proportionality, U, has the units Btu/(hr) (sq ft) (deg F) and is called the *over-all coefficient of heat transfer*. It is a complex function of exchanger design and conditions of operation, including the characteristics of both fluids and their velocities relative to the exchanger wall. Given sufficient data, it is possible to calculate U, but exchanger manufacturers usually determine it experimentally and publish the data, which may relate the coefficient to fluid velocity.

In tubular exchangers the question arises whether A refers to the inside or outside surface of the tube. When steam is inside the tube, the outside area is always used for calculations, and vice versa. Otherwise a published coefficient should state (often by subscript, as U_o or U_i) which surface has been used in the calculation. If no statement is made, the outside surface is assumed.

Although the true temperature differential, ΔT, is often a very complex concept*, for most plating applications it may be very much simplified and taken as the mean (arithmetical) temperature difference between the two fluids. Thus, when one liquid is being used to cool another in a tube exchanger, ΔT may be taken to be the arithmetical mean between the temperature differential at one end of the exchanger and that at the other end. When calculating ΔT for external exchangers, it is common to assume that the liquid returning to the tank will be 10 F above (or 10 F below, for cooling) operating temperature.

Calculation of Heating Load

Although many heat transfer problems can be solved by the nomographs given in this chapter, some are not readily or wholly amenable to this treatment. When a certain volume of solution is to be heated in a given time, the usual nomographic approach is to draw the figure so that it gives the heat actually required to heat the solution, plus some arbitrary percentage allowance for heat losses to surroundings during the heat-up period. It is implicitly assumed that this quantity of heat is then also sufficient to maintain the solution at operating temperature. If the heat-up period does not exceed 3 to 4 hr, this assumption is usually justified, and the nomograph may be used with confidence. Under unusual conditions, however, the heating load for both heat-up and maintenance should be calculated (a simple enough operation) and the larger value used for design. Calculation is indicated whenever:

(a) The heat-up time is 4 hr or more.
(b) Maintenance losses will be high, due to:
   (1) A large tank operating at high temperature.
   (2) A substantial air velocity across the tank, e.g., a high exhaust rate.
   (3) A constant feed of cold water, as to a running hot rinse.
   (4) A large number of heavy parts are being processed at a high rate (in almost all cases this factor is insignificant).

The heat losses to surroundings at operating temperature may be obtained from Fig. 1 and a knowledge of the tank size. The data of Fig. 1 are for still air; over exhausted tanks the velocity of air across the water surface may average from 3 to 25 ft/sec. Fig. 2 gives a factor by which the surface losses should be multiplied to arrive at the approximate actual losses. During the heat-up period, the rate of heat loss may be taken as one-half of that at operating temperature in

*See Appendix, p. 593

still air (the tank exhaust is usually not operating during the heat-up period).

The heat required to raise the solution from room to operating temperature is given by:

$$\text{Btu/hr} = \frac{\text{cu ft of solution} \times 62.4^* \times \text{temperature rise °F}}{\text{Heat-up time, hr}} \quad \text{(Eq. 2)}$$

The heat removed by the metal being processed is:

$$\begin{aligned}\text{Btu/hr} = &\;\text{lb of metal/hr} \\ &\times \text{specific heat of metal} \\ &\times \text{temperature rise, °F}\end{aligned} \quad \text{(Eq. 3)}$$

The specific heats of the metals will be found in Table 5, Chapter 1.

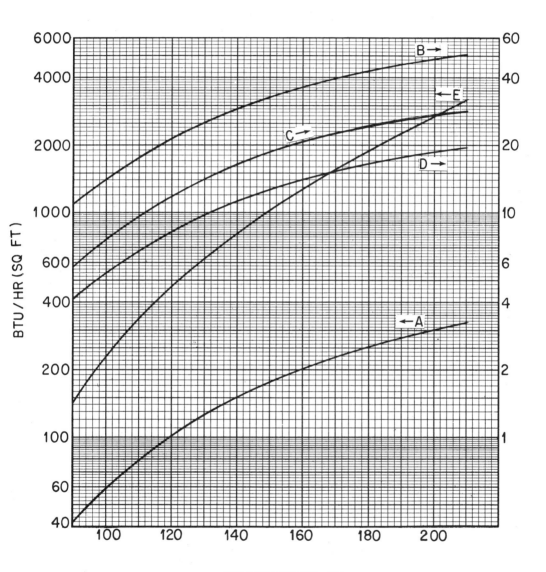

FIG. 1. Heat losses to still air at 60 F from uncovered water tanks.
Curve A—Heat loss through tank wall, bare steel—read left scale.
Curve B—Heat loss through tank wall, 1 in. insulation—read right scale.
Curve C—Heat loss through tank wall, 2 in. insulation—read right scale.
Curve D—Heat loss through tank wall, 3 in. insulation—read right scale.
Curve E—Heat loss from liquid surface to still air—read left scale.

If water is flowing into the tank, the heat required for it is:

Btu/hr = gpm of water flow
× temperature rise °F × 500* (Eq. 4)

The following simple example will illustrate the above calculations: An uninsulated hot rinse tank with exhaust, 2 ft × 4 ft × 4 ft deep is operated at 170 F, with 1 gpm of water feed at 70 F and a through-put of work amounting to 500 lb of steel parts per hour. If a 3-hr heat-up period is allowed, what is the heating load? Room temperature is 70 F.

Step 1. A hot rinse has an exhaust of a low velocity; at 5 fps and 170 F, Fig. 2 shows a factor of 2.4, and the heat losses at operating temperature from Fig. 1 are:

	Losses in Still Air Btu/hr	Losses, Exhausted Btu/hr
(a) 8 sq ft surface area × 1600 =	12,800 × 2.4 =	30,720
(b) 56 sq ft wall area × 230 =	12,880	12,880
	25,680	43,600

Step 2. For a 3-hr heat-up period, the heat input must be:

Btu/hr

(a) To heat solution: (2 × 4 × 4) ×
62.4 × (170 − 70)/3 =   66,560
To replace heat losses (still air):
25,680/3 × 2 =   4,280
                              ──────
                              70,840

Step 3. To maintain solution at temperature, the heat input is:

Btu/hr

(a) Heat lost to surroundings =   43,600
(b) To heat incoming water, from
    Eq. (3): 1 × 500 × 100 =   50,000
(c) To heat steel workpieces, from
    Eq. (2): 500 × 500 × 0.11 × 100 = 5,500
                                   ──────
                                   99,100

This is an example of more heat being required to maintain the solution than to heat it up; a little less than 100,000 Btu/hr must be supplied.

* Most plating solutions may be considered as water, weighing 62.4 lb/cu ft. The product of specific heat × specific gravity is taken as 1. In Eq. (4), 8.33 lb/gal × 60 min/hr = 500.

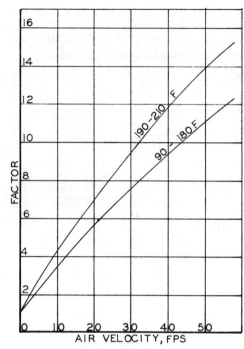

FIG. 2. Approximate factor for heat loss from water surfaces at various air velocities (fps). Multiply Btu/hr from Curve E, Fig. 1, by factor corresponding to appropriate air velocity and solution temperature range. (Adapted from "Heat Losses from Tanks, Vats and Kettles," S. J. Friedman, *Heating and Ventilating*, April 1948).

Calculation of Cooling Load

In plating, the usual reason for cooling is to extract the heat caused by passage of current through the solution. The most common operations requiring cooling are:

Operation	Result of Too High a Temperature
Chromium plating	Shift of bright range; loss of physical properties
Bright cyanide zinc plating	Loss of brightness
Anodizing (chromic)	Soft, porous coatings
Anodizing (sulfuric)	Coating redissolves too rapidly
Bright dipping	Etching, loss of luster
Electropolishing	Etching, loss of luster

In all the above operations except bright dipping, the cooling load is the removal of the excess electrical heat input and is readily calculated from the tank voltage and amperage:

3.4 × Volts × amperes = Btu/hr   (Eq. 5)

This simplified expression ignores the fact that the tank voltage is actually larger than the RI drop through the solution; furthermore because these are all room-temperature operations (except chromium plating and some electropolishing processes) the heat lost to the surroundings is neglected. This is good conservative practice, even in chromium plating and electropolishing at higher temperatures, because the usual average over-all coefficients of transfer do not take into account the viscous nature of these solutions. Where the cooling load is the chemical heat of reaction, as in bright dipping, its value cannot be calculated, but must be experimentally determined by measuring the temperature rise in a given interval, during which the production rate is at its maximum.

Specific Types of Heat Exchanger

Immersed Coils.* The most frequently used method of heating is to pass steam through pipe coils immersed in the solution (Fig. 3). The coils can also be used for cooling by passing water through them, but unless the solution is vigorously agitated, the heat transfer is extremely poor. This is due not only to the fact that the temperature differential in cooling is usually smaller than in heating, but also to the difference in convection currents. In heating, the rising convection currents (Fig. 4) assist considerably in heat transfer and distribution; whereas the downward currents set up by a cooling coil are much more sluggish and less effective.

Coils should be placed on the side walls of a tank rather than on the bottom, where they will interfere with cleaning, and in some solutions may become blanketed with sludge. In hot rinses, coils should be placed (Fig. 4) opposite the dam so that the movement of the surface water is toward the overflow, thus keeping the surface clean.

In electrified tanks, water and steam connections to coils (and to any other electrically conducting immersion heater) should contain an insulating section of hose at least 6 in. long, in both inlet and outlet lines, to avoid making an electrical ground through which stray currents will pass. Coils are usually mounted between the outside electrodes and tank walls, or at the end of the tank which is less desirable because it is difficult then to position the coils so as to minimize bipolar conditions, in which the coil becomes anodic in one section and cathodic in another.

The discharge traps from steam-heated exchangers in tanks containing concentrated solutions should either be connected to the sewer or, if steam conservation is important, to a condensate return line which is equipped with a conductivity control and alarm (see Chapter 37). Savings in heat and water achieved by returning the condensate to the boiler can be considerable, off-setting the modest cost of the conductivity alarm. Without the alarm, extensive damage to the boiler may occur in the event a coil leaks, thereby permitting solution to contaminate the condensate.

Some points of elementary importance are too frequently overlooked. Service mains, pipe coils and traps must be of sufficient size to handle the quantities of steam or water that will be required, at the operating pressure (see Table 28, Chapter 1). Tank coils must also be designed so that no condensate can be trapped. This cannot happen in the single-pipe coil of serpentine form, but if the header type of construction is used, the return

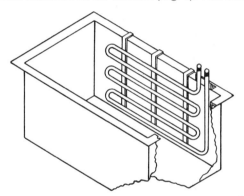

Fig. 3. Typical installation of tank coil. (*Courtesy of the Udylite Corporation*)

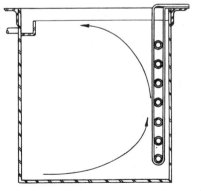

Fig. 4. Coil installed in tank with overflow. (*Courtesy of the Udylite Corporation*)

* See p. 586 for data on plastic coils.

header must rise from the lowest point in the bank of pipes so that all condensate is blown out by the steam.

In an installation with lead coils, care must be taken to keep the steam or water pressure low enough not to exceed the allowable fiber stresses in the pipe given by Fig. 5. An alloy of 6 per cent antimony-lead has considerably more strength below about 250 F but should not be used above this temperature. An alloy of lead with 0.04 to 0.1 per cent tellurium has also been used. It has strength equivalent to chemical lead and better resistance to sulfuric acid corrosion.

Embossed sheet metal coils (trade names "Platecoil," "Panelcoil," etc.) consist of two embossed sheets of 14- or 16-gauge metal welded together. The embossings form either a serpentine coil, Fig. 6, or a headed bank of pipes in various configurations, Fig. 7. Coils of this type are offered in 14-gauge mild steel and 16-gauge stainless steel (type 316). "Monel," Carpenter 20 alloy, nickel, "Ni-O-Nel" and "Hastelloy" B and C. Other weldable materials, including titanium, and other thicknesses may be supplied for special requirements. Plastic- or lead-coated exchangers are also available; in these there is a loss of about 25 per cent in heat transfer coefficient.

The construction of Fig. 6 is preferred for cooling applications, because for a given flow of cooling water, its velocity in the tube is higher than for the form in Fig. 7 and makes for better

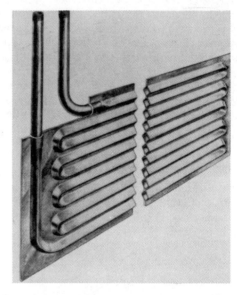

FIG. 6. Embossed coil, serpentine form. (*Courtesy of "Platecoil"*[R] *Division, Tranter Manufacturing, Inc.*)

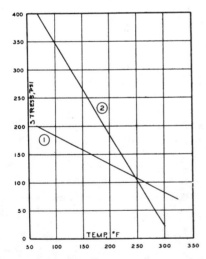

FIG. 5. Allowable fiber stress in extruded lead pipe at various temperatures.
Cure 1—Chemical lead
Curve 2—6% antimony lead or tellurium lead

Allowable pressure inside pipe, psi = $2ST/D$, where S is the allowable fiber stress from Fig. 5 and T and D are the pipe wall thickness and inside diameter, in., respectively. (*From data of the Lead Industries Association*)

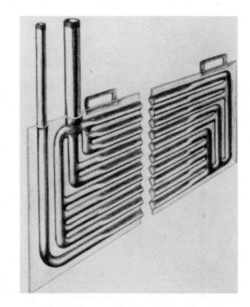

FIG. 7. Embossed coil, header form. (*Courtesy of "Platecoil" Division, Tranter Manufacturing, Inc.*)

heat transfer. The headed bank of pipes is appropriate for steam heating although the serpentine form is equally efficient.

Because of the increased radiation surface and closer "bends" an equivalent transfer can be obtained in a smaller space with the use of embossed rather than conventional pipe coils.

Electric Immersion Heaters. Although the cost of heating by electricity is in most areas considerably higher than by steam, electric immersion heaters are often chosen because of their many advantages. Temperature control is easy and precise; the heaters are portable and easily installed, and the utilization of electric energy is practically 100 per cent efficient. Because of this it is unnecessary to consider the over-all transfer coefficient or to use Eq. (1).

Electric heaters may be sheathed either in metal or fused quartz jackets. Quartz heaters are suitable for any acid bath except hydrofluoric acid and phosphatizing solutions. Metal-encased units transmit heat to the solution by conduction; the resultant high sheath temperature requires that the unit be so placed that it is a little distant from any sensitive surface such as a plastic tank lining. While quartz heaters transmit heat entirely by radiation and consequently have cool operating surfaces, the radiant heat itself may damage some materials if placed too close to them. Both types transmit more energy per unit area than steam pipes, i.e., they occupy less tank space, but mild agitation or solution circulation is desirable to prevent local boiling of the solution and in some cases increased scale formation. In solutions which tend to form scale, electric heaters should be examined frequently and cleaned, because a heavy scale not only interferes with efficient heating of the solution, but the heater may be damaged by excessive internal temperature.

Heaters of over 1 kw are commonly 230 or 460 v units; consequently a warning device (Chapter 37) should be installed to warn of possible shorting of the heater to the solution and injury to personnel.

Steam Jets. Although they utilize the heat of the steam most efficiently, steam jets are not widely used. Some dilution of the solution must be expected if the temperature of the solution is low enough (less than about 180 F) so that evaporation does not compensate for the formation of condensate. Where metal jets might be corroded, impervious graphite ("Karbate") jets are available.

Bayonet steam heaters are available in some materials such as graphite ("Karbate") and glass ("Pyrex") for use in small installations with limited space and low transfer area requirements (usually less than 5 sq ft).

Plate-type immersion heaters, up to about 50 sq ft transfer area, are also available in "Karbate" for larger installations. Since, like metal, graphite is an electrical conductor, they should be installed outside of the current path between electrodes. In solutions of specific gravity over 1.3, "Karbate" exchangers will float unless held in place by suitable means.

Gas-fired immersion heaters are not very widely used in the plating shop. They are chiefly used where very high temperatures must be maintained, as in some steel-blackening processes, or when it is desirable to avoid summer operation of a boiler. Gas may be burned in a submerged heating tube or, less commonly, fired directly against the tank bottom. Heat transfer occurs through a metal wall which must also be in contact with the solution because the temperatures reached preclude use of a plastic or rubber lining. This greatly limits the usefulness of gas-fired heaters.

External heat exchangers may be single-tube (Fig. 8), multi-tube (Fig. 9), or block type (Figs. 10 and 11). The former are more popular in plating installations than would be expected from their acceptance in general chemical engineering work. This is because by comparison, the quantity of heat to be transferred in plating operations is usually relatively much smaller, which makes it practical to use single-tube exchangers, that obviously take up much more floor space per unit of transfer area than the tube bundle design. They have, on the other hand, certain advantages. In cooling applications, the velocity of the cooling water past the tube may be higher than is possible in the multi-tube design, resulting in a better coefficient of heat transfer in cooling.

Single-tube exchangers are limited in the amount of transfer area they provide. For efficiency, the tube diameter must be relatively small so that reasonably high fluid velocities (preferably 6 to 10 ft/sec) may be attained, and the length cannot be more than about 20 ft as a practical matter, or in certain materials, 10 ft or less. It is therefore usually necessary to arrange a number of single-tube exchangers in a bank to realize the total transfer area needed.

Obviously, the fluid tubes may be arranged in

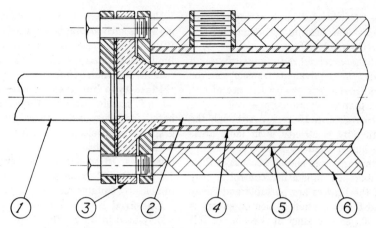

Fig. 8. Cross-section of concentric tube heat exchanger
(1) Flanged pipe or header (4) Steam baffle
(2) Core or tube (5) Outer tube
(3) End sealing flange (6) Insulation

(*Courtesy of Industrial Filter & Pump Manufacturing Co.*)

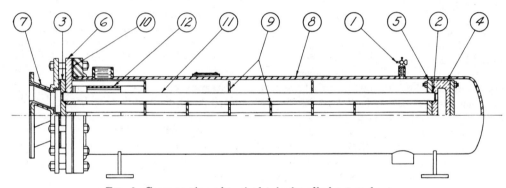

Fig. 9. Cross-section of typical tube-bundle heat exchanger
(1) Air vent cock (7) Bonnet
(2) Rubber ring (8) Shell
(3) Tube sheet (9) Transverse baffle assembly
(4) Floating head (10) Shell gasket
(5) Floating pressure plate (11) Tubes
(6) Stationary pressure plate (12) Impingement baffle

(*Courtesy of Industrial Filter & Pump Manufacturing Co.*)

series, parallel, or a number of different combinations of the two. For heating, it is only necessary to provide that the solution velocity in the tubes fall in the range recommended by the manufacturer. This is done by properly matching the pumping capacity with the series-parallel arrangement of the tubes. In cooling, the water side must also be considered. The water jackets in the bank may likewise be connected in various series-parallel combinations; furthermore the water may be in parallel or counter-flow to the solution. Better transfer is obtained with the cooling water in counter-flow to the solution, and a lower final solution temperature may be obtained with a given water temperature. If both the water and solution tubes are so arranged that the fluid velocities in them are within the range recommended by the manufacturer, then the over-all coefficient of heat transfer will be that given by the manufacturer in his rating of the unit. Lower velocities will result in a reduced coefficient and possible failure of the unit to perform satisfactorily. Higher velocities will result in increased pumping costs; the coefficient and leaving the exchanger. From a single set of readings, the transfer rate at any given time can easily be determined using Eq. (4), and from the record of weekly or monthly readings, excessive scaling or other cause of failure can be detected and corrected in time.

The materials of which tube exchangers of both types may be constructed include all the common construction metals, with or without organic coatings; the corrosion-resistant alloys; glass-lined steel; glass and graphite. Even though the coefficient of thermal conductivity of glass is substantially less than that of the other materials, the over-all coefficient of heat transfer for a well-designed glass exchanger (Fig. 12) does not suffer much by comparison with a metal or graphite exchanger (Table 1). There are several reasons for this. In the first place, improved methods of glass manufacture have permitted the thickness

FIG. 10. Block-type heat exchanger (graphite). (*Courtesy of Kearney Industries, Delanium Graphite Division*)

will tend to be increased, but less sharply because it is approaching some upper limit.

If it is desired to increase the capacity of an existing bank of exchangers, it is always best to add additional exchangers with appropriate pumps in *parallel* with the existing equipment. Simply to add parallel exchangers without pumps is to reduce the solution velocity, which is very damaging, and may actually result in less total transfer than before.

Adding new exchangers in *series*, even providing for the same pumping rate, will not increase the total transfer in proportion to the added area because the mean temperature differential between the two fluids will be lessened. This is especially important in cooling, and in one case, doubling the number of cooling exchangers in series increased the cooling transfer by less than 50 per cent even though the same flow rates were maintained; whereas if they had been placed in parallel, the transfer would have been doubled.

The over-all coefficients given by manufacturers will usually represent the behavior of equipment that has been in service for a reasonable period of time, and the tube surfaces are therefore moderately fouled. Under severe scaling conditions such as when very hard water is used for cooling, the coefficient may fall off very seriously. It is excellent practice, therefore, to provide flowmeters on each fluid line to detect serious changes in rate, and thermometers for at least one (preferably both) of the fluids entering

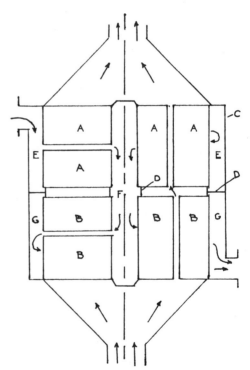

FIG. 11. Principle of "Polybloc" graphite exchanger.

A, B—Graphite blocks, indicating typical radial (shown on left side only) and axial (right side) passages
C—Shell
D—Baffles
E—First shell chamber for water or steam
F—Core
G—Second shell chamber for water or steam

Fig. 12. "Pyrex" heat exchangers operating on an electrogalvanizing solution. (*Courtesy Corning Glass Works*)

of "Pyrex" tube walls to be reduced to 0.030 to 0.060 in. without undue sacrifice of strength. Secondly, it has been said that the tube wall is actually only one of the five possible barriers to heat transfer. The smoothness of glass is a deterrent to scale formation and in addition, the coefficient of thermal expansion of the most common scaling compounds (sulfates and carbonates of calcium and magnesium) is so different from glass that any scale that forms tends to flake off.

"Pyrex" tube-bundle exchangers are not recommended for steam heating service unless the steam is in the tubes with the solution in the shell; steam may be used outside the tube in the concentric tube design, which usually has a steel jacket.

Multi-tube exchanger design differs in one important respect from the single-tube exchanger. It may be and often is constructed as a multiple-pass unit, with respect both to the tube- and shell-side fluids. In one popular design, there are two passes on the solution side so that the solution enters and leaves at the same end; in effect, it passes through a U-tube enclosed in a shell. In heating, with steam on the shell side, the path of the steam is of little importance. But in cooling, it is quite possible that for certain configurations of baffles directing the water flow, heat transfer will actually be in the wrong direction in some sections of the exchanger. A unit designed for heating should therefore never be employed for cooling without consulting the manufacturer.

Block-type exchangers are available only in graphite. The graphite is specially processed to make it very dense and nonporous. Such graphite has a high coefficient of thermal conductivity, approaching that of copper or red brass, and much better than steel and many other metals. The graphite is inert to solvents, alkalies, and all acids except strongly oxidizing acids, e.g., nitric above 10 per cent or chromic acid or very strong hydrofluoric (over 65 per cent) or sulfuric (over 96 per cent) acids.

There are two major exchanger designs. In one ("Delanium[7]" or Impervite[8] graphite exchanger) a block of graphite is made with numerous holes $3/8$ in. to $3/4$ in. in diameter in two directions at right angles to each other but not intersecting. By means of an outer shell, one of the two exchanging fluids is directed through each set of holes (Fig. 10). The fluid may be made to traverse the unit in one to four passes, depending on the design of the shell. For a given flow rate and the same number of passes, the velocity of the liquid in the "tubes" is the same in either direction. "Delanium"-type exchangers are presently available with transfer areas between 4 and 200 sq ft; for larger requirements two or more units are used. The other design (Impervite[8] "Polybloc[9]") features the use of identically molded cylindrical graphite blocks (A and B,

TABLE 1A. TYPICAL OVER-ALL COEFFICIENTS OF HEAT TRANSFER IN STEAM HEATING

Equipment	Solution Velocity (ft/sec)	U, Btu/(hr) (sq ft) (°F)	Special Conditions	Ref.
Metal coils*	—	120–250		—
	—	600	Boiling solution	1
Lead coils	—	60–110		—
Embossed coils	See Col. 4	100–200	Natural convection	2
	See Col. 4	150–275	Forced convection	2
	Natural convection	70–100	Hot water in coils	2
	Forced convection	110–160	Hot water in coils	2
	—	125		3
Graphite immersion heaters	—	80–150	No agitation	6
Single-tube exchangers				
Metal	—	150–400		1
"Duriron"	—	100		4
"Pyrex"	—	100–125		1
Graphite	4	350	Unfouled tube	6
Graphite	7	375	Unfouled tube	6
Graphite	9	415	Unfouled tube	6
Multi-tube exchangers	4–9	200–350		10
Metal	—	150–600		—
Graphite	6.5	330	Unfouled tubes	6
Graphite	8.5	350	Unfouled tubes	6
Block exchangers	4–9	200–350		10
(graphite)	7	200	See also Fig. 15.	6, 7, 9

* See Fig. 13 for graph at $U = 135$.

TABLE 1B. TYPICAL OVER-ALL COEFFICIENTS OF HEAT TRANSFER IN COOLING

Equipment	Water Velocity (ft/sec)	Solution Velocity (ft/sec)	U, Btu/(hr) (sq ft) (°F)	Special Conditions	Ref.
Metal coils	—	—	40–80		—
Lead coils	—	Agitated	30–60		—
Embossed coils	—	Natural convection	65–95		2
	—	Forced convection	105–155		2
	—	Still tank	35	Chilled water	3
	—	Still tank	20	Refrigerant	3
Graphite immersion	—	—	50–75		10
Single-tube exchangers					
Metal	—	—	150–300		1
"Duriron"	—	—	100		4
"Pyrex"	—	—	50–100	0.060 in. tube	5
Graphite	7	7	310	Unfouled tube	6
Multi-tube exchangers	3–7	4–9	130–250		10
Metal	—	—	130–250		
"Pyrex"	—	—	125–195	0.030 in. tube	5
Graphite	Varies	4–9	130–250		6, 10
Block graphite exchangers			See Fig. 15		

Fig. 11) with a central core and two sets of 5/16 to 3/8 in. diameter passages—radial and axial. The blocks may be stacked in multiples of 2 up to 20 blocks, and are encased in a shell, C, with suitable "Teflon" or graphite ring gaskets and baffles, D, which separate the fluids and direct them through the exchanger. The axial fluid, usually the corrosive or sensitive solution, flows axially straight through from one block to the next, while the other fluid flows from the first shell chamber, E, radially in through the first block to the central core, F, then radially out through

the second block to the second shell chamber, G, etc. For the same flow rates, the velocity of the radial fluid is from 1.5 to 2 times that of the axial, depending on the size of the unit. Three block sizes (6, 13 and 24 in. diameter) are offered, which make up into units with from 2 to over 500 sq ft of transfer area.

In either design, the shell may be made of or lined with suitable material so that corrosive liquids may be used on both sides, if desired.

Block graphite exchangers are characterized by quite high over-all coefficients of heat transfer. This, combined with the fact that because of their design the effective transfer surface per unit volume is very large (7 to 8 sq ft/cu ft of exchanger) this equipment provides high capacity in a small floor space.

Calculation of Transfer Area Required

The data given in this section in Tables 1A and 1B and Fig. 16 have for the most part been obtained from the equipment manufacturers and represent values used by them for equipment design. In most cases the values were obtained experimentally, but the interpretation of the test results will vary from one manufacturer to another. Sometimes tests are run on new equipment, and a scaled-down value deemed conservative is assigned for design purposes. Sometimes service tests are run on equipment fouled by considerable use. Thus, the data should be used for rough estimating purposes only, and definitely not for the purpose of comparing the performance of different pieces of equipment, or the products of different manufacturers.

Figure 13 may be used to estimate the linear feet of pipe coil required for the heating of still tanks by means of steam. The chart is based on an over-all heat transfer coefficient of 135 Btu/(sq ft)(hr)(°F). Although based on 1-in. steel pipe, the curve may also be used for other pipe sizes by multiplying by a factor which is the length of the desired pipe size having the same outside surface area as one foot of 1-in. pipe (0.344 sq ft). Several factors for common pipe sizes are shown in the figure. The chart also applies to coils of other metals except that for lead, due to its poor conductivity and thick wall, the indicated length should be multiplied by 1.5.

The square feet of embossed coil required may be obtained by multiplying the linear feet of 1-in. pipe as read from Fig. 13 by 0.344.

The transfer surface required for any steam-heated exchanger may be obtained from the

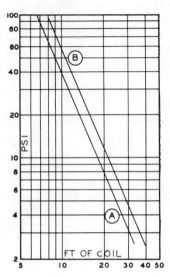

Fig. 13. Linear feet of immersed steel pipe coil required for steam heating.

Curve A—to raise temperature of 1 gal of water 100 F in 1 hr.
Curve B—to supply 100,000 Btu/hr.

For pipe size	Multiply feet by
½ in.	1.57
¾ in.	1.25
1¼ in.	0.79
1½ in.	0.69
2 in.	0.56

nomograph of Fig. 14, providing unusual conditions do not exist, as noted earlier. A value of U must be assumed or assigned with the help of Table 1A. Fig. 14 can be used with an error of usually less than 15 per cent, provided the rate of circulating solution in gpm is at least equal to, and preferably 3 to 6 times, the square feet of exchanger area. As a conservative practice, add 15 per cent to the area given by the nomograph to compensate for possible error.

Similarly, Fig. 15 will solve any cooling problem involving internal (immersed) exchangers or *counter-flow* external exchangers. The first operation (line 1) determines point A, the minimum flow rates of water and solution, assuming both flows are equal.* Typical values of U for use in the nomograph may be selected from Table 1B.

Figure 16 refers specifically to graphite block-

---

* If the solution flow, S, is k times the water flow, W, then the following minimum flow rates must be obtained if the heat transfer is to be thermodynamically possible: $S = kW = F[(k + 1)/2]$ where F is the value of the flow at point A from Fig. 14.

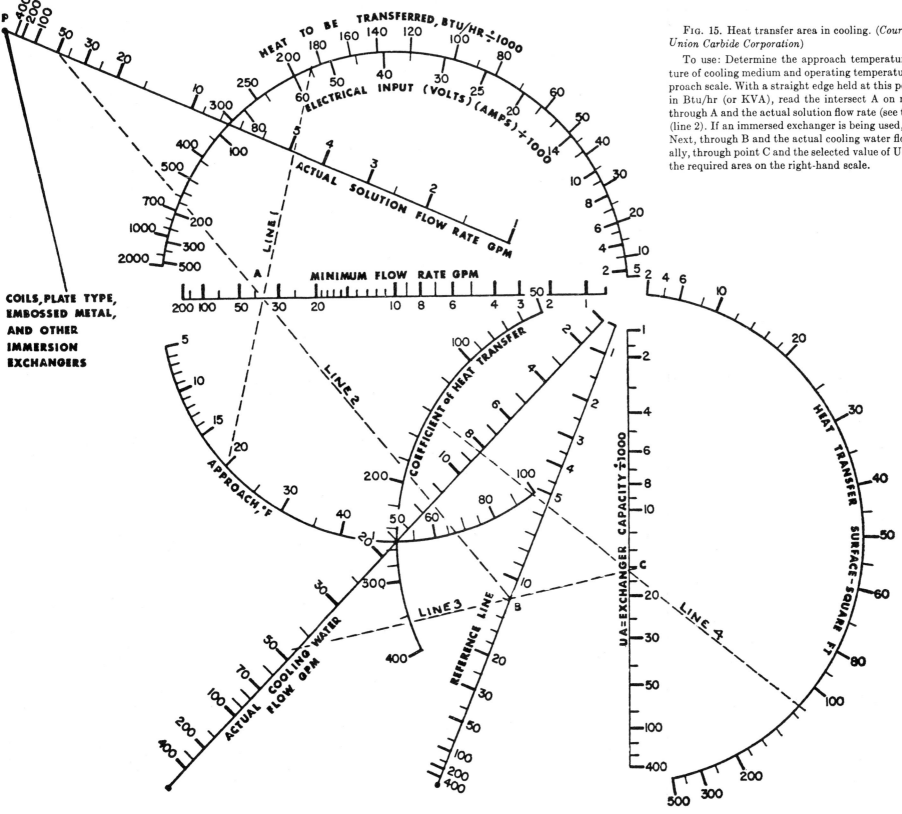

FIG. 15. Heat transfer area in cooling. (*Courtesy of National Carbon Co., Division of Union Carbide Corporation*)

To use: Determine the approach temperature (difference between design temperature of cooling medium and operating temperature of solution, °F) and locate on the approach scale. With a straight edge held at this point and at the appropriate cooling load in Btu/hr (or KVA), read the intersect A on minimum flow scale (line 1). Similarly, through A and the actual solution flow rate (see text) locate point B on the reference line (line 2). If an immersed exchanger is being used, line 2 is determined by points A and P. Next, through B and the actual cooling water flow, locate point C on the UA scale. Finally, through point C and the selected value of U (Table 1B, or manufacturer's data) read the required area on the right-hand scale.

HEATING AND COOLING EQUIPMENT 549

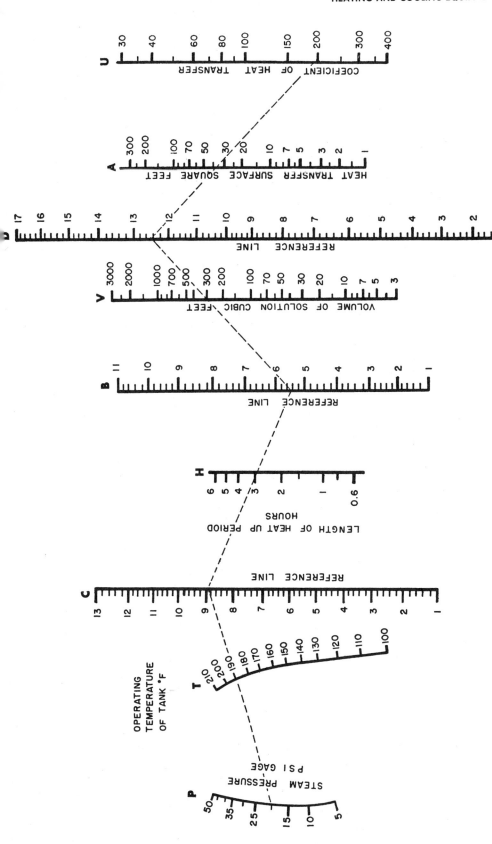

Fig. 14. Steam heated exchangers. Heating surface required to heat aqueous solutions from 60 F to operating temperature. To use, proceed from left to right through given data as shown (20 psi steam, 190 F operating temperature, 3-hr heat-up, 300 cu ft of solution, $U = 200$) to read the required surface on A (35 sq ft). (*Courtesy National Carbon Co., Division of Union Carbide Corporation*)

type exchangers and gives over-all coefficients as a function of liquid velocity. It is to be noted that the velocity abscissa scale refers to *solution* velocity for "Delanium" or Impervite exchangers and for steam heating with "Polybloc" exchangers, but to *water* velocity for cooling with "Polybloc" exchangers.

Calculation of Energy Requirements

Knowing the Btu/hr required for heating, the energy input is as follows:

$$\text{Lb steam/hr} = \frac{\text{Btu/hr}}{\text{Enthalpy}} \quad \text{(Eq. 6)}$$

(Read enthalpy from Table 2 or steam tables.)

$$\text{KW of electrical energy} = \frac{\text{Btu/hr}}{3412} \quad \text{(Eq. 7)}$$

Cu ft/hr of city gas

$$\text{(carburetted water gas)} = \frac{\text{Btu/hr}}{530} \quad \text{(Eq. 8)}$$

$$\text{Cu ft/hr natural gas} = \frac{\text{Btu/hr}}{1100} \quad \text{(Eq. 9)}$$

For cooling with refrigerated brine:

$$\text{Tons of refrigeration} = \frac{\text{Btu/hr}}{12,000} \quad \text{(Eq. 10)}$$

Engineering the Installation

While there is essentially nothing very complicated about the installation of heat transfer equipment, certain precautions are necessary, some of which have been mentioned: proper sizing of mains and headers, placing of immersion heaters in electrified tanks, and their electrical insulation from ground, etc. External heat exchangers are usually connected to the tank with hose, which insulates them electrically, but where plastic-lined steel pipe is used, the exchanger is not insulated, and if the exchanger tube is metal, corrosion of the tube can occur to the point of failure. In such cases one of the flanged joints in the pipe should be made up to be electrically insulating by the use of insulating gaskets and insulating sleeves in the bolt-holes.

Some of the most useful exchanger materials are relatively fragile, such as "Duriron," graphite, and glass. The manufacturer's instructions should be carefully followed. "Duriron" and graphite are liable to damage by impact or by being subjected to unequal strains, as when a joint is drawn up unevenly. Thermal shock should

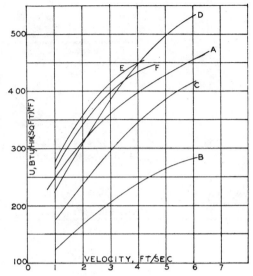

FIG. 16. Over-all heat transfer coefficients for graphite block exchangers.
Curve A—Polybloc 6 in. and 13 in.* size in steam heating
Curve B—Polybloc 6 in. and 13 in. size in cooling. Axial solution flow 1 fps.
Curve C—Polybloc 6 in. and 13 in. size in cooling. Axial solution flow 2 fps.
Curve D—Polybloc 6 in. and 13 in. size in cooling. Axial solution flow 3 fps.
Curve E—"Delanium," heating with 50 psi steam
Curve F—"Delanium," cooling with water at 6 fps.

* For 24 in. size, multiply by 0.87 for heating and 0.8 for cooling (*Courtesy of The Carbone Corporation and Kearney Industries, Delanium Graphite Division*).

TABLE 2. PROPERTIES OF STEAM

Gauge Pressure (psi)	T_s Temperature Sat. Vapor (F)	Enthalpy, Evaporation (Btu/lb)
5	227	960
10	239	952
15	250	945
20	259	939
25	267	934
30	274	927
40	287	919
50	298	911
60	307	904
70	316	897

be avoided by opening the cold water or steam valve just a crack at first and leaving it that way for some minutes before opening it fully. Modern glass technology has developed a glass that is highly resistant to impact, but it will fail by fatigue if subjected to a continued stress as by improper make-up of joints or support of pipe. Maximum pressure recommendations must also be strictly observed for the same reason. When "Pyrex" exchangers are to be used for steam heating, the steel-jacketed type is usually preferred; where the inlet steam impinges on the glass tube, a copper sleeve is provided to dissipate the heat and prevent the occurrence of local thermal stresses which in time can cause failure. Therefore the steam inlet and outlet must not be reversed.

Large quantities of cooling water may be required for room temperature operations having a large KVA input as in anodizing (high voltage) or electropolishing (high current) because in such cases, only a few degrees temperature rise can usually be obtained in the water. The economics of cooling with refrigeration may be worth investigating. The cost of the cooling water may be readily calculated from Eqs. (3) and (4) and the local cost of water. Refrigeration costs with power at 1.5¢/KWH will average about 10 to 20¢/hr for each ton of refrigeration (12,000 Btu/hr) or 2.8 to 5.7¢/hr for each KVA input to the tank, plus a small volume (1 to 2 gpm/ton) of cooling water for the condenser. If the refrigeration costs are lower, it is a simple matter to calculate whether the investment in refrigeration equipment will be amortized in a reasonable period. One type of refrigeration unit costs about $300–800/ton of refrigeration capacity in sizes of 10 tons and over; smaller units cost more per ton. The rating of a given unit (and hence its initial cost/ton) depends on the tank operating temperature. The costs cited are for 70 to 90 F operation, being lower at the higher temperature. This equipment is available in two types. The first simply recirculates chilled water or brine through coils in the tank or through an external heat exchanger, not provided by the refrigeration manufacturer. The other type[11] is completely self-contained and is designed to circulate the solution, which may be acid or alkaline, returning it at 10 F below the tank operating temperature. No coils or other exchange equipment are required.

Re-use of cooling water is an economy that is well-worth considering. For best results, this should be carefully planned, rather than blindly to divert the cooling effluent into a rinse tank. The rinsing requirements should be known and should be minimized as described in Chapter 34. The number of rinse tanks to be serviced by the cooling water should be such that their total demand is somewhat less than the average cooling water flow. The cooling water may be collected in an elevated tank with an overflow to carry off the surplus. This overflow could be used as part of the feed to still another rinse tank. Each rinse tank in the system would, through its conductivity control and a solenoid valve, demand water from the storage tank. Where the storage tank cannot be elevated, the stored water may be circulated by a pump at somewhat above the total rinsing rate through a line which returns to storage. The rinse tanks then draw upon this circulating water as required.

Immersed Plastic Coils. A relatively new design of plastic coils made of bundles of small-diameter (100 mil), thin-wall (10 mil) "Teflon"*-fluorocarbon-resin tubes (trade name "Supercoil"**), have been available for several years. (See Fig. 17.)

Lengths vary from 4 to 12 ft in three different models of 128, 168 and 280 tubes, arranged in 2,

FIG. 17. Typical installation of immersed plastic coil "Teflon" "Supercoil". (*Coutresy of the E. I. du Pont de Nemours & Co., Inc.*)

* Registered Trademark of E. I. du Pont de Nemours and Co., Inc., for its fluorocarbon resin.

** Information on removable end hardware and additional lengths of "Supercoils" are now available from suppliers.

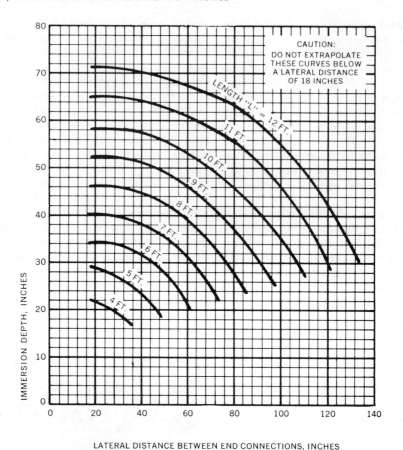

Fig. 18. Length selection chart for all plastic "Supercoil" models.

3 and 5 flattened braids, respectively. (See Fig. 18.) These give an available range of areas from 13 to 92 sq ft. Details are given in Table 3 and illustrated in Fig. 17.

According to the manufacturer, shell and tube heat exchangers containing small diameter tubes of "Teflon" and an earlier design of immersion coils of the same plastic material are available. Like these, the "Supercoils" will not corrode in any metal finishing solution and can be used in media up to 300 F. The latter are light weight, tough, flexible, easy to install and seldom need maintenance. Their high electrical resistance prevents short circuits, stray currents and related solution contamination. Any leak which might possibly occur can be quickly stopped by cutting the tube and heat sealing the ends.

Engineering data for sizing these coils have been worked out for steady-state cooling of air-agitated processing baths with average water temperatures of 90 to 50 F (see Figs. 20, 21 and 22).

The manufacturer has detailed the following sizing procedure for "Supercoils" for cooling chromium plating baths. The sizing procedure is as follows:

(1) Calculate the kilowatt heat input to the bath by multiplying the anode bar voltage by the amperage, and dividing this product by 1000.

(2) Read required surface area from Fig. 20. *If bath is not air-agitated, multiply the Fig. 20 area by 1.33.*

(3) Calculate the required cooling water flow by: Flow Rate, gpm = 0.45 × heat input, kilowatts.

(4) Refer to Fig. 19 and identify the smallest

TABLE 3. DATA ON 3 MODELS OF "SUPERCOILS"

Coil Length "L", ft	Immersed Surface,* sq ft		
	M 128H- 2-"L"	M 168S- 2-"L"	M 280- 2-"L"
	Number of Tubes**		
	128	168	280
4	13	17	31
5	16	21	39
6	20	26	47
7	23	31	54
8	27	35	62
9	30	40	69
10	34	45	77
11	37	49	85
12	41	54	92
Number of Braid Layers			
	2	3	5
Width of Coil, in.			
	3.5	3.5	3.5
End Connections	Reinforced Corrosion-Resistant Steam Hose with Clamp	Formed 316 S/S 1.0" NPT	Formed 316 S/S 1.5" NPT
Length, in.		End Con. 5.5 in.	End Con. 5.75 in.
"O" Ring Seals	None	Ethylene Propylene	Ethylene Propylene

* Includes allowance of 3 in. of exposed tubing at each end for connections.
** ("Teflon"), fluorocarbon resin, O.D. 100 mil; wall thickness 10 mil.

coil (fewest tubes) that has the surface computed in Step 2.

(5) Consult Fig. 19 to insure that the coil selected in Step 4 is long enough to be immersed to within about 1 ft of tank bottom. Select a longer coil, if necessary, to meet this criterion.

(6) Enter Fig. 21 with the cooling water flow rate found in Step 3 and read pressure drop for the heat exchanger selected in Step 4. If pressure drop is no greater than that specified in Step 4 (or Step 5, if applicable), selection is the coil required.

(7) If the pressure drop in Step 6 exceeds the maximum allowable, repeat Steps 4 through 6

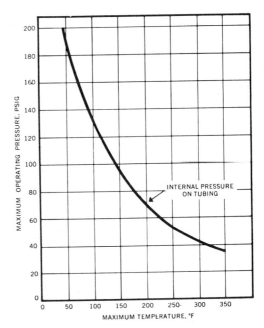

FIG. 19. Operating internal pressure limits for plastic coils of "Supercoil" design. Tubing—100 mil O.D. and 10 mil wall. Maximum steam pressure inside tubes—30 psi gauge

with the next larger tube count. Continue this process until the pressure drop obtained from Fig. 21 is no greater than the maximum allowable value.

(8) If the required heat transfer surface cannot be provided in a single coil without exceeding the allowable pressure drop, divide the net heat input in Step 1 by a whole number (2, 3, etc.) and repeat Steps 2 through 7. The whole number used is the number of coils required.

Figures 23 and 24 are to be used for determining heating requirements for unagitated baths, from an initial temperature of 60 F or higher to various operating temperatures. The procedure cannot be used for bath volumes less than 200 gallons.

To size coils for heating purposes, the manufacturer recommends proceeding as follows.

(1) Compute the volume of the tank contents in gallons. $(L \times W \times D)$ Ft $\times 7.5$ gal per cu ft = gallons.

(2) Enter Fig. 23 and read the required heat transfer surface (in square feet per 100 gallons). If the tank volume is between 200 gallons and 2,000 gallons, interpolate between the curves for these two volumes. If the volume is 2,000 gallons or more, use the appropriate 2,000-gallon curve.

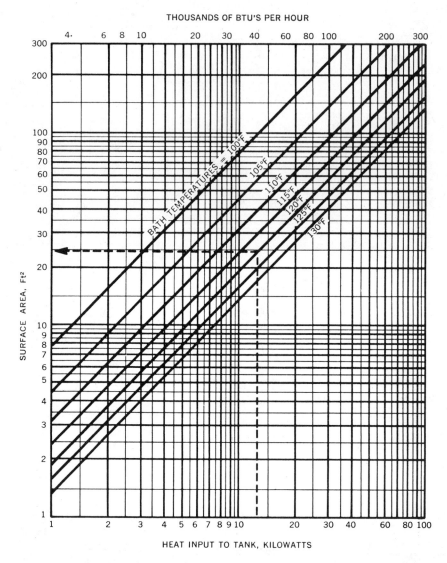

Fig. 20. Data for steady-state cooling of air-agitated baths.*

* Based on cooling water in and out of the tubing at 80 and 95 F, respectively. This rating method is to be used only for solutions with physical properties similar to water.

Multiply the Fig. 23 reading by the tank volume (in hundreds of gallons).

(3) If the heating time is different from 2 hr, multiply the Step 2 area by the appropriate correction factor shown on Fig. 23.

(4) Refer to Fig. 19, search for the smallest bundle (fewest tubes) that has the required surface, and read the length in feet of the shortest coil that provides the required surface.

(5) Consult Fig. 19 to insure that the coil selected in Step 4 is long enough to be immersed to within about 1 ft of the bottom of the tank. Select a longer bundle, if necessary, to meet this criterion.

(6) Compute the average bath temperature (add initial temperature to final temperature and divide by 2). Enter Fig. 24 at this temperature and read the unflooded length. If this length is greater than that of the coil selected in Step 4, the coil selected in Step 4 (or Step 5, if applicable) is the coil required. If the length in Fig. 24 is less than that of the Step 4 selection, repeat Steps 4 and 5 with a larger tube count. Continue this process until the Fig. 24 length is equal to or

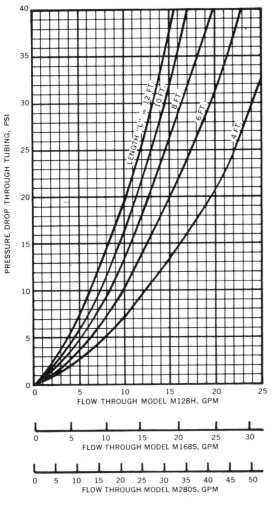

FIG. 21. Chart for pressure drop for flow of water through plastic "Supercoil" tubing.

greater than the Step 4 length. It is not necessary that the Step 5 length be less than the Step 4 length, but if the Step 5 length is greater than the Step 4 length, the Step 5 length represents the coil required.

(7) If the total surface obtained in Step 2 (or Step 3, if applicable) exceeds the maximum available in a single coil (or if Step 6 prevents use of a single coil that is long enough to provide the required surface), divide the Step 2 area by a whole number (2, 3, etc.) and repeat Steps 4 through 6. The whole number used as the divisor represents the number of coils required.

Automatic temperature controls for heating and cooling such controls have long been standardized and are well-known. The sensing element must be one which will not corrode in nor contaminate the solution; otherwise it should be placed in a glass tube filled with water (in hot solutions, care must be taken to keep water in this tube). With immersion exchangers, the sensitive element should be placed close to the heating or cooling unit in order to prevent overshooting. Simultaneously, in large tanks it is advisable to install an indicating thermometer remote from the heating unit, which will indicate whether or not a uniform tank temperature is being maintained. If not, agitation or some solution circulation should be employed.

Pumps for use with external exchangers are available in many corrosion-resistant materials; a suitable selection may be made by consulting Chapter 6. A number of exchangers are sold complete with pumps and hose. When pumps must be purchased separately, the pressure drop through the exchanger must be considered; these data will be available from the manufacturer.

The power for pumping cooling water or solution by any type of pump is given by:

$$P = \frac{fhs}{3960e} \quad \text{(Eq. 11)}$$

where P = horsepower input; h = total head on pump, in feet of the liquid being pumped; f = flow rate, gpm; s = specific gravity of liquid; and e = efficiency of pump, expressed as a decimal. The total head on the pump is the sum of the total static fluid head, the velocity head (often insignificant) and the friction losses, all expressed in feet of the liquid being handled. The total static head is the elevation at the discharge point, minus the elevation at the intake in feet. The velocity head is given by the expression:

$$\text{Velocity head} = \frac{V^2}{64.4}, \quad \text{(Eq. 12)}$$

where V = liquid velocity, ft/sec.

The friction loss in an exchanger may be obtained from the manufacturer's literature; it is usually expressed in psi, in which case it must be converted to feet of liquid by the expression:

$$\text{Feet of liquid} = \frac{2.31 \text{ psi}}{\text{sp gr of liquid}} \quad \text{(Eq. 13)}$$

The friction loss in pipes and fittings may be obtained from Tables 4 to 6, Chapter 5.

Whenever an installation requires both an external heat exchanger and a filter there is a strong

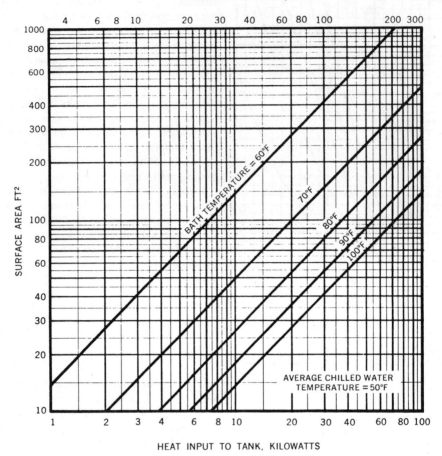

FIG. 22. Data for steady-state cooling of air-agitated baths* using *chilled* (50 F) water.

* Based on cooling water in and out of tubing at 45 and 55 F, respectively, $\Delta T = 10$ F. Based on this flow rate, gpm = 0.685 × heat load, kw. This rating method is to be used only for solutions with physical properties similar to water.

temptation to arrange them in series, thus saving piping. This can be a grave error. The exchanger has been designed for a certain constant flow rate of solution. On the other hand, a filter by its very nature will deliver at or slightly above its rated capacity when first put into operation, but this rate falls off rapidly and (what is more important) unpredictably in the first few hours of operation. If and when this rate becomes less than the design rate of the exchanger, there is an enormous loss in heat transfer because:

(a) The over-all heat transfer coefficient is sharply reduced.
(b) The rise (or fall) in solution temperature necessary to effect the required total heat transfer is increased.
(c) The temperature differential in the exchanger is reduced.

The simultaneous reduction of U and ΔT in Eq. (1) is obviously synergistic. An installation may be designed with an oversize filter to offset this situation, but entirely aside from the uncertainty about how much extra capacity should be provided, it is much more economical to arrange the filter and exchanger each in its own circuit. At the same time, this provides a uniform flow rate for the exchanger.

It is sometimes desirable, as in chromium plating, to use the same exchanger alternately for heating and cooling. If steam is admitted to an exchanger full of water, the resulting "hammer," due to water rushing in to fill voids left by col-

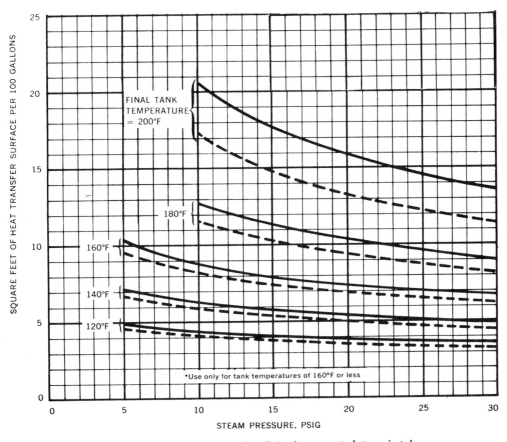

FIG. 23. Data for heating unagitated baths* using saturated steam in tubes.

Caution: Use only for steam pressures of 10 to 30 psi gauge.

Basis: Heating time of 2 hr. For heat-up times of 1, 3 and 4 hr multiply areas by factors 2.0, 0.7 and 0.6, respectively (these factors only applicable for temperatures up to 160 F).
Key: Solid-lines, 200 gal bath
　　　Dotted-lines, 2000 gal bath
This rating method is to be used only for solutions with physical properties similar to water.

lapsing steam bubbles, can pit a metal tube or break a fragile one. An external exchanger should therefore be provided with valves, either manual or motor-operated, to drain out the cooling water before the steam is admitted. This is not possible with immersed exchangers; in this case, the switch-over should be made slowly.

It is curious how many installations of external exchangers may be seen without adequate measuring devices, that is, flowmeters in each liquid line and thermometers in liquid lines at each end of the exchanger (for steam, a pressure gauge is adequate). With these instruments and a previous history of readings, the cause of inadequate heat transfer can be diagnosed in a few minutes.

Acknowledgments

The author gratefully acknowledges the cooperation of the companies in the list of references, who supplied much of the transfer data for this chapter. Special thanks are also due to W. M. Gaylord, Chemical Eng., National Carbon Co., Division of Union Carbide Corp., who provided the monographs for Figs. 14 and 15. Special thanks also are due Larry George, Udylite Corporation, for his cooperation, with that of W. van

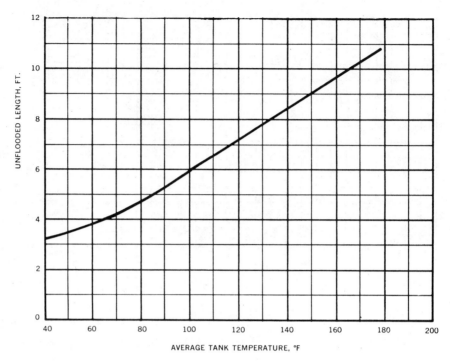

FIG. 24. Data on unflooded length of condensing steam* inside 100 mil O.D. tubes.

Caution: Use only for steam pressures of 10 to 30 psi gauge.

der Graaf, Fred Johns, and Frank Czeiner, E. I. du Pont de Nemours & Co., Inc., who supplied the data on the plastic immersion heating coils of their "Supercoil" design.

Appendix

The logarithmic mean temperature difference, ΔT_L, expresses accurately the effective temperature difference between two liquids flowing in an exchanger. If T_1 and T_2 are the entering and leaving temperatures of the hot liquid and t_1 and t_2 the same for the cold liquid, then:

$$\Delta T_L = \frac{T_1 - t_2 - T_2 - t_1}{2.3 \log \dfrac{T_1 - t_2}{T_2 - t_1}}$$

When $[(T_1 - t_2)/(T_2 - t_1)] < 2$, the simple arithmetical mean can be used with less than 4 per cent error. In most plating operations, this will be true.

References

1. Perry, J. H., "Chemical Engineers' Handbook," Third Ed., p. 481, New York, McGraw-Hill Book Co., Inc., 1950.
2. Platecoil Div., Tranter Manufacturing, Inc., Lansing 9, Mich.
3. Dean Products, Inc., Brooklyn, N. Y.
4. The Duriron Company, Dayton, Ohio.
5. Corning Glass Works, Corning, N. Y.
6. Heil Process Equipment Co., Cleveland, Ohio.
7. Kearney Industries, Delanium Graphite Div., South Plainfield, N. J.
8. Falls Industries, Inc., Solon, Ohio.
9. The Carbone Corporation, Boonton, N. J.
10. National Carbon Co., Division of Union Carbide Corp., New York, N. Y.
11. Mayer Refrigeration Engineers, Inc., Lincoln Park, N. J.
12. Heat Transfer Products Division, E. I. du Pont de Nemours & Co., Inc., Wilmington, Del.

22. DESIGN, CONSTRUCTION, INSULATION AND MAINTENANCE OF PLATING RACKS.

Herbert Tilton

Tilton Rack and Basket Corp., Fairfield, N.J.

Plating racks, their design and construction, are as important to good electroplating as is a rectifier, tank and solutions. They can be of any design, size, shape or form. As can be seen in Fig. 1, a plating rack has many different parts, depending on the part being plated and the type of plating. The tank size, the rectifier size and the solution all have an effect on the design and construction of the rack.

DESIGN

The most important feature of any rack is design; without good design, improper plating will result, production can suffer and the cost of plating the parts can be prohibitive. The racks should be designed for 100% efficiency in the plating operation for which they are intended. In designing racks, a number of features should be considered:

Is it going to be a manual or automated operation? This choice will determine rack size and weight. In a manual operation, the rack with parts should have a total weight of no more than 20 lb. It should be small enough so it can be carried easily and handled over tanks and bus bars so it does not contact the anodes to cause a short.

In an automated operation, the lift height and weight limitation of the machine is a factor. Design and construction of these racks depends on the type of machine—whether it is return or overhead.

What is the power source. It must be sufficient for the operation. It is not practical to design a rack to hold many pieces if the power supply is not large enough to provide adequate current for the number of parts to be plated (see Table 1).

What size shall the tanks be? Always design the plating rack around the smallest tank. Too many times, in both manual and automated operations, this is overlooked, and the rack is either too long or too wide to fit the operation. In automated operations, it is especially important to be aware of the presence of drain pipes, water lines, electrical conduits or air ducts that rise above tanks and cut down on machine lift.

What type of plating solution is to be used? The rack should be designed to complement the efficiency of the particular plating bath. In a chrome solution, it is necessary to put fewer parts on the rack and space them farther apart than in a silver solution. This is because of the different plating efficiency and throwing power characteristics of the baths.

After all these factors are considered, the type of rack should be determined. Is the best form a single spline with one straight spline, as in Fig. 2; a box rack with multiple splines tied together, as in Fig. 3; or a "T" rack, a single spline with multiple cross-bars, as in Fig. 4?

In choosing the right type of rack, then, the

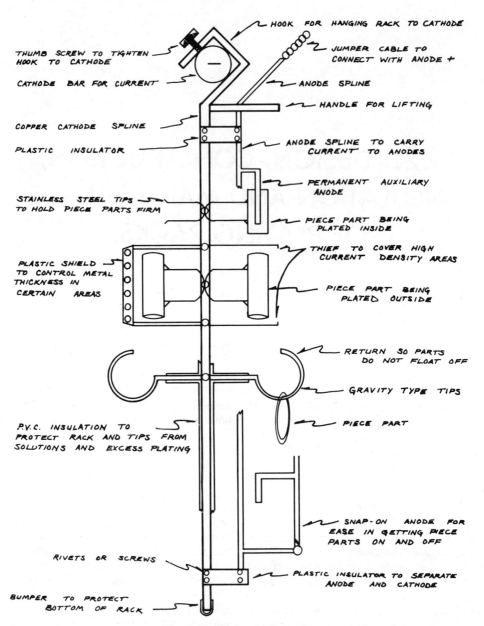

Fig. 1. Parts of a plating rack.

following factors must be taken into consideration:

1. Manual or automated operation
2. Power supply
3. Size of tanks
4. Type of solution.

There will be times when only through trial and error will the right type of rack be chosen.

Another consideration in design is that of dimensions. When designing the rack, there are three important dimensions that must be taken into account:

1. *The overall length* (see Fig. 5). This is the dimension from underneath the hook to the bottom of the rack. From the hook to the solution level to the tops of the parts being plated is called the free area of the rack. The working area is

TABLE 1. PLATING SOLUTIONS—CATHODE CURRENT DENSITIES

Plating Bath	A/Ft^2	Voltage
Brass	5-20	2-5
Cadmium	5-50	2-5
Chromium (decorative)	100-200	4-6
Chromium (hard)	200-500	5-12
Copper (sulfate)	15-50	1-4
Copper (fluoborate)	15-50	1-4
Copper (cyanide)	20-60	2-5
Gold (acid)	10-30	5-6
Nickel	20-100	4-8
Silver	5-30	0.5-2
Tin (fluoborate)	25-150	1-3
Tin (stannate)	30-100	4-6
Tin (sulfate)	10-40	1-4
Zinc (cyanide)	10-90	1.5-6
Zinc (low cyanide)	20-80	1.5-6
Zinc (acid—no cyanide)	20-80	1.5-6

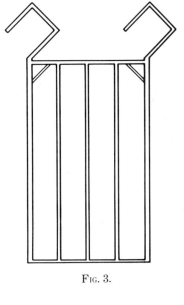

FIG. 3.

the overall length less the free area. This can vary depending on the anode length, heating coils, air agitation pipes or sludge.

2. *The width of the racks* (see Fig. 6). This means fitting one or more racks into the long tank dimension with adequate clearance between racks and tank walls. With a return type plating machine, the direction of travel should be considered as well as clearances for turning at the end of the machine. The width of the rack is extremely important, because racks tend to sway during transfer, and if too wide will catch on the tank walls at the entry station and jam the machine, or knock the rack off the carrier to the bottom of the tank.

3. *Thickness* (see Fig. 7). This is the area between anodes and/or cathodes which permits the rack to pass freely without contacting the other electrode.

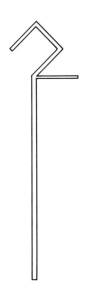

FIG. 2.

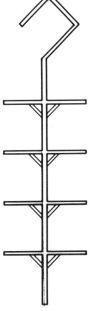

FIG. 4.

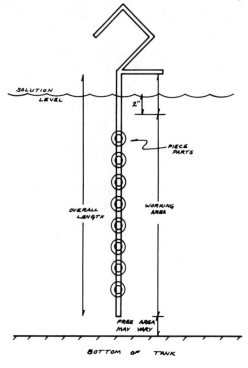

FIG. 5.

It is very important that, with all three dimensions, parts be included. Too many times this has not been done, with the result that rack and parts are damaged.

CONSTRUCTION

The material used most frequently for splines for plating racks is copper, because of its conductivity and its easy workability. It can be bent-shaped, drilled, soldered, sandblasted, insulated and baked to form a workable product. Other materials, including steel, brass, aluminum and stainless steel are available for manufacturing racks, but all have conductivity disadvantages compared to copper (see Table 2).

When constructing plating racks, the size of the copper should be sufficient to carry current from the cathode bar to the parts without overheating the spline or the tips (refer to Table 2). If box racks or "T" type racks are used, rack joints should be braced for rigidity and strength so as not to break down and increase resistance to current flow. Parts should be measured for area and this area multiplied by the number of pieces on the rack. Multiplying this area by the current density (A/ft^2) required by the particular plating bath for proper operation yields

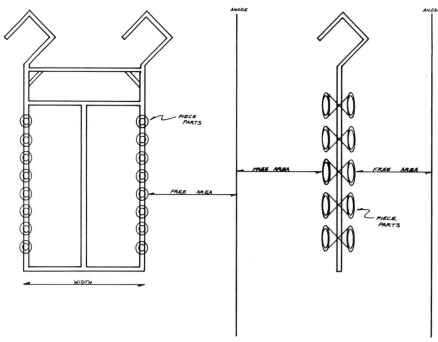

FIG. 6. FIG. 7.

Table 2. Relative Conductivity (Amps)

Size	Copper	Aluminum	Brass	Steel	Phosphorous Bronze	(300 Series) St. St.	Titanium
1 × 1 in.	1000	600	250	120	180	23	31
3/4 × 1 in.	750	450	185	90	135	17	23
1/2 × 1 in.	500	300	125	60	90	12	16
1/4 × 1 in.	250	150	63	30	45	6	8
1 in. dia.	785	470	196	94	141	18	24
3/4 in. dia.	445	265	111	53	80	10	14
1/2 in. dia.	200	120	50	24	36	5	6
1/4 in. dia.	50	30	13	6	9	1	1 1/2
3/16 in. dia.	28	16	7	3 1/2	5	5/8	7/8
5/32 in. dia.	20	12	5	2 1/2	3 5/8	1/2	5/8
1/8 in. dia.	12	6	3	1 1/2	2	1/4	3/8
3/32 in. dia.	7	4	1 3/4	1 7/8	1 1/4	3/16	1/4
1/16 in. dia.	3	1 3/4	3/4	1/2	1/2	1/16	3/32

the total amperage required. The proper size for the copper in the rack can then be determined. Hook shape on whatever type rack is used is also important to the plating operation. Hooks should be designed to minimize "walking" on agitated cathode bars, swaying or chattering during transfer on automatic plating machines and manufactured or cast to fit carriers or busbar.

The center spline should always hang directly under the cathode bar (see Fig. 8) so parts on either side of the spline are uniformly spaced between, and adequately separated from, the anodes. If uniform thickness of electroplate on both sides of the parts is important, the parts should hang directly under the cathode bar so one side is no closer to the anodes than the other (see Fig. 9). Yoke hooks (see Fig. 10) are utilized where two-point contact is necessary for current carrying capacity, or to prevent swaying of the rack. If strength is needed at this point, a stainless steel strap should be added over the hook (see Fig. 11).

Tips for plating racks are either pressure type to hold parts firmly, or gravity type where pieces just rest on the tips. The most desirable type will be determined by the weight of the part, the method of rack transfer, the plating solution and the current demand of the part. Tip material for plating racks can be either stainless steel, spring temper phosphorous bronze or a combination of both, according to application. Pri-

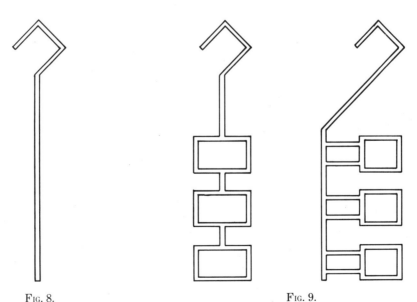

Fig. 8. Fig. 9.

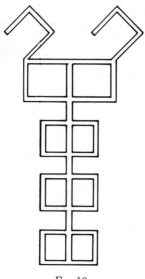

FIG. 10.

marily three hundred series full, hard, spring temper stainless steel is being used in the majority of plating racks manufactured today because of the ease of stripping excess plate from the tips. If the parts have a large plating area and a large cross-section is needed to carry current to the part, phosphorous bronze will be used because of its higher current carrying capacity. It cannot be stripped as readily as stainless steel, thereby posing a problem. In such

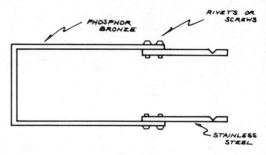

FIG. 12.

a case, a combination tip of P.V.C. coated phosphorous bronze, for current capacity, and stainless steel, for contact and ease of stripping, is used. This is also practical for gravity type tips (see Figs. 12 and 13).

Attachment of tips to splines is another area of importance in rack construction. If not constructed properly, tips may loosen under the insulation, causing skip plating or complete failure to plate. Rivets or screws are usually used for this connection. When rivets are used, they mesh into the copper for a more permanent construction. Screws have an advantage when it comes to maintenance, but with spring type contacts, if they are not properly secured, tips can work loose.

Welding of stainless steel tips is the ideal method of construction, but this must be done with great care or the temper can be taken out of the spring stainless.

For anodizing, titanium tips are frequently used because they are resistant to anodic oxidation and dissolution in the anodizing solutions. The suggestions given above for stainless steel tips apply equally well to titanium.

Tips should make contact in the most inconspicuous area of the part, such as crevices, rims, edges or holes. Even with the smallest contact area a rack mark or bare spot will usually show. Where it is not possible to make contact incon-

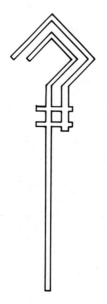

FIG. 11.

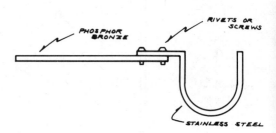

FIG. 13.

spicuous, rack tips should be as small as possible, and kept completely free of all excess metal before reracking. If not, large bare spots will occur, or gassing in the plating solution will take place at the point of contact, causing streaking of the plated part. Parts should be held so drainage is not a problem, to hold drag over from one solution to another at a minimum. Parts should also be positioned so as to avoid the accumulation of gas in pockets. With many shapes, some compromise may be necessary between these two requirements.

If the part must be completely and uniformly plated, the design of the rack and tips must be such that current can reach all surfaces of the parts (see Fig. 14). Current in all baths travels mainly in a straight line. Throwing power of the plating bath plays an important part here, too (see also Chapter 15).

If only the front side of the piece is important, the parts may be racked closer together, as in Fig. 15. Again, the throwing power of the bath is important.

When positioning parts on the racks, care must be taken that the parts are not too close to the anodes, causing excessive buildup in the high current density area. Current always concentrates at corners, points or projections. If the current density is too high, excessive metal will build up, or discoloration or burning will occur in this area. Thieves or robbers may have to

Fig. 15.

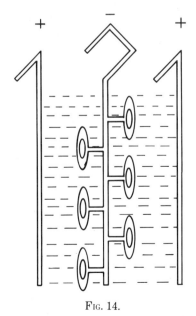

Fig. 14.

be used to avoid this (refer to Fig. 1). They can either be built into the rack, or simple wires can be attached to the cathode bars to restrict the buildup. Where possible, the rack should be designed so that the parts act as thieves for each other. Material for thieving is usually cold rolled steel, so it can be discarded after plating. If thinner plating is necessary on some area of the part, or if very deep recesses must be covered, a shielding device of some sort may be needed. This can either be built into the rack, or placed directly into the tank. Material for shields is usually a rigid plastic or cold rolled steel coated with plastic material. This can be used over and over again.

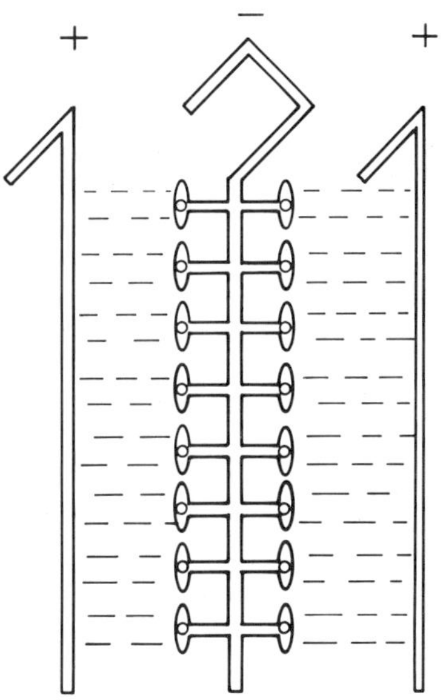

Anode or bipolar racks are a combination of two racks—a cathode section tied together with an anode section to make one common rack. It has an insulator between splines to prevent shorts while plating (refer to Fig. 1). The anode can be either built into the rack, or snapped into place after the parts have been loaded onto the rack. This type of rack may be necessary where plating is needed in deep crevices, internal diameters or blind holes, with plating baths that do not have the necessary throwing power to provide the required thickness without excessive buildup on other areas of the part. The anode section will have some form of jumper to make contact with the anode bar on the tank. Auxiliary anode material may be of any suitable material (see next page).

Auxiliary Anode Materials

Plating Bath	Material
Brass	Brass, steel
Cadmium	Cadmium, steel
Chromium (decorative)	Platinum clad titanium, lead, steel
Chromium (hard)	Lead, steel
Copper (cyanide)	Copper, steel
Copper (fluoborate)	Copper, Pt/Ti
Copper (sulfate)	Copper, lead
Gold (acid)	Pt/Ti, tantalum
Nickel	Pt/Ti, nickel
Silver	Silver, stainless steel
Tin (fluoborate)	Tin
Tin (stannate)	Tin
Tin (sulfate)	Tin, steel
Zinc (cyanide)	Zinc, steel

Auxiliary anodes, except for platinum clad titanium, and steel in cyanide solutions, dissolve in solution while plating, as in any plating operation. Anode racks should be designed and constructed with this in mind. Anodes themselves should be replaceable with ease so as not to disrupt the plating operation. Platinum clad titanium does not dissolve, but will wear. It is expensive, and not very practical for routine plating operations. It is most often used with precious metals, or for plating sophisticated and expensive parts where cost is a minor factor.

RACK INSULATION

All racks should be insulated after manufacturing. If not, they will plate on each cycle just as the parts do, causing many problems. Uncoated racks will steal current and metal from parts. Repeated cycles will cause the metal on the rack to delaminate and peel into the solution. They will trap solution, dragging it from one bath to another.

Polyvinyl chloride (PVC) is almost universally used on racks manufactured today. This is a plasticized PVC resin that can be manufactured in any color and is oven cured. Done properly, PVC coating will greatly extend the life of the rack. To apply the coating, the rack is first properly cleaned and coated with a suitable primer. This will assist in bonding the PVC to the rack. The rack should be given a double dip in the PVC resin to make sure there is adequate coverage and no pinholes in the coating. It is then baked at 380 F until cured. (If not cured properly, the coating will crack and peel, trapping solution and dragging it from one solution to the next.)

Rack maintenance is one of the most neglected problems in plating operations. Racks should be maintained at 100% efficiency, broken tips being replaced and repaired immediately. If ten tips on a 100-piece rack are not in use, 10% of production is lost. If broken tips are allowed to build up metal in the plating bath, they use unnecessary current, steal current from parts, carry solution from one tank to the next and make necessary additional filtration and purification. Racks with cracked and peeling insulation cause similar problems and should be stripped and recoated.

Racks must be treated as tools of the trade. They should be stripped regularly, and should never touch the floor.

23. MANUALLY OPERATED INSTALLATIONS

Lawrence J. Durney

Vice President for Technology
Frederick Gumm Chemical Co. Inc.
Lyndhurst, N.J.

Various methods of handling parts through treatment cycles are compared in Table 1. This chapter covers manually operated installations for carrying racked parts through treatment cycles (see methods 1 to 4 in Table 1). Chapter 24 deals with barrel plating of small parts in bulk, including both manually and automatically operated installations (see methods 1, 2, 3, and 5 in Table 1). Chapter 25 deals with semiautomatic and fully automatic machines, primarily for handling racked parts through treatment cycles (see methods 4, 5, and 6 in Table 1). Chapter 26 deals wholly with machines for continuous processing of steel-mill products through treatment cycles (see method 7 in Table 1).

ALL MANUAL OPERATION

The completely manually operated treatment cycle provides the maximum possible flexibility. Since treatment times are under the control of the operator, it is frequently the setup of choice for low to moderate production under variable conditions. Such situations may involve the application of a single type of plate to a variety of parts requiring variations in the pretreatment cycle; variations in deposit thickness, or in types of plate; or variations in post-treatment requirements.

Whichever situation dictates the choice, however, a careful analysis of tank requirements, number of operators, and traffic patterns are necessary to ensure an efficient operation. Operator travel distance should be minimized, and should always be productive. The most efficient arrangements for this type of layout are "U"-shaped.

A typical arrangement for small manually operated installations is given in Fig. 1. The simplest layout is shown in Fig. 1a. The operator moves counter-clockwise from tank 1 to the racking unracking station 10.

Replacing the main processing tank 5 with one or more semiautomatics, as shown in Fig. 1b, will increase production at the expense of some flexibility.

The double "U" configuration shown in Fig. 1c will provide even greater flexibility at some increase in space requirements. Such an arrangement is ideal for running highly varied pre-plate and post-plate cycles. The operator uses the inner or outer leg of the U, depending on the requirements of the part or the finish (see also suggested layouts in Chapter 17).

In an installation of this type, careful attention to all handling details is necessary to avoid operator fatigue and inefficiency. Figure 2 illustrates the use of a raised duck-board platform to make transfer of long racks less tiring.

COMBINATION HOIST AND MANUAL OPERATIONS

The above considerations also apply to methods of handling parts through treatment cycles involving the combined use of hoists and manual operation (methods 2 and 3 of Table 1).

TABLE 1. Comparison of Methods for Handling Parts Through Treatment Cycles

No.	Description	Limitation as to Parts		Control of Treatment and Transfer Times	Remarks
		Weight and Size	Handling		
1	Wholly manual operation	Limited to what one operator can handle	Racks, wires, trays or baskets	Dependent on operator	For barrels* or still tanks in any arrangement
2	Chain hoist and monorail, manually operated	Greater weight and size or large number in bulk	Racks, baskets, cylinders or barrels	Dependent on operator	For barrels or still tanks usually in straight line or sequence
3	Motor-driven hoist and monorail (or bridge crane) operated manually or automatically	Still greater weight, size or bulk work	Racks, baskets, cylinders or barrels	Dependent wholly or partially on operator	For barrels* or still tanks in either straight line or return path
4	Semiautomatic return-type machine, manual loading and unloading	Limited weight and size, little bulk work	Racks, hooks, wires, trays or baskets	Single treatment automatic, others depending on operator	For single plating tank only (Chapter 25)
5	Fully automatic return-type machine, manual or automatic loading and unloading	Wide range of weight and size, also bulk work	Racks, hooks, cylinders or barrels	All automatic	For all treatments in the cycle (Chapters 24 and 25)
6	Fully automatic, straight-line machine, manual or automatic loading, and unloading	Great variety in weight and size, but no bulk work	Racks or special fixtures	All automatic	For all treatments in the cycle (Chapter 25)
7	Fully automatic machine, automatic loading and unloading of steel mill products	Relatively wide range in weight and dimensions	Continuous for sheet, strip, wire or rigid conduit	All automatic	For all treatments in the cycle (Chapter 26)

*See Chapter 24.

However, the use of a hoist and monorail tends to restrict the arrangement of tanks to an orderly sequence and the position of the operator with respect to the height of the tank rim becomes less important. The tank arrangement may be either straight-line or a return ("U") type (see Figs. 3 and 4).

A simple bridge crane and hoist cannot conveniently be used to turn corners and therefore its use is ordinarily restricted to a straight-line arrangement vertical to the pair of crane rails. An exception would be a large bridge crane equipped for movement in any direction within a work area. Such an installation is costly and its use with any return (or "U") type tank arrangement would be restricted to a small production of relatively large items.

A straight-line arrangement can have the tanks positioned so that the horizontal work bar carrying the parts may be either (1) parallel or (2) at right angles to the direction of travel. If the tanks are single-cell units, they would be placed end-to-end for case (1) and side-by-side for case (2). The former is more commonly used for a return (or "U") type tank arrangement since the work bar travels from the loading station through the cycle of treatments and returns to the unloading and loading station. Case (2) is most commonly used for a straight-line tank arrangement. However, the latter re-

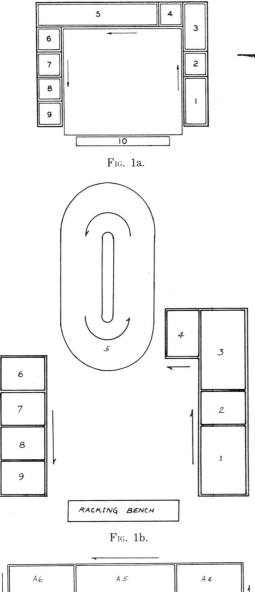

Fig. 1a.

Fig. 1b.

Fig. 1c.

Fig. 1. Typical tank arrangement for small manually operated installations.

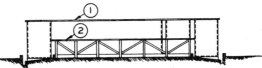

Fig. 2. Elevation view of Fig. 1 showing duckboard platform for proper operating level. (1) Tank rim. (2) Platform level.

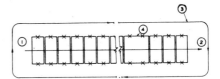

Fig. 3. Typical straight-line arrangement for use with hoist and monorail (or bridge crane).
(1) Loading zone
(2) Unloading zone
(3) Additional conveyor for returning rack and work bar to loading zone
(4) Illustrating a two-cell plating tank

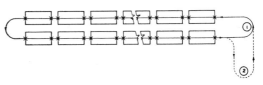

Fig. 4. Typical return "U" type tank arrangement for use with hoist and monorail.
(1) Loading and unloading zone.
(2) Dotted line indicating possible use of extension in other directions.

quires a separate conveyor to carry the work bar and rack from the unloading to the loading end. This requires additional floor space, but it also offers the opportunity for reloading the racks before reaching the loading end of the line.

The work bar also serves as the anode or cathode rod in anodic or cathodic treatments. Electrical contact is made by resting the bar on contact blocks at each end when the load is fully immersed (see Chapter 27). When tanks are placed side-by-side, these blocks are at the ends of the tank on each side of the line. When tanks are positioned end-to-end, the contact blocks are also at the tank ends. This places the contacts at the center line and below the monorail. Such an arrangement requires greater spacing between tanks to allow for bus connections and any agitation mechanism. The side-by-side tank positioning in a straight-line arrangement requires less space between tanks. It also enables one to use large tanks with two or more plating cells, if desired, instead of a single tank per cell. This simplifies the solution control and reduces the number of individual filters, pumps, heat exchangers, and purification installations. Some of these advantages can be obtained with the end-to-end tank positioning by piping two or more tanks together.

SIZING AN INSTALLATION

In order to properly size an installation, it is necessary to establish all the production requirements and the required finishing specifications. These will include:

(1) *Production.* The number of pieces required per day, week or month.

Example: Assume that a maximum of 1750 pieces per 8-hr shift of one design are required. Allowing for 1 hr of lost time per shift while starting up and shutting down, this is equivalent to 250 pieces/hr for 7 hr of operation. Also assume 40 hr/wk of 5 days of one 8-hr shift per day. The maximum production would then be 5 times 1750 or 8750 pieces/week.

(2) *Plating Specifications.* The following information must be given for each part to be plated.

(a) Metal or metals to be plated and the kind and quality of metal finish such as bright nickel, brush silver, etc.

(b) Coating thickness for each metal, usually stated as oz/ft^2 (for zinc) or the minimum thickness in inches or mils (for copper, nickel or chromium).

(c) A drawing giving the exact dimensions of the part, the location of any holes or screw threads, and clearly indicating (1) what are the significant surfaces requiring the specified coating, (2) what surfaces if any must be unplated (or subsequently machined) and (3) what surface may be regarded as "optional", i.e., not requiring any deposit.

(d) Other requirements such as accelerated corrosion or porosity test ratings (see Chapter 10).

Example (continued): Assume that the specification calls for a steel item to receive a protective coating of 1.5 mil minimum nickel. Further assume that the item is 5 ft long × 1 ft wide, is substantially a flat sheet and requires plating on one side only. The total area of one piece is therefore 5 ft times 1 ft × 2 (sides) or 10 ft^2 and the significant area (one side) is 5 ft^2.

It is impossible to prevent some metal from depositing on the back of such an item unless the surface is protected with a "stop-off" material; however this may be considered impractical and unnecessary in the case under discussion. The best that can be done to favor depositing most of the metal on the front significant surface or face is to mount the pieces back-to-back on opposite sides of the rack. One must experimentally determine the best arrangement in any case, but it may be assumed that by way of illustration 80% of the nickel is deposited on the face of the piece.

At an average current density for nickel plating of 40 amp/ft^2 a total of 80 amp is required for 1 ft^2 of both front and back surfaces. Since 80% of the total nickel is assumed in this case to be deposited on the front, the current flowing to this surface will be 80 amp × 0.8 or 64 amp, equivalent to a face current density of 64 amp/ft^2. Also the current required per piece would then be the total area of 10 ft^2 times the average current density of 40 amp/ft^2 or 400 amp.

It requires 18.7 amp hr/ft^2 (1120 amp min/ft^2) to deposit 1.0 mil of nickel at 100% cathode efficiency. Assuming an average efficiency of 95% it would therefore require (18.7/0.95) times 1.5 mil or 29.5 amp hr/ft^2 (1770) amp min/ft^2) to deposit 1.5 mil of nickel.

The time required to deposit 1.5 mil of nickel at a *face* current density of 64 amp/ft^2 is therefore 1770/64 or 28.6 min. However, the actual time required to obtain a minimum thickness of 1.5 mil of nickel at any point on the significant surface (or face in this case) will be longer. It will depend on how the pieces are racked and the many other factors influencing current and metal distribution as discussed elsewhere in this book. This can only be determined by actual plating test runs. If we assume for the case under consideration that the plating time must be increased by 45% to obtain the minimum specified nickel thickness on the significant surface of the item, the actual plating time would then be 28.6 times 1.45 or 40 min.

(3) *Racking Data.* The number of pieces that can be placed on one rack depneds on many factors (see Chapter 22), the most important of which are as follows:

(a) Part size and weight limitations.

(b) Tank or plating cell size and shape.

(c) Solution volume in relation to the amperes per gallon and ability to control the temperature.

(d) Available ampere and voltage capacity.

(e) Ability to plate successfully to the required minimum specification with the most favorable metal distribution.

Once this is established the amperes required per rack based on the assumed average current density can be determined.

Example (continued): Assume that 10 pieces can be racked horizontally at one time with 5 on each side and back-to-back. Under these conditions anodes would be placed around the outside of the loaded rack in the treatment tank in such a position as to favor uniform current and metal distribution.

The total area of the rack load would then be 10 pieces times 10 ft^2 or 100 ft^2.

The total current required per rack would be the average current density of 40 amp/ft^2 times the total area of the rack load of 100 ft^2 or 4000 amp/rack.

The tank voltage required depends on the anode to cathode spacing and other factors as discussed in Chapter 5.

(4) *Number of Plating Cells.* One loaded rack and its anode arrangement constitutes a plating cell. This may be an individual tank or a large tank containing two or more plating cells arranged side-by-side. The number of such cells required to give the desired production may be calculated from the number of pieces per rack, the plating time required to meet the specification, and the number of productive hours in a work day or week.

Example (continued): By racking 10 pieces at a time and plating 40 min one plating cell will produce 60/40 or 1.5 rack loads/cell/hr equivalent to 15 pieces/hr. Since there are assumed to be 7 production hours in each 8-hr shift, this is equivalent to 1.5 times 7 or 10.5 rack loads/cell/shift and 15 times 7 or 105 pieces/cell/shift. To obtain the assumed production of 1750 pieces/shift one would require 1750/105 or 16.7 plating cells (minimum). Assuming a factor of safety of 10% for downtime on plating cells (or for increased production) a total of 16.7/0.9 or 19 cells would be needed.

The total ampere capacity required is 4000 amp/cell times 19 cells or 76000 amp. Each cell requires 4000 amp or preferably 5000 or 6000 amp to ensure flexibility for future requirements.

In the above example a conservative figure of 40 amp/ft^2 was assumed for the average current density. The conditions were further assumed to give a face current density of 64 amp/ft^2. In many cases the face current density can be higher and if racking and other factors can be favorably changed to allow the use of higher values of average and face current densities without producing burnt or defective deposits, this would be done. By so doing the productive capacity of the installation could be further increased without additional tanks provided that the excess voltage and ampere capacity are available.

If the nickel must be buffed, it is necessary to deposit sufficient excess of nickel so that the minimum thickness after buffing will meet the 1.5 mil specified. This must be determined experimentally and the plating time and/or current density must be increased. This factor would ordinarily be included in experimentally arriving at the 45% increase in plating time required to meet the minimum thickness specifications as discussed earlier.

(5) *Cleaning Requirements.* The longest cleaning step in a manually operated plating installation may limit the production or may require an increase in the number of tanks to provide additional treatment cells.

In the previous example for nickel plating an average of 1.5 rack loads/cell/hour were processed, and there were 19 plating cells provided, which is equivalent to a maximum production of 28.5 rack loads per hour. It would therefore be possible to process each rack 60/28.5 or about 2 min in any cleaning step without delaying production, and only one cell for each step would be adequate as long as the treatment time did not exceed 2 min.

If a 4-min copper strike were required on the other hand, two tanks or plating cells would be needed. If the plating specification for nickel permitted a reduction of from 40 to 20 min, then the line production could be doubled. The maximum in a one-cell unit then would have to be reduced to 1 min, in which case a 2-min treatment would require two tanks or cells.

The importance of the above factors should be fully understood before attempting to settle the design factors for a plating line. This is especially so in considering future developments in relation to the flexibility that should be provided in any installation.

(6) *Transfer Time.* The time of transfer of a rack load of parts between treatments in an operating cycle must not exceed safe limits; otherwise the surface will dry, the metal will become stained or oxidized, and both the adhesion of the deposit and its appearance may be adversely affected. Consequently, automatic machines have an advantage over manually operated lines since the transfer times are designed to fall within a range of proven values for the various basis metals.

With manually operated lines there is less control of transfer time, and difficulties from this, as mentioned above, are more frequently experienced. The speed of travel from the last rinse of the cleaning cycle to the last plating tank in a straight-line or "U" type arrangement is the longest and therefore the most critical design factor. It will definitely limit the number of plating tanks which can safely be installed in one line. It should also be emphasized that hot, humid summer days will cause the most difficulty owing to the greater tendency for workpieces to dry and stain during transfer.

The illustrative example was selected to be sufficiently large to require a hoist line. The principles, however, will apply equally well to smaller parts and smaller production volumes.

Any plating installation, existing or proposed,

will benefit from a complete flow analysis to determine the optimum layout (see also Chapter 17).

A complete flow analysis consists of the following steps:

1. Layout of existing flow patterns.
2. Determination of problem points.
3. Rearrangement of equipment and/or procedures to eliminate problems and smooth flow.
4. Analysis of the cost/benefit ratio for the required change.

In order to carry out these steps, you will need:

1. A complete floor plan of the department to scale.
2. Transparent overlay paper.
3. Process sequences for all parts processed.
4. A supply of varied colored pencils.
5. A drawing board or table.

Follow these simple steps:

1. Fasten the floor plan to the drawing board, and cover it with an overlay.
2. If the floor plan does not show the arrangement of equipment, draw it in to scale. Make sure all storage areas are included. Ditto for inspection stations.
3. Select the highest volume part or finish; assign a color to it, and on the overlay draw in its progress through the department. Start with preprocessing storage. Number each handling step. Continue until it leaves the department.
4. Repeat using a different color for the part or finish with the second highest volume, etc.

Now check for the following points:

Backtracking. Do parts anywhere retrace their path to get to the next operation, or to get out of the department? If so, can the backtracking be eliminated by rearranging the equipment or providing another exit from the department? This generally reduces handling, and avoids congestion.

Poor storage planning. Extra storage space ahead of a machine or an operation sometimes pays for itself. If parts can be delivered directly to a localized storage area instead of to a generalized storage area, one handling will be eliminated. An added benefit may be greater flexibility in processing, since the operator will have a wider choice of parts to even out machine or operation flow.

Traffic congestion. Do too many flow patterns converge, or cross each other in one place? Can the patterns be altered to correct the problem?

Can any of the transfer paths be shortened?

These few rules can be applied in numerous ways to improve a layout. For example: An analysis showed that certain parts were being burnished in the burnishing area, and transferred wet to the nickel plating area, where they had to wait before entering the cycle. Locating a single burnishing barrel near the plating unit eliminated the transfer, and enabled the plater to coordinate the burnishing with the plating so parts no longer waited to enter the cycle. The result, in addition to the handling savings, was smoother flow and fewer rejects.

When the analysis is complete, tape a new overlay over the print, and draw in the changes and the new flow pattern. Analyze this layout closely. Have new problems been created? Can it be improved? Are there alternate approaches that should be considered? Continue to repeat the process until an acceptable layout has been produced and thoroughly critiqued.

Now based on this layout, calculate the cost of making the changes, including the cost of new equipment if any. Calculate the savings to be realized. Include an estimate of the savings due to reduced rejects, or increased production capacity. Calculate the time required for the savings to pay for the changes. Companies have different standards, but in general if the time to repay is five years or less, the changes are justified. Improved production flow or higher production capacity may justify the changes even in the absence of savings.

24. BARRELS

WILLIAM H. JACKSON

Formerly Chief Engineer, Barrel Department, The Udylite Corporation, Detroit, Mich.

AND

A. KENNETH GRAHAM

Consulting Engineer, Jenkintown, Pa.

Revised by Reginald K. Asher, Sr., CEF

Principal Staff Engineer/Scientist, Motorola, Inc., Phoenix, Ariz.

The rotary plating barrel is used to plate parts in bulk. As a rule such plating is cheaper than rack plating because all racking and unracking labor as well as rack costs and rack maintenance is eliminated. Required space for the total plating sequence is less, and so is the amount of labor to move the work from one operation to another.

Barrel plating also corresponds well with such other bulk-treatment operations as barrel polishing, barrel burnishing, barrel de-burring, barrel phosphating and the application of oxide coatings.

BARREL TYPES—SIZES AND CAPACITIES

Barrels can be classified as follows:

Horizontal barrels partially or completely immersed in a tank with plating solution (1) with perforated cylinder walls, perforated removable door and anodes on the outside of the cylinder (Figs. 1a and 1b) and (2) with solid cylinder walls but without door, and anodes suspended from the cylinder axis inside the cylinder (Fig. 2).

Oblique, or "45-degree," barrels open at the upper end (a) with solid walls that hold the plating solution and inside anodes (Fig. 3) and (b) partially immersed in a tank and having perforated walls and outside anodes (Fig. 4).

Type 1a, usually called the horizontal barrel is the one most commonly employed. The most popular size of cylinder is 30 in. long × 14 in. diameter (inside measurements). (On polygonal "cylinders" the "diameter" is the distance between opposite flat surfaces). It can handle up to half its volume or about one bushel (1.25 cu ft) of work. Another popular size is 36 in. long × 14 in. diameter; this holds a 20% larger load. Barrels up to 18 in. diameter (or larger) and up to 5 ft long are built, but they cause more abrasion of the deposited coatings than barrels of smaller diameter. Although there is a limit to the size of a part which should be barrel plated, larger barrels will plate larger pieces than will smaller barrels.

In barrel plating it is normally very important that the parts do not become mixed in the process; therefore because of the type or quantity of parts, it frequently becomes necessary to use horizontal barrels smaller than 14 in. diameter. Small portable barrels are used for plating of "odd lots" or for laboratory use; they are hung on the cathode rod in still tanks. Many sizes are available: a popular size is 6 in. diameter × 12

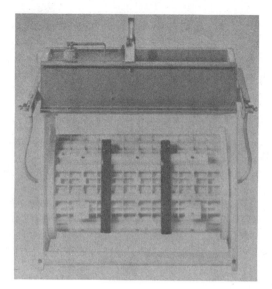

Fig. 1a. Horizontal honeycomb polypropylene mesh barrel. (*Courtesy of Sterling Systems.*)

Fig. 2. Imperforate horizontal barrel for chromium plating[2]. (*Courtesy of United Chromium Inc.*)

Fig. 1b. Perforate horizontal barrel. (*Courtesy of The Udylite Corporation*)

cylinders are removable from their support brackets, and one cylinder head is removable for loading and unloading. Two important features are inherent in the operation of this type of equipment: (a) the very gentle tumbling of the parts within the cylinders and (b) the excellent agitation of the solution, which definitely becomes replenished within the cylinders. From (a) it follows that frail parts may be processed and plated in bulk; whereas from (b) it can be seen that by regulation of the height of the

in. long. Smaller barrels, frequently of the oblique type, are also employed.

The so-called "cluster barrel," has put the plating of small lots on a commercial basis comparable to the production from larger barrels. It consists of one or more clusters of small, perforated horizontal cylinders mounted on a central insulated shaft which revolve with the shaft but not relative to it (Fig. 5). Cluster barrels are usually built to be interchangeable with the larger (regular) barrels in the same tanks. The cylinders are made in various sizes, and up to sixteen of them may be mounted on the center shaft; hence a different kind of part may be plated in each without mixing. The

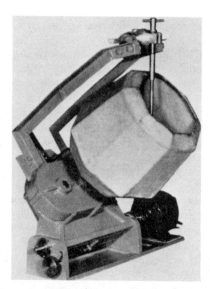

Fig. 3. Imperforate oblique barrel. (*Courtesy of Hanson-Van Winkle-Munning Co.*)

Fig. 4. Perforate oblique barrel. (*Courtesy of Daniels Plating Barrel and Supply Co.*)

solution in the tank (hence the amount of breakthrough by the cylinder when at the top of its circular path) any desired portion of the solution with the cylinder will be removed and immediately replenished during each revolution. It follows that any evolved hydrogen will be expelled during each revolution. Moreover, four cluster cylinders size 6.5 in. diameter ×

Fig. 5. Cluster barrel. The function of the handle (in front of the barrel) is to enable the operator to hold the cylinder with one hand, and at the same time retain the cylinder head in place, while he locates the cylinder in or removes it from its supports. The hooks of the handle fit around the cylinder hubs.

14 in. long will have a larger perforated area than will a 14 in. diameter × 24 in. long plating cylinder; hence they will plate faster. The scroll-type cylinder is a modification of the perforate horizontal cylinder, presumably developed to eliminate the necessity for removing the door panel. A United States patent was issued on this type of cylinder about 65 years ago.[1] The cylinder is revolved in one direction to keep the work in, and in the opposite direction to discharge it. Lately it has found use in England but has not been popular in the United States. The perforated horizontal barrel has also found use in fully automatic machines both in England[1] and the United States.

Barrels of type 1b are used exclusively for chromium plating with an inside insoluble anode.[2] Those now built in the United States are of the batch type. The 16- and 24-in. sizes will handle 3- and 10-lb loads, respectively, of such items as wood screws.[3] Some German and English makes include continuous units, in which a helical track on the inside of the cylinder guides the work from one cylinder end to the other while the cylinder rotates. Such equipment has also been built in the United States but found expensive relative to its productive capacity.[2]

Type 2a is the original type. Inasmuch as beer barrels and whiskey kegs used to be employed as the work containers, this piece of equipment was called a barrel, a name that now covers all rotary bulk-plating machines.

This type, as indicated, used to be available in both large and small sizes and for all kinds of work. Because of its general inefficiency, it is now used only for such small parts that fall or protrude through the perforations of other types and for small odd lots. The cylinder dimensions are of the order of 12 to 15 in. diameter and depth, and the capacity of the order of ½ peck (269 cu in.).

Type 2b was developed for use in fully automatic barrel platers and is still the type most commonly employed for that purpose. It is also used for hand-operated units. With dimensions of 6 to 17 in. inside diameter and 6 to 26 in. depth, it handles lots of 40 in.^{3}—0.75 ft^3, of work that is big enough not to get out through the perforations.

ROCKER-TYPE BULK PLATERS

Bulk plating may also be carried out in machines in which a rocking motion has been

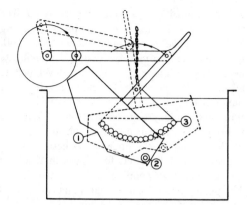

Fig. 6. Schematic view of oscillating cradle plater.
(1) Hump for turning the work
(2) Flexible cathode contact
(3) Basket for ball anodes

substituted for the rotary motion. Such units have a low capacity relative to their size and have never met with general acceptance. An example is the cradle plater of which a few were built in the United States (Fig. 6). The hump in the middle makes the work (the kind which tumbles freely) turn over as the container is swung from one side to the other. It was used (1) with rubber covering and a cathodic chain contact that moved with the work and (2) in bare steel, charged cathodically, for chromium plating. This type of machine will not be discussed further.

TYPES OF WORK THAT CAN BE BARREL PLATED

Whether to plate a lot of certain parts in a barrel or on a rack is one of the most important decisions to be made in the plating plant. It has a decided bearing both on the quality of the plate and on the cost of plating. The following factors must be considered:

Weight and Size of Part. Imperforate oblique barrels can handle parts of the size of a pin or the smallest bearing ball. In the usual horizontal perforated barrel one can plate parts weighing 1 lb. Parts weighing more than this are likely to crack cylinders which are made of solid rubber or plastics or to cause rapid wear on both cylinders and drives.

It usually does not pay to barrel plate parts having a volume larger than 25 in.[3]

Shape of Part. Parts that do not move freely relative to each other when the barrel rotates fail to come into the path of the current on the surface of the load with equal frequency. As a result, the average plate thickness will vary greatly from one part to another, and it becomes difficult to meet plate specifications.[3]

Parts *nest* (a) when they consist of large flat surfaces of thin sheet metal. Here the capillary attraction $F = \gamma A/\delta$ (where γ = surface tension, A = area of contact, δ = space between the sheets filled with liquid) pulls the parts together. They also nest (b) when they are shaped like cups that fit into each other, sometimes due to the friction between them, sometimes to surface tension, or to both. Finally parts nest (c) when they hook into each other purely mechanically. Examples are many types of wire goods such as corset stays and light stampings of complicated configuration. With some such parts it is possible at times to reduce the nesting tendency by use of low barrel speeds. In cases (a) and (b) the addition to the load of several 1-in. diameter, 6-in. long metal rods, with or without a plastic coating, sometimes causes the parts to break apart; also a higher peripheral speed and the use of additional tumbling ribs will help.

Stiff parts *jam* in the barrel when their length is equal to the barrel diameter or some other barrel dimension. At times two or several parts cooperate in the jamming. This difficulty may be overcome by reducing the barrel speed or using a barrel of different dimensions.

Kind of Coating. Certain lots of cast and malleable iron and carburized steel are difficult to plate in zinc cyanide solution because they have areas of low hydrogen overvoltage. This difficulty is particularly pronounced in barrels because it is difficult to reach "striking" cathode potentials with available voltage. Recently however, particularly with automatic equipment, potentials up to 24 volts, with voltage regulation, have been used. At times, coverage can be achieved by intermittent rotation, but best results are obtained by addition of sodium stannate or a cadmium salt.

Because the current density varies continuously and contact to the part is often broken, only parts of relatively simple shape that tumble well can be chromium plated. Chromium plating requires barrels of special construction.

It is impossible to apply nonconducting coatings in barrels because no points of electric contact remain after a short time; thus one cannot use a barrel to anodize aluminum or magnesium or to apply rubber or plastic coatings by electrophoresis.

TABLE 1. CONSTANTS IN THE THICKNESS EQUATION

Barrel Type	Constant a at Certainty of:			Constant b
	50%	95%	99.87%	
(a)	0.008	0.032	0.056	0.69
(b)	0.083	0.120	0.154	0.95
(c)	0.081	0.116	0.148	0.97
(d)	0.076	0.117	0.155	0.98

TABLE 2. REQUIRED BATCH THICKNESS AND PER CENT EXCESS FOR TWO INDIVIDUAL THICKNESSES, MIL

Barrel Type	Desired t_i mil	t_b (and % excess) at Certainty of:		
		50%	90%	99.87%
(a)	0.2	0.30 (50%)	0.34 (70%)	0.37 (85%)
	0.4	0.59 (47.5%)	0.63 (57.5%)	0.66 (65%)
(b)	0.2	0.30 (50%)	0.34 (70%)	0.37 (85%)
	0.4	0.51 (27.5%)	0.55 (37.5%)	0.58 (45%)
(c)	0.2	0.29 (45%)	0.33 (63%)	0.36 (80%)
	0.4	0.50 (25%)	0.53 (32.5%)	0.56 (40%)
(d)	0.2	0.28 (40%)	0.32 (60%)	0.36 (80%)
	0.4	0.49 (22.5%)	0.53 (32.5%)	0.57 (42.5%)

Plate Distribution. Even when the parts tumble well and electrical contact is maintained to the load, the variation in thickness from part to part is substantial. In order that the individual parts in a load meet a certain thickness requirement, an individual average thickness of t_i mil (average upon a certain area) the corresponding thickness on all parts of the load, the batch-average thickness t_b mil, must be bigger. One finds $t_i = -a + bt_b$ or $t_b = (t_i + a)/b$ where a and b are constants that depend on the equipment, solution and operating conditions as well as on the desired percentage of certainty, often 95 per cent, that the requirement be met. Table 1 shows the results of two investigations of threaded fasteners. One, (a) dealt with bright zinc plating at 70 to 90 F in 14 × 30 in. horizontal barrels at 6 rpm, the other (b) with cadmium plating at the same temperature in 21 in. deep × 21.5 in. diameter imperforate oblique barrels at 9 rpm, (c) fully immersed 14 × 30 in. horizontal barrels at 5 rpm, and (d) doorless scroll-type 16 × 22 in. horizontal barrels at 13 rpm. According to Table 1, in case (a) $t_i = 0.2$ mil and at 95 per cent certainty, the above equation takes the form

$$t_b = \frac{0.2 + 0.032}{0.69} = 0.34 \text{ mil.}$$

A number of such examples are shown in Table 2. The data in these tables lead to the conclusions that: (1) the horizontal barrels gave a somewhat better plate distribution than the oblique barrels, (2) the difference between the required average batch thickness for high quality plating (high certainty) and for low quality plating (low certainty) at a given individual average thickness is relatively small, (3) the plate tends to get more evenly distributed between the parts the higher the thickness, and (4) if other conditions during the two investigations were comparable, the cadmium solutions gave better plate distribution than the zinc.

Some of the most recent work on metal distribution during barrel plating was done by Craig and Harr[3] at Arizona State University. They carried out extensive tests using 4 × 4 in. and 6 × 6 in. barrels, using various solutions, current densities, parts and plating times. In general, they noted that there was no clear correlation between the spread factors for the various solutions and the critical anode current density, conductivity, cathode efficiency or Hull Cell tests. They found that the part-to-part spread in deposit thickness could be correlated by the equation

$$\sigma = K\bar{t}/\sqrt{\Theta}$$

Where Θ is plating time in minutes; $\bar{t}$ is the average thickness in μm; σ is the deviation in μm and K is the spread factor, independent of current and plating tank size.

They did find that the combination of the barrel geometry and part shape had definite effects. Parts which tumbled poorly due to sticking, tangling, or interaction with the danglers or other form of contacts, gave very poor mixing of the load and showed spread factors 50% greater than smoothly tumbling parts. Changes to correct the mixing problem restored the spread factor to the expected values.

They were able to quantitatively confirm that the spread in deposit thickness could be decreased by:

1. Decreasing the deposit thickness.
2. Increasing the plating time.
3. Increasing the rotational speed of the barrel.
4. Increasing the amount of barrel periphery open to solution transfer.
5. Decreasing the size of the load.
6. Decreasing the size of the barrel.

They also did extensive work on the effect of solution composition on spread factors for most of the common solutions and were able to determine distinct differences in performance.

The variation in plate thickness *over the surface of a single part* is likely to be greater in a barrel than on a rack. On one hand the throwing power is often reduced because the agitation lowers the concentration polarization at the cathode and, in zinc- and cadmium-cyanide solutions, because the temperature rises above the best value, thereby again reducing polarization. On the other hand the parts are not kept in the best position relative to the anode, and such strategems as shielding and thieving cannot be employed.

During barrel plating with the less abrasion-resistant metals, e.g., copper, zinc or cadmium, some of the coating is rubbed off high points. The plate distribution may be affected in a desirable manner, e.g., so that the top of machine-screw threads fit better into tapped holes, or adversely, so that corrosion protection suffers. The wear increases with heavier loading of the barrel.

Plate Appearance. Whereas broken or small, rounded surfaces can be bright-nickel plated in a barrel to an attractive appearance, it is impossible to produce a mirrorlike finish on flat or slightly curved surfaces of substantial size because the scratching that occurs is easily observed. Malleable coatings can be barrel burnished after barrel plating, but more brittle coatings such as bright nickel cannot.

THE PERFORATE HORIZONTAL BARREL

Construction Features

The barrel consists essentially of (1) the cylinder, rotatable around a horizontal axis, (2) a hanger from which the cylinder is suspended, (3) a tank that holds the electrolyte, carries anode rods with anodes, and supports the hanger and cylinder, a motor, and means to connect the barrel to a source of current, (4) a system of gearing which transmits the motive power to the cylinder, and (5) a system of contacts and conductors that connect the work to the cathode terminal. In addition, the barrel may be supplied with (6) means that lift the hanger-and-cylinder assembly.

Fully automatic equipment with cylinders of this type is discussed separately.

The Cylinders. The "cylinder" proper consists of a regular prism or a right circular cylinder supplied with two ends.

The prismatic polyhedral "cylinder" often has separate perforated sides or panels that fit into slotted ribs and separate ends or "heads" with slots and holes to receive the ends of the panels and the ribs. Such a cylinder is held together with bolts or perhaps tie rods, both of which should be insulated.

When the cylinder is made of acrylic resin, the sides or cylindrical surface, except for the door, may be made of one piece, mortised and cemented into or welded to the heads. Insulation-coated steel cylinders are usually of one-piece welded construction (except for the door) which has then been coated. Such constructions are free from solution-carrying joints except at doors and are especially suited to be moved through one solution after the other with a minimum of contamination of the solutions.

One of the panels or a corresponding wall portion of other types of cylinders is removable and serves as the *door*. A large door permits easy loading and unloading. The door opening should have close tolerances so that the door fits well and flat parts do not lodge between it and the door. The door should be supplied with a handle or knob for quick removal.

Much effort has gone into the design of *door-closing devices*. With cylinder-material that keeps its shape at the temperatures and load conditions that are encountered, the door may slide in from one side through the head opposite the drive, and any simple locking device will prevent it from sliding out. Usually the door is held by *clamps*. Preferably they should be maneuverable with one hand without undue effort. The clamps should be strong, wide enough, and of sufficient number to securely fasten the door. Thin doors of flexible material require especially effective clamping. The clamps, being in the path of the current, should be of nonconducting materials so that they do not become plated.

The *thickness of the panels or walls* and the door varies from $\frac{1}{4}$ to $\frac{1}{2}$ in. for large barrels to about $\frac{1}{8}$ in. for small barrels. One would expect that the thickness would be made proportional to the wear resistance of the panel material (see cylinder materials), but the opposite seems to be the case.

Panel and door perforations vary greatly. Hole sizes are usually ⅜, ⁵⁄₁₆, ¼, ³⁄₁₆, ⁵⁄₃₂, ⅛, ³⁄₃₂ and ⁵⁄₁₆ in. diameter for holes drilled in solid material. The user ordinarily chooses the largest holes that can be used for his parts in order to get the maximum percentage perforation (open area). While panels should be perforated as liberally as possible they should not be over-perforated because this will unduly weaken them. Generally speaking the open area should not exceed 30 per cent. On automatic barrel machines it is good to have matched cylinders, each having the same open area.

Herringbone-type Perforations (Fig. 7). This type of perforating is applied to plating cylinders to be used for plating commercial pins, nails and certain types of wire goods. It can be seen that the work will not go through the panels, and that the direction of cylinder rotation should be as indicated. Again, it can be seen that the hole size is a function of the panel thickness. For panels ½ in. thick, holes ⅛ in. diameter are suitable.

The thickness and percentage perforation of panels are important factors in the amount of current drawn. For example, in an hexagonal cylinder 14 in. in diameter, with panels 30 in. long, fully immersed, the voltage drop through the panels $(IR)_p$ can be estimated to be

$$(IR)_p = 0.042\, I\rho t/p$$

where

I = total current in amp
R = total resistance in ohm
ρ = specific resistivity of the solution in ohm cm
t = panel thickness in inch
p = percentage perforation on each panel

Thus in a zinc-cyanide solution where $I = 600$ amp at 12 volt and $\rho = 5$ ohm cm, $(IR)_p = 126t/p$. Therefore, if $t = ⅜$ in. and $p = 20\%$, $(IR)_p = 2.38$ volt, or nearly 20% of the total voltage drop. If the $(IR)_p$-drop were reduced to one-half of this value, the current drawn would be increased by something less than 10%, and the productivity of the barrel would increase by almost as much.

The maximum thickness of cylinder walls made out of coated steel is of the order of ¼

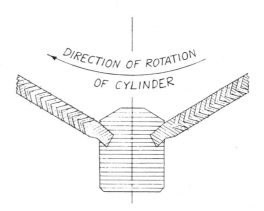

Fig. 7. Plating cylinder with herringbone perforations.

to ⅜ in. and tapers off to zero at the holes. For comparison with drilled solid panels, one has to calculate the average thickness of the imperforate area. The number of holes is given by the number of holes in the perforated or expanded metal base. The naturally formed holes are usually about ¼ in.; they may be enlarged somewhat by drilling or burning. The lower average wall thickness, which is the same as a larger-than-nominal hole diameter, reduces the resistance of the walls so that such cylinders draw almost as much current as others with good percentage perforation. It is evident, however, that the holes are too big for many small parts.

Cylindrical containers are often supplied with inside ribs to help in carrying the work pieces up the sides. In polyhedral cylinders, the sides serve a similar function. The larger the area over which the work is spread, the better is the current distribution and, thereby, the plate distribution during plating. (See also barrel speed.) The number of sides on a polyhedral cylinder is usually six or eight. The hexahedral cylinder is cheaper to make.

Using Fig. 8 as an illustration of a rotating horizontal hexahedral perforate barrel, it has been found that the fractional amount of the volume change of liquid within the immersed barrel during rotation about a substantially horizontal axis will vary with the percent of the diameter (d) of the circumscribing cylinder (y) to which the barrel is immersed in the solution. This is best illustrated in the figure in which the solid and dotted line figures of the hexahedral barrel (x) represents two critical positions during rotation, the dotted line figure A being that with a side B horizontal and the solid

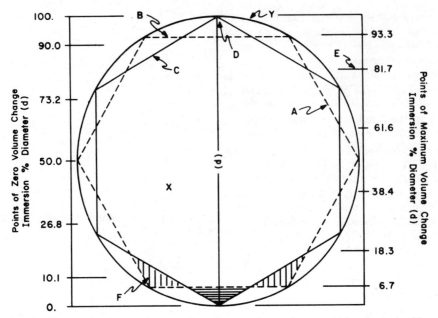

FIG. 8. Depth of immersion points of zero and maximum volume change within a horizontal hexahedral perforate barrel during rotation.

line figure C being that with an apex or corner D at the top. At various critical depths of immersion E, i.e., points of maximum volume change, either the dotted line figure A or the solid line figure C will represent a minimum or maximum contained volume during rotation. At each of these critical depths E, the contained volume always includes completely immersed triangles in cross section, (prisms in solid form) F, the maximum contained volume exceeding the minimum contained volume by one such completely immersed triangle F. The term "pumping action" (as used herein with respect to a polyhedral barrel of N sides) means the volume change of liquid within the barrel per $1/N$th revolution as per cent of the maximum contained volume.

It should be noted from Fig. 8 and Table 3 that pumping action during rotation of the hexahedral barrel becomes greater at the lower maximum point of immersion, with the advantage of better solution replacement and more

TABLE 3. DATA FOR STANDARD HORIZONTAL HEXAHEDRAL PERFERATE BARRELS

Immersion Depth at Points of Maximum P.A.,* % (d)**	Pumping Action* (P.A.) at Maxima, %	Immersion Range about Maxima for at Least 1% P.A., % (d)	Time at Maxima and 10 rpm‡ for P.A. to = Contained Volume, min	Immersion Depth at Points of Zero P.A., % (d)
				100
93.3	1.20	4.4 = (1.0)	1.45	
				90
81.7	1.32	8.0 = (1.83)	1.25	
				73.2
61.6	1.80	15.4 = (3.50)	0.93	
——————— axis of rotation ———————				50
38.4	3.47	19.6 = (4.45)	0.48	
				26.8

* P.A. = the volume change per ⅙ revolution as per cent maximum contained volume.
** d = diameter of the circumscribing cylinder of the hexahedral.
‡ Speed of rotation must be determined specifically for the kind of part being plated. Lower speeds may be required to avoid possible damage to external threads, surface finish, etc. (see page 620 under Cylinder Speeds).

rapid cooling within the barrel. However, the depth of immersion must be controlled to avoid operating at points of zero pumping action and, during plating, the work pieces should be totally immersed. In prior editions of this Handbook, it has been stated that polyhedral cylinders with separate wall panels have ribs on the outside that, during rotation, carry solution to the inside. This is limited to a relatively narrow range (about ±2%) of immersion depth about the 90% point for the hexahedral barrel, which is also a point of zero pumping action. Furthermore, a 12 to 16 in. size standard hexahedral barrel, when immersed with an average work load, will displace a volume of solution sufficient to raise its level by about 0.7 to 1.3 in. (or about 6% of the barrel diameter), respectively. This further complicates the control.

Circular cylinders have zero pumping action when rotated, regardless of the depth of immersion. Their use should therefore be discouraged, and especially if barrel plating to rigid deposit thickness specifications is required.

Hydrogen Explosions. Explosions have occurred in the horizontal polyhedral perforate barrel during plating. Presumably this would be more likely to happen when the cathode efficiency is low, as for example in bright cyanide zinc baths, and when the solution level is too low, and the evolved hydrogen mixes with the oxygen of the air and is ignited by a spark. This concept need not be limiting with all solutions and conditions as discussed later. Sparking may arise from the work tumbling, from the cathode contact, or, perhaps, from the gearing. The remedy commonly recommended is to raise the solution level so that the inside of the cylinder is filled with solution; then hydrogen cannot gather nor air get inside. Explosions may occur when many of the perforations are plugged, thus restricting the escape of hydrogen which must reach 4.1% by volume in air for explosions to take place. Any condition which lowers the cathode efficiency favors the evolution of hydrogen.

Unfortunately, when the horizontal polyhedral perforate barrel is fully immersed and rotated during plating, there is no pumping action. Under these conditions the metal in solution within the barrel has to supply the metal to be plated out. As the metal concentration decreases during plating, the cathode efficiency decreases and, without solution replacement within the barrel, the temperature rises, causing further variation in efficiency as well as the bright plating results. These disadvantages need not apply to all barrel plating. They can be overcome by plating at a controlled depth of immersion having good pumping action, while using carefully controlled bath compositions and operating conditions to give relatively high cathode efficiency. Barrel plating of parts to high reliability with respect to coating thickness is being done in this manner and without hydrogen explosions.

Further in respect to hydrogen explosions, the anodes should be inspected. If they are polarized on account of insufficient anode area or faulty solution composition, excessive oxygen evolution results.

For data on barrel nickel plating see Fig. 9.

Some manufacturers furnish plating barrels arranged for rapid circulation of solution through the cylinder (Fig. 10). With a pumping capacity of 25 gal/min (100 liter/min), the plating time for a 14 × 30 in. cylinder in cadmium is said to be shortened by some 20 to 25%, apparently by the increased agitation.[7]

The depth of immersion is also a factor in the amount of current drawn (Fig. 11). Other factors include the volume of the load and the tank voltage (Fig. 12). It can be seen in Fig. 12 that nothing is gained by increasing the volume of the load above 50% of the cylinder volume, which is the maximum at which the load tumbles properly.

Plating to Specifications. The factors discussed above have become much more important with the increasing demand for barrel plated parts requiring a very high degree of reliability. Specifications for control of minimum and maximum deposit thicknesses of high strength threaded fasteners, for example, have been very difficult to meet.

In this connection, the experience of one company[15] manufacturing high strength industrial and aircraft nuts and bolts is of great value. A large percentage of their fasteners require plating to thickness ranges of 0.2 to 0.4 mil, 0.3 to 0.5 mil, and 0.4 to 0.6 mil. Because the fasteners are made to a Class 3 tolerance, this presents a problem of deposit thickness control. One must plate to a minimum thickness on the shank, without building up the threads beyond the accepted gauge fit. This has been accomplished by very careful control of: (1) solution; (2) processing; and (3) final inspection, as outlined in Table 4.

Furthermore, the following barrel plating steps are required to successfully meet the spec-

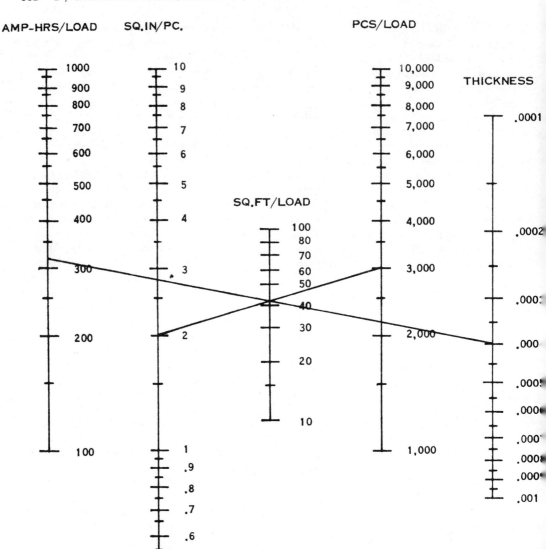

Fig. 9. Illustrative nomograph for barrel nickel plating (similar ones used for other metals). Other data required are as follows: Invoice No., Part No., Base No., Total pcs. in lot, Pieces per load, Amp hr per load.

ifications when plating these threaded fasteners:

1. Determine area of part (from area book or process sheet).
2. Weight-count for total pieces in production lot.
3. Divide lot into equal loads.
4. Determine area of load by use of nomograph (Fig. 9).
5. Determine ampere-hours for load from thickness required and area of load.
6. Set ampere-hour meter.
7. Follow plating procedure called out on process sheet.
8. Inspect work for blisters, thickness, pitch diameter and functional fit.
9. Bake within one hour after plating as indicated on process sheet.

A standard 14 × 30 in. size hexahedral perforate barrel has been used most successfully, although a 10 × 18 in. size barrel has also been used to advantage. The hangers for both sizes were specially made to enable one to control their immersion depth in the same plating tanks at a point of maximum pumping action during rotation corresponding to 81.7 per cent of the barrel diameter (see Fig. 8 and Table 3). This

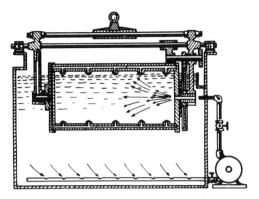

Fig. 10. Perforate horizontal plating barrel with external pump forcing solution through cylinder.[7] *(Courtesy of Hanson-Van Winkle-Munning Co.)*

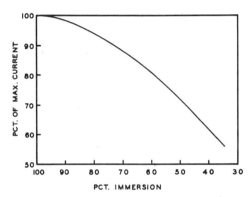

Fig. 11. Effect of cylinder immersion on amount of current drawn by perforate horizontal barrel in a cadmium cyanide solution.

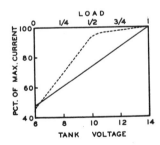

Fig. 12. Effect of load volume (broken line) and tank voltage (full line) on amount of current drawn in a cadmium cyanide solution by perforate horizontal barrel.

made possible closer control of bath composition, temperature, efficiency and deposit thickness. It has been used successfully, and to great advantage, with all acid and alkaline plating baths (including bright zinc cyanide) without any difficulty from hydrogen explosions (but always using the detailed controls).

The volume of the load for the 14 × 30 in. size

TABLE 4. BARREL PLATING PROCESS CONTROL

(1) *Solution Control*
  (a) All solutions analyzed at least once a week.
  (b) Some critical solutions analyzed two or three times weekly.
  (c) Baume and pH of nickel solution checked every eight hours.
  (d) Narrow limits on metal, total cyanide and other constituents.
  (e) Permanent record of solution analysis in numbered page, bound ledger book, consisting of solution analyzed, cc's of titration, factor for titrant, ounces/gallon (cc's × factor), recommended addition and initials of analyst.
  (f) Tank labels indicating type of solution; control limits of ingredients; the results and date of the last analysis; the recommended additions; name of analyst and addition maker.
  (g) An assigned person on each shift for weighing out and making additions to all processing tanks, i.e., cleaners, dips and plating solutions.
  (h) Permanent solution level indicator.
  (i) Continuous filtration.
  (j) Scheduled periodic solution purification.

(2) *Process Control*
  (a) Ammeters and volt meters standardized every six months; ampere-hour meters checked monthly. Post plate bake oven check weekly.
  (b) Temperature controls checked monthly.
  (c) Mandatory detailed plating procedures.
  (d) Controlled load size for optimum surface area.
  (e) Use of nomographs by operators to determine number of pieces per barrel load, and the ampere hours necessary for the required thickness.
  (f) All parts requiring post plate bake in the oven within one hour.

(3) *Post Plate Inspection*
  (a) Operator inspection for appearance, blisters, thickness dimensions, and gauge fit at plating line.
  (b) Inspection after post plate bake, prior to leaving Plating Department.
  (c) Final quality control inspection prior to shipment.

barrel is controlled between $\frac{1}{6}$ and $\frac{3}{4}$ of a ft^3. The maximum area per barrel load is 65 sq ft, but the optimum is 55 sq ft. Similar limits have been established for the 10 × 18 size barrel.

The standard limp-wire, dangler type barrel contacts are not used. The ball type contact has been replaced with a hexagonally shaped stainless steel plate. This change was made to prevent arcing due to the small contact surface area with the type of parts being plated.

Cylinder Materials. The most common cylinder materials are polypropylene, hard rubber, acrylic resins, phenolformaldehyde laminates, melamine-formaldehyde laminates and plastisol-covered steel. Usually walls, doors, and ends are made of the same material.

Properly compounded and cured hard rubber is used successfully in alkaline and acid solutions without contaminating them. Wheras older types began to deform at 57 C (135 F) and could not be used much above 65 C (150 F), new types are available which will withstand temperatures of 93 C (200 F) under load, and have softening points of 135 C (275 F) or higher. Cylinders made of such hard rubbers are good for "whole cycle" operations, and the wear resistance of these materials is well-known.

Acrylic resin ("Plexigals") is another thermoplastic that is available in high temperature form. This may be subjected to temperatures up to 82 C (180 F) depending on the weight and concentration of the load. It has excellent wear resistance and may be used in alkaline and acid baths.

Suitable grades of the thermosetting resins *phenol-formaldehyde* and *melamine-formaldehyde* find use as the binder in cloth laminates that are used to some extent as barrel materials. The former ("Bakelite," "Formica," "Micarta," etc.) are suitable only for acid solutions or low pH alkaline solutions up to 65 C (150 F), the latter ("Melmac," "Merlon," etc.) only for alkaline solutions and weakly acid solutions up to 87 C (190 F), the limitations being the solubility of the resin and consequent delamination of the material or contamination of the plating solution. The melamine cylinders are used throughout complete cleaning and plating cycles. Both types of material have only fair abrasion resistance.

Thin, flat workpieces stick to fresh, flat *cylinder surfaces* but not to worn surfaces. Some manufacturers, therefore, offer serrated or undulating panels, and others roughen the inside of the cylinder, e.g., by sandblasting.

Cylinders of expanded or perforated sheet steel may be covered with about $\frac{1}{8}$ in. coatings of vinyl plastisols that are resistant to most standard barrel-cycle solutions and temperatures. However, their wear resistance is not as good as cylinders made of hard rubber or acrylic resin of the usual thickness.

Drive. A ring gear is usually attached to one end of the cylinder; it should be strong, wear resistant and corrosion resistant in the solutions involved. Mild steel castings serve well in alkaline solutions of pH 8.5 or higher as long as they are substantially free of chlorides. If chlorides are present, the steel ring gear is likely to become bipolar and the anodic portion goes into solution. Cast iron or semisteel ring gears are used in alkaline solutions. For acid plating and for "whole cycle" operations involving an electrolytic pickel, the ring gear should preferably be made of a suitable plastic; such materials are available on the market. The foregoing remarks also apply to the pinion which drives the ring gear. V belts are also being used to both support and rotate the cylinder.

The *cylinder speed* should be high enough that (1) the load rises high on the cylinder side, exposing a large cathode area and thereby increasing somewhat the current drawn, (2) all pieces get into the path of the current at many and equal time intervals for good plate distribution from piece to piece, and (3) parts which stick together when wet separate more easily. The speed, however, must not be so high that (1) the parts which fall from the apex of their ascent fall back without making contact with those below that are still rising, (2) the wear on the plate and on the cylinder becomes excessive, or (3) delicate threads become nicked or fragile stampings distorted.

Most horizontal barrels of about 14-in. diameter are geared to a speed of about 6 rpm, and the gears are interchangeable for speeds of 3 rpm. It is believed, however, that in many cases considerably higher speeds should be employed for sake of plate uniformity, and that the resulting increased production and saving in plating metal would more than offset the cost of increased wear.

The cathode contacts and any current conductors in the cylinder, which lead current to the work, will be discussed later.

The Hanger. The hanger is designed for lightness and strength to carry the cylinder and its load without deflection. The horizontal member is frequently an inverted "T"-section, and for less heavy loads it may consist of three rods placed at the corners of a triangle.

On the ends of the horizontal member are usually two vertical *hanger arms*. These arms are, as a rule, formed like "T" or a cross. At each end of the horizontal bar of the "T" or cross is a *stud contact,* usually of round cross section.

The four stud contacts support the cylinder on the tank. The contacts are isolated from the horizontal hanger member and at times from the hanger arms. The two stud contacts on each hanger arm are interconnected electrically.

One of the two hanger arms carries a shaft with two *gears*. One of these gears meshes with the gear on the gear-motor or gear-box which is fastened to the tank. The other gear drives the large gear on the cylinder. An idling gear is sometimes placed between the last mentioned gears to obtain more complete immersion of the cylinder (Fig. 13). Because of no solution change within the barrel during rotation when fully immersed, this may be questioned.

In older constructions the insulated hanger arm itself carries the current from the stud contacts toward the cylinder (Fig. 14). The lower part of the hanger arm is then formed as a journal for a conducting shaft fastened to the cylinder end. Even with very elaborate shielding it is difficult to prevent current leakage to this conducting bearing. The trees formed tear the isolating sleeves and bushings; thus the trees grow at an ever increasing rate.

In newer constructions (Fig. 13) the lower end of the hanger arm carries a bushing, and the cylinder head the bearing sleeve. The current

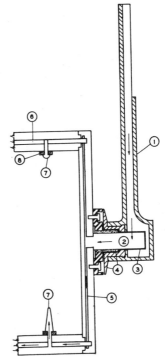

Fig. 14. Cross section of idler-end hanger arm of perforate horizontal barrel with cathode lead broken in cylinder supporting bearing.
(1) Hanger arm of hard rubber molded around conductor
(2) Shaft fastened to cylinder end
(3) Bearing surface
(4) Insulating sleeve
(5) Conducting spider in cylinder end
(6) Conductor in rails
(7) Button contacts of different designs
(8) Flexible insulator

Arrows show path of current from stud contacts to buttons.

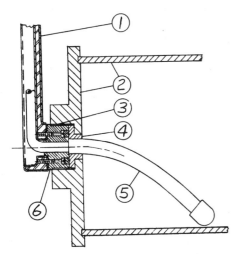

Fig. 13. Cross section of idle end hanger arm and bearing of perforate horizontal barrel with unbroken cathode lead from stud contact (not shown) to inside of cylinder.
(1) Hanger arm, plastisol covered
(2) Cylinder
(3) Bearing sleeve
(4) Arbor bushing (floating, but does not rotate)
(5) Insulated cathode lead and contact ball
(6) Collar bearing

goes through a separate insulated copper rod or cable from the stud contact to the inside of the cylinder without a break. The hanger arm has a "U" cross section with the rod or cable held inside the "U" to protect it from mechanical damage.

Cathode Contacts. A number of different cathode contacts are available. A type that seldom interferes with the tumbling of the work and provides positive contacts is the *button contact* (Fig. 14). The conducting shaft on the cylinder head is joined to a spider of copper in the head, which in turn contacts conductors in the rails, into which the buttons are screwed. A variety of button designs are available for different purposes. A drawback with a deeply immersed cylinder is that at least two rows of buttons are not covered with workpieces and rapidly build up with plate. Cleaning of the buttons is troublesome: their removal is time-consuming, and cleaning them in the cylinder is likely to damage the latter. Hence they often are not cleaned with sufficient frequency, and the plate grows sideways at an ever increasing rate and then out through the perforations. When that happens, serious trouble may ensue.

Other older constructions make use of a conducting centershaft that is covered by insulation except for areas about 6 in. apart. These areas are covered either by conducting sleeves from each of which hangs a contact consisting of a *chain with a ball at the end* or by *hairpin-shaped dangler* contacts. Both types suffer from buildup of plate at the centershaft, which, unless removed, freezes the sleeves to the shaft. Where the shaft rotates with the cylinder the chain then winds itself on the shaft or the hairpin danglers rotate with the barrel, thus destroying the contact to the work. In addition, unless the balls bury themselves in the load and keep the chains taut, little or no current is led to the work. *Star-shaped danglers* are better if the work can stand the impact as the "frozen" danglers rotate.

When the shaft does not rotate with the barrel, there is less danger, but as a rule these contacts are not very efficient and should be avoided.

An efficient cathode contact for use with conductors in the barrel head consists of three or four *bare metal rods* stretching from one head to the other in a symmetrical arrangement. They are arranged for easy removal through the head for cleaning.

The least troublesome contact is the modern flexible dangler (Fig. 15). It consists of rubber-insulated cable in which the successive layers of copper wire are stranded in opposite directions so that the same kind of cable can be used on both ends of the barrel without getting wound up and stiff or wound down and weak. The contact end of the cable is brazed into a metal sleeve that is threaded into a contact weight, and the metal sleeve, as well as the portions of the cable and of the weight nearest it, is covered with an insulating rubber sleeve. The weight can be chromium plated so that the plate comes off easily with a hammer blow after the weight has been removed for cleaning.

The cable may continue through the barrel end and up the hanger arm to the stud contact as in Fig. 13, and then it is necessary to clamp it tightly to prevent it from being forced out of the head. It may also end inside the barrel and be joined there to an insulated stiff copper rod, which then passes through the end and up the hanger arm. The joint must be insulated efficiently by means of a nonconducting sleeve. The cable may also be fastened to a continuation of the conducting shaft of Fig. 14 that protrudes inside the cylinder.

Other useful contacts include *discs* and *cones* screwed onto the incoming copper rods and sitting quite close to the barrel ends, or fastened to current-carrying members in the cylinder ends. Such discs and cones can be chromium plated and quickly removable for cleaning.

The flexible danglers have the greatest all-around usefulness. Only when the parts are fragile and become injured by the contact balls or so heavy that the contact balls do not bury themselves in the load, should one consider other types, especially the disc or the cone. Whatever contact is chosen, it must have ample current-carrying capacity.

The Tank

For a general discussion of tank construction and linings, of heating and cooling devices and of tank ventilation, see appropriate chapters.

Tank size. With modern solutions designed for high-current-density operation, barrel solutions are run at high voltages and high current concentration. Because it has been impossible to reduce solution resistivities in proportion, the heating effect $I^2R = EI$ has risen to high values. It has, therefore, become difficult to cool solutions that must operated at or slightly above room temperature.

Table 5[6] shows the amount of current drawn in the average 14 × 30 in. barrel with 22 to 25%

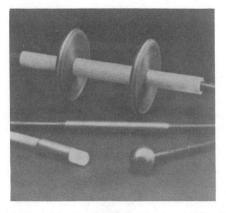

FIG. 15. Flexible danglers. *(Courtesy of Sterling Systems.)*

TABLE 5.[6] PERFORMANCE OF A 14 × 30 IN. BARREL

Solution	Amp	Heat Produced per Sec*			Time per 0.1 mil† average (min)
		joule	Btu	kcal	
Cadmium	500	5750	5.4	1.4	9
Zinc cyanide	600	6900	6.5	1.6	10½
Copper cyanide‡	280	2520	2.4	0.6	23
Brass	250	2880	2.7	0.7	25
Silver	300	3450	3.3	0.8	9
Stannate tin	300	3450	3.3	0.8	29
Watts-type nickel	200	2300	2.2	0.5	40

*Neglecting polarization at anode and cathode, which would reduce the listed values by as much as 10 to 20%.
†2.5 μ.
‡8–10 volt.

perforation, 60 sq ft load and 11 to 12 volts across the tank in common solutions.

In view of the large amount of cooling required for the zinc and cadmium solutions, they need large tanks so that the heat losses through the walls and free solution surface becomes large. In many localities the available water for the cooling coils is not very cold during the summer and vaporizing or refrigeration machines may be resorted to. However, up to about 200% more cooling effect may be obtained from a given volume of water passing through a cooling coil if the plating solution is pumped past the coil.

Plating barrels 14 × 30 in. operating in cyanide zinc, in regular production, are now drawing 900 to 1000 amp each at 16 to 18 volts.

Tanks for warm and hot solutions may well have somewhat smaller absolute size to conserve heat, though not so small that the solution composition varies rapidly. Tank working volumes of 200 to 300 gal are being used for 14 × 30 in. cylinders.

Multiple-Cylinder Tanks. Where production requires a number of barrels in the same solution, one may advantageously use one, proportionally larger tank for several cylinders, thereby reducing the number of solution analyses and corrections. However, if the solution is troublesome to operate and likely to require tedious treatments, it is better to have at least two units, so that one can be kept in operation at all times.

Double Stacking Cylinders. Plating barrels with two cylinders stacked vertically can be used in the electronics industry. These barrels give an excellent performance with electrolytic nickel deposition.

Anodes Rods and Anodes. When a plain steel tank is used for zinc or cadmium cyanide solution, the *anode rods* may be of steel immersed beneath the solution surface and in electrical contact with the tank. This assures positive anode contacts at all times. Stannate-tin, cyanide-copper and brass solutions, however, operate with polarized anodes and anodically charged steel would take part in the current transport. In these the tank must be neutral, and the anode rods must be above the solution level and insulated from the tank. To prevent plating on the tank sides, the anodes should be several inches away from them[8].

Special attention must be paid to the current-carrying capacity of anode rods in multiple-cylinder tanks. Frequently two rods, one for each cylinder, are used between cylinders.

For maximum current, the *anodes* should curve close to the cylinder to shorten the path of the current. For this purpose curved solid anodes may rest on insulated supports wedged into the tank, whereas curved ball anode holders may be tied together at their lower ends with steel wire.

Negative Contacts. Each of the four stud contacts on the hanger arms rests in a mating member, a so-called *contact saddle*, that is carefully insulated from the tank (Fig. 16). The four saddles are shimmed to fit the stud contacts. Two of the saddles may be shaped as in the figure to prevent the barrel from "jumping" when gears are worn; the other two saddles are plain V-blocks. At the bottom of the V is a drainage channel to reduce caking of solution.

The conical shape on each stud contact helps to seat the barrel in the saddles.

Spring contacts are also used. In that case a plain piece of bar is substituted for the stud

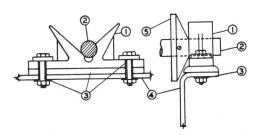

FIG. 16. Contact saddle (1) for stud contact, (2) on hanger arm of perforate horizontal barrel, (3) insulation, (4) tank, (5) conical shape fastened on stud contact for easy seating of barrel.

contact. The receiving member on the tank rim flares out at the top to facilitate locating the barrel. The position of this member and the spring pressure should be adjustable.

The saddles or spring contacts that carry current (on some barrels current enters only one end of the cylinder) are connected with bus bar, that ultimately leads to the negative pole of the current source. (The anode rods are, of course, similarly connected to the positive pole.)

Motor and Speed Reducer. Each cylinder requires from $\frac{1}{4}$ to $\frac{1}{2}$ hp, depending on the weight of the load that the cylinder is intended to carry.

The gear-motor or the motor with separate speed reducer is mounted upright on a shelf or sideways on a vertical support, which in turn is welded to the side of the tank. In any case, the driving unit should be adjustable in all directions and preferably rubber mounted or the like. It should be totally enclosed to prevent damage by solution dripping from passing barrels.

Multiple-barrel tanks are supplied with a separate motor for each cylinder or with one motor driving a shaft that in turn drives a separate gear-box for every cylinder. Long shafts are preferably center-driven. This system of connected gear-boxes is more difficult to adjust to smooth motion.

A push-button *starter* may be located on the tank in a convenient position.

Individual Hoist. When production is limited and only one or two barrels are used, they may be supplied with individual hoists fastened to the tank. The barrel is lifted on a chain attached to the horizontal hanger member by means of a hand-operated crank supplied with a ratchet stop.

The hoist construction varies widely. It should be substantial without interfering with the loading and unloading of the cylinder; the crank should be convenient to the operator's right hand; the ratchet should be heavy and the teeth on the ratchet wheel large.

Portable Barrels

Small horizontal barrels are often used for small odd lots of work. The gear-motor is mounted on the frame above the cylinder. The cylinder is driven by means of a vertical drive shaft with a friction wheel that is forced against the cylinder end by a belt running around the cylinder from a horizontal drive shaft, or by the same kind of gear system as the large units.

The negative contacts are formed into hooks instead of stud contacts, and the barrel is simply hung on the cathod rod of a still tank during plating.

When made of suitable materials, this barrel may be taken through the complete plating cycle, just as a big horizontal barrel.

LAYOUT AND AUXILIARY EQUIPMENT

The plating cycle for barrel plating includes the usual steps of cleaning, rinsing, acid dipping or equivalent, rinsing, plating, rinsing (cold and hot), and drying, and the usual variations thereof.

There are three principal methods of handling the work through the preplating steps, namely (1) in baskets, (2) in separate rotating barrels similar to the plating barrels, or in the plating barrels themselves, and (3) in multicompartment rotating drums.

Basket handling requires the smallest investment and is used almost exclusively when production is low or when the parts are fragile and easily damaged by tumbling. Nothing is gained by electrocleaning or electropickling parts in baskets; the current reaches only the outside layer of parts.

Handling in separate barrels is probably the most popular method for medium and large production. With the advent of nonconducting materials that resist all solutions and temperatures commonly encountered in the cycle, the plating barrel itself is often taken through the whole cycle. Contrary to all-metal cylinders, they can be electrified in any solution, which permits rapid and efficient cleaning and acid treatments.

Handling in multicompartment rotating drums is suitable only for large production. It is wholly automatic except for the loading and unloading of the plating barrels, and requires a minimum of labor, but it has not yet received the recognition it deserves. To this writer's knowledge, insufficient work has been done to adapt this type of barrel to the plating itself.

Cleaning and pickling by tumbling is up to 4 times more rapid than treatment in baskets, and more uniform results are obtained.

The basket should normally hold either a half or a full barrel working load; thus one should use either two 12 in. diameter × 12 in. high basket loads or one 16 in. diameter × 16 in. high basket load for a 14 × 30 in. barrel. These baskets provide room for the work to be shaken without its spilling.

TABLE 6. WIRE SCREEN SCHEDULE

Mesh*	Standard			1 Size Ex. Heavy		
	B and S Gauge	Dia mil†	Opening mil	B and S Gauge	Dia mil	Opening mil
2	11	92	408	10	105	395
3	13	72	261	12	80	253
4	14	63	187	13	72	178
5	15	54	146	14	63	137
6	17	47	120	15	54	113
8	18	41	84	17	47	78

Mesh*	2 Sizes Ex. Heavy			3 Sizes Ex. Heavy		
	B and S Gauge	Dia mil	Opening mil	B and S Gauge	Dia mil	Opening mil
2	9	120	380	8	135	365
3	11	92	241	10	105	228
4	12	80	170	11	92	158
5	13	72	128	12	80	120
6	14	63	104	13	72	95
8	15	54	71	14	63	62

* Mesh = Number of openings per sq in.
† 1 mil = 0.001 in. = 0.0254 mm.

FIG. 17. Well constructed (welded) basket. *(Courtesy of Hanson-Van Winkle-Munning Co.)*

The basket should be cylindrical and have a swingable handle for ease in loading and unloading. It should preferably be made of wire mesh or expanded metal for ease of circulation of liquid through the load. For the same purpose the mesh should be large, almost as large as the parts when convenient. Long basket life results from the use of heavy gauge wire and of a construction and size of supporting member that prevents distortion during the rough handling. A list of available wire screen used in basket manufacturing is given in Table 6[7]. Good basket construction is shown in Fig. 17.

Low-carbon steel is a satisfactory *basket material* for the usual alkali-acid cleaning cycle and for chromic acid bright dips for cadmium plate, as long as the acid is cold and the acid treatment is short. With hot acids or long treatments it pays to use "Monel" metal. The very corrosion-resistant "Nichrome" also resists solutions of high nitric-acid content. Plastic-coated baskets are suitable when nitric acid bright dips for zinc or cadmium plate are part of the cycle.

Figure 18 shows a simple *layout* for one barrel with individual hoist[6]. Often a 10 to 15 gal earthenware crock is substituted for the rubber- or plastic-lined acid tank. The basket is moved by hand from tank to tank, and emptied into the barrel cylinder as it sits in its tank by means of a simple sheet-metal hopper (Fig. 19). Following plating, the barrel is lifted by means of the hoist and the contents dumped onto a per-

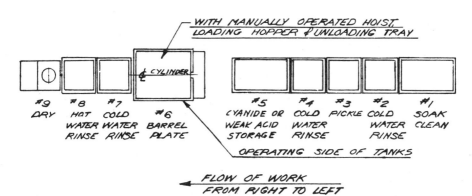

FIG. 18. Straight-line plant layout for one (or two) perforate horizontal barrel with individual hois

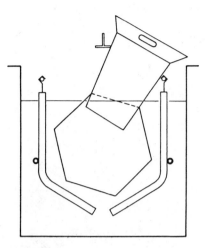

Fig. 19. Schematic cross section of perforate horizontal barrel with loading hopper.

forated chute (Fig. 20). This chute is emptied into two baskets, that are taken through the rinses, an eventual bright dip and the dryer.

A similar layout serves two barrels. Where more than two barrels are used, one should consider employing barrels without individual hoists, and transport the baskets as well as lift the barrels with pneumatic or electric hoists from monorails, but in that case one usually does better with barrel operation throughout.

Agitation of the basket is desirable in all tanks. For short operations, shaking by hand or raising and lowering the basket with a hoist from the ceiling will serve. For longer treatment time in the cleaner or during scale removal by pickling, the tank is preferably supplied with an oscillating work rod, advantageously one that moves the basket up and down.

The dryer should usually be of the centrifugal type and big enough to dry either one-half or the whole barrel load at a time. Such dryers should be heated; steam or electric heat is preferable. The speed of rotation is 400 to 750 rpm, the higher values for small units. In order that a centrifuge last, it should have good dynamic balance and either float or have strong and big enough bearings to take the side thrust of an unevenly distributed load. The direction of rotation should be reversible, and the machine should have a powerful foot-brake. The drying of some parts may be facilitated by sudden stoppage and reversal of rotation. That method, however, is not sufficiently effective to empty deeply cupshaped parts or such parts as lipstick or small shell cases or the blind holes in hollow rivets of water. They are best dried in a rotary dryer. Work of thin section that holds insufficient heat from the hot-water rinse requires a dryer with a hot-air blast.

Several centrifugal dryers on the market do *not* meet the foregoing requirements as to strength.

Some surfaces that stain easily, especially copper- and brass-plated surfaces, may best be dried with specially graded corn-cob meal. *Sawdust of hardwood* (not oak) may also serve. Such drying is best carried out in a tumbling barrel, either horizontal or oblique, that is arranged for separation of the cob meal or sawdust during rotation.

After use, the cob meal or sawdust must be dried thoroughly before it is reused, suitably by spreading it in a thin layer on a steam table. Great care must be taken in previous rinsing so that the drying medium does not become contaminated. With time such contamination will occur, and then the medium must be discarded or it will cause the work to stain when wet.

Barrel Handling

Figures 21 and 22 show typical layouts of two medium-size barrel plating plants[6], in which *all steps except the final rinsing and drying are handled in barrels*. In the first one, only one plating bath is used; in the second one, three different metals, e.g., zinc, cadmium and nickel are deposited.

Fig. 20. Unloading chute below perforate horizontal cylinder for discharge into baskets. *(Courtesy of Crown Rheostat and Supply Co.)*

Barrel Capacity

It should be noted that metal barrels may be almost filled with work, in contrast to plating

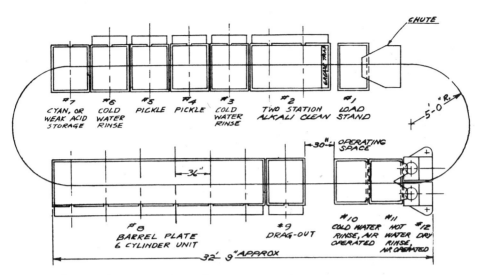

Fig. 21. Return-type plant layout for perforate horizontal barrels in one multicylinder tank (or in several tanks with the same plating solution). *(Courtesy of the Udylite Corporation)*

barrels which use d-c; thus, depending on the work, a 14-in. diameter metal barrel may hold almost twice as much, and an 18-in. diameter about three times as much as a 14-in. plating barrel of the same length. There is no objection to metal barrels of large size so long as they do not unduly damage the work, and that the hoist and monorail are strong enough. For convenience in loading the plating barrel, the capacity of a metal barrel should be a multiple of that of the plating barrel. Metal barrels may be made multiple compartment, usually two or three

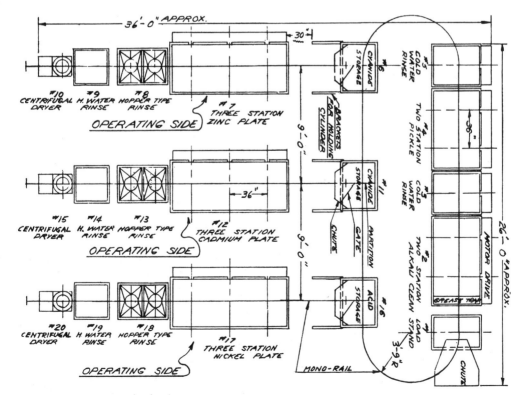

Fig. 22. Plant layout for perforate horizontal barrels in several multicylinder tanks with different plating solutions. Note common cleaning line. *(Courtesy of the Udylite Corporation)*

Fig. 23. Horizontal barrel line. (*Courtesy of Motorola Inc., Phoenix, Ariz.*)

compartments, then different kinds of work may be processed simultaneously without mixing. The loads which are put into these barrels may be pre-weighed, the weight being such as to correspond with an established surface area.

An example of a single manual hoist for a horizontal barrel plating line is given in Fig. 23. The maximum size barrel cylinder is 10 in. in diameter and 24 in. in length. Loading, cleaning, electroplating, rinsing, centrifuge drying (after a transfer to a drying basket) and packaging are done in this and similar production lines.

Construction of Barrels. The metal barrel is constructed substantially as a plating barrel except that it carries no d-c system. The heads are usually about $\frac{1}{4}$-in. thick, the perforated sides and door $\frac{1}{16}$ to $\frac{1}{8}$ in. Reinforcing ribs or stay-bolts give the barrel carrying capacity.

Barrel Materials

The horizontal bridge member of the metal barrel is usually of steel, the hanger arms, gears and cylinder of "Monel," which has high resistance to all common cleaning and pickling solutions. Other cylinder materials are stainless steel and copper-silicon-manganese ("Everdur") bronzes. All of these resist alkaline cleaning solutions very well, but whereas "Monel" and "Everdur," especially the latter, have high resistance to sulfuric and hydrochloric acids, stainless steel dissolves rather rapidly in the acids, unless they contain a substantial amount of copper salt or oxidizing agents. Stainless steel will serve in concentrated bright dips for copper and copper alloys. Hanger arms and gears may be of the aforesaid materials or of leaded phosphor bronze casting. The latter have good resistance to alkaline cleaners and sulfuric acid solutions, but are rapidly corroded in oxidizing acids and in commercial hydrochloric acid.

As has been mentioned previously, barrels made of certain plastics may be taken through the whole cycle. Such barrels can be employed for electrocleaning and electropickling when desired. When contamination or drag-out of the plating bath is feared, separate plastic barrels may be used for the cleaning-and-pickling part of the cycle, after which the work is transferred to other barrels for plating.

Loading Stand

The barrel may be loaded on a simple loading stand supplied with either a chute or a hopper (Fig. 24). The chute is operated by the regular hoist on the monorail or by means of a separate pneumatic lift.

Cleaning, Pickling, and Rinsing Tanks

These should all be motorized. A grease trap should be placed on one side of the cleaner tank (Fig. 25). At the lower end of the trap is an outlet pipe joined to a centrifugal pump connected to

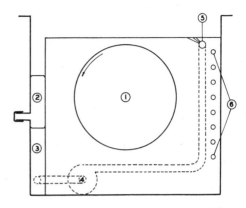

Fig. 25. Schematic cross section through barrel cleaning tank.
 (1) Cleaning cylinder
 (2) Grease in trap
 (3) Cleaning solution in trap
 (4) Centrifugal pump
 (5) Perforated pipe for forcing cleaning solution on surface into trap
 (6) Heating coils

a horizontal pipe near and across the side of the tank opposite the trap. This pipe, which is slightly above the solution line, has a number of holes in a slightly downward direction. As the solution is pumped through the holes, it washes the surface free of grease, which then cannot adhere to the barrel and work as they are removed through the solution surface. The grease trap is supplied with a drain cock half way up the trap so that the grease can be drained off at intervals. The pickling tank may also be furnished with a grease trap.

Cleaning tanks for plastic barrels should be supplied with anode rods, steel anodes, and automatic temperature control.

Transfer

The transfer from cleaning to plating barrel may conveniently be combined with a storage tank with weak cyanide solution for cyanide plating baths or weakly acid solution for acid plating baths. Such solutions will prevent rusting and other discoloration of the parts during transfer and during storage of the excess metal-barrel load that remains for the next plating barrel.

A suitable unit for a metal barrel of the same diameter as the plating barrel consists of a tank with a perforated chute that is divided in the middle. The spout of each half has a gate to hold back half of the load. To empty the chute it is

Fig. 24. Loading stands with chute (top) or hopper (bottom) for perforate horizontal barrels. (*Courtesy of the Udylite Corporation*)

FIG. 26. Storage tank for transfer of work from metal cleaning barrel to plating barrel. Self-contained pneumatic hoist is optional. (*Courtesy of the Udylite Corporation*)

FIG. 27. Hopper-type rinse tank for transfer of work from perforate horizontal barrel into baskets.[4] Note stand in bottom of tanks for locating of baskets and for free circulation of the rinse water.

raised either by means of the monorail hoist (Fig. 26) or by means of a separate hoisting device such as a pneumatic cylinder attached to the tank and chute.

Final Rinses and Bright Dips

After plating, the barrel is raised out of the plating solution and should be rotated (by hand on manually operated lines) for drainage. It may then either be emptied into baskets or another barrel (or retained in the plating barrel in some cases) in either the dragout station or the following first rinse station. The after-plating steps should comprise a dragout recovery station, a minimum of two counter-flow rinse stations and a hot water rinse station. If a bright dip is required, at least two counter-flow rinse stations should both precede and follow it prior to the final hot water rinse station.

The cheapest arrangement as far as equipment cost is concerned is to have a hopper swingably mounted on the dragout recovery or following first rinse tank or compartment (Fig. 27). This hopper has one or two openings that lead into one or two baskets depending on the size of the basket and the centrifugal dryer. The baskets are then taken by hand or hoist through the remaining tanks into the dryer.

One may also supply each tank and compartment with a perforated chute as in Fig. 25, dump the work out of the plating barrel into the first chute and transfer it from chute to chute and into the centrifugal dryer or dryers or the rotary dryer by means of the overhead hoist or an individual pneumatic lift on each tank (Fig. 28). Such pneumatic lifts may be tied together so that the operator need turn only one valve to set all chutes in operation, one after the other. The pneumatic-lift sequence may be timed and even arranged for moving the work up and down in the tanks.

Tanks and chutes are made of steel, except in bright-dip tanks and, when rust discoloration on the work is objected to, in the hot rinse tank. In the former case linings and coatings of rubber or plastic may be used; in the latter case the tank may be lead lined, the heating coils may be lead or stainless steel, and the chute may be stainless steel.

To prevent stains from plating solution residues on the work removal of all the plating solution should have been accomplished in the preceding cold rinses. A slight flow of water through the hot rinse is recommended to take care of inevitable contamination. If all staining is to be avoided, it may be necessary to use distilled or deionized water in the final rinse.

Dryers

Centrifugal dryers have already been discussed under *Basket Handling*.

Fig. 28. Pneumatically operated chute-type rinsing unit with two centrifugal dryers. (*Courtesy of Crown Rheostat and Supply Co.*)

The *rotary dryer* is becoming increasingly popular, especially in large production because of the labor saving obtained. It consists of an enclosed drum of a length 4 to 10 times its diameter, often slightly inclined from the entrance to exit end. To the inside of the drum is welded a helix that moves the work forward and frequently ribs that turn the work over. Hot air is forced through the drum countercurrent to the work. On the entrance end is a feed hopper, underneath the exit end are placed the receiving containers.

Because of the tumbling action, the rotary dryer is particularly suited to cupshaped articles and articles with deep closed holes and to flat pieces that tend to stick together when wet. On the other hand, precision parts that are easily bruised or distorted should be dried by centrifuge.

In determining the capacity of the drum dryer from the pitch of the helix and the speed of revolution, one should take into consideration that about a 2-min interval is required to prevent mixing of loads.

The occasional need for sawdust drying has been discussed under Basket Handling, Dryers. The rotary dryer has been adapted to this purpose. The first part of the drying cylinder is made of screen through which the excess water runs. The parts continue down the helix to a cylinder part made of solid metal where they are mixed and dried with corn-cob meal. The last part of the cylinder is again made of screen, which separates the moist meal from the dry parts. The meal is taken by an elevator up an inclined steam table where it dries, and returns through a chute to dry more work. While steam is mentioned as the source of heat for the above dryer, these machines are also available with gas heat. However, with direct gas heating, the products of combustion may tend to tarnish bright plated parts.

Hoists

Both pneumatic and electric hoists are used. The former requires that compressed air be already available, that headroom be ample and that the radius of action be limited to about 25 ft, the convenient length of air hose. Air hoists are particularly suited to basket handling. In use, they should be watched for oil drippage.[9]

On electric hoists, hooks and controls should be well insulated to prevent electrical shocks to operators.[9] Alternating-current motors are preferred as they give a constant lifting speed independent of the load.

A suitable lifting speed is 6 in./sec. The height of the lift should take the barrel well above the equipment in the line.

Hoists used only on straight-ways may be supplied with outrigger arms to facilitate moving them along the track; those used also on curved tracks are preferably moved by the hoisting cable or chain, as the outrigger arm multiplies the power required to move the hoist and gets in the way of cable or chain handling.

Monorails

Monorails are usually of the single I-beam or Coburn type with the individual wheels pivoted so that the wheel-carriage can turn curves. The track-hangers are spaced at distances depending on the load. Monorail curves should have a radius of not less than 4 to 5 ft. Monorail switches should be avoided wherever possible since they slow down the operator; transfers with storage tanks (see above) are used instead. Where switches cannot be avoided, as in the case of plastic barrels used for both cleaning line and several plating lines (Fig. 27), they should serve the lines with the lowest production.[6]

Bridge Cranes

Bridge cranes are heavy and slow to start and stop, and the hoist must be positioned from two directions to lift or lower a barrel. All this requires time and effort. Thus it is generally more efficient to lay out separate lines for different plating solutions and use monorails than to employ a bridge crane to bypass and not contaminate an impurity sensitive second plating solution in the same line. Only when space does not permit monorail arrangement should a bridge crane be installed.

EQUIPMENT REQUIREMENT FOR BASKET AND BARREL HANDLING

Plating Barrels

To calculate the equipment required one must know the amount of work to be plated per unit time: day, month or year. Four facts should be known about each part, namely, (1) number, (2) bulk volume, (3) area, and (4) weight. In addition one must know (5) the particular minimum thickness of plate that is required on significant surfaces of each part and (6) by test, experience, or from Table 2 of this chapter, the average thickness needed to meet the minimum thickness requirements.

The most popular sizes of plating barrels are the 14×30 in. and the 14×36 in.; the former holds 1.25 and the latter 1.50 cu ft of work approximately. The number of barrels required to process a given quantity of work per hour is $wt/60 \times l$, where w = lb of work to be processed per hour, t = time of cycle in minutes, and l = load in the barrel in pounds.

Example: Supposing 1000 lb or small parts are to be cadmium plated per hour to a batch average thickness of 0.3 mil in 14×30 in. barrels used through the whole cycle (except operations 11, 12, 13 and 14 when the work will be processed in baskets). The surface area is 0.6 sq ft/lb; the load 100 lb equals 60 sq ft, and the volume of load about 1.20 cu ft; and the barrel draws 500 amp (see Table 3). The cycle may be as follows:

Operation	Time, min (including transfer)
1 Load	1.5
2 Clean	1
3 Cold water rinse	1.5
4 Acid dip	3
5 Cold water rinse	1.5
6 Cyanide	1.5
7 Cadmium plate	28
8 Drag-out	1.5
9 Tumble rinse	1.5
10 Unload, into hopper rinse	1.5
11 Bridge dip	
12 Cold water rinse	
13 Hot water rinse	
14 Dry	
Total, 1 to 10 inclusive	46.5

Substituting the above values in the foregoing formula we have:

$$\frac{1000}{100} \times \frac{46.5}{60} = 7.5, \text{ say 8 barrels required.}$$

The number of plating stations will be:

$$\frac{1000}{100} \times \frac{28}{60} = 4.66, \text{ say 5 stations.}$$

In the above, the surface area in square feet could be substituted for the weight in pounds, so could the volume in cubic feet. By the same token, this formula may be used for any work container through any cycle.

Auxiliary Equipment

The usual cleaning cycle, including loading and unloading, takes from 10 to 30 min, depending on how much grease and scale must be removed and whether the work is tumbled and electrolyzed, simply tumbled, agitated in baskets, or simply immersed in baskets. Assuming that for the same kind of work the four different treatments require 10, 15, 15, and 25 min respectively, one plating barrel produced 6, one metal barrel $2 \times 4 = 8$, one agitated basket $1/2 \times 4 = 2$, and one nonagitated basket $1/2 \times 60/25 =$ 1.2 plating barrel loads per hour.

Thus if we were to feed 4 plating barrels with a plating time of 30 min, we must supply 8 plating loads, which requires $8/6 = 2$ plating barrels, $8/8 = 1$ metal barrel, $8/2 = 4$ agitated baskets or $8/1.2 = 7$ nonagitated baskets.

In this case, one station is sufficient in each tank of the cleaning cycle when barrel cleaning is used. As a rough but time-tested rule for use when an exact time cycle cannot be predicted, up to 5 plating barrels may be supplied with work from a single-station cleaner and pickle, 5 to 10 plating barrels from a 2-station cleaner and pickle and one metal barrel will prepare work for 2 to 3 plating barrels.

For basket cleaning, the cleaning and pickling tanks should be large enough to hold a total of 15 baskets. This number may be distributed according to the time of treatment in each tank. With space for 2 baskets in each rinse tank, there is room for at least one extra basket in every tank so that the operator simply can exchange baskets as he moves down the line. In such an installation the storage tank should be of ample size and hold at least 4 baskets.

Considering the cycle steps after plating, one

should always install a dragout recovery tank following the plating tank whether barrels or baskets are carried through the after-plating steps. In the past, this has not been regularly practiced. By so doing one can save chemicals, conserve water and reduce the size and cost of the waste treatment installation, provided the drag-out recovery and rinses are properly designed and operated.

The after-plating treatment tanks can handle 20 to 30 barrel loads per hour. Similarly, a centrifuge will dry 20 to 30 basket loads per hour, assuming that hot air is used for light parts. For the sake of labor saving, one frequently uses two centrifuges side by side, even if one would do the work.

Plant Layout

Reference is made to the layouts in Figs. 18, 21, 22 and 29.

In addition to what has been indicated in the foregoing, the following rules should be observed:

(1) Tanks should be laid out and used in such an order that drag-out from one solution will not contaminate another. Neither should it fall on electric motors or electric connections.

(2) Ample aisle space should be provided, at least 4.5 ft for one to two workmen, 5.5 ft for several.

(3) Electrical equipment, e.g., motors, rectifiers, and generators should be accessible and not jammed against walls.

(4) Generators and rectifiers should be at some distance from hot, moisture- or chemical-saturated points in the plant, but placed so that long runs of bus work with accompanying voltage drops be avoided.

(5) In large plants responsibilities should be divided among the workmen and reflected in the layout. Thus one man or group may handle a cleaning line, another a zinc line and a third a copper-nickel line.

MULTICOMPARTMENT ROTARY EQUIPMENT

This kind of equipment, like the rotary dryer and the imperforate horizontal chromium-plating barrel, consists essentially of horizontal cylindrical drums supplied on the inside with helical guides that move the work forward as the drums rotate. The drums are perforated and may be partially immersed in tanks holding the required solutions and rinse waters.

Three principal types are used and differ in the arrangement of transfer from one drum to another. Their construction is indicated in Fig. 30. At the top, one end of each drum takes the form of a cone, into which the helix continues. When the parts have climbed to the end of the cone they are discharged into a chute that drops them into the next drum. In this drawing the drums are mounted on a common center shaft by means of spiders, attached to the helix so that they do not interfere with the movement of the parts. The shaft runs in bearings between the drums, and the bearings are supported by floor stands.

In the middle of Fig. 30 is shown a second type in which both ends of the drum extend into cones. The conical ends are joined with short cylindrical portions. The helix continues through cylinders and cones. In this manner the work is brought from one drum to another gradually and without being dropped. The short cylindrical portions are supplied with trunnions that run on sets of wheels supported between the drums.

The third type, at the bottom of Fig. 30, has no conical sections; hence it requires very little space, and the work does not travel long distances under conditions when it is likely to dry. The work is scooped up at the end of each cylinder by means of pockets that empty it onto chutes leading into the next drum. This type of transfer, like the first one is likely to damage fragile parts and delicate threads.

Independently of the type of machine, the drive should be applied at the center of the unit to reduce torque. A common method is to place a large gear on a section between tanks. This gear is driven by means of a smaller gear attached to the shaft of a motor-speed reducer.

The drums may be hooded and ventilated as desired.

One common weakness of all these types is that they cannot be immersed very deeply, and the work is likely to be carried out of the solution during rotation, whereby the actual time of treatment is reduced. This difficulty can be overcome by supplemental spraying where there are no central shaft and spiders or any cylindrical portions between drums. Treatment solution is simply pumped out of the tank into the drum through a pipe with spray nozzles, and rinse water may be introduced by the same means. The nozzles are directed against the parts that rise out of the liquid in the tank.

The whole treatment may take place by *high pressure spraying only*. The perforated drums are then surrounded by imperforate drums that

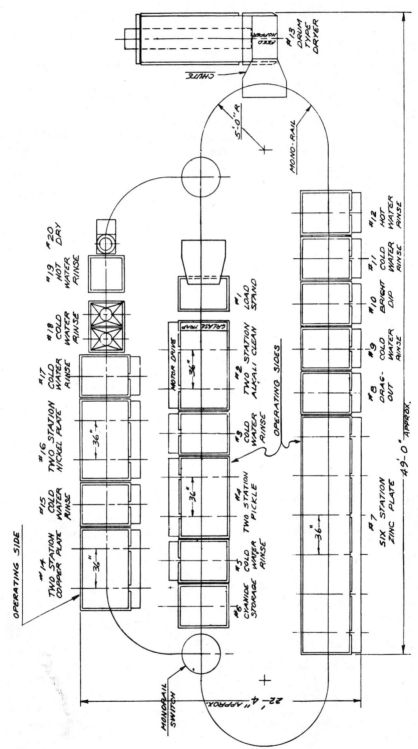

FIG. 29. Plant layout for two plating lines served by one cleaning line. Plating barrels used through both cleaning and plating lines. (*Courtesy of the Udylite Corporation*)

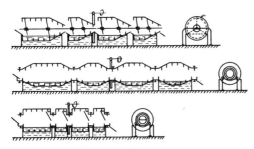

FIG. 30. Schematic representation of three types of multicompartment rotary equipment for, left to right, cleaning, rinsing, pickling, and rinsing prior to barrel plating.

serve to collect the spray and lead it back into sumps for respraying. These sumps are arranged with grease traps and also with heating means for hot treatment solutions.

The multicompartment rotary equipment may not only be used for pretreatment of the parts but also for after-treatment, including drying. It may also be employed for other treatments not connected with electroplating.

THE IMPERFORATE OBLIQUE BARREL

Whatever the shape or size of this type of barrel (Fig. 3), it is much less efficient than the usual horizontal barrel. Thus, (1) many parts do not tumble as well, (2) a small proportion of the total load is in the path of the current, (3) the anode area is small relative to the work area, which easily causes anode polarization with rapidly dropping metal content and considerable voltage losses, and (4) the volume of solution is small, which leads to overheating of such solutions that one wishes to run at room temperature or somewhat higher when any substantial voltage is used. Items (2) to (4) add up to a severe limitation on the amount of current that can be drawn, i.e., on the productivity of the barrel; hence this type of barrel does not compete with barrels with perforated cylinder, as indicated in the discussion of Barrel Types.

Construction Features

The Cylinder. The ratio of height to average diameter varies between 0.85 and 1.5 and the angle with the horizontal is 35 to 55°, usually close to 45°. The load volume is about 5 to 15% of the cylinder volume, depending on the cylinder shape and on how well the work tumbles. There is no evidence of barrels of larger height-to-diameter ratio running at smaller angles with the horizontal that approach the horizontal barrel in relative capacity or ability to tumble the work.

All sorts of shapes are in use: those that have a larger top diameter than bottom diameter and vice versa, those that have an inward flange, with reduced opening, to hold the solution instead of greater height, etc. No substantial advantages or disadvantages result from these details. For example, a cylinder that flares out toward the top will spread the work somewhat better, but the volume of solution is less than in a regular cylinder of the same area of wall plus bottom.

An hexagonal or octagonal cross section or 6 to 8 ribs on the inside assists in tumbling the work.

The cylinder is usually made of sheet steel, lined with plastic or rubber. It should be light but rigid when removed for unloading. For this purpose the top is often reinforced. Wooden cylinders are disappearing.

Current-Carrying Features. Both buttons in the cylinder bottom and flexible danglers through the top opening are used for cathode contacts. Reference is made to the comments under Horizontal Barrel. However, lack of cleaning of the buttons is possibly not as troublesome in these barrels. The buttons function as bottom ribs in spreading and tumbling the parts.

The anodes are either discs, circular or elliptical, or baskets with ball anodes. It should be recognized that the over-all immersed dimensions are more important than any increase in surface by corrugation because the bottoms of the corrugations supply little current relative to the tops until the tops polarize.

Naturally, the larger the anode, the smaller must be the bulk of the load in order that short-circuit or burning on protruding points be avoided. Successful plating depends to a considerable extent on a workable compromise between these dimensions.

When the same kind of parts are plated continually, it may pay to design a special anode that conforms to the load pattern during rotation. Such anodes require rigid suspension.

The anode and the flexible danglers are fastened to rods that extend above the cylinder. The rods are removably attached to current-connected members, fastened to but insulated from a support. In one elegant construction, that permits a relatively large anode area, the anode rod is a tube through which passes the flexible cable of the cathode contact.

The cathodic button contacts that go through

the cylinder bottom may continue into the flange of a keyed sleeve over the drive shaft. Brush or other connection is made to the sleeve, which is insulated from the shaft, but rotates with it.

The Drive. The flexible-dangler type of cylinder sits solidly in a cradle that is shaped to fit the cylinder, and this cradle is rotated by the drive shaft. The button-type of cylinder may be connected to the shaft by means of the aforesaid sleeve. It is important that the bearing or bearings for the shaft that carries the cylinder be sturdy enough to withstand the oblique load.

A number of constructions are available in *bench-type* barrels. In general they can be classified according to whether the cylinder angle is fixed or not. The former constructions are the simplest; the motor and speed-reducing means (or gear-motor) and the support for the anode rod and the eventual cathode rod are simply mounted in position on a base plate.

In the adjustable constructions the aforesaid plate may be hinged on top of a second plate and supplied with a hand screw for adjustment of the angle. Alternatively the gear-reducer shaft may carry the cylinder and the electrode support and swing around the horizontally extended motor shaft, which is journaled in a channel attached to the motor base plate (Fig. 3). None of these mechanisms is intended for use in emptying of the cylinder; as in the fixed-angle construction, the cylinder is lifted out of the cradle or off the drive shaft by its handle or rim.

Unless the base plate is to be bolted to a table or other support, it must extend forward underneath the cylinder when the motor and speed reducer are light; otherwise the loaded cylinder will tip forward and spill the solution.

Some barrels are tiltable for emptying (Fig. 31), the whole assembly fastened to a horizontal shaft with journals supported by a U-shaped floor or bench stand so that the motor-speed reducer assembly somewhat more than counterbalances the cylinder and its contents. The tilting is done by means of a lever or hand wheel, and a ratchet arrangement holds the barrel in position.

Often the driving gears or belt pulleys are interchangable so that two or more cylinder speeds may be had. The low speed is commonly 5 to 7 rpm for cylinders with an average diameter of 6 to 15 in., yes, even up to 30 in. This narrow range of low speed is used for plating, and no attention seems to have been paid to circumferential velocity. The higher speeds are employed for tumble cleaning (see below). For a discussion of the effect of speed, see Perforate Horizontal Barrels.

FIG. 31. Imperforate oblique barrel, tiltable for unloading. (*Courtesy of Rheostat and Supply Co.*)

The *motor size* for single units usually ranges from $1/6$ to $1/3$ hp, which is ample. Such motors are single-phase and may be connected to any convenient outlet.

Multiple Units. Up to 12 to 15 cylinders may be run from one motor and speed reducer. Each has its own clutch or equivalent device so that it can be stopped and started independently of the other cylinders.

Auxiliary Equipment

Solution Storage. Because the composition of the plating solution is seldom maintained in the oblique barrel, one must have a solution storage tank of a volume at least 5 to 10 times that of the barrel cylinder. The larger the volume, the less often one needs to analyze and correct the solution, and the better is the chance of uniform plating.

The cylinder is emptied, by hand or by tilting, through a screen-bottomed or perforated tray, resting in the upper part of the storage tank. Fine needles that would pass through a screen may be dumped onto a slightly inclined solid tray with perforations only at its low end. The tray should be small enough and supplied with handles so that it can be used for rinsing. The cylinder is usually loaded with work and with solution from the storage tank after it has been placed in plating position.

Cleaning Equipment. Odd or small lots of work are frequently *basket* cleaned. (See Horizontal Barrel.)

Work that because of its size is plated in oblique barrels is usually cleaned in *oblique barrels*. For this purpose one may conveniently use the multiple barrels mentioned above. The cleaning cylinder may be of the same construction as the plating cylinder, except for eventual button contacts. Tiltable barrels may be supplied with cleaning cylinders that have tight-fitting covers and that are operated as *horizontal barrels*.

Cleaning and pickling cylinders are supplied with storage tanks and with trays for separating the work and the solution. The tanks are heated as required.

Drying is usually with dry corn-cob meal or sawdust. (See Perforate Horizontal Barrels.)

THE PERFORATE OBLIQUE BARREL

General

The cylinder has a shape that is similar to that of the imperforate cylinder, but being perforated it will not handle as small parts.

Immersed in a tank of solution and with outside anodes, curving around the immersed part of the cylinder (Fig. 4), the cylinder itself need not be constructed for large solution volume or for large anode surface, and all attention can be directed toward effective tumbling and spreading of the work. In these respects, it is considerably better than the imperforate oblique cylinder although not as satisfactory as the horizontal perforate cylinder, especially for some types of work.

No difficulties are encountered with the temperature inside the cylinder as long as a part of the top opening is below the solution level. In this respect it is superior to the usual horizontal perforated barrel, not to speak of the oblique imperforate barrel. However, the bath temperature rises and must be controlled as in the perforate horizontal barrel.

Single barrels are used primarily for small and medium size lots and not for high production work. For plating of many small lots, several small cylinders may be operated in one tank.

This type of cylinder, however, finds its most important application in return-type fully automatics, to which it is well suited. (See Fully Automatic Barrels.)

Construction Features

The Cylinder. Two designs that are being widely used in fully automatics are shown in Fig. 32. The bottom one has two steps in the bottom

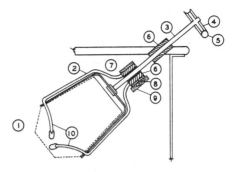

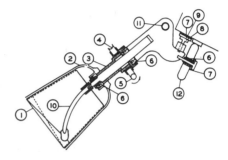

FIG. 32. Designs of perforate oblique barrels; top, arranged for hand operation as in Fig. 33; bottom, for fully automatic operation.

(1) Cylinder
(2) Bracket
(3) Drive shaft
(4) Gear
(5) Worm
(6) Bearing
(7) Insulation
(8) Contact
(9) Negative bus
(10) Flexible contact
(11) Bracket to carrier
(12) Wheel which rises on hump cams for transfer

that help turn the work. The one at the top has a fairly long conical section, which during the rotation in plating position simulates a horizontal cylinder. Both may be supplied with inside ribs for more efficient tumbling. The top of the cylinder is fastened to a 2- to 4-prong bracket, which at its center forms a sleeve. The sleeve is fastened to the drive shaft, which is concentric with the cylinder.

The same *materials* and the same *perforations* are used as for perforate horizontal cylinders. Not only the sides but also the bottom should be perforated.

Current-carrying Features. The *current* may be brought to the work from a bus by means of a contact shoe. This shoe is attached to a large bearing for a sleeve that is fastened to, but insulated from the drive shaft (Fig. 32, top). The whole weight of the loaded barrel thus rests on the bus. From the shoe the current passes through the bearing to the sleeve, and from there through flexible cables that are held to the outside of the cylinder until they enter it through holes. At the free ends of the cables may be contact balls.

In a somewhat different construction the sleeve discussed above may be extended into an insulated contact frame surrounding the cylinder. This frame leads the current to the bottom of the cylinder, which it enters by means of button contacts (Fig. 4).

A number of variants are available. It is desirable, however, that the current does not pass through the drive shaft to the tank or to the motor.

The Drive. On a shelf on the side of the tank is placed a gear-motor, that drives a worm. The drive shaft is fitted with a corresponding gear, which in operation is flexibly pressed against the worm.

The motor size for cylinders handling 40 cu in—0.75 cu ft of work is $\frac{1}{20}$ to $\frac{1}{3}$ hp.

The speed of rotation is the usual 5 to 7 rpm.

Loading and Unloading. Frequently a lever system, e.g., as shown in Fig. 33, is used to move the cylinder out of the solution and tip it over the edge of the tank for unloading. It should preferably be possible to interrupt the movement when the cylinder has just left the solution so that it may be turned by hand for drainage back into the tank. The parts are dumped into one or two baskets which stand in a tray that collects any remaining solution.

Button-contact cylinders that sit in a contact frame may be disengaged from the frame when the cylinder has reached an upright position (Fig. 4). Supplied with handles, the cylinder may be taken directly through rinsing and centrifugal drying or any other desired steps. If such steps include treatment in chromate solutions or any other solution that is likely to contaminate the plating bath, the cylinder should be free from joints, i.e., made of one piece. Such a cylinder may also be used for basket cleaning and pickling prior to plating.

Productivity. The two-step model in Fig. 32 with bottom and top diameters of about 19 and 13 in. and a height of 18 in. holds about 0.5 ft^3 of parts and draws about 50% of the currents listed in Table 5. The other model in the same figure with bottom, maximum and top diameters of about 10, 21, and 13 in. and a height of 26 in. holds about 0.75 ft^3 of parts and draws about 65 to 70% of the currents listed in Table 5.

The equipment requirements are calculated as for perforate horizontal barrels.

Cleaning and Pickling

Basket or tumble cleaning is the usual method. See last paragraph and discussion under Perforate Horizontal Barrels and Imperforate Oblique Barrels.

FULLY AUTOMATIC BARRELS

Both the liftarm- and the elevator-type of fully automatic are used to handle cylinders instead of racks. Both are operated by an indexing, or "stop-and-go," movement although in principle there is nothing to prevent both from being run continuously, except as noted below.

Perforate Oblique Cylinders in Liftarm-type Automatics

The reader is referred to the sections about the Perforate Oblique Barrel for discussions of the basic units that are combined here.

The barrel drive shaft, with the cylinder on one end and the gear toward the other, here runs in a bearing that continues as a bracket (Fig. 32, bottom). This bracket is attached to the carrier, which is suspended from the conveyor chain. To drive the cylinder the gear mates with rotating worms that extend horizontally along the two

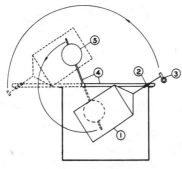

FIG. 33. Schematic view of perforate oblique barrel in plating position (full lines) and in unloading position (broken lines).
(1) Cylinder
(2) Bearing for drive shaft
(3) Gear and worm
(4) Lever system
(5) Counter weight
Arrows show motion for unloading

sides of the chassis. As in the rack-handling automatic, a hump-type cam is used to transfer the cylinders from one tank to another, and a delayed set-down mechanism may be employed when an especially short treatment time, shorter than the regular dwell period, is required in any particular tank.

The machine may be supplied with tanks for special treatments that are utilized for only a portion of the production. When such a treat-

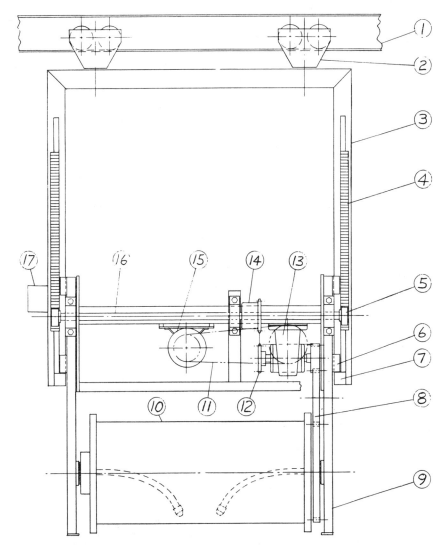

FIG. 34. Illustration of a carriage and barrel assembly of an automatic plating barrel. The electric motor revolves in one direction to lower and rotate the barrel. The clutch overrides when the barrel rotates in the *down* position. To elevate, the motor is reversed; this causes the clutch to grip the squaring shaft and rotate the pinions which mesh with the toothed racks. One reversing switch in control panel will control several barrels. Tanks not shown.

1. Monorail
2. Trolley or eyehook
3. Carriage frame
4. Toothed rack
5. Pinion
6. Barrel guide
7. Stop (for barrel in *down* position)
8. Gear train
9. Superstructure
10. Plating cylinder
11. V belt drive
12. Chain and sprocket drive
13. Speed-reducer (worm and worm wheel type, self-locking)
14. Over-running clutch
15. Electric motor
16. Squaring shaft
17. Motor switch, manually operated

ment is not desired, a skip bracket is inserted between two hump cams to prevent the cylinder from being lowered into a tank.

Drying is accomplished by blowing warm air into the cylinder as it rotates. For unloading, the cylinder rises on a hump-cam slightly above horizontal position and continues to rotate. The parts then fall via a chute into toteboxes, onto a conveyor belt, or the like.

For loading, the cylinder drops down into an upright position. It is loaded either by hand or automatically from a loading chute. The automatic loading may be combined with automatic weighing for equalization of the loads.

Perforate Horizontal Cylinders

These are now being successfully used on return-type automatic barrels, with the axes of the cylinders coinciding with the direction of travel around the machine. The advantage of this is that when the cylinders are lowered into the first station of the plating tank, they are not raised until they reach the last station; meanwhile they have been processed through a "trough" of anodes. If the cylinders were perpendicular to the line of travel, then either they would have to be raised to clear the anodes at each station or the anodes moved to clear the cylinders. Of course, in a "cell"-type machine, the cylinder (or cylinders) would stay in their respective stations until the predetermined processing time has elapsed; meantime the rest of the cylinders would bypass them.

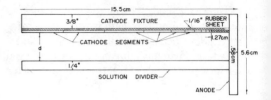
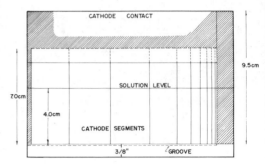

FIG. 35. Current distribution cell developed in AES research project #34.

FIG. 36. Automatic electroless nickel-gold plating line designed from data from AES research project #34. (*Courtesy of Motorola SG, Phoenix, Ariz.*)

In more recent developments of automatic barrels, the cylinder and superstructure assembly is mounted in a carriage in a manner which permits up and down movement, and a motor drive (or drives) is provided to rotate and elevate the cylinder. The carriage has trolleys which support it on a monorail and is provided with a means of moving it along the rail at the correct time intervals. Such machines are very flexible and accessible and permit the cylinders to be kept in the "up" position when not in use. They are also capable of "multiple transfer," i.e., with fewer cylinders than there are stations on the machine, the plating tank is kept full at all times and the cylinders in the rest of the cycle transfer twice or three (or more) times as often as those in the plating tank. The drying of the work is done off the machine, usually in a centrifuge, in a drum-type or in some other drying device. The door panel must be removed by hand for loading and unloading. (See Fig. 34.)

Both the perforate oblique cylinder type and the perforate horizontal cylinder type machines are built to operate on the "batch" principle to avoid the mixing of loads. Generally speaking, only a small percentage of the work processed in these machines consists of long "runs" of identical pieces. Most of it varies from several cylinder loads to just two or three; moreover single-cylinder loads of different pieces must be processed without mixing. Because the production from these machines is frequently very high, the problem of materials handling to and from them should receive serious study.

Calculation of Machine Size

The time of plating for each barrel load and the required number of barrels are calculated precisely as for horizontal barrels. Having thus obtained the number of arms over the plating tank of the automatic, one proceeds with the calculations of machine size just as if the automatic were rack-loaded.

References

1. Pottoff, I. L. U.S. Patent 806,835 (Dec. 12, 1905).
2. Dubpernell, G. and S. M. Martin. *Proc. Am. Electroplaters' Soc.* **34**:180-200 (1947).
3. Craig, S. E., Jr. and R. E. Harr. *Plating* **60**:617-621 (1973); 1101-1110 (December 1974).
4. Soderberg, G. *Proc. Am. Electroplaters' Soc.* **28**:154-160 (1940).
5(a). Geissman, W. and R. Carlson. *Proc. Am. Electroplaters' Soc.* **39**:153-165 (1952).
5(b). Wallbank, A. W. and D. N. Layton. *Trans. Inst. Metal Finishing* **32**:308-335 (1955). Data in Table 1 and 2 calculated by the present authors from figures in the reference.
6. Jackson, W. H. *Monthly Rev., Am. Electro platers' Soc.* **33**:1162-1173 (1946). The author has drawn freely on this excellent article, as well as on a private publication from Mr. Jackson, without making further reference to them.
7. Bulletin D-108. Hanson-Van WinkleMunning Co., Matawan, N. J.
8. Soderberg, G. *Monthly Rev., Am. Electroplaters' Soc.* **20**, No. 1:26-33 (Sept. 1933).
9. Saltonstall, R. B. *Plating* **36**:34-37 (1949).
10. Fink, C. G. and R. D. Eash. To Chemical Treatment Co., U.S. Patent 1,899,679 (Feb. 28, 1933).
11. Craig, S. E., Jr., R. E. Harr and S. Y. Wu. *Plating* **60**:1239-1245 (1973).
12. Casey, G. J., Jr. and R. K. Asher, Sr. *Plating and Surface Finishing*, pp. 56-58 (Feb. 1980).
13. Hannon, A. H. U.S. Patents 1,895,622 (Jan. 31, 1933); 2,148,552 (Feb. 28, 1939); 2,177,982, (Oct. 31, 1939); 2,187,079 (Jan. 16, 1940).
14. Graham, A. K. and H. L. Pinkerton. *Plating* **53**:1222-1229 (1966).
15. Private communication from John J. Laurilliard and Maurice Horoff, Precision Fastener Division, Standard Pressed Steel Company, with permission.
16. Casey, G. J., Sr. and R. K. Asher, Sr. *Plating and Surface Finishing*, pp. 51-54 (Aug. 1979).
17. Nobel, F. I., B. D. Ostrow, R. B. Kissler and D. W. Thomson. *Plating* **53**:1099 (1966).
18. Gissman, W. G. and R. A. Carlson. *Proc. Am. Electroplaters' Soc.* **39**:153 (1952).

25. SEMI- AND FULLY AUTOMATIC PLATING MACHINES

Martvig J. Moll

Formerly Manager of Conveyor Sales, Hanson-Van Winkle-Munning Company, Matawan, N. J.

Revised by Lawrence J. Durney

Automatic machines are used in the metal finishing industry to convey parts mechanically through a treatment cycle. Many parts to be treated are commonly held on racks although hooks, wires, trays, baskets, or barrels may be used depending on the kind of work to be processed. If the operation involves electroplating, the term "automatic electroplating machine" is commonly applied. However, such machines may also be used for anodizing of alloys of aluminum or magnesium and for carrying parts through treatment cycles not involving the use of electric current.

The fact that parts to be treated are handled automatically in these machines offers a number of advantages, as a result of which they have become universally used in the industry both with high or low volume of work, where direct savings in labor costs and/or other conditions result in over-all economies that justify the investments. The advantages of fully automatic processing compared to hand operation are as follows:

(1) Use less floor space (generally this is true only when production levels are considered in conjunction with labor involved). Depending on the transfer mechanism used, more floor space may be required, but will deliver higher production at lower unit labor cost.

(2) larger and deeper tanks are possible.

(3) larger and heavier rack sizes are possible. Gain depends on whether racks can be loaded and unloaded on the machine, or whether an automatic transfer mechanism is used to transfer to a secondary conveyor for unloading and reloading. If racks must be manually transferred to unloading and reloading stations, limitations on size and weight still exist, although these limits will be somewhat higher than if racks were to be processed through the cycle manually.

(4) Better control of dragout. This is mainly the result of controlled transfer time. This provides better and more complete drainage between tanks.

(5) Less danger of solution contamination. This gain is mainly dependent on (4) above. If a bypassing mechanism is used, or if a programmed hoist (*q.v.*) system which transfers over other tanks in the system is used, special care must be exercised to prevent contamination.

(6) Better rack life and control of insulation of contacts. This depends on the loading and unloading procedures used. See (3) above. (Not necessarily true if manual procedures are used.) Storage arrangements for idle racks also have a major influence.

(7) More efficient ventilation is possible.

(8) Lower installation cost for a given capacity. Depends on the equipment selected, modified by labor savings.

(9) Better process control. A function of controlled processing times only. Solution control and replenishment will be similar.

(10) Better production control at high level of capacity.

(11) Less costly and improved solution maintenance. Generally true only if proper ancillary equipment is installed, e.g., brightener feed pumps, continuous filtration and purification, etc. Since production is uniform, advantage can be taken of standard additions with reduced frequency of analysis.

(12) Better control of quality and specifications.

(13) Lower rejects. Generally true based on (4), (5), (9) and (11) above. May be offset by late discovery of defects; i.e., by the time a defect is discovered, a substantial number of additional parts will be "in the pipeline" and may well be defective also.

(14) More efficient use of labor and materials.

(15) Improved working environment.

Typical Application

Typical applications for fully automatic processing equipment may be classified as follows: (1) for parts to be handled in bulk as in barrel-type machines; (2) for electrified processing of parts, other than in bulk; (3) for nonelectrified processing of parts, other than in bulk.

Applications for parts to be handled in bulk, as in "barrel"-type machines, are covered in Chapter 24.

Applications for electrified processing of parts, other than in bulk, may include the following:

(a) Complete processing through the preplating or cleaning treatments, plating one or more metals, and final rinsing and drying operations.

(b) Similar to (a) except that an anodizing or electropolishing operation replaces the plating operations.

(c) Segregation of parts with a by-passing mechanism for selective plating combinations on one machine for either chromium, copper-nickel or nickel-chromium deposits.

(d) Complete cycle treatments including dual controls for deposits on outside and inside surfaces, or for deep recesses in which case an auxiliary anode may be used.

(e) Electrolytic alkali and acid treatments, with or without prior mechanical washing in combination, for purposes other than plating.

(f) Combinations of fully automatic and semiautomatic machines for specific treatment cycles.

Applications for nonelectrified processing of parts other than in bulk may include the following:

(a) Cleaning, pickling, and drying prior to painting.

(b) Cleaning, pickling, and applying a prime coat prior to vitreous enameling.

(c) Cleaning, pickling, phosphate coating, and drying.

(d) Cleaning, pickling, and blackening by any one of a number of processes, followed by rinsing and drying.

(e) Bright dipping of copper and copper alloys with sulfuric nitric acid dip, rinsing, and drying.

(f) Heating to stress relieve, anneal or harden in fused salt baths in combination with pickling, rinsing, and drying.

Types of Machines

The straight line is usually classified as a heavy-duty machine although it can be used to process light rack loads.

The return-type which covers the entire field of applications, both in weight and size of loads, can be classified as light, medium and heavy

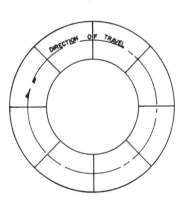

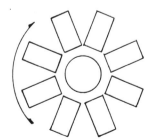

FIG. 1. Return rotary-type machine.

duty, with structural changes keeping pace with new conditions. The return type usually has all of the preplating and after-plating treatments mechanized, as well as the plating operations. There are conditions, where for specific reasons, the plating operation is done on one or more separate semiautomatic return-type plating tanks, working in conjunction with the return-type fully automatic machine where all preplating and after-plating treatments are mechanized.

A semiautomatic machine is a mechanism mounted over a plating tank only. It consists of a chain with hooks or carriers and with sprockets at both ends of a rectangular tank. Any point on the chain can be selected for load-unloading. This type machine is mentioned because of its usage in conjunction with precleaning, fully automatic machines. Usually the semiautomatic replaces "still plating tanks" manually operated, where all work is manually handled through all preplating and after-plating treatments.

There are but a few isolated conditions under which a semiautomatic machine might have value today. In most cases a small fully automatic machine serves to better advantage. For this reason, semiautomatic machines are not further discussed.

The rotary type is a circular machine, usually with a center lifting post raising and lowering hydraulically, having arms like the spokes of a wheel, where the post is the hub of the wheel. Each arm may be termed "a carrier," supporting a work load in an individual tank. Usually the size of the work load and of the individual tanks limits the number of arms or spokes that can be fitted into practical space conditions of 20 to 25 ft diameter. This may range from 6 to 8 or 10 arms. Its application is thus limited to short precleaning cycles using either bulk containers or racks and for specific prepickling processes requiring a minimum number of single-carrier treatment tanks within the range of the total carriers available on this type machine. It is usually not sufficiently flexible to cover the normal requirements of a plating cycle, although it has been used as an automatic bulk barrel plater under limiting conditions. Because the rotary does not cover the wide range of applications of return-type and straight-line machines, it is not discussed further.

There are small rotary return-type plating machines where a series of pins or single spine racks pivot on a chain in passing in and out of shallow treatment tanks. The racking and unracking is done employing only one operator and frequently only one part or a few pieces per rack. Its practical application is in a chrome plating cycle, or other short treatment cycle with small pieces and light loads.

The characteristics of the two major types of fully automatic machines are compared in Table 1.

Straight-Line Fully Automatic Machine. In the straight-line machine (Fig. 2) the plating racks are loaded on the work carriers at one end of the machine, then carried through the various steps of cleaning, rinsing, pickling, plating, rinsing, and drying operations, and finally

Table 1. Comparison of the Straight-line and Return-type Automatic Machines

Characteristics	Straight Line	Return Type		
		Light	Medium	Heavy
Production capacity	Large	Small	Large	Large
Rack size	Very large	Small	Large	Very large
Row of racks per carrier	1-2-3-4	1	1-2	1-2-3
Average weight of carrier load	100-2500 lb	10-30 lb	30-200 lb	200-1500 lb
Carrier spacing	30-96 in.	18-24 in.	18-48 in.	30-96 in.
Height of lift	36-96 in.	30-40 in.	30-72 in.	36-144 in.
Tank width	Very large	Small	Med. to large	Med. to large
Anode accessibility	Fair	Good	Good	Good
Machine accessibility	Fair	Good	Good	Good
Cycle revamp flexibility	Poor	Fair	Good	Good
Load and unload points	Separate	Same	Same	Same
Ventilation	Poor	Good	Good	Good
Headroom	High	Low	Low to high	Low to high

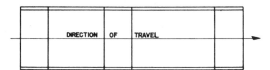

Fig. 2. Straight-line fully automatic machine.

unloaded at the opposite end of the machine. This type machine employs a straight bar-type carrier arm, with a pickup or lift mechanism at each end of the arm. The lift mechanism may be a pair of chains mechanically operated at each tank transfer point, or a single lifting mechanism on each side of the conveyor engaging all lifting carrier arms at the same time. This may be accomplished by use of a mechanical or hydraulic mechanism.

This type of machine may use two very wide lanes with two work loads per carrier, but more often three or four lanes and work loads per carrier are employed. Under the latter condition the tank width may vary from 10 to 20 ft. For parts of abnormal length, one single work load in either a basket or rack parallels the carrier arm and may be considered a single lane of work.

One or more carriers may be provided at the entering end for loading and an equivalent number at the exit end for unloading. The empty carriers are usually returned overhead to the entering end of the machine, in which case additional headroom is required. As an alternative requiring less headroom but more floor space, the empty carriers may be transferred mechanically to a separate monorail system by means of which they are returned to the loading end of the machine. When this practice is employed the empty racks are sometimes left on the carrier arm and can then be passed through a stripping cycle before returning to the loading point.

In one bumper application bars are racked back-to-back in a horizontal position and hung parallel to the carrier arm. The racks then pass through a series of single-station preplating treatments, then into single-station conforming anode plating cells, following a selective automatic by-passing movement over other plating cells, and then continue through the other afterplating treatments. Automatic loading and unloading is used, the weight of the loaded rack being about 1,000 lb.

The straight-line machine can be designed for very large racks and heavy loads and is generally considered a heavy-duty machine.

An important new development in the straight line fully automatic concept is the programmed hoist line. One or more traveling hoists are substituted for the carrier arm machine previously discussed. These hoists are controlled by a punched or magnetic tape fed through a tape reader. Barrels or "flight bars" on which racks are hung are picked up and transported from tank to tank. The main advantages of this system are that the dwell time in various tanks can be varied, movement through the tanks is not required, carriers can be "backtracked" (as, for example, to use a single rinse tank for multiple purposes) and tanks can easily be passed over to provide variations in finish. The number of hoists required will be determined by the complexity of the cycle. Hoists may also be programmed to return carriers to the loading station, allow variable drip times over different tanks, have variable dwell times in various tanks, etc. Most such systems allow for manual override of the tape control providing additional versatility.

The main disadvantage of the system is that changes in the cycle require preparation of a new tape at some expense and time delay. This is at least partially offset by the advantage that tapes for several different cycles may be prepared and stored. The machine may then be quickly altered from one cycle to another simply by inserting the proper tape into the tape reader.

The most recent development in this type of automatic system is the replacement of the tape control with a solid state read/write programmer. This permits rapid in-house changes in the program, interfacing with a computer or interfacing with a cassette type program storage unit.

There is little doubt that there will be continued application of solid state memory, microprocessor and computer control to this type of unit. Combining these new types of control with the increasingly sophisticated controls available for rectifiers, and solution control systems will result in a whole new generation of automated systems.

Straight-Line Trolly-Type Automatic Plating Machines

In recent years, where conditions permit, parts are now placed on plating racks by the operators in the manufacturing areas where final drilling, reaming, etc. are performed, rather than stacking in tote boxes, trays, etc.

Occasionally several separate areas perform this function prior to plating.

Straight-line plating machines are now being used, where these racks are suspended from the trolley of a "Power and Free" shop conveyor system. All trolleys converge by means of conventional switches and tracks to the loading end of the plating machine. At that point the trolly and work loads are released from the power chain of the shop system and ride on a track of the plating machine. The trolly functions in the same manner as the carrier arm of the conventional straight-line machine, advancing the work load from station to station. For the plating tank or any treatment that requires more than one station, the work load submerges at the first station and remains submerged until it reaches the exit station. At the unload end o the machine, the trolly again engages the shop power and free conveyor system and is directed automatically to the assembly or packiong area, where parts are removed from the plating rack.

This system eliminates tote boxes, trucking, separate racking areas and labor, permits storage of racked work for the next morning starting load, does away with bulk storage areas, minimum handling and keeps the work off the floor.

Each trolly is fitted with conventional yoke bars for suspension of the plating racks, either by the single-lane or double-lane systems, also collector shoes that engage low-voltage copper rails at electrified tanks of the plating machine.

Return-Type Fully Automatic Machine. On all return-type machines, loading and unloading zones are usually together. In all cases the work follows an elliptical path returning to the original loading point. Two parallel rows of tanks make up the straight away travel with sprockets or curved tracks moving this work through a 180 degree radius at both ends.

The loading and unloading zones are not always together. Sometimes a rack stripping treatment cycle is inserted between the unloading station and the loading station. Other derivations of the return type may have a complete processing cycle on one side, unloading at the opposite end, repeating an identical processing cycle on the opposite side of the straight away path. While loading and unloading zones are often at the one end of the machine (usually around the curved radius) a point on the straight away path is often used in preference and termed, "side loading and unloading."

As illustrated in Table 1, the return-type machine covers the entire range of industrial requirements. There are small assembled light-type machines suited for light loads, modest lifts and production, usually in the range of 60 to 100 racks per hour. The medium type covers the greater number of average requirements and has a wide range in the lift, work loads and current requirements. The heavy-duty type may be considered for exceptionally heavy rack loads of large dimensions where the pieces are racked directly on the machine or where a number of smaller racks may be suspended from one carrier. In the automotive field there are many requirements for the processing of molding, wide grille bars and other items that are logically suspended in a horizontal position, making it necessary to use wide spacing between the carrier bars. Where this construction is employed it not only provides for these large pieces but permits processing a number of smaller racks on the same carrier bar. The resulting heavier gross work loads per carrier and large current requirements necessitate a more rugged design than the light- or medium-type machines.

The return-type machine may also be (a) a continuous moving type, or (b) an intermittent type.

Continuous Moving Type. In this type the work load is in constant motion at all times. Any increase or decrease in the conveyor speed affects both the treatment time and the transfer time. Usually the carrier load is transferred between treatments by means of a cam or chain-operated device. When a chain-type transfer is used or a pickup device riding in a guide channel, the carriers index on and off a moving chain. Other types having a carrier arm fixed to the chain use a cam plat follower at all transfer points to raise and lower the arm from a hinge position at the chain.

Intermittent Type. In this type the carriers pause or "dwell" when the work is submerged in all treatments and are in motion only during the transfer of work from tank to tank, although there is one variation in which a drag chain is used to move carriers continuously while immersed in a large processing or plating tank. A

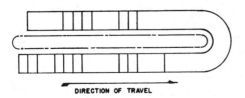

FIG. 3. Return U-type machine.

single lifting mechanism, actuated by mechanical or hydraulic means, engages all carrier arms or carrier hooks at transfer points, raising them in one operation through the required vertical lift. While in this elevated position, all carriers move forward horizontally through the selected carrier spacing. The lifting mechanism then reverses, lowering all carriers vertically to the rest position in the following tanks. When two or more carriers are required in any tank, as determined by the selected treatment time, the others remain "down" in their rest position and advance with the load submerged during the forward travel of the transfer operation. Counterweights are usually employed to offset the lifting load. There is another type using a rack and pinion to raise and lower the carrier arm at transfer points. One characteristic is common to all intermittent-type machines; the forward travel at all transfer points prevails only when the raised carriers are in the elevated position.

A constant transfer speed can be used with the intermittent-type of machine, predetermined for ideal conditions. The "dwell" or "immersion" time is usually varied by a timer mechanism for longer or shorter treatment times.

Return-type machines require a minimum floor space. The selection of a single-lane or two-lane machine is often determined by the existing space available. Single-lane machines are longer and narrower, double-lane machines are shorter and wider. They are very flexible in one respect, i.e., cycle changes may quickly be made by rearrangement of treatment and plating tanks. This often permits an existing machine to be "revamped" and utilized for an entirely different plating process.

Factors in Selecting A Machine

The following factors must be considered when selecting a fully automatic machine:

1. Production requirements
2. Specification
3. Work load and rack size
4. Carrier spacing
5. Number of lanes
6. Treatment cycle
7. Total lift at transfer points
8. Transfer time
9. Anode rod spacing
10. Current requirements per load

Production Requirements. When considering an automatic machine, a production study must be first made to determine specific requirements and their effect on the design and size of the proposed machine as a basis for selection. The following information is required:

(a) The size, shape, weight, and area of each part and the number to be processed per day, week, or month.

(b) Whether each part can be racked so that it will drain freely without difficulty due to air or gas pockets.

(c) The maximum dimension limitations as determined by the size and shape of the largest part to be processed.

(d) The relation between weight and area of individual parts, if a wide variation exists, as an aid to proper racking.

(e) The ideal spacing between parts for correct processing, recognizing the essential differences for processing deeply recessed parts as compared with shallow recessed parts in relation to fixed design factors such as anode spacing and carrier spacing.

Specifications. The specifications for the various parts to be processed must be clearly established. These should include:

(a) The basis metal for each part.

(b) The significant surfaces for each, i.e., those surfaces which must meet the specification.

(c) The nature of the treatment, i.e., nickel-chromium plating or copper-nickel-chromium plating.

(d) The minimum thickness of each plated coating or the time of treatment on a specific anodizing cycle.

(e) The salt-spray hours which each type of part must pass or similar requirements.

Work Load and Rack Size. The work load may be considered as the cubical space required for parts or workpieces as arranged on a loaded carrier. It largely determines the size and type of automatic machine best suited for the job requirements and may consist of any one of the following:

(a) A single part of unusual shape or size.

(b) A single rack containing a specified number of parts.

(c) Two or more racks not exceeding the overall space dimensions defining the work load.

(d) A cluster of parts strung on wires.

(e) A single tray or a stack of two or more trays.

(f) A basket, cylinder or other type container.

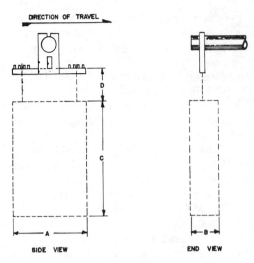

Fig. 4. Dimensions for work load and rack load.
   A Working width
   B Working thickness
   C Working length
   D Hook length

The dimensions of a work load are defined in various ways. The following terminology will be used in this chapter and is illustrated in Figs. 4 and 5.

A, Working width (sometimes called work length), is the space occupied by the work load in the horizontal direction of travel or forward motion.

B, Working thickness (sometimes called work depth), is the space occupied by the work load in

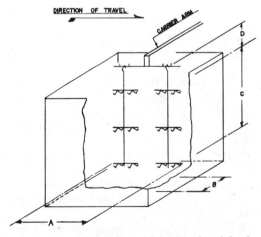

Fig. 5. Dimensions for work load and rack load.
   A Working width
   B Working thickness
   C Working length
   D Hook length

a single lane measured across the tank, i.e., at right angles to the direction of travel.

C, Working length (sometimes called work height), is the maximum vertical distance occupied by the work load when submerged in the solution, measured from the top to the bottom of the part or parts as racked.

D, Hook length, is the vertical distance from the top of the work load to the point of suspension on the carrier. This varies greatly with the individual design of an automatic.

Conventional electroplating factors as discussed in Chapter 22 should apply in determining the type and size of racks for use on an automatic machine. However, the following details should be carefully considered in order to take full advantage of automatic handling:

(1) A master rack size to accommodate a large variety of parts.

(2) A greater working length C than commonly used in manual operation.

(3) Segregation of parts of various shapes and sizes so that they may be racked in the most favorable manner with respect to the work load dimensions, weight, area, and current and metal distribution.

(4) The size and shape of the largest part to be processed which may determine the limiting rack size that can be used.

(5) The anode spacing, i.e., the distance between the work and the anodes, which will be influenced by the variables under (3) and (4) above.

(6) The arrangement and spacing of parts under (3) above which will influence the effective plating area by presenting the significant surfaces so as to receive the heaviest deposit while shielding the nonsignificant surfaces.

(7) The ability to rack and rerack parts without removing the racks from the machine as a factor influencing the rack size and design.

(8) If the racks are to be removed from the machine for loading and unloading, the necessary differences in rack size and design must be determined. Baskets, cylinders, and other type containers fall in this classification and in such cases automatic handling should also be considered.

A careful study of the production requirements will enable one to determine the above details. One should anticipate if possible what future changes involving size, shape, and weight of parts may develop and provide for these requirements in designing the machine.

It is sometimes possible to take care of the

major portion of the production requirements with a smaller working width A by processing odd-lot parts of abnormal size by manual methods. This may lower the cost of a machine by enabling one to reduce the carrier spacing and the length and width of the machine, but such a decision always involves compromises which must be carefully analyzed to obtain the most economical solution.

Carrier Spacing. The carrier spacing depends on the total work width A and although there is no accepted standard, the following values may apply for a vertical lift-type transfer on a machine using a 6-in. pitch chain or spacer bars. For work lengths or lifts approaching 48 in. or greater, a 6-in. spacing is not considered adequate.

When transferring from one treatment step to another, one has to lift the carrier and its work load over the ends of adjacent tanks or the steel plate separating the treatments when a two-compartment single tank is used. In many cases this does not exceed 1 in. If one assumes that the carrier spacing is the sum of the work width A plus 6 in. and deducts the above 1 in., the clearance between the edge of the work load and the end of the tank wall during the vertical up and down travel is 2.5 in. When a sloping transfer is used, the prevailing clearance is normally greater. While this clearance is adequate for average work loads, i.e., not too long or too wide, the following exceptions may be cited:

(1) Bent hooks or damaged racks may cause the work load to project beyond the working width A.

(2) The tendency for long pieces to swing or otherwise assume a position outside the dimension A.

(3) Elongation of the conveyor chain on some machines which may alter the clearance at transfer points.

(4) Faulty racking so that parts are positioned outside the working width A.

One should therefore allow ample clearance to avoid jamming of work loads at transfer points and costly shutdowns and repairs. The work width, A, plus 8 to 10 in. is a recommended safe carrier spacing.

Number of Lanes. As stated previously automatic machines may be constructed with one, two, three, or more lanes. The total rack or carrier work loads per hour, space conditions such as floor area and headroom, and many other factors influence the choice of single- or multiple-lane machines.

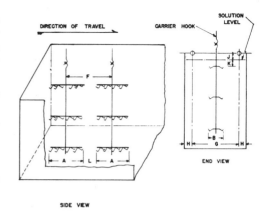

FIG. 6. Single-lane, single-hook, carrier.
F Carrier centers
G Distance between anode rods
H Distance of anode rod from tank wall
J Distance from solution level to top of tank
K Distance from solution level to top of work load

A *single-lane* or "single-file" machine is one in which only one work load per hook or carrier passes through the various treatments and succeeding work loads pass in "single file." For a plating tank this requires two anode rods only, one at each tank wall. As illustrated in Figs. 6 and 7 the dimensions of the single lane and all tanks are based on the dimensions of the work load and carrier centers, F, as follows.

G = the anode rod spacing.
H = the distance of the anode rod from the tank wall.
J = the distance from the solution level to the top of the tank.

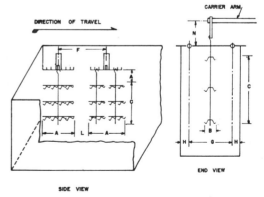

FIG. 7. Single-lane, horizontal work bar carrier.
A, B, C, F, G, and H same as in previous figures
L Minimum distance between work and adjacent carriers
N Distance of carrier center from the tank rim

K = the distance from the solution level to the top of the work load.
L = the minimum distance between work parts on adjacent carriers.
N = the vertical distance of the carrier center from the tank rim (optional).

In all cases the working width A as shown in Figs. 5 and 6 relate to the total work load per carrier, whether there is only one or more than one rack on the carrier yoke bar. In many production loads there will be more than one type or size of rack, both wide and narrow. The work width A per carrier should not be confused with the total rack requirements per hour when this includes more than one size rack or rack design.

A *double-lane* or "double-file" machine is one in which two work loads per carrier pass through the various treatments in adjacent single lanes as shown in Figs. 8 and 9. In a double-lane plating tank there are two outer anode rods adjacent to the sides of the tank and one or two center anode rods depending on the current requirements of the loaded racks. Figure 8 illustrates a horizontal work bar carrier for a double-lane machine in which P is the horizontal carrier arm distance between the lane center lines. This is identical with the anode rod spacing G except when double center anode rods are used. Figure 9 illustrates a vertical H-type work carrier for a double-lane machine.

Double lanes reduce the over-all length of the machine and require less carriers for a given production than a single-lane machine. They may be equipped with a rigid carrier arm, as

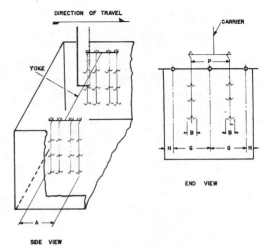

Fig. 9. Double-lane, vertical "H"-type work carcarrier.

A, B, G, H, and P same as in previous figures

illustrated in Fig. 8, consisting of a side-arm vertical lifting type or a T-bar type carrier arm. Return-type machines are frequently installed with double lanes.

A *triple-lane* or "triple-file" machine is one in which three rows of work are carried through three parallel lanes. A plating tank on such a machine is equipped with two outer anode rods along the tank sides and one additional anode rod between each lane. A triple-lane work bar is illustrated in Fig. 10.

A *multiple-lane* "straight-line" machine while more costly than a return type is used for high production loads. A triple-lane return-type machine may also be used provided the working thickness B is not too great.

The programmed hoist machines previously discussed eliminate many of these considera-

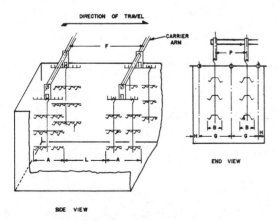

Fig. 8. Double-lane, horizontal work bar carrier.
A, B, F, G, H, and L same as in previous figures
P distance between lane center-lines

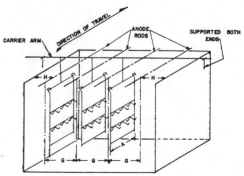

Fig. 10. Triple-lane work bar suspension.
A, G, and H same as in previous figures

tions. The essential factors are the number of racks, baskets or barrels per flight bar, and the number of flight bars per hour required to give the desired production.

Treatment Cycle. The number of steps in the cycle and the specified time and conditions for each treatment must be established in order to complete the design features for an automatic machine. One must make certain that the treatment cycle will adequately clean and process the parts according to the specifications, since any changes made after a machine is built are costly and time-consuming.

The size of a plating tank, for example, must be large enough to provide time to obtain the required thickness of deposit under the conditions specified for the plating process and must also enable one to process the required number of parts per hour. Assuming that a plating time of 40 min is required and that the parts are moving at a speed of 1 ft/min, the tank must then be a minimum of 40 ft in length, plus any additional clearance required at the entering and exit ends.

The number of parts processed per hour will depend on the number of parts per rack, the number of racks per carrier, and the number of carriers per hour. If under these conditions the required production can be met with one lane, the same production could be obtained with two lanes and a conveyor speed of ½ ft/min, in which case the tank length would be reduced to about 20 ft. In establishing such important design factors it is always well to provide for some excess capacity.

The length of each treatment tank must be similarly determined to give the required treatment time based on the chosen operating speed. It is well to provide excess time (additional tank length) so that the speed may be varied within limits and still enable one to meet required production. It is also well to provide for excess treatment time in the preplating cleaning treatments to ensure adequate flexibility since the condition of work coming to the machine can vary greatly from time to time with respect to the quantity and nature of the soil which must be removed.

Where combinations of more than one finish are provided on the machine such as brass, bronze, brass and chrome, bronze and chrome, nickel or nickel and chrome, it involves "hanging up," to bypass the plating operations not required for a specific work load. On these applications additional rinses or activating dips are usually required to compensate for the long exposure and drying of work that may result in the travel area bypassed.

Total Lift at Transfer Points. The total lift at transfer points is a function of the working length, the hook length, D, the distance from the top of the work load to the solution level, K, and the distance between the solution level and the top of the tank, J. The dimensions K plus J are usually 5 to 6 in. The hook length may vary widely depending on the construction of different machines. It is customary to allow an additional 3 in. or more to provide for clearance between the bottom of the work load and the top of the tanks when the load is at the highest point of transfer. For economical reasons the working length is usually made greater than is possible for manual operation and the added cost for greater lifts and deeper tanks does not as a rule increase greatly. When vertical mechanical agitation is used, the lift must be increased by as much as 2 in. or more.

Transfer Time. The transfer time must not be so fast that the racks or parts swing excessively, but it must not be so long that the parts become dry or stain excessively during transfer. Within these limits the time should be sufficiently long to allow proper drainage and to prevent drag-over of solution from tank to tank as far as possible. With very large work loads requiring high lifts and wide carrier spacing the permissible transfer speed is governed by mechanical limitations. While there is no fixed rule, many automatic machines operate with transfer times of 25 to 30 sec except when limited for mechanical reasons. The transfer time of a machine for acid bright dipping of copper alloys using a sulfuric-nitric acid dip is often 10 sec or less.

Anode Rod Spacing. In electroplating and cathodic cleaning treatment of parts the anode rod spacing corresponds to the cathode rod spacing in an anodic cleaning treatment, electropolishing or anodizing. This corresponds to the dimension, G, in Figs. 6 to 10, is a function of the working thickness, B, and is related to the shape of the parts being processed. A spacing of 6 in. represents good practice. Deeply recessed parts may require 8 to 12 in.; whereas tubular metal furniture requires 14 to 20-in. spacing. When horizontal mechanical agitation is used the anode rod spacing must be increased by an equivalent amount.

Current Requirements per Carrier Load. The current required per carrier load should be estimated for the processing step requiring the *maximum* current density as a basis for designing the current carrying capacity of the current carrying members and contacts on the automatic machine. The amperes per carrier load is the product of the total plating area and the average current density per unit area.

In order that this current will be available, a current source of sufficient voltage must be provided and the current carrying members on the machine must be adequately designed to avoid overheating and excessive voltage losses. A factor of safety must be provided in the current-carrying capacity of conductors in order that a machine will have the flexibility required for future variations of the work loads and process requirements. Proper maintenance and good housekeeping are important factors in this connection because dirty electrical contacts will cause increased voltage losses.

The amperes per gallon in a specific processing bath can sometimes influence the tank size and indirectly the design requirements of an automatic machine. Usually provision for cooling and circulation of the bath through an outside source will relieve critical situations of this nature without affecting the machine design.

General Characteristics

Parts that are usually large and identical in size and shape which lend themselves to a simple method of suspension or racking may, if the application permits, be synchronized with other mechanisms to transfer, automatically, the parts on and off racks or fixtures that are permanently mounted on the conveyor mechanism.

There are applications where bulk items are processed in trays or baskets through cleaning, pickling and phosphating cycles in conjunction with extrusions or press operations. These containers may be conveyed to and from press operations by various mechanical methods and arranged for automatically loading and unloading the containers at the processing conveyor.

Obviously, where the above auxiliary automatic procedures are considered, the rack, fixture, or container is sometimes modified to synchronize and simplify these auxiliary operations.

The transfer of plating rack loads automatically between an auxiliary chain and the plating conveyor may also be accomplished. This has merit where the racking and unracking operations are conducted in a remote area. Frequently the same objective can be accomplished by racking the work directly at the conveyor, without removing the racks from the plating machine.

Modern technology in anodizing cycles on aluminum, where specific treatments are bypassed depending on the alloy, or where aging of a bath increases the treatment time, or combinations of dyes are used for different colors are readily controlled by manual or automatic flags or trip pins provided on the plating machine. Occasionally the flag is mounted and positioned on the rack itself for a designated cycle, thereby permitting the work to be racked and removed on a remote monorail system and automatically transferred to the anodizing machine. Another method pre-sets the flag on the hook of a monorail, allowing the carrier engaging the work load at the automatic loader to receive an impulse that selects the required cycle. The technique of "selective cycling" has greatly expanded the use of automatic plating or processing machines.

Small full automatic conveyors are available that occupy little space and are usually shipped as a complete assembly. These machines are intended primarily for simple processing cycles, a small number of pieces per rack or carrier bar and very often require only one operator. When the processing or cycle requirements justify the use of bypass features, varied selective treatment cycles, mechanical agitation, or other such requirements, they very often fall into another more costly classification even though the units are still small in size and transported in a complete assembly.

There is, of course, a great range in the size and design of automatic machines depending on weight of work load, current required, carrier spacing, lift required, and many other factors.

When choosing a fully automatic machine the advantages insofar as specific mechanical design features are concerned should not be the major basis. It is far more important that the performance of the machine adequately meet the production and physical plant requirements. It is desirable that all the plant personnel concerned and the representative of the machine manufacturer jointly review the production requirements and specifications with respect to the proposed design features before making a final decision. A fully automatic machine may function well as a mechanical unit, but as a robot it is limited by the design features built into it.

Because of the great variety of plating and other processing requirements, no individual style or design of equipment will cover the entire field.

26. CONTINUOUS PLATING EQUIPMENT FOR STEEL MILL PRODUCTS*

A. Kenneth Graham

Consulting Engineer, Jenkintown, Pa.

AND

Samuel S. Johnston

Consultant, Steubenville, Ohio

Formerly Head Technician, Electrolytic Department, Weirton Steel Division,

National Steel Corporation, Weirton, W. Va.

Continuous Plating Equipment for Steel Mill Products

Heavy-duty equipment is required for the continuous electroplating of steel mill products, particularly for strip and wire. Such operations are customarily performed 24 hr per day and 7 days per week, minimally interrupted only as required for maintenance. Accordingly, equipment auxiliary to the electrodeposition processes is specially designed, is available from specialized manufacturers, and is of adequately rugged construction for this severe service.

The use of strip lines has been extended to include cleaning, pickling and phosphate coating applications for steel, as well as cleaning of aluminum.

Steel mill products may be considered to include sheet, strip, and wire. Sheet steel products involve separate lengths cut from strip. Strip is a continuous length of steel, usually welded successively to join multi-ton coils, in order to provide continuous operation of the strip plating lines.

---

* Richard M. Wick (deceased) contributed to the first edition.

Rods are produced commercially by hot rolling whereas wire is the product of cold drawing.

The Plating of Sheet

Electrodeposited coatings on sheet stock have been produced by a few companies. The sheets are processed horizontally, utilizing contact, drive and guide rolls, and are transferred successively through treatment steps. Since coiled steel strip (up to 100 mil thick and 4 to 5 ft in width) may now be processed continuously on strip lines, the economics of plating sheet may be questioned.

Reference should be made to the following discussion of strip plating for the quantitative relations between speed, current, coating weight and thickness, etc., that are involved.

Cycle steps may vary, but a typical cycle follows:

(1) Electro cleaner   (6) Scrubber rinse
(2) Cold spray   (7) Electroplate
(3) Hot acid   (8) Cold spray
(4) Cold spray   (9) Hot rinse
(5) Anodic acid   (10) Hot air dryer

The Plating of Strip

Continuous electrodeposition on strip steel was pioneered in the late 1930's. It was made possible by the development of the process for cold-rolling of strip steel, which was dependent on the development of the hot-strip mills and the continuous pickling of this product. The production metallurgical processes resulting in cold-rolled strip steel are described in the literature.[2,3] Following cold-rolling, the strip was cleaned, annealed, and temper-rolled prior to electroplating.

A description and analysis of tin and zinc plating of steel strip should provide the data required for plant engineering purposes as to the magnitude of the operations for a predetermined tonnage requirement. Such design factors are detailed, as well as the description of the equipment for production strip plating lines. These lines, modified in accordance with particular process requirements, likewise characterize strip plating lines for coatings other than tin or zinc, such as chromium, nickel, etc.

Some strip lines have been modified to interchangeably plate two metals, such as tin and chromium or tin and zinc. Interchangeably plating lead and various low-tin terne alloys, as well as tin and zinc, on one strip line may be considered where the demand for any one plated coating alone would not justify the cost of installing a line. However, this should only be done on a cell type plating line, engineered for the purpose intended, and providing separate plating cells, etc. (when necessary) with compatible materials of construction and auxiliaries.

Tin Plating of Strip Steel

The three original plating processes for production of electrotinplate were:

(1) Alkaline (sodium stannate, and subsequently the improved potassium bath).

(2) Two acid tin processes: (a) Ferrostan (polysulfonate electrolyte), and (b) Halogen (fluoride and chloride electrolyte).

In the alkaline processes, tin is deposited from its quadrivalent ion, while in the acid processes, tin is plated from its bivalent ion; hence, for equal cathode efficiency, twice the current input is required for the former process as compared to the latter for the same production rate. Also, the current density is more limited with the alkaline process, especially because of the tin anode polarization. The acid tin processes are, therefore, displacing the alkaline process, even though the latter gives a different and more favorable interfacilly layer of iron-tin alloy formation as plated than the former (an important quality factor). Also, as discussed below, the formation of an acceptable iron-tin alloy layer of controlled thickness is being achieved and improved by various techniques.

Since the development of the highly specialized continuous strip plating lines for production of electro-tinplate, many changes have taken place (See Ref. 18). These affect product quality and cost. They also require modification of processing cycles and equipment. Some of the changes are:

(1) Improvement in the chemical and physical properties of the steel.

(2) Reduction in the required tin-coating thickness.

(3) Introduction of "double-reduced" electrotinplate. The steel is reduced on a tandem mill from about 75 mil to about 10 mil, then cleaned and annealed. It is then further reduced by about a 40 per cent cold reduction with oil lubrication to final thickness. The product then requires an additional cleaning and scrub-rinsing in the strip plating line, prior to the usual acid preplating treatment.

(4) Applying an alkaline tin strike prior to an acid tin coating.

(5) Applying a thin (approximately 0.0025 mil) acid-tin strike, fusing at a carefully controlled temperature to convert to the iron-tin alloy, prior to applying an acid-tin coating of desired thickness and again melting.

Equipment for Processing Strip

The composite of strip-handling and processing equipment of the original electro-tinning lines comprised the following elements:

(1) Uncoilers	(11) Recovery rinse
(2) Welders	(12) Rinse
(3) Pinch rolls	(13) Reflow
(4) Looping tower or pit	(14) Drive bridle
(5) Drag bridle	(15) Chemical treatment
(6) Cleaner	(16) Rinse
(7) Rinse	(17) Dryer
(8) Acid treatment	(18) Oiling
(9) Rinse	(19) Looping tower or pit
(10) Tin plating	(20) Recoilers

Treatment Time. The basic relation between treatment time (θ), in seconds, length of strip (L) in feet, and speed of strip (S) in feet/minute is:

$$\theta = 60 \times L/S$$

(1) *Uncoilers.* Standard uncoilers are used, usually in duplicate. The expanding mandril type is preferred, rather than the cone type.

(2) *Welder.* The welding unit is used to connect successive coils. It comprises a set of squaring shears, a welder, and pinch rolls to allow the positioning of the strip ends during this operation.

(3) *Pinch Rolls.* These rolls control the speed and tension of the strip into the strip holding system.

(4) *Looping Tower or Pit.* The function of a looping tower (Fig. 1) or a pit (Fig. 2) is to hold a certain length of strip so that a time delay may be experienced without decreasing the strip speed in the electroplating step. The looping towers (with movable sets of top rolls) enables one to obtain constant speed in the electroplating operation.

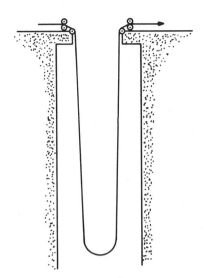

Fig. 2. Looping pit for strip plating line.

Provision may also be made for decreasing the strip speed while joining successive coils, whereby a looping pit of less capacity is feasible, although production efficiency is thereby sacrificed. In the more recent strip plating lines using a horizontal, straight-through, cell design (Fig. 6), a horizontal equivalent of looping is required for coiled steel of the heavier gages to enable one to operate continuously (see Fig. 3).

(5) *Drag Bridle.* This element of strip handling equipment is essential for the control of the strip tension entering the processing units.

(6) *Cleaner.* Acid process tin plating lines use alkaline electrolytic cleaners, but this is not true for all alkaline process lines. Electro-cleaning of steel prior to acid-tin plating consists of two stages, i.e., cathodic followed by anodic cleaning. Steel anodes in the cathodic stage polarize badly in continuous operation, reducing the current at full tank voltage to as low as 10 per cent of the rated value. Reversing the current at intervals of 20 to 60 min has been used as a means of overcoming this difficulty. Carbon (graphitized)

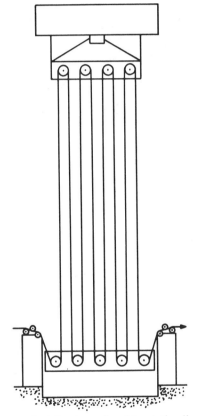

Fig. 1. Looping tower for strip plating line.

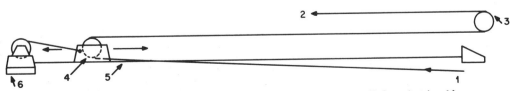

Fig. 3. Horizontal equivalent of looping tower or pit for plating line for coiled steel strip of heavy gages. Legend: 1—Strip from uncoiler, straightening rolls (for heavier gauges), and welder, 2—strip to plating line, 3—strip feed roll, 4—movable looping roll on carrier, 5—track for (4), 6—drive roll for (4).

anodes eliminate the difficulty. Their slow disintegration presents no serious problem.

(7) *Rinse.* Thorough rinsing after cleaning is important. The difficulty of rinsing alkaline films and particles from surfaces is well-known; therefore, resort is often made to the use of scrubber rinses. The alternative is spray rinsing at high water pressures, which involves substantial water volumes.

(8) *Acid Treatment.* The nature of the acid treatment depends on the requirements of the plating process. In the alkaline process, a simple sulfuric acid treatment from 3 to 18 sec is used. In the acid processes, an acid dip ending anodically, or an alternating polarity electrolytic acid treatment ending anodically, may be used.

(9) *Rinse.* The considerations of (7) above apply.

(10) *Tin Plating.* Both the alkaline and the acid "Ferrostan" processes operate with a skein-type plating cell in which the strip steel and the opposite tin anodes are in a vertical plane (Fig. 4). In the "Halogen" acid process, the strip steel is plated with tin on one side only in a series of individual cells; subsequently, the opposite side is plated after inversion afforded by the second tier of cells (Fig. 5). In this case, the reversal of direction of strip travel permits the inclusion of recovery and wash rinses in a third tier of cells. In the vertical pass alkaline lines, the tin anodes are single large castings, suspended both beneath the

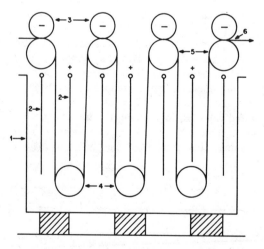

FIG. 4. Vertical-pass plating cell design for plating strip. 1—plating tank, 2—anodes, 3—contact rolls, 4—bottom "sink" rolls, 5—back-up rolls, and 6—strip.

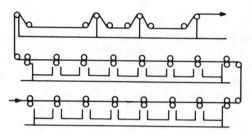

FIG. 5. Horizontal-pass, multi-tiered design for plating strip.

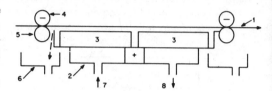

FIG. 6. Details of single plating-cell for horizontal-pass, multi-tiered, strip-plating line. 1—strip, 2—cell, 3—anodes, 4—contact roll, 5—backup roll, 6—overflow collection tray, 7—feed line from storage tank and 8—return line to storage tank. (Adapted from Ref. 6 with permission of the *American Electroplaters' Society*, and with minor modification).

rolls and between the rolls. These top mounted rolls are connected to the current source and serve both to position the strip and to complete the electrical contact; thus, the strip is simultaneously plated on both sides. In the vertical-pass acid "Ferrostan" cells, the arrangement is similar, but the anodes are sectional and smaller.

The horizontal-pass acid-plating configuration (Fig. 5) comprises a succession of small horizontal cells (Fig. 6). The plating solution is circulated continuously through each cell. Electrical contact is made to the strip by the top rolls at each end of the cell. The anodes, as shown, are positioned under the strip close to the strip travel path, and rest on an anodic center contact support and end supports within each cell. Figure 7 shows a patended[19] anode arrangement which maintains a constant anode to strip (cathode) spacing, facilitates spent anode replacement, and overcomes attack of the anode contact and support member.

Many continuous strip plating lines now include combinations of vertical and horizontal pass treatment cells, especially when exiting lines have been modified and available horizontal or vertical space has been a consideration.

A more recent design of a strip plating line uses a horizontal-pass, straight-through, cell configuration similar to that shown in Fig. 8. Some advantages of this design are: (1) the ability to process

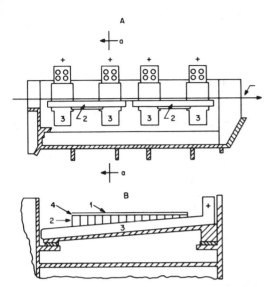

FIG. 7. Anode structure for continuous strip electroplating. A = schematic view of a plating cell. B = sectional view along the lines a–a. (1) strip, (2) anodes, (3) anode support and contact, and (4) constant spacing between strip and anodes as the latter are corroded away druing plating.

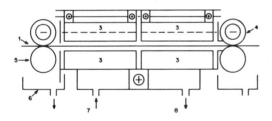

FIG. 8. Details of single plating-cell for horizontal-pass, straight-through, strip line. Anodes top and bottom. Legend: 1—strip, 2—cell, 3—anodes, 4—contact roll, 5—backup roll, 6—overflow collection tray, 7—feed line from storage tank, and 8—return line to storage tank.

steel strip about 6 to over 100 mil in thickness and up to 72 in. in width, and (2) the option of plating one side only or both sides simultaneously, with the same or different coating thickness on opposite sides.

A supplier of the latter design of the strip plating line reports that it is not economically feasible to have more than 18 cells in one line. The overall length of a plating cell is 7 ft with an effective (immersed) plating length of 6 ft. The distance between the rolls at each end is about 9 to 9.5 ft.

A contact roll 12 in. in diameter and 60 in. long, for example, can carry a maximum of 15,000 amperes without internal water cooling. Therefore, a single cell for plating a 60-in. wide strip, with a contact roll at each end, can operate at a maximum of 30,000 amperes. If there are several plating cells in line, the contact rolls between any two would carry 7500 amperes for each adjacent cell. Each cell, other than an end one, is then limited to 15,000 amperes, unless the contact roll is water cooled or a second contact roll is inserted between the adjacent cells. The design limitation of the current fed into one cell is the heating due to current passing through the available cross section of steel.

(11) *Recovery Rinse.* After passing through the plating cell, the strip is passed through a reclaim rinse in which the residual plating solution may be largely removed and collected for re-use.

(12) *Rinse.* The final rinse in clean water or condensate completes the removal of any remaining surface films on the tin-plated strip. To obtain full brightness, a concentration of 1.5 to 3.0 g/l of NH Cl in this rinse has proved beneficial.

(13) *Reflow or melting.* The bright, lustrous finish of electrolytic tinplate is obtained by fusing and quenching the melted tin coating under carefully controlled temperatures. This fusion is accomplished in one of three ways: (1) by gas heating; (2) by induction heating; or (3) by electrical resistance heating. In each case, the strip is quenched in water as soon as possible after the tin has melted. The gas fusion is accomplished in vertical towers equipped with special ceramic burners which radiate heat uniformly to the rapidly moving strip. The necessary control is obtained by varying the gas pressure or by moving the tower walls to alter the distance of the burner panels from the strip. The high-frequency induction method for fusing tinplate[6,7] involves passing the strip through a high-frequency field of magnetic flux. This method permits the use of short heating zones, and the energy input to the strip is controlled by varying the voltage to the induction coils. The electrical resistance heating method[5,7] operates by passing a heavy electrical current through the strip itself. The strip is run in a vertical position over a top contact roll and travels downward to a bottom contact roll, which is immersed in water. Operating with three contact rolls, the bottom two are grounded. Control is obtained by varying the applied voltage, and hence the current flowing through the strip, so that the

tin is melted just before it reaches the water quench.

(14) *Drive Bridle.* The drive bridle furnishes the motive power for passing the strip through the system. Formerly, in some installations, this equipment could be located farther from the plating lines. However, strip lines for plating tin or chromium now have as many as five bridle units, at critical locations, in order to reduce tension on the thinner gauge strip steel now being processed for the canning industries.

(15) *Chemical Treatment.* Following the fusion treatment, the bright tinplate is given one of several possible chemical treatments, usually involving chromic acid or a chromate, and is usually electrolytic.

(16) *Rinse.* Following the chemical treatment, the tinned strip is washed thoroughly with clean water, preferably deionized.

(17) *Dryer.* The washed strip is dried with steam or hot air.

(18) *Oiling.* The final processing step is the application to the tinplate of a controlled, very thin film of oil. The substance used is usually either cottonseed oil or, more commonly a substitute, dibutylsebacate. In emulsion oiling the surface of the strip is treated with an unstable oil-in-water emulsion. In this method the principal control elements are the oil concentration in the emulsion and the emulsion temperature. In the electrostatic oiling method vaporized oil is passed between grid electrodes and the strip, between which is a high potential gradient. Since the oil particles are charged, they are caused to deposit on the strip.

(19) *Looping Tower or Pit.* This unit performs the same function for the exit end of the line as the similar equipment described above under (4). In this case the loopers act as strip accumulators, permitting time for shifting the strip between the recoilers.

(20) *Recoilers.* Usually two recoilers are provided so that when one carries a full coil, the strip can be quickly shifted to the other. The full coil can then be removed for delivery to the inspection and shearing lines.

Inspection. Plated can stock is now shipped in coils to the large canners, instead of the formerly used sheets which were inspected at the steel mill. A continuous automatic inspection station for observing and detecting surface defects is now being used on some plating lines. This usually follows the fusion, quenching and drying operations.

Design Factors for Electrolytic Tin Plating of Strip.

Definitions:

Base Box. The unit of area is the base box, (BB), originally defined[16] as 112 sheets, each 20 in. × 14 in., which is 31,360 sq in. of tinplate. The area of a base box is 217.78 sq ft of sheet. Since both sides are tinplated, the total *surface* area of a base box is 435.56 sq ft.

Coating Weight. The coating weight, (w), of tin may be expressed as pounds per base box (lb/BB). The thickness of tin, (t), in inches, taking 7.28 as the specific gravity of tin, is:

$$t = 6.06 \times 10^{-5} \times (w) \qquad (2)$$

Example: When (w) is 1 lb/BB, (t) is 0.000061 in.

Base Weight. The weight of a base box of tinplate is the base weight (BW) and depends approximately on the thickness of the steel, (t), according to the relation:

$$(BW) = 9.107 \times 10^3 \times (t) \qquad (3)$$

or

$$(t) = 1.098 \times 10^{-4} \times (BW)$$

Example: When (BW) is 100 lb, (t) is 0.0110 in. steel thickness.

Nomenclature:

Symbol	Units	Description
BB	sq ft	base box or 217.78 sq ft of sheet
w	lb/BB	coating weight of tin
BW	lb	weight of one base box
θ	sec, min, hr, etc.	time
L	ft	length of strip
S	ft/min	strip speed
W	in.	width of strip
W^*	lb	weight of tin
I	amp	current
J	amp/sq ft	current density
t	in.	thickness
$N.T.$	ton	2000 lb
E_c	%	cathode efficiency, 100% = 100
E_a	%	anode efficiency, 100% = 100

Design Relations:

Production rate:

$$N.T. = 1.148 \times 10^{-5} \times S \times W \\ \times (BW) \times \theta \text{ (hr)} \qquad (4)$$

$$N.T. = 1.018 \times 10^{-1} \times S \times W \times t \text{ (steel)} \times \theta \text{ (hr)} \quad (5)$$

Efficiency (acid process)

$$E_c = \frac{4.097 \times 10^7 \times w \times (N.T./hr)}{(BW) \times I} \quad (6)$$

$$E_c = \frac{2.822 \times 10^3 \times w}{J \times \theta_{(min)}} \quad (7)$$

$$E_c = \frac{4.703 \times 10^2 \times w \times W \times S}{I} \quad (8)$$

$$E_c \text{ or } E_a = \frac{2.049 \times 10^4 \times (W^*)}{I \times \theta_{(hr)}} \quad (9)$$

$$E_a = \frac{7.787 \times 10^5 \times t(S_n)}{J \times \theta_{(hr)}} \quad (10)$$

* For the alkaline process, the constants of the equations for efficiency (E) are doubled.

whence,

$$t(S_n) = 1.284 \times 10^{-6} \times E \times J \times \theta \text{ (hr)} \quad (11)$$

Processing time:

$$\theta_{(sec)} = \frac{60 \times L}{S} \quad (12)$$

Strip length:

$$L = \frac{6 \times I}{W \times J} \quad (13)$$

Example:

To determine the strip speed required to produce tinplate at a rate of 15,000 $N.T.$/month, assuming 600 operating hr, 30 in. strip width and 100 lb base weight. Using Eq. (4).

$$S = \frac{15000}{1.148 \times 10^{-5} \times 30 \times 100 \times 600} = 726 \text{ ft/min}$$

To determine the current input required for this production by an acid tinplating process if the coating weight is ½ lb/BB and the plating efficiency is 95 per cent. Using Eq. (6):

$$I = \frac{4.097 \times 10^7 \times 0.5 \times (15000/600)}{95 \times 100}$$

$$= 53,900 \text{ amp}$$

The strip length being plated may be calculated, assuming a current density (J) of 200 amp/sq ft, from Eq. (13):

$$L = \frac{6 \times 53,900}{30 \times 200} = 54 \text{ ft}$$

The number of contact rolls required, assuming each carries equal current, is simply the quotient of the total plating current by the current per roll. If each roll carries 6000 amp, then:

$$\text{Number of rolls} = \frac{53,900}{6000} = 9$$

Allowing for entry and exit rolls carring only half the normal current, 10 rolls will be required. There are two plating passes per roll, hence 18 passes. The plating length per pass is therefore 54/18 = 3 ft, which is the effective anode length for the assumed conditions.

Electrolytic Zinc Plating of Strip

The horizontal-pass, cell-type lines used for tin plating strip (Figs. 5 and 8) have been adapted with little change to zinc plating of strip. At a current input of 60,000 amperes to the cells, a line speed of 160 ft/min is obtained at a current density of 200 amp/sq ft. Strip up to 38 in. wide may be zinc plated, the steel thickness ranging from 0.1 to 0.2 oz/sq ft, including both sides. The anodes are pure zinc, of which more than 16 tons are required for the cells. Except for the tin fusion equipment required for electro-tinning lines, there is no essential difference when zinc-coated strip is produced. The horizontal-pass, straight-through, cell-type line (Fig. 8) is also used for zinc plating especially for wide strip and for heavier gauges.

Design Factors for Electrolytic Zinc Plating of Strip

Definitions:

Area. It is noted that the area (A) refers to the total area of both sides of a sheet of steel, while the area (A^*) refers to one side only. The relation between the weight of the steel in lb/sq ft (W^* steel) to its thickness in inches (t steel) and area in sq ft (A^*), is:

$$(W^* \text{ steel}) = 40.73 \times (t) \times (A^*) \quad (14)$$

Coating Weight. The coating weight of zinc is expressed as ounces per square ft (w) distributed over both sides of the sheet. In terms of the coating thickness, (t_{zn}):

$$(w) = 1189 \times (t_{zn}) \quad (15)$$

and,

$$(t_{zn}) = 8.412 \times 10^{-4} \times (w) \quad (16)$$

Nomenclature:

Symbol	Units	Description
A	sq ft	total area of sheet, both sides
A^*	sq ft	area of sheet, one side
w	oz/sq ft	coating weight of zinc
θ	sec, min, hr, etc.	time
L	ft	length of strip
S	ft/min	strip speed $(S = L/\theta_{(min)})$
W	in.	width of strip
W^*	lb	weight of zinc on steel
I	amp	current
J	amp/sq ft	current density $(J = I/A)$
t	in.	thickness
$N.T.$	ton	2000 lb
E_c	%	cathode efficiency 100% = 100
E_a	%	anode efficiency 100% = 100

Design Relations:

Production rate

$$N.T. = 1.018 \times 10^{-1} \times S \times W_{(in)} \times t_{(steel)} \times \theta_{(hr)} \quad (17)$$

Efficiency (Zinc)

$$E_c = \frac{1.395 \times 10^5 \times w \times A^*}{I \times \theta_{(min)}} \quad (18)$$

$$E_c = \frac{1.162 \times 10^4 \times w \times W_{(in.)} \times L_{(ft)}}{I \times \theta_{(min)}} \quad (19)$$

$$E_c \text{ or } E_a = \frac{8.289 \times 10^7 \times t_{zn} \times A}{I \times \theta_{(min)}} \quad (20)$$

$$E_c \text{ or } E_a = \frac{1.382 \times 10^6 \times t_{zn}}{J \times \theta_{(hr)}} \quad (21)$$

whence,

$$t_{zn} = 7.239 \times 10^{-7} \times E \times J \times \theta_{(hr)} \quad (22)$$

$$E_c \text{ or } E_a = \frac{3.719 \times 10^4 \times (W^*)}{I \times \theta_{(hr)}} \quad (23)$$

Example: To determine the strip speed required to produce electrolytic zinc-plated strip at a rate of 10,000 N.T./month, assuming 600 operating hr, 0.030 in. strip thickness and 30 in. widths. Using Eq. (17):

$$S = \frac{10,000}{1.018 \times 10^{-1} \times 30 \times 0.030 \times 600}$$

$$= 182 \text{ ft./min}$$

To determine the current input required for this production if the coating weight is 0.12 oz/sq ft (t_{zn}) = 0.0001 in. and the electroplating efficiency is 90 per cent. Using Eq. (19):

$$I = \frac{1.162 \times 10^4 \times 0.12 \times 30 \times 182}{90}$$

$$= 84,600 \text{ amp}$$

Plating of Wire

The plant design for wire plating depends on the plating process to be employed. The substantial plant cost indicates immediately the need for high-current density plating processes since the current density that can be employed determines the physical size of the equipment. An estimated minimum current density for wire plating is 500 amp/ft, and a desirable average is 2000 amp/ft. The process used may operate either with soluble or insoluble anodes. Continuous recirculation of the cell electrolyte is commonly used, and continuous purification is beneficial or may be required for continuous operation.

In wire plating, a multiplicity of wires is processed in parallel through the cells. The sequence of operations of a typical line is listed below. The items marked with an asterisk are optional as required by the process or product specification.

(1) Welding
(2) Unreeling
(3) Heat treatment*
(4) Electrolytic cleaning*
(5) Rinse*
(6) Pickling
(7) Rinse
(8) Special acid treatment*
(9) Rinse*
(10) Electroplating
(11) Rinse
(12) Dry
(13) Recoil

Heat Treatment. The optional step (3) above has been applied successfully in a wire plating line. The annealing of low-carbon wire, and the special heat treatment (patenting) of high-carbon wire, have both been used in commerical production. When incorporated in a wire plating line either one is usually followed by a strong pickling step (6), which may be hot hydrochloric acid and which must be capable of removing completely the thin oxide coating formed after heat treatment.

Electrolytic Cleaning. The electrolytic cleaning step (4) is not always necessary. Wire fresh from the annealing furnace, for example, may require only acid treatments.

Special Acid Treatment. A special acid treatment, step (8), may be necessary depending on the electroplating process and the condition of the wire. This is a process-determined step. For example, nitric acid has been used in some cases and anodic electrolytic treatment in sulfuric acid in others.

Flat Wire or Narrow Strip

The electroplating of flat wire or narrow strip can be accomplished directly in wire plating lines where the work is moved through the cells substantially in a straight line. Widths up to an inch or more can be processed. Should the product packaging so require it, specialized unreeling and recoiling equipment may be required; otherwise that used for round wire may be adapted.

Electrolytic Zinc Plating of Wire

The electrogalvanizing of wire using insoluble anodes of silver-lead alloy has been described[9]. This is the Bethanizing process, originally developed by U. C. Tainton and operated by the Bethlehem Steel Company. Zinc oxide is leached from zinc calcine by means of sulfuric acid to form a neutral zinc sulfate solution. This is purified and used as a feed to the zinc electroplating cells, displacing electrolyte which is recycled for further leaching. This process operated at a current density up to 2000 amp/sq ft, which has subsequently been increased beyond 6000 amp/sq ft[11].

A soluble anode process for electroplating zinc on wire has been described[10] which is known as the Meaker process. It used a sulfate bath but at a pH between 2 and 4. Pure zinc anodes are used to maintain the zinc concentration, and small additions of sulfuric acid are made to maintain the pH between the desired limits. In this process a current density of at least 1000 amp/sq ft can be used so that it is suitable for continuous wire plating.

Equipment. The wire-handling equipment, comprising welding units for joining successive coils of wire, pay-off reels, guide sheaves and take-up blocks for recoiling the wire, is of standard design and commercially available. Figure 9 is a photograph of a wire-plating cell in which may be seen the multiple contacts that are used to complete the electrical circuit to each wire at successive intervals along the cell. Individually, these contacts are copper rods which ride on the wire and are insulated at all other points by porcelain, rubber or other suitable acid-resistant insulating materials. The contacts may ride directly on the wire and be individually insulated, or they may be inserted in a vertical hole in a ceramic contactor block so as to ride on the wire passing through a horizontal hole at right angles. The ganged contactors are connected through flexible leads to a header bar which in turn joins the main distribution bus from the generator. In the soluble anode process the anodes are zinc bars which rest on contacts below the level of the wire. The lead alloy anodes in the insoluble anode process are located between the contactor blocks and comprise a grid work of spaced vertical plates about 4 in. high, resting on a header bar and centered on the parallel wires.

Design Factors for Zinc Plating of Wire

Nomenclature:

Symbol	Units	Description
d	in.	wire diameter
E	—	electrochemical efficiency
I	amp	current
J	amp/sq ft	current density
L	ft	cell length (without contactors)
L'	ft	cell length (with contactors)
n	—	number of wires
p	—	number of cells
Q	lb/hr	production rate
R	—	coating ratio
p_c	—	sp gr of coating (Zn = 7.14)
p_s	g/cu cm	sp gr of steel = 7.83
S	ft/min	speed of wire travel
t	in.	coating thickness
w	oz/sq ft	coating weight
W	lb	weight of zinc

Design Relations:

Coating Ratio, (R)

$$R = \frac{\text{weight of zinc coating}}{\text{weight of steel wire}} \quad (24)$$

$$w = 1.629 \times 10^2 \times R \times d \quad (25)$$

$$R = \frac{6.138 \times 10^{-3} \times w}{d} \quad (26)$$

Coating Thickness, (t)

$$R = 4 \times \frac{p_c}{p_s} \times \frac{t}{d}\left(1 + \frac{t}{d}\right) \quad (27)$$

for zinc, $4 \times \frac{p_c}{p_s} = 4 \times \frac{7.14}{7.83} = 3.65$

$$t = \frac{R \times d}{R + 3.65} \text{ (approximately)} \quad (28)$$

Length and Area of Wire

$$\frac{\text{Length of steel wire}}{\text{pound}} = \frac{0.3751}{d^2}, \text{ft} \quad (29)$$

FIG. 9. Electrogalvanizing (Bethanizing process) plating cell installation. (*Courtesy of the Bethlehem Steel Company*)

$$\frac{\text{Area of steel wire}}{\text{pound}} = \frac{0.09820}{d}, \text{sq ft} \quad (30)$$

Production Rate

$$Q_{(\text{steel})} = 1.600 \times 10^2 \times n \times d^2 \times S, \text{lb/hr} \quad (31)$$

$$Q_{(\text{Product})} = Q_{(\text{steel})} \times (1 + R), \text{lb/hr} \quad (32)$$

$$Q_{(\text{Product})} = 6.453 \times 10^{-4} \times E \times I$$
$$\times \left(1 + 162.9 \times \frac{d}{w}\right), \text{lb/day} \quad (33)$$

Efficiency

$$E_c = \frac{3.651 \times 10^4 \times n \times w \times d \times S}{I} \quad (34)$$

whence,

$$S = 2.739 \times 10^{-5} \times \frac{E \times I}{n \times w \times d} \quad (35)$$

$$E_c = 1.395 \times 10^5 \times \frac{w \times S}{L \times J} \quad (36)$$

whence

$$S = 7.170 \times 10^{-6} \times \frac{E \times L \times J}{w} \quad (37)$$

$$E = 8.719 \times 10^2 \times \frac{Q(\text{steel}) \times w}{n \times d^2 \times L \times J} \quad (38)$$

$$E = 3.719 \times 10^4 \times \frac{(W^*)}{I \times \theta_{(\text{hr})}} \quad (39)$$

Current

$$I = 0.2618 \times n \times d \times L \times J \quad (40)$$

Example: To determine the parameters of an electroplating cell to produce 10 N.T./day of electrogalvanized wire, average diameter 0.12 in., diameter range from 0.18 to 0.08 in., average coating weight 1 oz/sq ft. The plating efficiency is assumed to be 90 per cent, and the average current density is 900 amp/sq ft.

The current required is:

$$I = \frac{10 \times 2000}{6.453 \times 10^{-4} \times 90 \times \frac{(1 + 162.9 \times 0.12)}{1}}$$

$$I = 16{,}760 \text{ amp}$$

The length of wire in the plating cell is:

$$nL = \frac{16760}{0.2618 \times 0.12 \times 900} = 593 \text{ wire ft}$$

For an 8-wire cell, $L = 74.1$ ft. If $L' = 1.2 \times$

L, the cell length with wire contactors is $1.2 \times 74.1 = 88.9$, or 90 ft.

The speed for the average gauge and coating weight may be obtained from either Eq. (37) or (35).

$$S = 7.170 \times 10^{-6} \times 90 \times 74.1 \times 900/1$$
$$= 43.0 \text{ ft/min.}$$

$$S = \frac{2.739 \times 10^{-5} \times 90 \times 16760}{8 \times 1 \times 0.12} = 43.0 \text{ ft/min}$$

and for 0.18 in. and 0.08 in.-diameter wire, the speed is 29 and 65 ft/min, respectively.

The current density range is calculated:

$$J = \frac{16,760}{0.2618 \times 8 \times 74.1 \times (d)} = \frac{108}{d},$$

so that it varies from 600 to 1350 amp/sq ft from largest to smallest diameter wire.

The average coating ratio is

$$R = \frac{1}{163 \times .12} = 0.0512$$

and the amount of zinc plated is:

$$W^* = \frac{10 \times 2000}{24} \times \frac{0.0512}{1.0512} = 40.6 \text{ lb of zinc}$$

plated per hour. This may be checked:

$$W^* = \frac{90 \times 16760 \times 1}{3.719 \times 10^4} = 40.6 \text{ lb, or } 975 \text{ lb of}$$

zinc plated per day, excluding losses.

The design equations permit the calculation of numerous other useful data and permit the construction of tables, etc. to suit the individual requirements. Analogous relations may readily be developed for metals other than zinc.

Electroplating Large Diameter Wire

One company operates, according to its patents[12], a helical coil method of plating wire in rod form, followed by subsequent consecutive redrawing and heat treatment to the required finished diameter[13]. The open coil is passed by external roller and comb guide members spirally along a rotating shaft, held down by a similar roller shaft. Electrical contact to the wire is accomplished by a roller contact on the rod opposed by a roller guide, functioning external to the plating bath. The internal anodes are semi-cylindrically disposed to the rod coil, and the external anodes are partly formed to the coil shape. Current densities up to approximately 500 amp/sq ft are said to be employable.

By this means, the necessary successive operations of cleaning, electroplating and rinsing may be performed continuously as required by the electrodeposition process.

The Plating of Rigid and EMT Conduit

The art of electroplating the external areas of tubular lengths has been resolved in usable manner by the development of suitable apparatus by L. E. Lancy[14]. In this method the tubular articles are disposed transversely and conveyed horizontally through the length of the electroplating bath. The work holders conduct the current, and no masking is required. The anode arrangement provides for successive feeding of ball anodes, utilizing open helical holders disposed to maintain uniform current density by varying anode-cathode distance to obtain a minimum at the mid-point of the tube. Through adjustment of the anode guide member angle, there may be obtained uniform current density, and therefore uniform coating weight of the (zinc) coating being electrodeposited. While this development is primarily intended to provide a specification grade bright zinc coating from a suitable cyanide zinc bath, it is evidently mechanically adaptable to other electrodeposits on similar work structures.

Standard EMT conduit, which has a light wall compared to the heavy wall rigid conduit, is being zinc plated from either acid or cyanide zinc baths, using automatic conveyor-type plating equipment of special design.[16, 17] The pipe is fed horizontally to the machine in cut lengths from a table from which it is engaged by the conveyor arms and fed successively through the pretreatment, rinsing, plating and drying steps and may be enameled and baked as a final post plating step.

Chromium Plating of Strip

A process for plating steel can stock with a thin chromium coating and applying a supplemental (oxide) conversion coating was originally developed by the Japanese (see Refs. 20, 21 and 22). It has created great interest in the U.S.A. It is now being used for beer can bottoms and is also being evaluated for can bodies. A special plastic cement, as well as a welding procedure, have been developed to replace the usual soldered seams employed for tinplate. The cemented seams are limited at high temperatures, hence the use of chromium for sanitary cans may be restricted. However, in the opinion of some, chromium-plated can stock may ultimately replace the use of 0.25 lb per base box grade of electro-tinplate.

Local Current Density Considerations. The vertical pass plating cell favored for chromium plating of strip is illustrated in Fig. 8. As the strip passes down under the bottom "sink" roll and back out at the top, the local current density varies from a maximum at the entering and exit end to none at the bottom. The deposition of chromium will, therefore, similarly vary from a maximum rate at the top to none at the bottom as the strip passes through. In between, deposition decreases through a zone of very thin chromium to one of low local current density with no chromium deposition. This is caused by the decreasing bath voltage locally available, a function of the increasing resistance as the strip length increases with depth, and also the fact that the efficiency of chromium deposition decreases rapidly with decreasing current density. Furthermore, substantial cathode polarization is required to deposit any chromium in the metallic state from a chromic acid electrolyte, and this requirement is not satisfied at the lower local current densities.

Anode Data. Because of the above limitations, grid anodes on each side of both the entering and exiting strip are limited in length. On some installations, the anodes are about 2 ft in length and parallel to the strip. Plastic shields between the anodes and strip which extend some 6 in. into the plating bath are sometimes used to reduce the otherwise very high current density at the solution level. Use of the short anodes (insulated on the back) also enables one to control both the anode and local cathode current densities at their desired values for satisfactory chromium deposition at the top entering and exiting ends of the strip.

Cast lead alloy anodes, containing 1 per cent each of tin and silver, give long use with a minimum formation of anode particles.

Process Data. Chromium is being plated from proprietary baths to thicknesses varying from 0.002 mil (0.05 micron) to as little as 0.0003 mil, now designated as 5 gm per sq ft of surface, i.e., on each side. This is equivalent to 2 per cent of the tin thickness of 0.25 lb per base box electro-tinplate.

One company[23] deposits only 0.0003 mil of chromium at current densities from 500 to 1000 amp per sq ft and a bath temperature of 115 F with an apparent cathode efficiency of about 25 per cent.

The preplating cleaning cycle is the same as for

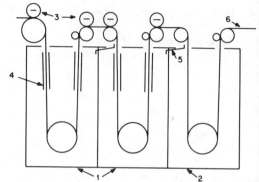

FIG. 10. Vertical-pass plating and rinsing cell design for chromium plating of strip. 1—plating cells, 2—rinse tank, 3—contact rolls, 4—special insoluble anode grids, 5—drip collection tray, and 6—strip.

tin plating. After chromium plating, the strip passes through a reclaim and rinse station into a chemical treatment station of a cell design similar to the plating cell (Fig. 10). The chemical treatment produces a coating which seals pores or cracks in the chromium coating and presumably gives a more uniform surface for the subsequent organic coating regularly applied to strip for beverage cans.

The chemical treatment solution may be a proprietary chromic acid solution as recommended by the Japanese.[20] A dilute plating bath composition also is used to give a thin coating[22], comprising a thin oxide layer over metallic chromium. The procedure for producing the coating is described in detail in the cited reference. The treatment is carried out cathodically and at 115 F, using about 150 to 200 coulombs per sq ft of surface, but in no case over 300. The coating produced may vary from 40 to 200 micrograms per sq ft, but there is considerable difference of opinion as to the best weight. No further treatment is required other than a thorough rinsing with deionized water, hot air drying and a final oiling of the surface.

Equipment. Each plating cell in one installation is connected to two 30 to 40 volt rectifiers, each of 15,000 amperes capacity, so that each contact roll has independent current control. Two plating cells are used, as shown in Figure 10, to deposit a 0.0003 mil chromium coating.

Each plating cell in the above installation is 4 ft long × 6 ft wide × 10 ft deep with a solution capacity of about 1,600 gal. The proprietary bath composition is maintained by pumping to and from a 10,000 gal chemical make-up tank at a rate to maintain the temperature in the plating tank (about 275 gpm). The solution from the latter tank is also pumped at the same rate through a heat exchanger to maintain a make-up solution temperature of 115 F. The bath temperature in the plating cell is thereby maintained at 110 F.

Continuous filtration is required to eliminate roughness from solid particles, some of which come from the insoluble lead alloy anodes. "Dacron" cloth screens have been used in front of the anodes to further reduce this source of roughness.

The plating tanks, cathodic chemical treatment tank and storage tanks must be installed with means of exhausting the spray and fumes. Means must also be provided to remove contaminants from the exhausted air to avoid environmental pollution.

Lead and Terne-Alloy Plating of Strip

Process for plating lead or (5 to 15% Sn) terne-alloy coatings, suitable for strip line applications, are now available. (See Chapter 6, Tables 12 and 13 and cited references.)

The baths for both type coatings are acid fluoborate electrolytes with hydroquinone as an addition agent. No production strip line for these coatings is known to have, as yet, been operated at the date of this publication. However, the evaluation of the properties and performance of these coatings (sponsored by the International Lead Zinc Research Organization) offer promise.

Performance of lead exposed in certain atmospheres, and of terne in certain can and automotive tests, is such that strip lines applications could well be realized in the future. Also, the combined cost of metal and power for plating zinc, tin, lead and terne indicate that lead and even low-tin terne alloys can be competitive with zinc at power costs within the range of 1 to 3 cents/KWH.

A suggested process flow sheet for strip plating steel with lead or low-tin terne alloys is as follows:

(1) Electric cleaner
(2) Scrub rinse
(3) Acid treatment
(4) Scrub rinse
(5) Deionized water spray
(6) Fluoboric acid dip
(7) Lead or terne plating
(8) Recovery plus deionized water spray
(9) Cold rinse
(10) Hot rinse
(11) Hot air dryer

Since lead and terne-alloy coatings may be applied to coiled steel up to 0.100 in. in thickness, the horizontal-pass, straight-through-type plating cell may be required. The fluoborate-hydroquinone bath for both lead and terne has very high limiting current densities and anode and cathode efficiencies of about 100 per cent, i.e., well above the practical average current density for strip of about 400 asf. The baths have resistivities of approximately 4 micro ohms-cm^3 and excellent stability. However, low-tin terne-alloy anodes, even though of high purity (See Chapter 6, Table 12 and references) will precipitate a thin tin film by immersion during idle periods. Since strip lines normally operate 3 shifts a day, 7 days a week (other than shut down periods for maintenance or repairs) this should not become a problem. Also, filtration must be provided for any strip plating bath.

Deionized water must be provided for make-up of the plating bath, the fluoborate acid dip preceeding it, and for the rinses before and after the above steps. Otherwise sulfate ions precipitate lead. Chloride ions can also cause trouble.

Plating of any single metal (such as lead, tin, zinc, etc.) on a strip line is simpler to operate and control, compared to the deposition of an alloy of only two metals, such as lead and tin. The thickness and alloy composition will vary with most of the bath composition and operating variables. In cell-type plating, where the voltage drop through the strip from both the entering and exiting ends of the cell to the middle of the cell will be appreciable, the local current density at both ends will be greater than at the midpoint. At the relatively high to low local current densities involved, a considerable change in alloy composition can result. While passing through one cell, there will be a variation of tin content from high to low and back to high. Passing through two or more such plating cells further minimizes the effect of this factor. Uniformly high (100%) cathode and anode efficiency over a range of temperature from 120 to 160 F also minimizes variations in alloy and bath composition. Fortunately, the terne-alloy composition varies little with strip surface velocity (from zero to 250 fpm) in the lower alloy baths and at current densities from 200 to 600 asf. (References 39a to 39c in Chapter 6 give complete details on these matters.)

A practical average current density for both lead and low-tin terne alloy deposition would be from 300–400 asf at 12 to 18 volts in the type plating cell under discussion. This will vary with the thickness of the strip to be processed, anode to cathode clearance, etc.

The plating time to deposit 0.001 in. (1 mil) of lead and low-tin terne alloys at 100% efficiency and an average current density of 250 asf, compared to tin and zinc, are as follows:

Metal or Alloy	Plating Time for 1 mil, sec	Relative Plating Time for 1 mil
Lead	100	1.00
(5% Sn) Terne	106	1.06
(14% Sn) Terne	133	1.33
Zinc	197	1.97
Tin	113	1.13

References

1. Private communication, courtesy of the Hanson-Van Winkle-Munning Co., Matawan, N. J.
2. United States Steel Co., Pittsburgh, Pa., "The Making, Shaping and Treating of Steel," 6th ed., 1951.
3. Ess, T. J., "The Modern Strip Mill," Association of Iron and Steel Engineers, Pittsburgh, Pa., 1941.
4. Cooper, W. B., "31st An. Proc. Am. Electroplaters' Soc.," 14–16 (1943); "Iron and Steel Engineering Year Book," 172–176 (1943).
5. Anon., *Can. Metals Met. Inds.*, **13**, 8–11 (1950).
6. Johnson, S., and Jenison, G. C., "33rd An. Proc. Am. Electroplaters' Soc.," 102–115 (1946).
7. Stoltz, G. E., Hutcheson, J. A., and Baker, R. M., "Iron and Steel Engineering Yearbook," 141–149 (1943).
8. Erbe, J. R., *Iron Age*, 70–72 (Nov. 16, 1944).
9. Winkler, L. H., *Wire and Wire Products*, **16**, 687–693, 712–715 (Nov. 1941).
10. Lyons, E. H., Jr., *Trans. Electrochem. Soc.*, **78**, 318–338 (1940).
11. Wick, R. M. (to Bethlehem Steel Co.), "Electrodeposition of Zinc," U.S. Patent 2,392,075 (Jan. 1, 1946).
12. (a) Camin, E. L. A., and Kenmore, H., "Electroplating Apparatus," U.S. Patent 2,495,695 (May 8, 1944).
    (b) Kenmore, H., and Manson, W. J., "Method and Apparatus for Continuously Electroplating Heavy Wire and Similar Strip Material," U.S. Patent 2,680,710 (Sept. 14, 1950).
13. Kenmore, H., and Durr, F. L., *Wire and Wire Products*, **23**, 135–138 (1948).
14. Lancy, L. E., "Electrolytic Treating Apparatus," U.S. Patent 2,601,535 (Aug. 9, 1948).
15. Hoare, W. E., Tin Research Institute, "Tinplate Handbook."
16. Potthoff, K. T., "Method and Apparatus for Treating Pipes, Bars, etc.," U.S. Patent 1,789,596 (Jan. 20, 1931).
17. Potthoff, K. T., "Methods and Apparatus for Treating Pipes," U.S. Patent 1,997,013 (Nov. 18, 1930).
18. Hoare, W. E., Hedges, E. S., and Barry, B. T. K., The Technology of Tinplate, St. Martin's Press, New York, 1965.
19. Johnston, S. S., Anode Structure for Continuous Strip Electroplating, U.S. Pat. No. 3,445,371 (May 20, 1969).
20. Kitamura, Y., Electrochemical Treatment of Metal Surfaces and the Products Thereof, U.S. Pat. No. 2,998,361 (Aug. 29, 1961).
21. Uehida, H., et al., Process for Electrolytic Chromium Plating Steel Strip Without a Bluish Tint While Using Two or More Plating Tanks, U.S. Pat. No. 3,316,160 (Apr. 25, 1967).
22. Fukuda, N., et al., "Process for Coating Tin-Free Steel With Layers of Metallic Chromium and Chromium Oxide," *J. Electrochem. Soc., Electrochem. Technology*, **116**, No. 3, pp. 398–402 (1969).
23. Allen, James E., "Tin Line Conversion to Chrome," Assoc. Iron & Steel Engineers Meeting, Pitts., Pa. (Oct. 7, 1969).

27. ANODE AND CATHODE ROD AND BUS SYSTEMS

A. Kenneth Graham

Consulting Engineer, Jenkintown, Pa.

AND

H. L. Pinkerton

Fellow Engineer, Process Engineering Department, Aerospace Division, Westinghouse Electric Corporation, Baltimore, Md.

Revised by Lawrence J. Durney

The widespread change from generators to rectifiers as the preferred power source for electroplating operations has materially altered the power distribution systems in plating rooms. It is now extremely advantageous to provide each electrified tank with its own power source, and to locate this source close to the tank it services. Since rectifiers generally are provided with voltage control devices and meters for monitoring voltage and amperage, the use of power wasting resistance type rheostats can be eliminated in most cases. Shorter busbar runs between tanks and the more conveniently located rectifiers reduce power losses even further, and/or permit the use of small conductor cross-section at an additional saving in installed cost.

In a few unusual cases, it may be desirable and/or necessary to operate two or more tanks from a single rectifier. For these cases, the use of a rheostat is necessary, and the proper arrangement is shown in Fig. 1.

In some installations, power requirements may dictate the use of more than one rectifier for a single tank. Suitable methods of connecting the rectifiers are shown in Figs. 2 and 3.

Figure 2 illustrates a typical arrangement for connecting two rectifiers to a single cathodic treatment tank. Each rectifier is connected to a separate anode rod but to a common cathode rod. The current from each rectifier must be approximately balanced; otherwise the cathode current density on the work on opposite sides will be different.

Figure 3 illustrates a more conventional arrangement for connecting two or more rectifiers to a single cathodic treatment tank in which the cathode rods are split so that each rectifier supplies current independently to both sides of the common cathode rod. This system is commonly used for large tanks on automatic plating machines.

Another special case in which continuous strip or wire can be alternately given cathodic and anodic treatment in one tank without direct current contact is illustrated in Fig. 4. This shows two direct sources, each separately connected to a pair of top and bottom anodes and cathodes. The strip passing between them serves as a series electrode, alternately becoming negative and positive. The spacing between adjacent anodes and cathodes must be somewhat greater

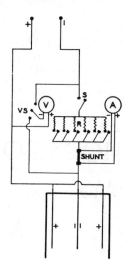

Fig. 1. Typical tank rheostat showing parallel resistances and contact switches, R, ammeter and shunt, A, voltmeter, V, switch for measuring tank or current source voltage, VS, and a master switch, S, which enables one to open the circuit without altering the setting of the individual resistor switches.

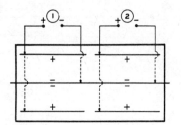

Fig. 3. More conventional arrangement for connecting two rectifiers to a single cathodic treatment tank.

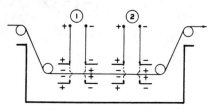

Fig. 4. Illustrating the use of two-current sources for alternate cathodic and anodic treatments of continuous wire or strip in a single tank without the use of a current contactor.

TABLE 1. TYPICAL CROSS SECTIONS FOR COPPER BUS

Rectangular	Round
0.25 x 3 in.	0.5 in. d.
0.25 x 4 in.	0.75 in. d.
0.25 x 6 in.	1.0 in. d.
0.5 x 4 in.	
0.5 x 6 in.	

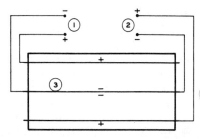

Fig. 2. Typical arrangement for connecting two rectifiers to a single cathodic treatment tank.

than the spacing between the electrodes and the strip to avoid direct feeding of current from anodes to cathodes.

Low Voltage d-c Bus

Engineering Data. Copper, brass, and aluminum have been used for connecting low voltage d-c sources to processing tanks. Copper is by far the most commonly employed because of its high conductivity.

Copper Bus. The manufacturers of copper bars for bus normally furnish high-purity, cold-rolled copper for this purpose with cross sections as shown in Table 1. Other shapes or sizes are employed in special locations. Flexible cables are used for connecting rigid bus to current-carrying moving parts such as an agitator. Woven copper braid is also used for this purpose.

The yearly cost of any bus installation is made up of two components: (a) the amortization of the investment represented by the installed cost of the bus, and (b) the cost of the inevitable lost power which is dissipated as heat in the bus. For a given amperage, as the cross section of the bus is increased, component (a) is increased in proportion, but component (b) is decreased. Since the installed cost of bus can often amount to an appreciable fraction of the complete installation cost for the plating department, it will be worthwhile to examine this subject a little more closely:

Let

B = bus cost, installed, \$/lb
D = current density in the bus, asi

I = operating current, amp
L = length of bus, ft*
C = cost or power, ¢/KWH
N = number of years to amortize bus cost
A = amortization, \$/yr
P = power cost, due to loss in bus, \$/yr
T = total cost, \$/yr

The power loss in watts in the bus is given by

$$\text{Power loss} = \frac{rI^2L}{\text{cross section of bus}} = rDIL \quad (1)$$

where r is the resistance (8.30×10^{-6} ohms at 25 C (67 F)) of a bar of copper 1 in. sq and 1 ft long, which weighs 3.86 lb. From these data it is a simple calculation to show that

$$A = \frac{3.86\ BIL}{DN} \quad (2)$$

and for an operating year of 250 24-hr days

$$P = 0.498 \times 10^{-6}\ CDIL \quad (3)$$

then

$$T = \frac{3.86\ BIL}{DL} + 0.498 \times 10^{-6}\ CDIL \quad (4)$$

Since C, B, I, N and L are constants fixed by the conditions of installation, the current density, D, is the only variable to be considered in designing. By a little calculus it can be shown that T (Eq. 4) will be a minimum when

$$D_{\text{opt}} = 2784 \left(\frac{B}{CN}\right)^{1/2} \quad (5)$$

The minimum total yearly cost, T', is then given by substituting the value of D_{opt} from Eq. (5) in Eq. (4)

$$T' = 2.772 \left(\frac{CB}{N}\right)^{1/2} IL \quad (6)$$

In passing, it may be observed that under minimum cost conditions $A = P$, i.e., the total cost is made up of equal amounts of amortization and power cost.

*Note: L is twice the actual run from current source to tank since there are two legs to the bus system.

In certain cases, notably in very long runs carrying high currents, consideration of voltage drop may override the economics of bus installation. For instance, suppose a 20,000 amp 6-volt current source is expected to deliver not less than 5.7 volts at a tank 50 ft away. The permissible voltage drop, E, is then 0.3 volt and the cross-section (from Table 17, Chapter 1) should be

$$\text{Cross-section, in.}^2 = 9.03 \times 10^{-6}\ LI/E$$
$$= 30.1\ \text{in.}^2$$

The bus current density must then not be over 660 asi regardless of the economics.

Installation Details. For large power requirements involving high amperage and low voltage, multiple copper strips of rectangular cross section are installed in a vertical position, as far as possible, with spacing provided between each strip to obtain maximum cooling and maintain the minimum temperature. The spacing is usually equal to the thickness of the copper, and this also facilitates joining as shown in Fig. 5.

In making joints the minimum length of contact area between strips is usually the width of the copper bus. The contact surfaces after cleaning are draw-filed to remove nicks, burrs, and roughness; then they are further finished with abrasive paper if necessary. Normally such surfaces will not be smooth enough to give the most perfect electrical contact between the surfaces. To correct this situation lead foil (0.008 to 0.010 in. in thickness) is commonly placed between each pair of contact surfaces. This serves as a

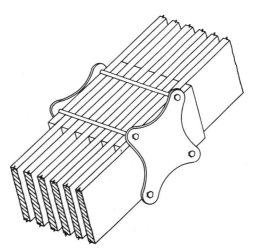

Fig. 5. Details of typical bus joint.

cushion of metal and gives very low contact resistance when the clamps are drawn up tight. After this is done, the edges at the joint should be coated with an acid-resistant paint. Joints prepared in this fashion have completely withstood oxidation or corrosion when examined after 10 to 20 years.

Silver plating of surfaces to be joined has been tried; however to obtain comparable results extra time must be spent in preparing the surfaces and the use of a soft lead foil offers a more full-proof procedure at less cost.

All bus work may be supported from the ceiling or walls by conventional brackets and must be suitably insulated against grounds with sheet plastic, rubber, or glass.

Insulated flexible cable is now extensively used instead of rigid copper bus of conventional design. The cable is available in a number of sizes, sometimes in two colors—black for the negative (cathode) conductor and red for the positive (anode) conductor. The insulation may be a polyvinyl chloride type passing both aging and heat requirements for service up to 90 C (194 F) and having a high degree of acid and alkali resistance. A specification for one size flexible cable is as follows:

Size	500 MCM
Current rating, amp	500
Number strands	37
Wall thickness, in.	5/64
Diameter (approx.), in.	0.98
Shipping weight (approx.), lb/M ft	2018

A 500 MCM copper conductor for d-c has a resistance of approximately 0.025 ohm/M ft at its operating temperature. When carrying 500 amp the voltage drop will be 1.25 volts/100 ft. For most installations this voltage drop is the maximum allowable from the current source to the tank; therefore if the distance is greater than 50 ft, it is necessary to parallel two or more conductors to limit the voltage drop.

Aside from voltage considerations, if the current requirement to a given tank is 1200 amp maximum, three cables in parallel would be required for both the positive and negative conductors.

When using flexible cable for low voltage d-c installations the problems encountered when using bus are largely eliminated. The number of joints usually can be restricted to the terminal ends only. The required supports and their insulation are greatly simplified, and the overall installed cost is favorable.

The Cadweld process for exothermic welding of the cable to copper conductors and/or fittings has largely been supplanted by swedging the connector onto the cable; filling the connector socket with solder (sweating) or both. While these methods do not provide the perfect connection of a weld, they have proved to be adequate, and the ease and economy of the procedure offsets the slight loss of contact effectiveness. These connections may then be bolted to the copper connectors (anode and cathode bars, rectifier terminals, etc.). Brazing is also sometimes used, particularly for connecting bar to flat stock, cable to bar stock, etc. Examples are shown in Fig. 6. Various clamps and bolted connections are also available for this purpose.

There are three basic types of supports for cables—racks, troughs and baskets. The necessary bends and fittings for all three types, the necessary hardware for hanging same from the ceiling or supporting same off a wall, and the necessary cable installation tools are all readily available from suppliers.

Aluminum Bus. From the standpoint of conductivity, the only bus metals competitive with copper are silver and aluminum. Silver is impractically expensive; whereas the installed cost of aluminum bus is about one-half that of equivalent copper bus. Millions of pounds of

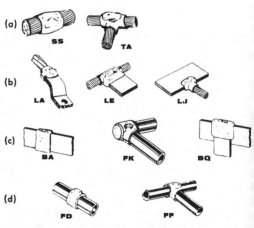

Fig. 6. Typical welded copper connections. (*Courtesy of Erico Products, Inc.*)
    (a) Cable to cable
    (b) Cable to lug or bus
    (c) Bus to bus
    (d) Bus to tube

aluminum bus are in use, with much of it in the electrochemical industry.

Bus may be designed on several bases: temperature rise, voltage drop or energy loss. In electroplating, the usual basis is voltage drop, whereby the cross section of aluminum bus should be 1.623 times that of copper, which means that 0.493 lb of aluminum will do the work of 1 lb of copper.

Bus joints may be bolted or clamped. Joint compounds are available, one of the best being Alcoa No. 2 Electrical Joint Compound which is a greaselike compound which chemically attacks aluminum oxide but not aluminum or copper. It should be applied to the joint area, which is then abraded with a wire brush or abrasive cloth. The joint is assembled without removing the compound, which is squeezed down to a monomolecular layer when the joint is drawn up; this layer has no measurable resistance. Joints thus prepared will operate for years without reformation of oxide, but if the joint is opened, the same procedure should be followed in making it up again. Bolts may be aluminum or steel. With aluminum bolts, flat washers are used, but because the coefficient expansion of aluminum is considerably greater than that of steel, Belleville washers are used with steel bolts in order to compensate for the differential expansion. As the number of bolts is increased from one to four, joint resistance is definitely decreased; there is no advantage in more than four bolts and in some cases additional bolts increase resistance owing to loss of contact area.

Bus bar clamps are more practical than bolts when a large number of bars are to be joined. Two types of clamps give satisfactory service: the curved back clamp and the spring washer clamp. The first has convex faces that will be parallel when the bolts are drawn down to the rated capacities of the clamp. The second type is more popular in the electrochemical industry. Large Belleville washers of rated capacity are incorporated as part of the clamps to distribute the bolting pressure evenly. When they are totally compressed, sufficient clamping pressure has been applied.

Welded or brazed connections are also used in permanent installations.

Aluminum bus has given many years of service over hot acid zinc baths and cyanide plating baths, but for the work rods in a still plating tank, for instance, copper may be preferred because it is easier to maintain a good contact with the rack. When joining aluminum and copper or any other metal, all parts should be coated before assembly with joint compound. In a corrosive atmosphere, it is essential to exclude moisture and electrolyte from the joint. After bolting, excess joint compound should be removed. The surface should be cleaned with a solvent and steel wool and a zinc-chromate primer (Mil Spec MIL-P-6879A) applied over the entire external area of the joint, including the hardware and as far as ½ in. beyond the joint. Finally the entire area should receive one or more coats of acid- and alkali-resistant paint.

Expansion joints should be provided in very long bus runs, i.e., over 50 to 75 ft.

ANODE AND CATHODE RODS

Anode rods are commonly made of round copper bar stock although bars of rectangular cross section are also used. They must have sufficient cross section to support the weight of the anodes and to carry the current. A design factor of 1000 amp/sq in. of area is satisfactory. Figures 7, 8, 9 and 10 illustrate various types of

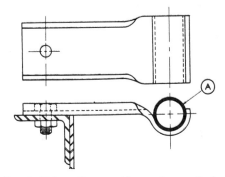

FIG. 7. Side wall support for anode or cathode rod showing an insulating sleeve (A).

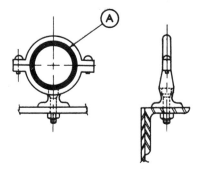

FIG. 8. Support for anode or cathode rod showing a sleeve insulator (A).

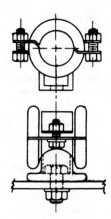

Fig. 9. Porcelain spool insulator support for anode or cathode rod.

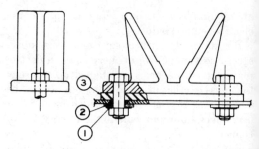

Fig. 11. Saddle type rod contact and insulated support. (1) Insulating washer (2) Insulating bushing, and (3) Insulating pad.

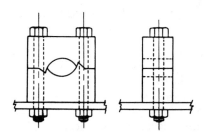

Fig. 10. Porcelain insulator support for anode or cathode rod.

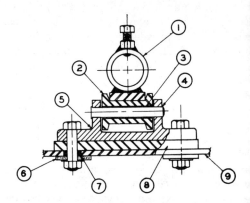

Fig. 12. Cathode rod roller for use with rod agitation. (1) Cathode rods protector shoe. (2) Roller. (3) Bearing. (4) Roller shaft. (5) Roller bracket. (6) Insulating washer. (7) Insulating bushing. (8) Insulating pad. (9) Tank rim.

insulated supports, which are usually mounted on the tank rim.

Cathode rods are similar to anode rods with respect to construction and installation. In plating and cathodic treatment tanks a long cathode rod may require added support to prevent sagging under the weight of the loaded racks. Pedestals are usually placed on the tank bottom to furnish this support. The anode rods are similarly supported in anodic treatment tanks.

The height of the rod support with respect to the tank rim and the solution operating level determines the length of hooks for the electrodes, i.e., the anodes or cathodes in cathodic or anodic treatments. The rods must also be so positioned that the electrodes will hang free without contacting the tank walls (especially in unlined tanks) or heating coils. The spacing between anode and cathode rods must also be carefully established to allow for proper spacing of the pieces being treated and for clearance when loading and unloading the tank.

Various insulating materials such as rubber, plastic, porcelain, and even wood may be used for mounting purposes. Figure 11 shows a saddle-type rod contact and support with a cutaway view to illustrate how it is insulated from the tank rim. This type depends on the weight to effect a positive contact.

A cathode rod roller for use with rod agitation such as is illustrated in Fig. 12 shows the same insulating details as in the previous figure.

All contacts between buses and the anode and cathode rods must be positive and of sufficient area to avoid over-heating. Two methods of connecting bus to the rods are illustrated in Figs. 13 and 14. When the cathode rod is used for agitation, there are two movements which are commonly used, namely, up and down and reciprocating horizontally. Both of these require a flexible connection to the bus. Woven wire copper braid or flexible cable are used for this purpose. Provision must be made for securely clamping or bolting the ends and to ensure ample contact area in making this type of connection. The current-carrying capacity of the flexi-

ANODE AND CATHODE ROD AND BUS SYSTEMS

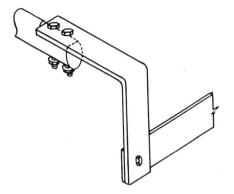

FIG. 13. Method of connecting a flat bus to a round rod.

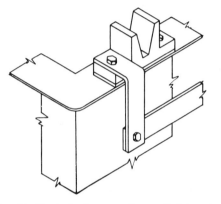

FIG. 14. Method of connecting a flat bus to a saddle contact.

ble connection also must be adequate to carry the current without over-heating. Spring actuated contact fingers or brushes are also used for making contact with sliding or rotating mechanisms.

Maintenance. No matter how well-designed and installed, an anode/cathode rod system can be a continuous source of problems unless properly maintained. This is especially true with solutions that are air agitated, or which gas heavily and thereby generate spray. Plating salts crystallized on insulating rod supports are conductive and can cause the insulators to become shorted to the tank.

Anode or cathode rods allowed to become oxidized, corroded or salt encrusted can lead to contact problems with anodes or racks. Such contact failure at the anode can be a source of improper current distribution, anode polarization resulting in excessive brightener consumption and excessive use of replenishing salts. At the cathode, effects include nonuniform plate distribution, failure to obtain required thicknesses and possibly loss of brightness and/or adhesion.

The required cleaning operations must be carried out with care. Copper is a contaminant for many plating solutions. The common practice of scrubbing the bars with a plater's brush dipped in plating solution, or hosing off the bars while adding water to the tank should be avoided except in the case of copper and copper alloy plating baths. With air agitated baths in particular, wrapping the anode's bars and anode contacts with transparent, flexible plastic sheet is quite helpful. Bolting the anode baskets to the anode bars also can be effective.

Acknowledgments

Thanks are due to Mr. John Davis of the Udylite Corporation, Dr. Robert Lloyd of the Westinghouse Electric Corporation, and Mr. W. Keith Switzer of Erico Products, Inc., for furnishing a number of the illustrations.

Reference

1. "Alcoa Aluminum Bus Conductor Handbook," Aluminum Company of America, Pittsburgh (1957).

28. EXHAUST SYSTEMS

ALLEN D. BRANDT

*Manager, Environmental Quality Control,
Bethlehem Steel Corporation, Bethlehem, Pa.*

AND

DAVID M. ANDERSON

*Assistant Manager, Environmental Quality Control,
Bethlehem Steel Corporation, Bethlehem, Pa.*

For reasons discussed briefly in Chapter 13, ventilation is necessary in electroplating and closely allied departments or shops. Not only the actual plating operations but also those frequently carried out preparatory to or subsequent to the plating may give rise to atmospheric contamination and require appropriate and adequate ventilation to avoid interference with vision, the creation of a nuisance, a housekeeping problem, or a health hazard. Since adequate ventilation is seldom inexpensive, it is advisable to keep to a minimum the amount and capacity of such equipment required by employing other means of controlling the air contamination to the extent that such measures are available and practicable. By way of example, inhibitors of various sorts are available for use in acid pickling baths, many of which decrease the amount of acid mist escaping into the air; plastic chips or balls, surface-foam type inhibitors, and surface-tension-depressive agents which decrease the mist escaping from plating and processing tanks are extensively used; and a great multitude of different degreasing or cleaning solvents are available so that one low in order of toxicity and volatility should be selected to meet operational requirements. In other words, it is well to give careful thought to other steps which might be taken advantageously with the view toward keeping the problem at a minimum, rather than to proceed blindly and then look to ventilation alone to accomplish the entire result. Not that it cannot be done by ventilation alone; it can, but it may be more expensive to do so.

Types of Ventilation

Industrial ventilation which has as its primary purpose the control of atmospheric contamination is of two main types; general or dilution, and local exhaust. The former type functions by causing sufficient air to pass through the room or shop to dilute the contaminant or contaminated air to a safe level by mixing with the relatively uncontaminated air entering the space under consideration from the outside. Thus, window fans, roof fans, central ventilating systems, etc., accomplish general ventilation. Local exhaust ventilation, on the other hand, functions by capturing the contaminant or grossly contaminated air as close to its source or point of release as possible and conveying it to the outside either directly or after passing through air pollution control devices where necessary. Slot-type exhaust hoods at tanks, hoods at grinding and buffing wheels, and canopy-type hoods above tanks are examples of local exhaust hoods. Even though the design, construction, and arrangement of the ventilating equipment indicate generally whether the effect produced is general or local exhaust ventilation, the determining criterion is the relationship between the concentration of air contaminant in the room air and that in the exhaust air before cleaning. Thus, if the concentration of contaminant in the exhaust air is essentially the same as or only slightly higher than in the room air, the ventilation is general; whereas, if the concentration of contaminant in the air exhausted is much greater than in the general room air, the ventilation is local exhaust.

As a general rule, local exhaust ventilation is preferred to general ventilation in the control of air contamination. In fact, there are many operations in the electroplating industry at which local exhaust must be used to prevent the creation of a health hazard. If the sources or areas of escape of the contaminant are definite and if the amount of contaminant escaping from such sources is considerable, no reasonable general ventilation rate will suffice to dilute the contaminant to a safe level before it reaches the breathing zone of the operators or attendants. Local exhaust ventilation, by acting in a wholly different manner, may be enployed to prevent the escape of huge amounts of contamination by capturing it at the points or areas of escape, and frequently comparatively low ventilation rates will suffice. For this reason, and possibly because the basic concepts of local exhaust ventilation are not generally understood, emphasis in this section will be placed on the principles governing the design of local exhaust systems. By the same token general ventilation will not be overlooked. It will be considered, but only in its proper perspective.

Fortunately, most operations involved in the electroplating industry may be lumped into one type insofar as ventilation is concerned, namely, open-tank type operations. They include alkali cleaning, solvent cleaning, pickling, plating, rinsing, electropolishing, anodizing and bright dipping, annealing, and others which are carried out in tanks, pots or vats. These will be considered as a whole and the miscellaneous additional operations requiring ventilation such as grinding, buffing, polishing and abrasive blasting will be considered separately.

Exhaust Systems for Tank-Type Operations. One of the early studies on ventilation for the control of health hazards from plating was reported in 1928[1]. Since then considerable research has been done and much has been written on the various aspects of ventilation of open-surface tanks[2-23]. Local exhaust may be applied to tanks, vats, or pots in any of three different ways, and the ventilation rate to accomplish adequate control of the contamination varies considerably one from the other. These different types of hoods are (1) enclosing hoods, (2) lateral exhaust hoods, and (3) canopy hoods. The characteristics of these different hoods follow.

Enclosing Hoods. Such hoods need not enclose the tank completely. Any hood which projects over the entire tank and encloses it on at least two sides falls in this category. Representative examples of enclosing hoods are shown in Figs. 1 and 2. It is the intent when employing hoods of this kind that the workers will have their heads outside the hoods at all times except possibly when repairs or adjustments are necessary. Control is effected by moving air through the hood at such rate that its velocity into all openings in the hood will prevent the escape of mist or gas. As a rule, an average air velocity into the hood openings* of 100 fpm is adequate to prevent escape of the contaminant. This rate has been employed commonly in the industry in the past in calculating the exhaust capacity required for enclosing hoods at tank operations, although it is frequently varied up or down as dictated by the experience of the designer. The rate of air flow through the openings necessary to accomplish adequate control (commonly referred to as the control velocity) is affected considerably by several factors, such as percentage of hood area that is open, activity or air disturbance inside the hood or at the openings, and relative harmfulness of mist or gas liberated by the enclosed operation. More recently these variations dictated by experience have been cataloged and condensed into very simple tables, making it possible for the inexperienced engineer to determine the capacity of the air-flow producing equipment for any existing or proposed open-surface tank operation[16, 18]. Since the procedure is applicable to lateral exhaust hoods and canopy hoods, it will be discussed later under the heading Exhaust Rate.

Lateral Exhaust Hoods. This type of hood is

---

* The entire open area at the hood through which the contaminant would escape into the room if air were not drawn through the hood by the exhaust fan.

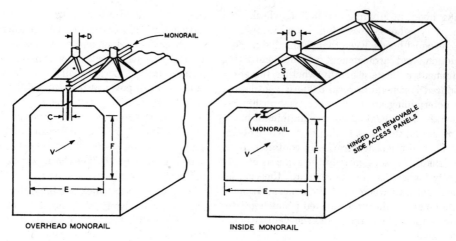

Fig. 1. Enclosing hoods. (*Courtesy of U.S.A. Standards Institute*)

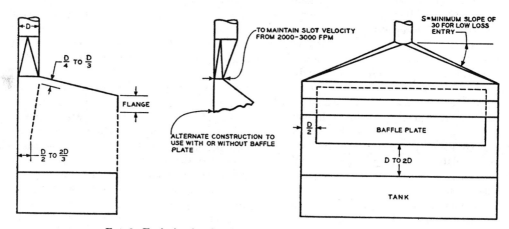

Fig. 2. Enclosing hoods. (*Courtesy of U.S.A. Standards Institute*)

more common than any other at open-surface tanks in plating and allied operations. It may take the form of slot-type hoods located along one or both long upper edges of the tank as illustrated in Fig. 3, or fishtail-type hoods usually along one long side of the tank as shown in Figs. 4 and 5, although they may be located along the middle as shown in Fig. 6, or along both sides if operating conditions permit.

At first the ventilation rate was specified on the basis of slot velocity into the hood. Since slot velocity in itself is meaningless, this basis was discarded early and rate per square foot of tank top came into general use. It is the one employed today and the "standard" value for plating, rinsing, and similar tanks is 120 cfm/sq ft and for degreasing tanks in which solvent loss by evaporation is an important consideration it is 50 cfm/sq ft. The ventilation rates used for design purposes by experienced engineers vary considerably from the "standard" bases to compensate for the influence of shape and size of tank, nearby air disturbances, toxicity of material in question, and rate of mist or gas release from the tank. Adequate design and satisfactory operation under these conditions depended heavily on the experience of the engineer. To improve this situation the different factors affecting the required exhaust rate were cataloged in such fashion as to reduce to a minimum the advantage of experience in calculating the capacity of the air-handling equipment[16,18]. This phase of the subject matter is treated later under Exhaust Rate, p. 680.

Even though the advantages of experience have been minimized, it is desirable that the influence of the factors cited above be recognized by all equipment designers so that the adverse influences may be avoided through selection, design,

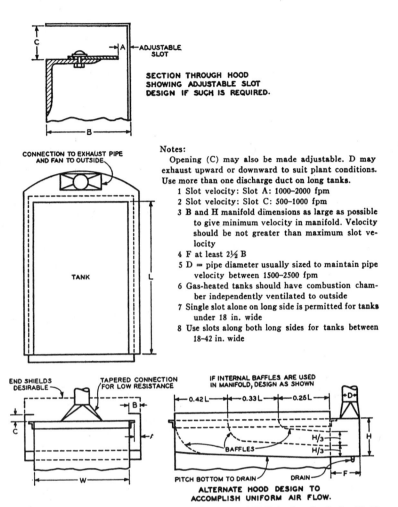

FIG. 3. Lateral exhaust hoods—slot type. (*Courtesy of U.S.A. Standards Institute*)

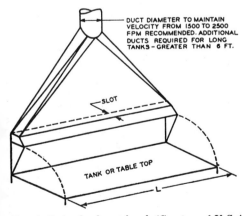

FIG. 4. Lateral exhaust hood. (*Courtesy of U.S.A. Standards Institute*)

and location of equipment and materials. The whole object of a lateral exhaust system at an open-surface tank is to cause all the contaminant rising from or being liberated at the surface of the tank, or from the surfaces of the parts or products removed from the liquid in the tank, to flow toward and enter the exhaust hood. It is actually unnecessary and impracticable to prevent every last bit of the contaminant from escaping into the workroom air since a limited amount will be diluted to a harmless concentration by the general ventilation in the room and by air flowing into the local exhaust system at the tank. However, if a very toxic liquid is used in the tank in question, the amount which may be be permitted to escape with impunity is much smaller than if the tank contains a relatively harmless liquid. Hence, a higher ventilation rate is required for relatively dangerous solutions than for the less dangerous ones. For the same reasons higher ventilation rates are required for tanks re-

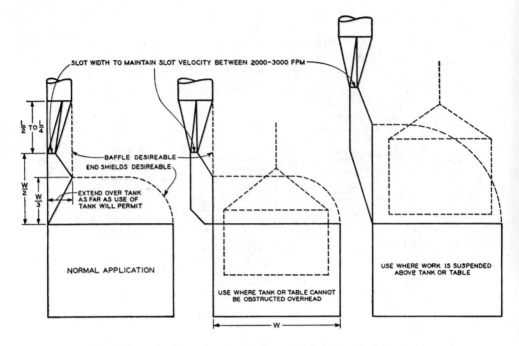

FIG. 5. Lateral exhaust hoods. (*Courtesy of U.S.A. Standards Institute*)

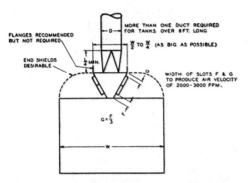

FIG. 6. Lateral exhaust hood along middle of tank. (*Courtesy of U.S.A. Standards Institute*)

leasing large amounts of mists and gases than for those from which little contaminant is liberated; higher rates are also required for those from which the contaminant escapes with considerable velocity (bubbles rising from a plating operation) than for those from which the contaminant is released with no significant velocity (evaporation from the surface of an unheated liquid). To capture all but a small portion of the contaminant released from the surface of the liquid in the tank it is only necessary that the velocity of the air moving toward the hood be adequate at all points above the tank to overcome the natural or artificial air disturbances which exist in all rooms. Lift trucks and ohter transport equipment, as well as people walking by at the rate of 3 to 6 mph (264 to 528 fpm), will temporarily upset the air flow pattern rather profoundly and will cause a spill of contaminated air into the workroom. For this reason the tanks should be located in large rooms and not close to open windows or doors, or to unit heaters. It is impracticable, and fortunately unnecessary, to provide exhaust ventilation at a rate adequate to overcome momentary air distrubances of this nature. Adequate control is accomplished if a positive air flow toward the hood of not less than 50 to 100 fpm is maintained at the points or areas above the liquid in the tank most remote from the hood. This requires a higher ventilation rate on the basis of cfm/sq ft for a short and wide tank than for a long and narrow one unless there are substantial shields or baffle at both ends of the tank (see Figs. 4, 5 and 6) a rather unusual arrangement, especially for slo hoods. This stems from the fact that for a tank which is long and narrow the air velocity at the point on the liquid surface most remote from the hood varies approximately directly with the distance from the hood; whereas for a square tank with a hood along only one edge the air velocity varies approximately as the square of the distance from the hood[14]. Consequently, in the case of tank not having large flanges at both ends higher exhaust rate is needed for a tank 2 x 2

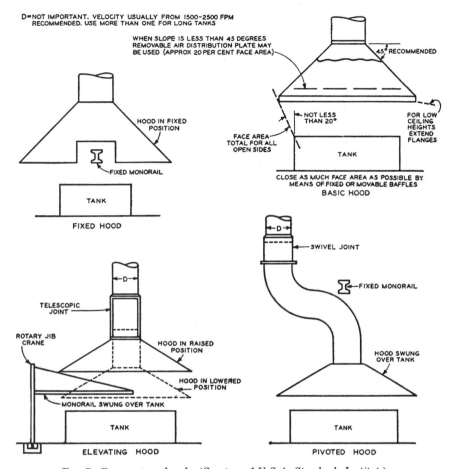

Fig. 7. Canopy-type hoods. (*Courtesy of U.S.A. Standards Institute*)

than for one 4 x 1 ft even though the surface area in each case is 4 sq ft.

Canopy Hoods. Hoods of this kind were very common in the past, but not at present. The usual type is illustrated in Fig. 7 (top right). Variations of canopy hoods arranged to accommodate certain operating fixtures are shown in Fig. 7 also. Such hoods should extend beyond the edges of the tanks in all directions. They are not recommended for operations which require workers to be in attendance at one or more sides of the tanks because the contaminated air rising toward the hood passes through the breathing zones of the workers. The mist and gas must be prevented from escaping into the workroom air by maintaining a curtain of air flowing into the hood through all openings or the entire open area between the top edges of the tank and the lower edges of the hood. Conventional design practice in the past has been to size the exhauster and ductwork on the basis of 140 cfm/sq ft opening. It has been the practice to vary this base value as dictated by experience to compensate or allow for unusual circumstances which would influence the effectiveness of control. The factors which may influence the ventilation requirements now have been cataloged and are presented in tables which permit satisfactory design by engineers with no experience in this field[16, 18] (see following discussion on Exhaust Rate).

Exhaust Rate. It has been shown earlier that the required ventilation rate to control adequately the nuisance or hazard associated with any operation in an open-surface tank varies considerably, depending on such factors as toxicity of the liquid in the tank or of the gas and mist rising therefrom, the rate at which mist or gas is released from the liquid in the tank, side and shape of the tank, and the type of hood used. These variables have been classified as to their effect on the required ventilation rate and summarized in Tables 1 to 5 in such fashion that if

TABLE 1. PROPERTIES OF MATERIALS WHICH GOVERN THE CAPTURE VELOCITY[1]

Substance	Threshold Limits (1968) (ppm)[2]	Threshold Limits (1968) (mg/cu. m)[3]	Flash Point (F)	Boiling Range (F)	Relative Evaporation Rate	Substance	Threshold Limits (1968) (ppm)[2]	Threshold Limits (1968) (mg/cu. m)[3]	Flash Point (F)	Boiling Range (F)	Relative Evaporation Rate
Acetaldehyde	200		−17	70	Fast	Lead		0.2	(solid)	—	—
Acetic acid	10		104	245	Medium	Manganese		5	(solid)	—	—
Acetic anhydride	5		121	284	Slow	Mercury (salt)		0.1	(solid)	—	—
Acetone	1000		0	134	Fast	Mesityl oxide	25		87	266	Medium
Acrolein	0.1		—	127	Fast	Methyl acetate	200		14	140	Fast
Acrylonitrile	20		32	173	Fast	Methyl alcohol	200		52	147	Fast
Ammonia	50		(gas)	—	—	Methyl bromide	20		none	40	Fast
Amyl acetate	100		92	290	Slow	Methyl butanone	100		—	199	Fast
Aniline	5		168	363	Nil	Methyl "Cellosolve"	25		105	255	Nil
Antimony		0.5	(solid)	—	—	Methyl "Cellosolve" acetate	25		132	289	Nil
Arsine	0.05		(gas)	—	—	Methyl chloroform	350		—	165	Fast
Benzene (Benzol)	25		12	176	Fast	Methylcyclohexanone	100		118	325	Nil
Bromine	0.1		(gas)	—	—	Methylethyl ketone	(see Butanone)				
n-Butanol	100		100	243	Slow	Methyl formate	100		−2	90	Fast
2-Butanone	200		30	176	Fast	Methyl isobutyl ketone	100		73	244	Medium
n-Butyl acetate	150		72	260	Medium	Monochlorobenzene	75		85	270	Medium
Butyl "Cellosolve"	50		141	340	Nil	Monofluorotrichloromethane	1000		none	75	Fast
Cadmium (salts)		0.2	(mist)	—	—	Mononitrotoluene	5		223	460	Nil
Carbon dioxide	5000		(gas)	—	—	Naphtha (coal tar)	100		100–110	300–400	Slow
Carbon disulfide	20		−22	114	Fast	Nickel carbonyl	0.001		140	110	Fast
Carbon monoxide	50		(gas)	—	—	Nitric acid		5	(mist)	—	—
Carbon tetrachloride	10		none	170	Fast	Nitrobenzene	1		190	412	Nil
"Cellosolve"	200		104	275	Slow	Nitrogen dioxide	5		(gas)	—	—
"Cellosolve" acetate	100		124	313	Slow	Octane	500		60	257	Medium
Chlorine	1		(gas)	—	—	Pentane	1000		−40	97	Fast
Chloroform	50		none	142	Fast	Pentanone	200		60	216	Medium
Chromic acid and chromate as CrO3		0.1	(mist)	—	—	Phosgene	0.1		(gas)	—	—
Cyanide as (CN)	—	5	(mist)	—	—	Phosphine	0.3		—	—	—
Cyclohexane	300		1	176	Fast	Phosphoric acid	—	1	(mist)	—	—
Cyclohexanol	50		154	322	Nil	Phosphorus trichloride	0.5		(mist)	—	—
Cyclohexanone	50		147	313	Nil	Isopropyl acetate	250		40	194	Fast
Cyclohexene	300		21	181	Fast	Isopropyl alcohol	400		53	181	Fast
o-Dichlorobenzene	50		151	354	Nil	Isopropyl ether	500		−18	156	Fast
1-1 Dichloroethane	100		39	134	Fast	Selenium		0.2	(solid)	—	—
1-2 Dichloroethane	50		70	181	Fast	Sodium hydroxide	—	2	(mist)	—	—
1-2 Dichloroethylene	200		43	141	Fast	Stibine	0.1		(gas)	—	—
Dichloroethyl ether	15		131	352	Nil	Stoddard solvent	500		100	300–400	Nil
Dichloromethane	500		none	104	Fast	Styrene monomer	100		90	295	Slow
Dimethylaniline	5		145	379	Nil	Sulfur chloride	1		(mist)	—	—
Dioxane	100		54	214	Medium	Sulfur dioxide	5		(gas)	—	—
Ethyl acetate	400		24	171	Fast	Sulfuric acid		1.0	(mist)	—	—
Ethyl alcohol	1000		55	173	Fast	Tellurium		0.1	(solid)	—	—
Ethyl benzene	100		59	277	Slow	1-1-2-2 Tetrachloroethane	5		none	295	Slow
Ethyl ether	400		−49	95	Fast						
Ethyl silicate	100		125	334	Slow						
Ethylene chlorohydrin	5		140	264	Medium						
Ethylene oxide	50		20	51	Fast	Tetrachloroethylene (perchloroethylene)	100		none	249	Medium
Formaldehyde	5		(gas)	—	—	Tin (salts)	—	2	(mist)	—	—
Formic acid	5		(gas)	—	—	Toluene (toluol)	200		40	232	Medium
Heptane	500		25	208	Medium	Toluidine	5		188	392	Nil
Hexane	500		−15	156	Fast	Trichloroethylene	100		none	189	Medium
Hydrogen chloride	5		(gas)	—	—	Turpentine	100		95	300	Slow
Hydrogen cyanide	10		(gas)	—	—	Vinyl chloride	500		20	57	Fast
Hydrogen fluoride	3		(gas)	—	—	Water	none		none	212	Slow
Hydrogen selenide	0.05		(gas)	—	—	Xylene (xylol)	100		63	291	Medium
Hydrogen sulfide	10		(gas)	—	—	Zinc chloride	—	1	(solid)	—	—
Iodine	0.1		(gas)	—	—						

[1] After Code Z9.1-1951 with permission of the U.S.A. Standards Institute.
[2] Parts of substance per million parts of air by volume.
[3] Milligrams of substance per cubic meter of air.

(a) the liquid in the tank, (b) the temperature at which the liquid will be maintained, (c) the type of hood to be used, and (d) the dimensions of the tank are known, the exhaust rate may be calculated from which the duct size and fan capacity may be derived.

Table 1 contains information on the toxicity (threshold limit), the combustibility (flash-

TABLE 2. HAZARD POTENTIAL OF MATERIALS
(According to Properties Given in Table 1[1])

Hazard Potential Class	Threshold Limits		Flash Point (F)
	(ppm)	(mg/cu m)	
A	0–100	0–0.10	under 100
B	101–500	0.11–0.50	100–200
C	over 500	over 0.50	over 200

[1] After Code Z9.1-1951 with permission of the U.S.A. Standards Institute.

TABLE 3. CONTAMINANT EVOLUTION RATE FROM TANKS
(According to Properties Given in Table 1, and Nature of Operation[1])

Contaminant Evolution Class	Temperature of Liquid in Tank (F)	Degrees below Boiling Point (F)	Relative Evaporation Rate	Gassing Rate[2]
a	over 200	0–20	Fast	High
b	150–200	21–50	Medium	Medium
c	94–149	51–100	Slow	Low
d	under 94	over 100	Nil	Nil

[1] After Code Z9.1-1951 with permission of the U.S.A. Standards Institute.

[2] For information on this factor see discussion under heading Exhaust Rate or refer to Tables 6, 7, and 8.

TABLE 4. REQUIRED MINIMUM CAPTURE VELOCITY[1], fpm

Class Combination (Tables 2 and 3)	Enclosing Hoods		Lateral Exhaust Hoods	Canopy Hoods	
	One Open Side	Two Open Sides		Three Open Sides	Four Open Sides
A-a, A-b, and B-a	75	100	100	125	175
A-c, B-b, and C-a	65	90	75	100	150
B-c and C-b	50	75	50	75	125
A-d, B-d, C-c, and C-d	Adequate general room ventilation sufficient				

[1] After Code Z9.1-1951 with permission of the U.S.A. Standards Institute.

point temperature), the boiling temperature, and the volatility of most liquids, gases, and solids which are used or encountered in open-surface tank operations in the electroplating industry. It will be noted that the threshold limit for those materials which occur in the air in the gaseous state are given in parts of substance per million parts of air (ppm) while those which are found in the air as mists or solids are given in milligrams of substance per cubic meter of air (mg/cu m).

These units have been chosen because they represent standard industrial hygiene practice.

From the data presented in Table 1 and some information as to the nature of the operation under consideration (see Tables 6, 7 and 8), the hazard potential and the contaminant evolution rate may be determined by referring to Tables 2 and 3. All materials used or encountered in open-surface tank operations are classified into three groups on the basis of those properties which determine the hazard potential. The ranges of the several properties as to hazard potential are shown in Table 2. Likewise all liquids used in cleaning, plating, and finishing tanks may be classified into four groups according to the rate of evolution of gas, vapor, or mist from the tank. The ranges of the properties which determine this classification are shown in Table 3. The gassing rate (Column 5 of Table 3) refers to the formation of gas bubbles beneath the surface of the liquid in the tank by chemical or electrochemical action. The bubbles usually are hydrogen or oxygen and form on the surface of the products being processed in the tank or on anodes or cathodes. As these bubbles rise to the surface, they escape into the air and may carry considerable liquid with them in the form of a mist. The rate of gassing is a function of the chemical or electrochemical activity in the tank and increases with the amount of work in the tank, the strength and temperature of the solution, and the current density in plating tanks. Consequently, this factor in Table 3 cannot be selected from the properties given in Table 1. The gassing rates for many

TABLE 5. REQUIRED VENTILATION RATE FOR LATERAL EXHAUST HOODS[1]
(Rate in cfm/sq ft of tank area for different tank proportions)

Required Capture Velocity (Table 4)	cfm/sq ft required to maintain capture velocity Ratio Tank Width / Tank Length				
	0.0–0.09	0.1–0.24	0.25–0.49	0.5–0.99	1.0–2.0
50	75	90	100	110	125
75	110	130	150	170	190
100	150	175	200	225	250

The above values are for lateral exhaust hoods which do not have the benefit of baffles to restrict the air flow from the ineffective areas. If adequate baffling as explained in the text is provided these values may be decreased 30 per cent.

[1] After Code Z9.1-1951 with permission of the U.S.A. Standards Institute.

TABLE 6. AIR-BORNE CONTAMINANTS RELEASED BY METALLIC SURFACE TREATMENT, PICKLING, ACID DIPPING, AND METAL CLEANING OPERATIONS*

Process	Type	Footnote	Component of Bath Which May Be Released to Atmosphere	Physical and Chemical Nature of Major Atmospheric Contaminant	Rate of Gassing	Usual Temp. Range (F)
Surface treatment	Anodizing aluminum		Chromic-sulfuric acids	Acid mist	Medium	95 and up
	Black magic		Conc. sol. alkaline oxidizing agents	Alkaline mist, steam	High	180–200
	Bonderizing	(a)	Boiling water	Water vapor & mist	Medium-high	140–212
	Chemical coloring		None	None	Nil	70–90
	Descaling	(b)	Nitric-sulfuric, hydrofluoric acids	Acid mist, hydrogen fluoride gas, water vapor & mist	Medium-high	70–150
	Dulite		Conc. sol. alkaline oxidizing agents	Alkaline mist, steam	High	180–200
	"Ebonol"		Conc. sol. alkaline oxidizing agents	Alkaline mist, steam	High	180–200
	Galvanic-anodized	(c)	Ammonium hydroxide	Ammonia gas, water vapor & mist	Low	140
	Hard-coating aluminum		Chromic-sulfuric acids	Acid mist	High	120–180
	"Jetal"		Conc. sol. alkaline oxidizing agents	Alkaline mist, steam	High	180–200
	Magcote	(d)	Sodium hydroxide	Alkaline mist, water vapor & mist	Low-medium	105–212
	Magnesium pre-dye dip		Ammonium hydroxide-ammonium acetate	Ammonia gas, water vapor & mist	Low	70–180
	Parkerizing	(a)	Boiling water	Water vapor & mist	Medium-high	140–212
	Zincate immersion	(e)	None	None	Nil	70–90
Pickling	Aluminum		Nitric acid	Nitrogen oxide gases	Medium	70–90
	Aluminum		Chromic, sulfuric acids	Acid mists	Low	140
	Aluminum		Sodium hydroxide	Alkaline mist	Medium	140
	Cast iron		Hydrofluoric-nitric acids	Hydrogen fluoride-nitrogen oxide gases	Medium-high	70–90
	Copper		Sulfuric acid	Acid, mist, water vapor & mist	Low-medium	125–175
	Copper	(f)	None	None	Nil	70–175
	"Duralumin"		Sodium fluoride, sulfuric acid	Hydrogen fluoride gas, acid mist	Low	70
	"Inconel"		Nitric, hydrofluoric-sulfuric acids	Nitrogen oxide, HF gases, acid mist, water vapor & mist	Medium-high	150–180
	Iron and steel		Hydrochloric-sulfuric acids	Hydrogen chloride gas-acid mist, water vapor & mist	Medium-high	70–190
	Magnesium		Chromic-sulfuric, nitric acids	Nitrogen oxide gases, acid mist, water vapor & mist	Medium	70–160
	"Monel"		Hydrochloric-sulfuric acids	Hydrogen chloride gas-acid mist, water vapor & mist	Medium-high	70–190
	Nickel		Hydrochloric-sulfuric acids	Hydrogen chloride gas-acid mist, water vapor & mist	Medium-high	70–190
	Nickel silver		Sulfuric acid	Acid mist, water vapor & mist	Low-medium	70–140
	Silver		Sodium cyanide	Cyanide mist, water vapor & mist	Low	70–210
	Stainless steel	(g)	Nitric, hydrofluoric, acids	Nitrogen oxide, hydrogen fluoride gases	Medium	125–180
	Stainless steel	(g) (h)	Hydrochloric-sulfuric acids	Hydrogen chloride gas-acid mist, water vapor & mist	Medium-high	70–190

TABLE 6.—Continued

Process	Type	Footnote	Component of Bath Which May Be Released to Atmosphere	Physical and Chemical Nature of Major Atmospheric Contaminant	Rate of Gassing	Usual Temp. Range (F)
	Stainless steel immunization		Nitric acid	Nitrogen oxide gases	Medium	70-120
	Stainless steel passivation		Nitric acid	Nitrogen oxide gases	Medium	70-120
Acid dipping	Aluminum bright dip		Phosphoric, nitric acids	Acid mist, water vapor & mist	High	200
	Aluminum bright dip		Nitric, sulfuric acids	Nitrogen oxide gases, acid mist	Medium-high	70-90
	Cadmium bright dip		None	None	Nil	70
	Copper bright dip		Nitric, sulfuric acids	Nitrogen oxide gases, acid mist	Medium-high	70-90
	Copper semibright dip		Sulfuric acid	Acid mist	Low-medium	70
	Copper alloys bright dip		Nitric, sulfuric acids	Nitrogen oxide gases, acid mist	Medium-high	70-90
	Copper matte dip		Nitric, sulfuric acids	Nitrogen oxide gases, acid mist	Medium-high	70-90
	Magnesium dip		Chromic acid	Acid mist, water vapor & mist	Medium	190-212
	Magnesium dip		Nitric, sulfuric acids	Nitrogen oxide gases, acid mist	Medium-high	70-90
	"Monel" dip		Nitric, sulfuric acids	Nitrogen oxide gases, acid mist	Medium-high	70-90
	Nickel and nickel alloys dip		Nitric, sulfuric acids	Nitrogen oxide gases, acid mist	Medium-high	70-90
	Silver dip		Nitric-sulfuric acids	Nitrogen oxide gases, acid mist	Medium-high	70-90
	Zinc and zinc alloys dip		Chromic, hydrochloric acids	Hydrogen chloride gas (If HCl attacks Zn)	Nil-Low	70-90
Metal cleaning	Alkaline cleaning	(i)	Alkaline sodium salts	Alkaline mist, water vapor & mist	Medium-high	160-210
	Degreasing		Trichloroethylene-perchloroethylene	Trichloroethylene-perchloroethylene vapors	High	188-250
	Emulsion cleaning		Petroleum-coal tar solvents	Petroleum-coal tar vapors	Low-medium	70-140
	Emulsion cleaning		Chlorinated hydrocarbons	Chlorinated hydrocarbon vapors	Medium	70-140

(a) Also aluminum seal, magnesium seal, magnesium dye set, dyeing anodized magnesium, magnesium alkaline dichromate soak, coloring anodized aluminum
(b) Stainless steel before electropolishing
(c) On magnesium
(d) Also manodyz, Dow-12
(e) On aluminum
(f) Sodium dichromate, sulfuric acid bath and ferrous sulfate, sulfuric acid bath
(g) Scale removal
(h) Scale loosening
(i) Soak and electrocleaning

* Based on Engineering Unit Plate No. 161, issued by the New York State Department of Labor, Division of Industrial Hygiene and Safety Standards, New York.

types of plating and allied operations are given in Tables 6, 7, and 8. If the operation under consideration is not found in these tables, a suitable gassing rate may be assigned by referring to a similar operation found in the table.

In using Tables 2 and 3 to find the key for selecting the required capture velocity from Table 4, the lowest letter (as regards order in alphabet) is chosen. For example, if for the substance under consideration the permissible limit from Table 1 is over 500 ppm and the flash point is under 100 F, letter A would apply, not C. The same rule applies in selecting the contamination evolution rate from Table 3.

These two letters in combination are then referred to Table 4 from which the required minimum capture velocity is found depending on the kind of hood used or to be used. For example, if the particular operation in question is classified as B from Table 2 and a from Table 3, the required capture velocity is 75 fpm if an enclosing hood is used with one side or a portion thereof open, 100 fpm if a lateral exhaust hood is used, and 175 fpm if a conventional canopy hood with four open sides is used. The exhaust rate in cfm for an enclosing hood and for a canopy hood is then the product of the resulting capture velocity in fpm and the open area in the hood or between the hood and tank in sq ft, as explained earlier. For lateral exhaust hoods the capture velocity is

TABLE 7. AIR-BORNE CONTAMINANTS RELEASED BY ELECTROPOLISHING, ELECTROPLATING AND ETCHING OPERATIONS*

Process	Type	Footnote	Component of Bath Which May Be Released to Atmosphere	Physical and Chemical Nature of Major Atmospheric Contaminant	Rate of Gassing	Usual Temp. Range (F)
Electropolishing	Aluminum	(a)	Sulfuric, hydrofluoric acids	Acid mist, hydrogen gas, water vapor & mist	Medium	140–200
	Brass, bronze	(a)	Phosphoric acid	Acid mist	Low	68
	Copper	(a)	Phosphoric acid	Acid mist	Low	68
	Iron	(a)	Sulfuric, hydrochloric, perchloric acids	Acid mist, hydrogen chloride gas, water vapor & mist	Medium	68–175
	"Monel"	(a)	Sulfuric acid	Acid mist, water vapor & mist	Medium	86–160
	Nickel	(a)	Sulfuric acid	Acid mist, water vapor & mist	Medium	86–160
	Stainless steel	(a)	Sulfuric, hydrofluoric, chromic acids	Acid mist, hydrogen fluoride gas, steam	Medium-high	70–300
	Steel	(a)	Sulfuric, hydrochloric, perchloric acids	Acid mist, hydrogen chloride gas, water vapor & mist	Medium	68–175
Strike solutions	Copper		Cyanide salts	Cyanide mist	Medium	70–90
	Silver		Cyanide salts	Cyanide mist	Medium	70–90
	Wood's nickel		Nickel chloride, hydrochloric acid	Hydrogen chloride gas, chloride mist	Medium	70–90
Etching	Aluminum		Sodium hydroxide-soda ash-trisodium phosphate	Alkaline mist, water vapor & mist	High	160–180
	Copper	(b)	Hydrochloric acid	Hydrogen chloride gas	Medium	70–90
	Copper	(c)	None	None	Nil	70
Electroless plating	Copper		Formaldehyde	Formaldehyde	High	75
Electroplating Alkaline	Platinum		Ammonium phosphate, ammonia gas	Ammonia gas	Medium	158–203
	Tin		Sodium stannate	Tin salt mist, water vapor & mist	Low	140–170
	Zinc	(d)	None	None	Nil-low	170–180
Electroplating Fluoborate	Cadmium		Fluoborate salts	Fluoborate mist, water vapor & mist	Low-medium	70–170
	Copper		Copper fluoborate	Fluoborate mist, water vapor & mist	Low-medium	70–170
	Indium		Fluoborate salts	Fluoborate mist, water vapor & mist	Low-medium	70–170
	Lead		Lead fluoborate-fluoboric, hydrofluoric acids	Fluoborate mist, hydrofluoric acid	Low	70–90
	Lead-tin alloy		Lead fluoborate, fluoboric acid	Fluoborate mist	Low	70–100
	Nickel		Nickel fluoborate	Nickel fluoborate mist	Low	100–170
	Tin		Stannous fluoborate, fluoboric acid	Fluoborate mist	Low	70–100
	Zinc		Fluoborate salts	Fluoborate mist, water vapor & mist	Low-medium	70–170
Electroplating Cyanide	Brass, bronze	(e) (f)	Cyanide salts, ammonium hydroxide	Cyanide mist, ammonia gas	Nil-low	60–100
	Black nickel	(f)	None	None	Nil	70–100
	Bright zinc	(f)	Cyanide salts, sodium hydroxide	Cyanide, alkaline mists	Low	70–120
	Cadmium	(f)	None	None	Nil	70–100
	Copper	(f) (g)	None	None	Low-medium	70–160
	Copper	(f) (h)	Cyanide salts, sodium hydroxide	Cyanide, alkaline mists, water vapor & mist	Low-medium	110–160
	Gold	(f)	Cyanide salts	Cyanide mist, steam	Nil-low	75–214
	Indium	(f)	Cyanide salts, sodium hydroxide	Cyanide, alkaline mists	Low	70–120
	Silver	(f)	None	None	Nil-low	72–120
	Tin-zinc alloy	(f)	Cyanide salts, potassium hydroxide	Cyanide, alkaline mists, water vapor & mist	Low-medium	120–140
	White alloy	(f) (i)	Cyanide salts, sodium stannate	Cyanide, alkaline mists	Low	120–150
	Zinc	(f) (j)	Cyanide salts, sodium hydroxide	Cyanide, alkaline mists	Low-medium	70–120
Electroplating Acid	Chromium		Chromic acid	Chromic acid mist	High	90–140
	Copper	(k)	Copper sulfate, sulfuric acid	Sulfuric acid mist	Nil-low	75–120
	Gold		Cyanide salts	Cyanide mist	Low-medium	70–160
	Indium	(l)	None	None	Nil	70–120
	Indium	(p) (q)	Sulfamic acid, sulfamate salts	Sulfamate mist	Low	70–90
	Iron		Chloride salts, hydrochloric acid	Hydrochloric acid mist, water vapor & mist	Medium	190–210
	Iron	(l)	None	None	Nil	70–120

TABLE 7.—Continued

Process	Type	Foot-note	Component of Bath Which May Be Released to Atmosphere	Physical and Chemical Nature of Major Atmospheric Contaminant	Rate of Gassing	Usual Temp. Range (F)
	Nickel	(d)	Ammonium fluoride, hydrofluoric acid	Hydrofluoric acid mist	Low	102
	Nickel and Black Nickel	(l) (m)	None	None	Nil	70–90
	Nickel	(l) (j)	Nickel sulfate	Nickel sulfate mist	Medium	70–90
	Nickel	(p) (q)	Nickel sulfamate	Sulfamate mist	Low	75–160
	Nickel	(l) (m)	None	None	Nil	70–120
	Palladium	(m) (n)	None	None	Nil	70–120
	Rhodium	(l) (o)	None	None	Nil	70–120
	Tin		Tin halide	Halide mist	Medium	135–150
	Tin	(l)	None	None	Nil	70–120
	Zinc		Zinc chloride	Zinc chloride mist	Low	75–120
	Zinc	(l)	None	None	Nil	70–120

(a) Arsine may be produced due to the presence of arsenic in the metal or polishing bath
(b) Dull finish
(c) Ferric chloride bath
(d) On magnesium
(e) Also copper-cadmium bronze
(f) HCN gas may be evolved due to the acidic action of CO_2 in the air at the surface of the bath
(g) Conventional cyanide bath
(h) Except conventional cyanide bath
(i) Albaloy, Spekwhite, Bonwhite (alloys of copper, tin, zinc)
(j) Using insoluble anodes
(k) Over 90 F
(l) Sulfate bath
(m) Chloride bath
(n) Nitrite bath
(o) Phosphate bath
(p) Sulfamate bath
(q) Air agitated

* Based on Engineering Unit Plate No. 162, issued by the New York State Department of Labor, Division of Industrial Hygiene and Safety Standards, New York.

converted to exhaust rate per square foot of tank area by reference to Table 5, and the total exhaust rate is the product of this value in cfm and the area of the tank (length x width in feet).

Baffling at Local Exhaust Hoods. Refernce is made in the footnote in Table 5 to the advantage to be gained by suitable baffling; a considerably lower exhaust rate being required for adequate control than without baffling. The reason for this very briefly is as follows. The reader should consult Reference 14 for more detail. Air approaches a suction opening essentially uniformly from all unrestricted directions. Hence a hood, for example, that illustrated in Fig. 3, will draw much of the air it receives from the area to the rear of it. Since the purpose of the hood is to create air flow above the tank in such a way as to cause the contaminant escaping from the entire liquid surface to be carried into the hood, that portion of the air which reaches the hood from the rear of it is relatively ineffective. By installing a baffle above the hood which extends upward a distance approximately equal to one half the width of the tank (the width of the tank if the tank has a hood along only one long side), all the air entering the hood must pass through the zone of contaminant escape. Furthermore the velocity of the air flow through the effective area is increased substantially by baffling because for a given quantity-rate of air flow into the hood the velocity of approach increases as the cross-sectional area of approach decreases. Consequently, the exhaust rate required to effect adequate control may be decreased as much as 50 per cent by complete baffling although the usual baffling results in a reduction of only about 30 per cent. While reference has been made only to a baffle installed at the hood, a tank having one long side against or very close to a wall is effectively baffled. Likewise, a tank having an exhaust hood located lengthwise along the middle of the tank is effectively baffled.

Push-Pull Systems of Tank Ventilation. Since air emerging from an opening under pressure is directional and "throws" a considerable distance if uninterrupted, it appears at first thought that the exhaust rate at tanks equipped with lateral exhaust hoods might be reduced a great deal if a curtain of air is blown toward the exhaust hood along one long side of the tank from a narrow slot in a hood or pipe located along the opposite long side of the tank. Such air stream could be of small quantity-rate and high velocity and thereby carry all the contaminant into close proximity to the exhaust hood with the result that it could be captured effectively with a minimum ventilation rate in the lateral exhaust hood. Such systems having various and sundry peculiarities have been tried time and again with few favorable reports. The merits and shortcomings of such systems will not be argued here, other than to point out that the high velocity air curtain increases rapidly in quantity-rate by virtue of the air induced or entrained so that the ultimate quantity-rate of contaminated air flowing toward the exhaust hood is many times the initial quantity-rate emerging from the slot or holes in the pipe[24].

TABLE 8. AIR-BORNE CONTAMINANTS RELEASED BY STRIPPING OPERATIONS*

Coating to Be Stripped	Base Metal (Footnote)	Component of Bath Which May Be Released to Atmosphere	Physical and Chemical Nature of Major Atmospheric Contaminant	Rate of Gassing	Usual Temp. Range (F)
Anodized coatings	1, 7	Chromic acid	Acid mist, water vapor & mist	Medium	120–200
Black oxide coatings	14	Hydrochloric acid	Hydrogen chloride gas	Low-medium	70–125
Brass and bronze	8, 14 (a)	Sodium hydroxide, sodium cyanide	Alkaline, cyanide mists	Low-medium	70–90
Cadmium	8, 14 (a)	Sodium hydroxide, sodium cyanide	Alkaline, cyanide mists	Low-medium	70–90
	2, 4, 14	Hydrochloric acid	Acid mist, hydrogen chloride gas	Low-medium	70–90
Chromium	7, 8, 14 (a)	Sodium hydroxide	Alkaline mist, water vapor & mist	Low	70–150
	2, 4, 8, 14	Hydrochloric acid	Hydrogen chloride gas	Medium	70–125
	2, 4, 8, 18 (a)	Sulfuric acid	Acid mist	Medium	70–90
Copper	8, 14	Sodium hydroxide, sodium cyanide	Alkaline, cyanide mists	Low-medium	70–90
	7, 12, 14 (b)	None	None	Nil	70–90
	14 (a)	Alkaline cyanide	Cyanide mist	Low-medium	70–160
	1	Nitric acid	Nitrogen oxide gases	High	70–120
	18 (a)	Sodium hydroxide-sodium sulfide	Alkaline mist, water vapor & mist	Medium	185–195
Gold	4, 5, 6, 8, 9, 14 (a)	Sodium hydroxide, sodium cyanide	Alkaline, cyanide mists	Low-medium	70–90
	4, 5, 18 (a)	Sulfuric acid	Acid mist	Low-medium	70–100
Lead	13 (c)	Acetic acid, hydrogen peroxide	Oxygen mist	Low	70–90
	14 (a), (c)	Sodium hydroxide	Alkaline mist, water vapor & mist	Low-medium	70–140
Nickel	2, 4	Sulfuric, nitric acids	Nitrogen oxide gases	Medium-high	70–90
	2, 4 (a)	Hydrochloric acid	Hydrogen chloride gas	Low	70–90
	2, 4, 14 (a)	Sulfuric acid	Acid mist	Low	70–90
	7	Hydrofluoric acid	Hydrogen fluoride gas	Low-medium	70–90
	14	Fuming nitric acid	Nitrogen oxide gases	High	70–90
	(a), (d)	Hot water	Water vapor & mist	Medium	200
	1, 18, 19 (a)	Sulfuric acid	Acid mist, water vapor & mist	Low-medium	70–150
Phosphate coatings	15	Chromic acid	Acid mist, water vapor & mist	Low	165
	16	Ammonium hydroxide	Ammonia gas	Low-medium	70–90
Rhodium	10	Sulfuric, hydrochloric acids	Acid mist, hydrogen chloride gas	Low-medium	70–100
Silver	1	Nitric acid	Nitrogen oxide gases	High	70–90
	2, 11	Sulfuric, nitric acids	Nitrogen oxide gases, water vapor & mist	High	180
	8, 14 (a)	Sodium hydroxide, sodium cyanide	Alkaline, cyanide mists	Low	70–90
	17 (a)	Sodium cyanide	Cyanide mist	Low	70–90
Tin	2, 3, 4	Ferric chloride, copper sulfate, acetic acid	Acid mist	Nil-low	70–90
	(a)	Sodium hydroxide	Alkaline mist	Low	70–90
	2, 4, 14	Hydrochloric acid	Hydrogen chloride gas	Low-medium	70–90
	14 (a)	Sodium hydroxide	Alkaline mist, water vapor & mist	Medium	70–200
Zinc	1	Nitric acid	Nitrogen oxide gases	High	70–90
	8, 14	Sodium hydroxide, sodium cyanide	Alkaline, cyanide mists	Low	70–90

1. Aluminum
2. Brass
3. Bronze
4. Copper
5. Copper alloys
6. Ferrous metals
7. Magnesium
8. Nickel
9. Nickel alloys
10. Nickel-plated brass
11. Nickel silver
12. Nonferrous metals
13. Silver
14. Steel
15. Steel (manganese-type coatings)
16. Steel (zinc-type coatings)
17. White Metal
18. Zinc
19. Zinc-base die castings
(a) Electrolytic process
(b) Refers only to steel (14) when chromic, sulfuric acids bath is used.
(c) Also lead alloys
(d) Sodium nitrate bath

* Based on Engineering Unit Plate No. 163, issued by the New York State Department of Labor, Division of Industrial Hygiene and Safety Standards, New York.

Anyone desiring to explore the possibility of using a push-pull system should consult the literature on this subject, including References 14, 15, 25 and 26.

General Ventilation. It will be noted from Table 4 that certain types of open-surface tank operations do not require local exhaust ventilation —appropriate general ventilation will suffice.

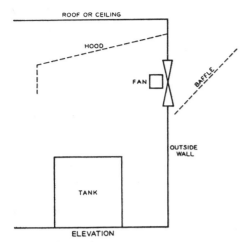

FIG. 8. General exhaust ventilation. (*Courtesy of U.S.A. Standards Institute*)

General ventilation may be accomplished by a central ventilating system, by roof fans, or by wall fans as shown diagrammatically in Fig. 8. It is difficult to state even in general terms what constitutes adequate general ventilation. Expressed in the common, even though rather meaningless, terms of air changes per hour, the general ventilation rate should be somewhere between 6 and 60 air changes per hour depending on the amount of mist and gas being released into the atmosphere. In some cases, the local exhaust systems at tanks requiring them produce enough general ventilation in the plating room to satisfy the requirements of tanks not needing local exhaust.

Make-up Air. Only too frequently local exhaust systems are installed at tanks needing them and roof or wall fans are provided to satisfy general ventilation requirements, but no thought is given to the entrance of make-up air into the room or building. If the plating and allied operations are housed in a relatively small space which is separated from the remainder of the building in which it is located by partitions or walls, as is often the case, the plating room may become air-bound because of wholly inadequate means for ingress of the air being exhausted by the ventilating systems. It is just as important to provide specific means for the air to enter as it is to provide the exhaust systems because the latter cannot operate otherwise. It is not entirely unusual to find rooms of this nature in which the air flow through low speed wall or roof fans is reversed by the negative pressure created in the tightly partitioned room by the exhausters in the local exhaust systems. This is particularly true in cold weather when doors and windows normally are closed tightly. The make-up air may need to be heated during the winter season in northern climates to avoid objectionable drafts over the workers. Also, it frequently is necessary to filter it to avoid plating defects. In those rare cases where make-up air is supplied to the plating room by a fan system, there should be an excess of exhaust capacity over supply so that air from the plating room which may contain considerable moisture and corrosive acids does not flow into the remainder of the building. The excess of exhaust capacity over supply should be in the order of 10 to 20 per cent, and even more if there is only partial partitioning of the plating room from the remainder of the building. However, if the presence of dust or other contaminants in the air entering the plating room is critical, the systems should be balanced and the supply air should be filtered.

Ventilation of Operations Other Than Open-Surface Tanks. Allied operations sometimes conducted as part of the electroplating process include abrasive blasting, grinding, buffing, and polishing. These operations seldom create a health hazard unless performed on rather toxic materials such as cadmium or lead. Abrasive blasting will create a serious health hazard if sand is used as the blasting medium. However, if steel grit or other material low in free silica is used, the safety hazard as well as the house-keeping problem far outweigh health considerations.

Abrasive Blasting. If blast cleaning is done, it should be carried out in properly ventilated equipment which has been designed for the purpose. These are many kinds of such equipment available commercially, including blasting rooms, blasting cabinets, blasting tables, and blasting barrels. The characteristics of the work to be done govern to a large extent which type of equipment is preferred. All such equipment requires proper ventilation. Most blasting rooms are so constructed that make-up air enters at or near the roof and is exhausted through a grill-type floor. The recommended minimum ventilation rate for blasting with friable abrasives (including sand) is 80 cfm/sq ft of cross sectional plan area. If a non-friable abrasive (such as steel shot) is used, and if the material being blasted is not coated with sand or a coating contains toxic components (such as lead), this rate may be cut in half. For rooms designed to be ventilated horizontally, the recommended minimum rate is 100 cfm/sq ft of cross sectional area in a plane at right angles to the direction of the air flow if a friable

TABLE 9. EXHAUST VENTILATION RATES FOR GRINDING, BUFFING, AND POLISHING WHEELS AND BELTS OF DIFFERENT SIZES[1]

Type of Equipment	Wheel or Belt Dimension (in.)	Hood Exhaust Outlet Diameter in Inches										
		3	3½	4	4½	5	5½	6	6½	7	8
		Exhaust Ventilation Rate in Cubic Feet per Minute per Outlet										
		220	300	390	500	610	740	880	1040	1200	1560
Grinding and cut-off wheels	Diameter	<9	—	9–16	16–19	19–24	—	24–30	—	30–36	—
	Width	1½	—	2	3	4	—	5	—	6	—
Buffing and polishing wheels	Diameter	—	<9	—	9–16	16–19	19–24	—	24–30	—	—
	Width	—	2	—	3	4	5	—	6	—	—
Horizontal single-spindle discs	Diameter	<12	—	12–19	—	19–30	—	30–36	—	—	—
Horizontal double-spindle discs	Diameter	—	—	—	—	<19	—	19–25 / 30–53 (2 outlets)	—	25–30	53–72 (4 outlets)
Vertical single-spindle discs (not covered)	Diameter	—	—	<20 (2 outlets)	—	—	20–30 (2 outlets)	30–53 (4 outlets)	—	53–72 (5 outlets)	
Vertical single-spindle discs (more than half covered)	Diameter	—	—	20–30 (2 outlets)	<20	—	—	30–53 (2 outlets)	—	—	53–72 (2 outlets)
Belts and straps	Width	<3	3–5	5–7	7–9	9–11	11–13	—	—	—	—

[1] After "Industrial Health Engineering" with permission of John Wiley & Sons, Inc. (Ref. 14).

abrasive is used and one-half this value if a nonfriable abrasive is used and the material being blasted is not coated with sand or a toxic coating. Abrasive blasting cabinets should be ventilated at such rates that the air flow into the cabinet through all openings (excluding curtains) is not less than 500 fpm. Rotary abrasive blasting tables should be enclosed or baffled as effectively as possible and should be ventilated at such rates as to produce a velocity of air into all openings of 250 fpm (based on free openings without baffles or curtains). Blasting barrels also should be ventilated at a rate capable of producing an air flow into the mill of 500 fpm through all openings.

Grinding, Buffing and Polishing. It is now rather common to find that commercial equipment of this kind incorporates hoods and connecting pipe outlets for exhaust ventilation as an integral part. This applies only to stationary machines, not portable ones. Consequently the stationary grinders, buffers and polishers may be ventilated very easily by providing exhausters of the proper size and the necessary piping or duct work. In Table 9 are summarized the recommended exhaust rates for different types and sizes of equipment. These rates are for hoods providing good enclosure, except where specified otherwise. Much thought has been given to dust control at portable grinding tools for a number of years. One satisfactory solution has taken the form of a ventilated work bench in which air is drawn down through the top

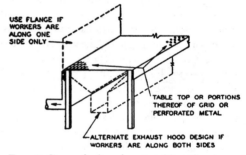

FIG. 9. Down-draft exhaust table. (*Courtesy of John Wiley & Sons*)

of the bench as illustrated in Fig. 9 or laterally across it to the rear similar to the arrangement shown in Fig. 4. Hoppers and suitable cleanout arrangements are required at downdraft tables to remove the dirt which accumulates. For downdraft tables the ventilation rate should be in the order of 200 cfm/sq ft of table top unless the working area is enclosed similar to the arrangement in Fig. 2, in which case this figure may be reduced, but it should not go below 100 cfm/sq ft of opening into the hood provided by the enclosure. If lateral exhaust to the rear of the table is employed, the capture velocity at the side of the table remote from the hood should be not less than 200 fpm. The exhaust rate can be computed by doubling the rate for a capture velocity of 100 fpm given in Table 5.

In recent years the so-called low-volume, high-velocity exhaust system approach has been applied to portable grinding tools with varying degrees of success. In this approach, dust control is achieved by exhausting air directly at the point of dust generation, using close-fitting, custom-made hoods. Exhaust volumes are small (usually less than 100 cfm), but system static pressures are high (5 in. of mercury or higher). A good description of these systems, including drawings, may be found in reference 23.

Drying Rooms or Tunnels. Where one of the preparatory or finishing operations is such as solvent degreasing, or dip or spray coating, it frequently is necessary to cause or permit the surfaces to become dry before the next operation is undertaken. In fully automatic plating departments this is sometimes accomplished by merely transporting the processed units about in loops on the conveying systems sufficiently to give them time to dry. This is not advisable generally because the evaporating liquid must be handled entirely by general ventilation which frequently requires an excessive ventilation rate. A common procedure is to convey the processed parts through ventilated, and sometimes heated, tunnels of such length that the parts being processed are dried when they emerge. Such tunnels can be enclosed entirely except for openings large enough to accommodate the entrance and exit of the parts, and require, therefore, only a minimum ventilation rate for very effective solvent control. As a rule, the exhaust rate is computed on the basis of 100 cfm/sq ft of opening into the tunnel, but the possibility of explosive vapor concentrations in the tunnel must not be overlooked where combustible liquids are being evaporated. This aspect of the problem may be resolved by estimating or measuring the rate at which liquid is carried into the tunnel and calculating the ventilation rate necessary to keep the vapor concentration in the tunnel below 20 per cent of the lower explosive limit. If this ventilation rate is higher than that calculated for contamination control, it should be used; otherwise the rate based on 100 cfm/sq ft of opening into the tunnel should be used. Table 10 gives the ventilation rates required to prevent explosion hazards for a number of combustible liquids frequently encountered as thinners in the plating industry. The necessary ventilation rate for any combustible liquid is given by

$$V = \frac{195.5 \times 10^3}{(\text{mol. wt}) \times (L.E.L.)}$$

TABLE 10. DILUTION RATE REQUIRED TO PREVENT EXPLOSION

Substance	Sp Gr	Mol. Wt.	Lower Explosive Limit (%)	Cu Ft of Air to Dilute 1 lb to Safe Concentration
Acetaldehyde	0.78	44	4.0	1,110
Acetic acid	1.05	60	5.4	600
Acetone	0.79	58	2.6	1,270
Acrylonitrile	0.80	53	3.0	1,210
Amyl acetate	0.88	130	1.1	1,365
Benzene	0.88	78	1.4	1,790
n-Butanol	0.81	74	1.4	1,890
n-Butyl acetate	0.88	116	1.4	1,200
Carbon disulfide	1.26	76	1.2	2,140
Cellosolve	0.93	90	2.6	835
Cellosolve acetate	0.98	132	1.7	865
Cyclohexane	0.78	84	1.3	1,775
Ethyl acetate	0.90	88	2.2	1,020
Ethyl alcohol	0.79	46	3.3	1,295
Ethylene oxide	0.89	44	3.0	1,480
Gasoline	0.75	86	1.3	1,750
Heptane	0.68	100	1.1	1,775
Hexane	0.66	86	1.2	1,820
Methyl acetate	0.93	74	3.2	825
Methyl alcohol	0.79	32	6.7	915
Methylethyl ketone	0.81	72	1.8	1,500
Methyl formate	0.98	60	4.5	725
Octane	0.71	114	1.0	1,630
Pentane	0.63	72	1.4	1,940
Isopropyl acetate	0.88	102	2.0	960
Isopropyl alcohol	0.79	60	2.0	1,630
Styrene	0.91	104	1.1	1,710
Toluene	0.87	92	1.3	1,675
Turpentine	<1	136	0.8	1,795
Vinyl chloride	0.97	63	4.0	775
Xylene	0.88	106	1.0	1,845

TABLE 11. MINIMUM DUCT WALL THICKNESS[1]

Diameter of Duct (in.)	U.S. Standard Sheet Steel Gage*	
	Gases, Mists and Fumes	Dusts
Up to 8"	20	16
8" to 18"	18	14
18" to 30"	16	12
Over 30"	14	10

* Use two gages heavier material for hoods enclosures, elbows and transitions.

[1] From "Steel Mill Ventilation" (reference 22), with permission of American Iron and Steel Institute.

where V is the volume of air in cubic feet required to dilute 1 lb of the liquid in question when in the gaseous state to a concentration equal to 20 per cent of the lower explosive limit; mol. wt is the molecular weight of the substance; and $L.E.L.$ is the lower explosive limit of the substance in per cent.

If the plating procedure is such that the parts are dripping as they leave the processing tanks, drainboards should be provided to keep the liquid from draining onto the floor and then contaminating the air through evaporation. In addition, local exhaust ventilation should be provided along the drainboard at about the same rate as it is provided at the tank, or the drainboard should be housed in a tunnel which should be ventilated as outlined in the preceding paragraph.

Design and Construction of Local Exhaust Systems

The subject will not be treated here except to emphasize a few details which frequently are overlooked. For detailed information the reader should consult References 14 and 19.

It is a mistake to fit the local exhaust system to some fan that happens to be available. All fans, but especially the exhaust fans in local exhaust systems, should be selected with great care. They must have the performance characteristics which will satisfy the demands of the hoods and piping system. As a rule, centrifugal fans are preferred although some axial flow fans have operating characteristics that are wholly satisfactory. Belt-driven fans are preferred to direct connected ones because they permit subsequent speed manipulations if such becomes necessary. Roof fans and wall fans, of course, are usually of the propeller type. Slow speed units should be chosen rather than high speed ones even in the face of much higher first cost because maintenance and repair are much lower and they produce much less noise—a constant source of annoyance to the workers.

A list of good practices for constructing and installing local exhaust systems is given below. If these are followed, the minimum requirements of most official agencies having jurisdiction will be met. However, it is always wise to consult the minimum requirements of the appropriate agency in your state to be sure that the system when installed will comply with them.

(1) Construct the hoods and ductwork of new materials, keeping the interior surfaces smooth and free from obstructions and making the joints tight as by soldering.

(2) Where the gases are not corrosive requiring special materials, use galvanized steel for the ducts and hoods. For corrosive gases and mists see section under "Special Materials of Construction."

(3) Use sheet metal thickness not less than specified in Table 11 for straight ducts, and two gages heavier for hoods, enclosures, elbows and transitions.

(4) Lap and rivet or spot weld on 3-in. centers the longitudinal duct seams.

(5) Lap girth joints in the direction of air flow using a 1-in. lap for ducts up to 19 in. in diameter and $1\frac{1}{4}$ in. for larger ones.

(6) Use long radius elbows (centerline radius preferably $2\frac{1}{2}$ pipe diameters but not less than $1\frac{1}{2}$ pipe diameters) with at least five sections per 90° elbow.

(7) Use tapered sections to accomplish changes in duct diameter and make the length of the taper not less than 5 in. for each 1-in. change in duct diameter.

(8) Connect branch to main in sides or top (not bottom) preferably near the large end of tapered transition section and at an included angle not greater than 45°, preferably 30° or less.

(9) Terminate all mains and sub-mains in a short dead end beyond the last branch and close it with a removable cap.

(10) Provide cleanouts in the ducts about every 12 ft and near bends and junctions.

(11) Use round ducts in preference to square or rectangular ones.

(12) Have a clearance of at least 6 in. between the duct and floor, wall or ceiling, and support the duct on 12- to 20-ft centers in such a manner that it will not fall if completely loaded with the contaminant.

(13) If dampers are needed to regulate or balance the air flow, use ruggedly constructed blast gates or other sliding dampers in preference to butterfly valves and provide for locking them in position after the system has been balanced.

(14) If necessary, equip terminal end of discharge with a drain-type stack[27]. Weather caps should not be used.

(15) Connect duct to inlet of exhauster by means of split-sleeve drawband at least one pipe diameter in length but not less than 5 in.

(16) If amount and type of contaminant in the air discharged to the outside are such that it may constitute a nuisance by way of air pollution,

include an appropriate air cleaning device in the ventilating system design. State or local air pollution control authorities should be consulted for requirements of applicable codes.

Special Materials of Construction

Many of the operations involved in the electroplating industry give rise to gases and mists which are very corrosive to the usual materials employed in the construction of hoods and ducts of ventilating systems. To eliminate the need for frequent replacement, special kinds of duct construction have been developed which have substantial resistance to corrosion[28,29]. Although ducts and hoods so made are more costly as a rule to install, they cost less per year than do others with shorter lives.

Certain plastic products resistant to chemicals and suitable for duct construction or for use in protective coatings are available. One which has been used for duct construction is an unplasticized extruded polyvinyl chloride which may be fabricated by low-temperature gas welding or by use of a special adhesive. Another is an extruded polyethylene plastic which also may be gas welded. A third is a polyester resin reinforced with glass fibers. Others are a furan resin reinforced with glass fabric and a phenolic-resin cast plastic. There are others which are similarly useful for duct and/or tank construction or treatment[30].

References

1. Bloomfield, J. J., and Blum, W., *Public Health Reports*, **43**, 2330 (1928).
2. Riley, E. C., and Goldman, F. H., *Public Health Reports*, **52**, 172 (1937).
3. Harris, W. B., et al., *Ind. Bull., N. Y. State Dept. Labor*, **18**, 132 (1939).
4. Witheridge, W. H., and Walworth, H. T., *J. Ind. Hyg. Toxicol.*, **22**, 175 (1940).
5. Battista, W. P., Hatch, T., and Greenburg, L., *Heating, Piping and Air Conditioning*, **13**, 81 (1941).
6. Brandt, A. D., *Heating, Piping and Air Conditioning*, **13**, 434 (1941).
7. Silverman, L., *J. Ind. Hyg. Toxicol.*, **23**, 187 (1941).
8. Silverman, L., *J. Ind. Hyg. Toxicol.*, **24**, 267 (1942).
9. Morse, K. M., and Goldberg, L., *Ind. Med.*, **12**, 706 (1943).
10. Brandt, A. D., *Heating, Piping and Air Conditioning*, **16**, 428 (1944).
11. Hirsch, M., *Heating, Piping and Air Conditioning*, **16**, 628 (1944).
12. Brandt, A. D., *Heating and Ventilating*, **42**, No. 3, 69 (1945).
13. Brandt, A. D., *Heating, Piping and Air Conditioning*, **17**, 237 (1945).
14. Brandt, A. D., "Industrial Health Engineering," New York, John Wiley & Sons, Inc., 1947.
15. Battista, W. P., *Heating, Piping and Air Conditioning*, **19**, 85 (1947).
16. U.S.A. Standards Institute, "Safety Code for Ventilation and Operation of Open-Surface Tanks," Z9.1, New York, 1951.
17. Stern, A. C., *Monthly Rev., N. Y. State Dept. Labor, Div. Ind. Hyg. and Safety Standards*, **30**, 45 (1951); **31**, 1 (1952).
18. Kingsley, I., *Monthly Rev., N. Y. State Dept. Labor, Div. Ind. Hyg. and Safety Standards*, **32**, 11 (1953).
19. U.S.A. Standards Institute, "Fundamentals Governing the Design and Operation of Local Exhaust Systems," Z9.2, New York, 1960.
20. Hemeon, W. C. L., "Plant and Process Ventilation," 2nd ed., New York, The Industrial Press, 1963.
21. Powell, C. H., and Hosey, A. D., Eds., "The Industrial Environment, Its Evaluation and Control—Syllabus," Public Health Service Publication No. 614, Washington, 1965.
22. American Iron and Steel Institute, "Steel Mill Ventilation," New York, 1965.
23. American Conference of Governmental Industrial Hygienists, "Industrial Ventilation," 11th ed., Lansing, Mich., 1968.
24. Madison, R. D., and Elliot, W. R., *Heating, Piping, Air Conditioning*, **18**, 108 (1946).
25. Williams, C. I., *Factory Management and Maintenance*, **96**, No. 7, 76 (1938).
26. Parker, J. H., *Air Engineering*, **9**, No. 7, 26 (1967).
27. Clarke, J. H., *Heating, Piping, Air Conditioning*, **35**, No. 10, 111 (1963).
28. McWilliams, J. W., *Standardization*, **22**, 111 (1951).
29. Seymour, R. B., and Erich, E. A., *Metal Finishing*, **50**, No. 6, 117 (1952).
30. "Modern Plastics Encyclopedia, 1968," New York, McGraw-Hill, Inc., 1967.

29. ELECTRODE MATERIALS AND DESIGN

H. L. PINKERTON

Fellow Engineer, Process Engineering Department, Aerospace Division, Westinghouse Electric Corporation, Baltimore, Md.

Revised by Lawrence J. Durney

The anode is the positive electrode in a plating bath: it conducts the current into the solution and, by its shape and position relative to the cathode, influences the distribution of current over the cathode surface. Anodes may be soluble or insoluble, or both may be used in combination. In most plating operations a soluble anode supplies the metal which is deposited on the cathode; the solution is merely the means by which the metal is carried from anode to cathode. Insoluble electrodes are used either as cathodes or anodes in electrocleaning, as cathodes in electropolishing, and as anodes in certain plating operations, the most important of which is chromium plating.

SOLUBLE ANODES

Characteristics

An ideal soluble anode has the following desirable characteristics:

1. Corrodes smoothly and evenly under the influence of the current.
2. Produces, in corroding, a minimum quantity of sludge and metallic particles.
3. Corrodes with a high anode efficiency under normal operating conditions.
4. Has a high limiting current density.
5. Has a low rate of solution in the bath (without current).
6. Introduces no objectionable amount of impurities into the bath.

The metallurgical history of the anode, its purity, and the presence of certain added constituents are all factors in determining the anode characteristics above described.

Metallurgical Variations

Cast Anodes. These are usually cast in chill or permanent molds, which may introduce traces of iron. Sand-cast anodes are no longer used to any extent as they tend to be rough and to corrode unevenly. The grain structure of a cast anode is generally coarse and is variable from anode to anode as well as within one anode. The maximum length obtainable is limited by casting difficulties and structural weaknesses.

Rolled Anodes. Anodes formed by rolling a cast billet into the desired cross section will show a finer grain structure in two dimensions with crystals elongated in the direction of rolling. Traces of iron may be introduced from the rolls. Corrosion characteristics are improved over the plain cast anode. Much greater lengths are obtainable: in nickel an elliptical bar anode 22 ft long is available.

Electrolytic Anodes. Electrodeposited sheets of various thicknesses and sizes can be used as anodes. They have the finest grain size and highest purity of all commercial anode forms. The dissolution characteristics are usually not too desirable: (1) anodic corrosion may be nonuniform because of irregularities in the grain structure, (2) solid metal particle formation may be relatively high, (3) the anode efficiency is often low, and (4) anode polarization is greater for some metals.

For many metals, considerable progress has been made in overcoming problems of this nature. Additionally, the widespread use of titanium anode baskets, which can be readily bagged to retain particulate matter, has allowed additional freedom in their use. Electrolytic nickel is routinely supplied sheared to 1×1 in. $\times \frac{1}{2}$ in., or as rounds approximately 1 in. diameter by $\frac{3}{8}$ in. thick for such use.

Extruded Anodes. In some metals elliptical bar anodes can be cold extruded, affording refinement of grain size. Some "ball" anodes are made by shearing extruded cylindrical rods to approximately the size and shape of spherical ball anodes.

Ball Anode. As originally introduced, this anode form was chill cast, but it is now also made by shearing forged, extruded, rolled or continuous cast sections. They are used in bare steel cages, or for acid solutions in titanium baskets. Their advantages include (1) little or no scrap loss and (2) easy maintenance of constant anode area. In some alkaline or cyanide solutions with a tendency for the metal content to increase, they make it possible to balance the ratio of inert steel to soluble anode area so metal concentration remains constant.

Anode Suspension

Contacts. The most satisfactory electrical contact between two pieces of metal is a point contact designed so that there is sufficient metal near the contact to carry the current and to remove any heat generated by contact resistance. A line contact is less satisfactory and a surface-to-surface contact is the poorest. This is because contact resistance is a direct function of intimacy of contact, i.e., pressure per unit area, which is highest for the point contact under a given total load. In addition, the point, and to a lesser extent the line contact will cut through films of corrosion and foreign matter often present on the surfaces, and any slight movement of such contacts will tend to maintain clean surfaces. Because of its small cross section, the point contact has a high resistivity per unit length; however, the length, in direction of current travel, is so short that the added resistance in the conductor is negligible.

Anode Hooks. As discussed in Chapter 27, anode bars are often round copper rods of sufficient cross section to carry the current. Where greater strength is required, there may be used a pipe of equal cross-sectional metal area or a rectangle with the long side vertical.

Contact is usually made between the anode and anode bar by means of hooks. Hooks for use with the various anode rod types are shown in Fig. 1, A–D. Figures 1A and 1B give point contacts and are preferred. Bar anodes are usually supported by screwing the hook into a tapped hole at one end of the anode. Less frequently, a cast bar anode may be supported by casting the anode around the shank of the hook, which may extend into the anode any distance up to its full length. Plate or sheet anodes may be supported by S-hooks, preferably of form 1A, through holes in the corners or in lugs at the corners. For all these types of contact the solution level is usually maintained below the top of the anode; otherwise anode corrosion will cause the hook to loosen and pull out before the anode is consumed. The factor of bath contamination must also be considered (see Undersolution contacts). Plate anodes may also be supported by bolting between two bus bars (Fig. 2). Because this would produce excessive scrap when using soluble anodes, the method is normally used for insoluble anodes. Sheet-lead anodes are best supported in this fashion.

Anode Baskets. Most production installations now use anode baskets with ball, sheared, pellet or button type anode materials rather than solid anodes. Alkaline solutions will use steel baskets with the advantages previously described. Acid solutions will generally use titanium. These may be fashioned either as the

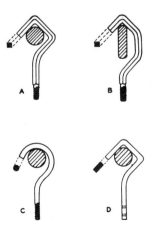

FIG. 1. Anode hooks.
(A), (B) Preferred types, illustrating knife edge or point contact
(C) Approaching line contact
(D) Line contact

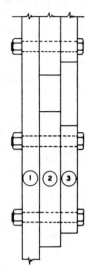

FIG. 2. Bolted contact for sheet anodes.
(1) Copper bus
(2) Sheet anode
(3) Copper bus (continuous or interrupted)

standard ball cage (Fig. 3) or as a rectangular basket. Common sizes are 2½ × 6 in. or 3 × 12 in., with the length as required by the tank working area. Titanium baskets can be used in most acid baths, including sulfamate and fluoborate types. Recently, however, they have been found unsuitable for certain types of acid tin. Titanium develops a nonconductive oxide which provides a protective film to prevent the basket from corroding. This film is thin enough so that it is easily broken by anode material placed in the basket, thereby establishing contact, but the basket itself draws no current. These baskets, therefore, cannot be used to control metal ratios in the manner described for steel baskets.

Experimentally, there has also been some success with the use of stainless steel baskets in acid copper solutions. In at least one case, stainless steel baskets are still in operation after two years of use.

Under-Solution Contacts

In alkaline solutions contacts are frequently made under the solution using hooks of steel or other metal not attacked by the solution.

Ball-anode containers of bare steel constitute the simplest under-solution contact. One form is shown in Fig. 3A. Plate anodes are sometimes suspended entirely beneath the solution by insoluble hooks, but this is normally limited to thin sheets such as silver because with heavy sheets the holes corrode excessively and the hooks tear out. Large plate anodes are occasionally supported as in Fig. 3B from holes near the middle of the anode. A rubber washer insulates the bolt head from the anode or the bolt head is covered with plastic; otherwise corrosion is accelerated by the contact of dissimilar metals and the bolt head pulls through. The springs serve to maintain contact pressure as the anode becomes thinner.

In acid solutions under-solution contacts are rare; however they are sometimes necessary in electroforming and other special applications such as in continuous plating of steel-mill products. The utmost care is required in designing such contacts to avoid solution contamination. The type depicted in Fig. 4A has been successfully used for positioning a large number of standard elliptical anodes under a nickel solution in which iron contamination had to be held below 15 ppm and copper below 2 ppm. The type shown in Fig. 4B can be used for installations of single anodes; 4C is not strictly an under-solution contact but is useful in violently agitated solutions or where the anode is nearly all immersed to produce minimum scrap. The molded plastic cap fits standard anodes and hooks and is available from supply houses.

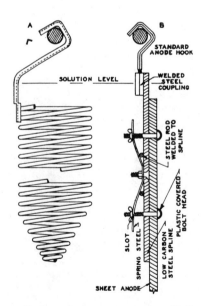

FIG. 3. Under-solution contacts for alkaline solutions.
(A) Steel cage for ball anodes
(B) Contact hook for large sheet anodes

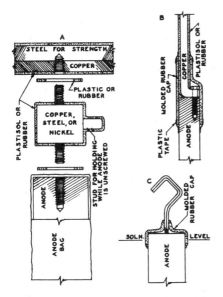

Fig. 4. Under-solution contacts for acid solutions.
(A) Submerged bus connection for a number of anodes
(B) Submerged bus connection for single anode
(C) Molded plastic anode cap

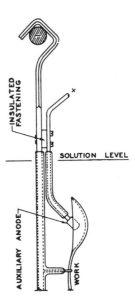

Fig. 5. Bipolar rack.

Standard Sizes and Shapes

Hooks. Standard hooks of types shown in Fig. 1A and 1B for bar anodes are made of steel, copper, nickel, or "Monel" metal from $5/16$, $3/8$, or $7/16$ in. sq stock, and threaded $5/16$—18, $3/8$—16, or $7/16$—14. They are available in lengths from 3 to 9 in., usually in 1-in. increments. Special sizes can be supplied to order. Hooks of $3/8$ in. sq titanium have recently been introduced.

Anodes. Elliptical bar anodes have a cross section about 1 to $1\frac{1}{4}$ in. $\times$ 3 to $3\frac{1}{4}$ in. and can be obtained cut to exact specified lengths up to the maximum available length for the particular anode material. They are usually sold drilled and tapped for a specified hook size. Curved bar anodes are also available and are especially useful in barrel plating, giving larger anode area and more even current distribution along the anode and on the work in the barrel.

Electrolytic sheet anodes can usually be obtained in any size up to 36 in. sq, in thicknesses from about $3/8$ to $5/8$ in., drilled as required for S-hooks.

Ball anodes average about $1\frac{1}{4}$ in. in diameter.

Auxiliary Anodes

Auxiliary anodes are specially designed anode forms to assist in such difficult problems as plating into deeply recessed parts. They may be either soluble or insoluble; they are connected to the positive bus but usually carry only a portion of the total current. In some cases the auxiliary anode circuit is isolated with separate control, but this is not always necessary. Auxiliary anodes are quite commonly used in industrial chromium plating and in electroforming.

Bipolar Rack. This is a plating rack which (a) holds the workpiece (cathode) in a predetermined position and (b) in addition carries an auxiliary anode in a fixed position relative to the cathode (Fig. 5). The anode with its support is insulated from the rest of the rack. Cathode contact is made in the usual way through the main rack hook, whereas the anode is connected to the anode rod by a flexible wire and clip or through an auxiliary support hook.

Bipolar Anodes. A bipolar anode is an intermediate or series electrode, usually soluble, with no electrical connection with the anode circuit. Such an anode is cathodic to the real anode and anodic to the workpieces being plated. Several ingenious arrangements have been described[1] using bipolar anodes.

INSOLUBLE ANODES AND CATHODES

The sole function of an insoluble electrode is to complete the electrical circuit to the solution; hence it is merely necessary for it (a) to be a good conductor and (b) to be unattacked by the bath with or without current flowing. Insoluble electrodes are used as anodes in plating, as cathodes in electropolishing and anodizing, and as either anodes or cathodes in electrolytic treatments in acid or alkaline solutions. In all such operations there will be strong gassing at the electrode, usually liberating hydrogen at the cathode and oxygen at the anode; however an insoluble anode in a chloride-bearing electrolyte will release chlorine. An insoluble anode will tend to oxidize certain bath constituents as organic addition agents and cyanide. The irreversible reactions occurring at insoluble anodes will materially increase the cell voltage over that required for soluble anodes operating under the same conditions.

Steel, nickel, alloyed lead and carbon (graphite) serve as insoluble anodes under specific conditions (see under specific materials). Platinized titanium, on the other hand, is insoluble in all common plating baths and is used extensively in gold plating.

Specific Anode Materials

This section deals primarily with the chemical composition of anode materials, referring to American practice. Information on usual anode current density and conditions of operation may be obtained from the tables in Chapter 6.

Brass. The composition of brass anodes should be approximately the same as the Cu/Zn composition of the deposit; otherwise the metal ratio in the bath will become unbalanced and will require correction by metal salt additions. In general 70/30 or 80/20 anodes are used for brass plating and 90/10 anodes for bronze (see Chapter 6). According to Thews[2] English practice seems to favor the use of copper anodes only, maintaining the zinc content of the bath with zinc cyanide.

The most damaging impurity is lead, and those shops which cast their own anodes from scrap or have this done at a local foundry must be very careful to exclude leaded or so-called "free-machining" brass from the melt.

The limiting current density for brass anodes under the best conditions is quite low: they may polarize excessively at anode current densities in excess of 8 to 10 asf; consequently a large anode area is required and brass anodes are available cast in corrugated or multi-edged form. Cast anodes perform best at anode current densities of 5 asf or less. Cold-rolled anodes corrode faster than anodes which have not been cold-worked and are recommended for installations where the anode area may be limited as in automatic and barrel-plating tanks.

There is little corrosion of brass anodes when the current is off.

The use of separate copper and zinc anodes will upset the Cu:Zn ratio in the bath and may cause roughness, owing to the deposition of copper on the zinc.

Suitable specifications for maximum percentage of impurities in brass anodes are:

Nickel .005   Tin   .005   Lead   .005
Arsenic .005   Antimony .005   Iron   .010

Cadmium.   Typical analysis:

Cd—99.95–99.97%   As—None–0.001%
Pb—0.008–0.03%   Zn—None–0.001%
Fe—0.005–0.008%   Tl—None
Cu—0.002–0.01%

The most objectionable impurities are antimony, arsenic, lead, tin, silver, and thallium. One U.S. government specification requires a minimum of 99.9 per cent cadmium, with not more than 0.05 per cent silver plus lead, plus tin and not more than 0.005 per cent of arsenic plus antimony plus thallium. All cadmium anodes are cast.

Commercial cadmium plating is largely done from a cyanide bath. Under normal operating conditions the rate of dissolution corresponds to over 100 per cent anode efficiency owing to chemical solubility. For this reason, no more anodes are placed in the bath than are required to carry the current without excessive polarization. Ball anodes in steel cages are preferred by many platers: these should be removed during periods of idleness because contact with the steel accelerates chemical corrosion when the current is off.

Because at a given current density the oxygen overvoltage on steel in a cyanide cadmium bath is higher than the potential of a dissolving cadmium anode, but lower than the oxygen overvoltage on a completely polarized cadmium anode[18], cadmium anodes in parallel with steel anodess never reach the completely polarized, insoluble state. However, it is also evident that if the cadmium anode is dissolving normally, extremely little current will be carried by the steel anode unless it is in a separate circuit at a higher potential. Unless so connected, therefore, steel anodes will not function to control cadmium build-up. When so connected, the steel anodes will oxidize, thereby forming a light rust which must be removed at intervals. It may be interesting to note here that except for this high oxygen overvoltage on steel, plain steel tanks could not be used for cyanide plating solutions, as otherwise stray currents would reach unmanageable magnitudes[19]. The limiting anode current density varies from 20 to 30 asf, is increased by an increase in free cyanide in the bath, and decreases as carbonate builds-up. Impurities of an oxidizing nature such as chromates or nitrates cause severe anode polarization even when present in very small quantity.

Carbon (Graphite). Sheet carbon electrodes are sometimes used as insoluble anodes in black nickel plating or sulfate-nickel baths and (rarely) in electrocleaning or as cathodes in stripping processes. They are fragile and exhibit a tendency to disintegrate under the action of anodic current especially when oxygen is evolved. Chlorine evolution is not detrimental to the anodes.

Copper. Table 1 illustrates several forms of commercially available copper. Of these, only the electrolytic and the rolled (usually rolled electrolytic) have been widely used as anodes.

Roughness in copper plating, both from acid and cyanide baths, has long been associated with anode performance. This in turn has been assumed to be related to the anode composition and perhaps also to its metallurgical history.

In acid sulfate baths it has been postulated[14] that the disproportionation of cuprous sulfate, known[15] to be formed in the electrolysis of these baths, results in the formation of very finely divided copper in the solution, thus causing roughness.

$$Cu_2SO_4 \rightarrow CuSO_4 + Cu^0$$

This explanation of the origin of roughness cannot, of course, apply to cyanide baths. Another explanation is that roughness is caused by very small crystal fragments dislodged from the anode by preferential electrolytic attack at the grain boundaries. Whatever its cause, it seems well established that roughness is influenced by the composition of the anode.

Nevers and his co-workers[14] found that in acid sulfate baths fire-refined anodes produced much smoother plates than the purer forms. A dark gelatinous film that was fairly adherent covered

TABLE 1. TYPICAL ANALYSES (IN PER CENT) OF VARIOUS FORMS OF COPPER

	OFHC	Electrolytic	Commercial Cast or Rolled	Commercial Phosphorized	Fire Refined
Copper	99.99+	99.97–99.98	99.9–99.94		>99.88
Oxygen	none	(f)	.05–.07		.042(c)
Sulfur	.002(a)	.001–.003(b)	.0015(b)		.004(c)
Phosphorus	<.0005	none	none	0.01–0.1	
Selenium				d	.0025(c)
Tellurium				d	.0013(c)
Silver	.001	.0007	(e)	d	.0095(c)
Arsenic				d	.0049(c)
Antimony	<.0005	<.0005	.0015		<.003
Iron	.001	none			
Nickel	.0008	.0003	.0014–.0025		<.05
Bismuth	<.0001	<.0001	.0001		<.003
Tin	.0001	.0001	.0001–.0002		
Zinc	<.0003	<.0003	.0015		
Manganese	.00005	.00005			
Lead	.0003	.0001	.001–.002		<.004
Reference	13	13	13	U.S. Pat. 2,689,216	ASTM B216-49

(a) As sulfide.
(b) As sulfate.
(c) Typical analysis, not part of Specification B216.
(d) Optional addition of .0005 to .01% of at least one of these elements.
(e) Usually included in copper analysis.
(f) Cast electrolytic copper will usually contain 0.3 to 0.4% oxygen, unless deoxidized with phosphorus when the oxygen content may be reduced to about 0.04%.

the anode, and there was practically no anode sludge and no appreciable build-up of copper in the solution. They attributed this phenomenon to the presence of silver, selenium, tellurium and arsenic in the anode. They later patented[16] the addition of 0.01 to 0.1% of phosphorus to the anode, which produces an adherent black film on the anode and prevents roughness. Additional advantages are claimed for the simultaneous presence of the other four elements.

Safranek and Faust[13] made an extensive study of various forms of copper anodes in acid sulfate and several cyanide electrolytes. They did not study phosphorized anodes in the acid bath, but stated that if chlorides were absent, all the anodes produced smooth deposits (at 1 mil, but nodular at 10 mil) but with 4 mg/l HCl in the bath, which tends to increase the production of anode sludge; only the OFHC (oxygen-free, high-conductivity) anode produced smooth deposits. In the cyanide baths in which phosphorized anodes were also studied, only the OFHC anodes consistently produced deposits with little or no roughness; furthermore with these anodes the solutions remained clear and free of copper particles. The use of PR in a potassium cyanide bath had no observable effect on roughness* but did decrease anode sludge formation and raised the current density at which polarization occurred under their conditions from 20 to 30 asf. Incidental results obtained in their study showed that in an air-agitated proprietary high-speed sodium cyanide bath, there was no anode polarization with any of the anode types at 80 asf and the anodes remained smooth and bright. They also showed that grain size had no effect on the production of roughness.

OFHC anodes have been used successfully in production in an air-agitated bright cyanide copper bath[24] without the use of bags or diaphragms.

The special anode types just discussed are a rather new concept; today most copper anodes are made from electrolytic copper and are either rolled (sheet or bar), extruded, or in the case of ball anodes, forged. Cast anodes produce much more anode sludge and consequently are not favored. Some electrolytic sheet anodes are used in cyanide baths.

Anode bags may be used in acid baths to minimize the production of roughness, but they are usually not wholly successful owing perhaps to the mechanism of cuprous ion formation suggested above. In cyanide baths, anode bags aggravate the tendency of the anode to polarize excessively at rather low current densities; therefore they are seldom used.

In acid baths, the limiting anode current density is so high (over 160 asf in air-agitated baths[13]) that for practical purposes it may be neglected. In cyanide baths copper anodes usually perform satisfactorily only up to about 20 asf.

Insoluble iron anodes may be used in cyanide baths to suppress anode polarization[2], but some ferrocyanide is introduced into the bath[3], which tends to reduce anode efficiency and causes dark, rough deposits[17] at about 1.5 to 2 g/l of iron.

Sometimes sheet copper anodes are hung from hooks in acid sulfate baths, and lead is poured around the hook and contact hole to prevent the hook from dissolving or pulling out when the anode is completely submerged to lessen scrap production.

Gold. Soluble high-purity rolled-gold anodes may be used in either the cyanide or acid baths.

In the rarely used acid solution gold anodes will passivate if the current density is too high or if insufficient free chloride ion is present.

In the usual potassium cyanide bath a gold anode usually dissolves satisfactorily under normal operating conditions. When present in the gold anode as impurities, lead, silver, bismuth, and arsenic, in that order, induce passivity as do certain impurities in the solution, notably ferrocyanide. It has been reported[2,4] that a gold anode is covered with a protecting layer of insoluble sodium gold cyanide when electrolyzed in sodium cyanide solutions and that even traces of sodium in potassium cyanide baths cause the formation of this film; however this restriction cannot be too serious since sodium cyanide baths have long been successfully operated with gold anodes. Because the behavior of gold anodes is likely to appear capricious, many cyanide solutions, especially in small installations, are operated with insoluble anodes[4]. The bath investment is then smaller and the possibility of theft

---

*It should be emphasized that the authors were concerned with "shelf roughness" caused by particles in the bath settling on a horizontal surface. This does not dispute the undoubted fact that PR in cyanide copper baths permits the production of smoother deposits at greater thicknesses than is possible with DC current by altering the habit of crystal growth.

is eliminated. Stainless steel is usually preferred as an insoluble anode, but hard carbon, "Nichrome," and "Duriron" have also been used.

Iron. Anodes for iron plating are high-purity forms such as low-carbon "Armco" ingot iron, wrought iron, or Swedish iron, usually in the form of chill-cast or rolled slabs. The iron dissolves chemically in these highly acid baths. Bags of "Vinyon," "Dynel," or blue African asbestos are a necessity to prevent roughness.

Lead. "Chemical Grade" lead is used for soluble anodes in lead plating. The ASTM (B29-49) specification for this material is:

Pb, min.—99.90%
Ag—0.002-0.020%   Zn, max.—0.001%
Cu—0.040-0.080%   Fe, max.—0.002%
As + Sb + Sn, max.— Bi, max—0.005%
  0.002%

Sheet lead may be used as anode: cast forms are also available with rectangular, multi-edged, or corrugated cross sections and with positive-contact hooks cast integrally with the anode or soldered into a hole in the top. Sheet lead is best supported by bolting between two bus bars.

Insoluble lead anodes are universally used in chromium plating and also find other applications in acid copper and sulfate nickel plating and as electrodes in electropickling, electropolishing, and anodizing.

In chromium plating pure lead anodes are attacked by the bath when the current is off; therefore the more resistant alloys with about 6 per cent antimony or tin are used. A small loss due to chromate formation cannot be avoided, and the anodes do not last indefinitely. The rate of chromate formation is lessened, and the desirable reoxidation of trivalent to hexavalent chromium in the bath is favored if the anode has a film of brown lead peroxide on its surface. This may be formed by anodic treatment in a sulfuric acid electrolyte overnight or for several days. It will be converted to chromate when the chromium bath is idle but will reform, with some loss, when the bath is again used. The reoxidation of trivalent to hexavalent chromium is also favored by a large anode area, and many special anode cross sections, some of them patented, are available. These may be ridged, ribbed, corrugated, or multi-edged. One study[26] has indicated that round cross sections are preferable to flat or oval shapes in that the entire anode surface is active. It was also shown that there was little difference in performance between rod and tubular anodes of the same outside diameter; the latter represented a saving in weight of 25 to 40% for the 1½ to 2 in. sizes. Anodes made by casting lead around a steel or copper core are also available. These have added strength, rigidity, and conductivity and are necessary when very long anodes are used and in other special applications. Lead-plated copper rods serving as anodes in the chromium plating of gun barrels has been described[23]. Flame-surfacing the lead anode prior to use allegedly improves its resistance to attack by the chromium bath[5]. A case alloy lead anode, containing 1 per cent each of tin and silver, is reported to perform best in a continuous strip plating line for chromium (see Chapter 27).

Although lead peroxide has a conductivity approaching that of metals, too heavy or irregular a coating may interfere with proper current distribution, especially when closely conforming anodes are used as in hard chromium plating. It is therefore customary to clean the anodes regularly, which is usually done by means of acid dips and scratch-brushing. A less tedious method has been patented[20] involving cathodic treatment in an alkaline pyrophosphate solution said to reduce the peroxide to lead.

Lead anodes are frequently called upon to carry heavy currents, and they must be thick enough to do so without overheating, which warps them and causes them to corrode excessively.

Nickel. Typical analyses of the various forms of nickel anodes available are shown in Table 2. The anode polarization of nickel increases and the evenness of corrosion in a given solution decreases with increasing purity of the anode and with the amount of mechanical work, i.e., rolling or forging performed on it. However, oxide additions, as in the depolarized anodes, increase the activity markedly. Rolled depolarized high-purity anodes corrode evenly and well in all modern nickel solutions.

Cast-carbon and rolled-carbon anodes do not corrode satisfactorily in solutions above pH 4.6[6]. The carbon and silica contents of these anodes form a film on the anode surface which retains loose nickel particles long enough to permit them to dissolve[7]; otherwise an intolerable amount of sludge containing much valuable undissolved nickel would be formed. This was the principal objection to the low-carbon, low-

TABLE 2. TYPICAL ANALYSES OF NICKEL ANODES

	Cast Carbon (%)	Rolled Carbon (%)	Cast Depolarized* (%)	Rolled Depolarized (%)	Electrolytic (%)
Ni, plus Co†	99.6	99.3	99.6	99.5	99.95
Al	Trace			0.01	
Cu	Trace	0.02	Up to 0.01	0.05	0.03
Fe	0.10	0.07	0.20	0.08	0.04
Mg	0.04	—	—	—	Nil
Mn	Nil	—	Nil	0.005	—
S	—	0.01	0.01	0.005	Trace
Si	0.05	0.25	0.10	0.02	Trace
O	Nil	Nil	0.2	0.25	Nil
C	0.21	0.25	Up to 0.01	Up to 0.01	Trace

* Rarely encountered.

† In the past, cobalt content averaged about 0.7 per cent, but because of recent cobalt shortages, more has been recovered resulting in a lower content, about 0.2 per cent, in nickel anodes.

purity, so-called "99 per cent cast nickel" anode, which because of its higher content of impurities corroded actively, but unevenly, at any pH. Electrolytic anodes, especially if rolled and annealed, are said to corrode satisfactorily[8] in solutions of low pH and high chloride content. Improper annealing will usually result in spalling and exfoliation during electrolysis. Improved performance can be obtained by breaking electrolytic sheet into small pieces for use in an anode saver basket, thus lowering the anode current density[9].

A process has been patented[10] for welding nickel anodes together for special purposes. Unlike most such welds, this weld corrodes at a slightly slower rate than the parent metal and has proved satisfactory in practice, even in low pH nickel baths.

The use of titanium anode hooks in acid solutions permits complete immersion of the anode in operation, and thus reduces the production of scrap. As with titanium baskets, the hook draws no current; it does not corrode nor influence the rate of corrosion of the nickel in contact with it. The hooks may be used repeatedly. The relatively higher electrical resistivity of titanium will increase the tank voltage somewhat, but the effect is small (of the order of 0.1 volt for a current of 100A/anode).

Silver. Rolled silver sheet for anodes is available 99.95(+) to 99.99(+) fine. In the lower purity anodes traces of lead, iron, bismuth, manganese, tellurium, selenium, sulfur, and antimony can cause the conditions of "black anodes" in which an insoluble black scum forms on the anode. This dissolves slowly in periods of idleness or with low anode current density in solutions of high free cyanide. Annealing and quenching decrease the grain size of the silver and tend to prevent the damaging impurities from concentrating in the grain boundaries in which condition they do the most harm. Fine grain size is therefore desirable in anodes of this purity but does not seem to be a factor in the very pure grades.

In decorative silver plating the anode current density is rarely allowed to exceed 10 asf, corresponding to a ratio of anode to cathode area of at least 1 to 1. As is usual in cyanide solutions, the current density at which the anode tends to polarize decreases with decreasing free cyanide or increasing carbonate concentration in the bath. The addition of nitrate and hydroxide anions, customary in industrial high-speed silver plating, as for bearings, greatly increases the permissible anode current density, as much as 500 asf being used in some cases without encountering black anodes.

It has been shown[28] that high purity (99.97 fine) silver anodes will flake or "shed" tiny particles under certain conditions. The particles are very fine and can only be detected in solution by the Tyndall effect (directing a concentrated beam of light through a beaker of the solution). Rough deposits may result, and the anode will corrode unevenly and be rough. Flaking can be caused by organic contamination of the bath and is aggravated by a low-anode current density and in high-concentration, high-temperature baths. Under these conditions, large grain anodes flake more than those with small grain size, and anodes of lower purity are even less satisfactory.

Steel. The principal use of steel electrodes is in electrocleaning. In many installations, these are simply steel sheets bolted to the anode or cathode bar as the case may be. Perforated sheet or heavy expanded metal should be used to allow free passage of solution and to improve current distribution. Periodic cleaning of these steel electrodes is usually necessary to remove encrustations which may interfere with the passage of current. Commercial electrodes of the expanded metal type, and various modifications are available.

Tin. High-purity tin is used for anodes;

"Straits Tin" is frequently specified, particularly the grades under the designations Banka, Billiton, or Penang. A typical analysis is:

Sn—99.940%   Cu—0.004%
Sb—0.007%   Fe—0.007%
Pb—0.015%   Ag—Nil
As—0.014%   S—0.003%
Bi—0.002%

Lead is the most damaging impurity because it not only interferes with plating operations but is undesirable in the many food applications to which tin plate is put.

In acid-tin baths the anode corrodes evenly under normal conditions, dissolving as Sn (II). In the alkaline stannate bath the tin must dissolve as Sn (IV) instead of Sn (II) for satisfactory operation, and to achieve this the anode must be polarized under special conditions[21]. If a clean tin anode is placed in an alkaline stannate solution and operated at a low current density, the tin will dissolve as stannite which disproportionates to stannate and tin, and the anode will be gray, covered with a loose black smut. As the current density is increased, a point is reached (the "critical" current density) at which the potential rises suddenly, usually causing the tank current to fall back at the same time. Further increase in current causes only a slow rise in voltage, and now the anode is not dissolving at all and is covered with a dense, shiny black, adherent film. If the current is now reduced below the critical current density, a yellow-green film is formed and is the indication that tin is dissolving as stannate. If the current is reduced too far, the yellow-green film becomes very thin and may show interference colors (iridescence), indicating that the region is being approached in which tin will again dissolve as stannite.

The actual value of the critical current density will vary, being higher at higher temperatures and higher free alkali content. It is most easily recognized by following voltmeter and ammeter readings during the filming process or by actual inspection of the anode (the low current density loose black scum must not be confused with the adherent black film). The anode must not be operated above the critical current density for longer than is required to form the black film; otherwise the anode becomes permanently passive and the black film will not redissolve when the current is lowered, but must be removed mechanically or with a mineral acid such as hydrochloric. A voltmeter is of help in maintaining correct anode conditions. The properly filmed anode contributed about 2 to 2½ volts to the bath voltage, which will be at least 4 if not 6 or 7 volts. Thus, when the bath voltage drops a volt or so below its usual value, defilmed anodes can be expected; however it must be remembered that if only one of a number of anodes is defilmed (as by low current density caused by a poor contact) plating trouble will be experienced even though the tank voltage may not be affected.

Anodes containing 1 per cent aluminum will operate satisfactorily[25] at 1.5 to 3 times the anode current density useful for pure tin and in addition may be properly filmed at current densities in the high end of the operating range rather than well above it.

Tin anodes are available in sheet or slab form, in the usual elliptical bar shape, or as balls.

Zinc. Three grades of zinc anodes are used: prime Western, intermediate, and high purity. The first two grades are not suitable for cyanide-zinc baths, especially of the bright type. Typical analyses of these grades of zinc are:

	Prime Western (%)	Intermediate (%)	Special High Grade (Horsehead) (%)
Zn	98.50	99.500	99.990
Pb	1.40	0.150	0.003
Fe	0.08	0.005	0.001
Cd		0.080	0.003

In acid-zinc plating slabs or cast oval anodes of prime Western of intermediate grades of zinc are commonly used. Lead will give rise to discolored deposits and if this is important, purer anodes should be used such as the good grade of intermediate zinc illustrated:

Zn—99.81%   Cd—0.08%
Pb—0.08%   Cu—0.01%
Fe—0.01%

In cyanide-zinc baths all the heavy metals are objectionable, especially if the deposits are to be bright-dipped.

In both acid and alkaline baths zinc tends to dissolve at more than 100 per cent anode efficiency, owing to chemical solubility. The addition of aluminum, or aluminum and mercury, reduces the anode efficiency and slows chemical attack in the idle bath[11]. This expedient is most effective in acid baths and when the anode contains about 0.3 per cent mercury and 0.5 per cent aluminum. In cyanide baths, particularly bright baths, the mercury is objectionable. The anodic efficiency of zinc in cyanide baths can be controlled by adding magnesium[12]: in general 0.18 per cent is recommended for still-plating and 0.05 per cent in barrel-plating. The use of steel anodes in conjunction with zinc anodes in a cyanide bath will reduce the amount of zinc taken into solution during electrolysis, but the iron-zinc couple increases the chemical solubility in the idle bath about tenfold[12]. It is usually recommended, therefore, that ball anodes in steel baskets be removed whenever the bath is not in use for any appreciable period such as overnight; however this is a laborious procedure and is rarely done in practice.

High purity zinc anodes are available as cast or extruded elliptical anodes and as cast balls or extruded cylinders.

Alloy Anodes. Brass anodes have already been discussed. While it is convenient to be able to use alloy anodes of the same composition as is done in brass plating, this is not always possible. For instance, in tin-nickel plating the deposit may contain 35 per cent nickel, but anodes of this composition consist of two intermetallic compounds, Ni_3Sn_2 and Ni_3Sn_4, of which the former dissolves preferentially. Consequently, anodes of 27 to 28 per cent Ni (i.e., entirely Ni_3Sn_4), a combination of these with nickel anodes, or else separate anodes[22] are used. These latter may be operated from a single anode rod provided a small fixed resistor is incorporated in the hook of the tin anode since the difference in anode potential is slight (about 0.1 volt). Otherwise separate current sources would be required.

In general, alloy anodes should be a single-phase solid solution because usually two phases will have different dissolution rates at the same potential. This may result in disintegration of the anode.

The use of separate anodes must be considered carefully in view of undesirable side-effects that may occur, especially when the current is turned off (see Brass, above). For instance, in tin-alloy plating from stannate baths, the tin anodes should be removed to reduce the possibility of reduction to stannite[22]; whereas in tin-nickel they are removed because otherwise a spongy deposit of tin and nickel forms on them.

Fluoborate baths for depositing terne-alloy coatings of from 5 to 15 per cent in tin content (Chapter 6, Table 12) have been developed for high-speed plating of strip. They require alloy anodes of the above compositions and of high purity, i.e., "straits Tin" and 99.99 per cent lead. The anode efficiency is about 100 per cent, but tin deposits by immersion on the anodes during long non-plating periods. It then flakes off when plating is resumed, rather than dissolving anodically. Fortunately, in the modern straight-through cell-type plating line (see p. 000) the solution can be quickly drained from the plating cell during the very infrequent idle periods, thus eliminating the difficulty.

ANODE BAGS

No soluble anode corrodes without the formation of more or less anode sludge; therefore loose-fitting bags are provided which slip over the anode and are tied around the hook above the anode and out of the solution. Bags should hang 2 to 4 in. below the bottom of the anode so that there is a pocket for falling sludge to collect in without insulating the end of the anode. The material of the bags should be woven closely enough to retain sludge without excessively restraining the passage of solution. If the latter occurs, the metal concentration will build-up in the solution inside the bag and polarize the anode. In cases where such a compromise is impossible, a tank diaphragm should be used instead of anode bags (see Chapter 19).

The life expectancy of an anode bag depends on its material and on the solution in which it is used, but in most cases it will be several times the life of an anode. The bag should be emptied of sludge and washed before reusing, preferably by inserting a hose or pipe inside the bag. Aside from accidental damage, bags usually fail first either at the bottom because of heavy sludge accumulations or at the air-solution interface where acidic attack is greatest and where a dried crust of salts may cause cracking and rupture of the fibers when the bag is flexed. Some suppliers furnish bags with synthetic resin-dipped ends to counteract this weakness. In violently agitated solutions the flapping motion given to the anode bags imposes more stringent requirements on the filtering and strength characteristics of the bag than are called for in the same solution without agitation.

Textile materials are available in many different weaves, weave counts, and weights; those given in Table 3 merely represent an example

TABLE 3. CHOICE OF ANODE BAG MATERIAL

Material‡	Plain Cotton Drill	Double Flannel	Muslin Filter	Nylon	"Vinyon"	"Dynel"
No. of plies	1	2*	3†	1	1	1
Weave	Twill	Single nap	Plain	Plain	Chain twill	Plain
Weave count	72 x 48	42 x 44	64 x 64		68 x 42	68 x 46
Weight, oz/sq yd	6.76	4.79	4.57		6.60	5.25
Approximate cost per inch finished length for standard anodes	1½¢	4¢	3½¢	2½¢	4½¢	5¢
Recommended for:	(a) Nickel plating except fluoborate (b) Nearly neutral solutions	Nickel with depolarized anodes and vigorous agitation	High pH nickel	Alkaline cyanide including not agitated solutions	(a) Alkaline and acid baths, including fluoborate (b) Temp. up to 250F (c) Agitated solutions	(a) Acid solutions, including fluoborate, up to 200F (b) Alkaline solutions, room temperature
Not recommended for:	Strongly acid or alkaline solutions	Strongly acid or alkaline solutions	(a) Strongly acid or alkaline solutions (b) Agitated solutions	Below pH 3.5		(a) Hot alkaline solutions (b) Vigorous agitation

* Two piles of single nap flannel, nap inside.
† Two plies of muslin with inner liner of filter paper.
‡ Blue African Asbestos is another anode material. It is satisfactory in iron baths. See Text, Iron Anodes.

of commercial usage. Anode bag material normally is not sized, but lubricants, etc., may be used in weaving, and new bags should invariably be washed thoroughly before placing them in use and preferably presoaked at the approximate pH to be employed, and then rinsed thoroughly.

References

1. Savage, F. K., *Monthly Rev., Am. Electroplaters' Soc.*, **29**, 301-308 (1942).
2. Thews, E. R., *Metal Finishing*, **46**, (3), 68-75; ibid. (4), 61-66, 75; ibid. (5), 74-79 (1948).
3. Blum, Wm., and Hogaboom, G. B., "Principles of Electroplating and Electroforming," 3d Ed., p. 298, New York, McGraw-Hill Book Co. (1949).
4. Weisberg, L., and Graham A. K., *Trans. Electrochem. Soc.*, **80**, 514-515 (1941).
5. Eckhardt, R., *Metal Finishing*, **38**, 313-314 (1940).
6. Pinner W. L., and Borchert, L. C., *Monthly Rev., Am. Electroplaters' Soc.*, **25**, 909-910 (1938).
7. id., *Proc. Am. Electroplaters' Soc.*, **26**, 84-90 (1938).
8. Pinner, W. L., "Method of Nickel Plating Using Electrolytic Nickel Anodes," U.S. Patent 2,358,995 (assigned to Houdaille-Hershey Corp.), (Sept. 26, 1944).
9. Amundsen, P. L., "Electrolytic Nickel Anodes for Nickel Deposition," U.S. Patent 2,371,123 (assigned to Parker-Wolverine Co.), Mar. 13, 1945.
10. Roehl, E. J. (assignor to the International Nickel Co.), "Nickel Plating," U.S. Patent 2,504,239 (Apr. 18, 1950).
11. Hogaboom, G. B., "Zinc Aluminum Mercury Anode," U.S. Patent 1,887,841, No. 15, (1932), and Graham, A. K., "Zinc Aluminum Anode," U.S. Patent 1,888,202 (Nov. 15, 1932).
12. Hull, R. O., *Proc. Am. Electroplaters' Soc.*, **28**, 29-32 (1940).
13. Safranek, W. H., and Faust, C. L., *Plating*, **42**, 1541-1546 (1955).
14. Nevers, R. P., Hungerford, R. L., and Palmer, E. W., *41st Proc. Amer. Electroplaters' Soc.*, 243-248 (1954).
15. Blum, W., and Hogaboom, G. B., "Principles of Electroplating and Electroforming," 3d Ed. p. 291, New York (1949).
16. Nevers, R. P., and Palmer, E. W., "Electrodeposition of Copper," U.S. Patent 2,689,216 (Sept. 14, 1954).
17. Wagner, R. M., and Beckwith, M. M., *26th Proc. Amer. Electroplaters' Soc.*, 147-167 (1938).
18. Soderberg, K. G., *J. Electrodepositors' Tech. Soc.*, **13**, (9), 1-12 (1937).
19. Soderberg, K. G., *Monthly Rev. Amer. Electroplaters' Soc.*, **20**, (1), 26 (1933).
20. Hyner, J., U.S. Patent 2,456,281 (Dec. 14, 1948).
21. Lowenheim, F. A., *Metal Finishing*, **49**, (3), 60-64 (1951).
22. Fishlock, D. J., *Metal Finishing*, **51**, (2), 48-51, 57; (3), 55-59 (1959).
23. Heiser, F. A., and Hyde, J. E., *Plating*, **46**, 385-388 (1959).
24. Swalheim, D. A., paper presented at Philadelphia Branch meeting Amer. Electroplaters' Soc. (1960).
25. Lowenheim, F. A., *Trans. Electrochem. Soc.*, **96**, 214-225 (1949), and U.S. Patent 2,458,912 (Jan. 11, 1949).
26. Friedberg, H. R., *Plating*, **46**, 834-836 (1959).
27. Available from Plating Products, Inc., Kokomo, Indiana and elsewhere.
28. Benham, R. R., *Trans. Inst. Metal Finishing*, **36**, 22-26 (1958-1959).

30. RECTIFIERS

Peter Cambria

Chief Engineer, Rapid Electric Co., Brookfield, Conn.

What is a rectifier? In the electronics industry, a rectifier is a device which conducts the flow of current in one direction and impedes the flow in the opposite direction, and therefore is used to convert alternating current to direct current. In the electroplating industry, a rectifier is the entire power supply which converts the alternating voltage supplied by the utility company to a low-voltage direct current (D.C.) suitable for electroplating or anodizing. It includes not only the uni-directional conversion element but the transformer which steps the voltage down from the high voltage supplied by the utility company to a low voltage suitable for electroplating, as well as all periphery equipment associated with the rectifier.

We will first break the rectifier down into its component parts and discuss the operation of each. Next, we will talk about the different types of control available, highlighting advantages and disadvantages of each. We will discuss wave shapes and power requirements and cover some trouble-shooting of rectifiers. Since they are now essentially obsolete, we will not discuss motor generators or copper oxide, vacuum tube, germanium or selenium type rectifiers.

BASIC ELECTRICITY

Before proceeding into discussion of rectifiers, a brief review of basic electricity is in order. Figure 1 shows a battery which provides an electromotive force measured in volts, connected by electrical conductors to an ammeter which measures the flow of electric current, in amperes, and then to a light bulb which will glow when current flows through it. The light bulb is connected to a variable resistance which will impede the flow of current depending upon the resistance setting. The resistance is measured in ohms. The relationship is governed by "Ohm's Law," which can be expressed as

Voltage(E) = current (I) × resistance (R); i.e.,
$$E = IR$$
or $I = E/R$
or $R = E/I$.

The amount of power (P) delivered to the light bulb is equal to

$$P = I^2 R$$
or $P = E^2/R$
or $P = E \times I$

where P is expressed in watts.

In the circuit shown, if the battery voltage is 12 volts, the variable resistance is set for 5 ohms and the resistance of the light bulb (which also enters into the formula) is 1 ohm, the current through the circuit is: $I = E/R$ or $I = 12/(5 + 1)$ or 2 amperes. The voltage across the bulb is $I \times R = 2$ volts, and the power delivered to the light bulb is $P = E \times I = 2 \times 2$ or 4 watts.

The circuit shown is one in which the current flows from the positive terminal of the battery, through the ammeter, light bulb and variable resistor and back to the negative terminal of the battery. Since the battery voltage polarity is always the same, the current flows in only one direction (from plus to minus). This is known as direct current and is the type of current required for electrodeposition of a metal from a solution.

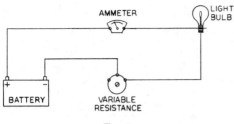

Fig. 1

Utility companies do not supply direct current. Their method of generating electricity supplies a current which alternately goes positive and negative. This is called alternating current or A.C., while the uni-directional current is called D.C. The rate at which the current alternates is known as the frequency, measured in hertz (Hz) or cycles per second. Most power generated in the United States is at a frequency of 60 Hz, while in many foreign countries the generated power is at a 50 Hz frequency. Since electroplating requires a direct current, the utility company power must be converted to D.C. This function is performed by the "rectifier."

RECTIFIER COMPONENT'S PARTS

The component, common to all types of rectifiers, is the main power transformer. This has the function of converting the high-voltage A.C. supplied by the utility company to a low voltage suitable for plating. The transformer is made up of turns of copper wire wound around a magnetic steel core. It generally has at least two windings. The windings connected to the high-voltage side are known as the primary windings. The windings connected to the rectifying elements are known as the secondary windings. The number of primary windings depend upon the number of input phases, and the number of secondary windings depend upon circuit configuration. The voltage which appears on the secondary winding is proportional to the voltage at the primary and the number of turns on the primary and secondary. It follows the relationship:

$$E_s = \frac{(E_p)(N_s)}{N_p}$$

where E_s = secondary voltage, N_s = secondary number of turns, E_p = primary voltage and N_p = primary number of turns.
The secondary current follows the relationship:

$$I_s = \frac{(I_p)(N_p)}{N_s}$$

where I_s = secondary current and I_p = primary current.

Transformers are rated in kilovolt-amps (KVA), or thousands of volt-amps. To determine the primary, or input, A.C. or line current of a rectifier is a simple matter if the KVA of the transformer is known. For a single-phase rectifier, divide the KVA by the line voltage, and multiply the result by 1000. For a three-phase unit, multiply the KVA by 1000 and divide by the line voltage times the square root of three (1.732).

With the power company voltage stepped down to a voltage useful for plating, the low-voltage A.C. must be converted into the direct current (D.C.) required for electroplating. There are two devices which will perform this function and which are used in modern rectifiers. They are the silicon diode and the silicon thyristor or SCR (silicon controlled rectifier). We will first discuss rectifiers which contain silicon diodes and then go into greater detail on SCR type rectifiers.

A silicon diode is a device which is made up of a silicon wafer grown under controlled, clean room conditions and exhibits a semiconducting characteristic; i.e., it will allow the flow of current in one direction and impede the flow of current in the opposite direction. The main advantage of the silicon diode over other devices which have been used in the past is its ability to operate at extremely high temperatures. A silicon diode can operate with junction temperatures as high as 150 C (300 F). Its high-temperature characteristics, along with its low forward voltage drop, allow a device to be made which is relatively small and can handle high currents. Another advantage over previous types of rectifying devices is that, when operated under proper conditions, the device has no aging associated with it, and hence has no known end of life.

Figure 2 illustrates how a diode converts A.C. to D.C. in a single-phase, full-wave bridge configuration. We will assume conventional current flow; i.e., current flows out of the positive terminal of the source back to the negative terminal. Current flow through a diode is from anode to cathode. The A.C. generator (utility) supplies a voltage (sine wave) which is alternately going

RECTIFIERS

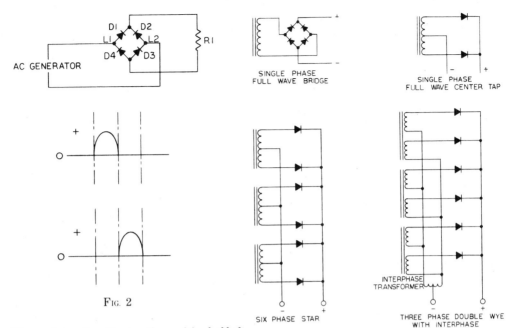

Fig. 2

Fig. 3

positive and negative. During the positive half of the cycle, L1 is positive with respect to L2, and current flows into the anode of diode D1, out the cathode, through the load resistor, R1, into the anode of diode D3, out of the cathode and back to the generator. During this half cycle, the voltage waveshape across resistor R1 is shown in Fig. 2. The generator voltage then reverses (L1 becomes negative with respect to L2; hence L2 is positive with respect to L1), and current flows into D2, through R1, into D4 back to L1. During this portion, the voltage across R1 is as shown in Fig. 2. Since the current flow through the load R1 was always in the same direction, the A.C. voltage has now been converted to a D.C. voltage (by definition, a voltage which is of single polarity). The voltage may not be of the same waveshape as a battery or D.C. generator would produce, but it is still uni-directional and hence suitable for plating. A voltage of this waveshape is generally termed "pulsating D.C." and has a high ripple content. This pulsating D.C. may not be optimal in the plating of some metals; therefore, a means must be devised to smooth out this pulsating D.C. to make the waveshape approach that of a battery. The amount of ripple depends upon circuit configuration and rectification means and can be reduced or "filtered" to a desirable level. This aspect will be discussed later.

Numerous rectification circuits are available and are shown along with circuit parameters in ANSI (American National Standards Institute) publication C34.2. The most common circuits used in plating rectifiers are the single-phase, full-wave bridge, the single-phase, full-wave centertap, the six-phase half wave (six-phase star), or the three-phase double wye with interphase (see Fig. 3).

The most common rectifier sizes are in the range of 500 to 3000 amps. These generally require a three-phase input and use the six-phase star configuration. Most manufacturers will use diodes which are rated typically at 300 amps average. In the six-phase star configuration the average current through each secondary leg of the transformer is one-sixth of the D.C. current. If a 1000 ampere rectifier was chosen with one diode per leg, each diode would be carrying one sixth of 1000 amperes or 167 amperes average. A silicon diode has a typical forward voltage drop of approximately three quarters to one volt. If, for the purpose of this discussion, we assume that it is one volt, this means that at 167 amperes, times one volt, there are 167 watts dissipated in the diode. If one can picture the size of a light bulb, which is 167 watts, and how hot that bulb would get, you can imagine that the diode, which is approximately a quarter of the size of that same light bulb, dissipating that many watts, is going to get extremely hot. This means that the diode must be cooled in some manner. The first step is to mount the diode on a metal heat sink, to provide a larger area for dissipating the heat. This heat sink generally is made from copper or aluminum for best thermal transfer. The added area provided by the heat sink still may not be adequate to prevent the diode from overheating and destroying itself without additional means of cooling. The two most common ways are the use of forced air and direct water cooling. Forced air cooling is the most economical, since it not only cools the diode but also the main transformer, allowing for a more economical transformer design. If the rectifier is to be placed in an extremely dirty and corrosive atmosphere, it is beneficial to seal the unit and use direct water cooling to remove the heat.

Since we now have a means of stepping down the high utility company voltage to a low voltage and converting it to D.C. for plating, we next need a means for varying this voltage so as to be able to control the amount of plating to be done. To plate different metals and substrates requires different current densities and hence different voltages. The way these voltages are varied are numerous and each way has its own unique advantages and disadvantages. The most common means used in modern-day rectifiers are the following.

Tap Switch Control

The simplest means of controlling the D.C. voltage is the tap switch method. This method offers two different ways in which to control the D.C. voltage—the extended primary method and the auto transformer method.

Extended Primary Method. Referring to the section on transformers, we find the formula for secondary volts as being equal to the primary volts times the secondary turns and divided by the primary turns. It can be seen that if the primary turns are increased, the secondary voltage is decreased. This is what is done in the extended primary method. More turns are added to the primary winding and at various points throughout this extended primary section, a connection is made. This connection is known as a tap on the transformer. A typical transformer would have eight per phase. These taps are now connected to one switch per phase and the common connection on the switch is tied to the high voltage input line. By changing the position of the switches, different taps on the transformer can be selected and hence different output voltages. By use of eight taps per phase, 22 discrete voltage steps of control are available. Caution must be taken with this type of control to assure that there is never an imbalance between switches of more than one position. As an example, if switch #1 and switch #2 were set on position #4, switch #3 must be on either position #3, 4 or 5. Any greater variation than that can cause severe unbalanced currents in both the transformer and diodes, thereby causing rectifier failure.

A disadvantage of this system is that the output can only be adjusted to some minimum voltage, not zero. Typically, the voltage range is approximately six to one; i.e., a 12-volt unit would be adjustable down to 2 volts. Another disadvantage is that smooth control is not obtained. In the case of the aforementioned 12-volt unit, the resolution would be approximately ½ volt per step.

Advantages of this type control are: (1) low cost, (2) ripple voltage of less than 5% as long as all three taps are on the same position and (3) simplicity of design, which reflects in low maintenance on the rectifier.

Auto Transformer Method. An auto transformer is a single winding transformer with multiple taps on it. As opposed to the extended primary type, the taps are all at points below the full-voltage tap. The line voltage is impressed on the full winding and the tap switch is connected to the taps which now select some lower voltage than the line voltage (including zero voltage). The wiper of the tap switch is then connected to the primary of the main step down plating type transformer. By varying the tap switch, the voltage to the primary is varied and hence the D.C. output voltage. The only advantage to this system over the extended primary is that control to zero volts can be obtained, while the disadvantage is somewhat higher cost.

Variable Auto Transformer (V.A.T.)

The next type of control we will discuss is the variable auto transformer type known under the trade names of "Powerstat," "Variac" and "Adjust-avolt." The V.A.T. is almost identical to the tap switch auto transformer type, with the exception that the voltage from the V.A.T., rather than being changed with taps, is changed by the means of a brush which slides across the turns of the transformer in a continuous motion from one turn to the next. It may be considered as the tap type auto transformer with a very large number of taps to allow smooth control of the voltage which is applied to the main transformer. Another advantage of this type of control, is that three single-phase V.A.T.s are assembled on a common shaft so all three phases are adjustable simultaneously thereby preventing any possible unbalance between phases. Disadvantages of these types of control are: (1) higher cost than the tap switch, (2) size is limited, (3) moving parts such as brushes which are subject to wear and require frequent maintenance and (4) exposed windings on the coil of the V.A.T. which could be troublesome in a corrosive atmosphere.

The V.A.T. type of control is also available in a motor driven version. Operation is the same as the manual one, except that control is accomplished by a servo type motor, which is given a command to either raise or lower its voltage by the operator pushing the appropriate button. This version allows ratings to be increased to much higher levels by ganging many sections of auto transformers, driven by a single motor through a gear train. Typical traverse time (from minimum to maximum) of this type is from 15 seconds to one minute. Since the V.A.T. is electrically driven, it now becomes possible to locate the rectifier in a remote area and out of the corrosive atmosphere, thereby increasing reliability and reducing maintenance cost. With the addition of a solid state servo amplifier and voltage or current regulator, it becomes possible to have constant voltage or constant current control. This is a feature which could not be done with a tap switch or manual V.A.T. type control. In modern automatic plating lines, it is almost mandatory that either constant voltage or constant current be used. Another version of the motor-driven V.A.T. is one in which it is immersed in a tank of insulating fluid. A trade name of one such type is the "Rapitrol." This has all the advantages of the regular type with an additional advantage. The brushes and exposed areas of the auto transformers are now sealed in a tank of fluid. This protects them from the atmosphere and they receive additional cooling and lubrication from the fluid, allowing them to be used at a higher rating.

All the aforementioned types of control had one thing in common, and that was the ripple content of the output voltage. The ripple content, in percent, is defined as the ratio of RMS voltage (A.C. component) to the D.C. average voltage times 100. If not specified otherwise, the content is given at maximum voltage and current. It may also be specified over a certain voltage range at maximum current, but still as a percentage of maximum. A third way of specifying ripple would be as a percentage of the output voltage setting at maximum current. In any case, all of the aforementioned types of control would yield a ripple of content of less than 5% of set voltages. The importance of ripple will be discussed in a later section.

Saturable Core Reactor

This type of control is one which until recently was the most effective means of achieving automatic control in high current ratings and at a reasonable cost. In recent years the cost of the reactor type has gone up substantially, while costs of solid state types have gone down. The saturable core reactor is a magnetic device which could be considered analogous to a large dam. The dam has flood gates which, depending upon how much they are opened or closed, will control the amount of water flow. The gates are controlled by a small motor which in turn is run by some electrical signal received from the operator. The reactor has gates also, which are connected between the input voltage and the transformer primary. The amount of current which flows into the transformer is dependent upon how much these gates are open (how much the core is saturated). The amount they are open is determined by its control or D.C. windings (the motor) which gets a signal from a small D.C. power supply controlled by the operator. Without going into great detail on the theory of saturable reactors, it is sufficient to say that the amount of voltage out of the rectifier is directly proportional to the amount of D.C. control current. With this type of control, it is also possible to have either constant voltage control or constant current control. The major advantage of this type of rectifier over all previously mentioned types is that the entire rectifier is solid state/magnetic and therefore very reliable, rugged and requires very little maintenance. The only moving parts would be the control potentiometer, the A.C. contactor and if the uit is air cooled, the fan. These parts are generally considered expendable and are relatively inexpensive to replace or repair. Another advantage of this type of control is that it could be used directly connected to high-voltage input lines and in fact has been used successfully on input lines as high as 15,000 volts, although in most plating shops, this is a rarity.

The drawbacks to this type of control are: (1) the minimum voltage out of the rectifier is something greater than zero (generally about 5% of rated voltage) and (2) the rectifier must always have some load connected to it, since without any load, the output would go to maximum even with the absence of a control signal. This could be harmful on plating tanks where live entry is required, since arcing would occur. This could be overcome by the addition of a dummy load or a magnetic preloading device which is available.

Thyristor

This type of control is one which is the latest in the state of the art and most commonly used when automatic control is required. This control is known as the "SCR" type, although the proper name is thyristor. Two types of SCR control are available, primary control or secondary control. Before examining these types of control, a brief explanation of the SCR is in order.

The SCR (silicon controlled rectifier) is a device which is similar to a silicon diode in that it will allow a current flow in one direction and impede the flow of current in the opposite direction. The major difference between the SCR and a diode is that, with a diode, current will flow whenever the anode voltage is positive with respect to the cathode, while with the SCR, a positive voltage on the anode is not only required but also a signal at the gate lead to "turn on the SCR." The signal at the gate lead need only be in the milliampere range to allow large currents to flow from anode to cathode, hence providing very large current gains.

Once the SCR has been switched "on" with a signal on the gate, it remains "on" (in a conduction state) even though the gate signal is removed. To turn it off, the anode current must be reduced to below the "holding current" of the SCR. For all practical purposes, this value can be considered to be zero. When the SCR is connected to an A.C. source, this condition occurs every cycle as the voltage reverses from positive to negative.

Referring to Fig. 4, we see that, depending upon which part of the positive half of the A.C. voltage the SCR is switched on, the amount of voltage impressed across the load can be varied.

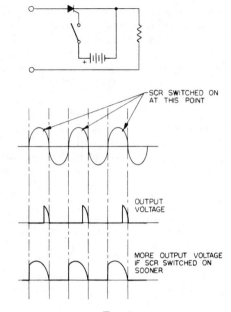

FIG. 4

Switching on early in the cycle produces a large amount of output voltage, while switching on later in the cycle produces a small amount of voltage. This method of varying the output is known as "phase control" and the device which controls the "turn on" point of the SCR is the "trigger" or firing circuit.

Referring now to Fig. 5, we see two SCRs connected in a "back-to-back" or "anti-parallel" configuration. If the forward SCR is triggered during the positive half of the cycle, and the reverse SCR is triggered on in the negative half of the cycle and this triggering is repeated every cycle, output across the load will be a varying A.C. voltage capable of controlling large amounts of current with only a very low-current trigger signal. If the tap switch, variable autotransformer or saturable core reactor previously discussed were replaced with a "back-to-back" configuration of SCRs, a primary controlled rectifier is made. The output of the SCR would be connected to the primary of the step-down transformer, and would control the voltage applied to it, hence controlling the secondary voltage. This voltage would then be rectified by the diodes and the output of the rectifier would be a D.C. replica of the A.C. voltage out of the SCRs.

In the primary method of SCR control, the SCRs were used strictly to control the voltage to the step-down transformer. On the secondary side of the transformer, silicon diodes were used

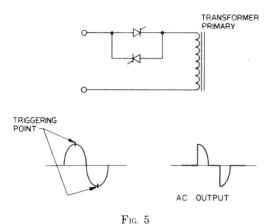

Fig. 5

Comparison Between Primary and Secondary SCR Control

In the primary SCR type control, the current through the SCR is small with respect to the D.C. or secondary currents. Referring to the section on transformers will explain why. In the primary circuit, the SCR is subjected to the line voltage. If the line voltage happens to be 480 volts, which is a typical three-phase input voltage, the peak voltage is 1.41 times 480 volts, or approximately 700 volts. SCRs are made with voltage ratings up to about 3000 volts, which is rare and very expensive. Typically 1600 volts is the economical maximum available. This means that the safety margin between the voltage applied to the SCR in the reverse direction and its voltage rating is something between 2 and 2½ times. On the surface, this seems fine; however, industrial power lines are subject to many high-voltage transients or "spikes." These spikes are very high-frequency in nature and sometimes achieve magnitudes in the order of 3000 volts on a 480-volt power line. The spikes can be caused by motors being started, power factor correction capacitors being switched on or off the line and resonating with components in the circuit, outside disturbances interfering with the power line and many other factors. If one of these spikes reaches the SCRs they would immediately break over in the reverse direction and be destroyed. Therefore, with primary SCR control, great pains must be taken to absorb these spikes and assure that the SCR does not see them. This can be done by putting inductors which present a high impedance to spikes in series with the SCRs. "Snubbing networks" can also be connected across the SCRs. These cause the spikes to be shunted around the SCRs. These protection networks are designed by assuming certain frequencies for these spikes. Since they can be caused by many different means, these assumptions do not cover all cases; therefore, these networks are not completely effective and failures still occur. When SCRs were first developed, they were only available in low-current ratings and the only way to make a high-current rectifier was by means of primary SCR control, or to parallel many, many devices in the secondary. In the past 5 to 10 years, the state of the art of SCR manufacturing has grown to the point where high-current SCRs are no longer a rarity. By going to the flat type of construction as opposed to a stud type package, the cooling can

to perform the rectification functions. If we now remove the SCRs from the primary of the transformer and feed the line voltage directly into the primary, remove the diodes and in their place put SCRs, the SCRs will be used to perform the function for which they were developed; i.e., control and rectify. As far as the plater is concerned, the output of the rectifier is the same whether primary or secondary SCR control is used. However, it is important that he know the advantages and disadvantages of each system. In order to explain the advantages and disadvantages, an understanding of the SCR ratings is important. Two very important ratings which the SCR manufacturer puts on the device are the reverse voltage breakdown rating, and the forward current rating. The reverse voltage is one that, if exceeded, will instantly cause the SCR to break-over in the reverse direction and destroy itself. The silicon diodes used in a rectifier also have a reverse voltage rating but it is often neglected since they are used in the low voltage side. The safety margin, i.e., the rating of the diode, is generally 10 times more than the reverse voltage impressed upon the circuit. The other rating (forward current rating) is a little more complex in the way it is described, because it is a function of temperature as well as current. The higher the temperature, the less the maximum allowable current. This indicates that if the rated current is exceeded it will not destroy itself instantly, since it must heat up first and exceed the rated temperature. The SCR also has a surge current rating on it, which means that it will allow a very high current to pass for a very short time, usually one cycle of line frequency. This very high current can be in the order of ten times the rated current.

now be achieved from both sides of the device and thereby allow much higher current ratings for the same silicon wafer. SCRs are now available which can handle well in excess of 3000 amps. By using high-current SCRs and putting them in the secondary side of the transformer, the peak voltage that the SCR sees is, on a typical 12-volt unit, approximately 17 or 18 volts. The minimum size SCR used by most manufacturers of secondary control is 200 volts. This allows a safety margin of well over 10 times as compared to less than 2½ times in the primary type control. The same 3000 volt spike mentioned previously would be reduced by the turns ratio of the transformer. In a rectifier which is connected to a line voltage of 480 volts and has a secondary of 12 volts, the turns ratio is 40 to 1; therefore, the 3000-volt spike would now be reduced to 75 volts. Since the SCR has a voltage rating of 200 volts, there is no problem with this spike. With secondary SCRs, controls are presently made using just a single SCR per leg, or 6 SCRs per rectifier, up to approximately 7500 amps D.C. Above that rating, the SCRs are then paralleled in equal increments, i.e., 15,000 amps would be 12 SCRs, 22,500 amps with 18 SCRs, etc. Actually, SCRs are made which would allow rectifiers to be made up to 15,000 amps while using only one SCR per leg; however, it is a matter of economics at which point paralleling of devices is used, since SCRs become extremely expensive in higher current ratings. The fusing to protect these higher current SCRs becomes almost as expensive as the SCRs themselves.

In the primary SCR control, aside from all the suppression networks required, rectification devices (diodes) are still required in the secondary. With the secondary SCR control, all that is required for the power circuit is the transformer and SCRs. Both methods of control require means for triggering the SCRs, contactors for starting, regulators to control and regulate the voltage or current, and a means to cool the rectifier. The major difference between the two systems is the number of parts required and the ease of protection. Because of fewer parts, the reliability of secondary SCR control far exceeds that of primary SCR control. Efficiency of both systems is approximately equal. Output waveshape is the same, for the same output voltage and current. Output ripple is the same. The cost of both systems is also comparable.

Since the SCR system (either primary or secondary type) has many advantages over all

Fig. 6. A 4000-amp, 12-volt thyristor rectifier with remote control. (*Photo courtesy of Rapid Electric Co.*)

other types mentioned, we will first discuss the disadvantages. The major disadvantage is output ripple. As mentioned in a previous section, the output ripple on tap switch and variable autotransformer type rectifiers is 5% of setting throughout the range. On the SCR type as well as the reactor type, the ripple is 5% only at maximum output. As the output voltage is reduced, the percent of maximum voltage becomes progressively worse until approximately 50% output, at which point it begins to improve slightly.

If low ripple is necessary for a particular plating operation, and an SCR rectifier is desired, a ripple filter may be added to the rectifier. This usually consists of a large inductor or choke. When low ripple is desired for loads below 10% of rated, capacitors are also required. One way of reducing the amount of filtering required is to operate the rectifier close to its maximum voltage rating. The most obvious way to achieve this is to purchase a rectifier for the specific voltage required. This is not always practical since the rectifier may be used with many different types of loads as well as plating solutions. When this is the case, the rectifier should be purchased with taps which will allow the maximum output voltage to be changed for different loads and/or solutions. Although this is one of the major disadvantages it is not an insurmountable problem. The other disadvan-

TYPE OF RECTIFIER	INITIAL INVESTMENT	REMOTE CONTROL	AUTOMATIC CONTROL AVAILABLE	OUTPUT RIPPLE	RELIABILITY
Tap switch (ext. prim)	Low	No	No	5%	Good
Tap Switch (Auto tran.)	Low	No	No	5%	Good
Variable (Auto tran.) (Manual)	Low to moderate	No	No	5%	Good
Variable (Auto tran.) (Motor driven)	Moderate	Yes	Yes but slow response	5%	Fair
Saturable core reactor	High	Yes	Yes	5% approx. approx. 30%	Excellent
Thyristor (primary)	Moderate to high	Yes	Yes	5% to approx. 30%	Fair
Thyristor (Secondary)	Moderate to high	Yes	Yes	5% to approx. 30%	Excellent

Fig 7. Comparison Chart.

tage to the SCR type control is cost. This is only a disadvantage when compared to a tap switch or manual V.A.T. Its cost is comparable to a motor driven V.A.T. and less expensive than a "Rapitrol" or saturable core reactor.

The advantages are numerous. Smooth control from zero to rated current or voltage is available. Remote control is possible. Voltage control is possible with or without load in the tank. Means of starting and stopping the rectifier output without operation of the main contactor is available. Regulation (ability to maintain current or voltage) is possible. In modern rectifiers, a major reason for using an SCR controlled rectifier is its ability to be interfaced with computers for total automatic control.

Most all of the different types of rectifiers available to the plater at this writing have been discussed, with one exception, and this is, in low current rating, there are available precision rectifiers which are generally used in laboratories and on precious metal plating. These rectifiers are the transistorized type and since their application is limited they will not be discussed in this chapter other than to mention their availability. Figure 7 shows a tabulated comparison of all the rectifiers discussed.

Regulation

In the previous section, such terms as *constant voltage, constant current, average current density* and *regulation* were used. In this section we will explain these terms as well as others which are commonly used and required in plating.

In the plating industry *constant voltage* is referred to as "AVS," which stands for automatic voltage stabilization. This means that, once set, the voltage out of the rectifier will remain constant as the work load is changed. If we refer back to the section of basic electricity and Ohm's Law, we find that if the voltage remains constant, in order for the current to change, the resistance must change. This is done by adding or removing work from the tank. In plating tanks, the total resistance is not just the work which is added or removed. The total resistance encompasses the connections, the conductivity of the plating solution as well as the resistance of the bus bar connecting the rectifier to the plating tank. Therefore, doubling the surface area of the part may not necessarily double the current because the resistance is not necessarily half. If the only resistance in the circuit were the work piece, then doubling the surface area of the work piece would be equivalent to halving the resistance, thereby doubling the current. This is not the case in the plating tank. What this means is that operating the rectifier in the constant voltage mode, and adding more work in the tank, the current increases somewhat but the current density or amps/ft^2 of work do not remain constant. In many shops, each time work is added or removed, the operator resets the voltage control to the point which will give him the proper amperage for the square footage of work in the tank. Once it is in the tank, with the voltage remaining constant, it is assumed the resistance will no longer change and therefore the current, and hence current density, will remain constant.

In many plating applications, the thickness of deposit is very critical. The thickness is a function of the current density at which it is plated multiplied by the time for which it is plated. Many platers prefer to operate using a constant current. Knowing the square footage and not changing it, plating with a constant current will assure that after a certain length of time, a known thickness will appear on the part, regardless of solution conductivity changing, connection points changing the effective resistance of the circuit, line voltage fluctuations and the like. In this case, the current remains constant for all of these variations, but the voltage out of the rectifier will change, allowing the current to remain constant. This is the "ACC" or automatic constant current mode of operation.

The term *regulation* refers to the amount that either the output voltage or output current will change when either the load or the line voltage changes. Typically, SCR units have regulation of either voltage or current of plus or minus 1% of rated. This means that a 12-volt unit operated in the AVS mode will change no more than 0.12 volts for a change in load from no load to full load or line voltage change of plus or minus 10%, or an ambient temperature change of 25 to 40 C. If it were a 1000-amp rectifier being operated in the ACC mode, the current would change no more than 10 amps under the same variable conditions.

A word of caution is in order when a rectifier is to be operated in the constant current mode. The regulator will correct the output voltage to provide the amount of current set by the operator. As the load resistance is increased, the voltage will increase to provide the current called for until the rated output voltage of the rectifier is reached. If load is removed from the tank under those conditions, burning could occur on the last parts removed. Precautions must be taken to ensure that the current control is turned to zero or the rectifier output is shut off when removing or putting work into the tank when operating in ACC mode. Another application when "ACC" is not desirable is in barrel plating, since the load is constantly making and breaking as the barrel is rotated.

Obviously, if the load on the rectifier is high in resistance, sufficient voltage may not be available to obtain the rated current. This is an important consideration when sizing the voltage of a rectifier.

Average current density control is a means of maintaining the same amps/ft^2 or amps/in.^{2}, as the surface area of the work to be plated is changed. Take the case of a printed circuit board which is to be copper plated. Assume that this board has 1 ft^2 of surface area and it is desired to plate at a current density of 100 amps/ft^2. If this in on an automatic line where boards are continually being put in or taken out of the tank, there could be times when only one board is in the tank and other times when there are 10 boards in the tank. Average current density (ACD) control will automatically change the voltage output to produce the proper current for the amount of square footage in the tank. This is done on an averaging basis and also makes some assumptions. The assumptions are that first, as additional square footage is added in the tank, the current will automatically increase. The second assumption is that the amount of correction is not more than 50% of the actual current required. In other words, if one board were in the tank and drawing 100 amps, and the second board entering the tank increases the current to 110 amps, it would require a correction of 90 amps or 90% to bring it up to the 200-amp level. This is out of the range of any ACD adjustment. The third assumption is that as each additional board enters the tank, it requires the same amount of correction as the previous board. As an example, if the addition of a second board brought the current to 170 amps, the correction required was 30%. When each additional board entered the tank, it too required 30% correction. The circuit is adjusted based on a known minimum and maximum amount of work to be used in the tank and assumes that in between the two extremes it is a linear function.

Plating rectifiers may contain numerous other

features depending upon a plater's needs. Some of the most popular additions to rectifiers are:

Ampere hour meter. An ampere hour meter measures the amount of amps at which a part is being plated, and multiplies it by the time producing a digital readout of the product of the two parameters. Since the ampere hours determine the thickness of plate, the use of the ampere hour meter is a good indication of how much metal is used, and thus can be utilized for inventory control. It may also be connected to sound an alarm or shut off the rectifier after a predetermined amount of ampere hours is reached. Use of the ampere hour meter assures the plater that regardless of load conditions, each part will always be plated with the same thickness. This is especially important in non-regulated units where the current could fluctuate for a number of reasons. The same result could be achieved with a regulated unit which is being operated in the ACC mode without the use of an ampere hour meter. Since the current is being held constant, the ampere hours are strictly a function of time and a simple timer could perform the same function.

Ramp start. Another very useful option, especially if one is anodizing, is to have ramp start. The ramp start ensures that the current will be applied to the part in a gradual manner. In anodizing, at the beginning of the cycle, the resistance of the part to be anodized is virtually a short circuit. As the part builds up an anodic coating, the resistance increases. By applying the current gradually, sometimes over a period of many minutes, the part is allowed to develop some resistance to limit the current for anodizing. The time of the ramp start is generally adjustable over a wide range to allow the operator to select the setting which suits his needs best.

Reversing. Another option which is sometimes desirable on a rectifier is the ability to reverse the polarity of the output. This is sometimes used at the beginning of a plating cycle for cleaning of a part by etching it prior to plating. The part is inserted in the tank, the power supply selected for the reverse mode of operation and started. The part is then etched for a short period of time and then the switch is activated to revert the power supply to the forward mode of operation for plating. This can be accomplished either mechanically or statically. Mechanical (either electrical or pneumatic) type reversing switches are commonly used up to approximately 3000 to 4000 amps. However, above that rating, they become cumbersome, difficult to maintain and expensive. Static reversing can be accomplished only with secondary SCR control. This is done by using two sets of SCRs, each connected with opposite polarity to each other. When reverse polarity is desired, the forward set of SCRs is shut off and the reverse set switched on. Static reversing is practical to high current levels—30,000 amperes and even higher.

Numerous other options are available for rectifiers. The rectifier manufacturer should be consulted as to which would be proper for a particular application.

Efficiency

The efficiency of a rectifier is a measure of the amount of power being put to use in the plating process. This power is measured in watts or kilowatts (1000 watts). The utility company charges the user based on the number of kilowatts consumed times the number of hours, (kilowatt-hours). The efficiency of a rectifier is equal to the power output of a rectifier divided by the input power. The differences between these two figures are power losses. The output power is equal to the (D.C. volts) $\times$ (D.C. amps). Every rectifier has power losses which do not contribute to the plating process, such as the power to operate the main contactor and the fan motor. These items are using power even when the rectifier is turned to zero output and not doing any plating. Therefore, it can be seen that the efficiency is usually highest at maximum output, since these items will reflect only a small percentage of the overall power consumed at that operating level.

As an example, a 1000-amp, 12-volt D.C. unit requires an input power of 13,800 watts. The output power is 1000×12, or 12,000 watts and efficiency is $(12,000/13,800) \times 100$ or 87%. In this case, the losses within the unit are 1800 watts. The silicon diodes develop approximately 1 volt with 166 amps flowing through them, which is equal to 166 watts times six diodes equals 1000 watts. The main transformer is approximately 97% efficient, which means that it contains 3% losses or 360 watts and the remaining 440 watts are needed to operate the fan motor and main contactor. If this were compared to a 1000-amp, 24-volt rectifier, we see that the losses from the diodes remain the same, since they are a function of the amps through them and not the output volts. The fan motor and contactor require approximately the same amount of power. The only difference is the main transformer whose losses would be equal to 720 watts. Total losses now equal 1000 + 720 + 440 or 2160 watts. Output is 1000 × 24 or 24,000 watts. Input power is output power plus losses, or 26,160 watts, and efficiency is now $(24,000/26,160) \times 100$ or 91.7%. Therefore it can safely be said that the higher the D.C. output voltage of the rectifier, the higher the efficiency.

What do these losses mean, and where do they go? Losses are energy which is used and does not directly affect the plating function. The energy in the control transformer is used to energize the contactor and start the rectifier; however, this in no way improves the plating or speeds up the plating process. The fan motor is used to remove heat from the rectifier, but again, this does not contribute to the plating process. The remainder of the losses from "SCR's" and transformer are used up in heat. The heat is then removed from the rectifier by the fan and dispersed into the outside air and it performs no useful function and is additional wasted energy. In these times of rising power costs, people are becoming very energy conscious and do not like to waste energy. They have devised means of ducting the heat from the rectifier into areas which require heating in the wintertime, thereby saving on heating costs. In a similar manner, water cooled units can use the output water which is heated from the losses in the rectifier.

MAINTENANCE OF RECTIFIERS

Modern rectifiers, as mentioned previously, use silicon devices having no known end of life. The transformer, when operated within its design rating and properly maintained, can also have an indefinite lifetime. They both require very little maintenance. It is, however, imperative that this small amount of maintenance be performed religiously if continued trouble-free operation is desired. One of the major causes of failure in rectifiers is lack of proper cooling of both the semiconductors and the transformer. In the design of a rectifier, a fan or fans are selected to provide enough air flow to maintain the components at temperatures within their ratings. In the case of water-cooled units, the water paths are designed to perform this function. Most plating shops have an atmosphere which is detrimental to the components in the rectifier. This means that the rectifier must be shut down and thoroughly cleaned. This is most important on fan-cooled units where the output screen for the fan, as well as spaces between heat sinks and windings of transformers tend to clog up. Once this happens, the heat dissipation capabilities are reduced and in a short time a rectifier failure will occur. Vacuum cleaning is a common maintenance procedure. Start with once a month and adjust the frequency based upon experience. Another common cause of problems which lead to rectifier failure is loose connections. With the rectifier constantly being turned on and off and being run at high currents, there is continued expansion and contraction of the conductors. This has a tendency to loosen many connections. These loose connections represent a high resistance and, at the currents encountered in plating, can cause appreciable voltage drops at these junction points. These drops will begin to heat to the point where the connection could melt. A regular preventative maintenance item should be the tightening of all high-current connection points. This applies to connections external to the rectifier as well. As an example of how a loose connection can cause a catastrophic failure, we will examine the case of the same 1000-amp rectifier previously discussed. If the loose connection occurred at one of the diodes, the circuit would eventually become a very high resistance. The remaining 5 diodes which had previously been carrying 167 amps will now each be carrying 200 amps. This might still be within the limits of the diodes but the transformer would now be delivering current from only 5 of the 6 phases and it would tend to run hotter. This additional heat would then blow on the diodes prior to exiting from the rectifier. The extra heat on the diode from the hotter transformer and from the additional currents in the diodes will eventually cause another diode failure and the process will cascade to the point where all the diodes and possibly the transformer would be destroyed—all due to one loose connection!

The third and last preventative maintenance task is to ensure that all moving parts are performing properly and, if necessary, repaired. Those moving parts consist of the contactor, the fan in fan-cooled units, the tap switch in tap switch units and the brushes in the variable autotransformer type units; and in automatic control type units, the control potentiometers.

The contacts of a contactor should be inspected to determine if pitting has occurred and, if so, be replaced.

The fan motors should be inspected to ensure that there is no binding, and the bearings, if not the permanently lubricated type, should be lubricated.

Some of the tap switches are difficult to inspect since they are sealed and the only way these could be inspected is by feel to determine if they can be rotated properly without any binding.

The critical part on the variable autotransformer is the brush. If it is worn down too far, it must be replaced and the exposed windings must be inspected and kept clean. The final item is the control potentiometer on automatic control units. These can easily be checked with an ohmeter to determine if the resistance change from zero to full resistance is smooth.

As we can see, preventative maintenance on modern rectifiers is neither difficult, very time consuming nor expensive. Most of the moving parts are relatively inexpensive compared to the cost of a new rectifier.

As a recap, the three major items of preventative maintenance are:

1. Cleanliness of the rectifier.
2. Tightness of all power connections.
3. Inspection or replacement of all moving parts.

The schedule for such maintenance must be determined on an individual basis since it will vary with the amount of usage and, more importantly, with the physical location of the rectifier. In the case of a rectifier, the old adage about an "ounce of prevention" definitely holds true.

TROUBLE-SHOOTING RECTIFIERS

Trouble-shooting a rectifier should normally be done by a qualified electrician or serviceman. However, with a knowledge of the equipment and a little care and common sense, an interested layman can perform many of the trouble-shooting functions and perhaps save excessive repair costs. The rectifier manufacturer's service department will be happy to give directions and advice by phone.

The two most important pieces of equipment for trouble-shooting are: (1) a good multimeter and (2) a clamp-on type A.C. ammeter. The size of the clamp-on ammeter is determined by the size of the rectifier to be serviced. Generally, one with a maximum scale of 300 amps is sufficient for rectifiers up to approximately 3000 amps. Larger rectifiers will require a clamp-on ammeter with a full-scale capability of 1000 amps. A third piece of equipment which would be helpful, but not mandatory, is a digital voltmeter.

Before getting into the trouble-shooting section, there is one point which must be stressed. If is the cause of the most common complaint (by far) heard about rectifiers and, ironically, it has nothing to do with the rectifier. In order to stress this point, an analogy will be used. Referring to Fig. 8, we see a tank full of water. At the bottom of this tank is an outlet pipe. At the end of this outlet drain pipe is a valve. Inserted in this outlet pipe before the valve is a meter which measures pressure and another meter which measures water flow. If the tank were filled with water and the valve completely closed, the pressure gauge would read a pressure depending upon the level of water in the tank, but the flow gauge would not read at all since the valve is closed and no water is flowing. If the valve is now opened, water will begin to flow and the flow gauge will register an amount dependent upon the amount which the valve is opened. We will now relate the tank of water to the rectifier. The pressure gauge is analogous to the voltmeter, the flow gauge to the ammeter and valve is equivalent to the total resistance in the circuit (including connections, workload and conductivity of solutions). If there is no work in the tank, and the rectifier is turned on and set to some output voltage, the voltmeter will register, and the ammeter will read zero. As work is inserted in the tank, the ammeter will begin to read depending upon the amount of work in the tank and the other circuit resis-

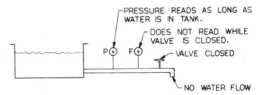

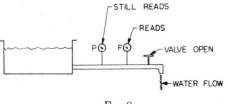

Fig. 8

tances. The main point that is being stressed is that the rectifier produces a voltage, and the amount of current which flows in the circuit is strictly a function of the resistance of the load circuit connected to it, in the same manner in which the water in the tank produced a pressure reading, but no water flowed until the resistance was decreased by opening the valve.

Returning to the major complaint registered about a rectifier, it is that "the rectifier is not working properly because it shows a voltage reading but no amperage reading." As stated previously, the rectifier only delivers a voltage, and current is determined by the load resistance. The problem usually is found to be a poor connection at the tank or at the work creating a high resistance. The main point to be remembered is that if the rectifier is putting out a voltage, look elsewhere for the problem.

Trouble-shooting a rectifier, if performed in a logical manner, can be very simple in most cases. One of the greatest aids can be the plating operator. Questioning the operator as to what symptoms occurred prior to breakdown, what output voltages and currents were being used prior to breakdown, whether the controls had to be set in any abnormal way, etc., can be a clue as to where to look for the trouble. If the rectifier cannot be started (fan not turning—contactor not energizing), one would not begin by checking diodes. The obvious place to start would be the control circuit (after checking to be certain that there is A.C. voltage at the input terminals). Checking the control circuit would be done by measuring voltage at the control transformer, next at the protective devices (thermal cutouts, D.C. overloads, fan overloads, etc.), next at the start and stop buttons and finally at the contactor itself. Locating the point at which the voltage is lost will then identify the defective part. If the proper voltage appears at the coil of the main contactor, and it will not energize, the problem is more likely than not a defective contactor. If the problem turned out to be one of the protective devices, it should be investigated further to determine the reason. If it was the D.C. overload, it probably was due to too much work in the tank or running at too high a current density. If the fan overload was the reason, check for cleanliness of the output screening. A clogged screen could overload the fan motor. If one of the thermal devices tripped out, investigation may take a little longer. The first thing to determine is if it is a semiconductor (diode/SCR) thermal or a transformer thermal.

A diode/SCR thermal shut-down will cool down in a matter of minutes and the rectifier will be able to be restarted. A transformer thermal will usually take about 15 to 30 minutes to cool down. Once it has been determined which thermal is tripping, the rectifier should be restarted and run at a reduced current for making further checks. First measure the three line currents and compare them for balance. There should be no more than 10% unbalance between any three lines. If the line currents are balanced, the reason for the thermal cutouts is most likely due to improper cooling, due to some blockage of air or water flow. If the line currents are unbalanced, check the voltage at the primary of the transformer. Unbalanced voltage at the primary is usually caused by a faulty controlling device in the primary circuit (tap switch, variable autotransformer, saturable reactor or SCRs in the primary). If the voltage at the primary of the transformer is balanced, further checks must be made on the secondary side. Check all secondary transformer voltages. Improper voltages indicate a faulty transformer. If proper voltages exist, the final check is to measure all diodes or SCR currents. At this point, if all other checks gave proper results, it more than likely will be a blown diode/SCR or diode/SCR fuse which caused an imbalance and, hence, thermal cutout.

The same method of trouble-shooting would apply if the symptom were something else, such as no output, no control, partial control, etc. The main point in this case would be to locate the controlling device and replace it. If the problem turned out to be in one of the electronic control cards, it is expected that the card would be replaced rather than trouble-shooting the card itself. Most instruction manuals will give a chart of voltages and/or resistances to be checked external to the card to determine if it is faulty.

Pulse Plating Rectifier

Prior to concluding this chapter on rectifiers, mention must be made of rectifiers for pulse plating. The art of pulse plating was defined as early as 1934 by J. R. Winkler, Jr. in his patent #1,951,893, titled "Electrodeposition of Metal Alloys." Its use has been stymied in most applications due to the unavailability of a power supply of sufficient current capacity, and at a reasonable cost, for industries other than the precious metal industry.

Recent advancements in state of the art technology now make it possible to have pulsing rectifiers as high as 30,000 amperes and in voltages of sufficient capacity for many hard coat anodizing applications, or of low enough voltage for plating printed circuit boards.

Advantages of pulse plating or anodizing are numerous and many papers have been written describing such advantages, but some of the more prominent advantages are faster plating, better distribution, less power consumption, less material usage, fewer rejects, reduced waste treatment costs and better quality parts.

It is this writer's opinion that in the not-too-distant future most plating will be performed with the assistance of pulsing.

This chapter was written with the intention of giving the plater a little more knowledge about equipment which is a part of his life. It is hoped that, with this knowledge, the task of specifying, operating, maintaining and troubleshooting the rectifier will be eased and he can devote more of his time to his primary task of electroplating.

31. PULSE PLATING

Robert Duva

Consultant, Catholyte Inc.

Pulse plating can be defined as current interrupted electroplating. Interrupted current is direct current applied for a specific time period and then returned to ground zero for another specific time period. Pulse plating is useful in certain cases where improved characteristics of electrodeposits are desired. Waveform, frequency and duty cycle of the pulsed current contribute to the changes made in electrodeposits. Although it is a useful tool, pulse plating should not be considered as a universal means to obtain electroplate improvements. Pulse plating has its advantages and disadvantages, as will be described here.

Electroplating solutions can be considered to have three segments:

1. The electrolyte or source of ions
2. The catholyte or cathode double layer, where electrodeposition occurs
3. The anolyte or anode double layer, where other electrolytic reactions occur.

Many electroplating systems utilize anionic complexes for metal deposition. Because these anions are electronegative, they will be repulsed from the negatively charged cathode, reducing the concentration of metal ions in the catholyte. Additionally, reduction to metal from these ions further causes a negative shift in metal ion concentration in the catholyte. A sound electroplating system is one in which the concentration of metal ions in the catholyte is in equilibrium with the metal ions in the bulk electrolyte. Increased agitation of the electrolyte is the most common way to permit higher diffusion rates of metal ions into the catholyte and thus approach this ideal state of ionic equilibrium. The basis for all sound electroplating systems is this replenishment of fresh metal ions in the catholyte as fast as their predecessors have been reduced to metal. Pulse plating can enhance this equilibrium because replenishment of metal ions into the catholyte occurs when the current is at ground zero. The frequency of the pulse, the duty cycle used and the waveform chosen will determine the rate of electroplating within the parameters of any given electrolyte system.

PULSE GENERATION

Pulse plating is generally performed with any of four generic waveforms; i.e., square, asymmetric sine, split sine and periodic reverse square wave (Fig. 1). Many variations of these basic waveforms can be used—each with its own advantage. Square waves (Fig. 1a) can be generated at very high frequencies (to 10 KHz) and these generators are usually the most expensive. Periodic reverse square waves (Fig. 1d) offer the advantage of catholyte metal concentration replenishment when the cathode is an anode for a short period of time. This can be disadvantageous in that dissolution of deposited metal can occur, inducing pits or cavities in the coating. However, it can also contribute to smoothing of the electrodeposit by removing asperities. This same statement can be made for asymmetric sine waves (Fig. 1b), which can only operate at low frequency (50 to 60 Hz) and which are the least expensive. Split sine wave systems (Fig. 1c) produce current flow which is always cathodic and are generally used where the electroplated metal would dissolve in the electrolyte matrix.

Although all of these waveforms can be produced and are being utilized, the square wave is

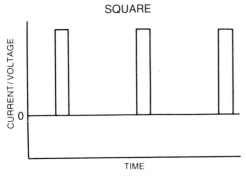

FIG. 1a

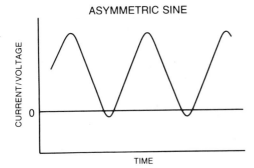

FIG. 1b

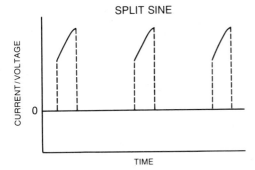

FIG. 1c

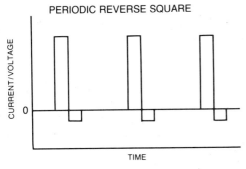

FIG. 1d

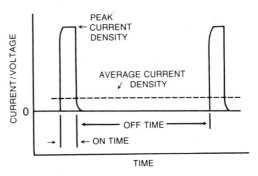

FIG. 2

the form generally accepted as the standard. Figure 2 depicts a typical oscilloscope trace of a square waveform attainable with existing pulse generation equipment. It is not a "pure" square wave in that there is gradual curvature to both the rising current flow and the descending current flow. The curvature on the rising current flow (or slow rise time) is intentionally induced electronically so that instantaneous rise time is avoided and current/voltage overshoot (spiking) does not occur (Fig. 3). When spiking occurs (it can be as high as ten times the peak current/voltage), electrochemical destruction of the electrolyte is imminent. In addition, when spiking is seen, a reverse spike (back EMF) also appears on the down side of the wave (Fig. 3). This back EMF changes the cathode to an anode for a short time at high current/voltage permitting unwanted electrochemical reactions to occur which may be detrimental to the electrodeposit.

The electrolyte chosen, the peak current density and the duty cycle (the ratio of on-time to off-time) will determine the physical characteristics of the electrodeposits. Applying some actual numbers to Fig. 2, if the on-time is 10% of the off-time, and the peak current density is ten times

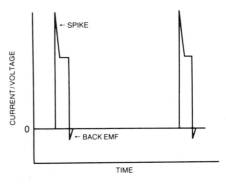

FIG. 3

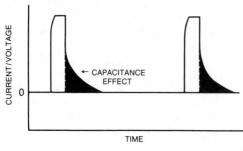

Fig. 4

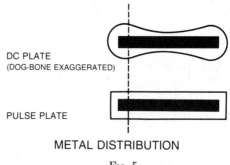

DC PLATE (DOG-BONE EXAGGERATED)

PULSE PLATE

METAL DISTRIBUTION

Fig. 5

the normal current density for the electrolyte, the average current density would be identical to the normal current density utilized with that electrolyte. It has been found that following this concept, there is no change in "real" time of deposition for any given electrolyte and that the physical properties of the electrodeposit can be greatly enhanced. One advantage of using the square wave is that one can build into the electrical system a wave form that would suit the electrolyte being used. Figure 4 is typical of what can be seen when an improper peak current density or duty cycle is chosen. If the peak current density is too high or the duty cycle is too long, the ion concentration in the catholyte will be reduced to approximately zero and a capacitance effect will occur in the double layer and show up as a very large curve on the descending current flow, as depicted in Fig. 4. When these large capacitance effects are seen, the resulting electrodeposit is normally granular, spongy and not cohesive. A simple adjustment of either the peak current density or the duty cycle will eliminate this large capacitance effect and bring the system back into the realm of producing a useful electrodeposit. Although this may seem to be an empirical means of attaining an end result, it has been found that it is functional 90% of the time.

METAL DISTRIBUTION

One of the primary advantages of pulse plating is that metal distribution over the parts being plated is extremely uniform. Figure 5 depicts a typical connector bar that has been plated with normal direct current. The dog-bone effect has been exaggerated; however, it is representative of what occurs. Pulse plating will level out the metal distribution across the part that is being plated. Because pulse plating is high-frequency cyclic DC plating, there is never a chance for the voltage to become static across the surface of the part. Because this voltage gradient is always dynamic, there is less chance for high or low current density effects to be noted over the part being plated. These high and low current density effects can be equated to ionic concentration differences across the part. It was stated earlier that the equilibrium that can be obtained utilizing pulse plating to maintain ionic concentrations uniform across the surface of the part with the bulk solution will promote good metal distribution as long as the pulsed cycle does not starve the catholyte and produce a capacitance effect, as shown in Fig. 4. However, this improvement in metal distribution can be detrimental to the manufacturing of the parts in question. For example, assume that the thickness of the coating on the part being manufactured is measured at a certain distance from the end of the part, as depicted by the dotted line in Fig. 5. With better metal distribution, the dog-boning effect that would occur at the end of the piece would no longer be there. This means that there would be an increase in the total quantity of metal required on the part being plated in order to satisfy the specification of a minimum thickness at that specific spot on the part. This could be a disadvantage since cost factors would increase in the utilization of pulse plating. It has been shown that a nominal 10 to 15% increase in the use of metal is required when pulse plating has been used where minimum thickness requirements are specified.

PHYSICAL PROPERTY CHANGES

Much work has been done utilizing pulse plating on gold electrolytes, and a considerable amount of data has been presented relative to the physical property differences in these gold coatings. Although a discussion of changes in density, wear resistance, crystalline structure, internal stress and corrosion resistance will be put forth, it is noted that all of these physical properties are interrelated as a function of the electrodeposit produced with pulse plating.

Density

Cobalt or nickel hardened golds have been used for over 20 years by the connector industry because they are wear resistant and exhibit good corrosion resistance. The nominal density of a hard gold is approximately 17.3 to 17.7 g/cm^3 as produced with normal DC plate. Pulse plating of cobalt or nickel hardened gold with a recommended pulse plating cycle has shown to increase the density of the electrodeposit to 18.5 to 18.7 g/cm^3. Viewing these two electrodeposits, one can see that the structure of the DC plated part in Fig. 6 is different than the structure of the pulse plated part in Fig. 7. Figure 6 depicts the normal structure for a cobalt hardened gold electrodeposit. Figure 7 shows a fine-grained equiaxed structure. Analysis of both of these deposits show the same quantity of cobalt and the same quantity of carbonaceous matter present; however, the density has been increased by approximately 6 to 7% through pulse plating. Because the electrodeposit has a fine grained equiaxed structure, the mean free path from the

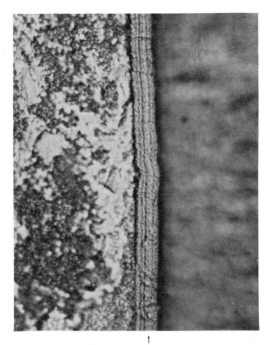

↑
Gold

FIG. 7. Pulse 1400x.

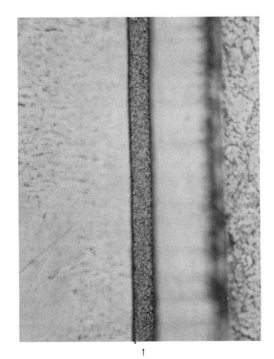

↑
Gold

FIG. 6. DC 1400x.

surface of the gold to the substrate is lengthened and corrosion testing will indicate a more corrosion resistant deposit at equivalent thicknesses. Increasing deposit density via pulse plating would mean an increase in the actual weight of the gold deposited for a specified thickness measured microscopically or, conversely, less thickness required for equivalent corrosion resistance. When a specification requires a minimum thickness, it would mean that an increase in the consumption of gold by 6% would be needed to meet that minimum thickness requirement because of the increase in density. However, it has been shown that with an increased density, the thickness required to withstand corrosion testing can be as great as 50% less than the normal gold thickness required when plated with standard DC. This can be a considerable reduction in gold consumption. In like manner, high-purity soft gold has been shown to increase in density through pulse plating. This improves corrosion resistance and end use requirements in the semiconductor industry, where high-temperature diffusion testing and discoloration are specified for high-purity gold. Densities of high-purity golds when electroplated with standard DC are nominally 19.0 to 19.1; when pulse plated, their density can be as high as 19.20 to

19.24, approaching the theoretical density of wrought gold.

Wear Resistance

Cobalt/nickel hardened golds are chosen for their wear resistance in the manufacture of connectors and contacts. A typical wear resistance test for hard gold uses the crossed wire method wherein the wires are plated with a nominal 2½ μm of gold. These plated wires are worn against each other with a 200 gram-force load.

A typical specification states that a cobalt/nickel hardened gold would withstand 2000 passes in such a test before wear through to the substrate is noted. Utilizing pulse plating, an increase of 50% has been determined in the wear resistance of these hard golds. Figure 8 depicts the chatter marks seen after standard DC plated gold has been worn through to the substrate. It occurs at about 2000 passes. Figure 9 is a pulse plated cobalt hardened gold, where the number of passes has already exceeded 3000 and the coating is just beginning to wear through to the substrate. This is a significant improvement in wear resistance and can be equated to the fact that the DC plated hard gold has a tensile stress

Wear Track↑

Fig. 9. 3000 pulse.

of 40,000 psi, whereas pulse plated cobalt hardened gold has a lower tensile stress in the range of 20,000 to 25,000 psi. Where wear is a primary functional criterion, a reduction by 25 to 30% in the thickness of gold coating can be realized utilizing pulse plating rather than DC plating. Although all of the foregoing has been based upon actual data relative to gold plating and gold electrodeposits, it can be said that, in general, the same statement could hold true for any electrodeposited metal.

As has been stated earlier, there is no set standard pattern to follow when utilizing pulse plating. Each metal alloy system must be evaluated on its own merit so that an empirical formula can be arrived at to obtain the kind of process cycle required in order to improve the physical properties of the electrodeposits. This is one of the main disadvantages of pulse plating. Each metal or alloy system must be investigated within its own merits without reliance upon data obtained from any other metal or alloy system.

COST CONSIDERATIONS

When equating cost to operations in pulse plating, a number of things must be considered. On the negative side, the cost of pulse generating

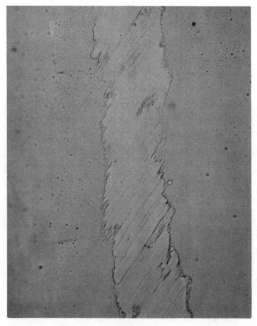

Wear Tracer↑

Fig. 8. 2000 DC

equipment (square wave) is nominally five to ten times that for similar peak current DC equipment. Sine-wave generators can cost considerably less, but sine-wave generators lack the high-frequency capability that is usually sought after with the square wave units. Also on the negative side is the matter of the thickness of electrodeposit as it relates to the end cost of the electrodepited coating. Noted earlier was the effect of metal distribution and the increase of 10 to 15% in the consumption of metal. This was related to specifications requiring a minimum thickness. The same increase in consumption is true when the density of the deposit is considered.

If the density increases by 5 to 6% and the thickness of the deposit via specification must be a minimum (which can increase consumption by 15 to 18%), a total increase in the cost of the part being plated using pulsed current would be about 20 to 25%.

On the positive side, where minimum thickness requirements can be waived and end use requirements are the rule, it has been shown that the excellent wear and corrosion resistance of pulse plated electrodeposits can significantly reduce metal consumption. From a corrosion standpoint, a reduction in 50% in nominal thickness of a deposit can be had and still obtain the same corrosion resistance. Relative to wear resistance, a 30% reduction in the thickness of a deposit has been shown to produce the same wear results.

Because the use of pulse plating is generally empirical at the present time, the following is a list of criteria which should be used to explore the feasibility of the use of pulse plating.

1. Define the characteristics of the deposits that must be improved by pulse plating.
2. Determine the kind of analyses that are to be performed on the deposits (corrosion resistance, wear resistance, stress, etc.).
3. Perform tests with both direct and pulsed current on like parts and compare results.
4. From these results, determine the optimum conditions for the specific applications of pulse plating to the specific part being tested.

By this determination, all of the above can demonstrate to the user the advantages and the disadvantages of pulse plating—based upon the characteristics of the deposit and the end results that he wishes to obtain utilizing pulse plating.

Bibliography

1. Knoedler, A., "Pulse Plating Processes in Gold Plating: Status and Prospects," *Galvanotechnik*, **72**, *11*:1167-1174 (1981).
2. Puippe, J. C. and N. Ibl. "Influence of Charge and Discharge of Electric Double Layer in Pulse Plating," *J. Appl. Electrochem.* **10**, *6*:774-784 (1980).
3. Wan, C. C., C. S. Tong and B. Y. Wu. "Effect of Pulsed Current on Heavy Gold Plating," *K'o Hsueh Fa Chan Yueh K'an* **8**, *9*:782-797 (1980).
4. Stimetz, C. J. and M. F. Stevenson. "Pulse Plating of Nickel Deposits," *Ann. Tech. Conf.* **67**. Sess. L: L1-15, AES (1980).
5. Morrissey, R. J. and A. M. Weisberg. "Further Studies on Porosity in Gold Electrodeposits," *Trans. Inst. Metal Finish.* **58**, *3*:97-103 (1980).
6. Wang, Y. Y., C. S. Tung and C. C. Wan. "Application of Pulsed Current to Gold Plating," *Metal Finish.* **78**, *9*:21-25 (1980).
7. Fluehmann, W. F., F. H. Reid, P. A. Mauesli and S. G. Steinemann. "Effect of Pulsed Current Plating of Structure and Properties of Gold-Cobalt Electrodeposits," *Plat. and Surf. Finish.* **67**, *6*:62-65 (1980).
8. Knoedler, A., "Metal Electrodeposition with the Pulse Plating Method," *Draht* **31**, *5*:363-364 (1980).
9. Dini, J. W. and H. R. Johnson. "The Properties of Gold Deposits Produced by D.C., Pulse and Asymmetric AC plating," *Gold Bull.* **13**, *1*:31-34 (1980).
10. Ibl, N. "Some Theoretical Aspects of Pulsed Electrolysis," *Surf. Technol.* **10**, *2*:81-104 (1980).
11. Stimetz, C. J. and J. J. Hren. "Pulse Plating of Hard Gold Electrodeposits," *Ann. Tech. Conf. AES* **66**, Sess. G:G1-14 (1979).
12. Bhar, T. N. and G. M. Lamb. "Effect of Processing Parameters on the Pulse Electroplated Gold Coating," *J. Electrochem. Soc.* **126**, *9*:1514-1515 (1979).
13. Leaman, F. H. and B. A. Kownacki. "Pulsed Versus Direct Current for Continuous Plating of Contacts for Electrical Connectors with Nickel Underplating and Cobalt Hardened Gold," *Int. Pulse Plat. Symp., AES Paper 13*, 19 pp. (1979).
14. Kleinkathofer, W., T. Muramaki and C. J. Raub. "Deposition of Nickel by Pulse Plating," *Int. Pulse Plat. Symp., AES Paper 5*, 7 pp. (1979).
15. Sun, T. P., C. C. Wan and Y. M. Shy. "Plating with Pulsed and Periodic-Reverse Current. Effect on Nickel Structure and Hardness," *Metal Finish.* **77**, *5*:33-38 (1979).

16. Duva, R. "Pulse Plating: the Pros and Cons of Practical Application," *Finish. Highlights* **11**, *1*:40–42 (1979).
17. Raub, C. J. and A. Knoedler. "Pulse Plated Gold," *Plat. and Surf. Finish.* **65**, *9*:32–34 (1978).
18. Rehrig, D. L., H. Leidheiser, Jr. and M. R. Notis. "The Influence of Current Waveform on the Morphology of Pulse Electrodeposited Gold," *Plat. and Surf. Finish.* **64**, *12*:40–44 (1977).
19. Cheh, H. Y., P. C. Andricacos and H. B. Linford. "Application of Pulsed Plating Techniques to Metal Deposition. Part III. Macrothrowing Power of Copper, Silver and Gold Deposition," *Plat. and Surf. Finish.* **64**, *7*:42–44 (1977).
20. Cheh, H. Y., H. B. Linford and C. C. Wan. "Application of Pulsed Plating Techniques to Metal Deposition. Part II. Pulsed Plating of Copper," *Plat. and Surf. Finish.*, **64**, *5*:66–67 (1977).
21. Raub, C. J. and A. Knoedler. "The Electrodeposition of Gold by Pulse Plating. Improvements in the Properties of Deposits," *Gold Bull.* **10**, *2*:38–44 (1977).

32. RINSING

H. L. Pinkerton

Fellow Engineer, Process Engineering Department, Aerospace Division, Westinghouse Electric Corporation, Baltimore, Md.

AND

A. Kenneth Graham

Consulting Engineer, Jenkintown, Pa.

It is recognized as axiomatic that high-quality plating is possible only when a clean cathode is plated in a solution of controlled purity. The part played by rinsing in securing these two fundamental conditions is too often overlooked. Poor rinsing will defeat the object of every other step in the plating cycle. It will cause stained, spotted, blistered or peeling work and contaminated solutions. Properly designed multiple counter-flow rinses will conserve water and reduce both the waste treatment installation and operating costs.

Rinsing is essentially an operation of dilution; its object is to dilute the dissolved chemicals on the surface of the work to the point where they are insignificant, not only in their effect on the quality of work being processed, but also with respect to ultimate solution contamination in the continuous operation of a plating line over a long period of time. Efficient rinsing is obtained when this object is accomplished with the minimum use of water—and this requires the use of properly designed and operated multiple counter-flow rinses.

Rinsing Theory

It is important to be able to calculate the concentration of contaminants in a rinse tank because this determines the effectiveness of the rinsing operation. The rinsed work will leave the tank with a solution of this concentration upon it, and the solution going to waste will be of this composition.

Several authors have proposed equations to describe the build-up of contaminants in a rinse tank[1, 2, 3, 4, 10]. In all cases, it has been necessary to make the principal assumption that mixing of drag-in and rinse water is complete and practically instantaneous. Good practice will approach this ideal as closely as possible by incorporating the following features (Fig. 1):

(1) Vigorous agitation of the rinse water with air.

(2) Introduction of fresh water at the bottom of the tank.

(3) Placing the overflow weir at the opposite end of the tank from the point at which water is introduced.

The principal equations describing various rinsing operations will be given using the following nomenclature:

V = volume of rinse tank, gal
Q = rate of fresh water flow, gpm
M = minutes interval between rinsing operations
D = volume of drag-over on rack and work per rinsing operation, gal
n = number of rinsing operations in t min
C_0 = concentration of contaminant solution being dragged in, oz/gal
C_t = concentration of contaminant in rinse tank after t min, oz/gal
C_e = equilibrium concentration of contaminant in rinse tank, oz/gal
C_r = equilibrium concentration in last rinse of r rinses in series, oz/gal
S = *average* rate of water use in sprays
(Any other consistent system of units may be used).

Case I—drag-in to a non-running reclaim rinse (or to a process tank).

$$C_e = C_0 \qquad (1)$$

$$C_t = C_0\left[1 - \left(\frac{V}{V+D}\right)^n\right] \qquad (2)$$

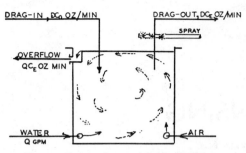

FIG. 1. Simple rinse, with ideal mixing.

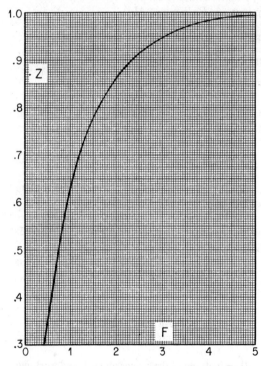

FIG. 2. Exponential curve, $Z = 1 - e^{-F}$. To use:
(a) for non-running reclaim rinse set $F = \dfrac{nD}{V}$, then $Z = \dfrac{C_t}{C_0}$
(b) for running rinse set $F = \dfrac{Qt}{V}$, then $Z = \dfrac{C_t}{C_e}$

Case II—drag-in to a running rinse.

$$C_e = \left(\frac{D}{QM + D}\right) C_0 \qquad (3)$$

$$C_t = C_e \left[1 - \left(\frac{V-D}{V+D}\right)^n e^{-Qt/V}\right] \qquad (4)$$

Case III—rinses in series. There are r rinses, with separate water supplies, $Q_1, Q_2, \cdots Q_r$.

$$C_r = \left(\frac{D}{Q_1M + D}\right)$$
$$\cdot \left(\frac{D}{Q_2M + D}\right) \cdots \left(\frac{D}{Q_rM + D}\right) C_0 \qquad (5)$$

Case III(a)—as Case III, but each tank is supplied with Q gpm.

$$C_r = \left(\frac{D}{QM + D}\right)^r C_0 \qquad (5a)$$

Case IV—multiple countercurrent cascade rinses. The rth rinse is supplied with fresh water at rate Q.

Let $\dfrac{QM}{D} = Y \quad (Y > 1)$

Then $C_r = \left[\dfrac{Y - 1}{Y^{(r+1)} - 1}\right] C_0 \qquad (6)$

or alternately,

$$C_r = \frac{1}{1 + Y + Y^2 + \cdots Y^r} C_0 \qquad (6a)$$

Case V—single running rinse with spray.

$$C_e = \left[\frac{D}{M(Q + S) + D}\right] C_0 \qquad (7)$$

(see text)

Figure 2 is the curve for the exponential equation:

$$Z = 1 - e^{-F}$$

This curve may be used as an approximate* representation of Eqs. (2) and (4) for the instantaneous concentration after n rinsing operations (i.e., at t minutes) in single rinses.

For a non-running rinse, if the value nD/V be used as the abscissa, F, then Z gives the value of C_t/C_0. Similarly, in a single running rinse, if Qt/V be used as F, then Z gives the approximate value of C_t/C_e, i.e., the fraction of equilibrium concentration which has been attained. By the reverse operation, Fig. 2 may be used to calculate the approximate time to establish a given

* The approximation is quite accurate for Eq. (2), especially for values of D/V less than .001 (in most cases D/V will be much smaller than this) and reasonably accurate for Eq. (4), up to values of $n = 500$ (above this value, equilibrium is not approached quite as rapidly as the curve indicates).

fraction of equilibrium concentration in the rinse. For instance, a concentration of 95 per cent of equilibrium corresponds to a value of Qt/V of 3. Stated another way, by the time, t, when the total water flow has amounted to three times the rinse tank volume, the concentration will be 95 per cent of equilibrium. At $Qt/V = 5$, equilibrium is substantially (over 95 per cent) complete. Translated into a practical example, a 200-gal rinse tank with a water feed of 2 gpm will attain about 95 per cent of equilibrium in 5 hr.

Figure 3 is a graphical representation of the equilibrium Eqs. (3), (5a) and (6). For multiple rinses, these curves are plotted assuming the flow to all rinses is equal, and Q is the *total* flow to the rinsing system. If this had not been done, curves 2 and 4, or 3 and 5 would have been almost indistinguishable. With respect to the rinse with a spray (Case V), it is important to note that while Eq. (7) gives the equilibrium concentration in the tank, this is greater than the concentration on the work after leaving the spray, because this arrangement is in effect a multiple cascade rinse. It is not possible to calculate what further dilution is accomplished by the spray, but in practical work the author has always simply assumed that the spray dilutes the equilibrium concentration on the work to one-half its withdrawal value, which should usually be well on the conservative side.

Rinsing Criterion

The fraction C_e/C_0 (or in a multiple rinse C_r/C_0) expresses the degree of dilution effected by the rinsing. The reciprocal of this fraction is a convenient criterion of rinsing quality[13] (i.e., 1000 to 1 dilution is better rinsing than 500 to 1). It will further be observed from the equilibrium equations that in all cases the criterion C_0/C_r may be equated to $(QM/D)^r$ with less than 1 per cent error whenever QM/D is 100 or more, which is often the case. This gives a very quick and convenient method of estimating and designing rinsing installations.

Drag-out

The importance of the drag-out volume, D, is apparent from the above equations. Soderberg[5] has shown that the volume of drag-out of solution per unit area of a flat surface is decreased by:

(1) Hanging the surface as nearly vertical as possible.

(2) Racking with the longer dimension of the work horizontal.

(3) Racking so that the lower edge is tilted from the horizontal, and the run-off bead is at a corner rather than a whole edge.

(4) Withdrawing from the solution slowly.

(5) Draining over the tank, up to 1 or 2 min or until the piece stops dripping.

He further showed that when the transfer interval from tank to tank is *fixed* at some time shorter than complete draining time (as in an automatic conveyor), drag-out is less when the work is withdrawn slowly and transferred rapidly than when the reverse is done.

Table 1 shows the drag-out to be expected per unit area for various classes of work as determined by Soderberg. These data are useful in rinsing calculations or in estimating expected solution losses and the corresponding load on waste treatment processes.

In the discussion following Soderberg's paper[5] it was brought out that solution viscosity, in the range of the viscosities of plating and cleaning baths, does not vary enough to effect the drag-out volume.* There is less drag-out from hot solutions, and a minor reduction in drag-out volume with solutions of low surface tension. No effects due to surface roughness were found in comparing drag-out by cold-rolled steel with that by hot-rolled and pickled stock. Kushner[7] properly considered that there are two phases controlling the drag-out volume, the withdrawal phase and the draining phase. In the former, the amount withdrawn by a flat vertical piece is a function of the square root of the product of withdrawal velocity and *kinematic* viscosity of the solution. From dimensional analysis and Soderberg's experimental figures he deduces the expression for f, the thickness (cm) of the film as withdrawn:

$$f = 0.02\sqrt{UV} \qquad (8)$$

where U and V are velocity and kinematic viscosity, also in c.g.s. units. Kushner further shows that the reduction in drag-out from hot solution noticed by Soderberg can be predicted from the reduction in viscosity.

* In a private communication, J. B. Kushner states that in many solutions used today the viscosity has been shown to be a factor. For example, the much more viscous 53 oz/gal chromic acid bath has a higher drag-out than the 33 oz/gal bath by a factor of nearly four even though the concentration is not even double.

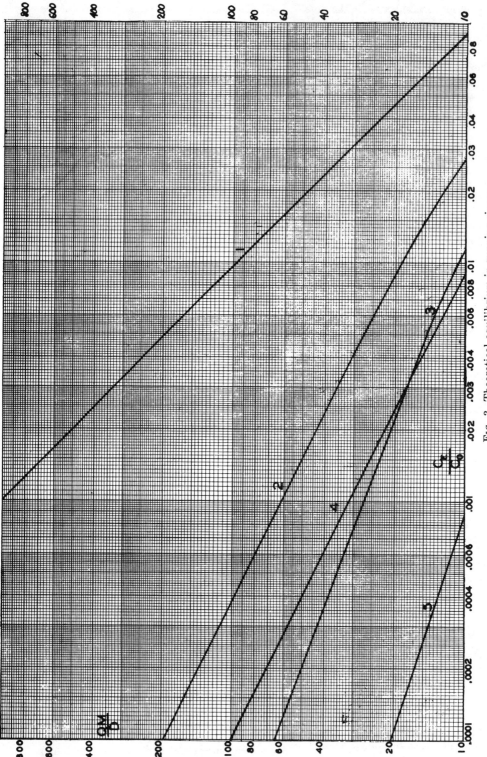

Fig. 3. Theoretical equilibrium in running rinses.

Curve 1. Single rinse.
Curve 2. Two rinses in series, with separate equal flows of fresh water.
Curve 3. Three rinses in series, with separate equal flows of fresh water.
Curve 4. Two countercurrent rinses, fresh water feed to second rinse.
Curve 5. Three countercurrent rinses, fresh water feed to third rinse.

Eq. (8) may also be written:

$$W = 0.02 A \sqrt{\frac{ap}{td}} \qquad (8a)$$

where W is the volume withdrawn in cm³; A is the area of the piece, cm²; a is the vertical length of the piece, cm, p is the viscosity, in poise; t is the time in seconds and d the density of the solution. The first four of Soderberg's five conclusions given above can be deduced from Eq. (8a).

The draining phase, during which the film thickness is reduced, is controlled by the same variables as in Eq. (8a) and in addition by surface tension. Drainage per unit area is increased by an increase in solution density and draining time, as well as by a decrease in the vertical length of the piece and in the viscosity and surface tension of the solution. Thus, all the common factors operate in the same direction in both phases since net drag-out is the withdrawal minus the drainage.

Effect of Tank Volume

It will be observed from Eqs. (1) through (7) that theoretically the rinse tank volume, V, merely determines how quickly equilibrium is reached, but does not influence the *average* equilibrium concentration. In the past, there has been some controversy on this point, which, it is hoped, is clarified in the appendix to this chapter. As a practical matter, it is silly to spend money for a larger rinse tank than is necessary comfortably to contain the work; furthermore the favorable effect of vigorous agitation is more easily attained in a small tank.

Effect of Agitation

Inasmuch as rinsing is a dilution process, it is advantageous if the solution on the work can be diluted with *all* the water in the rinse tank. All rinsing calculations are based on this assumption. If the mixing is inadequate, some of the fresh water will short-circuit to the overflow, and the concentration in the main body and in the drag-out will be indeterminately high. This explains why it is good practice to agitate the rinse water with air as violently as possible without undue splashing or dislodging work from the tank. Design data for air agitation of plating baths are given in Chapters 3 and 37, but the cfm of air used for rinse agitation may be somewhat greater than that used for plating solutions. The entering fresh water, except what may be used for sprays over the tank, should always be introduced at the bottom of the tank. The air should be so introduced (Fig. 1) as to give a rolling turbulence which keeps the surface of the rinse water clean by moving it toward the overflow. The air must be oil-free and should be supplied by a low-pressure blower.

TABLE 1. DRAG-OUT PER UNIT AREA

	Drag-out (gal/1000 sq ft)
Vertical parts, well drained	0.4*
Vertical parts, poorly drained	2.0
Vertical parts, very poorly drained	4.0
Horizontal parts, well drained	0.8
Horizontal parts, very poorly drained	10.0
Cup-shaped parts, very poorly drained	8.0 to 24.0 or more

* This figure may be taken as the absolute minimum for drag-out on a vertical sheet.

There should be an overflow weir along one whole end or side of the tank. A pipe outlet is wholly unsatisfactory. To provide for the periodic dumping and cleaning of the tank, every rinse tank should be provided with a bottom outlet valve.

Different substances vary in the ease with which they can be rinsed from metal surfaces. Dilute solutions of the mineral acids and their salts are very free-rinsing, but concentrated acids, especially certain very viscous electropolishing solutions, are difficult to rinse; moreover, agitation of the workpiece in addition to solution agitation is advantageous. Cyanides and alkaline cleaners are not very free-rinsing; consequently it is common practice to use a warm rinse (110 to 120 F) following a cleaner. Fog sprays are sometimes used over alkaline cleaners to dilute the alkali and to cool the work so that the cleaner is prevented from drying or crystallizing upon it. Surface-active agents are tenacious by their very nature and usually require multiple rinsing to remove them.

Racking

The importance of minimizing drag-out has been demonstrated. Workpieces should be racked so that major surfaces are vertical or nearly so, and so that no cup-shaped recesses are formed. If this is not possible, drain holes may sometimes be provided, the rack may be tilted as it is withdrawn, or an air-jet blow-off may be used. In general, the longest dimension of the work should be horizontal, not vertical; if possible parts should be racked not to drip on other parts hung

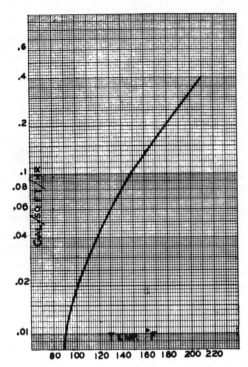

FIG. 4. Evaporation from water surfaces, gal/sq ft/hr in relatively still air (approximate).

TABLE 2. TOTAL RINSE WATER VOLUMES REQUIRED TO MAINTAIN EQUILIBRIUM CONCENTRATION C_e IN FINAL RINSE AT $0.001 \times$ DRAG-IN CONCENTRATION C_0

(In all cases drag-in volume taken as 0.01 gal each minute)

Type of rinse	Single	Series		Counter-flow	
No. of rinses	1	2	3	2	3
Total feed water Q, gpm	10.	0.61	0.27	0.31	0.1

lower on the rack. Rack insulation must be maintained in good condition to prevent trapping of solution behind fissures and cracks. Uninsulated racks should be stripped regularly since rough, treed deposits present a tremendous area, and much solution can be held on them by capillary action. Rack withdrawal from the solution should be slow, and draining time should be allowed over the tank if possible. In manual operations a draining rod can occasionally be provided above the tank; whereas in some automatic conveyors withdrawal and transfer time can be favorably regulated separately from the average conveyor speed or plating time. Sloping drain boards of suitable noncontaminating material should be provided at the exit end of each tank. In effect, these extend the draining period for the work and return as much of the drippings as possible to the tank of origin.

When work is being removed from a tank of heated solution, it is good practice to drench it with a fog spray while it is still over the processing tank. This affords a reduction in drag-out loss, and the water used serves to compensate for evaporation. Figure 4 shows the approximate volume of water evaporated per hour from the surface of solutions at various temperatures. In special cases air jets may be used over the process tank for the purpose of removing solution trapped in recesses, but this is not always practicable. Recovery of drag-out losses is discussed in a later section.

Multiple Rinsing

The advantages of multiple rinsing are apparent from inspection of Fig. 3. If, for instance, during the rinsing interval M, the ratio Y of water-feed volume QM to the drag-in volume D is 10 in a single rinse, the theoretical equilibrium concentration in this rinse will be 9 per cent of the drag-in concentration ($C_e/C_0 = 0.09$). But if two or three rinses with separate water feeds are used in series with the same *total* water usage, the equilibrium concentration in the last rinse will now be 2.8 or 1.2 per cent. If the series rinses are converted to counterflow rinses with the same water usage, the final concentrations will be 0.9 or 0.085 per cent, respectively. Suppose it is desirable to hold the maximum concentration in the final rinse to below 0.1 per cent of the drag-in concentration. The relative water quantities required, determined from Fig. 3, are shown in Table 2.

Thus, it can be seen that two counterflow rinses are about as effective as three in series with separate feeds and are many times as effective as a single rinse. For manual operations it is customary to design a counterflow rinse as a simple cascade because in this case it is not essential that the solution levels in all the rinses be the same. Frequently it is possible to convert a single rinse to a double or multiple counterflow rinse by welding one or more dams across the tank. When counterflow rinsing is desired on a fully automatic machine, pumps are preferably used to transfer the rinse water from one tank to another. In this way correct solution levels can be maintained, and the water feed to each rinse (except what may be used in sprays) can be entered at the bottom of the tank, which is desirable for good mixing.

Spray Rinsing

Spray rinsing can be very effective because even if only a relatively small volume of water is actually useful in diluting the contaminant on the work, the volume of this contaminant is usually still smaller and is being constantly diluted while it is under the action of the spray. This is of great value in outdated installations with only single rinse stations between processing steps. By using over-tank or "stripping" sprays on the work pieces, as they are raised above the rinse, the equivalent of an extra one-half counterflow rinse is obtained. If, in addition, deionized water is used in the sprays, it is possible to greatly reduce the dragover of unwanted ions (hard water type, etc.) into the following processing bath.

It is very important in spray rinsing to provide for maximum effectiveness of the water used. The sprays should be actuated only when work is in the spraying position, and properly designed spray nozzles should be used to effect uniform distribution of the water over the work and to control the volume of water. The frequently used device of one or more perforated pipes is wholly inadequate in this regard.

The most efficient spray will not reach into all recesses in the work and rack; a dip rinse should therefore also be used. An arrangement frequently employed is to position sprays above a dip rinse so as to impinge on the workpieces as they *leave* the rinse tank. These create in effect a counterflow rinse of high efficiency, especially if all the fresh water feed to the tank can be in the form of spray. To provide adequate water feed, however, it is usually necessary to supply part of the water directly to the rinse tank since the spray volume that can be used is limited by considerations of splashing, knocking work off the rack, etc. Many automatic machines are provided with sprays of this nature which are tripped automatically while the work is rising through the spray zone.

Water Economy

Several means for achieving economy of water have already been mentioned:
(1) Multiple countercurrent rinsing.
(2) Spray rinsing.
(3) Spray-and-dip rinsing.

Additional water may be saved if sprays are fed by water pumped from a succeeding rinse tank. Many plants have cooling water (which must be of suitable quality) available from process coolers, cooling coils, air conditioners, or air compressors, which is available for rinsing purposes before being discharged to the sewer. It is possible to reuse rinse water by pumping it to another rinse if care is taken that no undesirable reactions occur, especially precipitation. It is thus feasible to use water from the nickel-plating rinse in the acid rinse prior to nickel plating; however this should not be used in the rinse following alkaline cleaning where nickel hydroxides, etc., would be precipitated on the work. Both water and steam are conserved if a jet condenser is used to condense the steam from coils in a heated process tank, such as an alkaline cleaner, and to feed hot water to the following rinse.

Special Techniques

Ion Exchange. When the presence of hard-water chemicals is undesirable in a processing solution, the solution may be made up with deionized water. It is important in this case to recognize that the preceding rinse should also be made up and fed with deionized water; otherwise the carry-over and build-up of hard-water chemicals in the process tank will cause trouble. In some critical operations rinsing with deionized water may be desirable to prevent staining or other defects. It has been suggested[8] that it is more economical to deionize the rinse effluent and recycle it than to deionize the raw water feed and send the effluent to waste. A little calculation will show, however, that this recommendation is likely to be sound only when a single rinse is being used. When multiple rinsing is used efficiently, especially in cascade with effective controls, it is definitely more economical to deionize the quite small volume of raw water that will be needed. In such cases, one would only deionize the rinse effluent if there were some other good reason for doing so such as recovery of drag-out chemicals. Ion exchange for recovery purposes is more fully discussed in Chapter 11.

Rinsing in Ultracleaning Processes[12]. Some metal cleaning processes such as the cleaning of certain electronic components, typified by transistors, require the use of ultrapure water, sometimes called "intrinsic" water, for rinsing. Ionic and nonionic contaminants as well as microorganisms must be excluded. To produce this water, which may have a conductivity as low as 0.1 micromho/cm, it is first distilled, treated with activated carbon, passed through a mixed-bed ion exchange column, filtered through a fine mechanical filter having a pore size well below 1 micron and finally sterilized by heat or ultra-

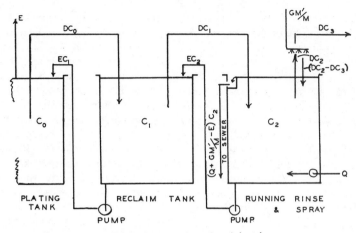

FIG. 5. Schematic operation of reclaim rinse.

violet light. Such high-purity water is contaminated rapidly because gases, notably carbon dioxide, are absorbed from the air; therefore either the water is used as soon as produced, or steps are taken to provide an inert atmosphere in the rinsing device. The used water may be recycled through the purification cycle, omitting the step of distillation. The conductivity of the effluent water from the rinsing operation may be monitored by a conductivity meter, which may be arranged to signal when water of sufficient purity is being discharged, indicating that the work is satisfactorily clean.

Reclaim Tanks. Non-running reclaim rinses have long been used as a means for recovering part of the drag-out. In manual operations, it is usual for the operator to replace evaporation losses, E, in the main tank from the reclaim tank whenever he notices that it is needed. In larger installations, the return may be made by a pump under level control. In either case,* the fraction, F, of the drag-out which is recoverable may be calculated by

$$F = \frac{E}{E + D} \quad (9)$$

When solution is returned intermittently, it is relatively insignificant whether the interval between returns are long or short, and the size of the reclaim tank is immaterial. A reclaim arrangement is shown schematically in Fig. 5, which has the added feature that the reclaim tank is refilled from the following running rinse, a refinement that becomes more economical with the use of multiple counterflow rinses.

In many circumstances it is economical to concentrate the reclaim solution before returning it to the process tank, in order to reclaim a greater fraction of the losses. Figure 6 shows the arrangement and design calculations for an actual large automobile bumper-plating operation[6], which has been quite successful in recovering chromic acid. The tanks preceding and following chromium plating are operated empty. In the tank preceding plating, the work is blown with air jets to reduce drag-in, which otherwise would be unusually large because of a cup-shaped recess in the part. In the tank following plating, the entering work is first rinsed by an intermittent spray; then as the work leaves, the diluted drag-out remaining on the parts is reduced in volume by blowing again with air jets. (It should be remarked that because of the cost and mechanical difficulty of installation, air jets should only be considered in cases such as this where severe cupping causes a very large drag-out.) The bumpers are then dipped in a slowly running rinse. The sprays of the preceding empty tank are fed from this rinse, and only enough water is added to replace that used by the sprays.

The recovered solution from the empty tank following plating is fed continuously through a heat exchanger to a packed column type concentrator having a bed of berl saddles (or equivalent) over which the heated solution trickles while air is introduced at the bottom.

With this system, the actual recovery rate has approached 95 per cent of the drag-out, and the consumption of chromic acid dropped from 18 lb/1000 sq ft of plating area to from 2 to 5 lb/1000 sq ft (a deposit of chromium 15 millionths of an in. thick corresponds to 1.05 lb chromic acid/1000 sq ft). This arrangement proved economical because:

---
* Eq. (9) is accurate for continuous return and a close approximation for intermittent return.

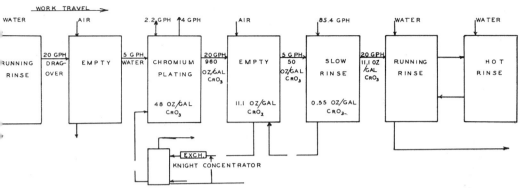

Fig. 6. Chromic acid recovery system (p = pumps).

(1) The unusually high drag-out rate caused a very high chromic acid consumption.

(2) Waste treatment cost savings were considerable as a result of (1) (see Ref. 15).

The successful operation of an evaporative recovery process for cyanides using a four-stage countercurrent rinse is described in detail by Weisberg and Quinlan[14]. The same authors describe the recovery of a chromate conversion solution containing 150 g/l chromic acid and other acids by a combination of ion exchange and a minimum amount of evaporation. The cost of evaporation in the cyanide recovery system amounted to about 18¢/lb of sodium cyanide. This is very favorable compared to the 82¢/lb, and it would have cost to destroy it (Chapter 11); there is a further credit in that the other bath components were recovered as well.

Controls. Where multiple counterflow rinsing is installed for the purpose of conserving water and/or reducing the volume of wastes produced, some form of automatic control should always be provided and then only small water flow rates are required for good rinsing (see Table 2, for example).

The safest and surest control is a conductivity meter[9] with an automatic valve to control the water flow and maintain the designed concentration level of contaminant in the rinse. A very inexpensive but less precise method is to install a flow control valve in the feed line. Such valves, available in capacities of ¼ to 10 gpm, will maintain the rated flow over quite large fluctuations in line pressure. These devices are discussed in more detail in Chapter 37.

The conductivity control will save much more water than the flow-restricting valve because water will only flow to the rinse while it is in use and only in the amount required. It also gives assurance that the desired degree of dilution is actually occurring; whereas the flow control valve must be sized on the basis of estimated drag-out and calculated water flow required.

When either type of control is used, a filling valve should also be provided to allow for quick refilling of the rinse after it is dumped. Provision should also be made that the filling valve will not inadvertently be left open when operation is resumed. One way to do this is to place the valve where a long filling hose will be necessary, which must be removed before resuming work. Another expedient, which has also been used to prevent operators from overflowing process tanks with make-up water, is to place the valve a short distance outside the tank with a short length of hose which must be manually held to direct the stream of water into the tank, thus necessitating the presence of the operator until the tank is full.

Design of Rinsing Installations

The purpose of designing a rinsing installation may be to ensure that rinsing is adequate, to conserve water, to reduce the volume of waste produced, to save chemicals, to reduce the size and cost of a waste treatment facility, or any combination of these reasons. The application of the principles of rinsing to an actual installation is well illustrated by Kushner[13] and by Hanson and Zabban[11], who also show the steps in the designing process:

(1) Determine the drag-over per unit time from Table 1 and the known area to be processed per hour. Alternatively, the drag-over may be determined experimentally by rinsing a number of racks of work in a non-running rinse and analyzing for one of the constituents being dragged in. In barrel rinsing, this is the only feasible means of determining drag-over with any degree of accuracy.

(2) Set the allowable concentration of con-

taminants in each final rinse. The authors[11] used the following, which in the absence of any compelling reason to use a higher or lower concentration, may be taken as reasonable:

	oz/gal	(ppm)
Rinse after nickel (or copper or any other heavy metal)	0.005	(37)
Rinse after cyanide	0.005	(37)
Rinse after chromium	0.002	(15)
Rinse after acid dip	0.1	(750)
Rinse after alkaline cleaner	0.1	(750)
Rinse after acid dip prior to chromium plating	0.002	(15)

(3) From these data and the composition of the process bath being rinsed, calculate the rinsing criterion C_0/C_r.

(4) Use Fig. 2, or the approximate form of Eq. (6) solved for Q:

$$Q = \frac{D}{M} \sqrt[n]{\frac{C_0}{C_r}} \qquad [6(b)]$$

to calculate the water required for 1, 2 or 3 countercurrent cascade rinses for each rinsing operation. Select the most economical number of rinses to use in each case, considering the added equipment and (if any) operating costs caused by the use of additional rinses balanced against water savings and reduced waste volume.

(5) To estimate the waste flow, total the flows to all rinses and multiply by a safety factor (two is suggested).

Hexahedral Barrel Operations

The majority of barrel plated work, especially when plating to rigid thickness specifications, is done in horizontal, hexahedral perforate barrels. As discussed in Chapter 25, the after plating treatment steps, including dragout recovery and rinsing, may be carried out in a hexahedral barrel or in baskets, into which the work is transferred immediately after leaving the plating tank. In the latter case, the operations are then completed manually. In the former case, the barrel may be processed manually or automatically on a hoist operated line, but automatically on a fully automatic line.

When transferring barrel work to baskets for after plating treatment, the baskets should be in a dragout reclaim rather than a running rinse tank, as is commonly done. The cold rinses should be at least double counterflow type with a final hot rinse before centrifuging and drying the work pieces. The baskets should be raised above each station to drain, while shaking with caution so as not to damage the work pieces. This should be repeated twice before transferring to the next station.

When manually rinsing work pieces in a hexahedral barrel between any processing steps, such as bright dips after the plating operation, a minimum of two counterflow cold rinse stations should be used. When this is carried out on a hoist line, a common procedure is for the operator to totally immerse the barrel, then to raise it above the rinse to drain while rotating. This should be done at least twice at each rinse station Unfortunately, where this practice is employed, the operator may do it too rapidly or only once. He may not drain or rotate the barrel above the rinse. Contaminated rinse water is then dragged over into the next rinse station or treatment tank, dripping and splashing on the way. There is then no adequate rinsing, but assured contamination of the following rinse or processing solution. This of course is exaggerated by piece work.

Rinsing work in horizontal hexahedral perforate barrels when immersed and rotated in rinse stations, on either automatic or manually operated hoist lines, involves a number of factors which have not previously been either properly understood or applied. The plating tanks are normally designed so that the barrel will be fully immersed during plating, to avoid hydrogen explosions and to favor drawing the maximum amount of current. (See Chapter 25, p. 000). Other processing tanks, including rinses, usually are similarly designed and operated with the barrel fully immersed and, in many cases, with only single rinse stations.

Application of the theoretical principles of rinsing discussed in this chapter have been largely neglected in barrel plating operations. This is in spite of the more concentrated plating baths generally used and the greater dragout (because of difficulty in draining), as well as the failure to recognize how to apply the rinsing principles to barrel operations due to the differences from rack operations. The most significant difference is the rapidity with which the quilibrium concentration in a given rinse can be reached on the work pieces in rack operations and the difficulty in reaching the equivalent equilibrium concentration in barrel operations. This applies to rinsing of barrel plated work in either baskets or barrels.[15]

TABLE 3. DATA FOR STANDARD (14 IN.) HORIZONTAL HEXAHEDRAL PERFORATE BARREL*

Immersion Depth at Maximum Points** of P.A., % (d)	Pumping Action*** (P.A.) at Maxima, %	Immersion Range about Maxima for at least 1% P.A.		Time at Maxima and 10 rpm for P.A. to = Contained Volume, min.	Immersion Depth at Points of Zero P.A., % (d)
		% (d)	inch		
					100.
93.3	1.20	4.4	0.7	1.45	—
					90.
81.7	1.32	8.0	1.29	1.25	—
					73.2
61.6	1.80	15.4	2.48	0.93	—
— — — — — — — — — —		— axis of rotation —		— — — — — — — —	50.
38.4	3.47	19.6	3.16	0.48	—
					26.8

* (d) = 16.1 in.; the diameter of the circumscribing cylinder of the hexahedral.
** Data for maximum points of 26.8 and 10.1% omitted as of no practical value.
*** P.A. = pumping action, i.e., volume change per 1/6 revolution as per cent of maximum contained volume.

Effect of Immersion Depth

The effect of the depth of immersion of a horizontal, polyhedral, perforate plating barrel on the pumping action during rotation is of great significance in both rinsing and dragout recovery, a rinsing operation. As shown in Table 3, there are depths of immersion for hexahedral barrels at which the pumping action during rotation may be a maximum or zero. Therefore, unless the depth of immersion is controlled during rinsing, the efficiency cannot be controlled. A totally immersed barrel has zero pumping action. Also, the rinsing time (wherein the sum of the volume changes in a rotating hexahedral barrel equals the maximum contained volume in the barrel) at a rate of 10 rpm, for example, is about 0.48 min for the 38.4 per cent immersion, about 0.93 min for the 61.6 per cent immersion, and about 1.25 min for the 81.7 per cent immersion depths, corresponding to points of maximum pumping action. The time required for the greatest efficiency of rinsing is less at the maximum points of less immersion depth. Also a retention time of 1.5 to 2.5 min per rinse station is possible on automatic lines processing 30 to 20 barrel loads per hour, respectively.

However, the immersion depth and rate of rotation for rinsing must be related to the character of the work pieces in the load as to possible damage during rotation where the work pieces might tumble outside the liquid. In view of these interrelated factors and the fact that barrels are normally loaded between about 25 per cent and 50 per cent of their volume, the preferred depths of immersion for rinsing in a rotating horizontal hexahedral perforate barrel, as shown in Table 3, are at the 61.6 or 38.4 per cent maximum points of immersion. About the 61.6 per cent maximum point is a range of 15.4 per cent of the diameter for at least 1 per cent pumping action, or a range of 2.48 in. for a standard 14-in. size barrel. When the barrel load will permit, the 38.4 per cent maximum point of immersion with a range about this point of 19.6 per cent of the diameter for at least 1 per cent pumping action (or a range of 3.16 in. for a standard 14-in. size barrel) is preferred, and especially in dragout reclaim rinsing. In either case, the time for efficient rinsing is most favorable.

The immersion depth should be controlled at or as close as possible to the points for maximum pumping action. Even if an installation is so designed, standard 12, 14 and 16-in. size hexahedral barrels with a 25 per cent volume work load of set screws, for example, when immersed, will displace a volume of liquid sufficient to raise the level of liquid approximately 0.7, 1.0 and 1.3 in. respectively (see Tables 4 and 5). It is therefore desirable to have the weir overflow location, with respect to the axis of rotation of the barrel, correspond to the desired maximum point of immersion. Also, the weir (or weirs) should be of sufficient size to provide quick removal of the excess displaced liquid, so as to return liquid level to the operating level of the weir as quickly as possible (see Tables 5 and 6 for design data).

TABLE 4. STANDARDIZATION OF HEXAHEDRAL BARREL SIZES AND TANK SIZES PLUS VOLUME DISPLACEMENT DATA

Barrel Length, in.	L, in.	Standard Rinse Tank Data W, in.	D, in.	Volume, gal/in.	Displacement,* gal Barrel Size, 12 in.	14 in.	16 in.
30	48	36	30	7.5	5.0	7.25	9.2
36	54	36	30	8.4	6.0	8.6	11.0
42	60	36	30	9.3	7.4	10.0	12.8
Average displacement, inches					0.7	1.0	1.3
Head difference between counterflow rinse stations, inches**					1.50	2.0	2.25

* ½ volume barrel members + ¼ total contained volume (for barrel loaded to 50% with work pieces that take up 50% of occupied volume).
** To prevent back-flow from one to following rinse station when barrel is immersed.

TABLE 5. HEXAHEDRAL BARREL VOLUME DISPLACEMENT* TO OVERFLOW WEIR SIZE FOR RINSING AT CONTROLLED DEPTH OF IMMERSION

Tank Length, in.	Weir Size in.	Data Volume, gal	Ratio Barrel Displacement to Weir Volume Barrel Size, in. 12	14	16
48	3 × 3	1.87	2.7	3.9	4.9
	4 × 4	3.37	1.5	2.2	2.7
54	3 × 3	2.10	1.9	4.1	5.2
	4 × 4	3.74	1.6	2.3	2.9
60	3 × 3	2.34	3.2	4.3	5.5
	4 × 4	4.15	1.7	2.4	3.1
Average Displacement, Liquid Level Rise, inches			0.7	1.0	1.3

* Based on ½ volume of barrel members plus ¼ total contained volume.

Effect of Different Size Barrels on One Line

A complication can arise when the next larger or smaller size barrel (a 12- or 16-in. size, for example) is substituted in a plating line designed for a 14-in. size hexahedral barrel. Assume that the overflow weir in a rinse tank is located so that the solution level will correspond to the 93.3 per cent maximum point for the immersed 14-in. size barrel. A 12-in. size barrel must have its axis of rotation suspended above that of the 14-in. barrel to use the same weir for its 93.3 per cent immersion point (as shown in Fig. 7). The 16-in. size barrel must have its axis suspended lower than the 14-in. barrel to accomplish the same result.

However, if the three barrels are all immersed so that their axes of rotation (or 50 per cent points) are at the same depth as that of the 14-in. size barrel for which the line was designed, the situation is quite different, as shown in Fig. 8. The 12 in. barrel is totally immersed below the weir at the 93.3 per cent point of the 14-in. size barrel and would have zero pumping action. The 16-in. barrel would be so high above the weir (see cross X for the 16 in. barrel) that it would be operating with an immersion depth close to its 90 per cent point of zero pumping action.

TABLE 6. WEIR DISCHARGE PIPE DATA FOR HEXAHEDRAL BARREL RINSING AT CONTROLLED DEPTHS OF IMMERSION

Weir Discharge Pipes d, in.	No.	Flow in Pipe at 1 fps Velocity,* gpm	Time to Discharge Displacement Volume, sec Barrel Sizes, in. 12	14	16
Barrel Length = 30 in., Tank Length = 48 in.					
3	1	22	14	20	25
	2	44	7	11	13
4	1	38	8	11	15
	2	76	4	6	7
Barrel Length = 36 in., Tank Length = 54 in.					
3	1	22	16	23	30
	2	44	8	12	15
4	1	38	10	14	17
	2	76	5	7	9
Barrel Length = 42 in., Tank Length = 60 in.					
3	1	22	20	27	35
	2	44	10	14	18
4	1	38	11	16	20
	2	76	6	8	10

* Velocity of flow assumed for basis of comparison.

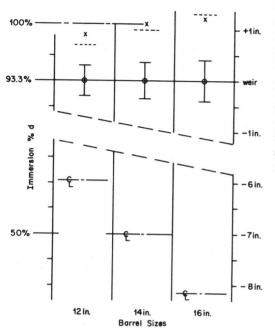

FIG. 7.* Illustrating three standard sizes of hexahedral barrels (12, 14 and 16 in.) immersed to a controlled depth at 93.3 per cent of their diameters, one of the six maxima immersion points for favorable pumping action and all at the same level as the overflow weir, requiring their axes of rotation at different depths of immersion.

* Note. Small circle is location of maximum point as per cent diameter of a specific barrel. Vertical line through small circle is range as per cent diameter for at least 1 per cent pumping action. Small cross is the location of the 100 per cent immersion point of a specific barrel. Fine dash horizontal line is the location of the displaced liquid level by a specific barrel when immersed with a typical load of work pieces. The centerline of the axis of rotation of a specific barrel is designated by the usual ℄.

On the other hand, the favorable maximum points of immersion for good rinsing, i.e., 61.6 and 38.4 per cent, for a 12, 14 and 16-in. size barrel, for example, are free from the above difficulties when used on a line designed with the weirs located for either of the above maximum points of immersion for a 14-in. size barrel, provided all three barrels have the same immersion depth for their axis of rotation. This assumes that the distance between the suspension point and axis of rotation of the several barrels will enable one to do this.* However, as shown in Fig. 9 the 61.6 and 38.4 per cent maximum points for the 12 and 16-in. barrels (shown by the small circles) are quite close to the corresponding points (or the weir locations) for the 14-in. size barrel. The range of immersion about the maximum points for the three sizes of barrels (as shown by the vertical lines drawn through the circles at their respective maximum points) are safely below their 73.2 and 50 per cent immersion points of zero pumping action, as shown by the small crosses above their respective maximum points. Also, the height to which the

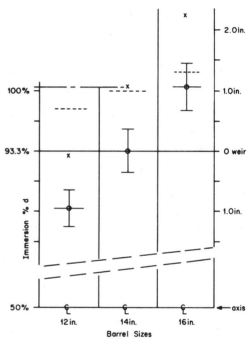

FIG. 8.* Illustrating three standard sizes of hexahedral barrels (12, 14 and 16 in.) immersed with their axes of rotation at the same depth** and with the overflow weir located at the 93.3 per cent maximum point for favorable pumping action of the 14 in. size barrel.

* Note. Small circle is location of maximum point as per cent diameter of a specific barrel. Vertical line through small circle is range as per cent diameter for at least 1 per cent pumping action. Small cross is the location of the 100 per cent immersion point of a specific barrel. Fine dash horizontal line is the location of the displaced liquid level by a specific barrel when immersed with a typical load of work pieces. The centerline of the axis of rotation of a specific barrel is designated by the usual ℄.

** Note. The suspension arms for the different size barrels have to be made the same length to accomplish this.

* Unfortunately barrel designs, as yet, are not standardized in this respect.

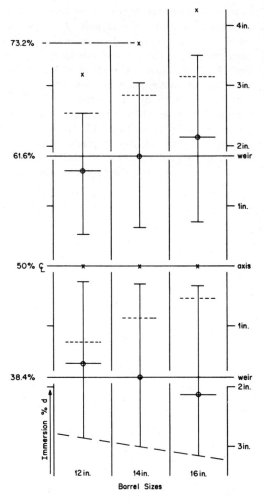

FIG. 9.* Illustrating three standard sizes of hexahedral barrels (12, 14 and 16 in.) immersed with their axes of rotation at the same depth** and showing two overflow weirs at the two maximum points of the 14-in. size barrel for favorable pumping action located immediately above and below the axis of rotation, i.e., at 61.6 and 38.4 per cent of the diameter of the 14-in. size barrel

* Note. Small circle is location of maximum point as per cent diameter of a specific barrel. Vertical line through small circle is range as per cent diameter for at least 1 per cent pumping action. Small cross is the location of the 100 per cent immersion point of a specific barrel. Fine dash horizontal line is the location of the displaced liquid level by a specific barrel when immersed with a typical load of work pieces. The centerline of the axis of rotation of a specific barrel is designated by the usual ℄ .

** Note. The suspension arms for the different size barrels have to be made the same length to accomplish this.

solution level is raised above the weirs, due to volume displacement when the barrels with a comparable work load are immersed, is shown by the fine dotted lines. In all cases, they are well within a range of pumping action of at least 1 per cent. With proper size overflow weirs, the solution level will quickly drop to the weir level, giving a maximum pumping action during rotation.

Comparison of Rack and Barrel Operation

The commonly used cyanide barrel plating solutions vary from 2.5 to about 11 oz/gal of metal and 4.5 to about 9 oz/gal of total (CN). In one plant the average dragout from a silver cyanide solution with an 18 × 30 inch horizontal hexagonal perforate barrel loaded with set screws was reported[15] to be 0.7 gal per load compared to a dragout of 0.015 gal per rack load. One rack was processed every 4 min. One barrel was processed every 6 min. Thus, the amount of contaminant carried out by the barrels was 30 times as much per minute as that carried out in rack plating. Assuming a rinsing efficiency** of 1000 as being good for the rack operation and of 5000 as being good for the barrel operation, the comparative water requirements were as shown below.

Rinses	Rack gal/min*	Barrel gal/min*
Single	3.3	583.
Double counterflow	0.1	8.25
Triple counterflow	—	2.0

* 4 gal/min of water was the maximum available per rinse station and was frequently much less during a working day.

Furthermore, these water data will only apply to the above barrel if it is rotated in each rinse at the required depth of immersion to give the desired movement of liquid in and out of the barrel during the time that it is immersed or immersed in and out of the solution, as discussed previously.

Barrel Rinsing and Dragout Recovery

Figure 10(a) shows a double counterflow barrel rinse tank with a horizontal hexagonal perforate barrel immersed to a controlled depth in the first compartment (while rotating) and the concentrated rinse water flowing over the weir and discharging to the drain or waste treatment as may

** The dilution of dragout chemicals on the work surface leaving a rinse.

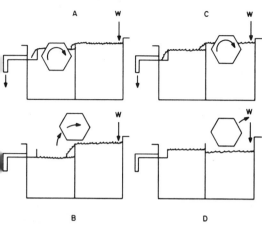

FIG. 10. Illustrating operation of a double-counterflow, hexahedral-barrel rinse with controlled depth of immersion equivalent to the 61.6 per cent point of maximum pumping action with water flowing into the second rinse station and out of the first station. Note the necessity of lowering the suspension of the barrel in the first station with this space saving design of counterflow rinse tank.

be required. In (b) the barrel is being transferred from the first to the second compartment. In (c) the rinsing procedure described in (a) is repeated and in (d) the barrel has been raised and is being transferred to the following cycle step. Note that water (w) in this case is flowing into the last compartment, preferably at a rate controlled by a conductivity cell, as may be required.

Immersing to a favorable depth is also important if a dragout recovery tank is used with the horizontal polyhedral perforate barrel in which rinsing in a reclaim solution is involved. Figure 11 shows one type of barrel dragout recovery operation in which the hexagonal barrel is: (a) first immersed to a controlled depth for a favorable pumping action while rotating in the reclaimed liquid for the desired time; (b) then raised above the liquid and rotated while draining; (c) then sprayed with a controlled volume of rinse water while still rotating; and (d) finally rotated to drain prior to transferring to the following rinse tank[15].

Figure 12 shows a two-tank barrel, dragout recovery operation followed by a double counter-

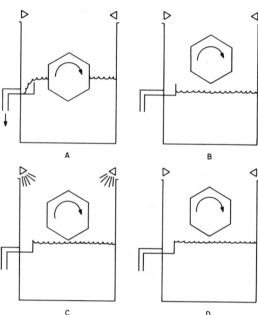

FIG. 11. One recommended design of dragout recovery tank for hexahedral barrels and method of operation with controlled depth of immersion for favorable pumping action during rotation while rinsing in reclaim solution.[15]

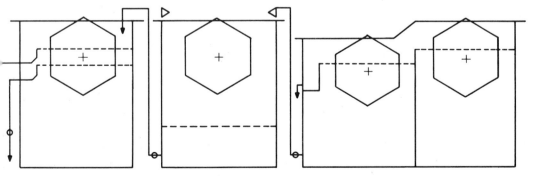

FIG. 12. An alternative two-tank, hexahedral-barrel, dragout-recovery installation, with a double counterflow rinse tank. There are two optional overflow weirs in the first dragout tank at the 61.6 and 38.4 per cent maximum points of immersion for favorable pumping action during rotation, a controlled volume of spray rinse in the second dragout tank and controlled depth of immersion in the double counterflow rinse stations.

flow rinse tank. In the first dragout tank, there are two weirs, corresponding to the most favorable 61.6 and 38.4 per cent maximum points of immersion for the hexahedral barrel. Either weir can be used interchangeably by closing and opening the related valves (shown as circles in the discharge lines). The reclaimed solution is either sent to a concentrator, if required, or directly to the processing tank from which it originated. The barrel is then transferred to a drain tank where it is sprayed with a controlled volume of used rinse water (while rotating) and then further drained prior to transferral to the first station of the double counterflow rinse tank. The barrels are shown immersed to a favorable controlled depth of immersion in both rinse stations. The reclaim spray solution from the second reclaim tank is pumped at the same controlled rate as spray to the first reclaim tank.

Since most alkaline barrel plating baths are operated at relatively low temperatures with very limited evaporation loss, any efficient rinsing and dragout recovery installation will still require concentration of the reclaim solution, in order to recover most of its volume.

Hexagonal Barrel Rinsing Design Considerations. Calculations for operating a single rinse tank (or station) for controlled depth of immersion using more than one size barrel, on a line designed for one size, are as follows:

Definitions:

d = diameter of circumscribing cylinder, neglecting extension of ribs.

D = depth of tank (standard size = 30 in.).

C = distance to centerline of axis of rotation of barrel when immersed from top support (or tank flange).

A = distance from top support (or tank flange) to the top of barrel with the diameter (d) from peak to peak in a vertical position.

B = distance from the bottom of tank to the bottom of the barrel with the diameter (d) from peak to peak in a vertical position.

Based on a standard 14-in. size hexagonal barrel, the data are:

d = 16.1 in.
A (by choice) = 5 in.
$C = A + d/2$ = 13.55 in.
$B = D - (C + d/2)$ = 21.6 in.

If the next smaller standard size (12 in.) barrel is immersed for rinsing at a controlled depth of immersion, the axis of rotation should be at the same depth as that of the 14-in. size barrel. There C is 13.55 in.; d is 13.8 in.; $A = C - d/2$ or 21.6 in., and $B = D - (C + d/2)$ or 9.55 in.

If the next largest standard size (16 in.) barrel is immersed for rinsing at a controlled depth of immersion, the axis of rotation should be at the same depth as that of the 14-in. size barrel. Then C is 13.55 in.; d is 18.4 in.; $A = C - d/2$ or 9.55 in.; and $B = D - (C + d/2)$ or 7.85 in.

Provided the barrels are so constructed that the supporting brackets are designed to immerse their axes of rotation at the same centerline, the advantages of efficient barrel rinsing in dragout recovery and water rinse stations can be realized. (For its importance in plating, see Chapter 25.) This is especially applicable to automatic barrel plating lines where the dwell time per station is favorable and the uncertainties of manual operation are not involved.

The above design need not limit the depth of immersion in the plating tank if provision is made to accommodate the largest barrel in the tank. The solution level can then be varied to plate at any controlled depth of immersion desired.

Appendix

There are several possible mathematical approaches to the rinsing problem:

(a) Set up the mathematical expressions describing what happens each time a rack is dipped into a running rinse and determine the limiting value of this infinite series[2].

(b) Set up the basic differential expression for conditions at any time in a rinse tank and integrate[10].

(c) Set up a material balance under equilibrium conditions[3].

Method (a) gives at equilibrium the limiting concentration of contaminants at the *end* of each rinsing cycle and is capable of expressing the concentration at the *end* of any given cycle before equilibrium is reached. Method (b) defines the *average* concentration during the rinsing cycle after n rinsing operations or at equilibrium. Method (c) define the *average* concentration during the rinsing cycle at equilibrium only.

In this chapter all equations dealing with *equilibrium* conditions have been developed by the simple method (c), and for non-equilibrium conditions by method (a). For equilibrium conditions, method (c) is preferable to method (a) for several reasons. Rinsing calculations are made generally with one of two purposes in mind:

(1) To determine the rinse water flow necessary to achieve a given dilution of contaminants on the work, or

(2) To estimate the concentration of a given contaminant in the water going to waste.

In case (1) above, a rack of work is usually withdrawn from the rinse before the middle of the rinsing cycle rather than at the end, and the calculated average concentration during the cycle therefore more nearly represents actual conditions. Method (a) would lead to the use of too little rinse water to achieve the purpose. In case (2), the concentration of contaminant in the waste is exactly the average concentration during the rinsing cycle; the concentration calculated for the end of the cycle will be too low. For *non-equilibrium* conditions, calculation by method (a) is considered sufficiently accurate.

Method (a) also purports to show that the tank volume, V, is one of the controlling variables; in the other two methods, V does not appear in the equations for equilibrium conditions. It has been shown that the tank volume is significant only in determining the time required to reach equilibrium and to a minor extent in satisfying the assumption of instant mixing. It does, however, determine how great is the fluctuation in concentration from its maximum just after dipping a rack and the minimum just before the next rack is dipped; therefore V appears in all equations derived by method (a).

As the interval between rinsing operations increases, the disparity between the two methods is emphasized. In a symposium on rinsing, presented in 1954 by Herrmann, Hendel and Marino before the New York Branch, American Electroplaters' Society, this point was very well illustrated experimentally. The authors used a 10-min interval between rinses and analyzed their rinse at the *end* of each cycle. As would be expected, their results agreed very well with calculations made by method (a) and differed widely from the other methods. If they had collected the overflow and analyzed an average sample, the results would have been very different.

By a variation of method (a)[3] it is possible to set up an expression for conditions at the *beginning* of a rinsing cycle; nevertheless this is open to the same sort of objections.

Tallmadge and his collaborators have added greatly to the theory of rinsing and dragout recovery (see references 16 to 22). Clarke has also published on the "Theory of Recirculating and Chemical Rinsing"[23] and "Vapor Rinsing."[24]

Chemical rinsing is an adaptation of Lancy's Integrated Waste Treatment technique.[25, 26] Vapor rinsing is based on condensation of steam to displace electrolytes dragged out on the work surface. While there are some acknowledged limitations in applying these techniques, experience may well establish specific cases where they can be used to advantage in production plating.

Appendix II
Fog Nozzles—Their Advantages and Proper Use*

What is a fog nozzle? A fog nozzle is a specially designed water spray nozzle which uses water pressure, or a combination of water and air pressure, to develop a very finely divided atomized mist of water droplets. Because of the very fine atomization attained, fog nozzles use only small amounts of water. Fog nozzles must not be confused with spray nozzles, which use much higher volumes of water.

What advantages do they have? Because of the fine atomization and low water usage, fog nozzles can often be used directly over a heated working tank without introducing enough water to cause the tank to overflow. In some cases they can effectively replace dragout rinses. Specifically, their advantages are:

(1) They reduce the temperature of the parts being withdrawn and thereby reduce or prevent drying on. This effect is particularly noticeable with massive parts that retain a lot of heat.

(2) They reduce the viscosity of the solution retained on the part, giving better run-off and reduced dragout. This reduces the load on the waste disposal system and may replace or eliminate the need for a dragout rinse.

(3) A corollary of (2). By prediluting the film of solution on the parts, better and more rapid rinsing is obtained in the rinse tank.

Typical Results of Properly Installed Systems

Case 1. Completely eliminated rejects due to partial drying of heavy-gauge parts during transfer from cleaner to rinse.

Case 2. Reduced dragout from nickel tank by over 50%, saving 12,000 lb of nickel salts per year.

*By Lawrence J. Durney.

Case 3. Reduced dragout, and, more important, greatly reduced fuming from a hot sulfuric acid pickle during transfer by cooling parts and washing air free of fumes.

To work well, however, a fog nozzle installation must be properly designed. Use the procedure which follows to design your system. Remember that good maintenance is a key requirement for successful operation. Install a good filter in the water line ahead of the nozzles and institute and maintain a nozzle cleaning program on a regular schedule to ensure optimum performance.

Design and Installation of Fog Nozzle Systems. Before attempting to design a fog nozzle system, the following information should be on hand:

1. Dimensions of rack to be fogged.
2. Dimensions of tank, particularly the distance from the most probable position of the fog nozzles to the nearest part on the rack.
3. Evaporation from the tank; i.e., the amount of water added hourly.
4. Approximate available water pressure.

With this information available, the following chart can be consulted to select the desired nozzles.

This elevation and angle should preferably be such that the bottom of the cone just touches the surface of the tank.

On automatic machines, it is possible to cam the valve for the fog installation to the machine in such a way that the nozzles are in operation only during the lift portion of the cycle. This will materially reduce the rate of water addition to the tank. On manual installations, the same effect can be obtained by using a foot treadle valve.

To apply this information, follow this example: Machine is a double-line elevator type; rack width is 26 in. and tank width is 42 in. (Notice that all the required information is not given. However, it is possible to assume some values, and calculate others.)

1. The distance to the middle of the tank (center anode bar) is 21 in.
2. If the thickness of the rack and work is 6 in., the distance from the nozzles to the work will be 7½ in.

At this distance, a 1.5 nozzle will have a pattern of 9 in. at 40 psi, or 11 in. at 80 psi. Therefore, three of these nozzles evenly spaced will adequately cover the face of the rack.

The installation should consist of 12 1.5 nozzles, placed three each on the front and back of the tank, and six in the middle of the tank with

Designation	Type Nozzle	Approximate Spray Angle	Factor	40 psi	Water Use in Gal/hr at 60 psi	80 psi
Narrow angle	0.6	35°	0.7			0.85
	1	40°–60°	1.0–1.2	1.0	1.2	1.4
	1.5	60°–70°	1.2–1.5	1.5		2.1
Wide angle	2 W	165°	11	2.0	2.5	2.8
	3 W	150°–160°	9–11	3.0	3.7	4.3
Multi-head	1–7 N2 at 3 ft from nozzle 65% will be in a 30 in. cone balance in a 54 in. cone			13.8	17.4	19.8

NOTE: Also specify whether male or female type fitting is required. Data is based on nozzles manufactured by Spraying Systems Co.

To determine the spray pattern of any nozzle, multiply the factor by the distance from the nozzle to the work in inches to find the diameter of the spray pattern in inches.

Ideally, the nozzles should be elevated above the tank slightly and aimed downward at a slight angle to prevent wide dispersion of the fog.

three facing each rack. The three on the front of the tank should be elevated approximately 12 in. The sets in the middle and on the back of the tanks should be elevated as much as possible without interfering with the carrier arm.

Water consumption under maximum pressure will be 2.1 gal/hr per nozzle, or approxinately 25 gal/hr for the installation. Since this may be more than the tank can accept, the valve should be cammed to the machine so the nozzles are on only during the lift portion of the cycle. Assuming a standard cycle of 10 seconds for lift, 10 seconds transfer and 10 seconds descent, with no dwell time, this would reduce the water consumption by 60%, making the rate of water addition to the tank 10 gal/hr, which for many heated tanks will be acceptable.

References

1. Kushner, J. B., *Monthly Rev., Am. Electroplater's Soc.*, **29**, 751-770 (1942).
2. id., *Plating*, **36**, 798-801, 866, 915-918, 1110, 1130 (1949).
3. Pinkerton, H. L., *Plating*, **39**, 1016-1017, 1031 (1952).
4. Mohler, J. B., and Sternisha, J., *Metal Finishing*, **44**, 59-67, 99-100 (1946).
5. Soderberg, K. G., *Proc. Am. Electroplaters' Soc.*, **24**, 233-249 (1936).
6. Keller, F. R., Cupps, C. C., and Shaw, R. E., *Plating*, **59**, 152-154 (1952).
7. Kushner, J. B., *Metal Finishing*, **49**, (11), 59-61, 64 and (12), 58-61, 67 (1951); *ibid.*, **50**, (1), 60-64, 74 (1952).
8. Bueltman, C., and Mindler, A. B., *Plating*, **42**, 1012-1018 (1955).
9. Mohler, J. B., *Plating*, **43**, 732-737 (1956).
10. McDonald, N., *Plating*, **36**, 1202 (1949); letter to the editor, with reply by J. B. Kushner.
11. Hanson, N. H., and Zabban, W., *Plating*, **46**, 909-918 (1959).
12. Koontz, D. E., and Sullivan, M. V., *ASTM Special Tech. Publication No. 246*, 183-194 (1959), published by ASTM, Philadelphia, Pa.; Pritchett, P. P., *ibid.*, pp. 205-213.
13. Kushner, J. M., *Metal Finishing*, **47**, (12), 52-58, 67 (1949).
14. Weisberg, L., and Quinlan, E. J., *Plating*, **42**, 1006-1011 (1955).
15. Graham, A. K., and Pinkerton, H. L., *Plating*, **53**, 1222-1229 (1966).
16. Tallmadge, J. A., and Buffham, B. A., "Rinsing Effectiveness in Metal Finishing", *J. Water Pollution Control Federation*, **33**, 817-828 (1961).
17. Tallmadge, J. A., Buffham, B. A., and Barbolini, R. R., *A.I.Ch.E.J.*, **8**, 648-653 (1962).
18. Tallmadge, J. A., and Barbolini, R. R., *J. Water Pollution Control Federation*, **38**, 1461-1471 (1966).
19. Tallmadge, J. A., "Entrainment of Liquid Films, Drainage Withdrawal and Removal", *Ind. Eng. Chem.*, **59**, No. 11, 19-34 (1967).
20. Tallmadge, J. A., Proceedings Second Mid-Atlantic Industrial Waste Conference, Drexel Institute of Technology (1967).
21. Tallmadge, J. A., and Sik, U. L., *A.I.Ch.E.J.*, **15**, 521-526 (1969).
22. Tallmadge, J. A., "Rinsing Mass Transfer in Plating Processes", Symposium of Fluid Flow in Electrochemical Processes, Drexel University, April (1970).
23. Clarke, M., *Trans. Inst. Metal Finishing*, **46**, 201-208 (1968).
24. Clarke, A., and Ashburn, R. J., *Trans. Inst. Metal Finishing*, **47**, 18-26 (1969).
25. Lancy, L. E., *Metal Finishing*, **49**, 56-59 (1951).
26. Lancy, L. E., and Hanson, H. F., *Plating*, **39**, 250-254 (1952).

33. DRYING PRACTICES AND EQUIPMENT

H. L. PINKERTON

Fellow Engineer, Process Engineering Department, Aerospace Division, Westinghouse Electric Corporation, Baltimore, Md.

Rapid and thorough drying of processed metal parts is important to prevent stains or watermarks on pieces plated for appearance; to remove all residual moisture from crevices or recesses, which could be a source of corrosion; or to prevent rusting of unplated pieces before further processing.

The extent to which refinements in drying practice must be carried will be determined by the degree to which any or all of the above objects must be accomplished; by the shape, weight, and orientation of the piece as it is lifted from the final rinse; and to a lesser extent by the metal and the nature of its coating.

Drying Processes

Atmospheric Drying. In atmospheric drying, the piece itself must supply all the heat of evaporation of the surface film of water. For successful drying at a reasonable speed, therefore, there is for each basis metal some minimum weight-to-surface relationship which must obtain; otherwise the piece cools before evaporation is complete. The following tabulation is useful in estimating whether successful atmospheric drying can be expected or not. The calculation is based on evaporating 1 gal/1000 sq ft of water from a sheet metal surface with no cup-shaped recesses, as the piece cools from 190 to 140 F. This amount of surface water corresponds to fairly well-drained vertical parts (Table 1, Chapter 34). The surrounding atmosphere is assumed to be dry enough to accept the water without approaching saturation.

Basis Metal	Minimum Thickness of Sheet for Successful Atmospheric Drying (in.)
Steel	0.070
Brass	0.088
Zinc	0.104
Aluminum	0.110
Magnesium	0.156

Forced Drying with Air Jet. A suitable directed air jet is useful in blowing off excess water from cup-shaped parts, or removing the heavy draining bead from long vertical parts. A fast-moving air stream will also speed up the atmospheric drying process which is often helpful in avoiding stains on sensitive metals such as copper, zinc, and their alloys. Unless the air movement is forceful enough actually to blow off some of the surface water as such, the use of an unheated air stream will not reduce the minimum metal thickness required for good atmospheric drying.

Forced Drying with Heated Air. This type of drying is usually accomplished in an oven immediately following a final hot rinse. It is more effective if heated air is blown directly against the work.

In most commercial dryers of this kind it is common to allow 2 to 4 lb of steam (or equivalent in other forms of heat)/lb of water evaporated from unheated surfaces. For a rough approximation of the amount of steam required to evaporate water from metal surfaces brought from hot rinses at 190 F into such a dryer, the following equation may be used:

$$h = k \left(\frac{w - 6cr}{120} \right) \qquad (1)$$

where h is the pounds of steam required per square foot of total surface area, w is the surface water in gal/1000 sq ft of total surface area, c is the specific heat of the base metal, and r is the ratio of the weight of pieces in pounds to the total surface in square feet. The factor k varies from 2 to 4 depending on the efficiency of the dryer.

For the rare case where wet metal at room temperature must be dried by a hot air blast:

$$E = a\left(1 + \frac{V}{4}\right)(t - t') \qquad (2)$$

E is the evaporation rate in gal/1000 sq ft/min, V is the velocity of the air stream in feet per second, and $(t - t')$ is the difference between the dry and wet bulb temperatures of the air in °F. The factor a is 0.002 for air flow parallel to horizontal surfaces and 0.004 for air flow impinging on verical surfaces.

Drying by Absorption. The ancient practice of drying parts by manually rubbing with hot sawdust is still followed in some shops. It has many disadvantages in the modern plating plant, chief among which are the high labor cost and the difficulty of preventing the dust from being blown about the room. Machines are available which overcome both these objections. Sawdust drying has the advantage that all the surface film of solution is removed by absorption—not merely the water, as in evaporation processes. There is also a burnishing action, especially in the mechanical dryers, which may remove stains already formed.

As the manual method is usually practiced, a 4- to 6-in. layer of sawdust is held in a metal box of suitable size, which is provided with a water bath for heating it evenly through the bottom. Steam is the usual source of heat although direct firing with gas or other means may be used. The absorption medium is a vegetable fiber meal or hardwood sawdust (maple, beech, or boxwood) and should be screened to eliminate fine particles smaller than about 60 mesh. Pieces to be dried are unracked after hot rinsing, placed in the box and covered with warm sawdust. The latter is rubbed over them, and the parts are moved about to contact fresh sawdust until they are dried, after which they are shaken to remove the sawdust. On some parts it is necessary to use an air blast to remove sawdust from recesses and crannies. This aggravates the problem of flying dust; therefore the sawdust box should be positioned as far as possible from operating solutions. This also helps to prevent contamination of the sawdust by splashes of concentrated solutions. The sawdust must be discarded and replaced at regular intervals since it builds up in concentration of absorbed chemicals. This is especially important when the same drier is used for different finishes. Any cyanide residue from brass plating, for example, would quickly spoil an oxidized brass or bronze finish.

The heat requirements for sawdust drying are somewhat less than for oven drying.

Mechanical sawdust dryers, which are described in more detail in the equipment section, are designed for handling small parts in bulk, a job for which the manual method is not well suited. Further, the sawdust is retained in the machine, minimizing the danger of dust in the atmosphere.

Drying with Dewatering Agents. Certain organic liquids or mixtures have the property of preferentially displacing water from a metal surface. These solutions can be compounded from relatively volatile components so that after the water is displaced, the organic film readily evaporates, leaving a dry clean surface; furthermore, nonvolatile ingredients can be incorporated in the mixture so that a film is left which has the properties desired for a particular application. Thus a heavy film can be left on steel parts to protect them from rusting in storage, or a light, paint-accepting film can be used to protect and seal phosphated surfaces for increased paint coverage. These films can be removed, if required, by simple vapor degreasing. Numerous proprietary dewatering agents are on the market.

The solutions are operated at room temperature in plain steel containers. Cold-rinsed work is dipped into the dewatering bath and preferably agitated to assist in removal of the water film. The time for complete dewatering is longer for large or complicated parts, but usually up to 1 min is allowed. Entry and withdrawal should be slow.

It is best to use a specially designed tank to hold the dewatering agent. A cover should be provided to prevent undue evaporation of the solvent when the tank is not in use. The displaced water sinks to the bottom of the tank and means must be provided for drawing this off. Because water usually reacts undesirably, if slowly, with these mixtures, it is best to design the tank so that the area of the collected water in contact with the solution is as small as possible. In small tanks this can be done by making the bottom conical (or tetrahedral) and in larger tanks by sloping the bottom to a small sump. The slope must be fairly sharp because the water motion is slow. There should be a grid or screen above the water level to catch fallen parts. It is also convenient to provide means for continuous automatic removal of the water as it collects. This may be done easily by providing an open riser pipe from the collecting sump (or bottom of the cone in small tanks) just long enough so that when filled with water, it just balances the full head of organic liquid in the tank. Thus, let l be the inches of organic liquid above the desired position of the liquid-water interface to the operating level, a the inches of pipe above the same point, and s the specific gravity of the dewatering solution, then:

$$a = ls \qquad (3)$$

Advantages. The advantage of using dewatering agents may be summarized as follows:

(1) All the surface film is removed, including substances dissolved in the water.

(2) Heat requirements for drying are reduced. No prior hot rinse is necessary or even desirable, and less heat, if any, is needed to evaporate the solvent.

(3) Water can be removed from very difficult crevices such as riveted joints, provided the piece is properly held or turned in the solution and sufficient time is allowed for the water to escape.

(4) Difficult staining problems can often be overcome because of the three points above.

(5) Special films can be deposited to meet unusual requirements.

Disadvantages. The following are the disadvantages of using dewatering agents:

(1) The solutions are relatively expensive, and drag-out and evaporation losses may be high. Drag-out alone will run from ½ to about 6 gal/1000 sq ft of treated surface.

(2) There is a certain fire hazard as the solutions are moderately flammable. (Flash points usually range from 100 to 150 F).

(3) A special tank with cover must be provided, together with means for agitating and/or slowly tumbling the work.

Drying with Infrared Radiant Heat. Infrared radiation has the advantage of being absorbed directly by the metal under radiation; thus the heat is applied efficiently and directly at the point where it is needed. The installation is costly, and in most localities electrical energy is more expensive than other means of heating. Drying by infrared heating is therefore not very widely used. It is probably best adapted to drying of continuous strip or sheet where its efficiency in transferring heat to the metal, combined with the continuous use factor, may operate to make it more economical than other sources of heat.

Air Used for Drying Processes. However it may be used in the drying process (air jet, hot-air blast, sawdust blow-off, etc.), the air must be dry, clean, and free from solid particles and oil. Air from a conventional compressor may be used but must be well filtered. Air for blowers in ovens may be partly recirculated for heat conservation, but sufficient fresh make-up air should be taken from a dry area in the room to keep the dew-point temperature of the air well below the operating oven temperature. This is more important when the air is not heated (note Eq. 2).

Prevention of Stains and Watermarks

Stains occur when some component of the surface water film reacts during the drying period with the metal to form a discoloration. Watermarks are usually caused when the surface water film containing a small amount of dissolved salts is concentrated by the evaporation during drying and shrinks to a few droplets. These then evaporate and visible spots or rings of salts are left, which may be difficult or impossible to remove except mechanically by rubbing or buffing.

Staining may be due to inadequate previous rinsing so that active ingredients of the processing bath build up in the final rinse. Assuming adequate rinsing prior to the hot rinse, stains or watermarks may arise from the water used to feed the final dip. If this is the case and if the spots are very slight, an increase in water flow to the hot rinse may remedy the difficulty because the dissolved constituents in the water feed rise to an equilibrium concentration in the tank given by the equation:

$$e = \frac{Q}{O} f \qquad (4)$$

where f and e are concentrations of water chemicals in the feed and at equilibrium in the tank, respectively, and Q and O are flow rates of feed and overflow. The rate of feed to a hot rinse is usually kept low to conserve on heat requirements, and evaporation can represent a substantial portion of this flow. A better answer to the staining or spotting problem is to use a wetting agent in the final rinse since the weight of solution retained on a given surface is a direct linear function of the surface tension of the solution. By means of a wetting agent, the surface tension of water can easily be halved. Soap is sometimes used for this purpose, to the extent of about 0.1 to 0.2 g/l, to reduce surface tension and to leave a slight protective film on the piece, provided such a film is not objectionable.

Where requirements are very critical, a final rinse feed is used of purity approaching that of distilled water. Clean boiler condensate, if available in sufficient quantity, is acceptable; but deionized water is more frequently used. The latter is generally recommended in color-anodizing processes, for instance.

Special cases warrant the use of dewatering agents, which were previously discussed.

Mechanical Drying Equipment

Centrifuges. Small parts, if suitable in shape and weight, can be very quickly and efficiently dried in bulk by centrifuging at moderate speeds. The centrifuge basket is perforated, and the parts are usually held in a wire mesh basket. By placing the basketful of pieces in the centrifuge and spinning them while still warm from the hot rinse, parts too light for normal atmospheric drying can often be dried successfully. Centrifuges are available with covers carrying a duct for a hot-air blast, the duct being designed so the cover can be swiveled open for loading and unloading. Steam or electric heat is used.

Centrifuges are frequently provided with two-speed motors and should always be capable of being reversed. In operation the centrifuge is spun for 1 or 2 min in each direction to reduce the possibility of some parts retaining water during the centrifuging. Flat parts such as washers are difficult to dry because large surface areas may contact each other and trap moisture. Several reversals of the centrifuge may be necessary to shift the load and expose new areas for drying. Some flat, lightweight parts may be dried better in a mechanical sawdust tumbler or by a dewatering operation.

Drying Machines. Continuous machines are available for drying; rinsing and drying; or washing, rinsing and drying. There are two usual types: one for small parts in which the work is carried through the machine in a perforated horizontal revolving cylinder with an internal helical rib for advancing the work; the other for larger parts in which the work is carried by an open-mesh metal conveyor belt. There are also specially designed dryers for work which must be held on a rack in a certain position, or for handling work on racks directly from the plating or other processing machine. Dryers of the latter type are usually built as an integral part of the processing machine.

In the small-parts machine, the revolving cylinder is perforated at least in the dryer section, and a hot-air blast is introduced at the exit end. A forced exhaust is preferred to ensure speedy removal of wet air. These machines are frequently provided with a loading chute into which a plating or processing barrel-load can be dumped and a discharge chute to a conveyor or to a table for inspection and sorting. The rinse-and-dry machine is also made with two drums in tandem for cold rinsing, followed by a hot rinse and dryer.

Larger parts which are not processed in bulk but on racks are not usually unracked for drying, and the tunnel-type dryer with chain conveyor is used for most work on racks. Therefore the horizontal conveyor-belt type of machine usually has a wash-rinse-dry cycle such as would be used for unplated parts in preparation for painting, for which racking is not required.

Sawdust Drying Equipment. The sawdust box for manual drying has been described. When used for bulk drying of small parts, the sawdust and pieces must be separated by screening. Mechanical separators are available, the most common being an oscillating screen or riddle, although other means have been resorted to, including blowing the sawdust to a collector bin.

Steam-heated sawdust tumbling barrels are often used. The simplest form is the oblique tilting barrel, similar to an ordinary burnishing barrel but fitted with a steam jacket. The steam enters, and condensate leaves through trunnion stuffing box connections in the base of the barrel. The barrel may be arranged to empty into a mechanical sawdust separator.

A larger type of batch tumbling dryer consists of a horizontal barrel which can be rotated in one direction for operating or reversed to unload. When rotating in the tumbling direction, sawdust and pieces are held in the tumbling compartment. A screen permits a portion of the sawdust to escape, and this is carried over a steam- or gas-heated conveyor and returned to the drum. When the barrel is reversed to discharge, the sawdust is automatically separated by a screen in the discharge section and returned over the heated conveyor to the working section. Wet pieces are charged through a chute directly to the working section, and the dried load is discharged at the opposite end of the machine. This unit and the continuous sawdust dryer described below have the advantage that the sawdust handling is all within the unit; consequently a minimum of dust escapes into the plating room atmosphere.

The continuous sawdust dryer is an adaptation of the continuous small-parts dryer, having a rotating horizontal cylinder with helical internal screw to move the work and sawdust through the unit. The cylinder, however, is not perforated except in the separator section near the discharge end. Here the sawdust drops out and is conveyed over a heater to be recharged hot and dry to the working end of the drum. The units are commonly steam heated, and the evaporated moisture is withdrawn up a stack by an exhaust fan.

Sawdust Substitutes. Several waste-product materials have been used as substitutes for hardwood sawdust. The processed bagasse of sugar canes, an almost pure cellulose fiber, will absorb 3 to 3.5 times its own weight of moisture, compared to an absorption for the average maple sawdust of about 1.8 times its weight. This may be used alone for drying or in combination with a meal made from the treated cobs of Argentine flint corn. The latter material is as hard as or harder than maple sawdust, thus contributing to the burnishing qualities of the mixture, and it will absorb about twice its own weight of water. Materials such as these also have an advantage over sawdust in their freedom from acids, pitch, or gum that may occur in inferior grades of sawdust, thereby spoiling the work.

Drying Ovens. Ovens may be of the box-type for batch drying; however far more common is the tunnel type for continuous drying on a conveyor. Because an almost infinite variety of parts are processed on racks, oven design will vary almost as widely; nevertheless there are two major varieties.

In one design, steam coils are placed on the side walls and bottom of the tunnel. Preferably, high-pressure steam is used to maintain an oven temperature of 220 F or higher. A mild current of room air is induced by a forced-draft exhaust duct usually at the entrance end. In large ovens, auxiliary recirculating fans may be used to speed drying, and blow-off jets are commonly used on certain parts. These may be placed in the first section of the tunnel or preferably, if conditions permit, over the hot rinse or in the transfer zone between hot rinse and dryer.

The other major design type uses a preheated blast obtained by blowing air across an electric heating unit or a bank of steam pipes, as is done in a unit space heater. A portion of the air stream may be recirculated, as has been mentioned, and provision must be made to withdraw moisture-laden air. The choice between the two design types will be largely dictated by the nature of the work and the conveyor design.

Drying ovens should be insulated for heat conservation, and the tunnel openings should be no larger than required for the passage of the racks.

Where large openings are unavoidable, an "air seal" can be used to diminish heat losses. This consists of a curtain of fast moving air directed downward across the opening. This air is picked up by a slot at the bottom and returned to the air seal blower. One or more separate recirculating blowers are used to distribute the confined hot air in the oven; still another blower provides the necessary exhausting. A fresh air supply finds its way across the air seal, and the volume of exhausted air should be carefully regulated to carry away the moisture evaporated, although it should not be so large as to reduce the oven temperature or to increase the heat requirements unduly.

Automatic controls are rarely provided on drying ovens. When a difficult drying problem exists, a study of hygroscopic conditions in the oven and exhaust stack will help to indicate what changes in air distribution or heat supply may be required.

Tunnel ovens are commonly designed for a 3 to 5 min drying time. Heat losses may be estimated roughly as follows:

For ovens and ductwork insulated with a high-grade material such as 85 per cent magnesia block, the loss through 1 in. of insulation is close to 1 Btu/hr/° F of temperature difference and is inversely proportional to the thickness of insulation (above $\frac{1}{2}$ in.). This loss figure is for average construction of this type of dryer, where usually little attention is paid to eliminating "through metal" losses at joints, flanges or bolts, etc. A carefully constructed oven will show about half this loss. Uninsulated balls and open ends will lose about 3 Btu/hr/° F of temperature difference. If there is an air seal, open-end losses will be less, but for a conservative figure 3 may still be used.

Accessories. *Air heaters* are usually placed in the ductwork of the recirculating air system of the oven. For best heat transfer the heaters, whether steam, pipes or tubular or finned-tubular electric heaters, should be installed transverse to the air flow, with adjacent rows staggered. Equation (1) may be used to estimate the heat requirements for the work of drying, to which must be added losses through oven and duct walls, open end losses, the heat to bring entering air to oven temperature, and any heat carried out by the work and racks, if their temperature is raised above entering temperature. To summarize, the approximate heat requirements are:

(a) Heat to evaporate water = $8.33\, k\, (w - 6cr)$ Btu/hr

(b) Heat loss, oven walls and duct = $S(t - 70)/i$ Btu/hr

(c) Heat loss, open ends = $3A(t - 70)$ Btu/hr

(d) Heat supplied to entering fresh air = $1.1\, Q(t - 70)$ Btu/hr

(e) Heat to raise temperature of work and racks = $0.12\, P(t - 70)$ Btu/hr.

New symbols used are:

S = surface area of oven walls and ducts, sq ft
t = working temperature of the oven, F
Q = volume of fresh air, cfm at 70 F
P = weight of work and racks, lb/hr
A = area of open ends of oven, sq ft
i = thickness of oven insulation, in. (½ in. minimum)

The above equations are only for estimating total heat requirements as a first approximation of dryer cost and for rough design. Careful designing requires the consideration of other factors such as choice of dryer operating temperature for optimum cost, the most advantageous fraction of air to be recirculated, the prevention of condensation on the walls of the dryer, etc.

Blowers may be of any type which delivers large quantities of air at low pressure. Except for the presence of moisture, there is no particular corrosion problem.

Nozzles for a direct air blast are available commercially. Other types of "nozzles" may consist of specially designed slots or openings in the recirculation system to direct this air most effectively against the work.

Infrared equipment consists of banks of infrared lamps, usually of 250 or 500 watts each. The practice, sometimes observed, of merely arranging one such bank on each side of the work lane without other enclosure is wasteful. In modern infrared ovens the lamps are set in prefabricated solid panels of a reflective insulator such as porcelain. The oven is built of these panels and enclosed as far as possible with other similar material; thus the electrical connections are protected from oven heat and most of the radiant heat and the rather considerable sensible heat of the air is held within the oven.

34. FILTRATION

RICHARD W. CRAIN

Vice-President, Sales, Industrial Filter and Pump Manufacturing Co., Cicero, Ill.

When applied to liquids, the term *filtration* is defined as the separation of particulate solids from a liquid. This is accomplished by passing the contaminant laden (prefilt) solution through a filter medium such as cloth (textile or wire), paper, nonwoven fabric or porous tubes. In practice, with electroplating solutions, the actual amount of solids (dirt) is relatively small when compared to the liquid volume. The filter's purpose is primarily one of "polishing." The cake volume is relatively small except when carrying out a periodic bath purification treatment. In almost all cases, the resultant cake has no value and must be disposed of as dictated by cognizant regulatory authorities.

Filtration may be carried out as either a batch, intermittent, or continuous operation. A batch filtration is performed by pumping the solution through a filter into a separate tank, then transferring back to the plating tank. This insures that every "drop" of the bath has been filtered.

Intermittent filtration is not so common but is done by operating the filter only when required. The filter is used when a plating problem develops.

The most common filtration is, of course, continuous filtration, accomplished by pumping solution through the filter at all times when plating is underway (sometimes even when the solution is not being worked). In all cases, the objective of filtration is either:

1. Prevention of deposit roughness, or
2. Removal of solids after periodic bath purification.

Deposit roughness is created by airbourne dirt, the contaminant being dragged in with the parts, or particles from the anodes migrating into the solution. Periodic bath purification is necessary either for the removal of powdered carbon which has been used to adsorb organic impurities, or precipitated hydroxides from high-pH treatment. In all cases, in order to achieve the objective, a good average flow rate is needed through the filter. Normally, with continuous filtration it would be at least one bath turnover/hour. The plater, solution supplier and filter manufacturer usually disagree on what number of turnovers/hour is required but it ranges from one to six times. The "bottom line" is whatever produces an effectively plated part.

Filter Media

Filter media used are widely varied and dependent upon the filter septum configuration. They consist of various weaves of textile cloth or wire mesh, discs of paper or nonwoven fabric, tubes or "socks" of paper or nonwoven fabric, porous carbon, stone or plastic and wound yarn cartridges. The solids to be removed from plating baths are usually slimy compressible precipitates of the heavy metals, airbourne dirt and some finely divided activated carbon. The initially high flow rate is rapidly decreased by the removal of these solids unless pressure is increased. Increasing pressure tends to pack the compressible solids tighter into the media so that the flow rate will continue to decrease.

In the case of surface type filters (plate, leaf, sock, pleated cartridge), the generated filter cake actually becomes the effective filter medium on the surface of the original medium. The use of filter cartridges (depth type) is somewhat different in that the particles migrate into the cartridges. With depth filtration, in order to achieve longer intervals between filter cleaning, it is necessary to employ a cartridge as coarse as possible but fine enough to do the job. In other words, the particles should migrate into the depth of the filter cartridge but not pass through. In both depth and surface filtration, the use of filter aids can be helpful.

Filter Aids

Filter aids are solid materials used to form a porous cake which is relatively incompressible and more permeable, thus extending the filter cycle. Some examples of filter aid used are diatomaceous earth, perlite, asbestos, clay or alpha cellulose.

They are commonly applied by making a precoat slurry with a portion of the plating solution (or with water) in a precoat tank provided for this purpose. Precoat application is then batchwise, returning the filtrate to the slurry tank until the filter aid forms a layer (precoat) on the original filter septum and the filtrate becomes crystal clear. Effectively, the precoat becomes the filter medium. Bath contaminants are removed and build up on the precoat, thereby forming the filter cake. The precoat also allows for easier cake removal from the original filter media. Filter aids are also sometimes used on the surface of depth cartridges in order to extend their life or allow for their reuse (must be washed off). The weights of various filter aids recommended for precoat vary from ¼ oz (cellulose or asbestos) to 2 oz (diatomaceous earth) per square foot of filter area, resulting in approximately ⅛-in. thickness on the media surface.

Care must be taken when using filter aids that they are completely compatible with the solution. Obviously, diatomaceous earth, perlite and asbestos are silicious in nature and should not be used when filtering fluorides, fluoborate, or fluosilicate electrolytes. Likewise, cellulosic material should not be used in highly alkaline solutions. Perhaps the best approach is to refer to a supplier's list of recommendations.

Adsorbents

Activated carbon is the principal adsorbent used in the plating industry. Both granular (10 × 10 mesh size) and powder (100 mesh) are used. The carbon is either added to the bath as a step in chemical purification treatment to remove organic contamination or by "slip streaming" a portion of the solution constantly through a separate chamber (see discussions of Purification, Chapter 6).

The average dosage of finely divided powdered carbon is 2 to 4 lb/100 gal solution. To aid in rapid settling of the carbon and subsequent batch filtration, filter aid is frequently added with or after the carbon. The usual amount for this purpose is 25 to 50% of the weight of the powdered carbon used. For continuous filtration of many plating baths, it's common practice to use a thin layer of carbon mixed with an equal amount of filter aid by weight following the conventional precoat of filter aid. Activated clay such as bentonite and Superfiltrol have also been used in this way.

TYPES OF FILTERS

Vertical Leaf (Vertical Chamber)

One design of a vertical filter is illustrated in Fig. 1. It consists of a set of rigid filter leaves which have a raised surface so as to allow for internal drainage. These leaves are covered with a textile bag (filter medium). The filter leaves are locked together, in this case, to form a horizontal top manifold. With other styles, they plug into a bottom discharge manifold with an "O" ring seal. The solution enters the chamber at

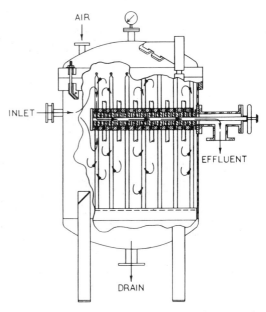

FIG. 1. Vertical leaf pressure filter. (*Courtesy of Industrial Filter and Pump Manufacturing Co.*)

the side, or bottom, passes through the filter leaves and is discharged through the outlet manifold. The air vent enables one to relieve air from the filter chamber when it is first put into operation or to periodically purge it when the air builds up in the dome of the filter.

The same connection is also used to introduce air for drying the cake. The contaminant (cake) builds up on the outside of the filter bags, which are supported by the filter leaves for the duration of the cycle. Then, for cleaning, the liquid is displaced with air. The cake may be released by an internal sluicing mechanism or by taking the leaves out and washing the bags.

Vertical Leaf (Horizontal Chamber)

The cake accummulates in the same way with this style. This unit is generally used for very large plating baths and is suitable for a dry cake (vibration) discharge, as shown in Fig. 2. Filters of this type are used to remove precipitated metallic hydroxides from a waste treatment system which often is associated with electroplating or metal finishing operations (see Fig. 2a).

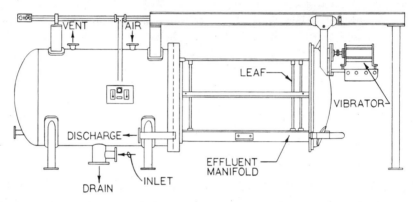

FIG. 2. Horizontal tank, vertical leaf pressure filter. (*Courtesy of Industrial Filter and Pump Manufacturing Co.*)

FIG. 2a. Horizontal tank, vertical leaf pressure filter. (*Courtesy of Industrial Filter and Pump Manufacturing Co.*)

TABLE 1. VERTICAL PRESSURE LEAF, TYPICAL FLOW RATES*

Type of Solution	Average Rate (gal/ft²/hr)
Watts, bright or high chloride type nickel	35
Acid copper sulfate, copper or nickel fluoborate, acid zinc or chromic acid	35–50
Acid tin, silver cyanide	25–35
Copper cyanide	25–40
Brass	25
Alkaline tin	15–25
Zinc or cadmium cyanide	15

*Using a precoat averaging 2 oz/ft² of diatomaceous earth or ¼ oz/ft² of cellulosic filter aid.

TABLE 2. VERTICAL PRESSURE LEAF, TYPICAL SIZES PLUS CAPABILITIES*

Filter Area (ft²)	Cake Space (ft³)	Filtration Rate** (gal/hr)
10	0.43	350
24	1.0	850
60	2.5	2000
95	4.0	3300
130	5.4	4500
170	7.1	6000
230	9.6	8000
304	12.7	10,800
425	17.7	15,000

*For rubber-lined steel (acid solution).
**Based on a rate of 35 gal/ft²/hr.

Horizontal Plate Filter

The horizontal plate filter is illustrated in Fig. 3. The solution to be filtered enters the chamber at the side and flows through the filter media (paper or nonwoven fabric) onto the perforated horizontal plate, then to the center core or manifold, it is discharged through a bottom outlet. An air vent is used as previously described. The cake forms on the top of the horizontal plates, having one advantage over a vertical leaf type, by eliminating the tendency for the cake to "sough off" from the filter media when the pump is stopped or flow is otherwise interrupted. However, the plate filter is more difficult to clean because the plates must be removed for cleaning and set aside, one at a time, then redressed with media before putting it back into operation (see Fig. 3a). A spare set of predressed plates can minimize cleaning downtime.

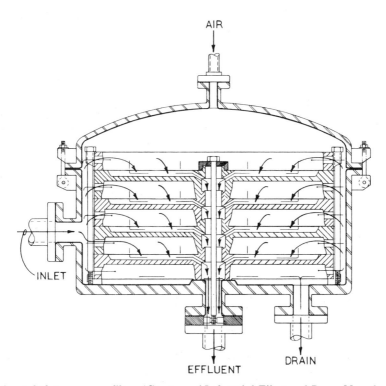

FIG. 3. Horizontal plate pressure filter. (*Courtesy of Industrial Filter and Pump Manufacturing Co.*)

FIG. 3a. Horizontal plate pressure filter. (*Courtesy of Baker Brothers.*)

Tubular Sock Type Filters

These units normally are rubber lined or plastic with a tube sheet at the top (see Fig. 4) to position and secure the tubes. The filter tube may either be a perforated metal or plastic or a polypropylene bag (see Fig. 4a). A nonwoven fabric "sock" or tube is inserted into the tube or bag, which then accommodates a solution flow from the top, through the "sock" to the chamber and out. The solids are retained in the "sock" and are easily disposed of as a "sausage" form to a proper receptacle.

All liquid in the filter chamber is filtered. No unfiltered heel (solution) remains at cleaning time (which could contribute to a pollution system load or solution loss).

Porous-Tube Pressure Filter

This type of filter, supplied by a number of manufacturers, accommodates several kinds of tubular filter media. The filter shown in Fig. 5 (also see Table 5) utilizes porous ceramic tubes mounted in a tube sheet. Filter aid is used to precoat the tubes, preventing them from becoming permanently clogged with finely divided solids. The contaminated solution, pumped in at the bottom of the chamber, is forced through the precoat and the walls of the supporting porous tubes. The filtered solution flows upward through the tube interiors and out of the upper chamber. A backwashing tube cleaning mode is accomplished by suddenly releasing the pres-

TABLE 3. HORIZONTAL PLATE PRESSURE, FILTER DATA

Typical Sizes and Capacities		
Filter Area (ft^2)	Cake Space (ft^3)	Filtration Rate* (gal/hr)
3.1	0.15	185
9.2	0.45	550
18.5	0.87	1110
59.4	2.5	3600
102.6	4.1	6000
151.2	6.5	9000

*Based on a rate of 60 gal/ft^2/hr.

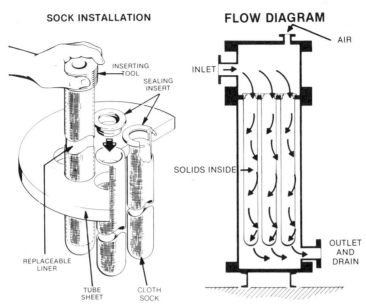

FIG. 4. Vertical tank "sock" type pressure filter. (*Courtesy of Industrial Filter and Pump Manufacturing Co.*)

FIG. 4a. Vertical tank tube type filter. (*Courtesy of Camac Industries.*)

TABLE 4. SOCK FILTER, TYPICAL SIZES AND CAPACITIES*

Filter Area (ft²)	Cake Space (ft³)	Filtration Rate* (gal/hr)
9.6	0.48	1700
16.8	0.84	3000
21.6	1.08	3800
36.0	1.5	6400
43.2	2.1	7800
72.0	3.2	12,900

*Based on 175 GPH/ft².

sure in the lower chamber and by opening a large, quick-opening backwash valve. This allows for a rapid backflow, cleaning the tubes with filtered liquid.

This type of tube filter has many variations: polypropylene with seamless sleeves or bags and wedge wire, as well as a spring suspension for vibratory (dry cake) cleaning.

Cartridge Filters

Cartridge filters of the wound type are available in the 3 to 100 range of micron particle rentention. Because of the variety of porosities available, they are sometimes best suited to handle higher dirt load conditions. A depth type cartridge consists of a series of layers that are

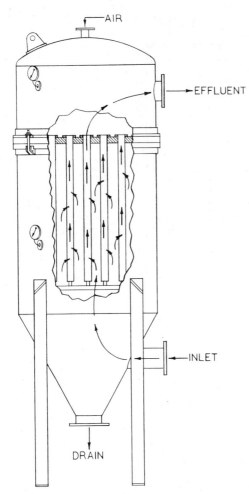

FIG. 5. Vertical tank, tube type back flushing filter. (*Courtesy of Industrial Filter and Pump Manufacturing Co.*)

TABLE 5. TYPICAL POROUS, TUBE FILTER DATA

Number of Tubes	Cake Space (ft³)	Filter Area (ft²)
3	0.34	8
7	0.82	19
21	2.4	57
38	4.3	102
54	6.2	146
111	12.9	300

formed by winding a twisted yarn around a core to form a diamond opening (see Fig. 6). The fibers are stretched across the diamond opening to become the filter media. Succeeding layers lock previous brushed fibers in place and since there is the same number of diamond openings on each layer, the openings become larger due to the increase in circumference. During filtration, the larger particles are retained on the outer layers of the cartridge where the openings are large, while the smaller particles are retained selectively by the smaller openings on succeeding inner layers. High flow rates are claimed for filter cartridges of this type of 100 GPH for each 10-in. long × 2½-in. diameter cartridge. When the cartridges become clogged, they are normally replaced with new ones. As may be seen from Table 6, the units are available in multiple (Fig. 6a) and single cartridge configurations.

Belt Filters

This equipment consists of a rectangular filter chamber of suitable materials of construction with rubber gasketed "doors" which seal the ends of the chamber. These are actuated by air cylinders or screw jack devices. The solution is pumped into the chamber through a top center inlet and forced through micron rated filter media (see Fig. 7). Filtration continues until the media becomes clogged, actuating a pressure response switch. At this point the pump stops, an automatic air blow-down takes place which evacuates the chamber through the media back to the process tank. Blow-down continues for several minutes, drying the solids filtered from the solution. The doors then open and fresh media is indexed from a roll at one end of the machine, while the spent media is rewound into a device at the other end. The dirt is scraped off into a special "tote box" for disposal. The chamber doors then close and filtration resumes. The indexing, self-cleaning cycle usually takes 3 to 4 minutes. Filtration cycles vary according to the amount of solids to be collected. They may be from ¼ hr to 10 hr. The functions are controlled by an electronic control panel. Because the media is indexed into the chamber as needed, the solution is always being filtered through fresh media.

ASME Code

At the present time 36 states and 30 major cities have pressure vessel laws. In almost all cases pressure vessels must be manufactured in accordance with the ASME Code, Section VIII, Div. 1, and bear an code stamp. The only vessels which are not within the scope of this division are:

1. Vessels with a capacity of 120 gal or less for containing water.

FIG. 6. Wound multiple cartridge filter. (*Courtesy of Serfilco, Ltd.*)

2. Vessels having an operating pressure not exceeding 15 psi.
3. Vessels having an inside diameter not exceeding 6 in. with no limitation on length or pressure.

A good source for determining where the ASME Code is required or where pressure vessel laws apply is:

Uniform Boiler and Pressure Vessel Laws Society
P.O. Box 512
Oceanside, NY 11572
Telephone (516) 356-5485

This organization publishes a synopsis of boiler and pressure vessel laws, rules and regulations by states, cities, counties and provinces (U.S. and Canada). The current trend is toward requiring the ASME Code and code stamp.

Materials of Construction

The list in Table 7 is limited to materials coming in direct contact with the processing solutions as they pass through filters, pumps, slurry tanks, valves, pipe and fittings.

Filter Cloths

Many of the filters used in the plating industry require cloth as part of the supporting filter medium. Materials commonly used for filter cloth are listed below:

Materials	Relative Cost	Applications
Cotton	1.0	Mild alkaline and acid solutions
"Nylon"	1.25	Alkaline and mild acid solutions
"Dacron"	1.3	Acid solutions (except concentrated sulfuric and nitric) and mild alkalies
"Dynel"	1.7	Strong acids and alkalies
"Polyprophylene"	1.4	Strong acids and alkalies
"Teflon"	10	All applications, including oxidizing acids
Glass	2.0	Strong acid solutions, except fluoride type

Cotton duck is the least costly fabric and is

Fig. 6a. Single cartridge "in-tank" type filter. (*Courtesy of Sethco Division, Met-Pro Corp.*)

Table 6. Typical Wound Cartridge Filter Data

Number of Cartridges (10-in.)	Solids Capacity/ Equivalent Depth Area (3.5 ft²/tube)	When used w/Precoat, ft²	Approximate Flow Rate Range (gal/hr)
1	3.5	0.6	50–75
3	10.5	1.8	150–225
6	21.0	3.6	300–450
24	84.0	14.4	1200–1800
175	612.0	105.0	8750–13,000
300	1050.0	180.0	15,000–22,500
445	1557.0	267.0	22,250–33,375

used extensively where applicable. The various synthetic fibers are used in the more destructive environments or where the higher cost can be justified by longer service life. Only glass and Teflon fabrics will withstand hot concentrated solutions of strong oxidizing acids such as chromic acid. Glass fabric is more subject to mechanical damage than the synthetic materials. Some of the synthetic fabrics have temperature limitations which must be considered. All filter cloths should be washed thoroughly before use in contact with plating solutions.

The details of cloth construction such as the type of weave, the threads per inch, the weight per square yard, and the methods of desizing and shrinking vary greatly. These data and the flow rates for specific cloths and applications may usually be obtained from the supplier.

FILTER INSTALLATION

The essential components of a typical pressure filter installation are illustrated in a schematic layout in Fig. 8. This includes a filter, a pump, pipe and valves, a small slurry tank and a

FILTRATION 725

FIG. 7. Indexing belt type filter. (*Courtesy of Summit Scientific Corp.*)

TABLE 7. MATERIALS OF CONSTRUCTION

Materials	Uses
Steel, low carbon	Tanks, pumps, chambers, pipe fittings
Cast iron	Pumps, valves, fittings
Stainless steel*	Tanks, pumps, chambers, valves, fittings
High-silicon cast iron	Pumps, pipe, fittings
Hastelloy C	Pumps, fittings
Titanium	Pump sleeves
Glass	Pipe, cloth filter bags
Chemical stoneware	Tanks, porous tubes
Hard rubber**	Pumps, pipe, fittings, linings
Rubber (approved compositions)**	Linings, hose
Cotton	Cloth filter bags
Plastics (approved compositions)**	Linings, hose, pipe, cloth filter bags, pumps, chambers

*Limited to special alloys
**Compatibility tests are necessary in the absence of data on both contamination and effect on deposits.

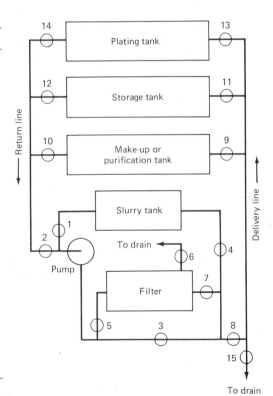

FIG. 8. Typical pressure filter installation showing operating valves.

plating tank. For precoating, filter aid is made up as a slurry in the addition tank and then fed into the suction line of the filter pump by cracking the discharge valve. The filter aid must be kept in suspension with some form of agitation. An external heat exchanger for heating or cooling is frequently inserted between the filter and the receiving tank. A disadvantage of this arrangement is that the flow will vary with the normal flow variation through the filter.

A positive displacement type of metering pump is occasionally connected between the filter pump and the filter for continuous feeding of filter aid slurry. This will assist in maintaining a more even flow. When using a diaphragm in a plating tank, the suction line of the pump should be connected so as to draw solution from both the anolyte and catholyte, unless two separate filters are used. In either case, the filtrate is always returned to the cathode side of the diaphragm. The flow must be controlled by valves in the suction lines (see Chapter 20).

PUMPS

Centrifugal pumps are the most commonly used type for filtering plating baths and related processing solutions. Pumps provided with liquid water seals between the packing glands and the impeller are preferred, since this feature reduces both the tendency to contaminate the processing solution with air or lubricant and the loss of the solution. Centrifugal pumps are available which feature magnetic coupling between the motor and the impeller, thus eliminating seals. The materials of construction depend on the type of plating bath or the processing solution, such as acid dips or electropolishing baths (see Chapter 3, Parts D and G, and Chapter 6). Unless a self-priming type of pump is used, the pump should be located so that the suction line is under a positive head of solution.

The following steps will serve as a guide when installing a pump:

1. Set the unit on firm foundation so the shaft will be level.
2. Check the name plate on the motor for proper voltage.
3. Connect to the power source.
4. Rotate the shaft of the pump by hand to make sure it is free (if impeller rubs against the suction cover or volute, adjust the clearance according to the manufacturer's instructions).
5. Check the direction of rotation of the pump which is usually indicated by an arrow in the volute, suction cover, or neck of the pump.
6. Lubricate according to instructions.
7. Adjust stuffing box gland while pump is running with solution in the pump to avoid possible breakage. (Never run it dry.)
8. Run the pump with the delivery line open until all adjustments have been made.
9. Follow strictly all maintenance instructions of the pump manufacturer.

VALVES

The valves in a filter system are a vital part of the installation. They must close or open a line positively and also operate freely to enable one to control the rate of flow or the pressure by adjustment while filtering. The valve must, therefore, seat properly and be kept in good condition; moreover, the packing gland on valve stems must be properly packed and adjusted. Globe or gate valves are commonly used. Flanged diaphragm valves are preferred by many, especially for the larger sizes and when rubber or plastic-lined pipe and fittings are required. Brass inserts in steel valves should be avoided.

Most filter installations are equipped with a standard set of eight valves (numbers 1 to 8, Fig. 8) and their manipulation for various operations is described below:

1. For open pumping, i.e., bypassing the filter and transferring solution from one tank to another, or for circulation within one tank, open valves 2, 3, and 8; close valves 1, 4, 5, 7 and 15.
2. For mixing in the slurry tank, open valves 1, 3 and 4; close valves 2, 5, 7 and 8.
3. For precoating the filter with filter aid (and carbon), open valves 1, 4, 5 and 7; close valves 2, 3, 6 and 8.
4. For filtering, open valves 2, 5, 7, and 8; close valves 1, 3, 4, 6 and 15.
5. To filter from the make-up or purification tank to the solution storage tank, set valves as described in (1) above and with valves 10 and 11 open and valves 9, 12, 13, and 14 closed.

When a single filter is used for filtering two or more types of solution (for example, bright and semi-bright nickel), every precaution must be taken to prevent mixing one solution with the other.

FILTRATION TROUBLE-SHOOTING

A filter installation requires periodic checking for satisfactory performance and maintenance and the same details require attention when filtration difficulties are experienced. The following list is offered as a guide:

1. All components of equipment must be checked for leaks.
2. Valves must perform properly.
3. Hose lines must not collapse under suction or pressure. Hose clamps must be tight and the open discharge end secured so that it will not slip and cause loss of solution.
4. All discharge lines must be below the solution level to avoid air entrapment.
5. A screen must be inserted in the suction line ahead of the pump to protect it against breakage, and this must be kept clean to prevent clogging of the suction line.
6. Intake and discharge openings from and to a tank must be located to provide full agitation and complete turnover of the solution.
7. Provision must be made for easy sampling of the filtrate for analysis and/or inspection as to clarity.

The pump performance should be specifically checked for the following:

1. Direction of rotation.
2. Free rotation (by hand).

If there is no solution flow from the discharge line:

a. The suction line may not be adequately primed with solution.
b. The speed may be too low.
c. The suction lift may be too high.
d. The discharge head and pipe friction losses may be too high.
e. The impeller or suction line may be plugged.
f. The packing gland may be sucking air.
g. There may be an open valve or an air leak in the suction line.

If solution flow from the pump is inadequate:

a. Check (a), (b), (c) and (d) above.
b. If a foot valve is used, it may be not be immersed to sufficient depth.

If there is insufficient pressure in the discharge line:

a. The speed may be too low.
b. Air may be entrapped in the line.
c. The impeller may be worn or damaged.

If the pump performance is variable:

a. There may be a leak in the suction line.
b. Some mechanical defect such as a bent shaft, binding of rotating part, a worn bearing or misalignment between the pump and the motor may be responsible.

GLOSSARY

Several terms commonly used in filtration should be defined:

Admix or bodyfeed. Filter aid fed in small quantities (usually one part/part of contaminant solid) on a continuous basis during a filter cycle.

Element. A unit in a filter such as a tube, leaf or plate.

Filtrate. The liquid that has passed through the filter element, used synonymously with effluent.

Heel. The unfiltered liquid left inside the filter after the filtration cycle.

Polishing. Filtering very small quantities of solids from a liquid in the range of 1 to 50 ppm.

Prefilt. Liquid to be filtered.

Septum. A permeable materials used to support the filter medium.

Acknowledgments

The author is indebted to the companies listed in the references for their cooperation in supplying materials for this chapter.

References

1. Industrial Filter and Pump Manufacturing Co., Chicago, Illinois.
2. Serfilco, Ltd., Glenview, Illinois.
3. Filterite, Technetics Division, Brunswick Corp., Timonium, Maryland.
4. Sethco Division, Met-Pro Corp., Freeport, L.I., New York.
5. Camac Industries, Fairfield, New Jersey.
6. Cuno Division, AMF, Inc., Meridan, Connecticut.
7. Baker Bros., Systems Engineering Corp., Stoughten, Massachusetts
8. Commercial Filters Division, Kennecott Corp., Lebanon, Indiana.
9. Summit Scientific Corp., Fairfield, New Jersey.

35. AUXILIARY EQUIPMENT FOR THE PLATING ROOM

H. L. Pinkerton

Fellow Engineer, Process Engineering Department, Aerospace Division, Westinghouse Electric Corporation, Baltimore, Md.

Revised by Lawrence J. Durney

Current Control Devices

Ammeters are essential in all electrolytic operations to measure the current flowing to the process tank. For the currents usually met with in plating, the meters will be of the external shunt type. The shunt is simply a conductor with a definite known resistance, to the ends of which the ammeter leads are attached. The combined resistance of the ammeter coil plus its leads bears some fixed relation to the shunt resistance; therefore a fixed fraction of the total current passes through the ammeter coils, and the meter can be calibrated to read the total current flowing.

It is important to realize that the leads are an integral part of the instrument's calibration. If the leads are damaged or changed in length or other leads are substituted, the meter must be recalibrated. The capacity of the meter should be only slightly above the maximum current to be used in the tank, and the scale should be large enough to permit accurate readings to be made over the range of current values to be used.

Both ammeters and voltmeters should be checked periodically with standard instruments to be sure they maintain their accuracy.

Voltmeters are useful in spotting irregularities in operation, as something is amiss if an unusual voltage is obtained when the current is flowing at its normal value.

Since it is usually desired to know the potential between the work and the anodes, the voltmeter is best installed between the anode and cathode work rods. If a tank rheostat is used, it is important that the voltage drop across the rheostat is not included in the measurement. Figure 1 illustrates correct and incorrect installations of voltmeter and ammeter. In the "incorrect" diagram ammeter A_1 reads the total current to two tanks, and A_2 reads the current to only one anode rod. Voltmeter V_1 gives only the general line voltage, and V_2 includes the drop across the rheostat.

The more general use of individual rectifiers located close to the operating tanks has changed many aspects of current distribution, monitoring and control. Voltmeters and ammeters are usually an integral part of the rectifier package and provision is usually made for the installation of the ammeter shunt and leads, and the voltmeter leads immediately adjacent to the rectifier. Where only a single tank is connected to the rectifier, this poses no problem other than a correction in settings of the rectifier to correct the applied voltage for the expected losses in the bussing system.

Where the rectifier is provided with a remote control, the meters are usually mounted on the control panel and this panel mounted conveniently close to the tank being serviced by the rectifier. The ammeter shunt and leads and the voltmeter leads are then connected in the circuit

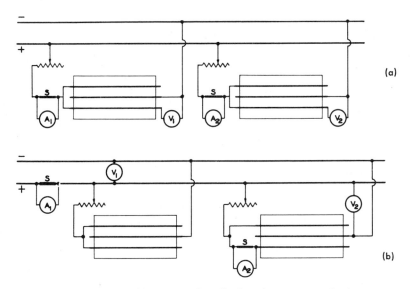

Fig. 1. (a) Correct and (b) incorrect installation of ammeter and voltmeter.

close to the tank and no voltage correction is necessary.

Rheostats are required when more than one tank is being operated from a single current source, and the current must be controlled separately in each tank. Plating rheostats are usually of the parallel type, consisting of a number of coils of resistance wire in parallel, with appropriate switches for control. (See Fig. 1, Chapter 27). The rheostat is installed in series with the tank; thus as more switches are closed, more parallel resistances are added and the total current to the tank is increased. Rheostats are designed to have a specific voltage drop (commonly 1, 2, 3, 4 or 5 volts) *when their full rated current is flowing*. In selecting a rheostat one must therefore know not only the maximum total current required but also the line voltage at the tank as well as the tank voltage required for the maximum current. If the line voltage is 5 at the tank and 4 volts are necessary to produce the desired maximum current, a rheostat with a 1-volt drop is selected.

The resistance coils are so chosen that the reciprocals of their individual resistances bear a definite relationship to each other. Then *at the rated line voltage* each coil might be expected to permit a predetermined number of amperes to pass, and these ampere ratings are frequently marked on each coil switch. Unfortunately the actual amperes passed will be the same as the rated amperage only when the conditions are exactly the same as those assumed in designing the coil.

Rheostats are supplied mounted on a panel, usually with built-in ammeters and voltmeters. The latter may be provided with a selector switch to read either line or tank voltage. Occasionally a short circuit switch to bypass the rheostat entirely is also supplied. The resistance coils are mounted on the back, in the open, with plenty of space to allow for air circulation. Care must be taken to install the rheostat so that nothing interferes with dissipation of the rather considerable heat that is generated.

A rheostat with the coils in series and switches to short-circuit all but the main coil might also be used, but it would be less flexible (with four coils, such a rheostat would give a maximum of 8 steps, while a similar parallel rheostat may provide up to 16 steps). Furthermore, the requirement that the main coil of lowest resistance must be sufficiently large to carry the full rated load by itself tremendously increases the cost.

When a number of tanks are served from one current source, the current to a given tank will fluctuate to some extent when other tanks are cut in and out of service. When this may be troublesome as in precision plating, it is preferable to provide one current source for each tank. This will increase the plant investment somewhat; however, the cost will be more or less offset by saving the power lost in rheostats.

The problem may also be overcome by utilizing a rectifier with automatic voltage control to eliminate the changes in line voltage as the load varies. The approach is somewhat limited by the capacity of the control to cope with major changes in the load (see Chapter 30). Losses of power in the rheostats will of course still occur.

Operation Timers

Ampere-hour meters[1] are integrating instruments reading directly in ampere-hours. Like ammeters, they are connected across an external shunt by means of calibrated leads. Some instruments are designed to signal after a given number of ampere-hours has been passed. They are useful in controlling the amount of metal deposited in precision plating or to indicate the necessity for addition of chemicals to the bath.

Electronic *ampere-hour recorders* are available[2], which may be connected to a remote recorder-controller. The latter may be arranged to signal at specified ampere-hour intervals, and it may also control solenoid or motor-operated valves to make predetermined additions of chemicals to the solution.

The use of ampere-hour controls for maintaining brightener levels has proven particularly effective. Consistent quality is maintained, usually with a substantial reduction in brightener consumption. A small brightener pump is cycled by the ampere-hour meter at set intervals to add brightener to the system. The amount of brightener added on each cycle is controlled by the duty timer for the pump and by controlling the concentration of the brightener in the pump reservoir.

When production is sufficiently consistent so amperage levels may vary only slightly, similar results can be obtained by a simple pulse pump. This pump is locked into some other meaningful circuit such as the rectifier, or the machine drive circuit, so the pump only operates when production is in process.

Simple *signalling timers* of various designs are used to control manual processing times.

Clocks with large sweep-second hands are useful for timing manual operations of short duration such as bright-dipping, strike plating, etc.

Dummy clocks are occasionally placed at manual plating tanks. When work is placed in the tank, the workman sets the hands to show when the work will be finished.

Temperature Controls

Automatic temperature controls are well-known and widely used to maintain the desired working temperature of processing solutions, without requiring the operator's attention. Complete dependence should not be placed on the automatic control, however. An indicating device should be incorporated in the control or installed separately so that the operator can occasionally check on the proper functioning of the control.

The location of the temperature sensing element is particularly important. When solution agitation is adequate to ensure uniform heating, location of the element as close as possible to the heating element produces the most sensitive control. In the absence of such agitation, the element should be located as closely as possible to the critical working area of the tank. Under these conditions, the possibility of temperature stratification, particularly when starting up after a shutdown, must be considered. Some supplementary agitation should be provided, or production delayed until a satisfactorily uniform temperature is attained.

It must also be remembered that a metallic bulb must not be placed in the current path as it may then be corroded away. The bulb must be made of or coated with a material compatible with the solution.

Control valves for steam heating are usually of the simple, self-contained type. Air-operated controls are usually used when both steam heating and water cooling are necessary, as is common in chromium plating baths.

Electrical heaters require special attention. In addition to the temperature control, they should be equipped with a low level protective device to cut off the heater in the event of low solution levels. Heaters not protected in this manner have started serious fires by igniting tank linings when solution levels have fallen low enough to expose the main heating area of the element.

A simple *warning device* should be used whenever *electric immersion heaters* are utilized in a solution which is insulated from ground to give warning in the event that a heater becomes shorted out to the solution. This can be very dangerous because such heaters are commonly 220 volt and a workman, who may be well-grounded by the usually wet floor, would be in danger of receiving the full voltage when he touches a rack to lift it out of the tank. The warning device consists of a light bulb (or several in series) with one contact grounded and the other in contact with the solution. The voltage of the bulb should correspond with the heater so that it will light, at least partially, when there is a dangerous short in the heater. If there is low-voltage D.C. being used in the tank, it will be insufficient to light the bulb or cause significant loss of D.C. current.

The use of the newly developed ground sensing relays are helpful in avoiding this type of incident. They are, however, considerably more expensive than the simple device described above.

Level Controllers

The solution level must be held within definite limits in all tanks. Manual control often leaves a lot to be desired, especially on heated solutions which may evaporate below the working level without being noticed in time. Also, manual addition of make-up water too often results, because of forgetfulness, in overflowing the tank and loss of the solution.

The increased use of dragout reclaim tanks has brought about a return to manual control in many instances, since the reclaimed solution can only be transferred back to the operating tank when sufficient evaporation has taken place. The electrical devices described below can be connected to the transfer pump instead of to a water valve, and thereby used to automate this function as well. Then this is done, however, a level controller should be installed on the dragout tank to replenish the volume pumped out. This controller should be equipped with a lockout relay to prevent it from functioning until the transfer pump has shut off so dilution of the dragout rinse does not occur during transfer.

A device which can serve very well to control the solution level in a heated tank by controlled addition of make-up water is shown in Fig. 2. The solution level (1) in tank (2) is controlled by the device shown, comprising a tee (3), an immersed pipe of small diameter (4), a side overflow pipe and discharge hose (5), an open discharge water feed pipe and control valve (6). The constantly overflowing water level is at (7). The pressure, p, of the water column at the bottom of pipe (4) is balanced by the pressure, p_2, of the denser solution in the tank. The solution level (1) is controlled by the immersed length of pipe (4).

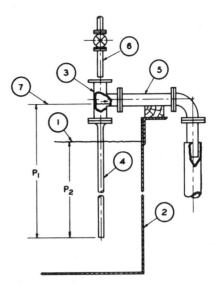

FIG. 2. Automatic level controller, differential head type.
(1) Solution level
(2) Tank
(3) Tee
(4) Inlet pipe
(5) Overflow discharge
(6) Control valve
(7) Overflow level

The pressure, p, equals the head, h, times the density, ρ. Since $p_1 = p_2$, $h_1\rho_1 = h_2\rho_2$. If $h_1 = 30$ in., $\rho_1 = 62.4$ lb/ft^3, and $\rho_2 = 68.8$ lb/ft^3, the solution level, h_2, will be 27.3 in. above the bottom of pipe (4) and 2.7 in below the water discharge level (7). In order to prevent pipe (4) from becoming filled with plating solution, it must be of sufficiently small diameter or provided with a bottom cap with only a small opening to minimize diffusion of the solution into the water.

Another device which has been used is described[3] by Couch and Brenner (Fig. 3). This is somewhat similar to the device of Fig. 2, except that the control of the water input is effected by a float (1) carrying a small cup of mercury (2). As the level rises, the mercury seals off the small water inlet tube (3). The main tube has a smooth bore and is, for example, 1 in. inside diameter and the float is ¾ in. in diameter.

Of the float-actuated controls, the simplest is the float switch, in which the buoyancy of the float either directly or through a lever system mechanically actuates a valve or an electric switch. In the corrosive atmosphere of a plating room, such controls frequently become inopera-

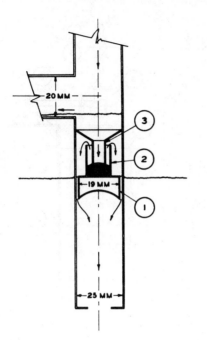

Fig. 3. Automatic level controller, float type.
(1) Float
(2) Mercury cup
(3) Inlet tube
(4) Main tube

tive, which is worse than no control at all. An improvement consists of a float, extension rod and guide (all of which may be made of corrosion-resistant material) and a sensing element at the upper end of the extension rod, which may be enclosed in a corrosion-resistant cabinet. The sensing element may be of the on-off or proportioning type. Typical of the former are a spring-loaded limit switch, or a magnetically[4], electrically or electronically operated switch. The Levelimiter[5] is a variant of this type, in which the float is toroidal (doughnut-shaped), contains a magnet, and is corrosion resistant. A closed corrosion-resistant tube passes through the hole of the doughnut float. Inside the tube, a second magnet is suspended on a metal tape. As the second magnet follows the movement of the float, the tape transmits the indication to the controller, which is arranged to take appropriate action.

Another major class of level controls is based on measuring, in one way or another, the hydrostatic pressure[6,7] at the bottom of an open-ended tube immersed in the solution. Since the tube may be made of any suitable material, there need be no problem of corrosion or contamination.

Still another class of controls is characterized by the use of one or more conductive probes[2,8] which actuate suitable relays when the liquid makes or breaks contact with them. This is a simple on-off control which enables the liquid level to be maintained between two probe settings.

A more sophisticated type of control depends on changes in liquid level changing the capacitance of a condenser[9] forming part of a high-frequency resonating electronic circuit. The change in liquid level detunes the circuit, whereby appropriate action may be taken to bring it back into resonance. The control can be arranged to operate on dielectrics as well as electrolytes and in some cases entirely out of contact with the solution.

Water Flow Controls

In the past decade, the substantial increase in the number of plating plants practicing waste treatment and also the increasing use of deionizers have focused attention on water conservation. Any water conservation program must incorporate flow controls.

For controlling the flow of water to rinses, the *conductivity control* is ideal, as has been pointed out in Chapter 32. Essentially, this control consists of a rugged conductivity cell which is in a balanced bridge circuit with means for varying the balancing voltage over the desired range. This is used for setting the control point of the instrument. The bridge circuit is connected to a relay and an automatic water valve in such a way that the valve is closed when the conductivity of the controlled solution is below the set-point and open when it is above.

It is a simple matter to set the instrument to control the flow of water to a rinse, once it has been established what degree of dilution constitutes satisfactory rinsing. Suppose a 1000:1 dilution is considered adequate after a nickel plating operation. A sample of the nickel solution is diluted with 1000 parts of the same water to be used in rinsing. The conductivity cell is immersed in the dilute solution and the balancing circuit adjusted until the relay just closes the valve. Too high a dilution ratio such as may be achieved in double or triple cascade rinses may make the control insensitive because of interference by the conductivity of the rinse water. In such cases the control cell can be placed in an earlier rinse where the concentration is higher and by calculation (see Chapter 32) arrange for the same final dilution.

At least one available instrument[10] of this type provides for controlling up to twelve rinses, each with its own cell connected to a central programming timer which monitors each cell every 2 min. The rather modest cost of this instrument is usually quickly paid off in water savings, especially when additional treatment expense, either of the feed water or waste, is involved.

Conductivity cells can also be used to analyze the solution in a rinse tank. This is the reverse of the operation of setting the control point to some specified concentration, i.e., conductivity described above. With the control point set to the conductivity of the rinse to be analyzed, place the cell in a known volume of clean rinse water and add from a burette solution from the process tank, the analysis of which will be known, until the meter is actuated, thus indicating that the concentration of the rinse tank has been reached. It is then a simple matter to calculate the composition of the rinse solution.

Flow control valves are useful when it is desirable to set the water flow rate at some predetermined value regardless of fluctuations in line pressure. The simpler example of this type consists simply of a pipe coupling of suitable size which contains a flexible diaphragm, somewhat conically shaped, facing upstream with a hole in the center. As the line pressure increases, the diaphragm moves to close the opening, thus maintaining a constant flow over quite wide pressure fluctuations. The valves are factory set to give the specified flow rate, which in the case of one manufacturer[11] may be chosen from 16 rates between $\frac{1}{4}$ and 10 gpm.

A larger and more complicated type of control valve consists of an adjustable annular metal orifice which is set by means of a calibrated dial. A weighted sleeve is arranged to rise and fall with changes in line pressure causing an appropriate change in the inlet port size, thus maintaining a constant pressure differential across the calibrated orifice. One manufacturer[12] of this type offers four valves, each adjustable over about a tenfold change in flow rate and covering the range from 0.1 to 100 gpm.

Flowmeters are useful where it is desirable to know the rate of flow (or to be sure that solution is flowing). They are most useful in heat exchanger applications, especially when operated in conjunction with a filter, as is discussed later. The Rotameter[13] or Flowrator[5] type of instrument is most convenient, and the construction is such that compatible materials can readily be used.

Agitators

Air Agitation. Low pressure air from a blower is preferred for agitation since it is less contaminated than air from a compressor. Even then precautions must be taken to avoid oil contamination and to remove solids.

For an acid copper bath W. Innis* recommends a minimum of 2 cfm of air/ft^2 of solution surface at a delivered pressure of 1 psi for each 21 in. depth of solution.

The air can best be distributed through a 1-in. diameter pipe placed from 1 to 3 in. off the bottom of the tank directly under the work to be plated. Staggered holes ($\frac{1}{16}$ to $\frac{3}{16}$ in. in diameter) are drilled 1 to 5 in. apart along two center lines on the underside of the pipe at an angle of 35 to 40° from the vertical. With this arrangement the air shoots out toward the bottom of the tank, thus providing the most efficient agitation. Such a pipe will use about 1 cfm of air per lineal foot of perforated length. Air volumes frequently recommended are 1 to 2.5 cfm/ft^2 of surface area for plating tanks and 3 to 4 cfm/ft^2 for rinsing. The supply pipe should have an interior cross section between 1 and 1$\frac{1}{2}$ times the total area of the holes.

The material of construction for the air distributing pipe depends on the composition of the solution, its corrosive action and its susceptibility to contamination.

Quite generally, the air distribution systems are now constructed of plastic pipe, both for ease of construction and resistance to corrosion. Since these systems can easily float out of place, they must be carefully anchored to avoid movement.

Low pressure blowers for the delivery of oil-free agitation air are available in capacities ranging from 10 cfm upward, at pressures of 1 to 8 psi, which adequately covers the requirements of any plating installation. One type[18] is a multistage unit consisting of a series of impellers rotating on a shaft within a steel or cast iron housing. The impellers may be aluminum or steel, and clearances within the unit are such that no lubrication is required, thereby assuring

*Private Communication

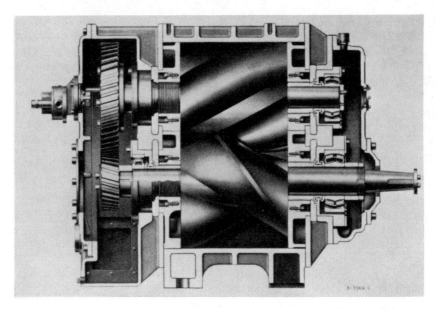

Fig. 4. Cross-section of Axi-compressor. (*Courtesy of Ingersoll-Rand Co.*)

the delivery of clean oil-free air. Another type[19] consists of a helical-grooved rotor (top, Fig. 4) driven through a set of timing gears by the main shaft carrying the two-lobed main rotor below it. The timing gears are located in a separate oil-tight compartment, and no lubrication is required in the main chamber. Although either type of blower will not introduce oil or dust into the air, it is important that the air intake be located where it can pick up only clean air.

Use of glass fiber or similar filters on the intakes are important. These must be properly maintained, being cleaned or replaced on a schedule dictated by the contamination of the pickup area. Under no circumstances should the blower be operated when these filters are not in place, since significant particulate contamination of the plating bath can occur surprisingly quickly.

Rod agitators are frequently used in manual plating tanks. The cathode rod or rods are mounted on rollers and reciprocated by a motor-driven cam at speeds usually between 6 and 20 fpm. Electrical contact to the rod is made by flexible cable or flat copper braid. The speed of agitation is limited because of (a) inability to maintain good contact of the work on the rack, (b) excessive swinging or "walking" of the rack on the bar, and (c) other mechanical factors. A combination of air and rod agitation is sometimes used.

Solution agitation by recirculating through a *pump* is rarely employed. There is the danger of undesirable impingement effects, and the degree of agitation obtained is usually very small and not as effective as other means. For the same relative speed, moving the cathode through the solution is more effective in disturbing the cathode film than moving the solution past the cathode. Solution agitation by any method imposes stricter requirements on solution clarity and freedom from suspended solids, if roughness is to be avoided.

Automatic conveyors and barrels, by their very nature, supply more or less agitation to the work.

Auxiliary Tanks

Mixing and storage tanks should be provided for all plating solutions. In most cases one tank can serve both for mixing and initial purification and for temperary storage in case the plating tank must be emptied for some reason. Means for pumping the solution to storage and filtering it back to the operating tank must be available. The storage tank should be at least sufficient in capacity to contain the largest operating solution and should be lined with the same material as the plating tank.

Because of space requirements, storage tanks are frequently placed outdoors. If this is done, they should at least be roofed over and means for heating provided. Components of many solutions, e.g., boric acid in nickel baths, will crystallize out at temperatures below 21 C (70 F). Sometimes advantage is taken of this fact, as in the removal of excess sodium carbonate from cyanide baths by storing them outside over the weekend in cold weather.

Electrolytic purification tanks are needed when continuous low current density electrolysis is used to remove metallic contamination from plating solutions. The tanks may be separate units or may be integral with the plating tank, and although many different designs are possible, Case[14] recommends that certain basic features are desirable:

(1) The cathode current density must be closely controlled; yet it should vary over the range known to be most effective for the specific bath and metal contaminants present. This is done by using a cathode with recurrent variations in recesses and high points. Corrugated black iron sheet is most widely used although a louvre type of cathode (like a venetian blind) has also been recommended.

(2) Agitation is essential to speedy removal of foreign metals, and air agitation should be used if the type of bath permits. A high rate of circulation and narrow passages in the purification tank increase agitation and assist the process.

(3) The rate of solution circulation should be from one-third to one-half the total solution volume per hour.

(4) For *continuous* purification, the maximum purification current likely to be needed is 0.05 amp/gal of solution. The maximum cathode area is based on this and on the optimum current density for purification.

For *batch* purification, the usual criterion for cleaning up a contaminated solution is 5 amp-hr/gal of solution. The cathode area is then based on the optimum current density for purification and on the time available for the purification operation.

(5) Anode-to-cathode spacing is close, averaging about 4 in., to keep the tank small and to increase the agitation effect; consequently only 1 to 2 volts will usually be needed to operate the tank.

(6) Positive contact must be made between the cathodes and the bus by using bolted contacts or the equivalent. This is essential to prevent local potential differences between cathodes and subsequent redissolving of the impure deposit or even of the corrugated sheet itself. Therefore, cathodes should not remain in the tank for extended periods with the current off.

(7) The solution should preferably be returned from the purification tank to the plating tank through a filter. This avoids contaminating the main solution with any solid particles which may detach themselves from the impure cathode deposits.

(8) If it is desired to use less than the maximum electrolytic purification, an appropriate number of cathodes are removed and the current is correspondingly lowered.

The electrolytic purification tank is a convenient place to dissolve additions, control pH, etc. If this is done in a separate section of the tank ahead of the purification section, the additions are thus subjected to purification before they get into the main solution.

Acid storage tanks must be provided when acid is received in bulk. They are often located in an elevated position at the receiving station, from which acid is fed by gravity to the points of use in the plant through pipelines. Materials of construction will be chosen on the basis of the acid being handled.

Special fixtures and piping for receiving the acid from the bulk carrier will be required and should be installed according to the vendor's instructions. Tank cars are usually emptied by air pressure; and tank trucks are emptied by air or by a pump carried on the vehicle. The storage tank must be at least of sufficient capacity to receive one shipment.

Newer government regulations have placed additional restrictions and requirements on such installations, e.g., in some instances it may be necessary to dike the area around the storage tank to prevent the spread of its contents in the event of tank failure. The impact of these regulations should be thoroughly investigated when considering such an installation. Additionally, special care must be taken to set up and monitor unloading and transfer operations and procedures. Serious accidents have resulted from incompatible materials being mixed by error during unloading and/or transfer.[20]

Filter and Heat Exchanger in Series

Most experienced plating engineers will hesitate to operate a heat exchanger on a plating solution in a circuit by itself without a filter following it because of the danger of encountering roughness arising from particles either formed in the heat exchanger or picked up along the circuit. There is also some saving in piping cost, especially in large installations with long runs when using a common circuit. This practice can lead to some serious difficulties in temperature control because a heat exchanger is designed for a certain flow rate; whereas by its nature the flow through a filter is variable and unpredictable. The filter manufacturer will usually rate his filters operating on water or clear solution with a precoat, if used, that is carefully prepared and is of the minimum thickness that will perform satisfactorily. With heavier cake thicknesses and continued operation, especially when the material being filtered out is compressible or tends to blind or plug the cake, the actual flow rate will fall well below the design rate because the rated pressure drop will be exceeded early in the operating life of the filter. This means that usually the flow rate in the exchanger-filter system is below the design rate and heat transfer suffers accordingly.

Suppose the exchanger is operating as a cooler designed to extract a stated number of Btu/hr from a solution flowing through it at the design rate, using cooling water at a fixed entering temperature, the flow rate of the water being under automatic temperature control. The effect of decreased solution flow rate on the heat transfer in the exchanger is threefold:

(1) Because the weight of solution traversing the exchanger per hour has been reduced, the exit temperature of the solution must be lower to effect the same total heat transfer per hour.

(2) This lower exit temperature for the solution reduces the temperature differential between water and solution which reduces the heat transfer rate proportionately.

(3) The solution velocity through the exchanger tube has been reduced, and this adversely affects the film coefficient of heat transfer on the solution side.

These three effects are synergistic, and the result is a serious loss of efficiency in cooling; in fact, in some cases the first effect alone may require that the exit temperature of solution be lower than the entering water temperature, an obvious impossibility. (Identical arguments hold for an exchanger operated in heating.) The solution to this difficulty is not always obvious; in fact the "obvious" answer is often wrong.

The addition of more exchangers will usually be ineffectual as is pointed out in Chapter 22. A booster pump to restore the flow rate to its proper value may work, but there is grave danger that the increased pressure on the filter cake will pack it so tightly that soon the situation will be less favorable than before.

The proper solution to this problem is to overdesign somewhat for the filter installation, preferably using several small filters rather than one or two large ones and to shut each filter down when (a) a flow meter in the line shows dangerously low flow or (b) the pressure drop across it exceeds some value to be recommended by the manufacturer. The first control is preferable; indeed it is well, especially in cooling problems, to measure the flow of solution and coolant as well as to provide for reading the entrance and exit temperatures of each. This gives a constant, immediate check on the operation of the exchanger, thereby eliminating any conjecturing.

An integrating Btu meter is available[17] to indicate directly the Btu added to or removed from a circulating liquid stream in any specified period.

Pumps, Piping and Accessories

Pumps, piping, fittings and hose are chosen on the basis of required operating characteristics and corrosion resistance, as in any chemical installation. When handling plating solutions, an additional requirement is imposed, that is, that the materials of construction must not contaminate the solution.

Pump packing, especially, must be carefully chosen, and it is good practice to specify a pump having a lantern ring in the packing gland, which is kept under water pressure. This serves to lubricate the packing and minimize leakage of solution; but more importantly, it prevents air from being sucked into the pump casing. Such entrained air, churned into the solution by the action of the pump, can be the source of serious pitting difficulties.

Magnetically coupled pumps are frequently used, since no packing seals are needed. The pump chamber and impeller can be made of suitably chemically resistant material. No leakage, air entrainment or packing contamination can occur.

A *diaphragm-type of valve* is usually chosen because the diaphragm can be of a non-contaminating material.

Hose is generally used for filter connections. The suction hose should be of the reinforced type to prevent collapsing in use.

Check valves are usually recommended on filter discharges to prevent back pressure from dislodging the cake when the filter is shut off. In lined piping systems, however, check valves are apt to be uncertain in operation.

Siphon breakers should be installed in any city water lines which dip into tanks of solution to prevent possible back-siphoning of solution into the water system should a temporary suction occur in the water lines. Some health codes make this provision a requirement.

Miscellaneous

Platform scales are necessary for weighing chemical additions and are useful in taking anode inventory. Dispensing scales of smaller capacity may be needed for weighing small amounts of addition agents, etc.

A *tank magnet* consisting of a powerful permanent magnet on a long handle is very desirable for recovering magnetic articles dropped into tanks. The magnet and handle can be covered with rubber or plastic, if necessary.

A *circuit tester* consisting of a volt-ohm-ammeter is an invaluable aid in the plating shop. It can be used to locate short circuits, stray currents, accidental grounds, and broken connections. Poor connections in bus runs, anode-to-rod contacts, etc., can be located. Racks with auxiliary anodes can be checked for partial or complete failure of insulation; moreover the instrument is extremely useful in routine electrical maintenance.

A *warning device* to signal the failure of d-c current can save rejects, especially in large automatic plating installations, where such a failure may go unnoticed until rejects start coming from the machine, by which time a great deal of work will be spoiled. Each electrified tank can be provided with such an alarm, consisting simply of a single-pole single-throw normally closed relay with a d-c coil of the same voltage rating as that applied to the tank. The coil is wired in parallel with the tank so that failure of the d-c source closes the contacts of the warning system. The a-c power to the warning system should be wired through the starting switch of the conveyor so that there will be a signal given if the conveyor is started without also activating all the d-c sources on the machine.

A *warning signal* for stray currents has been described[15], consisting simply of two low-voltage light bulbs. One terminal of each bulb socket is connected to a common ground. The other terminals are connected respectively to the positive and negative tank bus. When both lights glow evenly, there probably is no difficulty. When one light dims or goes out, a stray current is indicated. In the authors'[15] company, it was an ironclad rule, which allegedly paid dividends over the years, to halt operations until the stray current was found and eliminated even though there was no indication of any effect on plating quality. Of course, it one light burns out, a false indication is obtained, but this is simple to check before halting operations. The authors make no claim that this device will detect all stray currents; however it is reasonable to suppose that the most damaging ones would be indicated.

A *warning signal* to detect steam condensate contamination in the discharge from heating coils or exchangers is a necessity in plants where the condensate is returned to the boiler; otherwise the boiler may be severely damaged in the event of leakage. Even where the condensate is wasted, such an alarm may prevent the accidental siphoning of a valuable solution into the waste line by providing early indications of a leak. In its simplest form this device consists of a conductivity cell located in the main condensate return line, connected to a suitable sensing relay and alarm. When the conductivity of the condensate return exceeds some preset value (usually about 15 micro mhos/cm^3) the alarm is actuated. The relay may also be arranged to actuate a motorized three-way valve, diverting the condensate to waste until the condition is corrected.

Since the conductivity cell is located in the main return line, the alarm will not indicate which of a possibly large number of steam coils is leaking. It is useful, therefore, to provide a spare conductivity cell and electrode at the central control point and also sampling cocks in the discharge line from each individual steam coil. Thus the offending coil can quickly be located by sampling its discharge and testing with the spare cell.

A unit is commercially available[10] which consists of five or ten conductivity cells and a manual cell selector switch. One cell is mounted in the common steam return header; the others in individual return lines. The selector switch is normally positioned on the cell in the common header so that contamination anywhere will set off an alarm and/or divert the condensate to waste. By rotating the selector switch, the offending line can quickly be identified.

Automatic pH control is not given a more prominent position because it is only justifiable in quite large installations which are provided with circulating solution systems, and the best arrangement will depend on the physical layout of the plant. The vendors of the pH equipment can give valuable assistance in the design of such control systems. A proper installation will be capable of holding the pH within 0.02 to 0.05 unit, which is far preferable to manually controlled operations with only periodic corrections. Savage[16] has described one installation which has the advantage of considerable simplicity.

Safety equipment includes such items as carboy tilters or acid siphons for emptying carboys of acid, and specially designed hand trucks or fork lifts for handling full carboys. Ordinary hand or platform trucks do not provide for the safe anchorage of carboys in transit and should never be used for this purpose. Wherever workers are exposed to the hazard of splashing chemicals on their persons, a safety shower should be provided nearby. This is best designed to deluge the person as soon as he steps on the platform beneath the shower head.

Other auxiliary equipment items such as filters, dryers, deionizers, etc., are of such importance that they have been considered at length in earlier chapters.

References

Note: Some references indicate a possible source for the less common items discussed in the chapter. It is not meant to imply that the source cited is the only one available, nor even preferable to competitive suppliers.

1. Sangamo Electric Co., Springfield, Ill.
2. F. D. Pace, Grand Rapids, Mich.
3. Couch, D. E., and Brenner, A., *Anal. Chem.*, **24**, 922 (1952).
4. Magnetrol Inc., Chicago, Ill.
5. Fischer & Porter Co., Hatboro, Pa.
6. Uehling Instrument Co., Paterson, N. J.
7. Foxboro Co., Foxboro, Mass.
8. Photoswitch Div., Electronics Corp. of America, Cambridge, Mass.
9. Wheelco Instruments, Chicago, Ill.
10. Solubridge, Industrial Instruments, Inc., Cedar Grove, Essex County, N. J.
11. Dole Valve Co., Morton Grove, Ill.
12. W. A Kates Co., Deerfield, Ill.
13. Schutte and Koerting Co., Philadelphia, Pa.
14. Case, B. C., *34th Proc. Am. Electroplaters' Soc.*, 228–248 (1947).
15. Borchert, L. C., and Kinnerman, R. B., *Plating*, **36**, 456–461 (1949).
16. Savage, F. K., *39th Proc. Am. Electroplaters' Soc.*, 204–205 (1952).
17. Pollux Btu Meter, Air Conditioning Equipment Corp., New York City.
18. (a) U.S. Hoffman Machinery Corp., New York, N. Y.
    (b) Lamson Corp., Billmyre Blower Div., Syracuse, N. Y.
19. Ingersoll-Rand Co., Philadelphia, Pa.
20. *Industrial Fin.* (September 1982), pp. 9, 16.

36. GENERAL MAINTENANCE

Richard B. Saltonstall

Formerly Technical Director, The Udylite Corporation, Detroit, Mich.

and

Leslie C. Borchert

*Manufacturing Process Manager, Plating and Forming Department,
Houdaille Industries, Inc., Huntington, W. Va.*

Solutions and fumes encountered in plating plants are often highly corrosive in nature. While this fact presents many maintenance problems, it also increases the necessity for a sound operating plan of maintenance, which should include both building structures and equipment.

Proper execution of such a plan will produce three major benefits:

(1) Prevent or greatly decrease the frequency of costly production shutdowns for major repairs.

(2) Greatly prolong the life of expensive equipment and facilities.

(3) Have a beneficial psychological effect on personnel.

Maintenance of specific items of equipment is covered elsewhere in this volume under the sections dealing with these items. This chapter, therefore, deals only with maintenance of building structures and general service equipment.

Maintenance of Building Structures

Walls. Vitreous glazed tile or glass block walls present the least problems. Occasional washing will restore them to their original surface condition and appearance. Other types of walls should be painted, and the choice of colors is important. Light colors are preferable, and the use of white is increasing. However, white or light colors are not too practical near the floor where soiling or splashing occurs and black or dark-colored dadoes are helpful in these areas.

Walls of cinder block, concrete block, and common brick should be coated with a grade of paint which has good resistance to acids and alkalies. This also applies to wood, plaster, plasterboard, "Celotex," etc., and these should be repaired when mechanically damaged.

Occasional washing will decrease the frequency of repainting.

Doors and other means of exit should fit well, and door and other hardware should be kept functioning. These precautions will eliminate undesirable dust and dirt from outdoors and other sections of the plant.

Ceilings. Ceilings, as well as walls, should be light in color. Smooth surfaces are easily cleaned and coated and do not present lodging places for dust and dirt which may later become dislodged and fall into processing tanks. It is, therefore, desirable that wooden floor beams and cross braces be covered with a suitable ceiling material. If there is danger of dust and dirt falling from the floor above, the ceilings should be dust tight.

Use of good quality coatings having high resistance to moisture and acid fumes is of even greater importance on ceilings than on walls since ceiling coatings, if they flake or become powdery, may fall directly into processing solutions.

Exposed Structural Parts. Metal beams, posts, braces, and catwalks are usually made of steel and must be protected by suitable organic coatings both from the standpoint of their preservation and as a means of preventing contamination of processing solutions by the products of their deterioration. Such items are usually given at least a priming coat before they are erected. If not, or if they have been neglected to the point where rust has formed, all rust must be removed prior to coating in order to secure satisfactory adhesion.

When cleaning dust off beams, cleaning floors, or doing any mechanical work above or around processing solutions, the latter must be covered to prevent contamination. Vacuum systems are preferable for such cleaning.

Floors. Floors in most modern plating establishments are protected by a final layer of mastic, which prevents attack on the underlying material by water and processing solutions when unavoidably spilled. Any breaks in this surface layer should be repaired immediately. It is poor practice to use floors for general spillways, but adequate slope of all floor areas toward drains is important. Generous use of curbings to direct large volumes of waste liquids directly to drains will aid in keeping most floor areas dry. The use of false floors constructed of wood slats or metal gratings in areas unavoidably wet and over spillways and trenches is common practice and safety demands that these be kept in good repair.

General neatness and good housekeeping are advantageous with respect to both safety and psychological effect on employees. Plating racks not in use should never be piled on the floor, but hung on rods. This not only prevents costly damage to the racks but adds measurably to general neatness and facilitates proper cleaning of floor areas. Cleaning floors is preferably accomplished with a stream of water from a hose or by the use of vacuum equipment, as previously mentioned. Sweeping dry areas raises dust which can cause dirt contamination of processing solutions.

Load characteristics of floors, while not entirely a problem of maintenance, should be considered when any rearrangement or addition of tanks and equipment is contemplated.

In many instances it is necessary to install processing tanks and equipment in a pit below floor level. The sides and bottoms of pits are considered as parts of the floor area and should be inspected regularly and repaired as necessary. Pits must not be used for discarding or hiding waste and rubbish.

Windows. The principal items of window maintenance are protection of mullions and frames, replacing broken glass, and washing. The general principles of coating metal window parts for protection are the same as those outlined for other exposed metal construction. Dust and dirt blowing into a plating room through broken windows may cause processing difficulties, and such windows are unsightly. Fumes from processing operations, if not properly exhausted, may require very frequent washing of windows. The frequency of washing to maintain desired cleanliness is therefore dependent to a large extent on the efficiency of the area ventilating system and exhaust systems for processing tanks.

Lights. Adequate lighting is of considerable importance, and regular cleaning of lighting globes, tubes, and reflectors is necessary to maintain proper illumination. Light bulbs or tubes, which have failed completely or become dim from long use, should be replaced promptly. Safety and morale are both improved by good lighting, well distributed over all working areas.

Maintenance of General Service Equipment

Pipelines. Exterior surfaces of steel pipelines, like other iron surfaces, should be protected against rusting by suitable coatings, especially if they are located over processing solutions or where they are unavoidably splashed with liquids. This is frequently done with various colored coatings which, in addition to giving protection, may be used to identify lines handling different materials.

Pipelines made of corrosion-resistant metals and plastics are frequently brittle or fragile; they should be adequately supported and protected from mechanical damage wherever such damage is likely to occur. Color-marking or coating of such pipelines is used for identification and aids in maintenance.

Overhead pipelines carrying cold water should be insulated to prevent condensation and dripping. Those conducting steam and hot liquids should be insulated to minimize heat losses. Maintenance of such insulation is important.

All pipelines connecting heaters, cooling coils, air agitation coils, filter, or other auxiliary equipment with electrified processing tanks must be properly insulated to prevent stray electrical currents. Insulating couplings are obtainable for this purpose but their conductivity should be checked occasionally and, if necessary, they should be removed and cleaned. Short lengths of high pressure steam hose with clamps, which bridge intentional breaks in pipelines, are frequently used for insulators and require periodic inspection and maintenance.

Hoses, when not in use, should be coiled or hung neatly in designated locations for protection against kinking and breakage as well as general orderliness. Fast-acting couplings and connections are recommended for hoses which are moved frequently or used intermittently.

Drains. All floor drains should be equipped with strainers to prevent the entrance of solids which may lead to clogging. Such strainers are much more permanent if constructed of corrosion-resistant material such as high silicon iron.

A regular program of drain inspection, with special emphasis on drains located in relatively inaccessible places, can prevent clogging and damaging floods. This is very important in the case of open trenches carrying segregated plant wastes since clogging may cause mixing of different kinds of waste, and acid wastes backing up in pits affect maintenance of concrete and metal surfaces.

Heating Equipment. Steel parts of unit space heaters should be coated for protection against rust or corrosion. This is usually done by the manufacturer, but regular inspection and re-coating are necessary. Cast iron or steel radiators should be similarly maintained. All these items should be kept free of dust and dirt, preferably by vacuum cleaning.

To obtain maximum heat transfer the heating or cooling surfaces of heat exchangers or heating coils must be periodically cleaned to remove scale, slime, or films which frequently form in service. The metals for such items are chosen to be corrosion resistant to the processing solutions and obviously cannot be given organic protective coatings without impairing the heat transfer.

Meters and Controls. All indicating and controlling meters require regular inspections and maintenance and periodic calibration for satisfactory performance. The importance of this cannot be overemphasized.

Process Pumps. Centrifugal pumps are used almost exclusively for handling processing solutions in the plating industry. Packing and lubrication are the important maintenance items. When correctly packed, a centrifugal pump in operation should leak very slowly, i.e., one drop about every 20 sec. When this condition exists, it is reasonably certain that the packing is not tight enough to score the shaft. Many pump manufacturers recommend tightening the packing gland with a wrench, backing it off, and making the final setting finger tight. Routine inspection and maintenance are necessary.

Choice of packing is of considerable importance. Plating solutions will be contaminated by continuous circulation through a pump packed with a material harmful to the solution. Some solutions are more prone to leaks around the packing glands than others, and this tendency can be minimized by careful selection of the packing material. If the pump shaft is scored, or if the directly coupled motor is out of line, the pump cannot be properly packed. Recommendations of the manufacturer should be followed for packing and lubrication. Threaded components, which must be unscrewed occasionally, should be kept oiled to prevent seizing.

Some pumps are built to operate with water or liquid seals, that is, with a small stream of water or liquid continuously flowing over the packing gland. Maintenance of such pumps includes making certain that the liquid is flowing at all times while the pump is in operation.

Ventilating Equipment. Preventing corrosion, cleaning, and checking electrical insulation are the principal factors of the maintenance program.

Duct work frequently presents a severe corrosion problem on the interior surfaces, depending on the fumes and spray being exhausted through it. Little can be done in the way of maintenance after such corrosion starts, and the use of materials of construction or interior coatings having satisfactory resistance to attack is fundamentally a construction problem. The exterior surfaces of steel ventilating ducts should be coated to prevent rusting. This is especially important because ducts are usually constructed of light gauge sheet and will perforate quickly if not protected.

Slots in draft boxes must be kept open and clear of dried salts. These slots are designed to draw off a definite volume of air, and the system cannot operate properly if the slots are clogged. This is largely a matter of good housekeeping and cleanliness.

Exhaust blowers and motors are often neglected when they are located in difficultly accessible places. Regular lubrication of this equipment is important. Blower rotors tend to become "loaded" with dried salts and must be removed and cleaned occasionally to guarantee satisfactory operation.

Metal ductwork, like metal pipelines, must be electrically insulated when used with electrified processing tanks. This is best accomplished by the use of a nonconducting section of impregnated fabric or plastic adjacent to the draft box. Such sections and other forms of duct insulation should be checked frequently for leaks and other types of failure.

Occasionally exhaust systems are improperly designed with the result that spray removed from one tank may condense on duct walls and drip into another processing tank. If the condensed spray is corrosive, it may also increase maintenance of metal or other surfaces with which it comes in contact.

Dust collectors and fume separators are frequently used in preference to systems which exhaust dust and fumes into the outside atmosphere. These must be cleaned and lubricated regularly.

Valves and Gaskets. Certain types of plug valves require slight lubrication to keep them from seizing. Valves requiring packing should be inspected regularly for leaks and repacked when necessary. A pipe "tee" adjacent to a valve with a plug in the other end of the straight section of the "tee" facilitates cleaning of a plugged valve without removing it. For an open discharge bottom tank outlet a flat plate bolted to the outlet flange of a flanged valve prevents possible leakage of valves and also enables one to free the valve seat which frequently tends to freeze up with crystallized salts.

Electrically operated solenoid valves are in wide use in plating rooms for automatically controlling the flow of water, steam, and processing liquids. In the type in which a packing gland is incorporated the glands should not be so tight that the valve is rendered inoperable.

Gaskets should be replaced when damaged, and the gasket and the surfaces between which it is compressed should be smooth and free of foreign particles.

Electrical Conductors. Conductors for alternating current are usually installed in steel conduit and require little attention other than coating the outside surfaces of the conduit and fittings to prevent corrosion. Low-voltage conductors, on the other hand, usually consist of joined lengths of bare copper bus bar. These should be kept clear of any pipe, bare wire, plating racks, and other items which might cause short circuits, grounds, or stray currents. Bus bar joints occasionally become corroded, and this is evidenced by heating. When heating is detected, the joint should be disconnected and cleaned.

Moist salts should not be permitted to accumulate on any insulation used around an electrified processing tank.

Cranes and Hoists. Lubrication, corrosion prevention and checking of electrical insulation are the principal maintenance operations.

Lubrication is necessary, but excessive use of grease or oil can cause serious difficulties if lubricants drop into processing solutions.

Exposed steel surfaces should be protected with an organic coating, both to protect the equipment and to prevent the formation of rust which will eventually become detached and fall into solutions. Air hoists should be continually watched to prevent oil drippage.

Hooks and controls on electric hoists and cranes should be well insulated electrically to prevent stray currents to racks and possible electrical shock to operators.

Work Containers. Work containers may vary from steel tote pans or wooden kegs for small parts, to wood or metal crates for larger ones. Cleanliness is of prime importance. When delicate or highly buffed articles are processed, plastisol-coated metal crates or frames for holding items apart from each other may be used to advantage.

Cleaning facilities available in all plating plants simplify the problem of keeping bare metal work containers free of shop dirt.

Safety and Fire Equipment. These items are perhaps the least frequently used of any and because of this are most subject to neglect. Such neglect, however, may prove to be disastrous when emergencies arise. The equipment should, therefore, be inspected and tested regularly.

Agitation Equipment. Because it is unavoidably exposed to vapors, fumes and splashing of corrosive solutions, mechanical agitation equipment requires frequent special attention if it is to be maintained in good operating condition. Oscillating work-rods and sections of automatic machines usually function on rollers. Frequent lubrication of the bearings of these rollers and other frictional surfaces is essential if they are to function properly. Once the bearing of a roller is allowed to freeze, the roller very quickly wears flat on one side, requiring replacement. The use of steel or hard corrosion-resistant alloy shoes between copper work rods and rollers is advisable. Flexible cables used for conducting current to moving work carriers should be inspected frequently for breaks.

Air-agitation equipment also requires special attention. Proper lubrication of blowers and motors supplying low pressure air is important. Air filters should be cleaned or replaced at regular intervals, depending on the quantity of air-borne contamination which they are required to remove. Any breaks in the air distribution lines must be repaired immediately if full efficiency is to be maintained. Air-exit holes in air-agitation spiders frequently tend to become clogged with solids. This tendency can be decreased or eliminated by adding necessary water through the spiders, and by periodically shutting off the air for a time long enough to permit plating solution to fill the spiders and risers, then continuing pressure. The solution being forced through the holes helps to clear the holes, but temperature limitations of the materials of construction, as well as possible contamination from the steam, must be taken into consideration. Finally, air lines and spiders should be carefully inspected each time the solution is emptied. Any shifting of lines should be corrected promptly if a satisfactory agitation pattern is to be maintained.

37. ELECTROCHEMICAL MACHINING

Charles L. Faust

Formerly Chief, Electrochemical Engineering Division, Columbus Laboratories, Battelle Memorial Institute, Columbus, Ohio

Introduction

Electrochemical Machining, referred to as ECM, is an anodic process for removing metal to form a predetermined contour, shape, cavity, hole, or slot. In principle, the anodic dissolution is the same as for a plating anode. In practice, the conditions for ECM differ greatly from those for plating, and the anode is the workpiece. For practical machining of metals, the anodic dissolution of the workpiece must proceed at rates 10 to 100 times or more greater than a plating anode dissolves. Furthermore, the current distribution on the workpiece must conform to the metal dissolution pattern to effect the desired contour within the specified dimensional accuracy limits.

For practical attainment of the stock-removal rate and contour requirements, the cathode has the same contour in mirror image, is a tube of suitable contour, or is a rotating disk. The anode (workpiece) and the cathode are maintained 0.010 to 0.020 in. apart. This spacing or gap is filled with the electrolyte flowing at rates of 10 fps or more. The gap must be maintained during operation by advance of the cathode or tool or the workpiece at the same rate metal is dissolved from the workpiece.

Thus, at a metal-removal rate generally on order of 0.010 to 0.050 in. per minute, the workpiece (or more usually the tool) must advance 0.010 to 0.050 in. per minute. This tool movement must be accurate, and the electrolyte flow between the tool and workpiece must be uniformly turbulent. These conditions are met by special fixtures that confine the electrolyte (dispatched to the region between the tool and workpiece) by forced flow from a pump. The narrow channel formed by the gap between the wool and workpiece is the electrolysis zone. Although this zone holds only a small fraction of a gallon, in a minute of operation many gallons pass the workpiece surface and remove the products of anodic dissolution and the hydrogen produced at the cathode.

Figure 1 shows a typical fixture and cathode for producing a forging-die cavity by ECM. The tool, with a contour that is essentially the mirror image on one side of a turbine blade, is advanced through the Lucite box in which it fits snugly. In operation, the box is placed on the steel block against which it is held securely enough to resist being moved by the hydrostatic pressure of the electrolyte flowing between the tool and the workpiece block. Then when set up, the tool is advanced so that its nearest point is the desired distance (0.010 to 0.020 in.) away from the workpiece surface. This is done by connecting the shaft (visible through the Lucite box and at a right angle to the face of the tool) to the ram of an ECM machine, such as the commercial model shown in Fig. 2.

The quick fastening nipple for connecting the electrolyte flow tubing is visible at the upper center face of the Lucite box. From this nipple, electrolyte flows turbulently through the gap and out

ELECTROCHEMICAL MACHINING 745

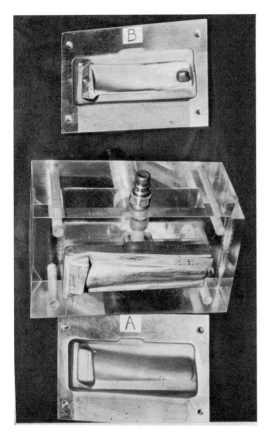

FIG. 1. Center. Lucite fixture for aligning tool and distributing electrolyte across the tool face. Tool for Cavity A. Different tool same fixture for Cavity B—the die to match A for forging a turbine blade. (*Courtesy of Sifco Industries*)

the opposite side of the fixture. Thus, the electrolyte flows across the face of the workpiece and of the tool, and exits at A.

The exit path is restricted to a narrow slot on the order of 0.01 to 0.02 in. high × width, the same as the tool length × flow length for the width of the Lucite fixture.

When the electrolyte is flowing, the current is turned on. In order to maintain the turbulent electrolyte flow in the confined space of the gap region, it is pumped at pressures of 50 psi or greater, depending on the length of path of flow, i.e., the width of the turbine blade in the example of Fig. 1. Because of this hydrostatic pressure and the need for accurate advance of the tool, a rugged machine is required. This distinguishes ECM from plating and completes the triumverate, electrochemistry, hydrodynamics, and fixture engineering required for successful design and operation.

Indeed, the crux of ECM success lies in the ability to design the fixtures for proper electrolyte flow and current distribution in the confined region of the gap. When the right fixture is available and is installed on a proper machine, the operation responds to pushing a button. The machine responds to the set time, maintains the current and electrolyte flow, and prevents sparking.

Under the condition of such close spacing of tool and workpiece, an aberration may occur during operation to cause local sparking. Such sparking may damage the expensive tool or the workpiece, or both, beyond repair. Therefore, ECM machines have spark detect and prevention units that operate in a few milliseconds to prevent the detrimental sparking. The commercial machine includes the d-c power source and electrolyte system. Clearly the ECM machines are far too sophisticated to be put together on a do-it-yourself basis. Thus, the knowledge and education needed for utilizing ECM lies in the skill required to design and build the fixtures for each type of job.

TYPICAL APPLICATIONS

Because the fixtures are relatively costly, ECM is most useful for production runs of many identical items—all products will be identical because there is no wear on the tool. However, metal hardness, per se, is not a factor. So ECM is used for shaping metal that is difficult to cut or is stress damaged by conventional chip-by-chip removal methods.

Typical ECM applications are: shaping, such as making a cavity as in Fig. 1; shaping turbine blades and other products, starting with a rod, an as-forged or cast part; drilling holes of round or other cross-sectional shape; trepanning; broaching, cutting off as to provide billets for forging or producing short lengths of extrusions without burrs; electrodeburring of parts machined by chip-by-chip methods; and sharpening tools. ECM is especially suited for hard-to-machine metals such as high-temperature aerospace alloys of nickel, cobalt, chromium, iron, tungsten, molybdenum, high-strength steels, carbides, silicides, etc.

ECM OPERATIONS

The first requirement is to decide on the size range of items to be put through ECM operations.

Fig. 2. Machine and System for ECM. (*Courtesy of Ex-Cell-O Corporation*)

This will determine the size of machine or machines to be purchased. If no commercial machine is large enough to handle the largest sizes, they must be eliminated from ECM or a special machine must be designed and built. The following text assumes that available machines can be used and points out the factors on which fixture and operation are based. The design of custom machines is outside the scope of this chapter.

Tool Design

Tools are designed for operation with one of two schemes for electrolyte flow: flow past the surface or flow through the tool. For cavity sinking and shaping operations, the flow-past mode has advantages and is the scheme for the tool and fixture of Fig. 1. This tool design is limited in use to the length of flow path across the face of the workpiece and entrance and exit flow channels in the fixture. The limiting length is determined by the gap width, rate of electrolyte flow at available pump capacity, temperature gradient from entrance to exit, and the type of electrolyte that determines the viscosity, which is affected by the hydrogen and by the anode product precipitated in the gap region. Turbine blades as large as 10 × 36 in. are successfully shaped with flow-past tools.

If the tool were designed for the flow-through method, it would be hollow to conduct electrolyte through itself, via holes in its face, and to deliver electrolyte into the ECM zone. Electrolyte exit may be through another system of holes in the tool or out as in the flow-past mode. A possible advantage of this mode is a larger part dimension, because flow past occurs in regions between the holes through which the electrolyte enters and exits. A limitation may result due to the raised spots on the workpiece surface opposite the holes in the tool. These spots must be removed by a subsequent operation.

Hole drilling is a special version of the flow-through mode. Usually, the electrolyte flows through the hollow-tube cathode tool, which is advanced into the cavity as drilling proceeds. Electrolysis proceeds at the tip of the tool and possibly from the sides. The side cutting may be useful and designed into the tool, depending on the kind of electrolyte, or the sides may be insulated so that the only side cutting is the small amount from current distribution at the tip. Usually, an insulated tool (both inside and outside) would be used.

Trepanning will be recognized as a special version of hole drilling in which the tool has such a large internal diameter that a slug remains. In hole drilling, the slug is cut away. It is obvious that the cross section shape of the tool can be varied to make holes or trepan cuts of special cross section shape.

Broaching is accomplished by electrolyte flow past the surface. The tool is tapered, similar to tapered plug, past which electrolyte flows in the gap region about 0.010 to 0.020 in. between tool and bore surface. As the broach tool is advanced into the bore, the workpiece dissolves, in the region opposite the tool, such that when the tool has passed a given point, essentially all the "cutting" has been done. Metal removal may be 0.020 in. per minute and the broach can advance 2 in. or more per minute.

Shaping, as for finishing turbine blades, would use two tools, each of which is a mirror image of the blade face. The tools advance simultaneously from opposite directions toward the workpiece positioned between the tools.

Tool Fixture

The tool fixture is designed so as to align the tool for advance by the ECM machine and confine the electrolyte flow to the gap region only, as required for the particular operation and tool configuration. In Fig. 1, the Lucite "box" is the fixture that guides both the tool and the electrolyte. When the Lucite fixture is positioned on the work, electrolyte flows into the slot in the fixture that distributes the electrolyte for uniform flow between the tool and the workpiece.

At the start, "cutting" (anodic dissolution) begins at the higher current density at the point on the tool of nearest approach to the workpiece. As the tool advances, other regions on the tool close in and the cutting increases as more surface area and current come into action. Thus, the anodic action is uniform on the workpiece face and side walls of the cavity. Steady state is reached when all points of the tool surface are equidistant from the workpiece surface. If ECM is continued, the cavity is deepened, but the contour at the bottom does not change. ECM is stopped when correct depth is reached, and the tool is backed out.

All the motions, electric current, and electrolyte flow are controlled by the machine, and every piece is identical with the others made in the same tool fixture. Therefore, the principal skills required for successful use of ECM are tool and fixture designing. No operator skill is needed when the tooling and fixturing are correct.

Electrolyte Flow System

Generally, the electrolyte system is a part of the machine. The important specifications concern the pump and electrolyte-tank capacities. These are selected on the basis of maximum size of items to be machined. In turn this size determines the maximum amperage to be provided. Now the mathematical relationships of the several variables enter the discussion.

Flow rate of electrolyte, current density, length of flow path, gap size, total area of ECM at steady state are interrelated. Total surface area is fixed by workpiece to be produced. Then, the machining engineer must set up his conditions by use of equations, such as those in the section on Operating Conditions. These equations reveal the electrolyte flow rate, in relation to the surface area and current density, to achieve the selected ECM rate within the capacity of the machine and flow system.

A flow system for a 1000-ampere ECM machine is shown in Fig. 3.

Operating Conditions

ECM operating conditions are directly related to the final size and shape of the workpiece, Faraday's Law, and hydrodynamic factors that support anodic dissolution at a practical cutting rate.

Machinists refer to cutting rates in cubic inches per minute. This provides a direct tie-in with Faraday's Law, because (volume) × (density of the metal) = weight. Faraday's Law states that the weight dissolved anodically is directly proportional to the amp-sec of electric current that passes. Thus, the operating conditions can be described by the relationships between: equivalent weight and density of the workpiece metal; current density; current efficiency; and time.

$$\text{Volume removed} = \frac{(CD)(Mew)(t)(CE)(A)}{(dm)(100)(F)} \quad (1)$$

FIG. 3. Electrolyte handling system for 1000-ampere ECM machine. It contains electrolyte tank, filter, heat exchanger, pump, and flow meters. (*Courtesy of Ex-Cell-O Corporation*)

In this equation, the terms are:

CD = current density (amp/in.^{2})
Mew = equivalent weight workpiece (oz)
t = time (min)
CE = current efficiency, per cent
dm = density (oz/in.^{3})
F = Faraday amp-min
A = area (in.)^2

Therefore, volume removed is in cubic inch (in.^{3}).

Although machinists refer to cubic inch/minute of stock removal, a handier term for ECM is feed rate, FR, as inch/minute. This is because the tool (or maybe the workpiece) advances in the direction of cutting at the same rate that stock is removed in the direction the tool advances. Thus, FR and penetration rate are the same.

If t in Equation (1) is 1 min, the stock removal by ECM becomes volume/min or in.^{3}/min. On the basis of unit area of 1 sq in.,

$$\text{Volume removed} = (\text{area})(FR)$$

and

$$FR = \frac{\text{Volume}}{\text{Area}}$$

by substitution Equation (1):

$$FR = \frac{(Mew)(CD)(t)(CE)(A)}{(dm)(100)(F)(A)} \quad (2)$$

The term (t) is shown as numeral 1, because FR is in./min. Equation (2) can be written as

$$FR = (km)(CD)(CE)$$

$$km = \frac{(Mew)}{(dm)(F)}$$

A value can be calculated for km for each metal and a chart can be made, as given in Fig. 4 for showing FR at various values of CD. For example, the km value for iron dissolving with a valence of 2 is 13.4×10^{-5} in.^{3}/min/amp.

For alloys, equivalent weights are calculated by adding the equivalent weights of each element multiplied by the percentage of that element in the alloy.

$$MAew = \frac{\Sigma \, (Mew)(\text{per cent})}{100}$$

Thus, significant variables for ECM can be related to each other by equations so, that by a series of calculations, numerical values can be obtained with sufficient accuracy for those variables in the intended ECM operations. Typical equations relating them follow.

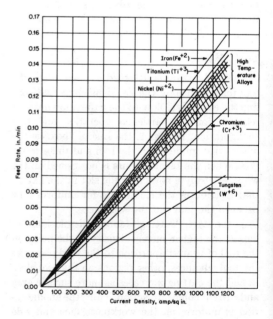

FIG. 4. Feed rate versus current density at 100 per cent current efficiency for various alloys and metals.

The term km obviously can be used in an equation for calculating volume of stock removed,

$$\text{Volume of stock} = (km)(t)(CD)(A)(CE) \quad (4)$$

Values for km for common metals in ECM are given in Table 1.

Thus far, the discussion recognizes no limit to current density that can be used. However, electrochemists know that a limiting current density does exist, because of polarization effects due to: mass transport of ions by diffusion, migration, and convection; activation; electron transfer; etc. Including terms for all the polarization parameters would result in an unwieldy and unusable equation for shop use. To simplify the situation, one may assume that under conditions of electrolysis that are controlled only by mass transport of ions, the current distribution is proportional to the distance (gap, G, in ECM) from anode to cathode—the limiting current density is given by Fick's Law. A further assumption is that the concentration gradient is linear across the diffusion layer at the workpiece surface (i.e., the anode film).

$$CD_L = \frac{(D)(n)(F)(\Delta C)}{\delta} + (\alpha)(CD_L) \quad (5)$$

CD_L = limiting current density, asi
n = valence of dissolving workpiece
F = Faraday, amp–min
ΔC = concentration gradient of dissolving metal across anode film
δ = thickness of anode film (inch)
α = migration coefficient
D = diffusion coefficient (inch²–min).

TABLE 1. VALUES FOR km AT 100 PER CENT CURRENCY EFFICIENCY

Metal	Valence	Density d_m, lb/in.³	km in.³/amp/min
Al	3	0.098	1.25×10^{-4}
Cb	3	0.310	
Cr	2	0.260	0.92×10^{-4}
Co	2	0.322	1.25×10^{-4}
Cu	2	0.324	1.34×10^{-4}
Fe	2	0.284	0.90×10^{-4}
Mn	2	0.270	1.39×10^{-4}
Mo	3	0.369	1.18×10^{-4}
Ni	2	0.322	1.25×10^{-4}
Ti	3	0.163	1.34×10^{-4}
W	6	0.697	1.70×10^{-4}
V	3	0.220	1.06×10^{-4}

If the assumptions are correct, CD_L can be large enough for practical ECM, provided δ can be made an order of magnitude (10 times or more smaller) than its value in even vigorously stirred plating or electropolishing operations. To achieve such a smaller δ requires forced movement of ions in ECM well beyond the rate attained by diffusion, migration, or convection. This brings hydrodynamics into the picture, along with controlled flow under uniformly turbulent conditions for experimentally determining whether the assumptions are correct. More importantly, can ECM be accomplished? For this text, it is sufficient to say that the experiments were successful and ECM can be done at current densities up to 1000 amp/in.² and higher in certain applications. Where Area and Total Amperage are small, current densities have been reported at 10,000 amp/in.²

Now, the quantity of electrolyte becomes important. Fresh solution must be provided in the electrolysis zone to take up the metal removed by anodic dissolution at the rate corresponding to the high current density, to remove the products from electrolysis zone, and to dissipate the heat of electrolysis. Flow volume normally will be at least 1 gpm per 400 amperes. As near the minimum as possible is desired so as to minimize the volume to be handled in effluent clarification.

Therefore, the volume of electrolyte may be related to the: current density; voltage; gap; length and width of flow path; and viscosity and conductivity of the electrolyte. Thus, another series of equations are needed for establishing conditions for ECM[1]. The first one relates the applied voltage, the current density, and the gap:

$$E_A = Ee + (\rho)(G)(CD) \quad (6)$$

E_A = voltage cross the ECM cell
Ee = sum of anode and cathode voltages
ρ = electrolyte specific resistance (ohm–inch)
G = gap, (inch).

The gpm, Q, of electrolyte is given by

$$Q = (4.175 \times 10^{-2})(V_T)(w)(G)(L) \quad (7)$$

Reynolds number is a convenient term for being indicative of turbulent flow which is essential for uniform electrolyte mass transfer required for ECM, and can be related to gpm as follows:

$$N_{Re} = \frac{(0.0534)(Q)}{(w)(\nu)} \quad (8)$$

w = width of flow channel in the gap (inch)
ν = kinematic viscosity of the electrolyte (ft²/sec).

Under turbulent flow conditions that exist in the ECM zone, the relatively stationary layer of electrolyte at the workpiece surface is the same order of thickness as in the anode film, δ in Equation (5).

Gallons per minute of electrolyte flow also can be related to the electrolysis conditions of ECM as follows:

$$Q = \frac{(0.0049)(E_A)(I)}{(de)(C_p)(\Delta T)} \qquad (9)$$

E_A = voltage across ECM cell.
I = total current, (amperes).
de = density of electrolyte.
C_p = specific heat of electrolyte.
ΔT = temperature rise in the electrolyte as result of passing through ECM zone.

The constant, 0.0049, in Equation (9) is derived from the fact that 1 BTU = 0.293 whr. Thus, $\frac{0.293}{60} = 0.0049$ BTU/min.

When the Reynolds number exceeds about 4000, the flow will be turbulent and the number can be estimated from the following equations:

$$N_{Re} = \frac{(1.67 \times 10^{-4})(G)(de)(V)}{\mu} \qquad (10)$$

or:

$$N_{Re} = \frac{(27)(E_A)(I)}{(C_p)(\Delta T)(\mu)} \qquad (11)$$

Thus, with this pair of equations, the Reynolds number and flow velocity can be estimated for turbulent flow in the ECM zone. In Equations (10) and (11),

μ = viscosity, (centipoise).
V = velocity of electrolyte flow (ft/sec).
de = density of electrolyte containing ECM products; i.e., hydrogen gas and sludge.
C_P = specific heat of electrolyte (BTU/lb/F).
ΔT = temperature rise from entrance to exit in ECM zone (F).
I = total current, (amperes).
E_A = voltage across ECM zone (volt).

For the term, de, the hydrogen is the cathode product in ECM and the sludge is the precipitate of the metals dissolved from the workpiece. In commonly used electrolytes of NaCl and NaNO$_3$, the dissolved metals precipitate as hydroxides or basic salts while still in the ECM zone. This affects the electrolyte density and conductivity values.

The temperature rise to point L downstream in the ECM zone, ΔT, when electrolyte flow is sufficient to carry out the heat is given by equating the right-hand terms of Equations (13) and (14). Then,

$$\Delta T = \frac{(136.7)(E_A)(CD)(L)}{(C_p)(de)(V)(G)} \qquad (12)$$

The terms are the same as those in Equations (10) and (11) and can be related to heat input and heat removal as follows. Heat input in

$$\text{Btu/sec} = (E_A)(CD)(L)(w)\left[\frac{3.47 \text{ Btu/whr}}{3600 \text{ sec/hr}}\right] \qquad (13)$$

Heat removed in

$$\text{Btu/sec} = (C_p)(\Delta T)(\rho_e)(V_T)(w)(G)\left(\frac{0.001}{144}\right) \qquad (14)$$

In order to achieve turbulent flow, a certain velocity of electrolyte flow will be required. This is related to other variables as follows:

$$V_T = \sqrt{\left(\frac{4.63}{C_F}\right)\left(\frac{1}{de}\right)\left(\frac{\Delta P}{L}\right)(G)} \qquad (15)$$

C_F = coefficient of friction.
de = density of electrolyte + hydrogen gas + sludge (oz/in.^{3}).
L = inch downstream from point of entry of electrolyte into the ECM zone.
$\frac{\Delta P}{L}$ = pressure drop/unit length (lb/in.^{2}).
V_T = turbulent flow velocity ft/sec.

The quantity of electrolyte flow to provide the velocity required by Equation (14) is given by:

$$Q \text{ gal/min} = (4.175 \times 10^{-2})(V_T)(w)(G)(L) \qquad (16)$$

or

$$Q = \frac{(N_{Re})(w)(\nu)}{(0.0534)} \qquad (17)$$

The pressure required for moving the electrolyte through the gap is ΔP in Equation (15). Thus, the pump pressure for maintaining Q of Equation (17) must equal line pressure drop, plus ΔP, plus any pressure restriction at the exit from the gap. Also, for approximation, the gap and pressure are related to the Reynolds number as follows:

$$\sqrt{(G)^3} = \frac{N_{Re}}{(K_T)\left(\sqrt{\frac{\Delta P}{L}}\right)} \quad (18)$$

$$K_T = 0.535\left(\sqrt{\frac{1}{C_F}}\right)\left(\frac{de}{\mu^2}\right) \quad (19)$$

C_F = friction factor, (order of 0.02 for sodium chloride electrolyte).

The sludge formed is an important factor in ECM and the amount can be estimated as follows: Volume of sludge, in.³/min:

$$S_V = \left(\frac{S_{mw}}{M_{mw}}\right)\left(\frac{dm}{de}\right)(M_V)\left[\frac{(CD)(A)(t)}{(n)(F)}\right] \quad (20)$$

When $A = 1$ in.² and $t = 1$ minute in Equation (18), $S_v =$ in.³ of sludge formed in the dry state per minute of ECM operation. This will be the minimum volume of sludge, which generally is voluminous in the ECM zone. In settling tanks or filter cakes, the wet volume will depend on the conditions under which these operations are carried out.

In ECM, hydrogen forms at the tool (the cathode). Assuming that electrolyte entering the ECM zone has zero hydrogen content, the amount of hydrogen formed is given by Equation (1) in which the terms Mew and dm are replaced by Hew and d_H, i.e., equivalent weight and density of hydrogen. Thus, the volume of electrolyte flowing must be sufficient to sweep out the hydrogen so that its effect on conductivity of the electrolyte is negligible. The actual volume of hydrogen will be the amount calculated from Equation (1), less the amount dissolved in the electrolyte, as corrected for temperature and pressure. Generally, the amount dissolved can be neglected. Therefore, the actual volume increase from point of electrolyte entry to point L, downstream in the ECM zone, may be approximated by:

$$\Delta H_V \quad (21)$$
$$= \left[\frac{(CD)(w)(L)}{F}\right]\left(\frac{Hew}{dh}\right)\left(\frac{14.7}{P + 14.7}\right)\left(\frac{460 + T_L}{492}\right)$$

H_V = volume of hydrogen (in.³)
P = pressure on ECM zone (lb/in.²).

The value of P may be calculated by employing several hydrodynamic equations. However, for practical ECM, pressure in the gap may be determined by gages placed where the electrolyte enters and leaves the ECM zone.

This is because the electrical conductivity of the electrolyte is decreased by the presence of hydrogen bubbles. Because temperature rise increases the conductivity, the decrease due to any hydrogen present is offsetting. The conductivity of hydrogen-containing electrolyte can be approximated by an empirical form of the Bruggeman equation:

$$\rho e = \frac{\rho}{(1 - f)^{3/2}}$$

ρe = specific resistance of the electrolyte-hydrogen mixture.
ρ = specific resistance of hydrogen-free electrolyte.
f = hydrogen void fraction.
Hc = amount of hydrogen, in.³ H_2/in.³ electrolyte.

$$f = \frac{Hc}{Hc + 1}$$

Thus,

$$\rho e = (\rho)(Hc + 1)^{3/2}. \quad (22)$$

The hydrogen volume changes with temperature and with pressure, which affect the value of Hc according to the gas laws and is approximated by Equation (21)

$$Hc = \frac{H_V}{(G)(w)(L)}$$

Therefore,

$$\rho e = (\rho)\left[\frac{(H_V) + (G)(w)(L)}{(G)(w)(L)}\right]. \quad (23)$$

ELECTROYLTE COMPOSITIONS

The most commonly used electrolytes for iron, nickel, and cobalt alloys are not harmful in contact with operators and are relatively noncorroding to machines made of mild steel and cast iron. In ECM, the electrolyte supports anodic dissolution of metals and alloys at 100 per cent current efficiency at current densities of 10 to 500 or more amperes per square inch. Solutions of sodium chloride, sodium nitrate, and sodium chloride-sodium nitrate in the concentration ranges of 1 to 3 lbs/gal meet these criteria. Whereas steels and high-temperature alloys of nickel, cobalt, and chromium dissolve very slowly (if at all) as anodes in such solutions, at usual electroplating current densities, under ECM conditions, they dissolve at nearly 100 per cent efficiency.

Chloride electrolytes are more efficient than nitrate electrolytes. For example, in chloride, mild steel dissolves at a rate 1.35×10^{-4} in.^{3}/amp-min and in nitrate 0.85×10^{-4} in.^{3}/amp-min. Unless the sides of the tool are insulated, overcutting at side regions not part of the main tool surface contour are greater in chloride than in nitrate electrolytes, and are least of all reported for perchlorate electrolyte[2] when it is applicable. Thus, passivating-type electrolytes require less tool correction or insulation, but use up more machine time.

Other metals such as tungsten and molybdenum alloys need alkaline electrolytes, such as NaOH solution. On titanium, NaCl electrolytes can be used. However, on some of its alloys, fluoride-containing electrolytes may be required.

When machine parts exposed to ECM electrolyte are constructed of chemically resistant materials, such as stainless steel and/or structural plastics, more active types of electrolytes are possibilities for special metals. At this time, the ECM user must develop his own electrolytes. If he has this interest, he must anticipate it when purchasing a machine and decide on the structural materials.

SURFACE EFFECTS

Smoothness and physical nature of the surface may be important to the end use of an ECM product. Smoothness can be in the range of 10 to 100 rms microinches by Profilometer measurement. In this range, the actual value is determined by ECM rate and grain size, phase structure, and chemical composition of the workpiece. Usually, smoothness is best at the highest ECM rate and for fine-grained, single-phase metals.

Multiphase metals present the problem of adjusting ECM conditions so that all phases dissolve at the same rate. There is no rule of thumb to be followed, except that passivating-type electrolytes are least affected by multiphases. Optimum ECM conditions are established empirically and, as in the design of tool and fixture, no substitute exists for experience.

Because it operates without mechanical work on the surface of a metal, no compressive surface stresses are introduced by ECM. Accordingly, metals which gain in fatigue strength from such stresses will appear to be weaker under fatigue loading after ECM. If this condition is critical, strength can be restored by mechanically cold working the surface by an operation such as vapor blasting or shot peening.

Some high-temperature alloys perform better at operating conditions after ECM. This is because stressed surfaces recrystallize at lower temperatures, which results in lower stress rupture strength. Being free from a stressed surface layer, the ECM material retains strength at higher temperature. Such effects are discussed in Surface Metallurgy.

In ECM as in any anodic treatment, some alloys tend to intergranular attack, i.e., the grain boundary material dissolves at a slightly more rapid rate than the grains. The net result is a crystal pattern appearance and a product that may be weak under fatigue stress. This is because of the notch effect caused by the etched grain boundaries. A similar tendency to notch weakness may prevail if surface pits result from ECM. Passivating-type electrolytes are least susceptible to these actions. If such detrimental effects exist for all ECM conditions, with a particular alloy, its composition and/or heat treatment may have to be changed.[3]

PLATING OUT

As electrodepositors will wonder, why doesn't the metal dissolved from the workpiece "plate out" on the tool? The main reason is that practically all the metal precipitates immediately, as a hydroxide or hydrated salt, while in the electrolysis zone. Any plating out is undesirable, because any such action will change the tool face, which in turn would change the contour and adversely affect the contour of the workpiece. Under some conditions, the precipitated particles will plate out by electrophoresis—this must be prevented by keeping the precipitated sludge at a minimum.

Even in the best controlled ECM systems, the possibility is ever present that a particle of sludge, or other insoluble material, will cause a spark over between the tool and workpiece. Such an event is likely to damage the expensive tool and/or workpiece and must be prevented. This is done by a spark detect and control unit on the ECM machine. The unit can anticipate the onset of a spark and prevent it from occurring.

EFFLUENT AND WASTE HANDLING

In ECM, the machining chips are insoluble precipitate of hydroxides and/or insoluble salts of the metals dissolved from the workpiece. This precipitate increases the viscosity and decreases the conductivity of the electrolyte. Therefore, it must be removed by filtering and/or settling, so that the clear electrolyte can be reused.

Because the precipitate is difficult to filter, as is characteristic of hydroxides, centrifuges are much more suitable than other means. The equipment cost is relatively high and maintenance may be expensive. Alternatively, settling in essentially nonmoving electrolyte in tanks or in lagoons is effective. Settling is usually so slow as to require large-volume retention, because sludge is produced at a high rate by ECM. In volume, the sludge is many times that of the metal removed. Ultimately, the time comes when the settled sludge must be removed to a dump or be dewatered for disposal or use in a metal recovery process. Settling involves relatively low first cost and is inexpensive to operate, but requires considerable floor or land space. Flotation also provides good separation and is inexpensive to operate. However, equipment cost is high.

The method to be used will depend on the size of the ECM operation and the amount of electrolyte to be handled. For small operations, filtration may be the best method. A large filtration area will be needed, but initial cost is low—however, it may be expensive to operate and maintain.

Present practice discards the sludge, and ECM would benefit greatly from improved methods for separating sludge from electrolyte in the liquid effluent.

BECOMING AN ECM USER

Becoming a user of ECM can be likened to becoming a piano player. The ECM machine is analogous to the piano. Both are designed and expertly built to do a job. The aforementioned equations are the sheet music of ECM. The user must learn to follow them and employ them for designing his tool-fixture units and for selecting his operating conditions.

Side cutting or overcutting is a situation to be reckoned with in ECM. Such effects can be compensated by trial and error to adjust a tool based on a mirror image design. The adjustments usually are slight and quickly made by an experienced ECM designer. Equations can be set up for initial approximation of tool adjustments for overcutting. Such equations and their use are briefly described elsewhere and are also useful for calculating the cavity contour generated by an ECM tool.

It is evident that no single equation relates all the factors. Therefore, a series of calculations must be made to find the combination of conditions that matches feed rate to total current, voltage, and flow rate of electrolyte that is attainable with the electrolyte circulation system and d-c source. As with any process, there is no substitute for experience in designing the tool and fixture for the job. When this is done properly and they are assembled on the machine, the operations are automatically controlled by the machine.

Glossary of Terms

I	total ECM current
CD	current density (amp/in.^{2})
CD_L	limiting current density (amp/in.^{2})
C_F	coefficient of friction
C_P	specific heat of electrolyte (Btu/lb/F)
Mew	metal equivalent weight (lb)
$MAew$	calculated equivalent weight of alloy (lb)
Maw	atomic weight of workpiece (lb)
dm	density metal (lb/in.^{3})
de	density electrolyte including sludge and hydrogen (lb/in.^{3})
ds	density of sludge (lb/in.^{3})
d_H	density of hydrogen (lb/in.^{3})
D	diffusion coefficient of metal ions (in.^{2}-min)
t	time (min)
F	Faraday (amp–min)
A	area of workpiece (in.^{2})
FR	feed rate (in./min)
n	valence of anode dissolution of workpiece
ΔC	concentration gradient of dissolving metal across anode film (oz/gal)
δ	thickness of anode film (in.)
α	migration of coefficient of metal ions
E_A	voltage across ECM cell (volt)
Ee	sum of anode and cathode voltages (volt)
f	void fraction of hydrogen
G	gap (in.)
Hc	amount of hydrogen (in.^{3} H$_2$ per in.^{3} electrolyte)
Hew	equivalent weight of hydrogen (lb/in.^{3})
Q	electrolyte flow (gal/min)
V_T	turbulent electrolyte flow rate (ft/sec)
V	velocity electrolyte flow (ft/sec)
w	width of flow channel through gap, G (in.)
L	length of flow path in the ECM region (in.)
N_{Re}	Reynolds number
ν	kinematic viscosity (ft^2/sec)
μ	viscosity (centipoise)
ΔT	temperature rise (F) in the electrolyte as result of ECM
S_V	sludge volume (in.^{3})
S_{mw}	sludge molecular weight (lb)
M_V	volume of workpiece dissolved (in.^{3})
ΔH	increase in volume hydrogen due to ECM (in.^{3})
P	pressure on ECM zone (lb/in.^{2})
T_L	temperature at point L downstream in ECM zone

ρ	specific resistance of hydrogen-free electrolyte (ohm–in.)
ρ_e	specific resistance of the electrolyte-hydrogen mixture (ohm–in.)
CE	current efficiency for anodic dissolution of workpiece
A	workpiece surface area (in.^{2})
K_m	electrochemical equivalent of workpiece
ΔP	pressure drop to distance L from entrance of electrolyte into G (lb/in.^{2})
ΔC	concentration gradient of dissolved workpiece metal across δ
Mmw	molecular weight of workpiece (lb)
Hv	volume of hydrogen generated at tool surface (in.^{3}).

References

1. Faust, C. L., Gurklis, J. A., and Cross, J. A., "Electrochemical Machining and Electrochemical Grinding," International Conference on Manufacturing Technology, pages 341–416, American Society of Tool and Manufacturing Engineers (1967).
2. Opitz, H., and Heitman, H., page 397, International Conference on Manufacturing Technology, American Society of Tool and Manufacturing Engineers (1967).
3. Faust, Charles L., "Surface Metallurgy," *Metal Progress*, **101** (May, 1957).

Additional Reading

Allison, Charles, "How the Electrolyte Influences the ECM Process," Creative Manufacturing Seminars (1963–64), American Society of Tool and Manufacturing Engineers, SP 64–79.

Baldwin, G. L., Brown, D. C., and Gulati, J. L., "Electrochemical Machining—Nomograms for Prediction of Process Parameters," *The Engineer*, 5 (Feb. 23, 1968).

Faust, Charles L., and Snavely, Cloyd A., "New Electrochemical Process for Tough Metals," *The Iron Age* (Nov. 3, 1960).

Faust, Charles L., "Electrochemical Machining of Metals," *Trans. Inst. Metal Finishing*, **41** (1964).

Faust, Charles L., "Review and Comparison of Nonmechanical Metal Working Processes," *Machine Design*, 164 (Aug. 13, 1964).

Galloway, D. F., "Development and Application of Electrochemical Machining as a Production Process," page 383, International Conference on Manufacturing Technology, American Society of Tool and Manufacturing Engineers (1967).

Throop, James W., "Influence of Electrolyte on ECM Equilibrium Gap Equation," page 359, International Conference on Manufacturing Technology, American Society of Tool and Manufacturing Engineers (1967).

38. ELECTRODEPOSITION OF ORGANIC COATINGS

George E. F. Brewer

Coating Consultant Birmingham, Mi.

It has been found that paint films can be deposited through the action of an electric direct-current impressed on certain specially formulated water-borne paints. This coating process was introduced in the early 1960s and is now in world-wide use, known as electrodeposition of coatings, electropainting, electrocoating, etc.

The rapid acceptance of electrocoating seems to be due to the resulting high corrosion protection, low cost, low energy requirement and little emission of solid, liquid or gaseous wastes.

Several thousand articles have been written about the new coating process, and many of these are accessible from quotations in the seven sources listed at the end of this chapter.

HISTORICAL DEVELOPMENT

The fact that colloidal particles of clay in aqueous dispersion migrate towards the anode in an electric D.C. field was discovered by Reuss in 1808, and named "electrophoresis." Reuss also observed that the water rises in a tube which surrounds the cathode, a phenomenon named "electroosmosis."

In 1859, Quincke referred to observations of electrophoresis made by Faraday and others, and observed that in aqueous dispersion most substances migrate towards the anode, while they migrate towards the cathode in turpentine.

In 1892, Picton and Linder observed that "Magdala Red" dissolved in alcohol is "repelled from the positive electrode." Magdala red is a diamine and therefore—from the present point of view—forms cations. Picton and Linder (1905) seem to have been the first investigators to be interested in an electrophoretic deposit: ferrous hydroxide from a dispersion in alcohol. The earliest patent on electrodeposition has been granted in 1919 to Davey, General Electric Co., for a process of making and applying "Japans" as insulation of wires, describing anodic deposition of bituminous materials in the presence of bases, ammonia being preferred.

In 1923, Klein, Anode Rubber Co., received a patent for the anodic deposition of rubber from latex.

From 1936 to 1943, Clayton and co-workers, Crosse & Blackwell Co., received various patents for the electrophoretic coating of food can interiors with beeswax emulsions, etc.

All of these electrodeposition processes worked with naturally occurring substances and essentially neutral resin dispersions. These processes seem to have been abandoned before 1950.

The technology of acidic, synthetic resins began to appear in the literature in the 1930s but these were not used for electrodeposition until the parameters of an effective process for painting by electrodeposition were defined by Ford Motor Co. The invention is probably best summarized as the utilization of synthetic, water dispersable and electrodepositable, film forming macro-ions.

PRINCIPLES

Metal Preparation

Any conventionally used metal preparation can be applied, such as degreasing, cleaning or phosphating. The obtained corrosion protection increases with the quality of the metal preparation, being highest for seven-stage zinc phosphate.

Paint Composition

An electric charge is exhibited by colloidal size particles in aqueous dispersions. Carbon, metals, sulfur, cellulose, etc. exhibit negative electric charges due to adsorption of hydroxyl ions. Electric charges are also exhibited by emulsified colloidal particles. In this case the electric charges are provided by a surface layer of ionized organic substances; for instance, arylsulfonic acids (Ar-OSO_3^-).

Water insoluble organic acidic resins (RCOOH) of suitable molecular weight (oligomers) can be dispersed in water through the addition of bases by formation of colloidal size anions ($RCOO^-$) and, similarly, aminated resins can be colloidally dispersed by the action of acids to form cations (R_3NH^+). The invention of electrocoating is probably best summarized as the utilization of synthetic, water dispersable and electrodepositable, film forming macro-ions.

The process of the formation of ions from oligomers may be looked upon as being eventually reversed by the process of electrodeposition, as shown in Table 1.

A somewhat different class of electrodepositable oligomers are the "Onium bases," like ammonium type (R_4N^+), phosphonium (R_4P^+) and particularly the sulfonium bases (R_3S^+). These do not require the use of external solubilizers, and the ionic group of some of them is irreversibly decomposed during the position and/or bake, leaving a neutral resin in the final film.

In all cases, the resinous moiety "R" may have the chemical composition typical for almost any of the known film formers, such as acrylics, alkyds, epoxies, phenolics, polyesters, etc. Thus, there is a large variety of formulations possible and actually used. The electrodeposited, cured films show essentially the same solidities as films of similar composition applied by any other method. Yet the corrosion protection afforded by electrodeposited films is much higher than that received from differently applied coatings of similar composition, due to the extremely uniform film thickness, even in highly recessed areas.

Electrodeposition

The outstanding ability to deposit paint evenly is caused by high electric resistance of the electrodeposited film. The electrophoretic deposition of film starts at the electrically most exposed edges. The first deposit reduces the flow of electricity and reduces the deposition rate, while electroosmosis forces the liquid phase to leave the coating, thus further increasing the electrical resistance. The electric current seeks the path of least resistance, causing electrodeposition on still available uncoated areas, until all surfaces of the object are evenly coated. This coat consists of approximately 95 weight % of nonvolatiles (resin, pigment, etc.) plus 5 weight % water and other volatile paint components.

Rinse

Electropainting is essentially a dip process, and sizeable quantities of paint bath therefore adhere to the merchandise and are lifted from the tank. It is lucrative to rinse the paint bath droplets back into the tank. The freshly deposited coat, particularly when blown dry, allows the application of properly chosen spray coats.

Schematic Process

If we consider the conventional metal cleaning and preparation, the electrodeposition, the rinse and the conventional cure or bake, we are faced with the four basic steps shown in Fig. 1.

Cure or Bake

The bake cycles range from ambient air drying to forced air drying to bake, 100 to 200 C (200–400 F) for 10 to 40 minutes being used frequently.

TABLE 1

	Water Insoluble Oligomer	+	External Solubilizer	Dispersion ⇌ Deposition	Dispersed Macro-Ions	+	Counter Ions	
Anodic	RCOOH	+	BOH aq	⇌	$RCOO^-$	+	B^+	+ aq
Cathodic	R_3N	+	HX aq	⇌	R_3NH^+	+	X^-	+ aq

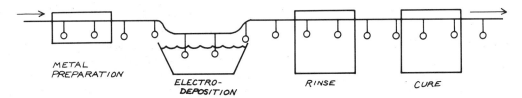

Fig. 1. Four basic steps of electrocoating.

Advantages

- High corrosion protection due to uniform film thickness even in extremely recessed areas which cannot be covered by other painting process. The ability to extend protective films into cavities, recesses and even capillaries is called "throwing power" and is one of the outstanding features of electrocoating.
- Much reduced cost of painting when compared with spray painting, dip and powder coating, with savings ranging from 50 to 10%.
- Electrocoating is one of the least polluting coatings, next to powder coat and radiation cure.
- Energy consumption is low since some drying and baking operations can be eliminated in metal preparation and for multi-coat systems.
- Coating efficiency is independent of size, quantity, and shape of merchandise going through the tank.
- Good edge protection, absence of runs, sags and wedging.
- Transfer efficiencies of up to 99%.

Disadvantages

- Sizeable capital investment.
- Cannot be applied as a second coat generally.
- Each color requires a separate dip tank.
- Coat does not hide metal defects.

APPLICATIONS

Any sufficiently conductive surface will accept electrocoats. Thus, ferrous and nonferrous metals, as well as generally nonconductive materials, can be coated, provided that their composition or an organic or metallic surface coat provides conductivity. However, if different metals are coated simultaneously, their film thicknesses may be slightly different. For instance, solder filled, sanded grooves receive usually a lesser film when compared with the adjoining steel.

The application possibilities can be divided into seven main groups:

Prime Coat

This is the largest outlet for electrocoating paints. The ability to coat either to a low or moderately high film thickness without runs or sags, to coat successfully in box sections and recesses, combined with the ability to protect capillary recesses such as butted joints, flanges, etc., has opened up new possibilities in corrosion protection for all kinds of merchandise.

Finish Coat

Electrodeposition as a one-coat system can show great advantages over conventional dipping and spraying. It is, however, in this field that the greatest limitations are felt. Some resins discolor during the heat cure (bake). Anodically deposited resins contain some anodically dissolved iron, but cathodic resins are almost free from this defect. Electrodeposited paints tend to follow the surface profile of metal, and the hiding of metal defects is incomplete, thus requiring some sanding.

High-Speed Coating of Coil Stock

Speeds of up to 1000 ft/min are sometimes used on coil stock, due to extremely uniform film thickness and the absence of runs, sags and tears.

Two-Coat Systems

These are used in special cases. A top coat is electrodeposited over an electrically conductive (at least to some extent) prime coat, which may be applied electrophoretically or by conventional methods. Sufficient conductivity in a paint is achieved by the introduction of conduc-

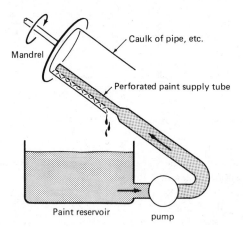

Fig. 2. Electroflow coating of symmetrical objects.

tive constituents such as zinc, aluminum, graphite, etc.

Flow Coating of Symmetrical Objects

This can be accomplished through rotating the electrically grounded merchandise by means of a mandrel while introducing paint through a stationary perforated tube closely located to the surface to be coated (see Fig. 2). The inside of cans, pipes, etc., may be coated by this process.

Bulk Coating

Fasteners, etc, are electropainted in peforated drums, or by use of a system of conveyors (Fig. 3).

Ceramic Coats

Glass powders (ceramic frits) are frequently deposited through the action of D.C. current. Some of these frits carry electric charges, particularly when dispersed in organic solvents like alcohols. In other cases, the frit does not carry an electric charge, but is suspended in a waterborne dispersion of a film forming electrodepositable partially ionized resin which envelops the frit particles. Thus, the resin is within the group of those used for electrocoating, and carries the frit to the electrode, where both resin and frit deposit. Ordinarily resins are expected to provide corrosion resistance and other durable properties. The resins used for the deposition of frit have one main requirement: They have to volatilize at a temperature below the softening point of the frit.

REQUIRED FACILITIES

Electropainting departments are designed to move the merchandise by means of either continuous moving conveyors or by intermittent motion which locates a batch of merchandise above treatment areas, lowers it, and then raises it after completion of the manufacturing step. Continuous motion seems to be preferred for high-volume production. Batch type operation may save space and energy and allows a certain amount of flexibility of production through skipping of certain steps for certain batches or selection of, say, one color out of several, etc. Figure 1 shows a schematic drawing of an electrocoating installation.

Pretreatment

Good pretreatment is as necessary for electrocoating, as for any other coating system. In fact, the unique opportunity to coat the insides of recessed areas requires more attention to metal preparation in these areas. Correct phosphate or chromate conversion coating adds protection

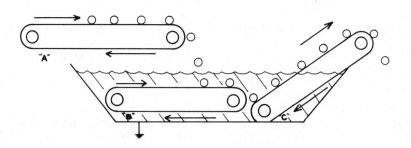

A- Feeding   B- Coating   C-Exiting

Fig. 3. Bulk electrocoating of fasteners.

against corrosion. Zinc phosphate is particularly effective. The use of a passivating rinse is important, and the introduction of ionized salts into the paint tank has to be avoided. A final rinse of deionized water should be employed.

Optional Drying Unit

Merchandise can be entered wet into the electrocoating tank, which saves energy and space. A blow-off section is sometimes provided. Care is taken to avoid carryover of large quantities of water because of paint dilution. Partially dry sections of work pieces are avoided because of flash rusting.

Cooling Channel

Pretreated merchandise should be cooled to a temperature which will not add a large burden to the cooling equipment which keeps the electrodip paint at a predetermined temperature.

Electrical Test and Connect Section

Electrical pick-up arrangements are usually built into the hangers. Large D.C. current loads, as required by automotive bodies, are supplied by cables which are clipped on manually. Electrocoating paints have enough "throwing power" to coat the inside of most merchandise adequately; however, auxiliary counter electrodes (see Fig. 4) are required for the coating of the insides of tanks, etc. These are usually inserted and connected manually. In this case, an automatic test for short circuiting may be provided.

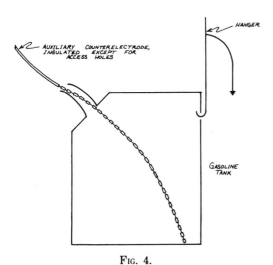

Fig. 4.

Dip Tank

Tanks are designed to provide clearance around the merchandise and allow for the required processing time, ranging from a few seconds for coil stock to approximately three minutes for the largest work pieces. The tank will also provide for agitation, filtration, cooling, heating and electrical needs, and usually accommodates the first rinse stage. Continuous motion conveyors require a tank area enclosure which prevents access to electrically dangerous areas and usually includes such devices as interlocked doors. Batch type coating tanks should have closed, electrically grounded covers in place before the start of the coating operation.

Electric Circuits

The simplest and earliest circuit for anodically depositing materials was the use of the entire cold rolled steel tank as cathode (see Fig. 5). In this case, the tank and the complete plumbing system, etc. are electrically grounded and the electric insulation is separating the grounded upper part of the hook from the electrically hot lower part of the hook, which receives energy through a shoe and busbar arrangement.

The entire tank can also be used as counterelectrode for anodic deposition by use of insulating tank supports and insulating links in all pipe connections (see Fig. 6).

Theoretically, the tank can also be used as counter electrode (anode) or positive electrode for electrodeposition of cathodic materials; however, due to anodic dissolution and acidic environment, the tank, etc., would have to be built from, or lined with, stainless steel or other corrosion resistant materials. The majority of all tanks are therefore designed as shown in Fig. 7. Here the tank is lined with an insulating coating, and the inlet and outlet pipes are connected to the tank by plastic links. The merchandise is grounded. The electrically hot electrodes are inserted into the tank, and separated from the coating bath by a membrane, if so required. When the positive side of the power source is grounded, anodic paints are deposited on the merchandise. Conversely, when the negative side of the power source is grounded, cathodic paints are deposited on the merchandise.

Rinse Deck

The electrically deposited film has been chemically reconverted from the ionized water

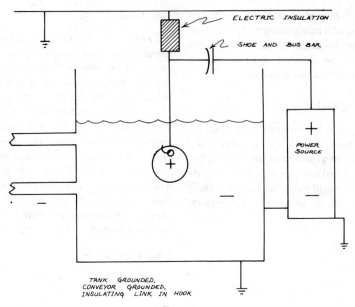

TANK GROUNDED,
CONVEYOR GROUNDED,
INSULATING LINK IN HOOK

FIG. 5.

dispersed form into a water insoluble oligomer (see Table 1), practically free from volatile components. This film is covered by a usually foamy or creamy layer of dispersed paint of higher concentration than the bath (see Fig. 8), the concentration gradient being the result of electrical migration towards the work piece. Most electrocoating installations use rinse decks (see Fig. 1) to recover the paint and the bath fluid which have been lifted out of the tank.

Removal of Electrics

When manual electric connections have been made, or when inserted counter-electrodes have been used, these are removed and readied for the next use.

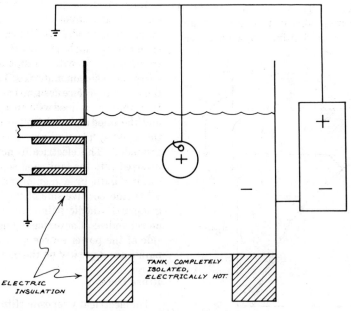

FIG. 6.

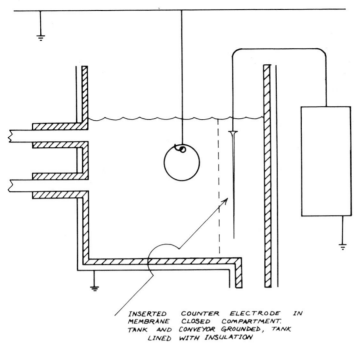

INSERTED MEMBRANE COUNTER ELECTRODE IN CLOSED COMPARTMENT. TANK AND CONVEYOR GROUNDED, TANK LINED WITH INSULATION

FIG. 7.

Blow-Off Section

Depending upon the complexity of the components which are being painted and the type of paint being used, it may be necessary to blow off the remaining drops of water. The blow equipment can vary from manual, compressed air to a stationary "collar," using ambient or heated air.

Application of Second Coat (Wet-on-Wet)

Top coat or sandable surfacer is sometimes applied to all or certain areas of merchandise by use of spray paints—a desirable, energy-saving practice.

Cure of Electrocoats

This may range from air drying to forced air drying to heat curing at temperatures varying from 93 to 190 C (200 to 375 F) for varying lengths of time.

Stripping of Hangers

Cleaning of hooks may be necessary only infrequently, since the hangers essentially carry only one coat, which rarely interferes with electric contact.

ACCESSORY EQUIPMENT

Power Source

The electrical equipment must provide the necessary D.C. current at voltages ranging from 50 to 500 volts at a ripple factor not exceeding 5% over the full range. Some coating tanks for coil stock, etc., run at up to 1000 volts.

The applied voltage depends upon the kind of paint used and the condition of the bath, etc., and is manually set to give the desired film thickness.

The amperage is determined by the quantity of merchandise going through the tank per unit of time. As a rule of thumb, 1 to 3 amp/ft^2 of work are required.

Current Requirement. Electrocoating paints exhibit electrical equivalent weights ranging from approximately 1000 to 2000; that is, one Faraday (F) (96,500 coulombs) deposits from

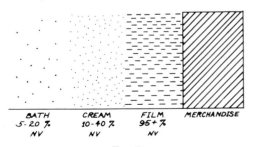

FIG. 8.

1000 to 2000 g. Compared with metal plating, the current requirement is small, since 1 F deposits 29 g of nickel. Furthermore, the volume of 1 g of paint is almost 10 times as large when compared with most heavy metals.

Knowing the apparent electrical equivalent weight, the density of the paint solids, the desired film thickness, and the area to be painted, we can predict the current requirement from Eq. 1.

$$\text{Coulombs} = \frac{F}{eq} \times \frac{A \times t}{K} \times 1000 \times d_{NV} \quad (1)$$

where

$F = 96{,}500$ coul
$eq =$ apparent electric equivalent weight in g
$A =$ area painted in m^2
$t =$ film thickness in μm
$K = 1000$ m^2 μm/l
$d_{NV} =$ density of paint Non-Volatiles in g/ml

The amperage load starts with an initial peak (dead entry), which gradually diminishes as the electrically resistant coating envelops the merchandise (Fig. 9). Of course, if the merchandise is gradually lowered into the electrically energized bath (live entry), the peakload of current is materially reduced (Fig. 10, curve I), further reduction of the peakload is accomplished through automatic amperage limitation (Fig. 10, curve II). The area under all three curves (coulombs) is identical.

Voltage Regulation. Regulating transformers or thryistors are used for voltage regulation. The former is a mechanical induction or contact system, while the later is a solid state device.

Rectification from A.C. to D.C. Silicon diodes are used, while the thyristor is its own rectifier.

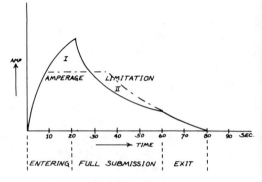

Fig. 10. Live entry.

Safety Devices. The main safety features are internal protection of the power source against short circuits by electronically distinguishing between normal and accidental current rise, regulated current wind-up on start and wind-down on conveyor stoppage, electrical leak detection and interlocking doors or trip switches protecting the access to the dip enclosure.

Distribution of Electricity. The electrically hot polarity is usually connected to stationary tanks or electrodes inserted into the tanks. The electrically grounded polarity does not cause problems except that many ground straps in different locations should be used. If the current requirement is small, say, under 50 amps, the grounded conveyor may carry this load to the hooks and goods. Yet, preferably, the hanger should rub against a grounded bus. Larger current loads should be transmitted via a pick-up shoe on each hanger which rubs against a bus bar. The largest electrodip-tanks transfer the goods to a special conveyor for the electrodip and then transfer the goods back to the factory conveyor. The special conveyor carries pick-up

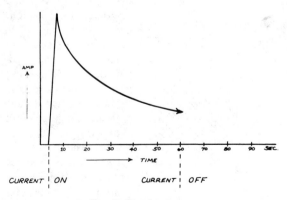

Fig. 9. Dead entry

shoes, on each hanger and on each of the cables for clip-on connection, contacting busbars to ensure full electrical grounding.

Busbar connections for internal counter-electrodes are also provided in the unusual cases when such is required.

Batch type vertical immersion tanks need a much simplified electric distribution system since contact has to be made only during the vertical downward motion of the merchandise and during the immersion time.

Paint Bath Maintenance

Level Control. Deionized water is added to the bath as required. The allowable upper and lower limits are guarded by an automatic alarm system which may also automatically shut off the addition of deionized water to avoid overflowing the tank.

Paint Agitation. The agitational system is based on the use of pumps, agitators, draft tubes, line shafts, and eductor nozzles, selected to deal adequately with the needs inside the dip tank and in the external parts.

The main requirements are:

- Maintenance of a homogenous mixture to avoid settling of solid paint components.
- Transport of fresh solids to the vicinity of the work to replace deposited solids.
- Entry and exit areas of the tank free of air bubbles, scum, etc., to avoid film defects.
- Distribution of cooled, heated, filtered or enriched paint as it returns to the tank from external circuits.

The average circulation rate seems to be suggested as six volume changes per hour.

Paint Heating and Cooling System. Paint heating is rarely necessary but provision should be made for it. Care should be taken to keep the contact temperature within the acceptable limit.

Bath cooling has to be applied to keep below the specified maximum temperature since practically all the electrical energy for electrodeposition plus the mechanical energy for pumping and agitation is converted into heat. Heat loss due to evaporation and radiation are small. Depending on the geographical location and the size of the installation, the following cooling media are employed: ambient air on cooling towers; plant water in heat exchangers; and (the only infallible method) refrigeration.

Paint Filtration. Strainers are built into weirs, etc., to remove larger pieces of refuse which may have fallen into the tank. Pressure filters with pore sizes from 5 to 70μ are employed. Many of the filters are of the self-cleaning or self-advancing type.

Ultrafiltration of Bath. Comparatively large quantities of paint solids are lifted from the paint tank, causing losses of 10 to 50% of the paint solids (see Fig. 8). It would be simple to recover the still dispersed paint by a water rinse, were it not for the fact that the tank is already full of paint. Thus, to allow the return of rinse fluid with its recovered paint, it is necessary to remove liquid from the tank. This is accomplished by ultrafiltration (see Fig. 11). Ultrafilters consist of porous plates, tubes or capillary tubes covered on one side with semipermeable membranes, selected to pass dissolved and dispersed substances of up to approximately 1000 molecular weight. Paint is circulated under pressure (approximately 20 lb/in^2) on one side of the membrane, while the permeate (ultrafiltrate) is collected on the other side. The ultrafilter is dimensioned to provide the needed quantity of permeate; as a rule of thumb, approximately 1 g/min/100 ft^2 of merchandise painted/min. Ultrafiltrate, deionized water, etc., are spawning ground for bacteria, but these clear liquids can be kept sterile by use of ultraviolet light units. Colored liquids require addition of bacteriostats.

Ultrafiltrate Rinse. The ultrafiltrate (permeate) is used on two or three rinse decks in counter-flow to the merchandise. This results in 70 to 90% recovery of the lifted out paint. Figure 11 shows the ultrafiltrate entering and being used on rinse deck #3; then being collected in a tank, from where it is pumped to rinse deck #2; then being collected in a tank and pumped to rinse deck #1, which is often located directly over a "drip" section of the electrocoating tank. Thus, the virgin ultrafiltrate is used on merchandise which has already passed through two rinses.

The ultrafiltrate rinses are usually followed by a deionized water rinse. This rinse water is usually discarded.

Ultrafiltrate Discard. Ultrafiltrate is sometimes discarded to purge electrocoating baths from contaminants like chlorides, phosphates, chromates, etc.. At the same time tanks are freed from low molecular weight organic substances (100 to 500 molecular weight) which sometimes are present in concentrations as high as 1 weight % of ultrafiltrate.

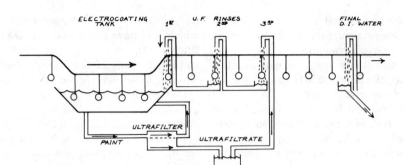

Fig. 11.

Permeate also contains amine solubilizer and co-solvents such as butyl cellosolve, which are lost through ultrafiltrate discard. In addition, the discard of ultrafiltrate is environmentally undesirable, and governmental effluent regulations in certain areas demand a costly clean-up. Reverse osmosis is therefore sometimes used to separate water, co-solvents and amines for reuse, while ionized inorganic and 100 to 500 molecular weight organic substances are concentrated in the retentate for discard.

Solubilizer Balance. Water dispersed film forming resins (R) are electrodepositable when carrying ionizable groups like —COOH —N, etc. The macro ions—(R—COO$^-$ or R$_3$NH$^+$), in these two cases—formed through the addition of external solubilizers (bases or acids) are formed in such quantities that only partial ionization occurs—enough to ensure a stable dispersion and not so much that film resolubilization occurs on lifting freshly coated merchandise out of the coating tank. Partially neutralized resins which envelope pigments, co-solvents and water form a dispersed system, exhibiting a certain electrical equivalent weight; for example, 2000 (that is, 96,500 coul deposit 2000 g, or the consumption of electricity is said to be 48 coul/g). If no resolubilization of the paint film has occurred, than 1 eq (or 1000 milliequivalents; meq) of solubilizer (acid or base) is present for every 2000 g of paint nonvolatiles. This is usually reported as 48 meq/100 g NV. Let us now assume that 20% of the paint solids in the tank are coated onto merchandise and lifted out of the tank.

Provided that no solubilizer is leaving, the tank would now contain 50 meq per 80 g NV, or 62.5 meq/100 g NV. Consequently, the electrical equivalent weight would be reduced from 2000 to 1600, with several accompanying disadvantages: higher current consumption, higher cooling requirement, increased resolubilization of film, etc.; thus, it is necessary to hold the solubilizer concentration within narrow, specified limits.

Solubilizer Discard. This is accomplished by one of several means:

1. Surrounding the counter-electrode by a membrane compartment (see Fig. 7). Theoretically, for every electrical equivalent weight of paint deposited, an electrical equivalent weight of counterion has to contact the counter electrode. Thus, the counterions can be flushed out of the membrane compartment.

2. Very early electrocoating tanks used deionizers to trap the cationic counter-ions of anodically depositing paints. After a certain period the deionizing resins have to be regenerated.

3. Theoretically, volatile external solubilizers such as ammonia, acetic acid, etc., can be removed via vapor phase. Also, substances like oxalic acid would be removed as gas on tact with the counter-electrode.

4. A more widely practiced counter-ion removal is the discard of ultrafiltrate.

Solubilizer Reuse. While a certain minimum amount of neutralization is necessary for a sufficient stability of a paint/pigment dispersion, a limited stability, but sufficient for transport from paint manufacturer to the user and storage on the user's premises, can be accomplished by use of small quantities of external solubilizer. Such under-neutralized paints are then mixed with solubilizer-rich bath to receive the needed complement of solubilizer (see Fig. 12).

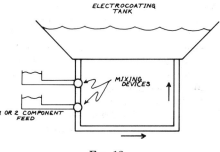

Fig. 12.

Exhaustion Method. An interesting variation of the electrocoating process has been developed for the painting of merchandise in very small tanks with low paint consumption, as is the case of fasteners. Here the cost of monitoring the tank for replenishment is comparatively high, while the cost of discarding the complete tank fill, or shipping it back to the paint manufacturer, is low. Special electrocoating paints are, therefore, placed into the tank at approximately 12% solids and 35% neutralization. The solids are gradually coated onto merchandise, until only 4% solids at about 70% neutralization remain. The tank is then refilled with 12% solid material.

One or Two Component Feed. Quite frequently electrocoating paints are delivered in two packages: a resin and a pigment slurry. These can be used to advantage in several ways. First, the resin portion can be underneutralized, and can be used to accomplish solubilizer balance. Secondly, the pigment can be protected against a comparatively high acid concentration of the resin portion. Third, the pigment concentration in the bath can be adjusted at will. Actually, the solubilizer is a third component which is sometimes added as needed.

Bath Stability. Electrocoating tanks are heavily agitated to prevent pigment settling, etc: thus high sheer stress resistance is required. Furthermore, like in all dip processes, the paint is subject to a long residence period in the tank, requiring high resistance against saponification and oxidation. To describe the residence of paint in the tank, the term "turnover period" is used.

Turnover Period. Suppose a 20,000-gal tank contains 10% nonvolatiles, or approximately 16,000 lb NV. Paint solids are coated onto merchandise and are leaving the tank, while feed materials maintain the solids concentration at 10% NV. When 16,000 lb NV have been added, the first turnover is complete. At that time, by the law of averages, more than half the original paint is still in the tank. Experience has shown that electrocoating baths should have a pumping stability exceeding four turnover periods.

REASONS TO ADOPT ELECTROPAINTING

Highest Corrosion Protection

If salt spray resistance exceeding 300 hr. is required for cold rolled steel merchandise, electrodeposition is the simplest and cheapest way of meeting the specification. Likewise, if salt spray resistance of more than 500 hr. is required, a combination of zinc coatings and electropainting is the best possible answer.

Compliance with U.S. Environmental Regulations

Solvent emission regulations in most of the 50 states limit the solvent contents of industrial paints for the coating of "Large Appliances" or "General Metal" to 3.0 lb volatile organic compounds (VOC) per gallon of paint minus water (approximately 58 volumes of nonvolatiles in 42 volumes VOC, possibly plus water or certain chlorinated solvents). A typical situation for a medium sized metal finishing department is the fact that approximately twice as much solvent vapor is emitted than the allowable amount. Typically, the solvent emission from a medium sized prime coating operation is, in round figures, 800 lb VOC per shift and an equal amount for top coating for a total of 1600 lb VOC/shift. Government regulations limit the emission to 500 lb for each operation or 1000 lb VOC/shift. This situation is often complicated by the fact that several, or even many, color coats are involved. With color matching and other solidities, it is extremely desirable to keep the currently used colors. According to the law in most localities, this can be done, provided that the total emission does not exceed the allowable 1000 lb/shift; in other words, if the prime coat does not emit more than 200 lb VOC/shift. Some electrocoating primers emit 1.2 lb VOC/gal paint minus water (82 vol % NV) or, in the above example, 150 lb VOC/shift, thus resulting in compliance, without change of color coats.

References

1. Raney, M. W. *Electrodeposition and Radiation Curing of Coatings.* Park Ridge, N.J.: Noyes Data Corp., 1970.

2. Yeates, R. L. *Electropainting.* Teddington (England): Robert Draper Ltd., 1970.
3. Brewer, G. E. F. and R. D. Hamilton. *Paint for Electrocoating, ASTM Gardner-Sward Paint Testing Manual, 13th Edition.* Philadelphia, Pa., 1972, pp. 486–489.
4. Brewer. G. E. F. (Editor). "Electrodeposition of Coatings," *Advances in Chemistry, Vol. 119.* Washington D.C.: American Chem. Soc., 1973.
5. Machu, W. *Handbook of Electropainting Technology.* Ayr (Scotland): Electrochemical Publications, Ltd., KA7 1XB, 1978.
6. Brewer, G. E. F. (Chairman). "Symposium on Newer Development in Electrocoating." *Org. Coatings and Plastics Chem.* 45:1-22; 45:92-113 (Aug. 1981).
7. Chandler, R H. "Advances in Electrophoretic Painting," *Bi- or Tri-annual Abstracts since 1966* Braintree, Essex. (England).

39. HIGH SPEED ELECTROPLATING

A. Applications

W. H. SAFRANEK

Technical Editor, Plating and Surface Finishing

INTRODUCTION

The electrodeposition of metals and alloys at a rate above 15 μm/min (0.0006 in./min) is called *high speed* or *fast rate plating*, which is achieved by combining a high cathode current density with a high rate of cathode or solution motion. To use a current density of 0.3 A/cm^2 (280 A/ft^2) or more, the cathode movement or solution flow should be greater than 1 meter/sec (3.3 ft/sec) in order to replenish the supply of metallic ions at cathode surfaces fast enough to avoid defective deposits that contain undesirable inclusions of oxides or salts or become excessively porous or powdery.

The impingement of jets of solution at cathode surfaces is a common technique for sustaining a high current density and a fast deposition rate, but rapid cathode motion and solution flow are capable of accomodating higher current densities above 3 A/cm^2 (2800 A/ft^2). Procedures involving ultrasonic agitation, voltage pulsing or mechanical abrasion have not been as effective as jet or solution flow.[1]

Solution temperatures are sometimes above normal during high speed plating, primarily to reduce cooling requirements that are a consequence of the heat generated in high-current-density, high-voltage cells; solutions have been specially formulated for some purposes, but the usual compositions for conventional, low current-density plating are suitable for high speed plating, as a rule, except for the baths containing a high concentration of free cyanide and a low metal concentration.

STEEL STRIP PLATING

Steel strip is plated continuously at a high rate with tin, zinc or chromium while the strip is advanced through horizontal or vertical cells. A line speed of 2.5 to 3 meter/sec (8.25 to 9.9 ft/sec) is customary for depositing tin at 0.6 A/cm^2 (560 A/ft^2) or zinc at 0.6 to 1.5 A/cm^2 (560 to 1400 A/ft^2). The first tin plating units, which deposited 0.75 μm (30 μin.) in 2 or 3 sec, were installed in the early 1940s.

Alkaline stannate solutions were used in some facilities in the 1940s, but most of these have been replaced with one of two types of solutions. One (the U.S. Steel Ferrostan process) consists of stannous sulfate, phenol sulfonic acid and small amounts of grain refining agents.[2] The other (E.I. DuPont's Halogen bath) contains stannous chloride, stannous fluoride, potassium bifluoride, sodium chloride and 1-2 g/liter of addition agents for grain refinement purposes. It is maintained at a pH of about 3 and a temperature of 65C (150 F). Both processes have been described comprehensively.[2]

A system including a series of vertical loops has been adopted for plating zinc in England and Germany.[3] A strip speed of 1.67 meter/sec (5.5 ft/sec) is combined with a current density of 0.6 A/cm^2 (560 A/ft^2) to achieve a deposition rate of 16 μm/min (0.00063 in./min), using a zinc sulfate solution and soluble zinc anodes. A facility in Germany for plating lead-tin (terne) alloy containing 7 to 8% tin uses similar equipment and a current density of 0.4 A/cm^2 (375 A/ft^2) to attain a deposition rate of 17 µm/min (0.00067 in./min).

A faster rate of zinc deposition is claimed for a horizontal system in the U.S., which uses a high solution flow rate in the opposite direction to the strip travel.[4] At a current density of 1.0 A/cm^2 (930 A/ft^2), the deposition rate is 27 µm/min (0.00106 in./min). Zinc is replenished by dissolving zinc oxide in an off-line tank containing solution that is pumped to the plating cells. With insoluble lead-antimony anodes, the tank voltage is about 10 V.

The largest facility for strip plating uses 18 rotating drums, each with a diameter of 2.4 meter (8 ft) for producing 250,000 metric ton/yr of steel, most of which is coated on one side with 15 µm (0.0006 in.) of zinc.[5] This facility is located in Gary, IN. The drums are rotated at a surface speed of 3 meter/sec (10 ft/sec) and zinc is plated at a rate of 40.6 µm/min (0.0016 in./min), using a current density of 1.5 A/cm (1500 A/ft^2). Unlike other strip plating facilities for zinc, which use sulfate solutions, the Gary plant uses a chloride bath.

Several strip plating facilities for producing container stock have been converted from tin to a two-layer film consisting of a thin layer (0.075 µm or 0.3 µin.) of chromium with a weight of about 54 mg/meter^2 (5 mg/ft^2) and another thin layer of chromic oxide with a weight of about 11 mg/meter^2 (1 mg/ft^2) also applied cathodically in a chromic acid electrolyte. The product—called tin-free steel—was developed in Japan[6-8] and later produced by several U.S. steel companies. It has been widely adopted for the manufacture of containers, using adhesives in place of solder.

To deposit chromium, a current density as high as 1.0 A/cm^2 (940 A/ft^2) is applied in some facilities. However, a lower current density of only 0.2 A/cm^2 (185 A/ft^2) is used in at least one plant.[9] A solution containing 125 to 175 g/L of chromic acid and small amounts of sulfate and fluosilicate ions is customary.

CATHODE ROTATION

Cathode rotation is another effective way to increase the transport of metallic ions to cathode surfaces and has been used for several commercial applications. The technique was used in the 1940s, for example, to electroform nickel fountain pen caps on stainless steel mandrels rotated at a peripheral speed of 0.25 meter/sec (0.83 ft/sec), while the current density was adjusted to 0.43 A/cm^2 (400 A/ft^2).[10] A small concentration of benzene sulfonamide was added to a Watts-type bath to refine the grain structure and harden the 0.2-mm-thick nickel deposit.

Chromium-plated aluminum mandrels with a diameter of 51 cm (20 in.) are rotated at a fast rate while endless nickel belts are electroformed with a wall thickness of about 0.15 mm (0.006 in.). To fabricate the belts, which are used in office copiers, the cathode current density ranges from 0.3 to 0.5 A/cm^2 (280 to 465 A/ft^2), which deposits nickel at the rate of 6 to 10 µm/min (0.00024 to 0.0004 in./min). In order to control stress and other properties in the desired ranges, a nickel sulfamate bath containing small concentrations of chloride ions and saccharin was adopted. Cooling the mandrel from the plating bath temperature of about 52 C (125 F) to room temperature in a rinse tank reduces its circumference about 0.25 mm (0.01 in.) and facilitates parting.[11-13]

Piston rings are plated at a fast rate with chromium while stacked on an arbor rotated on its vertical axis.[14] Surface speed ranges from 1.0 to 1.5 meter/sec (3.3 to 5 ft/sec). A 125 to 150-µm-thick deposit is applied in about 10 min to obtain a rate of 12.5 to 15 µm/min (0.0005 to 0.0006 in./min) using a current density of 4 to 5 A/cm^2 (3700 to 4650 A/ft^2). At these current densities, the cathode efficiency of the 2.5 M chromic acid bath with a chromic to sulfuric acid ratio of 100 : 1 is 45 to 52% when the temperature is adjusted to 50 C (122 F). With these conditions, a crack-free or nearly crack-free structure with a hardness of 720 to 840 kg/mm^2 is obtained.[15]

Insoluble lead alloy anodes around the rotating arbor are moved in and out of concentric rings to maintain a constant anode-to-cathode spacing when production schedules require a change in the size of the piston rings. Ring diameters vary from 6.5 to 16 cm (2.6 to 6.3 in.).

SOLUTION FLOW

Many high speed plating facilities installed in the 1970s were designed to rely on fast solution flow to achieve high cathode efficiencies at high current densities. The effectiveness of rapid solution flow was documented in 1948[16] but materials and designs for long lasting pumps were not sufficiently advanced to promote widespread use of the technique for another 20 years. Also, the need for conserving space and minimizing capital outlays for equipment became more pronounced in the 1970s.

Successful results depend on a flow rate above

TABLE 1. RATE OF HIGH-SPEED PLATING[a]

Metal	Solution[b]	Current Density A/dm²	Current Density A/ft²	Deposition Rate μm/min	Deposition Rate in./min
Chromium	3.0 M CrO_3	6.2	5800	20	0.0008
Cobalt	2.5 M $Co(BF_4)_2$	6.2	5800	125	0.049
Copper	2.0 M $CuSO_4$ or $Cu(BF_4)_2$	3.1	2800	75	0.003
Gold	0.05 M AuCN (citrate)	0.3	280	18	0.0007
Iron	2.3 M $Fe(NH_2SO_3)_2$	6.9	6400	150	0.0059
Lead	2.5 M $Pb(BF_4)_2$	1.6	1500	100	0.0039
Nickel	2.0 M $NiSO_4$ or $NiCl_2$	3.0	2800	60	0.0024
Tin[c]	1.0 M $Sn(BF_4)_2$	3.0	2800	120	0.0055
Cadmium[d]	1.0 M $Cd(BF_4)_2$	3.0	2800	120	0.0047
Zinc	2.0 M $ZnSO_4$	3.0	2800	80	0.0032

[a] Based on data for depositing 50 to 75-μm-thick coatings or foil, except for tin and cadmium.
[b] Metal salt solutions containing no organic addition agents.
[c] Data for 1 to 1.5-μm-thick deposits on steel.
[d] Data for 13 μm-thick coating.

1 meter/sec (3.3 ft/sec) but a rate of 2 to 3 meter/sec is frequently adopted in order to insure ion transport that is fast enough for the use of high current densities to deposit dense, high-purity metal. Above 1 meter/sec, solution flow is usually turbulent depending on the viscosity of the solution and the roughness of surfaces defining the flow path. Laminar flow at less than 1 meter/sec is insufficient.

The current densities and corresponding deposition rates attainable with a flow above 1 meter/sec are shown in Table 1, which also cites the solutions used to achieve these results.[1] The rates reported for copper, nickel and several other metals range from 60 to >100 μm/min (0.0024 to >0.004 in./min) which are 30 to 50 times faster than the conventional rates for plating in similar solutions. Table 1 shows a rate of 18 μm/min (0.0007 in./min) for thick deposits of gold with the use of dilute (0.05 M) solutions of AuCN. For thin deposits (1 to 2.6 μm) higher rates up to 120 μm/min (0.0047 in./min) can be reached by using more concentrated solutions and higher current densities.[17]

The cost effectiveness of high speed plating using fast solution flow is dependent upon the dimension of the gap between anodes and cathodes. To avoid excessive energy costs, this gap should be narrowed as much as possible. A high dc voltage and a large pump demanding considerable energy is required for a wide gap. A gap of 1 cm, or less, is recommended to limit the voltage to 9 V or less, when copper or nickel is plated at 2 to 3 A/cm² (1850 to 2800 A/ft²).[18] A smaller gap of only 3 mm was successfully adopted for electroforming flexible copper circuits at a current density of 1 to 2 A/cm² (925 to 1850 A/ft²).[19] Table 2 shows voltage data for

TABLE 2. OPERATING CONDITIONS FOR CHROMIUM HSP ON STEEL*

Chromic Acid Concentration, g/L	Cell Potential, V	Cathode Current Density, A/cm² At 68 C (155 F)	Cathode Current Density, A/cm² At 85 C (185 F)	Cathode Current Efficiency, % At 68 C (155 F)	Cathode Current Efficiency, % At 85 C (185 F)
50	5	1.7	1.8	22	17
50	7.5	3.8	4.0	30	22
50	10	5.4	5.6	45	28
50	12.5	7.1	7.1	52	33
300	5	3.7	3.9	18	12
300	6	4.8	5.2	19	14
300	7	6.2	6.4	26	17
300	8	7.4	7.9	32	18
300	9	8.9	9.1	40	20

*Data for a 2.5 mm anode-to-cathode gap and a flow rate of 4 meter/sec.[20]

high-speed chromium plating with an anode-to-cathode gap of 2.5 mm.[20]

With a small anode-to-cathode spacing, insoluble anodes are usually adopted in order to facilitate the maintenance of a constant gap and a constant flow rate because large variations in the gap and flow rate could result in defective deposits. Lead alloy anodes or lead-sheathed steel are satisfactory for chromium, copper, nickel, and zinc plating. Platinum-clad titanium is customary for gold and is preferred for some high-speed nickel plating facilities.

Operation with insoluble anodes requires metal replenishment in the form of a soluble compound. Chromic acid is the logical choice for chromium. Nickel carbonate not only supplies nickel ions but also raises the pH of the solution, compensating for the decrease that occurs when nickel baths are operated with insoluble anodes. Zinc oxide additions to sulfate baths supplies zinc ions and neutralizes the sulfuric acid generated at the insoluble anodes.

To replenish the copper consumed for electroforming flexible circuitry, circulation of the copper sulfate–sulfuric acid solution through a bed of clean copper scrap was proposed,[19] which would be helpful for keeping the cost of the high-speed fabrication process below that of circuit production conventionally by the substractive process. The rate of copper dissolution can be increased by maintaining a large copper surface area and a relatively high concentration of sulfuric acid, heating the solution, and/or using air or the oxygen generated at the insoluble anodes to speed up the oxidation of copper.

Several reel-to-reel high-speed plating facilities have incorporated techniques to recover dragout and eliminate waste discharge. Air wipes and the use of counterflow rinsing help to minimize rinsewater volume. In the process reported for electroforming printed circuits,[19] all rinsewater is pumped to the plating cell. An evaporative cooling system not only cools the solution but also evaporates sufficient water to accommodate the counterflowing rinsewater.

Unlike the solutions used for plating at conventional rates with little or no agitation, those used for high-speed plating with fast-speed flow do not require wetting agents to prevent pitting.[1] Because the fast-rate deposits tend to have a finer grain structure than conventional deposits, grain refining agents are seldom necessary, at least for chromium, copper, nickel and zinc, to control the density and surface smoothness of the high-rate deposits. In special cases, however, an addition agent is sometimes desirable for maintaining an advantageous property or characteristic. For electroforming nickel belts, for example, saccharin was added to the bath to promote compressive stress, which was important to facilitate belt separation from the mandrel.[11-13] For gold plating lead frames used in semiconductor devices, the reliability of wire bonds was maximized by adopting addition agents that affected surface topology.[21]

CATHODE EFFICIENCY

With either fast solution flow or rapid cathode motion, high cathode efficiencies above 95% are normal for plating most metals at high current densities. This was confirmed for copper, lead, nickel, tin, zinc, and high-purity gold.[1] The efficiencies for cobalt-hardened gold and chromium are lower but significantly higher than the customary efficiencies for these metals when plated conventionally at low current densities.

Data from two sources[20,22] confirm that high current densities favor high cathode efficiencies for high-speed chromium plating with fast solution flow. Efficiency increases to as much as 50 to 55% when current density is raised to the range of 5 to 9 A/cm^2 (4700 to 8400 A/ft^2).[22] However, efficiency was significantly reduced by an increase in temperature from 50 to 75 C[15] (122 to 167 F) or from 68 to 85 C (155 to 185 F).[20] A dilute solution (50 g/L CrO_3—6.5 oz/gal) resulted in a higher efficiency than a concentrated bath (300 g/L CrO_3—40 oz/gal) as Table 2 shows.[20]

PROPERTY DATA

High-speed deposits obtained with turbulent solution flow exhibit about the same properties as fast-rate deposits produced with rapid cathode motion.[1] This relationship was established for chromium plated on the bore surfaces of aluminum cylinders, using rapid solution flow, compared with chromium deposited on the outside of rods rotated at a surface speed equal to the rate of solution flow. (1.5 meter/sec—5 ft/sec). The equivalency of the properties of high-speed deposits produced with rapid cathode motion and fast fluid flow was confirmed for copper plated on rotating rods and in flow cells on flat cathodes.

With a fixed solution flow rate, the tensile properties of high-speed copper and nickel deposits depend on the current density.[22] Increasing the current density increased the yield and tensile strengths of copper but lowered them for nickel. Even so, the lowest tensile strength reported for high-speed copper (294 N/mm^2 or 43,000 psi from a fluoborate solution and 333 N/mm^2 or 48,000 psi from a sulfate bath) was greater than the maximum reported for conventional copper with no bath additives (255 N/mm^2 or 37,000 psi).[22] Similarly, the lowest values reported for high-speed nickel deposited in sulfate and sulfate-chloride solutions (883 and 785 N/mm^2 or 128,000 and 107,000 psi, respectively) were greater than the maximum for conventional nickel obtained with no addition agents. The higher values for the high-speed deposits were attributed to their finer grain structures.

Significant differences in the structure and properties of high-speed chromium deposits resulting from temperature changes have been reported.[15] In the case of chromium deposited at 3.7 A/cm^2 (3500 A/ft^2) in a 2.5 M CrO$_3$ solution containing 0.025 M H$_2$SO$_4$, a shift from 50 to 75 C (122 to 167 F) increased the hardness of the deposits from the range of 720 to 840 Knoop to the range of 920 to 1030 Knoop. Unlike conventional chromium obtained in 50 to 55 C (122 to 130 F) baths at a current density of 0.3 to 0.4 A/cm^2 (280 to 375 A/ft^2), the fast-rate deposits produced at 50 C (122 F) contained few or no cracks. With a temperature shift to 75 C (167 F), high-speed deposits exhibited the same crack structure (150 cracks/cm) characteristic of conventional chromium.

References

1. Safranek, W. H. and Layer, C. H., *Trans. Inst. Metal Finishing*, **53**, 121 (1975).
2. Hoare, W. E., Hedges, E. S., and Berry, B. T. K., *The Technology of Tinplate*. New York: St. Martins Press, 1965.
3. Mouthen, B., Proceedings of the Third Continuous Plating symposium. Winter Park, FL: Am. Electroplaters' Soc., April, 1980.
4. White, A. W., loc. cit.
5. Higgs, R. F., Private Communication. U.S. Steel Corp, Gary, IN.
6. Y. Kitamura, U.S. Patent 2,998,361 (Aug. 29, 1961).
7. H. Uehida, et al., U.S. Patent 3,316,160 (April 25, 1967).
8. Fukuda, N., et al., *J. Electrochem. Soc., Electrochem. Tech.*, **116**, No 3, 398 (1969).
9. Austin, L. W., Proceedings of the Third Continuous Plating Symposium. Winter Park, FL: Am. Electroplaters' Soc. (April, 1980).
10. Safranek, W. H., Dahle, F. B., and Faust, C. L., *Plating*, **35**, 39 (1948).
11. Bailey, R. E., MCIC Report 74-17. Columbus, OH: Metals and Ceramics Info. Center, April 1974.
12. E. M. Wallin, U.S. Patent 3,799,859 (March 26, 1974).
13. E. M. Wallin and C. H. Ling, U.S. Patent 3,876,510 (April 8, 1975).
14. S. Susuki, K. Yada, I. Yaguchi and H. Karasawa, U.S. Patent 4,080,268 (March 21, 1978).
15. Safranek, W. H., National Bur. Std. Special Pub. 452, 56 (1976).
16. Wesley, W. A., Sellers, W. W., and Roehl, E. J., *Proc. Am. Electroplaters' Soc.* **36**, 79 (1949).
17. Duva, R. Proc. of the 68th A. Electroplaters' Soc. Tech. Conf., Session C (June, 1981).
18. Safranek, W. H., *Plating and Surf. Fin.*, **69**, 48 (April, 1982).
19. Schaer, G. R. and Wada, T., *Plating and Surf. Fin.*, **68**, 52 (July, 1981).
20. Hoare, J. P., LaBoda, M. A., and Holden, A. H., *Plating and Surf. Fin.*, **69**,101 (May 1982).
21. Endicott, D. W., and Casey, G. J. Jr., *Plating and Surf. Fin.* **67**, 58 (March, 1980).
22. Safranek, W. H., Chapter 23, *The Properties of Electrodeposited Metals and Alloys*. New York: Am. Elsevier Pub. Co.,, 1974.

B. Principles

ROBERT DUVA

Consultant, Catholyte, Inc.

High-speed electroplating can be defined as any electroplating system which will produce a functional electrodeposit with no current density limitation. Theoretically, this implies that there is no limit to the thickness of an electrodeposit per unit time—a concept that is impractical. The basis for all sound electroplating systems is the replenishment of metal ions in the cathode film as fast as their predecessors have reduced to metal. Without this "instantaneous replenishment" to maintain constant concentration of metal ions in the cathode film (or concentration equilibrium with the bulk electrolyte), the applied current density must be reduced accordingly to permit production of sound electrodeposits.

Four factors must be explored so that high-speed electrodeposition can be attained:

1. Metal concentration
2. Agitation
3. Operating temperature
4. Cell geometry.

Metal Concentration

Most electroplating systems utilize anionic complexes for metal deposition. Because these complexes are anions, they will be repelled from the electronegative cathode. Diffusion of these ions into the catholyte must be depended upon so that electrodeposition can occur. Increasing the metal ion concentration in the bulk electrolyte will increase the probability of more rapid diffusion of these ions into the cathode film which in turn allows the use of higher current densities. Limits to concentration are set by the solubilities of all components within the system.

Agitation

High metal ion concentration alone in an electrolyte will not permit high current density operation. Diffusion of ions into the cathode film, the key factor for maintenance of high metal ion concentration in that cathode film, must be assisted by bringing those ions into close proximity of the catholyte. Solution movement, cathode rocker motion, fluidized bed and ultrasonics are some means to achieve agitation—the best mode or modes are determined by the physical and chemical systems employed. Agitation increases the probability of ion concentration equilibrium between the catholyte and the bulk electrolyte. This determines the highest current density permissible for the metal concentration of the electrolyte in use.

Operating Temperature

Increasing the operating temperature of an electrolyte can be classed as "micro-agitation" in that, as temperature increases, ionic activity increases. As ionic activity increases, the probability increases of ion diffusion into the catholyte of those ions in close proximity to it. The higher the temperature, the higher the activity; thus, the higher the diffusion rate. Upper temperature limitations are determined by the physical and chemical stability of the systems employed.

Cell Geometry

In order to properly obtain high current densities for high-speed plating, the equipment used to handle the electroplating bath must be well engineered. As discussed previously, metal ion diffusion into the catholyte is the most important consideration and will dictate the mode or modes of required agitation for the specific parts being electroplated. For example, in continuous wire or strip plating, auxiliary solution agitation must be used to eliminate the "doughnut hole" of ion starved electrolyte in the vicinity of the cathode caused by wire or strip movement thru the solution. The current carrying capacity of the cathode will determine rectifier sizing, anode and cathode bussing, type, size and number of cathode contacts and tank capacity. For example, it is better engineering design to use three 1-m cells with six cathode contacts than to use one 3-m cell with two cathode contacts when high-speed electroplating fine wire or thin strip. Anode to cathode distance, anode to cathode ratio and anode conformity to the cathode must also be taken into account because high cell voltages can change current distribution across the cathode and adversely effect metal distribution and throwing or leveling power of the electrolyte. High voltages can also cause adverse electrochemical anode reactions which can decompose the electroplating bath.

Other physical/mechanical features to be considered are:

1. Filter capacity and particle size retention.
2. Over-flow weirs for solution cleanliness.
3. Fog-spray and air blow-offs to minimize bath dragout.
4. Mist or vapor suppressants to reduce electrolyte loss and minimize air pollution.
5. Thermal blankets for tanks and electrolytes to reduce heat losses.
6. Readily accessible and repairable/replaceable cathode contacts.

Disadvantages

There are advantages and disadvantages to high-speed electroplating. The advantages are obvious: more product per unit time and per unit space with lower inventories. However, disadvantages can greatly outweigh advantages and should be carefully evaluated prior to the installation of any high-speed electroplating system. Included are:

1. Initial equipment costs are high.
2. Equipment maintenance is more frequent than it is with conventional operations.
3. Electrolytes require more frequent analysis for adequate control.
4. Adverse anode reactions at high current densities can cause electrolyte decomposition and frequent replacement.
5. Functional testing of end product must be more rapid to keep pace with productivity.
6. If *anything* is wrong with the overall operation, the reject rate is as high as the rate of production of acceptable product—and when this occurs, product costs are tripled to meet manufacturing quotas.

Summary

High-speed electroplating can be carried out on *any* metal deposition system as long as there is an understanding of basic electrochemistry and material handling techniques. There is really no upper limit to cathode current density and subsequent electrodeposition speed as long as the guidelines above are taken into consideration.

INDEX

Abrasive belts
  advantages of, 68
  description of, 67
  use of, 67
Abrasive descaling, 165
Abrasives for polishing wheels, 64
Abrasives, polishing
  grain selection, table of, 65
  standard grain sizes, 64
Acid pickling, 176
Acid pickling, *see also* Pickling, acid
Acid solutions, specific gravity of,
  chromic, 39
  hydrochloric, 37
  sulfuric, 38
Acids
  commercial,
    approximate concentration, 175
    weight/concentration ratios, 48
    data on, table of, 298
  reaction with metal oxides, 160
Activated carbon for solvent recovery with vapor degreasers, 137
Adhesion, effect of surface condition on, 59
Adhesives for polishing wheels, 62-64
Adsorbants, 717
Agitation
  for electropolishing, 114
  of plating solutions, 227
Agitators
  air, blowers for, 733
  rod, 734
  solution, 733
Agricultural products, media, mass finishing, 95
Airborne contaminants, from finishing operations, 646-650
Alkaline cleaning, 147-159
Alkaline descalers, operation of, 154
Alkaline descaling, 176
Alloys
  aluminum
    bright dipping of, 168
    chemical polishing of, 168
  composition of, 16-22
    aluminum, 28
    copper, 28-30
    gold, 30
    lead, 30
    magnesium, 31
    nickel, 31
    non-ferrous, 28-32
    silver, 32
    stainless steel, 26
    tin, 32
    zinc, 32
  copper
    bright dipping of, 166
    chemical polishing of, 167
    oxide removal from, 165
  densities of, 16-22
  electromotive series, 49
  ferrous, bright dipping of, 169
  heat resisting, composition of, 27
  special, descaling of, 165
Aluminum
  deoxidizing of, 155, 169, 185
  etching cleaners for, 155
  non-etch cleaners for, 155
  plating on, *see* Plating, on aluminum
    anodic process for, 187
    stannate process for, 187
  surface data for flat products, 41
  zincate process for, 186
Aluminum alloys
  bright dipping of, 168
  chemical polishing of, 168
Aluminum cleaners, 155ff
Aluminum oxide abrasive
  standard grain sizes, 64
  grain selection, 66
Ammeters, 728
Ammonia, commercial, weight/concentration ratios, 48
Analytical procedures
  EDTA for metals, 303
  atomic absorption, for metals, 300ff
  cyanide destruct pretreatment, 302
  for anodizing solutions, 307
  for carbonate, 302
  for chromium solutions, 304
  for copper, 302
  for copper solutions, acid, 306

Analytical procedures (*Cont.*)
  for cyanide, 302
  for electroless solutions, 307
  for gold, 303
  for hydroxide, 302
  for nickel solutions, 304
  for silver, 303
  for tin, 303
  for zinc solutions, non-cyanide, 306
  general discussion, 292
  index of, 289ff
  miscellaneous, 308
  reagents for
    list of, 291
    preparation of, 293ff
Anode
  bags, 227, 666, 667
  baskets, 657
  contacts, under solution, 658
  hooks, 657
  materials
    alloys, 666
    brass, 660
    cadmium, 660
    carbon, 661
    copper, 661, 662
    gold, 662
    iron, 663
    lead, 663
    nickel, 663, 664
    silver, 664
    steel, 664
    tin, 664, 665
    zinc, 665
  rod systems, 631ff
  rods, 635
Anodes
  auxiliary, 659
  insoluble, 659
  metallurgical variations of, 656
  soluble, characteristics of, 656
  suspension of, 657
Anodic coatings
  MIL specs. for, 397
  chromic acid
    testing of, 407
    weight of, 405
  classification of, 396, 397
  designation system for, 407–409
  sulfuric acid
    coloring of, 400
    properties of, 403
    testing of, 403, 404
    weight of, 397
Anodizing; *see also* Anodic coatings
  chromic acid, 404ff
    equipment for, 406, 407
    process steps, 405, 406
    solution maintenance, 407

definition of, 396
magnesium, *see* Magnesium
racks for, *see* Racks, plating
solutions, analytical methods for, 307
sulfuric acid
  equipment for, 400ff
  of cast alloys, 402
  process for, 398–400
  racking for, 401
  solution maintenance, 402
  types of, 396
Atomic distribution in materials, 354
Automation of barrel tumbling, 80

Backstands, for polishing, 74
Barrel plating, 228
  auxiliary equipment for, 593ff
  plate distribution in, 577
  process control, 583
  suitable work for, 576
  to specification, 581, 583
Barrel rinsing and dragout recovery, 704
Barrel tumbling
  automation of, 80
  method, 80
  process, 79
    equipment for, 79
    speed of rotation, 80
Barrels, plating
  fully automatic, 602ff
  imperforate oblique, 599ff
    auxiliary equipment for, 600
    design features, 599
  metal, capacity of, 590
  perforate horizontal
    pumping action in, 580
    construction of, 578–581
    construction materials for, 584
    contacts in, 585
    cylinders for, 578
    drives for, 584
    fully automatic, 603, 605
    hangers for, 584
    hole size, 579
    hydrogen explosions in, 581
    solution exchange in, 579, 580
    tanks for, 586
    voltage drop, 579
  perforate oblique, 601ff
    design features, 601
    fully automatic, 602, 604
  portable, 588
  rocker type, 575
  types and sizes, 573–575
Barrels, tumbling
  horizontal, 80
  oblique, 79
Basis metal, effect on cleaning, 149

Belts abrasive
  advantages of, 68
  description of, 67
  use of, 67
Beryllium, bright dipping of, 170
Bias buffs, 70
Bipolar anodes, 472
Blends
  of solvents, for cleaning, 122
  of vapor degreasing solvents, 129
Boiling points, of vapor degreasing solvents, 127, 128
Bowl type vibratory finishing machines, 81
Brass, surface data for flat products, 41
Bright dipping, 166ff
  costs, 172
  equipment, 172
  of aluminum alloys, 168
  of beryllium, 170
  of cadmium, 170
  of copper alloys, 166
  of ferrous alloys, 169
  of lead, 170
  of monel, 169
  of nickel, 169
  of silver, 170
  of stainless steel, 169
  of thorium, 172
  of titanium, 170
  of uranium, 170
  of zinc, 170
  of zirconium, 170
Brightness vs. smoothness, 101
Brightness, effect of surface condition on, 60
Brushes
  use of, 69
  wire, 69
Brushing, definition of, 69
Buffing, 69-73
  compositions, bar, 70
  compound removers, operation of, 151
  compounds, airless spray guns for, 72
  compounds
    as soils, 148
    color, 71
    cut and color, 71
    cutdown, 71
    grease base, 71
    liquid
      advantages of, 73
      application, 72
    spray
      application, 72
      description, 72
  equipment, 74-78
    description of, 74
  lathes
    manual, 74
    semiautomatic, 74
  machines, automatic, 75
  wheels, description, 69
  wheelspeed, table of, 72
Buffing, "mush," 72
Buffs
  bias, 70
  finger, 70
  folded, 70
  loose, 70
  pocketed, 70
  sewed, 70
Bus bars
  size of, 37
  aluminum, 634
  copper, 632
  for low voltage d-c systems, 632
  installation details, 633
  installed cost of, 633
  maintenance of, 637

CBF (*see also* Centrifugal barrel finishing), 84-87
CD (*see also* Centrifugal disc finishing), 87-92
CVD, *see* Chemical vapor deposition
Cadmium
  analytical methods for, 303
  bright dipping of, 170
Capacity
  of rectangular tanks, 36
  of round tanks, 36
Capture velocity, factors affecting, 644
Carbon treatment, of plating solutions, 228
Castings, electropolishing of, 119
Cathode rod systems, 631ff
Cathode rods, 635
Cathodes, insoluble, 659
Cements for polishing wheels, 64
Centrifugal barrel finishing (CBF), 84-87
  advantages, 85
  applications of, 87
  capabilities, 86
  effect on fatigue life, 86
  equipment, 85, 86
  future of, 87
  operation, 84
Centrifugal disc finishing (CD), 87-92
  application of, 90, 91
  equipment, 89
  future of, 92
  parts/media ratios, 92
  process, 89
Ceramic bonded media, mass finishing, 96
Chemical compounds, 5-10
Chemical polishing
  costs, 172
  equipment, 172
  of aluminum alloys, 168
  of copper alloys, 167

INDEX

Chemical vapor deposition, *see* Coatings, vapor phase, chemical vapor deposition
Chemicals, common names, 11
Chromate coatings
  cadmium, requirements for, 391
  cycles for, 393
  definition of, 390
  metals treated, 392
  protective mechanism, 391
  resistivity of, 392
  solutions for, 392
  specifications for, 395
  testing of, 394
  troubleshooting for, 394
  types, 391
  zinc, requirements for, 391
Chromating solutions
  equipment for, 393
  safety with, 395
  waste disposal, 395
Chromic acid anodizing, *see* Anodizing
Chromic acid solutions, specific gravity of, 39
Chromium
  analytical methods for, 304
  hexavalent, reduction of, 215
Circuit board production, solvents in, 124
Cleaner
  life, 156
  maintenance, 156
  operation, 150ff
Cleaners
  alkaline
    descaling type, operation of, 154
    operation of, 151
  buffing compound removers, operation of, 151
  electro, operation of, 153
  emulsifiable solvent, operation of, 150
  etching, for aluminum, 155
  for aluminum, 155ff
  non-etch for aluminum, 155
  soak, operation of, 152
  spray, operation of, 151
Cleaning
  alkaline, 147-159
    mechanisms of, 151
    before phosphating, 369
    bulk, 176
  cost, 157
  displacement, mechanism of, 151
  effect of base metal on, 149
  effect of finish on, 150
  electrolytic, 176
  of magnesium, 410, 411
  of plastics, for plating, 202
  soak, 176
  solvent, *see* Solvent cleaning
  spray, 175
  tests for, 150
  ultrasonic, 152, 176
  waterbreak test for, 150
Cloth wheels, 61
Coating weights of metals, 4
Coatings
  chemical reduction
    non-catalytic, 423ff
      copper, 424
      sensitizing for, 425
      silver, 424, 479
  displacement
    copper, 421
    gold, 421
    limitations of, 420
    mechanism of, 421
    nickel, 422
    tin, 421
  electroless plating, 438ff
    arsenic, 439
    boron nickel alloys, 457
    chromium, 439
    cobalt, 439
    copper, 440-444
      bath constituents, 441, 442
      formulations, 443
      operation, 443
      uses of, 440
    gold, 444
    iron, 444
    nickel, 445-457
      applications, 454
      bath constituents, 446-448
      bath formulations, 450
      deposit characteristics, 449, 451, 452
      deposition rates, 448, 449
      equipment for, 455-457
      heat treatment of, 451
      preparation for, 453
      principles of, 445
      stripping of, 454
    principles of, 438
  vapor phase, 426ff
    chemical vapor deposition, 433
      applicability, 433
      equipment for, 434
      principles of, 433
      properties of, 434
    ion plating, 431, 432
      applicability, 432
      equipment for, 432
      principles of, 431
      properties of, 432
    plasma assisted CVD, 435, 436
      applicability, 435
      equipment for, 435
      principles of, 435
      properties of, 436
    plasma spraying, 436-437
      applicability, 436

equipment for, 437
principles of, 436
properties of, 437
sputtered, 428–431
  applicability, 429, 431
  equipment for, 430
  principles of, 428
  properties of, 430
vacuum evaporation, 426–428
  applicability, 426
  equipment for, 427
  principles, 426
  properties of, 428
Cobalt, analytical methods for, 303
Cold cleaning, solvent, control systems for, 144
Cold solvent cleaning, 121
Color buffing compounds, 71
Composition
  of aluminum alloys, 28
  of copper alloys, 28–30
  of gold alloys, 30
  of heat resisting alloys, 27
  of lead alloys, 30
  of magnesium alloys, 31
  of nickel alloys, 31
  of non-ferrous alloys, 28–32
  of silver alloys, 32
  of stainless steel alloys, 26
  of tin alloys, 32
  of zinc alloys, 32
Compositions
  buffing, bar, 70
  spray, buffing, description, 72
Compounds
  buffing, see Buffing compounds
  greaseless
    polishing, 68
    uses of, 71
  mass finishing, 94–101
    function of, 98
  satin finishing, 70
Compress wheels, 62
Condensate volume, of vapor degreasing solvents, 128
Condensing system, for vapor degreasers, 133
Contact plating, 422
  tin, 423
Control
  of chromating solutions, 392
  of phosphating solutions, 370
Control systems
  for cold solvent cleaning, 144
  for conveyorized vapor degreasers, 146
  for open top vapor degreasers, 145
Conversion factors, 33, 34
  some plating salts, 34, 35
Conveyor speed, for vapor degreasers, 133
Cooling of solutions, see Heat exchangers
Cooling, of electropolishing solutions, 115

Copper
  analytical methods for, 302, 306
  electroless, see Coatings, electroless plating, copper
  surface data for flat products, 42
Copper alloys
  bright dipping of, 166
  chemical polishing of, 167
  composition of, 28–30
  oxide removal from, 165
Copper sulfate, concentration vs. specific gravity in sulfuric acid, 39
Corrosion resistant alloys, oxide removal from, 164
Cost
  of bright dipping, 172
  of chemical polishing, 172
  of cleaning, 157
  of cyanide oxidation, 213
  of disposing of sludge, 217, 224
  of dragout, for solutions, table, 209
  of hexavalent chromium reduction, 215
  of metal cleaning, 141
  of neutralizing acids, 216
  of precipitating heavy metals, 217
  savings, by conservation, 209
Covering power, 470
Crack detection by electropolishing, 107
Crossrod conveyor vapor degreaser, 131
Current distribution, 461ff
  primary, 461–464
    control of, 464–467
  secondary, 467, 468
  systems, 631ff
Current requirements, of plating solutions, 227
Cut and color buffing compounds, 71
Cutdown buffing compounds, 71
Cyanide, analysis of, 302
Cyanide oxidation
  control of, 214
  cost of, 213
  equations for, 213
  operating procedures, 214
Cycle, preplating, definition of, 147
Cycles
  for metals, 174–202
  plating
    elements of, 174
    for aluminum alloys, 185ff
    for beryllium, 192
    for cast irons, 182ff
    for chromium, 192
    for copper alloys, 183ff
    for copper plate, 198
    for corrosion resistant alloys, 194
    for germanium, 193
    for high carbon steel, 179ff
    for lead alloys, 190
    for low alloy steel, 179ff
    for low carbon steel, 178ff

Cycles (*Cont.*)
  plating (*Cont.*)
    for magnesium, 188
    for molybdenum, 193
    for nickel plate, 198ff
    for nickel silver, 189
    for niobium, 195
    for plastics, 202
    for powdered metal parts, 191
    for silicon, 195
    for stainless steel, 181ff
    for thorium, 196
    for titanium alloys, 195
    for tungsten, 196
    for uranium, 197
    for zinc base die castings, 184ff
    for zirconium, 197

Deburring
  by electropolishing, 117
  definition of, 73
  methods of, 73
Defects
  in anodizing, 337
  in electrodeposited coatings
    due to base metal, 335
    due to plating process, 335
Deflocculation, mechanism of, 151
Degreaser, vapor
  boiling sump thermostat for, 130
  condensor water thermostat for, 130
  crossrod conveyor, 131
  description of, 126
  elevator, 132
  ferris wheel, 131
  monorail conveyor, 130
  safety vapor thermostat for, 129
  solvent spray thermostat for, 130
  vibratory, 131
  water separator for, 129
Degreasers, vapor
  condensing system for, 133
  conveyor and hoist speed, 133
  conveyorized, control systems for, 146
  freeboard, 133
  freeboard chillers for, 137
  heat supply for, 132
  location of, 134
  materials of construction, 132
  open top, control systems for, 145
  size of, 133
  solvent conservation in, 135
  solvent recovery
    still for, 136
    with activated carbon, 137
  startup procedure, 135
  ventilation of, 133
Degreasing equipment, vapor, selection of, 134
Degreasing solvents; choice of (*see also* Solvents, degreasing), 138ff; 175

Degreasing, vapor, 124ff
  and EPA, 143
  and OSHA, 142
  and RCRA, 143
  equipment for, 129ff
  hazard evaluation and control in, 144
  quality control of, 142
  solvents, condensate volume of, 128
  ultrasonic agitation in, 140
Deoxidizers, for aluminum, 155, 169
Deoxidizing of aluminum, 185
Deposit growth mode, 58
Descalers, alkaline, operation of, 154
Descaling
  abrasive, 165
  alkaline, 176
  special alloys, 165
Design for plating, 50–57
  illustrative example, 55, 56
  principles of, 51
Desmutting, of aluminum, 185
Dimensions and areas of steel pipe, 46
Diphase cold cleaning, 124
Displacement cleaning, mechanism of, 151
Donut type vibratory finishing machines, 81
Dragout
  determination of, 209
  reduction of, 207
  rinses, 207
Drains, in floors, construction of, 505
Drills, twist, diameters, 47
Dryers, for barrel plating, 594
Drying
  atmospheric, 710
  by absorption, 711
  forced, air jet, 710
  forced, heated air, 710
  mechanical equipment for
    centrifuges, 713
    drying machines, 713
    ovens, 714
    sawdust machines, 713
    substitutes for, 714
  stains, prevention of, 712
  with dewatering agents, 711
  with infra red heat, 712

ECM
  applications of, 745
  definition of, 744
  electrolyte flow system for, 747
  electrolytes for, 751
  glossary, 753
  operating conditions, 747–751
  surface effects of, 752
  tool design for, 746
  tool fixtures for, 747
  waste treatment for, 752
Electricity, basic, 669
Electrochemical equivalents, 3, 4

Electrochemical machining, *see* ECM
Electrocleaners, operation of, 153
Electrocoating, *see* Electropainting
Electrodeposits
  properties of, 364
  structure of, 361
Electrodialysis recovery, 211
Electroforming
  applications of, 475, 476, 486
  defined, 474
  deposit characteristics, 480-482
  mandrels for, 476-478
    design and fabrication, 478
    surface preparation of, 478-480
  origins of, 474
  process control, 482-485
  solutions for, 480-482
Electroforms, finishing of, 485
Electroless plating solutions, analytical methods for, 307
Electroless plating, *see* Coatings, electroless plating
Electrolytic pickling, 163, 165
Electromotive series for metals and alloys, 49
Electropainting
  advantages of, 757
  applications of, 757
  auxiliary equipment for, 761ff
  current demand in, 762
  disadvantages of, 757
  electrical circuits for, 759
  facilities needed for, 758ff
  history of, 753
  paint bath maintenance, 763
  power source for, 761
  principles of, 753
Electrophoresis, *see* Electropainting
Electropolished surface, comparison with mechanically finished surface, 101
Electropolishing
  advantages of, 105
  agitation, 114
  applications of, 107
  as a machining tool, 117
  commercial methods, 108
  crack detection by, 107
  deburring by, 117
  definition of, 100
  effect of surface condition on, 101
  effect on friction, 119
  equipment, 114
  hydrogen embrittlement in, 119
  limitations of, 105
  metals processed, 107
  of castings, 119
  partial patent list, 108
  passivation by, 117
  pollution control in, 119
  power supply for, 115
  pretreatments for, 116
  procedures for, 116
  racks for, 114
  results on some stainless steels, 109
  rinsing after, 115
  solution operation, 112, 113
  solutions
    for metallographic samples, 110, 111
    heating and cooling, 115
  surface roughness in, 117
  use in metallography, 107
  ventilation for, 116
Elements
  atomic weights of, 12-15
  densities of 12-15
  melting points of, 12-15
  physical properties of, 12-15
  symbols of, 12-15
  thermal properties of, 12-15
Elevator vapor degreaser, 132
Emery, abrasive grain selection, 66
Emulsifiable solvent cleaners, operation of, 150
Emulsification, mechanism of, 151
EPA and vapor degreasing, 143
Equipment
  buffing and polishing, 74-78
  continuous, for steel mill products, 617ff
  electropolishing, 114
  elevation, 514
  evaporative recovery, 210
  for bright dipping, 172
  for chemical polishing, 172
  for chemical vapor deposition, 434
  for chromating solutions, 393
  for chromic acid anodizing, 406, 407
  for cold solvent cleaning, 143
  for drying, *see* Drying, mechanical equipment for
  for electroless nickel plating, 455-457
  for electropainting, 758ff
  for heating and cooling, 537ff
  for ion plating, 432
  for phosphate coating
    immersion, 375-381
    spray, 382-389
  for plasma assisted CVD, 435
  for plasma spraying, 437
  for plating strip, 618
  for spin finishing, 93
  for spindle finishing, 93
  for sputtered coatings, 430
  for sulfuric acid anodizing, 400ff
  for vacuum evaporation coatings, 427
  for vapor degreasing, 129ff
  for vapor degreasing, selection of, 134
  for vibratory finishing, 80
  miscellaneous, for the plating room, 737
  pickling, materials for, 171
Etching cleaners, for aluminum, 155
Exhaust hoods
  baffling, 649

Exhaust hoods (*Cont.*)
  canopy, 643
  enclosing type, 639, 640
  lateral type, 640, 641
Exhaust rate, 643
Exhaust systems, for tank type operations, 639

Ferris wheel vapor degreaser, 131
Ferrous alloys, bright dipping of, 169
Filter aids, 717
Filter media, 716
Filters
  ASME code for, 722
  cloths for, 723
  for electropainting, 763
  installation of, 724, 725
  materials of construction, 723, 725
  pumps for, 726
  series connection with heat exchangers, 735
  types of, 717
    belt, 722
    cartridge, 721
    horizontal plate, 719
    porous tube, 720
    tubular sock, 720
    vertical leaf
      horizontal, 718
      vertical, 717
    valves for, 726
Filtration
  adsorbants for, 717
  definition of, 716
  glossary, 727
  of plating solutions, 228
  troubleshooting of, 727
Finger buffs, 70
Finish, effect on cleaning cycle, 150
Flammability
  of degreasing solvents, 126
  of solvents, table, 128
Flammable solvents, safety with, 123
Flat metal products, surface data for, 40
Flexible polishing, 68
Flexible wheels
  advantages of, 68
  method of coating, 68
Flocculation, polyelectrolytes for, 218
Floors, 503-515
  acid proof brick
    brick dimensions, 508
    construction of, 507
  chemical mortars for, resistance of, 508
  chemical resistance of, 504
  comparative costs, 513
  drains in, construction of, 505
  dust free, 513
  interliners for, 505
  resin topped, 511
  slope of, 509, 510
  stairwells in, 507

substructure for, 504
sulfur concrete, 513
supports on, for equipment, 506
trenches in, capacity of, 511
Flow, in pipes, 48
Fluoride, analytical methods for, in chromium solution, 305
Freeboard, for vapor degreasers, 133
Friction, effect of electropolishing on, 119

Glue, hide
  adhesive for polishing wheels, 62
  preparation of, 62
Gold
  analytical methods for, 303
  electroless plating, *see* Coatings, electroless plating
Gold alloys, composition of, 30
Grain, abrasive, polishing, table of selection, 65, 66
Greaseless compounds
  uses of, 71
  polishing, 68
Growth mode, effect of surface on, 59

Hanging wires, size of, 36
Hazard evaluation and control in vapor degreasing, 144ff
Headroom, 523
Heat exchanger calculations, 538ff
  cooling load, 540
  cooling, steady state, 554, 556
  energy requirements, 550
  heating load, 538, 539
  transfer area required, 548-550, 557, 558
Heat exchanger selection, 537
Heat exchangers
  coefficients for cooling, 547
  coefficients for steam heating, 547
  electric heaters, 543
  embossed sheet metal, 542
  external, 543-546
  gas fired, 543
  immersed coils, 541
  plastic coils, 551
  pumps for, 555
  steam jets, 543
  temperature controls for, 555
  types of, 541ff
Heat resisting alloys, composition of, 27
Heat supply, for vapor degreasers, 132
Heat treatment, 358
Heating of solutions, *see* Heat exchangers
Heating, of electropolishing solutions, 115
Hide glue
  for polishing wheels, 62
  preparation of, 62
Hoist speed, for vapor degreasers, 133
Horizontal tumbling barrels, 80

Hydrochloric acid
  solutions, specific gravity of, 37
  pickling with, 163
Hydrogen
  other effects of, 231
  removal of, from steel, 231
Hydrogen embrittlement, 179, 231, 232, 264, 361
  in electropolishing, 119
Hygiene, industrial, 341
  contingency plans, 348
  controlling legislation, 341
  controls for, 345
  employee education, 346
  first aid equipment, 347
  toxicology, 342

Immersion, coatings, see Coatings, displacement
Indium, analytical methods for, 303
Inhibitors, effect on sulfuric acid pickling, 163
Ion exchange recovery, 211
Ion plating, see Coatings, vapor phase, ion plating
Iron phosphate, see Phosphate coatings
Iron, electroless plating, see Coatings, electroless
    plating, iron

Lathes, buffing, manual, 73
Lathes, buffing, semiautomatic, 74
Lead alloys, composition of, 30
Lead, bright dipping of, 170
Leather wheels, 62
Life, of cleaners, 156
Lining, design for, 533, 534
Lining materials
  acrylates, 530
  epoxy, 530
  fluorocarbon, 530
  lead, 530
  polyesters, 530
  polyethylene, 529
  polypropylene, 529
  rubber, 528
  suitability of, table, 531
  thermoplastic, 529
  urethanes, 530
  vinyl, 529
Linings, 525–536
  applied liquid, 526, 527
  care of, 535
  drop-in, 526
  flame sprayed, 527
  fluidized bed, 527
  lead, 526
  organic, 525
  plastisol, 527
  preconditioning of, 534
  protective coatings, 527
  repair of, 534
  requirements of, 531–533
  sheathings for, 528
  sheet, 525

Liquid buffing compounds, description, 72
Liquid compounds
  application, 72
  buffing, advantages of, 73
Location, of vapor degreasers, 134
Loose buffs, 70

Machining and forming oils, as soils, 149
Machining, use of electropolishing for, 117
Magnesium
  cleaning of, 410, 411
  pickling of, 411–413
  plating on, see Plating, on magnesium
  surface treatments for, 410ff
    #1, 413
    #7, 414
    #9, 414
    #10, 413
    #17, 416
    #18, 415
    #19, 415
    #20, 413
    #23, 419
    H.A.E. anodize, 417
    caustic anodize, 417
    chrome pickle, 413
    dichromate, 414
    dilute chromic acid, 415
    fluoride anodize, 416
    galvanic anodize, 414
    phosphate, 415
    sealed chrome pickle, 413
    stannate immersion, 419
    table of, 418
Magnesium alloys, composition of, 31
Maintenance; see also individual equipment
    listings
  of agitation equipment, 742
  of building structures, 739
  of ceilings, 739
  of cleaners, 156
  of drains, 741
  of electrical conductors, 742
  of floors, 740
  of heating equipment, 741
  of piping, 740
  of pumps, 741
  of valves, 742
  of ventilating equipment, 741
  of windows, 740
Manganese phosphate, see Phosphate coatings
Manufactured abrasives, media, mass finishing,
    95
Mass finishing, 78–100
  advantages, 79
  capabilities, 79
  compounds, see Compounds, mass finishing
  future trends, 100
  media, shape and size considerations (see also
      Media, mass finishing), 96

Mass finishing (*Cont.*)
  operating considerations, 98-99
  processes
    definition, 79
    limitations, 79
    types, 79
Materials of construction, for vapor degreasers, 132
Materials science, 352
Mechanical finishing, surface damage from 101
Mechanical properties of materials, 356
Media, filter, 716
Media, mass finishing, 94
  agricultural products, 95
  ceramic bonded, 96
  choice of, 97
  manufactured abrasives, 95
  metallic, 97
  natural abrasives, 95
  plastic bonded, 96
  preforms, 96
  resin bonded, 96
  shape and size considerations, 96
  testing of, 97
Metal
  cleaning costs, 141
  cleaning solvents, table of, 122
  conductivity, for racks, table of, 115
  content of metal salts, 5-10
  distribution, 461ff
    in pulse plating, 686
  oxides, reaction with acids, 160
  types electropolished, 107
Metallic media, mass finishing, 97
Metallographic samples, electropolishing solutions for, 110, 111
Metallography, electropolishing in, 107
Metals, 16
  atomic absorption analytical method for, 300ff
  coating weights of, 4
  densities of, 16-22
  electrochemical equivalents, 3, 4
  electromotive series, 49
  oxides and scales on, 160
  processing and operating sequences, 124-202
Molecular weights of compounds, 5-10
Monel
  bright dipping of, 169
  surface data for flat metal products, 40
Monorail conveyor vapor degreaser, 130
Mush buffing, 72, 75

Natural abrasive, media, mass finishing, 95
Neutralization, of wastewater, control of, 217
Nickel
  analytical methods for, 303
  bright dipping of, 169
  electroless plating, *see* Coatings, electroless plating, nickel
  solutions, purification of, 229
  surface data for flat products, 40
Nickel alloys, composition of, 31
Nickel silver, surface data for flat products, 41
Nickel strike
  glycollate, for plating on aluminum, 187
  neutral, for plating on aluminum, 186
  Woods, 181, 194, 233
Non-etch cleaners for aluminum, 155
Non-ferrous alloys, composition of, 28-32
Non-ferrous metals, wire gage for, 44

OSHA and vapor degreasing, 142
Oblique tumbling barrels, 79
Organic coatings, electrodeposition of, *see* Electropainting
Oxide removal, 159-173
  from copper alloys, 165
  from corrosion resistant alloys, 164
  from low/medium alloy steel, 161
  from stainles steel, 164
  with oxidizing salts, 165
  with sodium hydride, 165
Oxides on metals, character of, 160
Oxidizing salts, oxide removal with, 165

PEL, definition, 342
Painting, of finishing plants, 514
Part shape, effect on plating, 51
Passivation, by electropolishing, 117
Paste wheels, for polishing, 68
Phase diagrams, 355
Phosphate coatings
  cleaning for, 369
  colored, 389
  cycle requirements of, 371, 372
  cycles for, discussion of steps, 372-375
  definition of, 366
  factors affecting formation, 369
  immersion processes, 375-381
  immersion, equipment for, 375-381
  on aluminum, 369, 372
  on cadmium, 372
  on zinc, 372
  selection of, 371
  specialized processes, 389
  spray processes, 381-389
  spray, equipment for, 382-389
  structure of, 366-369
  uses of, 366ff
Phosphating solutions
  control of, 370
  sludge and scale in, 370
Phosphoric acid, pickling with, 164
Pickling equipment, materials for, 171
Pickling
  acid, 176, 177
    cast irons, 177
    copper alloys, 177

high carbon steel, 177
low carbon steel, 177
stainless steel, 177
electrolytic, 163, 165
gas phase, 164
of magnesium, 411-413
with hydrochloric acid, 163
with phosphoric acid, 164
with sulfuric acid, 161ff
effect of concentration and temperature, 162
effect of inhibitors, 163
effect of iron content, 162
effect of scale breaking, 163
Pipe, clay, capacity of, 512
Pipes, flow in, 48
Plant layout
effects of, 493
examples of, 499-502
information needed for, 487
items to include, 497-499
Plant location
area selection, factors affecting, 494
check list for, 502
effects of, 493
within a facility, factors affecting, 496
Plasma assisted CVD, see Coatings, vapor phase, plasma assisted CVD
Plasma spraying, see Coatings, vapor phase, plasma spraying
Plastic bonded media, mass finishing, 96
Plastics, plating on, see Cycles, plating, for plastics
Plate distribution
design effect illustrated, 52
in barrel plating, 577
ratios to evaluate design, 52
Plated coatings
applications of (see also Plating solutions, composition of), 230
stripping of, 232ff
Plating
chromium, high speed, conditions for, 769
design for, 50-57
high speed, 767
agitation for
cathode rotation, 768
solution flow, 768
cathode efficiency, 770
deposition rates, 769
principles of, 772, 773
properties of plate, 770
on aluminum
zincate process for, 186
on magnesium, 188
standards and specifications (see also Specifications), 263ff
steel strip, 767
uses of in design, 50
Plating barrels, see Barrels, plating

Plating, cycles, see Cycles, plating
Plating installations
all manual, 567, 569
automatic
advantages of, 606
applications of, 607
machines for, see Plating machines
barrel, 573ff
cranes for, 595
dryers for, 594
equipment requirements for, 596
handling methods, 588
layout of, 588-592
multicompartment rotary, 597
combined hoist/manual, 568, 569
fully automatic, 606ff
handling methods, 567
manual, 567ff
sizing of, 570, 571
work flow analysis of, 572
Plating machines
automatic
cycle considerations, 615
number of lanes, 613, 614
return type, 610
selection of, 611
straight line, 608, 609
straight line trolley, 609
work load and rack size, 611, 612
types of, 607
Plating racks, see Racks, plating
Plating solutions
agitation of, 227
carbon treatment of, 228
chemical purification of, 229
composition of
brass, see Plating solutions, composition of, copper/zinc alloys
bronze, see Plating solutions, composition of, tin/copper alloys
cadmium
cyanide, 236
fluoborate, 238
chromium, 242
copper
acid, 236
cyanide, 240
fluoborate, 237
pyrophosphate, 238
copper/zinc alloys, 239
gold, 241
indium, 243
iron, 244
lead, fluoborate, 246
lead/tin alloys, 245
nickel, 250
black, 247
special baths, 248, 249
rhodium, 235

Plating solutions (*Cont.*)
  composition of (*Cont.*)
    silver, 252
    strikes before silver, 251
    tin
      acid, 256
      alkaline, 257
    tin/copper alloys, 253
    tin/nickel alloys, 254
    tin/zinc alloys, 255
    zinc
      acid, 237, 259
      alkaline non-cyanide, 258
      cyanide, 258
  current requirements, 227
  efficiency of, 227
  electrolytic purification of, 228
  materials of construction for tanks, 229, 230
  operating conditions, *see* Plating solutions, composition of
  operation of, general principles, 227
  purification of, 228
  temperature of, 227
  ventilation of, 230
Pocketed buffs, 70
Polishing, definition of, 61
  flexible, 68
  wheel speed for, 67
Polishing backstands, 74
Polishing compounds, greaseless, 68
Polishing equipment, 74-78
  description of, 74
Polishing machines, automatic, 75
Polishing wheels, *see* Wheels, polishing
Pollution control in electropolishing, 119
Polyelectrolytes, proper use of, 219
Porosity, effect of surface condition on, 60
Power supply, for electropolishing, 115
Preform media, mass finishing, 96
Preplating cycle definition of, 147
Pressed felt wheels 61
Pretreatment
  definition of, 175
  for electropolishing, 116
Pulse generation, for plating, 684, 682
Pulse plating, 684ff
  changes in deposit, 686-688
  cost considerations, 688
  metal distribution in, 686
Pumps, 736
Purification, electrolytic, tanks for, 735

Quality control, of vapor degreasing, 142

RCRA and vapor degreasing, 143
Racks, for electropolishing, 114
Racks, plating, 559ff
  auxiliary anodes, 565
    materials for, 566
  construction of, 562-566
  design currents for, 561
  design of, 559-562
  insulation of, 566
  materials for, 562, 563
    current capacity of, 563
  parts of, 560
Ratios, parts/media
  centrifugal disc finishing, 92
  vibratory finishing, 82
Recovery, chemical, 208ff
  electrodialysis, 211
  electrowinning, 211
  evaporative, 209
  ion exchange, 211
  ion migration, 211
  reverse osmosis, 210
Recovery, water, 208ff
Rectangular tanks, capacity of, 36
Rectifiers, 669ff
  component parts of, 670
  controls for
    SCR, 674
    saturable core reactor, 673
    tap switch, 672
    thyristor, 674
    variable auto transformer, 672
  efficiency of, 679
  for pulse plating, 682
  maintenance of, 680
  operation of, 671
  regulation of, 677
  trouble shooting of, 681
Regulations, for solvents, 142ff
Resin bonded media, mass finishing, 96
Reverse osmosis, 210
Rheostats, 729
Rinses
  dragout, 207
  reclaim, 207
Rinsing, 691ff
  after electropolishing, 115
  comparison of rack and barrel operation, 704
  counterflow, 207
  criterion, 693
  design of installations, 699
  effect of agitation, 695
  effect of dragout, 693
  effect of tank volume, 695
  flow controls for, 732
  fog nozzles, 707, 708
  in barrel installations, 700
    design considerations, 706
    effect of barrel size, 702
    effect of immersion depth, 701
  multiple, 696
  reclaim tanks, 698
  special techniques, 697
  spray, 208, 697
  theoretical equilibrium, 694
  theory of, 691-693
  water conservation in, 697

Rod and tube
  buffing, automatic, 76
  polishing, automatic, 76
Round metal products, surface data for, 43–45
Round tanks, capacity of, 36
Rust proofing compounds, as soils, 148

Safety, 341
  electrical warning devices, 730
  solvents, for cleaning, 122
  with chromating solutions, 395
  with flammable solvents, 123
Saponification, mechanism of, 151
Satin finishing compounds, 70
Sawdust, substitutes for, 714
Scale removal
  from stainless steel, 177
  mechanical, 176
Scale softening, on stainless steel, 177
Scales on metals, character of, 160
Separator, water, for vapor degreaser, 129
Sewed buffs, 70
Shape considerations, media, mass finishing, 96
Shorts, AC, 472
Silicon carbide abrasives, standard grain sizes, 64
Silver
  analytical methods for, 303
  bright dipping of, 170
Silver alloys, composition of, 32
Size considerations, media, mass finishing, 96
Size of bus bars, 37
Size, of vapor degreasers, 133
Sludge
  dewatering of, 221ff
  volume and disposal cost, 217, 224
Smoothness vs. brightness, 101
Smoothness vs. polishing method, 105
Smoothness, surface, steel treated by different methods, 106
Smuts, as soils, 149
Soak cleaners, operation of, 152
Sodium hydride, oxide removal with, 165
Soils
  buffing compounds, 148
  definition of, 148
  machining and forming oils, 149
  rust proofing compounds, 148
  smuts, 149
Solution level controllers, 730
Solvency, of vapor degreasing solvents, 127, 128
Solvent
  conservation in vapor degreasers, 135ff
  vapor degreasing, choice of, 138ff
Solvent blends for cleaning, 122
Solvent cleaning 121–147
  cold, 121
    equipment for, 123
  uses of, 121

Solvent recovery
  from degreasers with activated carbon, 137
  still for vapor degreasers, 136
Solvent regulations, 142ff
Solvents
  degreasing
    flammability of, 126
    toxicity of, 126
  flammability of, table, 128
  flammable, safety with, 123
  in circuit board production, 124
  metal cleaning, table of, 122
  safety, for cleaning, 122
  vapor degreasing
    boiling points of, 127, 128
    condensate, volume of, 128
    solvency of, 127, 128
    stability of, 128, 129
    table of, 127
  vapor density of, 127
Specific gravity
  of chromic acid solutions, 39
  of copper sulfate/sulfuric acid solutions, 39
  of hydrochloric acid solutions, 37
  of sulfuric acid solutions, 38
Specifications, finishing, 263ff
  for anodized aluminum, 267
  for anodized magnesium, 267
  for cadmium on steel, 268
  for chromium, 268
  for composite coatings, miscellaneous, 278
  for copper, 269
  for copper/nickel/chromium on steel, 275
  for gold, engineering use, 269
  for lead and lead/tin on steel, 270
  for nickel, 271
  for nickel plus chromium, 272–274
  for nickel/zinc alloy, 276
  for palladium, engineering use, 276
  for rhodium, engineering use, 276
  for silver, 277
  for silver, commercial flatware, 277
  for tin, 278
  for tin plate (sheet), 279
  for tin/lead alloys, 279
  for tin/nickel alloys, 280
  for zinc on steel, 281
  general requirements, 263, 264
  list of, 265ff
  nickel plus chromium, supplemental requirments, 276
  test methods, 266
Specifications
  for anodes, for plating, 282
  for chemicals, for plating, 283
Speed of rotation, barrel tumbling, 80
Spin finishing
  applications, 94
  description, 92
  equipment, 93

INDEX

Spindle finishing
  applications, 94
  description, 92
  equipment, 93
  future of, 94
Spray cleaners, operation of, 151
Spray compounds
  application, 72
  buffing
    advantages of, 73
    description of, 72
Spray guns, airless, for buffing compounds, 72
Sputtered coatings, see Coatings, vapor phase, sputtered
Stability, of vapor degreasing solvents, 128, 129
Stainless steel
  activation of, 181ff
  bright dipping of, 169
  composition of, 26
  electropolishing results, 109
  light reflection from, 105
  oxide removal from, 164
Standard screw threads, 46
Steam, saturated, temperature of, 35
Steel
  low/medium alloy, oxide removal from, 161
  surface data for flat products, 40, 41
  wire gage for, 43
Steel pipe, dimensions and areas, 46
Steels, compositions of, 23-25
Still, solvent recovery, for vapor degreasers, 136
Stray currents, 471
Stress
  compressive, by centrifugal barrel finishing, 86
  effect of surface condition on, 60
  internal, of electrodeposited metals, 483, 484
String wheels, 69
Strip
  chromium plating of, 627
  lead plating of, 629
  terne plating of, 629
  tin plating of, design factors, 622
  zinc plating of, 623
Structure of materials, 353
Sulfate, analytical methods for, in chromium solution, 305
Sulfuric acid, pickling with, 161ff
  effect of concentration and temperature, 162
  effect of inhibitors, 163
  effect of iron content, 162
  effect of scale breaking, 163
Sulfuric acid anodizing, see Anodizing
Sulfuric acid solutions, specific gravity of, 38
Sump, waste, construction of, 510
Surface
  base metal, nature of, 58
  damage from mechanical finishing, 101
  electropolished, vs. mechanically finished, 101
  nature of, 58

Surface condition
  effect on adhesion, 59
  effect on brightness, 60
  effect on deposit growth mode, 59
  effect on porosity, 60
  effect on stress, 60
Surface data
  for flat metal products, 40-42
  for round metal products, 43-45
Surface roughness, in electropolishing, 117
Surface smoothness, steel treated by different methods, 106
Surfaces, structure and properties of, 359

TLV's
  list of, 644
  definition, 342
Table
  electrochemical equivalents, 3, 4
  flammability of solvents, 128
  metal cleaning solvents, 122
  metal conductivity for racks, 115
  of cost for dragout, 209
  of tank lining materials, 531
  of tank sheathing materials, 528
  of treatments for magnesium, 418
  vapor degreasing solvents, 127
Tank linings, see Linings
Tanks
  auxiliary, 734
  construction materials for, 516
  dam overflows, 521
  desion features of, 518
  diaphragms for, 522
  drain valve mountings, 521
  for acid storage, 735
  for electrolytic purification, 735
  maintenance of, 524
  rectangular, capacity of, 36
  reinforcing of, 520
  round, capacity of, 36
  steel, construction of, 517
Temperature
  of saturated steam, 35
  of plating solutions, 227
Temperature controls, 730
Testing
  of anodic coatings, see Anodic coatings
  of electrodeposited coatings, 309
    cass test, 330
    CorrodKote test, 330
    corrosion resistance, 328
    EC test, 331
    STEP test, 331
    outdoor tests, 330
    salt spray, 330
    sulfur dioxide test, 331
    defects due to base metal, 335

defects due to plating process, 335
defects in anodizing, 337
for porosity, 310
inspection, 332
salt spray, 310
thickness, 311
  beta back scatter method, 323
  chemical methods, 314
  coulometric methods, 320
  eddy current method, 318
  interferometry, 313
  magnetic methods, 316
  microresistance method, 325
  microscopic-optical methods, 312
  weight methods, 319
  X-ray methods, 321
Tests, for cleaning effectiveness, 150
Thermostat
  boiling sump, for vapor degreaser, 130
  condensor water, for vapor degreaser, 130
  safety vapor, for vapor degreaser, 129
  solvent spray, for vapor degreaser, 130
Thickness measurement of coatings, see Testing
Thorium, bright dipping of, 172
Threads, standard screw, 46
Throwing power, 468
  control of, 469
Timers
  ampere hour meters, 730
  ampere hour recorders, 730
  clocks, 730
  dummy clocks 730
Tin, analytical methods for, 303
Tin alloys, composition of, 32
Titanium, bright dipping of, 170
Toxicity, of degreasing solvents, 126
Troubleshooting, 285ff
  of filters, 727
Tub type vibratory finishing machines, 80
Tumbling barrels, see Barrels, tumbling
Twist drill diameters, 47

Ultrasomic cleaning, 152
Ultrasonics
  for electroless nickel plating, 454
  in vapor degreasing, 140
Units and factors, 33
Uranium, bright dipping of, 170

Vacuum evaporated coatings, see Coatings, vapor phase, vacuum evaporation
Vapor degreasers, see Degreasers, vapor
Vapor degreasing, see Degreasing, vapor
Vapor degreasing solvents, see Solvents, vapor degreasing
Vapor density, of solvents, 127
Ventilation
  dilution ratio to avoid explosions, 653
  for electropolishing, 116
  of abrasive blasting, 651
  of grinding, buffing, polishing, 652
  of plating solutions, 230
  of vapor degreasers, 133
  types of, 638
Vibratory finishing
  benefits of, 80
  equipment, 80
    future of, 83
  machines
    bowl type, 81
    description of, 80
    donut type, 81
    tub type, 80
  process
    advantages, 82
    limitations, 82
    operation, 81
    parts/media ratios, 82
    typical applications, 82, 83
Voltmeters, 728

Wastewater
  collection basins, sizing of, 213
  pretreatment of, 216
  segregation and collection of, 212ff
Wastewater control, 206-225
Wastewater treatment
  cyanide oxidation, 213
  flocculation, 218
  hexavalent chromium reduction, 215
  neutralization, 216
    control of, 217
  sludge dewatering, 221ff
  solids management, 220
  special methods, 219ff
Water
  conservation of, 207
  supply
    correction of, 206
    effect on plating baths, 206
    for finishing, 206
Water flow controls, 732
Waterbreak test, for cleaning, 150
Watermarks, prevention of, 712
Wheel speeds
  buffing, table of, 72
  for polishing, 67
  table of, 67
Wheels
  buffing, description, 69
  cloth, 61
  compress, 62
  flexible
    advantages of, 68
    method of coating, 68
  leather, 62

Wheels (*Cont.*)
  future of, 94
    abrasives for, 64
  paste, for polishing, 68
  polishing, 61
    adhesives for, 62–64
    cements for, 64
    coating of, 63
    recoating of, 63
    rules for selection, 62
  pressed felt, 61
  string, 69
Wire
  plating of, 624
  zinc plating of, 625

Wire brushes, 69
Wire gages, 43–45

Zinc
  analytical methods for, 303, 306
  bright dipping of, 170
Zinc alloys, composition of, 32
Zinc coatings, suggested standards for, 54
Zinc phosphate, *see* Phosphate coatings
Zincate process
  for plating on aluminum, 186
  for plating on beryllium, 192
Zirconium, bright dipping of, 170